COMPUTER MODELS OF WATERSHED HYDROLOGY

COMPUTER MODELS OF WATERSHED HYDROLOGY

Editor

Vijay P. Singh

Department of Civil and Environmental Engineering
Louisiana State University
Baton Rouge, LA 70803-6405, U.S.A.

Water Resources Publications

For Information and Correspondence:

Water Resources Publications
P. O. Box 260026 • Highlands Ranch • Colorado 80126-0026 • USA

Computer Models of Watershed Hydrology

Editor

Vijay P. Singh

Department of Civil and Environmental Engineering
Louisiana State University
Baton Rouge, LA 70803-6405, U.S.A.

Cover Designed *by Water Resources Publications*

ISBN Number - 0-918334-91-8

U.S. Library of Congress Catalog Card Number - 94-61748

Manufactured in the United States of America

To

Hydrologic modelers around the globe

PREFACE

Since the pioneering development of the Stanford Watershed Model by N. H. Crawford and R. K. Linsley in 1966, there has been a flurry of activity in the area of computer modeling of watershed hydrology. This activity has been fueled by ever-expanding computing power and growing capability to observe, store, retrieve, and manage hydrologic data. The result is virtual revolution in watershed models. In the United States alone, nearly every concerned federal agency has its own computer model of watershed hydrology. For example, the U.S. Department of Agriculture, U.S. Army Corps of Engineers, U.S. Environmental Protection Agency, U.S. Geological Survey, U.S. National Weather Service, to name but a few, have their own computer models. At the state level, the various state agencies have either developed their own models or adapted some variation of the existing models. Similar trends are found in Australia, Canada, China, Denmark, Germany, Great Britain, India, Japan, Sweden, and so on.

When the Stanford Watershed Model (SWM) was first reported in the hydrologic literature, the environmental and water resources problems were not as complex as they are today. Consequently, the SWM and the models that followed it were not as comprehensive as the current-generation computer models are. Those days the availability of data and the computer capability to analyze them were limited. In the intervening period of nearly three decades, the environmental and water resources problems have become many times more complex, and the demands for more comprehensive solutions considerably more heightened. To meet these challenges, the present-generation computer models are hence more sophisticated. Not only do these models simulate watershed hydrology more accurately, but they also simulate other components of the environmental and water continuum on one hand and socio-economic continuum on the other. For example, they, in addition to hydrology, simulate water quality, ecology, risk and uncertainty, environmental impact, etc. Furthermore, they are equipped to take advantage of GIS capability for data-base management, and capture input data received directly from remote sensing and satellite technology. They produce results that help with their interpretation by yielding the range of values in conjunction with their risk that the decision maker can use in making real-life decisions. Some of these models are intelligent systems and interact with the modeler as well as the user. In essence, the models not only do hydrology but also social and management science. In the process, the modeler and the user are afforded an opportunity to interact with each other and learn from each other, and in turn, advance the state of the art of hydrology.

It is now well accepted that modeling seems to be the only resort to address complex environmental and water resources problems. More and more people are using models these days. If the present trend is any indication, then models will continue to find increasing use in the entire gamut of water resources development and management. The motivation for putting together this book stemmed from the desire to provide, under one cover, a comprehensive account of some of the world's popular computer models of watershed hydrology. To that end, the objective of the book was to include a variety of models that together spanned a range of characteristics. Some of those models should be representative of a large class of models, some comprehensive, some applicable to not only civil works but also to agricultural, range and forest, and nonpoint source pollution fields, some equipped with the GIS and remote sensing capability, and some a good cross-section of models around the world. It is hoped that the 26 models included in the book fulfill the intended objective. Because there are tens of models available these days, any number of combinations of the models could satisfy the intended objective equally well. Therefore, in any model selection, personal bias is unavoidable, and the models included in this work may reflect my personal bias. This model selection, in no way, implies an endorsement of the models included here or a rejection of those excluded. I personally have used and like very much some of the models not included here.

The subject matter of this book is divided into 29 chapters. Beginning with introductory remarks on watershed modeling in Chapter 1, model calibration and reliability estimation are presented in Chapters 2 and 3, respectively. The next ten chapter (4 to 13) present some of the popular models from around the world. These models are in the realm of civil engineering applications of watershed hydrology models. Some of the models are more comprehensive than others and some have the management capabilities. The next two models, presented in Chapters 14 and 15, are large-scale models and embody GIS and remote sensing technology. The models presented in Chapters 16 to 23 are more physically-based and distributed in nature, quite suited to nonpoint source pollution modeling, and to assess environmental impact of land use changes. The remaining 5 models presented in Chapters 24 to 29 are within the realm of agricultural and forestry applications. Nonpoint source pollution, erosion and impact on soil productivity, drainage design, etc., can be modeled by applying these models.

The book will be of interest to those who are engaged in practice of hydrology, civil engineering, agricultural engineering, environmental science, forest and range science, earth science, climatology, and watershed sciences. Graduate students, teachers engaged in graduate instruction, and researchers will find the book useful. The book should be of special appeal to hydrologic modelers and the model users.

The editor expresses his deep appreciation to all the contributors who enthusiastically embraced the idea of this book and who spent much time, effort, and resources to write their contributions. The book reflects their collective effort and dedication. There are hundreds of people who have long been engaged in developing and applying the computer models of watershed hydrology. Without their efforts we will not have the models that we take for granted these days. The book is dedicated to all of these people for their unselfish work and devotion to advance the science of hydrologic modeling.

Ms. Susan Sartwell, Word Processor, Department of Civil and Environmental Engineering at Louisiana State University, did a lot of typing with skill, care, and unfailing devotion. My wife, Anita, drafted many of the drawings. My children, Vinay and Arti, allowed me to spend long hours during evenings, weekends, and holidays, and were always supportive. To all of these individuals I owe a debt of gratitude. Without their support, this book would not have been completed.

Baton Rouge, Louisiana
January 1995

V. P. Singh

CONTRIBUTORS

J. G. Arnold
USDA-ARS
Grassland, Soil and Water Research Laboratory
808 East Blackland Road
Temple, TX 76502, U.S.A.

J. C. Bathurst
NERC Water Resource Systems Research Unit
Department of civil Engineering
University of Newcastle Upon Tyne
Newcastle Upon Tyne
Great Britain

S. Bergström
Swedish Meteorological & Hydrological Institute
Box 923
S-60176 Norrkoping, Sweden

K. J. Beven
Centre for Research on Environmental Systems
Institute of Environmental and Biological Sciences
University of Lancaster
Lancaster LA1 4YQ, England, U.K.

B. R. Bicknell
AQUA TERRA Consultants
2672 Bayshore Parkway, Suite 1001
Mountain View, CA 94043, U.S.A.

G. Blöschl
Centre for Resource and Environmental Studies
The Australian National University
Canberra City, ACT 2601
Australia

D. D. Bosch
USDA-ARS
North Central Soil Conservation Research Laboratory
North Iowa Avenue
Morris, MN 56267, U.S.A.

R. J. C. Burnash
U.S. Department of Commerce
National Weather Service
California-Nevada River Forecast Center
3539 Ridgeview Dr.
El Dorado Hills, CA 95762, U.S.A.

A. Calver
Institute of Hydrology
Maclean Building
Crowmarsh Gifford
Wallingford, Oxfordshire OX10 8BB
England

D. H. Carlson
Department of Rangeland Ecology and Management
Texas A&M University
College Station, TX 77843-2146, U.S.A.

A. S. Donigian, Jr.
AQUA-TERRA Consultants
2672 Bayshore Pkwy., Suite 1001
Mountain View, CA 94043-1011, U.S.A.

J. Dørge
River Hydraulics Division
Danish Hydraulic Institute
Agern Alle 5
DK 2970 Horsholm, Denmark

A. D. Feldman
Hydrologic Engineering Center
Corps of Engineers
Department of the Army
609 Second Street
Davis, CA 95616-4687, U.S.A.

J. Freer
Centre for Research on Environmental Systems and Statistics
Institute for Environmental and Biological Sciences
Lancaster University
Lancaster, LA1 4YQ, U.K.

Contributors

D. C. Goodrich
Agricultural Research Service
U.S. Department of Agriculture
2000 E. Allen Road
Tucson, AZ 85719, U.S.A.

R. B. Grayson
Centre for Environmental Applied Hydrology
University of Melbourne
Parkwill, Victoria 3051
Australia

V. K. Gupta
Department of Hydrology and Water Resources
The University of Arizona
Tucson, AZ 85721, U.S.A.

K. Havnø
River Hydraulics Division
Danish Hydraulic Institute
Agern Alle 5
DK 2970 Horsholm, Denmark

W. C. Huber
Department of Civil Engineering
Oregon State University
Corvallis, OR 97331-2302, U.S.A.

J. C. Imhoff
AQUA TERRA Consultants
2672 Bayshore Parkway, Suite 1001
Mountain View, CA 94043, U.S.A.

G. W. Kite
Hydrological Sciences Division
National Hydrology Research Institute
11 Innovation Blvd.
Saskatoon, Sasketchewan S7N 3H5
Canada

W. G. Knisel
Biological and Agricltural Engineering Department
University of Georgia, CPES
P.O. Box 748
Tifton, GA 31793, U.S.A.

R. Lamb
Centre for Research on Environmental Systems
Institute of Environmental and Biological Sciences
University of Lancaster
Lancaster LA1 4YQ, England, U.K.

E. M. Laurenson
Department of Civil Engineering
Monash University
Clayton, Victoria 3168
Australia

G. H. Leavesley
Water Resources Division
U.S. Geological Survey
MS 412, Denver Federal Center
Box 25046
Lakewood, CO 80225, USA

Xinren Liu
Department of Hydrology
Hohai University
Nanjing, China 210024

M. N. Madsen
River Hydraulics Division
Danish Hydraulic Institute
Agern Alle 5, DK 2970 Horsholm
Denmark

R. G. Mein
Department of Civil Engineering
Monash University
Clayton, Victoria 3168
Australia

C. S. Melching
Water Resources Division
U.S. Geological Survey
102 E. Main, 4th Floor
Urbana, IL 61801, U.S.A.

I. D. Moore
Centre for Resource and Environmental Studies
Australian National University
G.P.O. Box 4
Canberra, ACT 2601
Australia

P. E. O'Connell
Water Resource Systems Research Group
Department of Civil Engineering
University of New Castle Upon Tyne
New Castle Upon Tyne NE1 7RU, U.K.

C. A. Onstad
USDA-ARS
North Central Soil Conservation Research Laboratory
North Iowa Avenue
Morris, MN 56267, U.S.A.

M. C. Quick
Department of Civil Engineering
University of British Columbia
Vancouver 8, B.C., V6T 1Z4, Canada

P. Quinn
Centre for Research on Environmental Systems
Institute of Environmental and Biological Sciences
University of Lancaster
Lancaster LA1 4YQ, England, U.K.

A. Rango
Hydrology Laboratory
Room 104, Building 007
USDA-ARS-BARC-West
U.S. Department of Agriculture
Beltsville, MD 20705, U.S.A.

J. C. Refsgaard
Danish Hydraulic Institute
Agern Alle 5
DK2970 Horsholm, Denmark

R. Romanowicz
Centre for Research on Environmental Systems
Institute of Environmental and Biological Sciences
University of Lancaster
Lancaster LA1 4YQ, England, U.K.

V. P. Singh
Department of Civil and Environmental Engineering
Louisiana State University
Baton Rouge, LA 70803-6405, U.S.A.

R. E. Smith
U.S. Department of Agriculture
Agricultural Research Service
Water Management Research Unit
P.O. Box E, 301 S. Howes, Rm. 233
Fort Collins, CO 80522, U.S.A.

S. Sorooshian
Department of Hydrology and Water Resources
University of Arizona
Tucson, AZ 85721, U.S.A.

D. D. Speers
Hydrologic Engineering Branch
Water Management Division
U.S. Army Corps of Engineers
North Pacific Division
Portland District
P.O. Box 2870
Portland, OR 97208-2870, U.S.A.

L. G. Stannard
Water Resources Division
U.S. Geological Survey
MS 412, Denver Federal Center
Box 25046
Lakewood, CO 80225, U.S.A.

B. Storm
Danish Hydraulic Institute
Agern Alle 5, DK-2970 Horsholm
Denmark

M. Sugawara
6-13-30 Mianai Karasuyama
Setagaya - Ku
Tokyo 157, Japan

T. L. Thurow
Department of Rangeland Ecology and Management
Texas A&M University
College Station, TX 77843-2126, U.S.A.

C. L. Unkrich
Agricultural Research Service
U.S. Department of Agriculture
2000 E. Allen Road
Tucson, AZ 85719, U.S.A.

J. M. Wicks
Sir William Halcros & Partners Ltd.
Burderop Park
Swindon, Wiltshire SN4 0QD, U.K.

J. R. Wight
U.S. Department of Agriculture
Agricultural Research Service
Northwest Watershed Research Laboratory
800 Park Blvd., Plaza IV, Suite 105
Boise, ID 83712, U.S.A.

J. R. Williams
USDA-ARS
Grassland, Soil and Water Research Laboratory
808 East Blackland Rd.
Temple, TX 76502, U.S.A.

W. L. Wood
Institute of Hydrology
Maclean Building
Crowmarsh Gifford
Wallingford, Oxforshire OX10 8BB
England

D. A. Woolhiser
Engineering Research Center
Colorado State University
Fort Collins, CO 80523, U.S.A.

R. A. Young
USDA-ARS
North Central Soil Conservation Research Laboratory
North Iowa Avenue
Morris, MN 56267, U.S.A.

R. J. Zhao
Department of Hydrology
Hohai University
Nanjing, China 210024

TABLE OF CONTENTS

Chapter 1

WATERSHED MODELING

V. P. Singh

1.1. MODELING FOR INTEGRATED WATER MANAGEMENT

Air, soil, and water constitute the environmental continuum, and are vital components for sustaining life on earth. These components are interactive and interactions amongst them are complex. Stated another way, a management change imposed on one component of the environmental continuum has effects that propagate to other components, and some of these effects are unknown and cannot be quantified. This interactive nature requires that the environment is managed and protected as a cohesive whole (or a system). This can be accomplished through an integrated approach to environmental management that clearly has to be interdisciplinary. Such an approach may entail management of individual components with proper accounting of their interactions and dependencies. The management of air, soil, and water is to be integrated as well.

Worldwide, environmental systems are rarely managed as a cohesive whole. That is partly because air, soil, and water resources have been and continue to be managed by independent organizations having little interaction amongst them. For example, there are a host of organizations in the United States that have authorities to manage nation's water resources at the local, state, regional, and national level. The U.S. Bureau of Reclamation and the U.S. Army Corps of Engineers are two of the federal agencies with broad authority for water resources management. Then there are the departments of water resources, natural resources, or environmental quality at the provincial (state) level having authority to manage water resources in their individual states. At the local (county or parish) level, there are the water boards, commissions, and the like that manage local water resources.

The integrated water management should be accomplished within a spatial unit called watershed through the instrument of modeling. The integrated water management can be viewed as a three or more dimensional process centered around the need for water, the policy to meet the needs and the management to implement the policy. Water elements encompassing physical, chemical, and biological aspects of water quantity and quality may constitute the first dimension. Water uses, including agriculture, water supply, energy generation, industry, fish production, recreation, transportation, etc., may constitute the second dimension. Clearly, these water uses have to be accomplished following a well-defined management policy balancing the demand for water amongst different uses. The strategy to implement this policy may form the third dimension. The management must be dynamic and evolve with time, in response to changing needs and objectives. Thus, the fourth dimension may be formed by these needs and objectives.

The water elements, their interactions, and the effects on them of natural as well as external constraints, as shown in Fig. 1.1, constitute the foundation upon which the edifice of integrated water management is to be built. External constraints such as economic, demographic, transportation and other forms of development directly influence one or the other water elements. Likewise, climatic vagaries, climatic change, and climatic extremes, and a host of natural hazards are some of the natural causes that greatly influence the water elements and have significant impact on the integrated water management.

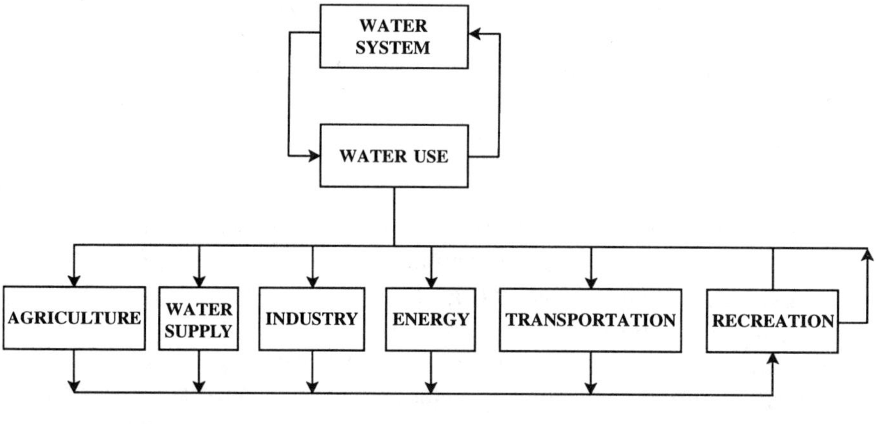

Fig. 1.1. Elements of a water system and their interactions.

The need for water is derived from a variety of activities in which it is used as shown in Fig. 1.2. These activities are vital for human society--its existence and development. Because usable water is limited in its availability, it has an economic value. Furthermore, different activities require water of differing quality. For example, for domestic use, water of high quality is needed. Clearly, all uses of water cannot be supported to the fullest extent and a management policy has to be developed that can prioritize water use following established criteria. There may be conflicts and interactions amongst different water uses, and these, in turn, interact with water elements. The management policy has to incorporate all of these considerations.

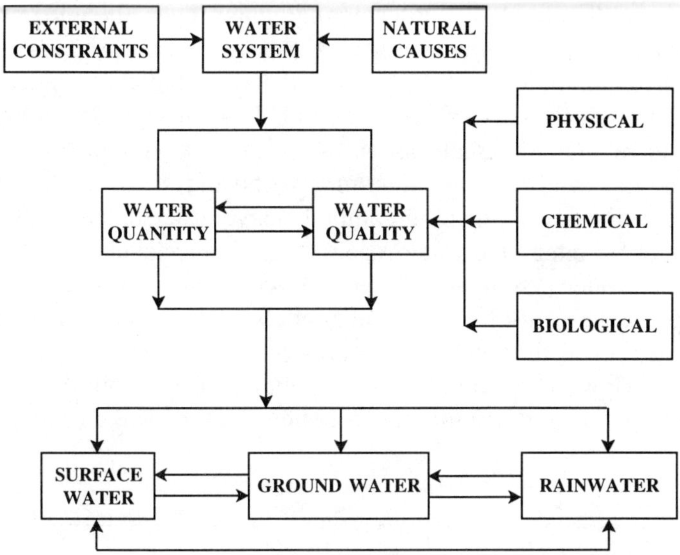

Fig. 1.2. Water use and its interaction with water system.

The criteria, as shown in Fig. 1.3, that form the foundation of the management policy must be developed following such considerations as cost effectiveness, economic efficiency, environmental impact, ecological and health considerations, socio-cultural aspects, to name but a few. The criteria must be practical, implementable, and must be acceptable to the

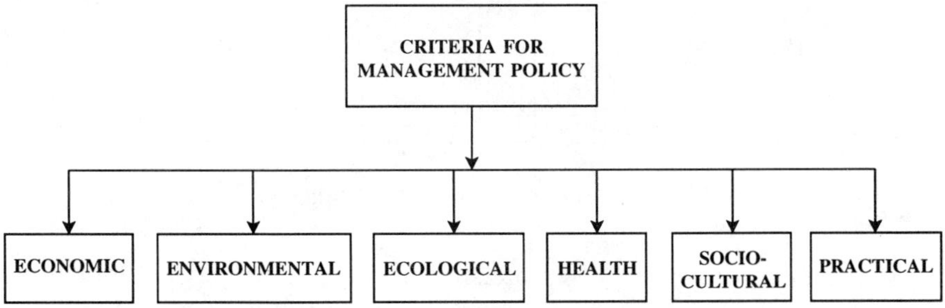

Fig. 1.3. Criteria for information of a management policy.

society at large. Furthermore, they must be capable of responding to changing needs of the society. The resulting policy must satisfy such interactive social goals as equity, efficiency, environmental quality, etc.

Once a management policy is established, a strategy, including administrative infrastructure, has to be employed to undertake the integrated water management as shown in Fig. 1.4. The components of integrated water management are interactive, and hence the administrative set up must be flexible and be responsive to changing goals. Thus, the integrated water management requires integration of the various components discussed earlier--physical, biological, chemical, social, economic, ecological, health, and environmental. This can be accomplished through development and application of mathematical models. The physical, chemical, biological, environmental, and ecological components and their models must be embedded in the development of comprehensive watershed models.

A model for integrated water management helps develop implementable solutions to water resources problems by combining into an optimization scheme all the essential component models. The model incorporates or accumulates all of the interactive forces or influences. Hence, it aids the decision-making process and keeps the policy results within the intersection of the social goals of the management policy and the legal constraints.

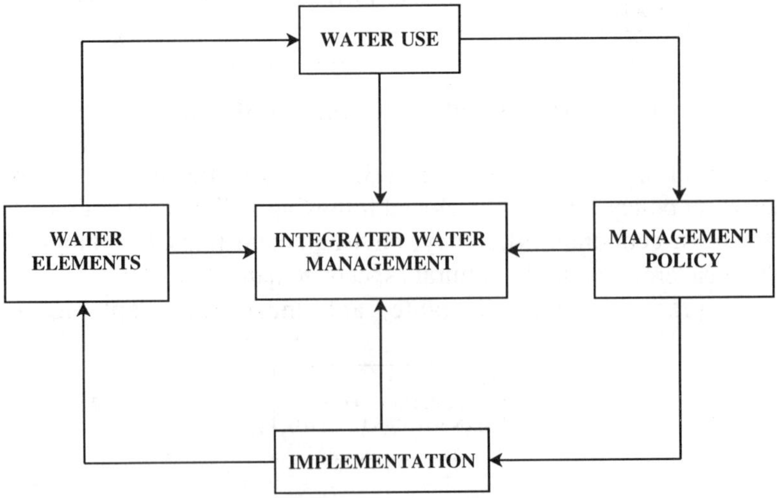

Fig. 1.4. Integrated water management.

Fundamental to integrated water management is watershed modeling. The watershed models abound in hydrologic literature (Singh, 1989), and the state of art of watershed modeling is reasonably advanced, especially when viewed in the context of practical application. However, these models have yet to become common planning or decision-making tools. To that end, two milestones will have to be achieved. First, these models will have to be transformed and packaged at the level of a common user who is not necessarily a hydrologist. For some models, this has already begun to occur. Second, these models will have to be integrated with social, economic and management models yielding information that is easily interpreted or understood by the user.

A majority of watershed models simulate watershed response either without consideration of water quality or inadequate consideration thereof. These models will have to be made more versatile and more comprehensive to include biological as well as chemical facets of watershed response. If these models are to be used for environmental or ecological modeling then this versatility becomes an essential attribute of the models. A decision maker wants to know as much about a watershed as he can, not just water-quantity information.

In the past few years, with revolution in information and communication technology, new vistas have unfolded that watershed models have not taken full advantage of. Examples include remote sensing and satellite technology (RSST), GIS, artificial intelligence, and risk and reliability analysis. The RSST is a powerful technology for data acquisition over extensive and even otherwise inaccessible areas. The GIS technology can be usefully employed for data management--data retrieval, manipulation, organization and the like--that may be entirely suitable for development of distributed models. Artificial intelligence allows one to make watershed models user-friendly and usable by nonhydrologists. Risk and reliability analysis allows for providing information on the accuracy of model results that the user wants to know about. It is this information that in many cases becomes crucial in selection of a model and determines the economic worth of data as well as of the model.

1.2. CLASSIFICATION OF WATERSHED MODELS

Since the development of the Stanford Watershed Model in 1966 by Crawford and Linsley, there has been a proliferation of watershed models (Renard, et al., 1982; James, et al., 1982; Singh, 1989). The models are of different types and were developed for different purposes. Nevertheless, many of the models share structural similarities, because their underlying assumptions are the same, and some of the models are distinctly different. The watershed models can be classified, as shown in Fig. 1.5, according to different criteria that may encompass (1) process description, (2) scale, and (3) technique of solution.

(a) Classification of models based on process description

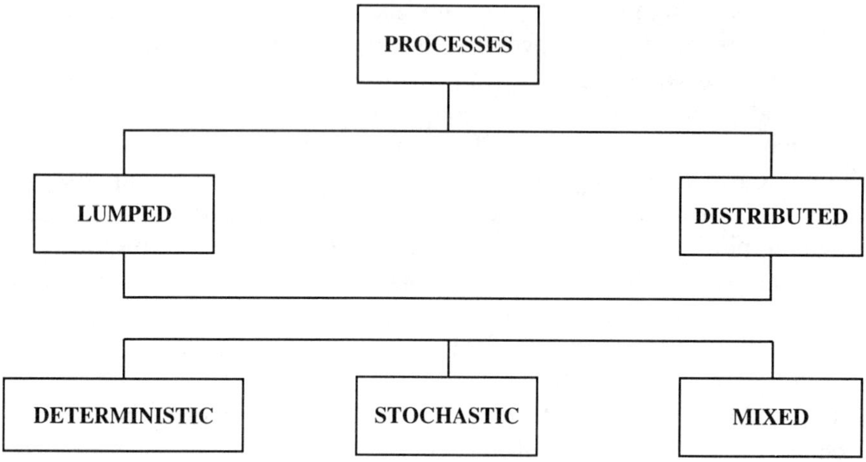

(b) Classification of models based on space and time scales

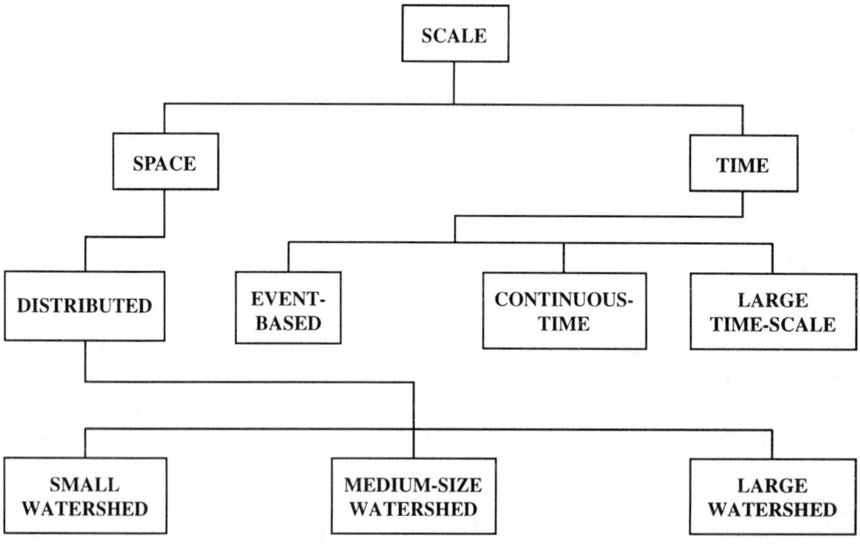

Fig. 1.5. Classification of watershed models.

(c) Classification of models based on solution technique

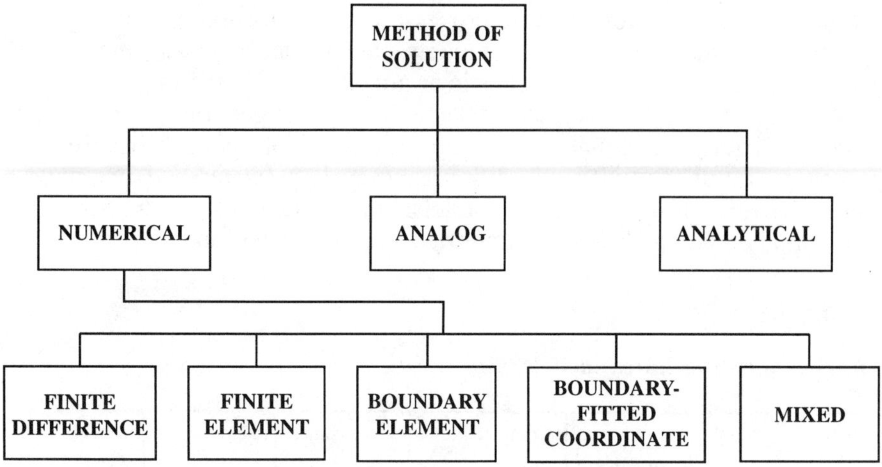

Fig. 1.5. (continued)

1.2.1. Process-Based Classification

A model, as shown in Fig. 1.6, has five components, including (1) system (watershed) geometry, (2) input, (3) governing laws, (4) initial and boundary conditions, and (5) output. Depending upon the type of the

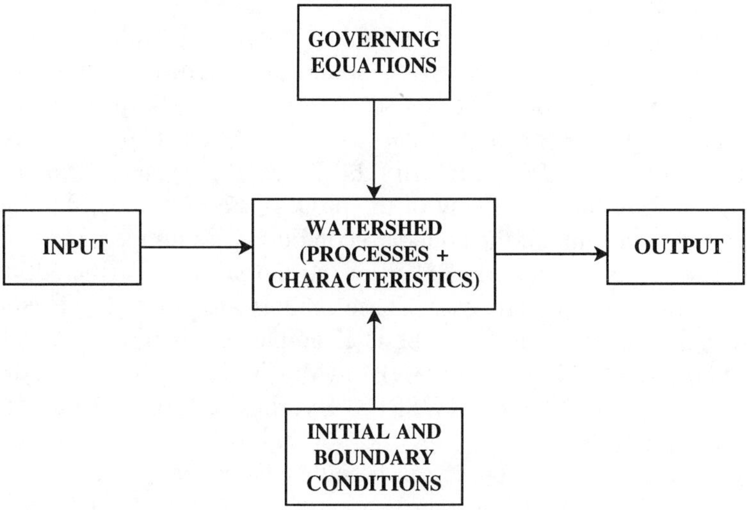

Fig. 1.6. Model components.

model, these components are variously combined. The processes include all of the hydrologic processes that contribute to the system (watershed) output. Based on the description of those processes, in conjunction with the system (watershed) characteristics, the models can be described as lumped or distributed, deterministic or stochastic or mixed.

A lumped model, as shown in Table 1.1, is, in general, expressed by ordinary differential equations taking no account of spatial variability of processes, input, boundary conditions and system (watershed) geometric characteristics. In most lumped models, some processes are described by differential equations based on simplified hydraulic laws, and

TABLE 1.1.

Lumped and distributed models.

Input	System Characteristics	Component Processes	Governing Equations	Output	Model Type
Lumped	Lumped	Lumped	ODE	Lumped	Lumped
Lumped	Lumped	Distributed	PDE	Distributed	Distributed
Distributed	Distributed	Distributed	PDE	Distributed	Distributed
Distributed	Lumped	Distributed	PDE	Distributed	Distributed

other processes are expressed by empirical algebraic equations. Examples of lumped models are HEC-1 (Hydrologic Engineering Center, 1981), HYMO (Williams and Hann, 1972), RORB (Laurenson and Mein, 1983), SSARR (U.S. Army Engineer, 1972), tank model (Sugawara, et al., 1984), amongst others.

Distributed models, as shown in Table 1.1, take an explicit account of spatial variability of processes, input, boundary conditions, and/or system (watershed) characteristics. Of course, in practice, a lack of data--field or experimental (laboratory)--prevents such a general formulation of distributed models. In a majority of cases the system (watershed) characteristics are lumped, many of the processes are lumped, the input is lumped, and even some of the boundary conditions are lumped, but some of the processes that are directly linked to the output are distributed, as for example, rainfall-runoff process. These models are not fully distributed; rather they are quasi-distributed at best. Examples of distributed models are SHE (Abbott, et al., 1986a, 1986b), IHM (Morris, 1980), SWMM (Metcalf and Eddy, Inc., et al., 1971), NWSRFS (Hydrologic Research Laboratory, 1972), amongst others.

The description of a process can be either deterministic, stochastic, or mixed. Depending upon the way the processes are described, the models can also be classified as deterministic, stochastic, or mixed (Singh, 1988). If all of the components of a model are deterministic, the watershed model is deterministic. If all of the components of the model are stochastic, the model is fully stochastic, and if only some components are stochastic then the model is quasi-stochastic. When the model components are described by a mix of deterministic and stochastic components, the model is a stochastic-deterministic or hybrid model. A vast majority of the models are deterministic, and virtually no model is fully stochastic. In some cases, only some parts of the model are described by the laws of probability, and other parts are fully deterministic. It is then fair to characterize them as quasi-deterministic or quasi-stochastic.

1.2.2. Time-Scale Based Classification

The watershed models can be classified based on the time scale of models. The time scale can be defined as a combination of two time-intervals (Diskin and Simon, 1979). One of the time intervals is used for input and internal computations. The second is the time interval used for the output and calibration of the model. Based on this description, the models can be distinguished as (a) continuous-time or event based, (b) daily, (c) monthly, and (4) yearly models. This classification is dictated by the interval of computation. If the model components are available at shorter time intervals, such as hourly interval or shorter, the model will be a short-interval model. Of course the choice of a time interval is often a function of model's intended use.

1.2.3. Space-Scale Based Classification

The spatial scale can be used as a criterion to classify models into small-watershed, medium-size watershed, and large watershed models. Of course, the definition of a small watershed is somewhat vague. Usually, the watersheds with area of 100 km^2 or less can be called small, those with area of 100 to 1,000 km^2 medium, and those with area of larger than 1,000 km^2 large. Such a classification is arbitrary, and is experimental rather than conceptual and is governed by data availability rather than physical meaning. The essential import is the concept of homogeneity and averaging of hydrological processes. For consideration of runoff generation on these watersheds, two phases can be considered: land phase and channel phase. Each phase has its own storage characteristics. Large watersheds have well-developed channel networks and channel phase, and thus, channel storage is dominant. Such watersheds are less sensitive to short duration, high intensity rainfalls. On the other hand, small watersheds have dominant land phase and overland flow, have relatively less conspicuous channel phase, and are highly sensitive to high-intensity, short-duration rainfalls.

If all three types of watersheds are spatially uniform, then they will behave hydrologically similarly. Under this condition, classification of watersheds serves no useful purpose and becomes moot. If a watershed is homogeneous with respect to its characteristics then it is called a small watershed. In reality, though, watersheds are seldom homogeneous, therefore, homogeneity is defined in an average or statistical sense. For example, for a subsurface environment if the average of logarithmic values of the hydraulic conductivity is the same, the aquifer may be considered homogeneous. Similarly, if the average system characteristics do not change, then the watershed may be considered small. For medium-size watersheds, the heterogeneity is present but it is in the medium range (± 2 standard deviation around the mean), whereas it is quite large for large watersheds. Even if a watershed is homogeneous in one characteristic, it may not necessarily be homogeneous in another characteristic. In that sense, both the soil and land use characteristics are important. Strictly speaking then no watershed may be small; all will be either large or medium.

1.2.4. Land-Use Based Classification

Based on land use, the watersheds may be classified (Singh, 1992) as (1) agricultural, (b)urban, (c) forest and range land, (d) desert, (e) mountainous, (f) coastal, (g) wetlands, and (h) mixed. In many cases, large or even medium-size watersheds have mixed land use. These watersheds behave hydrologically differently, indeed so differently that they have given rise to different branches of hydrology. For example, treatment of agricultural watersheds constitutes what is called agricultural hydrology. Similarly, there is urban hydrology for urban watersheds, coastal hydrology for coastal watersheds, forest hydrology for forest watersheds, desert hydrology for desert watersheds, wetland hydrology for marshes and wetlands, and mountain hydrology for mountainous watersheds. The soils, geology, and vegetation are different for these watersheds. As a result, the hydrologic processes are different in their evolution in these different watersheds, and hence their models are clearly different.

1.2.5. Model-Use Based Classification

Frequently, watershed models are classified on the basis of their intended use: (1) planning models, (2) management models, and (3) prediction models. A comprehensive watershed model can be employed to accomplish a considerable array of analytical tasks for planning and management of water resources. It offers the opportunity for systematic analysis of a management policy. On the other hand, it becomes an increasingly valuable predictor of river-basin response to management strategies. For planning purposes, the model defines geographical and temporal trends that may not be accurate at specific sites in the river basin. Thus, the primary use of a planning model is for analyses of river systems management strategies. The larger issues related to basin-wide trends and plans need to be analyzed to determine if there exist site interactions. Once these interactions are determined, a site-specific analysis can be performed with greater accuracy, perhaps with use of a prediction model. Probably the most powerful feature of a planning and management model is the ability it offers planners and managers to replace past management indices with improved economic performance measures of potential decisions.

1.3. SCALING AND ITS EFFECTS

Song and James (1992) reviewed five scales used in hydrologic simulation: (1) laboratory scale, (2) hillslope scale, (3) catchment scale, (4) basin scale, and (5) continental and global scale. The laboratory-scale models use hydrodynamic equations and are often one-dimensional. The hillslope-scale models combine surface and subsurface flows, may include preferential flow through soil pores, and may be two-dimensional or even three-dimensional. The catchment scale models add topography in simulation of surface runoff and geology in simulation of baseflow, and often divide larger catchments into smaller homogeneous parts. Basin-scale models employ storage and translatory routing schemes in combining catchment runoff. Continental- and global-scale models focus on atmospheric processes and their interactions with land-surface processes.

Scaling and its effects are tied to heterogeneity. At a smaller scale, heterogeneity is small and it is large at a larger scale. Clearly, it is the greatest at the continental or global scale and the smallest at the laboratory scale. This heterogeneity can be traced to three sources. First, weather, climate, topography, soils and geology are the principal sources of heterogeneities. The hydrologic response of a watershed is governed by local combinations of these factors. Second, the discontinuities or boundaries separating soil types (clay/sand), geologic formations (faults, fractures, folds, cleavages, etc), or land covers (agriculture/forest) induce heterogeneities. Hydrologic heterogeneity is further accentuated by the differences in the factors that control hydrologic processes of rainfall, interception, evaporation, infiltration, baseflow, and runoff. These factors may include physical properties such as trees, ridges in the ground surface, geometry of soil layers, and so on.

The hydrologic models use parameters to represent watersheds as a whole, whereas data on watershed characteristics are collected at a number of field locations and are difficult to transform into a collective measure. The question then arises as to the scale that is best suited for hydrologic simulation. As the spatial scale changes from a small area to a large area, the watershed response becomes less sensitive to spatial variations of input as well as system characteristics. On the other hand, greater heterogeneity suggests smaller optimal scales. This spatial extent, however, is constrained by the differences in physical, vegetative, and topographic features that can cause the pattern of hydrologic response to vary. This, then, leads to defining a scale as the size of a unit or a subwatershed within which the hydrologic response can be treated as homogeneous. This scale must not be too small to be dominated by local physical features, nor so large as to ignore significant hydrologic heterogeneity caused by spatial variabilities of watershed characteristics. The optimal scale in a watershed is determined by the collective working of multiple processes that generate hydrologic response and the availability of hydrologic data.

1.4. PERSPECTIVE ON MODEL BUILDING

Construction of a computer model of watershed hydrology may be driven by the need to solve a particular problem or the scientific pursuit of model building which may or may not be used for problem solving. Frequently, the type of a model to be built is dictated by the availability of data. Different models have different data needs. In general, distributed models require much more data than do lumped models. In most cases, needed data either do not exist or are not available in full. That is why regionalization and synthetic techniques are useful. Even if the needed data are available, problems remain with regard to incompleteness, inaccuracy, and inhomogeneity of data. Then, of course, storage, handling, retrieval, analysis, and manipulation of data have to be dealt with. If the volume of data required is large, data processing can be quite a sophisticated undertaking.

A watershed hydrology model is an assemblage of component models corresponding to different components of the hydrologic cycle. The time scale of model output (e.g., streamflow) greatly influences the type of the model or the details to be included in the model. For example, a monthly watershed model is quite different in its architecture and construction from, say, an hourly model. Thus, the elements of simulation differ with the model type. It remains an unresolved question as to the hydrologic laws operating at different time scales for different components of the hydrologic cycle. Solution of this question will greatly facilitate model construction and more clearly define the data needs.

After a model is developed, it needs calibration, verification, and an assessment of its reliability. Model calibration involves optimum parameter estimation subject to a specified error criterion. Usually the available data is split into two halves, one for calibration and the other for verification. The calibration set should encompass the full range of data variability that may be expected in verification. A range of fairly sophisticated parameter estimation techniques are available. However, it is not clear how a model's parameter changes with variation in the objective function. Furthermore, an error analysis of calibration results is to be performed. Through this analysis, it can be ascertained if the errors are due to errors in data, parameters, or model construction. Model verification and reliability of model results are essential to show if a model is any good.

1.5. REMOTE SENSING APPLICATION

A major problem in hydrology is lack of adequate data to quantitatively describe a hydrologic process accurately. To alleviate this problem, the remote sensing technology aids as follows (Schultz, 1988):

(1) It produces areal measurements in place of point measurements.

(2) All information is collected and stored at one place.

(3) It offers high resolution in space and/or time.

(4) Data are available in digital form.

(5) Data acquisition does not interfere with data observation.

(6) Data can be gathered for remote areas that are otherwise inaccessible.

(7) Once the remote sensing networks are installed, data measurement is relatively inexpensive.

The main drawback in obtaining remotely sensed data is in calibrating the electromagnetic signals in hydrologic terms.

Remote sensing and satellite data can be used for watershed modeling, and are useful, especially where data collection sites and cost-effectiveness of data collection are important on one hand, and increasing requirement for data on the other. Among the satellites with hydrologically useful data are NOAA series, TIROS N, SPOT, Landsat, GOES, GMS, and Meteosat.

Remote sensing permits obtaining data of a process from a location far away from the user or modeler. Sensing is taken to mean observation of the average value of a variable over some areal extent by examining the electromagnetic energy. Measurement techniques are either passive or active. Passive measurement techniques determine the amount of reflected sunlight or the amount of natural emissions from the target at various wave lengths. Active measurement techniques direct artificially generated signal at a target and measure the reflected signal. An example of a passive technique is air photography and that of an active technique is weather radar.

Several different remote sensing sensors provide hydrologically relevant information. These include aerial photography, scanning radiometers, spectrometers, and microwave radars. Since all information is obtained in the form of electromagnetic signals, it is important to determine the type of signal most suitable for a particular hydrologic variable. Furthermore, the needed resolution in space and time is obtained with certain types of sensors only.

Remote sensing data from satellites are useful in watershed modeling in two ways. First, satellite data can be used to better define soils and land covers over a watershed, which are needed to determine infiltration, evapotranspiration, and runoff. Second, remote sensing measures data over space rather than at a point, and can, therefore, be used to correct errors in input data based on point measurements as is with rainfall, evaporation, and the like. By far the most important data produced by remote sensing are on soil moisture and evapotranspiration through satellite thermal infrared images. These data can be employed to model water exchange between land surface and atmosphere (Engman, 1986), and to couple land-surface hydrology and atmospheric models (Otte, et al., 1985).

The remote sensing technology can be employed in a multitude of applications in hydrology. Radars are employed for rainfall measurement. Satellite data can be used for estimation of area and intensity of rainfall. Temperature estimates obtained from thermal infrared and soil moisture from microwave observations can be applied to develop mathematical models of evapotranspiration. Soil moisture is most adequately obtained through remote sensing. Inventories of water bodies such as lakes, dams, rivers, swamps, flooded areas, etc., are easily obtained by remote sensing imagery. Even groundwater resource evaluation can be assisted through the remote sensing information produced on geologic and geomorphologic characteristics. Snow and ice conditions are mapped with use of satellites. Water quality parameters such as algae, chlorophyll, and aquatic life parameters, as well as thermal pollution, oil spill, etc., can be observed by remote sensing techniques.

With the vastly improved capability to observe hydrologic data, remote sensing technology is finding its niche in a range of hydrologic applications, and these applications are becoming operational at an accelerated pace. Some of these applications are exemplified in real-time flood forecasting by a distributed model with radar rainfall measurements as input, computation of monthly runoff by a lumped model with NOAA infrared satellite data as input, model parameter estimation with multispectral Landsat satellite data, development of macroscale hydrological models coupled with atmospheric general circulation models with satellite data, and development of distributed hydrologic models based on a digital terrain model and a geographic information system with use of Landsat satellite imagery. These applications permit operationalization of space technology in such areas as weather forecasting, forecasting seasonal/short term snowmelt runoff, evolution of watershed management strategies for conservation planning, development of reporting services for drought assessment/forecasting, mapping of groundwater potential to support conjunctive use of surface and groundwaters, inventorying of coastal processes and marine process environment, environmental impact assessment of large-scale water resources projects, flood-damage assessment, and development of resource information matrix for irrigation development.

1.6. APPLICATION OF GIS

Geographic information systems (GIS) are computer-based tools that store, analyze, retrieve, manipulate and manage large amounts of spatial data. They are spatial data-base management techniques. Originally these tools were developed to ease cartography, but are being used these days for inventory analysis, estimation, planning, and modeling.

Watershed modeling requires accurate and representative data. Collection, storage, and manipulation of the required data can be cumbersome for three reasons:

(1) spatial nature of hydrologic data,

(2) large volume of data, and

(3) organization of data.

However, through the use of a geographic information system and the associated software, such data can be compiled and processed with relative use. The watershed and hydrology are inherently spatial. The physical characteristics of the watershed such as soils, land use and topography vary spatially. In distributed watershed models, these characteristics can be used directly. However, the watershed will have to be spatially characterized into homogeneous sub-areas. To that end, the traditional data file structure is less than troublefree. Clearly, a general-purpose spatial data analysis and manipulation tool will be desirable.

Distributed watershed models, especially numerical models, require large volumes of data. A hydrologic unit is characterized by many data, including those on topography, land use, soils, geology, climate, and so on. On the other hand, the various application programs may require the same data stored in different files, thus creating storage redundancy with traditional file structure system.

Hydrologic data are arranged in a variety of formats and stored on various media. It is often difficult and tedious to transform and manage these data. Furthermore, because many models have different data requirements, a collection program tailored to the demand of a particular model cannot be used for another with differing data requirements. Consequently, a separate collection program has to be developed and data transformation then has to be undertaken for each model. These data problems can be resolved through application of a GIS. Its ability to extract, overlay, and delineate watershed characteristics permits integration with watershed models. Design, calibration, modification, and comparison of these models can be significantly facilitated by use of the GIS. However, application of GIS technology to watershed modeling requires careful planning and extensive data manipulation, involving three major steps:

(1) spatial database construction,

(2) integration of spatial model layers, and

(3) the GIS and model interface.

With ever increasing volume of data available from many sources, the first step is time consuming. The second step is becoming less tedious due to advancing GIS capabilities in that a series of spatial overlaying and/or projecting procedures can be derived yielding the final modeling layer. The last step involves actual programming for interfacing GIS with watershed models.

Although large quantities of data can be efficiently manipulated in time by using a GIS, certain limitations need to be recognized. First, building the data base is expensive and time consuming. Also, appropriate data may be another problem, and making up for deficiencies in data is quite difficult with a GIS. Unless the same database can be used in a variety of analyses or models, the use of a GIS may not be cost-effective. However, with increasing complexity of investigations or models involving more data and more complex data processing, a GIS may indeed be cost effective. The PC-based GIS innovations have greatly cut down the cost and are more affordable now. A successful GIS application calls for data accuracy, data integrity (also called referential integrity), and multiple attributes capability. A GIS package should have a relational data base management system (DBMS) that provides both logic and physical data independence.

A GIS may be either raster-based or vector-based. A raster-based GIS is more storage efficient than a vector-based GIS. The larger the coverage the greater the storage space used by a vector-based GIS. The processing speed of a vector-based GIS is slower and can be an important consideration if extensive spatial processing is needed for a watershed model.

Construction of an appropriate GIS data base is the most difficult and expensive undertaking for watershed modeling. There exist large volumes of data on soils, geology, land use, vegetative cover, and so on in digitized form. However, there are problems with many of the existing digital data bases:

(1) They are in different formats,

(2) They use different coordinate (projection) systems,

(3) They are too coarse for use in 2-D or 3-D watershed models.

Integration of GIS with watershed modeling accomplishes a number of significant hydrologic functions: designing, calibrating, modifying, and comparing watershed models. The use of a GIS permits subdividing a watershed spatially into hydrologically homogeneous sub-areas. Depending upon the type of application requiring categorization of hydrologic properties many combinations of spatial overlay can be performed. With this technique, it is possible to delineate areas of varying soil loss rates (Ventura, et al., 1988), identify potential areas of non-point source agricultural pollution (Schmidt, 1987), map groundwater contamination susceptibility (Whittemore, et al., 1982), and so on. GIS improves the ability to incorporate spatial details beyond the existing capability of watershed models. With much better resolution of terrain--streams and drainage areas--the ability to delineate more appropriate grid layers for a finite element watershed model is enhanced.

1.7. EXPERT SYSTEMS IN HYDROLOGIC MODELING

A knowledge-based expert system is an interactive computer program that encodes judgment, experience, rules of thumb, intuition, empirical knowledge, and other forms of information or expertise to give knowledgeable advice on a specified problem. Such a program is knowledge based and is a domain-independent reasoning system. In other words, it can be applied to other numerical reasoning tasks in domains other than hydrology. The expertise about the problem area is specified in the form of a model of the decision-making process. A model of expertise in the form of judgment or experience is constituted by a set of statements in a stylized language. The program of reasoning reads a model or models from a disk file and creates an internal data structure that drives the consultation. Thus, changing the expertise in a knowledge-based system simply involves replacing one model by another. In this sense the structure of such systems is fundamentally different from most "traditional" programs.

Perhaps the first attempt to develop a knowledge-based computer consultation system, called HYDRO, for water resources problems was by Gaschnig, et al. (1981). They applied the HYDRO system to determine appropriate numerical values for various parameters that describe the physical characteristics of a watershed. The HYDRO system was intended to provide advice comparable to that of an expert hydrologist in selecting parameter values characteristic of the watershed under consideration. The parameter values so obtained served as input to the Hydrocomp HSPF Simulation program for evaluating various hydrological aspects of the watershed under consideration.

Although expert systems have vast potential, they have not yet found their niche in watershed hydrology. There is, however, every reason to believe that they will have commonplace application in water resources area in the not too distant future.

1.8. WHERE DO WE GO FROM HERE?

The future of watershed models will be shaped by increasing societal demand for integrated environmental management; growing need for globalization by incorporation of biological, chemical, and physical aspects of the hydrologic cycle; rapid advances in remote sensing and satellite technology, geographical information systems (GIS) database management systems (DBMS), and expert systems; enhanced role of models in planning and decision making; mounting pressure on transformation of models to user-friendly levels; and clearer statement of reliability and risk associated with model results.

Application of watershed models to environmental management will grow in the future. The models will be required to be practical tools--readily usable in planning and decision making. They will have to be interfaced with economic, social, political, administrative, and legal models. Thus, watershed models will become a component in the larger management modeling strategy. Furthermore, these models will become more global, not in the sense of spatial scale but in the sense of hydrologic details. Increasing fusion of biological and chemical courses in undergraduate curricula emphasizing hydrology is a healthy sign in that direction, and will help achieve this goal.

Watershed models will have to embrace rapid advances occurring in remote sensing and satellite technology, geographical information and data base management systems, and expert systems. With use of remote sensing and satellite technology, our ability to observe data over large and inaccessible areas is vastly improved, making it possible to develop truly distributed watershed models for both gaged and ungaged watersheds. Distributed models require large quantities of data which can be stored, retrieved, managed and manipulated with use of GIS and DBMS. This, of course is possible because of literally unlimited computing capability available these days and will be even more so in the future. If the watershed models have to become practical tools, then they have to be simple, easy to use with a clear statement as to what they can and what they cannot do, define the reliability with which they accomplish their intended functions, and require the user to posses the least of hydrologic training. Furthermore, the models will have to learn from the user as well as empirical experience. Many of these functions can be performed by use of expert systems in watershed modeling. The user is interested in what a model yields and how accurately it yields, not the biology, chemistry, physics and hydrology it is based on. The model results by themselves are useless unless they are associated with a clear statement of their reliability or accuracy.

The models will have to be described in simple terms such that interpretation of their results would not tax the ability of a user. They are designed to serve a practical end and their constituency is one of users. After all, computer models of watershed hydrology are to be used, not be confined to academic shelves. Thus, the model building will have to gravitate around the central theme of their eventual practical use in integrated environmental management. Although much progress has been made in computer modeling of watershed hydrology, there is still a long way to go before the models would be able to fully integrate rapidly occurring advances in information, computer, and space technology, and would become "household" tools.

REFERENCES

Abbott, M. B., Bathurst, J. C., Cunge, J. A., O'Connell, P. E. and Rasmussen, J., 1986a. An introduction to the European Hydrologic System - Systeme Hydrologique Europeen, "SHE" 1: history and philosophy of a physically based distributed modeling system. Journal of Hydrology, Vol. 87, pp. 45-59.

Abbott, M. B., Bathurst, J. C., Cunge, J. A. O'Connell, P. E., 1986b. An introduction to the European Hydrologic System - Systeme Hydrologique Europeen, "SHE" 2. structure of a physically based, distributed modeling system. Journal of Hydrology, Vol. 87, pp. 61-77.

Crawford, N. H. and Linsley, R. S., 1966. Digital simulation in hydrology: The Stanford Watershed Model IV. Technical Report No. 39, Department of Civil Engineering, Stanford University, Palo Alto, California.

Diskin, M. H. and Simon, E., 1979. The relationship between the time bases of simulation models and their structure. Water Resources Bulletin, Vol. 15, No. 6, pp. 1716-1732.

Engman, E. T., 1986. Hydrologic research before and after AgRISTARS. IEEE Transactions, Geoscience and Remote Sensing, Vol. GE-24, pp. 5-11.

Gaschning, J., Reboh, R. and Reiter, J., 1981. Development of a knowledge-based expert system for water resource problems. Final Report, SR1 Project 1619, SRI International, California.

Hydrologic Engineering Center, 1981. HEC-1, Flood Hydrograph Package - Users Manual. U.S. Army Corps of Engineers, Davis, California.

Hydrologic Research Laboratory, 1972. National Weather Service river forecast system: forecast procedures. Technical Memorandum NWS-HYDRO-14, National Weather Service, National Oceanic and Atmospheric Administration, U.S. Department of Commerce, Washington, D.C.

James, L. D., Bowles, D. S. and Hawkins, R. H., 1982. A taxonomy for evaluating surface water quantity model reliability. in Applied Modeling in Catchment Hydrology, edited by V. P. Singh, pp. 189-228, Water Resources Publication, Fort Collins, Colorado.

Laurenson, E. M. and Mein, R. G., 1983. RORB-version 3: Runoff routing program - user manual. 2nd edition, Department of Civil Engineering, Monash University, Monash, Australia.

Metcalf and Eddy, Inc., University of Florida and Water Resources Engineers, Inc., 1971. Storm water management model, Vol. 1, final report. Water Pollution Control Research Series 11024 DOC 07/71, U.S. Environmental Protection Agency, Washington, D.C.

Morris, E. M., 1980. Forecasting flood flows in grassy and forested basins using a deterministic distributed mathematical model. IAHS Publication 129, pp. 247-255, Wallingford, England.

Otte, C., Vidal-Madjar, D. and Girard, G., 1989. Remote sensing applications to hydrological modeling. Journal of Hydrology, Vol. 105, pp. 369-384.

Renard, K. G., Rawls, W. J. and Fogel, M. M., 1982. Currently available models. Chapter 13 in Hydrologic Modeling of Small Watersheds, ASAE No. 5, pp. 507-522, edited by C. T. Haan, H. P. Johnson and D. L. Brakensiek, American Society of Agricultural Engineers, St. Joseph, Michigan.

Schmidt, R. R., 1987. Groundwater contamination susceptibility in Wisconsin. Report 5, Wisconsin's Groundwater Management Plan, Wisconsin Department of Natural Resources, Madison.

Schultz, G. A., 1988. Remote sensing in hydrology. Journal of Hydrology, Vol. 100, pp.239-265.

Singh, V. P., 1988. Hydrologic Systems, Vol. 1. Rainfall-Runoff Modeling. Prentice-Hall, Inc., Englewood Cliffs, New Jersey.

Singh, V. P., 1989. Hydrologic Systems, Vol. 2. Watershed Modeling. Prentice Hall, Inc., Englewood Cliffs, New Jersey.

Singh, V. P., 1992. Elementary Hydrology. Prentice-Hall, Inc., Englewood Cliffs, New Jersey.

Song, Z. and James, L. D., 1992. An objective test for hydrologic scale. Water Resources Bulletin, Vol. 28, No. 5, pp. 833-844.

Sugawara, M., Watanabe, I., Ozaki, E. and Katsugama, Y., 1984. Tank model with snow component. Research Notes of the National Research Center for Disaster Prevention No. 65, Science and Technology, Ibaraki-Ken, Japan.

U.S. Army Engineer, 1972. Program description and user manual for SSARR model-streamflow synthesis and reservoir regulation model. Corps of Engineers, North Pacific Division, Department of the Army, Portland, Oregon.

Ventura, S. J., Chrisman, N. R., Conners, K., Gurda, R. F. and Martin, R. W., 1988. A land information system for erosion control planning. Journal of Soil and Water Conservation, Vol. 43, No. 3, pp. 230-233.

Whittenmore, D. O., Merchant, J. W., Whistler, J., McElwee, C. D. and Woods, J. J., 1987. Groundwater protection planning using the ERDAS GIS: Automation of DRASTIC and time-related capture zone. Proceedings, National Water and Well Association Focus Conference on Midwest Ground Water Issues, Dublin, Ohio.

Williams, J. R. and Hann, R. W., 1972. HYMO - a problem-oriented computer language for building hydrologic models. Water Research Laboratory, Vol. 8, No. 1, pp. 79-86.

Chapter 2

MODEL CALIBRATION

S. Sorooshian and V. K. Gupta

2.1. INTRODUCTION TO MODEL CALIBRATION

The reliance on models as hydrologic tools is increasing as hydrologists examine emerging problems and exploit new data sources. Models enable us to study very complex problems and to synthesize different kinds of information. However, model results are only as reliable as the model assumptions, inputs, and parameter estimates. Therefore, we face two problems. The first is to select a suitable model to represent the study site (watershed). The second is to then select values for the model parameters so that the model closely simulates the behavior of the study site. The process by which the parameters are selected is called model "calibration".

In this chapter, we discuss the process of model calibration from the perspective of the research that we and other researchers have conducted over the past decade and a half. Our emphasis here is directed towards the calibration of "conceptual" hydrologic models of streamflow. For a discussion of calibration of water quality models, please refer to the excellent review paper by Beck [1987]. Other relevant discussions can be found in Sorooshian [1983] and Sorooshian [1985].

2.1.1. Model Parameters

Many hydrologic models are based on conceptual representations of the physical processes that govern the flow of water through and over the soil. For example, the United States National Weather Service maintains and operates a system of such models of varying complexity, called the NWSRFS (National Weather Service River Forecast System). One of the more detailed of these models is called the SAC-SMA model (Sacramento Soil Moisture Accounting) and is used for riverflow forecasting; a representation of this model appears in Fig. 2.1. Such models typically have two types of parameters: "physical" parameters and "process" parameters.

(a) **Physical parameters:** Physical parameters represent physically measurable properties of the watershed. Examples are: the area of the watershed, the fraction of the watershed area that is impervious, the surface area of the streams and open water bodies, surface slopes, and so on.

(b) **Process parameters:** Process parameters represent watershed properties that are not directly measurable. Examples include: the average or "effective" depth of surface soil moisture storage, the effective lateral interflow rate, the coefficient of nonlinearity controlling rate of percolation to the groundwater storage, and so on.

2.1.2. Model Calibration

To calibrate a model, we must select values for the model parameters so that the model closely simulates the behavior of the study site. The process by which this is accomplished is called model "calibration". There are two parts to this process: parameter specification and parameter estimation.

(a) **Parameter specification:** Here, we use prior knowledge about the watershed properties and behavior to specify initial estimates for the parameters of the model. For "physical" parameters, estimates are made using measurements obtained from maps in the field. The parameters are then typically fixed at these measured values and not adjusted further unless determined to be in error. For "process parameters", estimates of the range (minimum and maximum values) of possible values for these parameters are determined based on judgement and understanding of the hydrology of the watershed. This uncertainty in the parameter estimates is then reduced by the process of parameter estimation described below.

(b) **Parameter estimation:** Here, we use various techniques designed to reduce the uncertainty in the estimates of the process parameters. A typical approach is to first select an initial estimate for the parameters, somewhere inside the ranges previously specified. The parameter values are then adjusted to more closely match the model behavior to that of the watershed. The process of adjustment can be done "manually" or using computer-based "automatic" methods, as discussed below.

2.2. MANUAL CALIBRATION

To calibrate a model, we must select some aspect of watershed behavior to which the model is to be matched; typically, we might select the streamflow hydrograph at one or more locations on the river. We then adjust the model

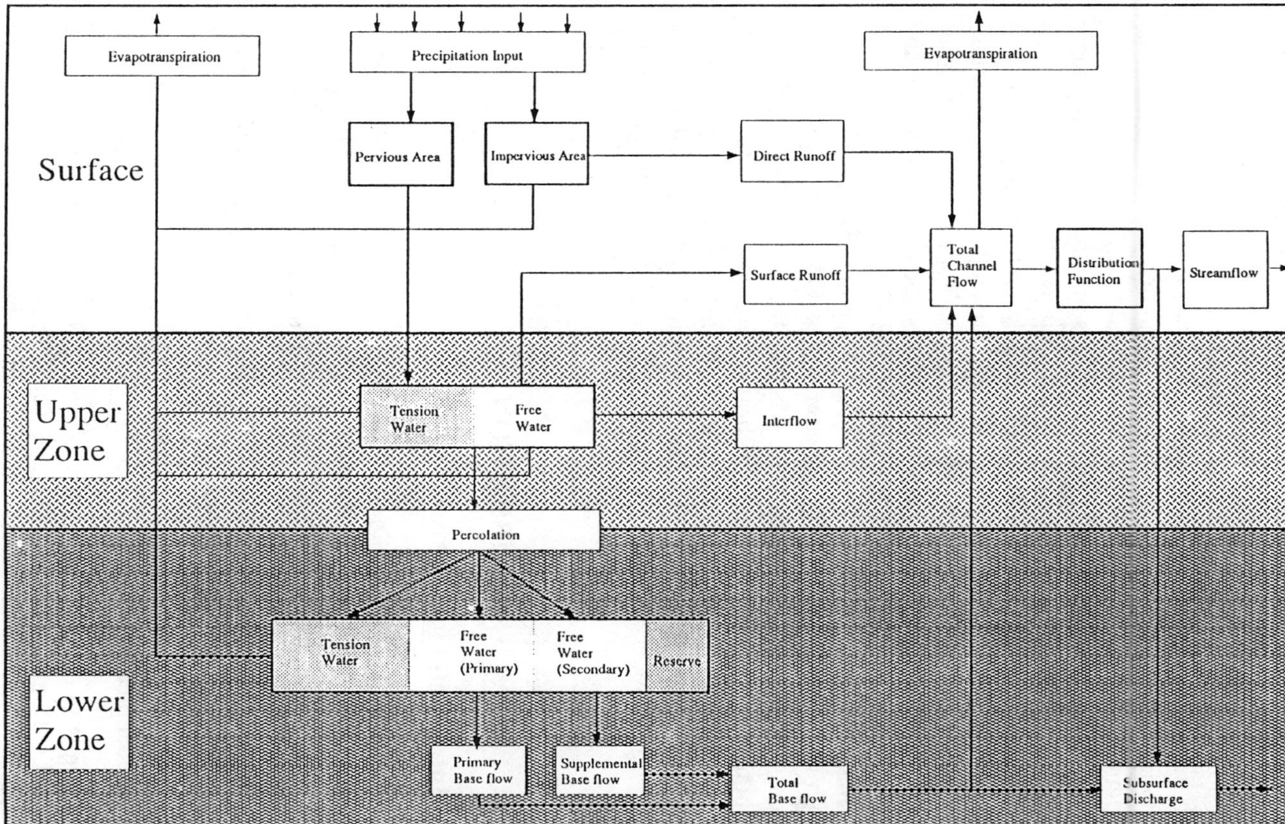

Fig. 2.1. Schematic diagram of the SMA-NWSRFS watershed model.

parameters to get the simulated streamflow hydrograph to resemble the observed hydrograph for some historical data period. In manual calibration, we use a trial-and-error process of parameter adjustments; after each parameter adjustment is made, the simulated and observed hydrographs are visually compared to see if the match has improved.

With training and a good deal of experience, it is possible to obtain very good model calibrations using the manual approach. However, for the inexperienced and untrained person, manual calibration can be a rather frustrating and time-consuming exercise. This is mainly because the logic by which the parameters should be adjusted to improve the match is difficult to determine (due to the compensating effects which the model parameters usually have on the model output). Recent developments in computer graphics have made the process of manual calibration somewhat less tedious by enabling the effects of a parameter adjustment to be rapidly observed and compared to previous parameter trials [Brazil, 1988].

The main weakness of manual calibration is that the absence of generally accepted objective measures of comparison makes it difficult to know when the process should be terminated--i.e., whether the "best" possible fit has been obtained. Because manual calibration involves a great deal of subjective judgement, different persons may obtain very different parameter values for the same watershed. This makes it difficult to explicitly assign measures of confidence to the calibrated model and to its simulations and predictions.

2.3. AUTOMATIC CALIBRATION

2.3.1. Introduction

The development of computer-based methods for "automatic" calibration of watershed models has been motivated by:

(a) the need to speed up the process of calibration;

(b) the fact that there are few model calibration experts available for each watershed model; and

(c) the need to assign some measure of objectivity and confidence to model predictions.

Early attempts to develop automatic calibration methods were reported by Dawdy and O'Donnell [1965], Nash and Sutcliffe [1970], and Ibbitt [1970], among others. These researchers set the stage by bringing the vast body of research on statistical regression and model fitting techniques to bear on the calibration problem.

Since these beginnings, a great deal of progress has been made. However, it is important to clearly state that *automatic calibration methods have not yet matured to the point that they can entirely replace manual methods.* Although quick to provide "solutions", automatic methods still require user expertise and are typically used in conjunction with a manual procedure.

A typical automatic parameter estimation procedure consists of four major elements: (1) objective function, (2) optimization algorithm, (3) termination criteria, and (4) calibration data, as discussed below. In addition, processes of verification (Section 2.4) and sensitivity analysis (Section 2.5) are necessary to establish confidence in the results.

2.3.2. Objective Function(s)

An objective function is an equation that is used to compute a numerical measure of the difference between the model-simulated output (typically the streamflow hydrograph) and the observed (measured) watershed output. The purpose of automatic model calibration is, therefore: "*to find those values of the model parameters that optimize (minimize or maximize, as appropriate) the numerical value of the objective function*".

2.3.2a Least Squares Methods

Drawing from statistical regression and model-fitting theory, the most commonly used objective function has been some form of the Weighted Least Squares (WLS) function:

$$F(\theta) = \sum_{t=1}^{n} w_t \bullet \left[q_t^{obs} - q_t(\theta) \right]^2 \qquad (2.1)$$

where:

q_t^{obs} = observed (measured) streamflow value at time t;

$q_t(\theta)$ = model simulated streamflow value at time t;

θ = vector of model parameters;

w_t = weight at time t; and

n = the number of data points to be matched.

The weights w_t indicate the importance to be given to fitting a particular hydrograph value. If the weights are all set equal to 1.0, the WLS function reduces to the familiar Simple Least Squares (SLS) function. Notice that the minimum value of the objective function $F(\theta)$ that can be attained is 0.0 (zero) if the model is able to perfectly reproduce the observed streamflow hydrograph. In general, however, a zero value is not attainable, and the purpose of automatic calibration is to find the value for θ which minimizes the value of the function.

In certain cases, additional sources of information regarding watershed behavior may be available and can be used in conjunction with the usual streamflow hydrograph data to improve model calibration. Kuczera [1982] investigated this possibility and suggested that use of data sources such as soil moisture and groundwater level data can improve the consistency and stability of the parameter estimates. This involves the use of a "generalized" least squares procedure which is described in detail in his paper.

2.3.2b Maximum Likelihood Methods

An alternative to least squares is the Maximum Likelihood approach. Diskin and Simon [1977] and Sorooshian and Dracup [1980] suggested that the choices of WLS, SLS, or other forms of objective function reported in the literature are typically based on subjective criteria. Similarly, Bard [1974] pointed out that if the objective function lacks a proper stochastic description of the data and the physical situation, the estimated parameters may not be close to their best "effective" values. The result is a phenomenon called "model divergence", in which the error variance of forecasted flows is much larger than the error variance of the flows simulated during calibration [Sage and Melsa, 1971].

According to Maximum Likelihood theory, use of the SLS function is equivalent to making the following two assumptions concerning the probability distribution of the errors:

(1) The joint probability of the errors over the available data record is Gaussian with mean zero, and

(2) The autocovariance matrix of the errors, denoted by Ω, has the form $\Omega = I.\sigma^2$, where σ^2 is the constant variance of the errors and I is the identity matrix. This is equivalent to stating that the errors are independent of each other.

Clarke [1973] emphasized that, because hydrologic data may often violate these assumptions, the effects can be quite serious. Therefore, Sorooshian and Dracup [1980] developed Maximum Likelihood based objective functions to properly account for the presence of either autocorrelation (nonindependence) or heteroscedasticity (changing variance) of the streamflow data errors. In many cases, the Maximum Likelihood criteria have proved to be superior to the SLS approach; this has been demonstrated by Sorooshian et al. [1982, 1983], Lemmer and Rao [1983], and Ibbitt and Hutchinson [1984], among others. A followup study by Sorooshian and Gupta [1983] revealed that among the reasons for the success of the ML objective functions were the superior convergence properties of the function response surfaces.

The most successful form of the Maximum Likelihood criteria has been one called HMLE (Heteroscedastic Maximum Likelihood Estimator) that accounts for nonstationary variance in the streamflow measurement errors [Sorooshian, 1978, 1981; Sorooshian and Dracup, 1980]. The HMLE estimator is the maximum likelihood, minimum variance, asymptotically unbiased estimator when the errors in the output data are Gaussian, zero mean, uncorrelated, and have variance related to the magnitude of the output (streamflows). Nonstationary variance errors are believed to be common in streamflow data. The HMLE is related to the SLS: it reduces to the SLS function in the special case that the streamflow measurement errors have equal variance.

The HMLE estimator has the form:

$$\min_{\theta,\lambda} HMLE = \frac{\frac{1}{n}\sum_{t=1}^{n} w_t \varepsilon_t^2}{\left[\prod_{t=1}^{n} w_t\right]^{1/n}} \tag{2.2}$$

where $\varepsilon_t = q_{t,obs} - q_t(\theta)$ is the model residual at time t; $q_{t,obs}$ and $q_t(\theta)$ are observed and simulated flows, respectively; w_t is the weight assigned to time t, computed as:

$$w_t = f_t^{2(\lambda - 1)} \tag{2.3}$$

where $f_t = q_{t,true}$ is the expected true flow at time t, n is the number of data points, and λ is the unknown transformation parameter which stabilizes the variance. The expected flow $q_{t,true}$ is approximated using either $q_{t,obs}$ or $q_t(\theta)$ [Sorooshian et al., 1983]. Fulton [1982] showed that the estimator can become unstable when $q_t(\theta)$ is used to approximate f_t, and recommends using $q_{t,obs}$. While this is the recommended procedure at present, it should be noted that use of $q_{t,obs}$ will cause some degree of bias in the estimate of λ [Gupta, 1984].

The HMLE is solved in two stages. First, given a set of model parameters, the residuals of the model are obtained. Next, an estimate of λ must be obtained; Sorooshian [1978] showed that this could be done by solving the implicit expression (4) to obtain an estimate of the parameter λ using an iterative numerical procedure:

$$\left[\sum_{t=1}^{n} \ln(f_t) \bullet \sum_{t=1}^{n} w_t \varepsilon_t^2\right] - n \bullet \left[\sum_{t=1}^{n} w_t \ln(f_t) \bullet \varepsilon_t^2\right] = 0 \tag{2.4}$$

The value of λ is substituted into eqs. (2.2) and (2.3) to compute the value of the HMLE function.

Duan [1991] devised an equivalent, but more stable procedure, for estimating , by rearranging eq. (2.4), as follows:

$$R = \frac{R_d}{R_h} - 1 = 0 \tag{2.5}$$

$$R_d = \sum_{t=1}^{n} w_t \varepsilon_t^2 \tag{2.6}$$

where:

$$R_h = \sum_{t=1}^{n} w_t \varepsilon_t^2 a_t \tag{2.7}$$

$$a_t = \frac{\ln f_t}{a_d} \tag{2.8}$$

$$a_d = \frac{1}{n} \sum_{t=1}^{n} \ln f_t \tag{2.9}$$

With this arrangement of terms, the HMLE function value can be computed as:

$$HMLE = \frac{\frac{1}{n} R_d}{\exp\left[\, 2(\lambda - 1)\, a_d \right]} \tag{2.10}$$

The revised procedure for estimating λ and computing HMLE, given $q_{t,obs}$ and $q_t(\phi)$, is as follows:

(1) Select $f_t = q_{t,obs}$ or $q_t(\phi)$ or $(\alpha\, q_{t,obs} + \beta\, q_t(\phi))$, where $\alpha + \beta = 1$, $\alpha,\ \beta \geq 0$, $t = 1,2,...,n$;

(2) Compute a_d, using eq. (2.9), and a_t (for t=1,2,...,n) using eq. (2.8).

(3) Use an iterative procedure (e.g., Golden Section, or Fibonacci Method) to estimate λ such that $R = 0$ in eq. (2.3). If the procedure requires an initial value, use $\lambda = 1$.

(4) Compute HMLE using eq. (2.10).

2.3.3. Optimization Algorithms

2.3.3a Introduction

The surface described by the objective function in the parameter space is called a "response surface". An optimization algorithm is a logical procedure that is used to search the response surface, constrained to the allowable ranges on the parameters, for the parameter values that optimize (maximize or minimize, as appropriate) the numerical value of the objective function. The procedure is typically implemented on a digital computer to enable a very rapid search. In the following discussion, we will assume that the objective function must be minimized.

Many different optimization strategies and computer codes are available (refer to the several excellent textbooks on optimization theory and methods). Virtually all these methods belong to a class of mathematical procedures called "*improvement strategies*". Such strategies begin with a guess of the solution, and then iteratively try to improve that guess by following a set of "*guessing rules*" (a sequence of educated guesses on which location to try next to improve the current solution) that make up the strategy. The most commonly used strategies are called Local Search methods.

2.3.3b Local Search Methods

Optimization strategies can be categorized as "local search" methods and "global search" methods. Local search methods are designed to efficiently find the minimum of unimodal functions--functions for which any strategy that seeks to continuously proceed downhill (a direction of improving function value) must eventually arrive at the location of the function minimum, *irrespective of where in the parameter space the search procedure is started*. To understand this process, it is helpful to imagine that you are a blind person standing on the side of a mountain and must find a way from your present location to the lowest point of the valley. You would have to:

(a) select a direction in which to move;

(b) move some distance in that direction; and then

(c) repeat the procedure over and over again until you are satisfied that no further improvement can be achieved.

Therefore, the procedure involves three main decisions: (a) which direction to move, (b) how far to move in that direction, and (c) how to decide that no further improvement is possible. Different local search strategies differ in the methods by which these decisions are made. Based on such differences, local search methods can be further classified as "*direct search*" methods and "*gradient*" methods. We discuss these methods in more detail below.

Direct search methods

A direct search optimization strategy uses only function value information in the decision process described above. A typical method is the following. Starting at the initial point, the strategy selects some direction and step size and evaluates the function at the new point. It may do this for more than one new point. Then, based on the differences in function values between the initial and new point(s), a prediction is made of which is the best direction to move to improve the function and how large a step should be taken in that direction. Each direct search strategy has its own "idea" about how this should be done in order to quickly solve the problem. A step is taken in the trial direction, and if the new point has a lower function value than the previous point, it replaces the old one, and the procedure is repeated. If the new point turns out to be worse than the previous point, the step size is reduced and another try is made at a new location. The search stops when the strategy is unable to find a direction in which improvement is possible.

Many direct search strategies have been tested for their effectiveness and efficiency in calibrating hydrologic models. Among the more popular methods have been the "Rosenbrock method" [Rosenbrock, 1960], the "Pattern Search method" [Hooke and Jeeves, 1961], and the "Simplex method" [Nelder and Mead, 1965] [see, for example, Dawdy and O'Donnell, 1965; Nash and Sutcliffe, 1970; Chapman, 1970; Ibbitt, 1970; Monro, 1971; Johnston and Pilgrim, 1976; Pickup, 1977; Sorooshian, 1978; Arfi, 1980;

Sorooshian et al., 1982, 1983]. Please note that the latter method is a nonlinear optimization approach and should not be confused with the Simplex method of linear programming.

Of the various direct search methods, the authors have a preference for the Simplex method because, in additional to its general efficiency, it has some characteristics that enable it to be somewhat more "globally" effective than the other strategies. The Simplex search strategy is presented below and is illustrated in Fig. 2.2.

(0) Select $n+1$ points (n = dimension) in the feasible parameter space and compute the function value at each point. This set of $n+1$ points is called a "simplex" Fig. 2.2a.

(1) Identify the point with the worst (largest) function value.

(2) Compute the centroid of the best n points of the simplex (i.e., exclude the worst point).

(3) Locate a new point by reflecting the worst point through the centroid (see Fig. 2.2a). If the function value at the reflection point is better than the worst point, go to Step 4; otherwise, go to Step 5.

(4) Locate a new point by expanding the reflection step by a factor of 2 (see Fig. 2.2b). If the function value at the expansion point is better than the reflection point, replace the worst point by the expansion point. If not, replace the worst point by the reflection point. Go to Step 7.

(5) Locate a new (contraction) point halfway between the worst point and the centroid (see Fig. 2.2c). If the function value at the contraction point is better than the worst point, replace the worst point by the contraction point, and Go to step 7. If not, go to Step 6.

(6) Shrink the simplex by moving each point (except the best point) to a location halfway between its current location and the best point (Fig. 2.2d).

(7) Repeat Steps 1-6 until the size of the simplex becomes smaller than some convergence criterion. The point with the best function value is selected as an estimate of the optimum.

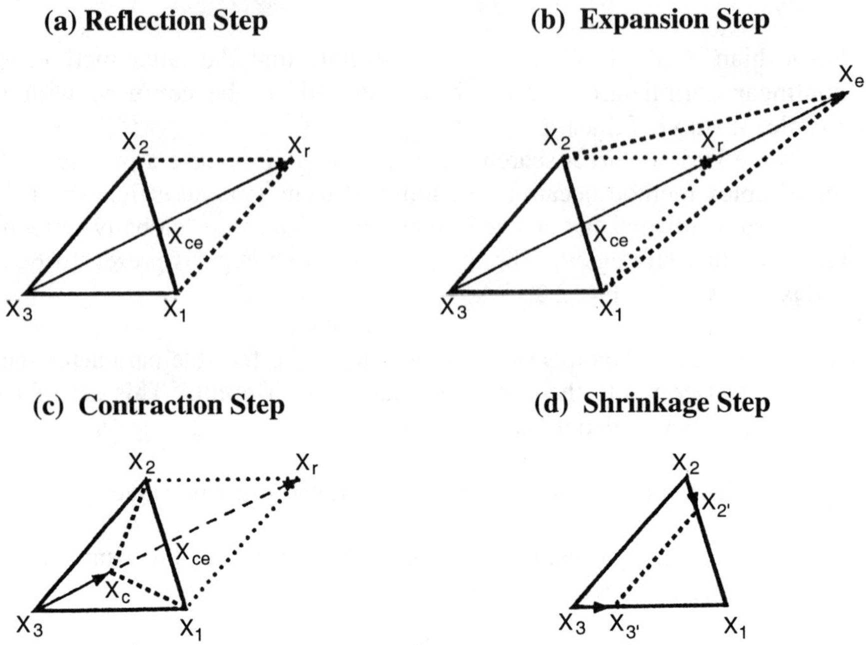

X_1 = the point with the lowest function value
X_2 = the point with the second lowest function value
X_3 = the point with the highest function value
X_{ce} = the centroid of the two best points X_1 and X_2
X_r = the reflection point
X_e = the expansion point
X_c = the contraction point
$X_{1'}$ and $X_{2'}$ = the shrinked points corresponding to X_1 and X_2

Fig. 2.2. The Simplex direct search algorithm.

Gradient methods

A gradient search optimization strategy uses information about both the function value and the function *gradient* in the decision processes listed above. Most gradient methods are based on the following equation:

$$\Theta_{I+1} = \Theta_I - \rho \,.\, A \,.\, \nabla\Theta_I \tag{2.11}$$

where Θ_I is the present (initial) point (parameter vector), $\nabla\Theta_I$ is the function gradient matrix at the present point, ρ is a step size parameter, A is a specially chosen square matrix, and Θ_{I+1} is the new point. It can be mathematically shown that, if the matrix A is any positive definite matrix, the vector in the direction from Θ_I to Θ_{I+1} will be a local improvement direction. Each gradient method uses different methods for selecting ρ and A. For example, the "*steepest descent*" approach uses the identity matrix (all diagonal values are one and off-diagonal values are zero); however, this method has been shown to be inefficient. In general, most gradient methods use some approximation to the "*Hessian*" matrix (matrix of second partial derivatives of the function with respect to the parameters). If the exact Hessian matrix were to be used, the optimization method would be the well-known "*Newton method*" which is the fastest way to solve a quadratic problem. The Newton method and several approximation methods can be found in books on nonlinear programming/optimization theory.

As with the direct search methods, if the new point has a lower function value than the previous point, the new point replaces the old one, and the procedure is repeated. If the new point turns out to be worse than the previous point, then the step size is reduced and another try is made at a new location. The search stops when the strategy is unable to find a direction in which improvement is possible. At this point, the gradient value will be numerically very close to zero.

Few gradient search strategies have been tested for their effectiveness and efficiency in calibrating hydrologic models. Although Gupta and Sorooshian [1985b] demonstrated a straightforward procedure for the explicit computations of the gradients, the programming effort involved in analytically computing the derivatives for many nonlinear hydrologic models with threshold structures is significant. Hendrickson et al. [1988] found that equally satisfactory estimates can be obtained by a simple finite difference method. However, their comparative tests of direct search and gradient optimization methods found that the gradient methods tested were not superior in their ability to find the minimum of the function for hydrologic model calibration problems.

2.3.3c The Limitations of Local Search Methods

The reason that local search methods have not generally given satisfactory results is that most practical problems involving calibration of nonlinear hydrologic models have response surfaces that are multi-modal--that is, there are several locations of the parameter space where the value of the function is a "local minimum". In such cases, the point where a local search algorithm terminates *will depend on the location where it is started.* As a result, it is impossible to know with certainty if the procedure has located the actual "global" minimum of the function.

The literature on calibration of hydrologic models indicates that researchers have rarely, if ever, been able to confidently state that the best (global) parameter set for their problem has been found. Typically, several different starting points were tried, and each search attempt terminated in a different location. Duan et al. [1992] recently discovered that conceptual rainfall-runoff models of the Sacramento model type have hundreds, if not thousands, of locally optimal solution points on their function response surfaces (see Section 2.4). If this finding is indeed true for other hydrologic models as well, then local search optimization strategies are inappropriate for calibration of such models. Instead, strategies designed for "global" search of the parameter space must be employed.

2.3.3d Global Search Methods

Global search strategies are designed to efficiently discover the minimum of multi-modal functions. Such strategies fall into three categories: *deterministic*, *stochastic*, or a *combination* of deterministic and stochastic. Deterministic strategies require that certain criteria related to the continuity of the function and its derivatives be satisfied to guarantee convergence to the global solution. These conditions are usually not met in the case of the calibration of nonlinear hydrologic models. To our knowledge, only stochastic and combination methods have been applied to the calibration of hydrologic models. For example, Ibbitt [1970] reported that a Random Search Method [Karnopp, 1963] was unable to obtain good estimates of the global optimum of the Stanford Watershed Model model, but could (in some cases) provide parameter values which were good starting points for a subsequent local search. Brazil and Krajewski [1987] investigated the use of Uniform Random Search and Adaptive Random Search (ARS) [Masri et al., 1978; Pronzato et al., 1984] for fine-tuning of the parameters of the SAC-SMA model (as part of a three-stage calibration process) and concluded that the ARS method was "an attractive alternative to nonrandom search techniques" [see also Brazil, 1988]. Wang [1991] reported that the genetic algorithm [Holland, 1975] with further tuning by a standard local search method was able to provide an "efficient and robust means for calibration of the Xinanjiang model".

Random search methods

Random (*stochastic*) search methods use the random number generators built into modern digital computers to randomly sample the parameter space in search of points with improved function values. The samples are generated according to some probability distribution applied to the feasible parameter space. In "pure" random search, the sampling is done using a uniform distribution. This assumes no prior knowledge of where in the feasible space the best parameter set exists. However, because pure random search does not make use of the function value information obtained during the search to guide the search, it is not very efficient. To understand why, assume that the volume of the total feasible parameter space is V and that the volume of the parameter space for which the function value is less than some amount f is V_f. Then the probability of successfully sampling a point that lies in the region represented by V_f is $P_f = V_f / V$. It is clear that because V_f becomes progressively smaller for smaller function values, the probability of success P_f, of getting a better function value with repeated sampling, becomes progressively smaller as the search progresses.

Because pure random search is not efficient, especially for problems with many parameters, other random search methods have been developed. These methods adaptively adjust the probability distribution used for sampling based on the function value information obtained during the search. The intention is to increase the probability of searching areas where lower function values have been found in the previous iterations. One such strategy is the Adaptive Random Search (ARS) proposed by Masri et al. [1978] and modified by Pronzato et al. [1984]. The ARS was used by Brazil and Krajewski [1987] as part of a procedure for calibrating the National Weather Services SMA-NWSRFS model. The ARS strategy is as follows:

(0) Choose a focal point (for example, this can be the "best" point obtained in the preliminary process of defining the parameter space, or it can be some arbitrary point such as the centroid of the feasible space);

(1) Generate a set of N points randomly distributed in the entire feasible space (for example, according to a uniform or normal distribution) and centered on the focal point. Store the location of the point with the best function value;

(2) Repeat Step (1) a pre-specified number of times, on the ith time using the initial parameter range divided by 10^i and centered on the focal point (Fig. 2.3) to restrict the search space. Each time, store the location of the point with the best function value;

(3) Compare all the stored points and determine the point with the best function value. Re-define this point to be the new focal point. Record in which range level this point was found; and

(4) Repeat Steps (1-3) until the best point is found in the smallest range level a user-specified successive number of times (say three). This point is chosen as the optimal parameter set.

Reports in the optimization literature indicate that the ARS strategy works well in practice. However, our research [Duan et al., 1992] found that the algorithm was neither effective nor efficient on a simple hydrologic model calibration problem. The best result we were able to obtain was a 30% success rate (30 out of 100 trials located the known region of the global optimum) after sampling the space 25,000 times.

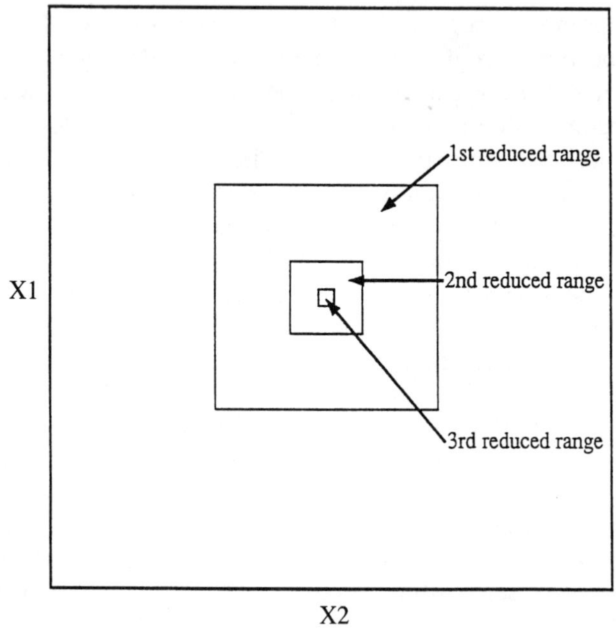

Fig. 2.3. The Adaptive Random Search algorithm.

Multi-start algorithms

A simple *combination* method for dealing with multiple optima has been suggested in the hydrologic literature [see e.g., Johnston and Pilgrim, 1976]. In this method, we run several trials of a local search (*deterministic*) optimization method from *randomly* selected starting points in the feasible space. The validity of this "multi-start" approach can be demonstrated by the following arguments. To have confidence in the results of any stochastic optimization procedure, we require that it have a relatively small failure probability on the problem of interest. Let us suppose that the failure probability of a local search optimization method is F (i.e., if we were to run 100 independent randomly initiated tests of the method, we would find that 100 x F of them would fail to locate the global optimum). Then, if we run the procedure r times from r independent randomly selected locations, the overall failure probability will decrease according to the equation $F(r)=F(1)^r$ and tend to zero as r becomes large. For example, if F is 0.65 (65 failures out of 100), then r equal to 12 will give a failure rate of less than 1 in 100.

The efficiency of any multi-start procedure varies nonlinearly with F, so that the number of restarts r required to achieve an overall failure probability of $F(r)$ is given by $r = \ln(F(r))/\ln(F(1))$. The curve of r versus F(1) is plotted in Fig. 2.4 for the cases of $F(r)$ equal to .01 (1 failure in 100) and .05 (5 failures in 100). Clearly, for single-start failure probabilities F(1) that are less than 0.8 (80 failures in 100), we do not require a very large number of restarts. However, as F(1) increases above 0.8 towards 1.0, the number of restarts required rapidly increases towards infinity, making the procedure impractical.

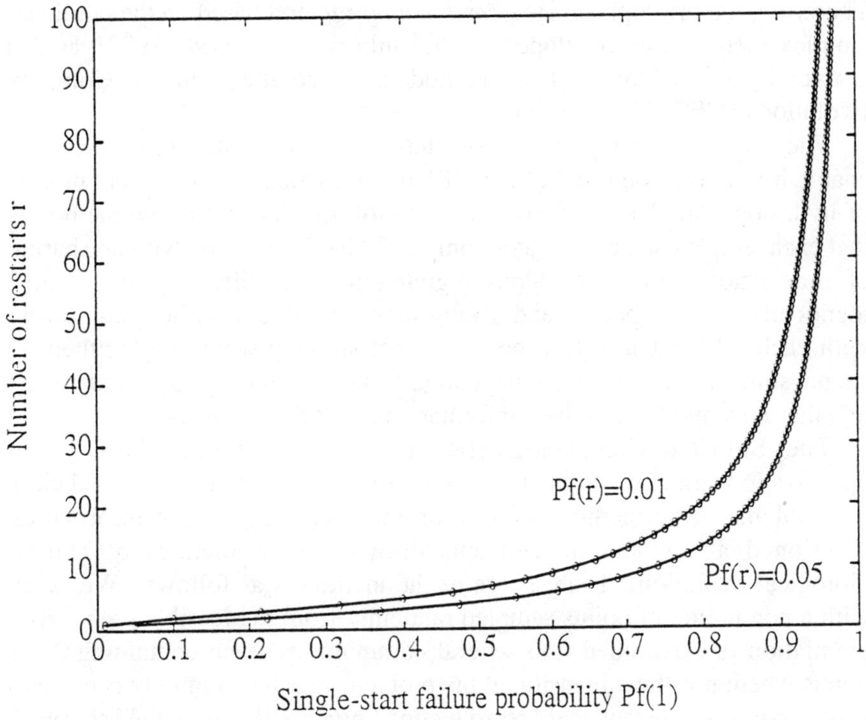

Fig. 2.4. Theoretical performance of any multi-start algorithm.

In Duan et al. [1992], it was demonstrated that a multi-start procedure based on the nonlinear simplex method (described earlier) worked well on a simple hydrologic watershed model. Whereas the ARS method was only able to achieve a 30% success rate with 25,000 function evaluations, the ordinary simplex procedure was found to have a single-start failure probability of approximately 65%, and the multi-start simplex (MSX) procedure achieved a failure rate of 1 in 100 (99% success) with only 12 restarts and approximately 10,500 function evaluations. Clearly, the MSX procedure is superior to the ARS method for such problems.

Shuffled complex algorithms

Although the MSX procedure performed well for the simple watershed model, it performed poorly in tests on the more complex National Weather Services SMA-NWSRFS flood forecast system model. We found the single-start failure probability of calibrating the 13 optimizable parameters of the model to be larger than 99% (all of the 100 independent trials of the algorithm failed to locate the global optimum) [Sorooshian et al., 1993]. Therefore, a more sophisticated *combination* method based on the nonlinear simplex method was developed by the authors [Duan et al., 1993], at The University of Arizona. This method is called the Shuffled Complex Evolution (SCE-UA) Algorithm.

The multi-start simplex strategy has certain desirable properties that enable it to overcome some of the difficulties encountered in the calibration of hydrologic models. However, a source of inefficiency in the method is that each simplex search operates completely independently, with no sharing of information. This is analogous to giving the same difficult problem to 12 identically capable people and asking them to solve it without conferring with each other. Clearly, a more efficient strategy would be for them to spend some time working independently or in small groups and to get together now and then to share information about their progress.

The Shuffled Complex Evolution (SCE-UA) method is a global optimization strategy designed to be effective and efficient for a broad class of problems. The method is based on the notion of sharing information mentioned above and on concepts drawn from principles of natural biological evolution. The essence of the method is as follows. We begin with a population of points sampled randomly from the feasible space. The population is partitioned into several communities, each containing 2n+1 points where n is the dimension of the problem. Each community is made to evolve based on a statistical "reproduction" process that uses the "simplex" geometric shape to direct the search in an improvement direction. At periodic stages in the evolution, the entire population is shuffled and points are reassigned to communities to ensure information sharing. As the search progresses, the entire population tends to converge towards the neighborhood of the global optimum, provided the initial population size is sufficiently large.

The SCE-UA strategy combines the strengths of the simplex procedure of Nelder and Mead [1965] with the concepts of controlled random search [Price, 1987], competitive evolution [Holland, 1975] and the newly developed concept of complex shuffling. The SCE-UA strategy is presented below and is illustrated in Fig. 2.5:

(0) **Initialize:** Select $p \geq 1$ and $m \geq n+1$, where p = number of complexes, m = number of points in each complex, and n = dimension of the problem. Compute the sample size $s = p \times m$.

(1) **Generate sample:** Sample s points $x_1,...,x_s$ in the feasible space Ω $\mathbf{R}^n$. Compute the function value f_i at each point x_i. In the absence of prior information, use a uniform sampling distribution.

(2) **Rank points:** Sort the s points in order of increasing function value. Store them in an array $D = \{x_i, f_i, i = 1,...,s\}$, so that $i = 1$ represents the point with the smallest function value.

(3) **Partition into complexes:** Partition D into p complexes $A^1,...,A^p$, each containing m points, such that $A^k = \{x_j^k, f_j^k \mid x_j^k = x_{k+p(j-1)}, f_j^k = f_{k+p(j-1)}, j = 1,...,m\}$.

(4) **Evolve each complex:** Evolve each complex A^k, $k = 1,...,p$, according to the Competitive Complex Evolution (CCE) algorithm outlined separately.

(5) **Shuffle complexes:** Replace $A^1,...,A^p$ into D, such that $D = \{A^k, k = 1,...,p\}$. Sort D in order of increasing function value.

(6) **Check convergence:** If the convergence criteria are satisfied, stop; otherwise, return to Step (3).

The Competitive Complex Evolution (CCE) algorithm required for the evolution of each complex in Step (4) of the Shuffled Complex Evolution method is presented below and is illustrated in Fig. 2.6:

(0) **Initialize:** Select q, α, and β, where $2 \leq q \leq m$, $\alpha \geq 1$, and $\beta \geq 1$.

(1) **Assign weights:** Assign a trapezoidal probability distribution to A^k, i.e.:

$$\rho_i = \frac{2(m+1-i)}{m(m+1)}, \quad i = 1,...,m \tag{2.12}$$

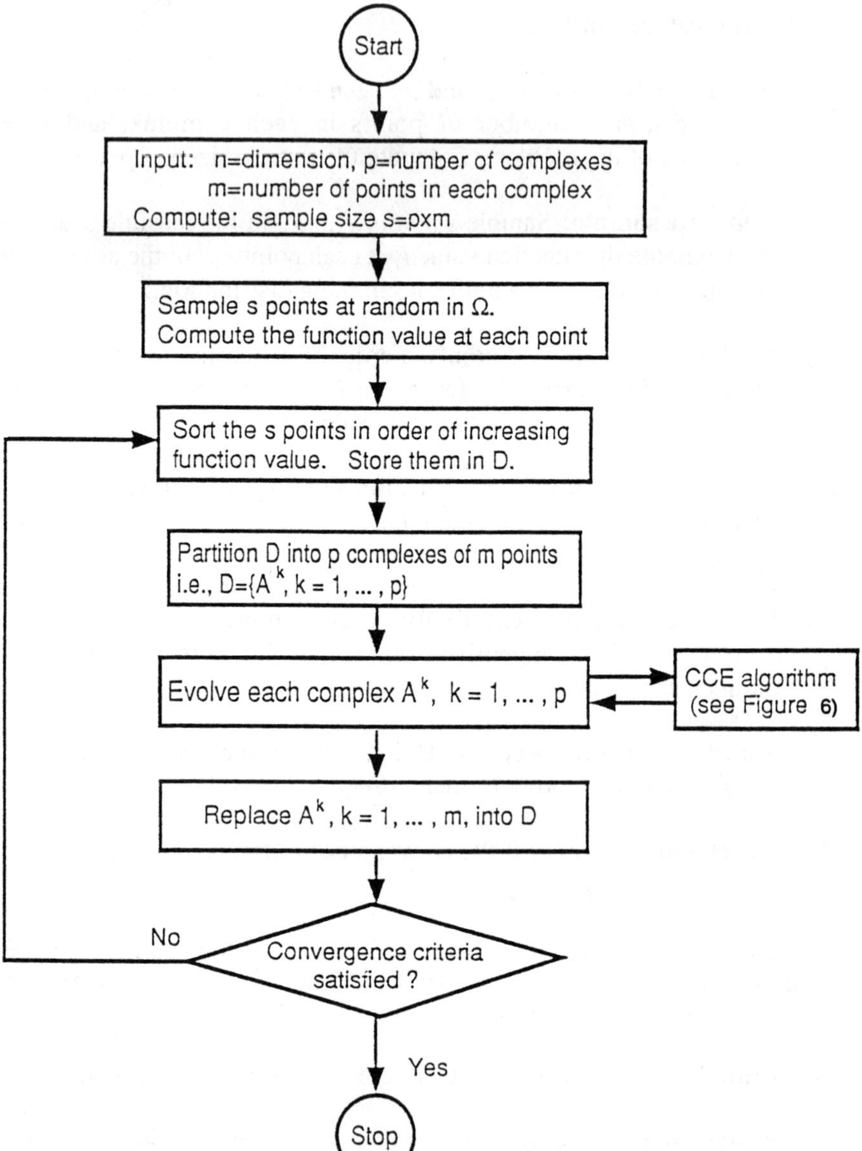

Fig. 2.5. Flow chart of the SCE-UA algorithm.

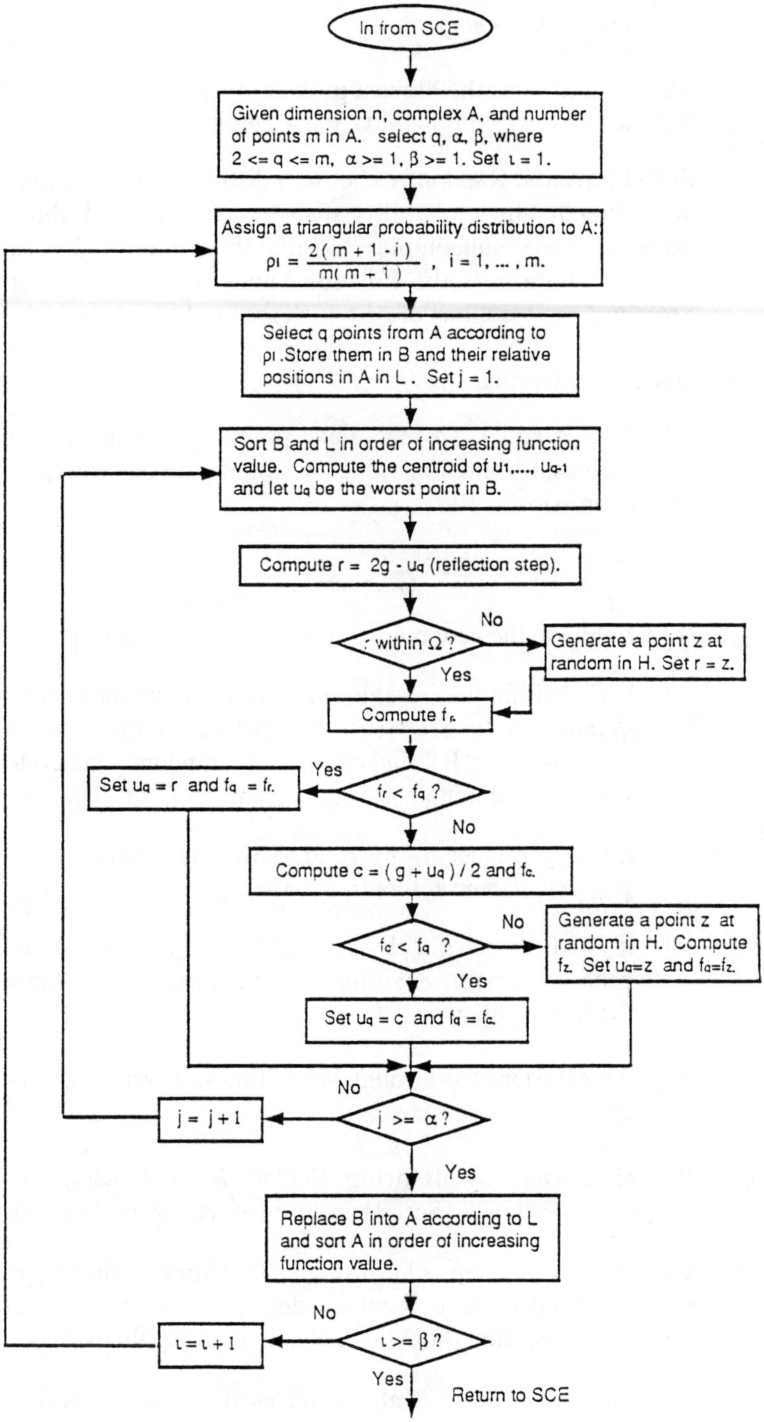

Fig. 2.6. Flow chart of the CCE portion of the SCE-UA algorithm.

The point x_1^k has the highest probability $\rho_1 = 2/m + 1$. The point x_m^k has the lowest probability $\rho_m = 2/m(m + 1)$.

(2) **Select parents:** Randomly choose q distinct points $u_1,...,u_q$ from A^k according to the probability distribution specified above (the q points define a "subcomplex"). Store them in array $B = \{u_i, v_i, i = 1,...,q\}$, where v_j is the function value associated with point u_j. Store in L the locations of A^k which are used to construct B.

(3) Generate offspring:

(a) Sort B and L so that the q points are arranged in order of increasing function value. Compute the centroid g using the expression:

$$g = \frac{1}{q-1} \sum_{j=1}^{q-1} u_j \tag{2.13}$$

(b) Compute the new point $r = 2g - u_q$ (reflection step).

(c) If r is within the feasible space Ω, compute the function value f_r, and go to Step (d); otherwise, compute the smallest hypercube $H \subset \mathbf{R}^n$ that contains A^k, randomly generate a point z within H, compute f_z, set $r = z$ and set $f_r = f_z$ *(mutation step).*

(d) If $f_r < f_q$, replace u_q by r, go to Step (f); otherwise, compute $c = (g + u_q)/2$ and f_c (contraction step).

(e) If $f_c < f_q$, replace u_q by c, go to Step (f); otherwise, randomly generate a point z within H and compute f_z (mutation step). Replace u_q by z.

(f) Repeat Steps (a) through (e) α times, where $\alpha \geq 1$ is a user-specified parameter.

(4) **Replace parents by offspring:** Replace B into A^k using the original locations stored in L. Sort A^k in order of increasing function value.

(5) **Iterate:** Repeat Steps (1) through (4) β times, where $\beta \geq 1$ is a user-specified parameter which determines how many offspring should be generated (how far each complex should evolve).

The version of the SCE-UA algorithm used in our research used the values $m = (2n+1)$, $q = (n+1)$, $\alpha = 1$, and $\beta = (2n+1)$. Hence, the only variable to be specified by the user is the number of complexes p.

The SCE-UA approach treats the global search as a process of natural evolution. The s sampled points constitute a population. The population is partitioned into several communities (complexes), each of which is permitted to evolve independently (i.e., search the space in different directions). After a certain number of generations, the communities are mixed and new communities are formed through a process of shuffling. This procedure enhances survivability by a sharing of the information (about the search space) gained independently by each community.

Each member of a community (complex) is a potential parent with the ability to participate in a process of reproduction. A subcomplex selected from the complex is like a pair of parents, except that a subcomplex may consist of more than two members. To ensure that the evolution process is competitive, we require that the probability that "better" parents contribute to the generation of offspring be higher than that of "worse" parents. The use of a triangular probability distribution ensures this competitiveness. Nelder and Mead's [1965] procedure is applied to each subcomplex to generate most of the offspring. This strategy uses the information contained in the subcomplex to direct the evolution in an improvement direction. In addition, offspring are introduced at random locations of the feasible space under certain conditions to ensure that the process of evolution does not get trapped by unpromising regions. This is analogous to mutation in response to stress that can occur in biological evolution. Each mutation also helps to increase the amount of information stored in the sample. Finally, each new offspring replaces the worst point of the current subcomplex. This ensures that every parent gets at least one chance to contribute to the reproduction process before being replaced or discarded. Thus, none of the information contained in the sample is ignored.

The processes of competitive evolution and complex shuffling inherent in the SCE-UA algorithm help to ensure that the information contained in the sample is efficiently and thoroughly exploited. They also help to ensure that the information set does not become degenerate. These properties endow the SCE-UA method with good global convergence properties over a broad range of problems. In other words, given a prespecified number of function evaluations (fixed level of efficiency), the SCE-UA method should have a high probability of succeeding in its objective of finding the global optimum.

In Duan et al. [1992] and Sorooshian et al. [1993], it was demonstrated that the performance of the SCE-UA method is far superior to that of the multi-start simplex (MSX) procedure. In particular, when tested on the National Weather Service SMA-NWSRFS flood forecast model, the MSX procedure was unable to locate the global solution even once in 100 random trials (Fig. 2.7), while the SCE-UA succeeded in 10 out of 10 attempts (Fig. 2.8). The SCE-UA method, therefore, appears to be the best method currently available for parameter estimation of conceptual watershed models.

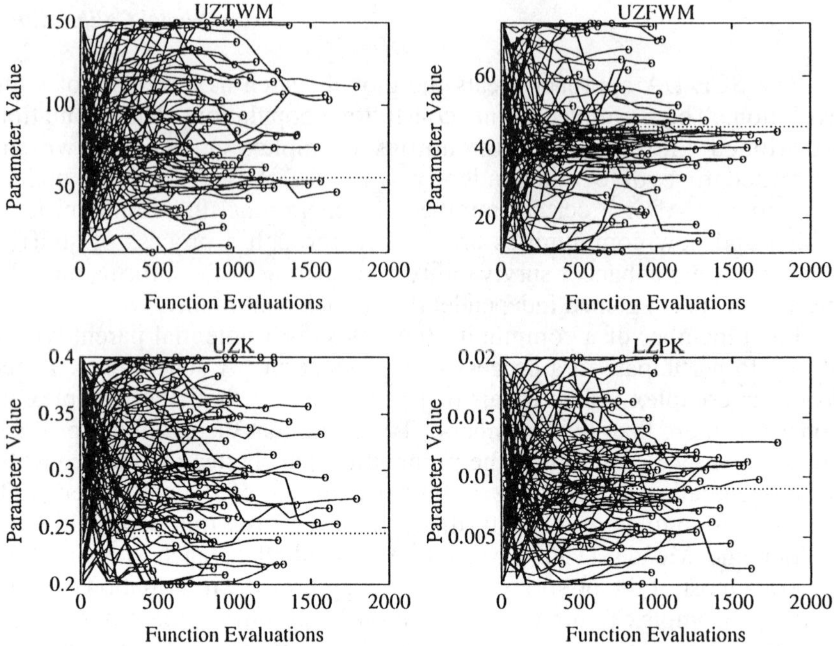

Fig. 2.7. Simplex algorithm convergence behavior (100 independent trials) for 4 parameters of the SMA-NWSRFS model (13 parameters were optimized) (source: Sorooshian et al., 1993).

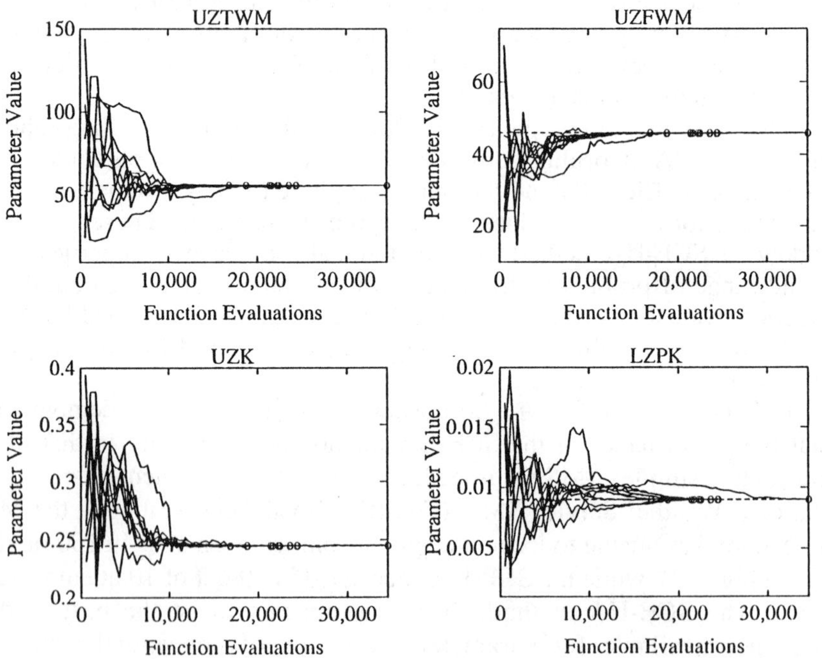

Fig. 2.8. SCE-UA algorithm convergence behavior (10 independent trials) for 4 parameters of the SMA-NWSRFS model (13 parameters were optimized) (source: Sorooshian et al., 1993).

2.3.4. Termination (Stopping) Criteria

The optimization strategies discussed earlier are all iterative procedures which search for the optimal parameter values by means of incremental improvement steps. Therefore, criteria are needed to determine when to stop the search. In principle, the solution exists at that point in the parameter space where the slope of the function response surface is zero and the function value is a minimum. In practice, it is virtually impossible to know when this point has been reached; hence, the criteria discussed below are more commonly used.

2.3.4a Function Convergence

One simple way to terminate the search is to stop when the algorithm is unable to appreciably improve the value of the function over one or more iterations. While this *can* indicate arrival at the location of an optimum, it could also mean only that a very flat region of the response surface has been reached. If precise detection of an optimum is not considered important, then function convergence can be a very useful stopping criterion. One typical implementation of this criterion is to stop when:

$$(f_{i-1} - f_i) \;/\; f_i \leq= \varepsilon_f \qquad (2.14)$$

where f_{i-1} and f_i are the function values at the i-1th and ith steps, respectively, and ε_f is the function convergence criterion (for example $\varepsilon_f = 10^{-3}$).

2.3.4b Parameter Convergence

Another way to terminate the search is to stop when the algorithm is unable to appreciably change the parameter values and simultaneously improve the function value over one or more iterations. While this *can* indicate arrival at an optimum, it could also mean only that a region of high parameter interaction (long narrow valley) on the response surface has been reached. In our experience, parameter convergence tends to be a more suitable criterion for calibration studies than function convergence. One typical implementation of this criterion is to stop when:

$$(\theta(j)_{i-1} - \theta(j)_i) \;/\; (\theta(j)_{max} - \theta(j)_{min}) \leq= \varepsilon_\theta \quad \text{for each } \theta(j) \qquad (2.15)$$

where $\theta(j)_{i-1}$ and $\theta(j)_i$ are the values of the jth parameter at the i-1th and ith steps, respectively, and ε_θ is the function convergence criterion (for example $\varepsilon_\theta = 10^{-3}$).

2.3.4c Maximum Iterations

If computer time is limited, and to ensure that the algorithm does not somehow enter an infinite loop, it is normal to terminate the search if a prespecified maximum number of iterations is exceeded, unless the parameter or function convergence criteria are met first. For random search methods, this is the normal way to terminate the search. It is not really possible to give guidelines on the value for this criterion, because it is both algorithm- and problem-dependent.

2.3.4d Limitations

None of these termination criteria are reliable in terms of indicating arrival at the global optimum, except in the most trivial cases where the function is convex and well behaved. However, in our experience, the parameter convergence criterion is most suitable for calibration studies; it stops the algorithm when the parameters are not changing significantly. The maximum iterations criterion is used as a backup to prevent waste of computer time; if the algorithm does not terminate within a "*reasonable*" number of iterations, the computer code may need to be examined for "*bugs*". The best way to establish proper values for the termination criteria is to run "synthetic data" test calibration studies [1] on the model.

2.3.5. Calibration Data

It is generally agreed that proper choice of the calibration data can do much to reduce the difficulties encountered during calibration of a hydrologic model. However, little is known objectively about what constitutes "good" calibration data. The critical issues here are *how much data* are necessary and sufficient for calibration and *what kind of data* will give the best results (most precisely specified parameter estimates).

2.3.5a Quantity of Data

It has been common practice to use as much data as were available for the calibration, after setting aside part of the data set for verification (see Section 2.4) of the results. However, the studies by Sorooshian et al. [1983] indicated that, if the calibration procedure was properly designed, only one year of daily data was sufficient to obtain conceptually realistic estimates for

---

1 A synthetic data study is one in which a user-chosen set of parameter values is assumed to be true, and a synthetic output hydrograph is created by running the model in simulation mode with the input data. The objective is then to start the optimization algorithm at some randomly selected parameter values and attempt to recover the "true" parameters (at which point, of course, a perfect match of the output hydrograph is obtained).

the SMA-NWSRFS model. Further, the use of longer data sets served only to *marginally* improve the parameter estimates. In general, from a statistical point of view, the data set used should be at least of length 20 times the number of parameters to be estimated (for example, if there are 10 parameters, then at least 200 streamflow data points shoud be used for computing the function). This is of course an approximate rule of thumb. Gupta and Sorooshian [1985a] showed that the standard error (j) of the estimate of parameter (j) decreases with sample size n approximately according to the formula:

$$\sigma(j) \; \alpha \; \frac{1}{\sqrt{n}} \tag{2.16}$$

Because the marginal improvement in 1/n becomes small after 500 to 1000 data points, this suggests that two to three years of calibration data should be sufficient, provided the data are of the *right kind*. This brings us to a discussion of data quality.

2.3.5b Quality of Data

From the viewpoint of model calibration, the *quality* of the data is dependent on the information (about the parameters) contained in the data and the noise (errors) in the data. Clearly, we wish the information content to be as large as possible and the noise to be as small as possible.

Informativeness

Qualitatively, an informative data set is one which contains or represents enough variability in watershed behavior that the different modes of operation of the hydrologic processes are properly represented. This maximizes the chances that the data contain sufficient information to properly identify the value of the model parameters (i.e., variations of the model parameters will result in detectable variations in the model output away from the observed hydrograph). For example, if the data selected are from a relatively dry year, certain runoff processes may not be activated; therefore, the model response will be insensitive to some of the model parameters that determine the partitioning of moisture between the various subsurface and overland flow components. However, if the data selected are from a year that is so wet that the watershed remains saturated most of the time, the model response may be insensitive to other subsurface flow controlling parameters.

The best choice seems to be a data set that contains a lot of "hydrologic variability". For example, Gupta and Sorooshian [1985b] showed that the identifiability of the upper zone threshold parameter of a typical watershed model depends not on how long the upper zone reservior overflows, but on *how many times* the reservior overflows. Thus, the more often the hydrologic regime switches from between dry, medium and wet modes, the more informative the data are likely to be.

Data errors

The presence of measurement and logging errors in the data causes the quality to deteriorate, thereby resulting in less confidence in the parameter estimates. Data errors come from many sources--machine misregistration, machine malfunction, methods for transfering the data to computer-ready media, approximations in rating curve transformations, shifts in the river bed, approximations in formulae used to convert field observations (e.g., river stage, precipitation at a gauge) to volume estimates, negligence, fraud, etc. While it has been suggested that the combined effects of these errors ought to tend towards a normal distribution, this has not been demonstrated to the authors' satisfaction. In practice, the data errors probably contain biases, particularly in the case of precipitation data which tend to underestimate the actual amount of rain falling on the watershed. Streamflow data are often the most reliable, but typically still contain approximately 10% error. In selecting data for model calibration, it is desirable that the data be carefully examined for obvious errors, such as periods where the hydrograph rises but no precipitation has been recorded. While errors in streamflow measurements can be mathematically handled by use of maximum likelihood and other statistical theories (see Section 2.3b), satistactory methods for proper handling of precipitation errors have yet to be developed.

2.4. VERIFICATION

Because the results of any calibration process are conditional on several factors--the calibration data, the objective function, and the optimization procedure--it is a good practice to conduct a verification test. The usual method is to test the performance of the calibrated model on a selected portion of the data that were not used for calibration. These verification data should contain hydrologic behavior of the kind under which the model is expected to provide streamflow estimates or forecasts. Such a test is often referred to as a "split-sample" test--part of the data record is used for calibration and the remaining part for verification. The purpose is to detect any biases that may have crept into parameter estimates due to imperfections in the calibration procedure. For example, it is not uncommon for the calibrated model to fit the calibration data very well, but then be found to not perform as well on the validation data period--a phenomenon known as "model divergence". When such behavior is observed, it may be necessary to critically examine the entire calibration procedure to determine if certain assumptions are not appropriate or valid and to thereby revise the procedure.

2.5. SENSITIVITY ANALYSIS

Once a model has been calibrated, it is useful to have some estimate of confidence in the calibrated model. One approach is to conduct a parameter sensitivity analysis--a procedure to determine which parameters can be considered to be well determined (sensitive) and which are poorly determined (insensitive). In principle, this requires obtaining an estimate of the "region of indifference" for the parameter estimates--i.e., the region around the best parameter estimates in which the function value varies from the best function value by only a small "indifference" value ε. This indifference region denotes the amount by which the parameter estimates can be in error (different from the best values) without significantly impacting on the closeness of the model simulated hydrograph to the observed data. By analyzing the indifference region, we can determine if there is high interdependence among two or more of the parameters (also called parameter correlation). High interdependence may indicate overparameterization or poor structural formulation of the model or perhaps inadequacy of the calibration data set.

As of yet, methods for obtaining the actual indifference region have not been developed. Sensitivity analyses are conducted using quadratic *approximations* to the response surface in the region of the best parameter values θ^*, using Taylor series expansions, to second order, of the objective function $F(\theta)$ [Sorooshian and Arfi, 1982, Gupta and Sorooshian, 1985a]. Using such an approach, the region of indifference can be approximately described by the expression [Gupta and Sorooshian, 1985a]:

$$\varepsilon(\theta,\theta^*) = [F(\theta) - F(\theta^*)] \simeq \frac{1}{2}[\theta - \theta^*]^T \, \nabla_\theta^2 F^* \, [\theta - \theta^*] \qquad (2.17)$$

where $\nabla_\theta{}^2F^*$ is the matrix of second derivatives of the function with respect to the parameter estimates, evaluated at the best parameter estimates θ^*.

Equation (2.17) describes a hyperellipse in the parameter space which bounds the region for which the difference between $F(\theta)$ and $F(\theta^*)$ is less ε. Figure 2.9 illustrates such a region for a two-parameter case. Note that each parameter θ_j can vary by an amount CP_j if all other parameters are held fixed at their best values θ^* and can vary by a larger amount P_j if all other parameters are simultaneously allowed to vary. The ratio of P_j to CP_j is a number larger than or equal to 1.0 and indicates the amount of multi-parameter correlation associated with parameter θ_j (a value of 1.0 indicates no correlation). This ratio can be computed using the equation:

$$\frac{P_j}{CP_j} = \frac{h_{jj}}{h_{jj} - g_j^T G_j^{-1} g_j} \qquad (2.18)$$

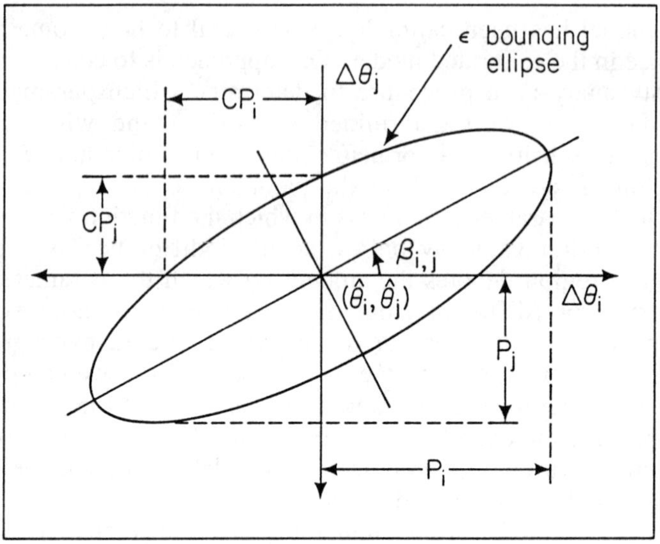

Fig. 2.9. Approximate indifference region (2 parameter case) (source: Gupta and Sorooshian, 1985).

where h_{jj} is the jj[th] element of $\nabla_\theta^2 F^*$, g_j is the vector obtained from the jth column of $\nabla_\theta^2 F^*$ by removing the jth element, and G_j is the matrix obtained by removing the jth row and column from $\nabla_\theta^2 F^*$.

2.6. STRUCTURAL PROBLEMS AND PARAMETER IDENTIFIABILITY

When calibrating a hydrologic model, it is important to be on the lookout for model structural problems that may result in calibration difficulties. Model structural problems are those that make it difficult or impossible to recover a unique global set of model parameters even under ideal test conditions using properly selected synthetic data. For example, we demonstrated the existence of such a problem in the structure of the SMA-NWSRFS flood forecast model [Sorooshian and Gupta, 1983; Gupta and Sorooshian, 1983].

The SMA-NWSRFS has a percolation equation which is considered to be the key element in the transfer of water within the model. It is a complex nonlinear function relating the capacities and contents of both the upper and lower soil zones and their free water depletion coefficients. The percolation process helps decide how much of the water in the upper zone contributes to the quickly reacting components of the storm hydrograph and how much is transferred to relatively longer term storage in the lower zone. Therefore, it is critical that the exact nature of the nonlinear relationship between the factors affecting percolation be identified accurately.

In attempting to calibrate the SMA-NWSRFS model, we discovered that an extended valley existed in the response surface of two of the parameters of the percolation equation (Fig. 2.10) [Sorooshian and Gupta, 1983]. Further analysis of the problem [Gupta and Sorooshian, 1983] indicated that the existence of the valley was not caused by poor data but was inherent to the structural formulation of the model. However, by examining the formulation of the percolation equation, we were able to show that a

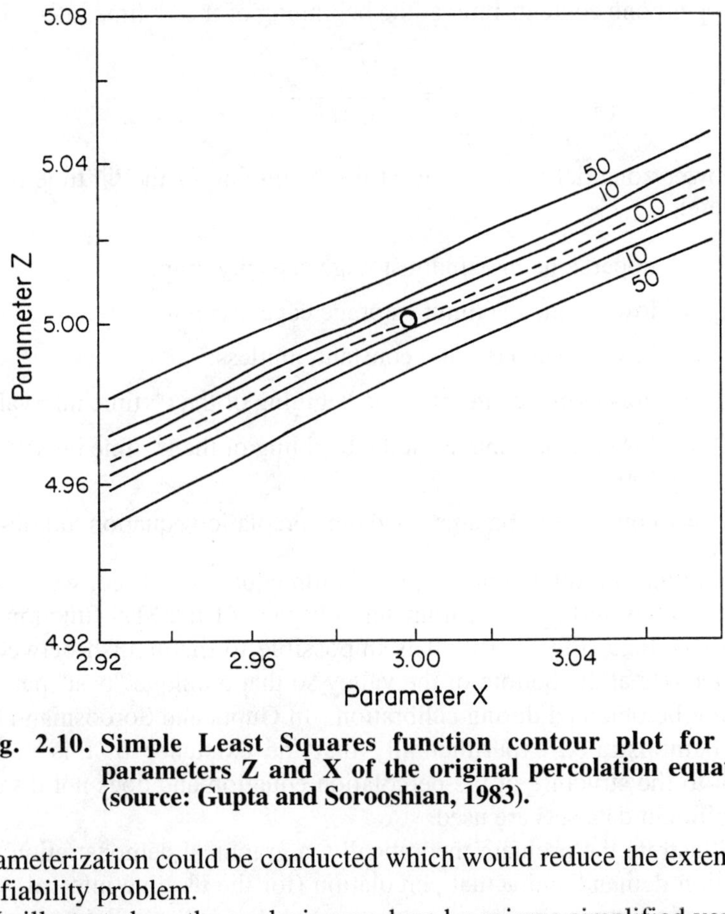

Fig. 2.10. Simple Least Squares function contour plot for the parameters Z and X of the original percolation equation (source: Gupta and Sorooshian, 1983).

reparameterization could be conducted which would reduce the extent of the identifiability problem.

We illustrate how the analysis was done by using a simplified version of the SMA-NWSRFS model, which we have named SIXPAR because it has only six calibration parameters. The percolation equation (eq. 2.19) is essentially of the same form as in the SMA-NWSRFS model, where PD_t is the percolation demand at time t:

$$PD_t = \beta \left[1 + Z(L_t)^X \right] U_t \tag{2.19}$$

where:

$$\beta = (LM)(LK) \tag{2.20}$$

is the minimum percolation demand when both upper and lower zones are full,

$$U_t = \frac{UC_t}{UM} \tag{2.21}$$

is the upper zone contents ratio at the beginning of the tth time interval,

$$L_t = \frac{LM - LC_t}{LM} \tag{2.22}$$

is the lower zone deficiency ratio at the beginning of the tth time interval, and where:

UM = upper zone maximum storage capacity, mm;

LM = lower zone maximum storage capacity, mm;

LK = lower zone recession constant, unitless;

UC_t = upper zone contents at the begining of the tth time interval, mm;

LC_t = lower zone contents at the begining of the tth time interval, mm; and

Z,X = constants to parameterize the percolation equation, unitless.

When this formulation of the percolation equation is used, we obtain an extended valley in the Z-X parameter subspace of the SLS function value response surface (Fig. 2.10). It is impossible to distinguish between the parameter sets at the bottom of the valley so that a unique "best" parameter set cannot be obtained during calibration. In Gupta and Sorooshian [1983], it was demonstrated experimentally that the existence of this valley is inherent to the structure of the percolation equation and does not disappear when different data sets are used.

To illustrate the problem theoretically, a graphical demonstration of the percolation demand and actual percolation (for the tth time interval) versus the change in lower zone deficiency ratio ($L_t = 0.0$ indicates lower zone full, $L_t = 1.0$ indicates lower zone empty) for the cases of upper zone full ($UC_t = UM$) and upper zone half full ($UC_t = UM/2$) is shown in Fig. 2.11. The curved lines represent the percolation demand for each of the cases, while the horizontal lines represent the amount of water available in the upper zone. The maximum percolation that can occur is equal to UM, which is the maximum capacity of the upper zone. Hence, percolation demand will

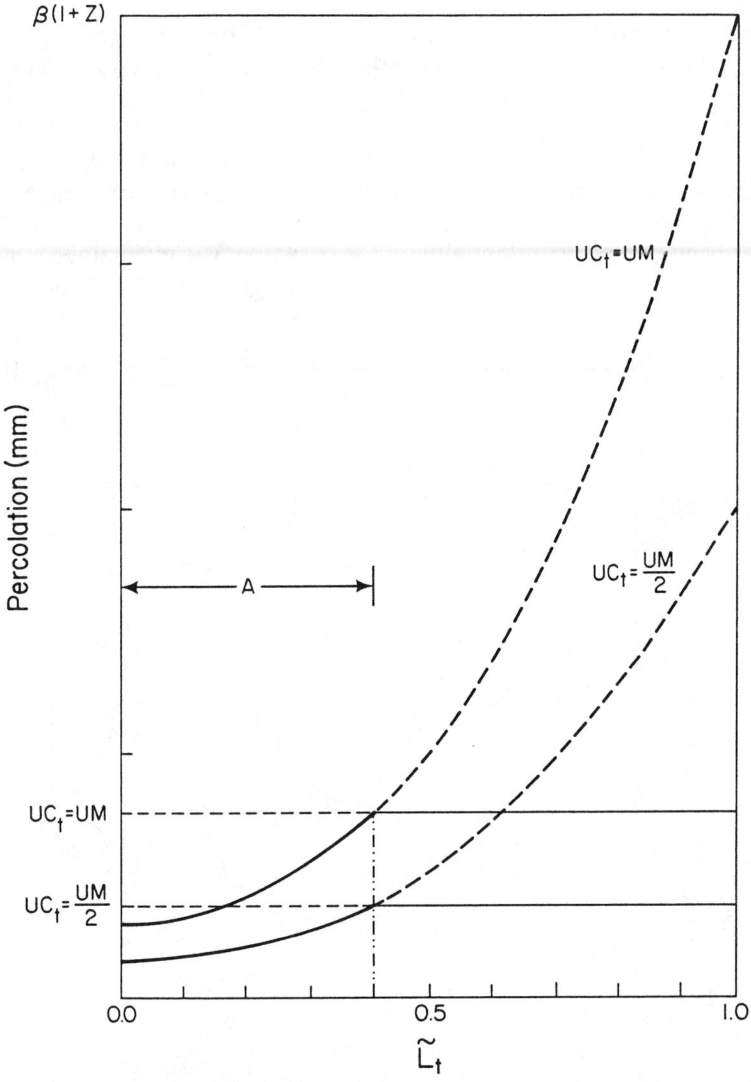

Fig. 2.11. Representation of percolation demand as a function of lower zone deficiency ratio at the tth time interval (source: Gupta and Sorooshian, 1983).

always be satisfied for L_t less than or equal to some value A (where $0 \leq A \leq 1$) and will never be satisfied for L_t greater than A. (Note that an important assumption here is that the maximum possible percolation demand [$\beta(1+Z)$] exceeds the size of the upper zone storage UM. This is considered to be realistic for most watersheds.) Therefore, A is a threshold. The implication of this is that, irrespective of the data selected for the estimation procedure, the output of the model is only sensitive to the parameters of the percolation equation (Z and X) during those time periods for which L_t is less than or equal to A. In addition, from Fig. 2.11, it can be seen that the value of A is significantly influenced by the magnitude of the upper zone UM. If UM is small, then A shifts to the left, making the parameters even less observable.

In Fig. 2.12, the manner in which this gives rise to the existence of the valley is demonstrated. Curve 1 indicates the demand curve corresponding to the true parameter set (say $\hat{Z}$, $\hat{X}$). Curve 2 corresponds to the case when Z has been reduced substantially to $\overset{*}{Z} < \hat{Z}$, while X is maintained equal to its true value $\hat{X}$. It is easy to see that fairly large changes in Z do not cause

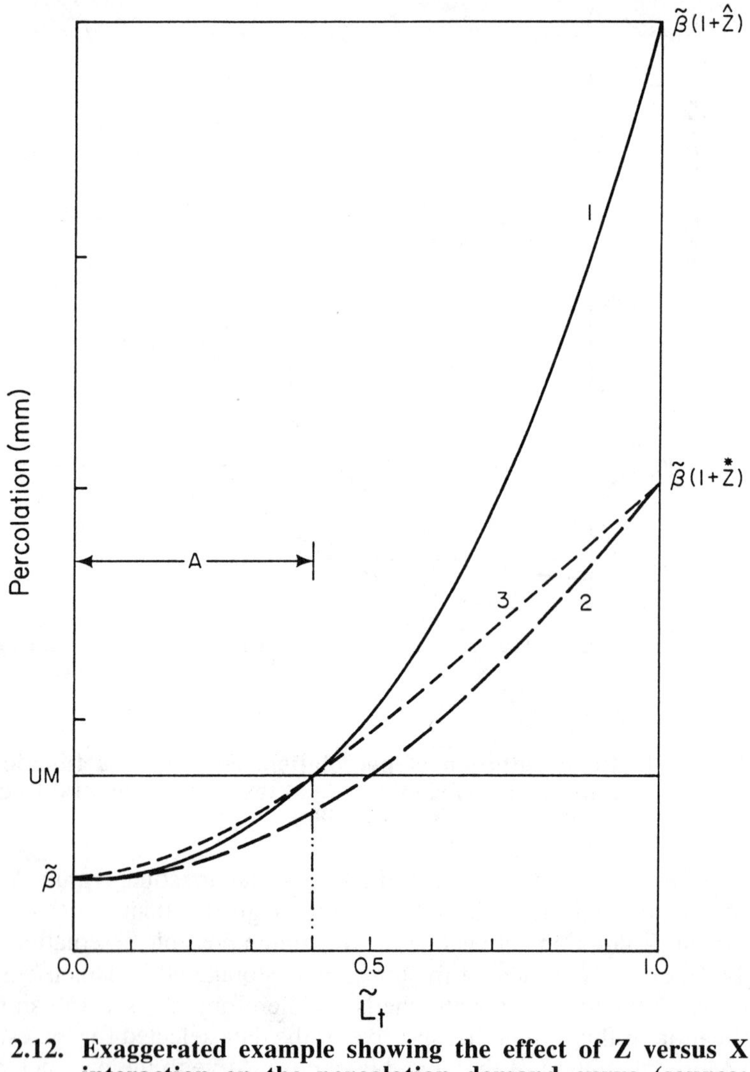

Fig. 2.12. Exaggerated example showing the effect of Z versus X interaction on the percolation demand curve (source: Gupta and Sorooshian, 1983).

correspondingly large deviations in the percolation curve for $L_t \leq A$. These deviations can be minimized by an appropriate compensating change in X (e.g., see curve 3, where $Z=\overset{*}{Z}$ and X has been reduced to $\overset{*}{Z}<\hat{X}$), thus making the errors in computed percolation almost neglegible. This will evidently give rise to the existence of a valley of virtually identical objective function values in the Z-X response surface.

To make the parameters identifiable, an appropriate reparameterization is required to make the response surface better conditioned. One possible (mathematical) approach is to conduct a linear transformation from the original cartesian coordinate system to the canonical coordinate system using an eigenvalue-eigenvector decomposition in the parameter space. An appropriate rescaling can then be employed. This approach, however, will result in a new (transformed) parameter set which may not possess any physical significance. In addition, this requires an accurate estimate of the Hessian matrix at each iteration of the search, a computational drawback. It would be more desirable to find a (linear or nonlinear) reparameterization of the equation that preserves the conceptual significance of the parameters. In the context of this problem, a parameter is required which is a function of both Z and X (and perhaps other parameters) and which represents the direction along the valley. This parameter should replace the relatively insensitive parameter Z.

From Fig. 2.11, it is clear that there is one "invariant" present which is unaffected by the amount of water available for percolation. This is the value A, which represents the value of L_t at which the percolation demand and available water curves intersect. This value has the favorable property of being scaled between 0.0 and 1.0, which means that a reasonable initial estimate for A can be made because it is bounded (Z, on the other hand, can theoretically vary between zero and infinity). This scaled property means that large changes in Z correspond to much smaller changes in A, thus making A a more sensitive parameter.

The relationship between the parameters A, Z, and X is now derived. The percolation demand PD_t will be equal to the upper zone contents UC_t when the lower zone deficiency ratio $L_t = A$. By substituting into eq. (2.19), we get:

$$UC_t = \beta \left[1 + Z(A)^X \right] \frac{UC_t}{UM} \tag{2.23}$$

By simplifying and rearranging eq. (2.23), A is obtained as a nonlinear function of Z, X, and other parameters:

$$A = [(UM - \beta) \ / \ (\beta Z)]^{1/x} \tag{2.24}$$

Similarly:

$$Z = \left[(UM - \beta) \ / \ (\beta A^X) \right] \tag{2.25}$$

By substituting for Z in eq. (2.19), we obtain relationships for the actual percolation PA_t:

$$PA_t = \left[\beta + (UM - \beta)(L_t / A)^X\right] U_t \quad \forall L_t <= A \tag{2.26}$$

and:

$$PA_t = UC_t \quad \forall L_t > A \tag{2.27}$$

Notice that, due to this reparameterization, there is no need to calculate a percolation demand value, as the actual value of percolation depends simply on whether L_t is greater or less than parameter A. When this form of the percolation equation is used in the model, we get the A-X parameter space SLS function response surface presented in Fig. 2.13. Notice that the contours are approximately elliptical in shape and display a unique optimum.

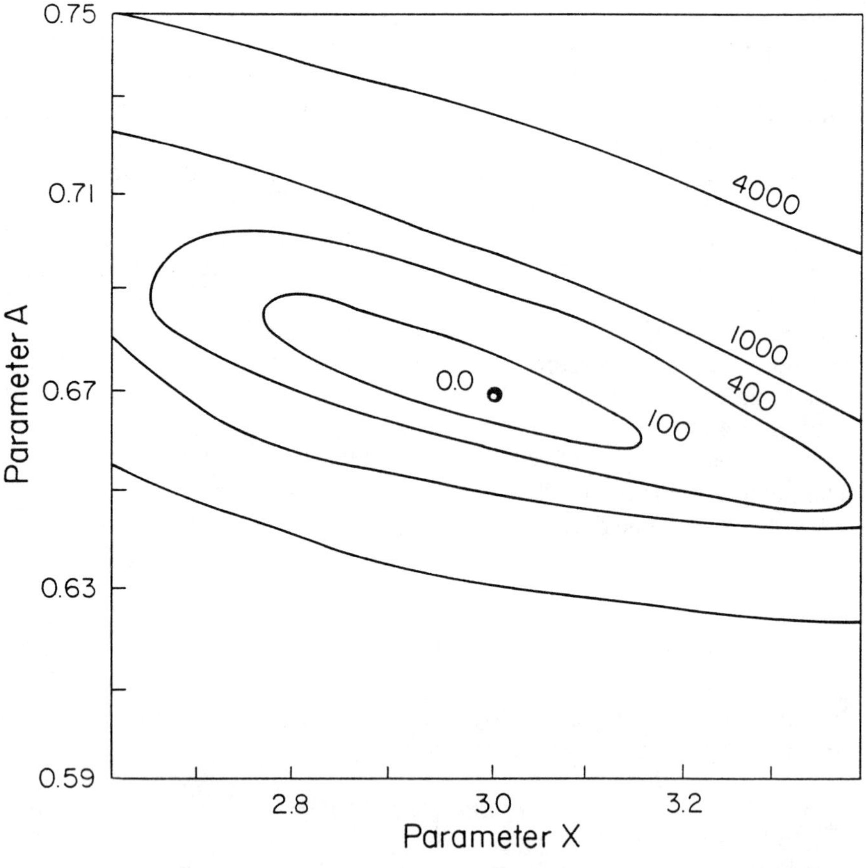

Fig. 2.13. Simple Least Squares function contour plot for the parameters A and X of the reparameterized percolation equation (source: Gupta and Sorooshian, 1983).

In the example presented here, it was possible to reparameterize the model equations to improve the extent of parameter identifiablity. Because many conceptual hydrologic models use a similar approach to modeling the percolation process, it is possible that they may be helped by some form of model restructuring. A procedure similar to the one used above might be adapted and employed to improve the structural identifiability of these other models.

2.7. CALIBRATION DIFFICULTIES

The successful application of a hydrologic watershed model model depends on how well the model is calibrated. In recent years, automated approaches to calibration, such as those discussed in this chapter, have received much attention, and several difficulties in the application of such methods have been reported [e.g. Ibbitt, 1970; Johnston and Pilgrim, 1976; Pickup, 1977; Larimore, 1981; Sorooshian and Gupta, 1983; Gan and Burges, 1990a,b]. These reports indicate that it is typically difficult, if not impossible, to obtain a unique set of optimal parameters for a watershed model using automatic calibration methods. While such problems may be partially attributable to limitations inherent in the calibration and verification data, the nonlinear structural characteristics typical of watershed models also lead to the existence of multiple optima (i.e., more than one solution). However, the local search optimization procedures in current use for automatic calibration are incapable of finding the globally optimal parameter estimates with any reasonable degree of confidence. This translates into uncertainty regarding the accuracy of the model forecasts. Furthermore, unless the best set of parameters associated with a given calibration data set can be found, it is difficult to determine how sensitive the parameter estimates (and hence the model forecasts) are to factors such as input and output data error, model error, quantity and quality of data, objective function used, and so on.

A comprehensive and detailed study of the reasons contributing to difficulties in calibrating hydrologic watershed models was recently conducted by the authors [see Duan et al., 1992]. We employed methods for detailed and exhaustive exploration of the response surface in conjunction with graphical display techniques to study the global properties of watershed model function response surfaces. Our studies demonstrated clearly the severity of the optimization problem encountered when attempting to calibrate a simple watershed using local search procedures. In particular, five major characteristics that complicate the optimization problem were identified, as follows:

(1) There are several major regions of attraction into which a search strategy may converge;

(2) Each major region of attraction contain numerous local minima (stationary points where the first derivatives are zero and the Hessian matrices are positive definite or positive semi-definite). These local optima occur both close to and at various distances from the best solution;

(3) The objective function surface in the multi-parameter space is not smooth and may not even be continuous. The derivatives are discontinuous and may vary in an unpredictable manner through the parameter space;

(4) The parameters exhibit varying degrees of sensitivity and a great deal of interaction and compensation. Much of the interaction is highly nonlinear; and

(5) The response surface near the true solution is often nonconvex.

Perhaps the most important of these characteristics is that the structure of multiple optima exists on at least two scales. At the "large" scale, we find that there is more than one broad "region of attraction" into which a search strategy may converge. However, at the "small" scale, each major region of attraction contains numerous local minima (stationary points where the first derivatives are zero and the Hessian matrices are positive definite or positive semi-definite). These minor optima occur both close to and at various distances from the best solution (Fig. 2.14a-d) and are not easily detectable using response surface mesh and contour plots.

The large number (hundreds to thousands) of minor optima is the most probable reason for the numerous reports in the literature of the inability to find unique "optimal" parameter values. In a local search procedure, when a line optimum or a stationary point is encountered, the optimization procedure will generally stop. Small step-size local searches will be unable to run the maze of minor optima, inevitably failing to reach the global optimum. Because many of the minor optima can be found quite far from the global optimum, the search may terminate without even finding an approximate solution.

In addition to the presence of optima at different scales, the objective function surface in the multi-parameter space is not smooth and has discontinuous derivatives (Fig. 2.15) that vary in an unpredictable manner through the parameter space. This helps explain why derivative-based local optimization methods have not performed well. Furthermore, it indicates that, for any global optimization to be successful, it must not depend on smooth and continuous derivatives either. Finally, the response surface in the region of the global optimum is not necessarily convex, exhibits widely varying degrees of sensitivity to the model parameters, and indicates the

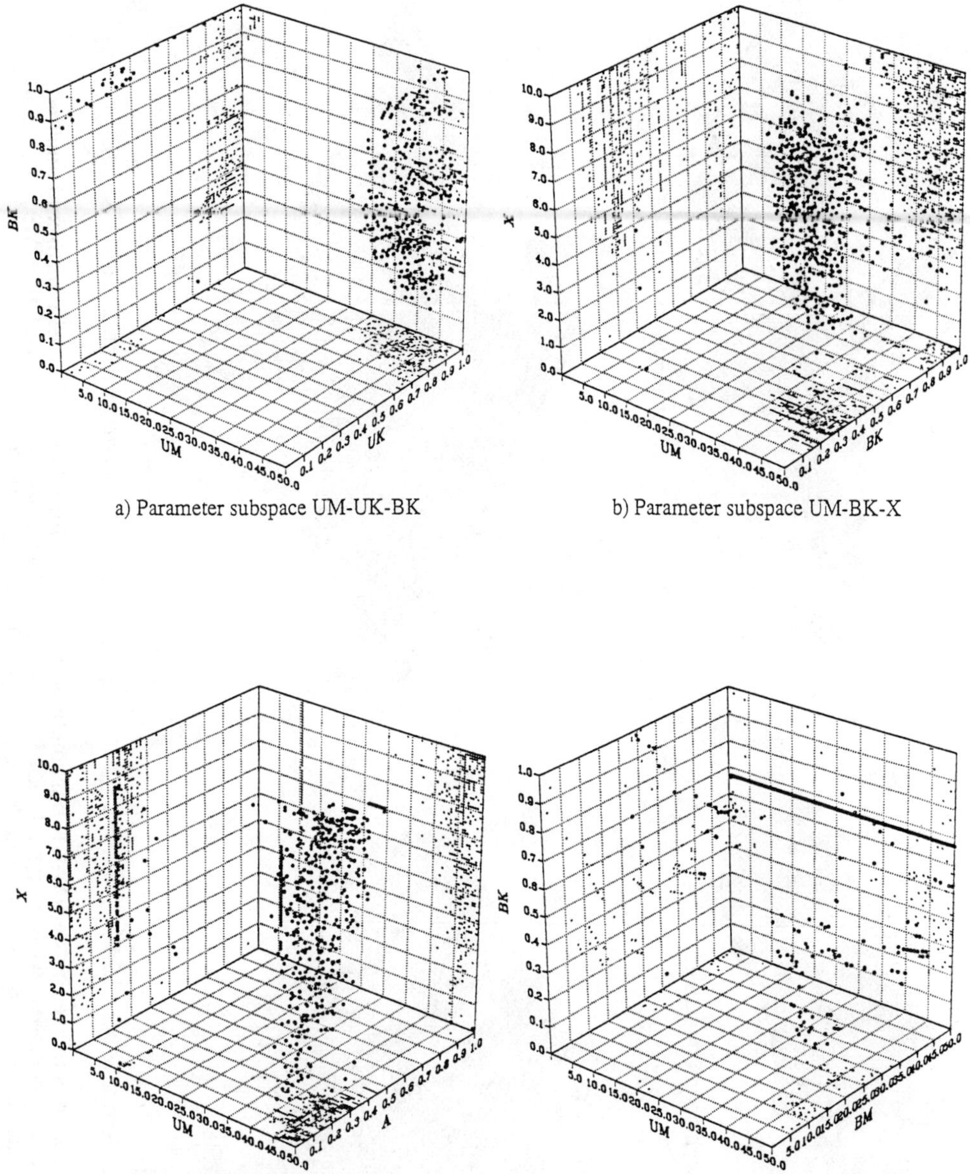

a) Parameter subspace UM-UK-BK

b) Parameter subspace UM-BK-X

c) Parameter subspace UM-A-X

d) Parameter subspace UM-BM-BK

Fig. 2.14. Locations of local optima in three-dimensional parameter subspaces: (a) parameter subspace UM-UK-BK; (b) parameter subspace UM-BK-X; (c) parameter subspace UM-A-X; (d) parameter subspace UM-BM-BK (source: Duan et al., 1992).

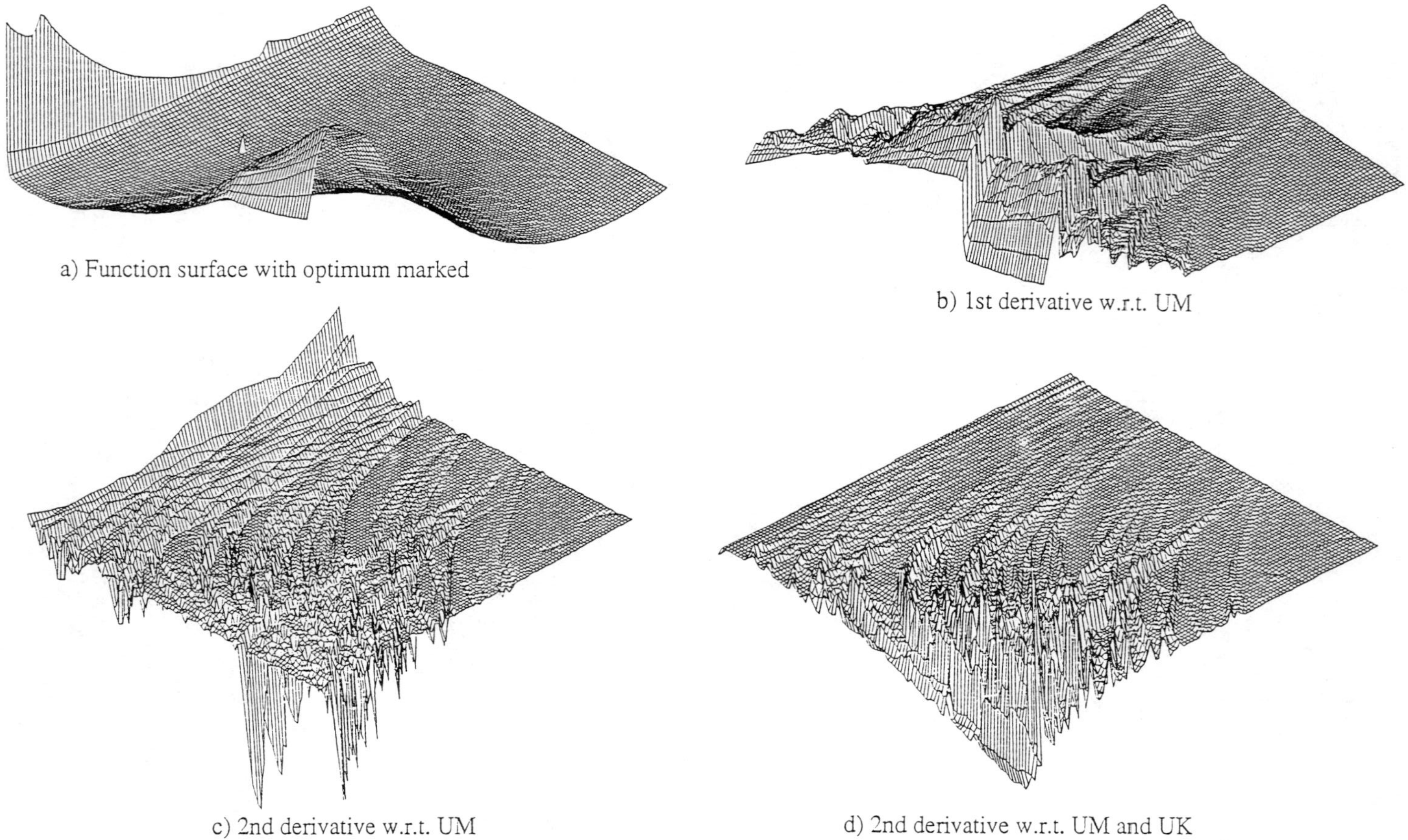

Fig. 2.15. **Derivative surfaces in the UM-UK parameter subspace: (a) function surface with optimum marked; (b) first derivative with respect to UM; (c) second derivative with respect to UM; (d) second derivative with respect to UM and UK (source: Duan et al., 1992).**

existence of a great deal of nonlinear parameter interaction and compensation.

The combination of the five features mentioned above makes the optimization problem difficult to solve. A successful optimization procedure must, therefore, be capable of dealing with these various difficulties. To deal with multiple regions of attraction, the algorithm must be globally based and possess global convergence properties. The algorithm must be able to avoid being trapped by minor optima, and it must not require the availability of explicit analytic expressions for the objective function in terms of its parameters or for the derivatives. It must be robust in the presence of parameter interaction and nonconvexity of the objective function surface. Finally, because watershed models usually involve a large number of parameters, the algorithm must be efficient in the presence of high dimensionality. The SCE-UA global optimization method described earlier in the chapter has been shown [Duan et al., 1992; Sorooshian et al., 1993] to have the capabilities described above and appears to be the most effective and efficient method for calibration of watershed models currently available.

ACKNOWLEDGMENTS

The authors gratefully acknowledge the support from the Engineering Division of the National Science Foundation and the U.S. National Weather Service's Office of Hydrology, which led to many of the results and findings reported in this chapter.

REFERENCES

Arfi, F., Response surface methodology as a parameter sensitivity analysis tool for rainfall-runoff models, M.S. Thesis, Systems Engineering Department, Case Western Reserve University, Cleveland, Ohio, 1980.

Bard, Y., Nonlinear Parameter Estimation, Academic Press, New York, 1974.

Beck, M.B., Water Quality Modeling: A Review of the Analysis of Uncertainty, *Water Resources Research*, 23(8), 1393-1442, August 1987.

Brazil, L.E., Multilevel Calibration Strategy for Complex Hydrologic Simulation Models, Ph.D. Dissertation, Dept. of Civil Engineering, Colorado State University, Fort Collins, Colorado, 1988.

Brazil, L.E. and W.F. Krajewski, Optimization of Complex Hydrologic Models Using Random Search Methods, Engineering Hydrology Proceedings, Hydraulics Division/ASCE/Williamsburg, Virginia, August 3-7, 726-731, 1987.

Chapman, T.G., Optimization of a rainfall-runoff model for an arid zone catchment, UNESCO Publ. 96, 126-143, United Nations Educational, Scientific, and Cultural Organization, Wallington, New Zealand, 1970.

Clarke, R.T., A review of some mathematical models used in hydrology, with observations on their calibration and use, *Journal of Hydrology*, 19(1), 1-20, 1973.

Dawdy, D.R. and T. O'Donnell, Mathematical models of catchment behavior, *Journal of Hydraulics Division*, American Society of Civil Engineers, 91(HY4), 113-137, 1965.

Diskin, M.H. and E. Simon, A procedure for the selection of objective functions for hydrologic simulation models, *Journal of Hydrology*, 34(1/2), 129-149, 1977.

Duan, Q., A global optimization strategy for efficient and effective calibration of hydrologic models, Ph.D Dissertation, Department of Hydrology and Water Resources, The University of Arizona, Tucson, Arizona, 1991.

Duan, Q., V.K. Gupta, and S. Sorooshian, A shuffled complex evolution approach for effective and efficient global minimization, *Journal of Optimization Theory and Applications*, 76(3), 501-521, March 1993.

Duan, Q., S. Sorooshian, and V.K. Gupta, Effective and efficient global optimization for conceptual rainfall-runoff models, *Water Resources Research*, 28(4), 1015-1031, April 1992.

Fulton, J., Discussion, modification, and extension of some maximum likelihood tech-niques for model calibration with application to rainfall-runoff models. M.S. Thesis, Case Western Reserve University, August 1982.

Gan, T.Y. and S.J. Burges, An assessment of a conceptual rainfall-runoff model's ability to represent the dynamics of small hypothetical catchments: 1. models, model properties, and experimental design, *Water Resources Research*, 26(7), July 1990a.

Gan, T.Y. and S.J. Burges, An assessment of a conceptual rainfall-runoff model's ability to represent the dynamics of small hypothetical catchments: 2. hydrologic responses for normal and extreme rainfall, *Water Resources Research*, 26(7), July 1990b.

Gupta, V.K., The identification of conceptual watershed models, Ph.D Dissertation, Case Western Reserve University, Cleveland, Ohio, January 1984.

Gupta, V.K. and S. Sorooshian, Uniqueness and observability of conceptual rainfall-runoff model parameters: the percolation process examined, *Water Resources Research*, 19(1), 269-276, February 1983.

Gupta, V.K. and S. Sorooshian, The automatic calibration of conceptual catchment models using derivative-based optimization algorithms, *Water Resources Research*, 21(4), 473-486, April 1985.

Gupta, V.K. and S. Sorooshian, The relationship between data and the precision of estimated parameters, *Journal of Hydrology*, 81, 57-77, 1985.

Hendrickson, J.D., S. Sorooshian, and L. Brazil, Comparison of Newton-type and direct search algorithms for calibration of conceptual rainfall-runoff models, *Water Resources Research*, Vol. 24(5), 691-700, 1988.

Holland, J.H., Adaptation in Natural and Artificial Systems, University of Michigan Press, Ann Arbor, Michigan, 1975.

Hooke, R. and T.A. Jeeves, Direct search solutions of numerical and statistical problems, *Journ. Assoc. Comput. Mach.*, Vol. 8(2), 212-229, 1961.

Ibbitt, R.P, Systematic parameter fitting for conceptual models of catchment hydrology, Ph.D. Dissertation, Imperial College of Science and Technology, University of London, London, England, 1970.

Ibbitt, R.P. and P.D. Hutchinson, Model Parameter Consistency and Fitting

Criteria, Proceedings of the IFAC 9th Triennial World Congress, Budapest, Hungary, Vol. 4, 153-157, July 2-6, 1984.

Johnston, P.R. and D. Pilgrim, Parameter optimization for watershed models, *Water Resources Research*, Vol. 12(3), 477-486, 1976.

Karnopp, D.C., Random search techniques for optimization problems, *Automatica*, Vol. 1, 111-121, 1963.

Kuczera, G., Combining site-specific and regional information: an empirical Bayes approach," *Water Resources Research*, Vol. 18, No. 2, 306-314, 1982.

Larimore, W.E., Applications of the Maximum Likelihood Parameter Identification to Hydrologic Forecasting, Report #TR-1480-3, TASC, 1 Jacob Way, Reading, Massachusetts, June 1981.

Lemmer, H.R. and A.R. Rao, Critical Duration Analysis and Parameter Estimation in ILLUDAS, Water Resources Research Center, Tech. Report #153, Purdue University, West Lafayette, Indiana, 1983.

Masri, S.F., G.A. Bekey, and F.B. Safford, An Adaptive Random Search Method for Identification of Large Scale Nonlinear Systems, Proceedings of 4th IFAC Symposium on Identification and System Parameter Estimation 1976, North-Holland, Amsterdam, 1978.

Monro, J.C., Direct Search Optimization in Mathematical Modeling and a Watershed Model Application, NOAA Tech Memo NWS HYDRO-12, U.S. Dept. of Commerce, Silver Spring, Maryland, 1971.

Nash, J.E. and J.V. Sutcliffe, River flow forecasting through conceptual models, part 1: a discussion of principles, *Journal of Hydrology*, 10(3), 282-290, 1970.

Nelder, J.A. and R. Mead, A simplex method for function minimization, *Computer Journal*, Vol. 7, 308-313, 1965.

Pickup, G., Testing the efficiencies of algorithms and strategies for automatic calibration of rainfall-runoff models, *Hydrogeological Science Bulletin*, 22(2), 257-274, 1977.

Price, W.L., Global optimization algorithms for a CAD workstation, *Journal of Optimization Theory and Applications*, Vol. 55(1), 133-146, 1987.

Pronzato, L., E. Walter, A. Venot, and J.F. Lebruchec, A general purpose global optimizer: implementation and applications, *Mathematics and Computers in Simulation*, Vol. 26, 412-422, 1984.

Rosenbrock, H.H., An automatic method for finding the greatest or least value of a function, *Computer Journal*, Vol. 3, 175-184, 1960.

Sage, A.D. and J.L. Melsa, Estimation Theory with Applications to Communication and Control, McGraw-Hill, New York, 1971.

Sorooshian, S., Considerations of Stochastic Properties in Parameter Estimation of Hydrologic Rainfall-Runoff Models, Ph.D. Dissertation, University of California-Los Angeles, 1978.

Sorooshian, S., Parameter estimation of rainfall-runoff models with heteroscedastic streamflow errors--noninformative data case, *Journal of Hydrology*, Vol. 52(1/2), 127-138, June 1981.

Sorooshian, S., Parameter Estimation of Rainfall-Runoff Models with Heteroscedastic Streamflow Errors: Noninformative Data Case, In: Y. Haimes and J. Knolles (eds.), Water and Related Land Resource Systems, IFAC Symposium, Cleveland, Ohio, May 28-31, 1980, Pergamon Press, New York, N.Y., pp. 477-485, 1981.

Sorooshian, S., Surface water hydrology: on-line estimation, *Reviews of Geophysics and Space Physics*, Vol. 21, No. 3, 706-721, April 1983.

Sorooshian, S., Synthesis of hydrologic and system sciences in the development of rainfall-runoff models, *Applied Mathematics and Computation*, 17, 279-298, 1985.

Sorooshian, S. and F. Arfi, Response surface parameter sensitivity analysis methods for post-calibration studies, *Water Resources Research*, 18(5), 1531-1538, October 1982.

Sorooshian, S. and J.A. Dracup, Stochastic parameter estimation procedures for hydrologic rainfall-runoff models: correlated and heteroscedastic error cases, *Water Resources Research*, 16(2), 430-442, April 1980.

Sorooshian, S., Q. Duan, and V.K. Gupta, "Calibration of the SMA-NWSRFS conceptual rainfall-runoff model using global optimization", *Water Resources Research*, 29(4), 1185-1194, April 1992

Sorooshian, S. and V.K. Gupta, Automatic calibration of conceptual rainfall-runoff models: the question of parameter observability and uniqueness, *Water Resources Research*, 19(1), 251-259, February 1983.

Sorooshian, S., V.K. Gupta, and J.L. Fulton, Parameter Estimation of Conceptual Rainfall-Runoff Models Assuming Autocorrelated Data Errors--A Case Study, In: V.P. Singh (Editor), Statistical Analysis of Rainfall and Runoff, Proceedings of the International Symposium on Rainfall-Runoff Modeling, Mississippi State, Mississippi, May 18-21, 1981, WRP, Littleton, Colorado, 491-504, 1982.

Sorooshian, S., V.K. Gupta, and J.L. Fulton, Evaluation of maximum likelihood parameter estimation techniques for conceptual rainfall-runoff models--influence of calibration data variability and length on model credibility, *Water Resources Research*, 19(1), 251-259, February 1983.

Wang, Q.J., The genetic algorithm and its application to calibrating conceptual rainfall-runoff models, *Water Resources Research*, 27(9), 2467-2471, 1991.

Chapter 3

RELIABILITY ESTIMATION

C. S. Melching

3.1. INTRODUCTION

The computer models of watershed hydrology described in this book are used for a wide variety of purposes including operational flow forecasting for flood warning or reservoir operation, design-flood estimation, recreation of historic flows, and generation of synthetic flows resulting from synthetic rainfall. Many of these purposes involve application of the model to storms and watershed conditions outside of the range of conditions for which the model has been successfully calibrated and verified (i.e., extrapolation). When using a model for extrapolation, hydrologists, engineers, and planners are very interested in the reliability of model output. Model-output reliability can also be used to assess model verification and the selection of appropriate models (Melching et al., 1991). Operational forecasts are greatly enhanced by estimates of the probability that discharges or volumes will exceed various levels (Georgakakos, 1986). Risk-cost design procedures require estimates of the probability that various target discharges, volumes, or stages will be exceeded. Finally, Beven (1989) has reviewed the limitations of the current generation of distributed, physical-simulation models of watershed hydrology and has suggested that a possible way forward requires a realistic estimation of prediction uncertainty. Therefore, estimation of the reliability of model output is important to hydrologic engineering and water-resources planning.

For water-quality modeling Beck (1987) noted that uncertainties affect primarily four problem areas that engineers and scientists must address:

1) uncertainty about model structure, i.e., what basic processes are involved, how they interact, and how these processes and interactions can be mathematically characterized in an efficient and parsimonious manner;

2) uncertainty in the model parameters, i.e., parameter identification and calibration problems;

3) uncertainty associated with estimates of the future behavior of the system, i.e., aggregation of uncertainties in model structure, model parameters, and in the definition of the design or decision scenario into overall estimation uncertainty; and

4) reduction of critical modeling uncertainties via carefully designed experiments and/or monitoring programs.

These problem areas are also highly important in modeling watershed hydrology. Furthermore, in modeling watershed hydrology the uncertainty in operational forecasts of rainfall-runoff events in progress is important. The forecast-uncertainty problem is similar to the estimation-uncertainty problem except that uncertainties in the real-time data replace the uncertainties in the design conditions (storm and antecedent) used for decision making.

The four problem areas are strongly interrelated. For example, evaluation of forecast and estimation uncertainty (area 3) assumes that the model structure (area 1) has been reasonably identified, and examines the propagation of model parameters (area 2) and data uncertainties through the model and the resulting forecast or estimation reliability. The purpose of this book is to describe a wide variety of computer models of watershed hydrology that are readily available to practicing hydrologists, hydrologic engineers, and water-resource managers. Thus, it is assumed that a model appropriate for application to the problem under consideration may be selected (if one exists) through careful examination of this book. In Chapter 2, Sorooshian and Gupta discuss in detail the effect of uncertainties on model calibration and ways to obtain unique and conceptually realistic parameters in the face of uncertainties; specific calibration procedures for many of the models presented in this book are discussed in the chapter describing the specific model. Thus, this chapter focuses on evaluation of the reliability of the forecasts and estimates obtained from computer models of watershed hydrology. Determination of the main sources of uncertainty in the forecast or estimate by the methods used to aggregate model structure, model parameters, and data uncertainties into an overall forecast or estimation reliability is also discussed.

The overall goal of this chapter is to provide the information needed to make a preliminary estimation of the reliability of watershed-model output. This chapter describes (1) sources of uncertainty and approaches to estimate the magnitude of the various uncertainties, (2) measures of output reliability, (3) methods of reliability analysis, (4) comparisons of these methods, and (5) a brief review of estimation of the key sources of uncertainty in output by use of reliability-analysis methods. Throughout the chapter, examples from the literature are cited to illustrate the techniques or points in question. Most of the examples cited involve applications of reliability-analysis methods to the computer models of watershed hydrology presented in detail in this book. To the best of this author's knowledge, some of the techniques discussed in this chapter have not been used or have had limited use in computer modeling of watershed hydrology, and, for these techniques, examples from the related fields of water-quality modeling, ground-water modeling, and hydraulic engineering are cited.

3.2. SOURCES AND MAGNITUDE OF UNCERTAINTIES

When using computer models of watershed hydrology to determine the runoff hydrograph resulting from a storm, snowmelt period, or both for the purposes of operational forecasting, hydraulic structure design, synthetic runoff data generation, etc., hydrologic modelers are faced with four types of uncertainty: (1) natural randomness, (2) data, (3) model parameters, and (4) model structure. Many researchers have examined the possible effects of these sources of uncertainty either alone or in combinations on watershed-model output. These will not be discussed here because the focus of this section is to provide suggestions as to how the magnitude of the uncertainty from the various sources may be estimated for the purpose of determining the overall reliability of watershed-model output.

The relationship between a typical hydrologic model application and an application of a hydrologic model combined with reliability analysis is shown in Fig. 3.1. The uncertainty information required by the reliability-analysis methods described in Section 3.4 includes the mean, variance, and, in some cases, the probability density function (PDF) or cumulative distribution function (CDF) of the basic variables describing the watershed system (i.e., data, model parameters, model-structure correction factors, etc.). If the basic variables are significantly correlated, the covariance matrix and, in some cases, joint PDF's are also needed. At first glance, these uncertainty information requirements might seem to add considerable complication to the process of watershed modeling. However, **it must be emphasized that much of this information is obtained in the normal process of calibration and verification of a selected hydrologic model for a given watershed, and development of useful uncertainty information usually requires small additional effort on the part of the hydrologic modeler**. For example, calibration over several events provides information on the range of values parameters may take, whereas physical parameters are selected from physically reasonable ranges; these ranges may be converted into standard deviations using simple assumptions. Such simple approaches are discussed in the following subsections.

Some of the approximations presented in the following are crude, and research is encouraged to find more rigorous methods. However, the

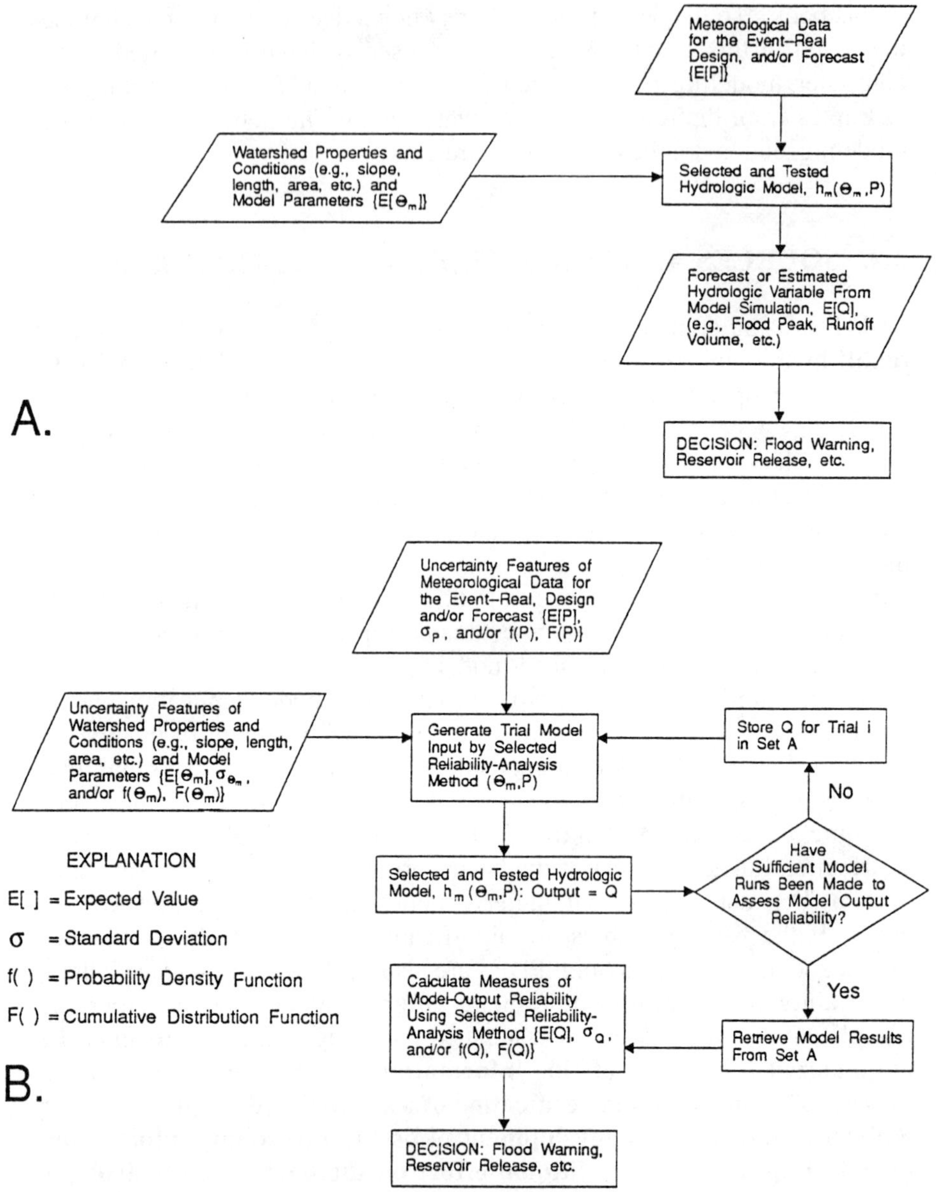

Fig. 3.1. Comparison of (A) a typical application of a computer model of watershed hydrology and (B) an application of a computer model of watershed hydrology combined with a reliability-analysis method.

underlying principle in any reliability analysis pointed out by Cornell (1972) must be remembered, namely that

> "It is important in engineering applications that we avoid the tendency to model only those probabilistic aspects that we think we know how to analyze. It is far better to have an approximate model of the whole problem than an exact model of only a portion of it."

3.2.1. Natural Uncertainties

Natural uncertainties include the random temporal and spatial fluctuations inherent in natural processes that almost always introduce a large amount of randomness into the physical process of runoff generation. Plate (1986) noted that even if a perfect model of the hydrologic cycle were available, hydrologists would still be faced with a large random residue caused by natural uncertainties.

The randomness of nature and the need to estimate flows over the design life or planning horizon of a project have been the driving forces in the development of stochastic hydrology. In one form or another all of stochastic hydrology assumes that the variability observed in past events is representative of events likely to occur during the planning horizon or design period. The analysis of natural uncertainties is typically limited to the magnitude of uncertainties seen in the data used to develop the watershed model. Hydrologic frequency analysis also assumes that hydrologic processes are stationary and independent in time allowing natural uncertainties to be considered as a purely random probabilistic process. Time-series analysis techniques assume that there are underlying processes causing variations--e.g., trends, periodicities, and correlations--in flow-data series around which natural randomness causes additional variation. The underlying processes are represented by linear or nonlinear relations (trends), harmonic analysis or seasonal standardization (periodicities), and autocorrelation and moving averages; and the natural randomness is represented by an error term which is independent, has a mean of zero, follows a designated PDF, and is additive.

The reliability-analysis techniques described in this chapter take an approach similar to that of time-series analysis in that the underlying process is described by the deterministic model and the additional, natural uncertainties are handled separately. That is, natural uncertainties affect the input data, model-parameter, and model-structure uncertainties. The uncertainties in the input data, model parameters, and model structure propagate through the model such that a probability distribution for the output from a deterministic model can be obtained. The influence of natural uncertainties on input data, model parameters, and model structure is discussed in detail in the following subsections.

3.2.2. Data Uncertainties

Nearly all computer models of watershed hydrology require data on some combination of precipitation, streamflow, evapotranspiration, and watershed morphology. The basic sources of error and uncertainty in each of these data types and methods to account for these uncertainties are discussed in detail in Melching (1987). The following discussion centers on the most significant sources of data uncertainty likely to influence typical applications of computer models of watershed hydrology. A comprehensive discussion of all uncertainties in the four data types listed above is beyond the scope of this chapter. Finally the effects of data uncertainties must be considered in different ways depending on:

(1) whether the model parameters are determined from calibration or from physical measurements and principles and

(2) whether the model is used to estimate real events (flood forecasting and recreation of historic flows) or to estimate synthetic events (design storms and generation of synthetic flows). These issues are considered separately in the following paragraphs.

For calibrated models, Dawdy et al. (1972) reported that the bias (systematic errors and uncertainty) in recorded rainfall at each gage location is compensated by the curve-fitting ability of the model to adjust parameters. Thus, the bias in the amount of recorded rainfall affects the resulting fitted-parameter values, rather than the accuracy of the fit of simulated to observed runoff. Troutman (1982a, 1983) also found that systematic errors and uncertainties in rainfall data because of variations in the rain-gage network used in modeling were transferred to the parameters of the model as bias in the parameters (i.e., deviations from their true values). However, due to the curve-fitting properties of the calibration routine, the performance of the model using erroneous data (i.e., a perturbation of assumed true data) and biased parameters is not greatly different from that using true (i.e., assumed hypothetical values) data and parameter values because the calibration, in effect, provides an adjustment factor for errors and bias. In fact for a simple regression model of rainfall volume to runoff volume, Troutman (1982b) found two noteworthy results. First, for cases where the rainfall-input error is equivalent to a random process (11 of 13 subsets of the rain gage network considered), even though the erroneous rainfall values have more variability than the true values, runoff computed from the model evaluated with the erroneous rainfall and adjusted parameters has less variability than that computed from the model evaluated with the true rainfall and parameters. Second, model predictions using the erroneous rainfall measurements and the corresponding biased parameter values are unbiased estimates of mean runoff. Troutman (1982a and b) further pointed out for more realistic (than simple regression models) nonlinear models, such as the U.S. Geological Survey Precipitation-Runoff Model System (Chapter 9), curve-fitting never quite eliminates the bias in mean-runoff estimation. However, if more than one parameter is adjusted, there is a good chance that the bias will be fairly small over a wide range of rainfall-input values.

A basic assumption of stochastic hydrology is that the calibration data are a representative sample of the range of flows to be estimated and the range of data uncertainties to be encountered. If this assumption is reasonably valid, then the data uncertainties may be assumed to be propagated into the calibrated model parameters and uncertainties. A key aspect of whether data is representative for cases of operational or real-time forecasting is that older data, collected before the establishment of a modern data-telemetry network, may be subject to errors of data handling, reading, and timing that are unlikely to occur or minimized with the telemetry system. Such older calibration data should be checked carefully for such errors, especially in the synchronization of the raingage and stream gage clocks. Jackson and Aron (1971) found that timing errors due to clock malfunctions and lack of synchronization were the most common cause of raingage unreliability. Laurenson and O'Donnell (1969) found that synchronization errors resulted in the greatest errors in unit-hydrograph determination. Since the data uncertainties are considered as part of the parameter uncertainties for calibrated models the details of this incorporation of uncertainties are discussed in Section 3.2.3. Model Parameter Uncertainties.

For synthetic rainfall events the uncertainties of the basic rainfall data have been incorporated into the stochastic or probabilistic structure of the procedures used to establish the design storms or synthetic rainfall series. Thus, the synthetic event uncertainties may be modeled by the confidence intervals on rainfall frequency analysis or the random term in temporal/spatial stochastic models of storm rainfall. For example, Tung (1987) used the mean-value first-order second-moment (MFOSM) method (Section 3.4.3) to estimate confidence bounds for the precipitation depths in the National Weather Service rainfall frequency atlas--U.S. Weather Bureau Technical Paper Number 40 (Hershfield, 1961). This leads to a theoretical dilemma for the case of estimating synthetic events with calibrated models because the data uncertainty structure absorbed by the model parameters is most likely different from the data uncertainty structure absorbed by the parameters of the sythetic-storm model. Thus, some rainfall data uncertainties could be double counted and others ignored when performing a reliability analysis considering uncertainties in the synthetic rain events and calibrated model parameters. Research to resolve this dilemma is needed.

On watersheds for which the streamflow record is too short to adequately calibrate a model or not available at all, flood forecasting and recreation of historic flows requires physical-simulation models for which all the parameters can be determined based on field measurements, physical principles, and past experience on similar catchments. Just as the model parameters for physical-simulation models must be considered and established from physical principles, so must the individual sources of data uncertainty be considered for a reliability analysis of estimates of runoff from actual storms using these models. Precipitation data are by far the dominant source of data uncertainty influencing the reliability of output from physical-simulation models. Thus, the remainder of this section will focus on the uncertainties in precipitation data and means to estimate the magnitude of these uncertainties in estimation of runoff hydrographs for larger storms.

There are eight primary uncertainties involved in using point-rainfall measurements to describe the true precipitation input for a watershed:

1) depth measurement error in the gage due to equipment malfunction,
2) representativeness of ground-level precipitation at the gaging point,
3) gage location, e.g., gages are in positions that consistently result in high or low readings relative to the areal-mean rainfall,
4) gage-network areal-mean rainfall versus true areal-mean rainfall,
5) rainfall spatial variability,
6) rainfall temporal variability,
7) lack of synchronization between time clocks for rain and stream gages, and
8) lack of synchronization between time clocks for the various rain gages in the watershed.

Measurement error due to malfunction is best detected by comparison of the storm rainfall totals and the general precipitation time series with the surrounding rain gages by traditional methods such as double-mass plots. While equipment malfunction is a major problem it need not affect a reliability analysis because data from a malfunctioning rain gage should be dropped from the modeling effort.

Ideally, in order to obtain the best estimate of storm runoff, knowledge of the rainfall intensity at each point in the basin and at each point in time is necessary. However, measurements of precipitation are always made at discrete intervals in time and at a limited number of points in a basin (although improvements in radar measurement of rainfall may offer great improvements of spatial rainfall measurements). Thus, there are uncertainties in the rainfall data due to unsampled temporal and areal variations in the true rainfall. Furthermore, Bras and Rodriguez-Iturbe (1976) among others have demonstrated that the natural "filter" characteristics of the watershed runoff process are insufficient to "damp out" the effects of the unaccounted for spatial and temporal variations in rainfall. Accounting for the data uncertainties due to spatial and temporal rainfall variability is not an easy task. Realistically, even if data from an extensive rain-gage network with readings every 1 minute are available, this fairly detailed reporting of spatial and temporal variability could not be used by the typical computer model of watershed hydrology. Almost all computer models of watershed hydrology, even so-called distributed models, "lump", to some extent, the input data and hydrologic characteristics of the watershed in order to make runoff estimation practical. Further, the temporal variations are also averaged for input to computer models of watershed hydrology partly for computational practicality and partly because the model structure is insensitive to more detailed information. Therefore, the uncertainty related to the inability to account for the true spatial and temporal variation in rainfall is in part a function of the model-structure uncertainties due to temporal and spatial "lumping" of the rainfall-runoff process.

For the case of establishing an operational-forecasting system on a previously ungaged watershed, proper network design should reduce the effects of poor and/or inadequate gage location and the use of modern telemetry systems should reduce time-synchronization problems. It should be noted that modern telemetry is not immune to system breakdown and breakdowns are more likely during the major floods of key interest. For the case of recreating historical flows, the synchronization and gage-location errors can be detected and removed by careful examination of the data relative to the nearby raingages.

Thus, for a reliability analysis of physical-simulation model output for estimation of runoff from actual storms the uncertainties requiring detailed consideration are those due to (1) representativeness of ground-level precipitation at the gaging point and (2) gage-network areal-mean rainfall versus true areal-mean rainfall. Sevruk (1982) performed a detailed review of the systematic errors which cause precipitation measured in the gage to be less than that actually reaching the ground. He proposed the following general model for correcting precipitation data to account for systematic and other errors

$$P_c = (1+k)(P_g + P_1 + P_2 + P_3 \pm P_4 - P_5) \pm P_r \tag{3.1}$$

where P_c is the corrected amount of precipitation; k accounts for wind deflection of precipitation around the gage orifice; P_g is the amount of precipitation measured at the gage; P_1 is the loss due to wetting on the internal walls of the collector; P_2 is the loss due to wetting in the container when it is emptied; P_3 is the loss due to evaporation from the container; P_4 is the effect of splash-in and/or splash-out; P_5 is the effect of blowing and drifting snow; and P_r represents random observational and instrumental errors. Clearly not all these factors are important for the different methods used to collect precipitation data. For example, rainfall data collected by a telemetered, calibrated tipping-bucket raingage during a major storm are only significantly affected by k, P_1, and P_r. Ranges of the magnitude of each of the systematic errors and the meteorological and instrumental factors affecting these errors are presented in Table 3.1.

TABLE 3.1.

Main components of the systematic error in precipitation measurement, magnitude and causitive meteorological and instrumental factors listed in order of general importance (after Sevruk, 1982).

Symbol	Component of error	Magnitude (percent)	Meteorological factors	Instrumental factors
k	Loss due to wind-field deformation above the gage orifice	2-10 10-50[1]	Wind speed at the gage rim during precipitation; the structure of precipitation	Shape, orifice area, and height of both the gage rim and collector
P_1+P_2	Losses from wetting on internal walls of the collector and in the container when emptied	2-10	Frequency, type, and amount of precipitation; the gage drying time; and the frequency of emptying the container	The same as above and the material, color, and age of both the gage collector and container
P_3	Loss due to evaporation from the container	0-4	Precipitation type; saturation deficit and wind speed at the gage during the period between the end of precipitation and the time of measurement	Orifice area and the isolation of the container; the color and in some cases the collector age, or the funnel type (rigid or removable)
P_4	Splash-out and splash-in	1-2	Rainfall intensity and wind speed	Shape and depth of the gage collector and the type of gage installation
P_5	Blowing and drifting snow		Intensity and duration of snow storm, wind speed, and the type of snow cover	Shape, orifice area, and depth of both the gage rim and collector

1 Magnitude for snow

The magnitude of the wind loss, k, may seem a bit low to readers familar with the work of Larson and Peck (1974) whose wind loss curves have been reproduced in several well known hydrology texts (e.g., Linsley et al., 1982, p. 60; Bras, 1990, p. 134). However, a second key variable which influences gage-catch accuracy along with wind speed is the size of the raindrops. Mueller and Kidder (1972) used wind-tunnel experiments with rain-gage models to define velocity patterns, and then used these velocity patterns and drag on raindrops to estimate gage-catch deficiences due to wind speed and drop size. For drops with diameters of 3 mm or greater, winds up to 15 m/s had virtually no effect on rain-gage catch. Laws and Parsons (1943) reported that high intensity (greater than 125 mm/hr) storms have median drop diameters greater than 3 mm and medium intensity (13 to 125 mm/hr) storms have median drop diameters between 2 and 3 mm. Thus, for medium to heavy storms of interest in flood forecasting and recreation of historic flows, under measurement of precipitation by 2 to 10 percent (as indicated in Table 3.1) is reasonable.

Estimates of corrections to rainfall data are part of the normal modeling process. The extra information required for a reliability analysis of model output is the distribution and magnitude of the random error term, P_r, in eq. (3.1). Mueller and Kidder (1972) estimated that the maximum probable error of their results is 15%. This could be an initial estimate of the portion of ΔP_r due to wind effects. The modeler must use personal judgement to estimate the portion of ΔP_r due to other sources.

The uncertainties in the relationship between the gage-network areal-mean rainfall and the true areal-mean rainfall may be approximated using data from dense rain-gage networks. If a dense rain-gage network is available for an area that is subject to climatic and topographic effects similar to those for the watershed being studied, then the uncertainty in estimating the areal-mean precipitation from a small number of gages may be approximated by the following procedure. Assuming that the "true" areal-mean precipitation is obtained from the complete (dense rain gage) network information, the uncertainty associated with areal-mean rainfall estimates for various numbers of rain gages may be approximated by comparing the averages estimated by different subsets of the total rain-gage network. This approach has been used often and successfully in hydrology (e.g., Horton, 1923; Light and Shands, 1947; Linsley and Kohler, 1951; McGuinness, 1963; and Huff, 1970).

The work of McGuiness (1963) is noteworthy because he developed a nomograph for estimating average error in areal-mean rainfall determination from typical gage networks applicable to 9 states in the midwestern U.S. He used data from (1) Coshocton, Ohio, as representative of small (less than 60 km^2) midwestern watersheds; (2) the Illinois networks for Panther Creek, Goose Creek, El Paso, and East Central Illinois as representative of medium sized (60 to 1000 km^2) midwestern watersheds; and (3) the Muskingum, Ohio, network as representative of large midwestern watersheds. From these data he prepared a nomogram relating average errors in areal-mean rainfall estimates to rainfall depth, gaging ratio, and a climatic/geographic factor (the 5-year, 24-hour rainfall depth from Hershfield, 1961). The nomograph was verified by comparing the error estimates with Linsley and Kohler's (1951) results for the Army-Navy-NACA Weather Bureau network near Wilmington, Ohio. Considering McGuiness' results it should be possible to develop similar regionalizations and to transfer uncertainty in areal-mean rainfall within homogeneous climatic/geographic regions.

3.2.3. Model-Parameters Uncertainties

For calibrated models the model-parameters uncertainties also incorporate the effects of data uncertainties because of the curve-fitting properties of the calibration process. Thus, for model applications where calibration is used, the standard deviations and distributions of the basic variables may be estimated directly from the results of model calibration for a number of storms that are representative of the storms to be estimated by the model for design, planning, operation, etc. When Melching et al. (1990, 1991) published this procedure several reviewers familiar with the many problems with calibration procedures were troubled by the fact that not only was calibration to be used to determine parameter "best estimates" but also to infer the uncertainties in the parameters. The difficulties of obtaining unique and conceptually realistic parameter values by model calibration are discussed in detail by Sorooshian and Gupta in Chapter 2. These problems, while formidable, are more related to model fitting than to the reliability analysis. That is, if a model cannot be adequately calibrated to the watershed being studied, the model should not be used operationally. The modeler should use every means to try to obtain unique and conceptually realistic parameter values for each storm including multiple starting points for calibration, examination of the objective-function surface, emphasizing the most sensitive parameters, model reformulation, etc. (see Chapter 2). The best parameter values obtained provide useful information regarding the parameter and propagated data uncertainties although there are identification uncertainties. In fact, Beck (1987, p. 1423) points out,

> 'In simple terms the results of the whole process of identification and the consequences of a lack of identifiability are crystallized in the a posteriori parameter estimation error variance-covariance matrix....Qualitatively, well (or poorly) estimated parameters will have relatively low (or high) error variances; parameters associated with redundant hypotheses should be highly uncertain; and parameters associated with multiple hypotheses, the collective effects of which cannot be disentangled, should yield significant off-diagonal covariances in the matrix.'

Finally, the reliability analysis either explicitly (weighting coefficients) or implicitly (multiple simulations) considers output sensitivity to the parameters, which partially compensates for some parameter-identification problems.

Inference of combined model-parameter and data uncertainties from model-calibration results has been used frequently for computer models of watershed hydrology. Kuczera (1988) determined the covariance matrix for the parameters of RORB (Chapter 5) directly from the nonlinear optimization package used to fit the model to a single storm. He determined the covariance matrix for the model parameters of the Soil Dryness Index streamflow yield model directly from the nonlinear optimization package used to fit the model to a 96-month period of runoff. Rogers et al. (1985) and Binley et al. (1991) used calibration results from 5 events (i.e., different events for the different catchments studied) to estimate the mean and standard deviation of model parameters for application of Rosenblueth's method to the Institute of Hydrology Distributed Model, IHDM (Chapter 17). Warwick and Wilson (1990) calibrated the Storage, Treatment, Overflow, and Runoff Model, STORM, (U.S. Army Corps of Engineers, 1977) for 11 storms on the watershed being studied to determine ranges for the model parameters. They found that the parameters appeared to be independent and assumed uniform distributions for the parameters over the ranges identified by the multiple-event calibration, and used this information for their Monte Carlo simulation of STORM water-quality output. Melching (1987, 1992a) and Melching et al. (1990, 1991) calibrated HEC-1 (Chapter 4) to 32 flood and near-flood events and RORB (Chapter 5) to 30 flood and near-flood events on a 1,600 km^2 agricultural watershed in Illinois. The expected value and covariance of each parameter was determined directly from the statistics of the calibrated storms. Figure 3.2 shows the histograms for the continuing-loss rate, CLR, and watershed delay time factor, C_1, used in RORB. The histograms were used to infer a lognormal distribution for CLR and a normal distribution for C_1. The hypothesized distributions were checked with the Kolmogorov-Smirnov test (Law and Kelton, 1990, p. 387-391) and skewness test for normality (Salas et al., 1980, p. 92), and then used for the reliability analysis of RORB. Similar analyses were used for the parameters of HEC-1.

The single-storm approach of Kuczera (1988) cannot pool information from several storms whereas the statistical analyses of results from several storms of Rogers et al. (1985), Binley et al. (1991), Warwick and Wilson (1990), and Melching (1987) provide a relatively simple pooling of uncertainty information on model parameters. Several more thorough methods of pooling results from several storms by Bayesian statistics have been developed. Bates and Townley (1988b) used the Maximum A Posteriori (MAP) estimation approach they developed (Bates and Townley, 1988a) to pool the model parameters and uncertainties for 8 calibration storms. The MAP approach combines prior information on the expected values and covariances of the model parameters with the results of ordinary or weighted least-squares fitting to obtain a posteriori expected values and covariance for the model parameters. In Bates and Townley's example the objective of calibration was to match the peak discharges from the 8 calibration events. Kuczera (1990a) also proposed a Bayesian approach to pooling parameter values between storms where data are of varying quality. He demonstrated his procedure for an application of RORB to several storms for an Australian watershed.

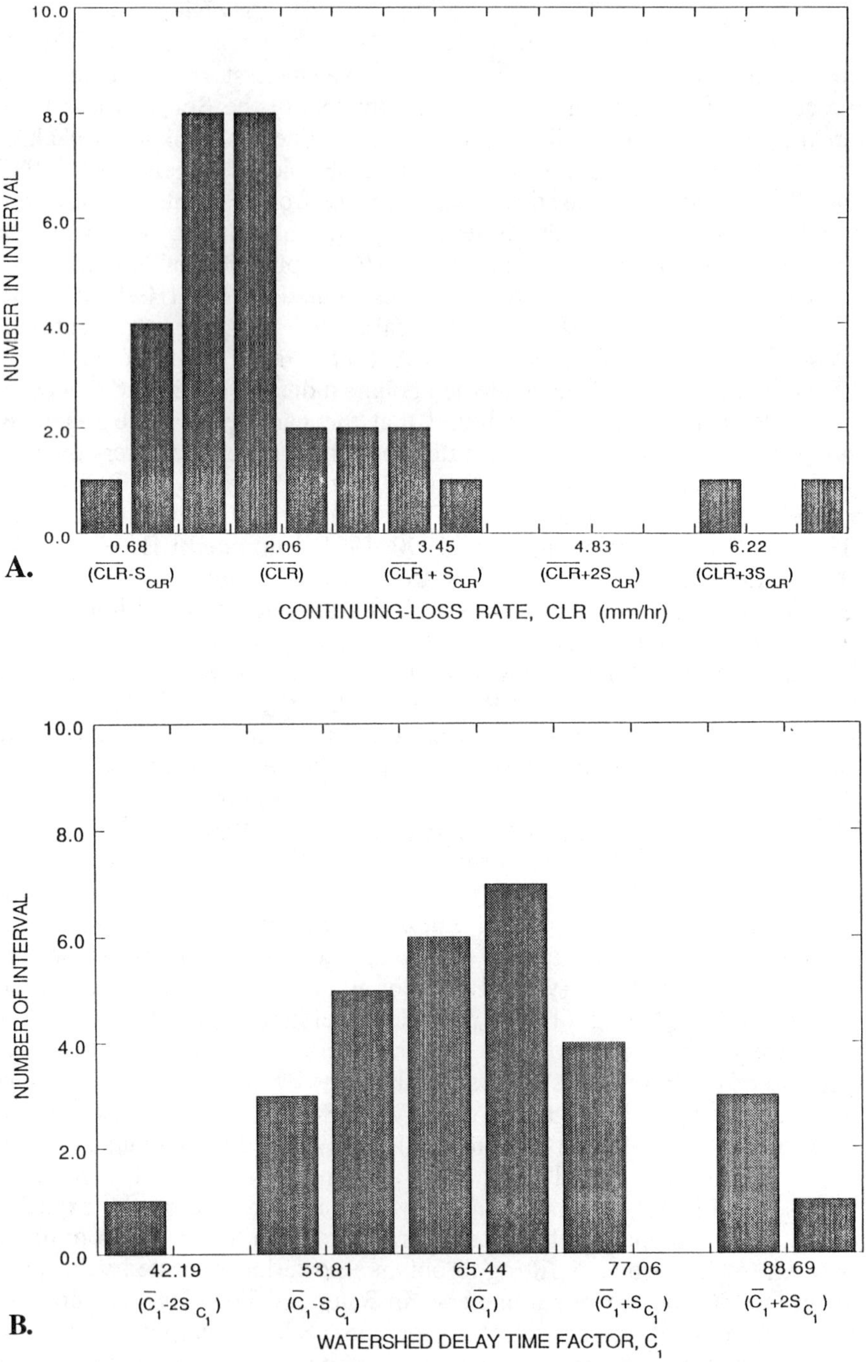

Fig. 3.2. **Histogram of optimal (A) continuing loss rate, CLR, and (B) watershed delay time factor, C_1, for RORB simulation of the rainfall-runoff process for the Vermilion River at Pontiac, Illinois. $\overline{CLR}$ = mean value of CLR, S_{CLR} = standard deviation of CLR, $\overline{C_1}$ = mean value of C_1, and S_{C_1} = standard deviation of C_1.**

Kuczera and Williams (1992) developed a procedure to consider model-parameters and rainfall-data uncertainties separately for calibrated models. Their procedure involves fitting a simple stochastic-transformed rainfall model to rain-gage data for a given storm on the watershed being studied. Kriging is used to calibrate and validate the rainfall model. The computer model of watershed hydrology is then calibrated using subareal-mean rainfall estimated by the rainfall model, which was calibrated to the individual rain-gage measurements, and the measured streamflow. As such this method offers an improved estimate of parameter uncertainty, however, rainfall uncertainties still affect and propagate to the model parameters. To determine the total uncertainty in estimated hydrographs, Monte Carlo simulation is used to generate random samples of (1) the model parameters from a multivariate normal distribution whose mean and covariance were determined during model calibration and (2) the subareal rainfall from the stochastic rainfall model. Kuczera and Williams (1992) demonstrated their method for design hydrograph estimation for a small catchment near Sydney, Australia. Kuczera and Williams also noted that while their procedure is general, it is, in practice, likely to be limited by the availability of sufficient data to identify and calibrate a rainfall model. Thus, they concluded the indirect assessment of the combined effects of model parameter and data uncertainties based on calibration to several storms (discussed above) is the most likely approach for practical applications, whereas their method is useful for assessing the relative importance of rainfall-data and model-parameters uncertainties.

For uncalibrated physical-simulation models the statistics of the parameters must be obtained from physical measurement, information in the literature, and transfer of calibration results from hydrologically similar watersheds. For model parameters, which are difficult to measure or for which there are insufficient measurements, physical constraints, literature review, and past experience with the model allow the modeler to estimate the "best" and probable range of parameter values. By assuming a triangular distribution over the range of parameter values with the apex at the "best" estimate, standard deviations of basic variables have been successfully approximated for several hydrologic applications (Tang and Yen, 1972; Tung and Mays, 1980). A uniform distribution over the parameter range may also be assumed if appropriate. For example, in their Monte Carlo analysis of potential water-quality constituent loads Warwick and Wilson (1990) assumed uniform distributions for the water-quality parameters in STORM (U.S. Army Corps of Engineers, 1977) over the typical ranges reported in the literature for these parameters. Distributional information may also be obtained from the literature. For example, many groundwater researchers have found hydraulic conductivity to be lognormally or geometrically distributed.

The work of Rawls et al. (1983) and Rawls and Brakensiek (1989) is perhaps the most complete source of uncertainty information for physically-based model parameters. They studied the relation between soil-water parameters of the Green and Ampt infiltration equation and the Brooks-Corey equation and soil-textural class, percent sand, percent clay, percent of organic matter, and other factors for approximately 1,200 soils (5,000 soil horizons) covering 34 states in the U.S. They developed tables and equations for prediciting infiltration parameters and their standard deviations on the basis of soil texture, grain-size distribution, etc. Goldman et al. (1990) used the relations developed by Rawls and his coworkers to estimate the mean and standard deviation of Green and Ampt infiltration equation parameters. They used a model combining Green and Ampt infiltration and kinematic-wave overland flow to estimate the results for 101 storms. Monte Carlo simulation was used to estimate the mean, standard deviation, skewness, and kurtosis of peak discharge and runoff volume for 101 storms assuming only the infiltration parameters are uncertain.

At this time (1995) a complete reliability analysis of a physical-simulation model considering uncertainties in the rainfall data and all model parameters has not been done. New research is encouraged to better develop the details of reliability analysis for these models because such research would clearly demonstrate the current quality of design-flood estimates for ungaged basins.

3.2.4. Model-Structure Uncertainties

Model-structure uncertainty stems from the inability to truly represent the watershed's physical runoff process in model simulation. The uncertainties introduced by the approximations used in the model affect the prediction of different runoff features--peak discharge, time to peak, runoff volume, etc.--in different ways and to different degrees. For example, the time to peak of a runoff event is much less sensitive to uncertainties in the abstraction model than is the runoff volume. Thus, it is difficult to consider and account for the total effect of model-structure uncertainty on the complete estimated runoff hydrograph. Therefore, it is necessary to focus on the most important outputs from the model for the problem being studied and to develop a simple, multiplicative correction factor, λ_m, between optimal model performance and the true value of the output of interest.

Garen and Burges (1981) pointed out that if a historical flow record representative of a wide range of watershed responses and of sufficient length to constitute a statistically significant sample is available, comparison of simulated and recorded flows might be extrapolated to estimate uncertainty in subsequent model estimates. For a physical-simulation model the variance between the measured discharge and the estimated discharge using the true input and model parameters mainly results from model-structure inadequacies. For calibrated models, the variance between the measured discharge and estimated discharge stems from model-structure inadequacies, input-data, and model-parameters uncertainties. Fortunately, the error compensation properties of the calibration process minimize the bias in the estimate resulting from error laden data and parameters allowing a reasonable estimate of the effects of model-structure inadequacies. Therefore, the expected value and standard deviation of the difference between measured and calibrated values for the runoff characteristic of interest taken over a wide range of storms provides a reasonable estimate of λ_m and the uncertainty. This approach was used by Melching (1987) and Melching et al. (1990, 1991) in their reliability analysis of peak discharge estimates made by HEC-1 (Chapter 4) and RORB (Chapter 5).

Generally, for physical-simulation models, calibration data are not available to allow approximation of model-structure uncertainty via comparison of estimated and measured runoff characteristics for the watershed being studied. However, calbration, verification, or other experience with the model on a hydrologically similar watershed may be used to approximate the model-structure uncertainty. Further, laboratory analysis of the component processes in the physical-simulation model can aid estimation of the total model-structure uncertainty. Transfer of model-structure uncertainty also applies to cases where a physical-simulation model is used to simulate runoff from a watershed that is changing, e.g., because of urbanization. Calibration for previous data for a changing watershed may be used to determine model-structure uncertainty effects, i.e., λ_m and the variance. At the very least, experienced modelers should be consulted and engineering judgement used to obtain rough estimates of λ_m and associated variance.

3.3. MEASURES OF OUTPUT RELIABILITY

The reliability of output from computer models of watershed hydrology is a function of the combined effect of data, model-parameters, and model-structure uncertainties discussed in Section 3.2 and measures of output reliability may be estimated by any of the 6 methods discussed in Section 3.4. A convenient way to consider the uncertainty in model output for a given storm is to assess the probability that the output variable of interest (peak discharge, volume, etc.) will exceed some limit state or hydrologic target level, T_H. The model performance may then be described by

$$Z = T_H - \lambda_m h_m(\theta_m) = T_H - H(\mathbf{X}) = g(\mathbf{X}) \tag{3.2}$$

where Z is the model-performance function; $h_m(\theta_m)$ is a function representing the model estimate of the hydrologic variable in question; θ_m is a vector containing the model input data and parameters; $H(\mathbf{X})$ is a function representing the corrected model output, $\lambda_m h_m(\theta_m)$; $g(\mathbf{X})$ is a functional form of the performance index or function; $\mathbf{X} = \theta_m \cup \lambda_m$, the vector of basic variables describing the model uncertainty, and $\cup$ is the union of sets.

The hydrologic target level is a dummy variable that is used in reliability analysis for estimation of the CDF of the model output. For a given target value T_H, the exceedance probability, P_E, (the complement of the CDF) is given by

$$P_E = \Pr(Z < 0) \tag{3.3}$$

where Pr(y) is the probability of event y occurring. For example, by setting T_H to 50 m^3/s, the probability that the peak discharge estimated by the model for the storm in question will be greater than or equal to 50 m^3/s can be obtained. Whether this probability reflects the actual likelihood that the storm will produce a peak discharge greater than or equal to 50 m^3/s depends on how well the model is suited to the watershed and how well the uncertainties in the basic variables (rainfall, model parameters, model-structure correction factor) have been estimated. The complete CDF of model output can be obtained by setting T_H to 100 m^3/s, 150 m^3/s, and so on. For certain cases the hydrologic target level may take on additional meaning. For example, T_H may denote the measured peak discharge for model verification or model calibration in the scheme proposed by Ibrahim and Liong (1992), or a hazardous flood discharge in flood-event forecasting.

The CDF and PDF of Z are the most complete measures of model-output reliability. The CDF of Z is obtained by varying T_H and calculating P_E. The PDF is then determined by numerical differentiation of the CDF. The CDF and PDF provide a quick visual assessment of model reliability and acceptability. The CDF and PDF may also be used to assess the probability of an acceptable percent deviation from the estimated value (the mean or the mode of the output probability distribution). For example, an acceptability criterion might be that 90% of the possible output values should be within ± 30% of the predicted value. Such a criterion could be quickly checked using the CDF or PDF of Z. The CDF may also be used in planning to assess risk-cost features of flood-protection measures, reservoir-operation rules, etc.

Often the complete reliability information provided by the CDF and PDF are not needed or are difficult to obtain because of the complexity of the watershed model being applied. Several other output reliability measures have been proposed that may be used in such cases. For watershed-model users, estimation of the standard deviation and confidence limits for the model output have great importance in model selection, model verification, and model use. Specific examples are presented in Sections 3.4 and 3.5.

The mean and standard deviation of the model output can be used to parameterize standard probability distributions (normal, lognormal, Gamma, etc.) that may be used to approximate the CDF and PDF of the model output so that the full probability distributions may be used in risk-cost design and planning. Tung and Mays (1980, 1981) used MFOSM, (Section 3.4.3) to obtain the mean and standard deviation of design discharge and hydraulic capacity for culverts (1980) and levees (1981). They then fit probability distributions for the design discharge and hydraulic capacity and integrated to find the failure risk. Tung and Hathhorn (1988) used MFOSM to estimate the mean and standard deviation of dissolved-oxygen (DO) deficit as estimated by the Streeter-Phelps (1925) equation. Distributions were then estimated for the DO deficit and compared to distributions estimated by Monte Carlo simulation (Section 3.4.1). This approach should be used with caution when the problem under study requires estimates of cases with very low P_E values because, as with hydrologic frequency analysis, the choice and parameterization of the probability distribution is primarily based on information for the central portion of the distribution and extrapolation to the tails of the distribution can be inaccurate. In such cases several distributions should be fit and the sensitivity of the planning decision relative to the choice of distribution should be considered.

3.4. RELIABILITY ANALYSIS METHODS

Ideally, the joint probability distribution should be obtained for all the significant sources of uncertainty affecting the reliability of output from the watershed model being used. As discussed in Section 3.2, however, the determination of probability distributions for the basic variables is quite difficult and involves several assumptions. Further, the multivariate combination and integration of the basic-variable distributions considering watershed model interactions between basic variables is impossible for all but the simplest (linear) models with identically distributed parameters. Thus, the aggregation of uncertainties in the basic variables of the watershed model into measures of overall model-output reliability must be done in an approximate manner. Six aggregation methods that have been used in water-resources engineering are described and their performance for a limited number of cases is compared in the following subsections.

3.4.1. Monte Carlo Simulation

In Monte Carlo simulation (MCS), probability distributions are proposed for the uncertain variables for the problem (system) being studied. Random values of each of the uncertain variables are generated according to their respective probability distributions and the model describing the system is executed. By repeating the random generation of variable values and model execution steps many times the statistics (mean, standard deviation, skewness, etc.) and an empirical probability distribution of the model output can be determined. Law and Kelton (1991, Chapters 7 and 8) provide a detailed discussion of how to generate random variables that follow a wide variety of probability distributions. The accuracy of the statistics and probability distribution obtained from MCS is a function of the number of simulations performed. For models with a large number of parameters and for which low probabilities (< 0.1) are of interest, tens of thousands of simulations are required. Rules for determining the number of simulations required for convergence are not available, and, thus, replication of the MCS runs for a given number of simulations is the only way to check convergence. The generation of correlated basic variables can also be a problem in MCS. Tung and Yen (1993) note that the generation of correlated normal or lognormal variables poses no difficulty; however, difficulty arises in the generation of a mixture of correlated non-normal and normal random variables. Chang et al, (1994) proposed a practical multivariate Monte Carlo simulation that preserves the marginal distributions of random variables and their correlation structure without requiring the complete joint distribution.

MCS is an extremely flexible and robust method capable of solving a great variety of problems. In fact, it may be the only method that can estimate the CDF and PDF of Z for cases with highly nonlinear and/or complex system relationships. Further, Beck (1985) notes that "when the computing power is available, there can, in general, be no strong argument against the use of Monte Carlo simulation." For complex, nonlinear models with many uncertain basic variables, however, the number of simulations (thus the computer time) necessary to achieve an accurate estimate may become prohibitive. For example, Brown and Barnwell (1987) report that for the QUAL2E multiple constituent (DO, nitrogen cycle, algae, etc.), steady-state, surface water-quality model 2,000 simulations are required to obtain accurate estimates of the output standard deviation.

MCS has been used to assess the output reliability of a wide variety of computer models of watershed hydrology. Many of these applications of MCS were to provide a check of the less computationally intensive reliability-analysis methods presented in the following subsections; these comparisons are discussed in Section 3.4.7. Warwick and Wilson (1990) applied MCS to STORM (U.S. Army Corps of Engineers, 1977) to ascertain probable ranges of water-quality estimations considering uncertainties in both hydrologic and water-quality parameters. The resulting 95% confidence intervals were compared with a limited water-quality data set to test model adequacy. Goldman et al. (1990) used MCS to determine the mean, standard deviation, skewness coefficient, and kurtosis of estimated volumes and peak discharges produced by a model using the Green and Ampt equation to estimate infiltration losses and the kinematic-wave method to route overland flow. Only the uncertainties in the infiltration model were considered in their analysis. Binley et al. (1991) used MCS to develop confidence limits on hydrographs estimated with IHDM (Chapter 17) to determine whether differences in runoff predicted for afforestation of an upland catchment in Wales are significant. Harlin and Kung (1992) used MCS to derive confidence limits on hydrographs estimated by the HBV hydrological model (Chapter 13) for the purpose of comparing several alternative calibration methods for two large catchments in Sweden.

Beven and Binley (1992) have developed an overall procedure for model calibration and reliability estimation employing MCS as a key component. Their procedure, called Generalized Likelihood Uncertainty Estimation (GLUE), is based on the premise that for distributed, physical-simulation models there is no single best parameter set that represents the watershed for a range of rainfall-runoff responses, but rather that a range of different sets of model parameter values may represent the rainfall-runoff process equally well. The method assigns likelihood weights to each set of parameters that reasonably reproduce measured system responses (runoff, soil moisture, etc.). These likelihood weights apply to the use of the parameter sets for design, verification, or synthetic storms or periods. For periods of estimation, forecast, or prediction, hydrographs are computed using each parameter set and output statistics and ranges are calculated as weighted averages (e.g., mean = sum of the normalized parameter-set likelihood weight times discharge value resulting from that parameter set).

The steps in the GLUE procedure are as follows.

1) Define the likelihood measure or set of likelihood measures. The likelihood measure is typically a standard model-calibration objective function, such as model efficiency, sum of squares to the -N power, maximum error, or sum of absolute errors. The likelihood function may consider a weighted combination of several model outputs (e.g., discharge and soil moisture at several locations in the watershed where measurements are available).

2) Determine an appropriate range for the initial distribution of parameter values. Previous experience with the model is helpful, but not necessary because the range and distribution for the parameter values may be refined after an initial run of the GLUE procedure.

3) Use MCS to generate a large number parameter sets and model outputs [Beven and Binley (1992) used 500 simulations for their analysis of the IHDM (Chapter 17)] and calculate the likelihood measure for each simulation and parameter set.

4) Choose a level of the likelihood function which divides acceptable simulations from unacceptable simulations.

5) Delete the unacceptable simulations, and normalize the likelihood values for the parameter sets that yield acceptable simulations such that the sum of the normalized likelihood values (i.e. likelihood weights) equals 1.

6) For design, forecast, or other predictions generate results for all the parameter sets that yielded acceptable calibration simulations and determine the weighted mean discharge and simulation probability bounds.

The GLUE procedure allows easy updating of the likelihood weights and range of parameter sets as additional data are collected and easy evaluation of the value of the additional data. The main short coming of the GLUE procedure is the dependency of the uncertainty estimation on the selection of the value separating acceptable and unacceptable simulations. If it is set too high, few parameter sets will be acceptable and the uncertainty bounds for predicted events will be unrealistically small; whereas if it is set too low, any parameter set will be acceptable and the uncertainty bounds for predicted events will be so large that the model will seem to be of little value. This subject requires further research. Beven and Binley (1992) illustrate the application and benefits of GLUE for a case of a small (3.9 km^2) upland watershed in Wales modeled with IHDM (Chapter 17).

Other important applications of MCS in watershed modeling include determining the key parameters in model design and building and attempts to determine the sensitivity of model output to given sources of uncertainties. Smith and Hebbert (1979) used MCS to analyze the effects of random spatial distribution of soil-infiltration properties on the hydrologic performance of small watershed areas as modeled by several infiltration models of differing complexity and spatial resolution. The goal of their study was to provide a working hypothesis for interpretation of watershed hydrologic response and a guide to meaningful infiltrometer sampling. Gardner et al. (1980) used MCS to study the reliability of daily values of water level and storage produced by a continuous model of marsh hydrology. MCS was mainly applied to verify the model. Many parameter sets were found that could match the water level and storage on one of four seasonally representative dates, but none of the 500 simulations could come within one standard deviation on all four seasonally representative dates. Takasao and Takara (1988) used MCS to compare simplified models on the basis of whether or not the models preserved the stochastic transformation properties of the runoff system represented by a more theoretically correct (ideal) model. They considered models of two types of hillslope runoff processes: (1) a surface-runoff (overland-flow) process, and (2) a surface-subsurface runoff process. Krajewski et al. (1991) used MCS to study the sensitivity of the runoff estimates made by a model combining the Soil Conservation Service curve-number model of precipitation excess and kinematic-wave routing of overland flow and streamflow to rainfall-input spatial and temporal sampling density. They found that the model results had a higher sensitivity to temporal resolution than to spatial resolution for a small (7.5 km^2) watershed, and that a unit hydrograph greatly underestimated peak discharges relative to the distributed models.

3.4.2. Latin Hypercube Simulation

Latin Hypercube Simulation, LHS, (McKay et al., 1979; McKay, 1988) is a stratified sampling approach that allows efficient estimation of the statistics of output. In LHS the probability distribution of each basic variable is subdivided into N ranges each with a probability of occurrence equal to 1/N. Random values of the basic variable are simulated such that each range is sampled only once. The order of the selection of the ranges is randomized and the model is executed N times with the random combination of basic variable values from each range for each basic variable. The output statistics, distributions, and correlations may then be approximated from the sample of N output values. McKay et al. (1979) have shown the stratified sampling procedure of LHS converges more quickly than the random sampling employed in MCS and more quickly than other stratified sampling procedures. As with MCS, accuracy is a function of the number of samples and no rules are available for choosing N. McKay (1988) suggests that N equal to twice the number of suspected important parameters might provide a good balance of accuracy and economy for models with large numbers of parameters, but ultimately each modeler must check for convergence for the model and problem in question.

For correlated basic variables McKay (1988) discusses several procedures to generate the stratified sample. The method with the broadest application is called the exact sampling method. For the case of two dependent basic variables, the marginal distribution of one of the basic variables can be stratified and sampled as in LHS, and the paired value of the codependent basic variable can be randomly sampled from the conditional distribution using the value of the first basic variable. These pairs remain together in the random assignment to model runs. For n-dependent basic variables, the conditional sampling can be extended, or the joint distribution can be sampled randomly in a stratified manner. The n-tuples remain together in the random assignment to model runs.

To the best of this author's knowledge, the application of LHS to HEC-1 (Chapter 4) for the purpose of comparing LHS to the MCS, MFOSM, and Rosenblueth methods presented in Section 3.4.7 is the first use of LHS for computer models of watershed hydrology. In water-quality modeling, Jaffe and Ferrara (1984) have used LHS to determine the key parameters in a 9-parameter model of sediment-water column interactions for hydrophobic pollutants. In hydraulic engineering, Yeh and Tung (1993) compared LHS to MFOSM and Harr's method for a model of the migration of gravel borrow pits in alluvial streams. They also used LHS to determine the key parameters in the 28-parameter model. Chang et al. (1992) compared LHS to MFOSM and Harr's method for the HEC2-SR model of water-surface profiles for streams with movable beds. These applications have shown that LHS is potentially a very powerful tool for reliability analysis that provides the flexibility of MCS without the high computational requirements. Research on the use of LHS in watershed hydrology is strongly encouraged.

3.4.3. Mean-Value First-Order Second-Moment Method

In the first-order reliability analysis methods, a Taylor series expansion of the performance function is truncated after the first order term

$$Z = g(\mathbf{X}_\mathrm{e}) + \sum_{i=1}^{p} (x_i - x_{ie})(\partial g / \partial x_i)_{\mathbf{Xe}} \tag{3.4}$$

where $\mathbf{X_e}$ is the vector representing the expansion point and p is the number of basic variables, x_i.

In MFOSM, the expansion point is at the mean value (or some other convenient central value) of the basic varibles. Thus, the expected value and variance of the performance fuction are

$$E[Z] \approx g(\mathbf{X}_\mathrm{m}) \tag{3.5}$$

$$Var(Z) = \sigma_Z{}^2 \approx \sum_{i=1}^{p} \sum_{j=1}^{p} (\partial g / \partial x_i)_{\mathbf{Xm}} (\partial g / \partial x_j)_{\mathbf{Xm}} E[(x_i - x_{mi})(x_j - x_{mj})] \tag{3.6}$$

where σ_Z is the standard deviation of Z and $\mathbf{X_m}$ is the vector of mean values of the basic variables. If the basic varibles are statistically independent, the variance of Z becomes

$$Var(Z) = \sigma_Z{}^2 \approx \sum_{i=1}^{p} \left[(\partial g/\partial x_i)_{\mathbf{Xm}} \sigma_i\right]^2 \tag{3.7}$$

where σ_i is the standard deviation of basic variable i. Similarly the mean and variance of the model output, Q, are

$$E[Q] \approx H(\mathbf{X}_\mathrm{m}) \tag{3.8}$$

$$Var(Q) = \sigma_Q^2 \approx \sum_{i=1}^{p} \sum_{j=1}^{p} (\partial H / \partial x_i)_{\mathbf{Xm}} (\partial H / \partial x_j)_{\mathbf{Xm}} E\left[(x_i - x_{mi})(x_j - x_{mj})\right] \tag{3.9}$$

If the basic variables are statistically independent, the variance of Q becomes

$$Var(Q) = \sigma_Q^2 \approx \sum_{i=1}^{p} \left[(\partial H / \partial x_i)_{\mathbf{Xm}} \sigma_i\right]^2 \tag{3.10}$$

In some cases the derivatives may be determined analytically, either directly or by adjoint sensitivity methods (e.g., LaVenue et al., 1989). In other cases it is best that the derivatives be determined numerically using finite difference methods. Melching (1987, 1992a) found that a forward difference wherein each basic variable value is increased 1% with the others held constant and the change in Z is calculated was adequate for examination of the HEC-1 (Chapter 4) and RORB (Chapter 5) watershed models. Other researchers have used backward differences and central differences with Δx values between 1 and 10% depending on the sensitivity of the model to the parameters being varied.

To estimate the CDF and PDF of Z, when using MFOSM, it is typically assumed that Z is normally distributed and that the system reliability can be defined in terms of a reliability index, β, defined as the reciprocal of the coefficient of variation of Z, that is,

$$\beta = E[Z]/\sigma_Z \tag{3.11}$$

The exceedance probability for the given hydrologic target level is then obtained as

$$P_E = 1 - \Phi(\beta) \tag{3.12}$$

where $\Phi(\beta)$ is the standard normal integral. The Gaussian (normal) assumption has several practical advantages. If the system performance is linear and the basic variables are normally distributed (this is a reasonable assumption for basic variables with small coefficients of variation, CV_i), Eq. (3.12) yields the exact exceedance probability. If the system performance function is nonlinear in a multiplicative sense and the basic variables are lognormally distributed, Eq. (3.12) yields a close approximation to the exact exceedance probability as long as the CV_i's of the basic variables are relatively small. Further, Yen et al. (1986) discuss cases where the Gaussian assumptions of Eq. (3.12) may be quite useful given limited information on the basic variable distributions.

The great advantage of MFOSM is its simplicity, requiring knowledge of only the first two statistical moments of the basic variables and simple sensitivity calculations about selected central values. However, when applied to engineering design problems, in general, including watershed modeling and design, MFOSM has several theoretical and/or conceptual problems (Melching, 1992a, Cheng, 1982). The main weakness of the MFOSM method is that it is assumed that a single linearization of the system performance function at the central values of the basic variables is representative of the statistical properties of system performance over the complete range of basic variables. For nonlinear systems this assumption becomes more and more inaccurate as the basic variables depart from the central values. For engineering design, MFOSM may be especially inaccurate because design failure should only happen because of extreme values of the basic variables describing the system. In MFOSM the assumption to use a normal distribution for the system performance function is only weakly supported, and no logical way can be used in MFOSM to include available information on basic variable probability distributions.

Several poignant examples are available in the water-resources literature to illustrate the nonlinearity effects on statistical properties of the system and on designed engineering systems.

Scavia et al. (1981) detected significant nonlinearity effects when comparing variance estimates from MFOSM and MCS for a eutrophication model of Saginaw Bay, Lake Huron, U.S. From a physical standpoint they reasoned that the discrepancy between the variance estimates was because MFOSM estimates measure variance with respect to a typical component of the algal population (i.e. the typical algal species), while MCS estimates consider variability over the entire algal population.

To circumvent nonlinearity problems, Kuczera (1988) applied Beale's (1960) nonlinearity coefficient to measure the relative nonlinearity of the system. He demonstrated the usefulness of Beale's nonlinearity coefficient for checking the validity of estimation confidence limits derived by MFOSM for two conceptual, nonlinear computer models of watershed hydrology--RORB (Chapter 5) and the Soil Dryness Index streamflow yield model. Bates (1988) also studied the utility of several measures of model statistical linearity applied to RORB including Beale's coefficient, Box's (1971) bias calculation, and the curvature measures of Bates and Watts (1980). He found that Beale's measure may underestimate nonlinearity, particularly in situations where the configuration of trial basic-variable vectors is constrained by basic variable feasibility considerations. Thus, Beale's measure should be used with caution. He also concluded that the curvature measures of Bates and Watts are probably the most useful measures in that they separate and quantify intrinsic and parameter-effects components of linearity. However, the accuracy of the curvature measures is reduced for cases when numerical differentiation must be used to estimate the curvature measures, and the curvature measures should be used with caution is such cases. Kuczera (1990b) recognized the problems with Beale's coefficient and recommended coupling Beale's coefficient with response-surface plots that illustrate the conditional relations between model parameters. Again RORB was used to illustrate the procedure.

LeVenue et al. (1989) clearly detected the nonlinearity problems associated with system failure at extreme values rather than central values. They wished to estimate the probability distribution of the time for a radioactive particle from a waste containment site to travel through a multilayer aquifer to an aquifer utilized for water supply considering uncertainty in the knowledge of aquifer characteristics. From MCS they found that travel paths different from the expected travel path led to faster travel times and MFOSM's sensitivity to the expected travel path did not adequately reflect the probabilities of the other travel paths.

Despite its short-commings MFOSM has been used quite successfully in hydrologic engineering. MFOSM has been directly applied to hydrologic design of storm sewers (Tang and Yen, 1972) and culverts (Yen et al., 1980), and indirectly applied to hydrologic design of culverts and levees by Tung and Mays (1980 and 1981, respectively) as described previously.

Garen and Burges (1981) have successfully applied MFOSM to a modified version of the Stanford Watershed Model (the forerunner of Hydrological Simulation Program Fortran described in Chapter 12) for estimation of design floods for a hypothetical catchment. MFOSM has been applied successfully to RORB (Chapter 5) by Melching (1987, 1992a and b) for a 1,600 km^2 watershed in Illinois, USA; Kuczera (1988) for a 90 km^2 and a 985 km^2 watershed in Australia; and by Bates and Townley (1988b) for a 153 km^2 watershed in Australia. MFOSM has been applied successfully to HEC-1 (Chapter 4) by Melching (1987, 1992a and b) for a 1,600 km^2 watershed in Illinois, USA. MFOSM has been applied sucessfully to the runoff block of HYSTEM-EXTRAN (a modification of the StormWater Management Model, SWMM-Chapter 22) by Lei and Schilling (1993) for a small urban watershed in Switzerland. The primary purpose of these applications was to test the utility of MFOSM relative to MCS and, thus, the results of these studies are discussed in detail in Section 3.4.7.

Several researchers (e.g., Bates and Townley, 1988b; Tung and Hathhorn, 1988) also discuss a second-order approximation of the expected value of Z given by

$$E[Z] \approx g(\mathbf{X}_{\mathrm{m}}) + 0.5 \sum_{i=1}^{p} \sum_{j=1}^{p} \left\{ (\partial^2 g / \partial x_i \partial x_j) COV(x_i, x_j) \right\} \tag{3.13}$$

where $COV(x_i,x_j)$ is the covariance between x_i and x_j. Some refer to the use of Eq. (3.13) for the expected value of Z and Eq. (3.6) or (3.7) for the variance of Z as a second-order reliability analysis or second-order approximation. This is both incorrect and misleading. It is incorrect because in structural reliability analysis a complete second-order approach has been developed (e.g., Lindgren, 1984; Breitung, 1984; and others) that provides second-order estimates of both the mean and variance at a nonlinear approximation of the failure surface (i.e., where Z = 0). The complete second-order approach is very accurate but quite computationally intensive. It is most useful for highly nonlinear problems (it is shown in Section 3.4.7 the nonlinearities in typical hydrologic model applications are rather small) where very low P_E values are of interest, as such it is probably not necessary for hydrologic modeling. Referring to the combination of Eqs. (3.13) and (3.6) or (3.7) as a second-order approach is misleading in that it implies a significant improvement over the MFOSM method. Such an improvement is not necessarily the case. Bates and Townley (1988b) found the "so-called" second-order approximation offered little improvement in the mean value relative to MCS for RORB. It has been this author's experience that the expected value of Z is generally well estimated by Eq. (3.5) (see Section 3.4.7) and the key to improving estimates offered by MFOSM is to provide a better estimate of the variance of Z. The Advanced First-Order Second-Moment (AFOSM) method described in the next section offers such an improvement in the estimate of the variance of Z.

3.4.4. Advanced First-Order Second-Moment Method

Hasofer and Lind (1974) proposed taking the Taylor series expansion at a likely point on the failure surface of the performance function (i.e., where Z= 0) as a method to maintain some of the simplicity of MFOSM and yet reduce its flaws. Rackwitz (1976) expanded on the ideas of Hasofer and Lind, and the following discussion follows the procedures proposed by Rackwitz. The expected value and variance of the performance function estimated by performing a Taylor series expansion at point $\mathbf{X}^*$ on the failure surface (i.e., $g(\mathbf{X}^*) = 0$) are

$$E[Z] \approx \sum_{i=1}^{p} C_i^*(x_{mi} - x_i^*) \tag{3.14}$$

$$Var(Z) = \sigma_z^2 \approx \sum_{i=1}^{p} \sum_{j=1}^{p} C_i^* C_j^* E\left[(x_i - x_i^*)(x_j - x_j^*)\right] \tag{3.15}$$

where C_i^* is $\partial g / \partial x_i$ evaluated at $\mathbf{X}^*$ and C_j^* is $\partial g / \partial x_i$ evaluated at $\mathbf{X}^*$.

For the case of stochastically independent basic variables the variance becomes

$$Var(Z) = \sigma_z^2 \approx \sum_{i=1}^{p} (C_i^* \sigma_i)^2 \tag{3.16}$$

For computational convenience it is often necessary to linearize the estimation of σ_Z as follows

$$\sigma_z \approx \sum_{i=1}^{p} \alpha_i C_i^* \sigma_i \tag{3.17}$$

in which

$$\alpha_i = (C_i^* \sigma_i) / \left[\sum_{j=1}^{p} (C_j^* \sigma_j)^2\right]^{1/2} \tag{3.18}$$

Thus, the reliability index, β, as estimated by AFOSM for the case of stochastically independent basic variables may be expressed as

$$\beta = \frac{g(\mathbf{X}^*) + \sum_{i=1}^{p} C_i^*(x_{mi} - x_i^*)}{\sum_{i=1}^{p} \alpha_i C_i^* \sigma_i} \tag{3.19}$$

where g(**X**) equals 0 at $\mathbf{X}^*$. Similar relations may be derived for the case of correlated basic variables.

For statistically linear models (linear failure surfaces) where all the basic variables are normally distributed, the exact exceedance probability is given by Eq. (3.12). For statistically nonlinear models with convex failure surfaces where all the basic variables are normally distributed, Hasofer (1974) showed that the exceedance probability is bounded as

$$1-\Phi(\beta) \leq P_E \leq 1-F_{\chi 2}(\beta^2) \tag{3.20}$$

where $F_{\chi 2}(\beta^2)$ is a chi-square distribution with p degrees of freedom. The conventional approach has been to set the exceedance probability equal to the lower bound of Eq. (3.20). The validity and accuracy of this assumption have been found acceptable for a number of nonlinear performance functions typical of those found in practical engineering problems (Ang and Tang, 1984, p.383). The quality of this approximation for water-resources models is illustrated in Section 3.4.7.

For most real modeling problems, not all the basic varibles are normally distributed. Rackwitz (1976) proposed a transformation in which the values of the CDF and PDF of the non-normal distributions are the same as those of the equivalent normal distributions at the failure point, i.e.

$$F_{xi}(x_i{*}) = \Phi[(x_i{*} - x_{miN})/\sigma_{iN}] \tag{3.21}$$

$$f_{xi}(x_i{*}) = \{f^N[(x_i{*} - x_{miN})/\sigma_{iN}]\}/\sigma_{iN} \tag{3.22}$$

where $F_{xi}(x_i{}^*)$ is the CDF of x_i at $x_i{}^*$, $f_{xi}(x_i{}^*)$ is the PDF of x_i at $x_i{}^*$, $f^N(\,)$ is the PDF of the standard normal distribution, x_{miN} is the mean of the equivalent normal distribution, and σ_{iN} is the standard deviation of the equivalent normal distribution. In order to do this, Rackwitz approximated the non-normal distribution by a first-order Taylor series expansion. Thus, the mean and standard deviation of the equivalent normal distribution become

$$x_{miN} = x_i{}^* - \Phi^{-1}[F_{xi}(x_i{}^*)]\,\sigma_N \tag{3.23}$$

$$\sigma_{iN} = f^N\{\Phi^{-1}[F_{xi}(x_i{}^*)]\}/f_{xi}(x_i{}^*) \tag{3.24}$$

Yen et al. (1986) present normal transformations of this type for several frequently used distributions. The accuracy of this transformation process for water resources models is illustrated in Section 3.4.7.

The formulations and procedures used in AFOSM are fairly simple and straightforward. The key to AFOSM is the determination of the failure point for the Taylor series expansion. Shinozuka (1983) has shown that for AFOSM the reliability index, β, (or its absolute value) is the shortest distance in standardized space between the system mean state (i.e., where all basic variables are at their mean values) and the failure surface. Thus, if the failure point is determined correctly it represents the most likely combination of basic variable values which produce the hydrologic target level. Because of β's shortest distance definition, optimization algorithms currently employed for constrained nonlinear optimization can be used to determine the failure point. For example, Cheng (1982) proposed using the generalized reduced-gradient algorithm to minimize the absolute value of Z with the constraints chosen such that the proper value of β is determined. Also, Shinozuka (1983) describes a Lagrange multiplier approach for direct minimization of β.

While such optimization approaches are attractive, iterative methods to determine the failure point are probably more easily adapted to reliability analysis of computer models of watershed hydrology. Melching (1987, 1992) has found that a modified version of Rackwitz' (1976) iteration scheme has worked well in applications to HEC-1 (Chapter 4) and RORB (Chapter 5). This iteration scheme is as follows.

1) Define the performance function g($\mathbf{X}$) for the model.

2) Input the statistics of each basic variable.

3) Assume initial trial values of β_t and $\mathbf{X}_t^*$, where the subscript t refers to the trial number for the interations. If information to guess the initial trial β_t and $\mathbf{X}_t^*$ is inadequate, the iteration may be started by assuming $\mathbf{X}_t^*$ using $\mathbf{X}_m$.

4) Evaluate the partial derivatives C_i for all i at the trial point $\mathbf{X}_t^*$.

5) Calculate x_{miNt} and σ_{iNt} for all i, according to Eqs. (3.23) and (3.24).

6) Compute α_{iNt} for all i, according to Eq. (3.18) substituting σ_{iNt} for σ_i.

7) Compute new trial values of $\mathbf{X}_t^*$ for all i, using the following equation derived from Eq. (3.19)

$$x_i{}^*{}_t = x_{miNt} - \alpha_{iNt}\beta_t\sigma_{iNt} \tag{3.25}$$

If β_t was not already assumed in the initial trial, calculate β_t using Eq. (3.19) and substituting x_{miNt}, σ_{iNt}, and α_{iNt} for x_{mi}, σ_i, and α_i, respectively.

8) Substitute new trial $\mathbf{X_t}^*$ obtained in step 7 into $g(\mathbf{X_t}^*)$ and solve for new trial β_t using Eq. (3.19) substituting x_{miNt}, σ_{iNt}, and α_{iNt} for x_{mi}, σ_i, and α_i, respectively.

9) Using the new trial value of β_t, recompute $\mathbf{X_t}^*$ using Eq. (3.25).

10) Repeat steps 4 to 9 until convergence is obtained.

Step 10 poses some questions in regard to acceptable convergence. Ideally one would like to have convergence to less than 1% for each basic variable. However, convergence of β_t to within a few percent between trials and of Z approaching 0 within 0.001 or 0.0001 times $g(\mathbf{X_m})$ is often the best that can be attained. Failure to converge in the basic variables results from discontinuities in C_i resulting from the use of threshold-type parameters. For example, for models using an initial loss-continuing loss rate abstraction scheme, such as HEC-1 (Chapter 4) and RORB (Chapter 5), a serious discontinuity with respect to the initial loss occurs when

$$P_j - bCL < IL < P_j \tag{3.26}$$

where P_j is the precipitation input up to and including time period j, CL is the continuing loss rate, IL is the initial loss, and b the portion of time period j over which the continuing loss rate applies. For IL values between the bounds of Eq. (3.26) the value of C_i equals 0 whereas outside these bounds the value of C_i may be greatly different from 0. Fortunately, Melching (1987, 1992) has found that convergence in β and of Z to zero is adequate for estimating the CDF of peak discharge estimated by HEC-1 and RORB.

AFOSM has been used by Melching et al. (1991) to compare the reliability of event peak-discharge estimates calculated by HEC-1 (Chapter 4) and RORB (Chapter 5). They found that in terms of output reliability the lumped, linear version of HEC-1 used performed as well as the nonlinear, quasi-distributed RORB model for a 1,600 km^2 agricultural watershed in central Illinois with sparse rainfall data. Ibrahim and Liong (1992) proposed a probability weighted calibration procedure based on AFOSM. In their procedure T_H is set equal to the measured peak discharge and the most likely parameter values from a range of reasonable parameter values are determined which match the peak by AFOSM procedures. They applied their procedure to three storms for a small urban watershed (6.11 km^2) in Singapore modeled with SWMM (Chapter 22).

3.4.5. Rosenblueth's Point Estimation Method

In 1975 Rosenblueth proposed a method for estimating the statistical moments of model output from the mean and covariance of the basic variables using a point estimation of a first-order Taylor series expansion as opposed to the derivative-based estimation in MFOSM and AFOSM. In a general case where there are p stochastic (uncertain) basic variables for a system, Rosenblueth (1975) demonstrated that the Nth moment of Q about the origin can be approximated via a point-probability estimate of the first-order Taylor series expansion as

$$E[(Q)^N] \approx [(q_{+++...p})(Q_{+++...p})^N + (q_{-++...p})(Q_{-++...p})^N + ... + (q_{---...p})(Q_{---...p})^N] \quad (3.27)$$

where $Q_{+++...p}$ is $H(x_{m1} + \sigma_1, x_{m2} + \sigma_2, x_{m3} + \sigma_3, ..., x_{mp} + \sigma_p)$, $Q_{---...p}$ is $H(x_{m1} - \sigma_1, x_{m2} - \sigma_2, x_{m3} - \sigma_3, ..., x_{mp} - \sigma_p)$, the various other + and - subscripts for Q refer to the mean of basic variable i + and -, respectively, one standard deviation (σ_i), and the function q is defined in terms of the correlation matrix ρ as

$$q_{i,j,...,p} = (1 + \sum_{g=1}^{p} \sum_{h=1}^{p} g' h' \partial_{g,h} \rho_{g,h}) / 2^p \quad (3.28)$$

where $\partial_{g,h}$ is 0 if $g \geq h$, $\partial_{g,h}$ is 1 if $g < h$, and g' and h' are -1 or +1 depending on the sign of the q function subscript for the basic variable being considered. For example

$$q_{+-+} = (1 - \rho_{12} + \rho_{13} - \rho_{23})/8$$

The procedure presented above is for symmetric random variables. Later, Rosenblueth (1981) expanded his two-point method to incorporate asymmetry of the random variables. Incorporating the skewness coefficient the q function can be computed as

$$q_{i,j,...,p} = \prod_{g=1}^{p} P_{g,\delta g} + \sum_{g=1}^{p-1} [\sum_{h=1}^{p} g' h' a_{g,h}] \quad (3.29)$$

where the subscript δg refers to the + or - subscript for the basic variable in the definition of q and

$$a_{g,h} = \frac{\rho_{g,h} / 2^p}{\{\prod_{g=1}^{p} [1 + (\upsilon_g / 2)^2]\}^{1/2}} \quad (3.30)$$

$$P_{g+} = y_{g-} / (y_{g+} + y_{g-}) \quad (3.31)$$

$$P_{g-} = 1 - P_{g+} \quad (3.32)$$

$$y_{g+} = \upsilon_g/2 + [1 + (\upsilon_g/2)^2]^{1/2} \quad (3.33)$$

$$y_{g-} = y_{g+} - \upsilon_g \quad (3.34)$$

and υ_g is the skewness coefficient. For asymmetrical basic variables the calculation $Q_{+++...p}$, $Q_{---...p}$, etc. in Eq. (3.27) is as follows:

$$Q_{+++...p} = g(x_{1+}, x_{2+}, x_{3+}, ..., x_{p+}) \quad (3.35)$$

$$Q_{---...p} = g(x_{1-}, x_{2-}, x_{3-}, ..., x_{p-}) \quad (3.36)$$

in which

$$x_{i+} = x_{mi} + \sigma_i y_{i+} \quad (3.37)$$

$$x_{i-} = x_{mi} - \sigma_i y_{i-} \quad (3.38)$$

Rogers et al. (1985) applied Rosenblueth's method to version 3 of IHDM (Chapter 17) for a small (0.9 km^2) forested watershed in Wales. They considered only the uncertainty in the soil hydraulic conductivity and Chezy's roughness coefficient for overland flow. Binley et al. (1981) applied Rosenblueth's method to version 4 of IHDM for a small (3.9 km^2) upland watershed in Wales. In this case the uncertainties in 4 parameters, soil hydraulic conductivity, saturated moisture content, capillary suction, and an overland-flow roughness parameter, were considered. They used the statistics of the model output to consider whether differences in runoff predicted by the physical simulation model for changing land-use conditions were truly significant. Melching (1992b) applied Rosenblueth's method to HEC-1 (Chapter 4) and RORB (Chapter 5) for the purpose of comparing Rosenblueth's method to MCS and MFOSM. These results are discussed in Section 3.4.7.

Rosenblueth's method requires 2^p model evaluations to estimate the statistical moments of model output. Thus, while Rosenblueth's method is quite efficient for problems with a small number of uncertain basic variables, its computational requirements are similar to those of MCS for p between 10 and 15 (i.e., 1,024 to 32,768 model evaluations).

3.4.6. Harr's Point-Estimation Method

To mitigate the problems experienced with Rosenblueth's method for cases with large numbers of basic variables, Harr (1989) proposed a modification that reduces the required number of model evaluations from 2^p to 2p. Yeh and Tung (1993) provide an excellent explanation of Harr's method that is followed in the steps presented below. In Harr's method, the correlation matrix of p stochastic input parameters, which is symmetric, is decomposed as follows

$$\boldsymbol{\rho} = \mathbf{V}\mathbf{L}\mathbf{V}^{\mathbf{T}} \tag{3.39}$$

where $\mathbf{V}$ is the eigenvector matrix ($\mathbf{v}_1, \mathbf{v}_2, ..., \mathbf{v}_p$) in which $\mathbf{v}_i$ are column vectors of eigenvectors; $\mathbf{L}$ is a diagonal matrix of corresponding eigenvalues ($\lambda_1, \lambda_2, ..., \lambda_p$); and the superscript T denotes the transpose of the matrix. The correlation matrix can be thought of geometrically as a hypersphere of radius $n^{1/2}$ centered at the expected values of the stochastic basic variables in the standardized coordinate system. Each eigenvector that passes through the origin of the hypersphere, intersects the sphere surface at two points. These 2p intersection points (for p basic variables) are used to estimate the statistical moments of the model output. Harr's (1989) point estimation method can be summarized by the following (Yeh and Tung, 1993):

1) Decompose the correlation matrix $\boldsymbol{\rho}$ of the basic variables into an eigenvector matrix $\mathbf{V}$ and corresponding eigenvalue matrix $\mathbf{L}$ as per Eq. (3.39);

2) Generate coordinates of the 2p intersecting points using

$$\mathbf{X}_{i\pm} = \mathbf{X}_{\mathrm{m}} \pm p^{1/2} \begin{bmatrix} \sigma_1 & . & . & . & . \\ . & - & . & . & 0 \\ . & . & - & . & . \\ 0 & . & . & - & . \\ . & . & . & . & \sigma_p \end{bmatrix} \mathbf{v}_i \qquad i=1,2,...,p \tag{3.40}$$

3) Compute $Q_{i\pm} = H(\mathbf{X}_{i\pm})$ and $Q_{i\pm}^2 = H^2(\mathbf{X}_{i\pm})$ for i = 1, 2, ..., p;

4) Compute the averaged model outputs for a given eigenvector i = 1, 2, ..., p using

$$Q_{mi} = (Q_{i+} + Q_{i-})/2 \quad \text{and } Q_{mi}^2 = (Q_{i+}^2 + Q_{i-}^2)/2$$

5) Compute the mean and variance of model output:

$$E[Q] = \left(\sum_{i=1}^{p} Q_{mi} \lambda_i \right) / \, p \tag{3.41}$$

$$E[Q^2] = \left(\sum_{i=1}^{p} Q_{mi}^{\;2} \lambda_i \right) / \, p \tag{3.42}$$

$$Var(Q) = E[Q^2] - E^2[Q] \tag{3.43}$$

To the best of this author's knowledge Harr's method has not been applied to a computer model of watershed hydrology. In hydraulic engineering, however, Yeh and Tung (1993) compared Harr's method to LHS and MFOSM for a model of the migration of gravel borrow pits in alluvial streams. Chang et al. (1992) compared Harr's method to LHS and MFOSM for the HEC2-SR model of water surface profiles for streams with movable beds. These applications have shown that Harr's method is a potentially useful tool for reliability analysis. Research on the use of Harr's method in watershed hydrology is encouraged.

3.4.7. Comparison of Reliability Analysis Methods

Several studies of the relative performance of the various reliability analysis methods have been reported in the literature. Most of these studies assume that the results of MCS for a large number of simulations is the standard against which the estimates from the simplified methods--LHS, MFOSM, AFOSM, Rosenblueth's method, and Harr's method--are compared to assess accuracy.

Melching (1992a) compared MCS, MFOSM, and AFOSM calculations of the CDFs of peak discharge estimates made by HEC-1 (Chapter 4) and RORB (Chapter 5) for a 1,600 km^2 agricultural watershed in Illinois, U.S. In this study both models were used in a forecasting or historic-event mode, that is, estimation of peak discharge assuming no knowledge of the true value. This was done because the purpose of the study was to compare the reliability-analysis methods and not to determine the utility of either model for flood forecasting or historic-event estimation. The statistics of the basic variables for the model applications were based on calibration of HEC-1 and RORB for 32 and 30 storms, respectively, on the example watershed. CDF and/or flood P_E values were calculated by and compared between the various methods for 16 storms not used in model calibration. The example events included several events with characteristics which violate the assumptions typically cited as necessary for MFOSM to yield good results: (1) significant nonlinearities in model performance, (2) large coefficients of variation of the most sensitive basic variables, and (3) non-normal distributions for the parameters of the abstraction models. AFOSM displayed generally good agreement with MCS for all storms from both models, and better agreement than MFOSM for the tails of the probability distributions. However, for the cases where the nonlinearity effects were not large MFOSM performed as accurately as and sometimes more accurately than AFOSM. The CDF comparison for a typical case of a storm event that causes significant nonlinearities is shown in Fig. 3.3, whereas the CDF comparison for a typical case of a storm that behaves fairly linearly is shown in Fig. 3.4.

Melching and Anmangandla (1992) compared the CDFs calculated by MCS, MFOSM, and AFOSM of DO concentration and DO deficit as

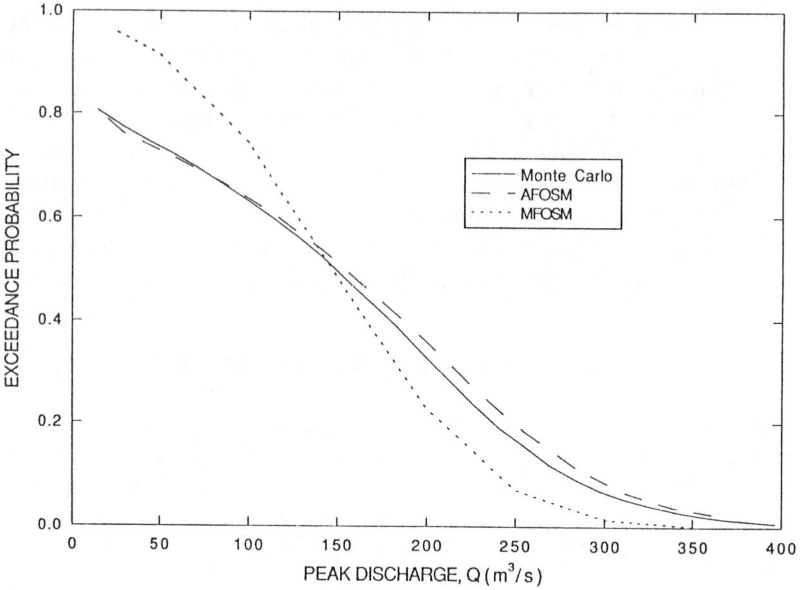

Fig. 3.3. Peak discharge exceedance probability estimated by Monte Carlo simulation, mean-value first-order second-moment (MFOSM), and advanced first-order second-moment (AFOSM) methods for the storm of July 13, 1957, calculated by RORB.

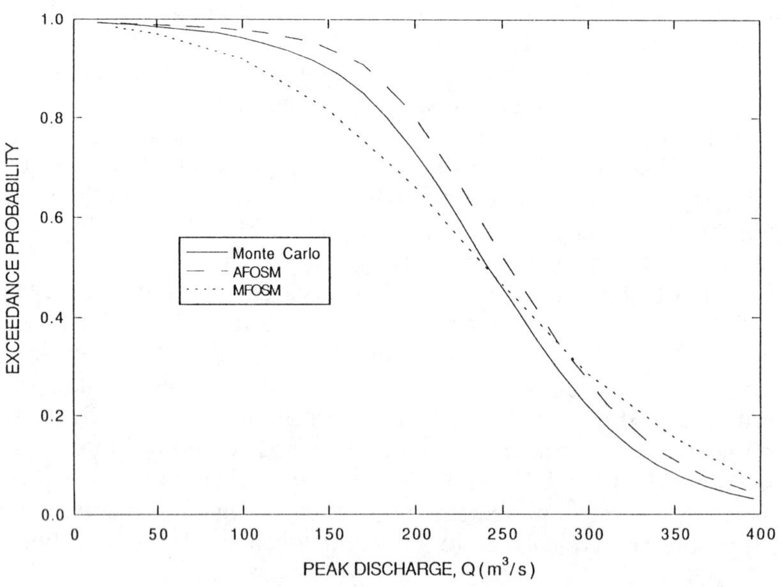

Fig. 3.4. Peak discharge exceedance probability estimated by Monte Carlo simulation, mean-value first-order second-moment (MFOSM), and advanced first-order second-moment (AFOSM) methods for the storm of May 4, 1965, calculated by RORB.

estimated by the Streeter-Phelps (1925) equation for two hypothetical streams. The basic variables were assumed to be normally, lognormally, Gamma, and uniformly distributed. In all cases, except that for uniformly distributed basic variables, the agreement between MCS and AFOSM was nearly exact, and that for the uniformly distributed case was acceptable for decision making. MFOSM agreed well with MCS in the central portions of the CDF and showed significant divergence only in the tails of the CDF.

For this chapter, the study of Melching (1992b) of the quality of mean and standard deviation estimates made by MCS, MFOSM, and Rosenblueth's method has been extended to include LHS estimates. The enhanced study compares estimates of the mean and standard deviation of estimated peak discharge produced by HEC-1 for 16 verification storms on a 1,600 km^2 agricultural watershed in Illinois, U.S. Harr's method has not been included in the enhanced study because it is believed that this method is most useful for cases with correlated basic variables. The original study (Melching, 1992b) also compared mean and standard deviation estimates made by MCS, MFOSM, and Rosenblueth's method for RORB simulation of the same verification data set. The results and conclusions of the comparison of reliability analysis methods for RORB were nearly identical to those for HEC-1.

For this study, the peak discharge estimate, Q_p, yielded by HEC-1 is as follows

$$Q_p = \lambda_H f(IL, CL, TC, R) = \lambda_H Q_{pH} \tag{3.44}$$

where f( ) is the HEC-1 function that estimates peak discharge; λ_H is a model correction factor that accounts for bias and error in the HEC-1 prediction of peak discharge, normally distributed with mean = 1.727-0.154 $\ell n\, Q_{pH}$ and standard deviation = 0.0472; IL is the initial loss in mm, lognormally distributed with log mean = 3.88 - 0.60 $\ell n\, Q_B$ and log standard deviation = 0.718; Q_B is the baseflow at the beginning of the storm (m^3/s); CL is the continuing loss rate in mm/hr, lognormally distributed with log mean = 0.591 and log standard deviation = 0.645; TC is the Clark time of concentration (hr), normally distributed with mean = 38.0 and standard deviation = 8.94; and R is the Clark watershed storage coefficient (hr), normally distributed with mean = 27.7 and standard deviation = 4.32. In this case there is no significant correlation among the parameters and all logarithms are taken with respect to base e. The distributions and statistics for the basic variables were determined by Melching (1987) from calibration of HEC-1 for 32 storms on the subject watershed.

To set the standard for comparison 10,000 MCS runs were made. For the LHS the probability distributions were subdivided into 50 intervals and 50 runs were made. LHS was also performed for 25 intervals and showed poor convergence in the standard deviation of model output, whereas 100 intervals did not show a great improvement in convergence over the results for 50 intervals. For each event the LHS was repeated 4 times to check convergence and the average value of the mean and standard deviation is reported below.

The mean values of peak discharge estimated by LHS, MFOSM, and Rosenblueth's method are compared to MCS for the 16 verification events in Fig. 3.5. The mean values estimated by each method are in very good agreement. Similar results have been obtained for MCS, MFOSM, and Rosenblueth estimates of mean values of peak discharge estimated by RORB (Melching, 1992b), MCS and MFOSM estimates of mean values of peak discharge estimated by RORB (Bates and Townley, 1988b), and Rosenblueth and MCS estimates of mean hydrographs for IHDM (Binley et al., 1991). Garen and Burges (1981) found that the agreement between MFOSM and MCS estimates of mean values of peak discharge and runoff volume produced by a modified version of the Stanford Watershed Model was good (within 15%) for 20, fair (within 15 to 30%) for 9, and poor (greater than 30% error) for 4 of 33 hypothetical combinations of storms, antecedent conditions, and parameter coefficients of variation. Causes for the variable quality of the estimates are difficult to discern from their results. In general, though it seems that any of the methods should be able to provide a good estimate of the mean value of the runoff characteristic of interest.

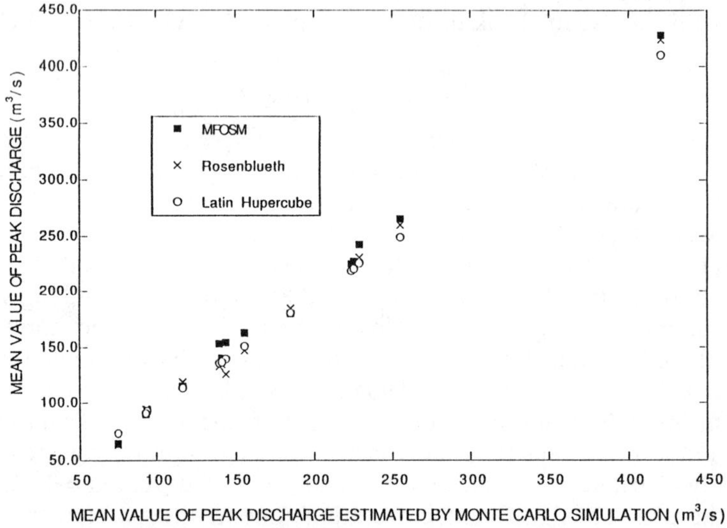

Fig. 3.5. Mean-value first-order second-moment (MFOSM), latin hypercube simulation, and Rosenblueth's method estimates of the mean value of peak discharge compared with estimates from Monte Carlo simulation for HEC-1 simulations of peak discharge for the Vermilion River at Pontiac, Illinois.

The standard deviations of peak discharge estimated by LHS, MFOSM, and Rosenblueth's method are compared to MCS for the 16 verification events in Table 3.2. The variation of standard deviation estimates between methods is much greater than that of mean value estimates. LHS provides the closest agreement to MCS with an average error of 7.5% and 10 of 16

TABLE 3.2.

Monte Carlo simulation (MCS), Latin Hypercube simulation (LHS), Mean-Value First-Order Second-Moment (MFOSM), and Rosenblueth's method estimates of the standard deviation of peak discharge calculated by HEC-1 for 16 verification storms on the watershed of the Vermilion River at Pontiac, Illinois, USA.

Storm Date	MCS (m^3/s)	LHS (m^3/s)	MFOSM (m^3/s)	Rosenblueth (m^3/s)
21/06/56	67.1	69.6	51.0	67.9
17/04/57	65.4	74.0	68.8	53.6
24/04/57	73.1	76.6	67.7	69.4
13/07/57	94.6	96.4	63.1	116.5
22/07/57	83.7	85.4	73.7	71.1
09/06/58	77.9	89.7	89.6	73.4
12/06/58	47.9	55.0	78.7	46.9
14/07/58	81.6	99.6	68.9	71.4
26/04/59	77.5	78.0	70.7	80.8
23/09/61	57.5	58.2	53.3	67.9
10/05/62	49.9	51.6	54.1	43.5
05/04/65	52.1	55.5	46.9	49.8
08/04/65	43.3	42.9	64.9	41.4
23/04/65	67.1	69.5	90.6	70.3
04/05/65	68.1	76.4	79.4	61.8
16/06/73	67.9	77.5	55.3	55.4

standard deviations estimated within 10%. For the comparison of Rosenblueth's method to MCS the average error is 9.9% and 9 of 16 standard deviations are estimated within 10%, while for the comparison between MFOSM and MCS the average error is 20.8% and 4 of 16 standard deviations are estimated within 10%. Similarly for RORB, Melching (1992b) found that for Rosenblueth's method the average error was 8.6% and 11 of 16 standard deviations were estimated within 10%, while for MFOSM the average error was 20.9% and 3 of 16 standard deviations were estimated within 10%. However, other researchers have reported much better agreement between output standard deviations estimated by MCS and MFOSM. Bates and Townley (1988b) found an average absolute error of 6.2% in MFOSM standard-deviation estimates for RORB calculations of 8 storms on an Australian watershed. Lei and Schilling (1993) show very good agreement between MCS and MFOSM standard-deviation estimates for HYSTEM-EXTRAN (a modification of SWMM - Chapter 22) calculations for 8 storms on a small urban watershed in Switzerland.

In the comparison of standard deviation estimates made for this chapter two points should be made clear. First, the number of model runs used is inversely proportional to the accuracy of the method, i.e., 5 were used in MFOSM, 16 were used in Rosenblueth's method, and 50 were used in LHS for evaluation of HEC-1. Second, some of the verification events had a very high CV for IL, which is the most important parameter for this application of HEC-1 (Melching et al., 1990). Large CV's of key basic variables are one of the known weaknesses of MFOSM. For example, Garen and Burges (1981) concluded that MFOSM provided reasonable estimates of standard deviation of model output for a modified version of the Stanford Watershed Model as long as the CV's of the most important parameters were less than 0.25. Thus, for more complicated models and smaller basic variable CV's, MFOSM's computational efficiency and accuracy might compare better relative to the other reliability-analysis methods. Finally, even though the standard deviations estimated by MFOSM were somewhat poor, the flood P_E values estimated by MFOSM compared well with those from MCS (Melching, 1992a).

The results in Table 3.2 also show that Rosenblueth's method tends to underestimate the standard deviation. The same was true for RORB (Melching, 1992b) and the Institute of Hydrology Distributed Model (Binley et al., 1991). Thus, while sampling the model performance at points representing plus and minus 1 standard deviation about the basic-variable mean values provides a more accurate estimate of the model output variance than does the local derivative at the central values, it still does not reflect the full range of variability in model output. Conversely, LHS consistently overestimates the standard deviation. This is because LHS forces extreme values of each basic variable to be sampled with the same frequency as the central values, thus, exploring a broader range of model outputs than would be generated by MCS for an equal number of runs.

Yeh and Tung (1993) compared LHS, MFOSM, and Harr's method for a model of the migration of gravel borrow pits in alluvial streams. They found good agreement between LHS and MFOSM in the estimated mean and standard deviation of pit depth 500 m dowstream from the initial pit location. Harr's method, however, overestimated the mean by 20% and the standard deviation by 46% relative to LHS and MFOSM results. Chang et al. (1992) compared LHS, MFOSM, and Harr's method for the HEC2-SR model of water-surface profiles for streams with movable beds. They found good agreement between the three methods for estimates of the standard deviation of the water-surface elevation, bed elevation, and sediment load at 2 points in the river studied over a 40-hour simulation interval.

In conclusion MFOSM provides adequate estimates of the mean, standard deviation, and central portion of the CDF of model output for cases where nonlinearity effects are not significant and basic variable CV's are not too large. MFOSM is a quick and computationally efficient method (p+1 model runs required), and modelers must be aware of the trade-off between accuracy and computational efficiency when using MFOSM. AFOSM provides adequate estimates of the CDF of model output with a considerable increase in computations. Experience with HEC-1 and RORB indicates that for target levels in the central portion of the CDF of model output 2 or 3 iterations (2p+2 or 3p+3 model runs) are required, whereas 5 to 7 iterations are required for target levels in the tails of the CDF of Z. Thus, approximately 500 hundred model runs could be required for a model with 10 basic variables and 14 target levels used to determine the CDF. Continued examination of the performance of MFOSM and AFOSM relative to MCS is encouraged.

Preliminary experience with LHS, Rosenblueth's method, and Harr's method shows that each method is potentially quite useful for reliability analysis of output from computer models of watershed hydrology. However, research is needed to define the strengths and weaknesses of applying these methods to computer models of watershed hydrology. For example, how many LHS model runs are necessary to obtain good results?

3.5. DETERMINATION OF KEY SOURCES OF UNCERTAINTY

The quality of estimates produced by a computer model of watershed hydrology may appear poor during model verification and yet the model structure may be adequate for the watershed and application studied. High levels of uncertainty in the input data and model parameters may cause poor model-verification results. For example, Melching et al. (1990) examined peak-discharge estimates made by HEC-1 (Chapter 4) for 17 verification storms on a 1,600 km^2 agricultural watershed in Illinois, U.S. They found that for several storms which were poorly estimated (greater than 40-percent error), the measured peak discharge was within the 95-percent confidence limits of the peak-discharge estimate, indicating HEC-1 provides good estimates from a stochastic point of view. Melching (1987) obtained similar results for RORB (Chapter 5) for the same example watershed and verification storms. For such cases, reducing uncertainties in the input data and model parameters may be the key to obtaining estimates that are good from both stochastic and practical points of view. Reliability-analysis methods may be used to determine the dominant sources of uncertainty affecting the reliability of model output and to provide guidance as to how improved information regarding these sources of uncertainty can increase the reliability of model output. Thus, reliability analysis can guide data collection so that model reliability and utility are enhanced at a reasonable data-collection cost.

A number of researchers (e.g., Gardner et al., 1981; Beck, 1987; Yeh and Tung, 1993) have shown that traditional sensitivity analysis, wherein basic variables are perturbed one at a time and the sensitivity of model output to the perturbation in each variable is observed, is not appropriate for determining the sources of uncertainty that most affect model output and, thus, require more detailed study. The sensitivity coefficient, $\partial g / \partial x_i$, does not account for the likelihood for the parameter to be different from its "best" value. Therefore, a highly sensitive parameter that is relatively well known may have much less influence on the reliability of model output than a much less sensitive parameter that is highly uncertain. The simulation based and first-order reliability-analysis methods described in Section 3.4 allow consideration of the combined effects of parameter sensitivity and parameter uncertainty in the determination of the key parameters affecting model-output reliability.

For the simulation-based methods (MCS and LHS), two approaches may be used to infer the key parameters. In the first approach the correlation between the parameter values and the model-output values is assessed, and the parameters with significant positive or negative correlations are determined to be important. The correlation analysis may be performed in 4 ways. The simple correlation coefficient (CC) and partial correlation coefficient (PCC) between parameter values and model-output values can be calculated to examine the linear relations between the model parameters and the model output. The CC and PCC can also be calculated for the ranked parameter values and ranked model-output values to examine possible nonlinear relations between model parameters and the model output. Gardner et al. (1981) note that if the modeler is interested in the underlying error-propagation properties of a model, the PCC is of greatest interest. In the field situation, all components of the system are subject to variability and are measured with error. These other sources of error cannot be controlled, and their influence on prediction reliability cannot be removed as in PCC analysis. Thus, in the design of field sampling programs, the ranking of parameters according to the CC is most relevant. Gardner et al. (1981) provide an example of CC and PCC analysis of model-parameter values and model-output values for a nonlinear stream-ecosystem model. Jaffe and Ferrara (1984) provide an example of rank correlation analysis for a model of sediment-water column interactions for hydrophobic pollutants. Yeh and Tung (1993) provide a comparison between standard correlation analysis and rank correlation analysis for a model of the migration of borrow pits in alluvial streams.

Simulation may also be used with Regional Sensitivity Analysis, RSA, (Hornberger and Spear, 1981) to determine the key basic variables. In RSA a reasonable range for model output is hypothesized. Simulations are performed and simulations where output fall in the reasonable range for model output are recorded as set A and simulations where output fall outside the reasonable range are recorded as set B. Empirical probability distributions are developed for model-parameter values for simulations in sets A and B. These empirical probability distributions for model parameter sets A and B are compared (e.g., using the Kolmogorov-Smirnov test; Law and Kelton, 1991, p. 387-391). If sets A and B are significantly different for a given parameter, that parameter significantly affects model-output reliability. Otherwise, that parameter is not important to model-output reliability. Hornberger et al. (1985) applied RSA to determine the important parameters for calibration of TOPMODEL (Chapter 18) for a small forested catchment in the Applachian Mountains of Virginia, U.S. Harlin and Kung (1992) applied RSA to the HBV hydrological model (Chapter 13) to determine the key parameters in design-flood estimation in Sweden. Beven and Binley (1992) applied RSA to the IHDM (Chapter 17) to determine key parameters for model calibration and uncertainty estimation.

In the first-order reliability-analysis methods the fraction of model-output variance contributed by each basic variable can be determined directly from Eq. (3.7) (or similar formulations for correlated basic variables). In general, modelers are interested in the variance relative to the best estimate produced by the model. Thus, for AFOSM the fraction of the variance evaluated at the mean values of the basic variables is generally preferred to the fraction of the variance evaluated at extreme failure surfaces. Melching et al. (1990) used MFOSM to determine the key parameters affecting HEC-1 (Chapter 4) and RORB (Chapter 5) peak-discharge estimates. The abstraction parameters dominated the uncertainty in HEC-1 and RORB peak-discharge estimates.

Yeh and Tung (1993) compared the results of correlation analysis, rank correlation analysis, and MFOSM in determining the key parameters for a model of the migration of borrow pits in alluvial streams. The correlation approaches and MFOSM gave considerably different lists of key parameters. The correlation approaches assess the global importance (i.e., importance over a wide range of possible basic variable values) of a basic variable (McKay, 1988; Yeh and Tung, 1993), whereas MFOSM assesses the local importance (about the best estimate of model output) of a basic variable. The choice of the most approriate method depends on the model use. If one is building or validating a model, global measures are probably of greater importance. If one is interested in fine tuning model output from models that are considered well identified and well calibrated, local measures are probably of greater importance.

3.6. SUMMARY

This chapter has presented a wide variety of methods that may be used to (1) estimate the reliability of output from and (2) determine the key sources of uncertainty for computer models of watershed hydrology. This chapter has also discussed the various sources (basic variables) of uncertainty and how the means, standard deviations, and distributions of the basic variables may be approximated for a number of model applications. It should be clear from the information presented that a modeler can obtain a reasonable preliminary estimate of output reliability for any of the computer models of watershed hydrology described in this book.

It should also be clear that much research is needed on reliability analysis of computer models of watershed hydrology. Many of the reliability-analysis methods have had only limited (or no) applications to computer models of watershed hydrology. Continued testing and determination of the utility of these methods are needed. Also the approaches to approximation of uncertainties in the basic variables are rather crude for several model applications. Further refinements in these approaches are needed, especially for the case of physical-simulation models applied to operational forecasting on previously ungaged catchments and in the assessment of model-structure uncertainties. Finally, the utility of reliability analysis for risk-cost design of hydraulic structures and flood-mitigation schemes also needs to be assessed. It is hoped that this chapter will provide stimulus to broader application of reliability analysis when computer models of watershed hydrology are used.

ACKNOWLEDGEMENTS

The author thanks Dr. William Kirby of the Office of Surface Water, U.S. Geological Survey, Reston, Virginia; Prof. Ben Yen of the University of Illinois at Urbana-Champaign; and Prof. Vijay Singh, the editor, for their detailed reviews of this chapter and the helpful comments that greatly improved the contents and readability of this chapter. The author also thanks Mr. Stephen Blanchard, Chief, Illinois District, U.S. Geological Survey, for granting the author time to work on this chapter.

REFERENCES

Ang, A. H.-S., and Tang, W. H., 1984, Probability Concepts in Engineering Planning and Design Vol. II: Decision, Risk, and Reliability, John Wiley & Sons Inc., New York.

Bates, B. C., 1988, 'Nonlinear, Discrete Flood Event Models, 2. Assessment of Statistical Nonlinearity,' Journal of Hydrology, **99**, 77-89.

Bates, B. C., and Townley, L. R., 1988a, 'Nonlinear, Discrete Flood Event Models, 1. Bayesian Estimation of Parameters,' Journal of Hydrology, **99**, 61-76.

Bates, B. C., and Townley, L. R., 1988b, 'Nonlinear, Discrete Flood Event Models, 3. Analysis of Prediction Uncertainty,' Journal of Hydrology, **99**, 91-101.

Bates, D. M., and Watts, D. G., 1980, 'Relative Curvature Measures of Nonlinearity,' Journal of the Royal Statistical Society, B, **42**(1), 1-25.

Beale, E. M. L., 1960, 'Confidence Regions in Nonlinear Estimation,' Journal of the Royal Statistical Society, B, **22**(1), 41-88.

Beck, M. B., 1985, 'Water Quality Management: A Review of the Development and Application of Mathematical Models,' Lecture Notes in Engineering 11, C. A. Brebbia and S. A. Orszag, eds., Springer-Verlag, New York.

Beck, M. B., 1987, 'Water Quality Modeling: A Review of the Analysis of Uncertainty,' Water Resources Research, **23**(5), 1393-1441.

Beven, K. J., 1989, 'Changing Ideas in Hydrology. The Case of Physically Based Models,' Journal of Hydrology, **105**, 157-172.

Beven, K., and Binley, A., 1992, 'The Future of Distributed Models: Model Calibration and Uncertainty Prediction,' Hydrological Processes, **6**, 279-298.

Binley, A. M., Beven, K. J., Calver, A., and Watts, L. G., 1991, 'Changing Responses in Hydrology: Assessing the Uncertainty in Physically Based Model Predictions,' Water Resources Research, **27**(6), 1253-1261.

Box, M. J., 1971, 'Bias in Nonlinear Estimation,' Journal of the Royal Statistical Society, B, **33**(2), 171-201.

Bras, R. L., 1990, Hydrology. An Introduction to Hydrologic Science, Addison-Wesley Publishing Co., Reading, Massachusetts.

Bras, R. L., and Rodriguez-Iturbe, I., 1976, 'Rainfall Network Design for Runoff Prediction,' Water Resources Research, **12**(6), 1197-1208.

Breitung, K., 1984, 'Asymptotic Approximations of Multinormal Integrals,' Journal of Engineering Mechanics, ASCE, **110**(3), 357-366.

Brown, L.C., and Barnwell, T. O., Jr., 1987, 'The Enhanced Stream Water Quality Models QUAL2E and QUAL2E-UNCAS: Documentation and User Manual,' Report EPA/600/3-87/007, U.S. Environmental Protection Agency, Athens, Georgia.

Chang, J.-H., Tung, Y.-K., and Yang, J.-C., 1994, 'Monte Carlo Simulation for Correlated Variables with Marginal Distributions,' Journal of Hydraulic Engineering, ASCE, **120**(3), 313-331.

Chang, J.-H., Tung, Y.-K., Yang, J.-C., and Yeh, K.-C., 1992, 'Uncertainty Analysis of a Computerized Sediment Transport Model,' Stochastic Hydraulics '92, Proceedings Sixth International Association for Hydraulic Research Symposium on Stochastic Hydraulics, Taipei, Taiwan, J.-T. Kuo and G.-F. Lin, eds., Water Resources Publications, Littleton, Colorado, 123-130.

Cheng, S.-T., 1982, 'Overtopping Risk Evaluation for an Existing Dam,' Ph.D. Thesis, Department of Civil Engineering, University of Illinois at Urbana-Champaign.

Cornell, C. A., 1972, 'First-Order Analysis of Model and Parameter Uncertainty,' Proceedings Intenational Symposium on Uncertainties in Hydrology and Water Resources Systems, University of Arizona, Tucson, Arizona, Vol. 3, 1245-1272.

Dawdy, D. R., Lichty, R. W., and Bergmann, J. M., 1972, 'A Rainfall-Runoff Simulation Model for Estimation of Flood Peaks from Small Drainage Basins,' U.S. Geological Survey Professional Paper 506-B, Washington, D.C.

Gardner, R. H., Huff, D. D., O'Neill, R. V., Mankin, J. B., Carney, J., and Jones, J., 1980, 'Application of Error Analysis to a Marsh Hydrology Model,' Water Resources Research, **16**(4), 659-664.

Gardner, R. H., O'Neill, R. V., Mankin, J. B., and Carney, J. H., 1981, 'A Comparison of Sensitivity and Error Analysis Based on a Stream Ecosystem Model,' Ecological Modeling, **12**, 173-190.

Garen, D. C., and Burges, S. J., 1981, 'Approximate Error Bounds for Simulated Hydrographs,' Journal of the Hydraulics Division, ASCE, **107**(HY11), 1519-1534.

Georgakakos, K. P., 1986, 'A Generalized Stochastic Hydrometeorolgical Model for Flood and Flash-Flood Forecasting, 1. Formulation,' Water Resources Research, **22**(13), 2083-2095.

Goldman, D. M., Marino, M. A., and Feldman, A. D., 1990, 'Runoff Prediction Uncertainty for Ungaged Watersheds,' Journal of Irrigation and Drainage Engineering, ASCE, **116**(6), 752-768.

Harlin, J., and Kung C.-S., 1992, 'Parameter Uncertainty and Simulation of Design Floods in Sweden,' Journal of Hydrology, **137**, 209-230.

Harr, M. E., 1989, 'Probability Estimates for Multivariate Analyses,' Applied Mathematical Modelling, **13**, 313-318.

Hasofer, A. M., and Lind, N. C., 1974, 'Exact Invariant Second-Moment Code Format,' Journal of the Engineering Mechanics Division, ASCE, **100**(EM1), 111-121.

Hasofer, A. M., 1974, 'Reliability Index and Failure Probability,' Journal of Structural Mechanics, **3**, 25-27.

Hershfield, D. M., 1961, 'Rainfall Frequency Atlas of the United States for Durations from 30 Minutes to 24 Hours and Return Periods from 1-100 Years,' U.S. Weather Bureau Technical Paper No. 40, Washington, D.C.

Hornberger, G. M., and Spear, R. C., 1981, 'An Approach to the Preliminary Analysis of Environmental Systems,' Journal of Environmental Management, **12**, 7-18.

Hornberger, G. M., Beven, K. J., Cosby, B. J., and Sappington, D. E., 1985, 'Shenandoah Watershed Study: Calibration of a Topography-Based, Variable Contributing Area Hydrological Model to a Small Forested Catchment,' Water Resources Research, **21**(12), 1841-1850.

Horton, R. E., 1923, 'Accuracy of Areal Rainfall Estimates,' Monthly Weather Review, **51**, 348-353.

Huff, F. A., 1970, 'Spatial Distribtuion of Rainfall Rates,' Water Resources Research, **6**(1), 254-260.

Ibrahim, Y., and Liong, S.-Y., 1992, 'Calibration Strategy for Urban Catchment Parameters,' Journal of Hydraulic Engineering, ASCE, **118**(11), 1550-1570.

Jackson, D. R., and Aron, G., 1971, 'Parameter Estimation in Hydrology: The State of the Art,' Water Resources Bulletin, **7**(3), 457-472.

Jaffe, P. R., and Ferrara, R. A., 1984, 'Modeling Sediment and Water Column Interactions for Hydrophobic Pollutants. Parameter Discrimination and Model Response to Input Uncertainty,' Water Research, **18**(9), 1169-1174.

Krajewski, W. F., Lakshmi, V., Georgakakos, K. P., and Jain, S. C., 1991, "A Monte Carlo Study of Rainfall Sampling Effect on a Distributed Catchment Model,' Water Resources Research, **27**(1), 119-128.

Kuczera, G., 1988, 'On the Validity of First-Order Prediction Limits for Conceptual Hydrologic Models,' Journal of Hydrology, **103**, 229-247.

Kuczera, G., 1990a, 'Estimation of Runoff-Routing Model Parameters Using Incompatible Storm Data,' Journal of Hydrology, **114**, 47-60.

Kuczera, G., 1990b, 'Assessing Hydrologic Model Nonlinearity Using Response Surface Plots,' Journal of Hydrology, **118**, 143-161.

Kuczera, G., and Williams, B. J., 1992, 'Effect of Rainfall Errors on Accuracy of Design Flood Estimates,' Water Resources Research, **28**(4), 1145-1153.

Larson, L. W., and Peck, E. L., 1974, 'Accuracy of Precipitation Measurements for Hydrological Modeling,' Water Resources Research, **10**(4), 857-863.

Laurenson, E. M., and O'Donnell, T., 1969, 'Data Error Effects in Unit Hydrograph Derivation,' Journal of the Hydraulics Division, ASCE, **95**(HY6), 1899-1917.

LaVenue, M., Andrews, R. W., and Rama Rao, B. S., 1989, 'Groundwater Travel Time Uncertainty Analysis Using Sensitivity Derivatives,' Water Resources Research, **25**(7), 1551-1566.

Law, A. M., and Kelton, W. D., 1991, Simulation Modeling and Analysis, McGraw-Hill, New York.

Laws, J. O., and Parsons, D. A., 1943, 'The Relation of Raindrop-Size to Intensity,' Transactions, American Geophysical Union, **24**(part II), 452-460.

Lei, J., and Schilling, W., 1993, "Propagation of Model Uncertainty,' in Proceedings Sixth International Conference on Urban Storm Drainage, Niagara Falls, Canada, J. Marsalek and H. Torno, eds., Seapoint Publishing, Victoria, British Columbia, Canada, 465-470.

Light, P., and Shands, A. L., 1947, 'The Reliability of Areal Rainfall Determination,' Thunderstorm Rainfall, Hydrometeorological Report 5, U.S. Weather Bureau, U.S. Army Corps of Engineers.

Lindgren, G., 1984, 'Extreme Ranks and Transformation of Variables or Extremes of Functions of Multivariate Gaussian Processes,' Stochastic Processes and Their Application, **17**, 285-312.

Linsley, R. K., Jr., and Kohler, M. A., 1951, 'Variations in Storm Rainfall Over Small Areas,' Transactions, American Geophysical Union, **32**(2), 245-250.

Linsley, R. K., Jr., Kohler, M. A., and Paulhus, J. L. H., 1982, Hydrology for Engineers, McGraw-Hill, New York.

McGuinness, J. L., 1963, 'Accuracy of Estimating Watershed Mean Rainfall,' Journal of Geophysical Research, **68**(16) 4763-4767.

McKay, M. D., 1988, 'Sensitivity and Uncertainty Analysis Using a Statistical Sample of Input Values,' Uncertainty Analysis, Y. Ronen, ed., CRC Press, Inc., Boca Raton, Florida, 145-186.

McKay, M. D., Beckman, R. J., and Conover, W. J., 1979, 'A Comparison of Three Methods for Selecting Values of Input Variables in the Analysis of Output from a Computer Code,' Technometrics, **21**(2), 239-245.

Melching, C. S., 1987, 'A Reliability Analysis on Flood Event Forecasting with Uncertainties,' Ph.D. Thesis, Department of Civil Engineering, University of Illinois at Urbana-Champaign.

Melching, C. S., 1992a, 'An Improved First-Order Reliability Approach for Assessing Uncertainties in Hydrologic Modeling,' Journal of Hydrology, **132**, 157-177.

Melching, C. S., 1992b, 'A Comparison of Methods for Estimating Variance of Water Resources Model Predictions,' Stochastic Hydraulics '92, Proceedings Sixth International Association for Hydraulic Research Symposium on Stochastic Hydraulics, Taipei, Taiwan, J.-T. Kuo and G.-F. Lin, eds., Water Resources Publications, Littleton, Colorado, 663-670.

Melching, C. S., and Anmangandla, S., 1992, 'Improved First-Order Uncertainty Method for Water-Quality Modeling,' Journal of Environmental Engineering, ASCE, **118**(5), 791-805.

Melching, C. S., Yen, B. C., and Wenzel, H. G., Jr., 1990, 'A Reliability Estimation in Modeling Watershed Runoff With Uncertainties,' Water Resources Research, **26**(10), 2275-2286.

Melching, C. S., Yen, B. C., and Wenzel, H. G., Jr., 1991, 'Output Reliability as Guide for Selection of Rainfall-Runoff Models,' Journal of Water Resources Planning and Management, ASCE, **117**(3), 383-398.

Mueller, C. C., and Kidder, E. H., 1972, 'Rain Gage Catch Variation Due to Airflow Disturbance Around a Standard Rain Gage,' Water Resources Research, **8**(4), 1077-1082.

Plate, E. J., 1986, 'Reliability Analysis in Hydraulic Design,' Stochastic and Risk Analysis in Hydraulic Engineering, B. C. Yen, ed., Water Resources Publications, Littleton, Colorado, 37-47.

Rackwitz, R., 1976, 'Practical Probabilistic Approach to Design,' Comite European du Beton, Paris, Bulletin No. 112.

Rawls, W. J., and Brakensiek, D. L., 1989, 'Estimation of Soil Water Retention and Hydraulic Properties,' Unsaturated Flow in Hydrologic Modeling Theory and Practice, H. J. Morel-Seytoux, ed., Kluwer Academic Publishers, Boston, Massachusetts, 275-300.

Rawls, W. J., Brakensiek, D. L., and Miller, N., 1983, 'Green-Ampt Infiltration Parameters from Soils Data,' Journal of Hydraulic Engineering, ASCE, **109**(1), 62-70.

Rogers, C. C. M., Beven, K. J., Morris, E. M., and Anderson, M. G., 1985, 'Sensitivity Analysis, Calibration and Predictive Uncertainty of the Institute of Hydrology Distributed Model,' Journal of Hydrology, **81**, 179-191.

Rosenblueth, E., 1975, 'Point Estimates for Probability Moments,' Proceedings National Academy of Sciences U.S.A., **72**(10), 3812-3814.

Rosenblueth, E., 1981, 'Two-Point Estimates in Probabilities,' Applied Mathematical Modelling, **5**, 329-335.

Salas, J. D., Delleur, J. W., Yevjevich, V., and Lane, W. L., 1980, Applied Modeling of Hydrologic Time Series, Water Resources Publications, Littleton, Colorado.

Scavia, D., Powers, W. F., Canale, R. P., and Moody, J. L., 1981, 'Comparison of First-Order Error Analysis and Monte Carlo Simulation in Time-Dependent Lake Eutrophication Models,' Water Resources Research, **17**(4), 1051-1059.

Sevruk, B., 1982, 'Methods of Correction for Systematic Error in Point Precipitation Measurement,' World Meteorological Organization, Operational Hydrology Report No. 21, WMO-No. 589, Geneva, Switzerland.

Shinozuka, M., 1983, 'Basic Analysis of Structural Safety,' Journal of Structural Engineering, ASCE, **109**(3), 721-740.

Smith, R.E., and Hebbert, R. H. B., 1979, 'A Monte Carlo Analysis of the Hydrologic Effects of Spatial Variability of Infiltration,' Water Resources Research, **15**(2), 419-429.

Streeter, H. W., and Phelps, E. B., 1925, 'A Study of the Pollution and Purification of the Ohio River, III. Factors Concerned in the Phenomena of Oxidation and Reaeration,' Public Health Bulletin 146, U.S. Public Health Service, Washington, D.C.

Takasao, T., and Takara, K., 1989, 'Evaluation of Rainfall-Runoff Models from the Stochastic Viewpoint,' Journal of Hydrology, **102**, 381-406.

Tang, W. H., and Yen, B. C., 1972, 'Hydrologic and Hydraulic Design Under Uncertainties,' Proceedings Intenational Symposium on Uncertainties in Hydrology and Water Resources Systems, University of Arizona, Tucson, Arizona, vol. 2, 868-882.

Troutman, B. M., 1982a, 'The Effect of Input Errors in Using Precipitation-Runoff Models for Runoff Prediction,' Statistical Analysis of Rainfall and Runoff, V. P. Singh, ed., Water Resources Publications, Littleton, Colorado, 305-314.

Troutman, B. M., 1982b, 'An Analysis of Input Errors in Precipitation-Runoff Models Using Regression with Errors in the Independent Variables,' Water Resources Research, **18**(4), 947-964.

Troutman, B. M., 1983, 'Runoff Prediction Errors and Bias in Parameter Estimation Induced by Spatial Variability of Precipitation,' Water Resources Research, **19**(3), 791-810.

Tung, Y.-K., 1987, 'Uncertainty of National Weather Service Rainfall Frequency Atlas,' Journal of Hydraulic Engineering, ASCE, **113**(2), 179-189.

Tung, Y.-K., and Hathhorn, W. E., 1988, 'Assessment of Probability Distribution of Dissolved Oxygen Deficit,' Journal of Environmental Engineering, ASCE, **114**(6), 1421-1435.

Tung, Y.-K., and Mays, L. W., 1980, 'Risk Analysis for Hydraulic Design,' Journal of the Hydraulics Division, ASCE, **106**(HY5), 893-913.

Tung, Y.-K., and Mays, L. W., 1981, 'Risk Models for Flood Levee Design,' Water Resources Research, **17**(4), 833-841.

Tung, Y.-K., and Yen, B. C., 1993, 'Some Recent Progress in Uncertainty Analysis for Hydraulic Design,' Reliability and Uncertainty Analyses in Hydraulic Design, B. C. Yen and Y.-K. Tung, eds., American Society of Civil Engineers, New York, 17-34.

U.S. Army Corps of Engineers, 1977, Storage, Treatment, Overflow, and Runoff Model, "STORM," User's Manual, Hydrologic Engineering Center, Davis, California.

Warwick, J. J., and Wilson, J. S., 1990, 'Estimating Uncertainty in Stormwater Runoff Computations,' Journal of Water Resources Planning and Management, ASCE, **116**(2), 187-204.

Yeh, K.-C., and Tung, Y.-K., 1993, 'Uncertainty and Sensitivity Analyses of Pit-Migration Model, Journal of Hydraulic Engineering, ASCE, **119**(2), 262-283.

Yen, B. C., Cheng, S.-T., and Melching, C. S., 1986, 'First Order Reliability Analysis,' Stochastic and Risk Analysis in Hydraulic Engineering, B. C. Yen, ed., Water Resources Publications, Littleton, Colorado, 1-36.

Yen, B. C., Cheng, S.-T., and Tang, W. H., 1980, 'Reliability of Hydraulic Design of Culverts,' Proceedings International Conference on Water Resources Development, International Association for Hydraulic Research Asian Pacific Division Second Congress, Vol. 2, Taipei, Taiwan, 991-1001.

Chapter 4

HEC-1 FLOOD HYDROGRAPH PACKAGE

A. D. Feldman

4.1. INTRODUCTION

4.1.1. History of HEC-1

The Hydrologic Engineering Center, HEC, developed several small computer programs for simulating the various parts of the precipitation-runoff, river-routing processes in the early 1960's. The small, individual programs were necessary because of the limited computer memory at that time. When computer memory became large enough in the mid to late 1960's, the several small programs were combined into a single package for river basin hydrology. The first version of the HEC-1 Flood Hydrograph Package was published in October 1968. It has been expanded and revised and published many times since then to incorporate improvements in hydrologic algorithms and advances in computer hardware and software. To simplify input requirements and to make the program output more meaningful and readable, a major revision was made in 1973. In 1981 the computational capabilities of the dam-break, project optimization, and kinematic wave special versions were combined into a single easy to use package. In late 1984 a microcomputer version (PC version) was developed. A menu capability was added to facilitate user interaction with the model; an interactive input developer, a data editor, and output display features were also added.

The current standard version, Version 4.0 (September 1990), represents improvements and expansions to the hydrologic simulation capabilities together with interfaces to the HEC Data Storage System (HEC-DSS). The DSS connection allows HEC-1 to interact with the input and output of many HEC and other models. New hydrologic capabilities in HEC-1 include Green and Ampt infiltration, Muskingum-Cunge flood routing, reservoir releases input, and improved numerical solution of kinematic wave equations. The Muskingum-Cunge routing may also be used for the collector and main channels in a kinematic wave land-surface runoff calculation. This new release also automatically performs numerical analysis stability checks for the kinematic wave and Muskingum-Cunge routings. The numerical stability check was added because many users did not check the validity of the time and distance steps used in the model.

In September 1991, version 4.0.1E of HEC-1 was released for use on extended memory PC's. A hydrograph-array size of 2,000 ordinates is available in that version. The increased array size reduces limitations encountered when simulating long storms using short time intervals. For example, simulation of the fixed-length (96-hour) standard project storm at 15-minute intervals requires 384 ordinates just for the storm; more time intervals would be required to simulate the full runoff hydrograph and route it through a river basin. The large-array version also allows greater flexibility in checking for numerical stability of simulation processes (e.g., kinematic runoff and routing computations). The large-array version uses an extended-memory operating system (386 or newer machines) with three megabytes total memory.

There are many different hydrologic simulation options in HEC-1. This evolves from the autonomy of the Corps field offices and different methods being preferred. Different geographic regions and/or the preferences of the hydrologic engineers have brought about the many methods. No one method for hydrologic analysis is prescribed by Corps headquarters; the only requirement is that the validity the selected method must be demonstrated.

A next generation of an HEC hydrologic modeling system is now under development. The new system will have broader capabilities than HEC-1, but will still primarily be a flood hydrology analysis model. This chapter describes the existing capabilities of HEC-1 and the developments underway for the next generation hydrologic simulation model.

4.1.2. Purpose of HEC-1

The main purpose of HEC-1 in its initial development was simulation of hydrologic processes. This capability would enable the Corps to analyze historical floods and design projects for flood damage reduction. HEC-1 was limited to flood event studies because flood protection was the Corps' major responsibility. Those flood events were typically major historical storms or hypothetical design storms. The simulation would begin with precipitation, then compute infiltration losses, and finally route the runoff throughout a river basin. HEC-1 soon expanded from just a hydrologic simulation tool; it increased its scope to include special hydrologic analyses and flood-control-project analyses that the hydrologic engineer typically performs using the hydrologic data derived from HEC-1. Hydrologic analysis tools include automated parameter estimation, multiple-storm simulation, precipitation depth-area relationship coordination, and flood frequency curve determination. The flood damage reduction project analysis capabilities include multiple-projects analysis, modified frequency curves, flood damage computation, and project-size optimization.

4.1.3. HEC Modeling Philosophy

HEC has developed and supported a full range of simulation models for understanding how water resources systems function. It is assumed that experienced professionals can deduce the appropriate solution to a problem given the insight provided by selective execution of the simulation models. This deduction process has historically been the dominant methodology for planning and operational decisions in the water resources community, and it continues. HEC has used hydrologic simulation methods in a variety of situations with notable success. Some optimization models are now being used to evaluate reservoir system operation.

4.1.4. Available HEC-1 Software, Documentation, and Support

HEC software, documentation, and technical support are available free of charge to U.S.A. Federal agencies. Others may obtain the software, documentation, and technical support from vendors who maintain a relationship with HEC so as to provide up-to-date information. Lists of domestic and foreign vendors are available from HEC; the software serviced and the type of services provided (distribution, hotline assistance, training, and/or engineering studies) are described on the listings.

Software documentation and HEC technical reports may be also obtained by nonFederal parties directly from HEC for a nominal charge. HEC's publications and computer-programs catalogs and vendor lists may be obtained free of charge directly from the HEC, 609 Second Street, Davis, California 95616, USA.

The currently available software and documentation related to HEC-1 are summarized in Table 4.1. The key publications describing HEC-1 and its use are the User's Manual and Training Document No. 32. Training Document No. 32 provides a detailed description of the HEC menu-system features including: executing the menu; creating data files; using the Corps editor; managing data files and file directories; executing HEC-1; executing the Data Storage System; and reviewing model results. The Corps editor,

TABLE 4.1.

HEC Software and Documentation Related to HEC-1

Software and User's Manual

Software release September 1990, Version 4.0. Includes: executable code for 640k PC; HEC operating shell/menu and editor; data storage system and graphics; and input data creation software.

Software release September 1991, Version 4.01. Same as above except uses 2.5 megabytes of extended memory for simulation of longer hydrographs and better testing of kinematic wave computation stability.

HEC-1 Flood Hydrograph Package User's Manual, September 1990, Computer Program Document No. 1. Includes description of theoretical basis, example applications, input description, output messages, and example uses of the HEC Data Storage System.

Installation Instructions for Microcomputer Version of HEC-1, March 1991.

Installation Instructions for Device Drivers, April 1991.

Training Documents

No. 10, Introduction and Application of Kinematic Wave Routing Techniques Using HEC-1, May 1979, updated 1993.

No. 15, Hydrologic Analysis of Ungaged Watersheds with HEC-1, April 1982.

No. 19, Application of Spatial Data Management Techniques to HEC-1 Rainfall-Runoff Studies, October 1983.

No. 32, Using HEC-1 on a Personal Computer, December 1991.

Technical Papers

No. 54, Adjustment of Peak Discharge Rates for Urbanization, September 1978.

No. 59, Testing of Several Runoff Models on an Urban Watershed, October 1978.

TABLE 4.1. Continued.

TABLE 4.1. Continued.

No. 62, Flood Hydrograph and Peak Flow Frequency Analysis, March 1979.

No. 70, Corps of Engineers Experience with Automatic Calibration of a Precipitation-Runoff Model, May 1980.

No. 100, Probable Maximum Flood Estimation - Eastern United States, June 1984.

No. 116, The HEC's Activities in Watershed modeling, July 1987.

No. 118, Real-Time Snow Simulation Model for the Monongahela River Basin, December 1987.

No. 121, Development, Calibration and Application of Runoff Forecasting Models for the Allegheny River Basin, June 1988.

No. 122, The Estimation of Rainfall for Flood Forecasting Using Radar and Rain Gage Data, September 1988.

Related Computer Programs

No. 45, HECDSS, User's Guide and Utility Program Manuals, December 1990.

No. 46, Probable Maximum Storm (Eastern United States) (HMR52) User's Manual, March 1984.

No. 56, Corps of Engineers Editor Users's Manual (COED), February 1987.

COED, provides HEC-1 specific help; most of the user's manual input data descriptions are provided via on-line help screens. The Data Storage System, described later, provides report quality custom tables and graphics.

4.1.5. A Suite of Hydrologic Engineering Software

This paper primarily addresses the HEC-1 hydrologic modeling activities of HEC. Not included in the paper is information about HEC's river hydraulics, reservoir operation, flood damage, and water quality models (Feldman, 1981). River hydraulics modeling receives a major emphasis like hydrology. In recent years, many improvements have been made to: HEC-2, the steady state water surface profile model; HEC-6, the sediment scour and deposition model; and UNET, the unsteady flow model. Detailed reservoir system operation for flood control, water supply, and hydropower may be accomplished with HEC-5. In the area of flood damage analysis, a Flood Damage Analysis Package provides a comprehensive set of tools to evaluate flood damage reduction, and a new program accounts for project benefits from individual flood event operations. HEC also maintains a set of river and reservoir water quality simulation models. The reservoir operation for water quality model, HEC-5Q, is a combination of the reservoir regulation and water quality models.

4.2. RIVER BASIN REPRESENTATION

4.2.1. Sizes and Types of Basins

HEC-1 is not limited to any watershed or river basin size except that it cannot simulate at time intervals less than one minute. It has been used for small urban areas ($1 km^2$ or less), and large river basins (100,000's km^2). The river basins may be of any complexity of subbasins, river reaches, reservoirs, etc., but they must be in converging dendritic form. HEC-1 is not a network model which will allow water to be reintroduced upstream; this can be accomplished through interfaces with other models as discussed later.

Within the smallest unit of watershed area, termed a subbasin, it is assumed that hydrologic processes occur uniformly. This is the so-called lumped watershed model. That is, the precipitation and infiltration are assumed to be uniform over the entire subbasin. There is a minor modification of this concept when using kinematic wave runoff procedures as discussed later.

The lumped-parameter assumption in HEC-1 is the same as is found in almost all hydrologic models. The assumption is not particularly limiting because the subbasins and river reaches can be made small enough to accommodate any practical spatial variation in the watershed characteristics.

4.2.2. Model Structure

The river network is defined by the order of the HEC-1 model building blocks which include subbasins, river reaches, confluences, reservoirs, diversions, and pumps. Each building block describes the hydrologic processes necessary to compute the hydrograph at the downstream outlet of the subbasin, river reach, reservoir or confluence. Diversions and pumps allow water to be withdrawn from a building block and saved on disk. Those diverted flows are labeled and may be retrieved at any subsequent building block; on retrieval they are treated like a subbasin runoff and may be routed and combined with other flows.

4.2.3. Philosophy of Use

HEC-1 offers many different hydrologic simulation options for each of the main hydrologic processes: precipitation; infiltration/interception; precipitation excess transformation to streamflow; river routing; and reservoir routing. As mentioned above, the simulation tools are a means to gain insight into the hydrologic function of a watershed or river basin. Use of more sophisticated methods does not mean better results by itself. The user must fully understand the physics of the hydrologic processes and how the model represents those processes before confidence in the results can be obtained. This is especially true in ungaged areas where hydrologic judgment is 90% of the application. HEC recommends starting with the simplest representation of the basin that will produce the desired information. Then, as the situation warrants, more complexity and sophistication can be added to the model.

4.2.4. Limitations of HEC-1

HEC-1 hydrologic simulation must be made at a constant computational time interval. This is only a limitation when some components of the model, i.e. building blocks, require more detail than others. For example, in a large river basin, it may be desired to simulate runoff in small urban areas in some detail. That detail would require the computational time interval to be much smaller than the other parts of the basin. Two solutions are possible: choose an average time interval that suffices for both the large and small components; or simulate the small areas separately. If the small areas are simulated separately at a shorter time interval, they may be interfaced with the larger river basin model through the HEC Data Storage System, DSS. The time-series input data to HEC-1 do not have to be the same as the computation time interval; HEC-1 interpolates the input time series to the prescribed computational time interval.

There is no recovery of precipitation loss rates (infiltration, interception, detention storage, etc.) during periods of no precipitation. That is, the soils, etc. do not dry out and recover their loss potential. Thus, HEC-1 is limited to flood event studies where the loss rates do not undergo recovery. The storm events may be very complicated in terms of precipitation patterns. If there are significant dry periods, then the storm should be separated into one or more flood events and the loss rates determined separately for each event.

This is usually not a significant limitation in flood studies where the losses are usually small with respect to the overall flood runoff volumes.

The river basin must be dendritic, i.e. a converging tree network. If flow reversals occur, then HEC-1 would need to be interfaced with an unsteady flow model; this may be accomplished through the data storage system as discussed later.

The hydrologic processes are considered to be average (lumped) over each time increment and spatial component of the model. This is a limitation common to almost all models. It is solved by making the simulation building blocks small enough to capture the desired variability.

The reservoirs in HEC-1 cannot be operated for downstream flow conditions. HEC-1 computes all hydrograph points in time for the current point in space, e.g. the reservoir. It does not know what is happening downstream. Thus, reservoir outflows are only a function of reservoir storage or stage. This limitation can also be overcome by use of the data storage system. The HEC-1 results are saved in DSS and the retrieved by HEC-5 which operates reservoirs to meet a variety of downstream requirements.

4.3. HYDROLOGIC SIMULATION COMPONENTS

HEC-1 views a river basin as a series of hydrologic building blocks representing the various parts of the precipitation-to-runoff process. Figure 4.1 shows a simple river basin and several of those building blocks. The hydrologic process building blocks are: watershed runoff; river routing; reservoir routing; confluences; and diversions. The HEC-1 User's Manual (HEC, 1990a) should be referred to for details about any of these simulation components. Each of these blocks will now be discussed.

4.3.1. Subbasin Runoff

The subbasin (also known as a watershed or catchment) runoff block involves four processes: precipitation; losses (interception, infiltration, and detention storage); excess-to-runoff transformation; and baseflow. HEC-1 contains several options for simulating each of these processes. The subbasin is the smallest areal element in the watershed conceptualization for which the precipitation-to-runoff process can be simulated. As noted before, it can be of any size only limited by the minimum of a one-minute computation interval. Each of the above processes is considered to occur uniformly over the subbasin.

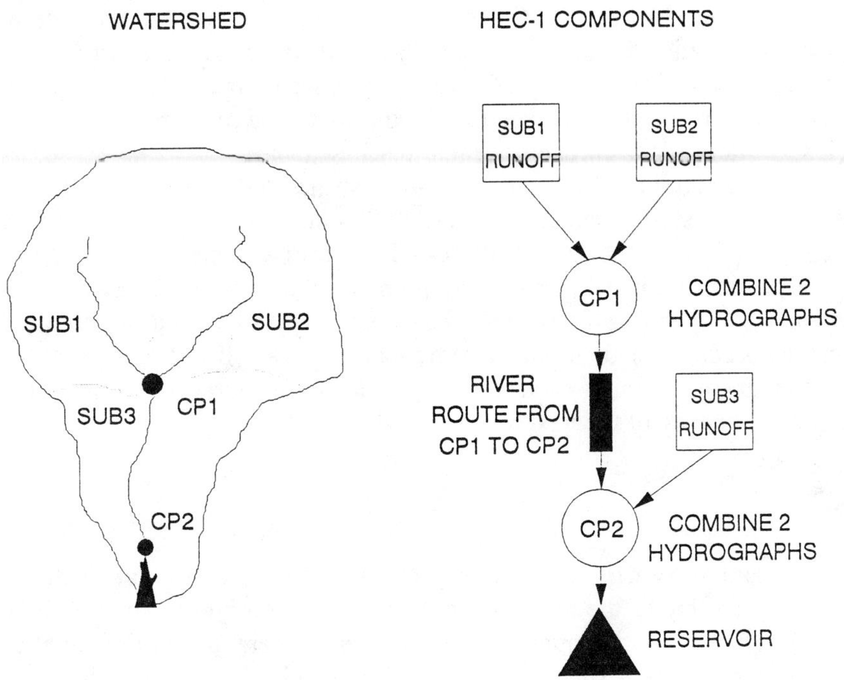

Fig. 4.1. HEC-1 Representation of Watershed Components.

Precipitation

For each subbasin, the average precipitation in each time interval must be determined. The precipitation will fall as rain or snow depending upon the air temperature (only if snowmelt computation is indicated). HEC-1 has the provision to use gaged precipitation or hypothetical storms. If subbasin average precipitation is already known, it may be input directly. If gages are used, the subbasin average precipitation must be calculated for each simulation time interval. The gages are weighted by Thiessen or other methods. The resulting time series of subbasin-average precipitation is then available for the loss calculations.

Hypothetical storms may also be used. These storms may be derived separately from HEC-1 and input, or they may be computed by HEC-1 by inputting storm statistics. Design storms such as the Standard Project or Probable Maximum can be calculated by HEC-1 but only for limited purposes. It is recommended that the design storm be computed by current criteria and then input to HEC-1 for runoff computation. An example of such a computation is the criteria of Hydrometeorological Report No. 52 (National Weather Service, 1982). HEC has a separate computer program, HMR52, to perform those calculations and provide the resulting storm input to HEC-1 via the data storage system as discussed later.

For frequency-based hypothetical storms, the storm statistics required are: storm area, storm frequency, and intensity-duration data. These data are typically available from NOAA atlases. The storm area is required because area adjustments must be made for point rainfall statistics. The storm frequency is required if the storm is frequency-based so that partial-duration adjustments can be made to the more frequent storms. HEC-1 then creates a balanced storm in which each intensity of a shorter duration is imbedded within the intensity of the next longer duration.

Snow

Snowfall and snowmelt can be simulated within each subbasin. Either a simple degree-day or the more comprehensive energy-budget method may be used. The degree-day approach requires air temperatures at the outlet of the subbasin for each time interval. An adiabatic lapse rate is input to compute the average air temperature in up to 10 elevation zones. Within each elevation zone, it is determined whether the precipitation falls as rain or snow. If snow, it is accumulated in a snow pack. If rain, it is subject to losses and runoff during each time interval. If the air temperature is warm enough, then melt occurs. The energy-budget method works in a similar manner except that the melt computation is also a function of wind speed, incoming solar radiation, and dew point. These data must also be provided for each time interval.

Losses

Losses refer to all rainfall or snowmelt that is not available for direct runoff due to interception, infiltration, and detention storage. These losses are not available for further use in the runoff simulation. Thus, the outflowing hydrograph from a subbasin is only the result of direct runoff and base flow. The loss rate methods available in HEC-1 are shown in Table 4.2.

The Initial & Constant method is commonly used for hypothetical storms or observed floods with poor definition of rainfall. The HEC Exponential method has logical parameters, but they are not readily related to watershed

TABLE 4.2.

HEC-1 Loss Rate Methods

Method	**Characteristics**
Initial & Constant	Satisfy an initial volume, then a constant rate
HEC Exponential	A logarithmic decay function based on accumulated losses
SCS Curve Number	Initial abstraction and infiltration per SCS
Holtan	Infiltration based on accumulated losses
Green & Ampt	Infiltration based on soil characteristics

characteristics, e.g. land use and soil type. The SCS method is easy to use because the Curve Number can be easily obtained from land use and soils maps; unfortunately, that "book" value may not be appropriate for historical storms of differing antecedent moisture conditions. The SCS now recommends calibration of the curve number to observed rainfall and runoff. The Holtan method is based on watershed characteristics, but its development stopped when the Green & Ampt method became popular. The Green & Ampt parameters may be obtained from soil characteristics and much work is currently being done to enhance its usage. As in any hydrologic model, it must be remembered that these loss characteristics are considered homogenous over the subbasin, while the real makeup of the subbasin is quite heterogeneous. Thus, what ever method is used must be considered to be "hydrologically representative" for the whole subbasin.

Excess-to-Runoff Transformation

The unit hydrograph and kinematic wave methods are available in HEC-1. For the unit graph, it may be input directly if the ordinates are known, or several analytical functions may be used. If the ordinates are input, they must be for a duration equal to the simulation time interval. The Clark, Snyder, or SCS analytical methods will compute a unit graph whose duration is equal to the simulation time interval. The SCS method is formulated as the single parameter LAG relationship using the SCS dimensionless unit graph shape. The Clark method uses the standard two parameters and a time-area curve. If the time-area curve is not known, HEC-1 supplies one with a typical elliptical shape. The Snyder method is simulated using the Clark method because Snyder, by itself, does not provide all ordinates of the unit graph.

Kinematic wave is a nonlinear runoff transformation function (HEC, 1993a). HEC-1's formulation of the kinematic wave method takes two forms: the standard runoff plane approach and an HEC storm-drain-subsystem approach for urban areas. The subsystem approach is described in the following hydrologic analysis section. Instability of the kinematic wave computation with respect to time and distance steps has been a frequent problem in kinematic wave solutions. The HEC representation of the method provides for automatic stability checks of time and distance intervals until the computation remains stable. Continuity and other diagnostic information is also provided.

Base Flow

HEC-1 has a rather simple representation of base flow because its primary interest is in simulating flood hydrographs where base flow is a relatively small contributor. Also, there is little agreement in the profession as to how base flow can be determined. Thus, a simple method is provided, or the user can determine the base flow by some other technique and directly input it to the program. In the later case, HEC-1's base flow can be set to zero.

The base flow capability consists of an starting flow, a recession threshold, and a logarithmic decay recession rate. The base flow begins with the starting flow which is recessed at the decay rate. Then the direct runoff is added to that recessed flow as the storm occurs. When the flood hydrograph falls to a recession threshold level, then the subsequent calculations are simply the recessed threshold value. If rainfall starts again, then rainfall excesses are again used. Usually the recession threshold value is small compared to the flood peak. The recession threshold value is chosen as the point on the falling limb of the hydrograph where the flows plot as a straight line on semi-log paper.

4.3.2. Channel Routing

Several methods for channel routing are available (as was the case with subbasin runoff methods). Channel infiltration may be computed in connection with any of the routing methods. The available channel routing methods are shown in Table 4.3.

TABLE 4.3.

HEC-1 River Channel Routing Methods

Method	Characteristics
Average-Lag	Averages flows over a period and lags the result
Muskingum	Classic method for upland, non-floodplain rivers
Muskingum-Cunge	Close approximation to full unsteady flow in most cases
Modified Puls	Uses floodplain storage-outflow characteristics developed from steady-flow backwater calculations.
Kinematic wave	Flood wave translation without attenuation

The Muskingum method is widely used in relatively steep channels with little floodplain storage. If floodplain storage has major attenuating effects on the flow, then the modified puls with the storage-outflow relationship determined with a program like HEC-2 is preferred. Modified Puls works well but it must be calibrated to the flood events in question; this is difficult in ungaged areas. The Muskingum-Cunge method has performed well for all but the flattest rivers and/or sharply rising hydrographs (Garbrecht and Brunner, 1991). It is a close approximation to the full unsteady flow computations and doesn't require calibration in the sense that Modified Puls does. It must be remembered that these hydrologic routings do not know what is happening downstream so it is not a full hydraulic solution considering flow resistances transmitted upstream.

4.3.3. Lake Routing

Flow through lakes is simulated with the same Modified Puls storage-outflow relationship as used for channels. The lake application is actually the more theoretically correct use of Modified Puls. It assumes a level pool (lake surface) whose outflow is only dependent on the storage (or elevation) of the lake. The channel usage is an approximation of channel flow as a series of reservoirs.

4.3.4. Confluences

Confluences are simply locations where one or more separate hydrographs are combined into a single hydrograph. They are treated as a separate building block because they compute a new hydrograph.

4.3.5. Reservoir Routing

Two methods are available for reservoir routing: Modified Puls and level pool. The modified puls is the same as described above for lakes. The level-pool routing allows the user to prescribe the following reservoir outlet characteristics: low-level outlet; spillway; and flow over the top of dam. A volume-elevation relationship (or surface area versus elevation) is also required. Low-level outflow is computed as orifice flow, and the spillway and top-of-dam flows are computed as weir flow. Continuity is maintained between the inflow, change in storage, and outflow. A level pool water surface is assumed throughout the reservoir. If the reservoir outflow rating curve is known or computed from other sources, it may be input directly instead of the outlet information.

The level-pool reservoir routing can also be used to represent other types of flow constrictions such as roadway embankments with culverts. This simulation is appropriate if the culvert is sufficiently submerged so that the orifice flow equations apply. If the open channel flow through the culverts is significant, then a separate rating curve for the outflow rating curve must be developed and input to HEC-1.

Reservoir releases may also be specified directly and HEC-1 will calculate the associated reservoir storage. This option is useful for quickly evaluating various reservoir release alternatives. If the reservoir storage is depleted or exceeded, then the release pattern is revised and simulated again.

4.3.6. Diversions

The diversion building block bifurcates the main-channel flow according to an input relationship of total inflow versus the flow to be diverted. There are several possible reasons for the diversion: natural separations in braided channels; flow constrictions which cause some of the flow to take a separate path; and inadequate channel capacities such as in urban storm drains. In some cases, the restricted flow will simply pond at that location and re-enter the stream when channel capacity permits. In other cases, the restricted flow will take a separate flow path, e.g. over the streets instead of the storm drains in an urban area. The flow which is not diverted continues down the main channel.

Because there is no general solution to these diversion possibilities, HEC-1 makes the user prescribe the path to be taken. The diverted flow is stored on disk for future retrieval in any subsequent subbasin building block, see Fig. 4.2. The diverted flow is retrieved in a subbasin building block and may be routed through reservoir storage or through river channels as appropriate. It may re-enter the flow system in any subsequent hydrograph computation.

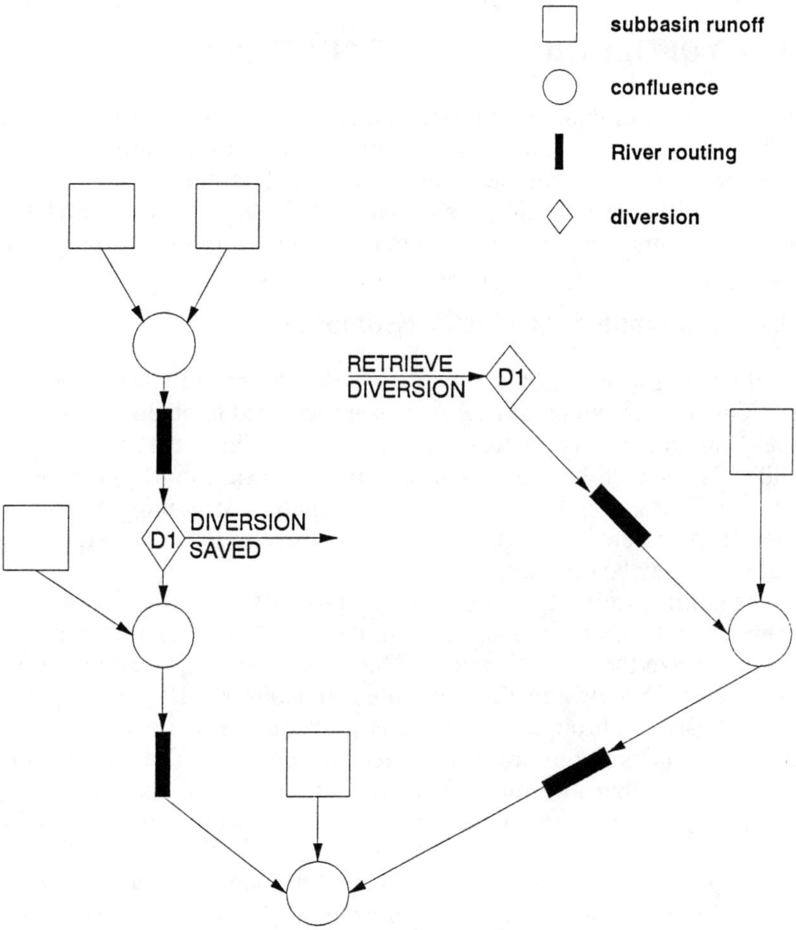

Fig. 4.2. Diversion Schematic.

4.3.7. Pumps

Pumps are not a natural watershed feature but are often part of man-made systems, especially in urban areas. Pumps may be used in urban areas where storm drains become too deep and the flow needs to be lifted to a higher elevation. Another common use is in areas behind levees of other obstructions become flooded and the flow must be pumped over the obstacle. The pump option provides for several pumps, each of which may be turned on and off at different water surface elevations. Pumping is accomplished in a storage (reservoir) routing building block. The pumped flows are saved like the diverted flows for retrieval in any subsequent building block.

4.4. HYDROLOGIC ANALYSIS FEATURES

HEC-1 is used in many hydrologic studies, and it has been found desirable to provide some extra automated features to assist the engineer with those analyses. This section describes automated features which assist the engineer with the hydrologic simulations. Many of these same features could be accomplished by repeated manual applications of the program.

4.4.1. Automated Parameter Estimation

In calibrating a hydrologic model to observed rainfall and runoff, many iterations of different parameter values are required to obtain a best estimate. HEC-1 automates this process using the unit gradient search method (HEC, 1980). The search may be performed for subbasin runoff parameters (loss rates and unit graphs) or for routing parameters. It is most often used for subbasin parameter estimation. Any or all of the loss rate and unit graph parameters may be optimized.

The unit gradient process systematically searches through each parameter to be optimized and determines how to change that parameter so as to minimize the error function. The error function is a weighted sum of the differences between the computed and observed hydrographs. The process searches through all parameters several times, always adjusting the parameter values to minimize the objective function. The process does not guarantee a global minimum; but this can be determined by starting the search process at different parameter values and observing the track and results.

The parameter estimation feature provides some checks for minimum and maximum values during the optimization. The user must verify that the results are reasonable, make good physical sense, and are consistent from one subbasin to another. The parameter estimation results are often used in a regional study to estimate parameters in ungaged basins.

4.4.2. Multi-Storm Simulation

This feature automatically takes ratios of a prescribed storm or flood and simulates it throughout the river basin. Ratios are used to evaluate the change in streamflow or project performance with respect to changes in the size of the flood. Ratios of storms are used where ever there is a change in land surface runoff characteristics, e.g. urbanization. Ratios of floods are used in hypothetical flood studies, e.g. ratios of the probable maximum flood which exceed a spillway capacity. This capability is often used in connection with the multi-Plan capability described in the Flood Control Projects Analysis section which follows.

4.4.3. Depth-Area Precipitation Simulation

The depth-area precipitation simulation option adjusts the average precipitation over an river basin to conform with depth-area statistics as the simulation proceeds downstream. This allows the user to compute, say, the 1% flood throughout a river basin without having to re-center the storm over each contributing area for each point of interest.

4.4.4. Urban Runoff

The kinematic wave land surface runoff capability typically available in urban models is usually a single runoff plane and collecting channel. That capability is also available in HEC-1, but it also has a powerful subsystem runoff feature. Figure 4.3 shows some of the flow features of an urban area;

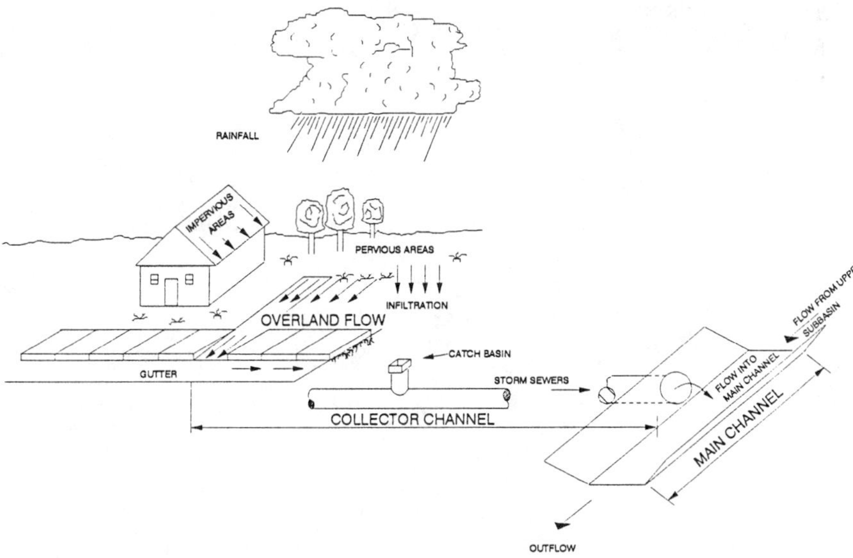

Fig. 4.3. Elements Used in Urban-Runoff Calculations.

note that there are two main types of runoff surfaces (impervious and pervious); two levels of collector channels; and the main channel. The subsystem feature depicts a typical runoff and storm drain system within a subbasin, see Fig. 4.4. The runoff is computed for the typical subsystem, and then that runoff is distributed along the main channel in proportion to

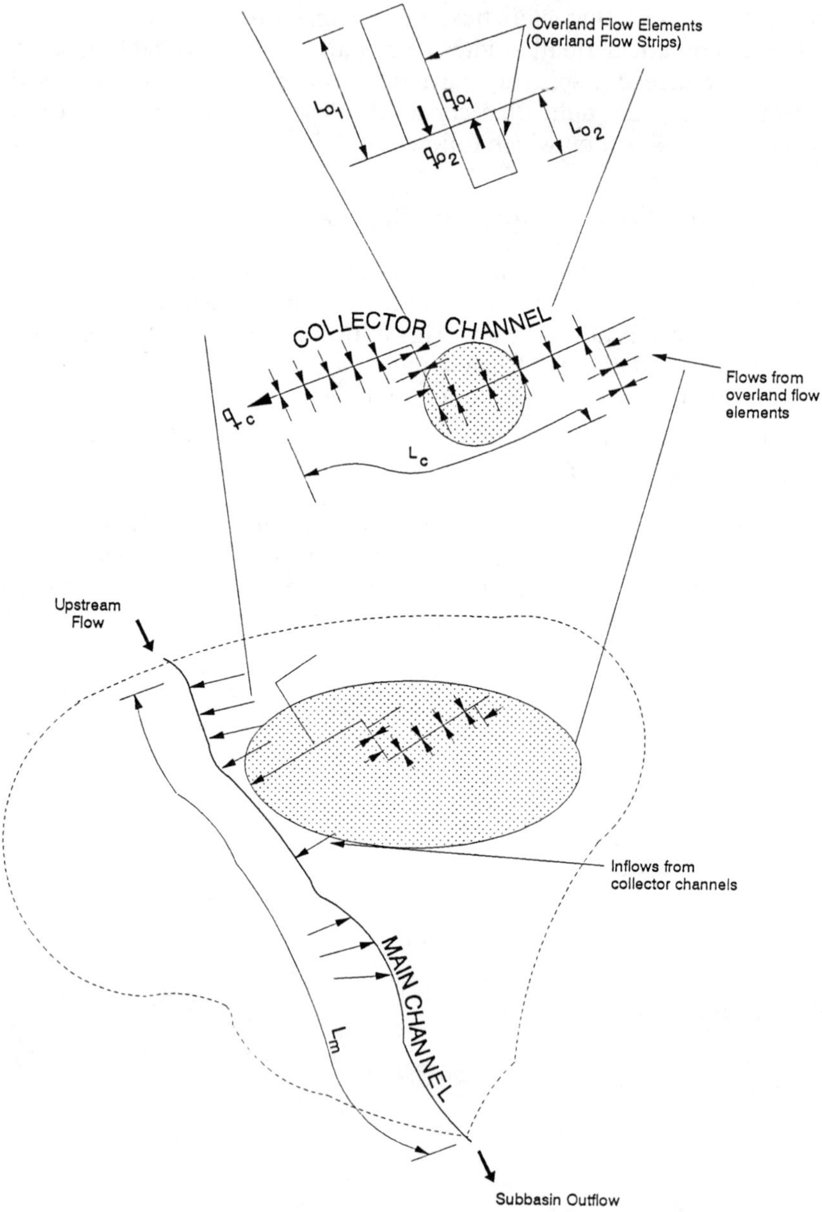

Fig. 4.4. Subbasin Typical Urban Collector System.

the size of the subsystem to the subbasin. This feature is particularly useful in cities where there are land development codes establishing consistent storm sewer systems.

Using the two runoff planes to represent pervious and impervious surfaces is advantageous to represent distributed runoff characteristics in a lumped hydrologic model. Although the two planes are average for the subsystem and subbasin, they do allow portrayal of the two main types of runoff within a subbasin. The pervious runoff surfaces have higher infiltration rates and are typically longer in transit to the collector channels. The impervious surfaces obviously have very little infiltration and are usually a short distance from the collector channels. Both of these surfaces contribute runoff to the first collector, typically a gutter in urban areas. The first collector is uniformly distributed into a second collector which may be a storm drain or intermediate sized channel. The outflow of the second collector is then uniformly distributed along the main channel in proportion to the size of the subsystem compared to the whole subbasin. Fractional contributions are used to ensure representation of the whole subbasin.

The same two-runoff-plane capability may also be important to simulation of natural watersheds. The idea of partial-area contributions of runoff can also be represented with the two planes. One plane can represent the faster, more efficient runoff areas using lower loss rates and shorter flow paths; the other plane can represent the farther away, slower contributing areas with higher loss rates. The areas of the two planes are fixed however, and could not change dynamically during a storm as the natural watershed might do.

If it is thought that the subsystem collector channel flow may exceed the channel capacity and the excess flow needs to be handled explicitly, then the subbasin must be further subdivided. If the excess flow just flows over the streets and ends up in the main channel in about the same manner, then the typical subsystem representation is adequate.

4.4.5. Dam-Safety and Dam-Breach Analysis

Two features were added to HEC-1 during the National Nonfederal Dam Safety Inspection Program: assessment of spillway adequacy, and dam breach simulation. Routing through a reservoir was described earlier in the hydrologic simulation section. Many nonfederal dams were found not to have engineering data and outflow rating curves. Thus the low-level outlet, spillway, and top-of-dam outflow computation capability was added as discussed earlier. Also, many dams were found to have nonlevel crests so additional algorithms were developed to simulate that type of overflow condition.

Spillway Adequacy

Assessing spillway adequacy is performed by routing ratios of hypothetical floods through a dam to see when the spillway capacity was exceeded. The hypothetical flood was usually the probable maximum flood, PMF. Before the new PMF criteria came out for the eastern U.S.A., the HEC-1 PMF calculation was used and ratios of the PMF were taken using the multi-flood option discussed earlier. For new PMF or other flood criteria, the probable maximum precipitation is computed separately and input to HEC-1 to compute the PMF. A special output summary was added to HEC-1 to analyze spillway capacity and dam overtopping.

Dam Breach

Together with the spillway capacity analysis, the impact of a dam breach also needed to be studied. The dam breach simulation capability of the National Weather Service's DAMBRK (1980) software was implemented in HEC-1 to meet this need. The water surface elevation which triggers the breach is specified along with the size and rate of breach development. An outflow submergence correction is made depending on backwater from an immediate downstream cross section input for that purpose. This differs from the DAMBRK model which transmits the flow resistance of all downstream reaches to the dam to determine the outflow water surface elevation for the submergence correction. The HEC-1 breach calculations are carried out at a much shorter time interval than the normal HEC-1 computations. After the breach hydrograph is computed, it is resampled at the normal HEC-1 simulation time interval in order to continue the routing downstream. Appropriate cautions and graphics are provided in HEC-1 to minimize user errors in resampling the breach hydrograph. The errors can be eliminated by making the HEC-1 simulation time interval shorter.

4.4.6. Flood Frequency Curve Determination in Gaged and Ungaged Areas

One of the most important hydrologic analyses in a flood control study is the determination of flow frequency curves at various sites. HEC-1 can assist with those computations by: computing frequency curves from hypothetical storm runoff; calibrating the runoff model to a given (adopted) frequency curve; and transferring frequency curves to ungaged locations. The hypothetical, frequency-based storm simulation was discussed earlier. For each storm frequency, the appropriate precipitation is input, antecedent conditions are assumed, and the peak runoff is assumed to be of the same frequency as the storm. This approach has well know limitations because of the unknown antecedent moisture frequency distribution.

If a flow frequency curve is known (adopted) for a location, then the HEC-1 model can be calibrated so that the 1% storm produces the 1% peak flow. For project analyses discussed in the next section, this is essential to represent existing conditions. Once the model is calibrated to produce the x% peak flow, that same model can be used to estimate the x% peak flow for ungaged areas which are hydrologically similar.

HEC-1 can also be used to translate hydrographs and peak flow from one land use condition (urbanization, storage projects, etc.) to another. This is often necessary when the historical annual peak flow series is nonstationary and must be adjusted to current conditions for project analysis. HEC-1 is calibrated to each land use condition and then a relationship between each past and current condition is developed for a range of floods. That relationship is used to translate peak flows between different land use conditions.

4.5. FLOOD DAMAGE REDUCTION PROJECT ANALYSIS

Analysis of flood damage reduction projects requires evaluation of the change from existing condition damage to the with-project damage. The change in damage may be due to land use changes such as urbanization or implementation of flood damage reduction projects such as reservoirs, levees, diversions, etc. HEC-1 has several features to facilitate these analyses.

4.5.1. Multi-Plan Analysis

Multi-plan analysis is a short-hand way to analyze different river basin developments in the same simulation. This eliminates making several data sets to analyze the different alternatives. It is especially convenient to analyze both the hydrologic and flood damage differences from existing to with-project conditions. A typical application is in the analysis of runoff from new urban developments. (The flood damage computation capability is discussed later in this section.) The increased urbanization causes increased flood runoff which must be counteracted by detention storage or some other on-site measure. HEC-1 can analyze this situation with three simultaneous plans: plan 1, existing watershed conditions; plan 2, future urban condition; and plan 3, future urbanization with detention storage. The impact on the flow frequency curve and flood damage can both be determined. The multi-plan option provides convenient output summaries to compare the runoff and damage for each plan of development.

4.5.2. Modified Frequency Curve Computation

Flood damage reduction project analysis requires the computation of a with-project flow or stage frequency curve. HEC-1 makes this computation through the combined use of the multi-storm and multi-plan analysis features discussed earlier. The multi-plan provides the alternative runoff conditions. The multi-storm provides analysis of floods across the full range of the frequency curve. To perform this analysis, a known or adopted frequency curve for existing conditions must bc input to HEC-1. HEC-1 computes the peak runoff for existing conditions (plan 1) for a range of flood sizes. The frequency of each storm ratio's peak flow is determined from the input frequency curve. That frequency is assigned to that storm ratio. Then the future, modified, with-project (plan 2) condition is simulated with the same storm ratios. A new peak flow is computed for each storm ratio and the frequency of the storm ratio is assumed to remain the same, hence the modified frequency curve is computed, see Fig. 4.5.

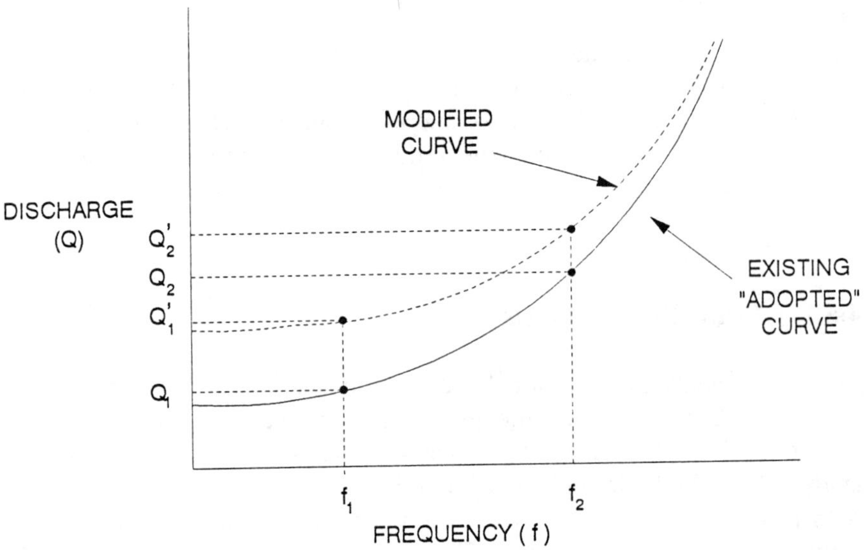

Q_i = PEAK DISCHARGE FOR STORM RATIO i

f_i = FREQUENCY FROM GIVEN EXISTING CURVE

Q'_i = PEAK DISCHARGE FROM STORM RATIO i UNDER MODIFIED (URBANIZED) WATERSHED CONDITION

Fig. 4.5. Modified Flow-Frequency Curve.

4.5.3. Flood Damage Computation

HEC-1 can compute expected annual flood damage or the flood damage associated with a single flood event. The expected annual computation is used for economic efficiency analysis of annualized flood damage reduction benefits versus annualized costs of the project. To compute flood damage, a stage- or flow-damage relationship is required for each damage center, e.g. town. If a stage-damage curve is used, a stage-flow rating curve must also be provided. For single flood events, HEC-1 simply looks up the damage associated with the peak flow for each flood (ratios and plans may be used).

For expected annual flood damage, the existing-condition frequency curve must be provided so that the modified, with-project frequency curve can be calculated. The damage-frequency curve is calculated from the flow-frequency and flow-damage relationships. Stages may also be used if a rating curve is provided. The damage-frequency curve is integrated (using the storm/flood ratios) to obtain the expected annual damage estimate. This analysis makes combined use of the multi-storm, multi-plan, and modified frequency curve capabilities of HEC-1.

4.5.4. Project-Size Optimization

The size of a flood control project may be optimized for economic efficiency by HEC-1. This capability combines the use of the expected annual flood damage computation described above with the automated parameter estimation search procedure described earlier. Additional information about how the cost of a flood control project varies with its size must be input. Several projects may be optimized at the same time.

The univariate search procedure analyzes alternative project sizes and adjusts the size so as to minimize the sum of the project cost plus expected annual damage (Davis, 1975). This is equivalent to maximizing the benefit-cost ratio. The search process works on each project sequentially and the iterates through all projects several times. This search procedure suffers the same limitations noted for the parameter estimation process, i.e. it cannot guarantee a global optimum. The search process will stop at a local optimum so the user must start the search process with different project sizes to determine if a global optimum has been reached. Also, the interplay between projects (e.g. reservoir and a levee) is only determined sequentially. The results do provide valuable information about the economic sensitivity of projects for planning purposes.

4.6. TYPICAL APPLICATIONS OF HEC-1

As noted in the previous discussion of HEC-1 capabilities, the software package has many more applications than pure hydrologic process simulation. The applications include special hydrologic system analyses and flood control project analyses. A summary of the different analyses is given in Table 4.4. Each of these applications has been discussed earlier in this chapter and is described in detail in the HEC-1 User's Manual and associated publications (see Table 4.1).

TABLE 4.4.

Applications of HEC-1

Application	Comments
Historical flood simulation	Several options for simulation of runoff from rainfall and snowmelt in river basins.
Hypothetical storm/flood simulation	Computes balanced storm from intensity duration statistics or accepts user-specified storm or flood.
Parameter estimation	Automatic loss rate and unit graph parameter estimation, also routing parameter estimation.
Urban hydrology simulation	Kinematic wave land surface runoff and typical storm drain subsystem simulation.
Multiple flood analysis	Automated storm- and flood-ratio simulation.
Dam safety	Spillway capacity analysis using ratios of hypothetical floods such as the PMF.
Dam breaches	Dam-breach simulation with outflow submergence correction and continued downstream simulation.
Precipitation depth-area simulation	Eliminates the need for multiple storm centerings and sizes to compute a frequency-based flood at several locations in a river basin.

TABLE 4.4. Continued.

TABLE 4.4. Continued.

Application	Comments
Modified frequency curve computation	Determines new flow frequency curve for changes in land use and addition/deletion of flood control projects.
Flood control projects	Can simulate reservoirs, levees, channel modifications, diversions, and pumps to control flows.
Flood control project optimization	Optimizes sizes of flood control projects for maximum economic efficiency.

4.7. INTERCONNECTION WITH OTHER MODELS

The HEC Data Storage System, HEC-DSS, (1990b) was created by a need to pass data from one analysis program to another, e.g. HEC-1 streamflows to HEC-5 reservoir operation. Also, there was no need for each program to have its own set of graphics routines, editors, etc. to aid in the program's use.

4.7.1. Data Storage System, HEC-DSS

HEC-DSS was developed to manage data storage and retrieval needs for water resource studies. The system enables efficient storage and retrieval of hydrologic and meteorologic time-series data. HEC-DSS consists of a library of subroutines that can be readily used with virtually any applications programs to enable retrieval and storage of information. At present approximately 20 applications programs have been adapted in this fashion. HEC-DSS has been used to connect EPA's Storm Water Management Model, SWMM, (Huber and Dickinson, 1988) to HEC software.

Approximately 17 DSS utility programs have been developed. A number of these programs are for data entry from such files as the USGS' WATSTORE data base or from the NWS precipitation data files. Other utility programs include a powerful graphics program, a general editor, and a program for performing mathematical transformations, DSSMATH. Macros, selection screens, and other user interface features combine with

DSS products to provide a set of tools whose application is limited only by the ingenuity of the user. HEC-DSS is depicted in Fig. 4.6.

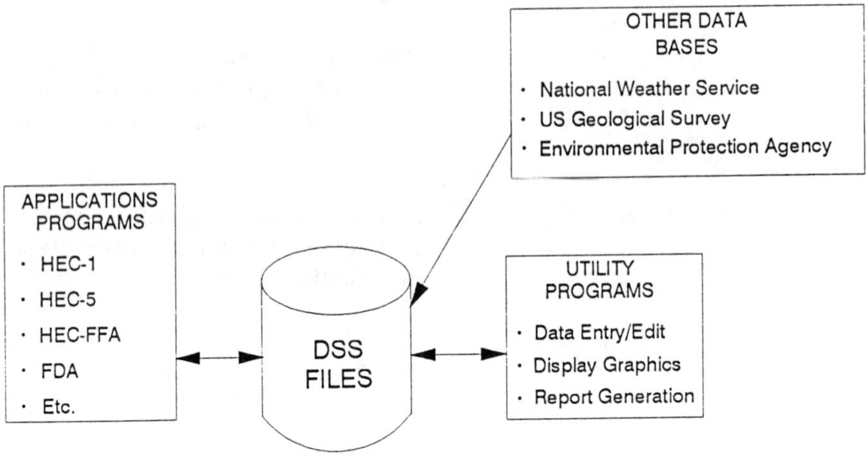

Fig. 4.6 HEC Data Storage System.

4.7.2. DSS Applications

Figure 4.6 identified a few models which can be connected via DSS. There are numerous hydrologic analyses and project evaluations which can be facilitated by DSS. Precipitation may be computed by several software packages including the HMR52 program (HEC, 1984) which computes the probable maximum precipitation for the eastern U.S.A. HMR52 creates an output DSS file for direct input to HEC-1. Other non-HEC software can easily be connected to DSS through direct fortran statements or translation of files to DSS format. In one Corps study, the EPA SWMM model software was revised to read and write from DSS; this was to enable input of HEC-1 land surface runoff into the SWMM EXTRAN storm sewer router. The EXTRAN output was then input to the HEC Interior Flood Hydrology (1992) model for further analysis.

For river routings where floodplain storage is significant, the modified puls storage-outflow relationship can be obtained from water surface profile programs like HEC-2 (HEC, 1990c). HEC-2 is set up to automatically compute multiple flow profiles and the associated storage in designated reaches. It then can output that storage-outflow relationship to DSS for use in the HEC-1 river routings.

In some river routings or urban area storm drains, the HEC-1 hydrologic routing assumptions are not adequate. Unsteady flow or flow in looped networks may exist that HEC-1 cannot simulate. For unsteady flow simulation, the land surface runoff from HEC-1 is stored in DSS and then retrieved by HEC's unsteady flow program, UNET (HEC, 1993b). UNET takes the HEC-1 hydrograph inputs from DSS and routes them through the river or storm drain system. The UNET results are also displayed by DSS graphics and reporting software.

For reservoir regulation studies, the HEC-5 (HEC, 1989) reservoir system simulation software is used. Again, the HEC-1 streamflows are saved in DSS and input to HEC-5. HEC-5 operates the reservoir system to minimize flooding at designated control points. The HEC-5 results are then also viewed via DSS.

Although HEC-1 can compute flood damage, there is more capable software available now in the HEC Flood Damage Analysis (FDA) Package, 1988. In this case, the FDA package can obtain the necessary frequency curves for with- and without-project conditions from HEC-1 via DSS.

4.8. THE NEXT GENERATION OF HEC-1

4.8.1. The HEC NexGen Project

In 1990, HEC embarked on a project, termed NexGen, to develop the next generation of its simulation models. The objectives of the new modeling capabilities were to provide the user with better means to visualize and understand the process being simulated, and to build more engineering expertise into the models themselves. With the recent advances in computer hardware, it was no longer necessary to have the computer programming constrained by the limitations of old machines. It was quite evident that the old batch processing format would not suffice and that interactive processing was necessary. The capabilities of modern workstations and PC's using the NT and UNIX operating systems offer a new level of processing power that could meet these next-generation software needs.

The intent of this next generation of models is to put the users inside the model and give them the tools to easily work with the data, simulation processes, and results. The user will enter data into a data base that is constructed in a logical engineering-analysis format, not a format for some computer input device. Output will also be stored in the data base for ready analysis. A graphical user interface will let the user view windows of the data, computations, and results for maximum understanding and analysis of the data and the physical processes. Spatial schematics of the system (e.g. river basin network) will be used to guide access to the data, computations, and results.

Four technical areas are being addressed in the current HEC NexGen effort: river hydraulics, watershed runoff, reservoir system, and flood damage analysis. The new models will have most of the capabilities of the existing HEC models in those areas plus new algorithms where appropriate.

For example, current river hydraulics modeling capabilities require reformatting the river geometry differently for each analysis - steady state, unsteady, and multi-dimensional. The new river analysis system of software will use consistent geometric representation of the river and floodplain for all applications. The simulation results will also be shown on the same geometric representations.

The ultimate goal is to have smarter models that automatically evaluate numerical stability (time and distance steps) and physical constraints of the process being simulated. The user will be advised of process-simulation problems, and alternative methods and analyses will be recommended where possible. Thus, more engineering expertise will be built into the models to enhance their application and interaction with the user. The NexGen project is a multi-year effort, and interim products will be made available as the work progresses.

4.8.2. New Hydrologic Modeling System Capabilities

The NexGen evolution of HEC-1 is termed the "Hydrologic Modeling System, HMS." HMS will contain most of the exiting HEC-1 capabilities, but also be a more generalized system to accommodate many other hydrologic simulation, analysis, and project analysis capabilities. Some of the planned new capabilities are shown in Table 4.5. Some of these features will be included in a beta version of the program. Pending experience with the beta version, other new capabilities will be added.

TABLE 4.5.

NexGen Hydrologic Modeling System Capabilities

Subject area	Description
Precipitation	Raster-based spatial rainfall input from radar or other spatial measurements. Gage weighing by inverse-distance squared method.
Soil moisture	Simple, continuous soil moisture accounting for long-term runoff simulation (more detailed accounting may be included if warranted).
Excess-to-runoff transformation	Diffusion routing for overland flow may replace kinematic wave. Spatially distributed unit graphs and kinematic wave/diffusion runoff.

TABLE 4.5. Continued

TABLE 4.5. **Continued**

Base flow	Simulation from soil moisture accounting.
River routing	Diffusion solution of unsteady flow equations.
Parameter estimation	More flexible regional optimization of parameters in several subbasins and routing reaches.
Flood frequency	Improved use of storm ratios and frequency estimation from given peak-discharge frequency curve.
Diversion	Variable diversion function of time and/or stage to simulate levee breakouts.
Spatially distributed processes	Increased use of GIS data and distributed processing.

4.8.3. GIS Hydrology

Some of HEC's earliest work in GIS hydrology involved development of a systematic methodology for automating the data preparation process. The raster-based organization chosen by HEC was called a grid cell data bank. Techniques for use of satellite data, for conversion of polygon data to grid format, and for use of commercially available software to manipulate and convert the data were developed. Parameters for HEC-1 and other hydrologic models were computed by a program called HYDPAR (HEC, 1985) which accessed the grid cell data. The grid cell data bank approach was formalized in the HEC Spatial Analysis Methodology (HEC-SAM). Remotely sensed land use and other hydrologic parameters were also incorporated in the SAM methodology. Later, HEC explored the use of triangular irregular elements, TINs, for representation of watershed characteristics. A program linking HEC-1 with the TIN was developed. Because of complex hardware, software, and study-management problems associated with the GIS approach, HEC has been less active in the evolution of this technology for the past decade.

Recent HEC efforts have included a review of GIS applications in hydrologic modeling (DeVantier and Feldman, 1993), and research into a method for combining the spatial GIS data with lineal hydrologic networks. A hybrid grid-network procedure for adapting GIS capabilities for hydrologic modeling is being investigated. Spatially distributed processes are represented on a grid and one-dimensional flow and transport occurs through an associated network. There is a duality between a grid and a network in that once the direction of flow on each grid cell is defined to a single neighboring grid cell, an implied flow network is created (Maidment, 1995). These ideas are being further investigated in HEC's NexGen and remote sensing/GIS projects. Digital elevation models, DEM, now widely available are being tested with algorithms to automatically capture the spatial processes of watershed runoff and link them to the river network. Current algorithms for raster or triangular element representation of DEM data are cumbersome with respect to accurate delineation of subbasins and rivers, especially in relatively flat terrain. Major improvements to hydrologic analysis should be forthcoming as georeferenced stream locations are included in the DEM watershed analysis.

4.9. CONCLUSION

Software like the HEC-1 Flood Hydrograph Package must be dynamic to keep up with changes in hydrologic simulation methods and new computer software and hardware. HEC-1 has succeeded in continuing to be a practical engineering tool and to keep up with advances in the hydrologic engineering and computer science industries. Much of this success is due to continued support and development of the software by the U.S. Army Corps of Engineers. The Corps recognized the need for continuing support several years ago and began a Numerical Model Maintenance program to complement its research and development program. Such support mechanisms are essential to the vitality of software.

HEC-1 has evolved from a series of small programs for simulation of parts of the precipitation-runoff process to a comprehensive hydrologic and project analysis software package. New hydrologic and hydraulic algorithms are added as they are proven in field applications. The computer-engineer interface is updated, or totally redesigned in the case of NexGen, as new software and hardware become available. HEC's philosophy is provide sound technical solutions to specific problems and then to generalize those solutions for use in many geographic areas. Those solutions are not just theoretically correct, but have stood the test of numerous field tests for technical validation and ease of use. The rapid advances in today's microprocessors and operating systems are being put to good use in the development on HEC's next generation of engineering software.

REFERENCES

Davis, Darryl W., 1975. Optimal Sizing of Urban Flood-Control Systems. Journal of Hydraulic Engineering, Vol. 101, HY8, American Society of Civil Engineers, New York, pp. 1077-1092.

DeVantier, Bruce A. and Arlen D. Feldman, 1993. Review of GIS Applications in Hydrologic Modeling. Journal of Water Resources Planning and Management, Vol. 119, No 2, American Society of Civil Engineers, New York, pp. 246-261.

Feldman, Arlen D., 1981. HEC Models for Water Resources System Simulation: Theory and Experience, in Advances in Hydroscience, edited by Ven Te Chow. Vol. 12, Academic Press, New York, pp. 297-423.

Garbrecht, Jurgen and Gary Brunner, 1991. Hydrologic Channel-Flow Routing for Compound Sections. Journal of Hydraulic Engineering, Vol. 117, No. 5, American Society of Civil Engineers, New York, pp. 629-642.

Huber, Wayne C. and R. E. Dickinson, 1988. Storm Water Management Model, Version 4: User's Manual. Environmental Protection Agency, Environmental Research Laboratory, Athens, GA.

Hydrologic Engineering Center, 1980. Corps of Engineers Experience with Automatic Calibration of a Precipitation-Runoff Model, Technical Paper No. 70, U.S. Army Corps of Engineers, Davis, CA.

_____, 1984. HMR52 Probable Maximum Precipitation (Eastern United States) User's Manual, Computer Program No. 46.

_____, 1985. HYDPAR Hydrologic Parameters, User's Manual. Computer Program Document No. 34.

_____, 1988. Flood Damage Analysis Package, User's Manual. Computer Program Document No. 59.

_____, 1989. HEC-5 Simulation of Flood Control and Conservation Systems, User's Manual. Computer Program Document No. 5.

_____, 1990a. HEC-1 Flood Hydrograph Package, User's Manual, Computer Program Document No. 1A.

_____, 1990b. HEC-DSS Data Storage System, User's Guide and Utility Program Manuals. Computer Program Document No 45.

_____, 1990c. HEC-2 Water Surface Profiles, User's Manual, Computer Program Document No 2.

_____, 1992. HEC-IFH Interior Flood Hydrology, User's Manual. Computer Program Document No 31.

_____, 1993a. Introduction and Application of Kinematic Wave Routing Techniques Using HEC-1. Training Document No. 10, originally 1979, updated 1993.

_____, 1993b. UNET One-Dimensional Unsteady Flow Through a Full Network of Open Channels, User's Manual. Computer Program Document No 66.

Maidment, David R., 1995. Watershed Data Structures and Digital Delineation of Watersheds and Stream Networks. A collection of research reports for the Hydrologic Engineering Center, Davis, CA.

National Weather Service, 1980. DAMBRK, The National Weather Service Dam-Breach Flood Forecasting Model, Office of Hydrology, National Oceanic and Atmospheric Administration, Silver Spring, Maryland.

_____, 1982. Hydrometeorological Report No. 52, Application of Probable Maximum Precipitation Estimates - United States East of the 103rd Meridian.

Chapter 5

RORB: HYDROGRAPH SYNTHESIS BY RUNOFF ROUTING

E.M. Laurenson and R.G. Mein

RORB is a runoff and streamflow routing program used to calculate hydrographs from rainfall and other channel inputs. It subtracts losses from rainfall to produce rainfall-excess and routes this through catchment storage to produce hydrographs. It can also be used to design retarding basins and to route floods through channel networks. The program is interactive, and can be used both for the calculation of design hydrographs and for model calibration. The model is areally distributed, nonlinear, and applicable to both urban and rural catchments. It provides for temporal and areal variation of rainfall and losses and can model flows at multiple gauging stations. In addition to normal channel storage, specific modeling can be provided for retarding basins, storage reservoirs, lakes or large flood plain storages. Base flow and other channel inflow and outflow processes, both concentrated and distributed, can be modeled. Conventional and unusual applications of the models are described.

5.1. INTRODUCTION

This chapter gives a brief description of RORB, a runoff-routing program used extensively in Australia for flood estimation and design. The package is also used in a number of overseas countries.

The first version of RORB was issued in 1975. Since then, it has undergone three major and many minor updates and revisions. The current release, Version (4.11) (Laurenson and Mein, 1992), is a user-friendly program with great flexibility in the modeling/synthesis of flood flows in river reaches and storages, on rural, urban, or part-rural part-urban catchments. Its generality allows RORB to mimic the operation of models like the Nash (1960), Clark (1945), Laurenson (1964) and Dooge (1959) and the Watershed Bounded Network Model (Boyd et al., 1979). All of these models, like RORB, are event models.

The Sections below give details of the sub-models used to simulate the hydrologic response of catchments, a description of the RORB program, and examples of applications.

5.2. THE RORB MODEL

The way in which RORB develops a stream hydrograph is shown in Fig. 5.1. The model of the catchment consists of two parts, a loss model and a catchment storage model. Inputs to the model can consist of rainfall on a catchment area or direct inflow to the channel system. In the former case, rainfall is operated on by a loss model to convert rainfall into rainfall-excess, which is then routed through the catchment storage

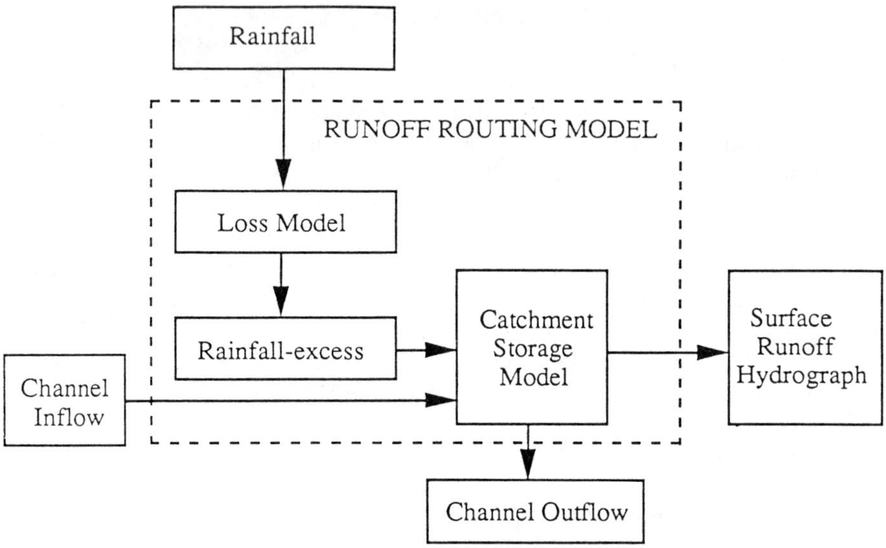

Fig. 5.1. Overall RORB Runoff Routing Model.

model to produce the surface runoff hydrograph. In the latter case, channel inflow enters the catchment storage model directly. In either case, diversions from the channel system can be modeled. Outputs are the calculated surface runoff hydrograph(s) at the required location(s), and/or channel outflows to effluent streams.

The sub-sections below give details of the modeling used to determine a catchment's response to a rainfall event.

5.2.1. Catchment Subdivision

RORB provides for spatial variability of rainfall and losses by subdividing the catchment into a number of sub-areas, typically five to fifteen. [This is also necessary for the modeling of the distributed nature of catchment storage, as described in Section 5.2.4] The subdivision is on watershed lines, so that hydrographs calculated at the downstream boundaries of sub-areas include all of the contributing area upstream. Figure 5.2 gives an example subdivision of a catchment for RORB modeling.

The adopted sub-areas are assumed to be homogeneous in hydrologic response, i.e. different response surfaces (e.g. urban, rural) should normally be assigned to separate sub-areas.

5.2.2. Rainfall Input

Storm rainfall is input to each sub-area at a node located at, or near, its centroid, but generally on a major stream. These sub-area nodes are also the points where rainfall-excess enters the storage. The depth and temporal pattern of rainfall can be the same on all sub-areas, or individually different, depending on the data available. Thus, temporal and spatial variation of rainfall over a catchment is provided for in the model; storm movement can also be simulated.

5.2.3. Calculation of Rainfall-Excess

Two alternative loss models are provided. Both provide for an initial loss, which is a threshold value of rainfall that must occur before any rainfall-excess is produced. The model of losses that occur after the initial loss has been satisfied may be either a constant loss rate or a loss directly proportional to rainfall intensity. Both loss models, therefore, contain two parameters, one determining the initial loss and the other the remaining loss. A more complicated loss model, the Green-Ampt model (e.g. Mein and Larson, 1973), providing for curvilinear time variation of loss rate, and containing three parameters, is to be included soon.

Modeling losses on urban areas. To provide for the modeling of impervious area runoff on urban or partly urban catchments, the fraction imperviousness for each sub-area of the catchment can be specified. [The same facility can be used on rural catchments, if desired.] A constant proportion of rainfall (90%) from impervious areas is assumed to run off; this is combined with the calculated runoff from pervious parts of the sub-area.

5.2.4. Modeling of Overland Flow and Reach Storage Effects

A simple and empirical model represents the effects of catchment storage, which includes both the detention storage on the surface of the catchment during overland flow and the channel storage contained in the stream network. All of this catchment storage is lumped together in any given model storage; channel storage is not modeled separately from overland

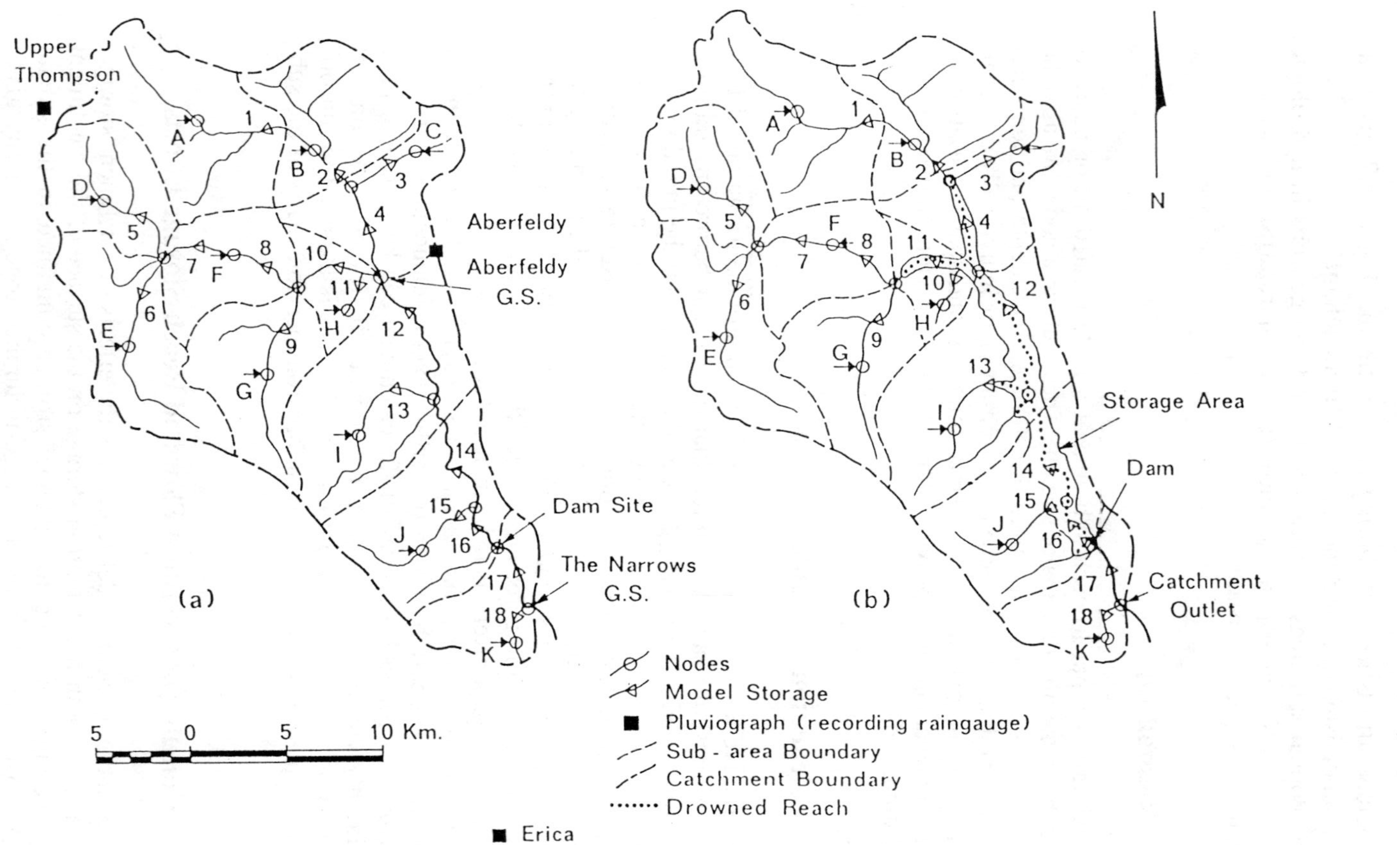

Fig. 5.2 Thomson River Catchment showing Catchment and Stream Subdivision, (a) Pre-dam; (b) Post-dam.

flow or sub-surface storage. The reason for this is that separate modeling of the surface and channel storage components has been found not to alter the calculated hydrograph significantly.

In any particular application, the catchment storage is represented by a series of model storages linked in the same way as the stream network on the catchment. This stream network is subdivided into a series of reaches, with each reach being represented by a model storage. The subdivision of the stream network can be independent of the subdivision of the catchment area into sub-areas, but it is convenient and sensible to undertake both subdivisions together. Thus, most model storages exist between nodes that represent the points where sub-area runoff enters the channel system and either similar nodes downstream or a downstream confluence of two streams. Figure 5.2 shows a typical set of model storages and sub-areas.

Each reach storage is defined by a storage-discharge relationship in the form of a power function:

$$S = 3600kQ^m \tag{5.1}$$

where

S is the volume of storage, m^3;

Q is the rate of outflow discharge from the storage, m^3/s;

k is an empirical coefficient;

m is an empirical exponent; and

3600 is the number of seconds in an hour.

This form of storage-discharge relationship is consistent with the observation that time parameters representing storage delay time can be reasonably approximated over the range of common observation by power functions, and also by theoretical modeling of idealised channel storage and corresponding channel discharge calculated by the Manning formula or some similar resistance equation (Mein et al., 1974).

In order to provide for differences between individual model storages, the coefficient k of Eq. (5.1) consists of two factors:

$$k = k_c k_r \tag{5.2}$$

where k_c is an empirical coefficient that is considered to be characteristic of the catchment as a whole, and

k_r represents the individual characteristics of a particular model storage, and is called the relative delay time. It is given by:

$$k_r = \frac{FL}{d_{av}} \tag{5.3}$$

where F is a factor dependent upon the type of channel,

L is the length of the actual channel represented by the model storage, and

d_{av} is the average flow distance of runoff from the catchment to the outlet.

The length of the reach is generally used as a surrogate for storage delay time, but other functions (e.g. $\text{length}/\sqrt{\text{slope}}$) may be substituted. The program then calculates d_{av} as the average value of this function between the catchment centroid and catchment outlet so that the relative delay times remain of the same order though the relative delay times of individual storages may change. Thus the k_c value is independent of the function used to represent storage delay time.

Modeling of non-natural reaches. The factor F can take one of four values, being:

- unity for a reach in its natural condition;
- zero for a reach drowned by a reservoir;
- $1/(3S_c^{0.25})$ if the reach has been excavated but not concrete lined; and
- $1/(9S_c^{0.5})$ if the reach is a concrete lined channel or a pipe, and S_c is the slope of the channel reach in percent (Mein and Laurenson, 1987).

The latter expressions are empirical ones derived from a few urban catchments and reflect the fact that, for artificial channels, roughness characteristics are relatively constant; the slope of the channel consequently has a significant effect on the flow velocity. [In natural streams, the effects of decreasing slope on flow velocity tend to be counteracted as a rule by the effects of decreasing relative roughness and increasing hydraulic radius] The value of zero for a drowned reach reflects the fact that waves pass through deep reservoirs as dynamic waves in times that are at least an order of magnitude less than the times taken by kinematic waves to pass down the same channel in its natural condition; this difference of an order of magnitude or more in wave speed justifies the approximation of zero travel time.

5.2.5. Modeling of Artificial and Natural Storages

The routing effects of significant artificial and/or natural storages on timing and attenuation of flow peaks downstream are usually major; and much greater than those found in routing through natural river reaches (see Section 5.2.4). With RORB, which bases its representation of a catchment on its topography and river network, such storages are included in the modeling sequence at points corresponding to their stream channel location [e.g. Thomson Dam in Fig. 5.2b].

Individual discharge-storage, or discharge-stage and stage-storage, relations for significant storages are used by the model, to route hydrographs. An extremely robust routing procedure was developed to cope with the discontinuities in the storage-discharge relationships characteristic of retarding basin storages (Laurenson, 1986). RORB also provides for interactive design of storage reservoirs and detention storages, by allowing users to change spillway and/or outlet pipe dimensions to meet the design criteria required.

5.2.6. Channel Inflows/Outflows

The program provides a good deal of flexibility to model channel inflows and outflows. Inflows/outflows can be:

- concentrated at a point (e.g. to simulate an input/output hydrograph to/from a reach);
- spread over a number of reaches (e.g. to simulate baseflow / transmission loss).

RORB allows for the modeling of streams which split into two or more channels, and provides for these channels to re-enter the stream system at points downstream. With this feature, looped systems can be handled.

5.2.7. Hydrograph Translation

Hydrographs can be translated forwards or backwards in time by any integral number of time steps. This feature can be helpful in the modeling of long reaches, to avoid the need to subdivide these into sub-reaches. It can also be helpful to deal with unexplained time shifts in data and/or catchment response such as often occur on catchments with distinct slope changes.

5.3. THE RORB PROGRAM

5.3.1. Input

Input to the RORB program is partly from a prepared data file on disk and partly from interactive use at a terminal. The file data describes the catchment configuration, and physical characteristics such as areas of sub-areas, lengths of stream reaches, types of streams, parameters of special storages; it also contains storm event data such as sub-area rainfall amounts, rainfall temporal patterns, and known hydrographs if there are any. This data file may be prepared by using a general purpose editor and following the instructions in the User Manual. Alternatively, the data file may be prepared by using an ancillary program, RORBDATA, which interactively prompts the user for the data required. Figure 5.3 shows a section of a data file required for the Thomson Dam catchment (Fig. 5.2b).

The interactive data required at the terminal include the global parameters of the model storages (k_c and m), the choice of loss model, initial loss, characteristics of retarding basin or dam outlets and spillways being designed, and control of the output.

5.3.2. Output

Most output from RORB goes to the screen and to a user-named log file. Separate graphics files of hydrographs may also be produced. The amount of output from a given problem can be governed in various ways by the user. The principal output of the program is calculated hydrograph data, which is always output to the screen and log, and parameter values, fit summaries, and error messages. Hydrograph data may be output for any node of the model. Other possible output includes hyetographs of rainfall and rainfall-excess, formatted listings of the data, indicators of progress in reading the data file, and storage-elevation-discharge tables for special storages. Figure 5.4 is an example of a hydrograph summary and output graph for the major storage shown in Fig. 5.2b.

```
   1,1,    3.000,     99                              ,Sub-area J, reach natural
   4                                                  ,Add in the last stored h'graph
   5,4,    3.000,     -99                             ,normal (reach) storage, reach drowned
   16                                                 ,Existing special storage
Thomson Dam
3
   0.000
   1                                                  ,Disch Reltn Flag; Drawdown; No. Spillways
   200.  ,            100.0000
   2.000,             -99                             ,Spillway crests, lengths; Weir coeff.
C Elevation (H) - Storage (S) data for this special storage
2
   2.2300E+07,    1.000,  200.,  -99                  ,H - S relation: S=a(H-Ho)**b; a, b, Ho
   5,1,    5.000,     -99                             ,normal (reach) storage, reach natural
3  ,Store the current hydrograph
   1,1,    2.500,     -99                             ,Sub-area K, reach natural
4                                                     ,Add in the last stored h'graph
   0                                                  ,0   Finished (Control vector)
C Sub-area Data
C
C Areas, km**2, of sub-areas A,B...
 62.000,  42.000,  44.000,  59.000, 41.000,  31.000, 44.000,  21.000, 80.000,  67.000
   28.000 ,-99                                        ,11 Sub-area(s)
C Impervious Area Flag
0  ,-99                                               ,No impervious area
A Design Storm
DESIGN
C Time: inc.(h), no. incs.; Rain: no. bursts, no. pluvis, areally uniform flag
   6.000,             20
   1,       1,0 ,     -99                             ,Areally uniform rainfall
C No. of time incs. (from initial time) to start and finish of each rain burst
   0,      8                                          ,  1 rainfall burst(s)
Design pattern
C Rainfalls (mm) for pluviograph  1
   7.00,  16.00,  41.00,  71.00,  71.00,  41.00,  16.00
   7.00 ,-99                                          ,End of data, pluvi.    1,   1 burst(s)
```

Fig. 5.3. Part of data file generated by RORBDATA showing the parameters for the Thomson Dam.

Results of routing through special storage Thomson Dam

Peak elevation = 202.17 m
Peak outflow = 641.83 m^3/s (spillway flow)
Peak storage = 4.85E+07 m^3

	Hydrograph	
	Outflow	Inflow
Peak discharge,m^3/s	642.	1233.
Time to peak,h	42.0	30.0
Volume,m^3	8.38E+07	9.03E+07
Time to centroid,h	53.1	32.7
Lag (c.m. to c.m.),h	29.1	8.7
Lag to peak,h	18.0	6.0

(a)

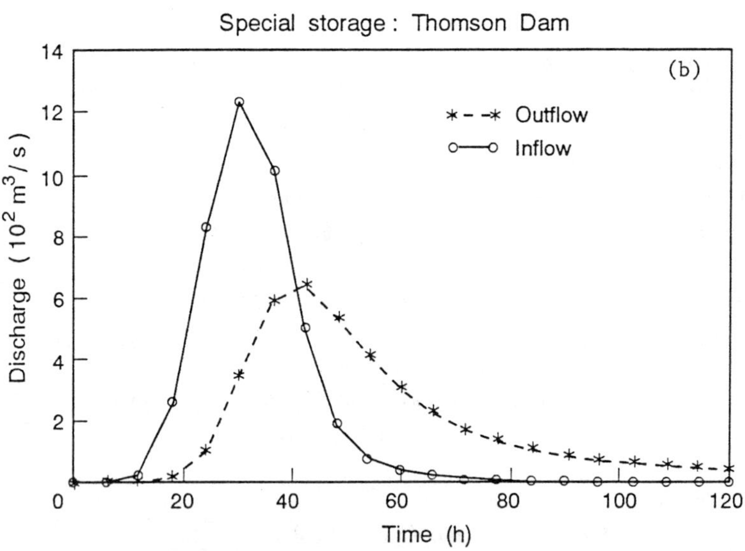

Fig 5.4. Sample output from RORB, in this case showing the results of routing a hydrograph through the Thomson Dam: (a) Extract from log file (also shown on screen); (b) graphical output to screen (and file).

5.3.3. Scope of Program Applications

Flexibility to enable RORB to be applied to a range of different flood problems and a range of different catchment models has been a key objective in the program development. It is achieved by the use of the sub-models described in Section 5.2, and combining these to represent the catchment being modeled. Thus, RORB can be applicd to:

- rural catchments, and to urban or part-urban catchments, by having facilities for altering both losses and travel times to reflect the influences of urbanised area or sub-areas;
- the design of retarding (detention) basins. In this application, the program calculates the required input hydrographs, provides for interactive sizing of the outlet pipes and the overflow spillway(s), and can, if desired, show the effect of a retarding basin on hydrographs at any downstream point;
- flood routing as distinct from runoff routing, where the input is an upstream hydrograph and the output the corresponding downstream hydrograph;
- converging, diverging or looping networks of channels.

The RORB program can be used in a calibration mode (i.e. determination of the model parameters in order to fit a recorded event), in a verification mode (for testing the validity of the model when applied to a particular catchment), or in a design mode for which the model parameters are known and the purpose is to calculate the hydrograph (or hydrographs at several different points) that will be produced by some design storm.

For ungauged catchments, information from derived regional relationships (e.g. The Institution of Engineers, Australia, 1987, Section 9.6.2) can be used for estimation of parameter values.

In urban applications, RORB can be applied to catchments that have pipe drains and/or artificial channels. However it does not design pipes or channels or any other hydraulic structure specifically, except in the case of retarding basin design.

The program has been written in such a way as to place very few specific limits on the catchment and stream characteristics or the amount of available data for the problems to which it is applied. Thus, there are no specific limits on such things as number of sub-areas, number of stream reaches, number of special storages, number of recording rain gauges, number of stream gauging stations, and so on.

5.4. UNUSUAL APPLICATIONS OF RORB

Originally, RORB was developed for "standard applications" such as the calculation of design hydrographs for the sizing of dam spillways (e.g. Laurenson and Mein, 1982), bridge waterways, and hydraulic structures. These and other features are noted in Section 5.3.3 above. It has been found, however, that the generality of RORB allows it to be adapted for a much wider range of problems.

Unusual applications of the program (Laurenson and Mein, 1993) have included:

- catchments in which an upland stream debouches on to a wide flood plain having distinctly different characteristics from the upper catchment. In these cases, the definition of special storages that represent the flood plain was necessary; this was done using the HEC2 program of the US Corps of Engineers;

- the design of excavated retarding basins in flat areas where the retarding basin storage was "balanced" on a main existing drainage line. Here, the design process consisted of determining a suitable area of excavated basin rather than designing the outlet works;

- the design of a retarding basin from which the outflow was pumped rather than being governed by gravity. It was tackled by developing a storage-discharge relation that incorporated the head-discharge characteristic of the pump (see also Laurenson and Mein, 1983);

- modeling of a complex network of channels on a large flood plain. The inflow consisted of a measured hydrograph and two outflow channels, but the application involved over 100 reaches, some 60-70 special storages, no sub-areas, three gauging stations, a number of spot measurements of discharge at particular points in the channel system, a flood duration of more than three months and a one-day timestep (Mitchell and Laurenson, 1983);

- modeling stormwater inflow/infiltration to wastewater sewers. Inflow behaviour is similar to stormwater runoff on urban areas, infiltration was modeled by a storage linked to a delay to account for the lag in response after rainfall (Mein and Apostolidis, 1993);

- catchments as small as 0.5 km^2 and larger than 700,000 km^2 (Kotwicki, 1986). There are really no size limits, either small or large, to the application of the program.

5.5. CONCLUSION

RORB is considered to be a simple yet effective modeling program. It requires little data for simple problems yet it can be applied with equal facility to large and complicated catchments for which there is a great deal of data. Regardless of the complexity of the problem, the model and its application retain their simplicity.

A key feature of the simplicity of RORB is its adoption of an empirical routing procedure based on a simple storage-discharge relationship. This is a great strength as well as a potential weakness of RORB. The weakness, however, is overcome by the inclusion of the special storage facility. The latter may be used with whatever complex calculations outside of the program are required to represent the storage-discharge relations of a particular storage or storages on the catchment.

Despite RORB's long development and the desire not to compromise its simplicity unnecessarily, further research is being undertaken in an effort to improve the model on which it is based and the user-friendliness of the program. As a result of this, both the capabilities of the program and its ease of use are likely to develop in the future. Already, developments in both these areas are in train.

REFERENCES

BOYD, M.G., PILGRIM, D.H. and CORDERY, I. (1979). 'A storage routing model based on catchment geomorphology'. *J. Hydrology*, **42**, 209-230.

CLARK, C.O. (1945). 'Storage and the Unit Hydrograph', *Amer. Soc. Civ. Engrs. Trans.*, **110**, 1416-1446.

DOOGE, J.C.I. (1959). 'A General Theory of the Unit Hydrograph'. *J. Geophys. Res.*, **64**, 241-256.

KOTWICKI, V. (1986). *Floods of Lake Eyre*, Engineering and Water Supply Dept., Adelaide, South Australia, 99pp.

LAURENSON, E.M. (1964). 'A Catchment Storage Model for Runoff Routing.' *J. Hydrology*, **2**, 141-163.

LAURENSON, E.M. (1986). 'Variable Time-step Nonlinear Flood Routing', *Proc.* Hydrosoft 86 - Hydraulic Engineering Software, Southampton, 61-72, Computational Mechanics, Springer-Verlag, Berlin.

LAURENSON, E.M. and MEIN, R.G. (1982). 'A Flexible Model for Spillway Design Flood Computation.' *Proc.* 14th Inter. Congress on Large Dams, Rio de Janeiro, Brazil, **3**, 1247-1261, International Commission on Large Dams.

LAURENSON, E.M. and MEIN, R.G. (1983). 'Hydraulic Design of Detention and Balancing Reservoirs.' *Proc.* Inter. Conf. on Hydraulic Aspects of Flood and Flood Control, London, 353-361, BHRA, Bedford, England.

LAURENSON, E.M. and MEIN, R.G. (1990) '*RORB Version 4 Runoff Routing Program: User Manual.*' Dept. of Civil Engineering, Monash University, 186pp.

LAURENSON, E.M. and MEIN, R.G. (1993). 'Unusual Applications of the RORB Program.' Engineering for Hydrology and Water Resources Conf., I.E. Aust., *Nat. Conf. Publ. 93/14*, 125-131.

MEIN, R.G. and APOSTOLIDIS, N. (1993). 'Application of a Simple Hydrologic Model for Sewer Inflow/infiltration.' *Proc.* 6th Inter. Conf. on Urban Storm Drainage. Niagara Falls, Ontario, Canada, 1994-1999.

MEIN, R.G. and LARSON, C.L. (1973) 'Modelling Infiltration during a Steady Rain'. *Water Resources Research*, **9**, 384-394. Also reply to Discussion, **9**, 1478.

MEIN, R.G. and LAURENSON, E.M. (1987). 'Adaptation of a Flood Estimation Model for Use on Urban Areas', in *Topics in Urban Drainage Hydraulics and Hydrology* (B.C. Yen, ed.), *Proc.* Tech. Session D, XXII IAHR Congress, Lausanne, 258-263.

MEIN, R.G., LAURENSON E.M. and McMAHON, T.A. (1974). 'Simple Nonlinear Model for Flood Estimation.' *J. Hyd. Div., ASCE*, **100**, HY11, 1507-1518.

MITCHELL, D.J. and LAURENSON, E.M. (1983). 'Flood Routing in a Complex Flood Plain.' Hydrology and Water Resources Symposium, I.E. Aust., *Nat. Conf. Publ. 83/13*, 315-318.

NASH, J.E. (1960). 'A Unit Hydrograph Study with Particular Reference to British Catchments'. Inst. Civ. Engrs. *Proc.*, **17**, 249-282.

THE INSTITUTION OF ENGINEERS, AUSTRALIA (1987). *Australian Rainfall and Runoff: A Guide to Flood Estimation*.

Chapter 6

TANK MODEL

M. Sugawara

6.1. WHAT IS THE TANK MODEL ?

6.1.1. The Tank Model for Humid Regions

The tank model is a very simple model, composed of four tanks laid vertically in series as shown in Fig. 6.1.

Precipitation is put into the top tank, and evaporation is subtracted from the top tank. If there is no water in the top tank, evaporation is subtracted from the second tank; if there is no water in both the top and the second tank, evaporation is subtracted from the third tank; and so on.

The outputs from the side outlets are the calculated runoffs. The output from the top tank is considered as surface runoff, output from the second tank as intermediate runoff, from the third tank as sub-base runoff and output from the fourth tank as baseflow. This may be considered to correspond to the zonal structure of underground water shown typically in Fig. 6.2.

In spite of its simple outlook, the behavior of the tank model is not so simple. If there is no precipitation for a long time, the top and the second tanks will empty and the tank model will look like Fig. 6.3a or 6.3b. Under such conditions, runoff is stable. In the case of Fig. 6.3a the discharge will decrease very slowly, and in the case of Fig. 6.3b the discharge will be nearly constant.

If there is a comparatively heavy rain of short duration under these conditions, the tank model will move to one of the states shown in Fig. 6.3c and 6.3d. In these cases, a high discharge of short duration will occur before the model returns to the stable state as before. In these cases, most of the discharge is surface runoff from the top tank and there is little or no runoff from the second tank.

If heavy precipitation occurs over a longer period then the tank model will take on the state shown in Fig. 6.3e. When the rain stops, the water in the top tank will run off quickly and the tank model will move to the state shown in Fig. 6.3f. Then, the output from the second tank will decrease slowly, forming the typical downward slope of the hydrograph following a large discharge.

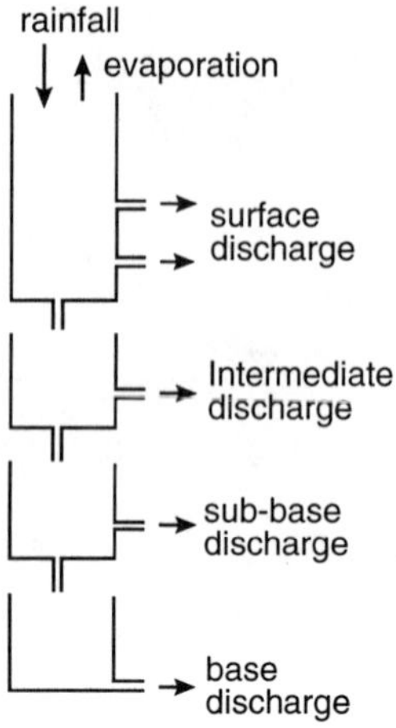

Fig. 6.1.

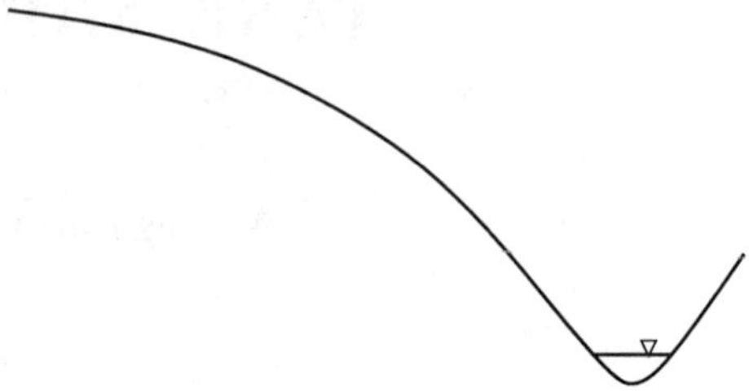

Fig. 6.2.

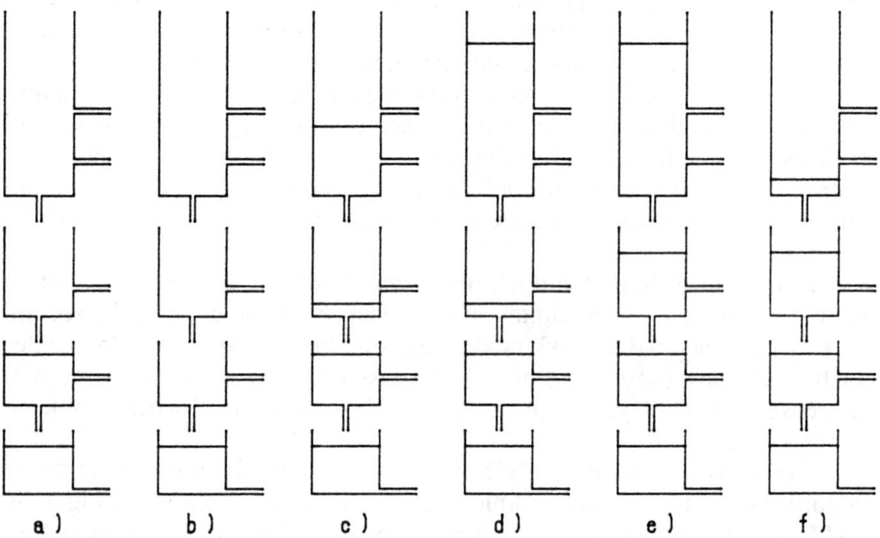

Fig. 6.3.

The tank model can represent such many types of hydrograph because of its non-linear structure caused by setting the side outlets somewhat above the bottom of each tank (except for the lowest tank).

The tank model described above is applied to analyse daily discharge from daily precipitation and evaporation inputs. The concept of initial loss of precipitation is not necessary, because its effect is included in the non-linear structure of the tank model.

For flood analysis the tank model shown in Fig. 6.4 is applied, where the inputs are usually hourly precipitation and the outputs are hourly discharge. This model contains only two tanks; the third and the fourth tanks are replaced by a constant discharge because the flows from lower tanks form a negligible part of the large flood discharge.

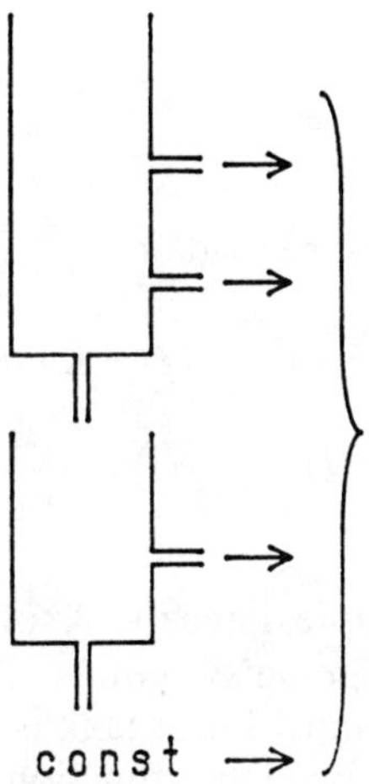

Fig. 6.4.

6.1.2. The Simple Linear Tank

If we move the side outlet(s) of each tank to the bottom of the tanks, we transform the model to one of the linear forms shown in Fig. 6.5a and Fig. 6.5b. Let us first consider a single linear tank of the form shown in Fig. 6.5b with input x(t) and output y(t) (Fig. 6.6). Then, if X(t) is the storage in the tank, the following equations hold:

$$\frac{d}{dt}X(t) = x(t) - y(t), \quad y(t) = k\,X(t)$$

Therefore

$$\frac{d}{dt}X(t) + k\,X(t) = x(t),$$

$$\left(\frac{d}{dt} = k\right)X(t) = x(t), \quad \left(\frac{d}{dt} + k\right)y(t) = k\,x(t).$$

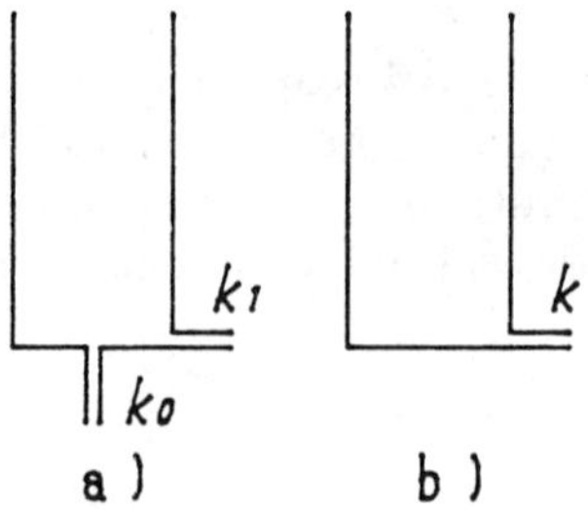

Fig. 6.5.

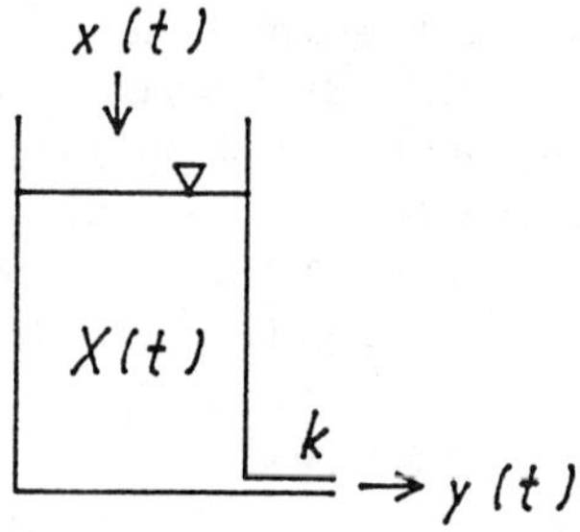

Fig. 6.6.

Or, using of differential operator D = d/dt

$$(D+K)X(t)=x(t), \quad (D+K)y(t)=k\,x(t)$$

$$X(t)=\frac{1}{D+k}x(t), \qquad y(t)=\frac{k}{D+k}x.$$

This shows, that the simple linear tank of Fig. 6.6 is the Heaviside operator k/(D+k), the first order lag system.

If the input at t = 0 to an empty linear tank is the δ function, i.e. x= $\delta(t)$, then the output will be the exponential function y(t) = k exp(-kt), as shown in Fig. 6.7. If we draw a tangent to this exponential curve at t = 0, then it will cut the horizontal axis(t-axis) at T = 1/k, which is the time constant of the operator k/(D+k) for a simple linear tank.

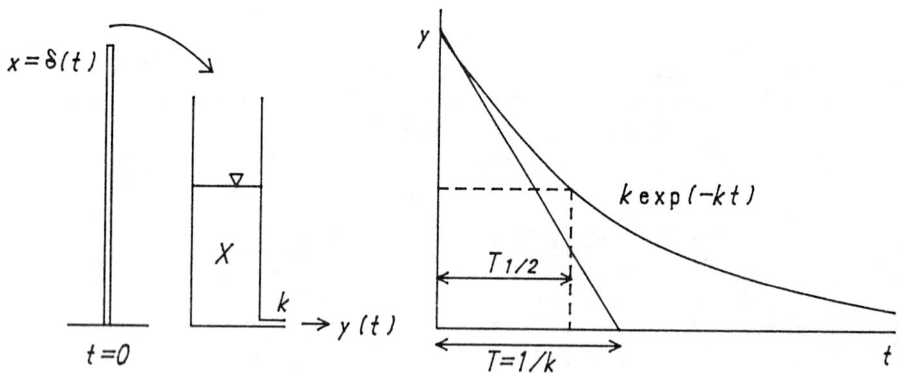

Fig. 6.7.

The meaning of the time constant T = 1/k is as follows: if the output y(t) is maintained at its initial value at t = 0, then the storage in the tank will disappear after T = 1/k. In reality, the output decreases as the storage decreases and after time T = 1/k the storage will becomes $1/\ell^{\bullet} =_{\bullet}$ 0.368. In practice, the half period $T_{1/2}$ is a rather convenient number as this is the time at which both storage and discharge decrease to half their original values. This is given by:

$$T_{1/2} = T \ln 2 = 0.6931\,T = 0.7\,T.$$

Consider now the non-linear tank shown in Fig. 6.8. When the water level is lower than the middle side outlet, discharge is controlled by the linear operator with a time constant T = $1/k_0$. When the water level is between the top and middle side outlets, discharge is controlled by the linear operator with time constant $1/(k_0+k_1)$, and when the water level is above the top outlet, the time constant is $1/(k_0+k_1+k_2)$. Such a tank structure has the property that the time constant becomes shorter as the storage becomes larger and should be a good representation of runoff phenomena.

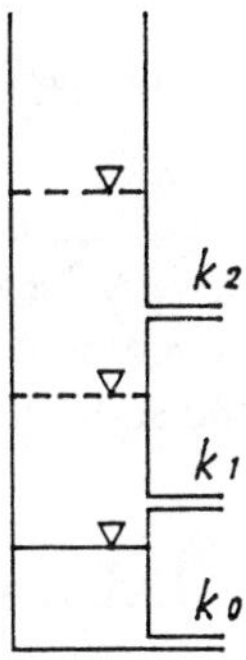

Fig. 6.8.

We can consider the time constant for each tank of the four-tank model by moving the side outlet or outlets to the bottom level. The top tank should have a time constant about a day or a few days, the second tank aweek or ten days or so, the third tank a few months or so, and the fourth tank should have a time constant of years. Roughly speaking, the time constants of the top, the second and the third tanks should have a ratio of $1{:}5{:}5^2$ or so. The time constant will also become longer as the catchment area becomes larger and will be proportional to the square root of the catchment area, i.e. to an equivalent diameter of the basin.

Regarding flood runoff, we have an approximate formula that the time constant T(hour) of the flood is given by

$$T = 0.15 \bullet A^{1/2},$$

where A(km^2) is the catchment area. We think that for flood analysis the time unit (T.U.) should be about one third of the time constant T, i.e.

$$T.U. = 0.15 \bullet A^{1/2}.$$

Using this formula, a basin of 10 km^2 would need precipitation and discharge data at a 10 minute interval, and a basin of 100 km^2 would need half-hourly data. This means that small experimental basins present difficult problems.

The formula T = 0.15 • $A^{1/2}$ was obtained mostly from examples of Japanese basins where the catchment areas are small and the ground surface slope is rather high. Therefore, the coefficient will probably be larger than 0.15 for basins in large continents, perhaps 0.2 ~ 0.3, or so. In the case of the Chang Jiang (the Yangtze River) at Yichang, the catchment area (Sichuang basin) is about half million sq. kms. (5 $10^5 km^2$), and the time constant of the flood is about 10 days (240 hours). If we apply T = C • $A^{1/2}$ to the Yangtze River at Yichang we get

$$C = 240/(5 \bullet 10^5)^{1/2} \bullet\!\!=_{\bullet} 0.34.$$

Even though the difference of the coefficient values between 0.15 and 0.34 is not small, when we consider the very large range of area from 10 to half a million km^2, it seems reasonable to say that the time constant of flood runoff is approximately proportional to the square root of the catchment area.

6.1.3. The Evidence for the Existence of Separated Storage of Groundwater

Despite giving good results for the calculation of river discharge from rainfall in many areas, the tank model has often been considered as only a black box model. However, there is important evidence that can give physical meaning to the tank model.

In Japan many stations measure crustal tilt for earthquake forecasting. The observations are often disturbed by noise, among which the largest is the effect of rain, as shown in Fig. 6.9 and Fig. 6.10. These data were observed in a tunnel at Nakaizu (position 138°59'48.4"E, 34°54'46.4"N; altitude 263m; lithology, tuffaceous sandstone). Figure 6.9 shows daily data and Fig. 6.10 shows hourly data. It is immediately apparent (particularly for the NS-component) that the curves of crustal tilt are very similar to hydrographs. By transforming the coordinate axis, the separation of the two types of tilt curve can be improved but this is not so important and so only original data are shown in the figures.

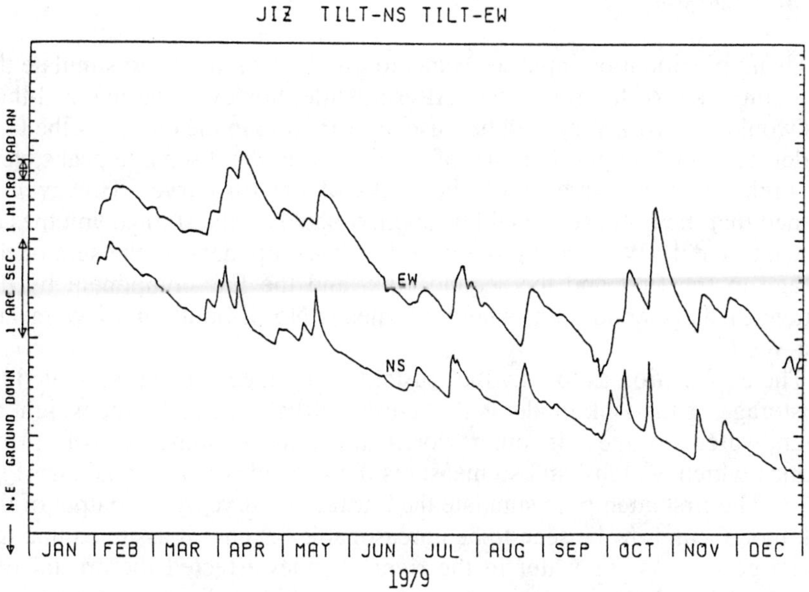

Fig. 6.9. Crustal tilt NS and EW-components at Nakaizu

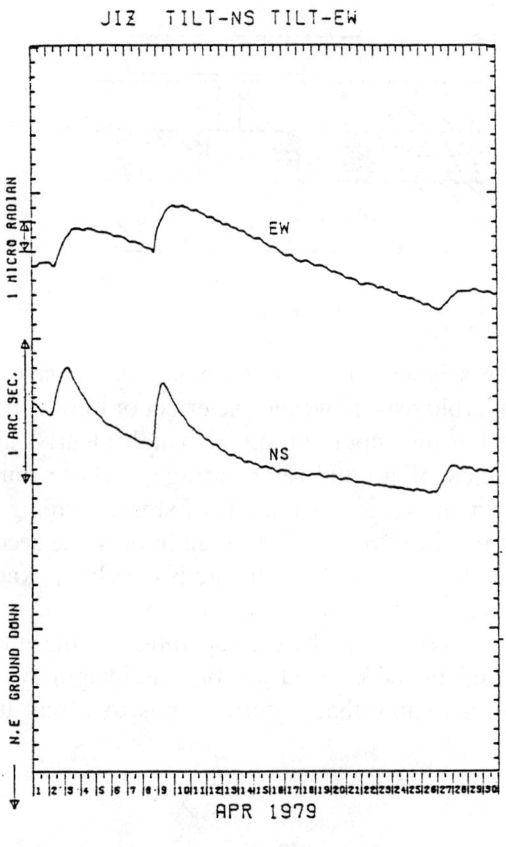

Fig. 6.10. Crustal tilt NS and EW-components at Nakaizu.

Using precipitation input we began to use the tank model to simulate the NS-component of the tilt curve. After a while, however, we realized that, this would not work very well because the variation in the output of the tank model was much larger than that of the tilt curve; the discharge peaks were too high when compared with the peaks of the tilt curve. However, it seemed that the tilt curves could be approximated by the storage volumes of the tank model. We can approximate the NS-component by the sum of the storage in the first and the second tank, and the EW-component by the storage of the second tank; where NS means North-South and EW means East-West.

The explanation as to why the crustal tilt curve can be approximated by the storage of the tank model is that crustal tilt is effected by the weight of groundwater storage. In other words, the crust is some sort of spring balance which 'weighs' and so measures the ground water storage (see Fig. 6.11). The first attempt to simulate the crustal tilt curve by the output of the tank model can now be seen to be unreasonable because it assumed that the discharge or mass of water in the river channel effected the crustal tilt. However, the mass of water in the river channel is negligibe in comparison with the groundwater storage.

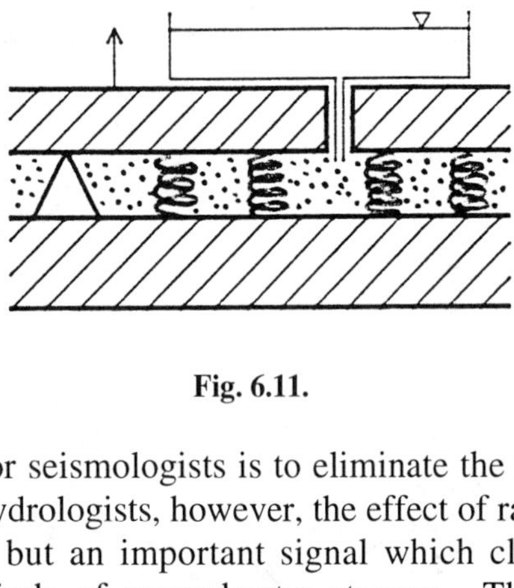

Fig. 6.11.

The main goal for seismologists is to eliminate the noise caused by the effect of rain. For hydrologists, however, the effect of rain on the crustal tilt curves is not noise but an important signal which clearly shows us the existence of two kinds of groundwater storage. There must be some unknown structure which 'weighs' two kinds of storage; storage in the first and the second tank for the NS tilt and storage in only the second tank for the EW tilt. Although the form of this structure is not clearly known, we can imagine such a structure, without any difficulty.

The most important point must be the separation of these two types of storage. For a long time the tank model was only an imaginary model in our minds, but now we can show that it corresponds to a real underground structure.

6.1.4. The Storage Type Model

It is reasonable to consider that runoff and infiltration from the ground surface layer are functions of the storage volume. Moreover, we can imagine that the relationship between runoff and storage must be of an accelerative type and that the relationship between infiltration and storage must be a saturation type, as shown in Fig. 6.12a. Such relationships can be simulated by a tank such as Fig. 6.12b. The two curves in Fig. 6.12a can be approximated by connected segments as shown in Fig. 6.13a and the corresponding functions can be simulated by a tank such as Fig. 6.13b. If we assume that infiltration is proportional to storage then the tank becomes as in Fig. 6.14.
This result lead the author to consider that all runoff phenomena could be simulated with the type of tank shown in Fig. 6.14 together with a simple linear tank for base discharge (see Fig. 6.15a). However, this was not

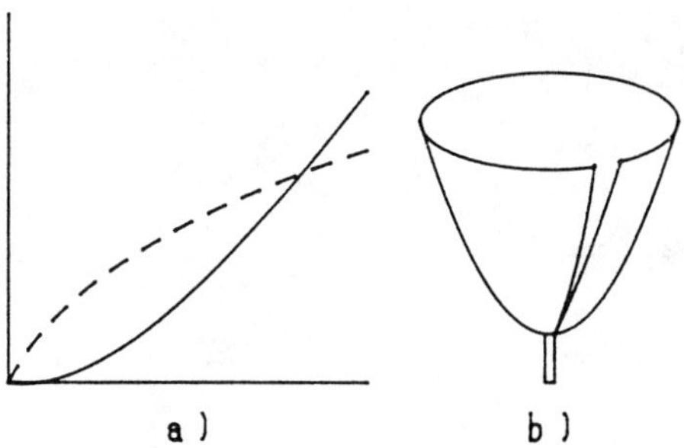

Fig. 6.12.

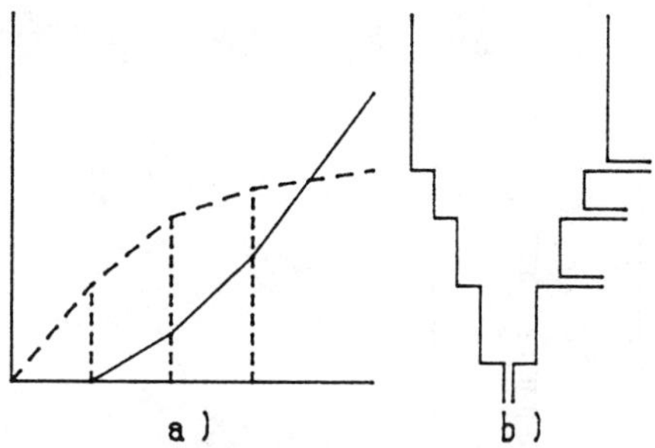

Fig. 6.13.

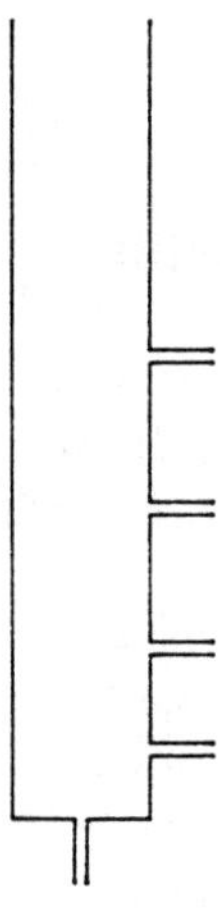

Fig. 6.14.

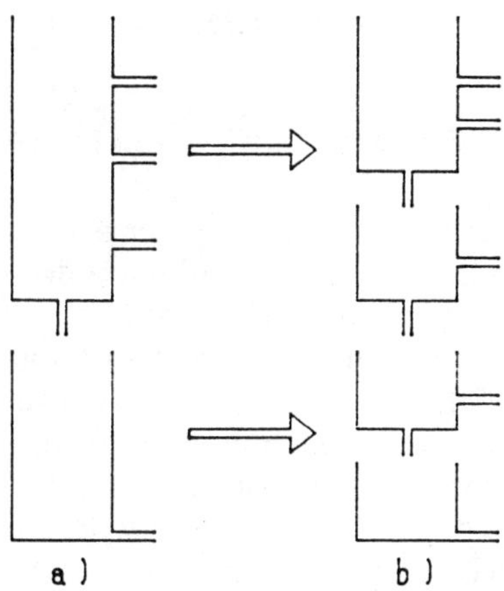

Fig. 6.15.

successful and to improve the performance we had to divide the upper storage type into two parts, one for surface runoff and another for intermediate runoff. Later, the lower tank was also divided into two tanks, one for sub-base runoff with a rather short time constant and one for stable base runoff with a long time constant (Fig. 6.15b).

In some cases, the tank with three side outlets is used as the top tank, and the tank with two side outlets is applied for the second tank. These are some sorts of variety of the usual tank model.

If we set many similar small-diameter evenly-spaced outlets on a tank, as shown in Fig. 6.16a, the relationship between runoff and storage is given by a parabola in the limiting case. In some cases, a tank of this type may be effective and the tank shown in Fig. 6.17a has been used as a top tank. In this case, when the storage is between H_1 and H_2, the relationship between runoff and storage is given by a parabola, and when the storage is larger than H_2, the relationship is given by the tangent of the parabola as shown in Fig. 6.17b.

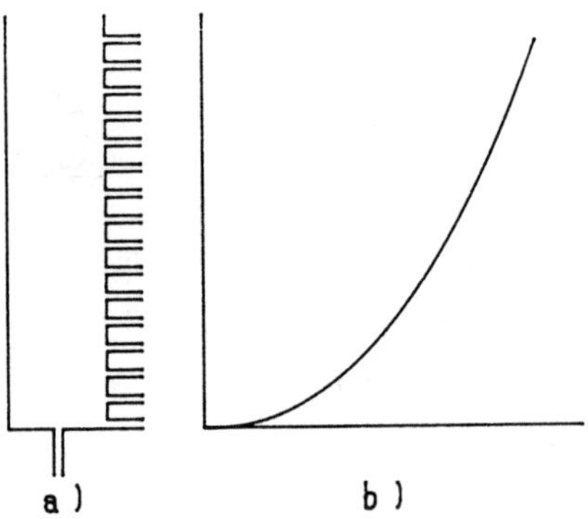

Fig. 6.16.

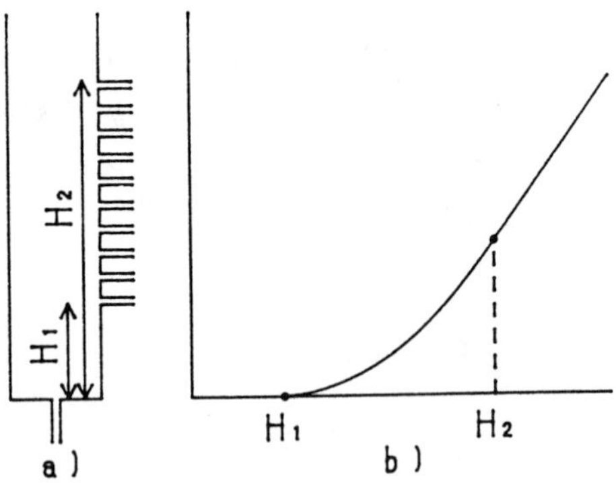

Fig. 6.17.

6.1.5. The Tank Model with Soil Moisture Structure

The ground surface layer is considered to have a soil moisture component. In humid regions without a dry season, such as Japan, the soil moisture is nearly always saturated and so the tank model of Fig. 6.1 can give good results without the need for a soil moisture component.

If, however we wish to consider the effect of soil moisture we must add a structure to the bottom of the top tank as shown in Fig. 6.18a. However, the effect of soil moisture is not so simple as shown in this model. If rain occurs on dry soil the moisture will first fill the space which is easy to occupy, i.e. free air space in the soil. Only then will water transfer gradually into the soil material itself.

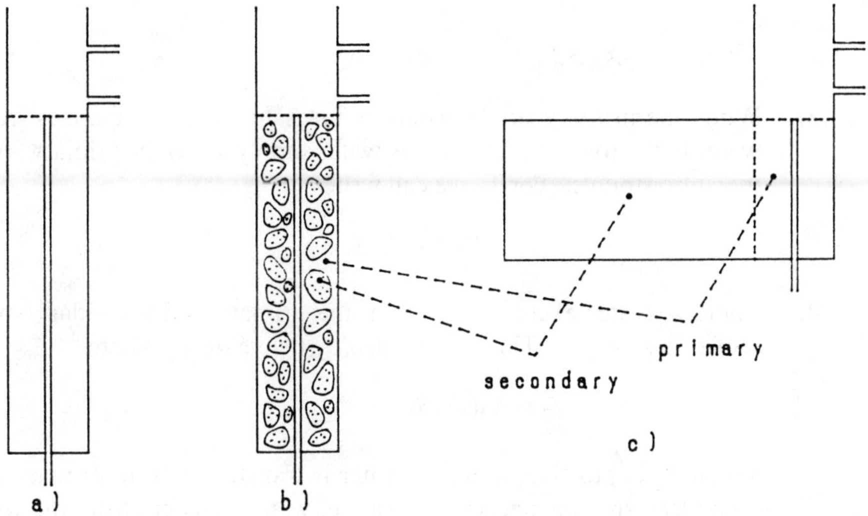

Fig. 6.18.

The first part is called the primary soil moisture and the latter the secondary soil moisture. They are shown symbolically in Fig. 6.18b. As this form of tank is somewhat inconvenient to use, we set the secondary soil moisture on the side of the tank as Fig. 6.18c shows.

The soil moisture structure attached to the tank model (see Fig. 6.19) is as follows :

1) Soil moisture storage has two components, the primary soil moisture storage X_P and the secondary soil moisture storage X_S, where the maximum capacity of each storage is S_1 and S_2, respectively.

2) The primary soil moisture storage and free water in the top tank together make a storage X_A in the top tank. Rainfall is added to X_A and evaporation is subtracted from X_A. When X_A is not greater than S_1, X_A is exclusively primary soil moisture storage and there is no free water, X_F, in the top tank; i.e.

$$X_P = X_A , \quad X_F = 0 : \quad \text{when} \quad X_A \leqq S_1 .$$

When X_A is larger than S_1, primary soil moisture is saturated and the excess part represents free water in the top tank; i.e.

$$X_P = S_1 ,$$

$$X_F = X_A - S_1:$$

when $\quad X_A > S_1$.

3) When the primary soil moisture is not saturated, and there is free water in the lower tanks, there is water supply T_1 to the primary soil moisture storage. For T_1 we can write

$$T_1 = K_1(1 - X_P/S_1).$$

4) There is water exchange between the primary and secondary soil moisture storages. This is regulated by the value T_2 where

$$T_2 = K_2(X_P/S_1 - X_S/S_2).$$

When T_2 is positive, it means water is transferred from primary to secondary storage, and vice versa; i.e. water transfers from the wet part to the dry part.

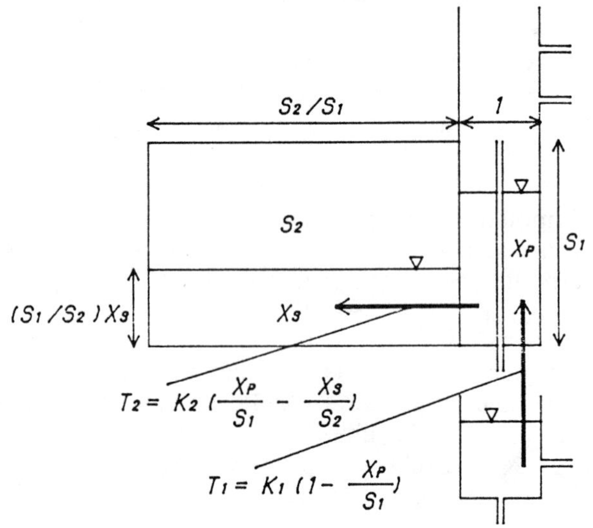

Fig. 6.19.

Now, we shall consider the behavior of the soil moisture storage under the condition that there is no rainfall but constant evaporation, E, there is no free water in the top tank but there is free water in the lower tanks.

Then, we can write the following equations:

$$\frac{dX_p}{dt} = -E + T_1 - T_2$$

$$= -E + K_1\left(1 - \frac{X_p}{S_1}\right) - K_2\left(\frac{X_p}{S_1} - \frac{X_s}{S_2}\right)$$

$$= -E + K_1 - \frac{K_1 + K_2}{S_1} X_p + \frac{K_2}{S_2} X_s ,$$

$$\frac{dX_s}{dt} = T_2 = K_2\left(\frac{X_p}{S_1} - \frac{X_s}{S_2}\right)$$

$$= \frac{K_2}{S_1} X_p - \frac{K_2}{S_2} X_s .$$

To make these equations homogeneous, we put $X_P = x_P + c_P$, $X_S = x_S + c_S$.. Then, to eliminate the constant terms, we must satisfy the following equations:

$$-(K_1 + K_2)(c_P / S_1) + K_2 c_S / S_2 = E - K_1$$

$$K_2 c_P / S_1 = K_2 c_S / S_2 .$$

By setting $c_P/S_1 = c_S/S_2 = c$, we obtain

$$-(K_1 + K_2)c + K_2 c = E - K_1$$

$$c = 1 - E / K_1 .$$

Therefore,

$$c_P = S_1 c = S_1 - (E / K_1) S_1 ,$$

$$c_S = S_2 c = S_2 - (E / K_1) S_2 .$$

The meaning of this result is simple and clear. If we determine new origins c_P and c_S, which are located $(E/K_1)S_1$ below the upper bound, as shown in Fig. 6.20, then the equations describing soil moisture storage can be rewritten in homogeneous form as:

$$\frac{dx_p}{dt} = -\frac{K_1 + K_2}{S_1} x_p + \frac{K_2}{S_2} x_s ,$$

$$\frac{dx_s}{dt} = \frac{K_2}{S_1} x_p - \frac{K_2}{S_2} x_s ,$$

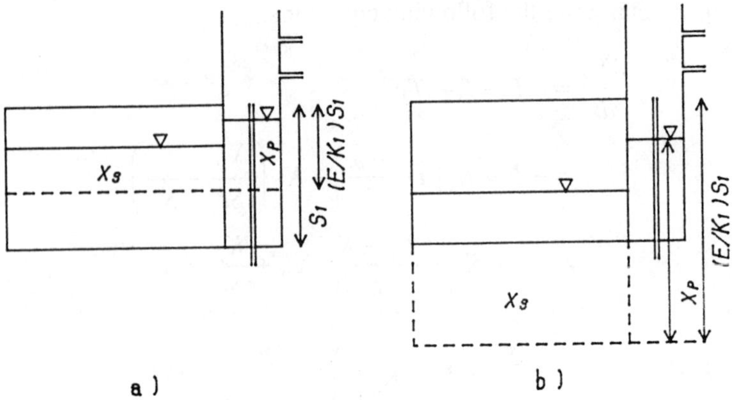

Fig. 6.20.

where x_P and x_S are primary and secondary soil moisture storage measured from the new origins c_S and c_P, respectively. When E is small ($E < K_1$) c_S and c_P lie relatively high, as shown in Fig. 6.20a, and when E is large ($E > K_1$) the new origins are low, as shown in Fig. 6.20b. Seasonal change in the level of these origins will play an important part in the behavior of soil moisture storage, as described later.

To solve the linear homogeneous equations, we put

$$x_p = A\exp(-kt), x_S = B\exp(-kt),$$

and obtain

$$-kA = -\frac{K_1 + K_2}{S_1}A + \frac{K_2}{S_2}B,$$

$$-kB = -\frac{K_2}{S_1}A - \frac{K_2}{S_2}B.$$

To have a meaningful solution other than A = B = 0, these equations must conform to the following characteristic equation:

$$\begin{vmatrix} -\frac{K_1 + K_2}{S_1}k & \frac{K_2}{S_2} \\ \frac{K_2}{S_1} & -\frac{K_2}{S_2} + k \end{vmatrix} = 0$$

$$\left(k - \frac{K_1 + K_2}{S_1}\right)\left(k - \frac{K_2}{S_2}\right) = \frac{{K_2}^2}{S_1 S_2},$$

$$k^2 - \left(\frac{K_1 + K_2}{S_1} + \frac{K_2}{S_2}\right)k + \frac{K_2{}^2}{S_1 S_2} = 0.$$

In our experience, S_2 is much larger than S_1, and K_2 is much larger than K_1. From Fig. 6.21, we can see, that the characteristic equation has two positive roots, k_1, k_2, where the large one satisfies the condition

$$k_1 > \frac{K_1 + K_2}{S_1} \text{ and the small satisfies } 0 < k_2 < \frac{K_2}{S_2}.$$

As k_1, k_2 satisfy equations

$$k_1 + k_2 = \frac{K_1 + K_2}{S_1} + \frac{K_2}{S_2}, \quad k_1 k_2 = \frac{K_1 K_2}{S_1 S_2},$$

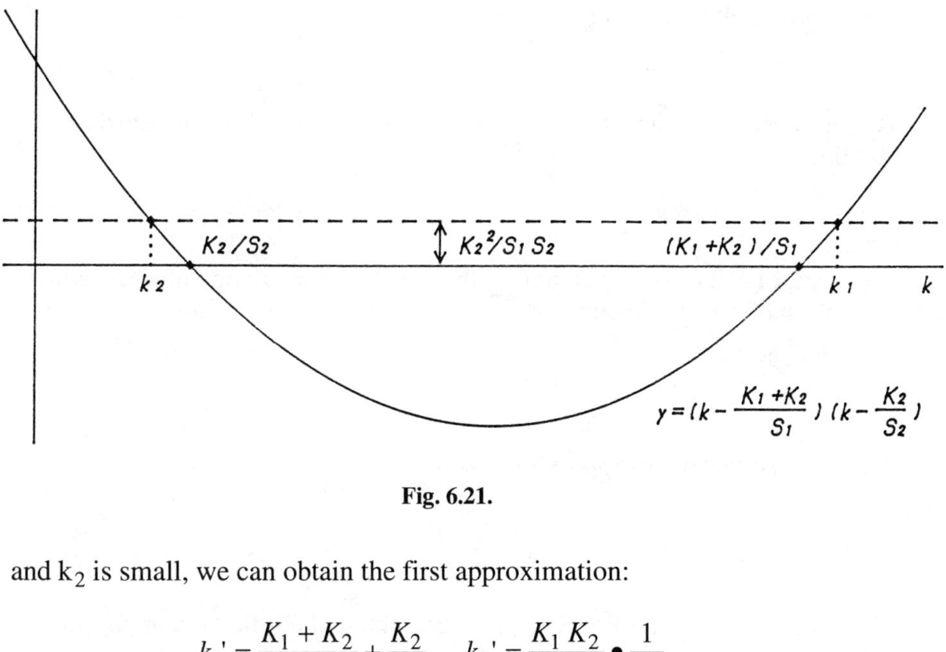

Fig. 6.21.

and k_2 is small, we can obtain the first approximation:

$$k_1' = \frac{K_1 + K_2}{S_1} + \frac{K_2}{S_2}, \quad k_2' = \frac{K_1 K_2}{S_1 S_2} \bullet \frac{1}{k_1'}.$$

Then, the second approximation is

$$k_1'' = \frac{K_1 + K_2}{S_1} + \frac{K_2}{S_2} - k_2', \quad k_2'' = \frac{K_1 K_2}{S_1 S_2} \bullet \frac{1}{k_1''}.$$

When $S_1 = 50$, $S_2 = 250$; $K_1 = 2$, $K_2 = 20$, the true values of k_1, k_2, and their first and second approximations are given as follows:

$$k_1 = 0.5137, \qquad k_1' = 0.52, \qquad k_1'' = 0.51385,$$
$$k_2 = 0.006228, \qquad k_2' = 0.00615, \qquad k_2'' = 0.006228.$$

To determine the vector (A,B), corresponding to the large characteristic value k_1, we use the first approximation

$$k_1' = \frac{K_1 + K_2}{S_1} + \frac{K_2}{S_2}$$

in the equation

$$-kA + -\frac{K_1 + K_2}{S_1} A + \frac{K_2}{S_2} B.$$

Then, we can obtain an approximate solution

$$A + B = 0.$$

By putting $A = C_1$ and $B = -C_1$ we can obtain the following approximate solution.

$$x_P = C_1 \exp(-k_1 t), \quad x_S = -C_1 \exp(-k_1 t).$$

The vector (A,B), corresponding to the small characteristic value k_2, which is very small, can be determined, by neglecting k in the equation

$$-kB = \frac{K_2}{S_1} A - \frac{K_2}{S_2} B.$$

Thus, we can obtain an approximate solution

$$\frac{A}{S_1} - \frac{B}{S_2} = 0.$$

If we set $A = S_1C_2$ and $B = S_2C_2$, we can obtain the following approximate solution;

$$x_P = C_2 S_1 \exp(-k_2 t), \quad x_S = C_2 S_2 \exp(-k_2 t).$$

Therefore, the general solution is of the form:

$$x_P = C_1 \exp(-k_1 t) + C_2 S_1 \exp(-k_2 t),$$
$$x_S = -C_1 \exp(-k_1 t) + C_2 S_2 \exp(-k_2 t).$$

The meaning of this general solution is simple and clear. If C1 = 0, we are left with the second component with a long time constant $T_2 = 1/k_2$ and both soil moisture storages are represented by

$$\frac{x_p}{S_1} = \frac{x_s}{S_2} = C_2 \exp(-k_2 t).$$

This means that both soil moisture storages have a common water content which decreases exponentially with the long time constant $T_2 = 1/k_2$. The first component is then a deviation from the common component and, itself, decreases exponentially with the short time constant $T_1 = 1/k_1$. We can also write this result in the following form:

$$\frac{x_p + x_s}{S_1 + S_2} = C_2 \exp(-k_2 t),$$

$$\frac{x_p}{S_1} - \frac{x_{p+} x_s}{S_1 + S_2} = \frac{C_1}{S_1} \exp(-k_1 t),$$

$$\frac{x_s}{S_2} - \frac{x_{p+} x_s}{S_1 + S_2} = -\frac{C_1}{S_2} \exp(-k_1 t).$$

These equations show that the relative wetness of the total soil moisture $(x_P+x_S)/(S_1+S_2)$ decreases exponentially with the long time constant $T_2 = 1/k_2$ and the deviation of the relative wetness of both primary and secondary soil moisture storages from the relative wetness of the total soil moisture decreases exponentially with the short time constant $T_1 = 1/k_1$.

If we assume that $S_1 = 50$, $S_2 = 250$, $K_1 = 2$ and $K_2 = 20$, both time constants can be estimated as

$$T_1 = 1/k_1 = 1.95(day), \;\; T_2 = 1/k_2 = 160.6(day).$$

Now, we have to notice that the points of origin for x_P and x_S are c_P and c_S, respectively, as in Fig. 6.20. On the Pacific Ocean side of Japan, winter is a rather dry season, evaporation E is smaller than K_1, and c_P and c_S values are high. Accordingly, even if the weather is dry, the soil moisture is high because of the high values of c_P and c_S.

On the other hand, under hot and dry conditions, e.g. $S_1 = 50$, $S_2 = 250$, $K_1 = 2$, $K_2 = 20$ and E = 6 (mm/day), the soil moisture will vanish when the relative wetness $(x_P+x_S)/(S_1+S_2) = 2/3$, as shown in Fig. 6.22. And as ln 2/3 = -0.4055, even if the soil moisture starts from 100%, it will disappear after 160.6 ·0.4055 = 65.1 days. To take another example, if E = 8 the soil

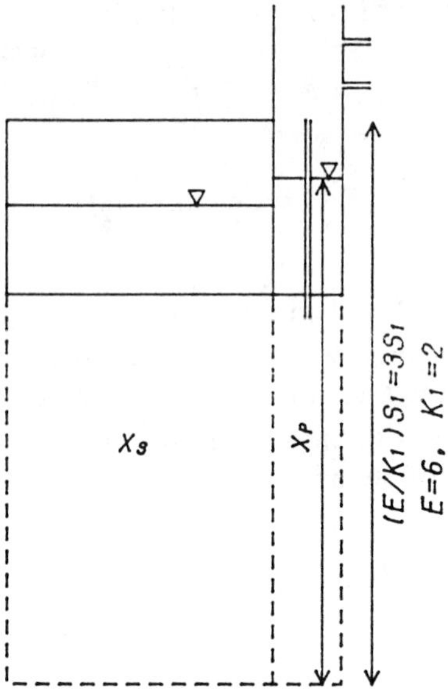

Fig. 6.22.

6.1.6 The Compound Tank Model for Arid Regions

moisture will disappear in 46 days from the saturation date. In conclusion, in spite of the long time constant T_2, the soil moisture will disappear completely in a relatively short period if the potential evaporation is high.

In arid regions where the annual potential evaporation is larger than the annual precipitation, mountainous areas become dry in the dry season, because groundwater moves downward by gravity. On the other hand, low areas along the river remain wet because they receive groundwater from higher areas. When the wet season returns and it rains, surface runoff occurs from the areas along the river, because they are still wet. In the dry mountain areas, however, rain water is absorbed as soil moisture and there is no surface runoff.

During the wet season the percentage of wet area increases with time and surface runoff increases. On the contrary, during the dry season the percentage of dry area increases with time and runoff decreases. As evaporation does not occur from dry areas the actual evaporation from the whole basin is smaller than the potential evaporation. Therefore, even if the annual potential evaporation is larger than the annual precipitation, there is the possibility of runoff.

Following these considerations we divide the basin into zones, for example, into the four zones shown in Fig. 6.23. Let the proportional areas of the zones be AR1, AR2, AR3 and AR4, where AR1+AR2+AR3+AR4 = 1 and apply a tank model of the same structure to each zone as in Fig. 6.24.

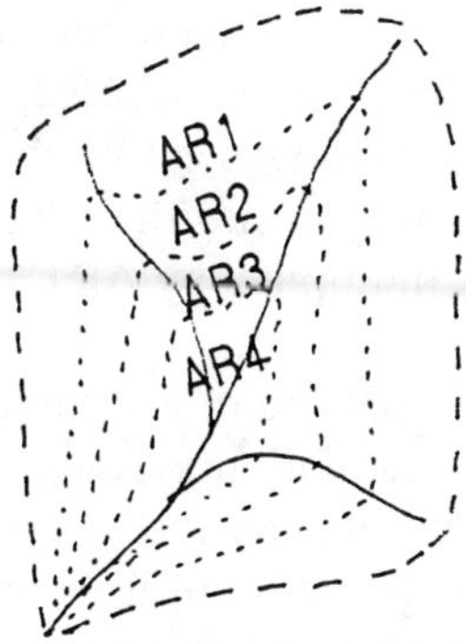

Fig. 6.23.

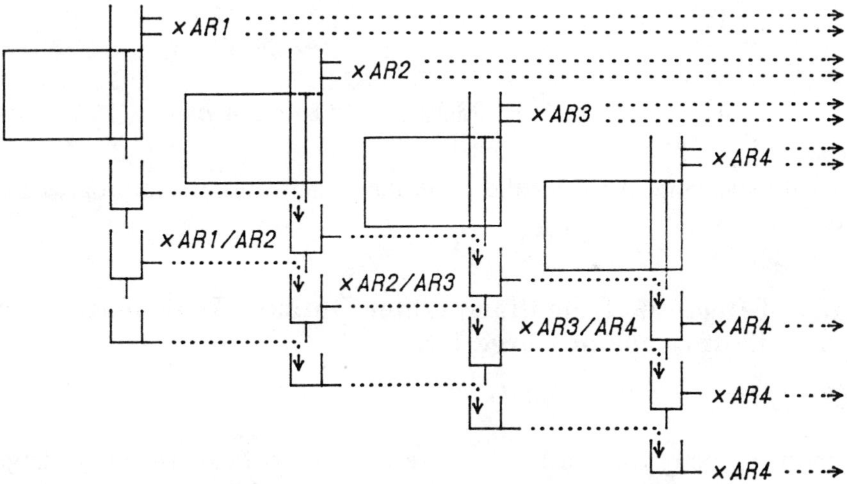

Fig. 6.24.

In this model we assume that, when surface runoff occurs in a zone, the lower zones are already wet and the surface runoff from the top tank of each zone will go directly to the river channel. For the second, third and fourth tanks, however, the output will go to the corresponding tank of the next lower zone, as Fig. 6.24 shows, because each tank corresponds to a groundwater layer (refer back to Fig. 6.2.

Consider the tank model for the first zone. The input to and output from the model (precipitation and evaporation) is measured and calculated in units of water depth, i.e. mm, but to transfer water to the next zone we have to convert to volume units by multiplying the outputs of each tank by AR1, the area of the first zone. The outputs from the tanks of the first zone are transferred as volumes into the corresponding tanks of the second zone and reconverted to depth units by dividing by AR2, the area of the second zone. Therefore, the outputs from the lower tanks of the first zone are multiplied by AR1/AR2 before they are put into the tanks of the second zone. The output from the top tank is multiplied by AR1 and goes directly to the river channel.

In the fourth zone, all outputs go directly to the river channel after being multiplied by AR4. The calculation system is shown in Fig. 6.24. The important factor in this type of tank model is the ratio of the zone areas, AR1 : AR2 : AR3 : AR4. If there are not enough data to determine this ratio, we have to determine it by trial and error method starting from a geometrical progression such as:

1	:	1	:	1	:		=	25%	:	25%	:	25%	:	25%
1.5^3	:	1.5^2	:	1.5	:	1	=	41.5%	:	27.7%	:	18.5%	:	12.3%
2^3	:	2^2	:	2	:	1	=	50%	:	25%	:	12.5%	:	6.25%
3^3	:	3^2	:	3	:	1	=	67.5%	:	22.5%	:	7.5%	:	2.5%
4^3	:	4^2	:	4	:	1	=	75.3%	:	18.8%	:	4.7%	:	1.2%

Roughly speaking, the drier the basin the larger ratio of the progression should be.

6.1.7. Effect of Irrigation Water Intake, Deformation of Hydrograph by Gorge, etc.

Effect of Irrigation Water Intake

In most Japanese river basins, there are wide paddy fields which use large volumes of irrigation water. In many cases, the irrigation water intake is nearly equal to the base discharge of the river because, over a long history, the paddy fields have been developed to the maximum possible extent considering the long-term river water supply. Therefore, if the effect of irrigation water is not considered, any runoff analysis would be meaningless. Fortunately, it is rather simple to calculate the effect of the irrigation water intake. Let the amount of irrigation water intake be Z; then Z is subtracted from the output Y of the tank model, and QE = Y-Z is the calculated discharge. However, irrigation water is not lost but infiltrates from paddy fields and adds to the groundwater, and so Z must be put back into the third tank as shown in Fig. 6.25. Using this simple method, good results have been obtained.

There is no need to subtract evaporation from irrigation water because evaporation has already been subtracted from the top tank. In most cases Z is determined by trial and error considering the local agricultural customs.

Deformation of the Hydrograph by Water Storage or Flooding Caused by Channel Restriction

If enough data are given about the relationship between the volume of stored or flooded water and the water level or discharge in a gorge or channel restriction then we can calculate the deformation of the hydrograph. If there are not enough data, we can estimate the hydrograph deformation using some sort of storage type tank such as shown in Fig. 6.26.

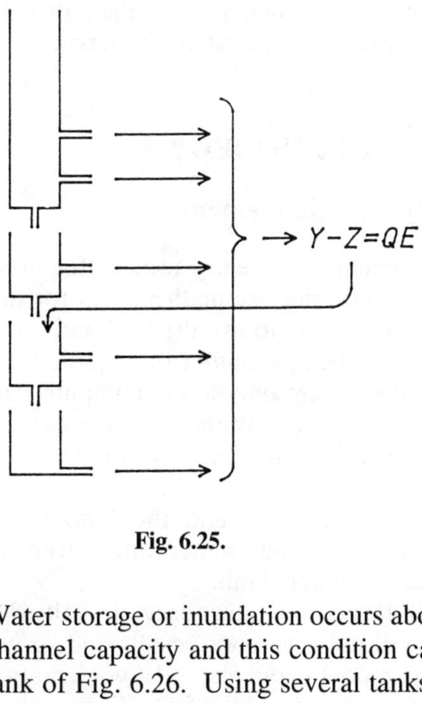

Fig. 6.25.

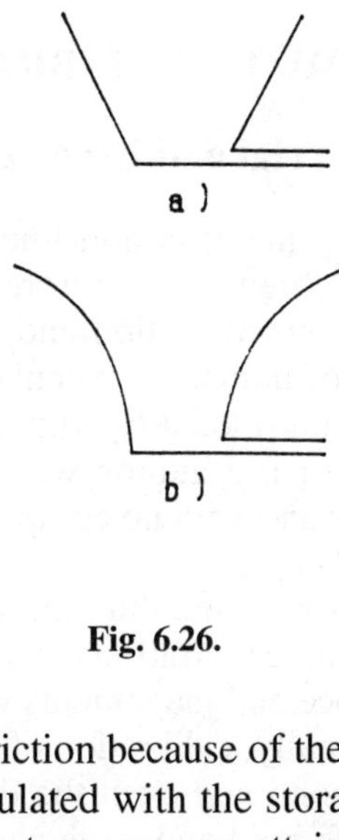

Fig. 6.26.

Water storage or inundation occurs above a restriction because of the limited channel capacity and this condition can be simulated with the storage type tank of Fig. 6.26. Using several tanks of such a type, we can attain a good result by trial and error.

Time Lag

Though the tank model itself can give some sort of time lag (the simple linear tank is equivalent to a kind of linear operator or first order lag) this time lag is often not enough. The output of the tank model Y(J), where J is the day number, may have a time lag when compared to Q(J), the observed discharge. To account for this it is necessary to introduce an artificial lag TL as follows:

$$QE(J) = (1 - D(TL)) \cdot Y(J + [TL]) + D(TL) \cdot Y(J + [TL] + 1),$$

where QE(J) is the calculated discharge, [TL] is the integer part of TL and D(TL) is the decimal part of TL.

As the discharge increases, the velocity also increases and, since time is inversely proportional to velocity, the time lag must decrease. However, since the change in velocity is usually small in comparison with the change in discharge, and the time lag itself is usually not large, we can often assume that the time lag is constant. However, for large basins with large time lags, we may have to consider the time lag as a function of discharge. In the case of the Yangtze River at Yichang the basin area is about half million sq. kms., the travel time from upstream points to Yichang is more than a week, and it is necessary to consider the time lag as the function of discharge.

6.2. HOW TO CALIBRATE THE TANK MODEL ?

6.2.1. Trial and Error Using Subjective Judgement

The tank model is non-linear and mathematics is nearly useless for non-linear problems. Therefore, the author could not use mathematics for the tank model calibration and the only solution was to use the trial and error method of numerical calculation. In 1951 when the author first applied the simple tank model for runoff analysis there were only a few computers in Japan and the author was not able to use one. Without mathematical solutions and with no computer, the numerical calculations necessitated long and hard labour.

However, the human mind is always curious and the labourious numerical calculations became not so boring but rather interesting as experience and judgements were built up in authors brain.

Gradually, calibration of the tank model became rather easy. Usually the first, second and even the third trials will not give good results and so we can make bold and large changes to the model parameters. However, after several trials the result should become fairly good and the fine adjustment of parameters can begin. After ten trials or so the result usually becomes very good.

However, another difficulty appeared. The author found difficulty in describing and explaining his tank model to others. Our language is based on common experiences; if two people have common experiences and knowledges about something, then they can talk and discuss the subject. However, if one person knows nothing about the subject, then it is difficult or impossible to discuss the subject because there is no common vocabulary.

Later, as the tank model became more well-known because of its good results, then the author had to find ways to describe and explain his method of working and his way of thinking. Some of these methods are:

The Initial Model

For the first trial, an initial tank model may be assumed; an example of such an initial model is shown in Fig. 6.27a. Or, we can derive an initial model to fit the basin by plotting the hydrograph in logarithmic scale, finding the peaks of the hydrograph and measuring the descending rates of discharge after the peaks. If the descending rate is r per day then the coefficients of the tank model shown in Fig. 6.27b may be determined as,

$$A0 = A1 = A2 = (1 - r)/3,$$

$$B0 = B1 = A0/5,$$

$$C0 = C1 = B0/5,$$

$$D1 = 0.001.$$

It is better to keep these coefficients as simple numbers, for example if A0 = 0.266 · · · ·, it is better to make it 0.25 or 0.3.

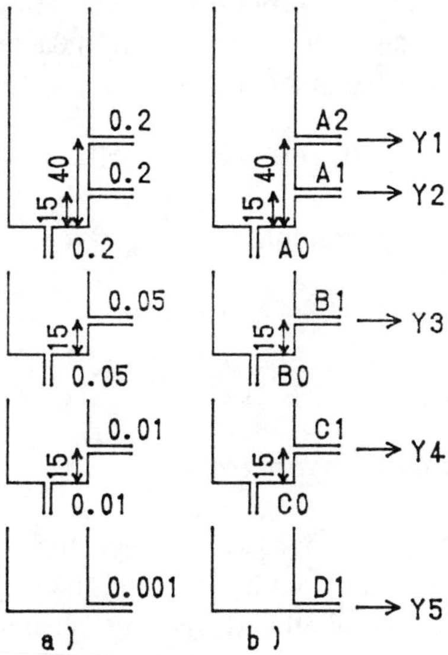

Fig. 6.27.

Find the Worst Point of the Running Model and Adjust the Parameter Values.

The entire hydrograph output of the running model must be plotted five times as Y5, Y4 + Y5, Y3 + Y4 + Y5, Y2 + Y3 + Y4 + Y5 and Y = Y1 + Y2 + Y3 + Y4 + Y5 (see Fig. 6.27b). Comparing the five calculated hydrograph components with the observed hydrograph, we can judge which component is the worst. In early trials, there may be many bad points and we have to select the worst one.

How to Make Some Component Larger or Smaller.

If the worst point is that the runoff from the top tank is too small, there may be two ways to correct for this. One way whould be to make A1 and A2 larger, and another way would be to make A0 smaller. However, the best way is to both make A1 and A2 larger, and to make A0 smaller, i.e. multiply A1 and A2 by k ($k > 1$) and divide A0 by k. If $0 < k < 1$, the output from the top tank will become smaller. The output from the second tank or the third tank can be adjusted in the same way.

In the case where judgement shows that the base discharge is too small, the method described above cannot work because the fourth tank has no bottom outlet. Therefore, we must increase the water supply to the fourth tank by making C0 larger. However this would decrease the runoff from the third tank and so it is better to supply more water to the third tank from the second tank. In such a case the adjustment to make the base runoff larger is made by increasing C0, B0 and A0 as:

$$C0' = k_1 C0, \quad B0' = k_2 B0, \quad A0' = k_3 A0,$$

where $k_1 > k_2 > k_3 > 1$, for example

$$k_1 = 1 + k, \quad k_2 = 1 + k/2, \quad k_3 = 1 + k/4,$$

$$\text{or } k_1 = k, \quad k_2 = k^{1/2}, \quad k_3 = k^{1/4}.$$

How to Adjust the Shape of Hydrograph.

The adjustment described above will change the volume of each runoff component. But the shape of the hydrograph may also be a problem. For example, the peak of the calculated hydrograph may be too steep or too smooth when compared with the observed one. If the peak is too steep, we must make A0, A1, A2 smaller, by multiplying them some constant k ($0 < k < 1$). In such a way, we can adjust the shape of hydrograph of each runoff component.

How to Determine the Positions of Side Outlets.

In Japan, experience shows that if it rains less than 15mm after about 15 dry days then there will be no change in river discharge. The position of the lower side outlet of the top tank, HA1 = 15, is determined from this experience.

Also we know that when it rains more than 50mm or so the discharge will increase greatly. The position of upper side outlet of the top tank, HA2 = 40, is determined in this way considering also the water loss from the top tank by infiltration and runoff during the rainfall.

In the tank model of Fig. 6.27, the side outlets of the second and third tank are set at HB = HC = 15. These are determined to be similar to HA1 = 15, but without such good reasoning, because HB and HC are not as effective as HA1. The effect of HB or HC appears when the runoff component from the second or the third tank vanishes under dry condition. Examples are shown in Fig. 6.28 of hydrographs from tank models with HC = 15, HC = 30 and HC = 100. These hydrographs were obtained under the following conditions.

a) There is no precipitation but evaporation of 2mm/day.

b) Both the top and the second tanks are empty.

c) The runoff and infiltration coefficient of the third tank are given as C0 = C1 = 0.01. The initial storage of the third tank is 80mm, 100mm and 200mm, respectively, corresponding to the case of HC = 15, HC = 30 and HC = 100.

d) The runoff coefficient of the fourth tank is 0.001 and its initial storage is 1,000mm.

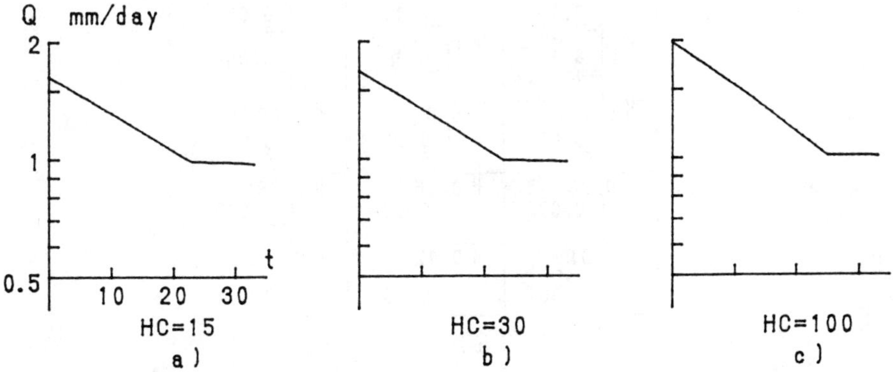

Fig. 6.28.

In some river basins where the observed hydrograph is similar to Fig. 6.28c, the side outlet of the third tank may be set at very high level.

In some cases the fourth tank may be replaced by the type shown in Fig. 6.29. In such a case, there is underground runoff from the basin and the baseflow disappears sometimes.

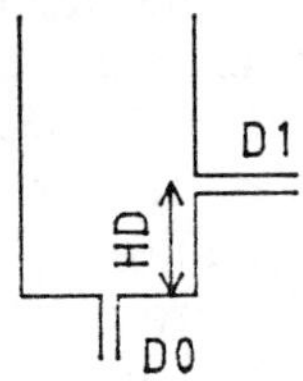

Fig. 6.29.

Volcanic Areas.

In regions where the land surface is covered by thick volcanic deposits the infiltration ability of the land surface is very high, the surface runoff is small and the base discharge is very large. The initial tank model of Fig. 6.27 is not suited for such regions and, instead, the tank models of Fig. 6.30 are recommended as the initial model.

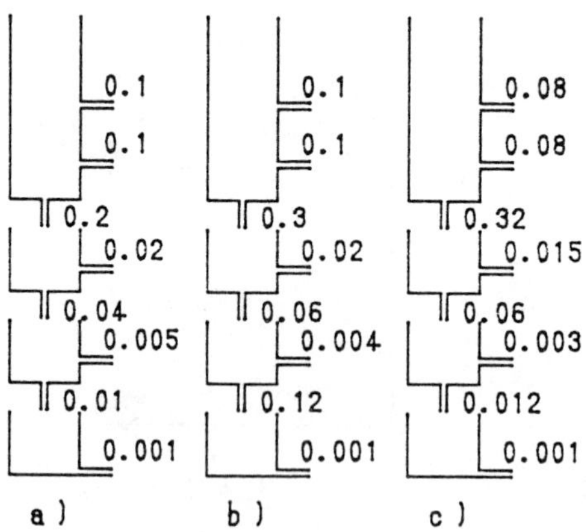

Fig. 6.30.

Some Sort of Balance and Harmony is Important for the Tank Model.

The coefficients of the initial tank models shown in Fig. 6.27 and Fig. 6.30, have the following relationships:

$$A0 : B0 : C0 = 1 : 1/5 : 1/5^2,$$

$$A1 / A0 = B1 / B0 = C1 / C0 = k,$$

where k = 1 for the initial tank model of Fig. 6.27 and k = 1/2, k = 1/3 and k = 1/4 for the initial tank models of Fig. 6.30a, b, and c, respectively. These initial tank models are rather simple and they have some sort of balance and harmony within the parameters. We think that such balance and harmony are important for the initial tank model.

By measuring the slope of the observed hydrograph after a peak flow (plotted in logarithmic scale) we can estimate the value of A0+A1+A2. Then, by assuming a value for k, we can obtain an initial tank model of the type shown in Fig. 6.30.

Even after starting from an initial model that has balance and harmony, however, the balance and harmony may be lost during the course of iterated trials and the results cannot become good, in some cases. When the balance and harmony have been lost in this way it is usually impossible to restore them again. The author considers that this process is somewhat similar resembles to writing poetry or painting a picture painting. In these cases it may be better to start again from a different initial model, which should be made considering the past unsuccessful trials. If the new initial model is appropriate, the result will improve after a few trials.

6.2.2. Automatic Calibration Program (Hydrograph Comparison Method)

Why the author wished to develop the automatic calibration program ?

In 1975, the author retired as director of the institute and had more free time for research. However, the following year, the author was warned by medical authorities that he might soon die from disease. This opinion turned out to be wrong but the scare convinced the author that it was his duty to develop an automatic method of calibrating the tank model so that knowledge of calibration techniques should not be lost with his death.

Once the work had begun it turned out to be rather easy. It was the type of problem which simulated the author's way of thinking and, after a few months, a program had been written which ran successfully and actually gave better results than the manual calibration!

The program is a trial and error method carried out by computer. A rough outline of the program is as follows:

1) Start with an initial model.

2) Divide the whole period into five subperiods, each subperiod to correspond to one of the five runoff components.

3) Compare the calculated hydrograph with the observed hydrograph for each subperiod, and define the criteria RQ(I) and RD(I) (I = 1,2,⋯,5) where RQ(I) is the criterion for volume and RD(I) is the criterion for the hydrograph shape of the I-th subperiod.

4) Adjust the coefficients of the tank model according to the criteria RQ(I) and RD(I).

5) The hydrograph derived by the adjusted tank model is compared with the observed and an evaluation criterion CR is calculated.

6) The next trial is made, using the adjusted model.

7) The automatic calibration procedure will usually finish after several iteration of a few trials and the model of the least criterion CR is considered to be the best model.

Division of the hydrograph into five sub-periods so that in the I-th subperiod the I-th runoff component is the most important.

Evaluation of the effectiveness of the I-th runoff component must be made in the period in which the I-th runoff component is most important. To accomplish this, the entire hydrograph is divided into five subperiod as follows:

Subperiod 1: Days on which Y1, the output from the upper outlet of the top tank, is most important belongs to subperiod 1, i.e. each day that

$$Y1 > CY,$$

where C is a constant (usually set to 0.1) belongs to subperiod 1. Similarly the rules for the remaining four subperiods are:

Subperiod 2:	When Y1 < CY	and Y1+Y2 > CY.
Subperiod 3:	When Y1+Y2 < CY	and Y1+Y2+Y3 > CY.
Subperiod 4:	When Y1+Y2+Y3 < CY	and Y1+Y2+Y3+Y4 > CY.
Subperiod 5:	Otherwise.	

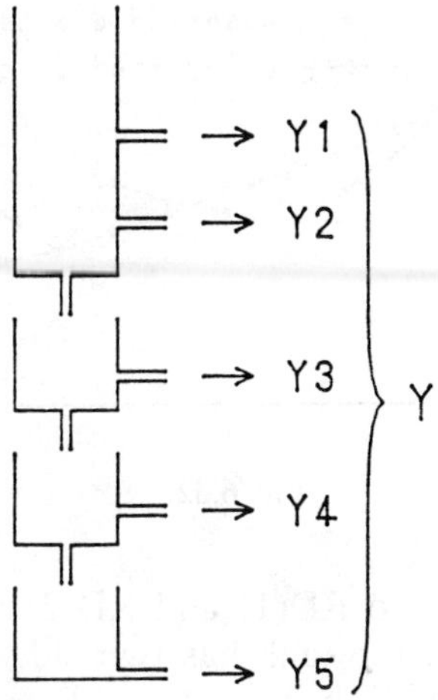

Fig. 6.31.

These rules were determined after some modification and improvements to simplify the programming.

Introduction of criteria RQ(I) and RD(I)

In each subperiod 1, 2, 3, 4 and 5, the discharge volume and the slope of the hydrograph (in logarithmic scale) of the calculated and observed discharges are compared using the following criteria:

$$RQ(I) = \sum_J QE(J) \,/\, \sum_J Q(J) \qquad (I = 1, 2, \ldots, 5)$$

$$RD(I) = \frac{\sum_J{}' \left(\log QE(J-1) - \log QE(J)\right)}{\sum_J{}' \left(\log Q(J-1) - \log Q(J)\right)} \qquad (I = 1, 2, \ldots, 5)$$

where Q is the observed discharge, QE is the calculated discharge, I is the index number of the subperiod, J is the day number, $\sum_J$ means the sum over the J days belonging to the subperiod I and $\sum_J{}'$ means the sum over the J days belonging to the subperiod I for which QE(J-1)-QE(J) is positive.

It should be noted that the conditions $\sum_J$ and $\sum_J{}'$ are determined only from the calculated discharge and not from the observed discharge. For example, in the definition of the criterion RD(I), the summation is carried out over the decreasing part of the calculated hydrograph. The observed hydrograph is not considered so that we avoid the random noise that will appear in it (Fig. 6.32).

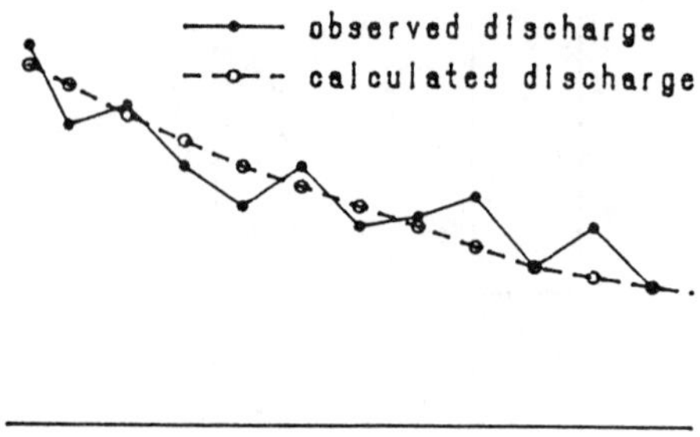

Fig. 6.32.

Later, an improvement to RD(1) and RD(2) was made based on the following reasoning: The top tank has two side outlets and one bottom outlet, and consequently, it has three coefficients A0, A1, and A2. However, there are four criteria RQ(1), RQ(2), RD(1) and RD(2) for the top tank, and so there is an 'excess' of criteria. To resolve this problem RD(1) is neglected (RD(1) = 1) and RD(2) is defined so that the summation Σ' is made over the J days belonging to both subperiods 1 and 2 for which QE(J-1)-QE(J) is positive.

Feedback formulae

For the top tank, adjustments to the runoff and infiltration coefficients are made as follows:

If RD(2) > 1 (< 1) then this means that the slope of the runoff component from the top tank is too large (small) and so the sum of the coefficients should be adjusted to

$$(AM0+AM1+AM2)=(A0+A1+A2)/RD(2),$$

where AM0, AM1 and AM2 are adjusted coefficients.

If RQ(1) > 1 (< 1) then this means that the runoff component from the upper outlet is too large (small) and so the ratio of the coefficients A2/A0 should be adjusted to

$$(AM2/AM0)=(A2/A0)/RQ(1).$$

Similarly, A1/A0 is adjusted to

$$(AM1/AM0)=(A1/A0)/RQ(2).$$

By solving these equations we can derive the following feedback formulae (written in FORTRAN form):

$$A=(A2/A0)/RQ(1), \quad B=(A1/A0)/RQ(2),$$

$$A0 = (A0 + A1 + A2) / (RD(2) \bullet (1 + A + B)$$

$$A1 = B \bullet A0, \qquad A2 = A \bullet A0.$$

Similarly, for the second tank, it is better to adjust the coefficients B0 and B1 by

$$(BM0 + BM1) = (B0 + B1) / RD(3),$$

$$(BM1 / BM0) = (B1 / B0) / RQ(3),$$

so that we can derive the following feedback formulae:

$$B = (B1 / B0) / RQ(3),$$

$$B0 = (B0 + B1) / (RD(3) \bullet (1 + B)$$

$$B1 = B \bullet B0.$$

In the same way, we can derive feedback formulae for the third tank.

$$C = (C1 / C0) / RQ(4),$$

$$C0 = (C0 + C1) / (RD(4) \bullet (1 + C)$$

$$C1 = C \bullet C0.$$

For the fourth tank, we cannot adjust the discharge amount by modifying the ratio of runoff and infiltration coefficients, because there is no bottom outlet. Therefore, we have to adjust it by modifying the water supply from the third tank, and, in consequence, from the second tank and the top tank, too. If RQ(5) > 1, the supply from the third tank should be made smaller by dividing C0 by RQ(5). If we decrease the value of C0, the runoff from the third tank will increase and we will have to reduce the supply from the second tank, i.e. we have to decrease the value of B0. In general this adjustment need not to be so large as the adjustment for the third tank. From these considerations, we can obtain the following adjustment formulae:

$$D1 = D1 / RD(5),$$

$$C0 = C0 / RQ(5),$$

$$B0 = B0 / (RQ(5))^{1/2},$$

$$A0 = A0 / (RQ(5))^{1/4}.$$

We would normally expect that, starting from an initial model and using the iterative feedback described above, the tank model would converge to a good fit. However, in order to make the feedback system converge, it may be necessary to include some additional modifications.

Some necessary modifications

RQ(I) and RD(I) should be limitted to the range (1/2, 2)

In some cases, some values of RQ(I) and RD(I) (especially RD(I)), may show values very different from 1. To avoid such extreme values we defined a maximum allowable range of 1/2 to 2, i.e. values of RQ(I) and RD(I) larger than 2 are replaced by 2, and values smaller than 1/2 are replaced by 1/2.

Effect of RD(I) must be halved

One of the reasons why the feedback system may not work well is the unreliability of RD(I) which is affected by large noise. We suppose that the subtraction included in the definition of RD(I) must be the main reason of its unreliability. To avoid this effect of noise, we halve the effect of RD(I) by replacing it by $(RD(I))^{1/2}$. At first, RD(I) was replaced by $(RD(I))^{1/4}$, but this was found to be too much.

RD(5) should be neglected

As the time constant of the fourth tank is very long, i.e. the slope of the calculated discharge in period 5 is very small, RD(5) is mainly governed by noise, and, accordingly, it is very unreliable. Therefore, the feedback of RD(5) is neglected by putting RD(5)=1.

The effect of random deviation of the time lag of peak discharge

RD(3) and RD(4) often show very unreliable values (i.e. negative values with large absolute values). This may occur when the peak calculated discharge has a one day lag when compared with the observed peak, as shown in Fig. 6.33a. One day difference between the peaks of observed and calculated hydrographs can easily occur because of the different measuring times of precipitation and discharge. For example, rain may occur just before or just after the normal measuring time of daily precipitation (usually 9 o'clock), and would then be recorded as either the precipitation of the foregoing day or of that specific day, only by chance. In the case of Fig. 6.33a, RQ(I) and RD(I) (especially RD(I)), are greatly affected when I = 3,4 or 5; and in the case of Fig. 6.33c, RD(I) for I = 1 or 2 is largely affected, too. To avoid cases in which day J belongs either to subperiod 1 or 2 and the previous day J-1 belongs to any of subperiods 3,4 or 5, the day J-1 is deleted from the definition of RQ(I) and RD(I) and the day J is deleted from the definition of RD(I).

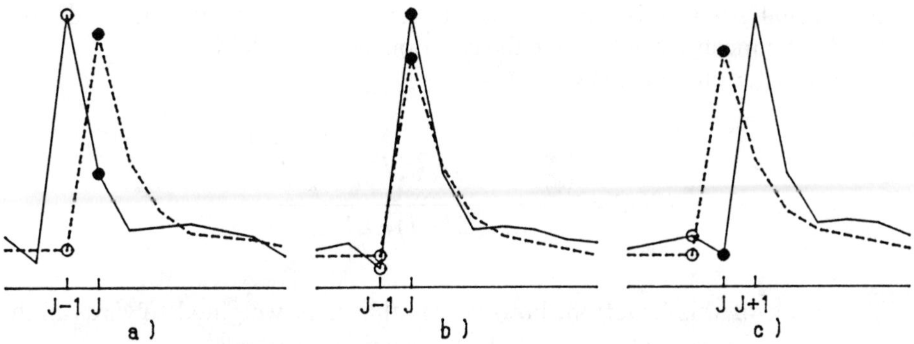

○ : **deleted in definition of RQ(I) and RD(I) where I=3 or 4 or 5**

● : **deleted in definition of RD(I) where I=1 or 2**

Fig. 6.33.

Values of RQ(I) and RD(I) which are close to 1 should not be used for feedback

In the early stages of developing the automatic calibration program, feedback procedures using RQ(I) and RD(I) did not give good results. This was probably caused by feedback of RQ(I) and RD(I) for values near to 1. Such criteria which are close to 1 include little information but are mostly composed of noise, and although the feedback effects of each of these criteria is not large individually, their total effect is greater and may lead the calibration astray.

In calibrating the tank model by the trial and error method there is one important principle; we must always concentrate our attention on only one specific point. We have to improve the weak points of the model one by one and not try to adjust many bad points at the same time. This principle is probably true for the automatic calibration method, too. If we apply many feedback procedures at the same time, they may interfere with each other and cause bad result.

After some trials we finally developed a method by which we select two criteria within RQ(I) and RD(I) which are most distant from 1 and use only these two criteria for feedback; i.e. the other criteria are eliminated by putting them equal to 1.

Evaluation criterion

When the author was calibrating the tank model by his trial and error method of subjective judgement, the best model was also determined subjectively. However, for the trial and error procedure by computer, an evaluation criterion is necessary to decide on the best tank model. While the most usual criterion is the mean square error, it is probably better to divide this by the mean value to make the criterion dimensionless.

The criterion obtained is

$$MSEQ = \frac{\left(\sum_J (QE(J) - Q(J))^2 / \sum_J 1\right)^{1/2}}{\sum_J Q(J) / \sum_J 1}.$$

The weak of this criterion, however, is that it is weighted towards high discharges, i.e. it can be thought of as a criterion for floods.

For the overall evaluation of the model it may be better to use the natural logarithm of discharge, as follows:

$$MSELQ = \left(\sum_J \left(\ln QE(J) - \ln Q\left(J'\right)\right)^2 / \sum_J 1\right)^{1/2}.$$

If d is small compared to x, the following relationship holds:

$$\ln(x+d) - \ln x = \ln(1 + d/x) = d/x.$$

Therefore, in the case of the error criterion:

$$\ln QE(J) - \ln Q(J) \quad (QE(J) - Q(J)) / Q(J).$$

This means that MSELQ is the mean square of the relative error and is probably better than MSEQ.

However, both MSEQ and MSELQ have a common weak point. In the case shown in Fig. 6.33a, QE shows a large error in days J-1 and J ; and in the case of Fig. 6.33c, QE shows large errors in days J and J+1. Since such cases occur frequently this means that both MSEQ and MSELQ are too high. Even if the difference between QE(J) and Q(J) is large in such cases, we can say that both hydrographs are similar except for the one day shift.

Compare QE(J) with Q(J-1), Q(J) and Q(J+1) and call the nearest one among these three Q(J'). Then a modified MSEQ may be defined as follows:

$$MSEQ = \frac{\left(\sum_J \left(QE(J) - Q\left(J'\right)\right)^2 / \sum_J 1\right)^{1/2}}{\sum_J Q(J) / \sum_J 1}.$$

In the same way MSELQ may be modified by searching ln Q(J-1), ln Q(J) and ln Q(J+1) for the nearest to ln QE(J), calling this ln Q(J'), and rewriting MSELQ as follows:

$$MSELQ = \left(\sum_J \left(\ln QE(J) - \ln Q\left(J^{'}\right) \right)^2 / \sum_J 1 \right)^{1/2}.$$

Finally, the evaluation criterion CR is defined by

$$CR = MSEQ + MSELQ.$$

Number of iteration of trials

In calibrating the tank model using the trial and error method of subjective judgement, if we make too many trials then the balance and harmony may be lost and the results may get worse. It is very interesting to note that the automatic calibration by computer also shows the same tendency. Since the computer program is a sort of simulation of the author's way of thinking, then it is also reasonable that the computer program should show this similar tendency.

There must be some reason why the number of iterations should be kept below a certain number. In the computer program, the tank model coefficients are adjusted by RQ(I)'s (I = 1,2,···,5) and RD(I)'s (I = 2,3,4). The two which are most distant from 1 within these eight RQ(I)'s and RD(I)'s, are selected, and these two are used to adjust the tank model coefficients. Therefore, after four iterations it is probable that all the effective RQ(I)'s and RD(I)'s will have been used for adjustment, the adjusted model will have become good, and all the RQ(I)'s and RD(I)'s will have become close to 1.

Generally speaking, each RQ(I) and RD(I) are composed of both signal and noise components. When they become close to 1 the signal component has mostly vanished and the remainder is noise. At this stage the RQ(I) and RD(I) are not usefull and may actually be harmful. Therefore, using such RQ(I)'s and RD(I)'s in the feedback will spoil the balance and harmony of the model. Once the balance is broken, the model can not become good again. It is some sort of divergence.

Uniqueness of the solution is probably nonsense

When the author started to develop the automatic calibration program he expected that there would be some final model, i.e. some definite solution, and that using the automatic calibration program we could approach this final best model. However, this was misplaced optimism. Starting from different initial models we will get different final models and it is very dificult to judge from the hydrographs of these different final models which one is the best.

The tank model is some sort of approximation of the runoff phenomena. There must therefore be many ways of approximating the result, i.e. there cannot be only one unique solution.

6.2.3. Automatic Calibration Program by Duration Curve Comparison Method

Why comparison of duration curve became necessary?

In 1975 and 1976, the author visited the Upper Nile Region around the Lake Victoria, and had the chance to analise the runoff of several basins. This region is located just under the equator and the rainfall is always local showers covering small areas. Therefore, even if there are several rainfall stations in the basin, this will not be enough to catch the rainfall over the basin. In some cases, nearly all stations may record heavy rain but the discharge from the basin is small. On the other hand, the raingauges may record very little but the discharge is large. Therefore, the calculated discharge from rainfall data may not show a good fit to the observed discharge.

Even under such bad conditions we were able to calibrate the tank model by comparing the shapes of the hydrographs, i.e. the slope of the discharge and the volumes of monthly discharge. Also, we were able to adjust the sub-base and the base runoff by comparing discharges in the dry season. However, it seemed to be very difficult to translate these ideas into a computer program.

Fortunately it occured to the author that he could use duration curves instead of hydrographs and it proved rather easy to develop an automatic calibration program using a duration curve comparison method.

The duration curve comparison method is much better than the hydrograph comparison method. It was first developed for basins in Upper Nile Region but it was applied to Japanese basins and gave better results than those obtained by the hydrograph comparison method. Now, the duration curve comparison method is used in every case.

Division of the duration curve into five sections corresponding to five subperiods

On the duration curve of calculated discharge derived from the working tank model, discharges for the days belonging to subperiod 1 are usually large and they gather in the left part of the curve and, conversely, discharges for the days belonging to subperiod 5 are usually small and they gather in the right part of the duration curve.

An example of such a distribution is shown in Fig. 6.34, where daily discharges for one year are ordered from the largest to the smallest, and each daily discharge is described by the number of the subperiod to which it belongs.

```
1111111111   1112122221   2222222212   2222222222   2222222222
2222222222   2222223223   2323322332   3333332222   2233222223
2333233333   3333333333   2333333333   2332222333   3322332332
3223333333   3333332333   3333333323   3323333333   3333343333
3323334432   3333433423   2433343234   4444343434   4434344333
4433344344   4444444444   4444434443   4344444444   4343443434
4444444444   4444444444   4444444444   4444444444   4444444444
4444444444   55555

1111111111   2221121222   2222121222   2222222222   2222222222
2222222222   2222222232   2222233322   3323223223   3333333233
3223333333   3222332333   2333333233   3233323333   2333332223
3223333333   2333343333   2433333333   3332432322   2243332443
4433444224   4423344443   3434444344   4344444444   2444424434
4343444443   4343444434   3344444444   4444443443   3443444444
4444434443   4234434434   3434444344   4454554354   4554555555
5555555555   55555
```

Fig. 6.34. Computed daily discharges of the River Sarugaishi at Taze (Japan) ordered and described by subperiod number (results for years 1969 and 1970)

After some consideration and hesitation we decided to use the following method: the duration curve is divided in the most simple way into five subsections, successively from the left, NQ(1) days, NQ(2) days,···, NQ(5) days, where NQ(I) is the number of days belonging to subperiod I.

Definition of RQ(I) and RD(I)

After the division of the duration curve into subsections the definition of the criteria RQ(I) and RD(I) is easy. Let Q(N) and QE(N) be the observed and calculated daily discharges in decreasing order, where N is the ordinal number of the day. Then, the five subsections are represented as follows, using the order number N:

subsection 1: $1 \leqq N \leqq NQ(1),$

subsection 2: $NQ(1) + 1 \leqq N \leqq NQ(1) + NQ(2),$

subsection 3: $\sum_{I=1}^{2} NQ(I) + 1 \leqq N \leqq \sum_{I=1}^{3} NQ(I),$

subsection 4: $\sum_{I=1}^{3} NQ(I) + 1 \leqq N \leqq \sum_{I=1}^{4} NQ(I),$

$$\text{subsection 5:} \quad \sum_{I=1}^{4} NQ(I) + 1 \leqq N \leqq \sum_{I=1}^{5} NQ(I).$$

We can define the criterion RQ(I) as the ratio of the sum of the calculated and observed discharges belonging to subsection I:

$$RQ(I) = \sum_{NY} \sum_{1} QE(N) / \sum_{NY} \sum_{1} Q(N) \quad (I = 1, 2, \ldots, 5),$$

where $\sum_{IL}$ is the sum over the subsection I, and $\sum_{NY}$ is the sum over years.

The criterion RD(I) can be defined as the ratio of inclination of calculated and observed duration curves in the subsection I, where the inclination is defined, for simplicity, by the difference of the mean value of duration curve in the left half and the right half of subsection I, respectively:

$$RD(I) = \frac{\sum_{NY} \left(\sum_{IL} QE(N) - \sum_{IR} QE(N) \right)}{\left(\sum_{NY} \left(\sum_{IL} Q(N) - \sum_{IR} Q(N) \right) \right)} \quad (I = 2, 3, 4),$$

where $\sum_{IL}$ and $\sum_{IR}$ mean the sum over the left half and the right half of the subsection I, respectively, and RD(2) is defined on the subsection, $1 \leqq N \leqq NQ(1) + NQ(2)$.

Feedback formulae

The feedback formulae are just the same as those used in the hydrograph comparison method. We knew that in the hydrograph comparison method we had to halve the effect of RD(I), because of its unreliability caused by the large noise effect. In the present case, RD(I) defined by a comparison of duration curves must be much more reliable because the duration curve is smooth and monotonously decreasing and, accordingly, we thought that they could be applied to the feedback formulae without dividing them by 2. Contrary to our expectation, we found that such a feedback formulae did not bring good results and we had to halve the effect of RD(I), just as we did before.

The feedback of RQ(I) probably corresponds to the displacement feedback in an automatic control system and the feedback of RD(I) probably corresponds to a velocity feedback. We also understand that, as a general principle, if we add a slight velocity feedback to a displacement feedback the total feedback system will usually work well. However, we can not understand the exact reason why the effect of RD(I) must be halved.

Evaluation criterion

For the duration curve comparison method we have to introduce new evaluation criteria as well as the ones previously used. Just as before, the mean square error of the duration curve of the calculated discharge or the natural logarithms of the calculated discharge are defined as follows:

$$MSEDC = \frac{\left(\sum_{NY}\sum_{N}(QE(N) - Q(N))^2 / \sum_{NY}\sum_{N} 1 \right)^{1/2}}{\left(\sum_{NY}\sum_{N} Q(N) / \sum_{NY}\sum_{N} 1 \right)}$$

$$MSEDC = \left(\sum_{NY}\sum_{N}(\ln QE(N) - \ln\ Q(N))^2 / \sum_{NY}\sum_{N} 1 \right)^{1/2}.$$

And we can introduce the evaluation criterion CRDC for the duration curve as

$$CRDC = (MSEDC + MSELDC)/2,$$

in comparison to the earlier criterion, now renamed CRHY, which is given by

$$CRHY = (MSEQ + MSELQ)/2.$$

The final evaluation criterion used in the duration curve comparison method is given by

$$CR = CRHY + CRDC$$

$$= (MSEQ + MSELQ)/2 + (MSEDC + MSELDC)/2.$$

This evaluation criterion seems to be fairly good because, by making the sum of four kinds of mean square error, the random deviation of all the components counteract each other and, consequently, the criterion becomes more reliable than each single criterion individually.

Some characteristics of the duration curve comparison method

The method is resistant to divergence

In this method RD(I)'s were defined under much better conditions than under the hydrograph comparison method and the result obtained was also better than that obtained by the hydrograph comparison method. Moreover, this method has the benefit that divergence hardly ever occurs.

The time lag problem is not important in this method

This method also has the benefit that we do not need to worry about the time lag. In the hydrograph comparison method it is important to determine an appropriate value of time lag and if the assumed time lag is inappropriate the criteria RD(I) become unreliable and, accordingly, the feedback procedures do not converge. Therefore, we have to determine the time lag first from the initial model which usually does not provide a good model for the object basin.

In the duration curve comparison method, such a troublesome problem does not exist at all. The criteria RQ(I) and RD(I) are defined by from the duration curves and so determination of the time lag is not necessary.

Goodness of fit of duration curves is probably more meaningful than goodness of fit of hydrographs

If we apply both methods to some basin and compare the resulting hydrographs, the hydrograph comparison method usually shows a slightly better result than the duration curve comparison method. This should be obvious because, in the hydrograph comparison method, the feedback procedures are made to give a good fit of the observed hydrograph, while, in the duration curve comparison method, the feedback procedures are made to give a good fit of the duration curves and usually, the evaluation criteria CRDC is smaller than CRHY, i.e. the duration curves fit better than the hydrographs.

We suppose that a good fit of duration curves is probably more meaningful than a good fit of hydrographs both from hydrological and practical points of view. It is clear, that for practical uses (say for hydroelectric power generation), a good fit of the duration curves is more meaningful than a good fit of the hydrographs. The hydrological point of view is not so obvious but can be developed as follows: There are many peaks in a hydrograph. In any particular storm rainfall data observed at some stations will show smaller values than the unknown real mean areal rainfall, and in some other storm the contrary case will occur. Therefore, in one case the observed peak discharge will be greater than the calculated one, and, in the other case, the observed peak will be smaller than the calculated one. Accordingly, to obtain a perfect fit between the observed and calculated hydrographs is impossible and to aim for a good fit of the duration curves seems to be more meaningful also from a hydrological point of view.

6.2.4. Calibration of other parameters than runoff-infiltration coefficients

Calibration of positions of side outlets

The automatic calibration program described above can calibrate runoff-infiltration coefficients A0, A1, A2, B0, B1, C0 and C1. However, the tank model has other coefficients than these: i.e. positions of side outlets, HA1, HA2, HB and HC; and parameters of soil moisture structure, S_1, S_2, K_1 and K_2.

The calibration of the positions of the side outlets is not so difficult. Parameters HA1 and HA2 relate to the behavior of the first and second runoff components Y1 and Y2, while HB relates to Y3 and HC relates to Y4. Therefore, within (HA1, HA2), HB and HC, there is not much interaction and we can calibrate them independently. For example, consider the case of HB and make several trials by changing its value. It is important to change the value of HB boldly with some large interval. For example, trials might be made, at first, setting HB as 10, 15 and 20, or 7.5, 15, 22.5. In each case, each trial is made by means of the automatic calibration program and the final model and criterion CR are obtained for each value of HB. In this way, three values of CR will be obtained for HB = h_1, h_2 and h_3 as shown in Fig. 6.35.

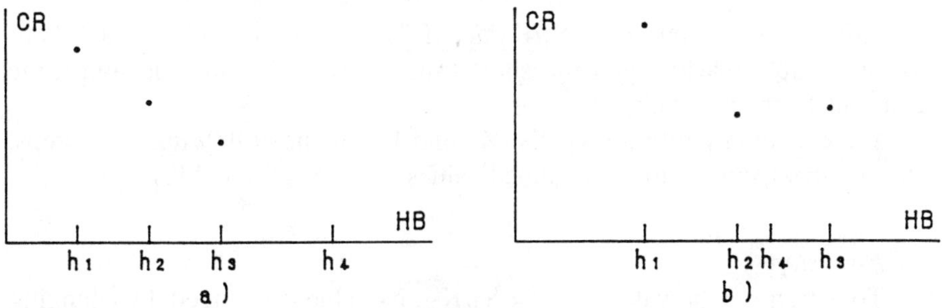

Fig. 6.35.

If the result obtained looks like Fig. 6.35a, we must make a fourth trial by putting HB = h_4, where the value of h_4 must be determined carefully and boldly. If the result obtained looks like Fig. 6.35b, we can determine h_4 without much difficulty, so that CR will attain a minimum at HB = h_4. In this way we can determine the value of HB after several trials. HC can be calibrated in the same way.

The top tank has two side outlets and so there are two parameters HA1 and HA2. We can consider that HA1 and HA2-HA1 will be nearly independent. In the initial model, HA1 = 15 and HA2 = 40 = HA1+25 and therefore we might make trials setting HA1 = 10, 15 and 20; HA2 = 10+25, 15+25 and 20+25. Next, the value of HA1 can be determined in the same way as the case of HB. After HA1 is determined, HA2-HA1 is changed to determine the optimum value and so HA1 and HA2 can be determined after ten trials or so.

Calibration of soil moisture structure

The soil moisture structure has four coefficients: S_1, S_2, K_1 and K_2 and therefore, if we wished to find the optimum values on a four dimensional mesh, the number of trials would become enormous. Therefore, we must find orthogonal factors and determine the optimum values one by one.

The soil moisture structure has two time constant T_1 and T_2. The short time constant T_1 shows how the difference between the relative wetness of primary and secondary soil moisture decreases with time as they approach the equibrium state. The long time constant T_2 shows how the total soil moisture decreases with time under drying condition. Therefore, T_1 and T_2 have entirely different hydrological effects, can be considered as independent factors and can be calibrated one by one.

Usually, S_2 is much greater than S_1, and K_2 is much greater than K_1 and so we can derive rough but simple approximate relations as follows:

$$T_1 = 1/k_1 \doteqdot S_1/K_2, \qquad T_2 = 1/k_2 \doteqdot S_2/K_1.$$

For simplicity, we put $T_1' = S_1/K_2$ and $T_2' = S_2/K_1$ so that T_1' and T_2' are values which nearly equal the short time constant T_1 and the long time constant T_2, respectively.

We can then calibrate S_1, S_2, K_1 and K_2 in the following four steps, starting from some initially assumed values of S_1, S_2, K_1 and K_2.

First step

To determine the value of $T_1' = S_1/K_2$, its value is changed, by changing the values of S_1 and K_2. For example, S_1 is changed to 0.95 S_1, S_1 and 1.05 S_1; and K_2 is changed to 1.05 K_2, K_2 and 0.95 K_2. This will change the value of T_1' about 10%. Trials are made, similar to the case of HB, and the value of T_1' is determined so as to give the minimum value of CR. As T_1' approximately equal to the short time constant T_1, which has important a hydrological meaning, it is important to determine the value of T_1' accurately.

Second step

Keeping the value of T_1' constant, the values of S_1 and K_2 are changed. For example, both are changed simultaneously ±10%. In this way, S_1 and K_2 are calibrated.

Third step

The value of T_2' is calibrated in the same way as in the first step.

Fourth step

The values of S_2 and K_1 are calibrated in the same way as in the second step.

6.3. SNOW MODEL AND SOME OTHER IMPORTANT FACTORS

6.3.1. Snow model

Division into zones

When winter comes, snow begins to deposit on high elevation area and then spreads to lower areas. When spring comes the snow deposit begins to melt first in low elevation areas and then moves up the elevation range of the basin. Therefore, it is necessary to divide the basin into elevation zones in order to calculate snow deposit and melt. The number of zones need not be too large; usually, it is sufficient to divide the basin into a few zones with equal elevation interval.

Temperature decrease with elevation

Snow deposit and melt are governed by air temperature, and so the rate of temperature decrease with elevation is one of the most important factors in the snow model. Usually, the temperature decrease per 1,000m in elevation is about 5°C~6°C, or so. In most cases, the precipitation and temperature station is located in the lowest zone and there are no stations in the higher zones. Therefore, we have to estimate the mean temperature of each zone from the value observed in the lowest zone. Under the assumption that the temperature decrease with elevation per 1,000m is 5.5°C, we can derive the mean temperature of the I-th zone T(I), from the observed temperature T at the station as,

$$T(I) = T + T0 - (I-1) \bullet TD,$$

where T0 is a correction term and TD is the temperature decrease per zone. These T0 and TD are initially given as,

$$T0 = (H_0 - H_1) \bullet 5.5 \bullet 10^{-3},$$

$$TD = H_d \bullet 5.5 \bullet 10^{-3},$$

where H_0 is the elevation of the station, H_1 is the middle elevation of the first zone and H_d is the elevation interval of zone.

Starting with initial values of T0 and TD we can calibrate them without much difficulty. In some basins, TD will show a seasonal change.

It is interesting to note that the calibrated value of T0 is usually much smaller than the initial value of T0 and is often negative. This means that the temperature at the recording station is usually higher than the temperature over most of the basin, after neglecting the effects of elevation. This must be due to the following two reasons: One, the station is, usually, set at a convenient place for observation, i.e. in or near a village or some place which is convenient to stay or visit. Such a place is usually warmer than the rest of the basin. The other reason might be that places where people live become warmer because of the effect of human activities.

Precipitation increase with elevation

Generally speaking, it rains more heavier in mountainous areas than on the plain. From a water balance based on a reasonable estimate of annual evaporation over the basin and the recorded annual runoff we can estimate the annual precipitation over the basin. Usually, it is about 120% ~ 130% or so of the annual precipitation observed at some place in the plain.

However, when the author tried to analyze a snowy basin in Japan for the first time, he was astonished by the fact that the discharge in the snowmelt season was very large compared with the total precipitation in winter as observed at several stations in the plain. This means that the precipitation increase with elevation must be very large.

Moreover, there was another curious fact, that the discharge in summer was not so large when compared with the rainfall observed in the plain. This means that the precipitation increase with elevation is not large in summer. So, there must be large seasonal change in the precipitation increase with elevation, i.e. it is large in winter and small in summer.

Such a large seasonal change can be considered reasonable under the following meteorological conditions. In Japan, the most snowy basins are located on the Japan Sea side. In winter, the seasonal north-west wind which blows over the warm Japan Sea brings heavy snowfall to the mountains. In such a case, the precipitation increase with elevation must be very large. In summer, the meteorological conditions are entirely different and rainfall occurs by typhoon, warm front, cold front, local shower, etc. In such cases, rainfall increase with elevation is small, because the wind is nearly independent of the mountain slope.

Assuming that the precipitation increase with elevation is linear and considering the seasonal change, the mean precipitation in the I-th zone is given by,

$$P(I) = (1 + P0 + (I-1) \bullet CP(M) \bullet PD) \bullet P,$$

where P is the observed precipitation, P0 is a correction term, PD is the precipitation increase per zone, CP(M) is the coefficient for the seasonal change and M is the month index.

In some case, it seems to be better to assume that the precipitation in the highest zone is the same as that in the next highest zone. For example, in the case of four zones the precipitation's are given as follows:

$$(1+P0)\bullet P, \qquad (1+P0+CP(M)\bullet PD)\bullet P,$$

$$(1+P0+2CP(M)\bullet PD)\bullet P, \quad (1+P0+2CP(M)\bullet PD)\bullet P.$$

However, as the area of the fourth zone is not large, the effect of this assumption is not so large.

Snowmelt constant

Snowmelt is mostly governed by air temperature and we may assume that the amount of snowmelt in a day is proportional to the mean air temperature T of the day; the coefficient may be called the snowmelt constant SM. Moreover, if it rains, we can assume that the temperature of the rain water is equal to the air temperature. Then, the snowmelt by rainwater is PT/80, where P is the daily precipitation.

Therefore, the maximum amount of daily snowmelt is given by

$$SM\bullet T+PT/80.$$

When the author analyzed Japanese snow basins he obtained good results by putting SM = 6. However, when he analyzed the snowy foreign basins it was found that SM = 4 gave better results. Japan is very humid region and humid air has much more energy than dry air. This is probably the reason that SM = 6 gave good results for Japanese snow basins.

Avalanche effect

In some cases it is very cold in high zones and the temperature remains under 0 C most of the year. Then, the snow deposit in high zones grow larger and larger and, in the model, might tend to infinity. To avoid this tendency, we have to include the movement of snow by gravity. There are many kinds of a movement, e.g. avalanche, glacier, etc., but we will call such a movement the avalanche effect, for simplicity.

A simple way to consider the avalanche effect is to assume that some part of the snow deposit in the I-th zone moves down to the (I-1)-th zone everyday and that this volume is proportional to the total volume of snow deposit in the I-th zone. We will name this coefficient the avalanche constant AV. Let P be the annual precipitation in the I-th zone and we assume that the daily precipitation is constant, P/365. Then, the snow deposit in the I-th zone will be in an equibrium state when the daily avalanche volume is equal to the daily precipitation, i.e.,

$$AV \bullet SD = P / 365,$$

where SD is the equibrium snow deposit volume in the I-th zone. If AV is 1%, then SD = P/3.65 = 0.274 P; if AV = 0.5%, then SD = P/1.825 = 0.548 P and if AV = 0.1%, then SD = P/0.365 = 2.74 P.

Though the real value of snow deposit in the I-th zone is unknown we can estimate its approximate value, from which we can estimate the approximate value of AV as follows:

$$AV = (P / 365) / SD.$$

This equation is correct for the highest zone. In the next zone, the input is precipitation and the avalanche from the highest zone, which is equal to the precipitation on the highest zone in the equibrium state. Therefore, the input of the next zone per unit area is

$$(S(I) \bullet P(I) + S(I-1) \bullet P(I-1)) / S(I-1),$$

where P(I) and P(I-1) are the annual precipitation in the I-th and the (I-1)-th zone, respectively, and S(I) and S(I-1) are the area of the I-th and the (I-1)-th zone, respectively. Therefore, the avalanche constant in the (I-1)-th zone is given as follows:

$$AV(I-1) = (((P(I) \bullet S(I) + P(I-1) \bullet S(I-1)) / S(I-1) \bullet 365)) / SD(I-1).$$

In the same way, we can derive the formulae for AV(I-2).

The value of AV is not so important for the calculation of discharge; its importance is to modify the accumulation of snow in high zones.

Runoff calculation by the tank model

To calculate runoff from the tank model is rather easy. Though the basin is divided into zones, only one tank model is applied to the whole basin. The precipitation as rain and the snowmelt water are summed up in each zone, the sum is multiplied by the area of each zone, they are summed up and the total is divided by the area of the basin. This is then put into the tank model to be turned into runoff.

6.3.2. Precipitation Factor CP(M)

In snowy basins, precipitation increase with elevation often shows a large seasonal change. Correspondingly, in most non-snowy basins, the precipitation factor CP(M) also shows a seasonal change, where M is the month-index, and sometimes this may be very large. As the effect of CP(M) on the calculated discharge is direct and large, it is very important to determine the appropriate values of CP(M) in runoff analysis.

Fortunately, calibration of CP(M) is not so difficult. We start from some initial tank model and CP(M) = 1 (M = 1,2,⋯,12). Then, we calibrate CP(M) by comparing the calculated and observed monthly discharges in parallel with the calibration of the tank model. It is better to adjust or change the initial tank model, if necessary. We can also make an automatic calibration program to determine CP(M).

In some special case, CP(M) shows very large value in some specific month. It seems to be curious, but probably from some specific meteorological condition.

6.3.3. Weights of precipitation stations

The Thiessen polygon method is illogical

Even though the Thiessen polygon method has been widely used for a long time, it is illogical and, in reality, cannot give results.

We can clearly see the defect of the method by considering two stations close to the boundary of a basin, as in Fig. 6.36. Using the Thiessen polygon method, the weight of A is very small and that of B very large. However, if the two stations are located close to each other, their weights should be almost equal. If the station A is moved to a A', the weight of A' becomes very large and that of B becomes very small. If the Thiessen polygon method were logical, such an absurdity would not occur.

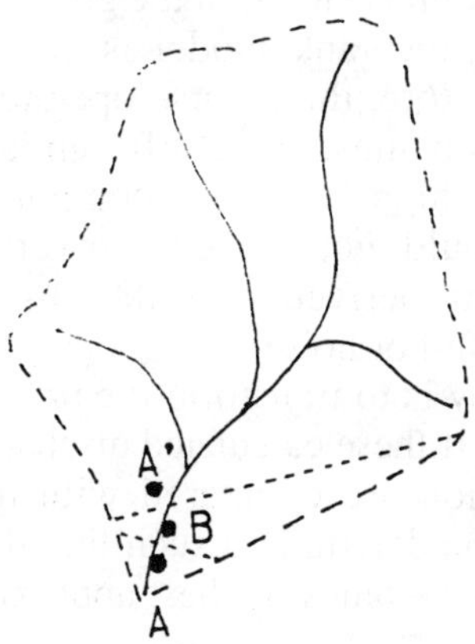

Fig. 6.36.

In principle, the weights of precipitation stations should be determined, not by their geometrical positions, but by the meteorological conditions. For example, in some cases, a station situated near the center of a basin may be almost useless for discharge forecasting, in spite of its large Thiessen weight. In such cases, due to micro-meteorological conditions, the rainfall in a small area around the station, may have a weak relationship with the areal precipitation over the whole basin.

The mean areal precipitation over the basin is unknown

When the basin is very small and there are no woods, no steep mountain, no lake, etc., that prevent the establishment of precipitation stations, then we can measure the approximate value of mean areal precipitation over the basin by setting many precipitation stations uniformly over the basin. It is some sort of sampling survey. However, in almost all cases, such a precipitation measurement is impossible, and so, to measure the areal precipitation over the basin is practically impossible.

As the mean areal precipitation over the basin is unknown, the problem of determining the weights of precipitation stations in order to get the best mean areal precipitation is meaningless.

When we have observed discharge

Even though the mean areal precipitation over the basin is unknown there is often observed discharge, which is some sort of areal integral of the precipitation over the basin. However, there is a large transformation between precipitation and discharge.

The problem then is to derive the mean areal precipitation over the basin from the observed discharge. To our regret, this problem seems to be impossible. The transformation from precipitation to discharge is an itegral-like operator. Most of the rain is stored in the ground before it is turned into discharge. For example, the tank model is some sort of a non-linear incomplete integral. Therefore, the inverse operator which will transform discharge into precipitation must be a differential-like operator, which means that it will amplify high frequency components and be practically useless. Even if we could get such an inverse operator, the areal precipitation over the basin derived from the observed discharge by this inverse operator would be full of noise.

The only remaining way is to transform the precipitation at each station into discharge time series. If these calculated discharge series corresponding to each precipitation station are compared with the observed discharge series, then weights may be determined such that the weighted sum of the calculated discharge series becomes the best approximation of the observed discharge series.

To derive the calculated discharge from precipitation some form of runoff model is necessary. So, we have to start the preliminary runoff analysis using some weights, usually, equal. If some precipitation station is found to be not representative in the course of runoff analysis, it is better to make the weight of this station small or neglect it. After the model has become good enough, the calibration of the weights using the derived runoff model can begin.

Determining the weights of precipitation stations by factor analysis

First, we make monthly or half-monthly sums of the observed discharge and the calculated discharge derived from the precipitation at the I-th station, where I = 1,2,...,N. Let the derived time series of monthly or half-monthly observed and calculated discharge be y(n), and $x_I(n)$ (I = 1,2,...,N), respectively.

Next, these time series are normalized by dividing them by the square root of the mean of square of each component: i.e. $(\sum_n y(n)^2/\sum_n 1)^{1/2}$ and $(\sum_n x_I(n)^2/\sum_n 1)^{1/2}$. Let the normalized time series be Y(n) and $X_I(n)$, respectively.

Then, the problem is to find the best approximation of Y(n) by the linear form $\{\sum_I w(I)X_I(n)\}$, where the coefficients w(I) is the weight of the precipitation station I. This problem can be solved easily by the method of least squares.

We put

$$D(n) = Y(n) - \sum_J^{NP} w(J)X_J(n) \quad (n = 1.2, ..., N).$$

where NP is the number of precipitation stations. Then, the problem is to determine w(I)'s which make $\sum_n D(n)^2$ minimum. To determine w(I), $\sum_n D(n)^2$ is differentiated partially with respect to w(I).

Putting $\frac{\partial}{\partial w(I)}\left(\sum_n D(n)^2\right) = 0$, we get the following equation:

$$\sum_J^{NP} A_{IJ} w(J) = B_I (I = 1, 2, ..., NP),$$

where

$$A_{IJ} = \sum_n X_I(n) X_J(n) / n, \quad B_I = \sum_n X_I(n) Y(n).$$

We had hoped that the solution of this equation would give good weights, but in most cases the solution was meaningless, showing positive and negative numbers with large absolute values. In most cases, the determinant of the matrix (A_{IJ}) is very close to zero, and the roots of the equation are given by the ratio of very small numbers, roughly speaking 0/0. Consequently, they are unreliable. The reason that the determinant $|A_{ij}|$ is very close to zero is that daily precipitation's at stations in or near the basin are usually similar to each other. In other words, they are highly correlated. Therefore, the daily runoffs derived from the precipitation at each station are also similar to each other, consequently, the time series of calculated monthly or half-monthly discharges. $x_I(n)$ (I = 1,2, ,NP) are similar to each other, as are the normalized time series $X_I(n)$. Hence, when we consider the time series $\{X_I(n)\}$ as a vector $\vec{X}_I$, the angle of intersection between any two of these vectors is small. Therefore, all A_{IJ} are close to 1, and, consequently, $\det(A_{IJ})$ shows very small value.

Moreover, if the working model has been calibrated well enough, the daily runoff derived from the precipitation at each station will be similar to the observed runoff. Accordingly, every vector $\vec{X}_I$, (I = 1,2, ,NP) is similar to $\vec{Y}$, i.e. the angle of intersection between $\vec{X}_I$, and $\vec{Y}$ is also small, and, consequently, B_I is close to 1.

To get a meaningful solution to such an equation the method of factor analysis was applied. The matrix (A_{IJ}) was transformed into diagonal form by orthogonal transformation. The derived diagonal elements are called the characteristic values of (A_{IJ}). By neglecting those characteristic values which are close to zero we can get a meaningful solution. Such a treatment is mathematics, not hydrology and so the author will stop the description here. It is not difficult to learn the method of factor analysis from some text book of statistical mathematics.

REFERENCES

Sugawara, M., Maruyama, F., 1952. Statistical method of predicting the runoff from rainfall. Proc. of the 2nd Jap. National Congress for App. Mech: 213-216.

Sugawara, M., 1961. On the analysis of runoff structure about several Japanese rivers. Jap. Jour. Geophys. 2(4):1-76.

Sugawara, M., 1967. The flood forecasting by a series storage type model. International Symp. on floods and their computation. Leningrad, USSR: 1-6.

Sugawara, M., et al., 1974. Tank model and its application to Bird Creek, Wollombi Brook, Bikin River, Kitsu River, Sanaga River and Nam Mune. Research note of the National Research Center for Disaster Prevention, No.11: 1-64.

Sugawara, M., 1979. Automatic calibration of the tank model. Hydrol. Sci. Bull. 24(3):375-388

Sugawara, M., et al., 1984. Tank model with snow component. Research note of the National Research Center for Disaster Prevention, No.65: 1-293.

Sugawara, M., 1993. On the weights of precipitation stations. Advances in theoretical hydrology, European Geophys. Soc. series on hydrological sciences, 1, Elsevier.

Chapter 7

THE XINANJIANG MODEL

R. J. Zhao and X. R. Liu

7.1. INTRODUCTION

The Xinanjiang model was developed in 1973 and published in 1980 (Zhao et al., 1980). Its main feature is the concept of runoff formation on repletion of storage, which means that runoff is not produced until the soil moisture content of the aeration zone reaches field capacity, and thereafter runoff equals the rainfall excess without further loss. This hypothesis was first proposed in China in the 1960s, and much subsequent experience supports its validity for humid and semi-humid regions. According to the original formulation, runoff so generated was separated into two components using Horton's concept of a final, constant, infiltration rate. Infiltrated water was assumed to go to the groundwater storage and the remainder to surface, or storm runoff. However, evidence of variability in the final infiltration rate, and in the unit hydrograph assumed to connect the storm runoff to the discharge from each sub-basin, suggested the necessity of a third component. Guided by the work of Kirkby (1978) an additional component, interflow, was provided in the model in 1980. The modified model is now successfully and widely used in China.

7.2. THE STRUCTURE OF THE XINANJIANG MODEL

7.2.1. The Concept of Runoff Formation on Repletion of Storage

Soil, or indeed any porous medium, possesses the ability of holding indefinitely against gravity a certain amount of water constituting a storage. This is sometimes called "field moisture capacity". By definition, water held in this storage cannot become runoff and the storage can be depleted only by evaporation or the transpiration of the vegetation. Hence evaporation becomes the controlling factor producing soil moisture deficiency.

In humid regions, soil moisture measurements show that the soil moisture can often reach field capacity within the entire soil cover. This implies that the replenishment in the next period of rainfall will be equal to the evaporation in the intervening period. This suggests a very simple relationship between rainfall and runoff viz. that before producing runoff the rainfall must satisfy the deficiency below field capacity which in turn is caused by the evaporation since field capacity last occurred and after that subsequent rainfall will all run off since soil has no more capability to hold them. This is the concept of runoff formation on repletion of storage which is the basic concept of Xinanjiang model.

7.2.2. Outline of the Xinanjiang Model Structure

The basin is divided into a set of sub-basins. The outflow from each sub-basin is first simulated and then routed down the channels to the main basin outlet. Based on the concept of runoff formation on the repletion of storage, the simulation of outflow from each sub-basin is consisted of four major parts:

1) the evapotranspiration which generates the deficit of the soil storage which is divided into three layers: upper, lower and deep;

2) the runoff production which produces the runoff according to the rainfall and soil storage deficit;

3) the runoff separation which divides the above so determined runoff into three components: surface, subsurface and groundwater;

4) the flow routing which transfers the local runoff to the outlet of each sub-basin forming the outflow of the sub-basin.

The flow chart of Xinanjiang model is shown in Fig. 7.1. All symbols inside the blocks are variables including inputs, outputs, state variables and internal variables while those outside the blocks are parameters.

7.2.3. The Variables

The inputs to the model are areal mean rainfall, P, and measured pan evaporation, EM. The outputs are the discharge, TQ, from the whole basin and the actual evapotranspiration, E, which includes three components EU, EL and ED.

The state variables are the areal mean tension water storage, W, and the areal mean free water storage, S. The areal mean tension water W has three components WU, WL, and WD in the upper, lower and deep layer,

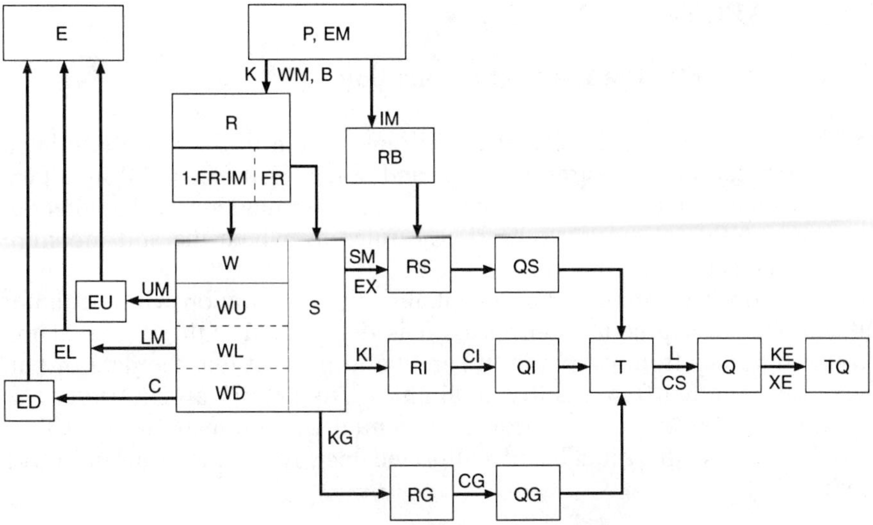

Fig. 7.1. Flow chart for the Xinanjiang model.

respectively. The FR is runoff contributing area factor which is related to W. The rest of the symbols inside the blocks are all internal variables. RB is the runoff directly from the small portion of impervious area. R is the runoff produced from the pervious area and divided into three components RS, RI, and RG referred to as surface runoff, interflow and groundwater runoff, respectively. The three components are further transferred into QS, QI, and QG and together form the total inflow to the channel network of the sub-basin. The outflow of the sub-basin is Q.

7.2.4. Parameters

The characters outside the blocks are all parameters. K is the ratio of potential evapotranspiration to pan evaporation if pan evaporation measurements are used as references.

WM and B are two parameters describing the tension water distribution. WM is the areal mean tension water capacity having components UM, LM and DM. B is the exponent of the tension water capacity distribution curve.

IM is the factor of impervious area.

SM and EX are similar to WM and B while they describe the free water capacity distribution.

KI and KG are coefficients relating to RI and RG.

CI, CG, L, CS, KE and XE are parameters for flow routing.

The meaning of all the parameters will be further discussed in later paragraphs.

7.3. EVAPOTRANSPIRATION

7.3.1. The Potential Evapotranspiration

In Xinanjiang model, the actual evapotranspiration of the basin is related both to potential evapotranspiration and soil moisture condition. Two problems are then raised, of which one is the determination of the potential evapotranspiration and the other is the rules by which the soil moisture governs the actual evaporation.

Direct determination of the potential evapotranspiration will encounter difficulties. For practical purposes, it is determined with the aid of the observed values from evaporation pan due to the fact that the data of pan evaporation are often readily available. An empirical coefficient K transferring the pan evaporation to potential evapotranspiration is thus introduced into the model and calibrated mainly by the trial-and-error method.

7.3.2. Three-Layer Soil Moisture Model

Traditionally, the actual evapotranspiration is treated to be proportional to the soil moisture. This is considered to be one layer soil model.

The main drawback of the one-layer computation lies in the fact that the distribution of moisture in the vertical profile can not be taken into account, which results in considerable error in the actual evaporation calculation, particularly when light rain falls in a long period of drought. This is because that the antecedent soil moisture content is already very low, thus the calculated value of evaporation by one-layer model will be very small. But in reality, the part of soil moisture replenished by new rainfall will be mainly distributed in the topsoil and, in addition, if there is a good vegetative cover, a certain portion of the rain will be intercepted by vegetation. The moisture which remains in the topsoil as well as on the branches and leaves of plants will evaporate at a rate comparable with the potential evapotranspiration, and, therefore, can no longer be taken the same thing as the antecedent soil moisture content. So that a top-layer is necessary to be introduced into the soil model for treading the new rain different from the old soil moisture.

Meanwhile, during dry season, still another error in the calculation of actual evaporation may occur especially in humid and well covered area. In this case, though the topsoil is dried out, the deeper layer may be still moist. The root system of vegetation will grow deeper to absorb the moisture there for its transpiration. Hence the rule of direct proportion no longer holds at this moment. It is necessary to further improve the calculation by putting forward a deeper layer to the model. Thus a three layer soil moisture model was so gradually developed. The rules of calculating the actual evapotranspiration are described below.

7.3.3. Calculation Rules of Evapotranspiration

The total areal mean soil moisture capacity of a basin, WM, is divided into three portions: UM the upper part, LM the lower part and DM the deeper part. WU, WL and WD are storage at any moment corresponding to the three layers. Evaporation occurs at the potential rate, equal to K times the pan evaporation rate in the case of using pan evaporation measurements,

$$EU = K \bullet EM \tag{7.1}$$

until the storage WU of the uppermost layer is exhausted. On exhaustion of the upper layer, any remaining potential evapotranspiration is applied to the lower layer, but the efficiency is modified by multiplication by the ratio of the actual storage WL to the capacity storage LM of that layer.

$$EL = (K \bullet EM - EU) \bullet WL \,/\, LM \tag{7.2}$$

When the lower layer storage WL is reduced to a specified proportion, C, of LM, evapotranspiration is assumed to continue, but at a rate ED given by

$$ED = C \bullet (K \bullet EM - EU) \tag{7.3}$$

7.4. RUNOFF PRODUCTION

The concept of runoff formation on repletion of storage holds true so far as one single point, in the sense of a very small area, of a basin is concerned. For an entire watershed, things are more complicated. The moisture deficit often varies from place to place. This non-uniform distribution effects the runoff production of a whole basin significantly. To solve the problem, in Xinanjiang model, a distribution of tension water capacity is suggested in statistical manner.

7.4.1. Distribution of Tension Water Capacity

To provide for a non-uniform distribution of tension water capacity throughout the basin or sub-basin whatever is concerned, a tension water capacity curve is introduced, see Fig. 7.2.

In Fig. 7.2, f/F represents the proportion of the pervious area of the basin whose tension water capacity is less than or equal to the value of the ordinate W'M. The tension water capacity at a point, W'M, varies from zero to a maximum MM according to the relationship

$$(1 - f \,/\, F) = (1 - W'M \,/\, MM)^B \bullet (1 - IM) \tag{7.4}$$

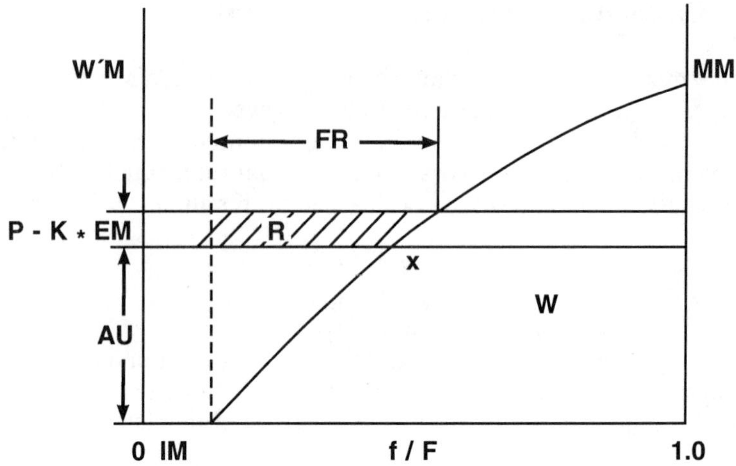

Fig. 7.2. The distribution of tension water capacity in the basin.

where MM and B are parameters.

The areal mean tension water capacity, WM, constitutes an alternative parameter to the maximum value MM. These are related through the parameter B. From eq. (7.4), by integration, it is easy to show that

$$MM = WM \bullet (1 + B) / (1 - IM) \tag{7.5}$$

The state of the basin, at any time, is assumed to be represented by a point x on the curved line of Fig. 7.2. The area to the right and below the point x is proportional to the areal mean tension water storage W (not capacity). This assumption implies that each point in the basin is either at capacity tension (points to the left of x) or at a constant tension (points to the right of x).

7.4.2. Runoff Production in a Basin

When rainfall exceeds evaporation, the ordinate of Fig. 7.2 is increased by the excess, x moves upwards along the curve and runoff is generated proportional to the area shown shaded to the left and above the point x in Fig. 7.2.

If, P - K • EM + AU is less than MM, then

$$R = P - K \bullet EM - WM + W + WM \bullet \left(1 - (P - K \bullet EM + AU) / MM\right)^{(1+B)} \tag{7.6}$$

otherwise

$$R = P - K \bullet EM - WM + W \tag{7.7}$$

7.4.3. State of a Basin

When evaporation exceeds rainfall, the tension moisture storage is reduced and the point x moves downwards along the curve of Fig. 7.2 to a level at which the areal mean tension water storage W (the area to the right and below the point x) assumes its appropriate value. It is perhaps worth noting that this implies a redistribution of water within the basin. If, initially, the tension water state of the catchment is represented by the curved line to the left and the horizontal line to the right of x in Fig. 7.2, a reduction in tension water storage at all points in the basin would be represented by a constant downward shift of the curved and horizontal lines. Instead, however, no reduction is imposed on points to the left and below x --these points remain at capacity level - - and a correspondingly greater reduction is imposed over the remainder of the basin, implying a greater reduction in the position of the horizontal line. This implies a redistribution of soil moisture during the drying period, with water flowing from the more elevated parts of the basin to the lower parts.

7.5. SEPARATION OF RUNOFF COMPONENTS

The runoff R, generated in a wet period in accordance with Fig. 7.2, should be further separated into its three components, RS surface runoff, RG ground water runoff, and RI interflow. To effect this, the concepts of free water storage and its distribution are developed.

7.5.1. Distribution of Free Water Capacity

It is assumed in Xinanjiang model that the free water storage and its capacity are non-uniformly distributed over the area, FR, on which the runoff is currently produced. The distribution curve is illustrated in Fig. 7.3. The free water storage capacity S'M are assumed to be distributed between zero and a point maximum MS in a parabolic manner, over FR.

$$(1-f/FR)=(1-S'M/MS)^{Ex} \tag{7.8}$$

where f is the portion of the basin area for which the free water storage capacity is less than or equal to S'M and Ex is a parameter.

It is further assumed that the current state of free water storage in the basin can be represented by a point (ordinate BU on the parabola of Fig. 7.3) implying that the portion of the basin to the left of that point is at capacity storage and to the right the storage is constant, below capacity level.

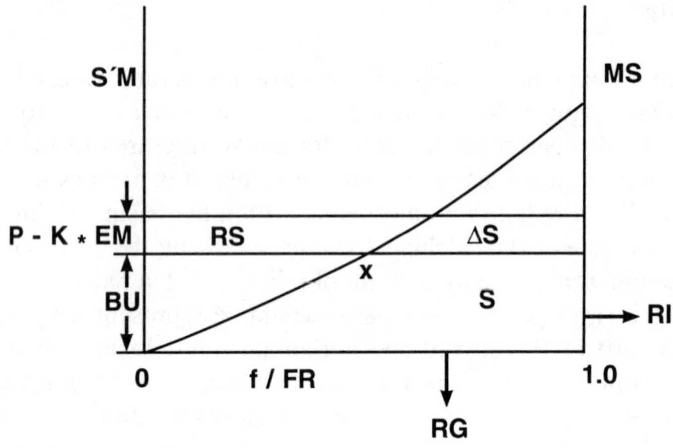

Fig. 7.3. The separation of runoff components.

The areal mean free water storage capacity SM may be used instead of MS as a parameter

$$SM = MS/(1+Ex) \tag{7.9}$$

By integration of S'M in eq. (7.8) and substitution of SM for MS from eq. (7.9), the equivalent free water storage S over the runoff producing area FR, can be found from the equation below

$$1-S/SM=(1-BU/MS)^{(1+Ex)} \tag{7.10}$$

7.5.2. Runoff Separation

The runoff R generated in accordance with Fig. 7.2 and expressed as the depth P - K • EM over the runoff producing area of the basin is applied by adding P - K • EM to BU in Fig. 7.3, yielding a contribution RS to surface runoff.

Algebraically, if BU + P - K • EM < MS, then

$$RS=\left(P-K\bullet EM-SM+S+SM\bullet(1-(P-K\bullet EM+BU)/MS)^{(1+Ex)}\right)\bullet FR \tag{7.11}$$

Otherwise,

$$RS=(P-K\bullet EM+S-SM)\bullet FR \tag{7.12}$$

The remainder of R becomes an addition, ΔS, to the free water storage S, which in turn contributes RI laterally to inflow and RG vertically to groundwater, according to the relations

$$RI = S \bullet KI \bullet FR \tag{7.13}$$

$$RG = S \bullet KG \bullet FR \tag{7.14}$$

where KI and KG are parameters.

7.6. FLOW CONCENTRATION

After the runoff R is produced and separated into three components RS, RI, and RG, a concentration processes begin. The three components may firstly take different ways from the place where they produced to the local streams, then flow together towards their sub-basin outlets, forming the outflow of that sub-basin, thereafter, move along the major channels and finally concentrate at the main outlet producing the total outflow of the whole basin. The entire process may be roughly divided into three stages. The first stage deals with mainly the water movements in the headwater area including the overland flow, subsurface flow and groundwater flow etc. The second stage involves basically the streamflow in tributaries and the third one concerns the flow routing along major rivers.

7.6.1. Flow Concentration in Sub-Basins

The surface flow is often several orders faster than the interflow and groundwater flow. Therefore, the surface runoff is treated to be unmodified passing over the hillslope and the headwater area into the channel systems as TS. While the interflow RI and the groundwater RG are routed through linear reservoirs representing interflow and groundwater storage respectively. Outflow TI and TG from these reservoirs, are determined by

$$TI(t) = TI(t-1) \bullet CI + RI(t) \bullet (1-CI) \tag{7.15}$$

$$TG(t) = TG(t-1) \bullet CG + RG(t) \bullet (1-CG) \tag{7.16}$$

where t is current time, t-1 is one time interval before. TI and TG are added to TS becoming the total sub-basin inflow T to the channel network.

The channel network routing within a sub-basin is represented by the convolution of T with an empirical unit hydrograph or by "lag and route" model with parameters L and CS. Here L is the lag time and CS is storage coefficient of the linear reservoir. The results of this convolution or routing are outflow Q from the sub-basin.

7.6.2. Flow Routing Along Main Rivers

Flow routing from the sub-basin outlets to the whole basin outlet may be achieved by applying any flood routing methods. Here in Xinanjiang model, the Muskingum successive reaches model is preferred to employ. KE and XE are Muskingum parameters for each sub-reaches.

7.7. DISCUSSION ON PARAMETERS

While all the parameters have clear physical meanings, their determination by measurement in the field is nevertheless impractical and recourse must be had to evaluation by some form of system identification. This is quite a difficult task, made more so by uneven sensitivity in the computed output to changes in the parameter values and dependence among the parameters so that the effect of changing the value of one may be offset by a corresponding change in another. Parameters for changes in which the output is insensitive, can be given fixed values in accordance with experience but dependence results in non-uniqueness and instability in optimized values.

In the model as described, there are 17 parameters when using lag and route method for sub-basin's and Muskingum method for main basin's flow routing, among which 15 parameters are for the sub-basin. The 15 parameters may be grouped as follows:

1) Evapotranspiration parameters K,UM,LM,C.
2) Runoff production parameters WM,B,IM.
3) Parameters of runoff separation SM,Ex,KG,KI.
4) Runoff concentration parameters CG,CI,CS,L.

Generally, the output is more sensitive to 7 parameters. They are K, SM, KG, KI, CG, CS and L.

7.7.1. Evapotranspiration Parameters

K is the ratio of potential evapotranspiration to pan evaporation. The output is particularly sensitive to this parameter which controls the water balance. K may be looked upon as the product of a number of factors. One of the factors is the ratio of pond evaporation to pan evaporation, K1, which is dependent on the type of pan, the installation of the instrument and environmental conditions of the site, etc. K1 can be determined by experiments and numerous data have been cumulated and analyzed.

The another one is the ratio of potential evapotranspiration to pond evaporation, K2, which is usually taken as 1.3-1.5 in summer and 1.0 in winter. Still another one is a coefficient, K3, converting the measured pan value to the basin mean value. K3 is dependent mainly on the elevation of the measuring site relative to the mean elevation of the whole area being considered.

WM, the areal mean tension water capacity, is the sum of UM in the upper layer, LM in the lower layer and DM in the deeper layer. DM is therefore completely dependent on the other three and need not be considered for optimization. WM is a measure of aridity, which varies from 80 mm in South China to 170 mm or more in North China. The model operation is generally insensitive to WM, provided its value is large enough to ensure that the computed areal mean soil moisture content W does not become negative. Parameters UM and LM are determined by experience. Typical values for UM are from 5 to 20 mm, for deforested to forested areas. LM, typically 60 to 90 mm, is taken as the range within which it is assumed that evapotranspiration is proportional to soil moisture content.

The coefficient of deep evapotranspiration, C, depends on the proportion of the basin area covered by vegetation with deep roots. It varies from 0.18 in South China to 0.08 in North China. The runoff is insensitive to C in humid areas and humid seasons, but quite sensitive in arid areas and dry seasons. The appropriate value of C is often dependent upon the sum UM+ LM but as this is usually kept to a fixed value of approximately 100 mm, the appropriate value of C can usually be determined.

7.7.2. Runoff Production Parameters

The distribution of tension water capacity completely determines the rainfall runoff relation according to the concept of runoff formation on repletion of storage. Three parameters IM,WM, and B describe the distribution curve as shown in Fig. 7.2. WM has been discussed above. IM is the ratio of the impervious to the total area of the basin. For natural basins in humid regions this is usually negligible, but in semi-humid or more arid regions the impervious area may be a significant portion of the runoff producing area of the basin and then could not be neglected.

The exponent of the tension water capacity curve, B, defines statistically the non-uniformity of the surface conditions mainly including the factors such as topography, geology, soil type and vegetative coverage. Experience indicates that for basins of area less than 10 km^2, B = 0.1 and for basins measured in thousands of square kilometers, B = 0.4 or even more.

7.7.3. Runoff Separation Parameters

The areal mean of the free water capacity of the surface soil layer, SM, represents the maximum possible deficit of free water storage. Surface runoff is sensitive to the value of this parameter. For thin soils, SM may be approximately 10 mm or less, increasing to 50 mm or more for thick and porous surface soils.

The exponent of the free water capacity curve, Ex, influences the development of the saturated area. Ex is dependent on SM but statistical analysis shows that the value of Ex lies in a range from 0.5 to 2.0, the best selection being between 1 and 1.5. It may be taken as a constant.

KG and KI are the outflow coefficients of the free water storage to groundwater and interflow. The sum of them determines the drainage rate from free water storage and their ratio determines the proportion going to interflow and groundwater flow, respectively. If the sum KG + KI increases and SM decreases while keeping the ratio KG/KI constant, the distribution of the runoff among the three components may be change little, if any. Insofar as the operation of the model is sensitive to these parameters, this dependence is undesirable. It may be avoided by fixing the sum KG + KI. As the recession duration of the upper interflow storage ordinarily lies between 2 and 3 days, it is suggested that the sum KG + KI may be taken as 0.7 - 0.8.

7.7.4. Flow Concentration Parameters

The daily recession constant of groundwater storage, CG, can be easily determined by the recession curve during the dry season and usually takes a value between 0.99 and 0.998. CI, the recession constant of the lower interflow storage, usually takes values between 0.5 and 0.9 in humid regions.

CS, the recession constant in the " lag and route " method for routing through the channel system within each sub-basin, is purely empirical. While this method of routing is not theoretically strong, it is easier to apply than the unit hydrograph. L is the corresponding " lag ", also of empirical value.

KE and XE are parameters of the Muskingum method which can be initially determined by hydraulic formulas. KE may be approximately take the travel time over a sub-reach. XE may be determined by the formula as below

$$XE = 0.5 \bullet \left(1 - \ell / LE\right) \tag{7.17}$$

where ℓ is the specific length defined by Kalinin (Kalinin and Milyukov, 1958), LE is the length of the subreach.

7.8. THE MODEL CALIBRATION

7.8.1. Principles of Calibration

The calibration of a conceptual model is a very difficult problem. A methodology of decomposition of system and degeneration of dimensionality has been developed in the calibration of Xinanjiang model. By decomposition, the model structure is analyzed hierarchically. Then the calibration is carried out from lower to higher hierarchy and use different objective functions for different levels. By degeneration, it means to reduce the number of parameters being optimized simultaneously at least to three or less. The reduction is based on the sensitivity analysis and structure constraint.

The foregoing mentioned classification of the parameters in four groups is indeed an hierarchy of Xinanjiang model structure and corresponds to different characteristics in optimization. Those of the first group are most stable and of the fourth group least so. Parameters within the same group tend to be mutually dependent. The parameter values in the higher groups have little effect on the optimized values in the lower groups, For example, variation of the parameter values of runoff concentration within a reasonable range, will not influence the optimum value of the parameter for runoff separation, and variation of the parameter value of runoff separation will not greatly influence the optimized value of the parameters of runoff production and evapotranspiration. Therefore, parameters of the lower numbered groups are optimized before those of the higher numbered groups and different objective functions are used for different groups.

This principles and process of parameter calibration has been successfully used in more than 70 basins in China.

7.8.2. Objective Functions

A comparison test has been made about the effect of objective function on the accuracy of parameter calibration. It shows that the results obtained by using absolute error as objective function are much more stable than that on least squares. For the four groups three objective functions are suggested:

1) The water balance error, being the difference between the total computed and the total measured flow.

$$\Delta R = \Sigma Q - \Sigma M \tag{7.18}$$

where ΣQ is the total computed and ΣM the total measured flow in the period of calibration.

2) The relative absolute error, being the sum of the absolute errors of prediction normalized by division by the sum of the measured flow

$$ABS = \Sigma abs(M - Q) / \Sigma M \tag{7.19}$$

3) The mean relative error

$$MRE = \Sigma abs(M - Q) / M / n \tag{7.20}$$

where n is the number of ordinates. In this objective function, the error occurring in low flow is emphasized by weighing with the reciprocal of its flow.

7.8.3. Calibration Procedures

The calibration proceeds as follows:

1) Assume initial values for all the parameters using reasonable values as far as possible.

2) Calibrate K alone, using daily data and the objective function of eq. (7.18). Normally at least 4 years of data are necessary for calibration. If in the course of running the model, it is observed that W becomes negative WM should be increased accordingly. Usually UM and LM may be chosen by experience and C may be adjusted after calibration of K.

3) By degeneration principles mentioned before, taking a fixed value (0.7 or 0.8) for the sum of KG and KI and fixing Ex at 1.5, the remaining parameters of the third group KG/KI and SM are optimized by a two-dimensional search using the objective function of eq. (7.20). Subsequently, the values of CG and CI may be adjusted.

4) Apply the model to some flood events only, and using very short time intervals with an adjusted value of SM, L and CS are optimized using the objective function of eq. (7.19). The adjustment to the previously optimized value of SM is necessary because, due to the unevenness of the distribution of rainfall during a 1-day period, SM for daily values is always less than the value appropriate to shorter time intervals.

7.9. THE MODEL SCALES

The Xinanjiang model is rather flexible both in time scale and spacial scale. The time scale of Xinanjiang model can be hourly for flood events forecasting, daily for continuous simulation and monthly for even longer term assessment. The Xinanjiang model may also be applied for a wide range of spacial scales from a small plot to a very large river basin since it contains sub-basin division in its structure. Therefore, the behaviors of the model when crossing a range of scales should be investigated and indeed it is an essential problem when dealing with regional and macroscale hydrology.

7.9.1. Time Scale

All parameters related to rates will change with the time scale that the model based on. Some of them are only pure converting with time interval. For example, CG, the coefficient of groundwater recession, for event-based model should be raised to the n-th power before applied to daily model, in which n is the ratio of 24-hour to the time interval used in event model. But some are really changed or even canceled as the time scale changing.

The flow chart shown in Fig. 7.2 describes the model structure basically for event based model. As the model scaling up to daily-base, the portion of model dealing with flow routing from the outlets of sub-basin to the outlet of main basin will often vanish out. This is because of the fact that, in common cases, the total travel time over major rivers in a basin is less than 24-hour, which means that the whole basin may be taken as one unit and no necessary to divide it into sub-basins in the sense of routing. Hence, in a daily model, KE and XE are not necessarily taken account.

As the time scaling up further to a monthly-base, the model structure may get even simple. In the case of monthly time scale, the difference between surface flow and interflow will not be able to identify. Therefore, KI and CI automatically have no meaning and can be diminished out from the model structure and only two components remain. One is the quick response with a time lag less than one month and the other is slow response with a lag longer than one month. In this case, the parameters of SM and KG will change accordingly. Since process of the free water percolating into local aquifer will mostly takes place within one month, KG in monthly model will be equal or nearly equal to 1. While SM in monthly model will be even less comparing with those in daily and event model, again because of the unevenness of the rainfall pattern in one month.

7.9.2. Space Scale

B, the power of the tension water capacity curve, describes statistically the non-uniform distribution of the tension water capacity of soil column by definition. In this sense, B would not be very sensible to the space scale as long as the landscape is statistically uniform within a basin. Nevertheless, B is calibrated by using rainfall and runoff data. In the case, it becomes an integral index representing all factors affecting the spacial distribution of runoff producing area. In view of this, B reflects mainly the pattern of land surface characteristics for smaller area, say less than hundred square kilometers, while for larger area, say over thousands of square kilometers, it represents the heterogeneity of the land surface as well as the spacial non-uniform distribution of rainfall.

In traditional hydrologic practice, with the spacial scales within thousands of square kilometers, B is slightly varies with the scale from 0.1 to 0.4 as mentioned before. However, when large scale hydrological modeling is dealt with, say over tens of thousands square kilometers, B increases significantly to 1 or 2 even more. In the later case, the non-uniform rainfall distribution might play a major role.

7.10. REGIONAL DISTRIBUTION OF THE PARAMETERS

Xinanjiang model is essentially based on a concept of storage controlling. The deficit of tension water storage determines the portion of rainfall which remains in the soil and the portion which runs off early or later. While the capacity of free water storage controls the division of runoff components. Free water storage with smaller capacity have less room for storing the runoff produced and draining it slowly later on. Correspondingly, in this situation, more runoff will flow immediately out of the places where they are produced and form the so called surface runoff component. On the contrary, large free water capacity will produce more slower components of runoff such as interflow or/and groundwater flow. Thus the characteristics of the two storages, tension water and free water, are the most vital features control the hydrologic behaviors of a basin. They are represented by the parameters WM and SM respectively in Xinanjiang model.

The value of WM in Xinanjiang model is actually represents the portion of tension water which is most active during the hydrologic process. That part of tension water which has never been evaporated plays no role in the hydrologic process and just like the dead storage in a reservoir. In view of this situation, WM depends not only on the soil layer itself but on the climate condition. Hence, generally, WM in humid region is smaller than that in arid region. But if there is thin soil cover or there is frozen soil then no matter how dry the climate is WM might be small and the region might be a easy producing area for runoff.

The SM represents the free water capacity which is sensitive for runoff components division. Smaller SM produces larger portion of surface runoff and less groundwater and vise versa. SM depends on the features of soil cover as well as the characteristics of the bedrock. For impervious bedrock and thin soil cover, SM might be very small and then most of runoff produced might go as surface runoff. In the case, the flood peak per unit area may be very high and the low flow is poor. While for pervious bedrock and thick soil cover, which corresponds to a high value of SM, situation might be quite different. In the case, surface runoff is a small portion in the total runoff and ground water is abundant. In reality, there are a range of situation between the two extremes combined by different soil cover and bedrock. SM is able to reflect this feature. An analysis based on more than 50 basins has proved that SM is really something related to the geological condition of the basin.

Apart from WM and SM, the potential evapotranspiration is also a vital element of the model. To avoid relying too much on the measured pan evaporation, a relation between solar radiation and potential evapotranspiration has been studied and proposed to take the place of empirical coefficient K.

Moreover, a study has also been carried out on the parameters of concentration L and CS associated with the morphology of river networks and useful results are obtained, by which the values of L and CS may be estimated by the morphology of river networks.

7.11. THE MODEL APPLICATION

The efficiency of the Xinanjiang model has been established by long use in China. The model has been applied successfully over a very large area including all of the agricultural, pastural and forested lands of China except the loess.

In China, the Xinanjiang model is currently used mainly for hydrological forecasting. Many kinds of software have been developed. Many large projects such as Gezhouba, Panjakou, Danjiangkou, Lubuge, Longyangxia and Ertan have involved the use of this model. Many forecasting systems such as Sanmenxia to Huayuankou on the Yellow River, the middle reach of the Huai River and the Yangtze Gorges have used the model, sometimes with adjustment in real time.

Use of the model has also spread to other fields of application such as water resources estimation, design flood and field drainage, water project programming, hydrological station planning, water quality accounting, etc. The model is being extended and developed to meet with miscellaneous surface conditions such as snow cover, karst, large plains, swamps, etc.

More recently, the model has found its usage in the macroscale hydrological modeling with its tension water capacity curve as a kind of parameterization and its regional distribution of parameters.

7.12. CONCLUSION

A most important task for us now is to strengthen the scientific basis of the technique, to prove the model structure, and to find relations between parameter values and natural conditions. Any conceptual model may be considered as a scientific hypothesis to be subjected to scientific examination. High accuracy alone is insufficient for many purposes. The final criterion must be whether the model structure and parameter values accord with the hydrological reality and the natural conditions. By the application of this criterion, the scientific status of modeling will be raised and scientific hydrology stimulated.

REFERENCES

Zhao, R..J., Zhuang, Y.-L., Fang, L.-R., Liu, X.-R.and Zhang, Q.-S., 1980. The Xinanjiang Model, Hydrological Forecasting Proceedings Oxford Symposium, IAHS 129, pp. 351-356.

Zhao, R.-J., 1992. The Xinanjiang model applied in China, Journal of Hydrology, 135 (1992), pp. 371-381

Zhao, R.-J., 1993. A non-linear system model for basin concentration, Journal of Hydrology, 142 (1993), pp. 477-482.

Chapter 8

THE UBC WATERSHED MODEL

M. C. Quick

8.1. INTRODUCTION

The UBC Watershed Model has been designed for estimating streamflow runoff using meteorological inputs of precipitation and temperature. Principally, the model is designed for mountain runoff prediction, in which runoff is generated by snowmelt, glacier melt, and rainfall. The model design uses elevation zones because the strongest gradients of behavior in mountain regions are the orographic temperature and precipitation gradients, and these gradients tend to follow similar patterns of behavior for each major storm and also exhibit systematic variation with variations in winter and summer weather conditions. For example, winter precipitation tends to increase more with elevation than for connective summer rain storms. Given continuous meteorological input data, the model will operate continuously, accumulating and depleting snowpacks and producing estimates of streamflows. For calibration and verification purposes, the model uses historical meteorological and streamflow records as reference data and calculates performance statistics of volume and hydrograph shape reconstitution.

The earliest version of the UBC Watershed Model dates back to 1963 when work started on the forecasting of floods in the Fraser River in British Columbia. Earlier, in 1948, a severe flood occurred which caused many millions of dollars damage. The BC Disaster Relief Fund gave help to many people who had suffered flood damage and loss, and residual funds were invested to support flood research at the University of British Columbia. This research effort resulted in the development of the UBC Model which was tested and found to be reasonably accurate during the high flood flows experienced in 1972 and 1974. The BC Water Resources Service then formed its own flood forecasting group and adapted the UBC Model for its own forecasting work. Arising from this work, from 1976 onwards, the model has been used by B.C. Hydro & Power Authority to forecast the Upper Columbia river to assist in reservoir operation for hydro-electric power generation. Further research has continued at UBC, resulting in the development of the present model which is being widely used in British Columbia and other areas, including the Himalayas in India and Pakistan.

The model incorporates a user-friendly interface and a semi-automatic calibration routine. In addition, there is easy and flexible access to graphical presentation of all aspects of watershed behavior and to statistical evaluations of model performance. Computing speed is fast and data can be accepted in a great variety of formats, but internally the model converts all files to binary, which again enhances computing speed, which is especially important for real-time operation and when carrying out semi-automatic calibration. In addition to streamflow, the model also provides information on the accumulation and depletion of snowpack, soil moisture budget, soil and groundwater storage values, contributions to runoff from various portions of the watershed, and surface and sub-surface components of runoff. These data can be viewed on the computer screen in color graphics, or can be printed or saved to a file.

The model will operate with any selected time step between 1 hour and 24 hours, and this selection is determined by the user based on data availability and by watershed response. The time step choice may also be influenced by the amount of reservoir or lake regulation in the system, because detailed flow variations will be damped out by the storage effects.

The model is designed so that many aspects of watershed behavior are pre-calibrated. For example, the snowmelt estimation routine has been pre-calibrated using data from the U.S. Corps of Engineers snowmelt study and other reliable data from mountainous watersheds. Similarly, evaporation and temperature lapse rate algorithms are also pre-calibrated. In addition, other parameters controlling precipitation gradients, partition into surface and sub-surface runoff and time distribution of runoff are all given default values which usually provide a good basis for the start of calibration.

8.2. AN OVERVIEW OF THE MODEL DESIGN

The model has seven major components which form a logical subdivision of the hydro-meteorological modelling and evaluation process, and these components are:

(1) **The meteorological sub-model**, which distributes the input data to all elevation zones of the watershed. This distribution process is the most important part of the total modelling process because (a) it controls the total volume of moisture which is input to the model, and (b) it specifies the variation of temperature with elevation, which controls whether precipitation falls as rain or snow and also controls the melting of the snowpacks and glaciers. As will be discussed later, this vital estimate of how precipitation and

temperature varies with elevation can sometimes be estimated directly from high elevation data, but such data is scarce and may only be available for a few special watersheds.

(2) **The soil moisture sub-model** controls the evaporation losses and the subdivision of the rainfall and snowmelt into the four components of runoff, which are fast, medium, slow and very slow components. The components can be thought of as surface, or near surface runoff, interflow, and upper and lower groundwater runoff. The model computes the soil moisture deficit and it is this deficit which controls the non-linear subdivision of rain and melt into the runoff components. The non-linear hydrologic response of the watershed is therefore determined by this soil moisture budgeting. There is also a flash runoff mechanism which operates when rainfall exceeds a certain threshold, and this produces increased fast runoff response.

(3) **The watershed routing sub-model** determines the time distribution of runoff. Each of the four components of runoff determined by the soil moisture sub-model are subjected to storage routing using either cascades or single linear reservoirs. Because these reservoirs are linear, conservation of mass is guaranteed and an accurate water budget is maintained. It has been found that this linear routing works well because the non-linear watershed response is handled by the soil moisture sub-model. Typically the time constants for this time distribution are of the order of one day or less for the fast component, five to ten days for the medium component, thirty to fifty days for the slow, upper groundwater component and one hundred to two hundred days for the very slow, lower groundwater component. It is because these time constraints are so distinctly different that it is possible to carry out a successful calibration of these components.

(4) **The output and evaluation sub-model** is designed to give flexible access to many aspects of the calculated watershed behavior. This information can be examined either in numerical format or graphically and, in addition, the data can be analyzed statistically. Part of this statistical output is used by the calibration sub-model.

(5) **The semi-automatic calibration sub-model** requires some user guidance to ensure that parameters are restricted to reasonable ranges. The calibration process is a constrained iterative search optimization which evaluates a maximum of four parameters at a time. Each of the three primary sub-models, meteorological, soil-moisture and time distribution, are sequentially optimized within

the constraints set by the operators. This calibration routine has proved to be a very powerful aid to the calibration process.

(6) **The updating sub-model** is based on a combination of feedback information from flow measurement and snowcover data from snowcourse or satellite. Both error sources must indicate a similar correction requirement before an adjustment is made. A stochastic updating procedure can be used externally to the model and this operates entirely from flow measurement information (Hamed & Quick 1991 a & b).

(7) **The routing sub-model**, based on the UBC Flow Model, combines watershed flows and routes these flows through a river, lake and reservoir system.

More information on the model design is given in the later section which describes the model algorithms.

8.3. USER INTERFACE

The User Interface gives simple menu driven access to all aspects of the model. For example, the editing of the program control file is menu controlled, and this avoids corruption of the file by inadvertent errors in typing. Calibration is also handled from the main menu and is semi-automatic, requiring some guidance from the operator, who must specify ranges for the parameters being optimized. The optimization will then run automatically and will update the model with the best set of parameters for that particular search, and will also display the model performance statistics for the top ten sets of parameter values.

There are numerous aspects of model operation controlled from the menu and these aspects are summarized below under the main menu headings:

UBC MODEL - gives access to the main program that runs the watershed model using the selected watershed control file (*.WAT) and produces a corresponding calibration file (*.CAL) which contains all the information on calculated flows and other aspects of the watershed status for each time interval; - it also provides access to the Optimization option for calibrating a watershed; - and finally the results of the optimization can be viewed.

GRAPHICS - any of the calculated watershed output information stored in the watershed calibration file (*.CAL) can be displayed graphically and compared with measured information; - a separate routine allows the user to compare the calculated snowpacks with measured snowcourse data.

STATISTICS - generates statistical information from the selected watershed calibration file and displays the output of the statistics file (*.STA) on screen.

EDIT WAT - lets the user edit the selected watershed control file (*.WAT), which specifies the physical characteristics of the watershed, the data inputs and all necessary information for the computer run.

UTILITY - provides the following utilities:

PRINT - prints selected information from the output files, watershed control files and statistics files, (*.CAL, *.WAT, *.STA).

CREATE.SNC - provides access to the snowcourse data files so that comparison can be made with the snow depths computed by the model for this same elevation in the watershed.

*.AES DATA ENTRY - allows manual input of meteorological data.

*.WSC DATA ENTRY - allows manual input of streamflow data.

*.SNC DATA ENTRY - allows manual input of snowcourse data.

DATA FILE CONVERSION - checks a new data file and/or converts it to the correct format.

8.4. USES OF THE WATERSHED MODEL

The Watershed Model was initially designed primarily for short term river flow forecasting, but, because the mountain runoff is generated from snow melt, the model was found to be very useful for estimating seasonal runoff volumes and the probable pattern of runoff. The accuracy of these forecasts depends upon:

- the current assessment of the snowpack, soil, and groundwater storage,
- the representativeness of the meteorological data to date, and
- the accuracy of the meteorological forecast.

When forecasting for hydro-electric generation, two types of forecast are required, the first being short term for immediate decisions, and the second being seasonal forecasts, which give estimates of seasonal volumes and also the probable time distribution of this seasonal runoff. Because the Watershed Model operates continuously if given continuous meteorological input, it is possible to use the model for either this short term or for the longer term forecasting by using projected patterns of weather data. This model capability can assess the range of possible outcomes for a whole season of snowmelt and rain runoff. This forecasting depends on an accurate assessment of the winter accumulation of snowpacks and this watershed status is established by the model, using data for the winter period and up to the present forecast date. Such seasonal forecasts can be upgraded with the recorded data as the season advances and the possible range of outcomes will gradually narrow. An extension of this mode of operation has been used to estimate complete years of data when the streamflow records have not existed, but for which meteorological data has been available.

8.5. DESIGN CONSTRAINTS

The Watershed Model was developed to describe and forecast watershed behavior in mountainous areas. This introduced several important design constraints because data in such regions are usually scarce, particularly at higher elevations. A major design consideration resulting from this was to provide the model with the ability to interpret meteorological data at a point in terms of basin-wide conditions. This requires assumptions to be made concerning temperature and precipitation distribution by elevation and by area. The next section describes the algorithms which carry out these functions. An important part of model calibration is the determination of suitable parameter values to describe these processes.

Another constraint is the availability of streamflow reference data used in the calibration process. For headwaters and major tributaries, independent streamflow data are usually available. For many smaller tributaries and for watersheds downstream of the headwaters no independent streamflow records may exist. To calibrate such watersheds it is necessary to use a streamflow routing model to separate out the streamflow contributions from each sub-area. These derived streamflow records will generally be lower in quality because distributed lateral inflows to a channel system may have to be treated as a lumped inflow and also measurement errors at the two reference stations will tend to interact to produce an enhanced random error component. The UBC Flow Model was developed to carry out this streamflow and lake routing and is available as sub-model.

A further design consideration was to provide the model with the ability to estimate snowpacks. Two data sources are available, precipitation measured at meteorological stations which are usually valley sites, and snowcourse measurements which are available at various elevations, from valley floor to higher snow fields. Snowcourse data are usually available only for certain dates during the snowmelt season, perhaps at two-week intervals. In addition, snowcourse data cannot be considered an absolute measure of snow at that elevation, but they are usually reasonably reliable indices of snow accumulation. In contrast to snowcourse data, precipitation measurements are available on a daily basis at all seasons of the year. For this reason, the Watershed Model uses daily precipitation inputs to estimate snow accumulation and rainfall amounts by elevation. The snowcourse data is used, external to the model, as calibration reference data. This measured snowcourse data is compared with the snowpacks calculated by the model, and these comparisons show a reasonably consistent pattern from year to year, perhaps not in absolute amounts, but they tend to show a consistent pattern from year to year with computed snow wedges. Snowpillow data is now available in many watersheds and this data is often available continuously from satellite data transmission. In the future, this data may be used to give a direct estimate of snowmelt inputs, but this use will depend on proven data reliability. At the present time, the existing snowmelt estimation method is still considered more reliable.

8.6. OVERVIEW OF SETUP PROCEDURES FOR THE WATERSHED MODEL

Before streamflow forecasting can take place three preliminary steps must be completed. These are:

- obtain historical streamflow and/or meteorological data for the region and enter them into the computer in the appropriate file format,
- create a description of the watershed and enter it into the computer (in the parameter file Template.WAT),
- calibrate system parameters to reflect the conditions existing in the watershed so that streamflow forecasts produced by the system using the description are consistent with historical data.

To develop a description of a watershed, the user must first examine topographic maps and aerial photographs of the region in order to determine the number and area of elevation bands, north-south orientation, percentage of forest cover, and so on. Once such issues are decided, the information is entered into the computer in the watershed control file. This process is menu driven, so that the operator is guided through the procedure.

Within the Watershed Model, streamflow information is contained in a file having the extension .WSC. The file name is entered by the user and should identify the watershed with which it is associated. Meteorological information, obtained from weather stations in the watershed, is stored in files with the extension .AES. Again, the file name is entered by the user and should distinguish one station from the other. The Watershed Model will handle data sets from a maximum of five meteorological stations and one streamflow (WSC).

When calibrating the Watershed Model, the model is supplied with a first estimate of parameters in the watershed control file and also with the meteorological data for the calibration period. The model is run to produce estimates of streamflow for each time step and these estimates are then compared statistically or graphically with the recorded streamflow. The calibration process is described more fully later, where both manual and semi-automatic calibration is discussed.

8.7. HARDWARE REQUIREMENTS

The Watershed Model has been developed for use on the IBM PC family of computers and compatibles, running the MS-DOS operating system. The following is a list of the minimum equipment required.

Computer: IBM 386- 33 MHz with a Math Co-Processor installed. Lesser machines will still run the model but will be slower.

Operating System: MS-DOS version 3.3 or higher.

Floppy Drive: 1.2 MB high density 5 1/4" floppy drive or 2 MB 3 1/2" floppy drive (for 1.44 formatted disks).

Hard Drive: Minimum 10 MB hard drive (20MB recommended) with at least 2 mb free after the installation of the UBC Watershed Model.

Graphics Card: Minimum IBM Enhanced Graphics adaptor (EGA), preferable with a color monitor. A monochrome monitor can be used if driven by a color graphics card or simulator.

Printer: LaserJet IIIP or any IBM compatible 9 pin dot-matrix printer.

8.8. THE USER MANUAL

This chapter is intended to give an overview of the watershed model and its use. More detailed information is presented in the User Manual which deals with the technical requirements for data formatting, watershed control file set-up, parameter ranges, etc. In addition, the software itself is largely self-documenting, so that the user is supplied with much of the information necessary for running the model.

8.9. SUMMARY OF WATERSHED MODEL STRUCTURE

The watershed Model is designed to operate on a variable time step which can range from one hour to 24 hours, depending on the availability of the meteorological data inputs of maximum and minimum temperatures and precipitation. From these inputs the Watershed Model calculates estimates of daily watershed outflows. Additional information is available on the accumulation and depletion of snowpack, the soil moisture status, the various soil and groundwater storage values and information on the contributions to runoff from various portions of the watershed and various surface and sub-surface components of runoff.

The watershed model is designed primarily for mountainous watersheds and calculates the total contribution from both snowmelt runoff and rainfall runoff. A separate calculation can also be made of runoff occurring from glacier covered areas. The basic structure of the model depends on a division of the watershed into a number of elevation bands. The elevation increment and the area for each band must be specified. The model has been used for watersheds ranging from a few square kilometers up to areas of several thousand square kilometers. The factors influencing choice of watershed size are the available streamflow reference data for calibration and the available meteorological data base.

In general, the meteorological data base is sparse for most of the mountainous regions modelled. In the majority of situations the meteorological data is from valley stations. As a result of these data constraints, an important aspect of the Watershed Model is the elevation distribution of data. Functional relationships are specified describing the variability of temperature lapse rates. The temperature lapse rate is a key relationship because it influences the precipitation distribution, and also is very significant in determining snowmelt rates at various elevations. Precipitation inputs are made functionally dependent on elevation and on temperature regime. This functional variation of precipitation automatically recognizes that precipitation undergoes greater orographic enhancement in winter than it does during warm summer rainstorms. The importance of these temperature and precipitation gradients is illustrated in the examples of watershed calibration. The general structure of the Watershed Model is indicated in the flow chart, Fig. 8.1.

The response of the watershed to snowmelt and rainfall is controlled by a soil moisture model. The soil moisture status of each area-elevation band

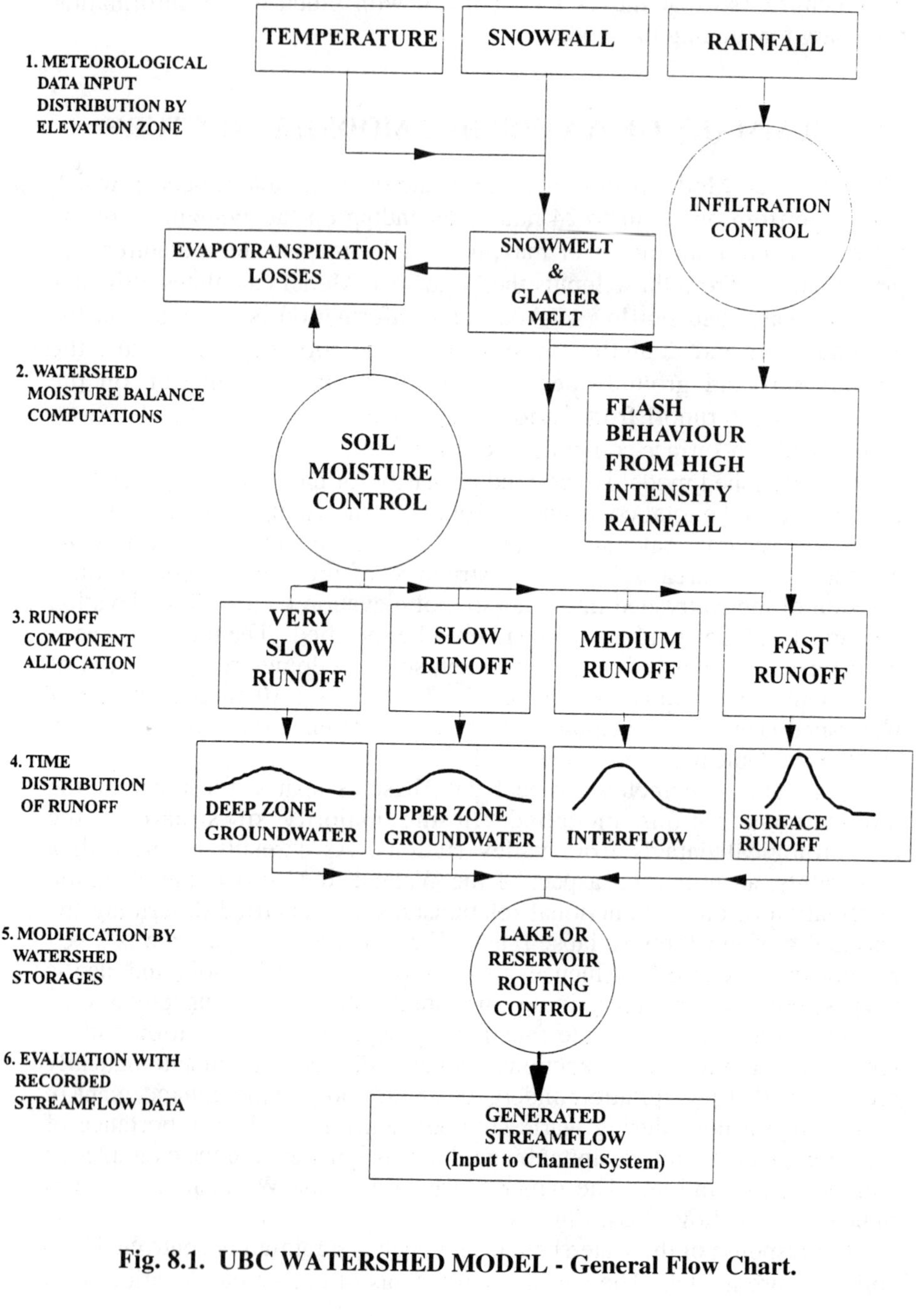

Fig. 8.1. UBC WATERSHED MODEL - General Flow Chart.

controls the subdivision of the total snowmelt and rain input into the various components of watershed runoff response. These components of runoff can be characterized as fast, medium, slow and very slow runoff, and they may be conceptually thought of as representing surface runoff, interflow, and superficial and deep groundwater components. The total snowmelt and rain input to each watershed band is subdivided on a priority basis, for example, a first priority is the satisfying of any soil moisture deficit which arises continuously because of evaporative demand. The various priorities of the components of runoff are described in the detailed outline of model behavior.

Each component of runoff undergoes delay before reaching the outflow point of the watershed. These delays, or time distribution runoff, are achieved by using unity hydrograph convolution. As explained later, the various delay processes, or time distribution processes, can be thought of in terms of cascades of linear reservoirs.

The following section outlines the various algorithms used to describe the processes involved in the runoff process.

8.10. METEOROLOGICAL DATA DISTRIBUTION ALGORITHMS

Meteorological data is available as point values at given elevations in a watershed. The model is designed to handle data from one up to a maximum of five stations. Before watershed response calculations can be made, these meteorological data at a point must be distributed to the mid-elevation points of elevation bands.

8.10.1. Temperature Lapse Rate

Lapse rates are known to be quite variable, ranging from high values of about the dry adiabatic lapse rate to low values representing inversion conditions. A complete and detailed representation of this lapse rate variability is not possible, but the main features of lapse rate can be represented as a function of daily temperature range.

The following major features of lapse rate variation are recognized in the temperature lapse rate algorithm:

(1) During continuous rainstorm conditions the lapse rate will approximate the saturated adiabatic rate. Under these conditions the daily temperature range will tend to be zero or very low.

(2) Under clear sky, dry weather conditions, the lapse rate during the warm part of the day will tend to the dry adiabatic rate. During the night, under these clear sky conditions, radiation cooling will cause the temperatures to fall to the dew point temperature, and this is particularly true for a moist air mass. As a result, night-time lapse rates under clear skies will tend to be quite low, and at times even zero lapse rates will occur.

Based on these considerations, two lapse rates are specified in the model, one for the maximum temperature and one for the minimum temperature. The lapse rate is calculated for each day using the daily temperature range (diurnal range) as an index. The general form of the equations used in the program are given below.

Maximum Temperature Lapse Rate - TXLAPS

$$TXLAPS = TZLAPS + (TLLXM - TZLAPS) * TD/TERM \tag{8.1}$$

Minimum Temperature Lapse Rate - TNLAPS

$$TNLAPS = TZLAPS - (TZLAPS - TLMN) * TD/TERM \tag{8.2}$$

where

$$TD = daily\ temperature\ range\ (TX - TN)$$

and

$$TZLAPS = TZ - (PP / PPM) * (TZ - TZP)$$

for the above,

PP	=	daily precipitation
TLXM	=	10#C/km
TLMN	=	0.5#C/km
TZ	=	6.4#C/km, that is, a reference lapse rate for rain-free conditions
TZP	=	3.2#C/km, that is, a reference lapse rate when PP ^ PM
PPM	=	10 mm/day

and the calibration parameter TERM equals the maximum temperature range under open sky conditions [selected from the data set (TX-TN)].

8.10.2. Precipitation Elevation Gradients

The enhancement of precipitation as a moist air mass is driven by wind across mountain barriers is an important aspect of mountain watershed modelling and is referred to as orographic precipitation.

Orographic precipitation is influenced by three main factors, the slope of the mountainside, the mountain barrier height and thirdly, the stability of the air mass. To generalize, the steeper the mountain side, the more rapid the increase in orographic precipitation. However, this increase in precipitation with elevation does not continue indefinitely, but tends to decrease substantially at about half the maximum mountain barrier height and shows no further increase above about two-thirds of the maximum mountain height.

The stability of the air mass depends on the relative values of the saturated adiabatic lapse rate and the prevailing lapse rate of the surrounding air mass. A useful index of stability can be developed by comparing the values of the saturated and dry adiabatic lapse rates. The saturated adiabatic lapse rate is considerably less than the dry adiabatic rate when temperatures are high, but becomes very similar to the dry adiabatic rate when temperatures are low. This variation of saturated adiabatic lapse rate with temperature will be used to account for the variation in orographic precipitation between winter snowfall, which shows a high orographic increase, and warm summer rain, which is relatively insensitive to orographic effects. Because the lapse rate is not usually available, a relationship between temperature and saturated lapse rate will be used.

The algorithms which describe the variation of precipitation with elevation are subdivided into two aspects to give the final orographic effect. The first algorithm describes the basic enhancement of precipitation with elevation barrier height if the temperature is zero degrees and the second algorithm modifies this basic distribution for variations in temperature.

8.10.3. Orographic Enhancement as Function of Elevation, Barrier Height and Temperature

The precipitation in any elevation band is calculated from the precipitation in the band immediately above or below using the equation

$$P_{I,J,L+1} = P_{I,J,L} * (1 + \mathbf{a}) \tag{8.3}$$

$P_{I,J,L}$ is the precipitation from meteorological station I for day J and elevation band L.

a is the precipitation enhancement factor defined on the next page.

The (1+**a**) factor produces a logarithmic increase in precipitation with elevation. The enhancement factor **a** is separately defined above and below a certain elevation, and is selected as a percentage. The multiplier is determined as:

$$multiplier = (1 + \mathbf{a})^{v\ elev}$$

where

a = gradient (%)

v elev = difference in elevation between the AES stations and the band

Three precipitation gradients can now be defined: P0GRADL, for bands below E0LMID; P0GRADM, for bands between E0LMID and E0LHI; and P0GRADU, for bands above E0LHI.

Modification of a with Temperature

The second algorithm can be used to modify the precipitation gradient for temperature when the band temperature differs from zero degrees, so that,

$$\mathbf{a} = \mathbf{a} - S * (T) \tag{8.4}$$

S is the model parameter A0STAB. The rationale for the use of the parameter S is derived as follows:

A measure of the stability of the air mass is $(L_D - L_S)/L_D$ where L_D is the dry adiabatic lapse rate. At higher air temperatures the greater difference between L_D and L_S indicates that warm moist air is much more unstable than moist cold air. Warm moist air is therefore likely to produce convective precipitation and does not rely so much on orographic lifting. In contrast, cold moist air has very little difference between L_S and L_D, so that convective rain is rather unlikely. Consequently, a cold moist air mass is very dependent on orographic lifting to produce precipitation. This argument leads to the defining of an orographic opportunity factor,

$$\text{Orographic opportunity} = 1 - \left(\left(L_D - L_S \right) / L_D \right) = L_S / L_D \tag{8.5}$$

If there is already considerable instability of the air mass, then there is a much reduced opportunity for precipitation to be enhanced by orographic lifting. The factor L_S/L_D will be proportional to the vertical flow rate induced by the orography and it will also be a measure of the condensation rate per unit volume of air. The reduction in orographic effect with temperature, S, will therefore depend on The gradient of $(L_S/L_D)^2$ as a function of temperature. A graph of $(L_S/L_D)^2$ versus temperature, reveals an almost linear variation between -30 and +20 degrees, and gives a value of S of 0.01, Fig. 8.2.

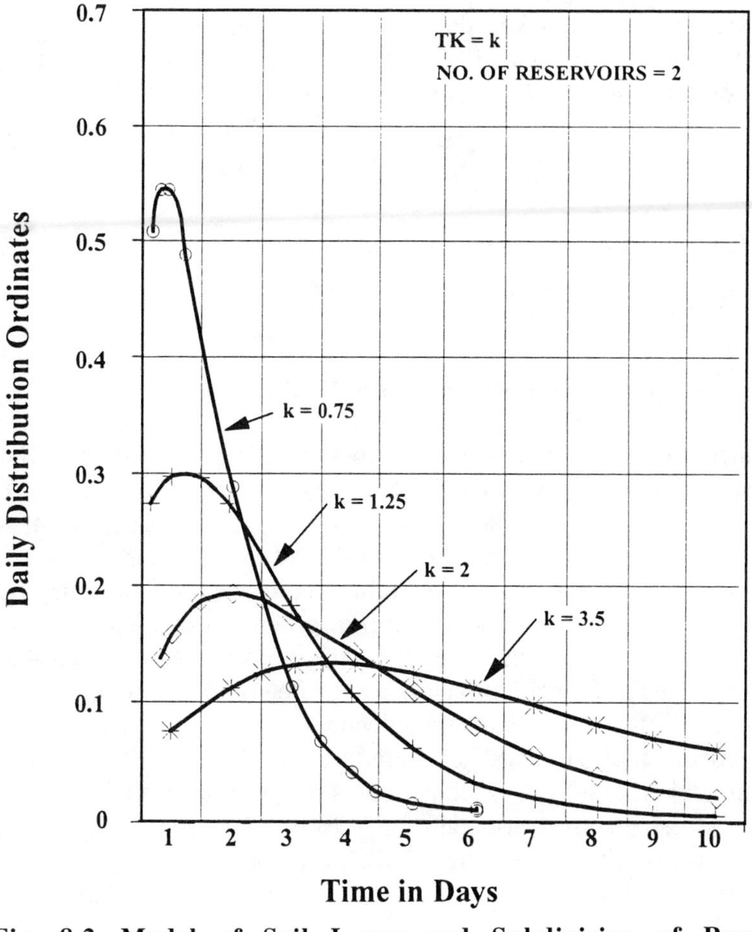

Fig. 8.2. Model of Soil Layer and Subdivision of Runoff Components.

8.10.4. Form of Precipitation

The model must distinguish between precipitation in the form of snow and precipitation falling as rain and this distinction must be made for each elevation band. Snow is stored until melted, whereas rain is immediately processed by the soil moisture model.

The form of precipitation is controlled by three logical statements and the temperature, T, used in these statements is normally the mean daily temperature in each band but it can be specified to be the maximum or the minimum daily temperature in each band.

If T < 0#C *all precipitation is* SNOW

If T > A0FORM *all precipitation is* RAIN.

A0FORM is specified in the parameter list. If it is not specified or is set below 0 degrees, it will take a default value of 2 degrees.

Between 0 degrees and A0FORM a proportion of the precipitation will be specified as rain, and this proportion is defined by FORMPP where,

$$FORMPP = \frac{T}{A0FORM} \quad 0 < FORMPP < 1 \tag{8.6}$$

Then rain, $RN = PP * FORMPP$

and snow, $SN = PP * (1 - FORMPP)$

8.10.5. Precipitation Representation Factors

Each meteorological station has two precipitation representation factors associated with it, one for snow, P0SREP, and one for rain, P0RREP. These factors are introduced because precipitation measured at a point is not always representative of the areal distribution of precipitation. For example, a meteorological station may be in a rain shadow situation, or it may be in a narrow valley where it is receiving precipitation which is more representative of the mountain side some hundreds of meters higher than the station. These representation factors can be determined by comparing long term volumes of runoff with computed values.

In general, snow measurements are more likely to be distorted by local exposure and orography. As temperatures rise, rain values tend to be more representative because warm rain tends to be more convective in origin. When temperatures are above 18#C, the local convective nature of the rain makes it more reasonable to introduce an area reduction factor. The program logic is:

If TX < O *use* POSREP

If 0 < TX < A0FORM *use linear interpolation between* POSREP *and* PORREP (8.7)

8.10.6. Evaporation

Evaporation estimation can be subdivided into three processes. In the first process, estimates are made of the daily potential evapo-transpiration for the reference meteorological station in the watershed (EVAP). In the second process, this EVAP value is distributed to each elevation mid-band level and is designated by PET. In the third Process, these PET values are used in conjunction with the calculated soil moisture deficit values to yield an actual evapotranspiration value for each band (EAT) and will be discussed in the soil moisture section.

$$EVAP = A0EDDF * V0EMOF * TX \tag{8.8}$$

where

A0EDDF is an evaporation constant = 0.133
V0EMOF is a factor which is specified as a monthly factor. It accounts for the seasonal variation of EVAP

$$PE(L) = EVAP - 0.90 * A0PELA \tag{8.9}$$

where A0PELA is the elevation range between the PET elevation and the EVAP elevation.

Tree cover may also be included in the formulation, as follows:

$$PET(L) = PE(L) * (1 + 0.5 * C0TREE) \tag{8.10}$$

where

C0TREE = fraction of tree-covered area.

The tree covered area can also be modified by a canopy factor, which represents the density of the tree cover. This canopy factor, C0CANY, multiplies the tree cover factor, C0TREE.

8.11. THE SOIL MOISTURE MODEL

The daily amounts of snowmelt and rain are subject to various delays and loss before eventually appearing as river flow at some watershed outflow point. The Watershed Model contains certain logical statements which decide how these snowmelt and rain inputs are subdivided between evaporation loss, and fast, medium, slow and very slow runoff.

The central control parameter for the subdivision of total watershed input is the SOIL MOISTURE DEFICIT. Note that rather than attempting to specify a total soil moisture capacity, the model operates from a lack of soil moisture. When this soil moisture deficit reaches zero, the watershed reaches its maximum runoff potential, except for FLASH runoff which will be discussed later.

A diagrammatic representation of the soil moisture model is shown in Fig. 8.3, and is described below in terms of the priorities of runoff.

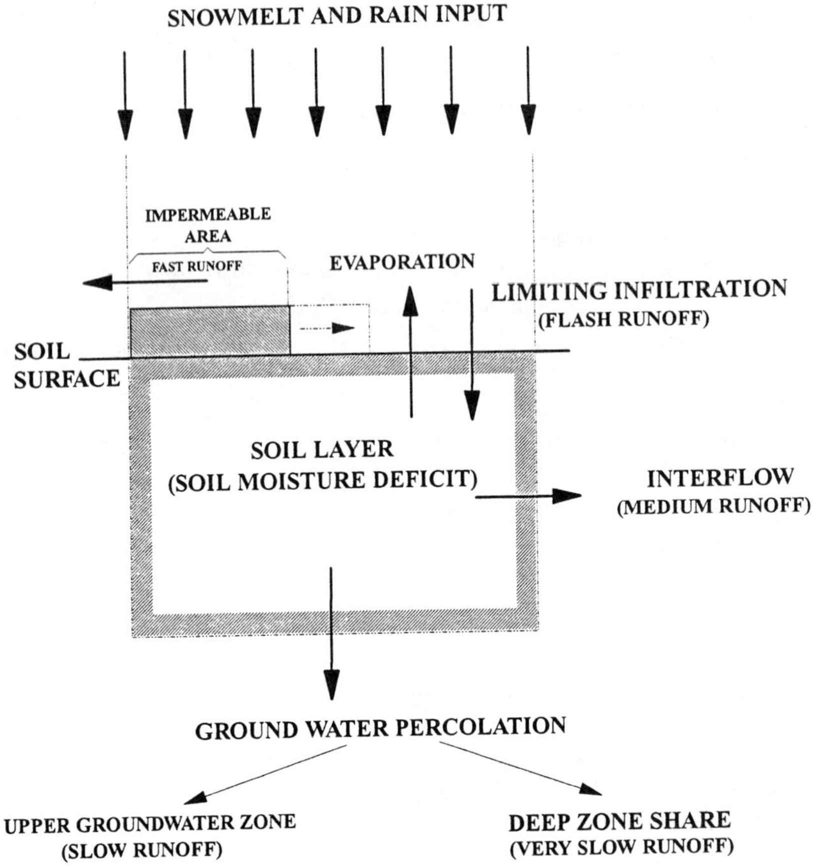

Fig. 8.3. Effect of Linear Storage Constant on Unit Hydrograph Distribution Ordinates.

8.11.1. First Priority: Impermeable Percentage - Fast Runoff Control

Part of each elevation band can be specified to be impermeable, so that any input of water to this area will enter the fast runoff component. Such runoff can be thought of in terms of surface runoff or very superficial percolation through coarse sediments. Usually such areas must be riparian in nature

The impermeable percentage of the watershed can be made to vary with soil moisture deficit. The algorithm which describes this process is:

$$Impermeable\ fraction = C0IMPA * 10 ** (-\ S0SOIL\ /\ P0AGEN) \quad (8.11)$$

C0IMPA is the maximum impermeable fraction when the soil is fully saturated.

S0SOIL is the soil moisture deficit in an elevation band.

P0AGEN is a constant which regulates how sensitive the impermeable area is to changes in soil moisture.

8.11.2. Second Priority: Soil Moisture and Actual Evapotranspiration

Before any further runoff can occur, other than fast runoff, the soil moisture deficit must be satisfied. While soil moisture deficits are being satisfied by incoming water from snowmelt and rain, there is also an evaporative demand which is continually building up a deficit.

Potential evapotranspiration has been described earlier. On any given day, in any given elevation band, there will exist a specified potential evapotranspiration. The soil moisture deficit which exists in that band will represent the actual evapotranspiration capability of that band. The algorithm describing this relationship is:

$$AET = PET * 10 ** \left(- S0SOIL \,/\, P0EGEN\right) \tag{8.12}$$

AET is the actual evapotranspiration.
PET is the potential evapotranspiration.
S0SOIL is the current value of the band soil moisture deficit.
P0EGEN is a specified constant which controls the rate at which BSD influences PET.

This actual evapotranspiration demand will only influence the area of the watershed which is not impermeable.

For each day a new value of soil moisture deficit is computed:

$$New\ value\ of\ S0SOIL = S0SOIL - PRN - BM + AET \tag{8.13}$$

where

PRN is rain input
BM is snowmelt input.

If the band soil moisture deficit reaches zero, any excess water inputs are subjected to further priorities.

8.11.3. Third Priority: Groundwater Percolation (P0PERC)

Groundwater percolation accepts any water excess up to a fixed limit (P0PERC). Any excess above this limit goes to the fourth priority, medium runoff.

Water which percolates to groundwater is assumed to be divided into specifiable fractions which go to the two groundwater components, the upper groundwater and the deep zone groundwater components. This subdivision of groundwater is controlled by P0DZSH, the deep zone share.

$$Upper\ groundwater\ zone\ recharge = (1 - P0DZSH) * P0PERC \quad (8.14)$$

$$Deep\ groundwater\ zone\ recharge = P0DZSH * P0PERC \quad (8.15)$$

8.11.4. Fourth Priority: Medium Runoff

Any excess moisture input now remaining is assigned to medium runoff, or interflow. Although this is the lowest priority, it is also frequently the most significant runoff component during active snowmelt and rain storms.

In the model, interflow is considered to be a large reservoir which receives inflows day by day during active snowmelt and rain. These inflows are the excesses remaining after satisfying soil moisture and groundwater abstractions. This reservoir releases a certain fraction each day, but the volume of water released does not immediately appear in the downstream channel system. Instead, this released water undergoes a convolution very similar to the fast component unit hydrograph. This release from an interflow storage reservoir and convolution before reaching the channel outflow point, produces a much more sluggish response for this medium runoff component.

8.12. WATERSHED ROUTING

Water allocated to each of the components of runoff, namely fast, medium, slow and very slow components, are subjected to a routing procedure which produces a time distribution of runoff. The routing procedure for each component is based on the same underlying concept, namely the linear storage reservoir. The fast and medium components of runoff are subjected to a cascade of reservoirs which is essentially identical to unit hydrograph convolution. The lower components of runoff simply use a single linear reservoir, thus avoiding the necessity to convolute for the final outflow.

8.12.1. Fast Runoff Routing

The Watershed Model calculates unit hydrograph ordinates for convolution of fast runoff. The calculation is based on a conceptual model of the runoff process developed by Nash in which inflows are passed through a cascade of linear reservoirs. The resulting outflow at a time t from a unit impulse of inflow is:

$$u(t) = \frac{1}{K^n} \bullet \frac{t^{n-1}}{(n-1)!} \bullet e^{-t/k} \tag{8.16}$$

K is the linear storage constant for each of the reservoirs in the cascade.

n is the number of linear reservoirs in the cascade (specified by N0FASR, N0FASS, N0GLAC).

t is the time after the water input has occurred.

A graphical representation of the time distribution of flow represented by Eq. (8.16) is shown in Fig. 8.4.

Rather than solving the above equation by using convolution, the Watershed Model uses a stepwise calculation of the outflow from each reservoir in the series. This method avoids convolution, which is much more demanding in storage of information, whereas the stepwise calculations of Eq. (8.17) require only the previous outflow and the latest inflow. Linear reservoir routing of the components of runoff is computed as follows:

$$\textit{First reservoir}: \ Q_1 = Q_1 + \left(1/(1+TK)\right) * \left(WB - Q_1\right) \tag{8.17}$$

$$\textit{mth reservoir}: \ Q_m = Q_m + \left(1/(1+TK)\right) * \left(Q_{m-1} - Q_m\right)$$

where

Q_m = outflow from the mth linear reservoir
TK = time constant of the linear reservoir (This is P0FSTK etc.)
WB = water input to the first linear reservoir.

8.12.2. Medium Runoff Routing

The medium speed runoff, often thought of as an interflow component, undergoes a two stage routing process. In the first stage the medium runoff for each day enters a linear storage reservoir with a release constant INTK. The daily release from this reservoir is therefore a constant percentage of the total storage on any given day. Such a release represents an exponential decay which is identical to Nash's result with the number of reservoirs equal to one.

$$QI = QI + \left(1/(1+INTK)\right) * \left(WI - QI\right) \tag{8.18}$$

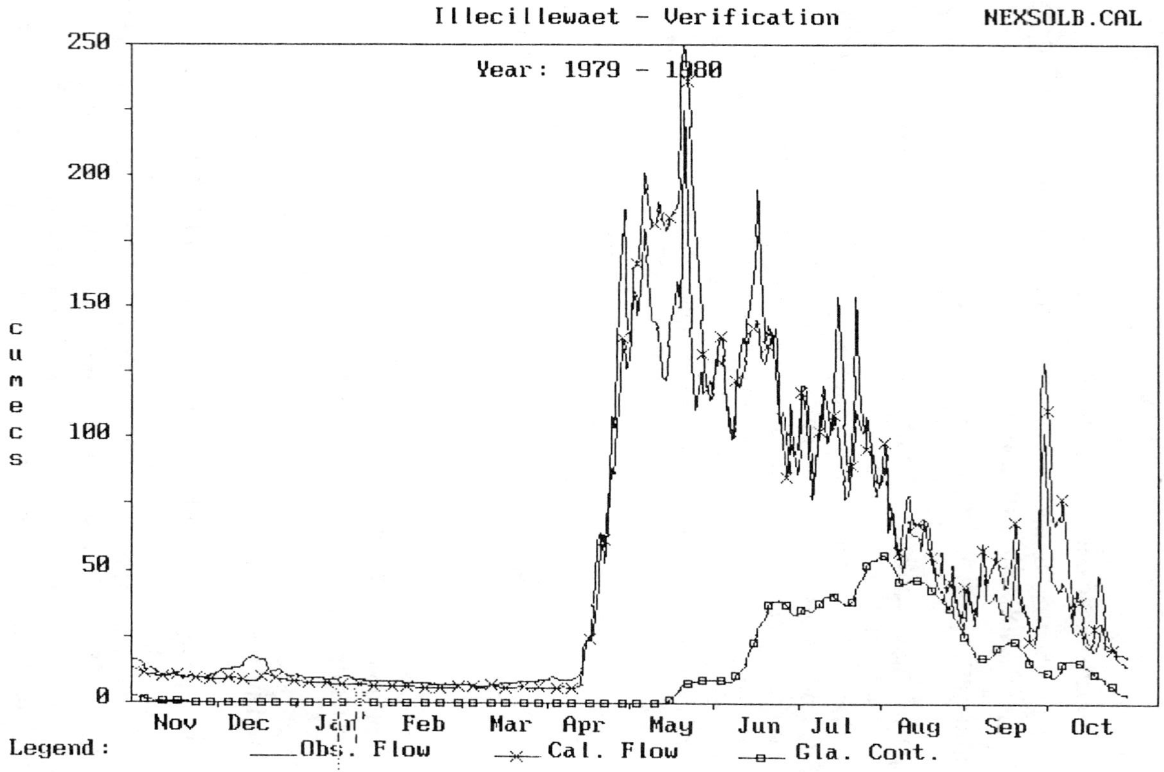

Fig. 8.4. Illedillewaet - Compute and observed flows for one of the verification years, 1979-80. The glacial contribution is shown.

where

WI is the water input to the interflow reservoir, and
INTK can be either POISTK for snow runoff or POIRTK for rain.

The daily release QI is then subjected to time distribution by convolution with a unit hydrograph which is similar to the fast unit hydrograph.

8.12.3. Slow Runoff Routing

The slow components of runoff, usually termed groundwater flow or base flow components, are divided by the soil moisture model logic into a slow, upper-zone component and a very slow, deep-zone component.

Both of these groundwater components are routed using a single linear reservoir which accumulates each day's inflow and releases a fixed percentage of the total storage each day.

The upper groundwater zone release fraction is controlled by P0UGTK. The deep groundwater release is controlled by P0DZTK.

8.13. SNOWMELT BUDGET

The Watershed Model accumulates precipitation falling as snow and then depletes these snowpacks according to the calculated melt rate. This snow accumulation and depletion is carried out separately in each area-elevation band. There are two ways in which this snow budgeting can be carried out, the first being a simple budget at the mid-band elevation, referred to as a block budget, and the second being a more complex wedge computation.

8.13.1. The Block Budget

From a computational viewpoint this is a straightforward calculation procedure in which snow is accumulated as if it were falling at the mid-elevation of each band. Snowmelt calculations are also made at this mid-band elevation.

8.13.2. The Wedge Budget

This analysis option recognizes that the final depletion of snow is a gradual recession of snowline from the bottom to the top of a band. In this option there is a smoother transition as snow is depleted completely from a band. It might be commented that the advantages of this more complex analysis are not as great as might be thought. However, if snow gradients are high, as can be the situation under some melt conditions, then the wedge budget does give a better representation of runoff behavior.

Essentially the wedge budget still uses the block budget method until the elevation band below is out of snow, At this point the wedge analysis takes over.

A set of calculations is still carried out at the mid-band elevation. When the snowpack is depleted to a value defined by S0PATS (usually of the order of 400 mm of water equivalent) a partial area of snowcover is computed as further melt occurs.

8.14. SNOWMELT

8.14.1. General

The UBC Watershed Model contains two options for the calculation of snowmelt:

(1) An energy budget approach which is simplified for use when there is only temperature data available. (The method is also easily usable if more detailed radiation, albedo and wind data are available.)

The choice of snowmelt method is governed by the parameter ROERGY.

If ROERGY = 0, the degree-day method is used.
If ROERGY = 1 (or any non-zero value), then the energy method is used.

(2) A degree-day approach in which two equations are used, one for forest melt and one for open area melt.

From the consideration of various energy exchange processes represented in the next section, it is clear that, in dense forests, the longwave heat exchange processes dominate and consequently a temperature index approach is reasonably valid. However, in open areas, an energy approach is superior because shortwave radiation and snowpack albedo are very important. Also the convective and condensation melt components are increased by turbulent mixing induced by wind and rugged topography. Temperature alone is not an adequate indication of melt because temperature varies with elevation and with incoming air mass, so that there is a weak and non-linear relationship between temperature and shortwave radiation. Albedo varies seasonally, but can rise considerably during periods of new snowfall, so that net solar radiation is a non-linear function of albedo. Cloud cover can greatly reduce the shortwave radiation and also has a strong influence on the longwave radiation balance. This raises the question of how well cloud cover can be estimated, especially its density, and to what extent the reduction in shortwave radiation is compensated by the increase in net longwave exchange. The conclusion from our extensive studies appears to be that open area melt is not a simple linear function of temperature, but relationships are put forward by which temperature may be used to estimate the various non-linear inputs which control the following simplified energy method for the calculation of snowmelt.

The energy exchange at the surface of a snowpack is made up of four major components:

(1) The shortwave radiation exchange, which consists of incoming solar radiation, and the reflected outgoing shortwave radiation. This shortwave component depends on the time of year, the site exposure, cloud cover and snow albedo.

(2) The longwave radiation exchange depends on black body radiation from the snow itself and from clouds and tree cover, and gray body radiation from the overlying air mass. Under clear skies, the net longwave is outgoing, or negative, unless air temperature exceeds about 20°C. Under cloudy conditions and also under tree cover, net longwave can be positive at temperatures above freezing.

(3) Convective heat transfer is produced by turbulent heat exchange between the air mass immediately above the pack. This heat transfer is dependent on both wind and air temperature and particularly on the stability of the air mass above the snowpack. A warm air mass above the cold snow surface tends to be stable, resisting any downward transport of heat to the snowpack, unless there is enough wind to produce turbulent mixing. This turbulent heat transfer is governed by the Richardson number, R_I, which is a measure of stability. The bulk Richardson number, defined below, is essentially the air temperature divided by the wind speed squared. As air temperature increases, and if the wind is only moderate, the stability can increase to the point where very little convective heat transfer can occur. Convective heat transfer is therefore self limiting and becomes quite small at higher air temperatures, unless there is very strong wind.

(4) Advective heat transfer, often termed condensation melt, is caused by the transport of moisture to and from the snowpack. Whether condensation occurs, cooling the pack, depends on the relative vapour pressures of the air and snow surface. Wind is once again an important factor and so is stability, as was discussed for convective transport. Advective heat transport can therefore produce snowmelt if the dewpoint temperature is above freezing, but, like the convective heat transport, becomes limited at higher temperatures by the stability of the warm air mass.

A simple set of equations has been developed which expresses the various snowmelt components in terms of millimeters of snowmelt per day, either negative or positive. Various simplifying assumptions have been used to write these equations, but they have been tested on high quality data sets from the U.S. Corps studies (1954) and are a considerable improvement on the calculation of snowmelt and glacier melt, especially for high elevation, open area regions in the mountains.

8.15. SIMPLIFIED ENERGY COMPONENTS

8.15.1. Net Shortwave - Energy Input

$$Melt = I_S\left(1 - C_L\right)\left(1 - A_L\right) mm, \qquad (8.19)$$

Where C_L is cloud cover, A_L is the albedo of the snowpack and I_S is the incident solar radiation, which varies seasonably and with latitude and with atmospheric conditions. At 35# North, I_S can be expressed in terms of millimeters of melt equivalent per day instead of Langleys per day,

$$I_S = 54 - 29 \cos 2pN / 365 \; mm / day \qquad (8.20)$$

The albedo and cloud cover reduce the potential melt values of Eq. (A-8.20) to the net values expressed by Eq. (A-8.19).

8.15.2. Longwave Radiation

Stefan's law can be expanded in terms of the temperature above freezing, T, so that the longwave black body radiant energy, I_L, can be expressed as a linear function of temperature plus small higher order terms,

$$I_L = r\,(273 + T)4 \text{ Langleys} \qquad (8.21)$$

$$= r \bullet 2734\,(1 + 4T/273 + 6T2/2732 + ...)$$

$$= \quad 661\ (1 + 0.015T + 0.0001T2 + \ldots)\ \text{Langleys/day}$$

$$= \quad 82.6\ (1 + 0.015T + \ldots)\ \text{mm/day}$$

Under clear sky conditions in an open, tree free area, the net longwave radiation received by the melting snowpack is the difference between the black body radiation of the 0#C snowpack and the incoming gray body radiation from the clear sky. Various estimated values for the gray body radiation are available, some of which are quoted in the U.S. Corps Snowmelt Report (1955). All of the equations show a small dependence on humidity of the atmosphere, such as equations due to Brunt and Angstrom. The simplest equation, from the Lake Hefner study, is

$$I_{LA} = rT4\left(0.749 + 0.0049\, e_a\right) \tag{8.22}$$

where I_{LA} is the atmospheric longwave radiation, and e_a is the vapour pressure in millibars. The dependence on vapour pressure is so small that it is reasonable to accept this value,

$$I_{LA} = 0.757\, rT4 \tag{8.23}$$

The net clear sky incoming longwave radiation, I_{LN}, can therefore be written, using the linearization of Eq. (8.3),

$$I_{LN} = 661\left(0.757\left(1 + 0.015T_A\right) - \left(1 + 0.015T_S\right)\right) Langleys\,/\,day \tag{8.24}$$

where T_A is mean air temperature and T_S is the snow surface temperature which is zero for a melting snowpack.

$$I_{LN} = 7.51T_A - 161 - 9.92T_S\ Langleys\,/\,day \tag{8.25}$$

In snowmelt equivalent,

$$I_{LN} = 0.94T_A - 20.1 - 1.24T_S\ mm\,/\,day \tag{8.26}$$

For a melting snowpack, T_S is zero, and I_{LN} does not become positive until T_A exceeds 21.4#C.

8.15.3. Net Longwave Under Cloudy Conditions

Clouds (temperature, T_C) act as black bodies so that under 100% cloud cover the net incoming longwave, I_{LNC}, is

$$\begin{aligned} I_{LNC} &= 9.92\left(T_C - T_S\right) Langleys\,/\,day \\ &= 1.24\left(T_C - T_S\right) mm\,/\,day \end{aligned} \tag{8.27}$$

The clear and cloudy sky equations are combined into an expression for the net longwave exchange, I_{LNT} so that for a partial cloud cover fraction, C_L and for a melting snowpack at 0#C,

$$I_{LNT} = (-20 + 0.94T_A)(1 - C_L) + 1.24\, T_C \bullet C_L \; mm/day \qquad (8.28)$$

8.15.4. Convective and Advective Heat Transfer

The U.S. Corps report (1955) and Anderson (1976) present equations for convective and advective heat transfer. Anderson refers to the question of air mass stability, but does not incorporate the results into a final relationship. The following equations are an approximate estimation of the new heat transfers which have been developed from the earlier work.

Under neutrally stable conditions the convective heat transfer, Q_C, is approximately,

$$Q_C = 0.113\,(p/101)\, T_A \bullet V \; mm/day \qquad (8.29)$$

in which p is the atmospheric pressure in kilopascals for the elevation being considered, T_A is the mean temperature and V is the wind speed in kilometers per hour.

Similarly, the advective, or condensation, melt Q_A under neutrally stable conditions is approximately,

$$Q_C = 0.44\, T_A \bullet V \; mm/day \qquad (8.30)$$

where T_d is the dew point temperature, which can be approximated by the minimum air temperature.

Both the convective and advective melt rates are reduced by a factor R_M which is a function of the bulk Richardson number R_I. A linearization of R_M, subject to the limitations given below, is

$$R_M = 1 - 7.7\, R_I \qquad (8.31)$$

where

$$R_I = 2g \bullet \frac{Z_A}{T_A + 273} \bullet \frac{T_A}{u_A^2} = 0.095\, \frac{T_A}{u_A^2} \qquad (8.32)$$

where V is the wind speed at a reference height Z.

This linearization is reasonable for R_I between +0.12 and -0.1. For positive air temperatures, the R_M factor is not less than zero, and for negative temperatures, R_M increases to about 1.8 and then slowly goes a little higher.

This R_M factor is based on idealized laboratory conditions. In steep mountain terrain, our own studies of snowmelt indicate that stability is greatly reduced by "terrain mixing" caused by large scale roughness and slope. Consequently R_M may increase by as much as a factor of 2.5 times, which is an additional "terrain mixing" factor.

8.15.5. Rainmelt

Snowmelt from rain, BMR, is computed as follows:

$$BMR = K * TM * RN \tag{8.33}$$

where K= a constant, RN = rainfall, and TM = mean air temperature.

This formula assumes that the rain falls at the mean air temperature and the K factor (mm/#C of rain) represents the heat content of the rain. This equation is also used in the energy method.

8.16. APPLICATION OF THE SIMPLIFIED ENERGY SNOW-MELT EQUATIONS

There are advantages to be gained from using energy equations for calculating snowmelt. The physical basis of the equation makes it possible to estimate snowmelt for forested and open conditions, for clear or cloudy weather, for various slope and aspects of mountainous watersheds and for changes in elevation. It is also possible to argue the impacts of changing forest cover or the snowmelt that would be experienced under extreme and unusual weather sequences. Some simplified energy relationships will be tested against measured point snowmelt data for both open area melt and forested melt.

For most high mountain regions data on radiation, cloud cover, etc., are not available, The factors which must be described are the cloud cover, albedo, wind, cloud temperature and dew point temperature. All other factors can be estimated from known physical behavior. As reasonable approximations, it has been found possible to represent these various factors with temperature-based estimates.

8.16.1. Cloud Cover

The cloud cover has been assumed to be related to the daily temperature range,

$$(1 - C_L) = (T_{MAX} - T_{MIN}) / D_R \tag{8.34}$$

in which C_L is the cloud cover fraction, T_{MAX}, T_{MIN} are the daily maximum and minimum temperatures, D_R is the daily temperature range for open sky conditions at a certain elevation in the watershed.

8.16.2. Wind Estimate

The wind tends to produce a decrease in daily temperature range, so that cloudy conditions are also windy conditions,

$$V_b = K_R \left(T_{MAX} - T_{MIN}\right)^{-1/2} - K_v \tag{8.35}$$

in which K_R and K_v vary slightly with elevation.

8.16.3. Albedo

The albedo is modelled by three relationships. Fresh snow is assumed to settle and change its albedo from a starting value of 0.9, which then decays by a factor of 0.9 per day until it reaches a settled value, A_{LS}, of 0.7, as described by Eq. (A-8.37).

$$A_{LS(j+1)} = 0.9 A_{LS(j)} \tag{8.36}$$

From then on, the albedo continues to decrease at an exponential decay rate, which is controlled by the calculated cumulative melt for the season, given by

$$A_{Lj} = A_{LS} \bullet \exp^{[-RM/KL]} \tag{8.37}$$

where A_{LS} is the settled value, usually 0.7, RM is the cumulative seasonal melt in millimeters and K_L is a constant, approximately of the order of the total seasonal melt, usually taken as 4000 mm.

When new snow occurs, if it is greater than 15 mm, the albedo is assumed to return to a value of 0.9. The albedo then decays again at the 0.9 rate, specified by Eq. (8.36), but continues down until it reaches the value computed on the day before the snow occurred, A_{Lj} from Eq. (8.37). The underlying recession described by Eq. (8.37) then takes over. This process allows a fairly rapid change of albedo from 0.9 to 0.7 or down to the calculated A_{Lj} value from Eq. (8.37), and then a slow recession as the season advances to albedo values of 0.4 or even 0.3 for very deep and aged snowpacks. Snowcover on a glacier is treated in the same manner, but the glacier, when it becomes free of snow, is assumed to have an albedo of 0.3.

These relationships for cloud cover, wind and albedo have been developed using SIHP data gathered in the Himalayas and also using U.S. Corps of Engineers data gathered in great detail for the Central Sierra Snow Studies (1952) as discussed by Quick (1987).

Activating the selection parameter ROERGY automatically implements either the degree day or the energy budget snowmelt equations given above. (The energy method is used if ROERGY = 0; the degree-day method is used if ROERGY = 1 or any non-zero value.)

The user must supply the following information:

(1) The temperature range which controls cloud cover specified by the parameters A0FOGY, A0SUNY, A0FOGX, where A0FOGX is the maximum range for a station, A0FOGY is the range below which complete cloud cover exists, and A0SUNY is the range above which the sky is assumed clear.

This temperature range depends on the elevation of the meteorological station and decreases as elevation increases. Typical values are specified in the calibration parameter files for the various watersheds.

(2) The percentage of forested area for each band.

(3) A general orientation index for the watershed, 0 for south facing or 1 for north facing.

8.16.4. Negative Melt Budget

If temperatures have been below freezing for any length of time, the snow becomes deeply frozen and requires ripening before melt can occur. These antecedent conditions are accounted for by keeping a running sum of antecedent mean temperatures, and allowing the total to decay by a certain fraction each day, specified by CCTX, the antecedent negative melt time constant. This decaying of the negative melt budget depends primarily on temperatures that have occurred over the previous 10 days or so. The memory of earlier cold temperature decays away. The ripening of the snowpack takes into account thermal input from rainfall, and further accounts for time delays induced by the water holding capacity of the snowpack.

The ripening of the snowpack is computed as follows:

$$CC = CC + \left(1/(1 + CCTK)\right) * \left(CCB - CC\right) \tag{8.38}$$

where

CCB = TM + KCC*RN

CC = cold content storage, and KCC is a constant.

KSS = latent heat contribution from freezing of rain falling on snowpack

A percentage of meltwater can be retained by the snowpack and this water holding capacity of the snowpack is expressed as a simple fraction of the cumulative snowfall.

8.17. RUNOFF FROM HIGH INTENSITY RAIN

There are two different controlling mechanisms for runoff from a watershed system. The runoff from moderate intensity rain and snowmelt events can be considered to be controlled by soil moisture levels and this is the normal operation of the UBC Watershed Model. The runoff from high intensity events is controlled by the rate at which water can infiltrate into the soil system and these infiltration rates are relatively independent of soil moisture levels.

For these high intensity rain events, some of the precipitation infiltrates into the soil system and is subjected to the normal soil moisture budgeting. The fraction of the rain which does not infiltrate is considered to contribute directly to the fast runoff component and is referred to as the FLASH SHARE.

It should be noted that even intense snowmelt rates do not appear to be adequate to produce FLASH runoff, even though 75 mm of snow may melt in the day. Apparently, this snowmelt is released to the soil system steadily over the whole day and the rate of release is below the infiltration capacity. In contrast, rain events of even less than 50 mm in the day can exhibit flash behavior. Examinations of rain intensity data suggests that rainfall events of this magnitude tend to occur in localized periods, or bursts, of high intensity rain, intensities which can be higher than the soil infiltration rate.

Comparing a rain event in which say, 50 mm of rain falls in the day with one in which 100 mm might fall, it appears that the greater rainfall may represent similar intensities occurring more frequently and over a greater period of the day.

These are the basic ideas underlying the FLASH SHARE behavior of the UBC Watershed Model. The relationships governing FLASH SHARE behavior are as follows:

(1) The portion of the total daily precipitation which is subjected to the normal soil moisture budget system is

$$(1 - MFR) * (1 - PMXIMP) * PRN \tag{8.39}$$

where

MFR	=	(MR-V0FLAS) /V0FLAX
MFR	=	FLASH SHARE fraction (MFR ^ 1.0)
PMXIMP	=	IMPERMEABLE percentage of the particular watershed elevation band
MR	=	total daily precipitation
VOFLAS	=	threshold value of total precipitation for flash runoff
VOFLAX	=	maximum value of total precipitation, which limits MFR range
PRN	=	rain inputs at each band.

(2) The portion of the total daily precipitation which flashes off is therefore, by difference,

$$(MFR) * (1 - PMXIMP) * PRN \tag{8.40}$$

(3) From the fraction of the watershed which is calibrated to be impermeable, the total daily precipitation input is assumed to go to fast runoff, and this would be true with or without the FLASH SHARE mechanism.

(4) When high intensity rain occurs at the same time as snowmelt, the snowmelt is added to the rain and subjected to FLASH SHARE.

(5) The FLASH SHARE mechanism only operates when specified in the WAT File facility of the UBC Watershed Model. In theory, with adequate precipitation data, the FLASH mechanism could be automatic. In practice, some point rainfalls, although intense, are so localized that there is limited watershed response of any kind. It has therefore proved more satisfactory to identify FLASH events by careful inspection of the river flow and meteorological data.

(6) It should be noted that FLASH is specified by only one parameter, FLASHR. The actual flash response is coupled to the normal watershed calibration by incorporating PMXIMP, the percentage of the watershed which behaves as if it were impermeable. With the usual gradient of impermeability which is observed to exist in mountainous watersheds, the run-off from lower elevations is slower, whereas the higher elevations, where PMXIMP is higher, automatically behave as much faster run-off regions.

(7) Rain inputs, PRN, for FLASH can be specified in two ways. Rain can be specified as the observed rain at a station and can be increased by the elevation parameter for rain. Alternatively, rain can be specified for each elevation band.

8.18. CALIBRATION OF THE WATERSHED

Calibration of a watershed involves adjusting parameter values in the watershed control file until the calculated volumes and patterns of runoff derived from this file match the historical streamflow records. The procedure usually involves many iterations because there are extensive non-linear interactions between the various groupings of parameters. Because of the iterative nature of the process, a semi-automatic calibration routine has been written which searches systematically for best-fit sets of parameters.

The optimization is designed to assist the user in the calibration of a watershed by removing the tedium of varying parameters to find the best model efficiency. The user is still required to have some knowledge of the hydrological processes involved in watershed runoff. Either by using the default values specified as normal starting values, or by using experience and judgment, the user can define reasonable ranges for the parameters and then let the computer do the work of selecting the best value.

The optimization module deals sequentially with three groups of parameters. The first group, **precipitation distribution**, adjusts the precipitation gradient parameters to distribute the point precipitation measurements over the watershed while maintaining the correct volume of outflow. Once the annual volumes of observed and estimated flow are reasonably close the optimization switches to the second group, **water distribution parameters**, which distributes the rain and snowmelt to groundwater, interflow, and surface runoff by controlling the soil moisture budget. When adjustment of these parameters shows no further improvement, the optimization proceeds with the third group of parameters, **the routing constants**, which control the time response of each component of flow, thereby controlling the length of time taken for water to pass through the watershed.

This optimization procedure can be run for any number of iterations, at the end of which the ten best efficiencies and the corresponding parameters are saved. The parameters corresponding to the best efficiency are then used to update the watershed control file which is then used with the next group of parameters. When all selected group are processed the resulting watershed control file will contain the best parameters for that optimization cycle. In general, the optimization cycle must be repeated a number of times, because of the non-linear interactions between the parameter groups. The results from the previous runs assist the operator to limit the parameter ranges, thereby speeding up the search process.

The operator can assist the semi-automatic calibration process by selecting suitable time periods when certain processes may dominate. For example, (p.5) the parameters controlling the time distribution of runoff subdivide the rain and snowmelt into fast, medium and slow runoff components. The slow component, groundwater, can best be analyzed by selecting a recession flow period, such as the end of summer when snowmelt has ceased, or after a large rain event when recessional flows dominate. These recessional flow periods may last for several months, which is consistent with the slow groundwater recession constants. On the other hand, the operator should select periods of intense rain or snowmelt. Because the fast response can be less than a day, it may be necessary to meter use of short-term data if such data is available. However, many watersheds have been successfully calibrated using daily data.

A further source of snow accumulation is from satellite images, which provide estimates of snow covered area, but not of snow depth. The watershed model also generates estimates of snow covered area on a daily basis, so that satellite snowcover information can be used to check these estimates and update the precipitation gradients in the model. At the present time there is only sparse satellite coverage, especially in the Pacific Northwest mountains, where cloud cover can be extensive. Radar-sat data should change this situation and in the future snow cover information may become a regular and direct model input. This is one more example of how model use is evolutionary and can adapt to new or improved sources of information.

The increase in precipitation as a function of elevation is probably the most important factor in watershed modelling for mountainous regions. Although this increase can be estimated from the total volumetric yield of the watershed, there is sometimes more direct evidence of this precipitation increase. For example, for watersheds where snowmelt is a major factor, there are often snowcourse measurements which can be used to estimate the precipitation gradient. To make this analysis, the model is used to simulate the accumulation and depletion of the snowpack at the elevations of the various snowcourses. This method has been found to be an excellent and reliable way of confirming precipitation gradients.

For rainfall, the increase with elevation can sometimes be estimated directly from high elevation measurement stations, but such stations are relatively scarce in many mountain regions. However, more of these high elevation stations are now being established, especially for watersheds which feed into hydro-electric systems, where there is sufficient economic value to justify the expense of remote stations which usually report the data by satellite link. In one example which will be quoted later, this high elevation data is available on a 3 hourly basis and is used to compute the precipitation gradient for each event. Although this data has proved of great value for rainfall, it has not yielded good information during snowfall because the gauge tends to freeze and cap over during snow. Work is continuing to try to find a better design of gauge which will be free of these capping problems.

This use of high elevation data is a good example of how model estimates of watershed behavior may gradually be replaced by actual measurements, especially as remote data gathering becomes more reliable and also less expensive.

8.18.1. Statistics

The Statistics package generates several statistical measures of model performance, the primary ones being the mean and total runoff statistics by the month and year and measures of model efficiency based on residual error values. The results of this analysis can be printed as a report using the Print option on the main menu or displayed on the screen using the View option. This report is central to stage one of the calibration process.

The Statistics Report provides the user a summary for the period selected and monthly values to indicate the model's performance. Comparison is made between the actual and the estimated total volume of runoff. The Coefficient of Efficiency is a measure of how well the estimated hydrograph compares to the actual hydrograph. Efficiency varies from negative values to plus one, which represents perfect agreement. A negative value indicates that the average flow is a better estimate, often called the naive forecast. The Coefficient of Determination is a measure of how well the regression of the observed flows compares to the actual flow.

8.18.2. View Watershed Graphics

The Graphics function displays information about watershed behavior in the form of graphs that can be viewed on the screen, printed, or sent to a file. The items available to be displayed and their units of measurement are:

- Observed flow (measured in cumecs)
- Calculated flow (cumecs)
- Difference between observed and calculated flow (cumecs)
- Snowmelt outflow (cumecs)
- Glacial contribution (cumecs)
- Rainfall outflow (cumecs)
- Groundwater (cumecs)
- Solar radiation (calculated - Langleys/day)
- Albedo (calculated as percentage)
- Area of snowcover (percentage of total area)
- Evapotranspiration loss potential (millimetres/day)
- Loss to et and soil (cumecs/day)
- Cloud cover (calculated - percentage)
- Snowpack water equivalent (millimetres)
- Temperature - maximum (˚C *10)
- Temperature - mean (˚C *10)
- Interception (precipitation * 10 mm)
- Snowfall (millimetres * 10)
- Rainfall (millimetres * 10)
- Snowmelt (millimetres * 10)

When you select items to display (the maximum is 4 items per graph), the scale used is the units of the top item tagged. For example, if you select

Observed and Calculated flow and Maximum and Mean temperatures, the scale units are cumecs. However, the values in the scale will be correct for the tagged items.

8.19. CASE STUDIES

8.19.1. Estimation of Precipitation Gradients

In mountain regions precipitation increases significantly with elevation. Assessing this change of precipitation with elevation is the most important part of the watershed model calibration process; because it controls the amount of moisture available for runoff, either as rainfall or as snowfall. Extensive studies of mountain precipitation indicate that precipitation can increase considerably on the lower mountain slopes, especially on the front range where incoming moist air first encounters a mountain barrier. The increase of precipitation generally gets less at higher elevations, and above half to two thirds of barrier height there is often no further increase and perhaps even a decrease in precipitation. Rain and snow show similar increases, although snow tends to increase more with elevation, mainly because of its slower fall rate, so that wind carries it to higher elevations.

A special situation, which produces a very steep gradient of precipitation, occurs on the lee side of a mountain range. High precipitation on the weather side of the range extends over to the lee side, but then decreases very sharply in this lee side subsidence zone. This is true for both rain and snow, but again, because of its low fall velocity, snow will show the greatest spill-over effect. There are many examples of this lee side behavior in mountain regions where considerable precipitation will occur at high elevations on the lee side, but then the precipitation will decrease drastically with decreasing elevation so that no precipitation at all may fall in the lee side valley regions. Consequently lee side valley precipitation data will usually be of little or limited use for estimating runoff potential from the higher mountain slopes. Under these conditions, precipitation measurements from the weather side of the mountains can be used to estimate lee side precipitation, although it will be necessary to reduce the amount of precipitation by some factor, which is usually reasonably constant, especially on a seasonal basis, so that precipitation estimation is still feasible even under these conditions. Clearly, the best data would be from high on the mountain slopes, and automatic data stations are increasing this possibility. These factors were taken into account when calibrating the following watersheds.

Illecillewaet River

The Illecillewaet River is typical of the rugged mountain streams in the Upper Columbia River basin. The flow is generated primarily from seasonal snowmelt which, because of the large depths of snowpack, continues from early April to late July in most years, and can extend even into August. Late season met is generated from the many glaciers, once they become snowfree. In addition to the sun and glacier melt, there are also summer rainstorms which can be very significant at times, especially when they occur while there are still extensive snowpacks. The Upper Columbia region contains a number large hydroelectric schemes both in Canada and the United States and the inflows to these various reservoirs are generated from steep mountain watersheds quite similar to the Illecillewaet. Therefore, a good understanding of the hydrological processes in watersheds like the Illecillewaet is an essential component of any stream flow forecasting and reservoir operating study.

The UBC Watershed Model has been used to study the Illecillewaet watershed using some years for calibration and others for verification. The verification result for 1979-80 is presented in Fig. 8.4 and Fig. 8.5 for the November to October period and is based on a calibration for a different series of years. Figure 8.4 shows the simulated flows calculated from the actual meterological data, and the recorded flows for the same period. Also shown is the model estimate of the glacial melt contribution. These flows are generated by calculating the snowpack using the watershed model and the recorded daily winter precipitation values. Also shown in Fig. 8.5 are the various calculated flow contributions, namely, the total flow, the fast snowmelt, the fast rainfall and the total groundwater flow. The model estimate of the groundwater contribution to runoff is seen to increase substantially as the season progresses and then undergoes long-term recession. It can be seen that snowmelt is the largest contributor. Glacier melt is a useful late summer contribution and rainstorms occasionally becomes significant. The statistics for this verification period give a Nash-Sutcliffe efficiency of 91% and a volume error of +2.9%.

For this region, the watershed model is used to estimate the seasonal volumes of snowmelt and glacier melt runoff and the probable seasonal distribution of these flows. For hydroelectric generation planning, it is very useful to have estimates of the possible range of flows which might result from various weather sequences during the spring and summer melt period. The watershed model is used to generate these estimates by using the known weather data for the preceding winter to generate the snowpacks at all elevations of the watershed. Then the probable melt season flows are estimated using various historical weather sequences recorded for a number

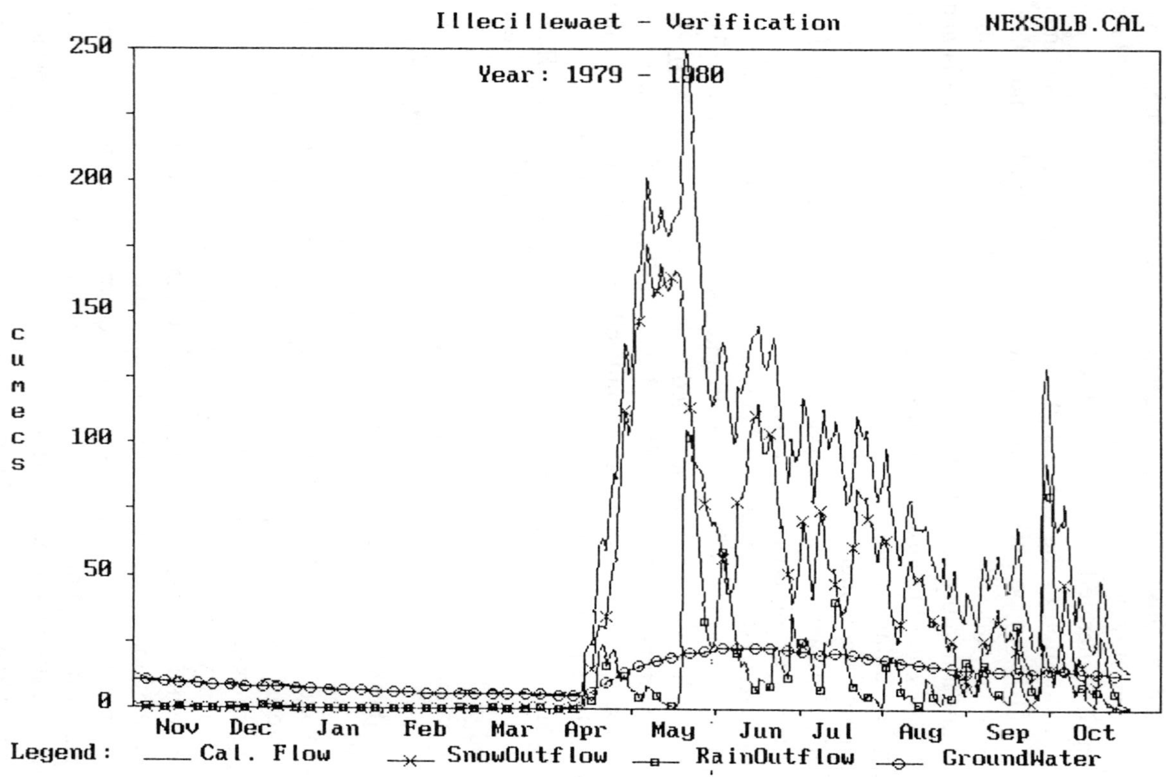

Fig. 8.5. Illedillewaet - The calculated component contributions for the same 1979-80 year: Total flow, fast snowmelt, fast rain and total groundwater contributions.

of years of record. The resulting flow estimates are used to give a statistical evaluation of both the volume of reservoir inflow and the time distribution of inflow. These estimates can then be updated as the season progresses, so that the final outcome becomes more deterministic.

This forecasting approach has now been used for about seventeen years and is used to generate the likely flow sequences for an optimization procedure which takes into account the various system constraints such as power demand, reservoir storage, flood control, downstream water demands, etc.

The Campbell River System

Forecasting for the Campbell River system has a quite different emphasis to the Illecillewaet because Campbell River is in the coastal mountains of Vancouver Island where the winter weather is dominated by large rainstorms interspersed with snowfall. These rainstorms produce a fast watershed response and require careful reservoir operation to prevent large floods downstream. The watershed response is very fast, of the order of three to four hours, but this response is somewhat slowed by the reservoir regulation, perhaps to some twelve hours or more. The system operators therefore have limited time to respond to a severe storm, even with good weather forecasting. Consequently, hourly data is used to generate flow estimates and this hourly data consists of the latest recorded information, together with the latest precipitation and temperature forecasts for the next few hours and even days, if available. This latest meterological data is received via satellite from data collection platforms at various locations in the watershed and is automatically fed into the watershed data files. The latest forecast information is entered manually and then the watershed model estimates are generated. These estimated flow sequences are routed through the reservoir system using an optimization program which seeks out the best operational strategy to avoid flood damage downstream and also to achieve good power generation benefits.

The nature of the Campbell River watershed response is illustrated in Fig. 8.6 which shows the measured and estimated daily flow sequences for the whole year, and also the estimated groundwater contribution. The statistics for this verification period show a model efficiency of 80% and a volume error of -1%. The more detailed flows for a particular flood event are shown in Fig. 8.7 which presents hourly flows. All the winter storm flows are primarily rain generated, but some of the largest events are rain or snow with associated snowmelt. The spring and early summer flows are generally lower and are mainly snowmelt events. These events are more easily forecast and do not usually present much hazard.

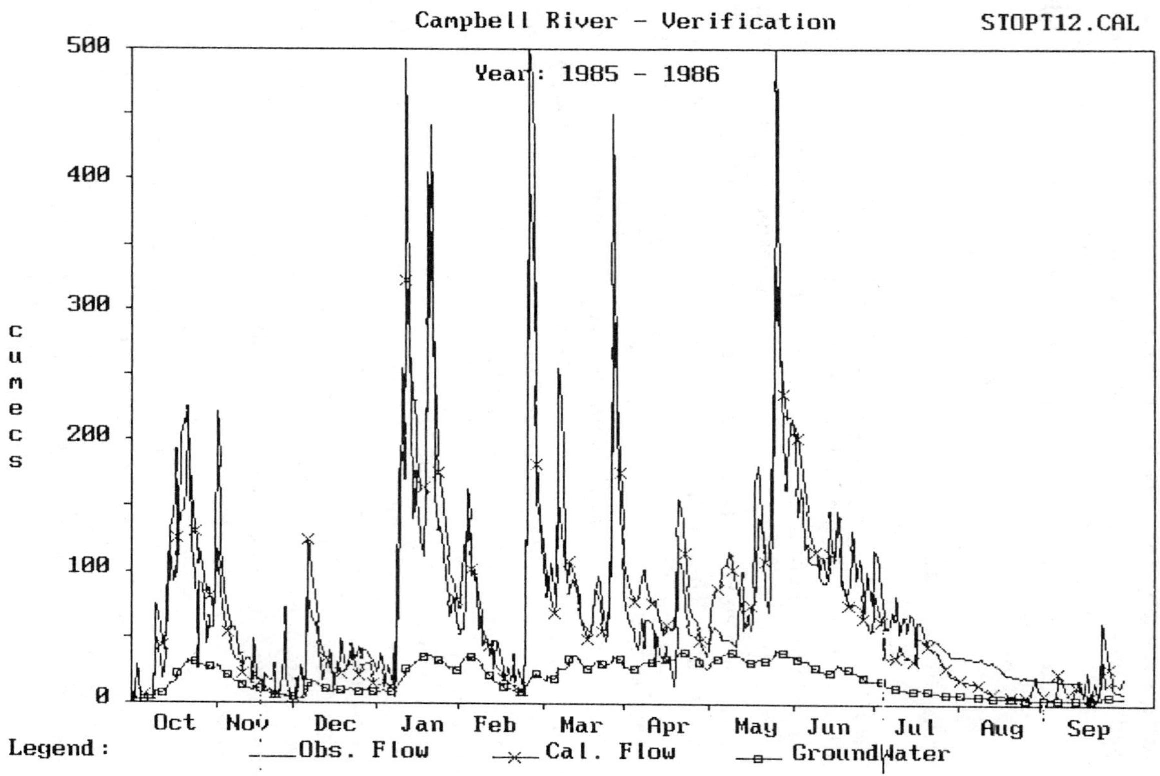

Fig. 8.6. Campbell River - Verification showing estimated and measured flows.

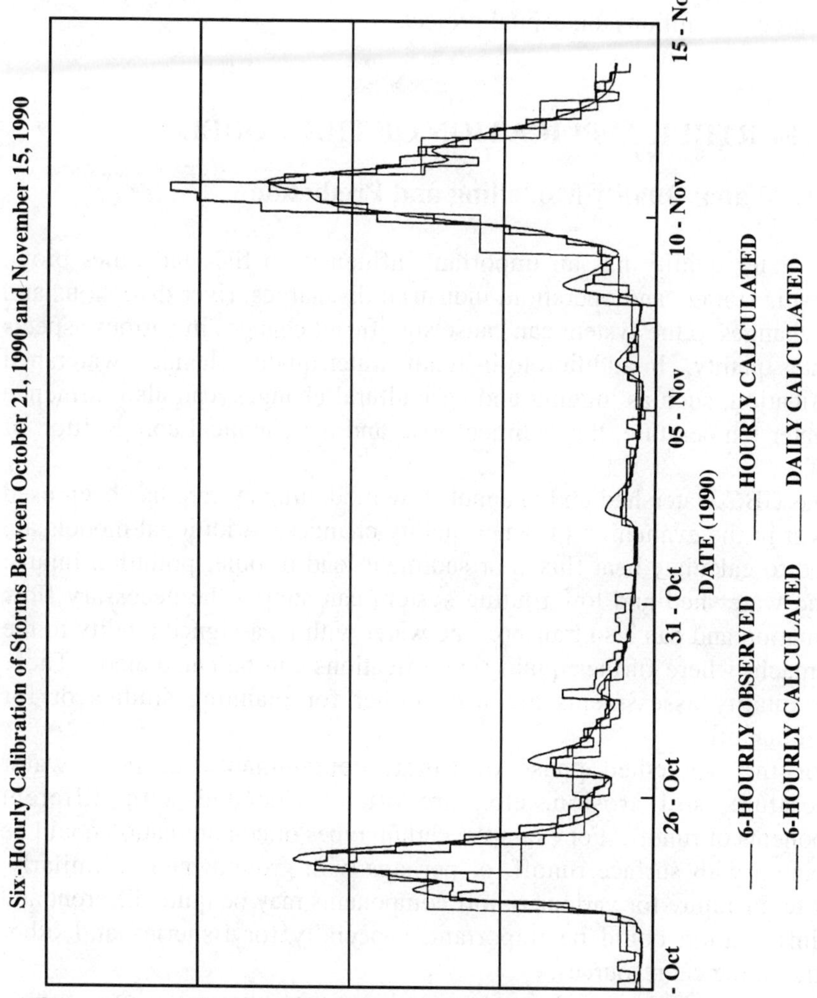

Fig. 8.7 Campbell River - Hourly estimated and measured flows.

The calibrated watershed model, proved to be doubly useful, because it not only provided a valuable forecasting tool for real-time use, but it also showed that the system could be operated more safely than had been thought, and large expenditures on additional spillway and downstream channel capacity were found to be unnecessary. This is an example where modest expenditures on a watershed model can save very large expenditures, of the order of millions, on capital projects.

8.20. FURTHER APPLICATION OF THE MODEL

8.20.1. Water Quality Modelling and Prediction

River water quality has an important influence on fish and other biota. Reservoir storage and operation, industrial discharges, river diversions and other changes to the system can cause significant changes in various aspects of water quality. In addition to instream water quality changes, watershed modification, such as logging and agricultural changes, can also influence the water temperature, the sediment load and the chemical composition of runoff.

The UBC watershed and channel flow modelling system has been used to assist in the evaluation of water quality changes. Additional models are needed to calculate heat fluxes or sediment load or other pollution inputs, but the watershed and low routing system can supply the necessary flow information and can also transport the water with its assigned quality to the next reach, where further quality modifications can be calculated. These water quality assessments are used either for planning studies or for operational use.

For the watershed phase of runoff contaminant transport, water temperature, soil erosion, etc., are often associated with different components of runoff. For example, certain types of contamination could be associated with surface runoff, or perhaps with groundwater. Similarly, water temperature for various runoff components may be quite different and this information could be important, especially for fisheries and other aquatic biological resources.

The watershed model is designed so that these various runoff components are accessible to the user either for each elevation zone, or for the total watershed. These flow components can then be imported to a user-selected sub model which would estimate the required water quality status.

It is anticipated that this water quality modelling will become an increasingly important part of the total process, and the watershed and flow modelling is a central element to this work.

8.21. AVAILABILITY OF THE MODEL

The UBC Watershed and Flow Routing System has been developed with public funding and is therefore available at nominal cost, which covers preparation of the computer diskettes and manuals, including sample input and output files.

Requests for software should be addressed to:

Dr. M.C. Quick
Mountain Hydrology Research Group
Department of Civil Engineering
University of British Columbia
Vancouver, B.C., Canada
V6T 1Z4

FAX: (604) 822-6901

ACKNOWLEDGMENTS

This model has been developed with funding from the B.C. Disaster Relief Fund administered by the Vancouver Foundation. More recently research grant funding has been received from the B.C. Science Council and the B.C. Hydro and Power Authority to improve the user interface, semi-automatic calibration and updating capability. The model versions developed up to 1987 were developed cooperatively with Anthony Pipes. Since that time various workers contributed to the latest version, the major contributors being Hamed Assaf, Edmund Yu, Dan Nixon, with additional help from Sakis Loukas, Rob Millar and Pratap Singh.

REFERENCES

The fundamental work by many researchers which has been used in the development of various algorithms in the UBC Watershed Model, is referenced in the papers cited below.

Assaf, H. and Quick, M.C., "Updating Hydrological Model Forecasts," Proceedings of the Canadian Society of Civil Engineers Annual Conference, Hamilton, Ontario, May 1990, pp. V-226 to V-245.

Assaf, H. and Quick, M.C., "Updating Snowpack Forecasts," Proceedings of the Canadian Society of Civil Engineers Annual Conference, Vancouver, B.C., May 1991, pp. 419-428.

Assaf, H. and Quick, M.C. "Updating Hydrological Model Forecasts," Canadian Journal of Civil Engineering, Vol. 18, No. 4, August 1991, pp. 663-674.

Loukas, A. and Quick, M.C., "Examination of the Hydrological Behaviour of a Small Steep Forested Watershed in Southwestern British Columbia," Proceedings of the Canadian Society of Civil Engineers Annual Conference, Vancouver, B.C., May 1991, pp. 376-386.

Loukas, A. and Quick, M.C., "Hydrologic Behaviour of a Mountainous Watershed," Canadian Journal of Civil Engineering, Vol. 20, No. 1, Feb. 1993, pp. 1-8.

Loukas, A. and Quick, M.C., "Distribution and Variability of Annual and Monthly Precipitation in Two Mountainous Watersheds in British Columbia," CSCE Eleventh Canadian Hydrotechnical Conference, Fredericton, New Brunswick, June 8-11, 1993, pp. 139-148.

Loukas, A. and Quick, M.C., "Spatial and Temporal Distribution of Storm Precipitation in Southwestern British Columbia," ASCE International Symposium on Engineering Hydrology, San Francisco, July 25-30, 1993.

Loukas, A. and Quick, M.C., "Rain Distribution in a Mountainous Watershed," Nordic Hydrology Journal, Vol. 2l4, No. 4, August 1993.

Quick, M.C., "River Flood Flows," Journal of the Hydraulics Division, ASCE, No. HY3, Vol. 9, May 1965, pp. 1-18.

Quick, M.C., "A Comparison of Measured and Theoretical Snowpack Temperatures," Journal of Hydrology, Vol. 5, No. 1, March 1967, pp. 1-20.

Quick, M.C., "Laboratory Simulation of Snowmelt," presented at the Western Snow Conference, Annual Proceedings of the Western Snow Conference, April 1970, pp. 37-43.

Quick, M.C., "Experiments with Physical Snowmelt Models," National Research Council Hydrology Symposium No. 8: Runoff from Snow and Ice, Quebec City, May 1971, pp. 131-150.

Quick, M.C., "Forecasting Runoff: Operational Practices," WMO, UNESCO, International Symposium on the Role of Snow and Ice in Hydrology, published by Canadian NRC, 1972, pp. 943-955.

Quick, M.C., "Computational Procedures in Streamflow Forecasting, Invited theme paper: Technical Workshop on Streamflow Forecasting. Atmospheric Environment Service, Toronto, November 1983, pp. 1-13.

Quick, M.C., "Snowmelt Hydrology and Applications to Runoff Forecasting," Invited Lecture for the 1986-87 National Lecture Tour sponsored by the Associate Committee on Hydrology of the National Research Council and the Hydrotechnical Committee of the Canadian Society of Civil Engineers. (To be published.) (This lecture was presented at twelve universities and cities cross Canada.)

Quick, M.C., "Snowmelt in High Unforested Mountain Regions," Invited Paper given at Indian Workshop on Himalayan Snowmelt, Manali, India, November 1988. Sponsored by Indian Water Commission and U.S. AID, and published in Proceedings.

Quick, M.C., "Design Flood Maximization: Re-examination of the Hydrological Assumptions and Data Reliability," (Invited Paper) Proceedings of the Workshop on Dam Safety and Design Floods in the Rocky Mountain Region, U.S. Geological Survey, Colorado Section, October 1990, pp. 1-32.

Quick, M.C. "Reliability of Flood Discharge Estimates," Canadian Journal of Civil Engineering, Vol. 18, No. 4, August 1991, pp. 624-630.

Quick, M.C., Loukas, A., Millar, R.G., Nixon, D. and Wu, E. - UBC Mountain Hydrology Group, "Application of the UBC Watershed Model to Two Climatically Different Watersheds," 27th Annual Canadian Meteorological and Oceanographic Society Congress, Fredericton, New Brunswick, June 1993.

Quick, M.C. and Pipes, A., "Daily and Seasonal forecasting with a Water Budget Model," Proceedings of the International Symposium on the Role of Snow and Ice in Hydrology, UNESCO, WMO, and NRC, Canada, Banff, September 1972, pp. 1017-1034.

Quick, M.D. and Pipes, A., "Nonlinear Channel Routing by Computer," Journal of the Hydraulics Division, ASCE, Vol. 101, No. HY6, September 1976, pp. 651-664.

Quick, M.C. and Pipes, A., "A Combined Snowmelt and Rainfall Runoff Model," Canadian Journal of Civil Engineering, Vol. 3, No. 3, September 1976, pp. 449-460.

Quick, M.C. and Pipes, A., "UBC Watershed Model," Hydrological Sciences Bulletin XXI,1,3, 1977, pp. 285-295.

Quick, M.C. and Pipes, A., "Modelling Large Scale Effects of Snowcover," Proceedings of he International Association of Hydrological Sciences at the International Union of Geophysics and Geodesy Conference, Vancouver, B.C., August 1987.

Quick, M.C. and Pipes, A., "Problems of High Mountain Snowmelt Runoff Simulation for Different Physiographic Regions," International Union of Geophysics, XIX General Assembly, Vancouver, B.C., Abstracts, Vol. 3, HW 1-42, August 1987, p. 954.

Quick, M.C. and Pipes, A., "High Mountain Runoff and Application to Runoff Forecasting," Proceedings of the Canadian Society of Civil Engineers Annual Conference, St. Johns, Newfoundland, June 1989, pp. 232-247.

Quick, M.C., Pipes, A. and Russell, S.O., "Simulating Snowmelt Hydrographs for the Fraser River System," presented at the Western Snow Conference. Annual Proceedings of the Western Snow Conference, 1970, April 1970, pp. 91-97.

Quick, M.C. and Singh, P., "Watershed Modelling in the Himalayan Region," International Symposium on Hydrology in Mountainous Areas, Shimla, India, May 28-30, 1992, pp. 201-233.

UBC Watershed Model Manual, Version 2.0, August 5, 1993, Mountain Hydrology Group, Dept. of Civil Engineering, Vancouver, B.C., V6T 1Z4.

Chapter 9

THE PRECIPITATION-RUNOFF MODELING SYSTEM - PRMS

G. H Leavesley and L. G. Stannard

9.1. INTRODUCTION

The Precipitation-Runoff Modeling System (PRMS) (Leavesley et al., 1983) is a modular-design, distributed-parameter, physical-process watershed model that was developed to evaluate the effects of various combinations of precipitation, climate, and land use on watershed response. Watershed response to normal and extreme rainfall and snowmelt can be simulated to evaluate changes in water-balance relations, flow regimes, flood peaks and volumes, soil-water relations, sediment yields, and ground-water recharge. Parameter-optimization and sensitivity-analysis capabilities are provided to fit selected model parameters and to evaluate their individual and joint effects on model output.

Selected model enhancements have been made and support programs have been coupled with PRMS to produce a watershed-modeling and data-management system that provides a range of hydrologic-simulation and data-analysis capabilities. Data-management and analysis components are provided by the U.S. Geological Survey's data-management program ANNIE (not an acronym) (Lumb et al., 1990). ANNIE is an interactive program that accepts and reformats data from a variety of sources for model application. ANNIE also includes a number of statistical, graphical, and data-management functions for analysis of measured and simulated data.

Hydrologic forecasting capabilities are provided by a modified version of the U.S. Weather Service's Extended Streamflow Prediction (ESP) program (Day, 1985). ESP applications include short-term and seasonal forecasting for floods and water supply, and evaluation of the effects of land-use and climate changes on hydrologic response.

The purpose of this paper is to provide an overview of PRMS, its system components, and system applications. More detailed discussions regarding individual components can be found in cited references for each component. The major components of the system, and the data transfers among these components, are depicted schematically in Fig. 9.1 and described hereafter.

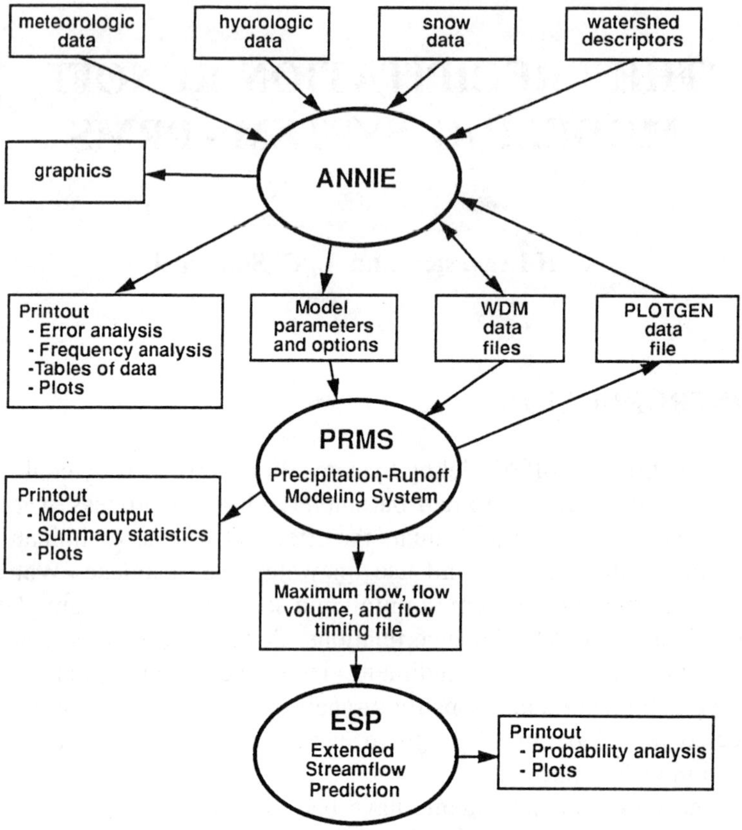

Fig. 9.1. Schematic diagram of the data transfers among PRMS.ANNIE. and ESP.

9.2. ANNIE

ANNIE is a modular-design interactive program that provides the data-management and analysis functions for PRMS. ANNIE is composed of software modules designed to help a user interactively create, check, and update input to hydrologic models, and to provide statistical and graphical tools to assist in the analysis of model input and output. Data from a variety of sources are reformatted by ANNIE to a system-compatible file structure. Meteorologic, hydrologic, and snow data are placed in a direct-access file structure called the Watershed Data Management (WDM) file. Data in the WDM file can be analyzed using a number of statistical, graphical, and data-management functions. Model- and watershed-specific data can be entered interactively into ANNIE to provide attribute information for selected data sets.

A WDM file is composed of one or more data sets, each of which is identified by a user-specified data-set number. Stored with each data set are a variety of user-selectable attributes that can be used to describe the type, geographic location, and other characteristics of the data set.

New data sets can be generated from existing data sets using a number of selectable procedures. One procedure generates a new data set by applying a mathematical operation to one, or selected combinations of two, data sets. Mathematical functions include log transformations, power transformations, weighted sums, and the addition or subtraction of, and multiplication or division by, constants. A second procedure generates a new data set by changing the time step of an existing data set. A disaggregate transformation expands a time series into shorter time steps. An aggregate transformation condenses a time series into longer time steps.

A plot component provides the graphic capabilities in ANNIE. Data sets in the WDM file, results of selected statistical analyses, and PRMS results placed in a PLOTGEN file can be displayed graphically on a monitor or on a pen plotter. Graph formats include time-series plots of one to five data sets, x-y plots of selected data sets, frequency plots, flow-duration plots, and scatter diagrams.

9.3. TIME AND SPACE IN PRMS

Watershed response can be simulated at both a daily and a storm time scale in PRMS. The daily mode simulates hydrologic components as daily average or total values. Streamflow is computed as a mean daily flow. The storm mode simulates selected hydrologic components at time intervals shorter than 1 day. The minimum time interval is 1 minute. The storm mode computes storm hydrographs and sediment yields for selected rainstorms. Sediment-modeling capabilities are provided only in the storm mode.

Distributed-parameter capabilities are provided by partitioning a watershed into units, using characteristics such as slope, aspect, elevation, vegetation type, soil type, and precipitation distribution. Each unit is assumed to be homogeneous with respect to its hydrologic response and to the characteristics listed above; each unit is called a hydrologic response unit (HRU). A water balance and an energy balance are computed daily for each HRU. The sum of the responses of all HRUs, weighted on a unit-area basis, produces the daily watershed response.

A second level of spatial partitioning is used for storm-mode computations. The watershed is conceptualized as a series of interconnected flow-plane and channel segments. An HRU can be considered the equivalent of a single flow plane, or it can be delineated into a number of flow planes.

9.4. PRMS DAILY-MODE COMPONENTS

The daily-mode components simulate the daily accretion, depletion, storage, and movement of water in a HRU. The rates and volumes of these processes are a function of HRU physical, hydrologic, and climatic characteristics. The conceptualized watershed system and its inputs are schematically depicted in Fig. 9.2. Brief descriptions of the components that comprise major hydrologic-process subsets of the watershed system and those that provide model-support capabilities are presented hereafter.

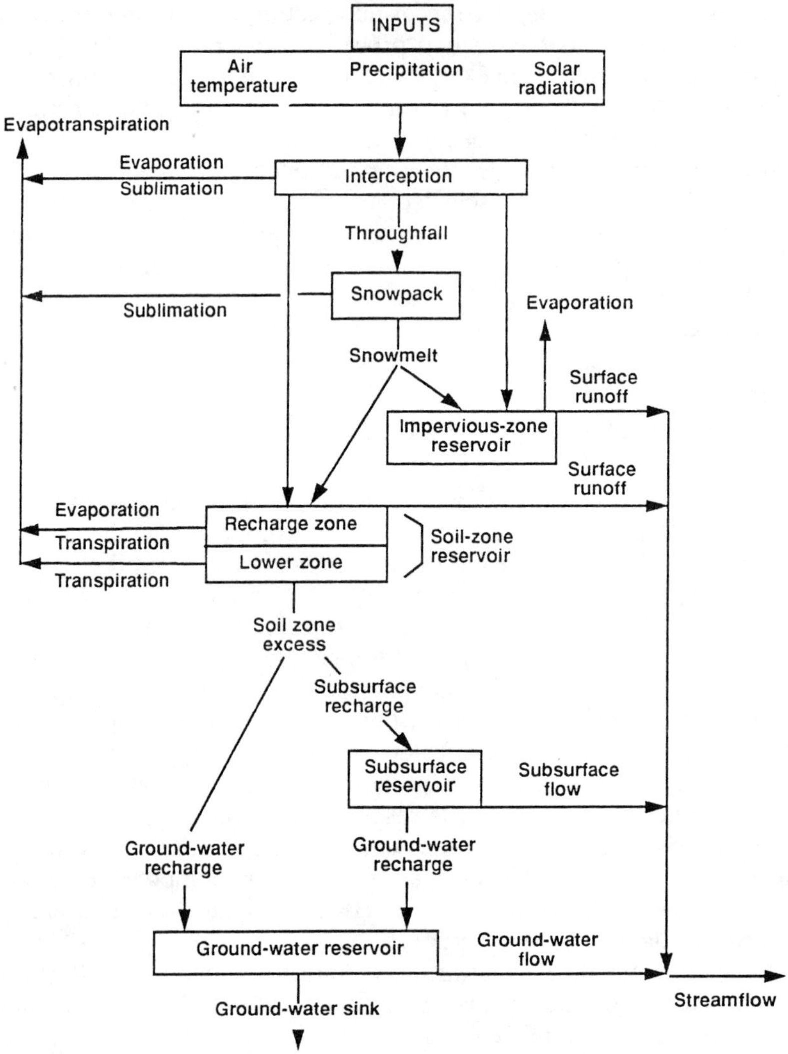

Fig. 9.2. Schematic diagram of the conceptual watershed system and its inputs (modified from Leavesley et al., 1983).

9.4.1. Climate Components

Model inputs are daily precipitation, maximum and minimum air temperature, and solar radiation. Precipitation in the form of rain, snow, or a mixture of both is decreased by interception and becomes net precipitation delivered to the watershed surface. The energy inputs of air temperature and solar radiation are used in the computation of evaporation, transpiration, sublimation, and snowmelt. Daily climate data are extrapolated to each HRU using a set of user-defined adjustment coefficients developed from regional climate data.

Temperature

Measured maximum (TMX) and minimum (TMN) daily air-temperature data are adjusted using monthly or daily lapse rates and the elevation difference between a climate station and each HRU. Up to five climate stations can be used. When using monthly lapse rates, a correction factor (TCRX) for adjusting TMX for each HRU is computed for each month (MO) by

$$TCRX(MO) = \left[TLX(MO) \bullet ELCR\right] - TXAJ \tag{9.1}$$

where

TLX is the lapse rate for maximum daily air temperature, in degrees per 1000 ft change in elevation for month MO,

ELCR is mean elevation of an HRU minus the elevation of the climate station associated with the HRU, in 1000's of ft, and

TXAJ is an average difference in maximum air temperature between a horizontal surface and the slope-aspect of an HRU.

A correction factor [TCRN(MO)] for adjusting TMN for each HRU is computed monthly with an equation of the same form as (9.1), using the monthly lapse rate for minimum daily air temperature [TLN(MO)] and a minimum temperature slope-aspect correction (TNAJ). An option is provided to compute TLX and TLN daily using the maximum and minimum daily air temperatures from a high-elevation and a low-elevation temperature station.

Adjusted daily maximum air temperature (TM) for each HRU is computed by

$$TM = TMX - TCRX(MO) \tag{9.2}$$

Adjusted daily minimum air temperature (TN) for each HRU is computed using an equation of the same form as eq. (9.2) and the variables TMN and TCRN(MO).

Precipitation

Total daily precipitation depth (PPT) received on an HRU is computed by

$$PPT = PDV \bullet PCOR \tag{9.3}$$

where

PDV is daily precipitation depth measured in the precipitation gage associated with the HRU, in inches (in.), and
PCOR is the correction factor for the HRU.

A maximum of five gages can be used to define the spatial and temporal variation of precipitation on a watershed. A PCOR value is defined for each HRU and accounts for the effects of elevation, spatial variation, topography, gage location, deficiencies in gage catch due to the effects of wind, and other factors. Different values of PCOR can be specified for summer and winter periods to account for different precipitation forms and for changes in storm characteristics. PCOR is set to parameters DRCOR for daily rainfall, DSCOR for daily snowfall, and UPCOR for storm rainfall.

In some geographic and climatic regions, the influence of elevation on precipitation volume varies seasonally with general storm patterns (Barry, 1972). Frontal origin storms may show a strong elevation effect, while convective storms show little elevation effect. If DRCOR and DSCOR include strong elevational influence, they can be overridden for a user-defined period where convective storms are dominant to account for these seasonal changes.

Precipitation form (rain, snow, or mixture of both) on each HRU is estimated from the HRU maximum (TM) and minimum (TN) daily air temperatures and their relation to a base temperature (BST) (Willen et al., 1971). Precipitation is all snow if TM is less than or equal to BST, and all rain if TN is greater than BST. If TM is greater than BST and TN is less than BST, then the part of the total precipitation occurring as rain (PRMX) is computed by

$$PRMX = \left(\frac{TM - BST}{TM - TN} \right) \bullet AJMX(MO) \tag{9.4}$$

where AJMX is a monthly mixture adjustment coefficient for month MO. For mixed events, rain is assumed to occur first.

The precipitation form estimation procedure can be overridden in two ways. One is the use of a parameter PAT. PAT is an air temperature value which when exceeded by TM, forces all precipitation to be rain, regardless of the TN value. This capability permits accounting for convective rainfall during spring, when TN may be below BST. PAT is user-specified for each month. The second override is for days when the form is known. Precipitation form can be specified in the model input. Also, snowpack volumes on each HRU can be adjusted using measured snowcourse data.

Shortwave Radiation

Daily shortwave radiation, expressed in langleys per day (ly/d), is used in snowmelt computations and can be used in evapotranspiration computations. Shortwave radiation is an optional input for areas with no snowmelt. Measured shortwave radiation (ORAD), from a horizontal surface, is adjusted to estimate daily shortwave radiation received on the slope-aspect combination of each HRU (SWRD) using a method described by Swift (1976). SWRD is computed by

$$SWRD = ORAD \bullet \frac{DRAD}{HORAD} \bullet \frac{1.}{COSSLP} \tag{9.5}$$

where

DRAD is daily potential solar radiation for the slope and aspect of an HRU (ly),
HORAD is daily potential solar radiation for a horizontal surface (ly), and
COSSLP is cosine of the slope inclination of an HRU.

DRAD and HORAD are linearly interpolated from a table of 13 potential solar radiation values that are calculated for each HRU slope-aspect combination and the horizontal surface. Because of the symmetry of the solar year, these 13 values represent the potential solar radiation for 24 dates throughout the year. These 13 values are computed using a combination of methods described in Frank and Lee (1966) and Swift (1976).

For periods of missing solar radiation data or where these data are unavailable, ORAD can be estimated using air temperature, precipitation, and computed potential solar-radiation data. One of two procedures can be selected. The first is a modification of the degree-day method described by Leaf and Brink (1973). This method was developed for a section of the Rocky Mountain region of the United States. It is most applicable to this region where predominantly clear skies prevail on days without precipitation.

The second procedure uses a relation between sky cover and daily range in air temperature demonstrated by Tangborn (1979) and a relation between solar radiation and sky cover developed by Thompson (1976). This procedure is applicable to more humid regions where extensive periods of cloud cover occur with and without precipitation.

9.4.2. Land Phase Components

Interception

Interception is computed as a function of vegetation cover density and the storage available on the predominant vegetation of an HRU. Net precipitation (PTN) on an HRU is computed by

$$PTN = [PPT \bullet (1. - COVDN)] + [PTF \bullet COVDN] \quad (9.6)$$

where

PPT is total precipitation received on an HRU (in.),
COVDN is seasonal cover density, and
PTF is precipitation falling through the canopy (in.).

COVDN is defined for summer (COVDNS) and winter (COVDNW). PTF is computed by

$$\begin{aligned} PTF &= PPT - (STOR - XIN) & PPT &> (STOR - XIN) \\ PTF &= 0. & PPT &\leq (STOR - XIN) \end{aligned} \quad (9.7)$$

where

STOR is maximum interception storage depth on vegetation (in.), and
XIN is the current depth of interception storage (in.).

STOR is defined by season and precipitation form; winter snow (SNST), winter rain (RNSTW), and summer rain (RNSTS). When rain falls on intercepted snow, STOR changes from SNST to the seasonal rain-storage value. Any snow in storage that exceeds the new rain-storage value is added to throughfall PTF. When snow falls on intercepted rain, STOR changes to SNST and currently available rain interception is used to satisfy SNST. Intercepted rain in excess of SNST is added to PTF.

Intercepted rain is assumed to evaporate at a free-water-surface rate (EVCAN). If pan evaporation data are used, EVCAN equals the pan loss rate. If potential evapotranspiration (PET) is computed from meteorologic variables, EVCAN is computed by

$$EVCAN = \frac{PET}{EVC(MO)} \tag{9.8}$$

where EVC is the evaporation-pan coefficient for month MO. Sublimation of intercepted snow is assumed to occur at a rate that is expressed as a user-defined percentage of PET. In addition, intercepted snow can be removed from the canopy by melting. An energy budget is computed for a 12-hour period (assumed 0600 to 1800). If the energy balance is negative or zero, no melt occurs, When the energy balance is positive, all melt is delivered to the snowpack or soil surface as net precipitation.

Soil-Moisture Accounting

The watershed system is conceptualized as a series of reservoirs, the responses of which combine to produce the total watershed response. The soil-zone reservoir represents that part of the soil mantle that can lose water through the processes of evaporation and transpiration. Average rooting depth of the predominant vegetation covering the soil surface defines the depth of this zone. The maximum available water-holding capacity of the soil-zone reservoir (SMAX) is the difference between field capacity and wilting point of the profile.

The soil-zone reservoir is treated as a two-layered system. The upper zone is termed the recharge zone and the remaining profile is the lower zone. The recharge zone has a user definable maximum available water-holding capacity (REMX) and losses from this zone occur from evaporation and transpiration; losses from the lower zone are assumed to occur only through transpiration. The maximum available water-holding capacity of the lower zone is the difference between SMAX and REMX.

Water storage in the soil-zone reservoir is increased by infiltration of rainfall and snowmelt and is decreased by evapotranspiration. Evapotranspiration losses occur at a rate that is a function of available soil-moisture storage and soil type. The attempt to satisfy potential evapotranspiration is made first from the recharge zone. Soil-moisture accretions must fill the recharge zone before water will move to the lower zone. When the soil-zone reservoir reaches SMAX, additional infiltration is routed to the subsurface and ground-water reservoirs.

The apportioning of water in excess of SMAX to the subsurface and ground-water reservoirs is done using a ground-water recharge parameter SEP. Daily ground-water reservoir inflow is equal to or less than SEP depending on the magnitude of soil-water excess. When soil-water excess is larger than SEP, the difference becomes subsurface reservoir inflow.

Evapotranspiration

Three procedures are available to compute potential evapotranspiration (PET). One procedure uses daily pan-evaporation data and a monthly pan-adjustment coefficient. PET (in./day) is computed by

$$PET = EPAN \bullet EVC(MO) \tag{9.9}$$

where

EPAN is daily pan evaporation (in.), and
EVC is a monthly pan-adjustment coefficient for month MO.

A second procedure computes PET as a function of daily mean air temperature and possible hours of sunshine (Hamon, 1961). PET is computed by

$$PET = CTS(MO) \bullet (DYL)^2 \bullet VDSAT \tag{9.10}$$

where

CTS is a coefficient for month MO,
DYL is possible hours of sunshine, in units of 12 hours, and
VDSAT is the saturated water vapor density (absolute humidity) at the daily mean air temperature, in g/m^3.

VDSAT is computed using empirical relations between VDSAT and saturated vapor pressure (Federer and Lash, 1978) and between saturated vapor pressure and mean daily air temperature (Murray, 1967). DYL varies with time of year and HRU slope and aspect. A table of 13 DYL values is computed for each HRU for the same 13 dates used in the solar radiation estimation procedure. Daily DYL values for each HRU are linearly interpolated from this table of 13 values.

The third procedure is a modified Jensen-Haise technique (Jensen et al., 1969) that computes PET using air temperature, solar radiation, and two coefficients that can be estimated using regional air-temperature, elevation, vapor-pressure, and vegetation data. PET is computed by

$$PET = CTS(MO) \bullet (TAVF - CTX) \bullet RIN \tag{9.11}$$

where

CTS is a coefficient for the month MO,
TAVF is the daily mean air temperature (° F),
CTX is a coefficient, and

RIN is daily solar radiation expressed in inches of evaporation potential.

Actual evapotranspiration (AET) is computed as a function of soil type, water currently available in the soil-zone reservoir, and the storage capacity of the soil-zone reservoir using an approach developed by Zahner (1967). When available water is nonlimiting, AET equals PET. PET is first satisfied from interception storage, retention storage on impervious surfaces, and evaporation/sublimation from snow surfaces. Remaining PET demand then is applied to the soil-zone storage. AET is computed separately for the recharge zone and the lower zone using the unsatisfied PET demand and the ratio of currently available water in the soil zone to its maximum available water-holding capacity. AET computed for the recharge zone is used first to satisfy PET; any remaining demand is attempted to be met by AET from the lower zone. Relations between AET and available soil moisture are defined for three soil types; sand, loam, and clay.

The active transpiration period is user defined by a beginning and ending month. The specific date of the start of transpiration is computed for each HRU using an HRU temperature parameter TST. The sum of the maximum daily air temperatures is accumulated for each HRU, starting on the first day of the month transpiration begins. When this sum exceeds TST, transpiration is assumed to begin. Transpiration is assumed to end on the first day of the month specified as the month transpiration ends.

Surface Runoff

Net precipitation reaches the snowpack or soil surface where it accumulates in the snowpack or is available for surface runoff and infiltration. Daily surface runoff from rainfall on pervious snow-free HRU's is computed using a contributing-area concept (Dickinson and Whiteley, 1970; Hewlett and Nutter, 1970). Net precipitation not becoming surface runoff infiltrates the soil surface. The percent of an HRU contributing to surface runoff (CAP) can be computed as either a linear or a nonlinear function of antecedent soil moisture and rainfall amount. In the linear case

$$CAP = SCN + \left[(SCX - SCN) \bullet \left\{ \frac{RECHR}{REMX} \right\} \right] \tag{9.12}$$

where

SCN is the minimum possible contributing area,
SCX is the maximum possible contributing area,
RECHR is the current available water in the soil recharge zone (in.), and
REMX is maximum storage capacity of the recharge zone (in.).

Surface runoff (SRO) then is computed by

$$SRO = CAP \bullet PTN \tag{9.13}$$

where PTN is the daily net precipitation.

The nonlinear scheme uses a moisture index (SMIDX) similar to that developed by Dickinson and Whiteley (1970) for estimating CAP. CAP is computed by

$$CAP = SCN \bullet 10.^{(SC1 \bullet SMIDX)} \tag{9.14}$$

where

SCN and SC1 are coefficients, and
SMIDX is the sum of the current available water in the soil zone (SMAV) plus one-half of the daily net precipitation (PTN) (in.).

Surface runoff from snowmelt is computed only on a daily basis. Snowmelt runoff from pervious areas is assumed to occur only when the soil zone of an HRU reaches field capacity. At field capacity, a daily maximum infiltration rate is assumed and user defined. Any daily snowmelt in excess of this maximum infiltration rate is assumed to become surface runoff. For impervious areas, snowmelt first satisfies available retention storage, and the remaining snowmelt becomes surface runoff.

Subsurface Flow

The subsurface reservoir simulates the relatively rapid component of flow that may occur in the saturated-unsaturated and ground-water zones during periods of rainfall and snowmelt. The subsurface reservoir can be defined as being linear or nonlinear. Outflow form the subsurface reservoir (RAS) is computed by

$$RAS = (RCF \bullet RES) + \left(RCP \bullet (RES)^2\right) \tag{9.15}$$

where

RCF and RCP are routing coefficients, and
RES is the storage volume in the subsurface reservoir (in.).

This is a nonlinear relationship. However, setting RCP equal to 0 makes the relationship linear. A subsurface reservoir can receive inflow from one or several HRU's. The number of subsurface reservoirs can range from one to the number of HRU's delineated.

Ground Water

The ground-water reservoir simulates the slower component of flow from the ground-water zone. It is conceptualized as a linear reservoir and is assumed to be the source of all baseflow (BAS). Inflow to the ground-water reservoir can be from both soil-water excess and one or more subsurface reservoirs. The vertical movement of water from a subsurface reservoir to a ground-water reservoir is computed as a function of the current volume of storage in subsurface reservoir and a linear routing coefficient. The shape of the baseflow recession of the simulated hydrograph will be effected by the relative proportion of ground-water recharge from the two source terms. Recharge from the soil zone occurs only on days when SMAX is exceeded while recharge from the subsurface reservoir occurs every day water is available in the subsurface reservoir. BAS is computed by

$$BAS = RCB \bullet GW \tag{9.16}$$

where

RCB is a routing coefficient, and
GW is the ground-water reservoir storage (acre-in.)

The movement of water through the ground-water reservoir to points outside the surface drainage boundary is treated using a ground-water sink which is computed as a function of GW and a linear routing coefficient.

9.4.3. Channel Reservoir Components

Channel reservoir components simulate the storage and routing response of channel reservoirs. Reservoir inflows are computed as the sum of the streamflow contributions from all HRU's and the parts of subsurface and ground-water reservoirs above the channel reservoir. Reservoir inflow also can include the outflow of up to 3 upstream channel reservoirs. Two types of routing procedures are available for simulating reservoir outflow. Both are based on the equation of continuity. One is a linear-storage routing procedure in which outflow (O) is computed as a linear function of storage (STO) by

$$O = RCS \bullet STO \tag{9.17}$$

where RCS is a routing parameter.

The second is a modified-Puls routing procedure (U.S. Soil Conservation Service, 1971). In this procedure the outflow (O) and inflow (I) terms are considered averages of O and I and the beginning and end of the computational time interval. The continuity equation for this approach is expressed as

$$\frac{2 \bullet STO_2}{\Delta t} + O_2 = (I_2 + I_1) + \frac{2 \bullet STO_1}{\Delta t} - O_1 \tag{9.18}$$

where the subscripts 1 and 2 specify start and end values for the computational time interval Δt. All terms on the right-hand side of equation (9.18) are known at the beginning of a routing period. An interpolation procedure described by Dawdy et al. (1978) is used to solvc for O_2.

9.4.4. Snow Components

The snow components simulate the initiation, accumulation, and depletion of a snowpack on each HRU. A snowpack is maintained and modified on both a water-equivalent basis and as a dynamic-heat reservoir. A snowpack water balance is computed daily and an energy balance is computed twice each day for 12-hour periods (designated day and night).

The energy-balance computations are a combination of equations and functional relationships taken or derived from several sources. The conceptual model for the snowpack system and its energy relations is one described by Obled and Rosse (1977). The snowpack is assumed to be a two-layered system. The surface layer consists of the upper 3-5 centimeters of the snowpack, and the bottom layer is the remaining snowpack. Heat transfer between the surface layer and the snowpack occurs by conduction when the temperature of the surface layer (Ts) is less than 0 °C. When Ts equals 0 °C, heat transfer occurs as conduction when the net energy balance at the air-snow interface is negative; but heat transfer occurs as mass transfer by surface melting when the net energy balance is positive. Heat transfer from precipitation occurs as a mass-transfer process. Conduction of heat from the soil surface to the snowpack is assumed to be negligible compared to the energy exchange at the air-snow interface. The conceptual snowpack system and the components of the snowpack energy balance are shown in Fig. 9.3.

Ts for the daylight period is computed as the mean of the maximum and mean-daily air temperature for an HRU. For the night period, it is computed as the mean of the minimum and mean-daily air temperature for an HRU. Ts equals 0 °C when these means exceed 0 °C. When Ts is less than 0 °C, the mean air temperature for the 12-hour period is assumed to integrate the effects of the radiation, latent heat, sensible heat, and diffusion processes expressed in the complete equation for the surface temperature. See Anderson (1968) for a complete discussion of the equation.

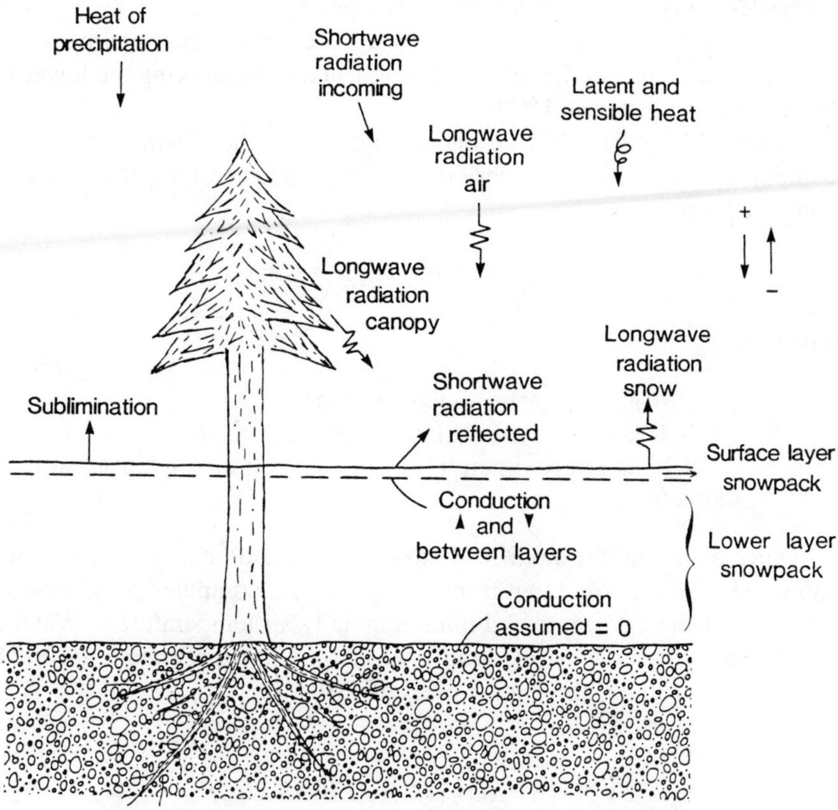

Fig. 9.3 Components of the snowpack energy balance.

The conduction of heat between the snow surface and the lower layer of the snowpack (QCOND) is computed by

$$QCOND = 2 \bullet DEN \bullet CS \bullet \sqrt{\frac{KEFF}{(DEN \times CS)} \bullet \frac{\Delta t}{\Pi}} \bullet (TS - PACT) \quad (9.19)$$

where

DEN is snowpack density (g/cm^3),
CS is the specific heat of ice in calories per gram per °C (cal/g/°C),
KEFF is the effective thermal conductivity of the snowpack (cal/s/cm/°C),
Δt is the time interval (s),
Π is pi, and
PACT is the temperature of the lower layer of the snowpack (°C).

KEFF is assumed equal to 0.0077 x DEN^2 (Anderson, 1968). DEN is computed daily using a procedure developed by Riley et al. (1973). PACT is recomputed each 12-hr period as a function of the snowpack water equivalent and the number of calories of heat needed to bring the lower layer to an isothermal state at 0 °C.

When Ts equals 0 °C, an energy balance at the air-snow interface is computed. For each 12-hr period the energy balance CALN in calories is computed by

$$CALN = SWN + LWN + CEN \tag{9.20}$$

where

SWN is net shortwave radiation (cal),
LWN is net longwave radiation (cal), and
CEN is an approximation of the convection-condensation energy terms (cal).

Daily precipitation modifies both the water and energy balances of the snowpack. The energy input from precipitation is computed as a function of the mean of the maximum and minimum daily air temperatures. Water and energy balance changes resulting from precipitation are made on a 24-hr basis and prior to the application of eq. (9.20).

SWN is computed by

$$SWN = SWRD \bullet (1.0 - ALB) \bullet TRNCF \tag{9.21}$$

where

SWRD is the computed shortwave radiation received on an HRU (cal),
ALB is the albedo of the snowpack surface, and
TRNCF is the transmission coefficient for the vegetation canopy over the snowpack.

ALB is computed as a function of the number of days since the last snowfall and whether the snowpack is in an accumulation or a melt phase (U.S. Army, 1956). A snowpack begins in an accumulation phase, but shifts to a melt phase when the temperature of the lower layer is isothermal at 0°C for 5 consecutive days or the Julian date exceeds a user defined forcing date. TRNCF is computed as a function of the winter cover density of the vegetation canopy over the snowpack. The relations between cover density and TRNCF were developed from relations presented by Miller (1959) and Vézina and Péch (1964).

LWN has two components, the longwave exchange between the air and the snowpack and the exchange between the vegetation canopy and the snowpack. LWN is computed by

$$LWN = (1.0 - COVDNW) \bullet \left[(EMIS \bullet AIRN) - SNON\right] + COVDNW \bullet (AIRN - SNON) \quad (9.22)$$

where

> COVDNW is the winter cover density of the predominant vegetation above the snowpack,
> EMIS is the emissivity of the air,
> AIRN is the longwave energy emitted from a perfect black body at the average air temperature for the 12-hr period (cal), and
> SNON is the longwave energy emitted from the snowpack surface at the temperature Ts for the 12-hr period (cal).

AIRN and SNON are computed using the Stefan-Boltzmann law. EMIS is a function of the moisture content of the air and ranges between 0.757 and 1.0 (U.S. Army, 1956). In the absence of humidity data, simplifying assumptions are made in the estimation of EMIS. During periods when frontal storms are dominant, EMIS is assumed to be 1.0 on days with precipitation and a user–defined parameter value for days without precipitation. During periods when convective storms are dominant, EMIS is computed separately for each 12-hr period as a function of precipitation occurrence and the ratio of observed to potential solar radiation.

The full equation for computing latent and sensible heat flux includes terms for temperature, vapor pressure, wind speed, and diffusivities of heat and vapor (U.S. Army, 1956). However, wind and vapor pressure or humidity data are generally not available. Therefore, the full equation is simplified to a temperature–index form to compute latent and sensible heat terms (CEN) by

$$CEN = CECN(MO) \bullet TAVC \quad (9.23)$$

where CECN is a parameter for month MO and TAVC is the mean air temperature for the 12-hr period (° C). CEN is computed only on days of rainfall or when the ratio of observed to potential solar radiation is less than or equal to 0.33. Areas of forest cover are assumed to receive only one-half of CEN computed by eq. (9.23).

When the energy balance CALN (eq. 9.20) is positive, there is a net gain of energy to the snowpack. This energy melts snow in the surface layer and the melt then transports heat into the snowpack by mass transfer. If the snowpack temperature is less than 0° C, the melt water is refrozen and decreases the cold content of the snowpack. When the snowpack becomes isothermal at 0° C, snowmelt is first used to satisfy the freewater holding capacity of the snowpack. Any remaining melt leaves the bottom of the snowpack to become infiltration or surface runoff. When melt reduces the snowpack water equivalent below a user-defined threshold, the snowcovered area of an HRU is decreased using the areal-depletion-curve approach developed by Anderson (1973). Up to 10 different delpetion curves may be user-defined.

Evaporation and sublimation from the snow surface are assumed to occur only when there is no transpiration from vegetation above the snowpack. Loss from the snow surface is computed as a percentage of the daily PET value. The daily percentage is a user-defined parameter.

9.5. STORM-MODE COMPONENTS

A watershed is configured into flow-plane and channel segments for storm-mode computations. An HRU can be considered a single flow plane or it can be divided into 2 or more flow planes to account for variations in slope and surface roughness. Cascading flow planes are not supported, so all flow planes discharge into channel segments. The watershed drainage network is characterized as a system of channel, reservoir, and junction segments that jointly describe the drainage pattern. Each channel segment can receive upstream inflow from as many as three other segments. In addition, each channel segment can receive lateral inflow from as many as two flow planes (left bank and right bank).

9.5.1. Surface Runoff

Storm precipitation is reduced by interception using eqs. (9.6) and (9.7) and resulting net precipitation is available for infiltration. On pervious areas, infiltration is computed using a variation of the Green and Ampt equation (Green and Ampt, 1911). Point infiltration capacity (FR) is computed by

$$FR = KSAT \bullet \left(1.0 + \frac{PS}{SMS}\right) \qquad (9.24)$$

where

KSAT is hydraulic conductivity of the transmission zone (in./hr),
PS is effective value of the product of capillary drive and moisture deficit (in.), and
SMS is the current value of accumulated infiltration (in.).

PS is varied linearly as a function of the ratio of the soil moisture quantity RECHR to REMX and user-defined parameters that represent the value of PS at field capacity (PSP) and the ratio of PSP at field capacity to PSP at wilting point. Point infiltration is converted to net infiltration over a flow plane using a procedure first presented by Crawford and Linsley (1966) which assumes that net infiltration varies linearly as a function of FR and rainfall rate.

Rainfall excess (net precipitation less net infiltration) is then routed as surface runoff over the flow planes into the channel segments using the kinematic-wave approximation to overland flow developed by Leclerc and Schaake (1973). The equation solved for each flow plane is

$$\frac{\partial h}{\partial t} + \frac{\partial q}{\partial x} = re \tag{9.25}$$

where

h is the depth of flow (ft),

q is the rate of flow per unit width ($ft^3/s/ft$),

re is the rate of rainfall excess inflow (ft/s),

t is time (s), and

x is the distance down plane (ft).

The relation between h and q is given as

$$q = \alpha h^m \tag{9.26}$$

where α and m are computed as functions of overland flow-plane characteristics using equations described by Dawdy et al. (1978).

9.5.2. Channel Flow

Channel flow is routed through the watershed channel system using the kinematic-wave approximation for channel flow described by Dawdy et al. (1978). The equation solved for each channel reach is

$$\frac{\partial A}{\partial t} + \frac{\partial Q}{\partial x} = q \tag{9.27}$$

where

A is the area of flow (ft^2),

Q is the flow rate (ft^3/s), and

q is the lateral inflow per unit length of channel ($ft^3/s/ft$).

The kinematic wave approximation relating Q to A is

$$Q = \alpha A^m \tag{9.28}$$

where α and m are computed from selected channel segment characteristics using equations described by Dawdy et al. (1978).

Routing through channel reservoirs can be computed using the same linear routing scheme or modified-Puls routing procedure described for the daily mode. Equations (9.17) and (9.18) are used but at shorter time steps.

9.5.3. Sediment

Sediment detachment and transport from flow planes is computed using a rill-interrill concept approach presented by Hjelmfelt et al. (1975) and is expressed as

$$\frac{\partial(ch)}{\partial t} + \frac{\partial(cq)}{\partial x} = ER + EF \tag{9.29}$$

where

c is sediment concentration in kilograms per ft^3 (Kg/ft^3),

h is flow depth (ft),

q is flow rate per unit width ($ft^3/s/ft$),

ER is the rainfall detachment rate of sediment ($Kg/ft^2/s$),

EF is the overland flow detachment rate of sediment ($Kg/ft^2/s$), and

x is distance down the flow plane (ft).

ER is assumed to occur in the interrill areas and is computed as a nonlinear function of rainfall rate and the mean depth of flow on the plane using a relationship described by Smith (1976). EF is assumed to occur in the rills and is computed as a linear function of the difference between sediment transport capacity at the current flow depth and the current sediment transport rate using a relationship described by Hjelmfelt et al. (1975).

Sediment delivered from a flow plane is currently transported as a conservative substance in the channel system; detachment and deposition are not included. The channel sediment-routing component will be upgraded in the future with the addition of detachment and deposition capabilities. The continuity of mass equation used to describe sediment transport is expressed as

$$\frac{\partial(cA)}{\partial t}+\frac{\partial(cQ)}{\partial x}=TLAT \tag{9.30}$$

where the equation variables are as defined above and TLAT is the lateral inflow rate of sediment per unit length of channel expressed in Kg/s/ft

9.6. OPTIMIZATION AND SENSITIVITY ANALYSIS COMPONENTS

Optimization components control the automatic adjustment of model parameters to obtain better agreement between observed and predicted runoff. A model parameter is broadly defined as a value that is used to represent a physical or hydrologic characteristic of a watershed, and is held constant during a simulation run. This definition produces a large number of parameters for optimization and sensitivity analysis; however, the availability of a large number of parameters is not meant to encourage optimization of all of them. The parameters are available primarily for development work in the theory of errors for the modeling system and to permit model users to evaluate the sensitivity of each parameter, and its interactions with others.

Two optimization techniques are available: one is the Rosenbrock technique (Rosenbrock, 1960); the second is a Gauss-Newton technique similar to the linearization method described by Draper and Smith (1966), but with the incorporation of the Box-Kanemasu interpolation modification (Beck and Arnold, 1977). One of four objective functions can be selected for use with either technique.

Spatially distributed parameters have an initial value assigned to each HRU, subsurface reservoir, or ground-water reservoir. Temporally distributed parameters have an initial value assigned for each time increment. One or any combination of parameters can be selected for optimization. For each iteration of a distributed parameter, all values of the parameter are moved in the same direction at the same time. The amount that each value is moved can be selected as the same magnitude or as the same percentage of the initial value. A major assumption in this fitting procedure is that the initial estimates of the values of a given distributed parameter are correct with regard to their relative differences in space or time.

An option in the fitting procedure allows the user to adjust different subsets of a distributed parameter independently. A special case of this option is that a distributed parameter may be adjusted independently on all HRU's.

Sensitivity-analysis components allow the user to determine the extent to which uncertainty in the parameters results in uncertainty in the predicted runoff. When sensitivity analysis is coupled with optimization, the user also can assess the magnitude of parameter standard errors and parameter intercorrelations. Discussions of sensitivity analysis and its interpretation are presented by Mein and Brown (1978) and Beck and Arnold (1977).

9.7. EXTENDED STREAMFLOW PREDICTION (ESP)

A modified version of the ESP program has been coupled to PRMS to provide forecasting capabilities. The ESP procedure uses historic or synthesized meteorologic data to forecast future streamflow, given the simulated hydrologic conditions for a watershed at a specified point in time. When historic data are used, the procedure assumes that past meteorological events are representative of future meteorological events. Alternative assumptions about future meteorological conditions can be made with the use of synthesized meteorological data.

To implement the ESP procedure, the user defines in PRMS a forecast period, the meteorological data set to be used, and the forecast variables of interest. The forecast period can vary from a few days to an entire water year. PRMS simulates to the start of the forecast period then stores all model variables that describe the state of the watershed system. A streamflow hydrograph for the forecast period then is simulated for each year in the meteorological data base. PRMS is reinitialized at the start of each iteration of the forecast period using the previously stored variables. The simulation results obtained from this iterative procedure are termed the conditional simulation.

Streamflow variables of interest are extracted from each forecast hydrograph and stored in an ESP file. The variables that can be extracted are the maximum and minimum daily flow values, flow volume, and the date that the flow decreases to less than a selected threshold value. One variable, or any combination of variables, can be selected for a given model run.

The ESP program reads the ESP file, performs a frequency analysis on the variables, and produces a probabilistic forecast for each of the variables in the file. The Log-Pearson Type III probability distribution is currently supported, but other distributions can be added. Frequency analysis results for the conditional simulation are presented in tabular form and in a printer plot.

A frequency analysis also can be performed with ESP using the historic streamflow record or the simulation results obtained from running PRMS using the historic or synthesized meteorologic data without resetting the initial conditions each year. The results of this continuous simulation are termed the historic simulation when historic data are used. A comparison of frequency-analysis results between the measured data and the historic simulation provides a measure of the adequacy of the PRMS calibration. Differences between the analyses reflect the effects of model and data errors on the historic simulation, and they also provide a measure of the bias in the conditional-simulation results.

A continuous simulation using synthesized meteorologic data is termed the synthetic simulation. Differences between the analyses for the historic simulation and the synthetic simulation provide a measure of the effect of the assumed meteorological changes in the synthesized record on streamflow.

The ESP component provides a valuable tool for a variety of hydrologic forecasting and analysis applications. These applications include short-term or seasonal forecasting for flooding and water supply, evaluation of the effects of land-use changes on streamflow characteristics under a variety of meteorologic and watershed conditions, and the evaluation of the potential effects of climatic changes on streamflow under existing or future watershed conditions.

9.8. SYSTEM APPLICATIONS

The modeling system as discussed above has been applied in a variety of climatic and physiographic regions. Reports on snowmelt-runoff applications of PRMS include Leavesley and Striffler (1979), Leavesley et al. (1981), Brendecke and Sweeten (1985), and WMO (1986). A comparison of PRMS with the Streamflow Synthesis and Reservoir Regulation (SSARR) model was presented by Brendecke et al. (1985). A frozen soil algorithm was developed for PRMS by Emerson (1991) to simulate this cold–region process

PRMS was used in a number of studies to investigate the effects of surface coal mining on basin hydrologic response. A summary of these applications on approximately 50 basins in selected regions of the United States was presented by Stannard and Kuhn (1989). More recent applications of PRMS have been to address the potential effects of climate change on basin hydrologic response. Climate change related investigations in Colorado using PRMS coupled with selected atmospheric models were reported by Leavesley et al. (1992) and Hay et al. (1993). Application of selected storm-mode components, including sediment detachment and transport, to tephra deposits from the eruption of Mount St. Helens, Washington, were reported by Leavesley et al. (1989). Other reports focusing on the sediment components of PRMS were presented by Carey and Simon (1984), Reed (1986), and Rankl (1987).

Model parameter estimation and evaluation techniques have been examined by several authors. Parameter estimation and calibration studies were reported for Colorado (Norris and Parker, 1985) and for Montana and Wyoming (Cary, 1991). The transferability of parameters to noncalibrated bains was examined by Kuhn and Parker (1992). A review and application of the parameter optimization and sensitivity analysis components was presented by Troutman (1985). Error analysis methodology was also examined by Rivera–Santos (1990).

A number of approaches have been developed to use GIS tools in the delineation of HRUs for PRMS applications. Leavesley and Stannard (1990a,b) presented a polygon HRU approach while Battaglin et al. (1993) presented a gridded HRU approach. Leavesley and Stannard (1990a) also coupled remotely sensed snowcovered–area data with the GIS tools to provide the ability to compare measured and simulated snowcovered area on selected dates. Flügel and Lüllwitz (1993) developed GIS procedures to compare modeling of micro– and mesoscale catchments in the United States and Germany.

9.9. THE NEXT GENERATION OF PRMS

One of the objectives in the development of PRMS was to provide a modular modeling system that could be used to address a variety of interdisciplinary environmental and water-resource problems. The modular concept enables the testing and development of a variety of modeling approaches that can incorporate knowledge from a broad range of scientific disciplines. Assessment of the initial PRMS modular structure identified a number of deficiencies for meeting these goals, so a new, more flexible, modular system was developed to provide a framework in which PRMS, and any other model or model component, can be incorporated.

The new system is called the Modular Modeling System (MMS) (Leavesley et al., 1992). MMS is an integrated system of Unix-based computer software that has been developed to provide the research and operational framework needed to support development, testing, and evaluation of physical-process algorithms and to facilitate integration of user-selected sets of algorithms into operational physical-process models. MMS uses a master library that contains compatible modules for simulating a variety of water, energy, and biogeochemical processes. A model is created by selectively coupling the most appropriate process algorithms from the library to create an "optimal" model for the desired application.

Where existing algorithms are not appropriate, new algorithms can be developed and incorporated in the library.

A geographic information system (GIS) interface has been developed for MMS to facilitate model development, application, and analysis. This interface permits application of a variety of GIS tools to characterize the topographic, hydrologic, and biologic features of a physical system for use in a variety of lumped- and distributed-parameter modeling approaches. MMS display capabilities permit visualization of the spatial distribution of model parameters and of the spatial and temporal variation of simulated state variables during a model run.

MMS provides a common framework in which to focus multidisciplinary research and operational efforts to provide improved understanding of complex water, energy, and biogeochemical processes. All the components of PRMS and ESP described above have been incorporated in MMS and the MMS modular library. Additional modules from a variety of other hydrologic and ecosystem models are currently being developed for incorporation into the modular library.

REFERENCES

Anderson, E.A., 1968, Development and testing of snow pack energy balance equations: Water Resour. Res., v. 4, no. 1, p. 19–38.

Anderson, E.A., 1973, National Weather Service river forecast system—Snow accumulation and ablation model: NOAA Tech. Memorandum NWS Hydro–17, U.S. Dept. of Commerce, Silver Spring, Maryland.

Barry, R.G., 1972, Climatic environment on the east slope of the Colorado front range: Boulder, University of Colorado, Institute of Arctic and Alpine Research Occasional Paper 3, 206 p.

Battaglin, W.A., Hay, L.E., Parker, R.S., and Leavesley, G.H., 1993, Applications of GIS for modeling the sensitivity of water resources to alterations in climate in the Gunnison River basin, Colorado: Water Res. Bull. v. 25, no. 6, p. 1021–1028.

Beck, J.V. and Arnold, K.J., 1977, Parameter estimation in engineering and science: New York, John Wiley and Sons, Inc., 340-349.

Brendecke, C.M., Laiho, D.R., and Holden, D.C., 1985, Comparison of two daily streamflow simulation models of an alpine watershed: J. Hydrol., v. 77, p. 171–186.

Brendecke, C.M., and Sweeten, J.G., 1985, A simulation model of Boulder's alpine water supply, in Proceedings of 53rd Annual Meeting of the Western Snow Conference: Boulder, Colorado, pp. 63-71.

Carey, W.P., and Simon, A., 1984, Physical basis and potential estimation techniques for soil erosion parameters in the Precipitation–Runoff Modeling System (PRMS): U.S. Geol. Surv. Water–Resour. Invest. Rep. 84–4218, 32 p.

Cary, L.E., 1991, Techniques for estimating selected parameters of the U.S. Geological Survey's Precipitation–Runoff Modeling System in eastern Montana and northeastern Wyoming: U.S. Geol. Surv. Water–Rcsour. Invest. Rep. 91–4068, 39 p.

Crawford, N.H., and Linsley, R.K., 1966, Digital simulation in hydrology: Stanford, Calif., Stanford Watershed Model IV, Tech. Rept. 39, Civil Eng. Dept., Stanford Univ., 210 p.

Dawdy, D.R., Schaake, J.C., Jr., and Alley, W.M.,1978, Distributed routing rainfall-runoff model: U.S. Geol. Surv. Water Resour. Invest. Rep. 78-90.

Day, G.N., 1985, Extended streamflow forecasting using NWSRFS: J. Water Resour. Plan. and Manag. Am. Soc. Civ. Eng., 111(2), 157-170.

Dickinson, W.T., and Whiteley, H.Q., 1970, Watershed areas contributing to runoff: IAHS Publ. no. 96, 1.12-1.28.

Draper, N.R., and Smith, H., 1966, Applied regression analysis: New York, John Wiley and Sons, Inc.

Emerson, D.G., 1991, Documentation of a heat and water transfer model for seasonally frozen soils with application to a precipitation–runoff model: U.S. Geol. Surv. Open–File Report 91–462, 97 p.

Federer, Anthony C., and Lash, Douglas, 1978, Brook—A hydrologic simulation model for eastern forests: Durham, New Hampshire, University of New Hampshire, Water Resources Research Center, Research Report No. 19, 84 p.

Flügel, W.A., and Lüllwitz, Th., 1993, Using a distributed hydrologic model with the aid of GIS for comparative hydrological modelling of micro– and mesoscale catchments in the USA and Germany, in Wilkinson, W.B., ed., Macroscale modelling of the hydrosphere: IAHS Pub. no. 214, p. 59–66.

Frank, E.C., and Lee, Richard, 1966, Potential solar beam irradiation on slopes: U.S. Department of Agriculture, Forest Service Research Paper RM–18, 116 p.

Green, W.H., and Ampt, G.A., 1911, Studies on soil physics, I-- Flow of air and water through soils: J. Agric. Res., 4, 1-24.

Hamon, W.R., 1961, Estimating potential evapotranspiration: J. Hydraul. Div. Am. Soc. Civ. Eng., 87(HY3), 107-120.

Hay, L.E., Battaglin, W.A., Parker, R.S., and Leavesley, G.H., 1993, Modeling the effects of climate change on water resources in the Gunnison River basin, Colorado, in Goodchild, M.F., Parks, B.O., and Steyaert, L.T., eds., Environmental modeling with GIS: Oxford Univ. Press, p. 173–181.

Hewlett, J.D., and Nutter, W.L., 1970, The varying source area of streamflow from upland basins, in Proceedings of Symposium on Interdisciplinary Aspects of Watershed Management: Montana State University, Bozeman, Montana, pp. 65-83.

Hjelmfelt, A.T., Piest, R.P., and Saxton, K.E., 1975, Mathematical modeling of erosion on upland areas in Proceedings of Congress of the 16th Int. Assoc. for Hydraul. Res.: Sao Paulo, Brazil, 2, pp. 40-47.

Jensen, M.E., Robb, D.C.N., and Franzoy, C.E., 1969, Scheduling irrigation using climate-crop-soil data, in Proceedings of the American Society of Engineers National Conference on Water Resoures Engineering: New Orleans, Louisiana, 20 pp.

Kuhn, G., and Parker, R.S., 1992, Transfer of watershed–model parameter values to noncalibrated basins in the Gunnison River basin, Colorado: AWRA 28th Annual Conference and Symposium, Managing water resources during global change, Reno, Nevada, November 1–5, 1992, p. 741–750.

Leaf, C.F., and Brink, G.E., 1973, Hydrologic simulation model of Colorado subalpine forest: U.S. Department of Agriculture, Forest Service Research Paper RM–107, 23 p.

Leavesley, G.H., Branson, M.D., and Hay, L.E., 1992, Using coupled atmospheric and hydrologic models to investigate the effects of climate change in mountainous regions: AWRA 28th Annual Conference and Symposium, Managing water resources during global change, Reno, Nevada, November 1–5, 1992, p. 691–700.

Leavesley, G.H., Lusby, G.C., and Lichty, R.W., 1989, Infiltration and erosion characteristics of selected tephra deposits from the 1980 eruption of Mount St. Helens, Washington: Hydrol. Sci. J., 34(3).

Leavesley, G.H., Lichty, R.W., Troutman, B.M., and Saindon, L.G., 1981, A precipitation-runoff modeling system for evaluating the hydrologic impacts of energy-resource development in Proceedings of 49th Annual Meeting of the Western Snow Conference: St. George, Utah, pp. 65-76.

Leavesley, G.H., Lichty, R.W., Troutman, B.M., and Saindon, L.G., 1983, Precipitation-runoff modeling system--User's manual: U.S. Geol. Surv. Water Resour. Invest. Rep. 83-4238.

Leavesley, G.H., Restrepo, P., Stannard, L.G., and Dixon, M., 1992, The modular hydrologic modeling system – MHMS: AWRA 18th Annual Conference and Symposium, Managing water resources during global change, Reno, Nevada, November 1–5, 1992, p. 263–264.

Leavesley, G.H., and Stannard, L.G., 1990a, Application of remotely sensed data in a distributed– parameter watershed model, in Kite, G.W., and Wankiewicz, eds., Proceedings of Workshop on Applications of Remote Sensing in Hydrology: National Hydrologic Research Centre, Environment Canada, p. 47–64.

Leavesley, G.H., and Stannard, L.G., 1990b, A modular watershed–modeling system for use in mountainous regions: Schweizer Ingenieur und Architekt, n. 18, p. 380–383.

Leavesley, G.H., and Striffler, W.D., 1979, A mountain watershed simualtion model, in Colbeck, S.C., and Ray, M., eds., Modeling of snow cover runoff: Hanover, New Hampshire, U.S. Army CRREL, 379-386.

Leclerc, G., and Schaake, J.C., Jr., Methodology for assessing the potential impact of urban development on urban runoff and the relative efficiency of runoff control alternatives, Mass. Instit. of Tech. Ralph M. Parsons Laboratory Report 167, 1973.

Lumb, A.M., Kittle, J.L., Jr., and Flynn, K.M., 1990, User's manual for ANNIE, a computer program for interactive hydrologic analyses and data management: U.S. Geol. Surv. Water Resour. Invest. Rep. 89–4080, 236 p.

Mein, R.G. and Brown, B.M., 1978, Sensitivity of optimized parameters in watershed models: Water Resour. Res. 14(2), 299-303.

Miller, D.H., 1959, Transmission of insolation through pine forest canopy as it effects the melting of snow: Mitteilungen der Schweizerischen Anstalt für das forstliche Versuchswesen, Versuchsw. Mitt., v. 35, p. 35–79.

Murray, F.W., 1967, On the computation of saturation vapor pressure: J. Applied Meteorology, v. 6, p. 203–204.

Obled, Ch. and Rosse, B., 1977, Mathematical models of a melting snowpack at an index plot: J. Hydrol. 32, 139-163.

Norris, J.M., and Parker, R.S., 1985, Calibration procedure for a daily flow model of small watersheds with snowmelt runoff in the Green River coal region of Colorado: U.S. Geol. Surv. Water–Resour. Invest. Rep. 83–4263, 32 p.

Rankl, J.G., 1987, Analysis of sediment production from two small semi arid basins in Wyoming: U.S. Geol. Surv. Water–Resour. Invest. Rep. 85–4314, 27 p.

Reed, L.A., 1986, Verification of the PRMS sediment–discharge model, in Proceedings of the Fourth Federal Interagency Sedimentation Conference: March 24–26, 1986, Las Vegas, v. II, p. 6–44 to 6–54.

Riley, J.P., Israelsen, E.K., and Eggleston, K.O., 1973, Some approaches to snowmelt prediction, in The role of snow and ice in hydrology: IAHS Pub. 107, p. 956–971.

Rivera–Santos, J., 1990, Parameter estimation in conceptual precipitation–runoff models with emphasis on error analysis, in Tropical Hydrology and Caribbean Water Resources: Proceedings of the International Symposium on Tropical Hydrology and Fourth Caribbean Islands Water Resources Congress, San Juan, Puerto Rico, American Water Resources Association, p. 91–100.

Rosenbrock, H.H., 1960, An automatic method of finding the greatest or least value of a function: Computer Journal, 3, 175-184.

Smith, R.E., 1976, Field test of a distributed watershed erosion/sedimentation model, in Soil erosion—Prediction and control: Ankeny, Iowa, Soil Conservation Society of America, p. 201–209.

Stannard, L.G., and Kuhn, G., in press, Watershed modeling, in Britton, L.J., Anderson, C.L., Goolsby, D.A., and Van Haveren, B. P., eds., Summary of the U.S. Geological Survey and U.S. Bureau of Land Management National Coal-Hydrology Program, 1974-84: U.S. Geol. Surv. Prof. Pap. 1464.

Swift, L.W., 1976, Algorithm for solar radiation on mountain slopes: Water Resour. Res. 12 (1), 108-112.

Tangborn, W.V., 1979, A model to predict short–term snowmelt runoff using synoptic observations of streamflow, temperature, and precipitation in Colbeck, S.C., and Ray, M., eds., Modeling of snow cover runoff: Hanover, New Hampshire, U.S. Army, CRREL, p. 414–426.

Thompson, E.S., 1976, Computation of solar radiation from sky cover: Water Resour. Res., v. 12, no. 5, p. 859–865.

Troutman, B.M., 1985, Errors and parameter estimation in precipitation-runoff modeling: Water Resour. Res., 21 (8), 1214-1222.

U.S. Army, 1956, Snow hydrology, U.S. Army Corps. of Engineers, Portland, Oregon, USA.

U.S. Soil Conservation Service, 1971, SCS National Engineering Handbook, Section 4--Hydrology, Washington, D.C., U.S. Government Printing Office.

Vézina, P.E., and Péch, G.Y., 1964, Solar radiation beneath conifer canopies in relation to crown closure: Forest Science v. 10, no. 4, p. 443–451.

Willen, D.W., Shumway, C.A., and Reid, J.E., 1971, Simulation of daily snow water equivalent and melt, in Proceedings of 39th Annual Meeting of the Western Snow Conference: Billings, Montana, pp. 1-8.

WMO, 1986, Intercomparison of models of snowmelt runoff: WMO Operational Hydrol. Rep. no. 23, WMO Publ. no. 646, WMO, Geneva.

Zahner, R., 1967, Refinement in empirical functions for realistic soil–moisture regimes under forest cover, in Sopper, W.E., and Lull, H.W., eds., International symposium of forest hyrology: New York, Pergamon Press, p. 261–274.

Chapter 10

THE NWS RIVER FORECAST SYSTEM - CATCHMENT MODELING

R. J. C. Burnash

In order to more effectively meet one of its primary missions, the protection of life and property from flood disasters, the National Weather Service has developed a computerized library of well tested, operational, hydrologic analysis procedures known as the National Weather Service River Forecast System or NWSRFS. Within the constraints of currently available computer systems, the NWSRFS now serves many of the needs of the River Forecast Centers which provide much of the hydrologic guidance for public warning purposes. In an attempt to improve the catchment modeling capability of this system, the National Weather Service conducted extensive tests of all catchment models which were then operational in the various River Forecast Centers which serve the NWS. These tests were later expanded under the auspices of the World Meteorological Organization to include procedures and catchments submitted by major hydrologic organizations throughout the world. As a result of these tests, the National Weather Service adopted the technology included in the Sacramento Catchment Model as the primary NWSRFS component for catchment modeling. The concepts which are involved in the Sacramento Catchment Model, its parameter structure, techniques for their determination, and requirements for improving future streamflow simulation capabilities are discussed.

10.1. NATIONAL WEATHER SERVICE HYDROLOGIC OPERATIONS

The National Weather Service is that agency of the United States Government which has been charged by Congress with the responsibility for providing public warnings of changes in river levels. This responsibility, which focuses on the protection of life and property, has led to the development of a nationwide hydrologic forecast program administered by the National Weather Service. A number of River Forecast Center Offices, distributed across the United States, collect timely hydrologic information from many diverse sources, and then assemble and analyze this information in support of the National Weather Service's hydrologic warning program.

An important factor in producing effective hydrologic warnings is the ability to maximize the lead time for responding to a definable hydrologic hazard. A principle requirement for meeting this objective is the ability to evaluate the runoff response of a wide variety of catchments to occurring and/or anticipated rainfall. As a portion of meeting its responsibility to enhance public safety, the National Weather Service has organized a variety of thoroughly tested rainfall-runoff procedures, discharge routing techniques, snow ablation methods, water supply analysis systems, and other hydrologic procedures into a centralized library of computerized forecasting techniques. This system, which includes many of the procedures which have proven their value for operational hydrologic forecasting, is known as the National Weather Service River Forecasting System, or NWSRFS.

The River Forecast Centers necessarily respond to their localized hydrologic warning responsibilities within a framework of constraints which has a significant geographic variation. These variations include substantially different data systems, a variety of interagency operating agreements, widely disparate hydrologic conditions, differing levels of cooperator involvement, and numerous variations in on site computer capabilities. Maximizing the effective lead time for hydrologic warnings, under the diverse conditions that exist, has required many of the River Forecast Centers to develop hydrologic procedures which are uniquely appropriate to local conditions and to the variety of operational environments which influence their activities.

Differences in operational computer capabilities, particularly those associated with localized real time operating systems versus the NWSRFS batch operating mode, have placed limitations on the interchange of procedures between those which are useful at some River Forecast Centers and those which could be included in the NWSRFS. Although the batch operating system utilized to implement the NWSRFS provides substantial capability for many NWS purposes, it also constrains the timeliness of many operations, limits the transfer of useful hydrologic procedures from the NWSRFS to many interested organizations within the United States, and complicates the transfer of hydrologic forecasting technology from the United States to other portions of the world. It is hoped that the continuing growth of microcomputer capabilities will lead to the transfer of many of the excellent procedures imbedded in the NWSRFS to a more appropriate real time operating system suitable for use on microcomputers. Such a change would encourage the inclusion of additional locally focused procedures, extend the applicability of existing NWSRFS routines, and would allow a broader portion of the hydrologic community to utilize those NWSRFS capabilities which would be of use to them.

10.2. RAINFALL RUNOFF PROCEDURES

Prior to the development of readily accessible computer systems, the primary NWS technique for determining runoff from rainfall was the Antecedent Index approach, Kohler and Linsley (1951). This system, which is still in use in portions of the United States, correlates prior precipitation and a second variable, generally week number, to determine a "runoff curve". The "runoff curve" provides a technique for determining the quantity of runoff which would be produced from storm rainfall totals. By comparing rainfall and runoff totals at fixed times throughout a storm, period runoff volumes can be computed. A unitgraph is then applied to the period runoff volumes, an estimate of base flow is added, and an estimate of catchment streamflow is obtained.

The Antecedent Index system was of greatest utility in areas which did not encounter protracted dry periods and where changes in base flow conditions did not create a serious problem in streamflow analysis. In areas where rainfall was infrequent and substantial drying of the soil could occur between runoff events, the Antecedent Index had difficulty in describing the degree of dryness and determining the amount of runoff produced from rainfall. In those regions where base flow variations are an important portion of the hydrograph, the Antecedent Index approach provided little guidance in describing the base flow changes. If storms occurred in close proximity to one another, the ability of the Antecedent Index to determine an appropriate "runoff curve" was substantially degraded. These continuing problems and the desire to obtain a more accurate and more complete definition of catchment response led to the search for a more comprehensive procedure for determining catchment runoff conditions.

In the 1960's computer capability became available for hydrologic analyses and the search for an improved rainfall-runoff methodology accelerated rapidly. At Stanford University where one of the first major computer systems was available, Professor Ray K. Linsley Jr. encouraged his students to investigate this problem. These efforts resulted in the "Stanford Model", Crawford and Linsley (1966). The potential for improved streamflow forecasting demonstrated by the "Stanford Model" encouraged many hydrologists to investigate further improvements in rainfall-runoff procedures made possible by the growing availability of computer systems.

One of the National Weather Service River Forecast Centers which operated in a climatological regime where the Antecedent Index approach was particularly deficient was the California-Nevada River Forecast Center located in Sacramento, California. In order to improve that area's operational capability for streamflow forecasting, personnel at the California-Nevada River Forecast Center attempted to construct an improved headwater forecasting model. This effort resulted in a "Generalized Streamflow Simulation System", Burnash et al (1973). In order to distinguish this effort from other simulation studies, this generalized system was frequently identified as the Sacramento Catchment Model.

The Sacramento Catchment Model underwent extensive development and testing on widely disparate catchments. The initial work focused on catchments in the California-Nevada area. When the model had been developed to a high level of capability in that region, tests were extended to catchments across the United States. Testing was further extended when the Sacramento Catchment Model was included in the international "Intercomparison of Conceptual Models Used in Hydrological Forecasting". This comparison, which was conducted under the auspices of the World Meteorological Organization (1975), utilized test catchments from diverse areas of the world. The comparison tested the ability of a variety of procedures to trace observed conditions and also to synthesize streamflow records for periods when the discharge data was masked. Upon review of the results of these tests and a careful examination of the Sacramento Catchment Model's logical structure, the National Weather Service decided to abandon a major component of the catchment modeling system which was then being utilized and to adopt the runoff determination capability of the Sacramento Catchment Model as a major component of the NWSRFS, Peck (1976).

10.3. DESIGN CONSIDERATIONS FOR A CATCHMENT MODEL

The primary purpose of the Sacramento Catchment Model is to provide an effective technique for determining catchment discharge. The model utilizes catchment precipitation and catchment evapotranspiration as basic inputs to the discharge determination process. These inputs are processed in accordance with a number of catchment characteristics which interact with mathematical representations of those hydrologic processes which govern catchment discharge. By utilizing a carefully structured conceptual design, coupled with an appropriate application of the model's features, the authors hoped to achieve an improved streamflow forecasting capability. They focused on a conceptual model in the belief that if there were a change in the catchment's morphology, a conceptual approach would provide a rational means of assessing the necessary changes in the characteristics used for modeling the catchment. This capability was later verified in actual forecast operations when major forest fires altered several catchment's runoff characteristics and it was possible to intelligently and successfully model the resulting changes before any hydrologic events took place.

The authors of the Sacramento Catchment Model were acutely aware that detailed modeling of the many physical processes involved in catchment hydrology would be too complex for general application to National Weather Service streamflow forecasting responsibilities. In addition, they recognized that operating a streamflow forecasting service within the limitations of generally available hydrologic data placed severe limitations on a model's structure. Responding to these constraints, they concentrated on developing a streamflow simulation system which would provide an effective integration of the primary physical processes and could operate without the mass of data and calculations which would be required to support a comprehensive mathematical simulation. In other words, the model had to be appropriate for regular operational application in support of the National Weather Service's public warning program.

The formulation which was adopted relied heavily on paralleling relevant physical processes which had been defined by soil scientists, botanists, and hydrologists. However, the realities of operational forecasting responsibilities and the data limitations associated with their operational flood warning mission, led the authors to utilize catchment integrations of processes which were known to be spatially diverse. In their effort to simplify the catchment representation, the authors attempted to limit the structure of their system so that it included only those processes which could be clearly identified by their contribution to streamflow simulation effectiveness. In doing this, they relied upon the historical discharge record as the arbiter of which parameters needed to be considered, which processes needed to be included, and which inputs were most effective in duplicating the recorded flows.

10.4. THE CATCHMENT WATER BALANCE

The authors of the Sacramento Catchment Model predicated their efforts on an expansion of the basic water balance equation. That is:

Runoff = Rainfall - Evapotranspiration - Changes in Soil Moisture

Thus, rainfall applied to a catchment must be utilized for evapotranspiration, stored in the catchment, or it must have exited the catchment either as channel flow or as subterranean drainage. As time passes, the moisture conditions associated with each component of the water balance must be continuously updated if the model is to meet its purpose of providing an effective estimate of streamflow.

It is generally assumed that measurements of catchment runoff, station rainfall, and pan evaporation can be utilized to index those hydrologic conditions which are needed to describe the catchment water balance. Unfortunately, there are many limitations to the effectiveness of each of these measurements. These limitations frequently influence the appropriateness of data which is necessary for the catchment modeling process and unless great care is taken, data problems can introduce distortions into the modeling process. None the less, there is no viable alternative to maximizing the use of available data as a necessary, but limiting, factor in any conceptual representation of a catchment's hydrological system. The completion of an effective simulation of streamflow requires a careful evaluation of all the significant components of the hydrologic cycle. These elements include catchment runoff, catchment precipitation, catchment evapotranspiration, as well as the catchment characteristics which are a portion of the hydrologic model.

10.4.1. Runoff

Catchment modeling represents an attempt to describe and quantify those processes which have produced historical flows and thereby provide a tool for determining future flows. Unfortunately, this effort may be influenced by discharge records which are less accurate than is desirable. In many areas, high flow records have been approximated with rating extensions which are based on measurements of discharge which are a fraction of the estimated peaks. All flow records, but particularly those which represent low flow conditions, may be corrupted by debris on the controls. Unstable stream beds can confuse the translation of gauge height records to discharge and the complexity of gauging sites with loop ratings can limit the accuracy of the recorded discharge. Despite these and many other problems associated with producing an effective record of stream discharge, the stream flow record is generally considered to be the most accurate element of a catchment's hydrologic history. However, the analyst should be aware that there is a real possibility that a carefully applied streamflow model may provide a better estimate of the stream's discharge than could be produced from the limited information which has guided some published determinations of flow. On many occasions, the author has utilized available data to demonstrate the lack of consistency between data sets and the inappropriateness of portions of previously published records. Such experiences prompt the warning that those portions of a catchment's hydrologic history which are not supported by all elements of the available data sets should be utilized with great care when defining a catchment's modeling characteristics.

10.4.2. Precipitation

The water balance characteristics of a conceptually consistent hydrologic model make it extremely important that the catchment precipitation estimate be as appropriate as possible. Definition of catchment precipitation is a complex problem which is made even more difficult in areas where precipitation data are sparse. Such areas constitute a substantial portion of the world and impose a genuine restriction on our ability to apply streamflow simulation techniques. However, even where rain gauges provide apparently adequate areal coverage, the gauges themselves are generally inadequate in their ability to "catch" true rainfall. Generally, true rain, rain which will affect the catchment's soil moisture system, is some value in excess of measured precipitation. In most cases, this discrepancy is due to the influence of wind over the orifice of the rain gauge. The World Meteorological Organization, Sevruk (1982), has indicated that as an average condition, rainfall measurements may be deficient by 10% with much greater errors being likely during particularly windy conditions. Orographic characteristics of a catchment may exacerbate this problem and make catchment rainfall evaluation even more complex.

An additional problem in determining the availability of moisture to the catchment occurs when snow is a significant portion of the catchment precipitation. Only as snow is transformed into moisture, available to the soil mantle, is it available to enter that portion of the moisture accounting system which is based upon the land phase of the hydrologic cycle. Due primarily to the efforts of Anderson (1973), the NWSRFS includes valuable routines which can be utilized to process the precipitation record and approximate the timing and volume of snowmelt reaching the land surface.

It is an interesting paradox that any system which translates point precipitation to a catchment average, introduces an error into simulation techniques. This occurs because the generation of runoff from rainfall is a nonlinear process while the weighting system which produces catchment rain from point measurements is a linear process. Even if the rain gauges are without error, a very unlikely condition, the technique for determining catchment precipitation can only be appropriate for a particular volume, timing and distribution of precipitation which occurs with a specific soil moisture condition. The problems created by this paradox are less severe in areas with relatively uniform rainfall and greatest where the rainfall has a marked spatial diversity. Thus, the "catchment precipitation paradox" precludes the exact determination of an effective catchment precipitation and places a genuine constraint on the capability of any streamflow simulation system even if it were perfect in every other detail.

Unfortunately, there is no effective alternative to catchment analysis as there is no mathematical process, applicable throughout a catchment, which can verifiably translate point rainfall to point runoff as neither value can be adequately measured. The ambiguities involved in deriving and integrating point runoff values introduce additional complications to catchment runoff determination. Indeed, attempting to resolve the "catchment precipitation paradox" by using point runoff for catchment forecasting, exceeds the descriptive capability of the available inputs and requires runoff integration techniques which are based upon a growing set of assumptions. It is operationally more efficient to start with a catchment as an analysis unit and to attempt to minimize the influences of the precipitation paradox by analyzing catchments which are large enough to integrate some of the effects of rainfall variability without being so large as to distort the concept of reasonable geologic and climatologic homogeneity.

10.4.3. Evapotranspiration

Detailed studies of evapotranspiration have demonstrated significant differences between measured evaporation at pan sites, natural water surfaces and field use of moisture by plants. Thus, the utilization of pan evaporation as an index to basin evapotranspiration generally requires a significant transformation of pan data. Inasmuch as evapotranspiration is frequently the dominant use of the moisture supplied by precipitation, and since it is one of the most difficult processes to evaluate in hydrologic analysis, evapotranspiration estimates are generally considered to be a significant source of error in streamflow simulation. Catchment evapotranspiration includes transpiration from leaf surfaces of water taken from the soil by plant roots, direct evaporation from the soil and from leaf and other surfaces wetted by rain, and direct evaporation from the surfaces of lakes, ponds and streams.

The processes of evaporation from a pan and transpiration from the surfaces of a plant leaf differ in several important aspects. An evaporation pan has considerable heat storage while a plant leaf has very little. Therefore, the temperature of a plant leaf responds more quickly to changes in air temperature and radiation. Soil temperatures, which influence both the growth of vegetation and the ability of roots to absorb soil moisture, change rather quickly at the soil surface, but much more slowly with depth. The growth of annual plants and the leaf development of deciduous perennials vary with the time of the year and past temperature and moisture conditions. Even for mature leaves, their condition and ability to transpire vary with season and past weather. The processes of water surface evaporation also differ from transpiration in that transpiration may be restricted by nocturnal or other closures of stomata, the small openings in leaf surfaces through which water vapor and other gases diffuse. Stomata closure may also be induced by excessive moisture stress in the leaf. This tends to reduce transpiration when external conditions are producing the highest evaporation rates.

In many areas more moisture is lost through evapotranspiration than is discharged by the streams. Potential evapotranspiration often exceeds rainfall on an annual basis, thus streamflow may result only from concentrations of precipitation. Due to the significance of evapotranspiration to headwater simulation, it is necessary to obtain a reasonable estimate of evapotranspiration. Inasmuch as the characteristics of the evaporation pan do vary significantly from those of vegetation, we can anticipate that any estimation of evapotranspiration demand based upon pan observations must reflect the differences between pan and catchment characteristics.

Catchment evapotranspiration demand is generally defined as that evapotranspiration which would occur if there were no deficiencies in the availability of moisture for evapotranspiration by that area's particular plant regime. Seasonal coefficients applied to pan data are for the purpose of dimensioning pan data to provide an estimate which approximates the potential seasonal evapotranspiration from that area's water surfaces and vegetation. When the area warms after a seasonal cold period where vegetative cover is largely annual or deciduous, pan evaporation may rise much more rapidly than catchment evapotranspiration demand. This is because evapotranspiration demand must wait for the growth of leaves and roots. Under these conditions, seasonal coefficients which are applied to pan data to transform it into an estimate of evapotranspiration may approach their smallest value when pan observations of springtime evaporation are increasing rapidly. When the catchment is undergoing seasonal cooling, pan evaporation values may decrease at a more rapid rate than is appropriate to evapotranspiration from fully developed leaf and root systems. However, if an early frost should prematurely kill much of the catchment's seasonal vegetation, actual evapotranspiration will drop to a very low value until the following spring. Unfortunately, the pan may return to fairly high values after the frost and its daily values will have little relationship to evapotranspiration. In general however, coefficients applied to pan data, to transform them into an estimate of catchment evapotranspiration demand, are likely to be larger in the fall than they are in the spring. This is because when the catchment is undergoing seasonal cooling, pan evaporation will generally decrease at a more rapid rate than is appropriate to evapotranspiration from fully developed leaf and root systems. At the same time, exposed water surfaces will generally have a larger thermal mass than the evaporation pan, contributing to the need for enlarging the coefficients. As a consequence, coefficients are likely to be larger during the annual cooling cycle than during a similar pan evaporation period in the spring. Unfortunately, this generalization is subject to modification if a severe frost should occur.

In the temperate latitudes of the Northern Hemisphere, the effective pattern of evapotranspiration demand generally is highest in July or August and reaches its lowest point in December, January, or February. Typical catchment evapotranspiration demand curves generally take the form shown in Fig. 10.1. However, variations in humidity, incoming radiation, vegetation, wind velocity, and other factors can influence the evapotranspiration demand in a particular catchment. Despite the differences between pan evaporation and catchment evapotranspiration, a carefully

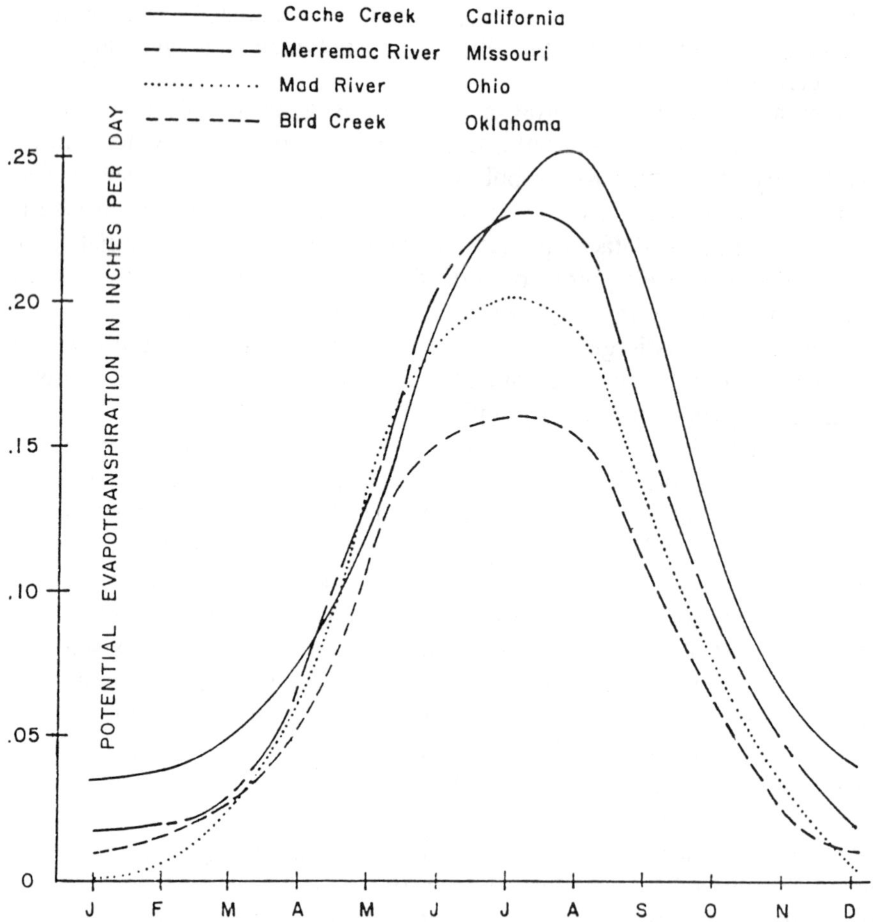

Fig. 10.1. Daily Average Evapotranspiration Demand.

maintained evaporation pan generally provides useful information on the annual amplitude of the evapotranspiration curve. If the observed pan values are modified on the basis of differences in pan versus plant behavior, the annual pan evaporation curve can provide a useful tool for approximating catchment conditions. Techniques for estimating potential evaporation from meteorological variables may suffer from similar deficiencies in their lack of a complete parallel to the transpiration process. Dimensioning problems in using such techniques can be similar to those encountered in utilizing pan data. These problems can be so severe that in many areas an average annual evapotranspiration curve appears to be as meaningful as any readily available discrete information.

10.4.4. Soil Moisture

The moisture accounting system utilized by the Sacramento Catchment Model is a carefully structured representation of the catchment's soil moisture storage system. It is based on using simple approximations of many of those soil moisture processes which have been reported in the hydrologic literature. The authors have organized these approximations in a manner which would allow the determination of many catchment characteristics from carefully selected portions of the catchment's hydrologic record. Inasmuch as many of the catchment characteristics are related to the soil moisture storage capabilities of the catchment, an intelligent application of the model starts with a good understanding of the three basic types of soil moisture which can potentially influence catchment runoff conditions. These soil moisture types are:

1) Hygroscopic Water

2) Tension Water and

3) Free Water

Hygroscopic Water

Hygroscopic water is that water which is so tightly bound to soil molecules that vegetation can not utilize it for transpiration purposes. Hygroscopic water can only be removed from the soil by extreme heating and will be replaced if any moisture is available. In most climates, the amount of surface heating is insufficient to produce any significant variation in the amount of hygroscopic water held within the soil. Consequently, an accounting system which discretely includes variations in a catchment's hygroscopic moisture is generally dealing with such small volumes, that they have little impact upon the effectiveness of the rainfall-runoff system.

In those climates which produce vapor pressure differences between the soil and the atmosphere which are large enough to involve a portion of the soil's hygroscopic moisture, the characteristics of such transfers are quite similar to those involving tension water. It should be noted that the magnitude of hygroscopic soil moisture change is not only a function of the soil type and lower level soil moisture, but is also a function of the dew point or temperature at which atmospheric moisture condenses and the wind velocity. The inclusion of hygroscopic moisture in an operational model would require a substantial increase in the types of data needed for catchment modeling and an expansion of the parameter set needed to define catchment characteristics. Inasmuch as hygroscopic moisture changes were not recognizable as influencing streamflow in any of the catchments used to develop and/or test the Sacramento Catchment Model, the authors chose to implicitly include any volume of time varying hygroscopic water as a portion of the catchment's tension water system.

Tension Water

Tension Water is that portion of the soil moisture system which can be separated from the soil and returned to the atmosphere through the normal processes of evapotranspiration. Normally, tension water is held in position against the force of gravity by the molecular attraction between soil particles and water molecules. However, pressure forces generated by plant and atmospheric conditions can transfer this water to the atmosphere. Whenever there is free water adjacent to soil particles with a tension water deficiency, the strong attraction resulting from the tension water deficiency will cause the adjacent free water to resupply the tension water deficiency. The potential volume of tension water within the soil is generally determined by an interrelationship between the area's climate, the catchment's soil types, and the characteristics of those plant forms which exist in the catchment.

Free Water

Free water is the third type of soil moisture which is involved in the catchment's soil moisture regime. Free water is water in the liquid state which is free to travel through or across the soil. Rainfall is applied to the catchment as free water. It is free water which moves into or across the soil. The free water which moves into the soil does so in response to the dryness of the soil and the soils permeability. Such water will percolate through the soil resupplying any soil moisture deficiencies it encounters. When free water reaches a soil particle which has lost, through evapotranspiration, any portion of the tension water which the soil is capable of retaining, then the deficient soil particle will absorb sufficient free water to eliminate that particle's tension water deficiency. Excess free water, that which is in excess of the tension water requirements at any depth, may be drawn by gravity to those deeper levels which have moisture deficiencies, or through a dominantly lateral course to a stream channel. When rainfall occurs at a rate which is greater than the rate at which the surface can transfer water to deeper depths, the excess rain will reach the stream channel as surface runoff.

The relative components of hygroscopic, tension, and free water which can be held within the soil, vary with soil types. For a given quantity of soil, the small particle sizes associated with clay soils will generally accommodate the highest proportion of tension water. The large particle sizes associated with sandy soils provide more storage space between soil grains, and such soils will generally accommodate the highest proportion of free water. Within a catchment, surface soil types can vary quite widely, and surface soils do not necessarily prevail to those depths which influence the catchment's rainfall runoff characteristics. As a result, surface manifestations of soil type may have little to do with defining the catchment's rainfall runoff characteristics.

10.4.5. Interception

There is another form of moisture storage which can influence a catchment's rainfall runoff regime. When rainfall occurs, a portion of the falling precipitation may remain suspended above the soil surface until trees and other vegetation which can intercept precipitation are sufficiently wet so that additional rainfall is transferred to the ground surface. That volume of water which is prevented from reaching the soil surface due to catchment vegetation, is frequently referred to as the interception volume. Inasmuch as interception represents a moisture use before soil moisture is affected, its influence on the catchment should be considered.

After each rainfall, as the catchment starts to dry, the intercepted water must be evaporated from the leaf surfaces so that the leaf surfaces are free to transpire. The intercepted water has temporarily interfered with the transpiration process. The evaporation of intercepted water from the vegetative surfaces substantially substitutes for moisture which the plants would otherwise have drawn from the soil and transpired. In a soil moisture accounting system it makes little difference, in determining the amount of free water which will be involved in the runoff processes, if the intercepted moisture is discretely stored and is then evaporated while transpiration is temporarily suppressed, or if the moisture which has been intercepted is lumped with the upper zone tension water and removed from that storage by evapotranspiration. Problems in the catchment moisture accounting which would arise from including the intercepted water as a portion of the tension water system would occur primarily if there were a substantial seasonal variation in the amount of water which is intercepted. For this reason, it is helpful to consider the seasonal variability of interception.

For many years I have observed a magnificent mature oak tree which appears to be well over a century in age. Most of the area beneath the oak is covered with vegetation, however portions of the area below the oak are traversed by concrete walks. Many times I have observed that when the tree was in full leaf, rainfall must exceed one and a half millimeters before the concrete areas below the tree receive any appreciable wetting. In contrast, during winter periods when the tree is bare of leaves, wetting of the concrete occurs as measurable precipitation begins. The interception capability of this mature tree is consistent with studies of interception in deciduous forests which suggest that the seasonal range of interception averages near two millimeters.

Interception in tropical rain forests has been reported as having a much larger potential, but such forests maintain a relatively consistent canopy and do not generally exhibit the seasonal range of interception encountered in areas where deciduous trees are dominant. Dense northern forests have an increasing proportion of nondeciduous trees. Such areas have been reported to exhibit an interception potential closer to 3 millimeters of rainfall per storm. However, in such forests the interception potential for rainfall has very little seasonal variation.

In temperate regions with deciduous vegetation, seasonal variations in the interception potential are generally the largest and because of this variability, likely to induce the greatest seasonal influence in evaluating how rainfall reaches the soil mantle. But considering only the seasonal variation in foilage exaggerates the magnitude of the interception problem. Those leaves which accumulate on the ground each fall, prevent a portion of the rain from reaching the soil. Until such time as the fallen leaves decompose, they will intercept a portion of the rain which they intercepted while they were on the tree. The result of all these influences is that the seasonal variability of interception on a catchment basis is quite small. In most areas, it is likely that catchment interception has a seasonal variation of little more than a millimeter in depth.

Because of the limited variability of the interception volume, and the parallels between the use of intercepted water and tension water, there is little likelihood that available data sets would provide sufficient discrimination to isolate the effect of intercepted water from that tension water storage which occurs in the upper portion of the soil mantle. For these reasons, the Sacramento Catchment Model, as a simplified representation of catchment processes, includes intercepted precipitation as a portion of the upper soil's tension water volume.

10.5. THE STRUCTURE OF THE SACRAMENTO CATCHMENT MODEL

Streamflow as computed by the Sacramento Catchment Model is the result of processing precipitation through an algorithm representing the uppermost soil mantle identified as the upper zone and a deeper portion of the soil mantle or lower zone. This algorithm computes runoff in five basic forms. These are 1) direct runoff from permanent and temporary impervious areas, 2) surface runoff due to precipitation occurring at a rate faster than percolation and interflow can take place when both upper zone storages are full, 3) interflow resulting from the lateral drainage of a temporary free water storage, 4) supplemental base flow, and 5) primary base flow.

10.5.1. Model Components

The model utilizes a set of storages of determinable capacities which are linked by processes which allow the system to approximate many of the soil moisture conditions which control the production of streamflow. The storages are filled if sufficient rain should occur, and their contents are depleted by vertical percolation, evapotranspiration and/or lateral drainage. The basic components of the system are presented in a schematic form in Fig. 10.2. Those elements of the modeling system which are depicted in

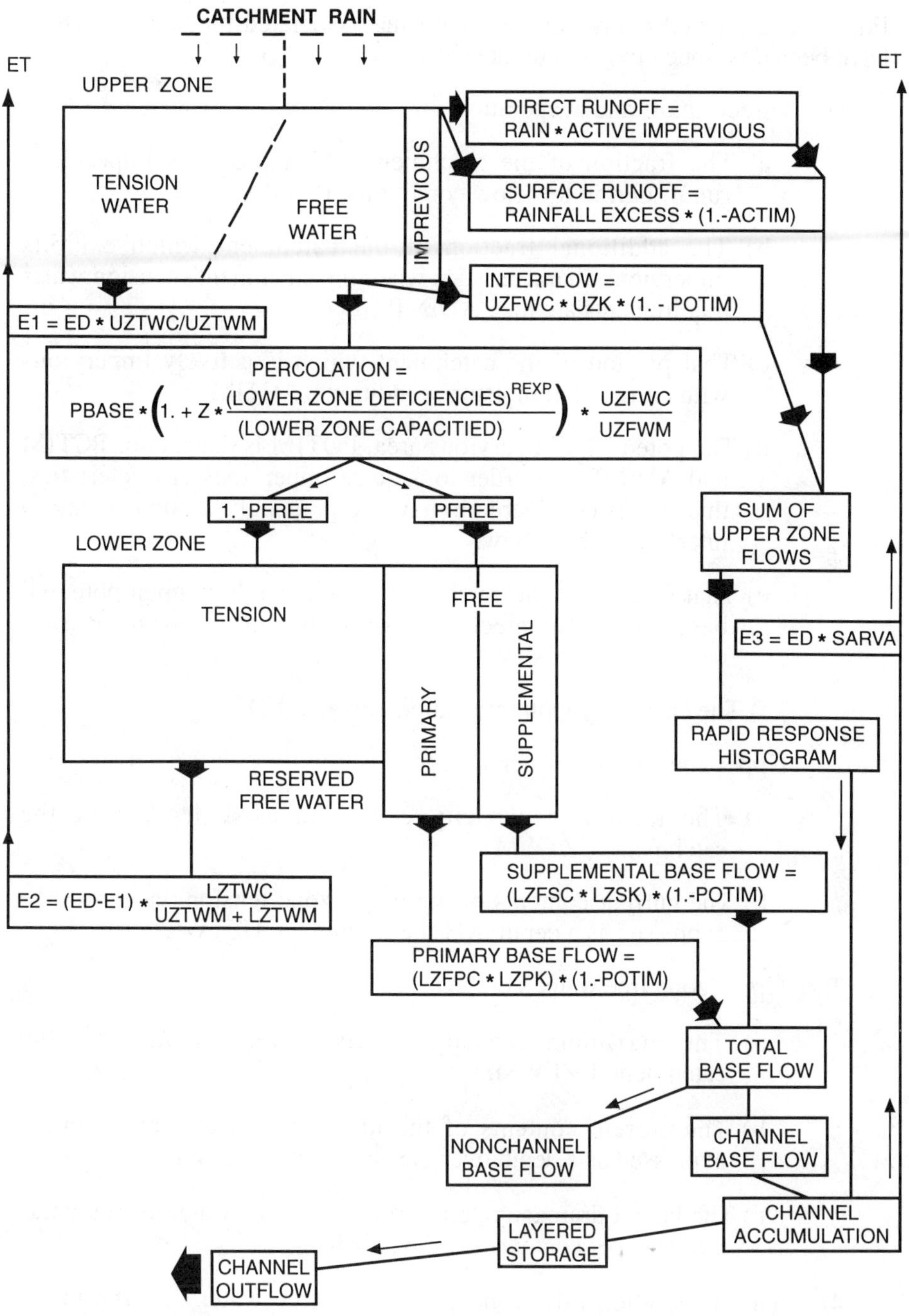

Fig. 10.2. Components of the Sacramento Catchment Model.

Fig. 10.2 are listed below together with the associated mnemonics which have been developed to assist in their identification:

1) Direct runoff and evaporation.

   a) The fraction of the catchment which produces impervious runoff during low flow conditions, PCTIM.

   b) The additional fraction of the catchment which exhibits impervious characteristics when the catchment's tension water requirements are met, ADIMP.

   c) That portion of the catchment which is actively impervious with current soil moisture conditions, ACTIM.

   d) The potentially impervious area, POTIM is the sum of PCTIM and ADIMP. In order to acquire impervious characteristics, this area is considered to have no potential for contributing to interflow or base flow.

   e) That fraction of the catchment which can draw upon potential base flow to meet evaporation and/or transpiration requirements, SARVA.

   f) The current evapotranspiration demand, ED.

2) Upper zone tension water storage

   a) The maximum capacity expressed as a depth over the catchment, UZTWM.

   b) The current contents of the upper zone tension water storage expressed as a depth over the catchment, UZTWC.

3) Upper zone free water storage

   a) The maximum capacity expressed as a depth over the catchment, UZFWM.

   b) The current contents of the upper zone free water storage expressed as a depth over the catchment, UZFWC.

   c) The lateral drainage rate of upper zone free water expressed as a fraction of contents per day, UZK.

4) The percolation rate from upper zone free water into the lower zone.

   a) The percolation capability during saturated conditions, PBASE.

b) The proportional increase in percolation from saturated to dry condition, Z.

c) An exponent determining the rate of change of the percolation rate with changing lower zone water contents, REXP.

5) Lower zone tension water storage

a) The maximum capacity expressed as a depth over the catchment, LZTWM.

b) The current contents expressed as a depth over the catchment, LZTWC.

6) Lower zone free water storage

a) Supplemental free water storage

1) The maximum capacity expressed as a depth over the catchment, LZFSM.

2) The current contents expressed as a depth over the catchment, LZFSC.

3) The lateral drainage rate expressed as a fraction of contents per day, LZSK

b) Primary free water storage

1) The maximum capacity expressed as a depth over the catchment, LZFPM.

2) The current contents expressed as a depth over the catchment, LZFPC.

3) The lateral drainage rate expressed as a fraction of contents per day, LZPK.

c) Direct percolation to Lower Zone Free Water, PFREE, the percentage of percolated water which is available to the lower free water aquifers before all lower zone tension water deficiencies are satisfied.

d) Ground water discharge not observable in the river channel

1) The ratio of non-channel subsurface outflow to channel base flow, SIDE.

2) The rate of flow which is conveyed by the channel's porous base and which must be available before channel flow can be observed, SSOUT.

e) The fraction of lower zone free water which is below the catchment's root zone and therefore incapable of resupplying a deficiency of lower zone tension, RSERV.

7) A non-dimensional histogram used to define the time distribution of rapidly responding runoff.

8) Channel storage characteristics to modify the timing of flow obtained from the normal histogram distribution.

a) The volume of flow for up to four layers of discharge which will be modified by the effects of varying channel storage. Note: the total volume encompassed by these layers must include all possible catchment flows.

b) The routing factor appropriate to each volume described above.

Rainfall occurring over the catchment is considered as falling on two types of surfaces which are a portion of the upper zone. These are, 1) a permeable portion of the soil mantle, and 2) a portion of the soil mantle covered by streams, lake surfaces, marshes, rocks, pavement, or other impervious material which is or becomes linked to the stream as soil moisture increases. The permeable area produces runoff when rainfall rates are sufficiently heavy, while the second area produces direct runoff from that portion of the catchment which is actively impervious.

Impervious Areas

It should be emphasized, that the area of the basin which is impervious, i.e., provides direct runoff, is not generally a constant area in this model. Although the model designates a portion of the catchment as having a permanently impervious characteristic, with sufficient wetting, it is possible for additional portions of the catchment to assume impervious characteristics. It has been observed in many catchments that coincidental with filling the tension water storages, an increasing fraction of the basin demonstrates impervious characteristics. This fraction, the additional impervious area, provides a useful representation of the filling of small reservoirs, marshes, and temporary seepage outflow areas which achieve impervious characteristics as the soil mantle becomes wetter.

Interflow Development

In the permeable portion of the basin, the water which arrives at the catchment's surface as rainfall is free to respond to the conditions which exist at its point of impact. The initial portion of each rainfall is generally claimed by that upper zone soil moisture storage identified as upper zone tension. This storage has such an attraction for moisture that it must be totally satisfied before any upper zone free water can accumulate for other purposes. Tension water is considered as that moisture storage which is held in place by the molecular attraction between soil particles and water. Upper zone tension represents that volume of precipitation which would be required under dry soil conditions to meet all those catchment moisture requirements which precede the development of interflow and percolation. When the upper zone tension requirements have been met, excess moisture is temporarily accumulated as upper zone free water.

Free water is that water which is not bound to soil particles. It is free to percolate to deeper portions of the soil mantle or to move laterally through the soil in response to gravitational and pressure forces. The upper zone free water storage provides water both for percolation to lower zones and for interflow. The demands imposed upon the upper zone free water vary with the amount of water available in both the upper zone and lower zone storages. Interflow, which takes place only after immediate percolation requirements have been met, is proportional to the residual upper zone free water volume. That is:

$$\text{INTERFLOW} = \text{UZK} \bullet \text{UZFWC}$$

where:

UZK is the upper zone free water storage depletion coefficient,

and

UZFWC is the residual volume of free water stored in the upper zone after immediate percolation requirements have been met. That is, the current upper zone free water contents.

Percolation

After upper zone tension water demands are satisfied, the next demand upon upper zone free water is for percolation. Percolation, the rate of vertical drainage to deeper soils, is controlled by the contents of the upper zone free water and the deficiency of lower zone moisture volumes. The preferred path for moisture in upper zone free water is considered to be downward as percolation. Horizontal flow in the form of interflow occurs only when the rate of precipitation exceeds the rate at which percolation can occur from the upper zone free water. When the precipitation rate exceeds both the percolation rate and the maximum interflow drainage capacity, then the upper zone free water capacity will be completely filled and the excess precipitation will result in surface runoff. Under this system, surface runoff is a highly rate dependent volume with the rate of runoff being determined by the rate of precipitation application and the degree of dryness of the different zones.

Percolation, the transfer of water from the upper soil to the lower soil, supplies moisture to three lower zone storages. These are, lower zone tension water storage, lower zone free water primary storage and lower zone free water supplemental storage. Lower zone tension water capacity is that depth of water held by the lower zone soil after saturation and drainage, and which is generally removed only by evapotranspiration. The two lower zone free water storages, primary and supplemental, represent those volumes which are available for drainage as base flow or subsurface outflow not appearing in the channel. The free water storages fill simultaneously from percolated water and drain independently at different rates, producing a variable ground water recession.

The percolation mechanics of the Sacramento Catchment Model have been designed to correspond with observed characteristics of the motion of moisture through the soil mantle, including the formation and transmission characteristics of the wetting front, as reported by Green et al. (1970), and Hanks et al. (1969). The mechanics of moisture transfers from the upper zone to those deeper soils which comprise the lower zone are based upon the computation of a lower zone percolation demand. When the lower zone is totally saturated, then the percolation into the lower zone is necessarily limited to the amount of water which can drain from the lower zone. The Sacramento Catchment Model represents the saturated drainage from the lower zone as the sum of the products of each of the two lower zone free water storages and its respective drainage rate. This limiting rate of drainage from the combined lower zone free water storages defines the Percolation BASE rate or PBASE where:

$$\text{PBASE} = ((\text{LZFPM} \bullet \text{LZPK}) + (\text{LZFSM} \bullet \text{LZSK}))$$

and,

LZFPM	=	The Maximum storage capacity of those Lower Zone Free water aquifers which provide the long term, ie: primary base flow to the catchment.
LZPK	=	The depletion coefficient of the Lower Zone free water's Primary storage.
LZFSM	=	The Maximum storage capacity of those faster draining Lower Zone Free water aquifers which Supplement the long term or primary base flow aquifers.
LZSK	=	The depletion coefficient of the Lower Zone free water's Supplemental storage.

When the soil is quite dry, the potential for percolation is much higher than the limiting value, PBASE. However, the upper limit of percolation may be expressed by a multiple of PBASE such that the:

$$\text{MAXIMUM PERCOLATION CAPACITY} = \text{PBASE} \bullet (1+Z).$$

If we further assume that percolation into the lower zone varies in proportion to the DEgree of WETness of the lower zone, a necessary condition if input into the lower zone is limited by outflow from the lower zone when the soils are saturated, then the Lower Zone Percolation Demand, LZPD, can be stated as:

$$\text{LZPD} = \text{PBASE} \bullet (1 + Z \bullet f(\text{DEWET})).$$

Where, the DEgree of WETness, DEWET is expressed as:

$$\text{DEWET} = 1 - \frac{\text{Sum of lower zone contents}}{\text{Sum of lower zone capacities}} .$$

However, numerous studies of the soil wetting process have concluded that the process of soil wetting varies exponentially with respect to the degree of soil moisture. This form can be obtained by modifying the degree of wetness by an exponent, REXP, such that:

$$\text{LZPD} = \text{PBASE} \bullet (1+ (Z \bullet \text{DEWET}^{\text{REXP}})).$$

The actual percolation to the lower zone is, however, not only related to moisture conditions in the lower zone, but it is also influenced by the availability of water in the upper zone. Since only free water in the upper zone can be utilized to meet percolation demands, we can complete our assessment of the percolation process by modifying the lower zone percolation demand on the basis of the availability of upper zone free water. Thus, the final determination of percolation to the lower zone, PERC is defined as:

$$\text{PERC} = \text{PBASE} \bullet (1+(Z \bullet \text{DEWET}^{\text{REXP}})) \bullet \frac{\text{Upper zone free water contents}}{\text{Upper zone free water capacity}} .$$

The sums of lower zone capacities and contents include both tension water and free water. Thus, percolation is defined by an interrelationship between soil drainage characteristics and soil moisture conditions.

The volume of moisture which is percolated to the lower zone is divided among three significant soil moisture storages. The first of these, lower zone tension, represents that volume of moisture in the lower zone soil which will be claimed by dry soil particles when moisture from a wetting front reaches that depth. Tension water deficiencies tend to absorb all percolating water until such deficiencies are satisfied. However, it has been repeatedly observed that nonuniformity of catchment conditions generally allows a fraction of the percolating water to reach the lower zone free water aquifers prior to the lower zone tension water being fully resupplied. The effect of these variations is approximated in the model by diverting a fraction of the percolated water into lower zone free water storages before lower zone tension water deficiencies are fully satisfied. An examination of Fig. 10.2 indicates that water percolating from the upper zone free water to the lower zone may go totally to tension water, or some fraction of the percolated water, PFREE, may be made available to the primary and supplementary storages. At any time that the lower zone tension storage becomes totally filled, continued percolation is divided between the two lower zone free water storages. At all times water made available to primary and supplemental storages is distributed between them in response to their relative deficiencies.

Infiltration

The relationship of percolation demand versus lower zone soil moisture deficiencies describes that portion of the infiltration curve which is constrained by lower zone soil moisture. Total infiltration is the result of supplying those soil moisture volumes needed for percolation to the lower zone, drainage of upper zone free water as interflow, and meeting those soil moisture deficiencies which may exist in upper zone tension water and upper zone free water. The developers of the Sacramento Catchment Model considered a limiting function to constrain the rate at which the upper zone storages could initially accept moisture. However, despite many years of looking at catchment records for indications of a demonstrable example of such a function, this author was unable to find records which were sufficient to guide the development of such a function. This may be analogous to the difficulties encountered by soil scientists in defining the upper limit for infiltration into dry soil. As a result, the upper limit for infiltration in that version of the Sacramento Catchment Model which was chosen for inclusion in the NWSRFS is implicitly constrained by the upper zone soil moisture conditions and the percolation potential of the lower zone. The determination of infiltration occurs through combining the resupplying of moisture deficiencies in the upper zone and the percolation of water to the lower zone. This provides a time varying infiltration which closely mimics the infiltration observed in field tests. Such a formulation provides a close parallel of infiltration described during continuous wetting of the soil, and also provides an effective emulation of the volumetric transfers which accompany partial drying and/or changes in the rate of rainfall.

Base Flow

Discharge from the lower zone free waters represents those volumes of catchment discharge which are generally identified as base flow. Base flow is produced by combining the drainages of two lower zone free water storages such that the potential base flow is equal to:

(VOLUME 1) • (DRAINAGE FACTOR 1) +

(VOLUME 2) • (DRAINAGE FACTOR 2).

Volume one is identified as the primary storage and Volume two as the supplementary storage. The approximation of base flow from two lower zone free water storages with fixed withdrawal functions allows the integration of observed recessions and the computation of the free water volumes with which they are associated. Primary base flow characteristics are generally estimated from daily flow values for those periods of the discharge record when the proportion of today's flow to yesterday's flow is remaining constant for an extended period. At such times, primary base flow characteristics can be described as:

$$K = (QPt/Q\ Po)^{1/t}$$

$$LZPK = 1\text{-}k$$

$$LZFPM = QPMAX/LZPK$$

where:

K	is the daily recession coefficient of primary base flow
t	is the number of days
QPo	is the discharge when the recession is occurring at the primary rate.
QPt	is the discharge t days after QPo when the recession is continuing at the primary rate.
QPMAX	is the maximum primary base flow which can be inferred by projecting the primary recession to the period when saturation of the soil is likely to have occurred.

In areas with long dry summers followed by significant cooling prior to rainfall, increases in channel flow can occur which are not accompanied by precipitation. Such increases occur when a larger portion of the recession flow is discharged from the catchment due to a decrease in evapotranspiration demand. A proper determination of base flow must provide sufficient water to accommodate the apparent discharge increase which occurs without recharge from rainfall. If rainfall occurs with sufficient frequency, other elements of flow may over ride the natural recession and obscure the features which define base flow. The analyst must be aware of those conditions which can influence reported discharge and take care that they do not invalidate the recession analysis. The major error in evaluating primary recession characteristics comes from the failure to consider the effects of evapotranspiration on the observed recession and/or selecting a portion of the recession curve where a single free water source is not a substantially dominant portion of the observed flow. In some cases, only extended drought periods will provide a significantly clear signal to adequately describe the drainage characteristics of the primary base flow aquifer.

Analysis of the supplemental portion of the recession is similar. However, it is complicated by the fact that supplemental flow never occurs without being accompanied by a component of primary base flow. None the less, the supplemental characteristics can be very closely approximated by careful selection and analysis of a plotted recession.

Figure 10.3 provides an example of a recession analysis which allows the derivation of both the primary and supplemental recession volumes and their associated drainage rates. The analysis of lower zone free water volumes and their drainage factors can be accomplished as follows:

1) Take a natural recession and add back the volume which was probably claimed by evapotranspiration, SARVA • Evapotranspiration Demand.

2) Find that portion of the modified recession a week or so after peak, which suggests a relatively stable slope. This indicates that supplemental flows are the dominant component of that recession period.

3) Project this slope back to when the aquifer system was filled to establish its maximum value, SMAX, and foreward to establish when this component becomes an insignificant portion of the total flow.

4) Analyze the primary recession slope from that time when supplemental has become insignificant to much later in the recession period. Project primary slope back to the period of recharge to establish PMAX.

5) Compute the average ratio of daily change in primary flow Qtda/Qyda.

6) Determine LZPK as equal to 1 - (the ratio above).

7) Compute LZFPM by dividing PMAX by LZPK.

8) Subtract primary flow from the recession volumes when supplemental dominates. Compute LZSK, ratio of daily change for supplemental flow.

9) Compute LZFSM, after subtracting primary flow, divide SMAX by LZSK.

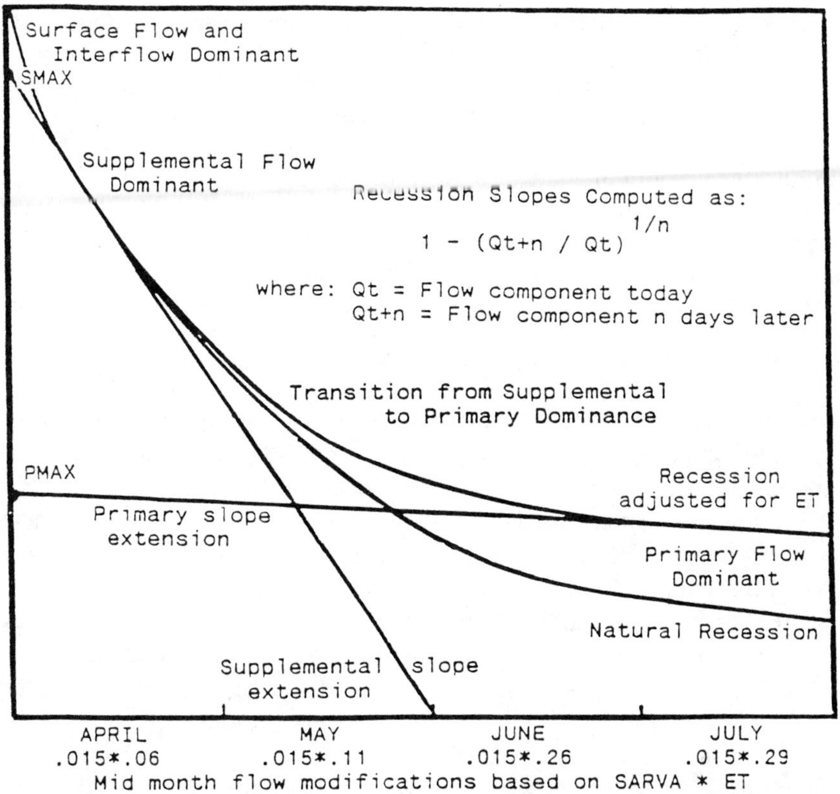

<u>Analysis of lower zone free water volumes and their drainage factors</u>

1) Take a natural recession and add back the volume which was probably claimed by evapotranspiration, SARVA * Evapotranspiration Demand.
2) Find that portion of the modified recession, a week or so after peak, which suggests a relatively stable slope. This indicates that supplemental flows are the dominant component of that recession period.
3) Project this slope back to when the aquifer system was filled to establish its maximum value, SMAX, and foreward to establish when this component becomes an insignificant portion of the total flow.
4) Analyze the primary recession slope from that time when supplemental has become insignificant to much later in the recession period. Project primary slope back to the period of recharge to establish PMAX.
5) Compute the average ratio of daily change in primary flow Qtda/Qyda
6) Determine LZPK as equal to 1 - (the ratio above).
7) Compute LZFPM by dividing PMAX by LZPK.
8) Subtract primary flow from the recession volumes when supplemental dominates. Compute LZSK, ratio of daily change for supplemental flow.
9) Compute LZFSM, after subtracting primary flow, divide SMAX by LZSK.

Fig. 10.3. Recession Anaylsis Utilizing Semi-Logarithmic Discharge Plot.

The early portion of the recession is also influenced by a third type of free water drainage which is identified as interflow. However, since interflow is produced from the same upper zone storage which is also providing water for percolation to the lower zone, only a portion of the volume which can be stored in upper zone free water is visible on the surface flow hydrograph. The limited depiction of upper zone free water on the observed recession, obscures an effective evaluation of its characteristics from direct hydrograph analysis. It is however, relatively easy to determine unique values for upper zone free water characteristics, the maximum storage size and the interflow drainage rate, using techniques which will be discussed in the topic dealing with the determination of model parameters.

Subsurface Discharge

If the natural boundary conditions should require all applied moisture to leave the catchment either as observable channel flow or through evapotranspiration, then the catchment's water balance could be completed with the features which have been reviewed. However, subsurface drainage may occur which is not readily apparent from the discharge record. In order to avoid the volumetric inconsistencies which would otherwise be imposed upon the catchment water balance, the Sacramento Catchment Model supplies two paths for unobserved water to drain from the catchment.

The first component of subsurface flow is identified as SIDE flow. SIDE flow is considered to be an extension of the lower zone free water aquifers which does not contribute to channel flow but which drains out of the catchment to a location which may not be known. The drainage characteristics of SIDE are considered to be the same as the characteristics which are appropriate to the observable portion of the base flow. This makes it possible to express SIDE as a ratio to the observed recession volumes. SIDE increases the total size of the lower zone storages without altering the volume which is observed in the channel. An important feature of SIDE, which substantially aids in its determination, is that SIDE increases the limiting percolation value, PBASE without increasing the subsequent volume of observable base flow.

The second type of subsurface flow is that volume which must be supplied to the channel to maintain sufficient moisture in the channel bed for visible flows to occur. This volume which is called subsurface outflow, or SSOUT represents the conveyance capability of the material in the base of the streambed. Although this factor can be quite significant in areas which have experienced glacial deposition of material in the streambed, and in a few other areas whose geology has produced permeable streambeds, in most areas it is not a factor. Fortunately, such flows provide a unique impact upon the recession and are relatively easy to determine.

Soil Moisture Responses to Evapotranspiration Demand

Evaporation from the area covered by surface water and/or phreatophyte vegetation is computed at the potential rate. Over other portions of the soil mantle, evapotranspiration use varies with both evapotranspiration demand and the volume and distribution of tension water storage. As the soil mantle dries from evapotranspiration, moisture is withdrawn from the upper zone at the potential rate multiplied by the proportional loading of the upper zone tension water storage. In the lower zone, evapotranspiration takes place at a rate determined by the unmet potential evapotranspiration demand times the ratio of the lower zone tension water content to total tension water capacity. The pore space which accommodates free water is considered to be contiguous with the soil particles which exert tension water demands. Consequently, if evapotranspiration should occur at such a rate that the ratio of contents to capacity for available free water should exceed the ratio of contents to capacity of tension water, then the free water would resupply any tension water deficiency with which the free water is in contact. This is accomplished by balancing the relative loadings of the available tension water and free water storages in order to maintain a soil moisture profile which is logically consistent. Depending upon basin conditions, some fraction of the lower zone free water may be considered to be below the root zone and, therefore, unavailable for such transfers. That free water which is not subject to evapotranspiration is considered to be reserved for drainage only.

10.6. VISUALIZING THE CATCHMENT MODEL

Many of the concepts which are involved in the catchment model which have been incorporated into the NWSRFS may be clarified by observing how the storages respond to a storm sequence. Figures 10.4 through 10.11 trace a typical seasonal response of soil moisture and discharge. The sequence shown in these figures traces the transition from a very dry condition, through a seasonal wetting cycle, and the return to dry weather.

In Fig. 10.4 the very dry catchment is producing a low level of base flow which is draining from a nearly empty aquifer which is identified as lower zone primary free water. The climax vegetation is clinging to life by tapping the dwindling supply of lower zone tension water. All other sources of moisture have been removed from the system by prior drainage and/or evapotranspiration.

Light rainfall begins to fall in the catchment, Fig. 10.5. That portion of the rain which falls on the permanently impervious portion of the catchment, PCTIM, produces immediate runoff. The rain which occurs on the permeable portion of the catchment is transferred through the upper zone free water to the upper zone tension water, where it will provide moisture

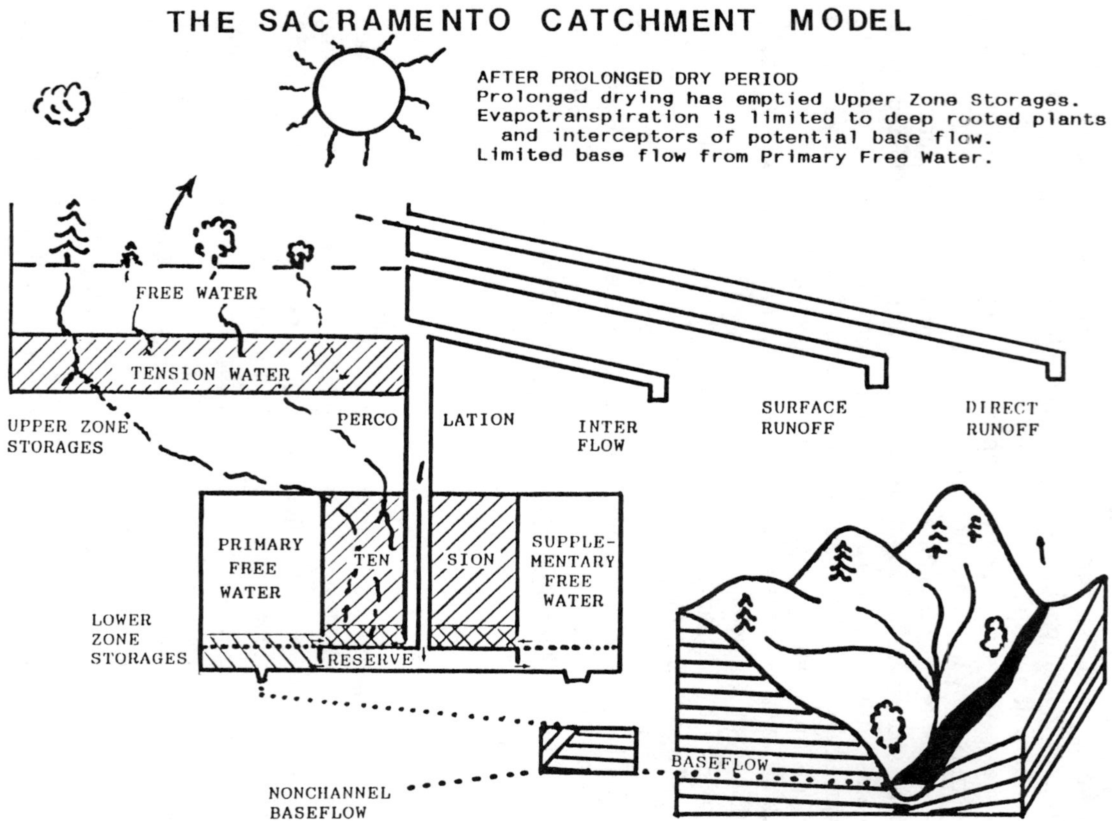

Fig. 10.4. The Sacramento Catchment Model after prolonged dry period.

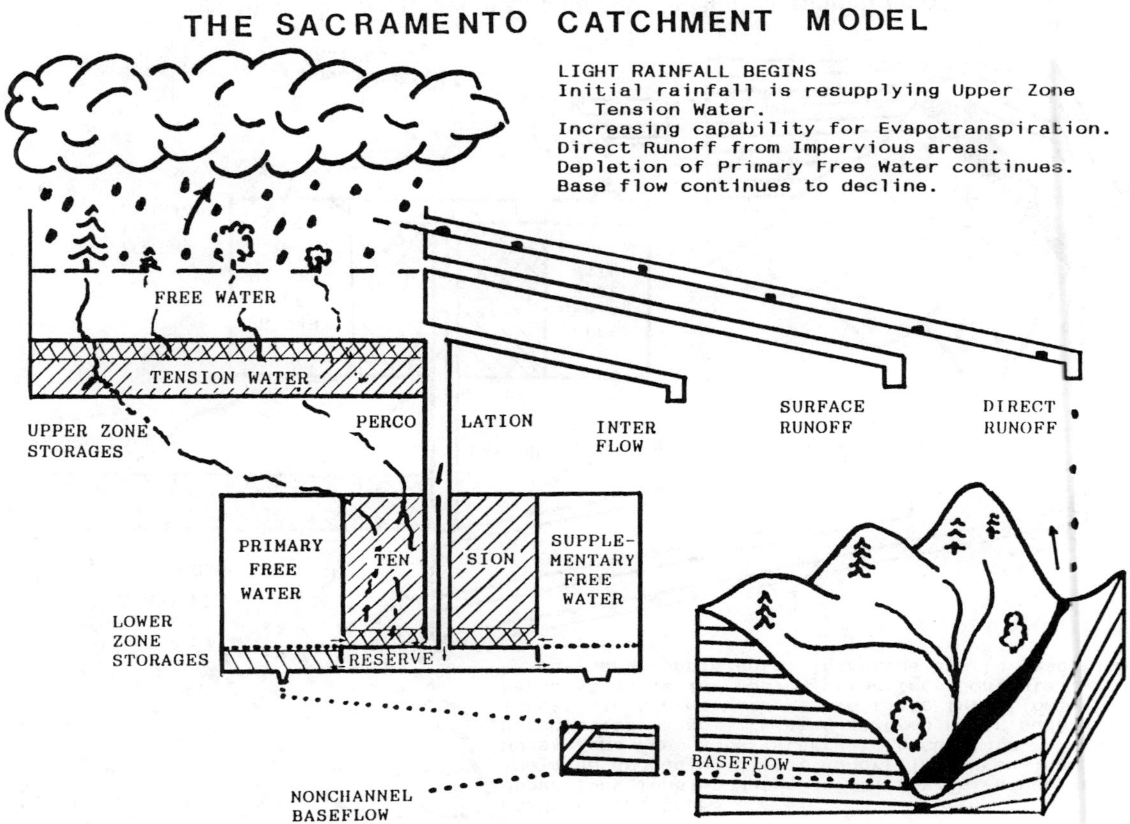

Fig. 10.5. The Sacramento Catchment Model after light rainfall begins.

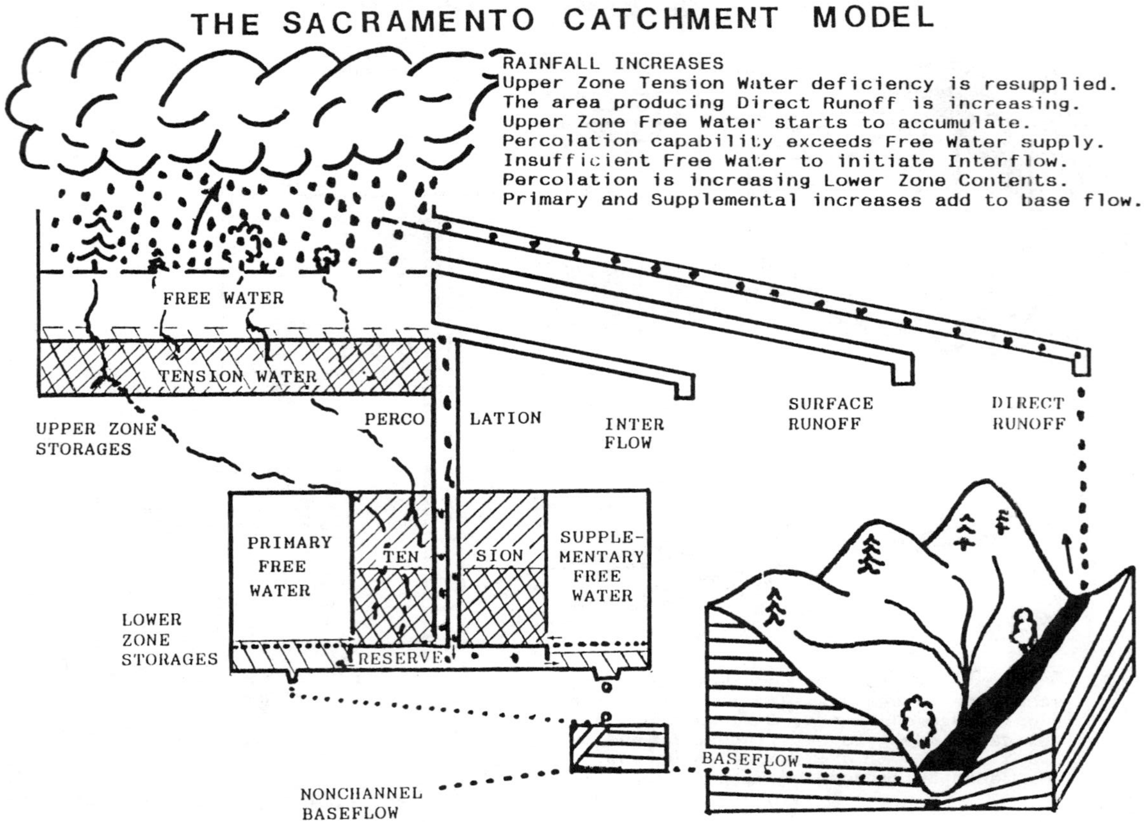

Fig. 10.6. The Sacramento Catchment Model after rainfall increases.

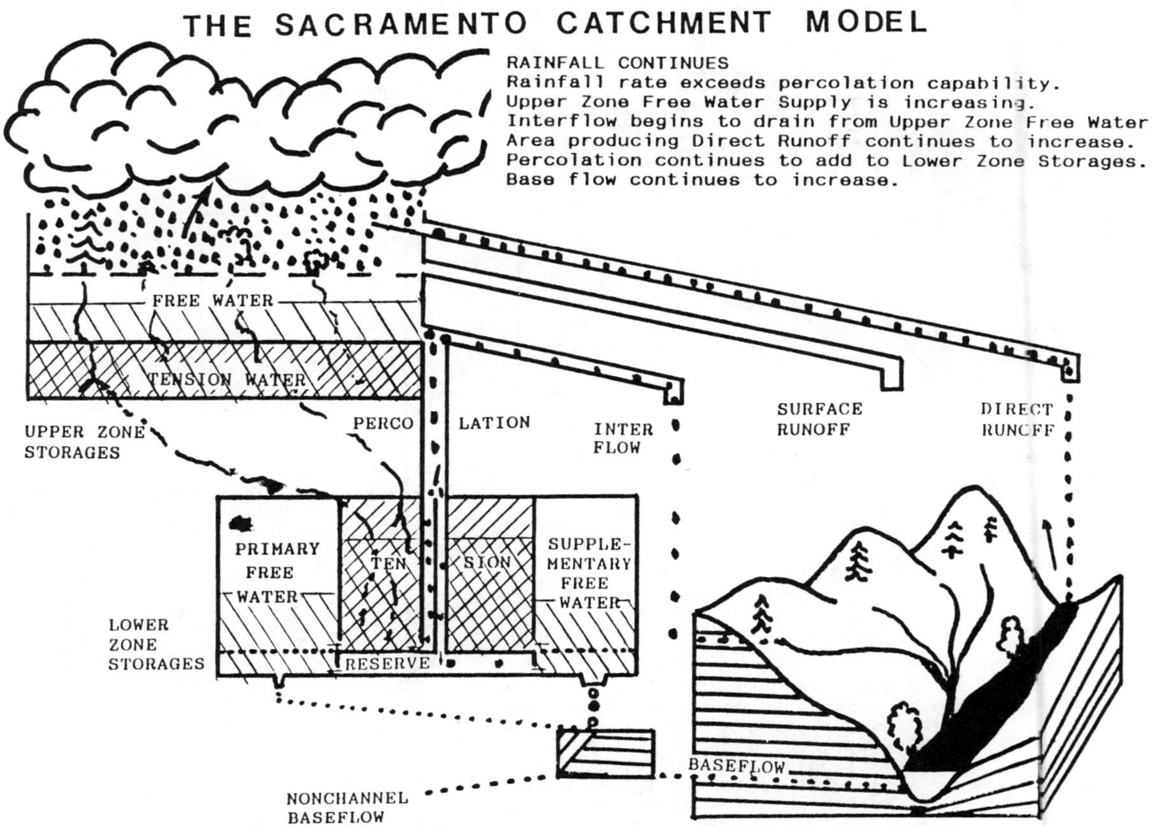

Fig. 10.7. The Sacramento Catchment Model rainfall continues.

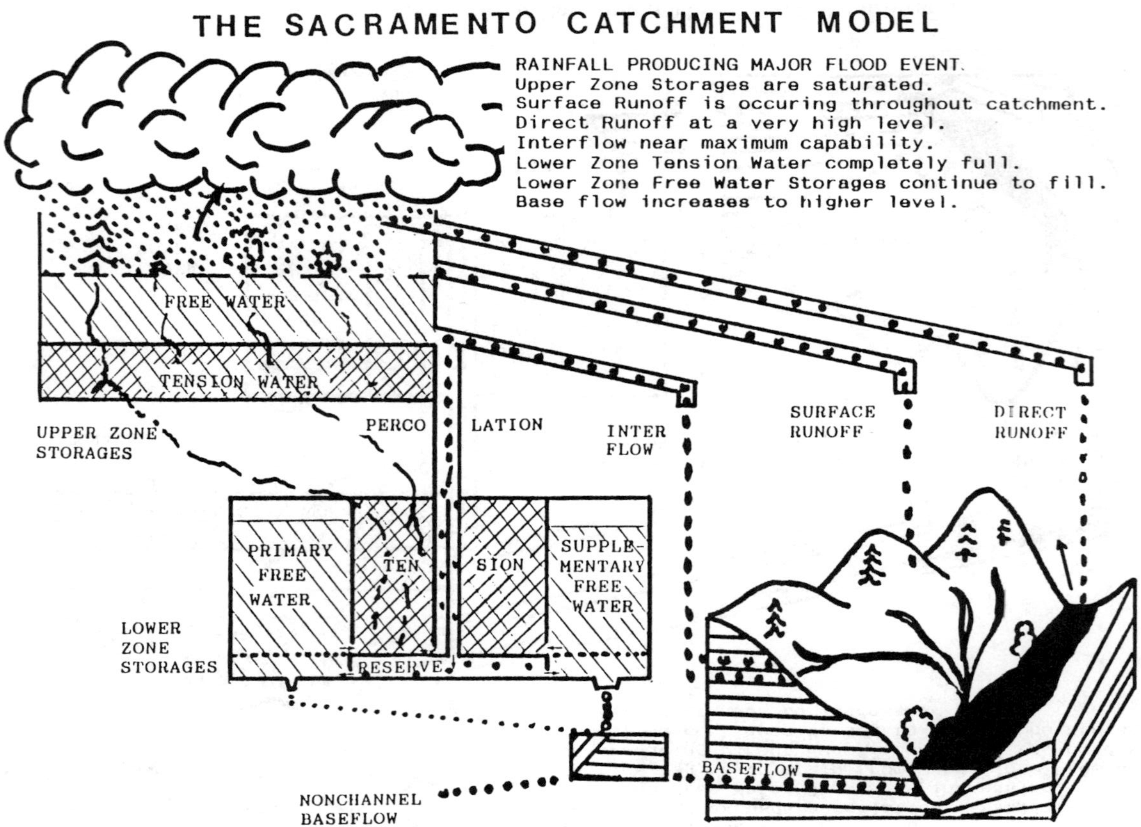

Fig. 10.8. The Sacramento Catchment Model rainfall producing major flood event.

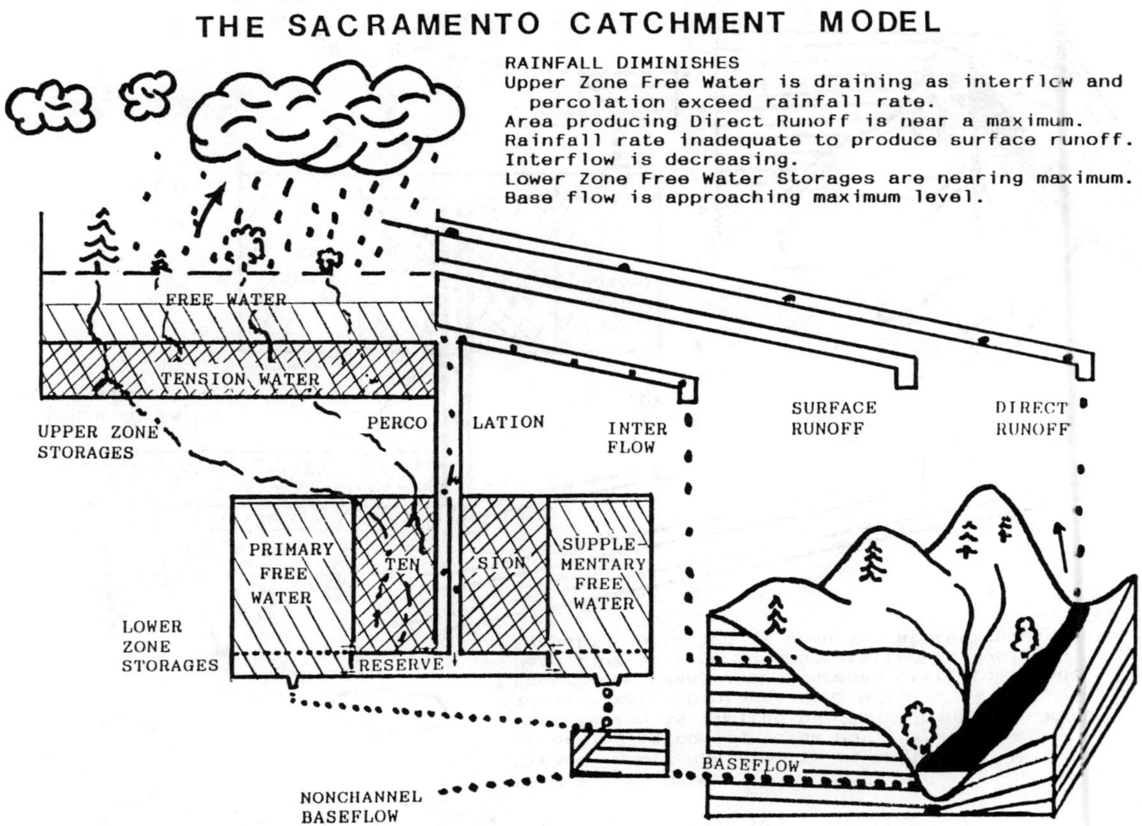

Fig. 10.9. The Sacramento Catchment Model rainfall diminishes.

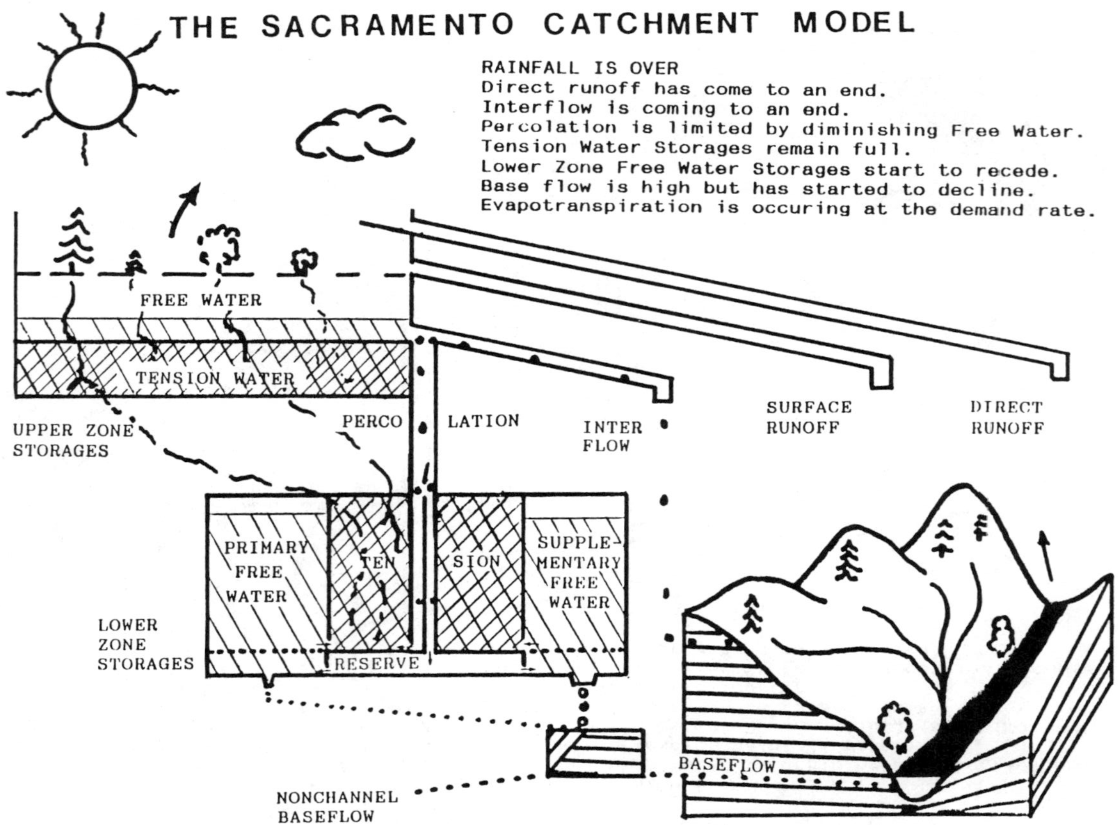

Fig. 10.10. The Sacramento Catchment Model rainfall is over.

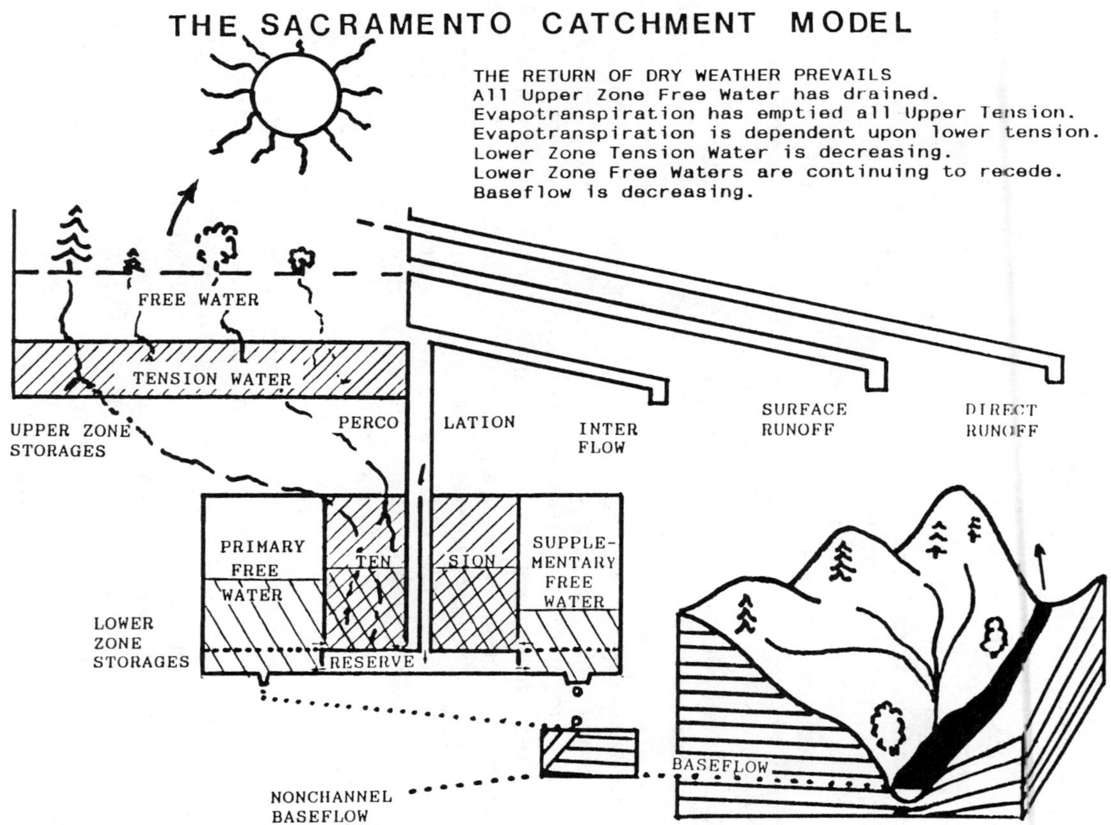

Fig. 10.11. The Sacramento Catchment Model - the return of dry weather prevails.

for the evapotranspiration process. No water has been conveyed to the base flow aquifers, so the availability of base flow continues to decline as the primary free water continues to be depleted. Depending upon the intensity of the rainfall, total channel flow may be the same as, higher than, or lower than the flow which took place in Fig. 10.4.

As the rainfall increases, Fig. 10.6, the upper zone tension water requirements are satisfied, and water begins to accumulate in the upper zone free water. The available free water is still quite small and as a consequence, percolation claims all of the available upper zone free water. That water which percolates to the lower zone is substantially claimed by the deficiency in lower zone tension water, but a small portion of the percolated water starts to recharge the lower zone free water aquifers. As a consequence, base flow starts to increase. Direct runoff from the impervious area of the catchment is increasing due to the increased rainfall rate, but also due to the increase in the effective impervious area which accompanies the increase in soil moisture. The increased direct runoff coupled with the base flow augmentation cause the river to rise.

The rainfall continues as shown in Fig. 10.7. Direct runoff from the growing impervious area increases. The rainfall is occurring at a rate which is greater than percolation can convey water to the lower soils and the upper zone free water supply is rapidly increasing. The availability of upper zone free water is now sufficient to produce a substantial amount of interflow. Inasmuch as the lower zone tension water deficiencies have not yet been met, percolation continues to be dominantly claimed by tension water requirements. There is, however, a portion of the percolated water which is augmenting the lower zone free water storages. The increasing volumes of free water in the lower zone aquifers are producing a continuing increase in base flow. The river has now risen significantly. However, the rise in river levels is still unaccompanied by surface runoff; in the permeable portions of the catchment, surface runoff has not yet occurred.

In Fig. 10.8, heavy rains are now producing a major flood event. The upper zone is completely saturated so that surface flow is now taking place in the permeable portion of the catchment. Direct runoff is occurring from all soils which have a capability of providing direct drainage to the channel. Interflow is occurring at close to its maximum capability. Percolation is being impeded by the increase in lower zone soil moisture. All lower zone tension water deficiencies have been met. The water which is being percolated to the lower zone is now going entirely to the free water aquifers. Discharge from the lower zone free water aquifers continues to increase. The entire system is approaching saturation and the river is extremely high.

After a time, the rainfall diminishes as shown in Fig. 10.9. Although the potentially impervious area is near its maximum size, the diminished rainfall rate is causing direct runoff to decrease. The rainfall has slackened to a rate which is allowing percolation and interflow to decrease the availability of upper zone free water. Interflow is diminishing in response to the decreased supply of upper zone free water. Due to the depletion of free water in the upper soil, a deficiency now exists and the pervious portion of the catchment is no longer producing surface runoff. The lower zone free water storages are claiming all of the water which is being percolated. Due to the near filling of the lower zone storages, and the diminishing availability of upper zone free water, percolation is being constrained at the same time as base flow is occurring at close to its maximum rate. In response to these conditions, the river has started to recede.

The rainfall ends, Fig. 10.10, and the drying process accelerates. Direct runoff, and surface runoff have both concluded and interflow is approaching an end. The available supply of upper zone free water is being consumed by percolation and interflow. In spite of the increased evapotranspiration demand, the upper zone tension water remains full due to the availability of free water in the upper soil. The lower zone tension water also remains full as the evapotranspiration demand is being met by the upper zone storages. The lower zone free waters are declining at a rate which exceeds the resupply by a rapidly diminishing percolation. Base flow remains very high, but it has begun to decline. The river system continues to recede.

The rainy period is over and dry weather has persisted for a considerable period. (See Fig. 10.11.) Evapotranspiration and drainage have substantially reduced the availabilty of soil moisture. Both upper zone storages are depleted. The resupplying of water to the various zones has long since terminated. Deep rooted vegetation is utilizing the moisture stored in the lower zone tension water, and drainage of the lower zone aquifers is diminishing the availability of base flow. The river system is continuing to recede.

10.7. CATCHMENT CHARACTERISTICS-DETERMINATION TECHNIQUES

The determination of appropriate values for the parameters in this system depends upon a good understanding of how the various characteristics interact to produce discharge and an appreciation of the information content which is appropriate to available data bases. It must always be remembered that historical data bases of catchment conditions are approximations. The problems of measuring station rainfall are compounded when observed values are combined to produce "catchment rainfall". As a result of the precipitation paradox, there can be no uniform rainfall weighting procedure which can produce an appropriate evaluation of catchment rainfall. This would be true even if it were possible to obtain perfect measurements of station rainfall. Determining the effective evapotranspiration demand on a catchment basis involves the dynamics of plants, water surfaces and climate in a manner which is likely to remain an approximation. Perhaps, if the value of increasing the effective evaluation of runoff becomes sufficiently high, improved approximations may become possible. For the time being, those approximations which exist will continue to frustrate the catchment modeler. There are, however, a few simple steps which can be taken to make the available data somewhat more useful for catchment modeling purposes. For these reasons, the first step in the modeling effort is to attempt to improve the descriptive capability of the precipitation and evapotranspiration estimates. For these purposes, the model includes a precipitation modifier which allows the scale of the composite precipitation record to be altered, and a monthly factor for improving the appropriateness of evapotranspiration estimates. These factors which are described below, are intended to improve the description of the catchment's water balance in order that the concepts involved in the structure of the Sacramento Catchment Model can be applied with the least degree of distortion.

10.7.1. Water Balance Input Modifiers

RAWT

Station rainfall weights used to compute catchment rainfall. In order to determine rainfall weights which are appropriate for describing the catchment's water balance, it is desirable to review the apparent rainfall requirement of the catchment. An estimate of the rainfall requirement can be obtained by examining the catchment water balance for particularly productive water years. In such years, there must be sufficient rainfall to produce the observed runoff, meet a substantial portion of the annual evapotranspiration demand, and if discharge levels change from the start of the water year to its conclusion, the change in the catchment's soil moisture volume must be included. By comparing a summary of reported station precipitation for such years with the estimate of required precipitation, a perspective on the catchment total which must be produced from the station values can be gained. If the catchment has relatively little in the way of terrain features which can induce orographic precipitation, and if all gauges are well protected from wind influences, a simple spatial weighting of the gauges may provide a sufficient estimate of catchment rainfall. However, in most cases the catchment will require about a 10% enhancement over measured rainfall in order to provide sufficient water to accommodate the water balance. In some areas which are exposed to severe winds, have significant orographic features, and/or have rain gauge locations which do not adequately express the spatial distribution of rainfall, the factors used to determine catchment rain may require more substantial modifications.

PCTPN

A set of twelve values for dimensioning mean daily evapotranspiration for each month.

1) If daily computations of evapotranspiration or daily measurements of pan evaporation are utilized, the values of PCTPN will represent the factors which will be applied at midmonth to dimension the computed evapotranspiration. Between the midmonth points, dimensioning values will be linearly interpolated to complete the daily evapotranspiration dimensioning.

2) If average evapotranspiration is to be used, the values of PCTPN will represent the monthly midpoint values of the mean evapotranspiration demand curve.

3) As a starting point, the evapotranspiration, ET, demand curve established by applying PCTPN values should be in approximate conformity with evapotranspiration curves derived for other catchment's in similar areas. The curve should have an annual amplitude similar to that of a representative pan curve for the general area. The spring ET values will, however, lag the pan values by a few weeks and fall values of ET will descend more slowly than pan values until frost occurs. As vegetation enters dormancy, ET values will drop precipitously. In catchments where pans freeze, evapotranspiration does not drop to zero, but will generally continue at about .1 millimeter per day.

10.7.2. Determination of Model Parameters

Determination of many of the model's parameters can be substantially aided by the preparation of a historical record for the catchment. The record should provide a listing of station and catchment rainfall which is accompanied by a plot of the discharge record utilizing a semi-logarithmic scale. Daily discharge is most conveniently expressed in the same units as the precipitation, that is as a depth of runoff. This provides for a more convenient determination of many catchment parameters and helps to identify those unfortunate conditions when reported discharge exceeds the volume of rainfall which was available to produce it. Such a listing not only assists in the identification of various components of flow, but it also helps to determine which periods of the hydrologic record are most appropriate for determining the characteristics of certain parameters which might otherwise be masked by the interaction of many catchment characteristics.

The model's parameters and their suggested derivation techniques include:

PCTIM

The percent impervious, a fraction of the catchment, contiguous with stream channels, which produces impervious runoff during low flow conditions.

This parameter should be determined by utilizing the catchment historical record in the following manner. During very low flow conditions select a time period when there has been an accumulated demand of 50 millimeters or more of evapotranspiration without intervening rainfall. This period should be followed by a rainfall event which does not exceed 30 millimeters. All runoff accompanying such an event will be from the permanently impervious area, PCTIM. Compute the fraction PCTIM as:

$$\text{PCTIM} = \frac{\text{the increment of runoff produced by the rainfall}}{\text{the quantity of rainfall which produced the runoff}}$$

Frequently, such runoff drains quite slowly due to the substantial friction encountered in the bed of the stream. Several events should be utilized to determine PCTIM. When averaging the resulting values, the effect of storm size on proportional error should favor a higher weighting for the larger rainfall events. The important factors which allow a storm to be utilized for this purpose are that the catchment is sufficiently dry so that only impervious runoff is included in the analysis, and the storm is sufficiently small so that all water falling on the permeable portion of the catchment will be claimed by the upper zone tension water deficiency. A careful analysis is a final analysis; the determination of PCTIM is complete.

ADIMP

The additional impervious area, that fraction of the catchment which becomes impervious as all tension water requirements are met.

In the catchment record, find an extremely wet period which produced major flows followed by at least ten days of dry weather. If the next rainfall event is not more than 60% of the intervening evapotranspiration, then the effective impervious area for that event, EIA, can be computed as; (storm runoff/storm rainfall). ADIMP is then computed as:

$$\text{ADIMP} = (\text{EIA-PCTIM})/((\text{TTC-IE})/\text{TTC})$$

where:

TTC is the total tension water capacity of the catchment's soil moisture system. A value for TTC of 300 millimeters is usually a sufficiently accurate approximation if the Total Tension Water Capacity is not yet known. However, if it is later determined that the total tension water profile differs significantly from 300 millimeters, a second approximation may be desirable.

IE is the sum of the intervening evapotranspiration during the dry period between the storms.

Under the conditions described, The upper zone tension water deficiency should be sufficiently large so that only the impervious area is producing runoff from the current storm. As a consequence, only that portion of the catchment which has assumed impervious characteristics should produce additional runoff. The computation of ADIMP should be made for a number of similar events and the median value of ADIMP determined. Due to uncertainties in the initial estimate of the tension water profile, a second estimate of this parameter may be appropriate as the model fitting proceeds.

SARVA

That fraction of the catchment which produces a reduction in streamflow due to evapotranspiration.

SARVA includes the evaporation losses from the exposed water surfaces of the streamflow system, evapotranspiration losses from phreatophyte vegetation along the channel, as well as evapotranspiration losses from any vegetation which deprives the channel of base flow. The magnitude of the SARVA area is demonstrated quite remarkably when there is a sudden change in the ability of plants to transpire and, as a consequence, there is a dramatic impact upon streamflow. Such changes may be induced by a killing frost, fire which destroys vegetation which meets evapotranspiration requirements by intercepting potential base flow, or activities of man which alter the catchment's morphology.

Such events are not a regular feature of most catchment histories. As a result, SARVA is generally determined by studying an extended recession which is composed of drainage from the lower zone primary aquifer and which continues through a substantial change in evapotranspiration demand. A time varying volume, computed by multiplying SARVA times the appropriate daily value of evapotranspiration, is added to the observed flow. That value of SARVA which produces a recession which most appropriately reflects withdrawal from a linear reservoir, a straight segment on the semi-log discharge plot, is the desired value. The determination of this parameter focuses on the selection of a dry period which undergoes significant changes in the evapotranspiration demand. If such a period is included in the historical record, SARVA can be discretely determined. In most catchments, the sensitivity of the simulation to SARVA is relatively low. If the records are inadequate to clearly define SARVA, it may be set to a value equal to PCTIM.

UZTWM

That depth of water which must be filled in permeable portions of the catchment before any upper zone free water becomes available for percolation and interflow.

The identification of UZTWM requires a protracted dry period which depletes all tension water in the upper soil. UZTWM can then be determined from that amount of rainfall which must fall before any runoff, other than that from the permanently impervious area, reaches the channel. In most areas, if the dry period has an evapotranspiration potential which exceeds the intervening rainfall total by approximately 200 millimeters, a positive determination of UZTWM can be made. In addition, any intervening rains must be relatively small in magnitude such that any soil moisture they contributed was likely to have been consumed by evapotranspiration prior to the onset of the storm used to determine UZTWM. Several events may be utilized with varying degrees of appropriateness. The factor we are attempting to determine is how much water can be applied to a dry soil before any free water accumulates in the upper zone to initiate interflow. Careful inspection of a sufficient historical data base will allow the direct determination of UZTWM by inspection.

UZFWM

Upper zone free water represents that depth of free water which must be available in the permeable portion of the catchment in order that vertical and horizontal drainage of the upper soil can take place at their maximum rates. This corresponds to the amount of free water in the upper soil which is required to maintain a wetting front at full potential, and to support the maximum level of interflow.

The size of UZFWM and the rate at which it drains, UZK, should be determined by isolating a storm event for analysis which clearly demonstrates the drainage characteristics of upper zone free water. The most effective event to portray this characteristic follows a saturation period which includes an extremely high recession which continues for two weeks or more without rain induced perturbations. The determination of UZFWM and UZK should be conducted by interacting with the model, making trial runs, after the lower zone free water storages have been evaluated analytically. Utilizing the lower zone free water conditions which have been previously determined, make successive model runs altering only UZFWM and UZK until the recession subsequent to the discharge of surface runoff is approximated as closely as possible.

If the soil moisture system is saturated, the influence of other parameters will be negated and only one set of values for UZFWM and UZK will maximize the ability to track the recession. Start the simulation a year or more in advance of the targeted recession so that there will be no residual influence from the initial state conditions of the various soil moisture storages. As long as the lower zone tension water is set sufficiently small so that the lower zone free water storages fill to a level which produces a match of their portion of the recession, the unique values which are appropriate to UZFWM and UZK can be positively determined.

UZK

The daily rate at which upper zone free water is discharged as interflow. See UZFWM.

PBASE

The percolation rate when all free water aquifers are filled. It is the maximum percolation which can occur when percolation to a saturated lower soil is limited to that which can replace soil moisture losses produced by base flow drainage.

PBASE is a value which is derived from the maximum discharge capability of the lower zone free water aquifers. PBASE is not directly provided by the analyst but is computed by the system as:

$$\text{PBASE} = (\text{LZFPM} \bullet \text{LZPK}) + (\text{LZFSM} \bullet \text{LZSK})$$

> The lower zone storages of LZFSM and LZFPM used to derive PBASE are those estimated by the investigator from hydrograph analysis and modified within the program to include the SIDE component.

Z or ZPERC

The proportional increase in percolation from saturated to dry condition is expressed by the term Z.

The value of Z is best determined by careful selection of that runoff condition which most clearly identifies its characteristics. This parameter should be determined subsequent to UZTWM, UZFWM, LZFPM, LZFSM, UZK, LZPK, LZSK, PCTIM, ADIMP, SIDE and estimating LZTWM. Z describes the percolation magnitude early in the wetting cycle, when the upper soil is approaching saturation and the lower soil is quite dry. Under these conditions the infiltration curve is relatively insensitive to REXP therefore the value of 1.5 can be temporarily utilized for that parameter. Select the first storm following extreme drying, which produces significant runoff in excess of the impervious components. Interact with the model, starting a year or more in advance of the target event to minimize the influence of the initial state conditions. Select that value for Z which modifies the percolation into the lower zone so that the storm runoff volume is correct.

Several events which occur under similar conditions may be utilized to improve the determination of Z. The state conditions for other parameters should be appropriate if a sufficient lead time precedes the analysis segment. The analyst should isolate on those storms which meet the characteristics for defining Z and temporarily disregard errors in those portions of the simulation which are being influenced by still undefined parameters.

REXP

The exponent in the percolation equation which guides the rate at which percolation demand changes from the dry condition, (Z+l)•PBASE, to the wet condition, PBASE.

In order to emphasize the effect of REXP on the percolation equation and consequently on the catchment hydrograph, we should interact with the model by concentrating on those events which occur as the lower zone storages are being filled. Subsequent to a dry period, when a series of storms is gradually increasing base flow, concentrate on those storms which increase base flow to about 35% of its maximum value. This will generally provide a period when variations in REXP will demonstrate the greatest sensitivity. The value of LZTWM will influence this evaluation so REXP should be considered concurrently with LZTWM, see LZTWM below. Select that value of REXP which produces the most effective evaluation of the targeted storms. Evaluating separate determinations of REXP from different flood seasons will assist in determining a generally appropriate value.

LZTWM

The lower zone tension water maximum storage capacity expressed as a depth over the catchment.

This parameter represents that tension water storage which supplies moisture to meet the evapotranspiration needs of those plants and trees which send their root systems many feet into the soil. Portions of the moisture in this storage may remain in place for many years. This feature which contributes to the long term memory of soil moisture provides an important indication of the degree of drought and the volume of water needed for a return to normal runoff conditions. However, the long term memory which contributes to the success of the model also requires a very consistent data base and an appropriate time period to evaluate.

The full magnitude of this storage is best defined when the record includes a protracted drought period with wet periods which include substantial surface flow and very high levels of base flow occurring both before and after the drought condition. As long as any deep rooted vegetation is surviving in the catchment, there must be residual moisture left in this storage, and when base flow reaches its highest levels, this storage must be full.

The most effective approach to determining the magnitude of LZTWM is to utilize computer runs for the entire period of record and examining those periods which have relatively long periods of saturation. After the other storages have been determined, look for a shift in the ability to evaluate discharge during a protracted period of wetness. If LZTWM has been established at too small a value, overforecasting of the simulated flows will occur and continue until late in the saturated sequence. The total volume of the overforecast will approximate the inadequacy in the size of LZTWM. If LZTWM is oversized, the early storms in the sequence will be underforecast with the later portion of the sequence being reasonably represented. The aggregate error of the underforecast will approximate the volume by which LZTWM must be reduced.

The errors induced in the simulation by errors in LZTWM tend to disappear late in the saturated sequence as long as LZTWM is not given such a large value that it can never fill. If this is done, the simulation will be replete with underforecasts and inadequate base flow. LZTWM must be set sufficiently small so that when observed base flow reaches its highest magnitude, LZTWM is completely filled. In contrast to errors produced by LZTWM, errors which are due to an improper application of SIDE are likely to be propagated throughout all events in each saturated sequence. See SIDE

LZTWM will respond to errors in catchment rainfall determination. If catchment rainfall is systematically underestimated the apparent value of LZTWM may be so small that it is clear there would not be sufficient soil moisture to support the area's vegetation. If catchment rainfall is overestimated, it may appear that the size of LZTWM needs to be increased each year.

The proper value of LZTWM is one which allows the contents of lower zone tension water to fill by the time base flow values reach their highest levels. Due to the long memory exhibited by water which has been stored in lower zone tension water, trial computer runs should establish an initial content for lower zone tension water which will allow it to fill when base flow first achieves its maximum rate. The median value which has been determined for LZTWM is approximately 150 millimeters, but there appears to be a significant variation which reflects the root depths from which the catchment's climax vegetation obtain water for evapotranspiration.

Lower Zone Free Water

The lower zone free water includes two lower zone base flow aquifers and their drainage constants, these are:

- LZFSM - The maximum capacity of the faster draining lower zone free water, supplemental storage, which drains at the daily rate expressed by LZSK. This capacity is expressed as a depth over the catchment.
- LZSK - The portion of water stored in lower zone free water supplemental storage which will be released each day.
- LZFPM - The maximum capacity of the slower draining lower zone free water, primary storage, which drains at the daily rate expressed by LZPK. This capacity is expressed as a depth over the catchment.
- LZPK - The portion of water stored in lower zone free water primary storage which will be released each day.

All values associated with lower zone free water are derived from hydrograph analysis. To the extent that the maximum level of base flow is represented in the historical record, the values of LZFSM, LZSK, LZFPM and LZPK can be determined by hydrograph analysis (see Fig. 10.3). The hydrograph utilized for the determination of the lower zone free water parameters should represent the highest level of base flow apparent from the record. If subsequent analysis reveals that a modification of these volumes is necessary, the volumetric additions to these storages must always be in proportion to their relative drainage factors. The computed values always reflect the base flow which is observed in the channel and if the historical record is adequate, they can be determined with a high degree of precision.

SIDE

Represents that portion of base flow which is draining to areas other than the observed channel. It is expressed as a ratio of the unobserved flow to the observed flow. When all of the catchment's saturated soils do not drain to the surface channel, SIDE allows the correct definition of PBASE, in order that the saturated percolation rate may be achieved.

The need for SIDE is most clearly demonstrated during those periods when the catchment is at or near saturation. Under these conditions if percolation, as limited by the analyzed size of the base flow aquifers, is regularly insufficient to remove enough water from the forecast hydrograph, and the magnitude of subsequent base flow is reasonable, then the rainfall has been overestimated, or runoff is leaving the catchment by bypassing the surface channel. If the scale of precipitation can not reasonably be diminished, then SIDE may be used to increase percolation without contributing to base flow. LZTWM which is inadequately sized will produce a similar manifestation when it fills prematurely. However, later in the saturated season the overforecast will continue if SIDE is required, but will no longer be apparent at the point where a properly sized LZTWM would have completed its filling process. Thus, several consecutive storms during the saturated period which show a change in the overforecast bias, can provide a means of discriminating between SIDE and LZTWM. If SIDE is required, all storms in the saturated sequence will be overforecast.

When SIDE is required, the limiting percolation computed from the recession characteristics, during saturated periods, will not be sufficient to satisfy the apparent soil moisture demands. As a result, although subsequent base flow will be effectively estimated by the applied model, too much water will appear to be available for surface runoff. This manifestation of the need for additional percolation is confirmed if it occurs when it is reasonable to assume by the magnitude of the base flow that the lower zone soils were saturated. Under these conditions, if the excessive runoff is not due to an overestimate of rainfall, the soil must accommodate a larger volume of water than has been computed from the historical recessions. If this problem generally appears to occur once in the rainfall sequence which is bringing the catchment to saturation, then the lower zone tension water has been evaluated at too small a magnitude.

However, if the overestimate of storm runoff regularly occurs in successive rises during a saturated period, then either the precipitation evaluation procedure is over estimating catchment rainfall or there are soils drained by aquifers that do not discharge to the catchment's surface stream channel. Since such drainage is not recorded in any readily available records the Sacramento Catchment Model assumes that those aquifers which drain to areas outside the catchment have soil drainage characteristics which parallel the soils which do produce observed flows.

This allows the free water drainages which produce non observed flows to be represented as an additional component of the lower zone free water storage volumes which have been integrated from the stream channel outflow hydrographs. This additional fraction of the lower zone free water system, SIDE, expands the size of the lower zone free water storages, but does not contribute to an increase in channel flow. The increased percolation made possible by enlarging the free water storages allows the soil to percolate more water, permits the surface runoff to be correctly evaluated, and maintains the integrity of that portion of the recession which is observable in the channel.

SSOUT

The subsurface outflow along the stream channel which must be provided to the stream before there is sufficient water to produce visible channel flow.

This volume is expressed as a rate of flow per unit area. It is only appropriate where the base of the channel is relatively porous. SSOUT occurs most frequently in sandy channels, those which have been partially filled with glacial debris, and others which can transfer water below the apparent base of the channel. If the observed discharge is corrected for the effects of evapotranspiration due to SARVA, it should represent a linear line segment for those periods when primary flow is the principle source of runoff. If instead of having this characteristic, the recession demonstrated on a semi-log plot curves downward with time, the stream bed is demanding a portion of the discharge which would otherwise produce observable base flow. That constant volume of flow which can be added to the observed recession to obtain a linear result on the semi-log plot is the volume of water required by the stream bed and is identified as SSOUT. Although SSOUT exists to some degree in most areas, it is generally so small that it can be set to zero and ignored. However in other areas, particularly those with well identified sub channel aquifers, SSOUT can be an important characteristic of the catchment system.

PFREE

Represents that fraction of the percolated water which is transmitted directly to the lower zone free water aquifers during the period when the lower zone tension water has a soil moisture deficiency.

This catchment characteristic represents that percent of the percolated water which directly replenishes the ground water aquifers without incurring any losses to tension water demands. This suggests that a portion of the ground water system is in direct contact with the upper soil mantle. PFREE is an expression of the areal variability which exists in the catchment. It is one of the most readily identified and least sensitive of the characteristics included in the Sacramento Catchment Model. From the same events utilized to identify Z, review the magnitude of the low flow recession following the runoff event. If PFREE is set to zero, there will be little if any, increase in base flow provided by the simulation as all percolated water is being captured by the lower zone tension water system. Slowly increase PFREE until the generated base flow corresponds to the observed. The range of values for PFREE which the author has encountered is from .00 to .40. A value of .5 would suggest a marsh like environment, and a value in excess of .99 would be physically impossible unless the entire catchment were a water surface. Generally, an effective value for PFREE will be about .12, indicating that 12% of the percolated water will be fed directly to the ground water aquifers.

RSERV

RESERVE, that fraction of the lower zone free water which is unavailable for transpiration purposes.

Generally, an appropriate value lies between zero and .40 with .20 being an acceptable value for most catchments. This parameter has very low sensitivity. In most cases, there is not ample evidence to justify manipulation of its value.

HISTOGRAM

The time distribution function for rapidly responding runoff. The nondimensional surface flow "histogram" utilized in this model is generally derived externally from the simulation process. Each ordinate represents the proportion of flow which will reach the channel outlet in successive time periods. This requires that the sum of the ordinates must equal 1. The base flow separation utilized in the Sacramento Catchment Model is based upon a significantly different assumption than that which was previously utilized in National Weather Service hydrological operations. The histogram should be derived after the determination of lower zone supplemental and primary characteristics have been completed. In this way the base flow separation can be more effectively determined. Aside from the difference in base flow separation, the histogram may be developed in a manner similar to a "unitgraph". However, whereas the "unitgraph" is attempting to describe the normal distribution of runoff, the histogram is attempting to define the shape of the more rapidly responding flows. As a consequence, the storms which are appropriate for determining the time distribution of the histogram are not likely to coincide with those appropriate for "unitgraph" determination. The utilization of a shape function which corresponds to those flows which reach the catchment outlet with the greatest efficiency allows the channel storage feature to provide the necessary increase in storage appropriate to varying flow volumes.

Channel Storage

A layered routing process. The channel storage mechanics are based upon a layered routing concept which expresses the efficiency with which up to four volumes of flow traverse the catchment. Each layer is routed by a technique which is parallel to the use of a Muskingum routing for each separate layer. The routing considers the outflow produced by the catchment model as inflow to a routing reach. The routing factors are determined from that volume of flow which would reach the outlet if there were no need for additional storage. When this feature is utilized, the volume associated with each discharge layer which has a significantly different storage attribute must be determined. Each layer must also be assigned a storage characteristic which describes that layer's efficiency.

As an example, a storage characteristic of 1. implies that the layer is 100% efficient and all flow encompassed by the accompanying volume would leave the catchment without further attenuation. Similarly, another layer volume, which would accompany flows in excess of that defined by the first layer might leave the catchment with an efficiency of .6. This would mean that the layer was only 60% efficient and that only 60% of the flow in that layer would exit the catchment in the time period defined by the other components of the model. In each subsequent time period, 60% of the residual flow from the first period which had not exited the catchment would be added to the other components of flow from other layers and from other time periods. The total of all the elements which influence each time period defines the final outflow shape.

The layered routing is invariably required for those small volumes of flow which are in contact with the bed of the stream. As flows increase, rocks, gravels, sand bars and other frictional features tend to be submerged and the channel efficiency increases. Efficiency may decrease once again as flood flows spill across the flood plain. The layered routing provides a simple and convenient way of improving the ability to model these variations in channel efficiency. There are, however, many hydrologists who have not availed themselves of this feature of the model. The author believes that this may be due to a focus on flood events. However, where water has a high value, and/or low flows are significant for water supply, dilution capability, pollution estimates, or many of the growing concerns about our water resources, a channel routing process can substantially improve the accuracy of flow estimation. Indeed, in any area where the basic histogram approach is inadequate due to variations in storage with different levels of flow, the layered routing can enhance the streamflow modeling capability.

10.7.3. Reviewing the Resulting Analysis

It should be noted that all of the parameters have been determined by focusing on those portions of the catchment history which contribute to isolating the influence of one or two parameters. This approach allows the analyst to maintain the consistency of the concepts upon which the model has been predicated and to achieve an operationally useful procedure which can be rationally altered if there is a change in the vegetation, the paved area, or any of a number of conditions which can be induced by man or by nature.

After the parameter values have been determined in the manner outlined above, it is worthwhile to review the initial estimate of the evapotranspiration curve. In order that the simulation capability may be as appropriate as possible, the initial state conditions (the starting soil moisture values) should be set so that they are consistent with the observed flow conditions at the start of the analysis. The entire record should then be simulated to assure that those assumptions which were utilized to produce the evapotranspiration demand are in phase with the catchment's response. An examination of the monthly errors for the entire period of record is likely to suggest that the initial assumption of the shape of the evapotranspiration demand curve has not been supported by the catchment analysis. Minor modifications of the evapotranspiration demand curve, which do not violate the information content appropriate to the source from which the curve was derived, are appropriate and will assist in providing a more effective procedure. Generally, such modifications may interact with the value of LZTWM but are unlikely to have significant influences on any of the other parameter determinations.

Finally, it is worthwhile to review the entire simulation capability to assure that those portions of the record which guided the determination of the parameters were representative of the data set. If any discrepancies are discovered, it is appropriate to review the rationale involved and to modify any inconsistent determinations.

10.7.4. Parameter Optimization

When the Sacramento Catchment Model was initially developed, it was hoped that computerized optimization procedures would allow the system to be rapidly applied to wide areas. This hope has not been fully realized as parameter detection, utilizing computerized optimization procedures, has generally been guided by optimization functions which are relatively insensitive to many of the model's features. This could occur because the data set was not stable, the historic sequence did not include an adequate sequence of events to exercise some of the model's characteristics, or the optimization function was not sensitive to the discrete functions associated with the proper use of particular parameters. These and other difficulties associated with designing an appropriate function for guiding automated optimization features have impeded the conceptually consistent application of the model.

The ability of automated procedures to produce parameter sets which provide a substantial level of streamflow simulation capability, but which contain parameters with conceptually invalid parameters, has demonstrated that the catchment modeling system can be applied at different levels. The model can be utilized as a black box approach to catchment hydrology, often inadvertently, by a hurried or unskilled modeler, or it can serve as a conceptually consistent approach in the hands of a skilled analyst. This suggests that if automated optimization procedures are to be utilized, they should be constrained so that parameter manipulation takes place within a range of values which retains conceptual consistency. Even then, the author remains concerned that the modified parameter set may be distorted by the lack of proportional sensitivity of the optimization function to all of the factors which influence the catchment's history. Sorooshian's interesting outlook on many of these problems is presented in Chapter 2 of this book.

The convenience associated with using automated optimization procedures has encouraged their application even though many of the results convert the model from a powerful conceptual tool to a distorted black box system. Inasmuch as inadequately constrained optimization procedures have thc capability of producing parameter sets which provide reasonable appearing simulations, but which can contain parameters with conceptually invalid values, it is important for the analyst to restrain parameter modification to a range which fit the conceptual basis on which the model has been formulated. This is particularly important if the catchment model is to be utilized to investigate or respond to, changes in catchment conditions such as the effects of logging, urbanization, wild fires, or climatological changes. For, if the parameter determinations are not conceptually consistent, the logical basis for altering them to describe changing conditions is lost.

An additional problem associated with automatic optimization procedures is that the analyst who may be applying the model may lose track of the mechanics behind the system, and as a result, fail to appreciate the limitations imposed by the data base used for parameter determination. The search for improved optimization techniques continues. However, this author has repeatedly demonstrated the capability of obtaining conceptually consistent, highly effective analyses utilizing the techniques described in this document. Once the data has been organized and the information is available to an interactive microcomputer system, a skilled analyst can complete an effective, conceptually consistent, catchment analysis in a few hours. Although an automated optimization procedure may be applied at this point in the catchment analysis, the author is not convinced that improving an optimization function necessarily produces a more appropriate catchment modeling capability.

The determination of model characteristics with automated optimization procedures requires a very similar amount of manpower to the interactive procedure which has been described, but with such procedures, much of the educational value which stems from carefully thinking through the concepts involved in the simulation process is lost. As a result, there is little transfer of the knowledge gained from fitting the system to its operational application. There is little doubt that an automated determination of the model's parameters will produce a higher level of that optimization function which guides it than can be achieved with a nonautomated approach. But there is great doubt as to whether or not automated procedures produce the most effective model for representing those processes which describe catchment behavior.

Nonetheless, the development of effective automated parameter optimization procedures is a goal which needs to be pursued. But the problems associated with automatic optimization of a streamflow model's parameters present a series of dilemmas. Should we develop procedures which optimize each concept in the model, or can we find a technique which derives a single optimization function which assures the maximum appropriateness of the separate processes? Do all parameters need to be optimized or only those whose characteristics are most obscure? There is also the need to consider factors beyond the degree of automated optimization which might be achieved. Ultimately, intelligent interactive analysis may contribute more to the development of effective hydrologists and analysis systems than can be achieved by accepting the determinations of computerized optimizations. Would such an approach ultimately represent a more appropriate goal for society?

10.8. COMMENTS ON STREAMFLOW SIMULATION CAPABILITIES

Since its origination, many people have proposed or implemented changes to the Sacramento Catchment Model. Some of these changes were for theoretical reasons, others were to accomplish particular objectives. That version reported on in this chapter is that which has demonstrated a substantial capability through extensive testing under rigorous guidelines and as a result, its principal components were incorporated into the catchment modeling program of the NWSRFS.

The author of this chapter has had the opportunity to work with the Sacramento Catchment Model utilizing data from every continent other than the polar regions. The results of these applications, which now number several hundred catchments, have consistently demonstrated that the effectiveness of the simulation is directly related to the appropriateness of the data set. If an adequately descriptive data set is available, excellent conceptually appropriate simulations can be produced. Such data sets may encompass five to fifteen years of data depending upon the hydrologic events included in the record. In those areas where data sets are inadequate, a surprising amount of information may be obtained by focusing on the concepts involved in the modeling process.

When the Sacramento Catchment Model is properly applied, the information content, involved in its multiple storages and sources of flow, has demonstrated many interesting additional capabilities beyond streamflow forecasting. These include such diverse but hydrologically related functions as the ability to provide "a marked improvement in water quality analysis", Barnes (1973); a tool for evaluating future fodder production, Burnash and Ferral (1973); and the "capability to simulate the soil moisture conditions which are appropriate to landslide occurrences" produced by soil saturation, Leader et al. (1983). All of these determinations were accomplished with the information imbedded in a catchment model which had been tuned for operational forecasts of streamflow. In addition, the information content and adaptability to changing environmental conditions are important attributes of a carefully conceived conceptual modeling systems.

A wide variety of procedures incorporating the logic imbedded in the model are now operational. These include the format utilized in the NWSRFS, but also a variety of modes which range from those which can function on microcomputers operating under an early version of a disk operating system to those suitable for use on advanced multi-tasking systems. Various versions of the catchment model have been prepared and applied to operational forecasting situations. These range from the use of data at twenty four hour intervals to systems appropriate to processing continuous real time data fields. Overall, the results of these numerous applications have been very encouraging. The demonstrated capabilities of the model are extremely high, but coupled with this capability are increased demands upon those who attempt to use it. Those problems which have been observed from extensive experience with these systems focus on two urgent needs. These are adequate data and hydrologists with sufficient training to properly apply and utilize the capabilities of the catchment modeling system.

Data problems are those common to much of hydrology: the lack of stable long term precipitation records which successfully integrate the effects of wind influences over the gauge orifice, accurate discharge records which allow catchment modeling to be applied to areas with reasonable climatologic and hydrologic homogeniety, evapotranspiration demand information which adequately describes catchment conditions, and sufficient information to allow the application of advanced snow analysis procedures. These factors are basic to defining reliable catchment water balance systems. Their solution may await the development of new technologies for measuring the many environmental elements which are required for a comprehensive description of hydrologic conditions.

The training of hydrologic personnel in the application and use of procedures is a complex problem which may involve budgets and the political determinations which control the distribution of scarce resources. However, one thing is clear. Any individual who attempts to design or apply a streamflow simulation system should understand the processes well enough so that when there are difficulties with that system's performance, whether they be from data deficiencies or catchment changes, rational decisions can be made as to why the difficulties exist and how they can be resolved. This was a fundamental consideration in the conceptual design of the Sacramento Catchment Model and it remains of great importance in the effective application of the system within the NWSRFS.

ACKNOWLEDGMENTS

The material included in this chapter, owes a great deal to my previous collaboration with R. L. Ferral and to many years of patience and assistance upon the part of my wife, Jeanne M. Burnash, I would like to thank both of them.

REFERENCES

Anderson, E. A., National Weather Service River Forecast System - Snow Accumulation and Ablation Model. NOAA Technical Memorandum NWS Hydro-17, U. S. National Weather Service, November, 1973.

Barnes, G.W., Water Quality Simulation Utilizing a Generalized Streamflow Simulation Model. Master of Science project, California State University, Sacramento, 1973.

Burnash, R.J.C. and R.L. Ferral, Generalized Hydrologic Modeling, a Key to Drought Analysis. Second International Symposium in Hydrology, Fort Collins, Colorado, 1972.

Burnash, R.J.C., R.L. Ferral and R.A. McGuire, A Generalized Streamflow Simulation System - Conceptual Modeling for Digital Computers. U. S. Department of Commerce, National Weather Service and State of California, Department of Water Resources, March, 1973.

Crawford, N.H. and R.K. Linsley, Digital Simulation in Hydrology - Stanford Watershed Model IV. Technical Report No. 39, Department of Civil Engineering, Stanford University, Stanford, California, 1966.

Green, D.W., H. Dabiri, C.F. Weinang and R.Prill, Numerical Modeling of Unsaturated Groundwater Flow and Comparison of the Model to a Field Experiment. Water Resources Research, Vol.6 No.3, 862-874, 1970.

Hanks, R.J., A. Klute and E. Bresler, A Numerical Method for Estimating Infiltration, Redistribution, Drainage and Evaporation of Water from Soil. Water Resources Research, Vol.5 No.5, 1064-1069, 1969.

Leader, D.C., R.J.C. Burnash and R.L. Ferral, An Incident of Serious Landslide Occurrences Related to Upper Zone Soil Wetness as Computed with the Sacramento Streamflow Model. International Technical Conference on Mitigation of Natural Hazards through Real-Time Data Collection Systems and Hydrological Forecasting. Sacramento, California, 1983.

Peck, E.L., Catchment Modeling and Initial Parameter Estimation for the National Weather Service River Forecast System. NOAA Technical Memorandum NWS Hydro-31, U. S. National Weather Service, June, 1976.

Kohler, M. A., and R. K. Linsley, Predicting the Runoff from Storm Rainfall. U. S. Weather Bureau Research Paper 34, 1951.

Sevruk, B., Methods of Correction for Systematic Error in Point Precipitation Measurement for Operational Use. Opcrational Hydrology Report No. 21, World Meteorological Organization, 1982.

World Meteorological Organization, Intercomparison of Conceptual Models Used in Operational Hydrological Forecasting. Operational Hydrology Report No.7, Secretariat of the World Meteorological Organization, WMO pub. no. 429, Geneva, Switzerland, 1975.

Chapter 11

SSARR MODEL

D. D. Speers

The Streamflow Synthesis and Reservoir Regulation (SSARR) computer model is described, including the program structure and the features of the watershed and river models. This program enables the user to simulate a river system from rain/snowmelt runoff to regulation of runoff through a reservoir system. Continuous simulation is possible at time increments varying from 0.1 hour to 24 hours. The program has been applied in a variety of hydrologic regimes throughout the world and in numerous applications. Several applications of the model are described as examples. Previously executed on mainframe computers, the program is now distributed for use on PC computers operating under DOS, and is also executed for real-time forecasting and large system studies on RISC-processor workstations under the UNIX operating system.

11.1. INTRODUCTION

The Streamflow Synthesis and Reservoir Regulation Model (SSARR) has been in the process of development and application since 1956. It was initially developed by the Corps of Engineers to provide hydrologic simulation for the planning, design, and management of water control works in the Columbia River basin. Further development in concert with the Northwest River Forecast Center of the National Weather Service led to provisions in the model for real-time river forecasting, and it has been used in this application for rivers in the Pacific Northwest for over 30 years. The model has been applied throughout the world in a variety of hydrologic regimes and in numerous applications. Rockwood (1982) has described the history of the program's usage and the fundamental principles inherent in its formulation.

11.2. OVERVIEW OF PROGRAM

The SSARR program provides the user the capability of developing a numerical model of the hydrology of a river basin system. This includes the response of a watershed to precipitation or snowmelt, the actions of the river network as water flows through it, and the effects of engineering structures such as diversions and reservoirs. The computational center for the program is in two main modules, the watershed model and the river/reservoir system model. These are interfaced to numerous utility programs and several databases, providing the user with a variety of optional configurations and applications in setting up the program. Figure 11.1 is a schematic of the program components, which are summarized below. More detailed descriptions of key components are contained in subsequent paragraphs in this chapter. A user manual - Corps of Engineers (1987) - is available for coverage of model theory and program usage.

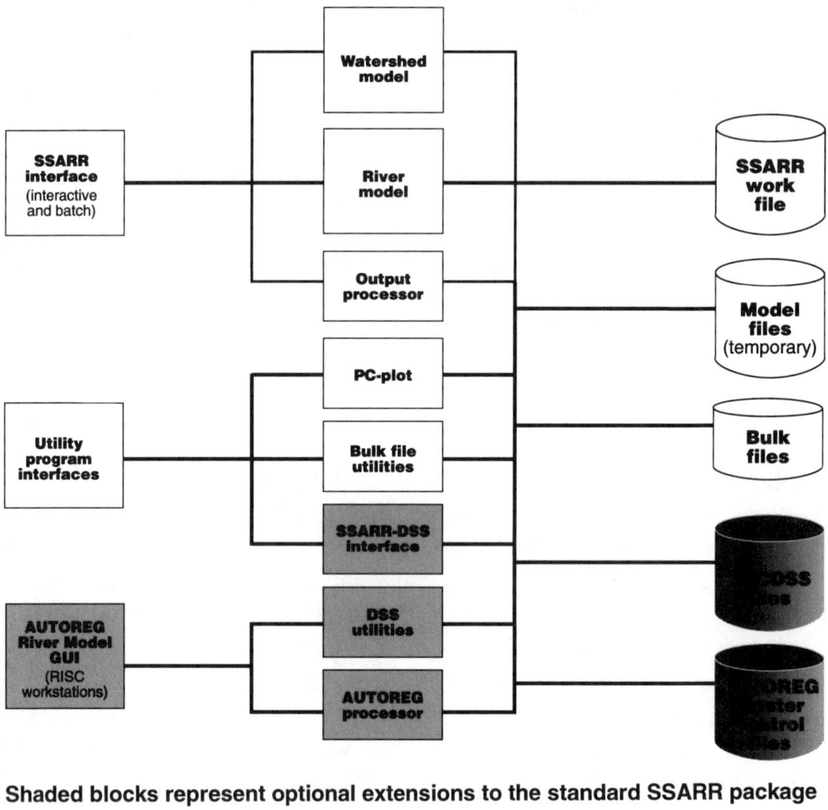

Fig. 11.1. Schematic Diagram of SSARR Program Components.

11.2.1. Watershed Model

This module simulates rainfall, snow accumulation and melt, interception, the state of the soil system as it affects runoff, and the translation of runoff in several components to the stream system. Evapotranspiration and long-term soil moisture storage routing is accounted for, making it possible to simulate continuously throughout a several-year period in low as well as high flow conditions. The computational time increment may be as short as 0.1 hour or as long as 24 hours.

11.2.2. River/Reservoir Model

In this module the routing of streamflow through a river network is simulated. Several variations in specifying channel routing methodology are available, including the provision to account for backwater effects from independent downstream sources. Diversions, overbank flows, local inflows, etc. are accounted for as are the effects of natural lakes. This module also contains algorithms for simulating reservoir operations, and a variety of operational specifications are permitted. Input to the river/reservoir model may be output from the watershed model (as in a forecasting application), or it may be a historic flow record (as in a reservoir system study application).

11.2.3. Databases

The SSARR model has several data files that are utilized in conjunction with the computational modules. The SSARR Work File is a permanent master file in which characteristics, initial conditions, data input and output are stored. It can be accessed interactively to display file contents. Several temporary utility files are also utilized. The user may wish to use the Bulk File, which is a binary file designed to hold large quantities of time-series data. The Watershed and River Models have provisions to access this file directly. In addition to these standard files that are associated with the SSARR program, the user may also interface to the HECDSS database through a utility program, using this as an input and output source. A suite of utility programs is associated with this database.

11.2.4. User Interface

Users of the SSARR watershed and river models utilize an interactive driver that prompts the user for input and provides help screens on request. Both free-form and fixed-format input is available. A program is available in the package which permits the color display of data. A special user interface program called AUTOREG is available on RISC-based workstations for preprocessing, and control when performing large scale reservoir system regulation studies with the River Model.

11.3. BASIC ROUTING TECHNIQUES

The propagation of flood impulses in both the watershed and river models is accomplished by storage routing in the SSARR program. The basic method can be classified as a "cascade of reservoirs" technique as illustrated in Fig. 11.2. In any lake or reservoir having a non-varying outlet, there is a unique curvilinear relationship between storage and outflow derived from two functions, the storage vs. elevation and the discharge vs. elevation functions. The slope of the storage-discharge function, dS/dQ, has the units of time; and, in fact the magnitude of the slope is a rough measure of the lag time involved between inflow and outflow. In the SSARR program this derivative is called "Time of Storage":

$$Ts = dS/dQ \tag{11.1}$$

where Ts = Time of Storage

S = Storage

Q = Outflow

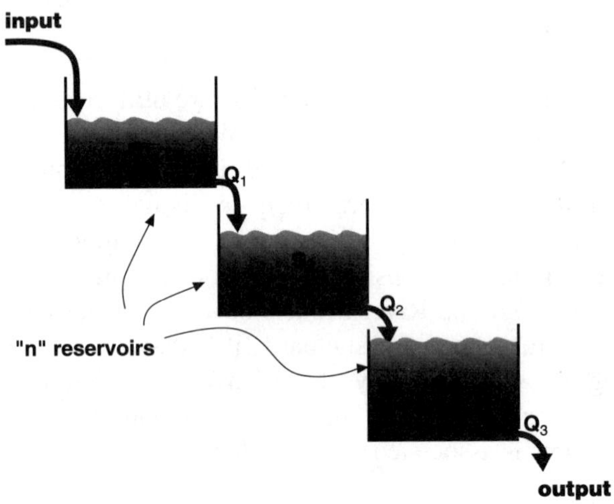

Fig. 11.2. Conceptual Representation of SSARR Routing Method.

By combining eq. (11.1) with the fundamental hydraulic equation for continuity, the resulting equation can be derived:

$$O_2 = O_1 + t(I_m - O_1)/(Ts + t/2) \tag{11.2}$$

where $O_{1,2}$ = Outflow at beginning and end of compute period

I_m = mean inflow, $(I_1+I_2)/2$

t = time interval

This equation is used in both watershed and channel routing in SSARR. The value t / (Ts + t/2) is evaluated for any time step and the solution for the next value of outflow, O_2, is a direct solution. This routing method offers several advantages including flexibility and ease in defining characteristics, and a convenient ability to preserve model state values.

11.3.1. Watershed Model Routing

In the watershed model, the transformation of rainfall or snowmelt to streamflow is accomplished directly by routing through increments of storage. In this case the storage vs. outflow function in the reservoir cascade is generally assumed to be linear; i.e., Ts is constant (a non-linear function may be specified if desired). To furnish the characteristics defining the routing the user needs only to specify the number of "reservoirs" (N) and the value of Ts for each reservoir. The total lag time between the maximum input and the resulting peak of the runoff component is approximately equal to the product of the two. For a given total lag time, a greater number of reservoirs with low Ts produces less attenuation than a lower N with higher Ts. Separate routing coefficients need to be determined for each component of runoff.

11.3.2. River Model Routing

The river model utilizes the same basic equation for channel routing except that Ts is defined by a non-linear equation

$$T_s = KTS/Q^n \tag{11.3}$$

where KTS = a constant specified by the user

Q = discharge

n = a constant specified by the user

It can be shown that the constant "n" is related to the cross-sectional shape of the channel, and that it typically varies between 0.2 and 0.5. With this a range of values for n, the equation results in a decreasing Time of Storage as discharge increases. As with the watershed equation, the value of KTS and therefore Ts can be estimated from the total lag time experienced in the channel being modeled. Provision is made in the river model for the user to specify Ts in curvilinear form as a function of Q. This permits departing from the standard function if desired, and it is especially useful in defining routing characteristics in overbank situations.

11.4. WATERSHED MODEL

An important part of the SSARR program is the watershed module, with which the user can construct a model of the runoff process. Meteorological variables such as precipitation and temperature are input with the derived catchment characteristics to simulate runoff into the river system.

11.4.1. Watershed Configuration

The SSARR program contains two basic options for configuring a watershed and accounting for the accumulation and depletion of a snowpack in a mountainous region; these are the "Snowband" and "Depletion Curve" options. The snowband approach is illustrated in Fig. 11.3. Here the basin is subdivided into as many as 20 elevation bands in which an independent accounting of temperature, snow quantity and condition, and soil moisture will be maintained. The second option treats the basin as one area in which precipitation is calculated as a lumped, basin-wide quantity (taking into account the temperature of the basin). In this option snow is not explicitly accumulated; however, its depletion is calculated by means of a pre-determined depletion curve that relates elevation to proportion of seasonal runoff.

The choice of which option to use is a matter of the application involved and the degree of complexity desired. The snow band option is the more complete and complex of the two, and would be used where detailed accounting of the snowpack is required. Continuous simulation for studies and forecasting applications is one obvious application area. The depletion curve model has been used for forecasting applications in the mountainous interior basins of the United States, wherein the melting of a large snowpack occurs during the spring and summer. The initial snow water equivalent is computed independently in these applications. For rain-only applications in basins with relatively flat topography either option could be used; however, the snowband option permits greater detail in simulation (e.g. interception, lower zone runoff component) so might be preferred even though snow is not actually involved in the hydrologic process. One capability not currently available in the snow band option, however, is the use of the energy budget equations - the temperature index method must be used with this option.

Regardless of the snow accounting option that is employed, the user can subdivide a basin being modeled into as many sub-watersheds as is desired. The limitation in this regard is usually that the lack of adequate precipitation and temperature data precludes making such subdivisions meaningful.

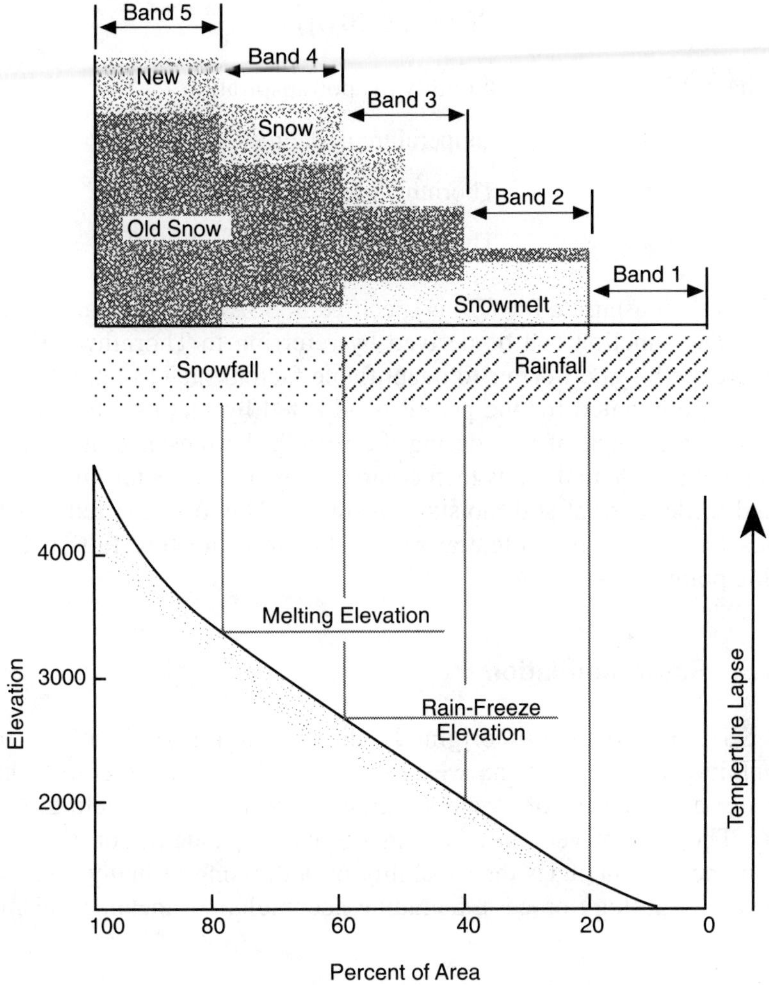

Fig. 11.3. Illustration of the SSARR Snowband Formulation.

11.4.2. Evapotranspiration

The calculation of evapotranspiration (ET) in the watershed simulation process enables the model to be applied on a continuous basis through dry as well as wet periods. Potential evapotranspiration (PET) may be furnished as an observed time series, or as a constant for each month. The most common option, however, is to have the model compute PET and ET as a function of air temperature and soil moisture index. The basic formulation is that proposed by Thornthwaite - described by Veihmeyer (1964) - in which PET is related to air temperature through the equation

$$PET = 1.6(10T/I)^{a} \tag{11.4}$$

where PET = monthly evapotranspiration

T = temperature in degrees centigrade

I = Thornthwaite's "Yearly Heat Index"

a = Thornthwaite's coefficient

The relationship PET = f(T) is specified in SSARR in tabular or equation form. The function can be derived by referring to Thornthwaite's tables which relate his coefficients to quantifiable factors such as latitude, season, etc. PET computed for the period is adjusted by factors which take into account the amount of rain during the period, the presence of snowcover, and elevation. Actual ET is then computed by a final factor DKE which is related to the current soil moisture quantity. This function reduces ET to values lower than the potential rate as the soil moisture falls below the wilting point.

11.4.3. Snow Simulation

The SSARR model was originally developed for applications in the mountainous regions of the western United States and therefore had to consider as a key factor in the simulation process the modeling of snow melt. The current version of the model has expanded upon the original design and now includes the capability of addressing virtually all facets of the snow hydrology process, including accumulation, metamorphism, and melt.

Overview of Process

Figure 11.4 is a schematic representation of the snow simulation process, beginning with precipitation input and ending with moisture input into the ground. This is shown for one elevation band in the snowband option. Temperature is computed at the appropriate elevation by applying the specified temperature lapse rate to a computed base temperature, which in

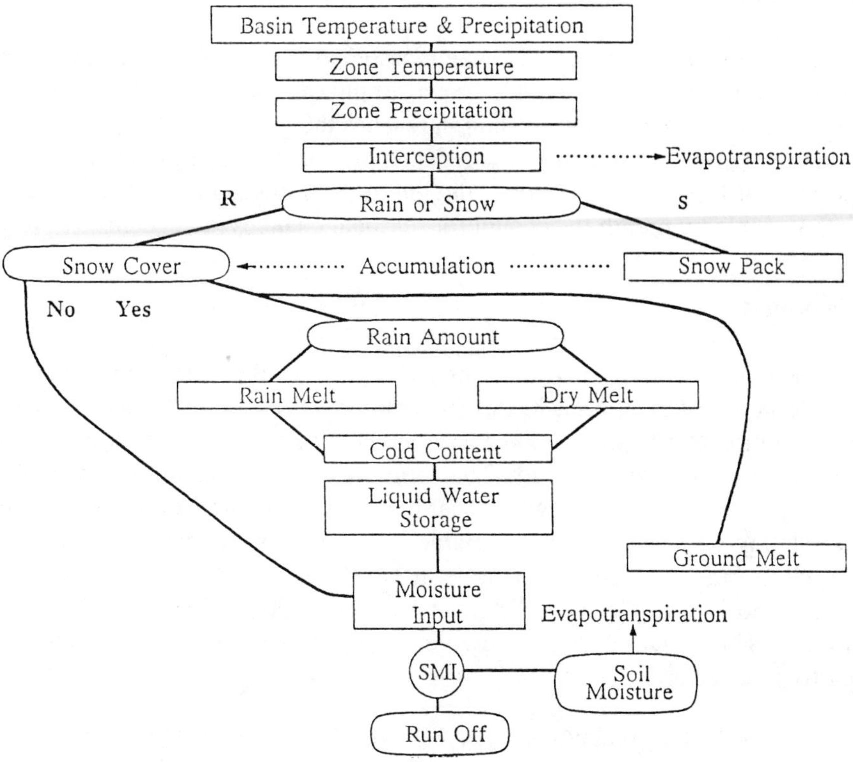

Fig. 11.4. Schematic Diagram of SSARR Snow Simulation.

turn is based upon one or more temperature time series furnished as input. Factors that control interception and the temperature at which precipitation falls as rain or snow are also furnished as a definition of the model.

Precipitation

Basin-average precipitation is computed from station precipitation by use of station weighting factors determined by the user, usually in the calibration process. Several stations can be weighted together in computing the mean basin quantity. Adjustments for gage-catch deficiency during snow events may be made through user-specified characteristics of the gage. When using the elevation band option, precipitation is distributed vertically to each band by means of a distribution function specified in the basin characteristics.

Interception

The relatively small loss of water associated with the interception of precipitation by trees is modeled by a simple overflow reservoir. Once the capacity is exceeded all precipitation reaches the ground. Any accumulation of water stored in the reservoir is reduced by evapotranspiration and thus is a permanent loss. The quantity of the interception reservoir may be specified as a variable by month to account for seasonal changes in vegetative cover.

Snowmelt

SSARR offers two basic methods of computing snowmelt: (1) the temperature index approach, in which snowmelt is computed as a function of air temperature; and, (2) the energy budget method, in which a number of meteorological are required. The temperature index solution, which is widely used in forecasting applications, relies upon a "melt rate factor" which can be specified as a variable. The energy budget formulation in SSARR draws upon the Cooperative Snow Investigations research conducted by the Corps of Engineers in the 1950s (Corps of Engineers, 1956). The equation used is the generalized equation for snowmelt in a partly forested area:

$$M_s = k_r(1\text{-}F)(0.0040I_i)(1\text{-}a) + k_w(0.0084v)(0.22T'_a + 0.78T'_d) + F(0.029T_a) \qquad (11.5)$$

where

M_s = Daily snowmelt in inches/day

T'_a = The difference between the air temperature measured at 10 feet and the snow surface temperature, in °F.

(Snow surface temperature is often assumed to be 32°F)

T'_d = The difference between the dewpoint temperature measured at 10 feet and the snow surface temperature, in °F

v = wind velocity at 50 feet above the snow in miles/hour

I_i = The observed or estimated insolation (solar radiation on horizontal surface) in langleys

a	=	The average snow surface albedo
k_r	=	The basin short-wave radiation melt factor. It depends upon the average exposure of the open areas to short-wave radiation in comparison with an unshielded horizontal surface (i.e., for a north-facing slope, k_r<1.0, but for a south-facing slope, k_r>1.0.
F	=	An estimated average basin forest canopy cover, effective in shading the area from solar radiation, expressed as a decimal fraction
k_w	=	The basin convection-condensation melt factor. It depends on the relative exposure of the area to wind

Since the data required for this (or any other) energy budget equation are typically not available in a real-time setting, the energy budget calculation of snowmelt is usually limited to theoretical applications such as the derivation of hypothetical design floods.

Snow Conditioning

A snowpack undergoes a metamorphosis as it progresses in time from its more crystalline freshly fallen state to a condition of being isothermal and primed for melt. In effect, this process results in a time delay between the onset of a melting potential until the meltwater actually enters the ground. It has practical significance particularly in continuous simulation applications, where for example there is need to simulate the effects of rain and/or warm temperatures occurring abruptly after a protracted cold spell that leaves the snowpack in an fresh "cold" condition. In the SSARR program the work of Anderson (1973) was utilized, wherein a continuous accounting is made of "cold content", an index of accumulated heat deficit in the snowpack. This deficit has to be made up by above freezing temperatures before snow is melted. A second factor, the "liquid water holding capacity" of the snow is also simulated by a single factor specified by the user.

11.4.4. Soil Moisture Accounting and Runoff Computation

Figure 11.5 depicts the process used in the SSARR program to maintain an accounting of soil moisture status, compute runoff, and distribute runoff to be routed through up to four zones of runoff transformation. The process involves several equations and relationships which are described briefly in this figure.

Soil Moisture Index

The runoff process centers around the continuous accounting of the Soil Moisture Index (SMI), which represents the weighted mean value of water stored in the basin's soil mantle that can be removed through evapotranspiration. Since the SMI is eventually subject to evapotranspiration, any increases in this index represents a permanent loss when considering the water balance for the basin as a whole. The updating of this parameter is accomplished through the following equation:

$$SMI_2 = SMI_1 + (MI\text{-}RGP) - (PH \bullet ETI)/24 \qquad (11.6)$$

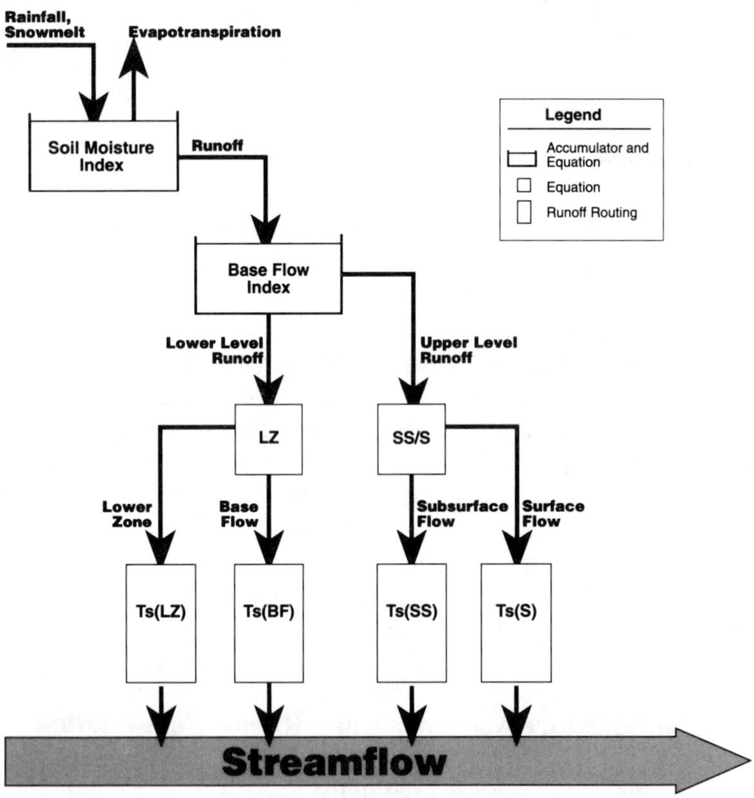

Fig. 11.5. Conceptual Diagram of Soil Moisture and Runoff Computation in SSARR.

where $SMI_{1,2}$ = Soil Moisture Index at beginning and end of period in inches

MI = Moisture input for the period in inches

RGP = Runoff generated for the period in inches

PH = Compute period length in hours

ETI = Evapotranspiration Index in inches/day

Runoff Calculation

Runoff - moisture input to the four transformation components - is calculated by means of a continuously calculated percent runoff factor (ROP) which in turn comes from a function, ROP = f (SMI, precipitation rate). Figure 11.6 is typical of this function. If the user desires, the precipitation rate parameter may be eliminated and a two-variable function used. This has been found to be satisfactory in most regions except those subject to cloudburst precipitation events.

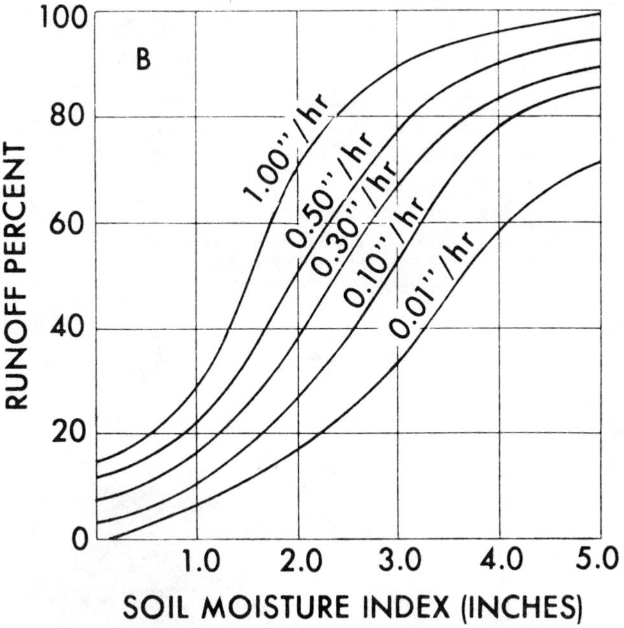

Fig. 11.6. Typical SSARR Runoff Percent Function.

Separation of Runoff Components

Three functions operate on the runoff generated by the model as shown on Fig. 11.5. Another continuously updated quantity, the baseflow infiltration index (BII), is a key to this process. This variable, which conceptually can be thought of as the relative wetness of the lower zones of the soil mantle, is updated through the following equation:

$$BII_2 = BII_1 + (24 \cdot RG - BII_1) \cdot (PH / (TSBII + 0.5 \cdot PH)) \quad (11.7)$$

where $BII_{1,2}$ = Base flow infiltration index in inches

RG = Runoff rate, inches/hour

PH = Compute period length in hours

TSBII = Time of storage for delay in base flow index, in hours

Input to the lowest zones is then determined by a decay function relating percent base flow to BII, BFP = f(BII). BFP is at its highest level for low values of BII, then decays rapidly to a relatively low percentage as BII increases. That portion of runoff that remains in the higher runoff zones is then assigned to either a surface or a subsurface component. This again is accomplished through a function that separates these two components depending upon the rate of moisture input as shown on Fig. 11.7. Here lower rates of input tend to be contained in the subsurface component, but as the input rate increases the subsurface component reaches a limiting value and a greater proportion of runoff becomes surface runoff.

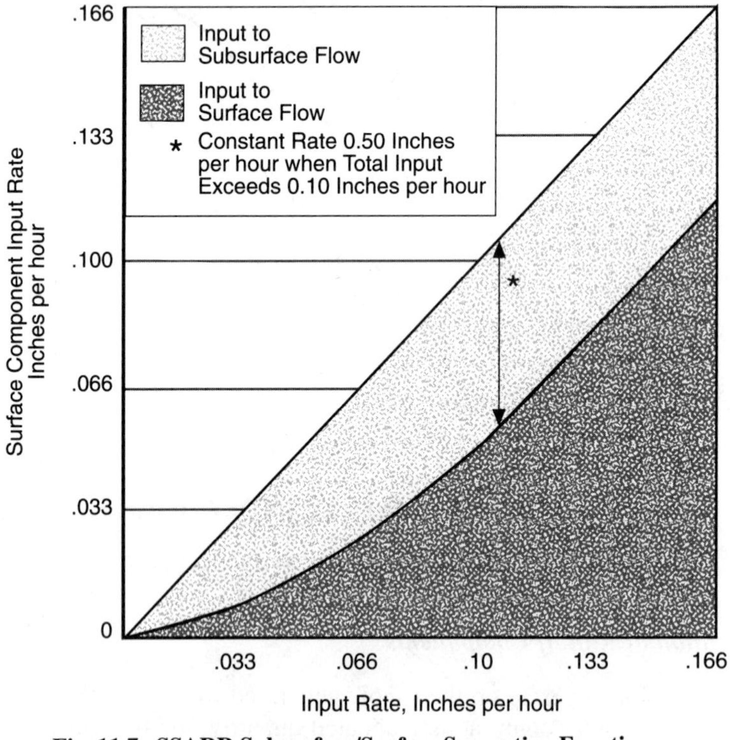

Fig. 11.7. SSARR Subsurface/Surface Separation Function.

Transformation to Runoff

The time delay of precipitation excess to streamflow input is computed independently for each of the four components using the cascade of linear reservoir technique described earlier. The two factors that define each component are given by the user as developed in the calibration process. Of course, an increasing time delay and degree of attenuation are specified in the successively lower zones of runoff. It is during this process that the SSARR program also converts the units of runoff from inches or millimeters per time increment to units of flow.

11.5. RIVER/RESERVOIR MODEL

A separate computational module within the SSARR program provides a variety of features for modeling the river system environment. The user may choose to execute both the watershed and river model in the same computer run, or can review the results of the watershed output (if used) before executing the river/reservoir system simulation.

11.5.1. River System Configuration

A river system model is configured by constructing a network of computational points, which can be any of the following:

1) A transfer point, at which tributaries are combined, diversions take place or flow is converted to stage.
2) A routing reach, in which streamflow is routed to the next downstream point.
3) A lake; in which outflow is related explicitly to lake elevation.
4) A reservoir, in which outflow can be specified by a reservoir operating code.

These nodes in the network are identified by the user by a unique alphanumeric identifier. The structure of the river system is then specified by simply stringing together a list of computational point identifiers in order so that each points to the next downstream station. A list so composed is called a "model" and is identified by an assigned name. Subset models can be included - for instance to execute only one tributary in the system, or to by-pass reservoirs to compute "unregulated conditions" - making it easy to execute alternative computer simulations within the same computer run.

11.5.2. Transfer Point Computations

A river system model typically contains many transfer point nodes for which a variety of computations may be performed. These are the vehicles for combining tributary flows from upstream points, or by providing for specified diversions from the river. A transfer point may refer to an equation or table which can be used to specify such relationships as a elevation-discharge rating table, a relationship defining diversion from the river as a function of river flow, an irrigation return flow function, etc.

11.5.3. Streamflow Routing

The basic methodology for streamflow routing in SSARR has been described, but the program allows several variations for specifying characteristics, to accommodate unusual channel configurations. The most simple and direct approach is to utilize eq. (11.1) which requires three factors for any routing reach. These can be estimated fairly closely knowing the approximate travel time involved. A second, more thorough method requires observed or inferred knowledge of the river. Here "time of storage" is specified in terms of flow through a table. This approach permits accounting for overbank storage wherein abrupt changes in attenuation and travel time may occur when the main channel of the river is overtopped in a extreme flood condition. A third method employs the backwater routing provision discussed below.

11.5.4. Backwater Routing

Often in a river system a physical situation occurs wherein a downstream feature backs water into the influence of an upstream station thereby affecting not only the stage-discharge relationship at the upstream point but the dynamics of the flood wave movement between the points. The controlling, independent downstream source might be a large tributary entering the river in question, a tidal estuary, or a river structure. The SSARR program has provisions for simulating such situations, wherein elevation and flow at the upstream point are related to elevation or flow at a downstream point through a three variable table of characteristics. These characteristics are developed based upon observed or computed steady-state water surface profiles. A succession of such relationships can be chained together over a river reach to provide elevation and flow simulation throughout the reach. This type of configuration is useful in flat-sloped channels having loop ratings, or any reach subject to downstream influences where detailed flow and elevation simulation is needed.

11.5.5. Reservoir Simulation

The SSARR program provides a comprehensive method of specifying the regulation of reservoirs whether it be short-term flood control regulation or seasonal regulation specifications over a period of several months. There are several operating modes possible, ranging from fixed outflows to specified rate of change of storage in the reservoir. The program automatically accommodates cases where the reservoir fills and needs to pass water over a spillway.

11.5.6. The Autoreg Module

To facilitate a more automated method of simulation of a large integrated reservoir system such as the Columbia River basin, the AUTOREG program was developed as a front-end preprocessor for SSARR. This program assists the user by calculating operating rule curves and other regulating guidance; pre-regulating reservoirs in accordance with reservoir and system rules, operating in a semi-automated mode to downstream control specifications, and displaying results in plotted and tabular form. AUTOREG uses the SSARR river/reservoir model as the computational engine for these calculations. At the date of this writing this program has been in operation only for the Columbia basin.

11.6. OPERATING THE SSARR PROGRAM

11.6.1. Operating for Study Applications

Hardware

The SSARR program is exported to users throughout the world most popularly configured for use on a Intel/PC computer. The export package comes with the program modules shown in Fig. 11.1, along with test input and documentation. Installation requires at least 2.5 megabytes of disk storage.

User Interface

An interactive menu system with help commands is provided for user-interface to the river and watershed models as shown on Fig. 11.1 and separate utility program menus are used to manage utilities and files. Large files of characteristics and data are prepared off-line with any editor program according to specified formatting, then input into the work file using interactive commands. Run controls and characteristics can be input and changed for successive runs using the interactive commands. For large, repetitive runs a DOS batch command file can be set up to facilitate the processing.

File System

Several file systems as shown in Fig. 11.1 are available for both study and forecasting applications. The Work File is the repository for characteristics, initial conditions, and most data files. It is accessible for review and editing with utility programs. For large data files the bulk data file is used. Calls to this file are made with special code designators in the SSARR input stream. For users of the HECDSS data storage system developed by the Corps of Engineers' Hydrologic Engineering Center, a SSARR-DSS interface program is available to convert input and output, allowing transfer of data back and forth. Users can then use the DSS utilities for display and manipulation of the SSARR input and output.

11.6.2. Operating for Forecasting Applications

The SSARR program has been used in real-time forecasting applications for many years, in a variety of hydrologic regimes, and over the years a number of special features have been developed for forecasting applications. Among these are the provisions for carrying forward the model state variables from one forecast date to the next; features for specifying input data efficiently using relative dating and departures from normal; and methods of computing "balances" conveniently. These reflect the experience of the National Weather Service Northwest River Forecast Center in Portland, Oregon, as well as the Corps of Engineers.

Hardware

The main difference between a forecasting and study application in considering computer hardware is that - at least in large forecasting operations - the preparation and execution of the forecast is done by more than one person in a team effort. Thus, a multi-user/multi-tasking operating system is typically required. For that reason a version of the SSARR is available for the UNIX operating system on RISC-based workstations. This replaces a version for IBM mainframe computers that was previously distributed. The SSARR program has been ported to UNIX environments on Hewlett-Packard, Intergraph, and Control Data Corporation workstations.

User Interface

Typical forecasting applications do not require massive amounts of data as do many study applications (the exception is "ESP" forecasting described later), but instead feature repetitive runs of relatively short duration. Thus, the interactive SSARR interface to the watershed and river/reservoir models are relied upon for data input, initial conditions, and forecast adjustments. A typical forecast requires first that a simulation be made of a "backup" time period preceding the current date of the forecast to verify model variables and assumed input. This can be done interactively, with output reviewed on the monitor. Once this is done the actual forecast run is made.

Automatic Adjustment Feature

The user may employ an automatic adjustment routine in the backup simulation to assist in determining to his satisfaction the state of the model preparatory to the forecast run. Observed streamflow is provided as input, along with observed meteorological data. The watershed model is executed, then it iterates through a series of adjustments in data to improve, if necessary, the agreement between computed and observed streamflow. This procedure assumes that the estimate of basin mean precipitation and temperature data is the most likely source of error; if however, the reconstitution is not satisfactory the user may have to manually adjust state variables (or perhaps make corrections in observed streamflow) before executing the forecast.

File System

Another important factor that distinguishes real-time forecasting from study applications is that a real-time data collection system and database is usually involved. For recent SSARR forecasting configurations the HECDSS system has been used for this processing, and therefore the connection with

HECDSS is required. As described above, the interface program facilitates this connection.

ESP Forecasting

The SSARR program is used to produce a type of long-term forecast denoted by the National Weather Service as "Extended Streamflow Prediction" or "ESP". In this procedure a continuous simulation model is run for an extended backup period, during which the model's state variables - soil moisture factors, snowpack, etc - are brought up to represent the state of the basin's condition at the forecast date. Then, the forecast is produced by simulating several historic years of meteorology each in turn beginning with the current model state. Statistical analysis of the resulting future sequences provides information for the forecast. The SSARR program includes provisions to conveniently process many historic data sequences as if they were forecasts.

11.7. MODEL CALIBRATION

The SSARR watershed and river models are calibrated manually, by comparing observed versus computed streamflow and making adjustments to characteristics to improve the fit. Calibration using the continuous simulation features of the model usually employs as many as 20-30 years of continuous data, so the data processing aspects of this effort are not insignificant. This process, however, can be carried out in an efficient manner by following a systematic calibration procedure and by utilizing statistical and plot capabilities of the model described below.

11.7.1. Watershed Calibration Procedure

Users of the SSARR program are encouraged to follow a tested, systematic sequence of steps in developing, calibrating, and testing a new watershed model. The procedure involves (1) obtaining, checking and entering hydrometeorological data into the database, and finding the most representative precipitation and temperature stations by independent correlation analysis; (2) setting up a preliminary model structure using rough estimates of characteristic factors and tables (default values are available for many factors); (3) executing preliminary runs in which long term water balance is emphasized. Alternative hydromet stations and station weightings need to be tried during this phase to find the best precipitation and temperature stations that will be representative of the basin's mean conditions; (4) once long-term (monthly and annual) volumes are approximately calibrated, adjusting the short-term characteristics (i.e, routing parameters) to obtain fit of hydrograph shape during runoff events. The tabular relationships that define the characteristics of a model should be intuitively reasonable, and consideration needs to be given to how these might vary regionally if a large model is being developed. Initial estimates can determined from the SSARR User Manual or from the example models furnished in the program package.

11.7.2. Watershed Calibration Aids

Since calibration of the watershed model involves long periods of record it is inefficient to rely upon tabular output from the model to review results. Output listings are available for each band in a distributed watershed formulation, as well as for a basin summary. These are optional, however, and are not normally printed unless a detailed understanding of the simulation is required in the final stages of calibration. Most users prefer to utilize summary statistics that are printed out to summarize the run. These include the root-mean square error statistic, and summary volume statistics. A plot summarizing cumulative error can be printed out to conveniently summarize the results of a calibration run. Plots of hydrographs, along with plots of internal computed time series such as soil moisture and runoff components are also invaluable, particularly in the later stages of calibration. Rockwood and Kuehl (1993) have described calibration procedures in detail.

11.7.3. River Model Calibration

Streamflow routing coefficients can easily be estimated using guidelines furnished in the User Manual, but it is preferable to verify these by comparing with observed data. Discharge hydrographs at upstream and downstream locations need be chosen so that they reflect a range of possible magnitudes, to ascertain the degree of non-linearity in the range of the coefficient. As discussed above, the program provides a high degree of flexibility in describing this relationship. Of course, ungaged "local inflow" needs to be considered in this process. One simple procedure that is effective in some situations is to route flow with trial coefficients and subtract the routed hydrograph from the observed downstream hydrograph (this is easily done in the river model). By inspection, one can compare and evaluate the routing coefficients by judging whether the computed local flow has a reasonable shape and magnitude. Other characteristics in a river model such as lake and reservoir relationships, discharge ratings, and backwater relationships are physically based and therefore their development is directly determined.

11.8. EXAMPLES OF APPLICATIONS

As described by Rockwood (1982) the SSARR model has been applied in a variety of settings, from small watersheds of only a few square miles to complex basins such as the Columbia River. The following are some selected examples of how the SSARR program has been employed by various users.

11.8.1. Long-Term Forecasting Using "ESP" Technique

For long-term forecasting in the Columbia River with the SSARR model the distributed, snow-band" formulation of the model is used so that the model can be operated continuously through the snow accumulation and snowmelt seasons. The model is initiated in the beginning of the water year when the basin is devoid of snow, soil moisture is at minimum levels, and streamflow is comprised almost entirely of base flow. As the winter season progresses in this region, snow aççumulates and streamflow generally remains at base flow levels. This is modeled with periodic executions of the model. Computed streamflow can be compared against observed data during this period, and calculated snow water equivalent in the model can be compared against observed snow readings. Adjustments in the model can be made if it is felt necessary.

Forecasts are typically made beginning in early January. Following the general ESP technique, the model is executed in a forecast mode by simulating possible future conditions, given the current state (snow, soil moisture, etc) of the model. This is done by using sequences of historic future precipitation and temperature. Early forecasts reflect considerable variance in the results since future conditions predominate in determining runoff. As the season progresses, however, the variance decreases. A sample of model output from a simulation made for the North Fork of the Clearwater basin (Dworshak dam inflow) in Idaho is shown in Fig. 11.8. This is one of two optional listings, in which the summary of seven elevation bands is summarized on one page. In this listing, the status of each of the bands is shown down the right side by codes indicating the

SSARR SNOWBAND MODEL - DWORSHAK - 1971-1980

DWR9 DWORSHAK RESERVOIR INFLOW (RUN 03/10/93)

	AREA	BASE-TEMP	ZONES
NOV-1970	2440.	33.5	7

DA	HR	PCPN	INT	SNOWLINE	WE	LR	TA	MR	RG	ET	ROP	BFP	SURF	SUBSF	BASEF	LOWERZ	TOTAL	OBS	FLAGS BY ZONE 1***5***10***15***20
4	240	.00	.00	6375.	.1	3.3	59.	.036	.00	.03	21.	89.	13	40	660	1185	1898	1385	D
5	240	.32	.27	6375.	.1	3.3	56.	.036	.01	.02	21.	89.	21	32	623	1177	1853	1510	A
6	240	.02	.02	6375.	.1	3.3	56.	.036	.00	.02	21.	89.	18	30	585	1169	1802	1865	D
7	240	.01	.01	6375.	.1	3.3	52.	.000	.00	.02	21.	89.	7	26	547	1161	1741	1960	S
8	240	.23	.06	6375.	.1	3.3	53.	.000	.03	.02	21.	90.	44	28	512	1153	1736	1835	A
9	240	.76	.03	5461.	.3	3.3	52.	.000	.12	.02	22.	88.	221	90	492	1144	1948	2400	AA
10	240	.06	.02	5132.	.3	3.3	51.	.000	.00	.02	23.	81.	183	187	502	1136	2009	2725	AAA
11	240	1.01	.02	5132.	.7	3.3	51.	.035	.14	.02	23.	85.	319	272	542	1128	2261	2685	QAA
12	240	.45	.02	5132.	.8	3.3	51.	.035	.06	.02	24.	79.	441	397	614	1120	2572	3230	QAA
13	240	.00	.00	5132.	.8	3.3	49.	.000	.00	.02	25.	81.	247	474	708	1112	2541	3235	SSS
14	240	.00	.00	5132.	.8	3.3	48.	.000	.00	.02	25.	85.	60	392	798	1104	2354	2625	SSS
15	240	.25	.06	5132.	.9	3.3	53.	.035	.04	.02	25.	88.	68	258	872	1096	2294	2320	DDA
16	240	.47	.02	5132.	.9	3.3	54.	.035	.11	.02	25.	86.	248	225	937	1088	2498	2855	DDA
17	240	.64	.03	5132.	1.1	3.3	51.	.035	.11	.02	27.	81.	441	331	1011	1080	2863	3395	AAA
18	240	.20	.02	4233.	1.2	3.3	46.	.000	.01	.02	29.	79.	301	444	1097	1073	2915	3195	AAAAA
19	240	.59	.02	4233.	1.7	3.3	48.	.035	.03	.02	29.	84.	161	426	1182	1065	2834	2970	QAAAA
20	240	.83	.02	4233.	2.3	3.3	48.	.035	.05	.02	29.	85.	191	353	1253	1058	2856	2810	QAAAA
21	240	.12	.02	2167.	2.4	3.3	31.	.000	.00	.02	29.	84.	125	301	1307	1051	2784	2390	AAAAAAA
22	240	.06	.02	2167.	2.4	3.3	28.	.000	.00	.02	29.	87.	37	225	1338	1044	2644	1875	AAAAAAA
23	240	1.92	.02	4233.	3.5	3.3	49.	.055	.31	.01	34.	88.	473	258	1368	1037	3135	2715	RRRCAAA
24	240	1.48	.01	4233.	3.4	3.3	56.	.061	.73	.02	45.	72.	3072	1108	1480	1030	6691	7520	RRRRR
25	240	.46	.02	3438.	3.8	3.3	44.	.000	.01	.02	47.	49.	2505	2468	1731	1024	7728	9210	AAAAAA
26	240	.13	.02	2167.	3.9	3.3	41.	.000	.00	.02	47.	63.	751	2716	2053	1019	6539	6110	AAAAAAA
27	240	.11	.02	3438.	3.9	3.3	45.	.035	.01	.02	47.	76.	107	1889	2353	1014	5362	4775	DCAAAAA
28	240	.00	.00	3438.	3.9	3.3	44.	.000	.00	.02	46.	82.	24	1066	2585	1009	4683	4220	SSSSSS
29	240	.25	.04	3438.	4.1	3.3	46.	.035	.02	.02	46.	86.	49	547	2733	1005	4334	3925	DAAAAA
30	240	.51	.02	3438.	4.4	3.3	47.	.035	.06	.02	46.	87.	169	315	2806	1001	4291	3865	DAAAAA

VOLUME - INCHES

PCPN	INT	RG	ET	SURF	SUBSF	BASEF	LOWERZ	TOTAL	OBS
10.88		1.89		.16		.53		1.42	
	.83		.61		.23		.50		1.43

MELT	MI	PET	LIQD	G-MLT
.30	5.82	.72	-.01	.07

EXPLANATION OF CODES

PCPN	Period precipitation, in
INT	Interception, in
SNOWLINE	Elevation of snowline, ft
WE	Water equivalent of snowpack, in
LR	Lapse rate, degrees F/1000 ft
TA	Air temperature at sea level, degrees F
MR	Melt rate, in/degrees F-day
RG	Runoff generated, melt + precip-int-loss in soil
ET	Evapotranspiration, in/day
SMI	Soil moisture index, in
ROP	Computed runoff percent
BFP	Computed baseflow percent
SURF	Surface flowrate, cfs
SUBSF	Subsurface flow rate, cfs
BASEF	Baseflow flowrate, cfs
LOWERZ	Lower zone flowrate, cfs
TOTAL	Total discharge, cfs
OBS	Observed discharge, cfs
FLAGS	Indicators of current snow activity on each band

- D Dry weather melt occurring
- R Rain melt occurring
- S Snow on band, no accumulation nor melt
- A Snow being accumulated
- L Dry melt being restricted by band transition
- Q Melt, but no RO because of liquid water deficiency
- C Melt, but no RO because of cold content

Fig. 11.8. Sample of basin summary output listing.

activity on the band. Figure 11.9 is a compilation of the ESP forecast runs. The forecast, made on April 1st, 1993, utilized 20 years of meteorological data. The April snow conditions were well below average in this year, as shown on the plot of ordered results which also shows the similar distribution but with historic snow conditions.

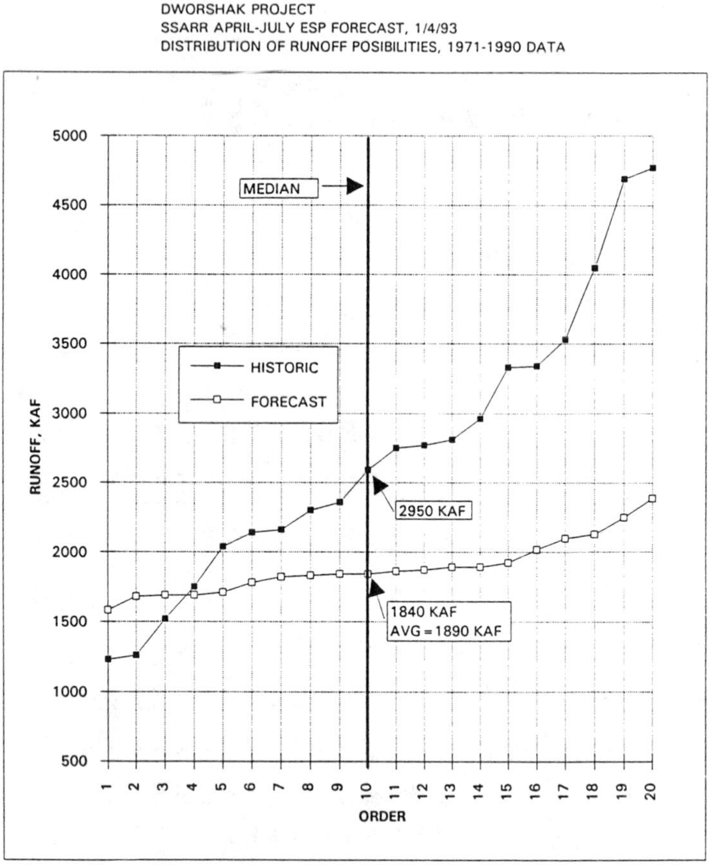

Fig. 11.9. Results of "ESP" Forecast with SSARR.

11.8.2. Probable Maximum Flood Derivation, Snow-Melt Environment

In dam design studies, or for studies reviewing the safety features of existing dams, it may be required to derive a inflow design hydrograph representing the probable maximum flood magnitude. The SSARR model has been used for numerous such studies, particularly those requiring the simulation of snowmelt as a significant portion of the runoff. In these studies it is desirable to utilize an energy budget equation, such as eq. (11.5) for the computation of snowmelt since such equations contain parameters that are quantifiable and can be maximized for the PMF derivation. Reference is made to the Corps of Engineers (1969) for a description of such a derivation.

11.8.3. Flood Control Analysis Using Hypothetical Floods

In river basins subject to snow runoff it is common to express flood control rule curves as a function of measured snow water equivalent (or forecasted residual runoff) in the basin. History has shown that rainfall occurring during the spring runoff can augment an otherwise insignificant snowmelt runoff to one that requires flood control operations. This possibility requires that a factor of safety be reflected in the rule curve to account for this spring rainfall. To quantify this factor of safety for several reservoirs in the Columbia River basin the SSARR model was used to generate hypothetical floods of rain-on-snow, then simulate the regulation of the flood by the reservoirs in question. The hypothetical flood was created by imposing a precipitation event of known magnitude and timing on several historic years having varying snow conditions. Figure 11.10 shows the results of such simulations (two of many), representing two rainfall patterns falling on a nominally low snowpack. The historic streamflow hydrograph is shown for comparison. With the historic snow condition representing the forecastable parameter, the magnitude of desirable factor of safety determined from each simulation could be related to snowpack/forecast to derive the generalized relationship or rule curve.

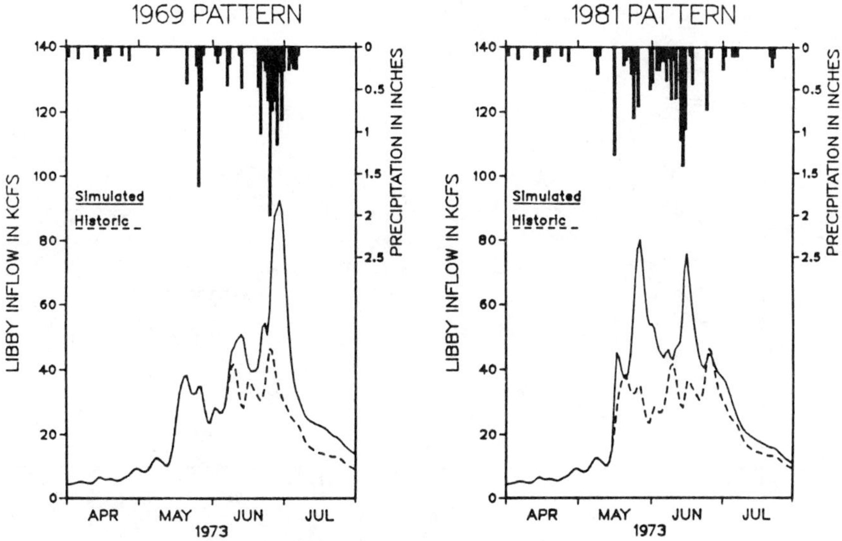

Fig. 11.10. Computation of Hypothetical Flood, Rain-on-Snow Simulation.

11.8.4. System Flood Control Analysis Using Historic Streamflow Data

Studies to establish or modify reservoir operating guidelines, and to derive flood frequency curves reflecting those guidelines, can make use of a data set of historical streamflows rather than to simulate streamflow using the watershed model. In the Columbia basin a 60-year study period of mean daily flows is used. These data have been first adjusted to remove historical effects of upstream reservoir operation, and they have been modified to reflect a consistent level of irrigation depletion. The SSARR model is used to compute the adjustments. They are then stored in the bulk data file for use in simulating future conditions with the river/reservoir model. Regulation runs have traditionally been performed a year at a time with the reservoir system simulation being carried out manually by a person experienced with the regulation of the system. Now, the AUTOREG program described earlier enables this activity to be carried out in a semi-automated fashion, with initial runs being executed with the entire period of record at once. Figure 11.11 is the hydrograph of streamflow for part of one year at the furthest downstream control point in the system, showing the results of the system's flood control regulation compared to the unregulated condition.

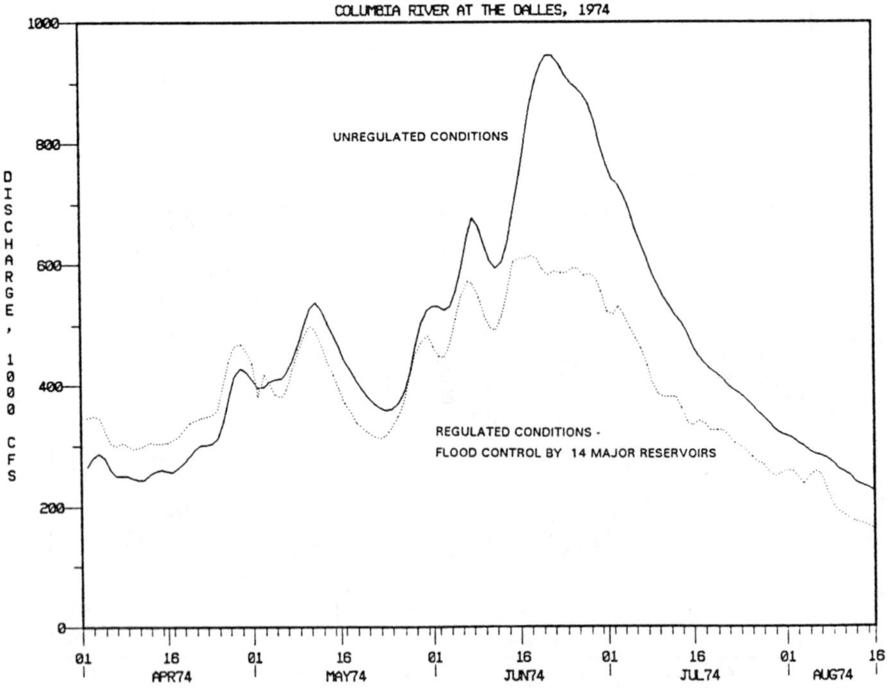

Fig. 11.11. Results of Flood Control System Simulation, Columbia Basin, 1974 Flood - Unregulated and Regulated Conditions.

11.8.5. Natural Lake System Analysis and Management

The SSARR model was used by the USDA Soil Conservation Service (1990) to analyze and provide a forecasting system for control of a series of lakes in northern Minnesota. The purpose was to improve the environment for the growing of wild rice by the White Earth Indian Reservation Tribe through the manipulation of manual control structures on the lakes. The SSARR program was linked with several custom made utility programs to facilitate routine usage for the specific task, and a management forecast system was set up using both the watershed and river models. The model consisted of 24 drainages varying from 1.32 to 31.7 square miles which provided runoff into 15 small lakes. Input is user-defined hypothetical or historical precipitation sequences starting from the date of the thawing of the lake ice cover and running throughout the summer. Figure 11.12 is a sample of the output - one run for an analysis examining alternatives of stoplog settings for a lake.

1

STREAMFLOW ROUTING
UPPER OTTERTAIL ABOVE ROCHERT

RUN DATE RUN NO. INITIAL DATE, HOUR
27 JUN 90 2400

	RICELK	RICELK	HTLAND	HTLAND
	RICE	RICE	HEIGHT OF LAND	HEIGHT OF LAND
DATE-PREC	FLOW CFS	ELEVATION FEET-MSL	FLOW CFS	ELEVATION FEET-MSL
27 JUN 90 0.00	182.	1,455.20	83.	1,454.00
28 JUN 90 2.50	218.	1,455.41	103.	1,454.17
29 JUN 90 2.50	258.	1,455.62	139.	1,454.46
30 JUN 90 0.93	293.	1,455.79	171.	1,454.72
1 JUL 90 0.00	306.	1,455.87	188.	1,454.85
2 JUL 90 0.00	306.	1,455.89	197.	1,454.92
3 JUL 90 1.03	309.	1,455.91	210.	1,455.02
4 JUL 90 0.00	310.	1,455.92	224.	1,455.11
5 JUL 90 0.09	299.	1,455.91	231.	1,455.15
6 JUL 90 0.00	287.	1,455.88	235.	1,455.18
7 JUL 90 0.31	277.	1,455.85	239.	1,455.20
8 JUL 90 0.00	265.	1,455.83	242.	1,455.22
9 JUL 90 1.33	264.	1,455.83	252.	1,455.28
10 JUL 90 0.00	261.	1,455.85	262.	1,455.34
11 JUL 90 0.00	248.	1,455.83	262.	1,455.34
12 JUL 90 0.66	247.	1,455.83	265.	1,455.35
13 JUL 90 0.52	248.	1,455.84	271.	1,455.39
14 JUL 90 0.00	242.	1,455.84	273.	1,455.40
15 JUL 90 0.00	234.	1,455.82	270.	1,455.38
16 JUL 90 0.00	227.	1,455.79	265.	1,455.35
17 JUL 90 0.04	219.	1,455.76	260.	1,455.33
18 JUL 90 0.00	211.	1,455.72	255.	1,455.30
19 JUL 90 0.11	203.	1,455.68	250.	1,455.27
20 JUL 90 0.00	192.	1,455.63	245.	1,455.24
21 JUL 90 0.03	182.	1,455.59	240.	1,455.20
22 JUL 90 0.00	174.	1,455.54	234.	1,455.16
23 JUL 90 0.00	165.	1,455.49	227.	1,455.12
24 JUL 90 0.00	157.	1,455.44	220.	1,455.08
25 JUL 90 0.00	150.	1,455.39	214.	1,455.04
26 JUL 90 0.00	141.	1,455.33	207.	1,454.99
27 JUL 90 0.00	133.	1,455.28	200.	1,454.95
28 JUL 90 0.00	123.	1,455.23	194.	1,454.90
29 JUL 90 0.00	115.	1,455.18	188.	1,454.85
30 JUL 90 0.00	108.	1,455.13	182.	1,454.80
31 JUL 90 0.00	102.	1,455.08	176.	1,454.76

Fig. 11.12. Sample Output of Minnesota Lake Simulation Application.

REFERENCES

Anderson, E. A. (1973). "National Weather Service River Forecast System -- Snow Accumulation and Ablation Model." NOAA Technical Memorandum NWS HYDRO-17, National Weather Service, Silver Spring, Maryland.

Corps of Engineers (1956). "Snow Hydrology." North Pacific Division, Portland, Oregon.

Corps of Engineers (1969). "Lower Columbia River Standard Project Flood and Probable Maximum Flood." North Pacific Division, Portland, Oregon.

Corps of Engineers (1987). "SSARR User's Manual." North Pacific Division, Portland, Oregon.

Rockwood, D.M. (1982). "Theory and Practice of the SSARR Model as Related to Analyzing and Forecasting the Response of Hydrologic Systems." Proceedings of the International Symposium on Rainfall-Runoff Modeling, May 1981, Mississippi State University.

Rockwood, D.M. and Kuehl, D.W. (1993). "Interactive Calibration for the SSARR Watershed Model." Proceedings of the Eastern/Western Snow Conference, Quebec City, Quebec.

Soil Conservation Service (1990). "SSARR Hydrologic Model; User's Manual, Upper Otter Tail River Basin." St Paul, Minnesota.

Veihmeyer, F.J. (1964). "Evapotranspiration." Handbook of Hydrology, ed V.T. Chow, pp 11-1 to 11-38.

Chapter 12

HYDROLOGICAL SIMULATION PROGRAM - FORTRAN (HSPF)

A. S. Donigian, Jr.
B. R. Bicknell
J. C. Imhoff

12.1. INTRODUCTION

In the mid-1970's, the U.S. EPA Environmental Research Laboratory in Athens, Georgia was in the beginning stages of model development and testing efforts that focused on tools and procedures for quantifying nonpoint sources (NPS) of pollution. Initiated by legislative mandates that required assessment of both urban and agricultural NPS contaminants, the Laboratory was supporting development and field testing of mathematical models (along with companion data collection programs) to be used to estimate these NPS loadings and, ultimately, to evaluate potential management and control alternatives. However, EPA scientists realized that although these field-scale models could provide loading values, they alone would not be sufficient to evaluate water quality impacts at the larger watershed, or regional scale. Thus began an extensive, comprehensive watershed model development effort to integrate the field-scale models with instream hydraulic and water quality process models, within a flexible, modular framework, to allow continuous simulation of complex watersheds with multiple land uses, both point and nonpoint contaminant sources, networked channels and drainage patterns, and lakes and reservoirs. The HSPF model produced by this development effort has been applied throughout North America and numerous countries and climatic regimes around the world; it enjoys the joint sponsorship of both the U.S. Environmental Protection Agency and the U.S. Geological Survey, and continues to undergo refinement and enhancement of its component simulation capabilities along with user support and code maintenance activities.

In this chapter we provide an overview of the HSPF model, along with background on its historical development, current ongoing and planned enhancements, selected examples of model applications to demonstrate the breadth of simulation capabilities, and projections of the possible paths for watershed model development activities over the next decade.

12.1.1. Overview

The Hydrological Simulation Program-FORTRAN, known as HSPF, is a mathematical model developed under EPA sponsorship for use on digital computers to simulate hydrologic and water quality processes in natural and man-made water systems. It is an analytical tool which has application in the planning, design, and operation of water resources systems. The model enables the use of probabilistic analysis in the fields of hydrology and water quality management. HSPF uses such information as the time history of rainfall, temperature, evaporation, and parameters related to land use patterns, soil characteristics, and agricultural practices to simulate the processes that occur in a watershed. The initial result of an HSPF simulation is a time history of the quantity and quality of water transported over the land surface and through various soil zones down to the groundwater aquifers. Runoff flow rate, sediment loads, nutrients, pesticides, toxic chemicals, and other quality constituent concentrations can be predicted. The model uses these results and stream channel information to simulate instream processes. From this information, HSPF produces a time history of water quantity and quality at any point in the watershed.

HSPF is a valuable tool to water resource planners. Because it is more comprehensive than most systems, it permits effective planning. Benefits to the user include:

- Flexibility in solving a wide range of water quantity and quality problems using a single model
- Convenient data management features that save time and money
- Modular program structure which facilitates program changes and additions for special applications

HSPF is currently one of the most comprehensive and flexible models of watershed hydrology and water quality available. It is the only available model that can simulate the continuous, dynamic event, or steady-state behavior of both hydrologic/hydraulic and water quality processes in a watershed, with an integrated linkage of surface, soil, and stream processes. The model is also unusual in its ability to represent the hydrologic regimes of a wide variety of streams and rivers with reasonable accuracy. It has been applied to such diverse climatic regimes as the tropical rain forests of the Caribbean, arid conditions of Saudi Arabia and the Southwestern U.S., the humid Eastern U.S. and Europe, and snow covered regions of Eastern

Canada. The potential applications and uses of the model are comparatively large including:

- Flood control planning and operations
- Hydropower studies
- River basin and watershed planning
- Storm drainage analyses
- Water quality planning and management
- Point and nonpoint source pollution analyses
- Soil erosion and sediment transport studies
- Evaluation of urban and agricultural best management practices (BMPs)
- Fate, transport, exposure assessment, and control of pesticides, nutrients, and toxic substances
- Time-series data storage, analysis, and display

HSPF is designed so that it can be applied to most watersheds using existing meteorologic and hydrologic data; soils and topographic information; and land use, drainage, and system (physical and man-made) characteristics. The inputs required by HSPF are not different than those needed by most other simpler models. The primary difference in data needs is that long, rather than short time-series records are preferred. Typical long time-series records include precipitation, waste discharges, and calibration data such as streamflow and constituent concentrations.

12.1.2. Background

HSPF is an extension and improvement of three previously developed models: 1) The EPA Agricultural Runoff Management Model - ARM (Donigian and Davis, 1978), 2) The EPA Nonpoint Source Runoff Model - NPS (Donigian and Crawford, 1979), and 3) The Hydrologic Simulation Program (HSP), including HSP Quality (Hydrocomp, 1977), a privately-developed proprietary program. In the late 70's EPA recognized that the continuous simulation approach contained in these models would be valuable in solving many complex water resource problems. Thus, a fairly large investment was devoted to developing a highly flexible non-proprietary FORTRAN program which contains the capabilities of these three models, plus many extensions.

HSPF incorporates the field-scale ARM and NPS models into a watershed-scale analysis framework that includes the capabilities needed to model fate and transport in one-dimensional stream channels. It is the only comprehensive model of watershed hydrology and water quality that allows the integrated simulation of land and soil contaminant runoff processes with instream hydraulic and sediment-chemical interactions.

HSPF was first released publicly in 1980, as Release No. 5 (Johanson et al., 1980), by the U.S. EPA Water Quality Modeling Center (now the Center for Exposure Assessment Modeling); since its initial release, the model has maintained a reputation as perhaps the most useful watershed-scale hydrology/water quality model that is available within the public domain. Fig. 12.1 is a timeline of the development activities of HSPF and its predecessor models beginning about 1971 and extending through the current time. The development of HSPF in the late 1970's represented an integration of a variety of EPA-sponsored model development and testing efforts associated with the ARM and NPS models, and the WEST (Watershed Erosion and Sediment Transport) model (Leytham and Johanson, 1979); the basic watershed modeling approach embodied in HSP was chosen, a highly modular code design and structure was developed, and all the individual models were redesigned and recoded into FORTRAN to make the resulting package widely useable and available to potential model users.

Throughout the 1980's and early 1990's, HSPF underwent a series of code and algorithm enhancements producing a continuing succession of new releases of the code, culminating in the recent release of Version No. 10 in 1993 (Bicknell et al., 1993). Table 12.1 lists some of the key enhancements and changes for the various HSPF releases since 1980, along with a preview of potential enhancements expected for the next release (Version No. 11) in the 1994-95 time period. The capabilities and changes for the various versions are discussed as part of the model capabilities overview in Section 12.2.

Since 1981, the U.S. Geological Survey has been developing software tools to facilitate watershed modeling by providing interactive capabilities for model input development, data storage and data analysis, and model output analysis including hydrologic calibration assistance. The ANNIE, WDM, Scenario Generator, and HSPEXP designations in Fig. 12.1 are USGS products that have greatly advanced and facilitated watershed model applications, not only for HSPF but also for many other USGS models. For example, the WDM (Watershed Data Management) file has effectively replaced the Time Series Store (TSS) file used in the earlier versions of HSPF due to the expanded data analysis and graphical capabilities of the ANNIE software (Lumb et al., 1990). These products are discussed in greater detail in Section 12.3.

HSPF applications since its initial release in 1980 have been world-wide and number in the hundreds; on the order of 50 current active applications

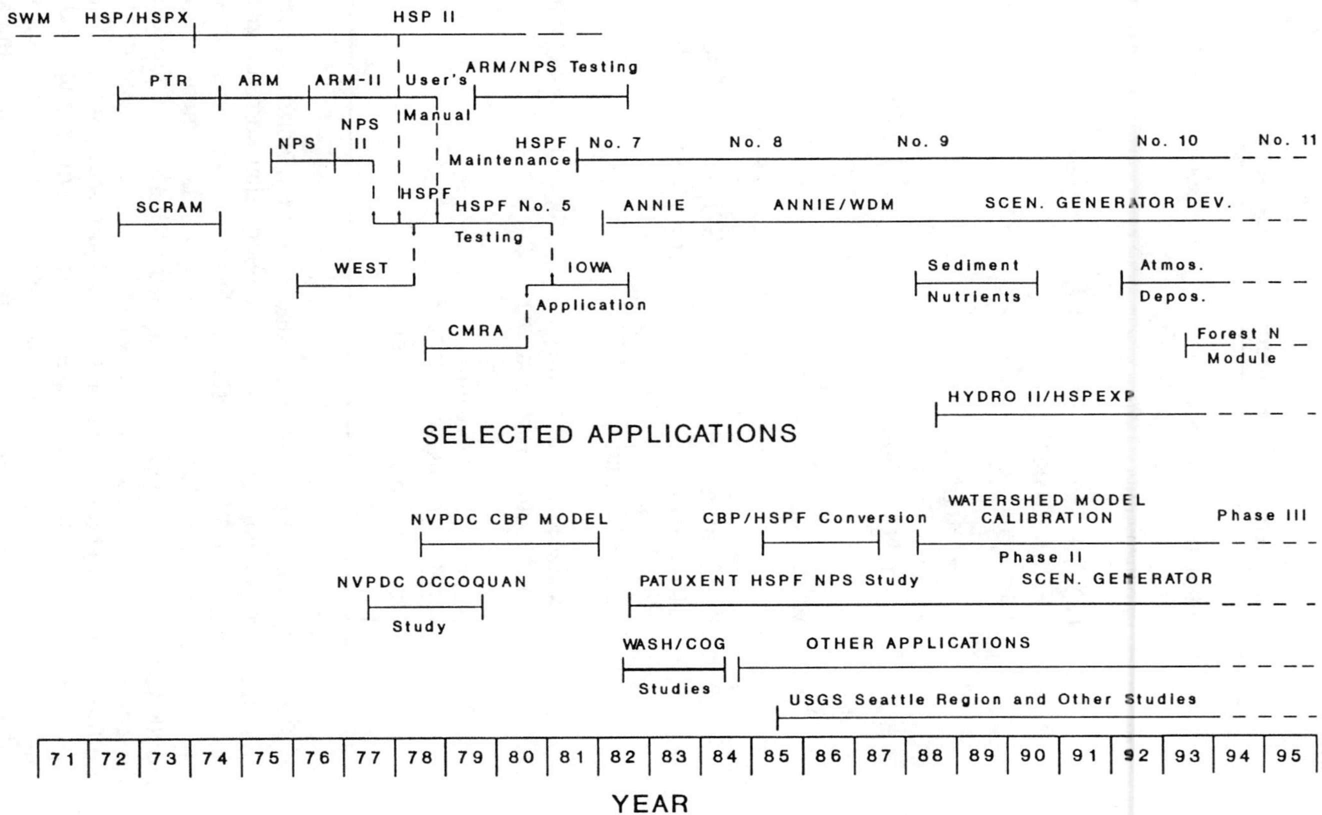

Fig. 12.1. HSPF and Related Model Development, Testing and Applications.

TABLE 12.1.

Historical Progression of HSPF Releases

Year	Version	Comments/Enhancements	Document
1980	5	Initial public release	Johanson et al. (1980)
	6	Performance and portability enhancements	
1981	7	GQUAL,SEDTRN, MUTSIN, GENER, DURANL enhancements	Johanson et al. (1981)
1984	8	Special Actions enhancements	Johanson et al. (1984)
		Initial PC version	Application Guide (Donigian et al., 1984)
1988	9	WDM implementation PC version distributed	CEAM publication
1993	10	Sediment-nutrient interactions Mass-Link/Schematic Acid-pH Module	HSPF Rel. 10 Manual (Bicknell et al., 1993)
1994(?)	11	Enhanced special actions Water regulation/accounting	Updated documentation
		Atmospheric deposition HSPF/DSS linkage (COE) Increased operations limit Forest Nitrogen Module	

continue around the world with the greatest concentration in North America. Fig. 12.1 includes a few applications on the timeline that correspond to studies discussed in Section 12.5; numerous studies have been completed or are continuing in the Pacific Northwest, Washington, D.C. metropolitan area, and the Chesapeake Bay region. Today the model serves as the focal point for cooperation and integration of watershed modeling and model support efforts between the Environmental Protection Agency and the U.S. Geological Survey. Over the years, development activities and model enhancements, along with these model applications, have continued to improve the model's capabilities and preserve its status as a state-of-the-art tool for watershed analysis.

12.1.3. Chapter Contents

The purpose of this chapter is neither to duplicate nor to summarize the materials available in the HSPF User's Manual (Bicknell et al., 1993). Interested readers and potential users are referred to the HSPF User's Manual for complete details of the model algorithms. Rather, our goals are to provide an overview of capabilities, to relate our experiences over the past 13 years in both applying and maintaining the model, and to project the directions that the HSPF development effort are likely to travel in the upcoming years.

To this end, the contents of the remainder of this chapter are divided into five major sections. First, we provide an overview of the major modules of HSPF; the overview reflects the current state of development of HSPF, as of early 1994. Next, we describe the most recent enhancements to the model implemented in the 1988 to 1993 time period. We focus our discussion on the new features available in Release 10.0, including improvements in database and input management, as well as improved algorithms for such processes as instream sediment/nutrient interactions and acid mine drainage. In addition, we summarize a stand-alone expert system for hydrologic calibration, using the HSPF water budget computations, that has been developed by the U.S. Geological Survey and made available to the public.

In the subsequent section we describe issues related to model distribution and hardware requirements, and we relate our experiences with the model maintenance and support activities which have been a key factor in HSPF's continued popularity.

Next, we provide summaries of selected applications that feature comprehensive and/or innovative approaches to using HSPF; we believe the examples that we have selected demonstrate the power of the model and the breadth and complexity of problems to which it can be successfully applied. The five examples of model application that are described include the Chesapeake Bay Watershed Model; the Patuxent Basin study and scenario generator; STREAM^{PC} , an interactive PC-based implementation of the STREAM methodology for exposure assessment of agricultural pesticide runoff; a regional parameter approach to developing runoff files for making urban management decisions in the Seattle area; and an interactive reservoir management system based on HSPF. We also provide a summary table of additional model applications during recent years to document the wide range of problem assessments that have been evaluated using HSPF.

Finally, we reveal our vision of the direction(s) that we believe HSPF will evolve in the coming years. The recent resurgence of government concern for nonpoint source issues and problems and the focus on watershed scale assessment and management, as catalyzed by various sections and amendments to the Clean Water Act in the United States, has renewed interest in nonpoint source and comprehensive watershed modeling. The comprehensive nature of HSPF, and its flexibility in allowing consideration of the combined impacts of both point and nonpoint source pollutants at the watershed scale, is leading to increased interest in model applications. To support this increased interest and potential usage, the final section identifies and explores areas of current model enhancements, and future research and model development. In addition to potential algorithm enhancements, we explore future needs for databases and user interaction capabilities that will be required to keep pace with changing software/hardware capabilities.

12.2. OVERVIEW OF CAPABILITIES

HSPF contains three application modules and five utility modules. The three application modules simulate the hydrologic/hydraulic and water quality components of the watershed. The utility modules are used to manipulate and analyze time-series data. Table 12.2 summarizes the constituents and capabilities of the HSPF modules. A brief description of each of the modules follows.

TABLE 12.2.

HSPF Application and Utility Modules

Application Modules

PERLND	IMPLND	RCHRES
Snow	Snow	Hydraulics
Water	Water	Conservatives
Sediment	Solids	Temperature
Soil temperature	Quality	Sediment
Quality		Nonconservatives
Pesticide		BOD/DO
Nitrogen		Nitrogen
Phosphorus		Phosphorus
Tracer		Carbon/pH
		Plankton

Utility Modules

COPY	PLTGEN	DISPLY
Data transfer	Plot data	Tabulate, summarize
DURANL	GENER	MUTSIN
Duration Analysis	Transform or combine time series data	Input sequential time series data

The three application modules within HSPF, and their primary functions, are as follows:

(1) PERLND - Simulates runoff and water quality constituents from pervious land areas in the watershed.

(2) IMPLND - Simulates impervious land area runoff and water quality.

(3) RCHRES - Simulates the movement of runoff water and its associated water quality constituents in stream channels and mixed reservoirs.

12.2.1. PERLND Module

As PERLND simulates the water quality and quantity processes that occur on pervious land areas, it is the most frequently used part of HSPF. To simulate these processes, PERLND models the movement of water along three paths: overland flow, interflow, and groundwater flow. Each of these three paths experiences differences in time delay and differences in interactions between water and its various dissolved constituents. A variety of storage zones are used to represent the processes which occur on the land surface and in the soil horizons. Snow accumulation and melt are also included in the PERLND module so that the complete range of physical processes affecting the generation of water and associated water quality constituents can be represented. Some of the many capabilities available in the PERLND module include the simulation of:

- water budget and runoff components
- snow accumulation and melt
- sediment production and removal
- nitrogen and phosphorus fate and runoff
- pesticide fate and runoff
- movement of a tracer chemical

Figure 12.2 defines the structure and contents of the PERLND module. The module features individual compartments for modeling air temperature as a function of elevation (ATEMP), snow accumulation and melting (SNOW), hydrologic water budget (PWATER), sediment production and removal (SEDMNT), soil temperature (PSTEMP), surface runoff water temperature and gas concentrations (PWTGAS), generalized water quality constituents (PQUAL), solute transport (MSTLAY), pesticides (PEST), nitrogen (NITR), phosphorus (PHOS), and conservatives (TRACER).

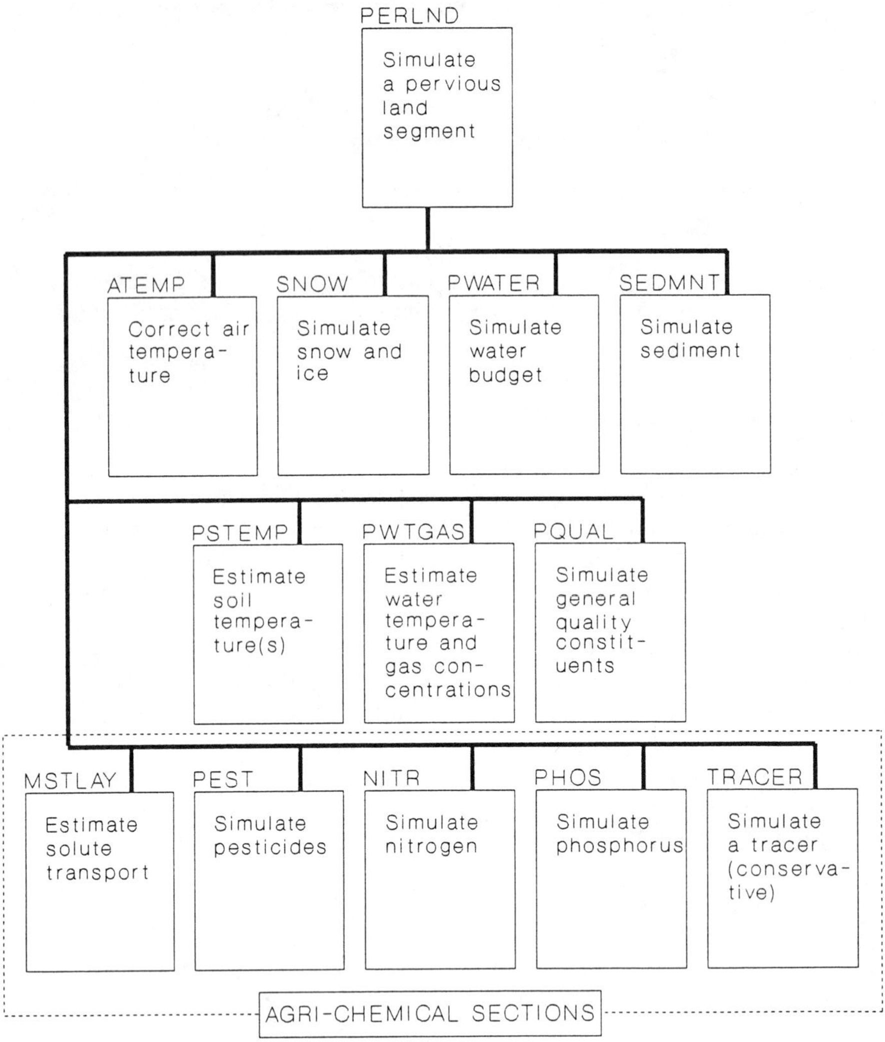

Fig. 12.2. PERLND Structure Chart.

ATEMP modifies the input air temperature values from the weather station record assigned to each land segment to account for elevation differences between the station and the land segment. SNOW uses meteorologic data (precipitation, air temperature, solar radiation, dewpoint, wind movement) to determine whether precipitation is rain or snow, to simulate an energy balance for the snowpack, and to determine the effect of heat fluxes on the snowpack. A combination of physical and empirical formulations are used to model such processes, fluxes and variables as snowfall accumulation; snowpack dullness; snowpack compaction; pack evaporation; atmospheric heat exchange; snow, ice and water occurrence in the pack; and melt from ground heat.

PWATER is used to calculate the water budget components resulting from precipitation on pervious land areas; as a result, it is the key component of the PERLND module. The basis of the water budget computations contained in HSPF is the Stanford Watershed Model (Crawford and Linsley, 1966). Like the SNOW code, the PWATER code uses both physical and empirical formulations to model the movement of water through the hydrologic cycle. PWATER considers such processes as evapotranspiration; surface detention; surface runoff; infiltration; shallow subsurface flow (interflow); baseflow; and percolation to deep groundwater. Lateral inflows to surface and shallow subsurface storages can also be modeled.

The equations used in the SEDMNT code to produce and remove sediment are based on the ARM and NPS models, and are modifications of soil and gully erosion equations developed by Negev (1967) and influenced by Meyer and Wischmeier (1969) and Onstad and Foster (1975). Many of the sediment model parameters are derived from the Universal Soil Loss Equation (Wischmeier and Smith, 1978). Removal of sediment by water is simulated as washoff of detached sediment in storage and scour of the soil matrix. Soil detachment is modeled as a function of rainfall, land cover, land management practices, and soil detachment properties. If the modeler so specifies, soil detachment can be incremented by lateral input from an upslope land segment and/or net external additions/removals caused by human activities or wind. Removal of detached sediment and scour of the soil matrix by surface flow are both modeled empirically as a function of surface water storage and surface water outflow.

PSTEMP simulates soil temperatures for the surface and subsurface layers of a land segment for use in computation of dissolved gas and agri-chemical concentrations. Surface layer soil temperatures are estimated using a regression equation as a function of air temperature. Subsurface soil temperatures can be estimated either with a regression approach, as a mean departure from air temperature plus a smoothing factor, or the modeler may specify fixed or time-variable values.

PWTGAS estimates the water temperature for surface, shallow subsurface (interflow) and groundwater outflows. The temperature of each outflow is considered to be the same as the soil temperature of the layer from which it originates. PWTGAS also computes the dissolved oxygen and carbon dioxide concentrations of overland flow using empirical formulations; concentrations are assumed to be at saturation.

PQUAL simulates generalized water quality constituents in the outflows (surface and subsurface) from a pervious land segment using simple relationships with water and/or sediment yield. The behavior of a constituent in surface outflow is considered more complex and dynamic than the behavior in subsurface flow. The code allows quantities in surface outflow to be simulated by one, or both, of two methods. Either (1) a constituent can be modeled using "potency factors" to indicate constituent strength relative to the sediment removal computed by SEDMNT, or (2) storage of a constituent on the land surface can be modeled, considering accumulation and depletion/removal, and a first-order washoff rate of the available constituent can be removed by overland flow, as computed by PWATER. In addition, both formulations can be used for representing the washoff behavior of particulate and dissolved components of an specific pollutant.

The remaining five code compartments in PERLND are used together to model detailed behavior of soil nutrients (i.e. nitrogen and phosphorus) and non-reactive tracer chemicals (e.g. chloride). These five code sections have been referred to as the AGCHEM module because their primary use to date has been for modeling the mass balance and runoff of agricultural chemicals. MSTLAY estimates the storages and fluxes of moisture in the four soil layers -- surface, upper, lower, groundwater -- which define soil layers used by the remaining four code compartments. MSTLAY is required because the moisture storages and fluxes computed by PWATER must be modified to effectively simulate solute transport through the soil. Estimates of solute flux are computed based on the assumption that the concentration of solute being transported is the same as that for storage; uniform flow through the layers and continuous mixing of solutes is also assumed. Leaching retardation factors are computed to modify the solute fluxes from the top three soil layers based on user-defined model parameters.

PEST simulates pesticide behavior in the soil and runoff from pervious land segments in three forms: dissolved, adsorbed and crystallized. The PEST code utilizes time series data generated by other compartments of PERLND (i.e., PWATER, SEDMNT, MSTLAY) to compute transport (runoff and leaching), adsorption/desorption and degradation. Pesticide transport is modeled as a function of water flow and/or association with transported sediment. Chemicals in solution move to, through, and from storages according to the fractions calculated in MSTLAY. Computations are performed that compute the movement of adsorbed pesticide associated with removal of sediment from the surface layer via scour and washoff. Adsorption/desorption is a function of both chemical and soil layer characteristics; several options for characterizing sorption are offered, including a first-order kinetic approach and the use of two different Freundlich isotherm methods. Degradation from all processes is modeled as a lumped rate in each of the four soil layers.

The NITR code section simulates the transport and soil reactions of nitrate, ammonia and organic nitrogen. Figure 12.3 shows the nitrogen and phosphorus transformations performed by the NITR and PHOS code sections for each soil layer. Nitrogen species are transported by the same methods used for pesticides. Nitrate and dissolved ammonium are

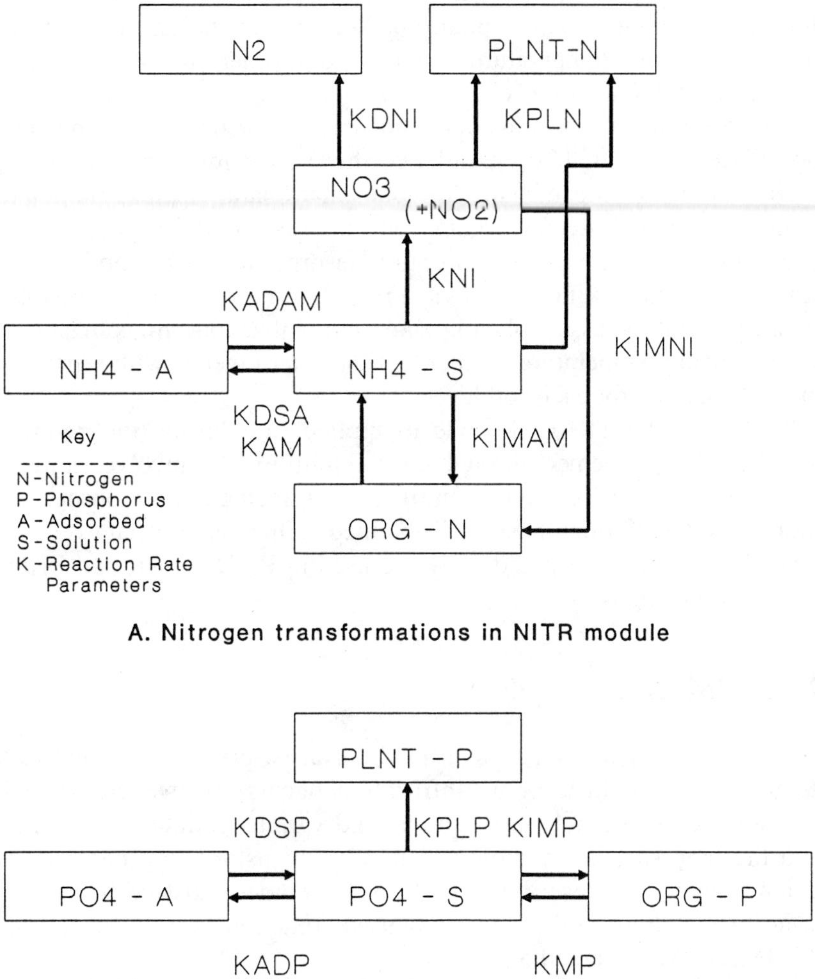

Fig. 12.3. Nutrient Transformations Simulated by the AGCHEM Module.

transported as a function of water flow; organic nitrogen and adsorbed ammonium are removed from the surface layer storage by association with sediment scour and washoff; nitrate and ammonium in the soil water are transported according to the fractions calculated in MSTLAY; and computations are performed that compute the movement of adsorbed organic nitrogen and ammonium associated with removal of sediment from the topsoil surface layer. First-order kinetics or a Freundlich isotherm can be used to model adsorption/desorption. Nitrogen transformation processes (denitrification, nitrification, plant uptake, immobilization, mineralization) are modeled using temperature-corrected, first order kinetics with separate rate constants defined for each soil layer.

PHOS simulates the transport and reaction of phosphate and organic phosphorus using methods parallel to those used for nitrogen species in NITR. Transport mechanisms for phosphate parallel those modeled for ammonium, and those for organic phosphorus parallel organic nitrogen. Like ammonium, phosphate adsorption/desorption can be modeled using either first-order kinetics or a Freundlich isotherm. Phosphorus transformation processes (plant uptake, immobilization, mineralization) are modeled using temperature-corrected, first order kinetics with separate rate constants defined for each soil layer.

Typically, the TRACER code is applied to chloride (or bromide) to calibrate solute movement through the soil profile. This involves adjustment of leaching retardation factors until good agreement with observed soil chloride concentrations has been obtained. Once appropriate retardation values have been determined, they are used in PEST, NITR and PHOS to simulate solute transport.

12.2.2. IMPLND Module

IMPLND is used for impervious land surfaces, primarily for urban land categories, where little or no infiltration occurs. However, some land processes do occur, and water, solids, and various pollutants are removed from the land surface by moving laterally downslope to a pervious area, stream channel, or reservoir. IMPLND includes most of the pollutant washoff capabilities of the commonly used urban runoff models, such as the STORM, SWMM, and NPS models.

Figure 12.4 defines the structure and contents of the IMPLND module. The module shares much of its code with PERLND, but is simplified since infiltration and other interactions with the subsurface cannot occur. The module features individual compartments for modeling air temperature as a function of elevation (ATEMP), snow accumulation and melting (SNOW), hydrologic water budget (IWATER), solids accumulation and removal (SOLIDS), surface runoff water temperature and gas concentrations (IWTGAS), and generalized water quality constituents (IQUAL). Since IMPLND is only used to model impervious areas, the code is not concerned with detailed simulation of agri-chemicals, and hence has no capabilities parallel to those contained in MSTLAY, PEST, NITR, PHOS or TRACER.

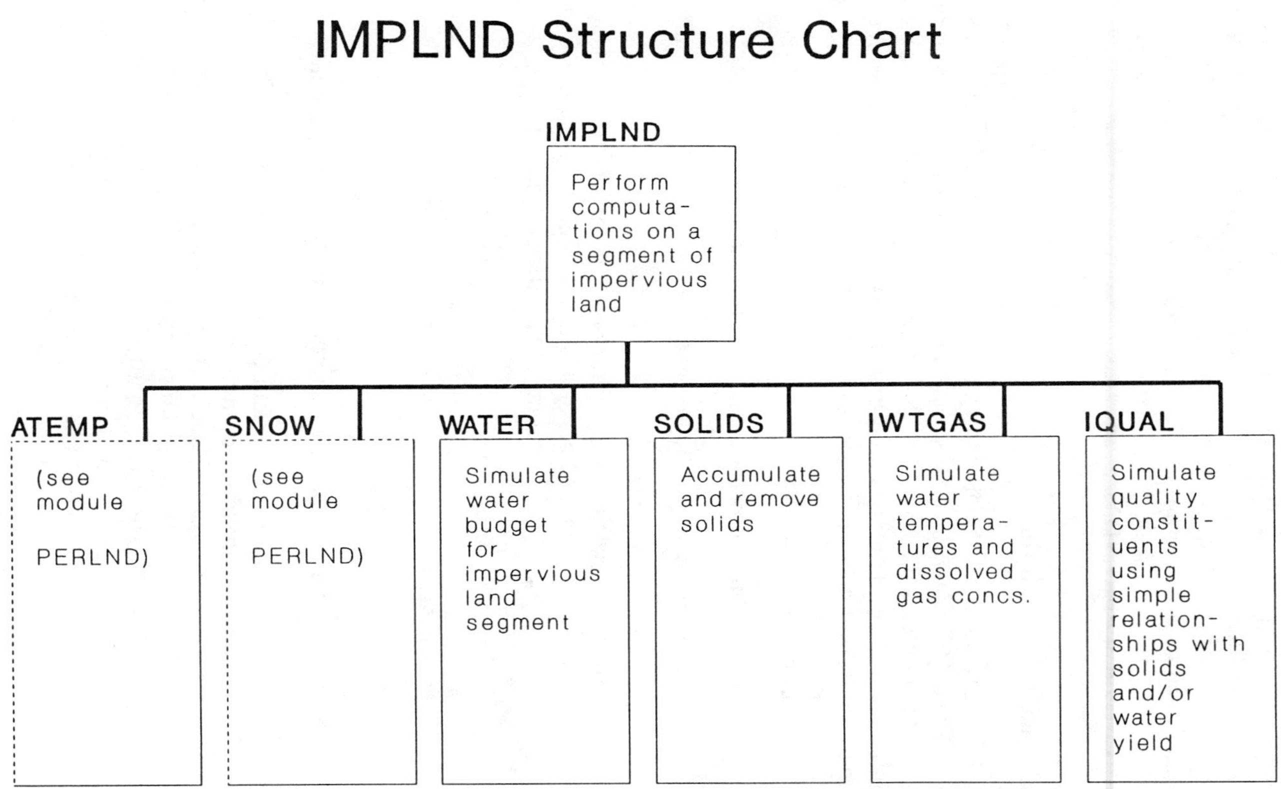

Fig. 12.4. IMPLND Structure Chart.

One difference between PERLND and IMPLND process representation is of note. In the SOLIDS code section, IMPLND offers the capability to model the accumulation and removal of urban solids by processes which are independent of storm events (e.g., street cleaning, decay, wind deposition or scour). To use this option, the modeler needs to assign monthly or constant rates of solids accumulation and removal, and to provide values for model parameters that define an empirical relationship between solids removal and impervious surface outflow by a method that parallels that used in the SEDMNT code of PERLND.

12.2.3. RCHRES Module

RCHRES is used to route runoff and water quality constituents simulated by PERLND and IMPLND through stream channel networks and reservoirs. The module simulates the processes that occur in a series of open or closed channel reaches or a completely mixed lake. Flow is modeled as unidirectional. A number of processes can be modeled, including:

- hydraulic behavior
- heat balance processes that determine water temperature
- inorganic sediment deposition, scour, and transport by particle size
- chemical partitioning, hydrolysis, volatilization, oxidation, biodegradation, and generalized first-order (e.g. radionuclides) decay, parent chemical/metabolite transformations
- DO and BOD balances
- inorganic nitrogen and phosphorus balances
- plankton populations
- pH, carbon dioxide, total inorganic carbon, and alkalinity

Figure 12.5 defines the structure and contents of the RCHRES module. The module features individual compartments for modeling hydraulics (HYDR), constituent advection (ADCALC), conservatives (CONS), water temperature (HTRCH), inorganic sediment (SEDTRN), generalized quality constituents (GQUAL), specific constituents involved in biochemical transformations (RQUAL), and acid mine drainage phenomena (ACIDPH). Because the ACIDPH code section has not been fully field-tested, it is not a documented feature of Release 10 and is not shown in Fig. 12.5 (see Section 12.3.3).

HYDR simulates the processes that occur in a single reach of an open channel or a completely mixed lake. Hydraulic behavior is modeled using the kinematic wave assumption. All inflows to a reach are assumed to enter at a single upstream point. The outflow of a reach may be distributed across several targets that might represent normal outflows, diversions, and multiple gates of a reservoir. In HSPF, outflows can be represented by either, or both, of two methods:

(1) Outflow can be modeled as a function of reach volume for situations where there is no control of flows or gate settings are only a function of water level.

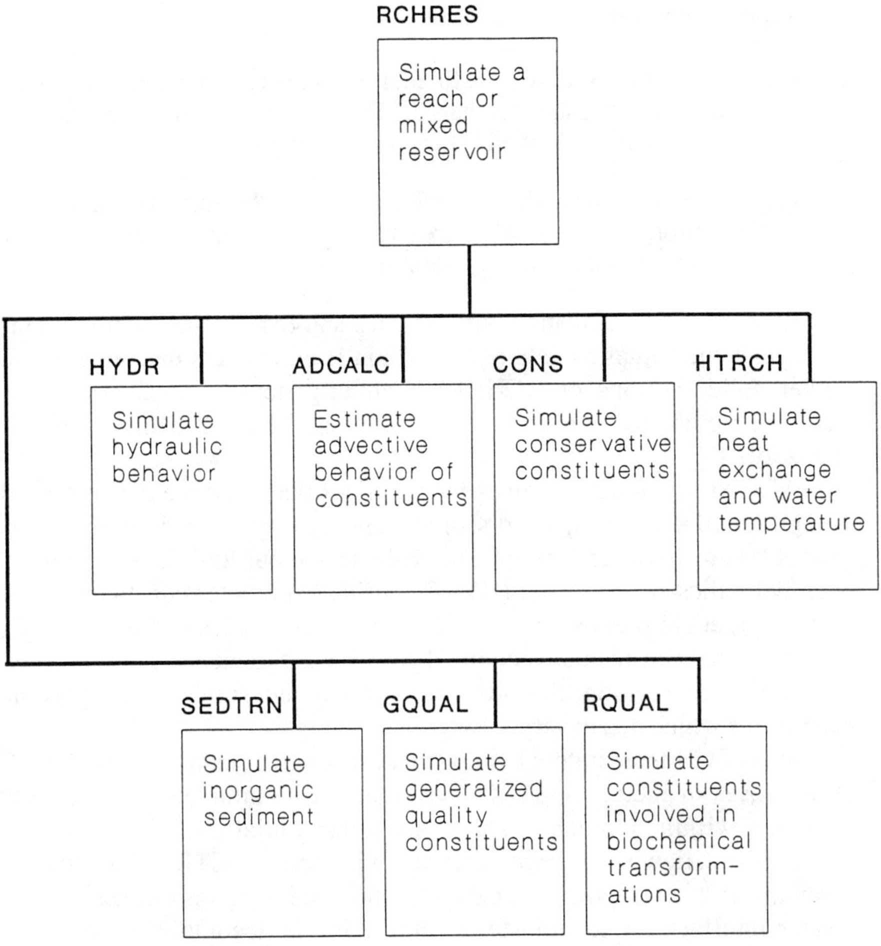

Fig. 12.5. RCHRES Structure Chart.

(2) Outflow can be modeled as a function of time to represent demands for municipal, industrial, or agricultural use. To do so, the modeler

must provide a time series of outflow values for the outflow target that is time-dependent and independent of reach volume.

If an outflow demand has both volume-dependent and time-dependent components, the modeler can, and must, specify how the components are combined to define the resulting outflow demand. HSPF allows the modeler to define the resulting demand in one of three manners: (1) as the minimum of the two components, (2) as the maximum of the two components, or (3) as the sum of the two components.

HSPF makes no assumptions regarding the shape of a reach; however, the following assumptions are made:

(1) There is a fixed, user-defined, relation between water depth, surface area, volume, and discharge. This is specified in a Function Table (or FTABLE) defined for each reach by the user.

(2) For any outflow demand with a volume-dependent component, the relation between the above variables is constant in time. (This assumption precludes modeling flow reversals.)

In addition to calculating outflow rates and reach water volumes, HYDR computes the values for additional hydraulic parameters that are used in the other code sections of RCHRES including depth, stage, surface area, average depth, top width, hydraulic radius, bed shear stress and shear velocity.

ADCALC calculates values for variables that are necessary to simulate longitudinal advection of dissolved and entrained constituents. These variables are all dependent upon the volume and outflow values computed in the hydraulics compartment (HYDR). CONS simulates constituents which, for all practical purpose, do not decay with time or leave the reach by any other mechanism than advection. Typical constituents that are modeled as conservatives include chlorides, total dissolved solids, and hydrophilic chemicals which decay very slowly.

In HTRCH temperature is simulated using a heat-balance approach. Five time series of meteorologic data are required to simulate the temperature balance within a reach. These are solar radiation, cloud cover, air temperature, dewpoint temperature and wind speed. HTRCH considers two major processes: heat transfer by advection and heat transfer across the air-water interface. Heat transfer by advection is accomplished by treating water temperature as a thermal concentration, and using the standard advection computations contained in CONS. The net transport of heat across the air-water interface is computed as the sum of a number of mechanisms, each of which is evaluated individually. Sources of heat that are computed include absorption of shortwave solar radiation, absorption of longwave radiation, and conduction-convection. Sinks of heat that are computed are emission of longwave radiation, conduction-convection, and evaporation.

The approach taken by the SEDTRN code compartment to compute transport of channel sediment is based on the SERATRA model developed by Battelle Laboratories (Onishi and Wise, 1979). Both noncohesive (sand) and cohesive (silt, clay) sediments are simulated in SEDTRN; migration of each sediment fraction between suspension in water and the bed is modeled by balancing deposition and scour computations. The code allows the modeler to compute the deposition or scour of noncohesive sediment by selecting one of three empirical formulations:

(1) A user-defined power function of streamflow velocity,

(2) A relationship (Toffaleti method) dependent upon median sand particle diameter, average stream velocity, reach hydraulic radius, reach slope, settling velocity for sand (user-specified), and water temperature,or

(3) A relationship (Colby method) dependent upon median sand particle diameter, average stream velocity, reach hydraulic radius, fine sediment load concentration, and water temperature.

The simulation of cohesive sediment transport consists of two steps. First, advective transport is calculated; then deposition and scour is calculated based on the calculated bed shear stress. To evaluate deposition, the modeler is required to provide values for settling velocity and critical shear stress for deposition for each fraction (silt, clay) of cohesive sediment that is modeled. To evaluate resuspension, or scour, the modeler must provide values for the erodibility coefficient and critical shear stress for scour for each fraction.

The focus of the GQUAL code development was to allow simulation of agricultural pesticides and other synthetic organic chemicals. Given the diversity of pesticides that might be modeled, the code provides the user with the capability to model any subset of the following generalized processes: advection of dissolved material; decay of dissolved material by hydrolysis, oxidation by free radical oxygen, photolysis, volatilization, biodegradation, and/or generalized first-order decay; production of one modeled constituent as a result of decay of another constituent; advection of adsorbed suspended material; deposition and scour of adsorbed material; and adsorption/desorption between dissolved and sediment-associated phases. Using the GQUAL section in conjunction with the sediment transport code (SEDTRN), adsorbed chemicals may settle or resuspend during each simulation time step, depending on hydrodynamic conditions.

Decomposition of adsorbed chemicals may be simulated, both in suspended materials and in the bed, by using a first-order, temperature-corrected decay formulation.

The RQUAL code provides detailed simulation of constituents involved in biochemical transformations. Included are dissolved oxygen, BOD, ammonia, nitrite, nitrate, phosphate, phytoplankton, benthic algae, zooplankton, refractory organics, and pH. The primary dissolved oxygen and biochemical oxygen demand balances are simulated with provisions for decay, settling, benthal sinks and sources, reaeration, and sinks and sources related to plankton. The primary nitrogen balance is modeled as sequential reactions from ammonia through nitrate. Ammonia volatilization, ammonification, denitrification, and ammonium adsorption/desorption interactions with suspended sediment fractions are also considered. Both ammonium and phosphate adsorption/desorption to suspended sediment fractions are modeled using an equilibrium, linear isotherm approach. Both nitrogen and phosphorus species are considered in modeling three types of plankton - phytoplankton, attached algae and zooplankton. Phytoplankton processes that are modeled include growth, respiration, sinking, zooplankton predation, and death; zooplankton processes include growth, respiration and death; and benthic algae processes modeled are growth, respiration and death. Hydrogen ion activity (pH) can be calculated by two independent code sections. The first, named PHCARB, is contained within the RQUAL section and computes pH by considering carbon dioxide, total organic carbon and alkalinity. In doing so, the code considers the effects on the carbon dioxide-bicarbonate system of carbon dioxide invasion, zooplankton respiration, BOD decay, net growth of algae, and benthal releases.

ACIDPH is a general module for performing user-defined instream chemical computations. Its intended application is primarily to model acid mine drainage and acid rain affected waters, where the pH computations in the PHCARB module, which are based solely on carbonate system equilibria, cannot adequately represent the processes which determine pH (see Section 12.3.4 for further details).

12.2.4. Utility Modules

The five utility modules are used to access, manipulate, and analyze time series information stored by the user in HSPF's TSS (Time Series Store) and WDM (Watershed Data Management) files. These time series, such as hourly precipitation, daily evaporation, daily streamflow, are used by the application modules and are often a valuable resource in the analysis of a watershed's characteristics. The five utility modules and their functions are as follows:

(1) COPY - Copies data in the TSS or WDM to another file. The user can change the time step of the time series during the COPY operation; for example, a five-minute rainfall record can be aggregated to an hourly time interval.

(2) PLTGEN - Generates a specially formatted ASCII file for subsequent data display on a plotter, or input to other software.

(3) DISPLY - Creates data display tables. Aggregated values, as well as summary information, can be generated.

(4) DURANL - Performs frequency, duration, and excursion analyses; computes statistics; and performs toxicity/lethality analysis. DURANL can be used to answer such questions as: "How often does dissolved oxygen stay below 4 mg/l for 4 consecutive hours?"

(5) GENER - Transforms one or two time series to produce a new, or different time series. GENER is a powerful tool that allows the user to perform any one of 22 optional transformations (e.g., absolute value, truncation, division, logarithm).

12.3. RECENT IMPROVEMENTS

Recently Release 10.0 of HSPF (Bicknell et al., 1993) was published and made available through the U.S. EPA. This most current release of HSPF reflects recent improvements in database and input management, as well as improved algorithms for such processes as instream sediment-nutrient interactions and acid mine drainage. In addition to improvements to the batch version of HSPF, a stand-alone expert system for hydrologic calibration using the HSPF water budget computations in the PERLND module has been developed by the U.S. Geological Survey and made available to the public.

12.3.1. ANNIE/WDM

The United States Geological Survey has sponsored the development of interactive pre- and post-processing capabilities to facilitate applications of HSPF and other hydrologic models. ANNIE (Lumb et al., 1990) is an interactive computer program written in FORTRAN and designed for portability to mainframe computers, minicomputers, and microcomputers (i.e., PC's and workstations). ANNIE helps users interactively store, retrieve, list, plot, check, and update spatial, parametric, and time-series data for hydrologic models and model analyses. A binary, direct-access file is used to store data in a logical, well-defined structure and is called a

Watershed Data Management (WDM) file. HSPF and a number of other hydrologic and water quality models and analysis tools developed by the USGS currently use either ANNIE or the WDM file, or both. The WDM file provides the user with a common data base for many applications, thus eliminating the need to reformat data from one application to another.

12.3.2. Software Enhancements

SCHEMATIC and MASS-LINK Blocks

In HSPF Release 10, two new blocks were added to the input file to facilitate the definition of the watershed network. The SCHEMATIC and MASS-LINK blocks allow the user to specify the basin structure and linkages in a more conceptual manner than is possible using the NETWORK block. Another advantage of these blocks for most model applications is that the NETWORK block will be much simpler.

The SCHEMATIC block contains the global specifications of the watershed structure, i.e., connections of land segments to stream reaches and reach-reach connections. A major advantage of this block is that it requires the user to input the area of a land segment tributary to a stream reach in a single location, instead of including the area in multiple lines in the NETWORK block. Each entry in the SCHEMATIC block refers to a table in the MASS-LINK block where all of the detailed material connections for that watershed linkage are defined.

The MASS-LINK block contains the specific time series or material quantities to be transferred from one watershed unit to another. This block also contains any required units conversion factors or other multiplication factors that may be needed, other than the area. Each MASS-LINK table contains a set of time series transfers that are associated with one or more of the watershed linkages in the SCHEMATIC block. The HSPF program combines the schematic linkages with the mass time series transfers and automatically generates all of the necessary time series connections and includes them in the NETWORK block.

FILES Block

A new section, called the FILES block, was added to the input file in Release 10 to facilitate interaction with all other input and output files. Previously, all I/O file definition in the input file was accomplished with file unit numbers, which were associated with actual disk file names using an external, operating system-specific "command" file such as a DOS batch file. The FILES block allows the user to accomplish the correspondence between file unit numbers and file names entirely within the input file, and reduces any confusion created by using separate command files.

12.3.3. Sediment-Nutrient Interactions

By the late 1980's, it became evident that the instream nutrient algorithms contained in the RQUAL module of Release 9.0 of HSPF had deficiencies that, under certain conditions, precluded effective modeling of nutrient loadings to endpoint receiving waters. Perhaps the process representation deficiency of greatest concern was the inability of HSPF to account for instream sediment-nutrient interactions such as adsorption/desorption and advection with sediment. To correct these model deficiencies, the EPA Chesapeake Bay Program sponsored a model enhancement project in 1990-1991.

The enhanced RQUAL module that resulted from the effort for the Chesapeake Bay Program is included in HSPF Release 10.0. The module now simulates the exchange of phosphate and ammonium between the dissolved state and adsorption on sediment suspended in the water column; adsorption/desorption in bed sediment is not modeled. The sorbents considered are the sand, silt and clay fractions of instream sediment that are simulated in the SEDTRN module. The adsorption/desorption process for each sediment fraction is represented with an equilibrium linear isotherm. The advective processes for phosphate and ammonium adsorbed to each sediment fraction are:

(1) Inflow to the reach of nutrient attached to suspended sediment.

(2) Migration from suspension to the bed as a result of deposition of the sediment to which the nutrient is adsorbed.

(3) Migration from the bed into suspension as a result of scour of the bed sediments to which the nutrient is adsorbed.

(4) Outflow from the reach of nutrient attached to suspended sediment.

The following modeling assumptions are used for inorganic nutrients in the bed:

(1) The phosphate and ammonia storages in the bed are infinite as long as the bed sediment storage is non-zero.

(2) Phosphate and ammonia that is deposited to the bed is assumed to be lost from the reach.

(3) The scoured sediment is assumed to have a constant (user-specified) adsorbed concentration for each of the nutrients. Thus, the scoured nutrient fluxes are limited only by the storage of sediment in the bed.

These enhancements provide a more realistic means of representing phosphate and ammonia behavior in streams and lakes, and allowed a more accurate estimation of nutrient loadings, especially for phosphate, as part of the Chesapeake Bay HSPF application (discussed in Section 12.5.1).

12.3.4. Acid Mine Drainage

To address the needs within the USGS for modeling acid mine drainage from Pennsylvania coal mines, a new module, called ACIDPH, was designed and implemented in Release 10 of HSPF. ACIDPH is a general module for performing user-defined instream chemical computations. Its intended application is primarily to model acid mine drainage and acid rain affected waters, where the pH computations in the PHCARB module, which are based solely on carbonate system cequilibria, cannot adequately represent the processes which determine pH.

The chemistry and methods used in the ACIDPH module were adapted from an earlier computer code called PHCALC (Gherini, 1984), and were specifically designed to represent the conditions exhibited by streams in the Pennsylvania coal mining regions. The code implemented in HSPF is comprised of computations that consider the effects on pH of aluminum and carbonate equilibria; extended alkalinity; and the possible effects of iron complexation and competition with aluminum. Three solution options were coded based on the PHCALC methods. These routines perform the following computations:

(1) Compute pH and extended alkalinity assuming Gibbsite ($Al(OH)_3$) saturation; includes interactions from aluminum and ferric complexes of fluoride, sulfate, and water.

(2) Compute pH and extended alkalinity assuming ferric hydroxide ($Fe(OH)_3$) saturation; includes interactions from aluminum and ferric complexes of fluoride, sulfate, and water.

(3) Compute pH from alkalinity assuming saturation of Al^{+3} and Fe^{+3} complexes of fluoride, sulfate, and water.

ACIDPH simulates the following seven chemical species: total aluminum (Al^{+3} and Al complexes), free Al^{+3}, total Fe (Fe^{+3} and Fe complexes), free Fe^{+3}, H^+, total inorganic carbon, and alkalinity. The CONS section of HSPF is used to represent sulfate and fluoride.

To date, the ACIDPH module has had limited testing and use. Its only application has been performed in the district office of the USGS at Pittsburgh, Pennsylvania. The applicability of the module to other mine drainage areas is uncertain, and may require the user to develop customized, situation-specific algorithms within the HSPF framework; the module has been designed to accommodate this need with a minimum of programming effort. Due to the minimal field-testing of the module, ACIDPH is currently an undocumented feature in HSPF Version 10.

12.3.5. HSPEXP

A stand-alone version of the HSPF land surface hydrologic computations has been developed as an expert system for calibrating watershed models for drainage basins. The resulting product, called HSPEXP (Lumb and Kittle, 1993), represents an effort to make the knowledge of experienced surface hydrology modelers available to general model users. The knowledge consists of the statistical representation of the observed hydrograph in terms of the system parameters that drive the precipitation-runoff process. The estimation procedure consists of a set of hierarchical rules designed to guide the calibration of the model through a systematic evaluation of the model parameters.

To date, the system has been tested on watersheds in Washington and Maryland, and the developers report encouraging results. In each instance, the system correctly identified the model parameters to be adjusted and the adjustments led to an improved calibration. It is anticipated that HSPEXP will serve as a major component in future calibration efforts for watershed models.

12.4. MODEL DISTRIBUTION, MAINTENANCE EXPERIENCE, USER SUPPORT / TRAINING, SOFTWARE / HARDWARE ISSUES

Since its initial release in 1980, HSPF has undergone a number of corrections, revisions, and enhancements; more than 300 copies of the code have been distributed to users throughout the world. To enable the widespread use of a complex model that has continued to evolve over many years, various methods of user support have been provided by both private parties and government agencies. The purpose of this section is to identify model distribution requirements and to share our experiences over the past thirteen years in attempting to support model users and improve their HSPF modeling opportunities.

12.4.1. Model Distribtuion and Hardware Requirements

Over the years, general distribution of HSPF has been the responsibility of the U.S. EPA Environmental Research Laboratory in Athens, Georgia. The model distribution activities are carried out by the Center for Exposure

Assessment Modeling (CEAM) at the Athens Laboratory. For the first couple of releases, this distribution focused primarily on IBM mainframe and various minicomputers such as Hewlett-Packard, DEC VAX, and Prime. Since 1988, this distribution has shifted largely to IBM-compatible microcomputers running the DOS operating system, and somewhat to Unix machines. The principal platform is now a 386 or 486 DOS-based microcomputer with extended memory; and the program is compiled using an extended-memory FORTRAN compiler. Distribution by the U.S. Geological Survey is generally limited to USGS District offices and their cooperators. The USGS version is distributed for use with the Unix operating system, and is targeted specifically at a Data General AViiON workstation.

In order to maintain consistency, both agencies attempt to distribute the same code; and, as of Version 10, they also distribute the ANNIE software which is necessary for development and maintenance of the WDM time series data used by HSPF.

12.4.2. Model Maintenance

Software maintenance of HSPF has been supported by the EPA Athens Laboratory in Georgia, and, since the late 1980's, cooperatively by the U.S. Geological Survey, Water Resources Division in Reston, Virginia. Since the initial release in 1980, almost all of the actual maintenance effort has been performed by two firms: Anderson-Nichols, Inc. and its successor, AQUA TERRA Consultants in Mountain View, California. These maintenance activities have included maintaining a list of software errors, correcting errors, implementing enhancements, adapting the code to new computer environments (hardware and operating system), testing, and providing new versions to EPA and USGS for distribution to users.

12.4.3. Training Workshops

Workshops and seminars on the theory, structure, application and operation of HSPF have been held on a periodic basis since 1981 sponsored by a variety of organizations. They have included one-day overviews of model capabilities and case-study applications, two-day sessions specifically focusing on hydrology and data management, and complete four-day intensive training workshops covering all model capabilities (i.e. hydrology and water quality) along with hands-on computer work sessions for model operation and calibration. The U.S. EPA has been the primary sponsor for most of the training workshops; however, the U.S. Geological Survey has also sponsored and co-sponsored (with EPA) a number of training sessions, and has held annual training sessions over the past few years on hydrology and data management using HSPF and ANNIE, at its National Training Center in Denver, Colorado.

Customized workshops designed for specific agencies have been held both in the U.S., for the Maryland Department of the Environment and the EPA Chesapeake Bay Program, and abroad for the City of Madrid (Spain), the Sydney Water Board (Australia), the Saudi Arabian Ministry of Agriculture (Riyadh, Saudi Arabia), and the Canada Systems Group (Toronto, Ontario). A number of private consulting firms have also presented training workshops.

After a hiatus of about 9 years, EPA sponsored its most recent complete four-day HSPF training workshop in Seattle, Washington in August 1993 at the specific request of local agencies who have enthusiastically adopted HSPF for urban stormwater drainage and management. The U.S. Geological Survey also held a complete hydrology/water quality workshop in February 1994 in Denver, and two additional workshops are planned by EPA for the summer of 1994.

12.4.4. Support for Model Users

Although a few current model users have successfully applied HSPF without any formal training and little or no user assistance or support, most potential users will greatly benefit from attending a training workshop (discussed above) and will need some degree of application assistance or support. The U.S. EPA Center for Exposure Assessment Modeling (CEAM), and its predecessor, the EPA Water Quality Modeling Center, located in Athens, GA has been the primary center for model distribution and user support activities for HSPF since its initial release. These activities have included code distribution, model updates, model 'bug' reporting and corrections, work-arounds, phone consultation, and varying degrees of model calibration and application assistance. Recent budget cutbacks have severely limited the extent of user support currently available from CEAM, in spite of an increasing demand for these services from local, regional, state, and federal agencies across the country. Currently, model support from CEAM is limited to code distribution, model updates, and limited phone consultation.

Since about 1988, the U.S. Geological Survey's Water Resources Division has provided support for both model enhancements and user assistance for HSPF in association with its ongoing development and support activities for ANNIE/WDM, HSPEXP, and other USGS software products. However, this user support is primarily for USGS district and regional offices, and associated cooperators. Also, a limited number of consulting firms have capabilities for providing HSPF application assistance and support functions under contract arrangements.

12.5. SELECTED APPLICATIONS

As was previously mentioned, applications of the HSPF model have numbered in the hundreds. Many applications have utilized a straight-forward approach and a limited subset of the model's capabilities to evaluate watershed problems that require a continuous simulation approach. Other applications have featured more comprehensive and/or innovative approaches to using HSPF that demonstrate the power of the model and the breadth and complexity of problems to which it can be successfully applied. Five examples of such model applications are described briefly below.

12.5.1. Chesapeake Bay Watershed Model

The EPA Chesapeake Bay Program uses the HSPF model as the framework for the Chesapeake Bay Watershed Model to determine total watershed contributions of flow, sediment and nutrients (and associated constituents such as water temperature, DO, BOD, etc.) to the tidal region of the Chesapeake Bay (Donigian et al., 1986a; 1991). The Watershed Model represents pollutant contributions from an area of more than 68,000 sq. mi., and provides input to drive a fully dynamic three-dimensional, hydrodynamic/water quality model of the Bay. The watershed drainage area is divided into land segments and stream channel segments; the land areas modeled include forest, agricultural cropland (conventional tillage, conservation tillage, hay), pasture, urban (pervious and impervious areas), and uncontrolled animal waste contributions. The stream channel simulation includes flow routing and oxygen and nutrient biochemical modeling in order to account for instream processes affecting nutrient delivery to the Bay.

At the onset of the study, buildup/washoff type algorithms were used for urban impervious areas, potency factors for all pervious areas, and constant (or seasonally variable) concentrations for all subsurface contributions and animal waste components. Subsequently, detailed process (i.e. agri-chemical modules) simulation was performed for cropland areas to better represent the impacts of agricultural BMPs. The Watershed Model is currently being used to evaluate nutrient management alternatives for attaining a 40 percent reduction in nutrient loads delivered to the Bay, as defined in the 1987 Chesapeake Bay Agreement signed by the governors of the member states, the EPA Administrator, and the District of Columbia. Figure 12.6 shows the Watershed Model estimates of Total Nitrogen loads to the Chesapeake Bay from the major subbasins under a range of alternative scenarios or conditions, including base conditions for 1984-87, progress made through 1991, load reductions required by the Bay Agreement, 'limit of technology' controls on both point and nonpoint sources, and expected loads for the year

2000 conditions. This type of analysis is being used to evaluate both the feasibility of the 40% reduction goal, the nature of the controls required, and the distribution of the needed load reductions among the Bay tributaries and the member states.

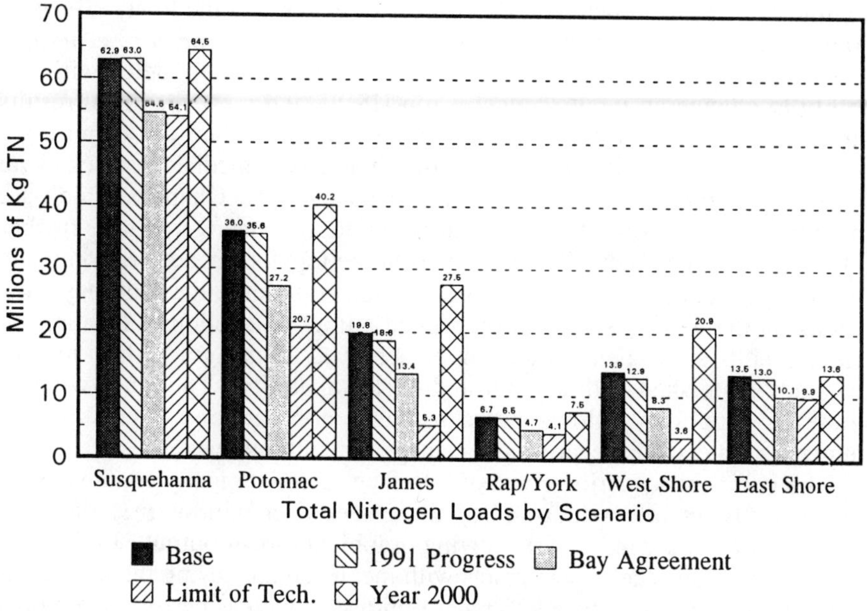

Fig. 12.6. Chesapeake TN Loads by Scenario.

12.5.2 Patuxent River Study and Scenario Generator

The Maryland Department of the Environment is conducting a study of the Patuxent River to quantify nonpoint source contributions and evaluate alternative means of improving downstream water quality in the Patuxent River Estuary. The study includes a multi-year monitoring program that involves observations of runoff quality at both field size (single land use) locations and instream, multi-land use sites. HSPF is being applied to calculate nonpoint loadings from the forest, agriculture, and urban land areas of the watershed, and the instream water quality throughout the river system. The Patuxent is a microcosm of the larger Chesapeake Bay, a complex watershed with multiple land uses, point and nonpoint sources, and reservoirs draining to a tidal estuary. Like the larger Chesapeake Bay study, both simple and complex nonpoint runoff algorithms are being used to represent the various land uses and effects of potential management practices.

Use of HSPF for applications such as the Patuxent study has traditionally involved using a text editor to build a series of input sequences that describe the physical conditions resulting from various management alternatives or scenarios. Each of these input sequences might comprise thousands of lines, and hence, the process of making changes has been time consuming and complex. In order to facilitate the process of evaluating the likely effects of potential management decisions in the Patuxent basin, a *scenario generator* was developed. The scenario generator provides the modeler with the ability to change an input sequence, run HSPF, and view graphical results interactively within a multi-window environment. When a user makes a change to an existing input sequence, the change is immediately checked via a set of programmed rules to see if it makes sense. Changes which are incomplete or inconsistent are indicated to the user, so that further refinement can be made. When changes to the input sequence are complete, the HSPF model is executed. After execution completes, the user may view results on the display screen or on printed output. Selection of locations at which results are to be reported can be achieved by using a built-in map of the study watershed with a 'point-and-click' capability. The multi-window graphics capability allows users to view comparisons of results for numerous watershed scenarios and management alternatives.

The Patuxent scenario generator interface screen typically contains three concurrently visible windows: 1) the interaction window, 2) the map window, and 3) a window for viewing graphical model output. In the user interface window, the user interacts with menu screens (using the keyboard) to select the various map and graphics options, as well as perform the model simulations. In the map window, the user operates a mouse to select locations for data display and zoom on specific areas. The graphics window is used to display the data plots that are selected in the interface window. The user can also interactively control the sizes and locations of the different windows on the screen.

The Patuxent basin scenario generator has proved itself a powerful tool for dealing with the literally thousands of input/output time series datasets that have been developed to represent possible watershed management scenarios (Lumb, 1993). Additional scenario generators are currently being developed for the USGS for use at other geographical locations, including the Carson-Truckee River basin in California and Nevada, and small urban basins in the Seattle, Washington area (Lumb, personal communication, 1994) .

12.5.3. STREAMPC

In 1993, STREAMPC was developed for the U.S. EPA Environmental Research Laboratory in Athens, Georgia, in response to the need for an automated screening procedure to rapidly evaluate pesticide impacts on a regional basis. It was based on the original STREAM methodology

(Donigian et al., 1986b; Donigian and Mulkey, 1992). The original STREAM system was developed by applying the HSPF model to various watersheds in agricultural regions across the United States, then performing sensitivity analyses on key pesticide parameters to generate cumulative frequency distributions of pesticide concentrations and loadings in each region for each of the major crops grown in that region. Users of STREAM were required to manually select the frequency curves appropriate for their situation of interest, perform graphical interpretation of the data (including interpolation or extrapolation), and make adjustments for application rate and percentage of area treated.

STREAM^{pc} (Hummel et al., 1993) incorporates the original STREAM methodology into an interactive software package, thus automating the analysis procedure. The package guides the user through the process of selecting the regions, crops, and pesticide of interest. A pesticide database, containing chemical property values for over 125 active ingredients, has been incorporated to assist the user in defining pesticide behavior. Additional input options allow the user to define up to six duration times for analysis and to define application rates and percentage of area treated.

The specified duration analyses are then performed, using the output from the original HSPF-STREAM simulations, for the user-specified scenarios. STREAM^{pc} generates the output results by performing any needed interpolations and reports them in either graphical or tabular format. Graphical output is in the form of exceedance probability plots with concentration (solution or bed) or daily loads reported. Tabular output is generated when the user specifies a particular exceedance probability or concentration/load value of interest.

Currently, STREAM^{pc} is limited to four major agricultural regions of the United States -- Eastern Cornbelt, Western Cornbelt, Southeast, Mississippi Delta -- and includes two primary production crops in each region. The system is flexible and expandable so that additional regions and/or crops can be added as resources become available.

12.5.4. Seattle Regional Parameter Approach

A "Runoff Files" implementation of HSPF has been developed as a hydrologic modeling tool for western King County, Washington (Jackson, personal communication, 1994). The method produces results (design flows, detention pond sizing, etc.) comparable to those obtained by direct use of the HSPF model, but with significantly less effort. This is achieved by providing the user with a set of time-series files of unit area land surface runoff (i.e., "runoff files") pre-simulated using HSPF for the range of land cover conditions and soil types found within several regions of King County. Estimation of design flows and design detention facilities is then accomplished by directly accessing and manipulating the runoff file data by means of supporting software.

The runoff files consist of runoff values generated from 8 representative years of a 40-year record of hourly meteorological data; the runoff files represent simulated unit area land surface runoff for six rainfall stations and eight land use classifications (forest till, pasture till, grass till, forest outwash, pasture outwash, grass outwash, wetlands, impervious areas). For each land use classification, a characteristic set of parameter values that satisfy the input requirements of the PERLND module of HSPF was developed. Taking advantage of the resulting 48 runoff files requires software which scales up and combines unit area runoff from a mix of land use types to represent the total runoff to specific design facility locations. King County has developed a software package, named KCRTS, with which to perform the manipulations necessary to use the runoff files to design stormwater detention facilities. The County has subsequently performed tests comparing the Runoff Files/KCRTS software implementation versus routing a full 40-year time-series of HSPF-simulated hourly flows through the facility. These analyses demonstrated that most designs using the simplified method met, or came close to meeting, their performance standards when tested using HSPF (Jackson, personal communication, 1994).

12.5.5. Interactive Water Resource and Reservoir Management Tools

Using the hydrology process algorithms of HSPF, Hydrocomp Inc. has developed a number of customized forecast/analysis models to simulate water resources within specific watersheds or geographic regions (Crawford, personal communication, 1994). These software packages enable interactive modeling of natural and anthropogenic hydrologic processes, including reservoirs, hydroelectric plants and irrigation projects. Process algorithms are identical to HSPF, but are coded in Pascal and C programming languages. The components of the water resources management systems include real-time data networks; historic hydrometeorologic databases; watershed, channel and reservoir characteristics; modeling software; and interactive graphical modeling using engineering workstations. The packages feature a graphical user interface (GUI) developed for the X-Windows environment, using Xt Intrinsics and the Motif Widget Set. The software runs on engineering workstations.

The increased speed and excellent graphics of engineering workstations, equivalent to having desktop "Cray" super computers, prompt re-examination of traditional methods used for operation of multi-purpose water resource reservoirs. The Hydrocomp forecast/analysis models enable advanced techniques for reservoir operation and design; the use of current watershed state and real-time data to forecast future water supplies; and the use of 'alternate futures', 'exceedance probabilities' and 'sensitivity analysis' for reservoir operation.

12.5.6. Other Recent Model Applications

Table 12.3 lists a number of recent and/or current HSPF applications to provide a brief summary of the many types of watershed hydrology and

TABLE 12.3.

Selected Recent and/or Current HSPF Applications

- Chesapeake Bay Watershed Model of nutrient loadings and management alternatives
- Seattle Metropolitan Area watershed and urban drainage studies
- Metropolitan Washington, DC urban nonpoint and water quality studies
- U.S. EPA Office of Pesticide Programs assessment of alachlor surface water concentrations
- Sydney Water Board (Australia) assessment of water supply quality and nonpoint pollution
- Maryland Department of the Environment Patuxent River nonpoint source study
- Numerous Florida applications for hydrologic assessments, nutrient loadings and water quality simulation
- Flooding and Water Resource Development studies for Saudi Arabian Ministry of Agriculture
- Upper Grande Ronde (OR) temperature TMDL
- Walnut Creek (IA) MSEA/MASTER surface water exposure assessment
- Minnesota River Nonpoint Source Assessment Project
- Water Management for the Humber River, South Africa

water quality assessments that can be, and have been, performed with HSPF. In addition to the studies mentioned above, the Washington metropolitan area, the Seattle metropolitan area, and Florida have been regions with numerous model applications by county and local agencies, often focusing on stormwater management. Foreign applications have included modeling studies in Australia for nonpoint and water quality assessment, in Saudi Arabia for flooding and water resources development, and in South Africa for water management. The Minnesota River Assessment Project is a comprehensive application with the ultimate goal of implementation of a watershed-based nonpoint source control program to improve water quality conditions related to low dissolved oxygen and high levels of turbidity, fecal coliform and ammonia. Current research-oriented applications include the Walnut Creek (Iowa) exposure assessment for surface/subsurface interactions related to agricultural BMPs, and the Upper Grande Ronde (Oregon) water temperature modeling study to evaluate impacts of riparian forest buffers on stream shading.

12.6. FUTURE DIRECTIONS

The recent resurgence of government concern for nonpoint source issues and problems, as catalyzed by various sections and amendments to the Clean Water Act in the United States, has renewed interest in nonpoint source modeling and assessments at the watershed scale. The comprehensive nature of HSPF, and its flexibility in allowing consideration of the combined impacts of both point and nonpoint source pollutants, is leading to increased interest in model applications. To support this increased interest and potential usage, this section identifies and explores areas of ongoing model and algorithm development, many of which will be implemented in the upcoming Version 11 of HSPF, and future research areas where additional development efforts are needed. In addition to potential algorithm enhancements, we explore future needs for databases and user interaction capabilities that will be required to keep pace with changing software/hardware capabilities.

12.6.1. Current Model Development Activities for HSPF Version 11

As shown in Table 12.1, HSPF has experienced a flurry of model development activity during the past five years, primarily related to needed improvements identified through selected model applications. This section describes the major improvements and added capabilities that users can expect in HSPF Version 11, planned for release in 1994-95.

Atmospheric Deposition

It has become apparent that atmospheric transport and deposition of nitrogen and other pollutants can contribute a significant fraction of the total load delivered to surface waters. Therefore, the ability to represent these inputs in water quality models is important. Currently, atmospheric deposition inputs are difficult or impossible to model with HSPF, especially to the land surface; and these inputs generally require extensive use of the Special Actions section. Consequently, the next release of the program will contain significant improvements in this area. These enhancements are also designed to facilitate linkage of the model with atmospheric transport models, such as EPA's Regional Acid Deposition Model (RADM) (Chang et al., 1987).

HSPF has been enhanced to accept atmospheric deposition inputs to all water quality constituent state variables, both on land surfaces and in water bodies. The deposition can consist of wet deposition, computed from user-specified concentrations in rainfall, and dry deposition, which is specified as areal loading rates. The input concentrations and loading rates can be in the form of time series or monthly averages. Also, the program's output contains summaries of the deposition inputs.

Plant Uptake and Forest N Cycling

The nitrogen (NITR) subroutines of the AGCHEM module are currently being enhanced to improve the ability to represent plant uptake for agricultural crops and nitrogen cycling in forested watersheds. The current first-order representation for plant uptake makes the model extremely sensitive to nutrient application rates (i.e. fertilizers and manure); to evaluate the impacts of changes in these application rates the user is required to re-calibrate the plant uptake rates whenever the application rates are significantly changed, such as under a nutrient reduction or management plan. To eliminate the need for re-calibration under these conditions, the plant uptake algorithms are being modified to be a function of crop needs and expected yields, in addition to available soil nitrogen. These changes will be incorporated and tested as part of the ongoing Chesapeake Bay Watershed Model refinement effort (Linker, personal communication, 1994), and are planned for inclusion in Version 11 of HSPF in late 1994.

The EPA Athens Laboratory is currently sponsoring a review of the AGCHEM module with specific focus on the applicability of the soil nitrogen (NITR) algorithms and structure to represent nitrogen cycling and discharge to streams from forested watersheds (Barnwell, personal communication, 1994). Because forests are a dominant fraction of watersheds in many parts of the country, a mass-balance, deterministic approach to modeling nitrogen on forest lands provides a more accurate means of determining nitrogen loads from various sources, including atmospheric deposition contributions (discussed above). In addition, improving estimates of nitrogen loads from forests will result in more accurate loads from other sources in the watershed, many of which are more controllable and affected by management alternatives. Scientists at Oak Ridge National Laboratory are currently reviewing the NITR subroutines and algorithms to identify appropriate changes to nitrogen storages, rate constants, and flux pathways for forested conditions. Some of the changes being considered include, (1) addition of nitrogen fixation by plants and microbes, (2) return of plant nitrogen to soil organic nitrogen, and (3) partitioning of organic nitrogen into labile, refractory, and soluble forms. Many of these improvements will also provide expanded capabilities applicable to non-forested areas (e.g. croplands, rangelands). The changes will be incorporated into the NITR subroutines by AQUA TERRA Consultants, and tested on primarily forested subbasins of the Chesapeake Bay drainage, and are planned for inclusion in HSPF Version 11.

Water Allocation/Ownership

In many river basins in the western U.S., the entire flow is allocated to various owners, such as municipal suppliers, farmers, ranchers, and industrial facilities, who can utilize their allocations at specific times of the year. In order to improve our ability to analyze such basins, the HSPF RCHRES module has been enhanced to keep track of the ownership of water within the reach network. This capability has been implemented as a set of water "categories" that represent ownership. A user can assign the ownership of water inflows and outflows from each stream segment. The ownership of outflowing water can be defined in the form of specified priorities or percentages, or they can be proportional to the current mixture in the stream segment. The initial application of this feature is the Carson-Truckee River system in California and Nevada, where the U.S. Geological Survey is developing allocation and water quality models of the basins.

Impacts of Riparian Forests on Instream Water Temperature

The EPA Office of Water, Assessment and Watershed Protection Division, is sponsoring a riparian characterization and water temperature modeling study of the Upper Grande Ronde (UGR) River in Northeastern Oregon (U.S. EPA, 1994). The focus of the study is to assess the causes of elevated summer water temperatures in the UGR River that impair its ability to sustain a healthy coldwater ecosystem for annual salmon runs and resident salmonid populations; reduced shading of the stream channel and riparian zone are the suspected causes. A TMDL (Total Maximum Daily Load) is being developed under authority of section 303(d) of the Clean Water Act to determine the temperature reductions needed to sustain a viable salmon habitat.

HSPF is being applied to the watershed, along with code enhancements to better represent the seasonal and directional aspects of both topographic and vegetal shading, to evaluate alternative management approaches for reducing the summer water temperatures. A GIS is being used to determine critical stream and riparian parameters -- such as stream direction and orientation; topographic variation; riparian zone forest density, height, crown, etc. -- for segments of the stream throughout the channel system (U.S. EPA, 1994). These parameters will then be used to evaluate the impacts of stream shading on water temperature, as impacted by diurnal shading variations, seasonal shading changes, and riparian forest characteristics. These changes in solar radiation impinging on the stream surface will then lead to changes in the energy input to the heat-balance equations used in the HTRCH section of HSPF and produce the resulting water temperature changes. Changes to existing forest buffers, or development of buffers where none existed, can then be evaluated as potential management alternatives for improving the coldwater ecosystem habitat. The study is a coordinated effort among the EPA Office of Water and the Athens Environmental Research Laboratory, the Oregon State Department of Environmental Quality, the Columbia River Tribes, the U.S. Forest Service, the Northwest Power Planning Council, and others.

Increased Limit on HSPF Operations

As HSPF has been used to model larger and more complex river basins, it has become obvious that the limit on the number of discrete operations (i.e., PERLND's, RCHRES's, PLTGEN's, etc.) in a single run must be increased. This limitation on the number of operations has forced simulations of basins that consist of many land segments and stream reaches to be divided into multiple runs that are executed sequentially from upstream to downstream. For example, the Susquehanna River portion of the Chesapeake Bay Watershed Model (Donigian et al. , 1991) consists of six separate runs. This results in significant computational overhead in the form of disk space and execution time to transfer the results between successive runs. It is expected that the current maximum of 75 operations in a run will be increased to 200 or more in Version 11 of the program.

Enhancements to SPECIAL ACTIONS

Addition of chemicals to the soil in HSPF requires use of the SPECIAL ACTIONS section in which each application of a chemical to a PERLND requires a separate instruction. In large basins containing multiple PERLND's representing different crops, soil types, and meteorologic regions, the number of these SPECIAL ACTIONS can easily grow into the thousands. Several enhancements have been added to the program to significantly reduce the number of SPECIAL ACTIONS instructions required in complex basins where agricultural operations are simulated. These enhanced capabilities are briefly discussed below.

Repeat -- Each SPECIAL ACTION can now be "repeated" at regular time intervals. This facilitates application of chemicals several times per year and each year of the simulation.

Distribution -- A SPECIAL ACTION can be "distributed" over time (equal time increments) with a user-defined pattern that is based on fractions of the total amount. This is useful in representing the activities of multiple farmers applying chemicals on different days when all of the farms are represented by a single PERLND.

User-defined -- Several SPECIAL ACTIONS can be combined as a single "user-defined" action which can be invoked multiple times for different PERLNDS and at different times. This reduces the number of actions required to represent incorporation of chemicals in two or more soil layers as a result of plowing, and application of multiple chemical species.

Conditional -- In addition to the enhancements designed to reduce the user-input requirements of SPECIAL ACTIONS, conditional SPECIAL ACTIONS are now possible in which an action can be dependent on the value of some other variable in the model. This can be useful for deferring agricultural operations that are dependent on rainfall or soil moisture, and for reservoir operations that are dependent on river flow or reservoir volume.

U.S. Army Corps of Engineers DSS File Interaction

Currently, the two optional HSPF time series database systems utilize the U.S. Geological Survey's Watershed Data Management (WDM) file or the Time Series Store (TSS) file. HSPF can read and write time series data directly to these files. The U.S. Army Corps of Engineers and other users of their software maintain large amounts of meteorologic and flow data using HEC Data Storage System (DSS) files; and many COE models and utility software packages have been designed to interact with these files. In order to facilitate use of HSPF by Corps of Engineers personnel, Version 11 of HSPF will contain the capability to read/write data to DSS files. The HSPF-DSS interface will be essentially identical to the interface used by the WDM and TSS files.

12.6.2. Future Algorithm/Computation Enhancements

The requirements of the process modeling capabilities of a model such as HSPF fall into five categories:

- Provide capability to define the physical setting of the modeling effort
- Represent important environmental state variables and processes
- Represent man-made effects on environmental state variables and processes
- Use process algorithms that utilize available data
- Provide a spectrum of analysis level capabilities

While HSPF provides capabilities to perform all these functions, there is room for significant improvement. In this section, possible enhancements to process algorithms and computational techniques are noted, first in general terms. Then specific capabilities of particular interest to the authors are elaborated.

Capability to Define Physical Setting of Modeling Effort

Developing and verifying model input that defines the physical setting of the modeling effort and specifies spatially-referenced materials properties can be confusing and time-consuming. A graphical approach to performing these tasks would be a highly useful enhancement to HSPF. Performance of these tasks is best performed within a GIS environment. GIS capabilities can be used effectively to accomplish the following tasks:

- Insert, delete and move rows and columns to define cells or segments
- Zoom, pan and window capabilities for detailed design
- Overlay a digitized map to aid in visualization or assign parameter values
- Rotate and offset a site schematic relative to a digitized base map
- Establish connectivity between cells or segments
- Compute segment-dependent data volumes and interfacial areas
- Assign parameter values (global, spatially variable, time dependent) depicting natural and anthropogenic conditions

- Assign boundary conditions
- Assign initial conditions

Important Environmental State Variables and Processes

In order to provide the basis for multi-stressor analysis of whole-ecosystem effects, many chemical and biological state variables and processes must be represented. While the majority of these state variables are already considered in the model, HSPF might be enhanced to include the following additional state variables:

- Selected additional biological variables (herbivorous fish, predatory fish)
- Selected habitat variables (% pools and riffles, streambank vegetation and shading, substrate character, turbidity)
- Selected ecosystem variables (elemental dynamics, energy dynamics, trophic dynamics, biodiversity, critical species (presence/abundance), genetic diversity, dispersal and migration, natural disturbance, ecosystem development)

Man-made Effects on Environmental State Variables and Processes

In addition to representing natural processes, modeling systems such as HSPF must provide process algorithms that represent the effects of man-induced sources or processes on environmental state variables. Models must include algorithms that can be used to represent any environmental disturbance that could influence the behavior of the natural watershed system. Examples of such phenomena include nutrient and pesticide application, tillage practices, crop harvest and residue practices, tile drainage, livestock grazing, feedlot runoff, highway drainage, urban development, stormwater detention structures, stream channelization, combined sewers, construction practices, mine drainage, silvicultural practices, municipal and industrial discharges, etc.

Many of these effects can be represented by adjusting values for parameters contained in existing HSPF algorithms; others may require development of new algorithms. We envision that considerable work will be done to develop additional sets of HSPF parameter value changes (i.e., model scenarios) that reflect our best understanding of the physical/chemical changes resulting from a particular modification or activity. This may be the most critical area of model development activity as it directly affects our ability to use models like HSPF for environmental management and decision-making.

Process Algorithms that Utilize Available Data

HSPF was developed prior to the proliferation of a new generation of data and data generation techniques that offer refined spatial detail for a number of parameters critical to watershed modeling. In some cases these new data are best used to support existing process algorithms that are solved for a higher resolution grid. However, the potential also exists to replace or enhance certain process algorithms to improve the simulation of natural processes by taking advantage of new data. For example, satellite data, GIS and digital elevation models (DEMs) have made it possible to compute the *aspect* (i.e., the direction toward which a slope faces) for watersheds or watershed segments at a high level of detail. The availability of techniques to reliably compute aspect invites the incorporation of improved process algorithms for snowmelt, soil temperature, and water temperature in areas of significant topographical relief.

The two technologies that offer the greatest body of new data that could be used to refine process algorithms are satellite remote sensing data and the transformation of remote sensing data, by use of GIS and related capabilities, to derive other useful data types.

The remote sensing data available from current and future satellites offer an opportunity to develop new process algorithms that could improve representation of precipitation, surface runoff, soil moisture, groundwater, and water quality variables including thermal pollution, erosion, sediment load, and trophic state of receiving waters. An immediate need of watershed-scale models are algorithms using radar imaging data to represent thunderstorms.

Future Modeling Research Areas

Every modeler has one's own views and opinions as to the most important areas of future modeling research. These are often the result of modeling applications and experiences where obvious deficiencies have been identified, and no satisfying resolution has been developed. Below we discuss a few of the areas we feel deserve attention in future model research and development, borne from our experiences with HSPF and other models.

Wetlands - Modeling of wetlands is an issue that arises in many watershed modeling studies, to varying degrees, and not only in the humid, coastal plain areas of the southeastern U.S. The beneficial effects of wetlands on flood retention, sediment filtration, and nutrient and toxics processing are well known, but not adequately understood. HSPF has been used to approximate the impacts of wetlands, and various code modifications have been proposed (Nichols and Timpe, 1985); however, these approaches have been primarily 'stop-gap' measures due to lack of resources and alternative models. Coordinated data collection and modeling research efforts (i.e. algorithm development) are needed to improve our ability to represent the complex water quality impacts of wetlands on the watershed system.

Fish - Fish share all zooplankton processes including growth, respiration, death and predation; additional important processes for fish include exposure to environmental stresses such as high temperatures, low dissolved oxygen, toxic chemicals, and sedimentation. Models of various fish species exist, but few are appropriate for inclusion within a comprehensive watershed modeling framework.

Habitat Suitability - As a group, habitat state variables (e.g., velocity, channel gradient, flow, depth, % pools and riffles, streambank vegetation and shading, substrate character, turbidity, salinity, pH, temperature, dissolved oxygen) characterize the physical or chemical setting in which biotic communities live. The physical state variables are tied to considerations of topographical relief, runoff, erosion, sedimentation, channel characteristics and thermal inputs. To a large extent, the habitat state variables that characterize the chemical setting need to be modeled irrespective of whether modeling goals include habitat analysis. A watershed modeling system, like HSPF, is ideally suited to include assessment of habitat variables.

Ecosystem Modeling - The goal of ecological modeling is to determine self-sustainability. To do this, modeling may focus on system elements or components (i.e., species), system structure/organization, system function (based on physical, chemical, and/or biological principles), system dynamics (material and energy transport), or the integration of one or more of these system characteristics, habitat features, and biotic communities. Relative to the other categories described above, habitat and ecological modeling are in their infancies; consequently, it is not possible to identify the important processes in a rigid manner. However, the need exists to integrate these areas into the watershed modeling arena to allow consideration of the full extent of human impacts on the watershed system and its component ecosystems.

12.6.2. User Interaction

User interaction, or more accurately the lack thereof, is a major weakness of HSPF and a significant stumbling block to many new model users. However, at the time of its initial release in 1980, interactive water resources and environmental models were non-existent, and many of the popular menu-driven, business and word processing programs (e.g. Lotus, WordPerfect) did not emerge until the mid-1980's. Although HSPF was developed as a state-of-the-art model, with respect to both software and model algorithms, advances in 'useability' features and interactive software packages quickly aged these aspects of the code. Today it remains a batch program, with interactive features reserved for its ancillary and related programs (i.e. ANNIE/WDM, HSPEXP, Scenario Generators), largely due to the lack of resources to bring it up to current 1990's standards of user interaction.

Many sophisticated users have developed a variety of software tools to automate specific functions needed to facilitate model application, data manipulation, and results analysis. Possible tools that could improve HSPF user interaction can be categorized into five categories of functionality:

- Data exploration/visualization
- Model support before processing
- Model support during processing
- Model support after processing
- Decision support

Potential enhancements for each of the categories are noted below.

Data Exploration/Visualization

- Basic display of data spatial and/or temporal variation (e.g., superimpose model results on location maps; time series plot of user-defined model variables; 2-dimensional and 3-dimensional plots, contour plots).
- Graphical comparison of different data types at common locations or times to make preliminary assessments of cause and effect.

The primary source of the list of capabilities to support modeling support activities (i.e., pre-processing, processing, post-processing) is the GEO-WAMS effort (DePinto, 1993).

Model Support Before Processing

- Site specification and problem definition. Provide an interactive graphical site schematic configuration and manipulation system with sophisticated capabilities (e.g., insert, delete and move rows and columns to define cells or segments; global and zonal modification; zoom, pan and window capabilities for detailed design; overlay of multiple information types).

- Graphical assignment of parameter values (global, spatially variable, time dependent)
- Hooks to retrieve raw data. Integrate custom programs to automate retrieval of data from commonly-used external databases (archival or commercial).
- Data transformation tools. Provide automated techniques for aggregating or disaggregating data. The modeling system should automatically transform data to meet process model requirements, and hence should be transparent to the user.
- Data derivation tools. When commonly accepted relationships of either an empirical or physical nature are available to estimate data needed by the component process models, the modeling system should include tools that enable automated value estimation based on available data.
- Computation of segment-dependent data volumes, interfacial areas
- Graphical prescription of initial conditions and boundary conditions

Model Support During Processing

- Spatial and temporal animation of results
- Model interrupt/restart capability

Model Support After Processing

- Basic display of primary variable spatial and/or temporal variation (e.g., superimpose model results on location maps; 3-dimensional plots, contour plots)
- Sensitivity analysis
- Segment/system mass balance diagrams
- Uncertainty analysis
- Data export for use in other modeling systems. The modeling system needs a full library of tools designed to expedite data export from the application data base.

Decision Support

- Improve on-line model capabilities to assist users developing model input: estimating parameter values, selecting appropriate distributions for parameters used in uncertainty analysis, developing physically correct and mathematically compatible configurations and discretization (spatial and temporal) for modeled areas.

- Improve on-line model capabilities to assist users in interpreting model output: enhanced interpretive graphics, interpretation of time series data statistical analysis results as a potential calibration aid.
- Limited expert systems for specific processes, state variables
- Additional scenario generators (urban and non-urban)

12.7. CLOSURE

In this chapter, we've attempted to summarize the historical development of HSPF, its current capabilities, selected model applications, near-term future enhancements, and potential areas for future model development and research. At the time of its initial release in 1980, the original model developers felt that HSPF would have about a five-year 'useful life' before it would need to be replaced with more advanced technology. Now, almost 15 years later, HSPF continues to enjoy a growing user community both in the U.S. and abroad, a renewal of interest from governmental agencies awakening to the need for watershed-scale assessments, and the joint sponsorship of the U.S. EPA and the U.S. Geological Survey. This longevity can be attributed to the foresight and vision of the original developers, who provided the framework and flexibility needed to withstand the test of time, and to the sponsoring agencies for their continuing code and user support. It remains unique in its comprehensive capabilities for watershed hydrology and water quality modeling, and its framework and structure is sufficiently flexible to accommodate many advances in process science and understanding.

Clearly, HSPF will continue to exist and evolve over the foreseeable future. Whether or not it provides the basis for the **'next generation'** of watershed models depends on the availability of continuing support and resources to adapt it to current standards of useability and interactions with current and future technology (i.e. GIS, remote sensing, computerized databases). The inertia inherent in the current HSPF user community and ongoing development efforts, in addition to its ability to satisfy the specific niche for comprehensive watershed-scale assessments, will sustain its evolution and popularity for many years, likely through the end of this century.

REFERENCES

Bicknell, B.R., J.C. Imhoff, J.L. Kittle Jr., A.S. Donigian, Jr. and R.C. Johanson. 1993. Hydrological Simulation Program - FORTRAN. User's Manual for Release 10. U.S. EPA Environmental Research Laboratory, Athens, GA.

Chang, J.S., R.A. Brost, I.S.A. Isaken, S. Madronich, P. Middleton, W.R. Stockwell, and C.J. Walcek. 1987. A Three-Dimensional Eulerian Acid Deposition Model: Physical Concepts and Formulation. J. Geophysical Research, 92(14):681-700.

Crawford, N.H. and R.K. Linsley. 1966. Digital Simulation on Hydrology: Stanford Watershed Model IV. Stanford University Technical Report No. 39, Stanford University, Palo Alto, CA.

DePinto, J. 1993. "Use of ARC/INFO in Developing a Modeling Support System for Great Lakes Watersheds," In Developing Dynamic Watershed Loadings Modeling Capabilities for Great Lakes Tributaries. The Proceedings of an EPA Workshop at Heidelberg College in Tiffin, Ohio, August 4-6, 1993. Office of Research and Development, U.S. Environmental Protection Agency, Athens, GA, (J.C. Imhoff, Ed)

Donigian, A.S. Jr. and H.H. Davis. 1978. User's Manual for Agricultural Runoff Management (ARM) Model. EPA-600/3-78-080. U.S. EPA Environmental Research Laboratory, Athens, GA.

Donigian, A.S. Jr. and N.H. Crawford. 1979. User's Manual for the Nonpoint Source (NPS) Model. Unpublished Report. U.S. EPA Environmental Research Laboratory, Athens, GA.

Donigian, A.S., Jr., J.C. Imhoff, B.R. Bicknell and J.L. Kittle, Jr. 1984. Application Guide for the Hydrological Simulation Program - FORTRAN EPA 600/3-84-066, Environmental Research Laboratory, U.S. EPA, Athens, GA. 30613.

Donigian, A.S. Jr., B.R. Bicknell and J.L. Kittle Jr. 1986a. Conversion of the Chesapeake Bay Basin Model to HSPF Operation. Prepared by AQUA TERRA Consultants for the Computer Sciences Corporation, Annapolis, MD and U.S. EPA Chesapeake Bay Program, Annapolis, MD.

Donigian, A.S. Jr., D.W. Meier and P.P. Jowise. 1986b. Stream Transport and Agricultural Runoff of Pesticides for Exposure Assessment: A Methodology. EPA-600/3-86-011. U.S. EPA Environmental Research Laboratory, Athens, GA.

Donigian, A.S. Jr., B.R. Bicknell, A.S. Patwardhan, L.C. Linker, D.Y. Alegre, C.H. Chang and R. Reynolds. 1991. Chesapeake Bay Program Watershed Model Application to Calculate Bay Nutrient Loadings. Prepared by AQUA TERRA Consultants for U.S. EPA Chesapeake Bay Program , Annapolis, MD.

Donigian, A.S., Jr., and L.A. Mulkey. 1992. STREAM, An Exposure Assessment Methodology for Agricultural Pesticide Runoff. In: Fate of Pesticides and Chemicals in the Environment. J.L. Schnoor (ed.). John Wiley & Sons, Inc. New York, New York. p. 297-330.

Gherini, S.A., C.W. Chen, L. Mok, R.A. Goldstein, R.J.M. Hudson, and G.F. Davis.1984. "The ILWAS Model: Formulation and Application." In: The Integrated Lake-Watershed Acidification Study. Chapter 4: Summary of Major Results. Prepared by Tetra Tech, Inc. for the Electric Power Research Institute, Palo Alto, CA.

Hummel, P.R., A.S. Donigian and B.R. Bicknell. 1993. STREAMpc - An Interactive PC-based Implementation of the STREAM Methodology for Exposure Assessment of Agricultural Pesticide Runoff. EPA Environmental Research Brief, U.S. Environmental Protection Agency, Environmental Research Laboratory, Athens, GA.

Hydrocomp, Inc. 1977. Hydrocomp Water Quality Operations Manual. Hydrocomp, Inc., Palo Alto, CA.

Johanson, R.C., J.C. Imhoff and H.H. Davis. 1980. User's Manual for the Hydrologic Simulation Program - FORTRAN (HSPF). EPA-600/9-80-105. U.S. EPA Environmental Research Laboratory, Athens, GA.

Johanson, R.C., J.C. Imhoff and H.H. Davis, J.L. Kittle, and A.S. Donigian, Jr. 1981. User's Manual for the Hydrologic Simulation Program - FORTRAN (HSPF): Version No. 7.0. Prepared for U.S. EPA Environmental Research Laboratory, Athens, GA. Prepared by Anderson-Nichols & Co. Inc., Palo Alto, CA.

Johanson, R.C., J.C. Imhoff, J.L. Kittle, Jr. and A.S. Donigian. 1984. Hydrological Simulation Program - FORTRAN (HSPF): Users Manual for Release 8.0, EPA-600/3-84-066, Environmental Research Laboratory, U.S. EPA, Athens, GA. 30613.

Leytham, K.M. and R.C. Johanson. 1979. Watershed Erosion and Sediment Transport Model. EPA-600/3-79-028. U.S EPA Environmental Research Laboratory. Athens, GA.

Lumb, A.M., J.L. Kittle, and K. M. Flynn. 1990. Users Manual for ANNIE, A Computer Program for Interactive Hydrologic Analyses and Data Management, Water-Resources Investigations Report 89-4080, U.S. Geological Survey, Reston, VA.

Lumb, A.M., and J.L. Kittle, Jr.. 1993. "Expert System for Calibration and Application of Watershed Models," In: Proceedings of the Federal Interagency Workshop on Hydrologic Modeling Demands for the 90's. U.S. Geological Survey Water Resources Investigation Report 93-4018. J.S. Burton, ed.

Lumb, A.M. 1993. "Integrating New Technologies to Support Watershed Management Decisions - A USGS Perspective," In: Developing Dynamic Watershed Loadings Modeling Capabilities for Great Lakes Tributaries. The Proceedings of an EPA Workshop at Heidelberg College in Tiffin, Ohio, August 4-6, 1993. Office of Research and Development, U.S. Environmental Protection Agency, Athens, GA. J.C. Imhoff, ed.

Meyer, L.D. and W.H. Wischmeier. 1969. Mathematical Simulation of the Process of Soil Erosion by Water. Trans. Am. Soc. Agric. Eng. 12(6):754-758, 762.

Negev, M. 1967. A Sediment Model on a Digital Computer. Department of Civil Engineering, Stanford University. Stanford, CA. Technical Report No. 76. 109 p.

Nichols, J.C. and M.P. Timpe. 1985. Use of HSPF to simulate Dynamics of Phosphorus in Floodplain Wetlands over a Wide Range of Hydrologic Regimes. In: Proceedings of Stormwater and Water Quality Model Users Group Meeting, January 31 - February 1, 1985. T.O. Barnwell, Jr., ed. EPA-600/9-85/016, Environmental Research Laboratory, Athens, GA.

Onishi, Y. and S.E. Wise. 1979. User's Manual for the Instream Sediment-Contaminant Transport Model, SERATRA. Draft Report prepared by Battelle Laboratories, Richland, Washington for U.S. EPA Environmental Research Laboratory, Athens, GA.

Onstad, C.A. and G.R. Foster. 1975. Erosion Modeling on a Watershed. Trans. Am. Soc Agric. Eng. 18(2):288-292.

U.S. EPA. 1994. Upper Grande Ronde River Riparian Characterization and Temperature Modeling Projects: Project Summary, January 1994. EPA 841-S-94-001. U.S. EPA, Office of Water, Washington, D.C.

Wischmeier, W.H. and D.D. Smith. 1978. Predicting Rainfall Erosion Losses - A Guide to Conservation Planning. U.S. Department of Agriculture, Agricultural Handbook No. 537. 58 p.

Chapter 13

THE HBV MODEL

S. Bergström

It is now more than 20 years since the first presentation of results from the HBV model. Since then the use of hydrological models has grown dramatically, and today they are standard tools for an increasing number of applications. The HBV model has followed this trend, and there are examples of applications in some 30 countries. The span of applications has also widened and covers today hydrological forecasting, spillway design, studies of effects of climate change, synoptic water balance mapping, simulations of groundwater response among others. The chapter describes the structure of the most widely used version of the HBV model and summarizes Swedish and international experiences from its application.

13.1. INTRODUCTION

When the first steps in hydrological modelling were taken at the Swedish Meteorological and Hydrological Institute (SMHI) in early 1972 no one could foresee the development. Today, more than 20 years later, we can look back at a tremendous expansion of the use of hydrological models. These models are now everyday tools for many of the Swedish hydrologists. The best known model is the HBV model which exists in a number of versions in Sweden and abroad. Applications have been found in some 30 countries (Bergström, 1992). The PULSE model is a modified HBV model, used for environmental applications, which also exists in several versions.

The SMHI is concentrating on applied research and the use of models for practical purposes. Therefore the models developed by the institute belong to the conceptual family. The following general, and rather simple, principles in model development were identified when the work started:

> The models must be based on a sound physical description but must not be so complex that they have higher data demand than can be met by our standard climatological and hydrological network.

The number of free model parameters (used for calibration) shall be kept to a minimum, and their sensitivity and interaction must be controlled.

The performance of the models must be controlled by field observations of many kinds. As far as possible the control shall be based on independent data.

It must be easy to understand and to use the models.

The work on the HBV model followed, to a high degree, the ideas lined out by Nash and Sutcliffe (1970) among others, who saw the risk that increasing computer capacities may result in too complex model formulations, unless the significance of model components is carefully checked. This philosophy is responsible for the relative simplicity of the HBV model. A lot of modifications of the model have been tested, but few have been showed to improve the results. The basic ideas behind the model are discussed in more detail by Bergström (1991).

The first successful run with an early version of the HBV hydrological model was carried out in the spring of 1972 (Bergström, 1972; Bergström and Forsman, 1973, Fig. 13.1).

The development process continued with an increasing number of test applications and introduction of a routine for snow accumulation and melt (Bergström, 1975). In the early summer of 1975 the first operational forecasts were carried out.

After twenty years the HBV model has become a standard tool for runoff simulations in the Nordic countries, and the number of applications in other countries is growing. Some of the applications abroad are carried out by, or in cooperation with, the Swedish Meteorological and Hydrological Institute using a standard computer code and the version six of the model (HBV-6). Such examples are the applications to five basins within the WMO project on intercomparison of models for snowmelt runoff (WMO, 1986) and a large number of applications to basins in Latin America (Häggström et al., 1990). Some other applications are carried out by modified codes or model versions which are totally rewritten. New standard codes with modified snow routines are, for example, in use in Norway, Finland and Switzerland. Many of the applications abroad can be considered as scientific tests of the model to judge its feasibility under specific conditions, but the number of operational applications is growing along with the development of user-friendly desk top computer systems.

The work with the HBV model has generated a number of useful by-products, as its general approach has proved to be a valuable tool for a large number of problems related to quantitative aspects of water resources. Its

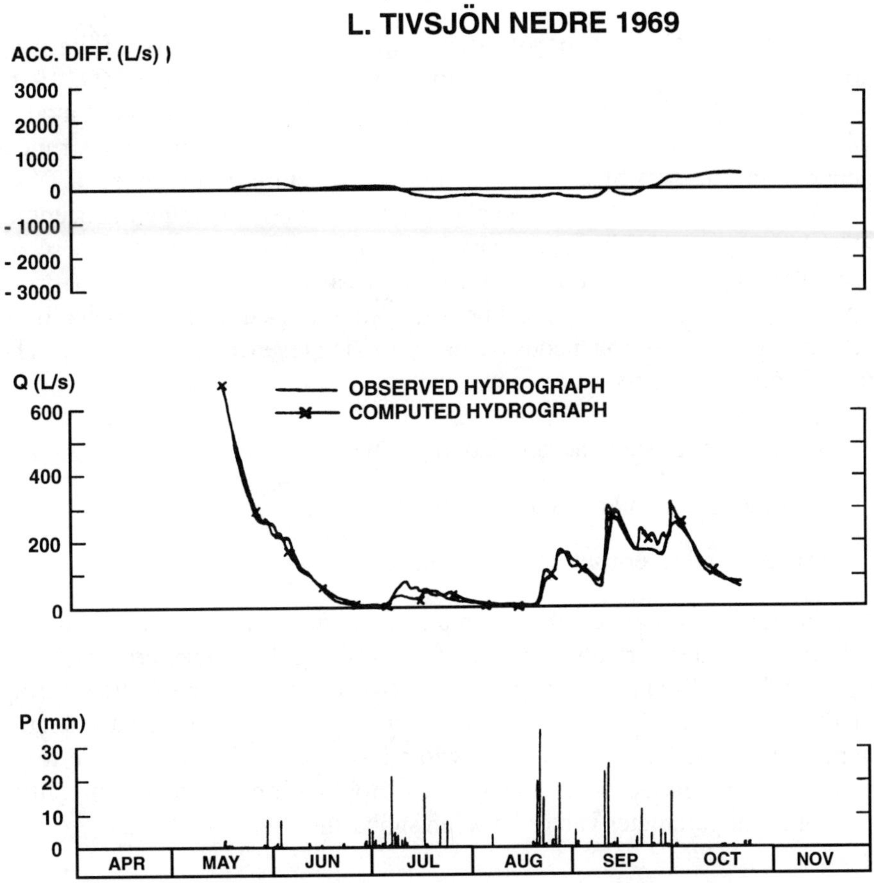

Fig. 13.1. The first successful application of the HBV-1 model carried out in 1972. Lilla Tivsjön research basin in Northern Sweden (14 km^2). ACC. DIFF. = accumulated volume error, Q = runoff and P = precipitation.

successor, the PULSE model, is used for hydrochemical simulations and simulations in ungauged catchments.

In the following sections a description of the structure of the most commonly used model version at the SMHI is given and the 20 years of experience, in Sweden and abroad, under a wide span of geographical and climatological conditions, are summarized. Work with the HBV model has been reported on numerous occasions and in a large number of scientific papers. The list of references is far from complete but covers some key titles and a span of applications.

13.2. THE HBV MODEL AND ITS PARAMETERS

There is a large number of hydrological runoff models of varying complexity available in the world today (see, for example, WMO, 1986; Nemec, 1986). The HBV model belongs to the second generation of models which is characterized by attempts to cover the most important runoff generating processes by as simple and robust structures as possible.

The HBV model can best be classified as a semi-distributed conceptual model. It uses subbasins as primary hydrological units and, within these, an area-elevation distribution and a crude classification of land use (forest, open, lakes) (Fig. 13.2). The subbasin option is used in geographically or climatologically heterogeneous basins or in the presence of large lakes. The model consists of three main components:

subroutines for snow accumulation and melt,

subroutines for soil moisture accounting,

response and river routing subroutines.

The model has a number of free parameters, values of which are found by calibration. There are also parameters describing the characteristics of the basin and its climate which, as far as possible, remain untouched during model calibration. The use of subbasins opens the possibility to have a large number of parameter values for a whole basin. It is, however, wise to be restrictive in this respect, and in most applications there is only little variability in parameter values between subbasins.

13.2.1. Data Requirements and Corrections

The SMHI version of the HBV model is usually run with daily time steps, but higher resolution can be used if data are available. Input data are precipitation and, in areas with snow, air temperature. Other versions of the model may require more input data for the snow routine.

The soil moisture accounting procedure requires data on the potential evapotranspiration. Normally monthly mean standard values are sufficient, but more detailed data can also be used. The source of these data may either be calculations according to the Penman formula or similar, or measurements by evaporimeters. In the latter case it is important to correct for systematic errors before entering the model.

Areal averages of the climatological data are computed separately for each subbasin by a simple weighing procedure where the weights are

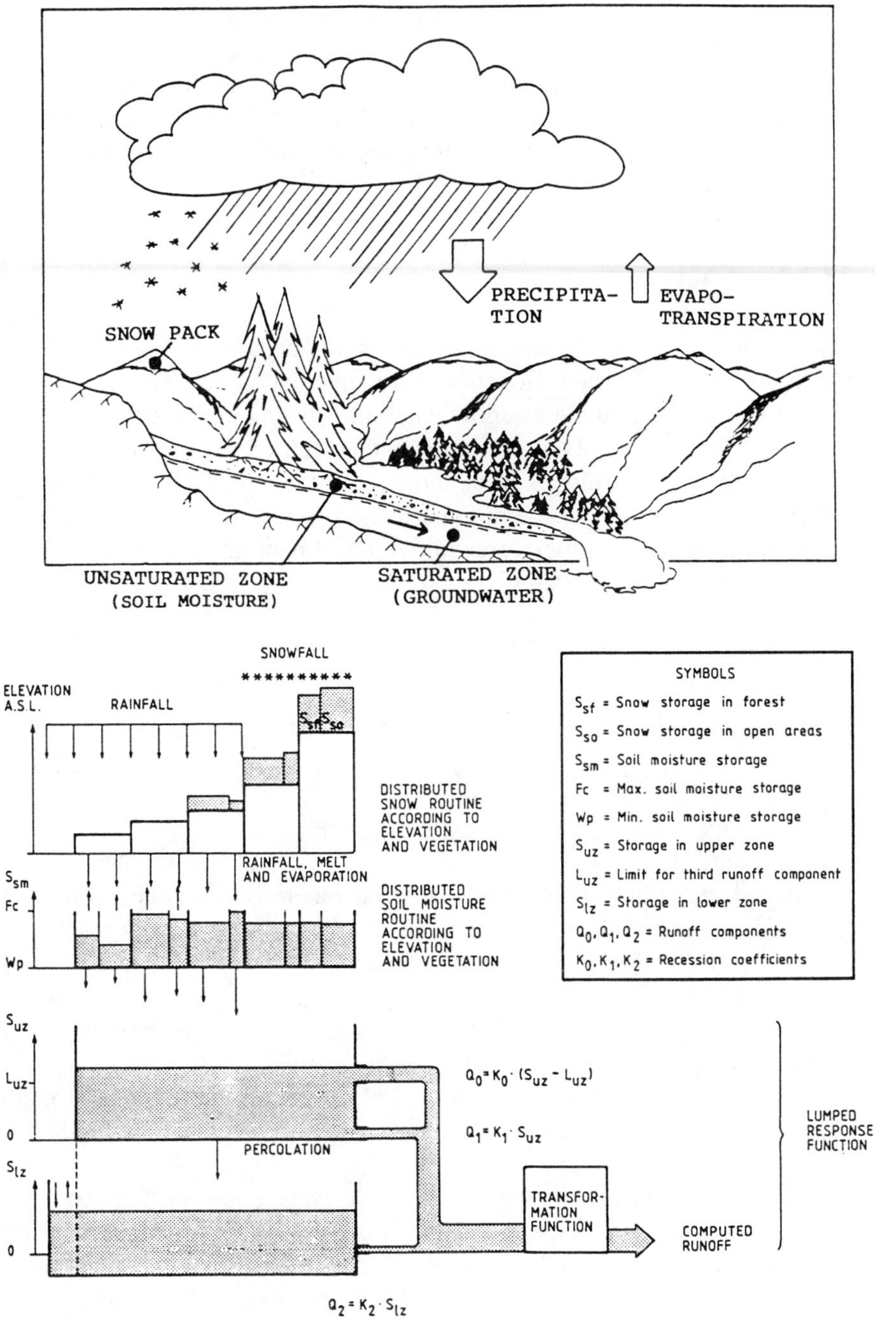

Fig. 13.2. **The general structure of the SMHI version of the HBV model when applied to one subbasin.**

determined by climatological and topographical considerations or by some geometric method like the Thiessen polygons. The climatological input is further corrected for elevation above sea level by constant lapse rates. The temperature lapse rate is usually set to -0.6 °C per 100 meter deviation from station level. The precipitation lapse rates are more site-specific and set by local climatological considerations.

13.2.2. Snow

The snow routine of the model controls snow accumulation and melt and works separately for each elevation and vegetation zone. The precipitation is assumed to accumulate as snow when the air temperature drops below a threshold value (TT). To account for undercatch of snow precipitation and winter evaporation, which is little known, snow accumulation is adjusted by a free parameter, C_{SF}, the snowfall correction factor.

Melt starts with temperatures above the threshold, TT, according to a simple degree-day expression:

$$MELT = C_{MELT} \bullet (T - TT) \tag{13.1}$$

where: MELT = snowmelt (mm/day)

C_{MELT} = degree-day factor (mm/°C • day)

TT = threshold temperature (°C).

The liquid water holding capacity of snow has to be exceeded before any runoff is generated. It is usually preset to 10 %. A refreezing coefficient, which is used to refreeze free water in the snow if snowmelt is interrupted, is fixed in the code.

Thus the snow routine of the HBV model has primarily three free parameters that have to be estimated by calibration: TT, C_{SF} and C_{MELT}. If a separation into vegetation zones is used, the number doubles. It is also common to use separate threshold temperatures for snow accumulation and melt.

The snow routine of the HBV model has been subject to major modifications in the Norwegian, Finnish and Swiss versions of the model. A statistical routine for redistribution of snow over the timber-line has, for example, been introduced by Killingtveit and Aam (1978), and several attempts have been made to introduce glacier-melt subroutines (see, for example, Braun and Aellen, 1990).

13.2.3. Soil Moisture.

The soil moisture accounting routine computes an index of the wetness of the entire basin and integrates interception and soil moisture storage. It is controlled by three free parameters, FC, BETA and LP, as shown in Fig. 13.3. FC is the maximum soil moisture storage in the basin and BETA determines the relative contribution to runoff from a millimeter of rain or snowmelt at a given soil moisture deficit. LP controls the shape of the reduction curve for potential evaporation. At soil moisture values below LP the actual evapotranspiration will be reduced, as shown in Fig. 13.3.

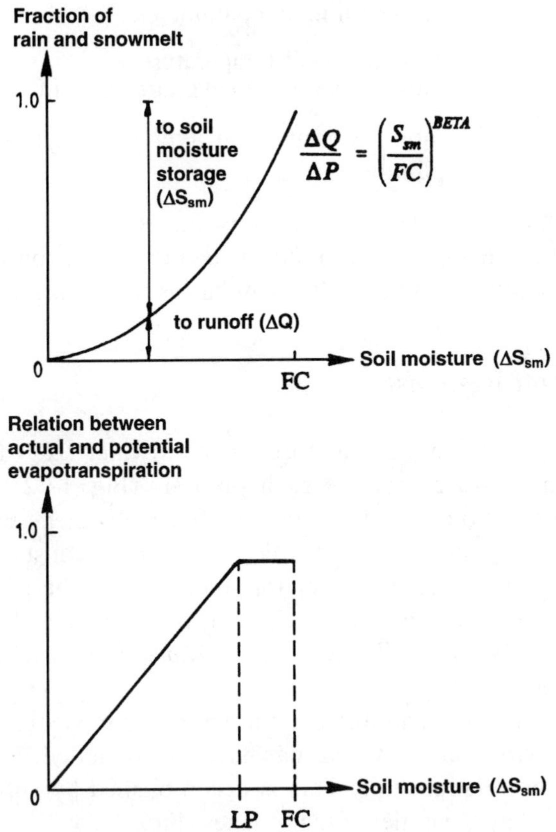

Fig. 13.3. Summary of the soil moisture accounting routine of the HBV model. FC, LP and BETA are empirical parameters.

The effect of the curve determining runoff generation is that the response is gradually increasing with increasing wetness. Thus the routine may be said to account for small-scale variabilities of the soil properties within the catchment.

Recently a modification of the evapotranspiration routine has been introduced in order to improve the model performance when the spring and summer is much colder or warmer than normal (Lindström and Bergström, 1992). This routine accounts for temperature anomalies by a correction which is based on mean daily air temperatures and long term averages according to eq. (13.2):

$$PE_A = \left(1 + C \bullet (T - T_M)\right) \bullet PE_M \tag{13.2}$$

where:

PE_A = adjusted potential evapotranspiration,

C = empirical model parameter,

T = daily mean air temperature,
T_M = monthly long term average temperature,

PE_M = monthly long term average potential evapotranspiration.

The adjusted potential evapotranspiration is limited to positive values and is not allowed to exceed twice the monthly average. The routine has so far given very encouraging results in the eight basins where it has been tested.

13.2.4. Runoff Response

The runoff response routine transforms excess water (ΔQ) from the soil moisture routine, to discharge for each subbasin (Fig. 13.2). The routine consists of two reservoirs with the following free parameters: three recession coefficients, K_o, K_1 and K_2, a threshold, UZL, and a constant percolation rate, PERC. Finally there is a filter for smoothing of the generated flow. This filter consists of a triangular weighting function (Fig. 13.4) with one free parameter, MAXBAS. There is also a Muskingum routing procedure available for flood routing.

Lakes in the subbasins are included in the lower model reservoir, but in later model versions lake routing can also be modelled explicitly by a storage discharge relationship. This is accomplished by subdivision into subbasins defined by the outlet of major lakes (Fig. 13.5).

The relative large number of free parameters in the response routine of the HBV model (six) has focused the risk for overparameterization. Work is therefore in progress on a routine with less parameters (Harlin, 1992). Experience has, however, shown that it is rather easy to assess the parameters after some training. The use of an explicit lake routing routine has also proved to simplify the calibration of the recession parameters of the model, as most of the damping is accounted for by the lakes (Bergström et al., 1985).

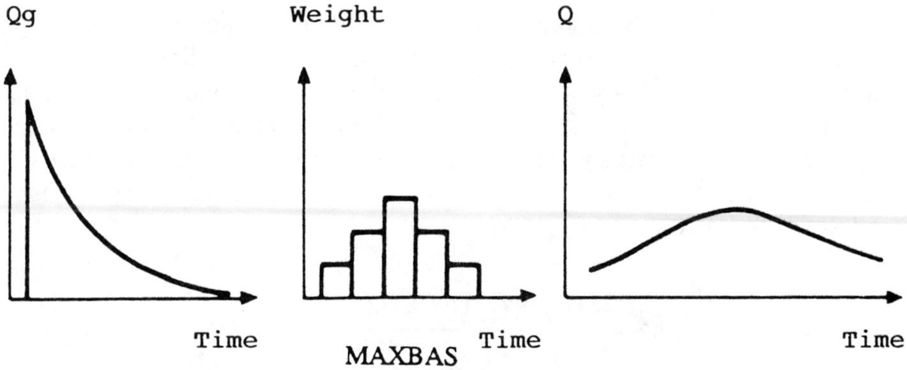

Fig. 13.4. The filter used to smooth generated runoff (Qg) to obtain discharge (Q). MAXBAS is a model parameter.

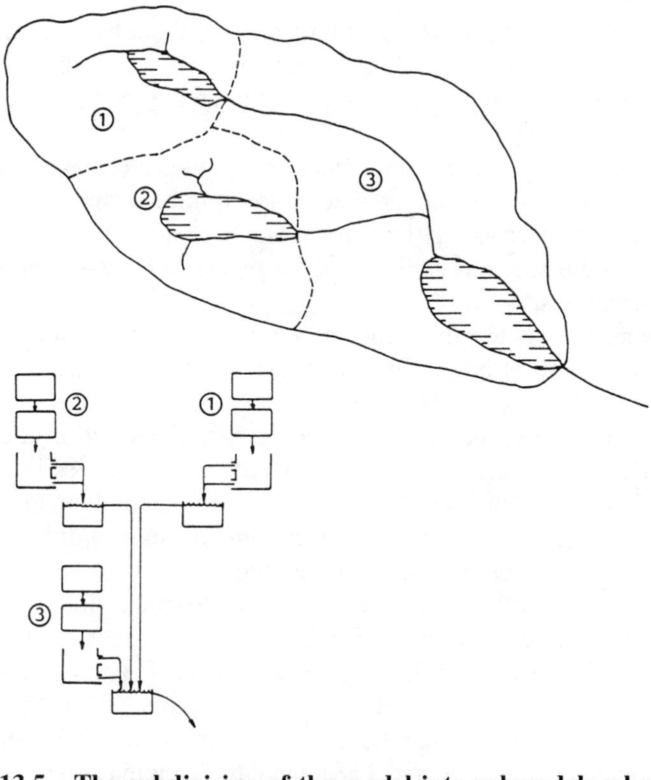

Fig. 13.5. The subdivision of the model into submodels when an explicit lake routing routine is used.

13.3. MODEL CALIBRATION

The HBV model, in its simplest form with only one subbasin and one type of vegetation, has altogether 12 free parameters. The calibration of the model is usually made by a manual trial and error technique, during which relevant parameter values are changed until an acceptable agreement with observations is obtained. The judgement of the performance is also supported by a statistical criterion, normally the R^2-value according to Nash and Sutcliffe (1970):

$$R^2 = \frac{\Sigma(\overline{Q}_o - Q_o)^2 - \Sigma(Q_c - Q_o)^2}{\Sigma(\overline{Q}_o - Q_o)^2} \qquad (13.3)$$

where: Q_o = observed runoff

$\overline{Q}_o$ = mean of observed runoff

Q_c = computed runoff

R^2 has a value of 1.0, if the simulation and the observations agree completely, and 0 if the model does not perform any better than the mean value of the runoff record. Negative values can be the result of poor model performance or poor data.

Another useful tool for the judgement of model performance is a graph of the accumulated difference between the simulated and the recorded runoff. This graph reveals any bias in the water balance and is particularly useful in the initial stage of calibration, for example for assessment of the snow-fall corrections.

Recently an automatic calibration procedure has been developed (Harlin, 1991). This method is based on the experience from a large number of manual calibrations and will successively replace these in future applications. It is, however, not likely that the visual inspection and manual interaction can be entirely replaced because of the difficulties in finding a statistical criterion which is as flexible and general as the human eye. The subjective manner in which the model is calibrated entails a difficulty when trying to generalize model parameters, as the results to some degree are depending on the individual who was responsible for the calibration.

It is not possible to specify the required length of records needed for a stable model calibration for all kinds of applications. The important thing is that the record includes a variety of hydrological events, so that the effect of all subroutines of the model can be discerned. Normally 5 to 10 years of records is sufficient when the model is applied to Scandinavian conditions.

Optimal values of the parameters from model calibrations in Sweden have been summarized by Bergström (1990). These values are specific for the SMHI model version and can be used as a guide for further applications.

13.4. PRESENTATION OF RESULTS

Due to the limitations of the statistical criteria it is of great importance to have a clear presentation of the model output and to be able to compare it visually with observations. It is also very useful to be able to present graphs of inputs, the main water balance components and the volume error, expressed as accumulated difference between the simulation and observations. The different versions of the HBV model system have different options for presentation of input and output. In Fig. 13.6 a standard presentation, which is fairly representative, is shown.

13.5. MAIN FIELDS OF APPLICATION OF THE HBV MODEL

Hydrological forecasting was the aim of the first operational applications of the HBV model. Since then the field of applications has widened and covers today real-time forecasting, control of data quality, extension of runoff records and filling in of gaps, design floods, synoptic water balance mapping, water balance studies, simulations of the effects of a changing climate and simulations of groundwater response. A successor to the HBV model, the PULSE model, is used for water quality studies and simulations in ungauged basins.

13.5.1. Real-Time Forecasting

In Scandinavia the need for flood warnings and reliable forecasts of the inflow to the reservoirs of the hydropower systems has stimulated the operational use of hydrological models. The forecasts are either short term, with a few days lead time, or long term, covering the whole snowmelt season (several months). For short term forecasting meteorological forecasts are often used as input to the models (see, for example, Björkenes, 1990). The long term forecasting is based on the present hydrological conditions, as described by the model, and a number of simulations with climatic data from the same season earlier years (Fig. 13.7). The output from the model is then subject to a statistical analysis which can be used to judge the risk of flooding or the probability of refilling of a reservoir.

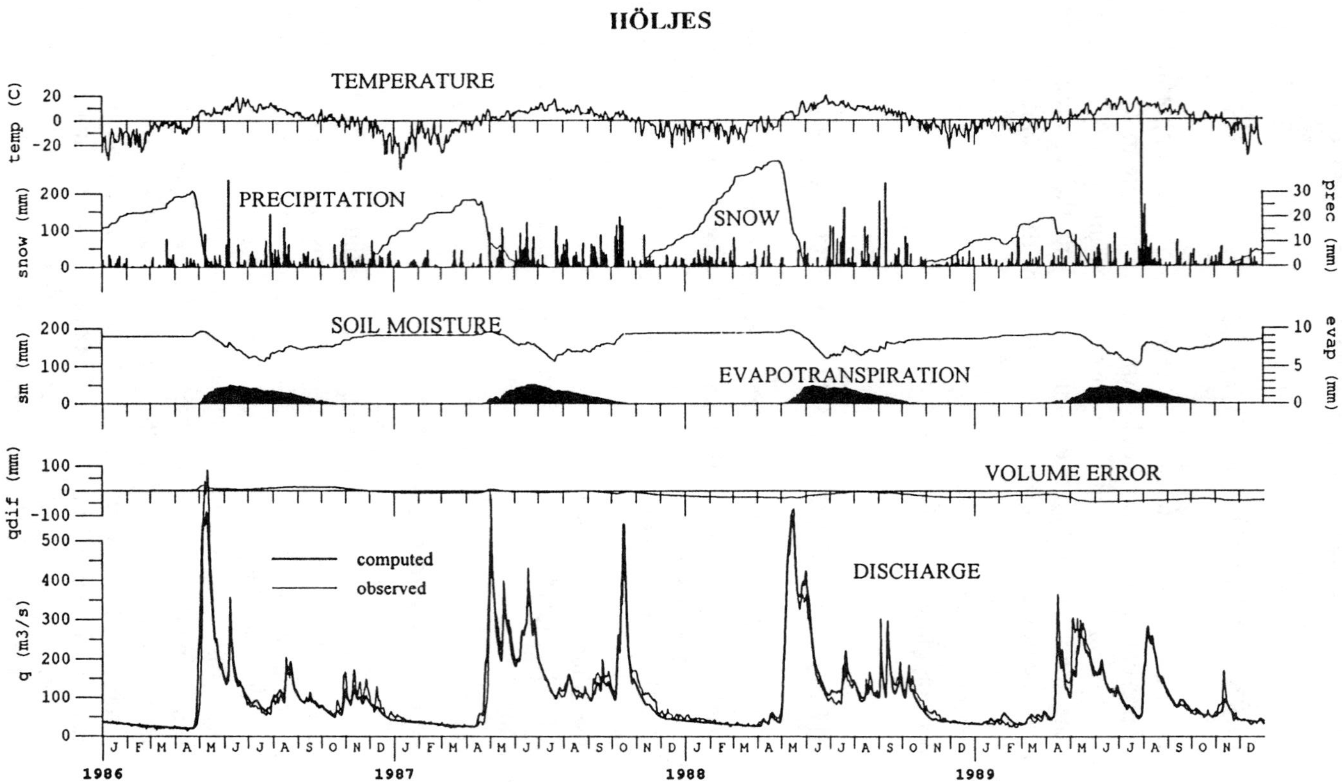

Fig. 13.6. Example of graphical presentation of output from a model simulation of inflow to the Höljes reservoir for a four year peiod.

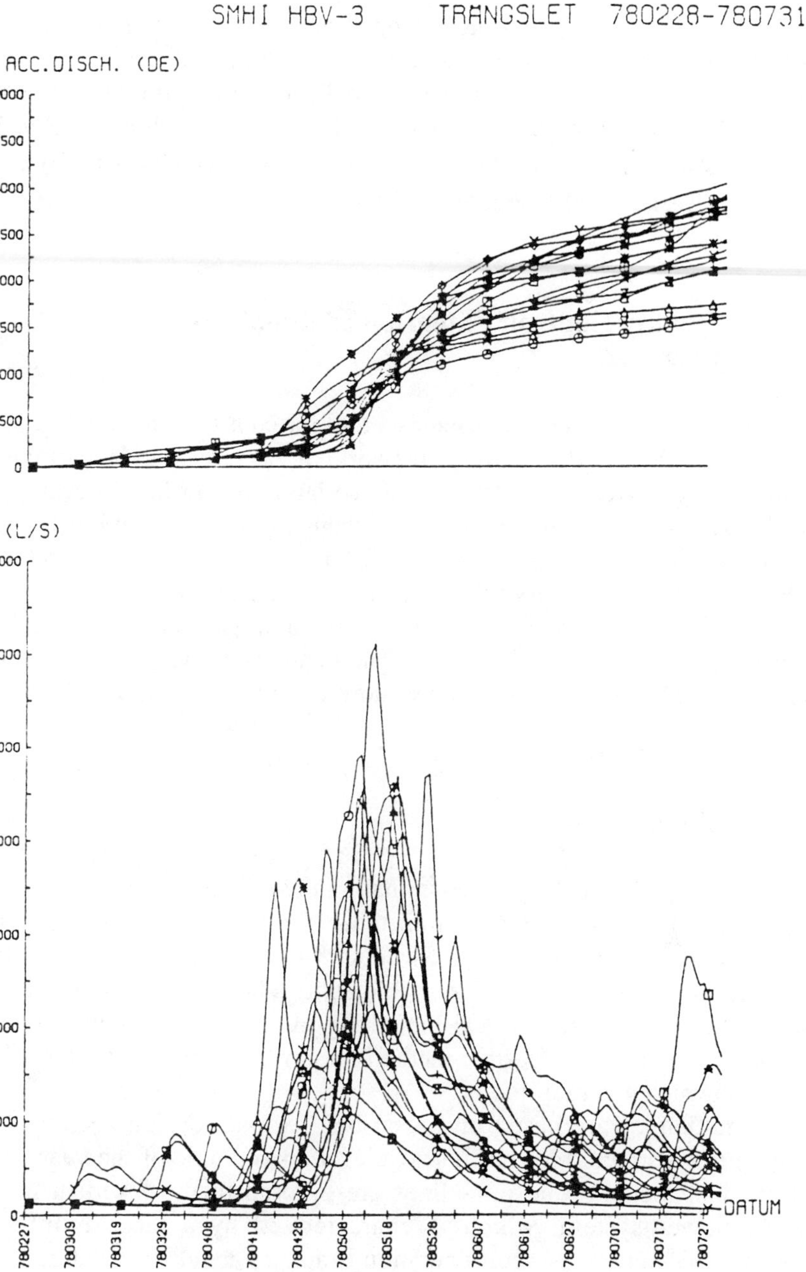

Fig. 13.7. Example of a long range hydrological forecast by the HBV model. The lower graphs show alternative discharge simulations and can be used to judge the risk for flooding. The upper graphs show the corresponding cumulative values and are useful for reservoir operation. (Trängslet basin, 4 483 km^2.)

Spring is the main forecasting season in Scandinavia, but models are also in operation during rainfloods in summer and autumn. The Swedish basins where the HBV model is available for real-time forecasting are shown in Fig. 13.8. These forecasts are produced either centralized by the Swedish Meteorological and Hydrological Institute or decentralized by river regulation enterprises or power companies.

13.5.2. Quality Control, Extension of Runoff Records and Filling in of Gaps

The HBV model is sometimes used as a tool for control of the quality of the runoff data of the Swedish national network. It has proved to be particularly useful for the correction of effects of ice-jamming on the records. The model output helps to decide whether a change in the observed water level has any relation to snowmelt or rain. There are also examples where the model has helped to identify inhomogeneities in the runoff records.

The extension of runoff records and filling in of gaps are straightforward applications of hydrological models. The methods are very useful in areas where the climatological records are more complete than the hydrological ones.

13.5.3. Design Floods

In 1990 new guidelines for the computation of spillway design floods were adopted in Sweden (Flödeskommittén, 1990; Norstedt et al., 1992; Bergström et al., 1992). The Swedish Committee on Design Flood Determination closely analysed possible approaches to the design problem and concluded that hydrological modelling in combination with reservoir simulation was the most feasible method for a multi-reservoir system with a mix of snowmelt and rain floods.

The guidelines are based on prescribed regional design sequences of precipitation, corrected for basin size, elevation and time of the year, and a hydrological model. The guidelines are tailored to the Swedish hydroelectric power system, which is characterized by a large number of reservoirs. Therefore reservoir operation strategies are also considered. For high hazard dams the most critical timing of all flood generating factors is found by an iterative simulation where the design precipitation is systematically inserted, at all possible locations, into a climatological record of ten years, and the response of the levels of the reservoirs is analysed.

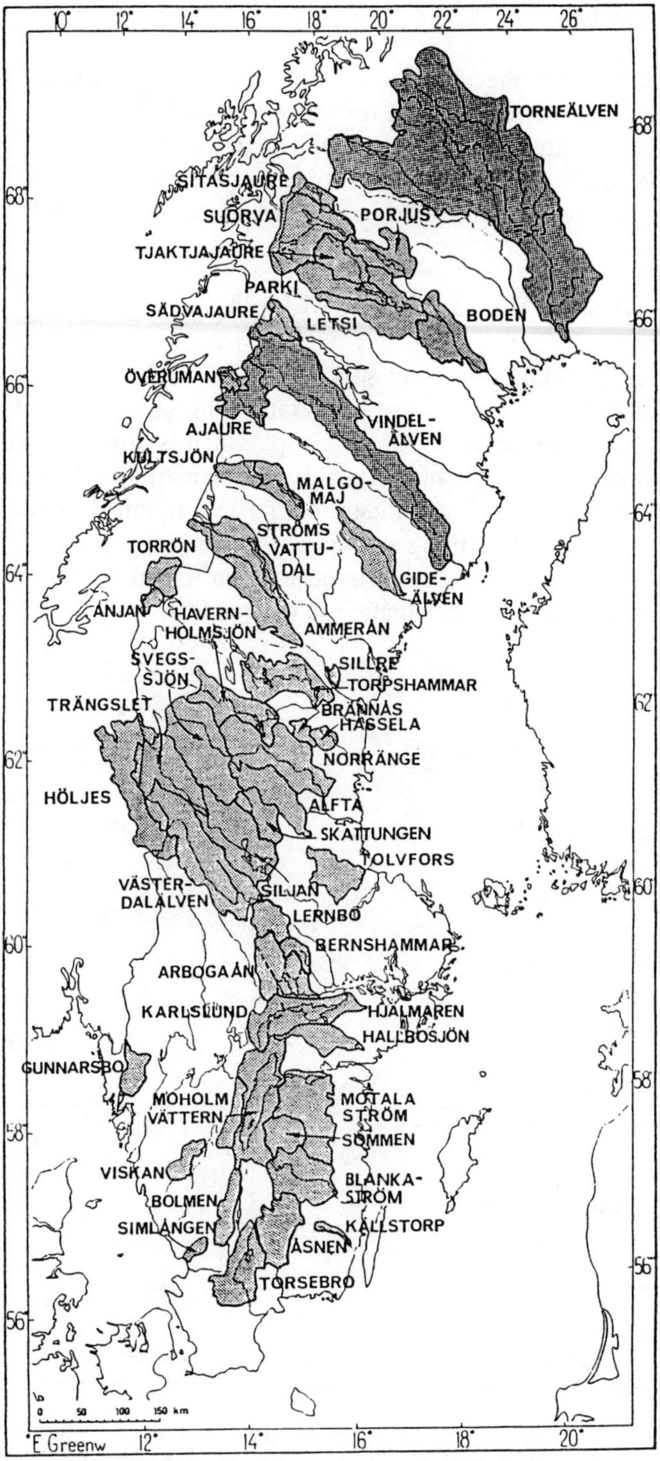

Fig. 13.8. Swedish basins where the HBV model is available for hydrological forecasting.

In order to meet the requirements of the new guidelines a computer system has been worked out by the Swedish Meteorological and Hydrological Institute, where the HBV model provides the hydrological input to the reservoirs (Fig. 13.9). In Norway a simplified version of the HBV model is in use for a similar purpose (Andersen et al., 1983).

13.5.4. Synoptic Water Balance Mapping

A number of research projects, or spin-off products, have been generated by the HBV model. One of the most important of these is the development of an operational synoptic water balance map, which is produced daily by the Swedish Meteorological and Hydrological Institute (Bergström and Sundqvist, 1982). This map illustrates the hydrological situation in the country by symbols describing snowpack, soil moisture deficit and runoff generation (Fig. 13.10). It has proved to be a useful tool to get a quick overview of the hydrological situation in the country.

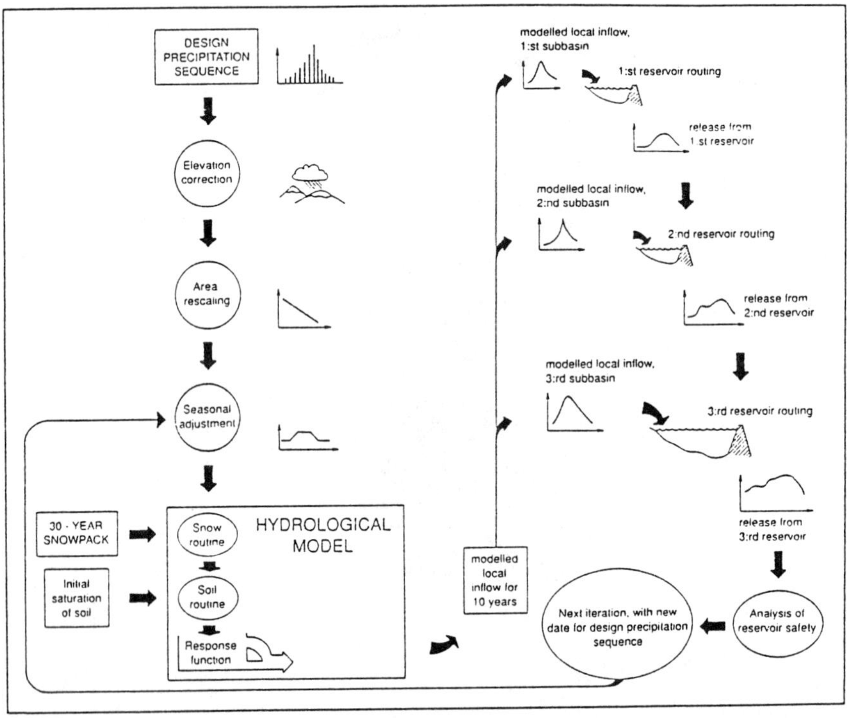

Fig. 13.9. Schematic presentation of the simulation of design floods by an iterative process according to the new Swediah guidelines.

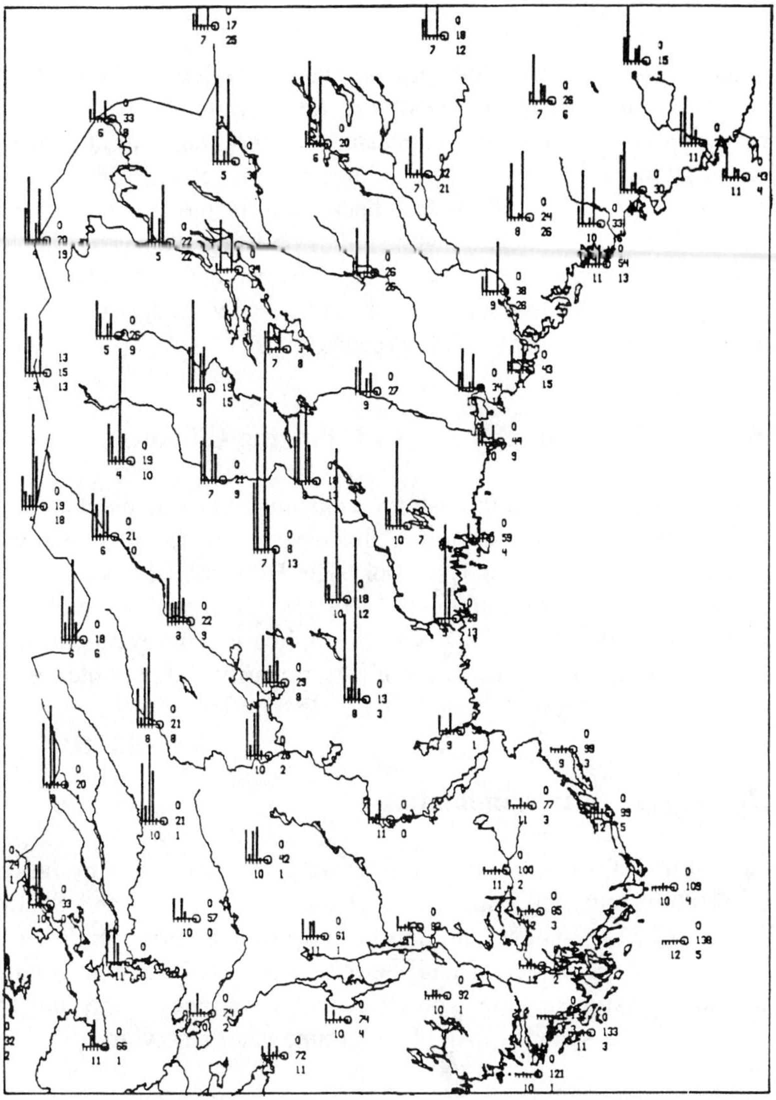

SP = *modelled snow water equivalent,*
SMD = *modelled soil moisture deficit,*
Q = *modelled runoff generation (5 days),*
P = *observed precipitation,*
T = *observed air temperature.*

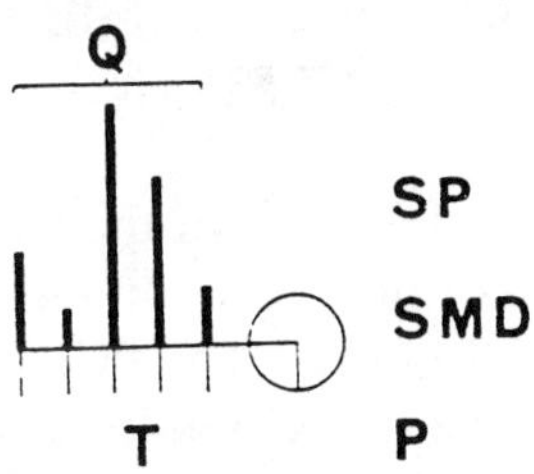

Fig. 13.10. Synoptic water balance map from a flood in September, 1985 in Central Sweden.

13.5.5. Water Balance Studies

In connection with the work on spillway design guidelines the HBV model was used in a nationwide study of the interaction and timing of flood generating processes. This study revealed the importance of critical timing and had great impact on the final guidelines (Brandt, et. al., 1987).

The modelling technique has also been used for the computation of soil moisture statistics in connection with studies on forest damage (Grahn et al., 1985) and for analyses of the effects of forest management on runoff (Brandt et. al., 1988). Jutman (1992) used the HBV model concept for the production of a new runoff map of Sweden.

13.5.6. Studies of the Effects of a Changing Climate

The growing concern about the risk for a changing climate and its effects on our water resources has started a discussion on the possible use of hydrological models as an analysis tool. The HBV model has been used in this respect in Norway (Saelthun et al., 1990) and in Finland (Vehviläinen and Lohvansuu, 1991). In this type of application the problem of local interpretation of climate scenarios, model stationarity and possible feed-back mechanisms by vegetation have to be carefully considered.

13.5.7. Groundwater Simulations

Another spin-off of the HBV model is a model for simulation of groundwater response to climatological input (Bergström and Sandberg, 1983, Fig. 13.11). These simulations required a modification of the saturated zone of the model. Thanks to the success in the attempts to model groundwater response, the synoptic water balance map has been complemented by a synoptic map of the groundwater reservoirs.

13.5.8. Water Quality and Simulations in Ungauged Basins, the PULSE Model

The modification of the HBV model used for groundwater simulations was the starting point for the development of a new model generation, the PULSE model. The PULSE model has most of its subroutines in common with the HBV model but has a better representation of the shallow groundwater and is thus more feasible for hydrochemical simulations. It is

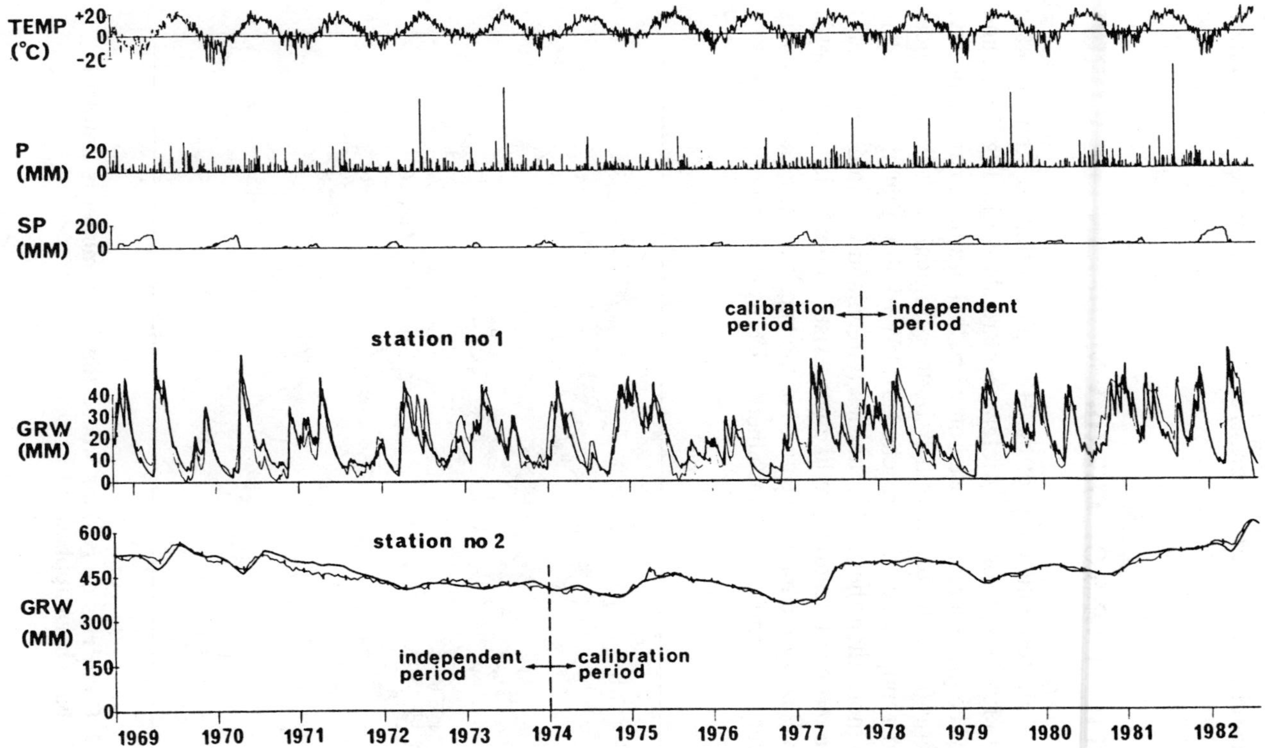

Fig. 13.11. Simulation of groundwater response in a shallow aquifer in till soil (Station No. 1) and a deep esker aquifer (Station No. 2) in central Sweden. (From Bergström and Sandberg, 1983) The graphs show the dynamics of the storage in the aquifers.

used for studies of short term variability in stream flow acidity (Bergström et al., 1985), modelling of transport of non-point source pollution (Brandt, 1990) and has also become widely used for the simulation of runoff in ungauged catchments after generalization of the model parameters (Johansson, 1986).

13.6. APPLICATIONS OF THE HBV MODEL IN THE WORLD

A summary of those applications of the HBV model that are known to the author today (1993) and some key references are shown in Fig. 13.12 and Table 13.1. The presentation is based on work that has been carried out by, or in cooperation with, the Swedish Meteorological and Hydrological Institute and reports that have reached us. We know that the model is available in more than these 31 countries (Greenland counted separately) and therefore more applications are likely to exist. In the table an attempt is also made to classify the main objectives of the applications although this is not always very straightforward since there exist overlapping objectives.

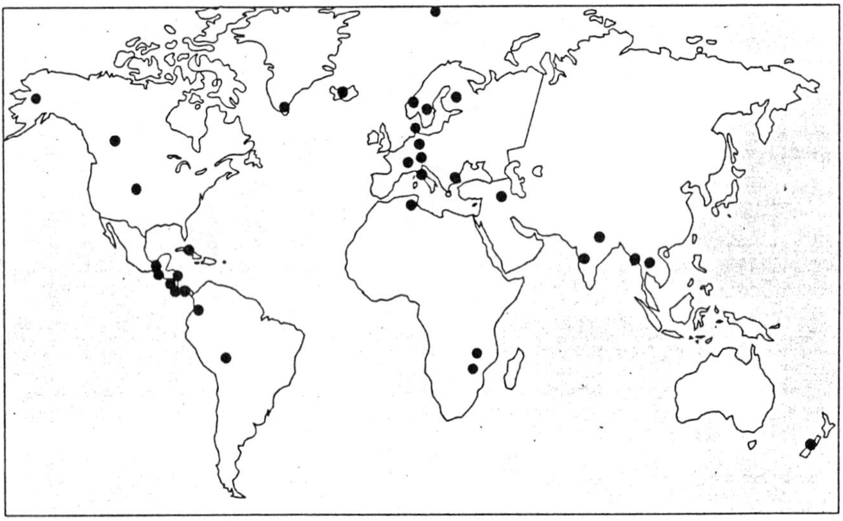

Fig. 13.12. Countries or regions where the HBV model is known to have been applied until 1993.

TABLE 13.1.

Known applications of the HBV model in the world.

	EUROPE
Sweden:	Some 100 basins for hydrological forecasting and computation of design floods (Bergström, 1976; Häggström, 1989; Bergström, 1990). Some 50 applications of the PULSE model for environmental studies and production of runoff records (Brandt, 1990).
Norway:	Some 50 basins for hydrological forecasting and studies of the effects of a changing climate. Norwegian version of the model (Killingtveit and Aam, 1978; Saelthun and Taksdal, 1988; Saelthun, 1990; Saelthun et al., 1990; Killingtveit et al., 1990). Scientific test of Norwegian version in Svalbard (Bruland, 1991).
Finland:	Some 20 basins for hydrological forecasting and studies of the effects of a changing climate, Finnish version (Vehviläinen, 1986 and 1992; Vehviläinen and Lohvansuu, 1991).
Denmark:	Scientific test of early version (Houmøller, 1976).
Iceland:	Scientific test of SMHI version (Bergström et al., 1982).
France:	WMO intercomparison of models of snowmelt runoff (WMO, 1986). WMO real-time intercomparison of hydrological models (WMO, 1987).
Italy:	Scientific test of SMHI version (Capovilla, 1990).
Poland:	WMO intercomparison of models of snowmelt runoff (WMO, 1986).
Bulgaria:	Scientific test of Norwegian version (Stanev and Sælthun, 1993).
Switzerland	WMO intercomparison of models of snowmelt runoff (WMO, 1986, SMHI version). Scientific test of the model in four basins (Renner and Braun, 1990; Braun and Aellen, 1990, Swiss version). Hydrological forecasting in the Rhine (Jensen and Braun, 1990).
	NORTH AMERICA
Canada:	WMO intercomparison of models of snowmelt runoff (WMO, 1986). WMO real-time intercomparison of hydrological models (WMO, 1987).
USA:	WMO intercomparison of models of snowmelt runoff (WMO, 1986). WMO real-time intercomparison of hydrological models (WMO, 1987). Scientific test of SMHI version (Hinzman and Kane, 1991).
Greenland:	Scientific test of Norwegian version (Bruland, 1991).

Table 13.1. Continued

Table 13.1. Continued

	LATIN AMERICA
Bolivia:	Hydrological forecasting, SMHI version (Johansson et al., 1987).
Colombia:	Hydrological forecasting, SMHI version (Häggström et al., 1988).
Costa Rica:	Hydrological forecasting, SMHI version (Johansson et al., 1985; Häggström et al., 1990).
	LATIN AMERICA
Cuba:	Hydrological forecasting, SMHI version (Rodriguez et al., 1991).
El Salvador:	Hydrological forecasting, SMHI version (Häggström et al., 1990).
Guatemala:	Hydrological forecasting, SMHI version (Häggström et al., 1990).
Honduras:	Hydrological forecasting, SMHI version (Häggström et al., 1990).
Nicaragua:	Hydrological forecasting, SMHI version (Häggström et al., 1990).
Panama:	Hydrological forecasting, SMHI version (Häggström et al., 1990; Espinosa, 1991).
	ASIA
Burma:	Scientific test of SMHI version (Gyaw and Persson, 1985).
India:	Scientific test of SMHI version (Bhatia et al., 1984).
Iraq:	Hydrological forecasting, SMHI version (unofficial report).
Nepal:	Scientific test of the Swiss version (Braun and Demierre, 1991). Water supply study, Norwegian version (Løvoll and Bellingmo, 1991).
Thailand:	Hydrological forecasting, SMHI version (Harlin, 1990).
	AFRICA
Tunisia:	Scientific test, local version (Berndtsson et al., 1985).
Malawi:	Hydrological forecasting, Norwegian version (Killingtveit and Brate, 1992).
Zimbabwe:	Siltation studies, Norwegian version (Bøe Olsen, N.R., 1987).
	OCEANIA
New Zealand:	Scientific test of local model version (Moore and Owens, 1984).

The applications in Table 13.1 cover a large number of basins in different climatological and geographical regions, and the basins are ranging in size from less then one up to 40 000 km^2. Thanks to the subbasin option there is no theoretical upper limit in size. In Fig. 13.13 is shown a sample of model simulations under the very different geographical and climatological conditions in Sweden, Poland, Iceland, India, Canada, Costa Rica and Bolivia for the same hydrological year of 1974-75. The graphs are taken from different applications and are therefore presented by different graphical systems. In all these plottings the computed runoff is represented by a thicker line than the observed runoff.

From Fig. 13.13 it is clear that the HBV model is very general and gives reasonable results under a wide span of geographical and climatological conditions. The differences are accounted for either by the set up of the model structure or its calibration. The parameter values of the model vary a lot between basins and regions, and it is not easy to relate these to physical

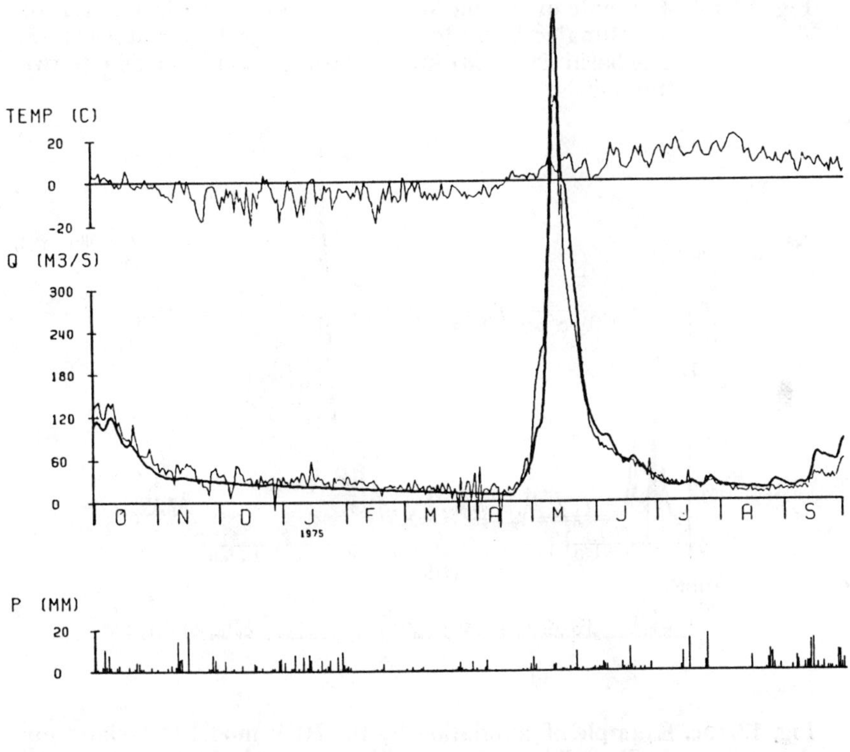

Fig. 13.13a. Example of simulation by the HBV model in Sweden for the Trängslet basin in the river Dalälven for the hydrological year of 1974-75. The basin size is 4 483 km^2 and the model is divided into two submodels.

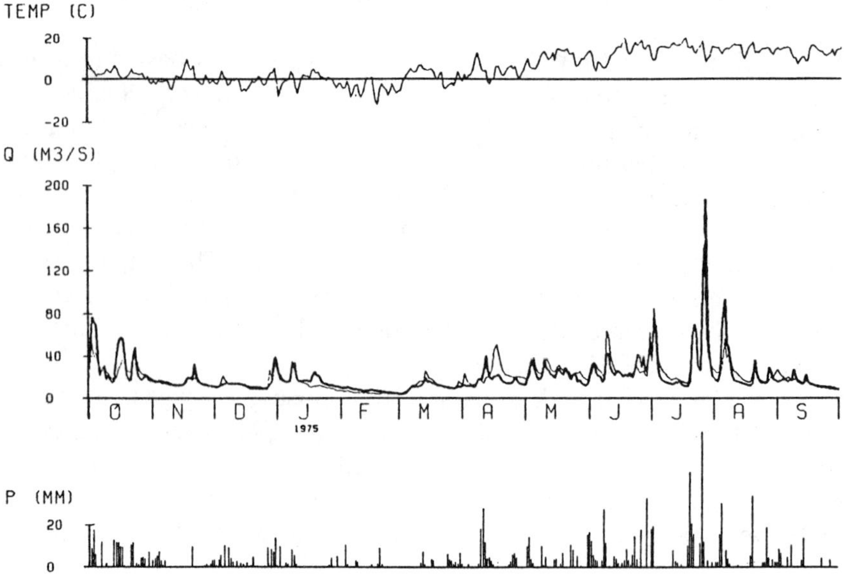

Fig. 13.13b. Example of simulation by the HBV model in Poland for the Dunajec basin for the hydrological year of 1974-75. The basin size is 681 km^2 and the model is divided into two submodels.

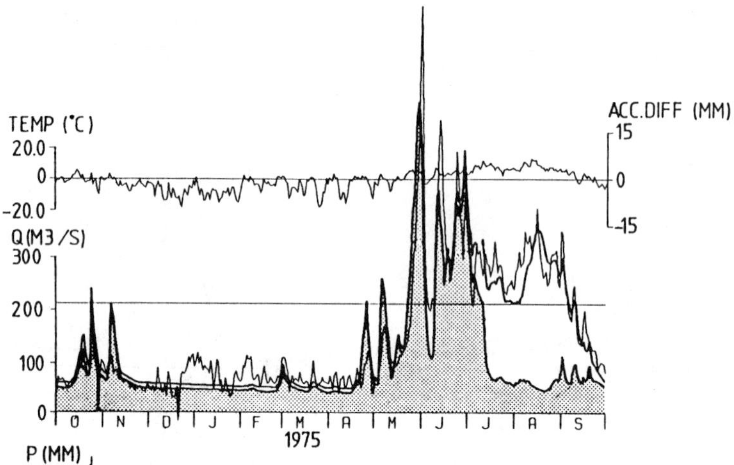

Fig. 13.13c. Example of simulation by the HBV model in Iceland for the Burfell basin in river Thjorsa for the hydrological year of 1974-75. The basin size is 2 890 km^2 and the model is divided into three submodels. One of these submodels represent glaciers. The shaded area represents simulated runoff from the part of the basin that is not covered by glaciers. (From Bergström et al., 1982.)

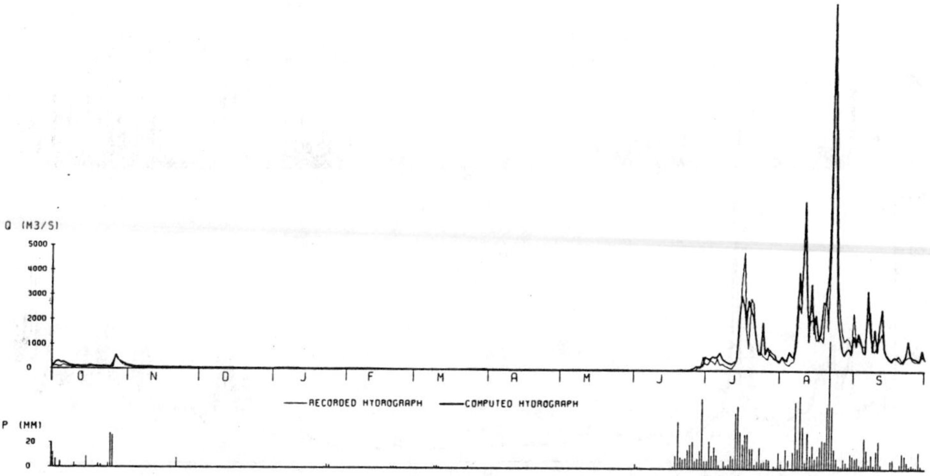

Fig. 13.13d. Example of simulation by the HBV model in India for the Upper Narmada in Jabalpur for the hydrological year of 1974-75. The basin size is 16 576 km^2 and the model is divided into five submodels. (From Bhatia et al., 1984.)

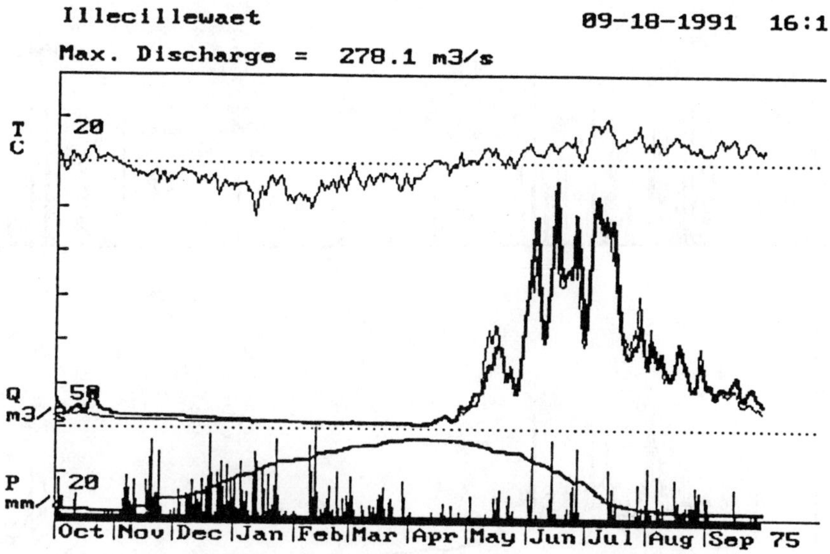

Fig. 13.13e. Example of simulation by the HBV model in Canada for the Illecillewaet basin in the Rocky Mountains for the hydrological year of 1974-75. The basin size is 1 135 km^2 and the model is divided into two submodels.

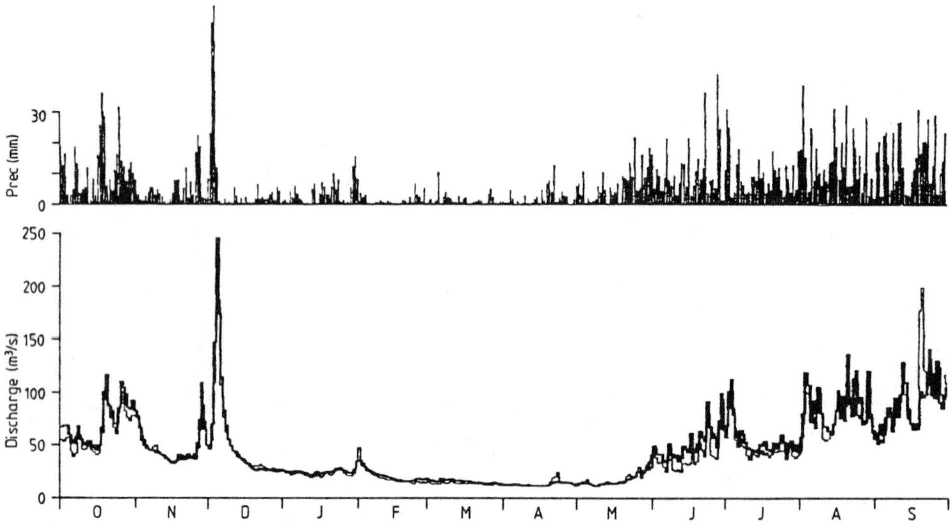

Fig. 13.13f. Example of simulation by the HBV model in Costa Rica for the Cachi basin for the hydrological year of 1974-75. The basin size is 785 km^2 and the model is divided into nine submodels. (Published with permission from Instituto Costarricense de Electricidad.)

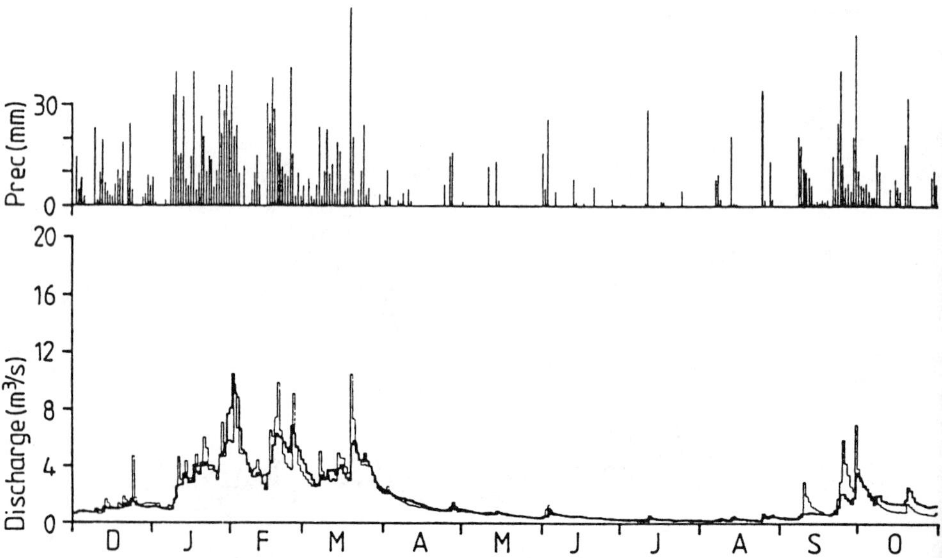

Fig. 13.13g. Example of simulation by the HBV model in Bolivia for the Malaga 3200 basin in Rio Malaga for the hydrological year of 1974-75. The basin size is 39 km^2 and the model is run without subdivision. (From Johansson et al., 1987.)

properties except for in a few special cases. The snow melt routine is empirical, and this is reflected in differences in the melt factor between regions. Normal values in Sweden are 2 - 3 mm/°C • day, while the corresponding values for Iceland are around 6 mm/°C • day. The difference is explained by the windy and moist climate in Iceland and thus the effective energy exchange at the snow surface.

The soil moisture routine has proved to be very efficient and flexible with its three free parameters. It is capable of describing soil moisture recovery after very long dry periods, such as in India, and it works well in the humid tropical climate of Central America as well as in the cold climate of Northern Scandinavia. A clear relationship between climate soil conditions and the maximum soil moisture storage can (FC) be found. FC has a value of 25 mm in Iceland where a soil cover is practically lacking, between 100 and 300 mm in Sweden and around 350 mm in the application in India.

13.7. CONCLUDING REMARKS

Thanks to the relative simplicity of the structure of the HBV model the demand for computer power is low and can be met by most modern desk top computers. The SMHI version of the model is normally operated in a PC-environment, but there are also main frame versions available. The code contains subprograms for input data control, model calibration, hydrological forecasting and computation of design floods in systems of reservoirs.

Although the HBV model has an uncomplicated structure and user-friendly computer systems are available, it is not recommended to apply it without a thorough insight into its structure. All models of this kind contain complex interactions and feed-back mechanisms, which require a dynamic way of thinking to be fully understood. It is also important to have a sound philosophy when choosing subbasins and climate stations and when correcting input data according to undercatch, elevation and other systematic errors. The introduction of an automatic calibration procedure replaces a lot of manual work but should not be used without a full understanding of the model. A final visual check of the calibration results must be made to avoid unpleasant surprises in future applications.

The large number of applications of the HBV model in the world has proved that this relatively simple conceptual model is surprisingly general. The key processes seem to be described in a reasonable way for the conditions under which they are applied. The model has also the advantage of limited demand on data coverage and computer facilities which makes it very versatile.

The low complexity of the model has made it easy to teach others how to use it and to install it in a number of countries. There are also several examples where scientists have been able to reproduce the model code just from a description in the literature.

The existence of different versions of the HBV model and different groups who apply it makes generalization of parameter values difficult. The parameter values are depending on the type of input data used, computational details in the model and, to some degree, on the calibration skill of the user.

The field of applications of the HBV model is growing and shows that the conceptual philosophy that it represents is very useful for the solution of a large number of problems related to water resources.

ACKNOWLEDGEMENTS

The author is grateful to all friends and colleagues, in Sweden and abroad, who have participated in the long process of model development and application. Thanks are also due to all those who have funded our research, development and applications during this period, in particular the Swedish Association of River Regulation Enterprises (VASO), a number of hydro-electric power companies and river regulation enterprises, the Swedish International Development Agency (SIDA), the Swedish Agency for Technical and Economic Co-operation (BITS), the Swedish Natural Science Research Council (NFR), the Swedish Environmental Protection Board (SNV) and the Swedish Meteorological and Hydrological Institute itself. Special thanks to Jorge Granados from Instituto Costarricense de Electricidad for model results from Costa Rica for 1974-75, to the staff of SMHI for comments on the manuscript and to Vera Kuylenstierna for word processing and editing.

REFERENCES

Andersen, L., Hjukse, T., Roald, L. and Saelthun, N.R. (1983). Hydrologisk modell for flomberegninger. (A hydrological model for simulation of floods, in Norwegian). Rapport nr 2-83, NVE, Vassdragsdirektoratet, Oslo.

Bergström, S. (1972). Utveckling och tillämpning av en digital avrinningsmodell. (Development and application of a digital runoff model, in Swedish). SMHI, Notiser och preliminära rapporter, serie HYDROLOGI, No. 22, Norrköping.

Bergström, S. (1975). The development of a snow routine for the HBV-2 model. Nordic Hydrology, 6, 73 - 92.

Bergström, S. (1976). Development and application of a conceptual runoff model for Scandinavian catchments. SMHI, Reports RHO, No. 7, Norrköping.

Bergström, S. (1990). Parametervärden för HBV-modellen i Sverige. (Parameter values for the HBV model in Sweden, in Swedish). SMHI, Hydrologi, No. 28, Norrköping.

Bergström, S. (1991). Principles and confidence in hydrological modelling. Nordic Hydrology, 22, 123 - 136.

Bergström, S. (1992). The HBV model - its structure and applications. SMHI, Reports RH, No. 4, Norrköping.

Bergström, S., Brandt, M. and Carlsson, B. (1985). Hydrologisk och hydrokemisk modellberäkning i sjörika skogsområden. (Hydrological and hydrochemical modelling in forested lake districts, in Swedish). Vatten 41:3.

Bergström, S., Carlsson, B., Sandberg, G. and Maxe, L. (1985). Integrated modelling of runoff, alkalinity and pH on a daily basis. Nordic Hydrology, 16, 89 - 104.

Bergström, S., Einarson, K. and Persson, M. (1982). Modellering av glaciäravrinning på Island. (Modelling glacier runoff in Iceland, in Swedish). Proceedings, Nordic Hydrological Conference, Förde, Norway.

Bergström, S. and Forsman, A. (1973). Development of a conceptual deterministic rainfall-runoff model. Nordic Hydrology, 4, 147 - 170.

Bergström, S., Harlin, J., and Lindström, G. (1992). Spillway design floods in Sweden. I. New guidelines. Hydrological Sciences Journal, 37, 5, 10/1992, 505 - 519.

Bergström, S. and Sandberg, G. (1983). Simulation of groundwater response by conceptual models.- Three case studies. Nordic Hydrology, 14. 71 - 84.

Bergström, S. and Sundqvist, B. (1982). Synoptic water balance mapping in Sweden. Proceedings from IAHS Workshop on new approaches in water balance computations, Hamburg, IAHS Publ. No. 148.

Berndtsson, R., Dahlblom, P., Hogland, W., Larson, M., Nguyen, V.L., and Niemczynowicz, J. (1985). Hydrological studies in Tunisia. Dept. of Water Resources Engineering, Lund University, Report No. 3107, Lund.

Bhatia, P.K., Bergström, S. and Persson, M. (1984). Application of the distributed HBV-6 model to the upper Narmada basin in India. SMHI, Reports RHO, No. 35, Norrköping.

Björkenes, A. (1990). Operational use of precipitation forecasts in the Arendal catchment area. Proceedings, Nordic Hydrological Conference in Kalmar 1990. NHP-Report No. 26, Norrköping.

Brandt, M. (1990). Human impacts and weather-dependent effects on water balance and water quality in some Swedish river basins. SMHI, Reports RH, No. 2, Norrköping.

Brandt, M., Bergström, S., Gardelin, M. (1988). Modelling the effects of clearcutting on runoff - Examples from Central Sweden. Ambio, 17, 5: 307 - 313.

Brandt, M., Bergström, S., Gardelin, M. and Lindström, G. (1987). Modellberäkning av extrem effektiv nederbörd. (Modelling extreme effective precipitation, in Swedish). SMHI, Hydrologi, nr 14, Norrköping.

Braun, L.N. and Aellen, M. (1990). Modelling discharge of glacierized basins assisted by direct measurements of glacier mass balance. Hydrology in Mountainous Regions. I. Hydrological Measurements, the Water Cycle. Proceedings of two Lausanne Symposia, August 1990. IAHS Publ. No. 193.

Braun, L.N. and Demierre (1991). Erste Anwendung des konzeptionellen Niederschlag-Abflussmodells HBV3-ETH im Langtang-Gebiet, Nepal. (First use of the rainfall-runoff model HBV3-ETH in Langtang basin, Nepal, in German). 1. Zwischenbericht Projekt "Aufbau eines Messdienstes für Schnee- und Gletscherhydrologie". Geogr. Inst. ETH, Abt. Hydrologie, Zürich.

Bruland, O. (1991). Vassbalanse og anlaupsmodeller i permafrostområder. (Water balance and runoff models in permafrost regions.) Division of Hydraulic and Sanitary Engineering, University of Trondheim/ The Norwegian Institute of Technology, Trondheim.

Bøe Olsen, N.R. (1987). Siltation and monitoring programme for dams in Zimbabwe. M. Sc. Thesis at the Norwegian Institute of Technology, Division of Hydraulic and Sanitary Engineering, Trondheim.

Capovilla, A. (1990). Applicazione sperimentale del modelo idrologico HBV al bacino del Boite. (Experimental application of the HBV hydrological model to the Boite basin, in Italian). Tesi di laurea, Universitá degli studi di Padova, facoltá di agrari, dipartimento territorio e sistemi agro-forestali, Padova.

Espinosa, D. (1991). Aplicacion del modelo hidrologico HBV a la cuenca del Rio Chiriqui Viejo. (Application of the HBV hydrological model to River Chiriqui Viejo, in Spanish). Instituto de recursos hidrologicos y electrificacion, departemento de hidrometeorologia, seccion de hidrologia, Panama.

Flödeskommittén (1990). Slutrapport från Flödeskommittén. (Final report from the Swedish committee on spillway design, in Swedish). Swedish State Power Board, Swedish Power Association, Swedish Meteorological and Hydrological Institute, Norrköping.

Grahn, G., Johansson, B. and Norlander, B. (1985). Beräknat markvattenunderskott i Simlångens avrinningsområde 1934-83. (Computed soil moisture deficit in the Simlången basin 1934-83, in Swedish). SMHI, HO 29, Norrköping.

Gyaw, O. and Persson, M. (1985). Application of the HBV model to pilot basins in Burma. SMHI, Reports HO, No. 31, Norrköping.

Harlin, J. (1990). Khao Laem dam and reservoir operation study, Part B: Hydrological forecasting. In: Final Report, Phase 1. SwedPower and Electricity Generating Authority of Thailand (EGAT).

Harlin, J. (1991). Development of a process oriented calibration Scheme for the HBV hydrological model. Nordic Hydrology, 22, 13 - 36.

Harlin, J. (1992). Hydrological modelling of extreme floods in Sweden. SMHI, Reports RH, No. 3, Norrköping.

Hinzman, L.D. and Kane, D.L. (1991). Snow hydrology of a Headwater arctic basin. 2. Conceptual analysis and computer modeling. Water Resources Research, Vol. 27, No. 6, 1111 - 1121.

Houmøller, O. (1976). En hydrologisk-geologisk undersøgelse af Giber å-området. Grundvandsinvindningens inflydelse på det hydrologiske kredsløb. (A hydrogeological investigation of the Giber river basin. The effect of groundwater use on the hydrological cycle. In Danish.) Aarhus University, Laboratory for Applied Geophysics, Aarhus, Denmark.

Häggström, M. (1989). Anpassning av HBV modellen till Torneälven. (Application of the HBV model to River Torneå, in Swedish). SMHI, Hydrologi, No. 26, Norrköping.

Häggström, M., Lindström, G., Cobos, C., Martínez, J.R., Merlos, L., Alonzo, R.D., Castillo, G., Sirias, C., Miranda, D., Granados, J., Alfaro, R., Robles, E., Rodríguez, M. and Moscote, R. (1990). Application of the HBV model for flood forecasting in six Central American rivers. SMHI, Hydrology, No. 27, Norrköping.

Häggström, M., Lindström, G., Sandoval, L.A., and Vega M.A. (1987). Application of the HBV model to the upper Rio Cauca. SMHI, Hydrology, No. 21, Norrköping.

Jensen, H. and Braun, L. (1990). Investigations on short-range forecasts for Rhine at Rheinfelden. ETH Zürich, Geographisches Institut, Abteilung Hydrologie. Report No. 100.34, Zürich.

Johansson, B. (1986). Vattenföringsberäkningar i Södermanlands län - ett försöksprojekt. (Discharge simulations in the county of Södermanland - a pilot study, in Swedish). SMHI, Hydrologi, No. 6, Norrköping.

Johansson, B., Persson, M., Aranibar, E., and Llobet, R. (1987). Application of the HBV model to Bolivian basins. SMHI, Hydrology, No. 10, Norrköping.

Johansson, B., Persson, M., Sandberg, G., and Robles, E. (1985). Aplicación del modelo HBV a la Cuenca del Lago de Arenal en Costa Rica. SMHI, HO 28, Norrköping.

Jutman, T. (1992). Production of a new runoff map of Sweden. Contribution to The Nordic Hydrological Conference in Alta, Norway.

Killingtveit, Å. and Aam, S. (1978). En fordelt modell for snöackumulering og -avsmältning. (A distributed model for snow accumulation and melt, in Norwegian). EFI - Institutt for Vassbygging, NTH, Trondheim, Norway.

Killingtveit, Å., Saelthun, N.R., Saether, B., Taksdal, S., and von Hirsch, R. (1990). Programmet till HBV-modellen (The program for the HBV model, in Norwegian). SINTEF, Norwegian Hydrotechnical Laboratory, Rapport STF 60 A90033, Trondheim.

Killingtveit, Å., and Brate, T.I. (1992). Flood forecasting and warning system for Lower Shire Valley in Malawi - Technical Documentation (3 reports). The Norwegian Institute of Technology, Division of Hydraulic and Sanitary Engineering, Trondheim.

Lindström, G., and Bergström, S. (1992). Improving the HBV and PULSE-models by use of temperature anomalies. Vannet i Norden, No. 1, 1992.

Løvoll, O., and Bellingmo, H. (1991). Water supply in Kathmandu-Lalitpur, Nepal. M. Sc. Thesis at the Norwegian Institute of Technology, Division of Hydraulic and Sanitary Engineering, Trondheim.

Moore, R.D., and Owens, I.F. (1984). A conceptual runoff model for a mountainous rain-on-snow environment, Craigieburn range, New Zealand. Journal of Hydrology (N.Z.), Vol. 23, No. 2.

Nash, J.E., and Sutcliffe, J.V. (1970). River flow forecasting through conceptual models. Part I. A discussion of principles. Journal of Hydrology, 10, 282 - 290.

Nemec, J. (1986). Hydrological forecasting. Design and operation of hydrological forecasting systems. Water science and technology library. D. Riedel Publishing Company, Dordrecht, Holland.

Norstedt, U., Brandesten, C.-O., Bergström, S., Harlin, J., and Lindström, G. (1992). Re-evaluation of hydrological dam safety in Sweden. International Water Power and Dam Construction, June 1992.

Renner, C.B., and Braun, L. (1990). Die Anwendung des Niederschlag-Abfluss Modells HBV3-ETH (V 3.0) auf verschiedene Einzugsgebiete in der Schweiz. (The application of the HBV3-ETH (V 3.0) rainfall-runoff model to different basins in Switzerland, in German). Geographisches Institut ETH, Berichte und Skripten Nr 40, Zürich.

Rodriguez, M., Johansson, B., Lindström, G., Planos, E., and Remón, A. (1991). Aplicacion del modelo HBV a la cuenca del Rio Cauto en Cuba. (Application of the HBV model to River Cauto in Cuba, in Spanish). SMHI, Hydrology, No. 31, Norrköping.

Saelthun, N.R. (1990). Forbedring av de hydrologiske rutinene i HBV-modellen. (Improvement of the hydrological routines of the HBV model, in Norwegian). Contribution to the Nordic Hydrological Conference in Kalmar 1990.

Saelthun, N.R., Bogen, J., Hartmann Flood, M., Laumann, T., Roald, L.A., Tvede, A.M., and Wold, B. (1990). Klimaendringer og vannressurser. (Climate change and water resources, in Norwegian). Bidrag till den interdepartementale klimautredningen. NVE Publikasjon nr V 30, Oslo.

Saelthun, N.R., and Taksdal, S. (1988). KARMEN, System for hydrological simulation and forecasting. VHD notat 24/88, Norwegian Water and Energy Administration, Oslo, Norway.

Stanev, K., and Sælthun, N.R. (1993). Snow melt and snow melt modelling. Literature review and model applications to Bulgarian catchments. Institute Report Series, No. 84, University of Oslo, Inst. for Geophysics, Oslo, Norway.

Vehviläinen, B. (1986). Modelling and forecasting snowmelt floods for operational forecasting in Finland. Proceedings from the IAHS symposium: Modelling Snowmelt-Induced Processes, Budapest, IAHS Publ. No. 155.

Vehviläinen, B. (1992). Snow cover models in operational watershed forecasting. The Water and Environment Research Institute, National Board of Waters and the Environment, Finland. Publ. No. 11, Helsinki.

Vehviläinen, B., and Lohvansuu, J. (1991). The effects of climate change on discharges and snow cover in Finland. Hydrological Sciences Journal, 36, 2, 4.

WMO (1986). Intercomparison of models of snowmelt runoff. Operational Hydrology Report No. 23, WMO-No. 646, WMO, Geneva.

WMO (1987). Real-time intercomparison of hydrological models. Report of the Vancouver Workshop, 1987. Technical Report to CHy No. 23, WMO/TD No. 255, WMO, Geneva.

Chapter 14

THE SNOWMELT RUNOFF MODEL (SRM)

A. Rango

The Snowmelt Runoff Model (SRM) is a simple degree-day model that requires remote sensing input in the form of basin or zonal snow cover extent. The model has been tested successfully on over 60 basins worldwide in the simulation and forecast modes. Model variables are derived from actual observations of temperature, precipitation, and snow covered area. Model parameters can either be derived from measurements or estimated by hydrological judgement taking into account the basin characteristics, physical laws, and theoretical or empirical relationships. To facilitate use of SRM, a microcomputer version of the program has been developed for IBM compatible personal computers. The program itself features user-oriented input and multiple self-help screens which allow the user to select the kind of data input employed and the output products desired.

14.1. INTRODUCTION

The Snowmelt-Runoff Model (SRM) is designed to simulate and forecast daily streamflow in mountain basins where snowmelt is a major runoff factor. SRM was developed by Martinec (1975) in small European basins. Thanks to the progress of satellite remote sensing of snow cover, SRM has been applied to larger and larger basins. The largest basin where SRM has been applied so far is about 122000 km^2. Runoff computations by SRM appear to be relatively easily understood. To date the model has been applied by various agencies, institutes and universities in about 60 basins situated in 19 different countries as listed in Table 14.1. About 30% of these applications have been performed by the model developers and 70% by independent users. Some of the localities are shown in Fig. 14.1. SRM also successfully underwent tests by the World Meteorological Organization with regard to runoff simulations (WMO, 1986) and to partially simulated conditions of real time runoff forecasts (WMO, 1992).

TABLE 14.1.

SRM Applications and Results

Country	Basin	Size (km^2)	Elevation Range (m a.s.l.)	Zones	Years (seasons)	R^{2*}	D_v(%)*
Germany	Lange Bramke (Harz)	0.76	540-700	1	1	N/A	N/A
Germany	Wintertal (Harz)	0.76	560-754	1	1	N/A	N/A
Czechoslo- vakia	Modry Dul (Krkonose)	2.65	1000-1554	1	2	0.96	1.7
Spain	Lago Mar (Pyrenees)	4.50	2234-3004	1	1	N/A	N/A
U.S.A.	W-3 (Appalachians)	8.42	346-695	1	10	0.81	8.8
Germany	Lainbachtal (Allgauer Alps)	18.7	670-1800	1	5	N/A	N/A
Switzerland	Rhone-Gletsch (Alps)	38.9	1755-3630	4	1	N/A	N/A
Switzerland	Dischma (Alps)	43.3	1668-3146	3	10	0.86	2.5
Japan	Sai (Japan Alps)	57.0	300-1600	3	3	0.86	N/A
United Kingdom	Feshie (Cairngorms)	106	350-1265	1	2	0.88	N/A
Switzerland	Sedrun (Alps)	108	1840-3210	3	2	0.79	1.9
U.S.A.	American Fork (Wasatch)	130	1820-3580	4	1	0.90	1.7
Switzerland	Landwasser-Frauenkirch (Alps)	183	1500-3146	3	1	N/A	N/A
India	Kulang (Himalyas)	205	2350-5000	N/A	N/A	N/A	N/A
Switzerland	Tavanasa (Alps)	215	1277-3210	4	2	0.82	3.1
U.S.A.	Dinwoody (Wyoming)	228	1981-4202	4	2	0.85	2.8
U.S.A.	Salt Creek (Utah)	248	1564-3620	5	1	N/A	2.6
Italy	Cordevole (Alps)	248	980-3342	3	1	0.89	4.6
India	Beas-Manali (Himalayas)	345	1900-6000	N/A	4	0.68	12
Norway	Laerdalselven (Lo Bru)	375	530-1720	5	1	0.86	5.2
Norway	Viveli (Hardangervidda)	386	880-1613	1	2	0.73	11.3
Japan	Okutadami (Mikuni)	422	782-2346	3	3	0.83	5.4
U.S.A.	Bull Lake Creek (Wyoming)	484	1790-4185	4	1	0.82	4.8
Switzerland	Tiefencastel (Alps)	529	837-3418	5	2	N/A	N/A
U.S.A.	South Fork (Colorado)	559	2506-3914	3	7	0.89	1.8
Argentina	Las Cuevas (Andes)	600	2500-7000	N/A	N/A	N/A	N/A
U.S.A.	Independence R. (Adirondacks)	618	261-702	1	1	0.81	5.0

continued . . .

Table 14.1. Continued

Country	Basin	Size (km^2)	Elevation Range (m a.s.l.)	Zones	Years (seasons)	R^{2*}	$D_v(\%)^*$
Poland	Dunajec (High Tatra)	700	577-2301	3	1	0.73	3.8
India	Saing (Himalayas)	705	1400-5500	N/A	N/A	N/A	N/A
U.S.A.	Conejos (Colorado)	730	2521-4017	3	7	0.87	1.1
Switzerland	Ilanz (Alps)	776	693-3614	5	2	N/A	N/A
Argentina	Lago Alumin (Andes)	911	1145-2496	N/A	N/A	N/A	N/A
China	Urumqi (Tien Shan)	924	1920-4000	N/A	N/A	N/A	N/A
India	Parbati (Himalyas)	1154	1500-6400	5	1	0.73	7.5
Canada	Illecillewaet (Rocky Mtns.)	1155	509-3150	4	4	0.86	7.0
India	Buntar (Himalyas)	1370	1200-5000	N/A	N/A	N/A	N/A
New Zealand	Hawea (S. Alps)	1500	300-2500	N/A	N/A	N/A	N/A
Switzerland	Ticino-Bellinzona (Alps)	1515	220-3402	N/A	N/A	N/A	N/A
U.S.A.	Spanish Fork (Utah)	1655	1484-3277	4	1	0.85	1.0
Argentina	Tupungato (Andes)	1800	2500-6000	8	1	0.63	6.4
Switzerland	Inn-Martina (Alps)	1943	1030-4049	2	1	0.82	4.3
Argentina	Chico (Tierra del Fuego)	2000	N/A	N/A	N/A	N/A	N/A
China	Gongnisi (Tien Shan)	2000	1776-3200	N/A	N/A	N/A	N/A
U.S.A.	Boise (Idaho)	2150	983-3124	3	3	0.84	3.3
France	Durance (Alps)	2170	786-4105	5	5	0.85	2.6
U.S.A.	Madison (Montana)	2344	1965-3234	2	2	0.89	1.5
Morocco	Tillouguit (Atlas)	2544	1050-3411	3	1	0.84	0.5
Austria	Salzach-St. Johann (Alps)	2600	570-3666	N/A	N/A	N/A	N/A
U.S.A.	Henry's Fork (Idaho)	2694	1553-3125	3	2	0.91	1.5
U.S.A.	Cache la Poudre (Colorado)	2732	1596-4133	3	1	N/A	N/A
U.S.A.	Sevier R (Kingston, Utah)	2929	1823-3260	4	1	0.75	5.1
Switzerland	Rhine-Felsberg (Alps)	3249	562-3425	5	2	0.80	1.2
U.S.A.	Rio Grande (Colorado)	3419	2432-4215	3	13	0.84	3.8
U.S.A.	Kings River (California)	4000	171-4341	7	5	0.82	3.2
India	Beas-Thalot (Himalayas)	5144	1100-6400	6	2	0.80	1.5
U.S.A.	Upper Yakima (Cascades)	5517	366-2121	5	1	0.92	2.8
Canada	Sturgeon (Ontario)	7000	N/A	N/A	N/A	N/A	N/A

continued . . .

Table 14.1. Continued

Country	Basin	Size (km^2)	Elevation Range (m a.s.l.)	Zones	Years (seasons)	R^{2*}	$D_v(\%)^*$
Argentina	Grande (Tierra del Fuego)	9050	N/A	N/A	N/A	N/A	N/A
Canada	Iskut (Coast)	9350	200-2556	5	N/A	N/A	N/A
U.S.A.	Sevier (Juab, Utah)	13380	1506-3719	4	1	0.93	4.0
U.S.A.	Snake River (Idaho)	14897	1524-4196	N/A	11	0.90	0.4
Pakistan	Kabul (Himalayas)	63657	305-7690	1	1	0.66	6.0
China	Yellow (Anyemaqen Shan)	121972	2500-5224	3	N/A	N/A	N/A

*The following accuracy criteria listed in the Table are defined as:

$$R^2 = 1 - \frac{\sum_{i=1}^{n}(Q_i - Q_i')^2}{\sum_{i=1}^{n}(Q_i - \overline{Q})^2} \qquad D_v = \frac{V_R - V_R'}{V_R} \bullet 100$$

where:

R^2 = a measure of model efficiency

Q_i = measured daily discharge

Q_i' = simulated daily discharge

$\overline{Q}$ = average daily discharge for the simulation year or simulation season

n = number of daily discharge values

D_v = percentage difference between the total measured and simulated runoff (%)

V_R = measured runoff volume

V_R' = simulated runoff volume

Approximate Coordinates

A Illecillewaet 57° N, 118.1 W
B Boise 43.4° N, 116.1° W
C Henry's Fork 44.5° N, 111.4° W
D Dinwoody 43° N, 109° W
E Kings River 37° N, 119.5° W
F Sevier 38° N, 112.5° W
G Spanish Fork 40.1° N, 111.4° W
H Rio Grande 37.4° N, 107.2° W
I Tupungato, 33° S, 70° W
J Lago Alumin 38.6° S, 71.1° W
K W-3 37° N, 82° W
L Lago Mar, Pyrenees 42.5° N, 2° E
M Feshie, Scotland 57.1° N, 3.7° W
N Durance 44.3° N, 6.2° E
O Felsburg 46.7° N, 9.5° E
P Lainbachtal 47.3° N, 11.05° E
Q Wintertal 51.7° N, 10.51 E
R Viveli 60.2° N, 7.1° E
S M. Dul 50.7° N, 15.3° E
T Dunajec 49.2° N, 19.5° E
U Kabul River 34.1° N, 71.2° E
V Beas 32.1° N, 77.1° E
W Sai River 36.5° N, 138° E
X Hawea 44.5° S, 169° E
Y Tillouguit 32°N, 5.8°W
Z Grande 53.8°S, 68.3°W
a Yellow 34.5°N, 99.5°W
b Gongnisi 43°N, 88°E
c Upper Yakima 47.3°N, 121°W
d Iskut 57°N, 131.3°W

Fig. 14.1. Selected locations where SRM has been tested.

14.2. RANGE OF CONDITIONS FOR MODEL APPLICATION

SRM can be applied in mountain basins of almost any size (so far from 0.76 to 122000 km^2) and any elevation range (for example, 305-7690 m a.s.l.). A model run starts with a known or estimated discharge value and can proceed for an unlimited number of days, as long as the input variables - temperature, precipitation and snow covered area - are provided. As a test, a 10-year period was computed without reference to measured discharges (Martinec and Rango, 1986).

In addition to the input variables, the area-elevation curve of the basin is required. If other basin characteristics are available (forested area, soil conditions, antecedent precipitation, and runoff data), they are of course useful for facilitating the determination of the model parameters.

SRM can be used for the following purposes:

1. Simulation of daily flows in a snowmelt season, in a year, or in a sequence of years. The results can be compared with the measured runoff in order to assess the performance of the model and to verify the values of the model parameters. Simulations can also serve to evaluate runoff patterns in ungauged basins and in a hypothetically changed climate.

2. Short term and seasonal runoff forecasts. The microcomputer program (Micro-SRM) includes a derivation of modified depletion curves which relate the snow covered areas to the cumulative snowmelt depths as computed by SRM.

These curves enable the snow coverage to be extrapolated automatically by the user several days ahead by temperature forecasts so that this input variable is available for discharge forecasts. The modified depletion curves can also be used to evaluate the snow reserves for seasonal runoff forecasts. The model performance may deteriorate if the forecasted air temperature and precipitation deviate from the actual values, but the inaccuracies can be reduced by periodic updating.

14.3. MODEL STRUCTURE

Each day, the water produced from snowmelt and from rainfall is computed, superimposed on the calculated recession flow and transformed into daily discharge from the basin according to eq. (14.1):

$$Q_{n+1} = \left[c_{Sn} \bullet a_n(T_n + \Delta T_n)S_n + c_{Rn} \bullet P_n\right]\frac{A_A \bullet 10000}{86400}(1 - k_{n+1}) + +Q_n \bullet k_{n+1} \quad (14.1)$$

where

Q	=	average daily discharge [m^3s^{-1}]
c	=	runoff coefficient expressing the losses as a ratio (runoff/precipitation), with c_S referring to snowmelt and c_R to rain
a	=	degree-day factor [$cm°C^{-1}d^{-1}$] indicating the snowmelt depth resulting from 1 degree-day
T	=	number of degree-days [°Cd]
ΔT	=	the adjustment by temperature lapse rate when extrapolating the temperature from the station to the average hypsometric elevation of the basin or zone [°Cd]
S	=	ratio of the snow covered area to the total area
P	=	precipitation contributing to runoff [cm]. A preselected threshold temperature, T_{CRIT}, determines whether this contribution is rainfall and immediate.
A	=	area of the basin or zone [km^2]
$\frac{10000}{86400}$	=	conversion from $cmkm^2d^{-1}$ to m^3s^{-1}
k	=	recession coefficient indicating the decline of discharge in a period without snowmelt or rainfall: $k = Q_{m+1}/Q_m$ (m, m + 1 are the sequence of days during a true recession flow period).
n	=	sequence of days during the discharge computation period. Equation (14.1) is written for a time lag between the daily temperature cycle and the resulting discharge cycle of 18 hours. In this case, the number of degree-days measured on the nth day corresponds to the discharge on the n + 1 day. Various lag times can be introduced by a subroutine.

T, S and P are variables to be measured or determined each day. c_R, c_S, lapse rate (to determine ΔT), T_{CRIT}, k, and the lag time are parameters which are characteristic for a given basin or, more generally, for a given climate. A guidance for determining these parameters will be given in a later section.

If the elevation range of the basin exceeds 500 m, it is recommended that the basin be subdivided into elevation zones of about 500 m each. For an elevation range of 1500 m and three elevation zones A,B,C, the model equation becomes

$$Q_{n+1} = \Big\{[c_{SAn} \bullet a_{An}(T_n + \Delta T_{An})S_{An} + c_{RAn} \bullet P_{An}]\frac{A_A \bullet 10000}{86400} +$$

$$[c_{SBn} \bullet a_{Bn}(T_n + \Delta T_{Bn})S_{Bn} + c_{RBn} \bullet P_{Bn}]\frac{A_B \bullet 10000}{86400} +$$

$$[c_{SCn} \bullet a_{Cn}(T_n + \Delta T_{Cn})S_{Cn} + c_{RCn} \bullet P_{Cn}]\frac{A_C \bullet 10000}{86400}\Big\}(1 - k_{n+1}) + Q_n k_{n+1} \quad (14.2)$$

The indices A,B and C refer to the respective elevation zones and a time lag of 18 hours is assumed. Other time lags can be selected and automatically taken into account as explained in the section 14.4.3.7.

In the simulation mode, SRM can function without updating. The discharge data serve only to evaluate the accuracy of simulation. In ungauged basins the simulation is started with a discharge estimated by analogy to a nearby gauged basin. In the forecasting mode, the model provides an option for updating by the actual discharge every 1-9 days.

Eqs. (14.1) and (14.2) are written for the metric system but an option for model operation in English units is also provided in the computer program.

14.4. NECESSARY DATA FOR RUNNING THE MODEL

14.4.1. Basin Characteristics

14.4.1.1. ***Basin and Zone Areas.*** The basin boundary is defined by the location of the streamgauge (or some arbitrary point on the streamcourse) and the watershed divide as identified on a topographic map. The basin boundary can be drawn at a variety of map scales. For the larger basins, a 1:250,000 scale map is adequate. After examining the elevation range between the streamgauge and the highest point in the basin (total basin relief), elevation zones can be delineated in intervals of about 500 m (or 1500 ft). In addition to drawing the basin and zone boundaries, several intermediate topographic contour lines should be highlighted for later use in constructing the area-elevation curve. Once the boundaries and the contours have been determined, the areas formed by these boundaries should be planimetered manually or automatically. Once the zones are defined, the various model variables and parameters are applied to each zone for the calculation of snowmelt runoff. To facilitate this application, the mean hypsometric elevation of the zone must be determined through use of an area-elevation curve.

14.4.1.2. ***Area-Elevation Curve.*** By using the zone boundaries plus other selected contour lines in the basin, the areas enclosed by various elevation contours can be determined by planimetering. These data can be plotted (area vs elevation) and an area-elevation (hypsometric) curve derived. This area-elevation curve can also be derived automatically if the user has access to digital elevation data and computer algorithms used in an image processing system.

14.4.2. Variables

14.4.2.1. ***Temperature and Degree-Days, T.*** In order to compute the daily snowmelt depths, the numbers of degree-days must be determined from temperature measurements or, in a forecasting mode, from temperature forecasts.

The program accepts either the daily mean temperature (option 0) or two temperature values on each day: T_{Max}, T_{Min} (option 1). The temperatures are extrapolated by the program from the base station elevation to the hypsometric mean elevations of the respective elevation zones. For option 1, the average temperature is computed in each zone as

$$\overline{T} = \frac{T_{Max} + T_{Min}}{2} \tag{14.3}$$

When using daily means (option 0) or when using T_{Max}, T_{Min} (option 1), it is recommended that negative temperature values (when they occur) be used in the calculation. In line with this recommendation, the original "effective minimum temperature" alternative (automatic change of negative temperatures to 0°C) was removed from the computer program beginning with Version 3.0. If the user still prefers this alternative, the occasional negative temperatures can be changed manually to 0°C when inputting the data to SRM.

Because the average temperatures refer to a 24 hour period starting always at 0600 hrs, they become degree-days T [° • d].

The altitude adjustment ΔT in Eq. (14.1) is computed as follows:

$$\Delta T = \gamma\left(h_{st} - \overline{h}\right) \bullet \frac{1}{100} \tag{14.4}$$

Where

γ = temperature lapse rate [°C per 100 m]

h_{st} = altitude of the temperature station [m]

$\overline{h}$ = hypsometric mean elevation of a zone [m]

Whenever the degree-day numbers ($T + \Delta T$ in Eq. 1) become negative, they are automatically set to zero so that no negative snowmelt is computed. The values of the temperature lapse rate are dealt with in Section 14.4.3.3.

The program accepts either temperature data from a single station (option 0, basin wide) or from several stations (option 1, by zone). With option 0, the altitude of the station is entered and temperature data are extrapolated to the hypsometric mean elevations of all zones using the lapse rate. If more stations are available, the user can prepare a single "synthetic station" and still use option 0 or, alternatively, use option 1. With option 1, the user may use separate stations for each elevation zone, however, the temperatures entered for each zone must have already been lapsed to the mean hypsometric elevation of the zone. Although SRM will take separate stations for each zone in this way, it is not required. The measurement of correct air temperatures is difficult, and therefore one good temperature station (even if located outside the basin) may be preferable to several less reliable stations.

In the forecast mode of the model, it is necessary to obtain representative temperature forecasts for the given region and altitude in order to extrapolate the expected numbers of degree-days for each elevation zone.

14.4.2.2. ***Precipitation, P.*** The evaluation of representative areal precipitation is particularly difficult in mountain basins. Also, quantitative precipitation forecasts are seldom available for the forecast mode, although current efforts in this field are improving this situation. Fortunately, snowmelt generally prevails over the rainfall component in the mountain basins. However, sharp runoff peaks from occasional heavy rainfalls must be given particular attention and the program includes a special treatment of such events (see Section 14.4.3.6.).

The program accepts either a single, basin-wide precipitation input (from one station or from a "synthetic station" combined from several stations) (option 0) or different precipitation inputs zone by zone (option 1). If the program is switched to option 1 and only one station happens to be available, for example in zone A, precipitation data entered for zone A must be copied to all other zones. Otherwise no precipitation from these zones is taken into account by the program.

A critical temperature (see Section 14.4.3.4.) is used to decide whether a precipitation event will be treated as rain ($T \geq T_{CRIT}$) or as new snow ($T < T_{CRIT}$). When the precipitation event is determined to be snow, its delayed effect on runoff is treated differently depending on whether it falls over the snow-covered or snow-free portion of the basin. The new snow that falls over the previously snow-covered area is assumed to become part of the seasonal snowpack and its effect is included in the normal depletion curve of the snow coverage. The new snow falling over the snow-free area is considered as precipitation to be added to snowmelt, with this effect delayed until the next warm day. This precipitation is stored by SRM and then melted as soon as a sufficient number of degree-days has occurred.

This may take place on the first day warm enough to produce snowmelt or on a series of days. The following example in Table 14.2 illustrates a case where 2.20 cm water equivalent of snow fell on day n and then was melted on the three successive days.

TABLE 14.2.

Calculation of the melt of new snow deposited on a snow-free area (P_n = 2.20 cm; T_{CRIT} = + 1.0°C)

Day	*a* (cm•°C^{-1}•d^{-1})	T (°C•d)	S	P (cm)	Melted Depth *a* • T(cm)	P Stored (cm)	P contributing to Runoff *a*•T•(1-S)(cm)
n	0.45	0	0.72	2.20	0	2.20	0
n+1	0.45	0.11	0.70	0	0.05	2.15	0.02
n+2	0.45	2.70	0.68	0	1.22	0.93	0.39
n+3	0.45	3.70	0.66	0	0.93	0	0.32

In this example, S is decreasing on consecutive days because it is interpolated previously from the snow-cover depletion curve. In reality it should remain constant as long as the seasonal snowpack is covered with new snow, however, the model currently uses the incremental decrease of S shown in Table 14.2.

Sharp peaks of discharge are typical of rainfall runoff as opposed to the relatively regular daily fluctuations of the snowmelt runoff. SRM has been adapted to better simulate these rainfall peaks whenever the average daily rainfall calculated over the whole basin equals or exceeds 6 cm. This threshold can be changed by the user. The procedure is described in the Section 14.4.3.6.1. in connection with the recession coefficient. In spite of these precautions, rainfall runoff peaks may cause problems because local rainstorms may be missed if the network of precipitation stations is not dense enough. Also, the timing of rainfalls within the 24 hour period is frequently unknown. Actually, this period (for P_n) usually lasts from 0800 hrs on the day n to 0800 on the day n+1 and is published as precipitation on the day n. In some cases, however, the same precipitation amount is ascribed to the day n+1 on which the measurement period ended. In such case precipitation data must be shifted by one day backwards.

14.4.2.3. *Snow Covered Area, S.* It is a typical feature of mountain basins that the areal extent of the seasonal snow cover gradually decreases during the snowmelt season. Depletion curves of the snow coverage can be interpolated from periodical snow cover mapping so that the daily values can be read off as an important input variable to SRM. The snow cover can be mapped by terrestrial observations (in very small basins), by aircraft photography (especially in a flood emergency), and most efficiently, by satellites. The minimum area which can be mapped with an adequate accuracy depends on the spatial resolution of the remote sensor. Examples are listed in Table 14.3.

TABLE 14.3.

Possibilities of remote sensing for snow cover mapping

Platform Sensor	Spatial resolution	Minimum area size	Repeat period
Aircraft orthophoto	3 m	1 km^2	flexible
Landsat			16-18 days
MSS	80 m	16-20 km^2	
TM	30 m	2.5 - 5 km^2	
NOAA			
AVHRR	1.1 km	100-500 km^2	12 hr
Meteosat			
Visible	2.5 km	500-1000 km^2	30 min.
SPOT	10-20 m	2-3 km^2	26 days
MOS	50 m	5-10 km^2	17 days

MSs = Multiple spectral scanner, TM = Thematic mapper,
NOAA = National Oceanic and Atmospheric Administration,
AVHRR = Advanced Very High Resolution Radiometer,
SPOT = System Probatoire d'Observation de la Terre (France)
MOS = Marine Observation Satellite (Japan)

Figure 14.2 shows the snow cover in the alpine basin Felsberg mapped from Landsat MSS 5 data. When the time interval between the subsequent useable satellite images becomes too long, e.g., due to visibility obscured by clouds, the depletion curves derived from the measured points may be distorted by occasional summer snowfalls. To avoid such errors, satellite images showing the short-lived snow cover from the summer snowfalls must be disregarded when deriving the depletion curves. In order to identify the new snow events, coincident precipitation and temperature data should be consulted. The transitory new snow is accounted for as stored precipitation eventually contributing to runoff as explained in the Section 14.4.2.2.

As an example, Fig. 14.3 shows depletion curves of the snow coverage derived for five elevation zones of the alpine basin Felsberg from the Landsat imagery (Baumgartner, 1987). Together with temperature and

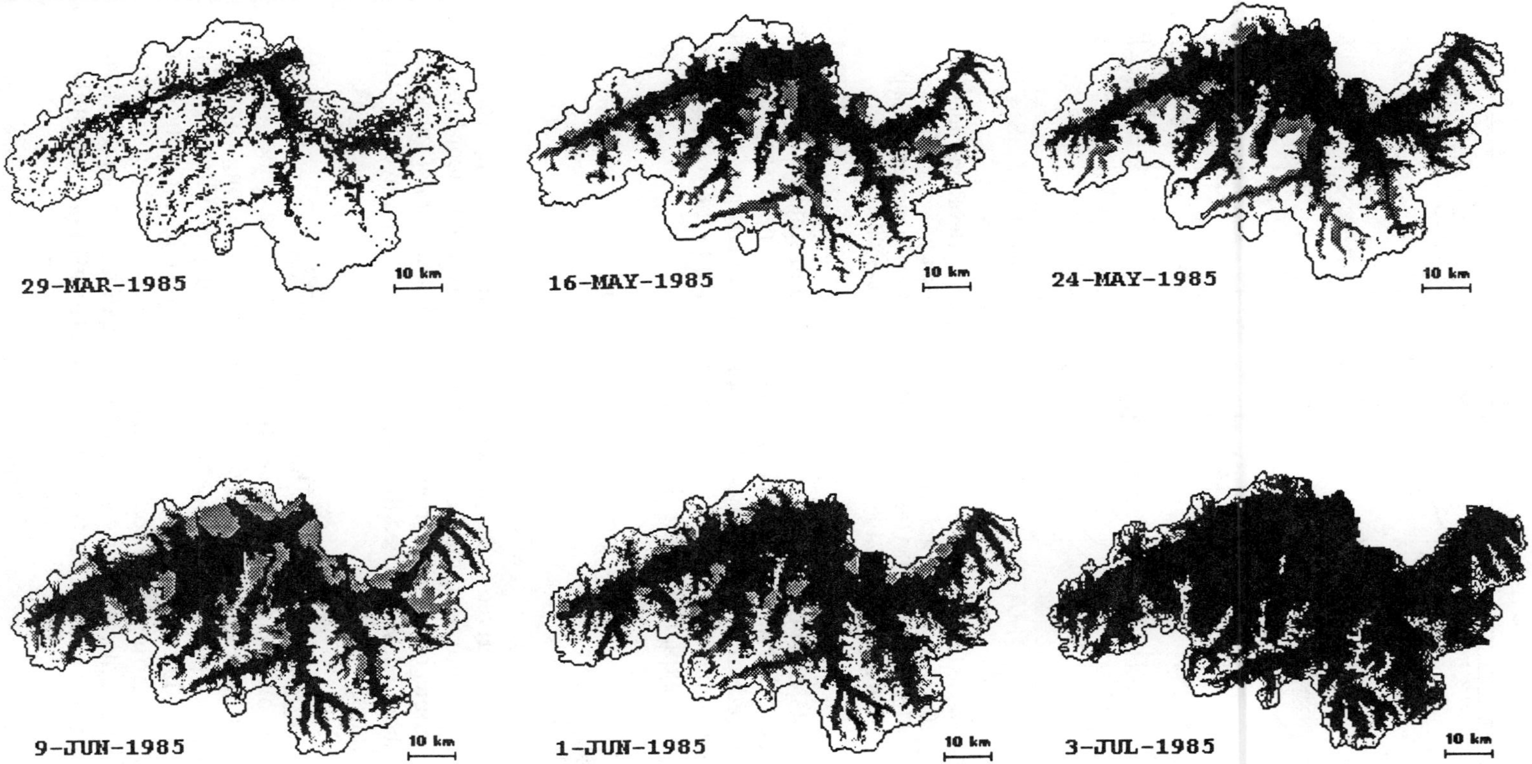

Fig. 14.2. Sequence of snow cover maps from Landsat MSS 5, Upper Rhine River at Felsberg, 3250 km^2, 560-3614 m a.s.l. Published by courtesy of the Institute for Communications Techniques, Institute of Technology, Zurich, Switzerland.

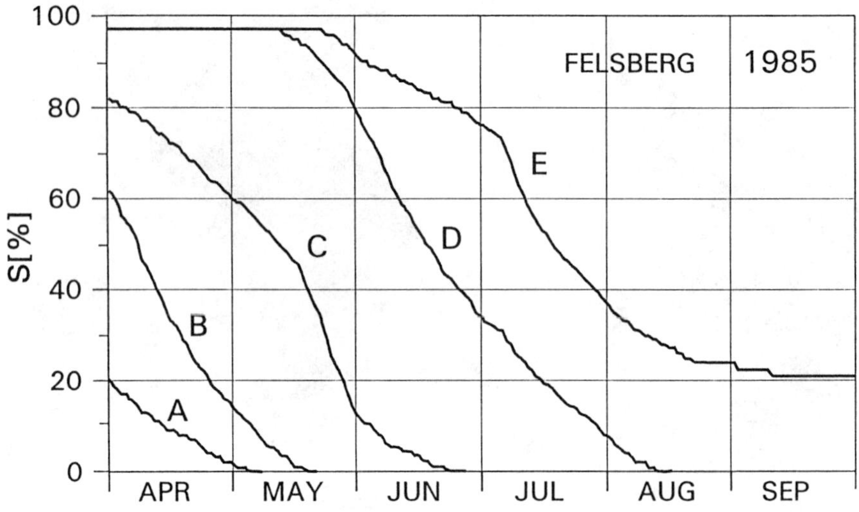

Fig. 14.3. Depletion curves of the snow coverages for 5 elevation zones of the basin Felsberg, derived from the Landsat imagery shown in Fig. 14.2. A: 560-1100 m a.s.l.; B: 1100-1600 m a.s.l.; C: 1600-3100 m a.s.l.; D: 2100-2600 m a.s.l.; E: 2600-3600 m a.s.l.

precipitation data, such depletion curves enable SRM to simulate runoff in a past year. For real-time runoff forecasts, however, it is necessary to know the snow covered area within days after a satellite overflight and also to extrapolate the depletion curves of the snow coverage to the future weeks. This procedure is explained in the Section 14.6.

14.4.3. Parameters

The SRM parameters are not calibrated or optimized by historical data. They can be either derived from measurement or estimated by hydrological judgement taking into account the basin characteristics, physical laws and theoretical relations or empirical regression relations. Occasional subsequent adjustments should never exceed the range of physically and hydrologically acceptable values.

14.4.3.1. ***Runoff Coefficient, c.*** This coefficient takes care of the losses, that is to say of the difference between the available water volume (snowmelt + rainfall) and the outflow from the basin. On a long term basis, it should correspond to the ratio of the measured precipitation to the measured runoff. In fact, comparison of historical precipitation and runoff ratios provide a starting point for the runoff coefficient values. However, these ratios are not always easily obtained in view of the precipitation catch deficit which particularly affects snowfall and of inadequate precipitation data from mountain regions. At the start of the snowmelt season, losses are usually very small because they are limited to evaporation from the snow surface, especially at high elevations. In the next stage, when some soil becomes exposed and vegetation grows, more losses must be expected due to evapotranspiration and interception. Towards the end of the snowmelt season, direct channel flow from the remaining snowfields and glaciers may prevail in some basins which leads to a decrease of losses and to an increase of the runoff coefficient. In addition, c is usually different for snowmelt and for rainfall. The computer program accepts separate values for snow, c_S, and rain, c_R, and allows for half-monthly (and, if required, daily) changes of values in each elevation zone.

Of the SRM parameters, the runoff coefficient appears to be the primary candidate for adjustment if a runoff simulation is not at once successful.

14.4.3.2. ***Degree-Day Factor, a.*** The degree-day factor *a* [$cm°C^{-1}d^{-1}$] converts the number of degree-days T[°Cd] into the daily snowmelt depth M [cm]:

$$M = a \cdot T \tag{14.5}$$

Degree-day ratios can be evaluated by comparing degree-day values with the daily decrease of the snow water equivalent which is measured by radioactive snow gauge, snow pillow or a snow lysimeter. Such measurements (Martinec, 1960) have shown a considerable variability of degree-day ratios from day to day. This is understandable because the degree-day method does not take specifically into account other components of the energy balance, notably the solar radiation, wind speed and the latent heat of condensation. If averaged for 3-5 days, however, the degree-day factor is more consistent and can represent the melting conditions. The effect of daily fluctuations of the degree day values on the runoff from a basin as computed by SRM is greatly reduced because the daily meltwater input is superimposed on the more constant recession flow (Eq. 14.1).

The degree-day method requires several precautions:

1. The degree-day factor is not a constant. It changes according to the changing snow properties during the snowmelt season.

2. If point values are applied to areal computations, the degree-day values must be determined for the hypsometric mean elevation of the snow cover in question and not, for example, for the altitude of the snow line.

3. If the snow cover is scattered, a correctly evaluated degree-day factor will produce less meltwater than if a 100 percent snow cover were assumed. A meltwater difference that arises from erroneous snow cover information should not be compensated for by

"optimizing" the degree-day factor. Instead, the correct areal extent of the snow cover should be determined and used.

4. In large area extrapolations, point measurements should be weighted depending on how well a specific station represents the hydrological characteristics of a given zone (Shafer, et al., 1981).

In the absence of detailed data, the degree-day factor can be obtained from an empirical relation (Martinec, 1960):

$$a = 1.1 \frac{\rho_s}{\rho_w} \qquad (14.6)$$

where

a = the degree-day factor [cm°C^{-1}d^{-1}]

ρ_s = density of snow

ρ_w = density of water

When the snow density increases, the albedo decreases and the liquid water content in snow increases. Thus the snow density is an index of the changing properties which favor the snowmelt.

Figure 14.4 illustrates the seasonal trend of the degree-day factor in the Alps and in the Rocky Mountains. Because the geographic latitude of a basin influences the solar radiation, it may be advisable to adjust the degree-day factors accordingly. In glacierized basins, the degree-day factor usually exceeds 0.6 cm °C^{-1}d^{-1} towards the end of the summer when ice becomes exposed (Kotlyakov and Krenke, 1982). The computer program accepts different degree-day factors for up to 8 elevation zones which are usually changed twice a month (although daily changes are possible).

Sometimes the occurrence of a large, late season snowfall will produce depressed a-values for several days due to the new low-density snow. The a-values in the model can be manually modified and inserted to reflect these unusual snowmelt conditions.

As is evident from Eq. (14.1), the degree-day method could be readily replaced by a more refined computation of snowmelt without changing the structure of SRM. Such refinement appeared to be imperative in a study of outflow from a snow lysimeter (Martinec, 1989), but is not considered to be expedient for hydrological basins until the necessary additional variables and their forecasts become available.

14.4.3.3. Temperature Lapse Rate, γ. If temperature stations at different altitudes are available, the lapse rate can be predetermined from historical data. Otherwise it must be evaluated by analogy from other basins or with

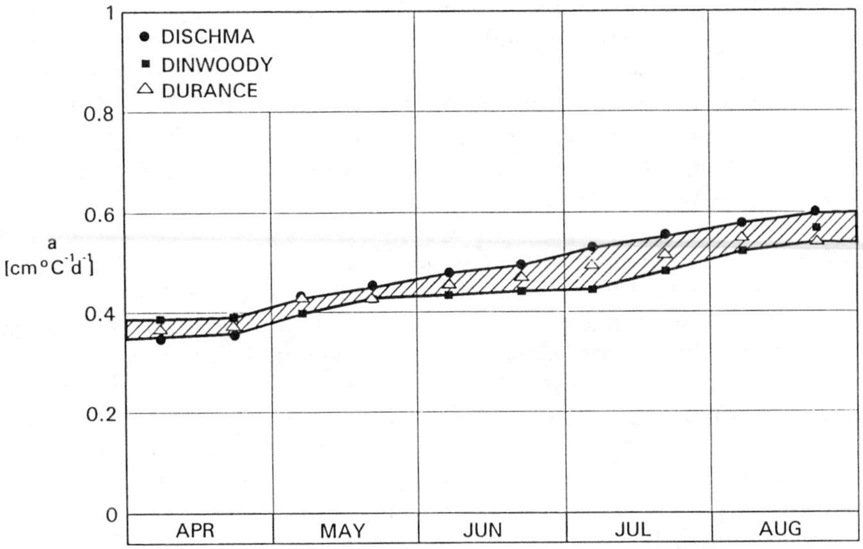

Fig. 14.4. Average degree-day ratio (a) used in runoff simulations by the SRM model in the basin Dischma (43.3 km^2, 1668-3146 m a.s.l., Switzerland, 10 years), Durance (2170 km^2, 786-4105 m. a.s.l., France, 5 years) and Dinwoody (228 km^2, 1981-4202 m. a.s.l., Wyoming, 2 years), (Martinec and Rango, 1986).

regard to climatic conditions. In SRM simulations, a lapse rate of 0.65°C per 100 m was usually employed. Slightly higher, seasonally changing values appeared to be adequate in the Rocky Mountains.

The computer program accepts either a single or basin wide lapse rate (option 0) or different rates for each zone (option 1). Seasonal variations can be accounted for by changing the values twice a month. The lapse rate can be also changed manually on selected days if a special meteorological situation (for example, an inversion) requires a different value.

If the temperature station is situated near the mean elevation of the basin, possible errors in the lapse rate are to some extent canceled out because the extrapolation of temperature takes place upwards as well as downwards. If, on the other hand, the temperature station is at a low altitude, SRM becomes sensitive to the lapse rate. For some basins in Wyoming, for example, the closest temperature station was more than 100 km away and 2000 m lower than the highest snow-covered parts of the basins. In such a case, an error in the lapse rate of only 0.05 °C per 100 m causes a deviation of 1 degree-day from the correct degree-day value which corresponds to an error of about 0.5 cm of the computed daily meltwater depth late in the snowmelt season.

Such situations sometimes necessitate an adjustment of the originally selected lapse rates taking into account the course of the depletion curves of snow coverage. If high temperatures result from extrapolation by a certain lapse rate but no change in the snow areal extent is observed, then probably no appreciable snowmelt is taking place at this altitude. The high temperatures result from a false lapse rate which must be increased or decreased, depending on whether the temperature station is lower or higher than the mean zone elevation.

14.4.3.4. ***Critical Temperature, T_{CRIT}.*** The critical temperature determines whether the measured or forecasted precipitation is rain or snow. Models which simulate the build-up of the snow cover first in order to simulate the runoff depend heavily on this parameter not only in the ablation period but particularly in the accumulation period. SRM needs the critical temperature only in the snowmelt season (unless a year round computer run is made) in order to decide whether precipitation immediately contributes to runoff (rain), or, if $T < T_{CRIT}$, whether snowfall took place. In this case, SRM automatically keeps the newly fallen snow in storage until it is melted on subsequent warm days, as explained in the Section 14.4.2.2.

When SRM was applied in the alpine basin Dischma, T_{CRIT} started at + 3°C in April at the beginning of snowmelt and diminished to + 0.75°C in July. This seasonal trend with a narrower range appears to be applicable in other basins. At certain times, SRM may not take notice of a sharp rainfall runoff peak because the corresponding precipitation is determined to be snow, the extrapolated temperature being just slightly below the critical temperature. In such cases the assignment of critical temperature and the temperature lapse rate values should be reviewed and logical adjustments made in order to change snow to rain. It is of course difficult to distinguish accurately between rain and snow because the temperature used is the daily mean while precipitation may occur at any time, day or night, i.e., in the warmer or colder portion of the daily temperature cycle.

As a possible refinement, formulas have been proposed (Higuchi et al., 1982) to determine the proportion of rain and snow in mixed precipitation conditions.

14.4.3.5. ***Rainfall Contributing Area, RCA.*** When precipitation is determined to be rain, it can be treated in two ways. In the initial situation (condition 0), it is assumed that rain falling on the snowpack early in the snowmelt season is retained by the snow which is usually dry and deep. Rainfall runoff is added to snowmelt runoff only from the snow-free area, that is to say the rainfall depth is reduced by the ratio snow-free area/zone area. At some later stage, the snow cover becomes ripe (the user must decide on which date) and the computer program should be switched to condition 1. Now, if rain falls on this snow cover, it is assumed that the same amount of water is released from the snowpack so that rain from the entire zone area is added to snowmelt. The melting effect of rain is neglected because the additional heat supplied by the liquid precipitation is considered to be small (Wilson, 1941).

14.4.3.6. ***Recession Coefficient, k.*** As is evident from Eq. (14.1), the recession coefficient is an important feature of SRM since (1-k) is the proportion of the daily meltwater production which immediately appears in the runoff. Analysis of historical discharge data is usually a good way to determine k. Figure 14.5 shows such an evaluation for the alpine basin Dischma (43.3 km^2, 1668-3146 m a.s.l.). Values of Q_m and Q_{m+1} from

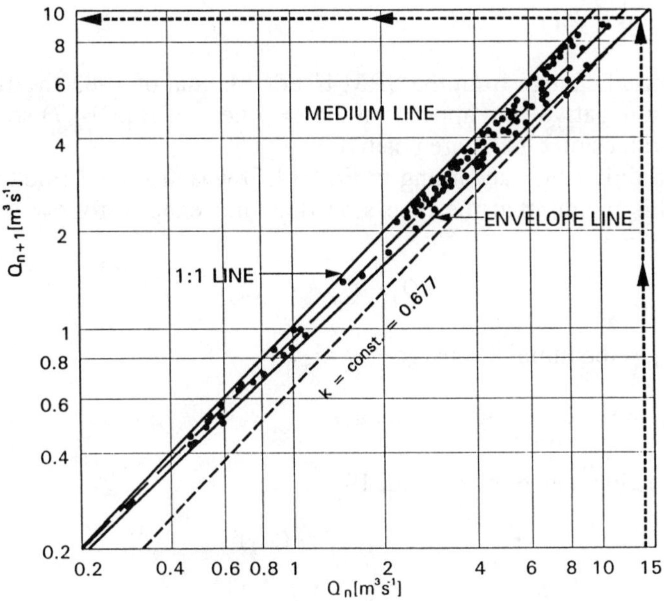

Fig. 14.5. Recession flow plot Q_n vs Q_{n+1} for the Dischma basin in Switzerland. Either the solid envelope line or the dashed medium line is used to determine k-values for computing the constants x and y in Eq. (14.7). (Martinec and Rango, 1986).

recession periods are plotted against each other and the envelope line of all points is considered to indicate the k-values. As can be read off, k = 0.677 for Q = 14 m^3s^{-1} and k = 0.85 for Q = 1 m^3s^{-1}. This means that k is not constant, but increases with the decreasing Q according to the equation

$$k_{n+1} = xQ_n^{-y} \tag{14.7}$$

where the constants x and y must be determined for a given basin by solving the equations:

$$k_1 = xQ_1^{-y}$$

$$k_2 = x Q_2^{-y}$$

$$\log k_1 = \log x - y \log Q_1 \qquad (14.8)$$

$$\log k_2 = \log x - y \log Q_2 \qquad (14.9)$$

In the given example,

$$\log 0.677 = \log x - y \log 14$$

$$\log 0.85 = \log x - y \log 1$$

$$x = 0.85, y = 0.086$$

As a formal change from the SRM User's Manual of 1983 (Martinec, et al., 1983), a negative sign appears in the exponent of Eq. (14.7) so that the numerical values of x and y are positive.

The variability of k according to Eq. (14.7) was also confirmed in other basins. This means that the recession does not exactly follow the usual equation

$$Q_n = Q_o k^n \qquad (14.10)$$

where Q_o = the initial discharge

Q_n = the discharge after n days

but the following equation (Jaccard, 1982)

$$Q_n = x^{1/y} \left(\frac{Q_o}{x^{1/y}} \right)^{(1-y)^n} \qquad (14.11)$$

where x and y are the constants of Eq. (14.7).

The envelope line in Fig. 14.5 and the resulting values of x and y must be determined for each basin. For ungauged basins and when historical discharge data are insufficient, x and y can be derived indirectly from the size of the basin as follows:

$$k_{Nn} = \left[x_M \left(\frac{\overline{Q_M}}{\overline{Q_N}} \cdot Q_{Nn-1} \right)^{-y_M} \right]^{4\sqrt{A_M / A_N}} \qquad (14.12)$$

where: x_M, y_M are the known constants for the basin M; $\overline{Q_M}, \overline{Q_N}$ are average discharge values from the basin M and the new basin N; and A_M, A_N are the areas of the respective basins.

Equation (14.12) indicates that recession coefficients are generally higher in large basins than in small basins. If the increase in k with size appears to be too large, the exponent may be replaced by $\sqrt[8]{A_M / A_N}$. Even if the envelope line in Fig. 14.5 can be reliably derived in a basin, it is possible that the resulting k-values may be too low to represent average conditions during the snowmelt season, especially in large basins. In such cases the SRM model will react too quickly to any change of the daily input. The simulated peaks would be too high and the simulated recession too fast. A quick improvement is possible by deriving a new x and y, not from the envelope line, but from an intermediate line between the envelope and the 1:1 line. This modification may especially be needed if the runoff simulation is extended to the whole year. Recession coefficients which may be right for the snowmelt period are usually too low for the winter months so that the simulated flows drop below the measured minimum values.

In very small basins, noticeable differences in the recession flow conditions and the k-values may occur from year to year. Figure 14.6 illustrates the range of k-values for varying basin size and for the mentioned alternatives of evaluation. For a very small mountain basin, the range includes the variations of k in different years (Martinec, 1970). For a larger alpine basin, the limits refer to the envelope line and to the medium line in

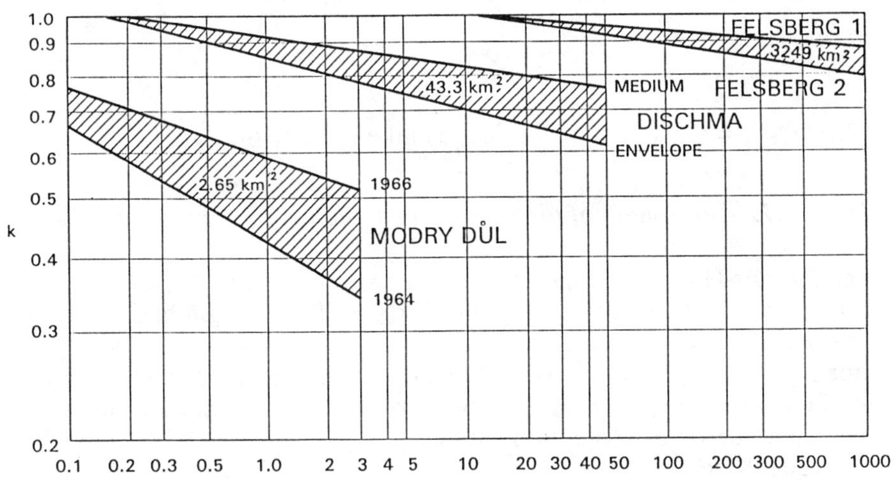

Fig. 14.6. Range of recession coefficients, k, related to discharge Q resulting from various evaluations: In Dischma, the range results from using either the envelope line or the medium line in Fig. 14.5. In Modry Dul, the relation varies in different years. In Felsberg, the relations (1) and (2) are derived from the size of the basin by two alternative formulas (Martinec and Rango, 1986).

Fig. 14.5. For the largest basin, the higher limit (1) of the indicated range was derived by substituting the constants x = 0.085, y = 0.086 derived for the Dischma basin into Eq. (14.12). The lower limit (2) was obtained by replacing $\sqrt[4]{A_M / A_N}$ in Eq. (14.12) with $\sqrt[8]{A_M / A_N}$.

Figure 14.6 shows that k can theoretically exceed 1 for very small discharges in large basins. This does not really happen because such small discharges do not occur there. Such a situation could be produced, however, by the user inadvertently taking the x and y values derived in a large basin and using them for a small basin without modification. In this case, if the daily snowmelt input exceeds the previous day's runoff, SRM computes a runoff decrease instead of increase. In order to avoid this error, the computer program prevents k from exceeding 0.99. However, it is advisable to avoid approaching this limiting situation by checking the x and y values with regard to the lowest flow to be expected. Recalling Eq. (14.11), it follows for n = ∞, because y > 0 and (1-y) < 1:

$$Q_\infty = x^{1/y}\left(\frac{Q_o}{x^{1/y}}\right)^{(<1)^\infty} = x^{1/y} \tag{14.13}$$

Therefore the values x and y should fulfill the condition

$$Q_{Min} > x^{1/y} \tag{14.14}$$

where Q_{Min} = the minimum discharge in the given basin.

14.4.3.6.1. Adjustment of the recession coefficient for heavy rainfalls

The formula (14.7) for computing the recession coefficient reflects the usual conditions characterizing the snowmelt runoff in the given basin. When a heavy rainfall occurs, the input is concentrated in a short time interval creating an abrupt rise and subsequent decline of the hydrograph. In order to simulate such events, the computer program automatically adjusts the recession coefficient whenever the daily rainfall averaged over the whole basin equals or exceeds 6 cm:

If $P\ (\text{Rain})_n \geq 6$ cm $\rightarrow$ $k_{n+1} = x\,(4\,Q_n)^{-y}$

$$k_{n+2} = x\,(4\,Q_{n+1})^{-y}$$

$$k_{n+3} = x\,(4\,Q_{n+2})^{-y}$$

$$k_{n+4} = x\,(4\,Q_{n+3})^{-y}$$

$$k_{n+5} = x\,(4\,Q_{n+4})^{-y} \tag{14.15}$$

after which it returns to the normal formula (14.7). In this way, k gets lower so that the basin response to input becomes faster. If the precipitation is recognized by T_{CRIT} as snow and not rain, the mechanism will not be activated.

The threshold value can be changed in order to activate the rainfall peak program for smaller rainfall amounts or to delay activation until higher rainfall amounts are reached.

14.4.3.7. ***Time Lag, L.*** The characteristic daily fluctuations of snowmelt runoff enable the time lag to be determined directly from the hydrographs of the past years. If, for example, the discharge starts rising each day around noon, it lags behind the rise of temperature by about 6 h. Consequently, temperatures measured on the nth day correspond to discharge between 1200 h on the nth day and 1200 h on the n+1 day. Discharge data, however, may only be published for midnight-to-midnight intervals and would need adjustments in order to be compared with the simulated values. Conversely, the simulated values can be adjusted (Shafer et al., 1981) to refer to the midnight-to-midnight periods. Figure 14.7 illustrates the procedure for

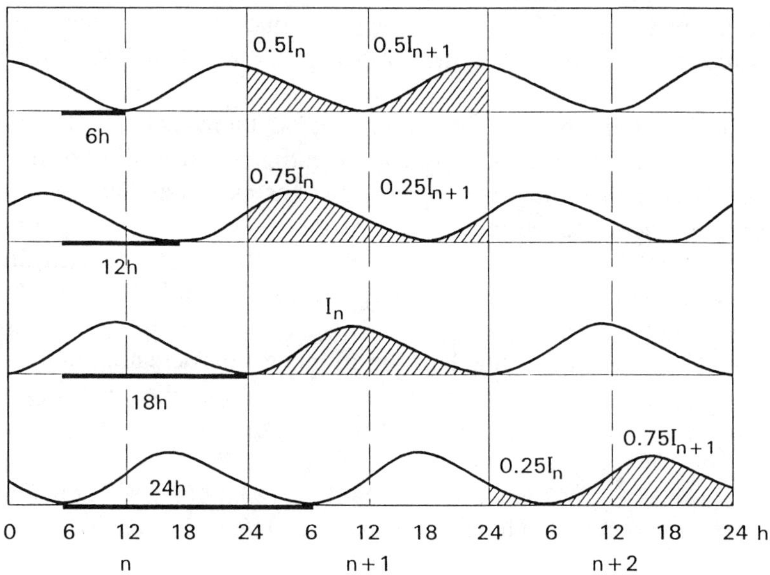

Fig. 14.7. Snowmelt hydrographs illustrating the conversion of computed runoff amounts for 24 hr periods to calendar day periods. The various time lags (bold lines) are taken into account by proportions of the daily inputs, I (Martinec and Rango, 1986).

different time lags. For L = 6 h, 50 percent of input computed for temperature and precipitation on the nth day (I_n) plus 50 percent of I_{n+1} results in the n+1 day's runoff after being processed by the SRM computer program:

$$L = 6\text{ h} \qquad 0.5I_n + 0.5I_{n+1} \rightarrow Q_{n+1} \tag{14.16}$$

similarly:

$$L = 12\text{ h} \qquad 0.75I_n + 0.25I_{n+1} \rightarrow Q_{n+1} \tag{14.17}$$

$$L = 18\text{ h} \qquad I_n \rightarrow Q_{n+1} \tag{14.18}$$

$$L = 24\text{ h} \qquad 0.25I_n + 0.75I_{n+1} \rightarrow Q_{n+2} \tag{14.19}$$

This procedure is preferable, at least in mountain basins smaller than 5000 km^2, to evaluations of L by calculating the velocity of overland flow and channel flow. It has been shown by environmental isotope tracer studies (Martinec, 1985) that overland flow is not a major part of the snowmelt runoff as previously believed. There is increasing evidence that a major part of meltwater infiltrates and quickly stimulates a corresponding outflow from the groundwater reservoir. With this runoff concept in mind, the seemingly oversimplified treatment of the time lag in the SRM model is better understood.

If the hydrographs are not available or if their shape is distorted by reservoir operations, the time lag can be estimated according to the basin size and by analogy with other comparable basins. Generally, the time lag in a basin increases as the snow line retreats.

In the WMO intercomparision test (WMO, 1986) most models calibrated the time lag. However, these results appear to be of little help to determine the proper values. Contradictory time lags have been calibrated by different models. However, if the time lags for all models participating in the WMO intercomparsion test are averaged for each basin, the resulting values support the expected relation between L and basin size:

> Basin W-3 (8.42 km^2): 3 h; Dischma (43.3 km^2): 7.2 h; Dunajec (680 km^2): 10.5 h; and Durance (2170 km^2): 12.4 h.

If there is some uncertainty, L (percentages in Eqs. 14.16-19) can be adjusted in order to improve the synchronization of the simulated and measured peaks of average daily flows. It should be noted that a similar effect results from an adjustment of the recession coefficient.

14.5. ASSESSMENT OF THE MODEL ACCURACY

14.5.1. Accuracy Criteria

The SRM computer program includes a graphical display of the computed hydrograph and of the measured runoff. A visual inspection shows at the first glance whether the simulation is successful or not. SRM additionally uses two well established accuracy criteria, namely, the coefficient of determination, R^2, and the volume difference, D_v.

The coefficient of determination is computed as follows:

$$R^2 = 1 - \frac{\sum_{i=1}^{n} (Q_i - Q_i')^2}{\sum_{i=1}^{n} (Q_i - \overline{Q})^2} \tag{14.20}$$

where

Q_i is the measured daily discharge

Q_i' is the computed daily discharge

$\overline{Q}$ is the average measured discharge of the given year or snowmelt season

n is the number of daily discharge values

Eq. (14.20) also corresponds to the Nash-Sutcliffe coefficient in which case $\overline{Q}$ is a long-term average measured discharge applied to the respective years or seasons.

The deviation of the runoff volumes, D_v, is computed as follows:

$$D_v\ [\%] = \frac{V_R - V_R'}{V_R} \bullet 100 \tag{14.21}$$

where

V_R is the measured yearly or seasonal runoff volume

V_R' is the computed yearly or seasonal runoff volume

Numerical accuracy criteria are never perfect, as illustrated by Fig. 14.8. From the visual judgement both simulations look good because the fundamental difference between two extreme years is well reproduced. However, $R^2 = 0.95$ in 1979 while it amounts only to 0.48 in 1977. In spite of this unfavorable value, the simulation (or forecast) in 1977 would

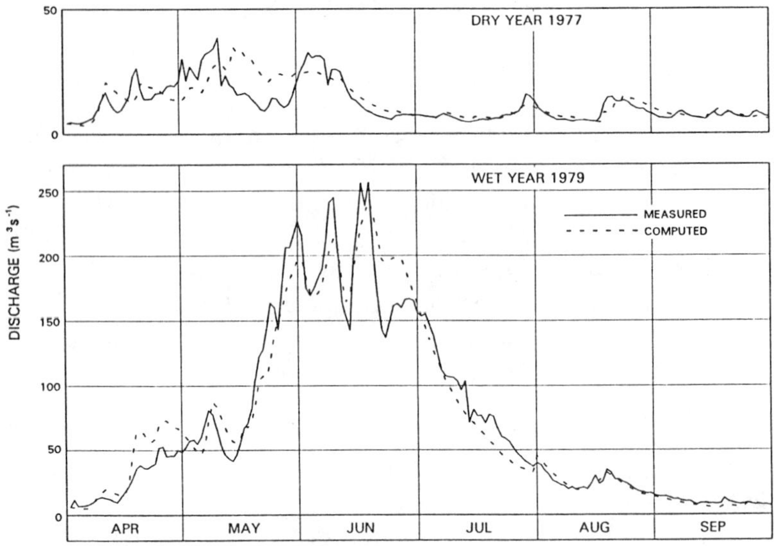

Fig. 14.8. Runoff siumultaions in the Rio Grande basin, Del Norte, Colorado (3419 km^2, 2432-4215 m a.s.l.) (Martinec and Rango, 1986).

certainly be useful for water management because it correctly reveals an extremely low runoff.

With $\overline{Q}$ as a long-term average substituted into Eq. (14.20) (Nash-Sutcliffe) instead of the average for the specific year, a much more favorable but deceptive value for R^2 of 0.97 results for the year 1977. This value is high because the long-term $\overline{Q}$ is much higher than any Q in 1977. Consequently, the daily deviations of the simulated runoff become relatively insignificant although in absolute terms they are not negligible.

In addition to these criteria which are automatically computed and displayed after each model run, the coefficient of gain from daily means, DG, could be computed by the user as follows:

$$DG = 1 - \frac{\sum_{i=1}^{n}(Q_i - Q_i')^2}{\sum_{i=1}^{n}(Q_i - \overline{Q}_i)^2} \tag{14.22}$$

where

Q_i is the measured daily discharge

Q_i' is the computed daily discharge

$\overline{Q}$ is the average measured discharge from the past years for each day of the period

n is the number of days

Thus, R^2 compares the performance of a model with "no model" (average discharge) and DG with a "seasonal model" (long term average runoff pattern). Negative values signal that the model performed worse than "no model" or worse than the "seasonal model".

14.5.1.1. ***Accuracy Criteria in Model Tests.*** The World Meteorological Organization (WMO) organized an international comparison of snowmelt runoff models in which hundreds of model runs were performed in six selected test basins. Figure 14.9 shows a summary of all numerical values of R^2, DG and D_v published by WMO (1986). Each prism refers to a tested model. The length along the x-axis corresponds to the arithmetic mean of all ($1-R^2$) values, length along the y-axis to the arithmetic mean of all (1-DG) values, and length along the z-axis to the arithmetic mean of all D_v values as achieved in the snowmelt seasons of 10 test years. Inaccurate

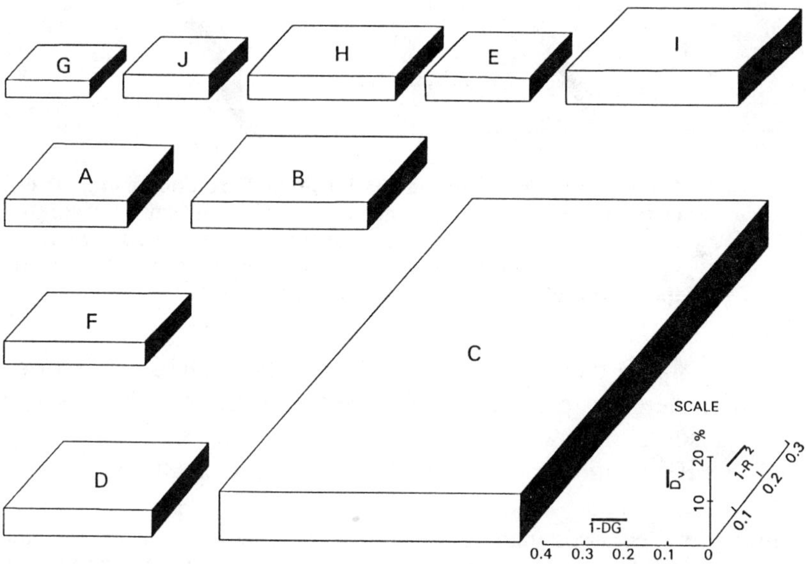

Fig. 14.9. Combined representation of model performance using three criteria: R^2, DG and D_v. The volumes of the prisms indicate the average inaccuracies of the tested models from all results for snowmelt seasons reported in the WMO project (Rango and Martinec, 1988).

results mean low values of R^2 and DG, thus longer dimensions along the x and y axis. Volume deviations prolong the dimension along the z-axis. Consequently, the volume of a prism is proportional to an overall average inaccuracy of a model.

In Fig. 14.10, the dimensions of the prisms are determined by the worst results of each model for R^2, DG and D_v, i.e., $(1\text{-}R^2)_{MAX}$, $(1\text{-}DG)_{MAX}$, and D_{vMAX}. All available data for the individual years and snowmelt season as

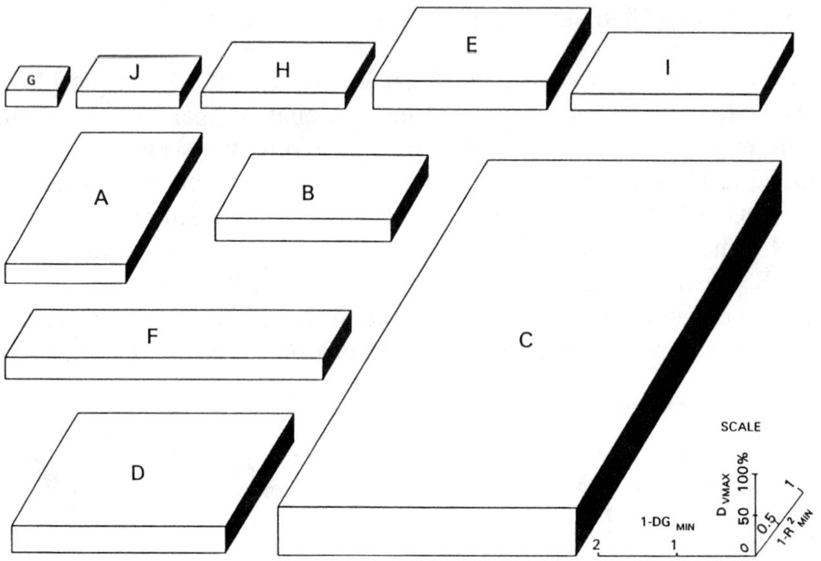

Fig. 14.10. Combined representation model of performance using three criteria: R^2, DG and D_v. The volumes of the prisms indicate the maximum inaccuracies of the tested models from all models results and listed for snowmelt seasons and individual years in the WMO tables. (Martinec and Rango, 1989a).

listed in the WMO tables (WMO, 1986) are thus contained within each prism. The differences between models are larger than in Fig. 14.9 which means that some calibrating models had difficulties in the years with unusual runoff conditions but improved their average results (Fig. 14.9) by the more normal years.

A more detailed assessment of accuracy criteria with regard to the needs of the model user is published elsewhere (Martinec and Rango, 1989a). For the WMO intercomparison, SRM produced the smallest prisms shown in Figs. 14.9 and 14.10.

14.5.1.2. ***Model Accuracy Outside the Snowmelt Season.*** SRM has been designed to compute runoff during the snowmelt season, but it can be run for the whole year if required. According to the mentioned WMO intercomparision test (WMO, 1986), about the same accuracy as for the snowmelt season can be achieved for the entire year in mountain basins with a low winter runoff. An example in Fig. 14.11 shows that SRM can even be run without updating for 10 years.

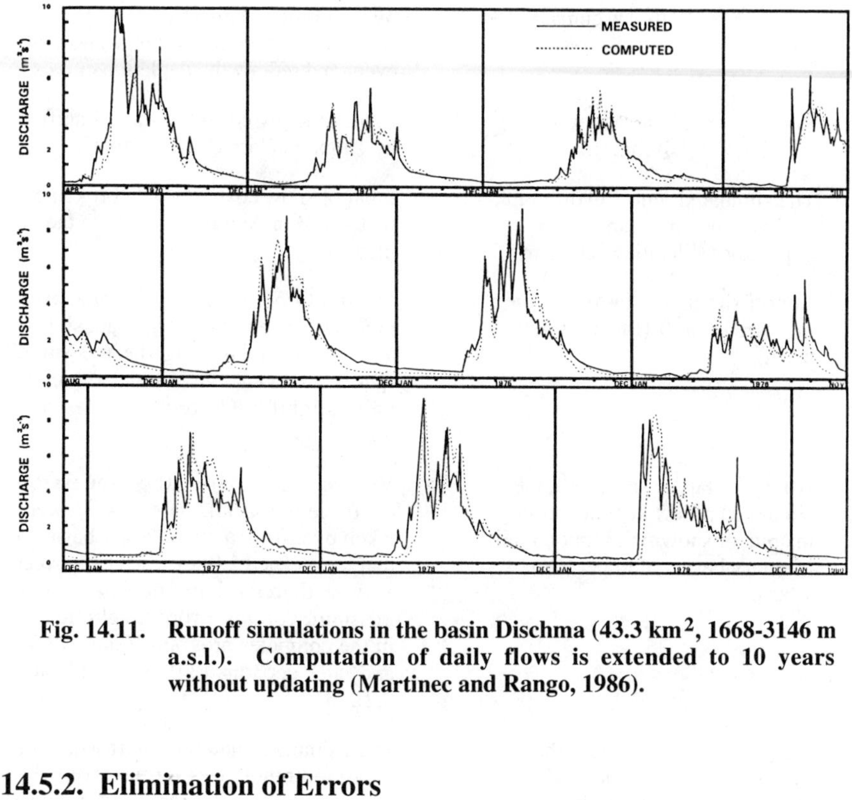

Fig. 14.11. Runoff simulations in the basin Dischma (43.3 km^2, 1668-3146 m a.s.l.). Computation of daily flows is extended to 10 years without updating (Martinec and Rango, 1986).

14.5.2. Elimination of Errors

It is not possible to give threshold values of accuracy criteria which would determine whether a model run is successful or whether something must be changed. With good data, a value like $R^2 = 0.80$ might still be improved. In unfavorable conditions, with incomplete data, a user may be satisfied even with lower R^2 values. Sometimes, however, the graphical display as well as the numerical criteria indicate that something went definitely wrong. Before adjusting one or the other parameter, it is recommended to check several probable sources of error first. Examples in Table 14.4 are based on the actual experience of various users.

TABLE 14.4.

Errors experienced by SRM users and their correction

Error:		Cause:
1.	Runoff simulation went too high (Dinwoody, 1976, Dischma, 1977).	Snow cover depletion curves distorted by summer snowfall: Eliminate satellite observation after a snowfall event and redraw the depletion curves.
2.	Runoff simulation suddenly deteriorated (Dischma).	Input of snow coverage data S shifted by 1 month: Correct the S-input.
3.	The simulated runoff hydrograph declined uncontrollably in September (Illecillewaet basin).	Input of snow coverage data broken off by the end of August: Complete the S-input.
4.	Runoff simulation far below the measured runoff (Illecillewaet basin).	Decimal point in the measured precipitation and discharge data shifted and thereby increasing the measured runoff 10 x, but the simulated runoff only slightly: Correct the decimal point.
5.	After the start of snowmelt, the simulated runoff kept decreasing in spite of snowmelt input (small tributary basins of upper Rhine, 1985).	Values x, y for the recession coefficient formula (Eq. 14.7) were taken over from a much larger basin so that k exceeded 1.0 for low Q: Correct x, y with regard to the basin size. Possibility of this error was eliminated in the computer program Version 2.01 and later versions (3.0, 3.1, 3.11, and 3.2).
6.	Frequent deviations from the measured runoff, periodical lows of the hydrograph not simulated.	SRM simulates natural runoff while the measured runoff was influenced by by storage (on weekends) and release from artificial reservoirs for by hydropower. These interventions must be corrected in order to compare simulated and measured runoff.
7.	Rainfall peaks inadequately simulated (Illecillewaet)	Rainfall input is concentrated to shorter periods than snowmelt input which accelerates the basin response. Program Versions 2.01 and later versions take this feature automatically into account whenever rainfall exceeds a preselected threshold.

TABLE 14.4. Continued.

TABLE 14.4. Continued.

8.	Rainfall peak inadequately simulated even with the special rainfall peak program (small tributary basins of upper Rhine 25 August 1985).	Temperature extrapolated from station 800 m a.s.l. to zone D 2380 m a.s.l. by 0.65°/100 m $\rightarrow$ T = + 0.43°C while T_{CRIT} = + 0.75°C. By decreasing T_{CRIT}, snowfall was converted to rainfall and the runoff peak was better simulated.
9.	Rainfall peak inadequately simulated because the daily amount was below the special program automatic threshold. Threshold was lowered but the peak simulation deteriorated further. (small tributary basins of upper Rhine, 1985).	Only 1 precipitation station, was available but the "By zone" option was switched on. Consequently precipitation input took place only in 1 zone. With 1 station, select option "basin wide".
10.	Runoff simulation went very high in a basin with a large elevation range of 7400 m. (Kabul River, 1976)	Compact snow cover was assumed above the snow line so that the snow covered area was exaggerated. Temperature was extrapolated by too low a lapse rate. Snowmelt was not computed separately for elevation zones but for the entire basin. This usually leads to over-estimation of meltwater input. Simulation was improved by re-evaluating the snow coverage and selecting a higher temperature lapse rate corresponding to the climate. With regard to the large elevation range, the basin should have been divided in several elevation zones.
11.	Distorted runoff simulation (outflow from a snow lysimeter).	Values of recession coefficient false: Negative y values used with the equation $k = xQ^{-y}$ so that the exponent became positive. Consequently, k increased with the increasing Q while it should have decreased. Use equation $k = xQ^{-y}$ and positive values of y, so that the exponent is always negative.
12.	Difficulties with the timing of rainfall-runoff peaks the stations were ascribed to (WMO test for simulated operational forecasts, Illecillewaet basin).	Precipitation data from one of the stations were ascribed to the date on which it was measured at 0800: Data had to be shifted one day backwards. See also Section 14.4.2.2.

14.6. OPERATION OF THE MODEL FOR REAL TIME FORECASTS

In order to be applied for real-time discharge forecasts, a model should be able to simulate the runoff not only in selected test basins with good data or in specially equipped experimental basins where a particular calibration model was developed, but also in basins where such forecasts are required by the user. SRM has relatively modest requirements for input variables (temperature, precipitation, and snow covered area) and therefore it was easily possible to shift the runoff simulations to the basins delivering water for hydroelectric schemes, as required by an electric company. Examples of such simulations are shown in Fig. 14.12.

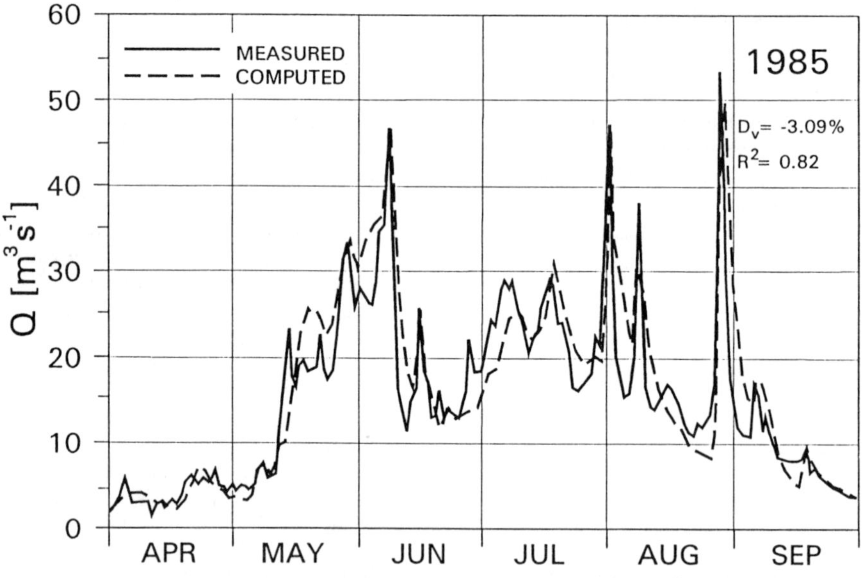

Fig. 14.12. Runoff simulation in the catchment area of the hydroelectric station Tavanasa (Swiss Alps, 215 km^2, 1277-3210 m a.s.l.).

As mentioned in Section 14.2., SRM can be used for short term (for example weekly) forecasts of daily flows as well as for longer time period forecasts such as monthly runoff volumes or seasonal runoff volumes. For short term forecasts, temperature, precipitation and snow covered area must be forecasted or predetermined for the coming days and substituted into the model. Temperature and even precipitation forecasts are becoming increasingly available from meteorological services, but the snow covered areas must be extrapolated by the model user.

14.6.1. Extrapolation of the Snow Coverage

The future course of the depletion curves of the snow coverage can be evaluated from the so-called modified depletion curves (MDC). These curves are automatically derived by SRM from the conventional curves (CDC) by replacing the time scale with cumulative daily snowmelt depths as computed by the model. The decline of the modified depletion curves depends on the initial accumulation of snow and not on the climatic conditions, as is the case with the conventional depletion curve.

Procedure for weekly forecasts of daily flows:

Assumption - A family of modified depletion curves has been derived from the past snow cover monitoring and temperature measurements in the given basin. Two of these curves representing the initial water equivalents $H_w = 20$ cm and $H_w = 60$ cm are plotted in Fig. 14.13.

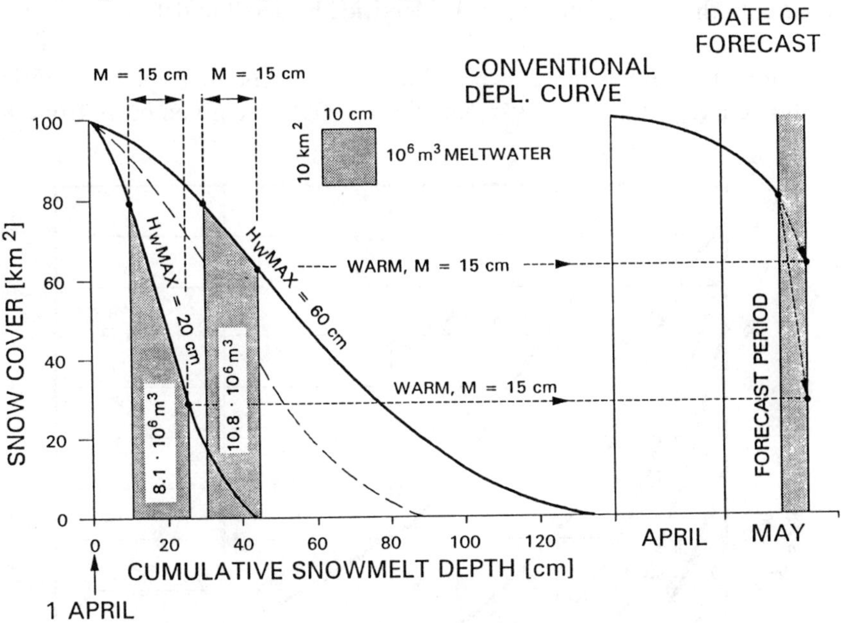

Fig. 14.13. Extrapolation of snow cover depletion curves in real time from modified depletion curves with the use of temperature forecasats. H_wMAX is the water equivalent of the snow cover at the beginning of the snowmelt season (Hall and Martinec, 1985).

Example 1 - Snow accumulation in the basin unknown, snow coverage measured by Landsat on 15 May, S = 80 percent, cumulative snowmelt depth (from degree-days and degree-day ratios) to date: 30 cm. Temperature forecast: 30 degree days for the next week, converted to meltwater depth M = 15 cm by a degree-day ratio $a = 0.5$ cm °C^{-1}d^{-1}. S = 80 percent and M = 30 cm indicate that the curve for H_w = 60 cm is applicable. The snow coverage will drop to 64 percent in 7 days. Extrapolated conventional depletion curve indicates values for day-to-day discharge computations.

Example 2 - As above, but the cumulative snowmelt depth to date is only 10 cm. Consequently, the curve for H_w = 20 cm is applicable and the snow coverage will drop to 33 percent in 7 days, which leads to a different extrapolation of the conventional depletion curve and to a different weekly total of forecasted daily runoff volumes. If the initial water equivalent is known, for example from SNOTEL (a system of data transmission using meteor paths for reflecting the signals and operated in the U.S.A.), the appropriate modified depletion curve can be selected at the start of the snowmelt season. Otherwise, the average curve (dashed line in Fig. 14.13) is used until the correct curve can be identified by satellite data.

Figure 14.14 shows that it is possible to derive a nomograph of modified depletion curves (Rango and van Katwijk, 1990a) for a given basin from the

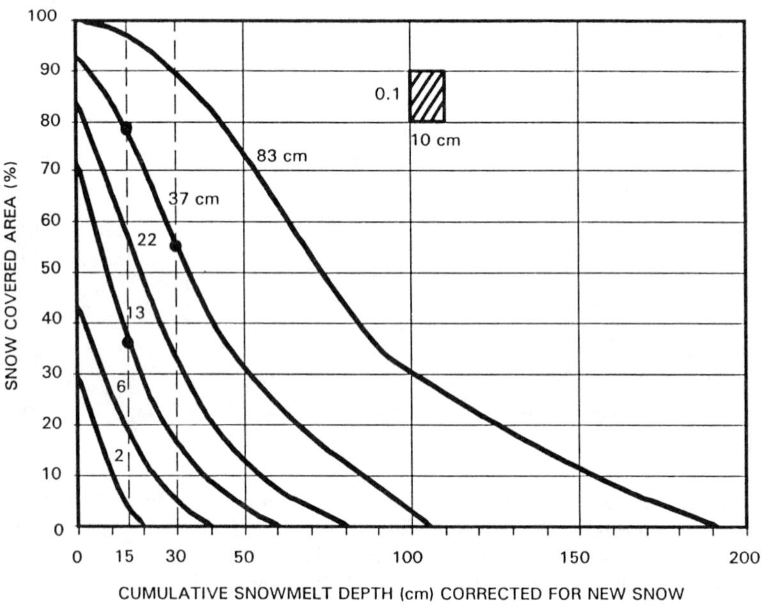

Fig. 14.14. Nomograph of modified depletion curves for the elevation zone B (1284 km^2, 2926-3353 m a.s.l.) of the Rio Grande basin. The curves are labelled with the areal average water equivalent (1 April) of the snow cover wihich they represent (Rango and van Katwijk, 1990a).

past years. As noticed in Fig. 14.13, the area below a modified depletion curve indicates the water volume stored in the snow cover if the y-axis scale is in km^2. If the y-axis scale is in percent snow coverage, it indicates the water equivalent of the snow cover as an areal average. Therefore each curve in Fig. 14.14 can be labelled by the water equivalent which it indicates. The rectangular shaded area means 0.1 • 10 cm = 1 cm. Because the area below the highest curve is 83 times larger, this curve indicates that at the beginning of computations of the cumulative snowmelt depth (usually on 1 April), the snow accumulation corresponded to the water equivalent of 83 cm. The values for each curve are automatically determined by the computer program.

The nomograph is used for real time forecasts as follows: In a current year, the snow covered area is monitored from the start of the snowmelt season and simultaneously the cumulative snowmelt depth is computed. The snow covered area must be evaluated as quickly as possible after each satellite overflight. The degree-days necessary for melting the temporary snow cover from intermittent snowfalls are disregarded. If, after some time, for example on 15 May, the snow coverage amounts to 80% and the cumulative snowmelt depth amounts to 15 cm, the modified depletion curve labelled by 37 cm is identified to be valid for that year. This curve can be used for extrapolating the snow covered area. For example, if another 15 cm will be melted in the next week according to temperature forecasts (total 30 cm), the snow coverage will drop to 55%. The snow covered areas thus extrapolated are used for real time forecasts of daily flows. The modified curve also indicates the water equivalent of snow (37 cm) at the start of the snowmelt season for seasonal runoff forecasts. If, in another future year, the cumulative computed snowmelt depth coincides for example with the snow coverage of only 36%, the curve labelled 13 cm is valid. The appropriate curve can thus be identified but with a certain time delay. If the initial water equivalent of the snow cover can be evaluated from point measurements, the proper curve can be selected at the start of the snowmelt season with no time delay.

The computer program also provides an option for plotting a modified depletion curve in which the totalized melt depth includes new snow that falls occasionally during the snowmelt period. While the new snow-excluded MDC is used to evaluate the water equivalent of the seasonal snow cover at the start of the snowmelt period, the new snow-included MDC can be used to evaluate the shifting of the conventional depletion curves by changed temperatures, as will be explained in the Section 14.7. The depletion curves of snow covered areas are dealt with in more detail in an earlier publication (Hall and Martinec, 1985). In the absence of temperature and precipitation forecasts, runoff forecasts can be issued on condition, for example, that long-term average values or extreme values (maxima, minima) will occur. It is also possible to use fictitious values as will be shown in the section dealing with climate change.

14.6.2. Updating

The model performance in the forecasting mode is naturally affected by the reduced accuracy and reliability of temperature and precipitation forecasts. The propagation of errors can be avoided by periodical updating. In the most recent versions of the computer program (Versions 3.11 and 3.2), the computed discharge is replaced every 1-9 days by the measured discharge which becomes known for the corresponding day so that each subsequent forecast period is computed by using a correct discharge value. Even without this updating, SRM prevents persistent large errors by a built-in self adjusting feature which is efficient if Eq. (14.7) is carefully assessed.

Further possibilities of updating will be made available to users when more experience in real time situations is accumulated. For example, it should be possible to adjust some parameters (e.g., the runoff coefficient) in the progress of the forecast, but only within hydrologically and physically acceptable limits. In any case, false forecasts of temperature and precipitation should be updated whenever a correction by new data is indicated. What is generally called "updating" can be thus divided into 3 categories:

1. Updating the computed discharge by the measured discharge when it becomes known, i.e., checking with the measured discharge to avoid carry-over of errors when the next forecast is issued.

2. Adjustment of model parameters in the process of forecast.

3. Correction of temperature, precipitation, and snow cover forecasts according to actual observations.

Short-term discharge forecasts can be updated as frequently as each day (Baumann et al., 1990).

14.7. RUNOFF SIMULATION FOR A FUTURE CHANGED CLIMATE

Because all parameters of SRM are predetermined and not calibrated, the model is suitable for computing snowmelt runoff in ungauged basins and, in response to a recently emerging problem, in a changed climate. If temperature can be forecasted several months ahead, predicted with a certain probability for an unspecified year, or just hypothetically assumed, SRM is applicable for computing the resulting runoff hydrograph. To this effect, it is necessary to derive the changed depletion curves of snow covered areas with the use of the modified depletion curves, MDC. Possible changes in precipitation may also influence the new runoff patterns (Rango and van Katwijk, 1990b). The opinion now prevails that the climate will become warmer and not colder. It is of course possible to simulate the effect of both options and to consider various changes of future precipitation.

If SRM is used only to predict snow covered areas in a changed climate (for example in mountain areas which are not hydrological basins), the computer program requires the following data:

Number of elevation zones, their areas, and mean hypsometric elevations
Present snow covered areas, daily values (S)
Daily average or Max/Min temperatures (T)
Precipitation, daily (P)
Degree-day factor (*a*)
Temperature lapse rate (γ)
Critical temperature (T_{CRIT})

The remaining SRM parameters c_S, c_R, RCA, k, L and the initial Q are required only for runoff computation.

The following example (Martinec and Rango, 1989b) explains the effect of a warmer climate on the snowmelt runoff in a mountain basin. It is assumed that the seasonal snow cover of 1 April 1983 will exist in a future year when temperatures will be higher by + 3˚C each day during the snowmelt period than temperatures of 1983. Figure 14.15 shows the MDC

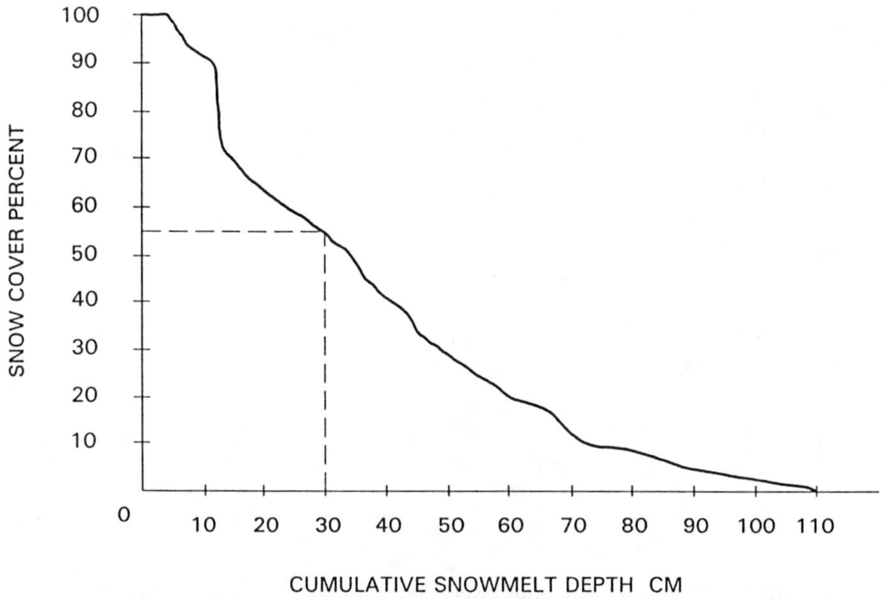

Fig. 14.15. Modified snow cover depletion curve (melt depth includes new snow) for the zone B (2926-3353 m a.s.l.) of the Rio Grande basin in 1983 (Martinec and Rango, 1986b).

derived by the computer program for the snowmelt season of 1983. The same occurrence of summer snowfalls is assumed so that the melt depth is obtained, for this purpose, by totalizing, starting on 1 April, the computed daily melt depths of the seasonal old snow cover as well as of the intermittent new snow. It can be read off that a totalized melt depth of 30 cm had the effect of reducing the snow covered area to 55%. The conventional depletion curve of snow covered areas in Fig. 14.16 shows that this occurred on 28 May 1983. With temperatures increased by + 3°C, the computed daily melt rates arc higher so that the total of 30 cm is reached already on 10 May of a future year, and with it the snow covered area of 55%. In this manner, the depletion curve of 1983 is shifted to earlier dates. The new depletion curve reflects the effect of a warmer climate and indicates snow covered areas for each day which are needed to compute the new runoff hydrograph. The climate-shifted depletion curves are automatically derived by SRM computer program Version 3.2 for any selected change of temperature and precipitation.

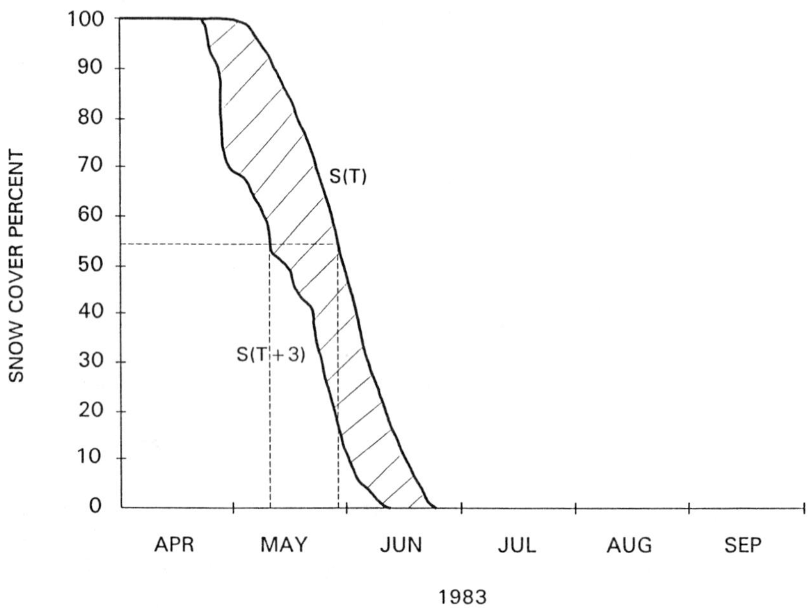

Fig. 14.16. Conventional snow cover depletion curve for zone B of the Rio Grande basin in 1983 and a predicted shifted curve for a temperature increase of +3°C (Martinec and Rango, 1989b).

The following symbols are used in this procedure, which is explained in more detail elsewhere (Rango and Martinec, 1993):

CDC	=	conventional depletion curve of snow covered area
CDC_{CLIM}	=	conventional depletion curve of snow covered area in the changed climate
MDC_{INCL}	=	modified depletion curve of snow covered area with new snow included. This curve indicates how much snow, including new snow falling during the snowmelt period, must be melted (in terms of calculated snowmelt depth) in order to decrease the snow covered area to a certain proportion of the total area and ultimately to zero. The shape of this curve depends on the initial accumulation of snow and on the amount of new snow.
MDC_{EXCL}	=	modified depletion curve of snow covered area with new snow excluded. This curve indicates how much of the initial seasonal snow cover must be melted (in terms of calculated cumulative snowmelt depth) in order to decrease the snow covered area (new snow excluded) to a certain proportion of the total area and ultimately to zero. The shape of this curve depends only on the accumulation of snow at the start of the snowmelt period, independent of subsequent snowfalls.
MDC_{CLIM}	=	modified depletion curve of snow covered area for the changed climate. This curve takes into account the amount of snowfalls changed by the new climate. If there is no change, it is identical with MDC_{INCL}.
$CDC_{CLIM(MA)}$	=	conventional depletion curve of snow covered area in the changed climate adjusted for the input to SRM runoff computation (Model Adjusted).

The adjustment of $CDC_{CLIM(MA)}$ is carried out by the computer program as follows: If CDC_{CLIM} shows more than one S-value on a date (the curve decreases vertically), the first S-value (the highest) is used. If there is no S-value on a date, the S-value of the previous day is used until there is a new S-value. Consequently, $CDC_{CLIM(MA)}$ has one S-value on each day as required for SRM runoff computation.

Due to the time shift by the changed climate, S-values may be missing towards the end of the computation period. The last available S-value is then repeated by the described adjustment so that $CDC_{CLIM(MA)}$ becomes a horizontal line. This part of the curve must not be used for runoff computation. The user can either derive realistic S-values to use instead or stop the runoff computation at this point.

If such a horizontal line occurs in the original CDC, it usually indicates a glacier within the elevation zone in question. In this case, the horizontal line which also appears in CDC_{CLIM} and $CDC_{CLIM(MA)}$ is valid and the S-values can be used for runoff computation.

The derivation of CDC_{CLIM} includes the following steps:

MDC_{INCL} and MDC_{EXCL} are derived in the same way as in previous SRM computer programs and displayed for visual check.

MDC_{CLIM} is derived from MDC_{EXCL} by adding melt depths for climate-adjusted new snow falls and displayed together with MDC_{INCL} and MDC_{EXCL}. Check: For warmer climate, MDC_{CLIM} should be to the left of MDC_{INCL}, for colder climate to the right. If all new snowfalls become rainfalls, $MDC_{CLIM} = MDC_{EXCL}$. If there is no change in the amount of snowfalls, $MDC_{CLIM} = MDC_{INCL}$.

CDC_{CLIM} is derived from MDC_{CLIM} by determining dates on which the cumulative snowmelt depths in the changed climate equal or exceed the cumulative snowmelt depths for MDC_{CLIM} as explained by the above example. CDC_{CLIM} is displayed together with CDC and should be to the left for a warmer climate. For unchanged temperature and increased precipitation it should be to the right because of increased new snowfalls.

The computer program takes care of the changed snow/rain conditions: with higher temperatures, some of the spring or summer snowfalls now become rainfalls. When these changes of input variables are also carried out in the remaining zones of the basin, the new runoff hydrograph is computed in the usual manner. Figure 14.17 shows that the warmer climate would increase the runoff and discharge peaks in the Rio Grande basin in the first half of the snowmelt season and reduce the runoff in the second half. The seasonal runoff volume appears to be slightly higher because the high flows are shifted to April and May when relatively high values of the runoff coefficient are assumed. However, the resulting conditions in the warmer climate may imply that adjustments of the runoff coefficients are necessary as well. So far, the effect of a future climate change is evaluated only for the duration of the snowmelt season. The existing snow accumulation at the

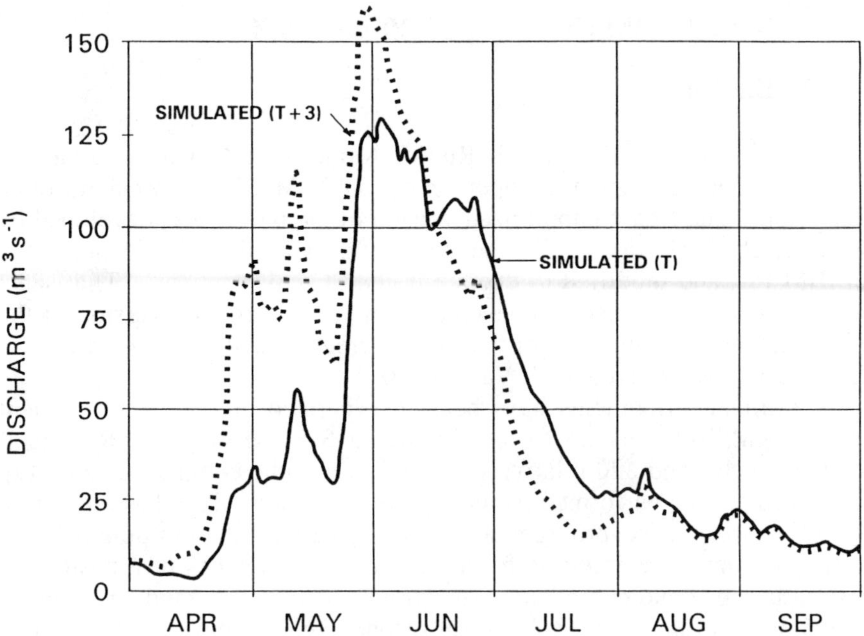

Fig. 14.17. Simulated change in runoff on the Rio Grande basin for 1983 as a result of a 3°C increase in temperature (T + 3) (Martinec and Rango, 1989b).

beginning of the snowmelt season is assumed to be unchanged. However, if the warmer climate would be extended to the winter months, the snow accumulation would be reduced. Consequently, a lower modified depletion curve MDC_{EXCL} would have to be used as a starting point to derive the snow covered areas as well as the redistribution of winter and summer runoff in the new climate. The adjustment of MDC_{EXCL} by the changed winter climate is in the research stage. New patterns can be simulated also by various assumptions regarding precipitation changes, but already the given example reveals some aspects of the climate change:

1. The redistribution of the seasonal runoff must be taken into account by water power and irrigation schemes.

2. A revision of design floods in certain areas may be indicated.

3. While the climate will change the snow covered areas, the resulting changed albedo may in turn influence the climate.

The current efforts concerning climate predictions open a new area of SRM applications.

14.8. MICRO-SRM COMPUTER PROGRAM

14.8.1. Background

The Martinec/Rango Snowmelt Runoff Model (SRM) was originally a Fortran model designed to operate on an IBM 370-series mainframe computer. The first computerized version of the model was developed by Martinec et al. (1983). In 1986 the model's Fortran code was downloaded to an IBM PC, and modified to operate in the PC environment. That same year a decision was made to develop a unique PC-oriented version of the model, taking full advantage of the PC's inherent capabilities. The results of that decision was Micro-SRM, Version 1.0.

Additional refinements have been incorporated in several subsequent Micro-SRM Versions up to and including 3.2. However, SRM itself remains unchanged and relatively simple, so that the computations by Eq. (14.1) can still be performed by any pocket calculator which has a function x^y. Of course the PC program automatically handles the multiple input of temperature and precipitation for up to 8 elevations zones of a basin, any desired lag time and complicated snow/rain situations. A model run for up to 365 days is finished within several seconds, the computed hydrograph is immediately displayed in comparison with the measured discharge and, if desired, quickly printed. Also, the achieved accuracy is automatically computed and displayed. A summary of parameter values can be displayed after each run so that adjustments can be made and their effect assessed.

SRM does not require numerous runs because calibration is not necessary. The ease with which the results are obtained should not lead to a replacement of the deterministic approach of SRM by a "try and see" philosophy. SRM is designed to operate with physically based estimates of parameters which should not require much change after the initial selection. As mentioned earlier, seemingly unsatisfactory results have been frequently improved not by adjusting the parameters but by correcting errors in data sets and in the input of variables.

14.8.2. Micro-SRM Availability

Micro-SRM has been distributed to over 100 individuals and institutions worldwide since 1985. The Hydrology Laboratory supports and distributes the model at no charge to any interested party. The distribution diskette, available on either $5^1/_4$" or $3^1/_2$" diskette, contains the executable code, supporting files, and example data files. To obtain the latest version of Micro-SRM, contact:

USDA-ARS, Hydrology Laboratory
Bldg. 007, Room 104, BARC-W
10300 Baltimore Avenue
Beltsville, MD 20705-2350 U.S.A.
(301) 504-7490

More detailed instructions for the use of the SRM model and a complete description of the SRM computer program are contained in the User's Manual (Martinec et al., 1992). An alternative source for obtaining SRM and related information is through Internet by accessing FTP file server hydrolab.arsusda.gov.

14.9. ACKNOWLEDGMENTS

The author would like to acknowledge the important and essential contributions of Dr. J. Martinec and Mr. R. Roberts in model development and computer programming, respectively.

REFERENCES

Baumgartner, M. F., 1987: 'Schneeschmelz-Abflusssimulationen basierend auf Scheeflaechenbestimmungen mit digitalen Landsat -MSS and NOAA-AVHRR Daten (snowmelt runoff simulations)', Remote Sensing Series, **Vol. 11**, Department of Geography, University of Zurich, Switzerland, 200 pp.

Higuchi, K., Ageta, Y., Yasunari T. and Inoue, J., 1982: 'Flow pattern of meltwater in mountain snow cover', Hydrological Aspects of Alpine and High Mountain Areas, (Proceedings of the Exeter Symposium), IAHS Publ. No. 138, 63-69.

Hall, D. K. and Martinec, J., 1985: Remote Sensing of Ice and Snow, Chapman & Hall Ltd., London - New York, 189 pp.

Jaccard, C., 1982: 'Recession coefficient formula', personal communication.

Kotlyakov, V. M. and Krenke, A. N., 1982: 'Investigations of the hydrological conditions of alpine regions by glaciological methods', Hydrological Aspects of Alpine and High-Mountain Areas, (Proceedings of the Exeter Symposium), IAHS Publ. No. 138, 31-42.

Martinec, J., 1960: 'The degree-day factor for snowmelt runoff forecasting', IUGG General Assembly of Helsinki, IAHS Commission of Surface Waters, IAHS Publ. No. 51, 468-477.

Martinec, J., 1970: 'Study of snowmelt runoff process in two representative watersheds with different elevation range', IAHS- Unesco Symposium, Wellington, N.Z., IAHS Publ. No. 96, 29-39.

Martinec, J., 1975: 'Snowmelt-Runoff Model for stream flow forecasts', Nordic Hydrology, **6**(3), 145-154.

Martinec, J., 1985: 'Time in hydrology'. In J. C. Rodda (ed.): Facets of Hydrology, **Vol. II**, John Wiley & Sons, London, 249-290.

Martinec, J., 1989: 'Hour-to-hour snowmelt rates and lysimeter outflow during an entire oblation period', Snow Cover and Glacier Variation (Proceedings of the Baltimore Symposium), IAHS Publication No. 193, 19-28.

Martinec, J. and Rango, A., 1986: 'Parameter values for snowmelt runoff modeling', Journal of Hydrology, **84**, 197-219.

Martinec, J. and Rango, A., 1989a: 'Merits of statistical criteria for the performance of hydrological models', Water Resources Bulletin, **25**(20), 421-432.

Martinec, J. and Rango, A., 1989b: 'Effects of climate change on snowmelt runoff patterns', Remote Sensing and Large-Scale Global Processes (Proceedings of the Baltimore Symposium), IAHS Publ. No. 186, 31-38.

Martinec, J., Rango, A. and Major, E., 1983: 'The Snowmelt-Runoff Model (SRM) User's Manual', NASA Reference Publication 1100, Washington, D.C., 118 pp.

Martinec, J., Rango, A. and Roberts, R., 1992: 'User's manual for the Snowmelt Runoff Model (SRM) (updated edition 1992, version 3.2)', Hydrology Laboratory Technical Report HL-17, Agricultural Research Service, Beltsville, MD, 70 pp.

Rango, A. and Martinec, J., 1988: 'Results from international intercomparisons of snowmelt runoff model performance', Proceedings of the 45th Annual Eastern Snow Conference, Lake Placid, New York, 121-128.

Rango, A. and Martinec, J., 1993: 'Areal extent of seasonal snow cover in a changed climate', Nordic Hydrology, 12 pp.

Rango, A. and van Katwijk, V., 1990a: 'Development and testing of a snowmelt runoff forecasting technique', Water Resources Bulletin, **26**(1), 135-144.

Rango, A. and van Katwijk, V., 1990b: 'Water supply implications of climate change in western North American basins', Proceedings of the Symposium on International and Transboundary Water Resources Issues, American Water Resources Association, Toronto, 577-586.

Shafer, B. A., Jones, E. B. and Frick, D. M., 1981: 'Snowmelt Runoff Simulations Using the Martinec-Rango Model on the South Fork Rio Grande and Conejos River in Colorado', AgRISTARS Report CP-G1-04072, Goddard Space Flight Center, Greenbelt, Maryland.

WMO, 1992: 'Simulated Real Time Intercomparison of Hydrological Models', Operational Hydrology Report No. 38, World Meteorological Organization, Geneva.

WMO, 1986: 'Intercomparison of Models of Snowmelt Runoff', Operational Hydrology Report No. 23, World Meteorological Organization, Geneva.

Wilson, W. T., 1941: 'An outline of the thermodynamics of snow-melt', Transactions Amer. Goephys. Union, Part 1, 182-195.

Chapter 15

THE SLURP MODEL

G. W. Kite

SLURP is a distributed watershed model in which the parameters are related to land cover characteristics. The model is designed to use satellite data for land cover information (Landsat, NOAA), for snow covered area (NOAA) and for snow water equivalent (DMSP) but can also operate without remotely sensed data. The model simulates the behaviour of a watershed by carrying out vertical water balances for each element of a matrix of land covers and sub-units of a watershed, and then routing runoffs between sub-units. The model operates on a micro computer and can simulate up to five years of daily data for up to ten land cover types and up to fifty watershed sub-units. SLURP has been used to investigate alternate climate scenarios in a mountain catchment and to interpret outputs from a general circulation model for a macro-scale basin. The model simulates areal average evapotranspiration, snowpack and soil moisture over each land cover within each response area of the catchment as well as streamflow hydrographs.

15.1. INTRODUCTION AND GRU CONCEPT

A hydrological model, like any other model, is intended to be a realistic representation of a physical system and the detail of its design will depend on the purpose of the modelling. For simulation or forecasting of a streamflow hydrograph one of the many simple lumped models will often suffice but, for scientific research, it may be necessary to use a more detailed, physically-based, fully-distributed model. SLURP, as a distributed conceptual model (Fig. 15.1) fits somewhere in the middle of a line linking lumped basin models at one end with fully-distributed physically-based models, such as the Système Hydrologique Européen (SHE) model (Abbott et al., 1986), at the other. SLURP - this was an acronym for **S**imple **LU**mped **R**eservoir **P**arametric in the original version of the model and the name has been retained as the model has developed through a semi-distributed version to the latest, distributed, version - was developed for

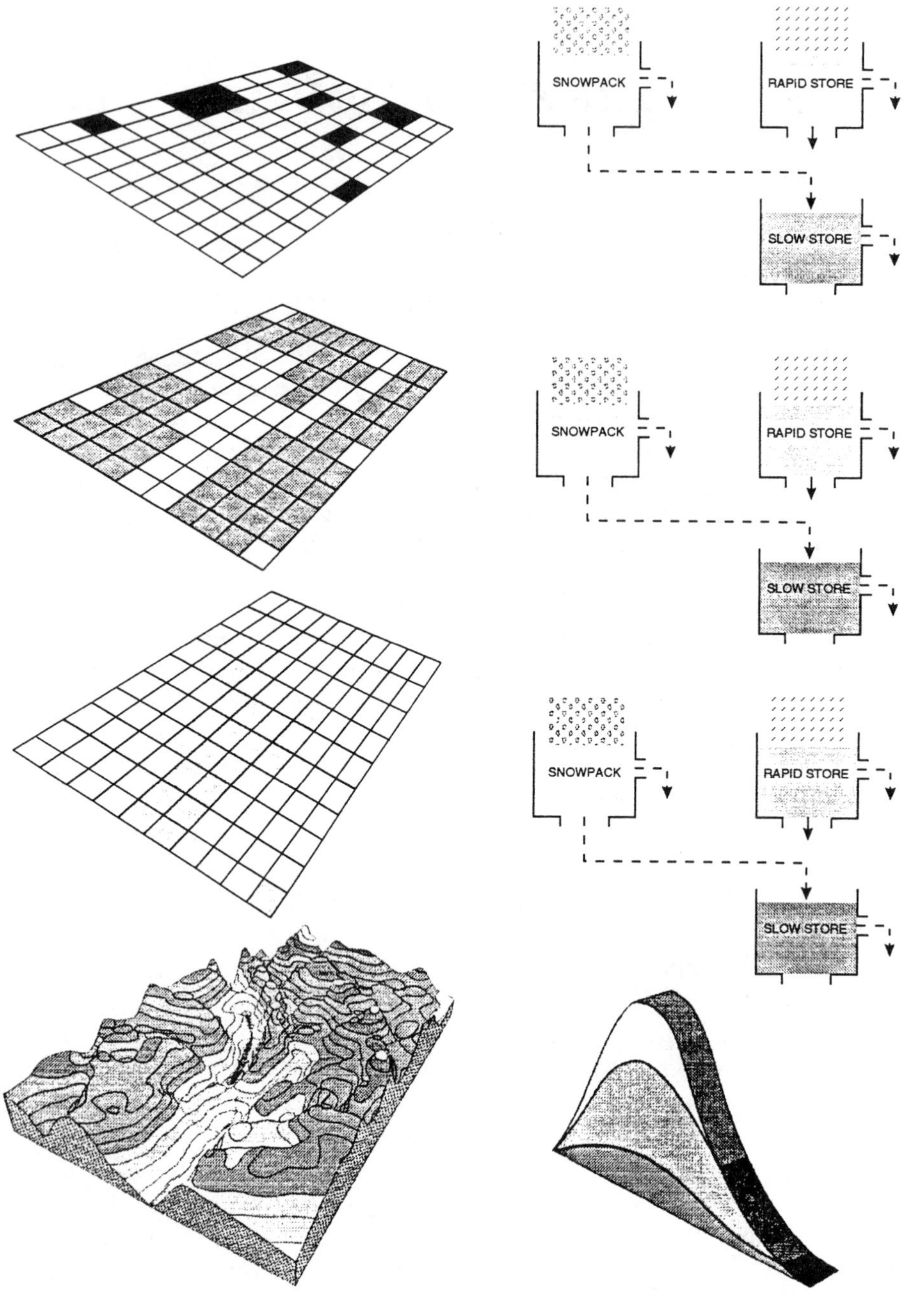

Fig. 15.1. Concept of SLURP watershed model.

meso-scale Canadian watersheds (Kite, 1978) as an alternative to the use of larger and more complicated models such as the U.S. Army Corps of Engineers' Streamflow Synthesis and Reservoir Routing model Reservoir routing model (Kite, 1975).

The major advantage of models in this middle ground is that they can incorporate the necessary physics while retaining simplicity of operation. The SLURP model is able to simulate the behaviour of a watershed at many points and in many variables but avoids the data and computation-hungry excesses of the fully-distributed models. In practical applications, the users of both lumped and physically-based fully-distributed models may also tend towards this middle ground. For example, users of lumped models may simulate the use of a distributed model by successively applying a lumped model to sub-watersheds and, from the other end, Jain et al. (1992) describe the use of areally-averaged variables in the fully-distributed SHE model.

The SLURP model divides a watershed into Grouped Response Units (GRUs) and into areas of different land covers. A GRU is simply a sub unit of the watershed and may be an individual grid square, a group of grid squares or a smaller watershed. The only constraint is that each GRU must have known land covers or land uses (Kouwen et al., 1990). Figure 15.2 shows the concept of collating land cover data from sub-units such as grid squares or satellite image pixels to GRUs. Land cover data are easily obtained from satellite and are used in SLURP as an indicator of climatic zone, vegetation type, soil characteristics and physiography. This embedded use of land cover information makes the SLURP model particularly useful for studies in which land cover is expected to change; for example in climatic change studies. The model has been used in this way to investigate methods of application of an alternative climate scenario to a mountain watershed.

The model may use up to fifty GRUs and up to ten land covers, potentially dividing a watershed into five hundred sub-units. Such division of watersheds or areas of interest is best done on a geographic information system (GIS) which can collate the areal data from different sources and present them in a form suitable for the model. If the model is to be calibrated, then at least some of the GRUs should have streamflow gauges at their outlets. To be sure of stability when calibrating, it is better if the number of GRUs equals or exceeds the number of land cover classes.

For each land cover type the model carries out a vertical water balance at each time interval using land cover roughnesses, infiltration rates and hydraulic conductivities. The resulting rapid and slow runoffs are routed within the GRU using physiographic data obtained from a GIS. The routed runoffs from each land cover are combined into a streamflow from the GRU and this is routed down the stream system to the next GRU.

A key concept in the development of the SLURP model is that it was designed to make maximum use of remotely-sensed data. As well as using

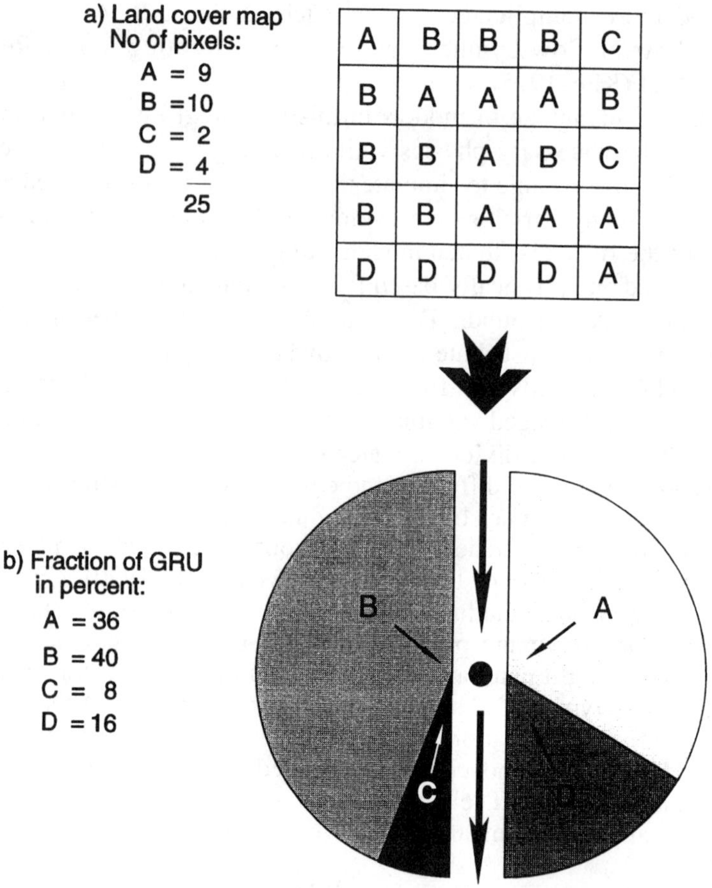

Fig. 15.2. Aggregation of land cover data into GRUs.

land cover information from Landsat or NOAA satellite, the SLURP model will also use NOAA AVHRR visible and infrared data to augment the calculation of snow extent and DMSP satellite data to compute average snow water equivalent over each GRU. This use of satellite data is particularly useful in applying the model to macro-scale watersheds where sufficient land-based data may not be available.

The SLURP model makes full use of the complementary relationship areal evapotranspiration (CRAE) model described by Morton (1983). This method is more realistic than the more usual computation of actual evaporation as a direct proportion of potential evaporation. CRAE provides the areal estimates required by the GRU modelling system and is very suitable for further development using measurements from satellite.

15.2. THE VERTICAL WATER BUDGET

Figure 15.3 shows a simplified flow chart of the vertical water balance which the SLURP model applies to each element of the matrix of GRUs and land covers. The structure contains three non-linear reservoirs or tanks, one for snowpack, one for a rapid response (may be considered as a combined surface storage and top soil layer storage), and one for a slow response (may be considered as groundwater). The three tanks have specified initial contents P(1), P(2), and P(3) mm; there is a maximum depression storage P(7) for the rapid response and a maximum field capacity for slow storage, P(8) mm. The vertical water balance operates at a daily time step.

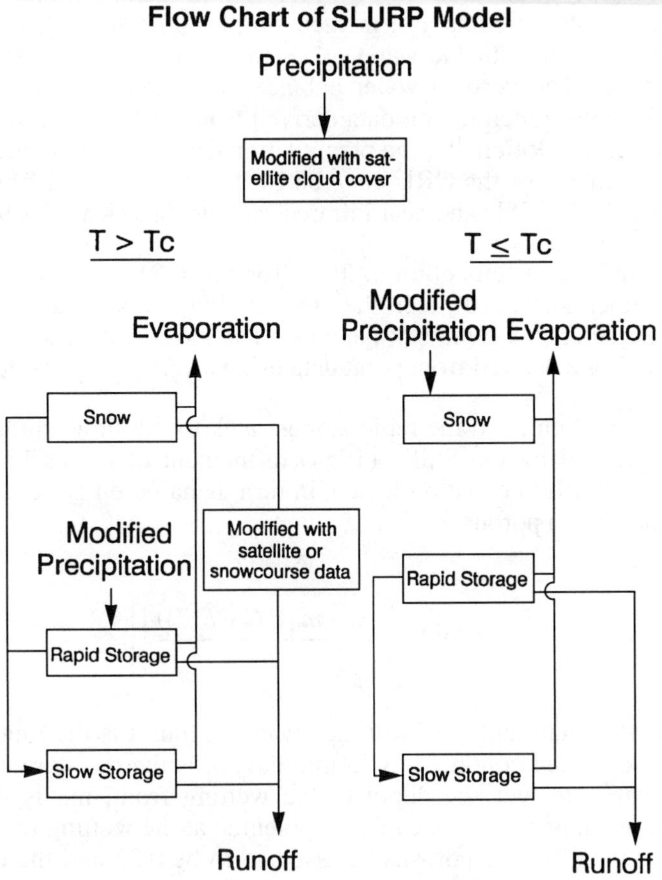

Fig. 15.3. Simplified flow chart of the vertical water balance applied to each land class within each GRU.

At a particular time step, the sequence of operations within the vertical water balance for a particular land cover and GRU would be as follows:

- Daily precipitation, mm, is read from file and may be multiplied by a correction factor P(9). The biggest problem currently facing hydrological modellers is undoubtedly that of obtaining distributed precipitation data. The most physically-sophisticated distributed model will give no better results than the simplest lumped model without good distributed precipitation data. Many so-called distributed models are, in fact, simply lumped models run many times over with the same parameters and the same precipitation input. The vertical water balance in SLURP assumes that each GRU has precipitation data derived from an (external) distribution scheme. Optionally, the precipitation data may be modified by the percentage of the GRU covered by cloud, as derived from NOAA AVHRR visible and near infrared satellite data (Kite, 1989a).

- If the mean temperature, T°C, for the GRU is above a critical temperature P(10), then the precipitation is assumed to be rainfall and enters the rapid storage tank. The mean temperatures for each GRU are derived from point data using a vertical lapse rate.

- Percolation from the rapid storage tank to the slow storage tank is governed by the Philip (1954) refinement of the early Green & Ampt (1911) equation which, in turn, is based on Darcy's equation for flow in porous media.

$$\frac{dF}{dt} = P(6)\left[1 + \left[\frac{(m - m_0)(C + P(7))}{F}\right]\right] \tag{15.1}$$

where F is the total depth of infiltrated water in mm, t is the time in days, P(6) is the saturated conductivity in mm/day, m is the moisture content of the soil averaged over the depth to the wetting front, m_0 is the initial moisture content and C is the capillary potential at the wetting front in mm. In practice, the effective porosity is assumed to be 0.33 and the change in moisture content is calculated as:

$$m - m_0 = 0.33\ [1 - D_s\ /\ P(8)] \tag{15.2}$$

where D_s is the depth of the slow tank, so that the percolation is

$$\frac{dF}{dt} = P(6)\left[1 + [0.33\,[1 - D_s\ /\ P(8)]\ [D_r + \phi]\ /\ D_s\right] \tag{15.3}$$

where D_r is the depth of the rapid tank and ϕ is the potential (suction) head defined as

$$\phi = 250[-\log(P(6)/86400)] + 100 \tag{15.4}$$

- If the contents of the rapid storage is greater than the depression storage then it is also depleted to runoff, Q_s, at a rate based on the Manning equation:

$$Q_s = (D_r - P(7))^{1.67} \sqrt{S_0}\, A/n \tag{15.5}$$

where S_0 is the average overland slope, A is the area of the land cover and n is a roughness parameter for overland flow.

- At the same time, any snowpack, S_p, is depleted to snowmelt, S_m, depending on a melt rate, P(11) mm/day/°C, and the temperature in degrees above P(10). Use of a simple snowmelt factor and temperature index instead of the more physically-based energy budget approach is justified where the seasonal depletion of snowpack is the only variable of interest (Tarboten et al., 1991). Vehviläinen (1991) concluded that there is no reason to use the complex and data-demanding energy balance models in place of the simpler temperature index for operational hydrological models.

$$S_m = P(11) S_p (T - P(10)) \tag{15.6}$$

Snowpack, in terms of snow water equivalent (SWE), may also be input directly to the model from snowcourse data or computed from passive microwave data (Slough & Kite, 1992). It was found that, for a mountain watershed, data from individual snowcourses are of little or no use in the model; point SWE measurements are not good estimates of GRU averages. SWE computed from passive microwave data, on the other hand, provide better estimates because they themselves are average values from large areas (e.g. 25km × 25km for data from the DMSP satellite for south-eastern British Columbia). In addition, the microwave data are available in all weathers and at frequent sampling intervals. Snowmelt is input directly to the slow storage.

- If, on the other hand, the daily mean temperature is below the critical temperature P(10), then the precipitation is assumed to be snowfall and is added to the snowpack storage, if any. There would be no snowmelt in this case, but the rapid storage would be allowed to infiltrate and to runoff as before.

- Snowmelt may, optionally, be modified by the areal extent of snowcover from NOAA AVHRR images (Kite, 1989a).

- Areal evapotranspiration is satisfied, if possible, first from the snowpack (if it exists), then from the rapid storage and, finally, from slow storage. Most hydrological models compute a point potential evapotranspiration E_p using the Penman (1948) or similar method and then calculate a point 'actual' evapotranspiration E_a as a positive linear function of the potential:

$$E_a = \alpha E_p \tag{15.7}$$

where α is a coefficient such as the Priestley-Taylor (1972) factor. A variation on this method replaces α with resistance factors such as those of Monteith (1981).

On the other hand, the SLURP model uses the complementary relationship areal evaporation (CRAE) concept described by Morton (1983) which first computes an areal wet-environment evaporation E_w and then calculates the areal 'actual' evapotranspiration E_a as a complementary function of the wet-environment value:

$$E_a = 2E_w - E_p \tag{15.8}$$

The CRAE model computes E_p by solving the energy balance and aerodynamic equations at equilibrium temperature using a modification of the Penman equation to replace the wind function with a vapour transfer coefficient. E_w is computed from an empirical equation using the slope of the saturation vapour pressure/temperature curve and net radiation. The version of the CRAE model built into SLURP uses mean daily temperature, dew-point temperature and hours of bright sunshine.

Figure 15.4 shows the relationships between E_a, E_w and E_p. The abscissa represents water supply and the ordinate indicates the magnitude of the various evaporation terms. The left hand side of the diagram shows that with no water supply there can be no actual evaporation resulting in a very high potential evaporation. Moving to the right along the x-axis, as the water supply increases, the actual evaporation increases and the potential evaporation decreases until, when the water supply reaches its limiting value, both values meet at the wet-environment evaporation.

Apart from its appeal to common sense, the advantages of using the CRAE concept are that i) areal rather than point values of evapotranspiration are derived which ties in well with the GRU concept of representative areas

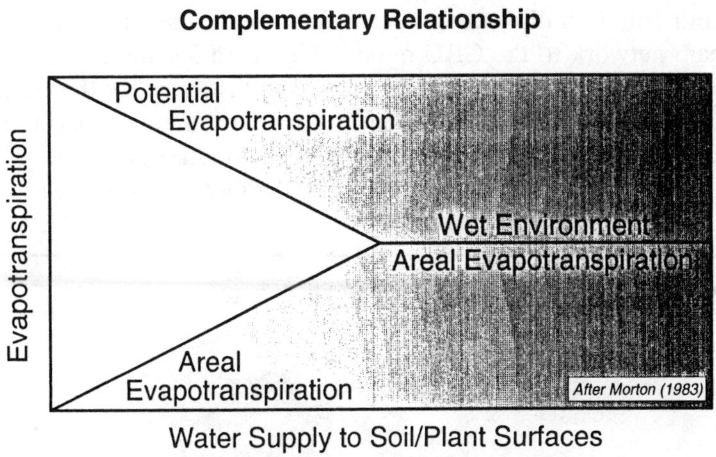

Fig. 15.4. Relationships between actual, potential and wet- environment evaporations in the CRAE model (Morton, 1983).

used in SLURP and that ii) the method is adaptable to the use of satellite data. This again ties in well with the development plans for SLURP.

- Slow storage contributes to GRU runoff, Q_S, at a rate depending on the contents of the tank, D_S; on the temperature ratio (T-P(10))/P(10), and on the conductivity for the slow store P(5):

$$Q_s = P(5) D_s^{[D_s / P(8)] + [T - P(10) + T_{diff}] / [2T_{diff}]} \tag{15.9}$$

The use of a temperature index incorporates the effects of changing viscosity and surface temperature (McKim, 1993), reflects the influence of melting and refreezing of groundwater in cold zones, and may serve as a simple model of the effects of permafrost on the annual cycle of runoff from cold areas.

The vertical component of the model is then applied to all the remaining land covers within the GRU. The concepts used in the vertical water balance are simple and use parameters which may be readily estimated. Application of other, more sophisticated models has demonstrated that, in practice, the use of field and laboratory measurements of infiltration and groundwater parameters may make no improvement to the model results (Jain, 1993).

15.3. ROUTING WITHIN A GRU

A GIS is used to distribute the runoff from each land cover over time. An analysis of the land cover data combined with a GRU streamflow network yields a distribution of distances both to the nearest stream and then along the stream network to the GRU outlet. Figure 15.5 shows, as an example, the sum, L, of the to-stream and the in-stream distances for the Great Slave GRU in the Mackenzie Basin. Figure 15.6 shows the distribution of these distances when associated with the five land classes. The GIS then calculates the change in elevation from each point to the GRU outlet and an

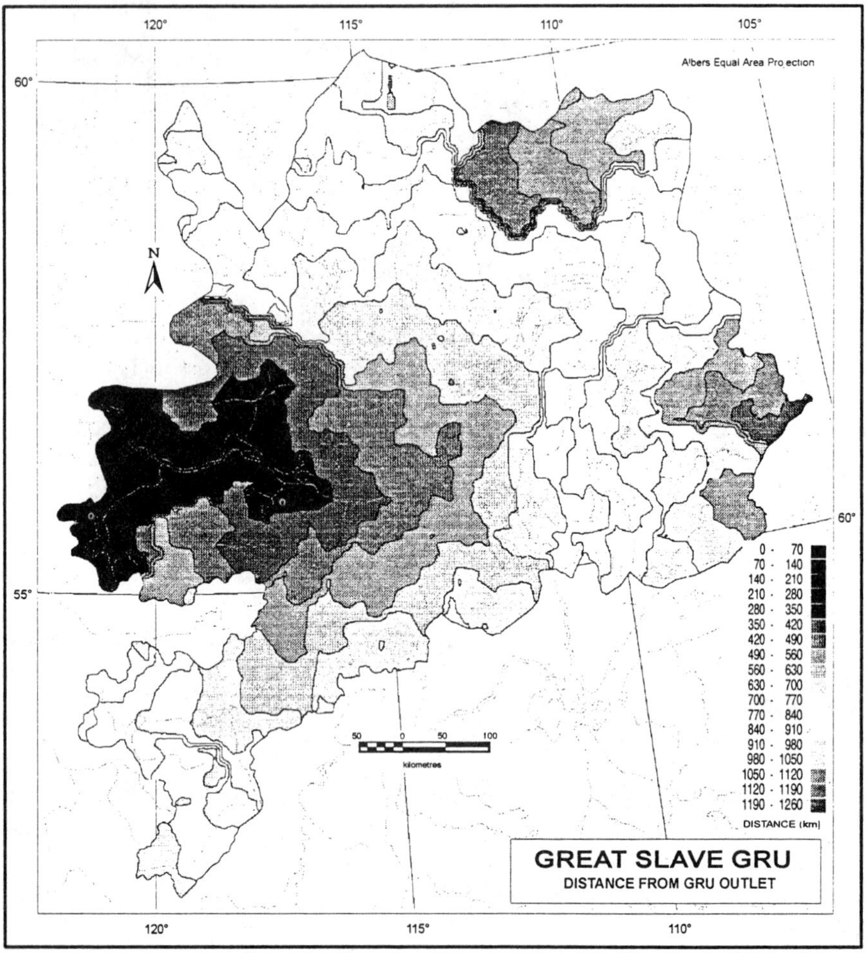

Fig. 15.5. Sum of distances to nearest stream and distances down the stream network to the GRU outlet for the Great Slave GRU, Mackenzie Basin, kilometres.

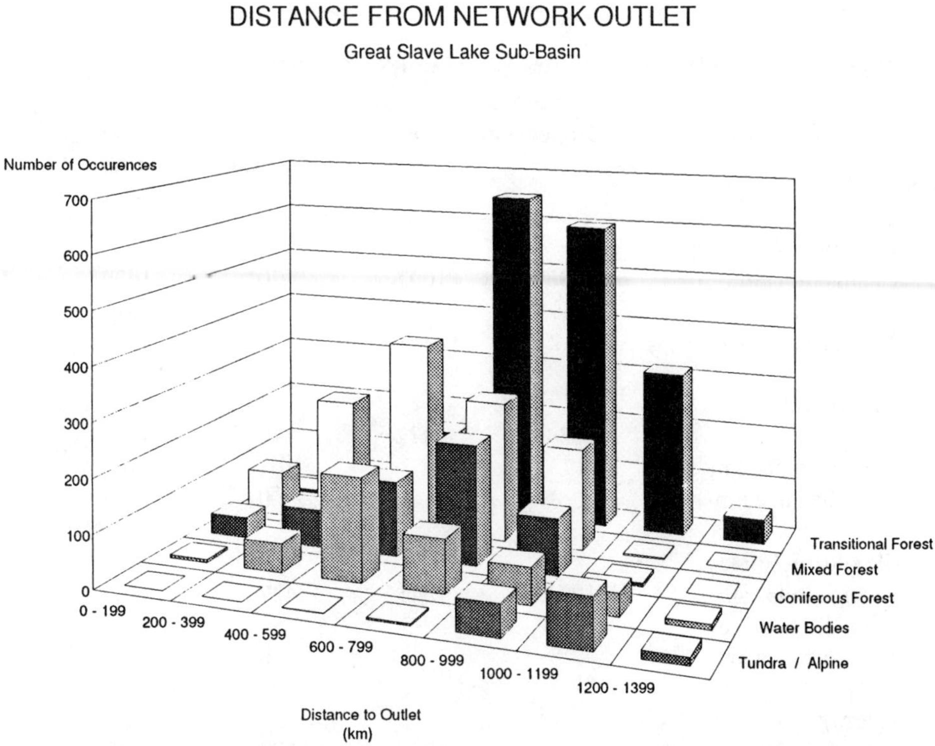

Fig. 15.6. Distribution of the distances to the GRU outlet for each land cover for the Great Slave GRU, Mackenzie Basin, kilometres.

average elevation change for the land cover, H. Table 15.1 shows similar data for the Kootenay Basin used as an example later in this chapter.

Naden (1993) assumed a constant velocity of 0.6m/s for networks of the order of 10,000 km^2 and Wyss et al. (1990) used a subsurface velocity of 2×10^{-3}m/s and a channel velocity of 0.6m/s for small New England catchments. Here, we use a GIS to calculate the changes in elevation between each pixel and the GRU outlet, compute an average velocity for each land cover, V m/s, using Manning's equation with a coefficient of roughness for each land cover, n, and assuming that the hydraulic radius, R, (equal to cross-sectional area divided by wetted perimeter) is unity for an infinitely wide shallow channel.

$$V = (1.49/n)R^{2/3}(H/L)^{1/2} \tag{15.10}$$

From the average velocity and the minimum and maximum distances to the GRU outlet, the minimum and maximum travel times, T_{min} and T_{max} for the land cover may be computed. These travel times are then used in a linear

TABLE 15.1.

Physiographic data and computed travel times Kootenay Basin.

Land cover GRU	Impervious	Forest	Crop/ grass	Average over all classes
Kootenay Crossing				
Change in H, m.	1737	1097	670	1168
Max. distance,km.	39	34	34	36
Min. distance,km.	2	0	1	1
Canal Flats				
Change in H, m.	2234	1472	710	1472
Max. distance,km.	173	167	161	167
Min. distance,km.	8	2	0	3
Skookumchuck				
Change in H, m.	1951	1341	427	1240
Max. distance,km.	98	95	89	94
Min. distance,km.	5	1	0	2
Average over all GRUs				
Change in H, m.	1974	1303	602	1293
Max. distance,km.	103	99	95	99
Min. distance,km.	5	1	0	2
Computed slope, roughness, computed velocity and computed travel times				
Slope	0.0366	0.02616	0.0127	
Roughness	0.23	1.00	0.70	
Velocity (m/s)	1.24	0.24	0.24	
Min. time (days)	0	0	0	
Max. time (days)	1	5	5	

smoothing filter to distribute the runoff from the land cover over time and are then weighted by the percentages of the GRU covered by the land cover, converted to m^3/s and added to the flows of the GRU.

15.4. ROUTING BETWEEN GRUS AND PROVISION FOR EXTERNAL ROUTING

After the runoffs from all the land covers within a GRU have been combined into a GRU streamflow, this flow is routed from one GRU to the next GRU down the stream system. In the SLURP model this routing may be carried out using a simple hydrological procedure or by accessing, from within the program, a user-written routine specifying any set of routing or regulation rules. The simple hydrological routing uses the following non-linear reservoir inflow-storage and storage-outflow relationships:

$$Q = \alpha S^{\beta} \tag{15.11}$$

and

$$S = (P / \alpha)^{1/\beta} \tag{15.12}$$

where P is the inflow in m^3/s , S m^3 is the storage, Q m^3/s is the outflow and α and β are parameters. The parameters are specified to give the degree of lag and attenuation required. For example, if $\alpha = 10$ and $\beta = 0.5$, then the lag will be 2 days and the peak will be attenuated by 10% while maintaining the correct volume.

Alternatively, flows may be routed from one GRU to the next using any other externally-specified method. This is necessary when the model is used, for example, to simulate watersheds containing reservoirs or lakes with defined regulation plans. The user prepares a computer program (in any language) to describe the transformation of inflow to outflow (which may also involve other data files), compiles the program, links in any other routines needed, and names the program SLRP_RUT.EXE. The SLURP model will then provide SLRP_RUT.EXE with the flow data from the correct GRU, SLRP_RUT will carry out the transformation, and SLURP will re-read the modified flows and continue with the simulation.

15.5. CALIBRATION OF THE PARAMETERS

In the ideal model, all parameters will be physically based and easily measured and calibration will be unnecessary. However, this ideal state is unlikely ever to be reached given the extreme variability in hydrological conditions across a watershed. The SHE model is probably one of the most physically realistic of current models, requiring vast numbers of distributed data; yet even SHE is calibrated in a practical application (Refsgaard et al., 1992). SLURP uses parameters which may be estimated directly for different land classes (e.g. Manning's n, infiltration rate, hydraulic conductivity, soil depths, etc) but, because the model is distributed and the land class parameters are applied over large areas, some form of calibration is often necessary. Once parameters have been derived for a particular set of land classes the model may be applied to other watersheds which contain the same land classes without any further calibration.

Table 15.2 lists the parameters of the model and shows which are measured, estimated or calibrated. The sensitivities of the various parameters are classed as **L**ow, **M**edium or **H**igh.

TABLE 15.2.

Parameters of the SLURP Watershed Model.

Parameter	How Derived	Sensitivity
Initial contents of snow store (mm)	Measured	L
Initial contents of rapid store (mm)	Estimated	L
Initial contents of slow store (mm)	Estimated	L
Manning overland roughness, n	Estimated then calibrated	M
Conductivity for slow store (mm/day)	Calibrated	M
Saturated conductivity (mm/day)	Estimated then calibrated	H
Depression storage (mm)	Calibrated	M
Field capacity for slow store (mm)	Calibrated	M
Precipitation multiplication factor	Calibrated	H
Rain/snow division temperature (°C)	Calibrated	H
Snowmelt rate (mm/°C/day)	Estimated then calibrated	H
Between-GRU routing coefficients, α, β	Specified	L

The goodness of fit of the SLURP model may be judged both graphically and from statistical criteria. After each year of model simulation the observed and computed GRU hydrographs will be displayed on screen (Fig. 15.7) and may be output to printer. The screen shows the daily rainfall in blue and snowfall in white along the top of the graph with daily temperatures shown below. The differences between the recorded and simulated streamflows are shown in yellow if the observed streamflow is greater and in purple if the computed streamflow is greater. Annual depths of precipitation, observed runoff, computed runoff and evapotranspiration in mm over the GRU are shown on the screen as are the sum of squares and the standard error of the daily flows.

The model generates statistics for the recorded and simulated streamflows for each GRU. Table 15.3 lists the statistics that correspond to the hydrographs shown in Fig. 15.7.

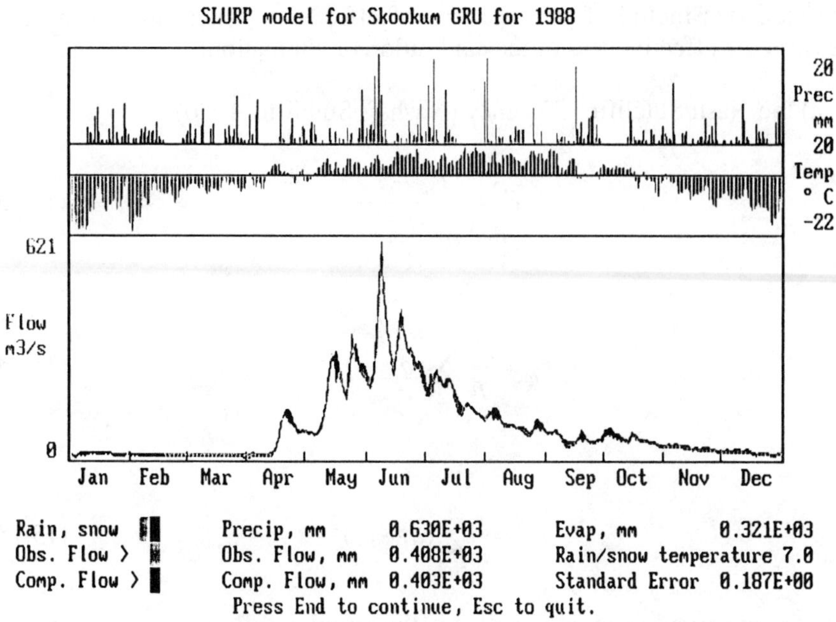

Fig. 15.7. **SLURP screen showing precipitation, temperature, recorded and simulated hydrographs, Kootenay Basin above Skookumchuck, 1988.**

TABLE 15.3.

SLURP watershed model statistics Kootenay Basin above Skookumchuck, 1988.

Mean computed flow, m^3/s	90.66
Mean observed flow, m^3/s	91.96
Mean error, m^3/s	-1.30
Ratio of mean error/observed mean	-0.01
Ratio of abs. error/observed mean	0.11
Coeff. of variation (observed)	1.09
Standard error, m^3/s	15.42
Largest daily error, m^3/s	84.99
Nash/Sutcliffe criterion	0.98
Garrick criterion	0.91
Previous-Day criterion	0.36
Volume deviation, %	1.41

The statistics include four measures of efficiency which may be used to measure the effectiveness of the calibration or simulation:

a) the Nash/Sutcliffe efficiency (Nash & Sutcliffe, 1970):

$$F^2 = \frac{F_m^2 - F_d^2}{F_m^2} \tag{15.13}$$

where

$$F_m^2 = \frac{1}{n}\sum_{i=1}^{n}(q_i - \bar{q})^2 \tag{15.14}$$

and

$$F_d^2 = \frac{1}{n}\sum_{i=1}^{n}(q_i - c_i)^2 \tag{15.15}$$

qi is the observed flow on day i, c_i is the simulated flow on day i and $\bar{q}$ is the mean of the observed flow. Martinec & Rango (1989) have discussed the implications of defining $\bar{q}_i$ as the long-term mean flow.

b) The Garrick efficiency measure (Garrick et al., 1978) in which F_m^2 in (5.13) above is replaced by:

$$F_m^2 = \frac{1}{n}\sum_{i=1}^{n}(q_i - \bar{q}_i)^2 \tag{15.16}$$

where $\bar{q}_i$ is the long term mean flow on day i. This is a more stringent criterion as it compares the model simulation on a particular day to the long term mean for that day. This criterion is also known as the Coefficient of Gain from Daily Means (WMO, 1986).

c) The Previous-Day criterion (Kite, 1991). Given no other information, one possible estimate of flow on day i might be the flow on the previous day, i-1, taking advantage of the persistence component of river and lake time series. A suitable measure of efficiency replaces F_m^2 in (5.13) above with:

$$F_m^2 = \frac{1}{n-1}\sum_{i=2}^{n}(q_i - q_{i-1})^2 \tag{15.17}$$

The sensitivity of the first three criteria to changes in the parameter values is very different. It is easy to calibrate the model to give a high value of the Nash/Sutcliffe statistic, less easy to improve the Garrick statistic and very difficult to obtain a calibration with a high value of the Previous-Day criterion. Figure 15.8 shows an example of the variation in (from the top) Nash/Sutcliffe, Garrick and Previous-Day as the values of two parameters are stepped across a grid. The three criteria are all displayed to the same scale and clearly show the relative sensitivity.

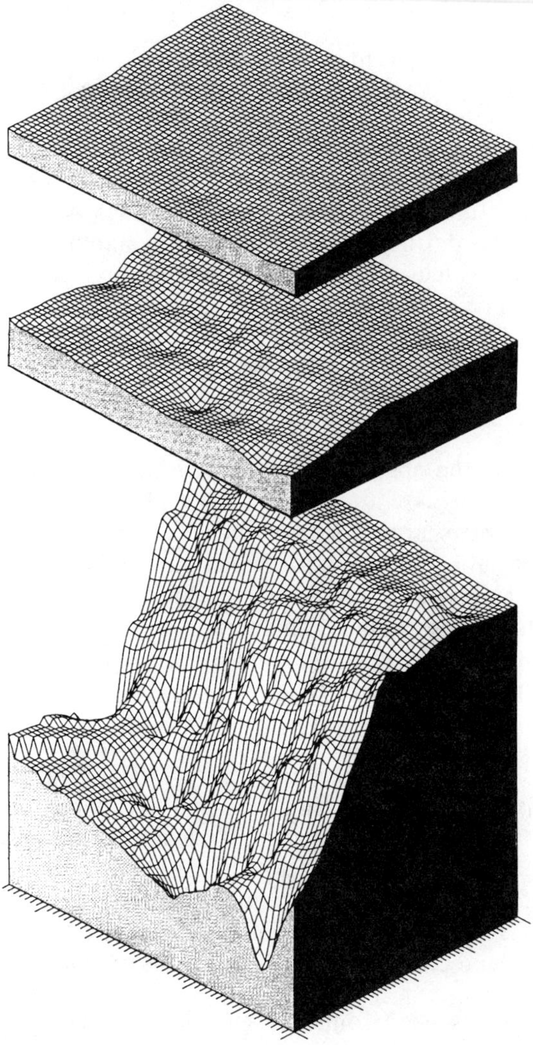

Fig. 15.8. Relative sensitivities of three performance criteria used in the SLURP watershed model as two parameters are changed in a regular grid pattern.

d) The fourth criterion computed in the model is the deviation of runoff volumes, D_v, (WMO, 1986):

$$D_v(\%) = \frac{V_m - V_c}{V_m} \bullet 100 \tag{15.18}$$

where V_m is the measured volume of runoff and V_c is the computed volume of runoff over the period of interest.

The model parameters may be calibrated in three ways:

i) by changing individual parameter values, making a series of simulations and comparing the criteria values;

ii) by running a grid search in which the model is automatically run with parameter values changing in turn by one-tenth of their respective allowable ranges until all values have been tried. The ranges are defined by the maximum and minimum allowable parameter values. The simulation with the lowest sum of squares error will be retained

$$SSQ = \frac{1}{n}\sum_{i=1}^{n}(q_i - c_i)^2 \tag{15.19}$$

where q_i is the observed flow on day i and c_i is the simulated flow on day i.

Following a grid search a screen such as Fig. 15.9 will be displayed showing the original and the new parameter values, the original and the new sum of squares, and a request as to whether the old parameter values are to be updated with the new values. This screen will be repeated for each land class. Note that in the example in Fig. 15.9 no improvement in the optimization function was achieved.

iii) The third calibration option is to use a hill-climbing optimization such as that programmed by Monro (1971). This procedure works in many cases but may find local and not global optima depending on the starting conditions (see, for example, the peaks and valleys in Fig. 15.8). If this option in SLURP is chosen, the user is requested to provide the maximum number of iterations to be used (Fig. 15.10). Following an optimization, Fig. 15.9 will be screened for each land class just as in the previous option.

In practice, it has been found best to use a combination of all three methods. First a number of simulations are made to determine the responses to a few extreme cases. Next, a grid search is made using a wide range for each parameter and, finally, an optimization is made within the most likely parameter ranges.

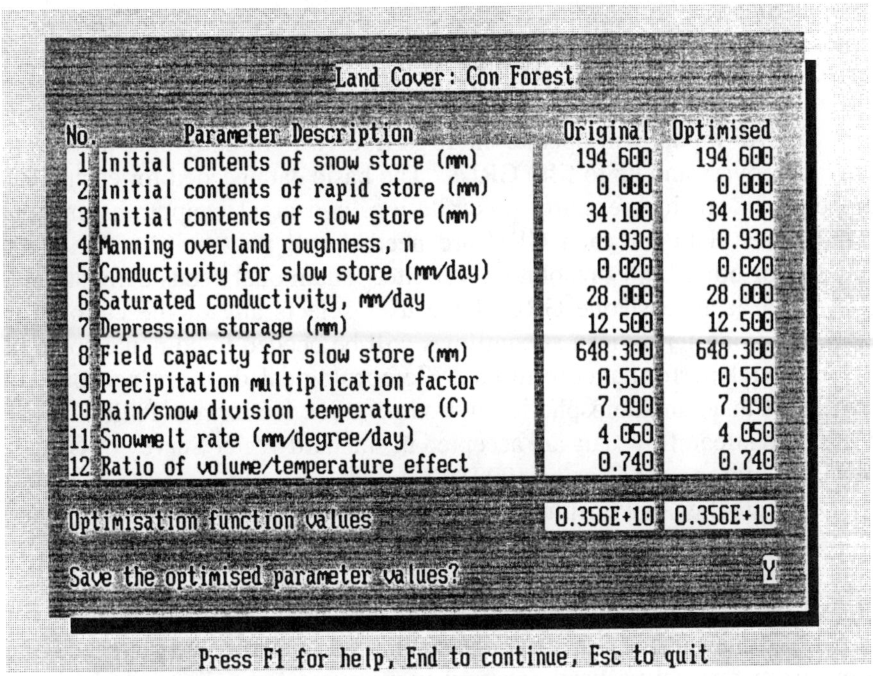

Fig. 15.9. SLURP screen after a parameter grid search showing old and new parameters and sum of square values.

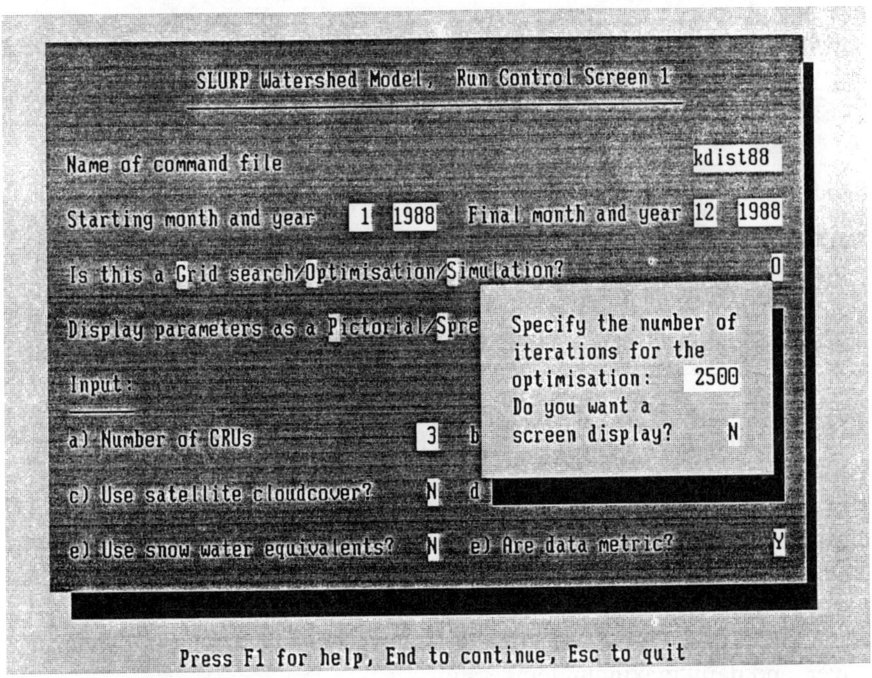

Fig. 15.10. SLURP screen showing the optimisation window on top of the data input screen.

15.6. DATA STRUCTURE

The SLURP model will operate on up to five years of daily data with up to ten land covers and up to fifty GRUs. The basic data needed for each GRU are daily mean temperature, precipitation, and areal evapotranspiration. Land cover data for each GRU are needed and may be obtained from Landsat or other high - resolution satellite. The model needs daily streamflow data at at least one GRU if the user selects any of the calibration options. Long-term mean streamflows will be needed to compute some of the optional performance criteria. Meteorological data are accepted in standard Canadian Atmospheric Environment Service format (Phillips et al., 1988) and streamflow data are accepted in standard Water Survey of Canada format (Environment Canada, 1990).

Daily areal evapotranspiration data are derived using the CRAE method (Morton, 1983) with dewpoint temperatures, mean air temperatures and hours of bright sunshine.

All data units should preferably be metric but provision is made for imperial dimensions. Other options such as snow and cloud cover may be input from remotely-sensed data and snow water equivalent may be input from snowcourses or from satellite (Table 15.4).

TABLE 15.4.

Satellite data used in the SLURP watershed model

Satellite & sensor	Resolution	Frequency	Cost $/image	Use in the model
Landsat MSS	80 m.	8/16 days	800	land cover classification
NOAA AVHRR	1 km	2 per day	25	i) snow and cloud cover ii) land cover classification
DMSP SSM/I	25 km	1 per day	20	snow water equivalent

Snow and cloud cover for the SLURP model have been estimated (Kite, 1989a) using daily visible and near infrared data (bands 1 and 2) from the NOAA satellite series. Figure 15.11 illustrates the multi-spectral classification used to distinguish between snow, cloud and background. Snow water equivalent may be estimated from passive microwave data from the DMSP SSM/I satellite using an algorithm developed by Slough & Kite (1992) using brightness temperature in the 19 and 37 GHz bands, land cover, and daily maximum temperature.

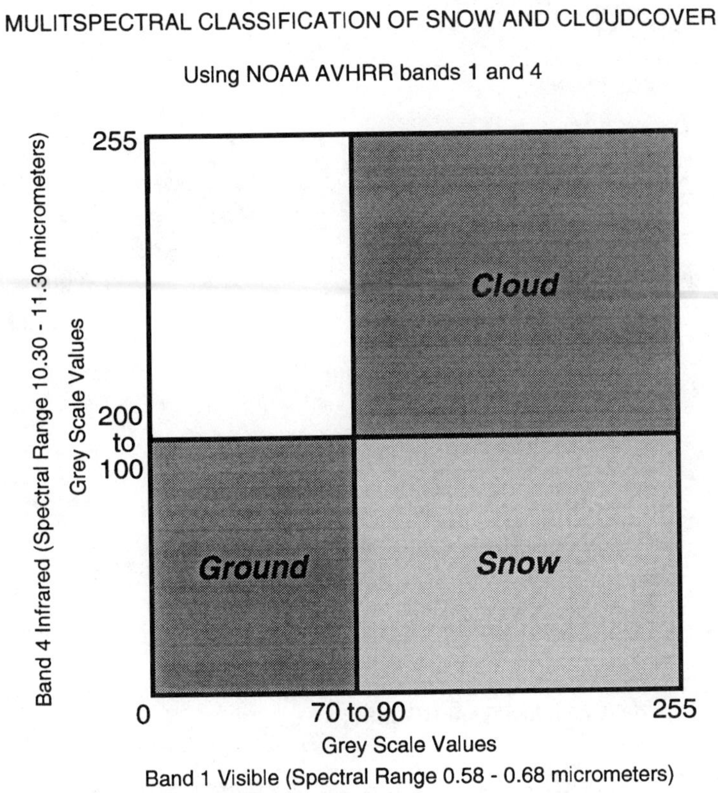

Fig. 15.11. Multispectral classification of snow and cloud cover using NOAA AVHRR bands 1 and 4.

Both the point data (temperature, precipitation, streamflow and other time series) and the spatially distributed data (cloud cover, snow cover, physiographic data, soil types, and land cover classifications) may be retrieved automatically by SLURP from a microcomputer database called HBASE (Kite, 1989b) for any period and any GRU for which data are available. Figures 15.12 and 15.13 show examples of the graphical display of time series and spatially distributed data in HBASE. Daily data from the climate stations are interpolated to grid points and then averaged over sub-watersheds. The interpolation and averaging process accounts for missing data.

SLURP uses a GIS to relate data on topography, land cover and stream distribution and to derive information to define the GRUs and for within-GRU and between-GRU flow routing. Table 15.5 lists the information derived from a GIS.

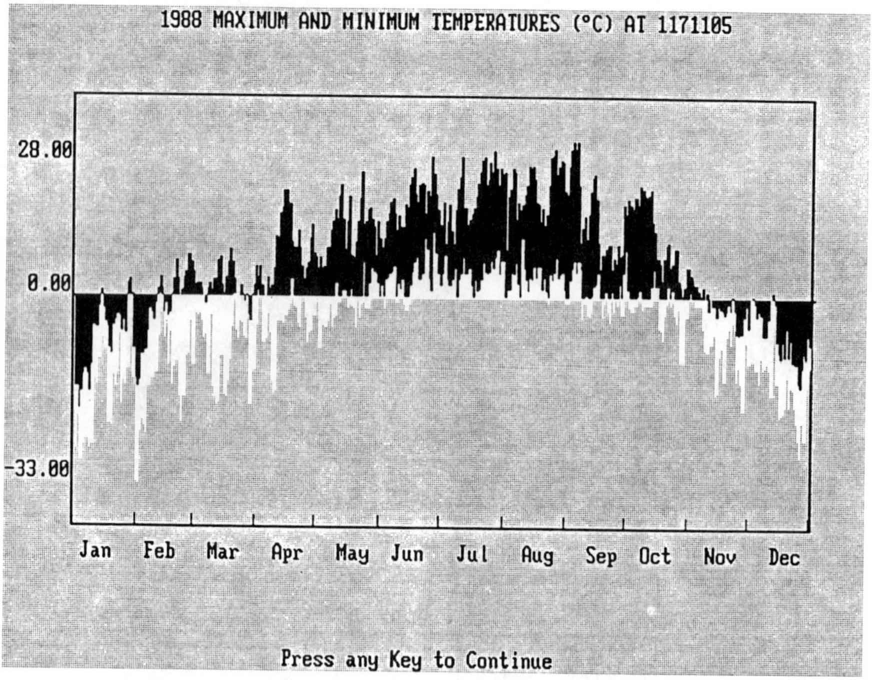

Fig. 15.12. **Daily maximum and minimum temperatures for station 1171105, 1988, graphed from HBASE.**

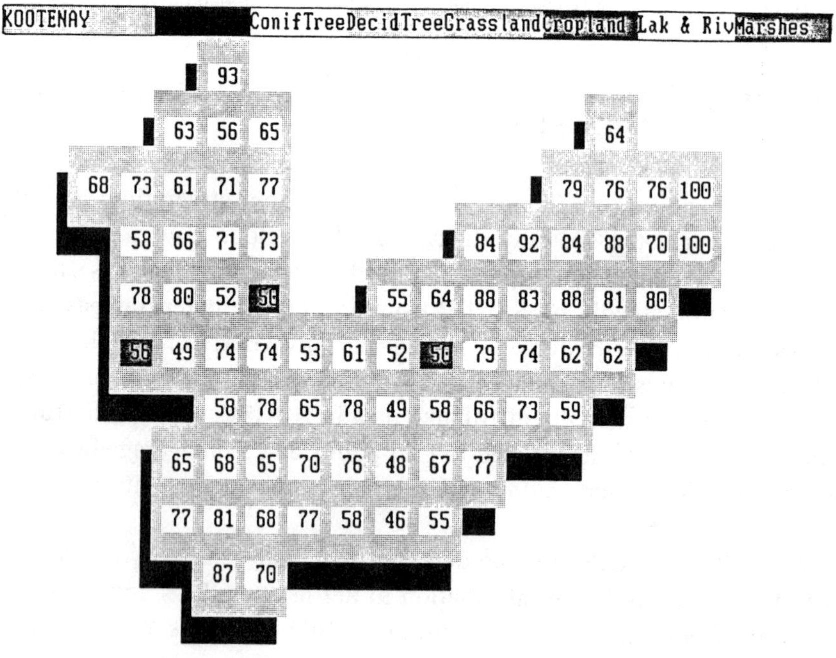

Fig. 15.13. **Percentage of the most significant land cover within each 12×12km grid square, Kootenay Basin above Skookumchuck, graphed from HBASE.**

TABLE 15.5.

Information from GIS used in the SLURP watershed model

Information
GRU boundary and area
land cover percentages within a GRU
distances to nearest stream
distances along the stream network
differences in elevation
weights used for climatic stations

15.7. THE COMPUTER PROGRAM

The computer program for SLURP is written in ProFortran with screens generated using Saywhat for interactive use under PC-DOS or equivalent. The main executable occupies 550K. When started, a menu will appear (Fig. 15.14) offering five options. The first option runs the watershed model.

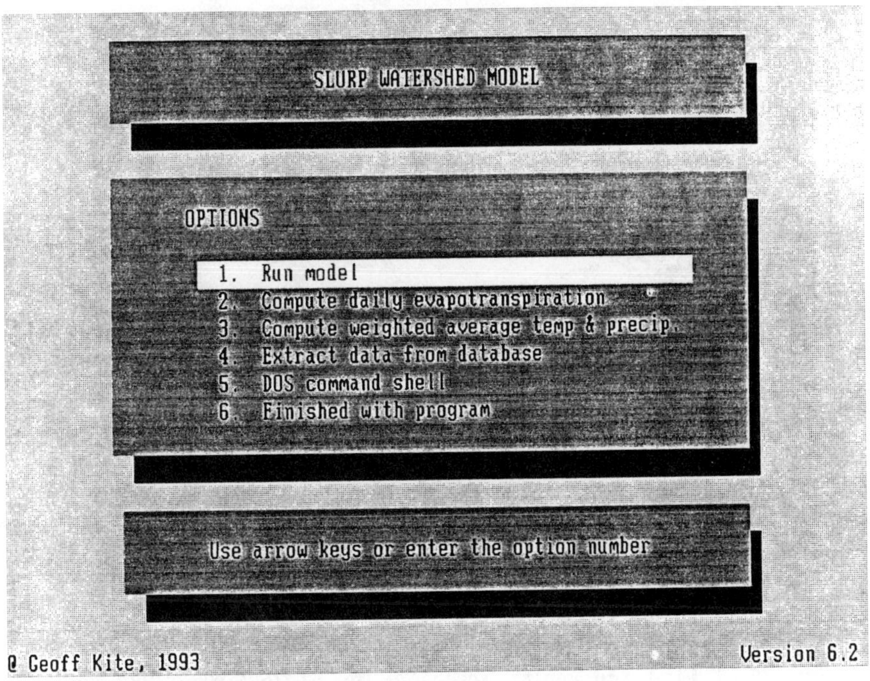

Fig. 15.14. Main menu of the SLURP watershed model.

A data entry screen will appear and the user should key in a name of up to eight characters to identify the watershed. The model will use this name with the added suffix '.CMD' to identify a command file which will keep track of all the data files used and keep a record of the options specified by the user. This file will be updated each time the program is run.

If the user has previously created a command file then the last set of parameters used will be written to the screen. If this is a new command file, a default set of parameters will be presented. In either case the user may modify any or all of the screen fields specifying the start and end of the simulation period, the number of land classes and GRUs to be used, and whether to use satellite data for cloud cover, snow cover and snow water equivalent (Fig. 15.15). The user also specifies at this time whether the model is to be in a calibration mode or in a simulation mode. If the model is to be calibrated then at least one of the GRUs must have recorded runoff data available.

The user then specifies on a second screen (Fig. 15.16) the types of output required (hydrographs, statistics) and where the outputs should be directed (screen, printer, disk file). Note that the user may specify computation of the full range of statistics used in the WMO model intercomparisons (WMO, 1986).

On the third data entry screen (Fig. 15.17) the user completes the matrix of percentages of each land cover class in each GRU. As the model can

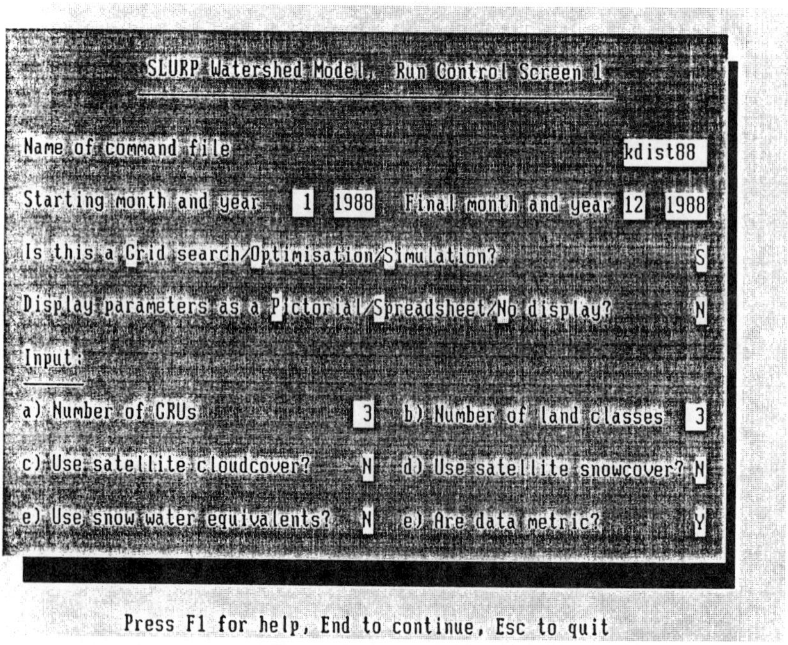

Fig. 15.15. First run control screen of the SLURP watershed model, specifying input options.

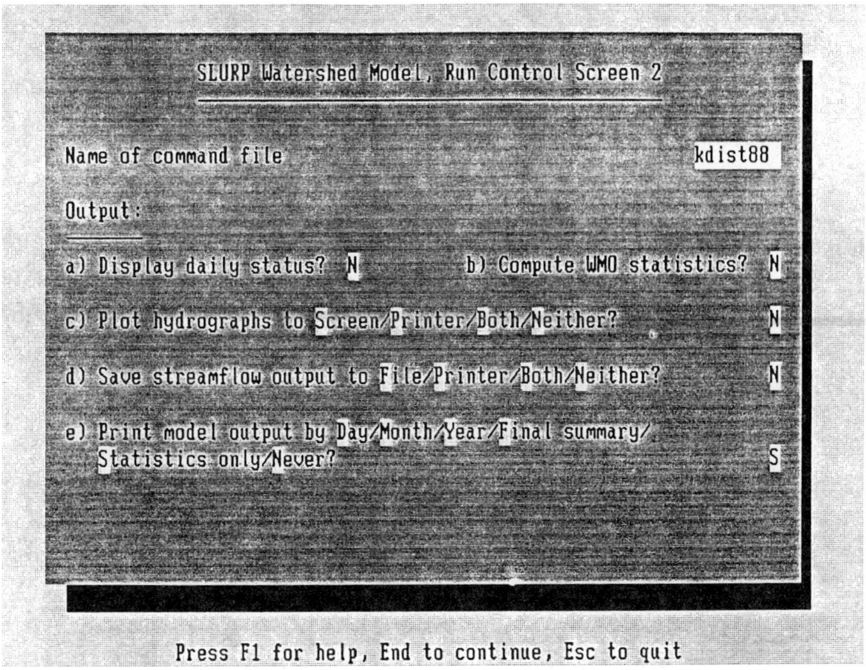

Fig. 15.16. Second run control screen of the SLURP watershed model, specifying output options.

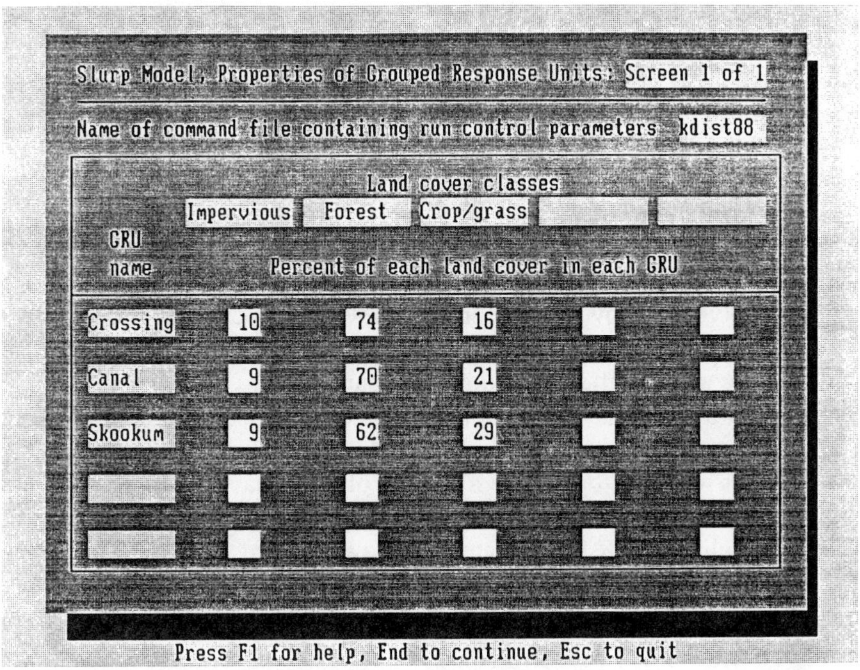

Fig. 15.17. SLURP watershed model data input screen specifying the percentages of each land cover in each GRU.

handle up to 50 GRUs and 10 land covers, if either the number of GRUs or land covers exceeds 5, this screen will be repeated as necessary. The areas of each GRU are also specified on this screen. All this information is stored in the '.CMD' file.

The next two screens provide information for the routing of runoff within each GRU and routing of flows between GRUs. The user estimates a surface roughness (as Manning's n) and (from the GIS analysis) provides average change in elevation and minimum and maximum distances to the GRU outlet for each land cover (Fig. 15.18a). The program then computes minimum and maximum travel times, in days, and displays them in the last two columns on the screen (Fig. 15.18b). The user is given the option to accept these calculated values or to go back and change the input data (for example, to try a different n value) until the results are satisfactory. The values of Manning's n from this screen are also used in the vertical water balance (see Section 15.2).

Figure 15.19 displays the input screen used to specify the between-GRU routing. The model needs to know where the flows from each GRU go to, whether recorded streamflows exist at the outlet of the GRUs and if hydrologic routing (specified by the parameters alpha and beta) or external routing is to be used (see Section 15.4).

Each land class has a separate parameter set which, depending on the options specified earlier, may be presented either pictorially or as a spreadsheet. If the pictorial option was selected (Fig. 15.20) then the screen will be repeated for each land class. Each parameter has three values; the top or right-hand value in each set is the maximum allowable value of the parameter, the centre value is the initial value to be used in simulation, and the bottom or left-hand value of each set is the minimum allowable value of the parameter.

If the spreadsheet parameter display option was selected (Fig. 15.21) then the minimum values will be displayed in the left-most column, the maximum allowable values will be in the right-most column, and the initial values for each land cover class will be in the middle columns. If there are more than five land covers then this screen will be repeated as necessary.

In either case any of the parameter values may be modified by the user and the new values will be stored in the command file.

If the parameters have already been defined in an existing command file, then the parameter-display step may be bypassed entirely.

The user may also opt to see a display of the behaviour of the simulated catchment (Fig. 15.22) showing the daily precipitation, evaporation, snowpack, changes in storages and the snowmelt and runoffs. The display cycles through each land cover and each GRU, all values are displayed proportional to their maximum values over the simulation period, and the display may be stopped at any time.

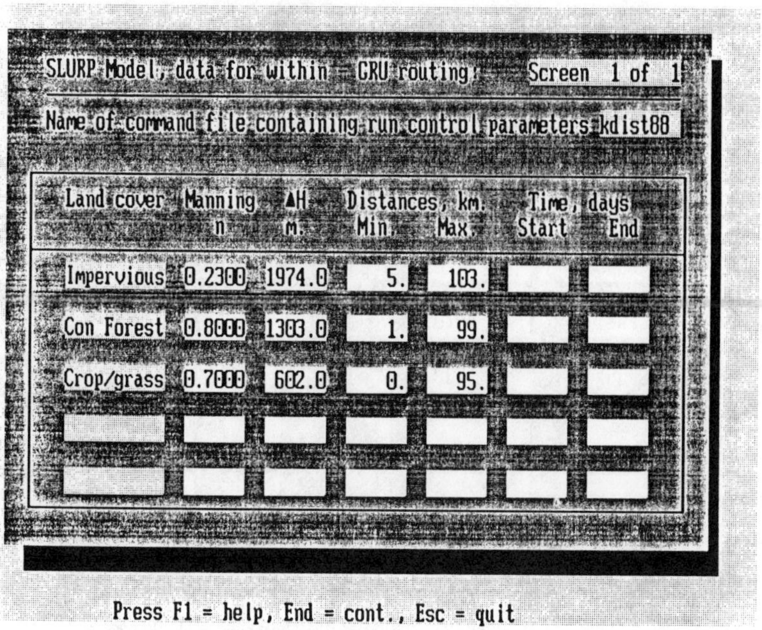

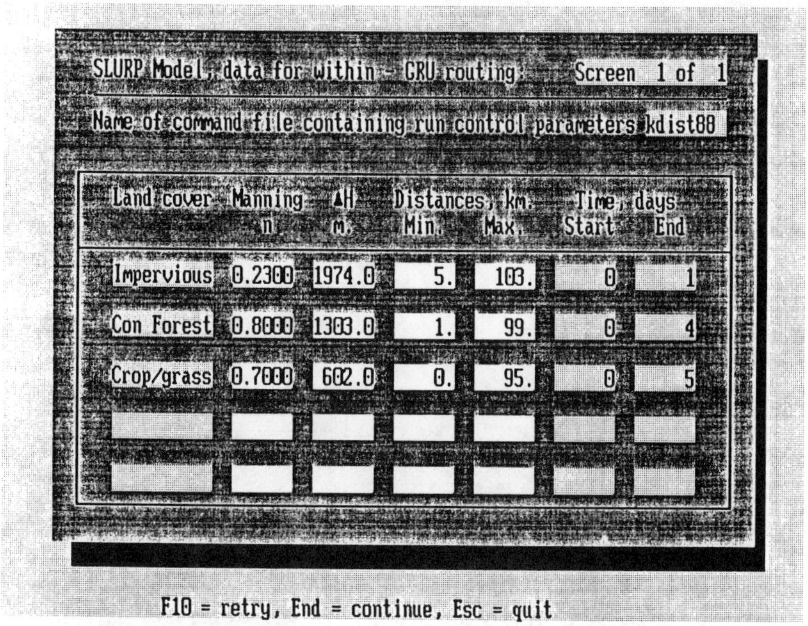

Fig. 15.18. **a) SLURP watershed model data input screen specifying surface roughness, change in elevation and distances to GRU outlet for within-GRU flow routing; b) showing the maximum and minimum travel times computed for each land cover.**

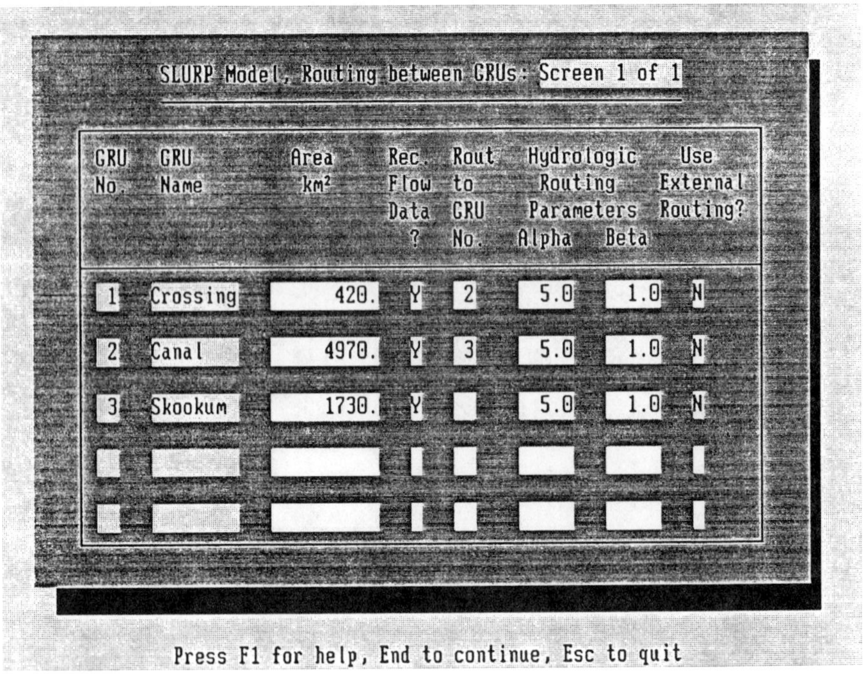

Fig. 15.19. SLURP watershed model data input screen specifying between-GRU flow routing matrix.

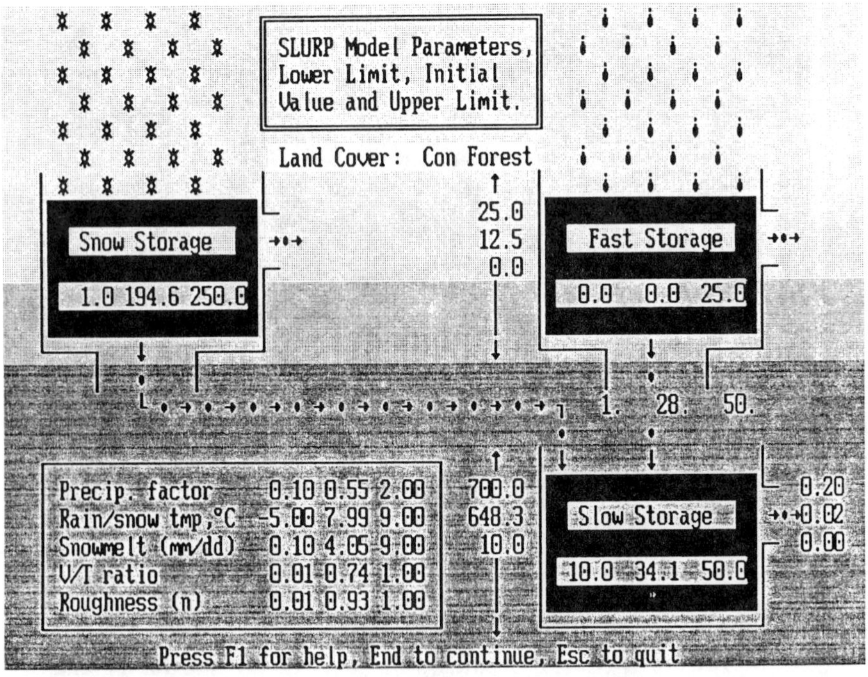

Fig. 15.20. SLURP watershed model parameter input screen; pictorial option.

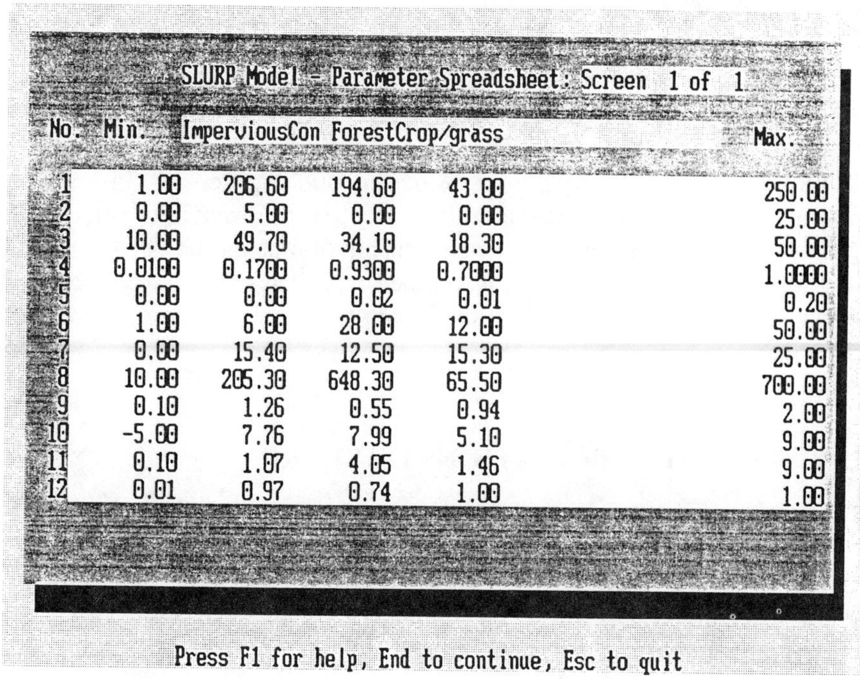

SLURP Model - Parameter Spreadsheet: Screen 1 of 1

No.	Min.	Impervious	Con Forest	Crop/grass	Max.
1	1.00	206.60	194.60	43.00	250.00
2	0.00	5.00	0.00	0.00	25.00
3	10.00	49.70	34.10	18.30	50.00
4	0.0100	0.1700	0.9300	0.7000	1.0000
5	0.00	0.00	0.02	0.01	0.20
6	1.00	6.00	28.00	12.00	50.00
7	0.00	15.40	12.50	15.30	25.00
8	10.00	205.30	648.30	65.50	700.00
9	0.10	1.26	0.55	0.94	2.00
10	-5.00	7.76	7.99	5.10	9.00
11	0.10	1.07	4.05	1.46	9.00
12	0.01	0.97	0.74	1.00	1.00

Press F1 for help, End to continue, Esc to quit

Fig. 15.21. SLURP watershed model parameter input screen; spreadsheet option.

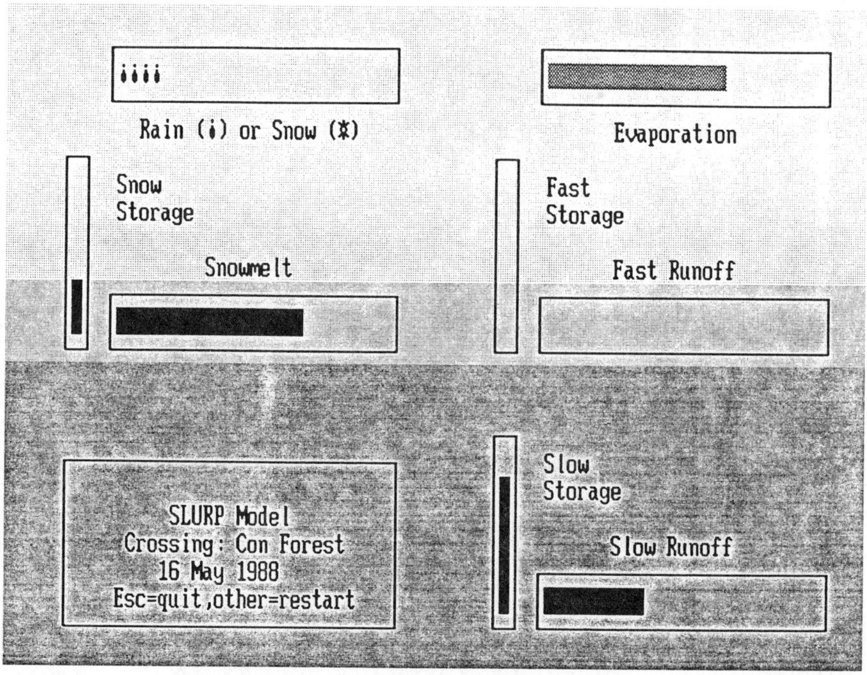

Fig. 15.22. SLURP watershed model screen displaying the behaviour of a grouped response unit.

The program will then cycle through the matrix of land covers and GRUs optionally displaying hydrographs (e.g. Fig. 15.7) and statistics (e.g. Table 15.3) for each GRU. After completion of the simulation, and any outputs required, the model recycles back to the opening screen (Fig. 15.15). Escaping back to the main menu (Fig. 15.14), the user may select other options to compute daily areal evapotranspiration using the CRAE method, compute daily precipitation and temperature for a GRU from recorded climatological data, or to extract more data from the hydrological database, HBASE.

15.8. RECENT APPLICATIONS OF SLURP

15.8.1. Investigation of Alternate Climate Scenarios in a Mountain Watershed

While the author has found no evidence to support the popular global warming hypothesis (Kite, 1993b) it is evident that the existing tools used to investigate the effects of alternative climate scenarios on water resources are inadequate. The SLURP model, because it relates the parameters to land cover information, can be used to study the effects of changes in land cover caused by climatic change.

General circulation models (GCMs) are tools derived from numerical weather prediction models which are used to investigate alternative climates including that resulting from a doubling of atmospheric greenhouse gases. They are based on the physical laws of conservation of energy and mass for a rotating sphere subject to an external heat source and include descriptions of both large-scale atmospheric processes and smaller-scale processes such as cloud distribution, precipitation generation, surface transfers of energy, moisture and momentum and land surface hydrology. The models operate at time intervals of about 20 minutes and at resolutions in the order of 2 or 3 degrees latitude and longitude.

The effects of these alternative climates on water resources have been estimated with conventional hydrological models which have generally assumed implicitly that the watershed itself would remain unchanged. However, it is more likely that a changing climate would alter the distribution and the transpiration efficiency of vegetation types within a watershed.

The SLURP model was calibrated and verified for the Kootenay River Basin in the Rocky Mountains of British Columbia. The changes in land cover associated with a doubled carbon dioxide scenario were estimated and the model was rerun incorporating the changed climate and the changed land cover (Kite, 1993a).

A Landsat MSS image of the Kootenay River Basin above Skookumchuck was classified (Kite, 1989a) into three land cover groups (bare ground & ice, coniferous forest, crop & grassland). The basin was divided into three GRUs consisting of groups of 12 × 12 km grid squares each containing land cover characteristics (Kite & Kouwen, 1992). The percentages of each of the three land covers were known for each grid square and were aggregated for each GRU.

Areal averages of recorded daily mean temperature, precipitation and evapotranspiration were derived for the period 1986-1990 for each GRU. Simulated climatic data for 1 × CO_2 and 2 × CO_2 scenarios were obtained from the second generation Canadian Climate Centre GCM (CCC, 1991). Mean monthly values of climatological variables were obtained for a 10-year control run (1 × CO_2) and a 10-year experiment run (2 × CO_2) at a grid point (116.25°W, 50.10°N) within the Kootenay Basin. The monthly differences between the two GCM runs were applied to the recorded daily 1986-1990 temperature, precipitation and evapotranspiration data to provide a second set of data for the hydrological model.

The distribution of land covers obtained by classifying Landsat images was modified by considering the effects of climatic change on ecoclimatic zones. The mean annual change in temperature for a doubling of CO_2 was converted to a vertical movement of vegetation zone and transferred to each GRU in the Kootenay Basin. Figure 15.23 shows an example of the changes in area within one GRU.

Daily areal evapotranspirations under the changed climate were calculated using the changed temperatures and were then modified to reflect an increased stomatal activity caused by the doubling of CO_2.

The hydrological model was calibrated using recorded streamflow for the period 1986-1990. Figure 15.24a shows the recorded and simulated hydrographs. The model was then run again using the modified climatic data and the changed land cover distribution but retaining the original parameter set (Fig. 15.24b).

The 2 × CO_2 results show a considerable increase in precipitation and a corresponding increase in mean streamflow. Evapotranspiration is decreased by about 10% in the 2 × CO_2 run because of the changed temperatures and land covers. Figure 15.24b shows that, although the maximum streamflow is virtually unchanged, the frequency of high flows has increased significantly in the 2 × CO_2 run. For example, mean daily flows above 300 m^3/s occurred only six times in the period 1986-1990 but 16 times under the 2 × CO_2 scenario.

The inter-annual distribution of hydrological variables is also affected by the climatological changes. Despite the increased precipitation, the monthly snowpacks were reduced because of the higher temperatures. Accordingly, the mean monthly streamflows are increased and the peak has been shifted from June to April. Table 15.6 compares the percent differences in monthly

CANAL FLATS SUB-BASIN
AREA-ELEVATION CURVES

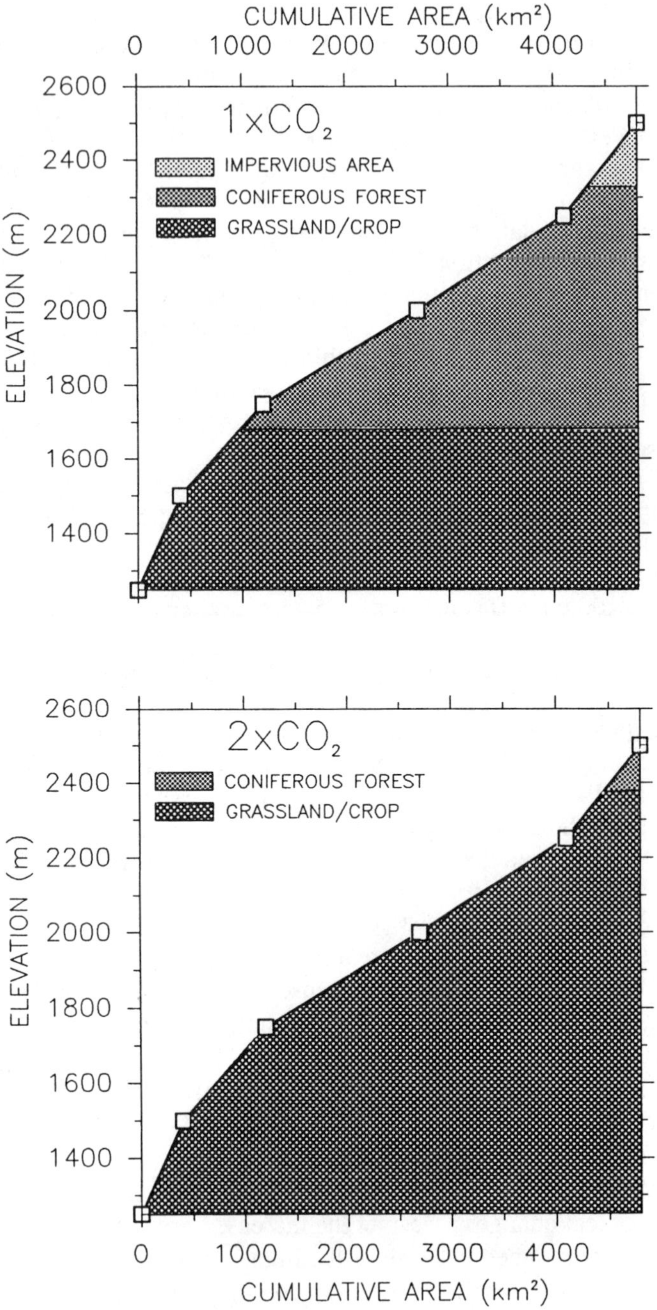

Fig. 15.23. Hypsometric distribution of land covers under 1 $\times$ CO_2 and 2 $\times CO_2$ scenarios, Kootenay Basin above Skookumchuck.

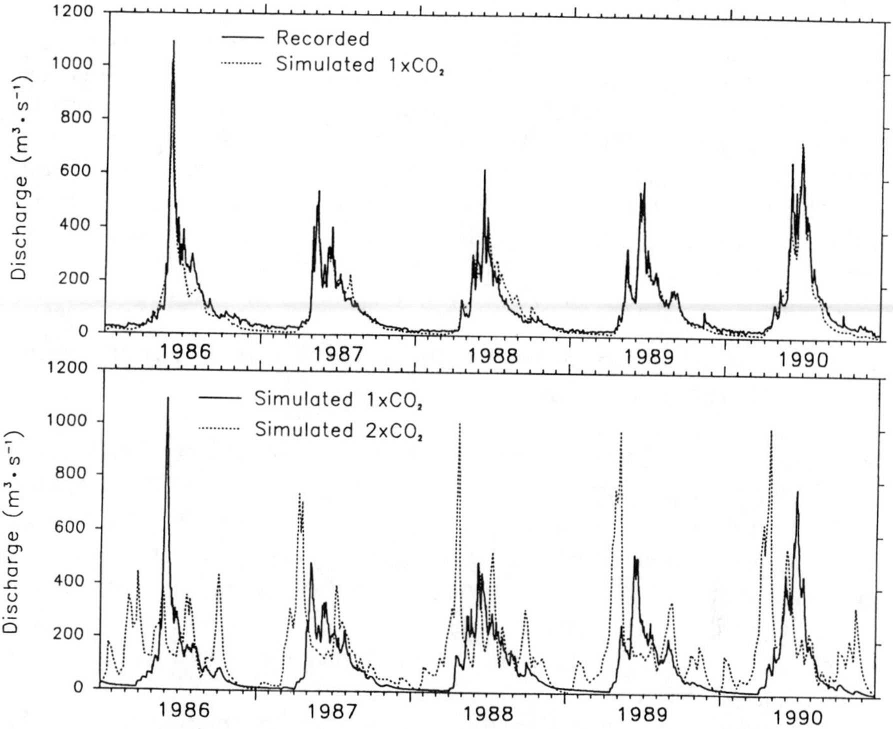

Fig. 15.24. Hydrographs, Kootenay River at Skookumchuck. a) Recorded and simulated hydrographs under 1 × CO_2 conditions. b) Simulated hydrographs under 1 × CO_2 and 2 × CO_2 scenarios.

TABLE 15.6.

Percent differences in mean monthly snow water equivalent between 1 × CO_2 and 2 × CO_2, Kootenay Basin, mm.

Month	GCM	SLURP model
January	-91	0
February	-97	-8
March	-99	-35
April	-100	-86
May	-100	-100
June	0	-100
July	0	0
August	0	0
September	-100	-100
October	63	-75
November	-96	-33
December	-98	-4
Annual average	-97	-33

snow water equivalent, averaged over the basin, between 1 × CO_2 and 2 × CO_2 as computed by both the hydrological model and the GCM. It is clear that the GCM shows a much greater decrease in snowpack than does the hydrological model. This is because each land cover in the hydrological model has a different snowmelt rate so that, as the land covers change from one climatic scenario to the other, the overall melt-rate changes. In the GCM, on the other hand, no such changes take place. SLURP also shows a more reasonable annual cycle in the snowpack than does the GCM.

15.8.2. Using Input Data from a GCM in a Macro-Scale Watershed

While the atmospheric components of GCMs are often very sophisticated and the land-phase components may be very detailed, most GCMs contain no lateral transfers of water within the land-phase. A GCM carries out a vertical water distribution at each grid point using precipitation, evapotranspiration and changing groundwater storages, but any water excess or overflow is simply discarded and takes no further part in the model computations. Such GCMs are modelling an incomplete hydrological cycle (Kite et al., 1993).

From a climatological viewpoint this loss of information may not be important. However, if the data from GCMs are to be used to investigate the possible effects of alternative climates then this shortcoming becomes important. The political impetus for climatic change studies is that such changes may affect our lifestyle and our lifestyle depends, to a large extent, on the lateral distribution of water. Also, measured data are needed to validate GCMs and while areal estimates of precipitation, evaporation and soil moisture are difficult to derive, river runoff data is accurate, readily available and represents an integral of the watershed activities.

A computer program was written to carry out a vertical water balance of published data from the Canadian Climate Centre's GCM II (CCC, 1991) at 12-hourly intervals over a 10-year test period at each GCM grid point with an area within the Mackenzie Basin. The program aggregated the 12-hourly water excesses from each relevant grid point to 10-year mean monthly combined water excess for the whole Mackenzie Basin. No time delays or attenuation were included and no consideration was made of the geography or physiography of the basin. (Fig. 15.25, short dashes).

The gridded GCM outputs were then used as distributed climatological inputs to SLURP. Daily precipitation, temperature and evaporation for the 10-year test period were computed for each of the five GRUs using data from surrounding GCM grid points. The resulting simulated flows were averaged to mean monthly values and are compared in Fig. 15.25 (long dashes) to the recorded data (solid line) and to the water excesses computed from the GCM (short dashes). The improvement due to the use of the hydrological model is dramatic.

The SLURP model was calibrated using recorded climatological and hydrometric data for five GRUs. Data from the NOAA satellite's AVHRR sensor were classified to give the land cover distribution for the Mackenzie Basin. The resulting 14 land cover classes were re-sampled to five major categories and the percentages of each category within each GRU were measured using the GIS (Fig. 15.26).

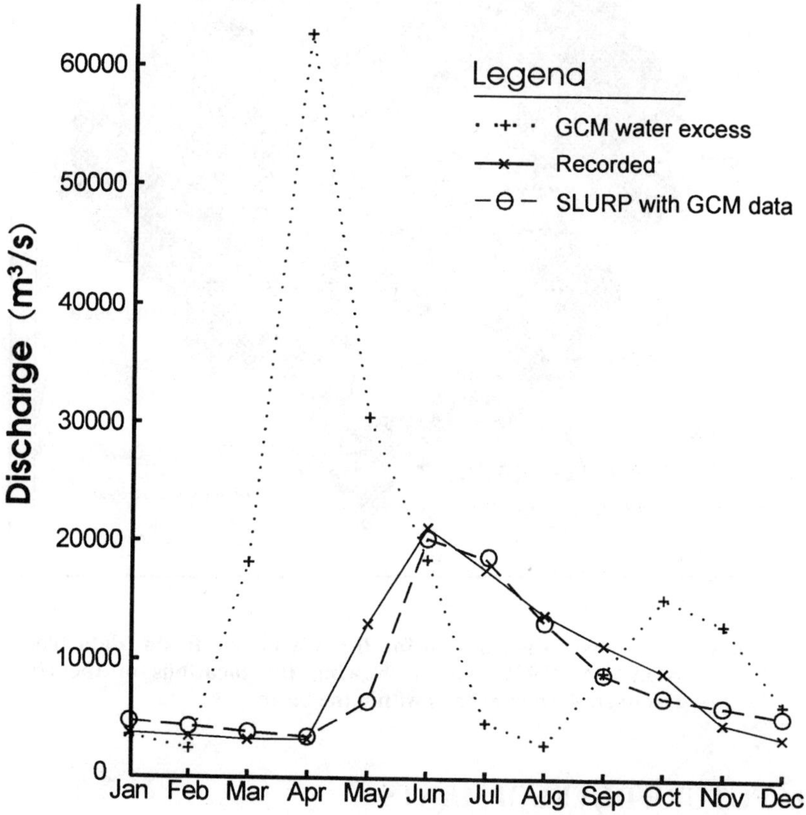

Fig. 15.25. Comparison of mean 1972-1990 observed monthly streamflow at station 10LC014, Mackenzie River above Arctic Red River, (solid line) with mean monthly streamflow simulated from a water balance of CCC GCMII data over a 10-year period (short dashes) and with mean monthly streamflow simulated by the SLURP hydrological model using distributed input data from the CCC GCMII (long dashes), m^3/s.

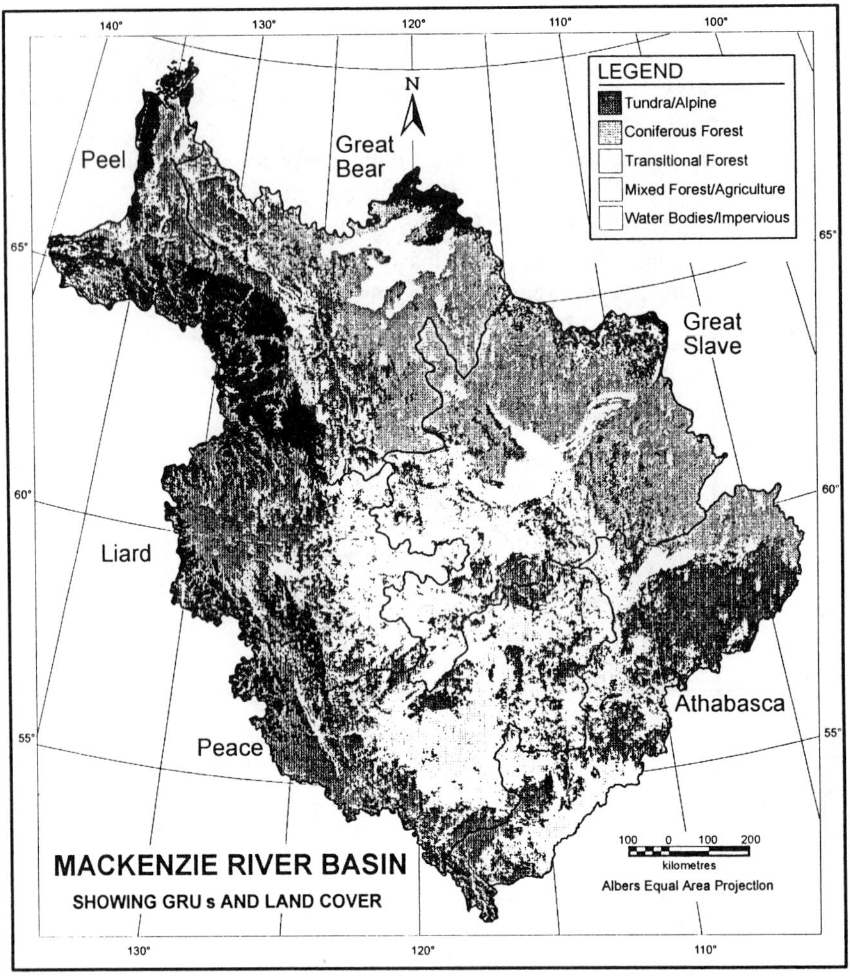

Fig. 15.26. Land cover distribution for the Mackenzie Basin (data from NOAA AVHRR sensor) showing the locations of the five Grouped Response Units within the basin.

15.9. FUTURE DEVELOPMENTS

15.9.1. Glacier and Permafrost Components

Glaciers are an important hydrological factor in many cold regions because of their storage capacities and modifying influence on streamflow. Work is in progress at the National Hydrology Research Institute (NHRI) to develop an advanced glacier component for the SLURP hydrological model based on remote sensing data (Landsat & SAR) and mass-balance data collected over many years for Canadian glaciers (Brugman, 1993).

The hydrologic response of permafrost basins is typified by increased surface runoff due to the impermeability of frozen soils and a consequent reduction in baseflow. While this is also true of seasonally frozen soils, the permafrost zones also have severely reduced storage capacities. As development of northern areas progresses, there is an increasing need for a hydrological model able to simulate the performance of permafrost basins. Research at NHRI is using data collected over many years of northern fieldwork to develop a permafrost component for the SLURP hydrological model. It is anticipated that the algorithms developed will be adaptable to other distributed models.

15.9.2. Evapotranspiration from Satellite

Research at the CCC (Bussières & Louie, 1989) and at the NHRI (Granger, 1993) is aimed at using data from GOES and NOAA satellites in the CRAE model. The work is based on the feedback relationship between surface temperature and vapour pressure deficit. Granger (1993) has demonstrated the method for a variety of land cover types including bare soil, wheat, prairie grass and a closed forest canopy. The present work will incorporate the algorithms in the SLURP hydrological model at a regional scale and will tie in with the use of the experimental global vegetation index (NOAA, 1992) in SLURP.

15.9.3. Hierarchical Hydrological Model

There is a clear need for hydrological models that will work interactively with GCMs. To adapt the existing type of hydrological model to a new role as a macro-scale surface-phase component of a global model will mean major changes. A macro-scale hydrological model must be applicable to large and diverse regions without recalibration and must be able to operate at a wide range of time scales. The model must include the effects of topography, land cover and local meteorological variations (for example, orographic precipitation effects). The spatial coverage of the model must range from the smallest hydrological element to the size of the grids used by GCMs. Experience in developing large-scale models (c.f. Becker & Nemec, 1987; Kouwen et al., 1990; Kite, 1991) indicates that, from many viewpoints, a grid size of 10-12km is appropriate for regional hydrological modelling. Data for land cover are available at a resolution of 30 m., or finer, but for practical purposes have usually been aggregated up to the ten to twelve kilometer scale (Ott et al., 1991; Kite 1991). Channel hydraulics for hydrological modelling may be performed adequately using a reach length of 10 km (Tao & Kouwen, 1989).

The time resolution of a macro-scale hydrological model depends on both the internal requirements of the model simulations and on the need to transfer data to and from a GCM. Channel routing is probably the most time-critical component of the hydrological model, varying from hourly routing using kinematic wave theory to daily water transfer using non-linear reservoirs and monthly water balance calculations using linear reservoirs. The model must include a mechanism for aggregating and disaggregating data to the typical GCM requirement of data at 15-20 minute intervals over a period of years.

To accommodate these complex requirements, the macro-scale hydrological model must be physically-based with appropriate representations of the land-phase processes. It is possible to use lumped models as, for example the current work on soil / vegetation / atmosphere / transfer schemes (SVATS), (Avissar & Verstraete, 1990). But, given the distributed nature of precipitation and the non-homogeneity of most basins, a realistic physically-based model should also be distributed. Each of the physically-based components of the distributed model should be verifiable individually. The key to the development of such a model is that the physical descriptions must be appropriate; there is no sense in trying to apply too detailed a model. The data needed will not be available at the macro scale and the micro-scale physics may not be applicable anyway (Becker & Nemec, 1987).

Most existing distributed hydrological models are based on a grid-square arrangement which allows for the definition of physical parameters in terms of fixed direction and the easy importation of data from remotely-sensed sources. However, using grid squares may mean unnecessary computation if many grid squares have the same properties and if flows are routed from square to square. The finer the grid size used, the more inhomogeneity will be found and the more computation will be needed. Given the uncertainties in meteorological inputs to hydrological models, a point will be reached at which this becomes unreasonable.

One possible approach is to develop a hierarchy of model components which may operate at different spatial and temporal scales but use common concepts and a common database to provide a continuum from the GCM scale to the scale of the hydrological/hydraulic application (Fig. 15.27). SLURP will be used as the intermediate level component acting as a watershed model in its own right and as a filter connecting the top and bottom levels of the HHM, feeding data on state variables and fluxes up and down the scale.

Top Level (input/output to GCM's)

Intermediate Level (watershed model)

Bottom Level (hydraulic routing)

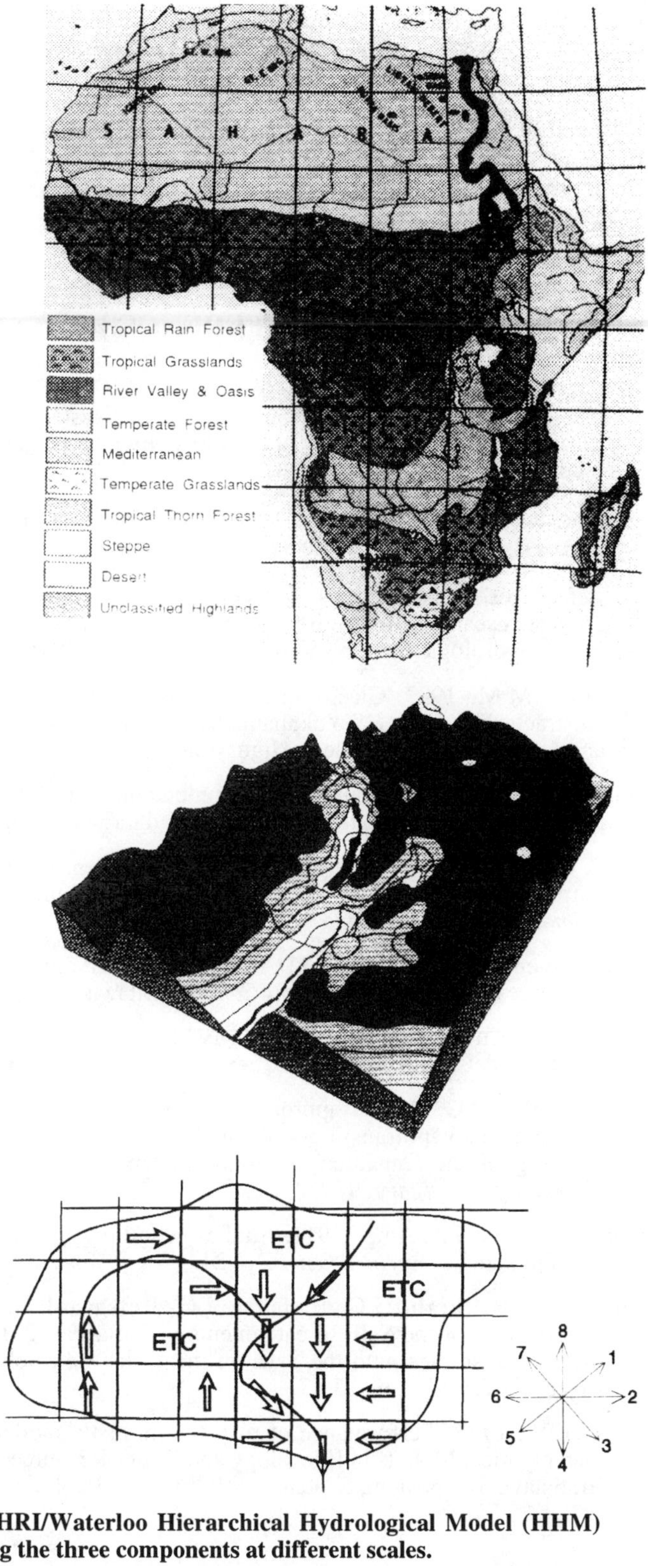

Fig. 15.27. The NHRI/Waterloo Hierarchical Hydrological Model (HHM) showing the three components at different scales.

ACKNOWLEDGMENTS

ProFortran is a trademark of Prospero Software, London, England, Saywhat is a trademark of Software Science, Burlingame, California and PC-DOS is a trademark of International Business Machines Corp.

REFERENCES

Abbott, M.B., J.C. Bathurst, J.A. Cunge, P.E. O'Connell & J. Rasmussen, 1986. 'An introduction to the European Hydrological system - Système Hydrologique Européen, 'SHE', 1: History and philosophy of a physically-based distributed modelling system'. *J. Hydrol.* 87,45-59.

Avissar, R., & M.M. Verstraete, 1990. 'The representation of continental surface processes in atmospheric models'. *Rev. Geophys.*, 28, 35-52.

Becker, A. & J. Nemec, 1987. 'Macroscale hydrologic models in support to climate research'. In: The influence of climate change and climatic variability on the hydrologic regime and water resources, *IAHS* Publ. no. 168, 431-445.

Brugman, M.M., 1993. 'Glacier mass balance applied to hydrologic model design'. Abstracts *IAMAP-IAHS* Yokahama Symposia, Workshop HW3: Role of Snow and Ice in the Water Cycle in High Mountainous Areas (abstract only).

Bussières, N. & P.Y.T. Louie, 1989. 'Implementation of an algorithm to estimate regional evapotranspiration using satellite data'. *AES*, Downsview, Ontario.

CCC, 1991. 'Application of the Canadian Climate Centre general circulation model output for regional climate impact studies; Guidelines for users'. *Canadian Climate Centre*, Downsview, 25pp.

Environment Canada, 1990. 'HYDAT CD-ROM version 3.0 user's manual'. Water Resources Branch, *Environment Canada*, Ottawa.

Garrick, M., Cunnane, C. & J.E. Nash, 1978. 'A criterion for efficiency of rainfall-runoff models'. *J. Hydrol.* 36, 375-381.

Granger, R., 1993. 'A new approach for the use of remotely-sensed data in the estimate of evapotranspiration'. In: Program and abstracts, 19th Scientific Meeting of the Canadian Geopyhsical Union, Banff, Alberta, May 1993, *University of Calgary*, 97.

Green, W.H. & G.A. Ampt, 1911. 'Studies in soil physics, 1. Flow of air and water through soils'. *J. Agric. Science*,4,1-24.

Jain, S.K., B. Storm, J.C. Bathurst, J.C. Refsgaard and R.D. Singh, 1992. 'Application of the SHE to catchments in India Part 2. Field experiments and simulation studies with the SHE on the Kolar subcatchment of the Narmada River', *J. Hydrol.*, 140, 25-47.

Kite, G.W., 1975. 'Performance of two deterministic models'. In: Application of mathematical Models in Hydrology and Water Resources Systems (Proc. of the Bratislava Symposium, September 1975) *IASH* Publ. no. 115, 136-142.

Kite, G.W., 1978. 'Development of a Hydrologic Model for a Canadian Watershed', *Canadian Journal of Civil Engineering*, 5, 1, 126-134.

Kite, G.W., 1989a. 'Using NOAA data for hydrological modeling'. In: Quantitative Remote Sensing: An Economic Tool for the Nineties. Proc. IGARSS'89, *IEEE*, 2, 553-558.

Kite, G.W., 1989b. 'Hydrological modelling with remotely sensed data'. Proc. 57th Annual *Western Snow Conference*. Ft. Collins, Co., April 18-20, 1-8.

Kite, G.W., 1991. 'A watershed model using satellite data applied to a mountain basin in Canada'. *J. Hydrol.*, 128, 157-169.

Kite, G.W., 1992. 'Applying a climatic change scenario to a semi-distributed watershed model'. Proc. 60th Annual *Western Snow Conference*. Jackson Hole, Wyoming, April 13-16, 113-120.

Kite, G.W., 1993a. 'Application of a land class hydrological model to climatic change', *Wat. Resour. Res.*, 29(7), 2377-2384.

Kite G.W, 1993b: 'Analyzing hydrometeorological time series for evidence of climatic change'. *Nordic Hydrology*, 24(2/3) 30-40.

Kite, G.W., Dalton, A. and K. Dion, 1993. 'Simulation of streamflow in a macro-scale watershed using GCM data', *Wat. Resour. Res.*, 30(5), 1547-1559.

Kite, G.W. and N. Kouwen, 1992. 'Watershed modeling using land classifications'. *Wat. Resour. Res.*, 28(12), 3193-3200.

Kite, G.W., E.D. Soulis & N. Kouwen, 1995: A hierarchical approach to the connection of hydrological and atmospherica general circulation models. In: Space/Time/Scale Variability and Interdependence for Various Hydrological Processes, *Cambridge University Press* (in press).

Kouwen, N., E.D. Soulis and A. Pietroniro, 1990. 'Enhancing rainfall-runoff modelling of mixed land-use/land-cover areas with remote sensing'. In: G.W. Kite and A. Wankiewicz (eds.) *Applications of Remote Sensing in Hydrology*, Proc. NHRI Symposium No. 5, NHRC, Saskatoon, 94-108.

Martinec, J. & A. Rango, 1989. 'Merits of statistical criteria for the performance of hydrological models'. *Wat. Resour. Bull.*, 25,2,421-432.

McKim, H.L., E.A. Cassel & P.J. LaPotin, 1993. 'Water resource modelling using remote sensing and object-oriented simulation'. *Hydrol. Proc.*, 7(2), 153-165.

Monro, J.C., 1971. 'Direct search optimization in mathematical modeling and a watershed application', *NOAA Tech. Memo. NWS HYDRO-12*, Natl. Oceanic and Atmos. Admin., U.S. Dep. of Commer., Washington, D.C.

Monteith, J.L., 1981. 'Evaporation and surface temperature', *Q. J. Roy. Meteor. Soc.*, 107, 1-27.

Morton, F.I, 1983. 'Operational estimates of areal evapotranspiration and their significance to the science and practice of hydrology'. *J. Hydrol.*, 66, 77-100.

Naden, P.S., 1993. 'A routing model for continental-scale hydrology'. In: Macroscale modelling of the hydrosphere (Proc. Yokahama Symposium, July 1993), *IASH* Publ. no. 214, 67-79.

Nash, J.E. & Sutcliffe, J.V., 1970. 'River flow forecasting through conceptual models; part I - a discussion of principles'. *J. Hydrol.* 10(3), 282-290.

NOAA, 1992. 'Global change data base, Vol. 2 - Experimental calibrated global vegetation index. *National Oceanic and Atmospheric Administration*, Boulder, Colorado.

Ott, M., Z. Su, Schumann, A.H. & G.A. Schultz, 1991. 'Development of a distributed hydrological model for flood forecasting and impact assessment of land-use change in the international Mosel River Basin'. In: *Hydrology for the Water management of Large River Basins* (Proc. of the Vienna Symposium, August, 1991), IAHS Publ. no. 201, 183-194.

Penman, H.L., 1948. 'Natural evaporation from open water, bare soil and grass'. Proc. *Roy. Soc.*,London, 193, 120-145.

Philip, J.R., 1954. 'An infiltration equation with physical significance'. *Soil Sci.*, 77(1), 153-157.

Phillips, D.W., D.W. Gullett & M.S. Webb, 1988. 'Handbook on climate data sources of the Atmospheric Environment Service'. *Canadian Climate Centre*, Downsview, Ontario.

Priestley, C.H.B. & R.J. Taylor, 1972. 'On the assessment of surface heat flux and evaporation using large-scale parameters', *Mon. Wea. Rev.*, 100, 81-92.

Refsgaard, J.C., S.M. Seth, J.C. Bathurst, M. Erlich, B. Storm, G.H. Jorgensen and S. Chandra, 1992. 'Application of the SHE to catchments in India, Part 1. General results', *J. Hydrol.*, 140, 1-23.

Slough, K. and G.W. Kite, 1992. 'Remote sensing estimates of snow water equivalent for hydrological modelling applications'. *Can. J. Wat. Res.*, 17(4), 1-8.

Tao, T. and N. Kouwen, 1989. 'Remote sensing and fully-distributed modelling for flood forecasting'. *J. of Water Resour. Plng. and Mgmt., ASCE,* 115 (6), 809-823.

Tarboten, D.G., Al-Adhami, M.J., and D.S. Bowles, 1991. 'A preliminary comparison of snowmelt models for erosion prediction'. Proc. 59th Annual *Western Snow Conference*. Juneau, Alaska, April 12-15, 79-90.

Vehviläinen, B., 1991. 'A physically based snowcover model'. In: D.S. Bowles and P.E. O'Connel (eds.) Recent Advances in the Modeling of Hydrologic Systems, NATO ASI Series, Vol. C345, *Kluwer Academic Publishers*, Dordrecht, 113-136.

WMO, 1986. 'Intercomparison of models of snowmelt runoff'. Operational Hydrology Report No. 23, *World Meteorological Organization*, Geneva.

Wyss, J., Williams, E.R. and R.L. Bras, 1990. 'Hydrologic modelling of New England river basins using radar rainfall data'. *J. Geophys. Res.*, 95, D3, 2143-2152.

Chapter 16

THE SHE/SHESED BASIN SCALE WATER FLOW AND SEDIMENT TRANSPORT MODELLING SYSTEM

J. C. Bathurst, J. M. Wicks and P. E. O'Connell

The SHE/SHESED combination is a response to the need for modelling systems able to represent river basins under a range of future conditions for the purposes of decision making. Developed at the University of Newcastle upon Tyne and based on the existing SHE hydrological modelling system, it is a physically based, spatially distributed modelling system for water flow and sediment transport, applicable at the basin scale. The flow component (i.e. the SHE) is an integrated surface and subsurface representation of the land phase of the hydrological cycle. The erosion and sediment transport component (SHESED) accounts for sediment yield derived from soil erosion by raindrop impact and overland flow. A decade of experience with SHE has shown the system to be at home in a range of environments. Data provision priorities for minimizing model output uncertainty, techniques for handling the large quantities of data and a validation methodology for testing model predictive capability have been developed. Experience shows hydrological expertise and an understanding of basin response mechanisms to be vital elements in model application.

16.1. INTRODUCTION

The SHE/SHESED combination is a physically based, spatially distributed modelling system for water flow and sediment transport, applicable at the scale of a river basin, which has been developed at the Department of Civil Engineering, University of Newcastle upon Tyne, UK. Compared with other types of model it has particular advantages in predicting the impacts of future land use and climate change, and in investigating optimum strategies for land management. It is based on the Système Hydrologique Européen (SHE) hydrological modelling system; this was developed initially by the Danish Hydraulic Institute, the Institute of Hydrology (UK) and SOGREAH

(France), although subsequent UK deployment of the SHE has been concentrated at the Water Resource Systems Research Unit (WRSRU), University of Newcastle upon Tyne, while activity in France has continued at the Laboratoire d'hydraulique de France, an affiliation of SOGREAH.

This chapter describes the background to the development of the SHE/SHESED combination, the system's process equations, data provision, validation technique and examples of applications. It is not intended as a discussion of the theoretical capabilities and limitations associated with physically based, spatially distributed models, which are addressed in more detail elsewhere (e.g. Beven, 1989; Bathurst and O'Connell, 1992). Instead the emphasis is on practical techniques for model application and the lessons learnt from a decade of experience with the SHE. In this way it is hoped to demonstrate the advances which have been achieved in physically based, spatially distributed modelling in general and with the SHE in particular.

The chapter also concentrates on developments at the WRSRU and it should be noted that other SHE-based developments and applications have been carried out by the SHE partners in Denmark and France.

SHESED has been referred to as SHESED-UK in other publications, with the qualification UK denoting the country of origin as agreed with the international SHE partners. However, the qualification is dropped for the rest of the chapter.

16.2. REQUIREMENT FOR PHYSICALLY BASED MODELLING

In recent years, considerable attention has been focused worldwide on the impacts of environmental change induced by human activity. Current issues in the hydrological domain include the effects of large scale deforestation and climate change on soil erosion and flow regime, and the dispersal of contaminants from agricultural and industrial activities. Strict regulations for environmental quality and curbs on agricultural and industrial activities are being introduced, usually with significant economic implications. At the same time, industrial and agricultural activity is essential for maintaining, or improving, our physical, material and economic well-being. A trade-off is therefore required between the level of activity (providing an economic return) and the maintenance of an acceptable environment.

Establishing such a trade-off is a decision-making process involving many elements, one of which is an appreciation of how environmental systems respond to imposed change. Of particular concern is the movement of water, sediments and contaminants through the land phase of the hydrological cycle. Mathematical modelling is increasingly relevant but needs to be based on a sound physical understanding of those processes which have a key role in determining the impacts of basin change. It must also provide information which contributes usefully to decisions concerning basin management. An effect of the increasing environmental pressures, and related energy constraints, is that the value of hydrological and related developmental predictions has increased rapidly. Further demands for water use, flood protection and basin development on the one hand, and further increases in energy costs and environmental penalties on the other, can only increase the value of such predictions even further. There is therefore an increasing requirement for a new approach to hydrological modelling to help optimize project planning. In particular there is a need for a means of assessing the impacts of different levels of project sophistication or different approaches to land management on basin hydrology, soil erosion and contaminant concentration, in advance of any changes being implemented.

The task of representing a river basin in a range of possible future altered states is beyond the traditional black box and conceptual hydrological models, which are essentially regression relationships relying on the availability of past rainfall and hydrological records for their calibration. Their parameters lack physical meaning, so cannot be specified in advance. Similarly, records derived for the past state of a basin cannot support model calibration for a range of future altered states. Such models, therefore, have a limited basis for representing the impacts of, for example, future land management options or climate change. Further, they are spatially lumped, so cannot readily account for spatial variations in basin characteristics, rainfall input and hydrological response.

By contrast, physically based, spatially distributed modelling systems have particular advantages for the study of basin change impacts and applications to basins with limited records. Their parameters have a physical meaning and can be measured in the field. Model calibration is therefore less dependent on the existence of past data records, while the ability to specify representative parameter values means that models can be set up for a future altered state of the basin. Basin response is represented on both a spatially and a temporally distributed basis and in terms of multiple variable outputs: i.e. rather than providing just one output variable, such as basin discharge, the models provide distributed predictions of all relevant hydrological, sediment and contaminant variables. Finally, they can make use of as much information and knowledge, qualitative and quantitative, as are available, particularly if relevant to the eventual decision-making process.

Disadvantages of physically based models include heavy computer requirements, the need to evaluate many parameters, and a complexity which implies a lengthy training period for new users.

16.3. BACKGROUND TO SHE/SHESED

The SHE is an integrated surface and subsurface representation of water movement through a river basin, incorporating the major elements of the land phase of the hydrological cycle. It is a general, physically based, spatially distributed modelling system: that is, it can be used to construct and run models of all or any part of the land phase of the hydrological cycle for any geographical area. Further details of the development and structure of the SHE are described in Abbott et al. (1986a, b).

The SHE is aimed at providing a strong European capability in hydrological modelling and its development was therefore aided financially by the Commission of the European Communities. The decision to develop it was taken in 1976 and a coordinated programme of work proceeded at the Danish Hydraulic Institute, the Institute of Hydrology and SOGREAH. The first production version became operational in 1982 but further extensive testing and development has continued to increase its reliability, running speed and general efficiency, and to broaden its scope.

Water movement is a basic driving mechanism for contaminant and sediment transport and the SHE has therefore provided a framework for a growing family of transport components. These make use of the hydrological information, such as overland and channel flows, soil moisture contents and groundwater flows, supplied by the SHE, to determine such quantities as concentrations and rates of transport, deposition and storage of sediments and contaminants on a spatially and temporally distributed basis. Different additions are being introduced at the three European centres as needed. In the UK, following the concentration of SHE activity at the WRSRU in 1985, the soil erosion and sediment yield component SHESED was developed (Wicks and Bathurst, in press). The combination of SHE and SHESED provides a general system for modelling water flow and sediment transport at the basin scale.

16.4. SHE STRUCTURE

The SHE and SHESED are physically based in the sense that the various flow and transport processes are modelled either by finite difference representations of the partial differential equations of mass, momentum and energy conservation, or by empirical equations derived from independent experimental research. The model parameters have a physical meaning and can be evaluated by measurement. Spatial distribution of catchment properties, precipitation input and catchment response is achieved in the horizontal through the representation of the catchment by an orthogonal grid network and in the vertical by a column of horizontal layers at each grid square (Fig. 16.1). The channel system is represented on the boundaries of the grid squares.

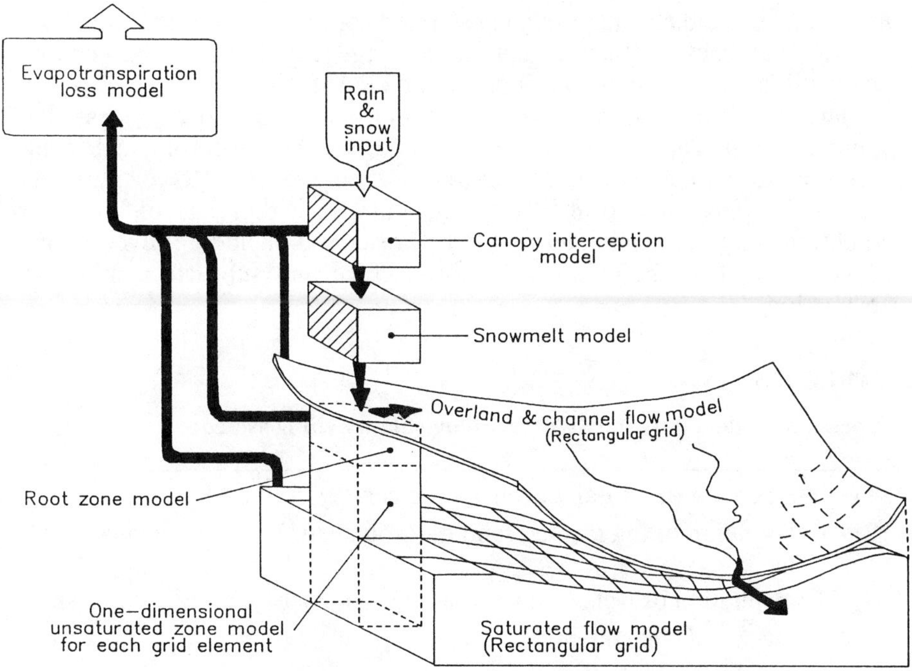

Fig. 16.1. Schematic diagram of the Système Hydrologique Européen (SHE) hydrological modelling system.

In water resources projects the level of complexity at which modelling can be carried out varies according to data availability, type of problem, scale of operation, required accuracy, computer facilities, economic criteria and other considerations. The SHE has therefore been designed as a flexible modelling system, encompassing several levels of complexity. This has been achieved through the use of structured programming, modular design and incorporation of various levels of sophistication for the process models. Within the model software each hydrological and transport process is allocated its own component and the simultaneous operation of all the components is controlled by the central FRAME component. Additional flexibility is assured through the ability of the SHE to use as few or as many data as are available, either from field measurements or from the scientific literature: the philosophy of model use with scarce data is discussed in Abbott et al. (1986a).

16.5. THE SHE HYDROLOGICAL MODELLING SYSTEM

16.5.1. System Design and Capability

The basis of the SHE is described in Abbott et al. (1986b). The system consists of subcomponents accounting for evapotranspiration and interception, overland and channel flow, unsaturated zone flow, saturated zone flow, snowmelt and channel/surface aquifer exchange (Fig. 16.1). It also contains an overall control subcomponent called the FRAME.

The capability of the SHE for modelling water flow processes is summarized in Table 16.1. The model is driven by meteorological inputs and provides inputs to the sediment transport component. However, it does not receive feedback from that component. A wide range of boundary conditions and source/sink terms can be prescribed, including several types of head and flow boundaries for the surface and subsurface and well abstraction.

TABLE 16.1

Processes modelled by the SHE hydrological modelling system

(1) Interception of rainfall on vegetation canopy
(2) Evaporation of intercepted rainfall, ground surface water and channel water
(3) Transpiration of water drawn from the root zone
(4) Snowpack development
(5) Snowmelt
(6) One-dimensional flow in the unsaturated zone
(7) Two-dimensional flow in the saturated zone
(8) Two-dimensional overland flow
(9) One-dimensional channel flow
(10) Saturated zone/channel interaction, including an allowance for an unsaturated zone below the channel
(11) Saturated zone/surface water interaction

The basic process equations used in the water flow component are as follows.

16.5.2. Interception

The interception subcomponent calculates net rainfall reaching the ground through the canopy, the amount of water stored on the canopy and evaporation from the canopy. Interception is modelled by an accounting procedure for canopy storage in which the rate of change of storage is calculated as

$$\frac{\partial C}{\partial t} = Q - ke^{b(C-S)} \tag{16.1}$$

where

$$Q = p_1p_2\,(P - E_pC/S) \text{ when } C < S$$

$$Q = p_1p_2\,(P - E_p) \text{ when } C \geq S$$

and C = depth of water on the canopy; S = canopy storage capacity, the minimum depth of water required to wet all canopy surfaces; P = rainfall rate; p_1 = proportion of ground in plan view hidden by vegetation; p_2 = ratio of total leaf area to area of ground covered by vegetation; $p_1p_2 = p_1p_2$ when $p_2 < 1$; $p_1p_2 = p_1$ when $p_2 \geq 1$ since p_1p_2 cannot exceed p_1; E_p = potential evaporation rate; k and b = drainage parameters; and t = time.

Equation 16.1 is solved by analytical integration. In general, the parameters C, S, k and b cannot be measured directly in the field but must be estimated indirectly from measurements of rainfall, net rainfall below the canopy and evapotranspiration. Changes in vegetation cover through time can be allowed for by varying S, p_1 and p_2 with time.

16.5.3. Evapotranspiration

The evapotranspiration subcomponent calculates actual evapotranspiration and translates it into a loss term for the unsaturated zone, describing uptake of water by plant roots and its transpiration in a response which is continuous as the canopy varies from a dry, through a partially wetted, to a fully wetted state. The basis of the model is the Penman-Monteith equation for actual evapotranspiration:

$$E_a = \frac{R_n\Delta + \dfrac{\rho c_p \delta_e}{r_a}}{\lambda\left[\Delta + \gamma\left(1 + \dfrac{r_c}{r_a}\right)\right]} \tag{16.2}$$

where E_a = actual evapotranspiration; R_n = net radiation; Δ = rate of increase with temperature of the saturation vapour pressure of water at air temperature; ρ = density of air; c_p = specific heat of air at constant pressure; δ_c = vapour pressure deficit of air; r_a = aerodynamic resistance to water vapour transport; λ = latent heat of vaporization of water; γ = psychrometric constant; and r_c = canopy resistance to water transport, usually determined as the stomatal resistance.

In view of the difficulties in evaluating r_c, three modes of operation are specified:

(1) r_c is a constant for each vegetation type;

(2) r_c varies as a function of soil moisture tension as well as vegetation type;

(3) r_c is zero, so that Eq. 16.2 gives E_p, the potential evapotranspiration rate; the actual evapotranspiration rate E_a is then calculated from a specified relationship between the ratio E_a/E_p and soil moisture tension; for this method, E_p can also be supplied directly from, for example, pan measurements.

Total actual evapotranspiration calculated for each grid square depends on how wet the canopy is and on the degree of ground coverage by the canopy:

$$E_t = p_1 p_2 E_p C/S + p_1 p_2 E_{at}(1 - C/S) + (1 - p_1 p_2) E_{as} \qquad (16.3)$$

where E_t = total evapotranspiration; E_p = potential evapotranspiration; E_{at}= total transpiration from uptake through the roots; and E_{as} = evaporation for bare soil.

16.5.4. Overland and Channel Flow

Surface water is routed as overland and channel flow as a function of topography, channel shape and flow resistance, using the diffusion wave approximation of the Saint Venant equations of continuity and momentum. Overland flow, generated by excess of rainfall over infiltration or by upward saturation of the unsaturated zone, is modelled in two dimensions as

$$\frac{\partial h}{\partial t} + \frac{\partial (uh)}{\partial x} + \frac{\partial (vh)}{\partial y} = q \qquad (16.4)$$

$$\frac{\partial h}{\partial x} = S_{ox} - S_{fx} \quad \text{x - direction} \qquad (16.5)$$

$$\frac{\partial h}{\partial y} = S_{oy} - S_{fy} \quad \text{y - direction} \qquad (16.6)$$

where h(x,y) = local water depth; t = time; x,y = horizontal cartesian coordinates; u(x,y), v(x,y) = flow velocities in the x and y directions respectively; q(x,y,t) = net precipitation minus infiltration; $S_{ox}(x,y)$, $S_{oy}(x,y)$ = ground slopes in the x and y directions respectively; $S_{fx}(x,y)$, $S_{fy}(x,y)$ = friction slopes in the x and y directions respectively. The Manning/Strickler formula is used for flow resistance and the equations are solved with an explicit finite difference procedure.

Channel flow is modelled in one dimension as

$$\frac{\partial A}{\partial t} + \frac{\partial (Au)}{\partial x} = q_L \tag{16.7}$$

$$\frac{\partial h}{\partial x} = S_{ox} - S_{fx} \tag{16.8}$$

where A(x) = flow cross-sectional area; S_{ox} = channel bed slope; $q_L(x)$ = source/sink term for evaporation, rainfall, lateral inflow and outflow, and stream/aquifer exchange. Again the Manning /Strickler formula is used for flow resistance and the equations are solved with an implicit finite difference scheme.

16.5.5. Unsaturated Zone Flow

The unsaturated zone flow subcomponent determines the soil moisture content and tension in the unsaturated zone, infiltration at the ground surface and an exchange with the saturated zone at the phreatic surface. The unsaturated zone is mostly nonuniform in its physical properties and the upper part, the root zone, exhibits considerable fluctuations in moisture content as a result of evapotranspiration and rainfall infiltration. Its lower boundary also varies as the phreatic surface elevation changes and there is scope within the model for the zone to disappear as the phreatic surface rises to the ground surface. The subcomponent plays a crucial role in the flow calculations since all the other flow subcomponents draw upon the boundary data which it supplies.

Based on the assumption that there is flow in the vertical only, the solution is obtained with the one-dimensional Richards equation:

$$C\frac{\partial \psi}{\partial t} = \frac{\partial}{\partial z}\left(K\frac{\partial \psi}{\partial z}\right) + \frac{\partial K}{\partial z} - S \tag{16.9}$$

where ψ = soil moisture tension or pressure head; t = time; z = vertical space coordinate (positive upwards); $C = \partial \theta / \partial \psi$ = soil water capacity; θ = volumetric water content; $K(\theta, z)$ = hydraulic conductivity; S(z) = source/sink term for root extraction and soil evaporation. Two functional relationships are needed to solve the equation. These are the relationship between unsaturated conductivity K and the volumetric moisture content θ and the relationship between soil tension ψ and the volumetric moisture content. The equation is then solved by an implicit finite difference scheme, which is also iterative to allow for the pronounced nonlinearity resulting from the strong dependence of K and C on θ or ψ.

16.5.6. Saturated Zone Flow

The saturated zone component determines the phreatic surface level and the flows in the saturated zone of an unconfined aquifer, stream/aquifer interactions, groundwater seepage at the ground surface, and artificial groundwater extraction. The variation through time of the phreatic surface level at each square is modelled by the nonlinear Boussinesq equation. This combines Darcy's law and the mass conservation of two-dimensional laminar flow in an anisotropic, heterogeneous aquifer, to give

$$S\frac{\partial h}{\partial t} = \frac{\partial}{\partial x}\left(K_x H \frac{\partial h}{\partial x}\right) + \frac{\partial}{\partial y}\left(K_y H \frac{\partial h}{\partial y}\right) + R \tag{16.10}$$

where S(x,y) = specific yield; h(x,y,t) = phreatic surface level; K_x(x,y), K_y(x,y) = saturated hydraulic conductivities in the x and y directions respectively; H(x,y,t) = saturated thickness; t = time; x,y = horizontal cartesian coordinates; and R(x,y,t) = instantaneous vertical recharge into the saturated zone. This can be expressed by

$$R = \sum q - \frac{\partial}{\partial t}\int_h^{gs} \theta dz \tag{16.11}$$

where Σq = transpiration + soil evaporation + infiltration + stream/aquifer exchange + external boundary flows; θ(z,t) = soil moisture content in the unsaturated zone; and gs = ground surface level.

Equation 16.10 is solved by a finite difference approximation using an implicit scheme. Allowance is made for the disappearance of the saturated zone and for the rise of the phreatic surface to the ground surface. Stream/aquifer interactions are simulated for all combinations of the phreatic surface in direct contact with or lying below flowing and dry channels. In the long term, the current unsaturated and saturated zone subcomponents will be replaced by a fully three-dimensional subsurface subcomponent capable of extending simulation to perched and confined aquifers.

16.5.7. Snowmelt

The snowmelt subcomponent uses snowpack and vegetation parameters, along with meteorological input data, to predict the transfer of moisture and the variation in snowpack thickness resulting from processes of precipitation onto the snowpack, snowmelt from the snowpack, lateral spatial variations in snowpack conditions, and interception and evapotranspiration in the presence of a snowpack and at air temperatures below freezing. The subcomponent is structured so that first the total heat flux to the snowpack is calculated, then the amount of melting engendered by this flux is determined and finally the meltwater is routed through the snowpack. Thus both energy and mass exchanges with the snowpack are modelled.

Depending on the availability of data or general requirements, the total heat flux can be determined using either an energy-budget or a temperature-based method. In the former, contributions from all the energy inputs and outputs are calculated for the pack as a whole, giving

$$H = R_n + C + V + I + G \quad (16.12)$$

where H = heat flux into the snowpack; R_n = net radiation; C = heat gained by convection from the air; V = heat gained from condensed vapour; I = heat gained from precipitation; and G = heat conducted from the underlying ground.

The temperature-based method is an empirical method used when available data are limited to air temperatures. The heat flux is calculated as

$$H = k\, T_{am}\, S\, \rho_w\, L_i \quad (16.13)$$

where k = degree-day factor; T_{am} = air temperature above the melting point; S = specific gravity of snow; ρ_w = density of water; and L_i = latent heat of fusion of ice.

A detailed description of the snowmelt subcomponent is given in Bathurst and Cooley (in press).

16.6. THE SHESED EROSION AND SEDIMENT TRANSPORT COMPONENT

16.6.1. Component Design and Capability

The basis of SHESED is described in Wicks (1988) and Wicks and Bathurst (in press); the component consists of subcomponents accounting for soil erosion by raindrop impact, leaf drip impact and overland flow, channel bed erosion by channel flow, and sediment transport by overland and channel flow.

The capability of SHESED for modelling soil erosion and sediment transport processes is summarized in Table 16.2. The component is driven by inputs from the SHE water flow simulations but feedback to the flow simulations is not modelled as the effects are unlikely to be significant at the sediment concentrations and relative scales of erosion and deposition likely to be considered.

The basic process equations used in the erosion and sediment transport component are as follows.

TABLE 16.2.

Processes modelled by the SHESED erosion and sediment transport component

(1) Soil erosion by raindrop impact, leaf drip impact and overland flow

(2) Two-dimensional total load convection in overland flow, including input to channels

(3) Deposition and resuspension of sediments in overland flow

(4) Channel bed erosion by channel flow

(5) One-dimensional convection of cohesive and noncohesive sediments in channel flow

(6) Deposition and resuspension of noncohesive sediments in channel flow

16.6.2. Soil Erosion

Soil detachment by raindrop and leaf drip impact (Fig. 16.2) is given by

$$D_r = k_r F_w (1 - C_g)\left[(1 - C_c) M_r + M_d\right] \qquad (16.14)$$

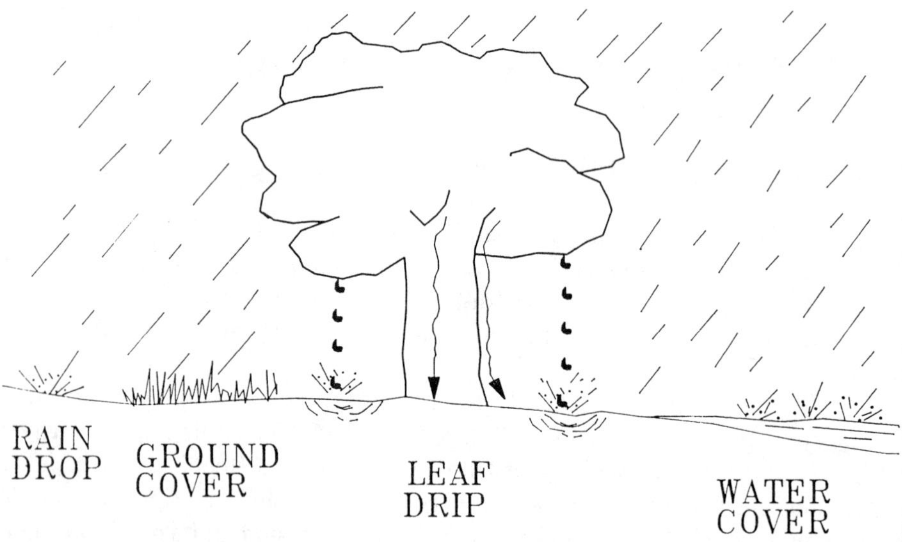

Fig. 16.2. Soil detachment by raindrop and leaf drip impact, modified by ground cover, canopy cover and surface water layer, as represented in the SHESED erosion and sediment transport component.

where D_r = the rate of detachment of material per unit area; k_r = raindrop impact soil erodibility coefficient; C_g = proportion of ground shielded by ground cover; C_c = proportion of ground shielded by canopy cover; M_r = momentum squared for rain falling directly to the ground; and M_d = momentum squared for leaf drip, spatially averaged. The effect of a surface water layer in reducing the energy imparted to the soil by raindrop impact is accounted for by F_w, where

$$F_w = 1 \qquad \text{if } h <= d_m \tag{16.15a}$$

$$F_w = e^{(1-h/d_m)} \qquad \text{if } h > d_m \tag{16.15b}$$

where h = water depth; and d_m = median raindrop diameter (determined as a function of rainfall intensity).

The momentum squared is calculated for the full distribution of raindrop sizes. As it is unlikely that the drop size distribution will be available for the storm under consideration, the model includes a default drop size distribution.

Soil detachment by overland flow is calculated as

$$D_f = k_f\left[\frac{\tau}{\tau_c} - 1\right] \qquad \text{for } \tau > \tau_c \tag{16.16a}$$

$$D_f = 0 \qquad \text{for } \tau <= \tau_c \tag{16.16b}$$

where D_f = rate of detachment of material per unit area; k_f = overland flow erodibility coefficient; τ = flow shear stress; and τ_c = critical shear stress for initiation of sediment movement.

As with other physically based erosion models, the two erodibility coefficients have not yet been quantitatively related to a measurable soil property and must be determined empirically. Similarly, Eq. 16.16 does not allow specifically for rill erosion so the effect of rilling is accounted for empirically by k_f.

16.6.3. Overland Flow Sediment Transport

In the sediment routing procedure, the eroded sediment is transported downslope by overland flow under the time- and space-varying conditions of erosion and overland flow characterizing the simulation period. The routing computations are based on the two-dimensional, mass (total sediment load) conservation equation:

$$\frac{\partial(ch)}{\partial t} + (1-\lambda)\frac{\partial z}{\partial t} + \frac{\partial g_x}{\partial x} + \frac{\partial g_y}{\partial y} = 0 \qquad (16.17)$$

where h = water depth; c = volumetric sediment concentration; λ = soil porosity; z = depth of loose soil; g_x, g_y = volumetric sediment transport rates per unit width in the x and y directions respectively; and t = time. The transport rate is then limited either by the amount of loose soil available or by the capacity transport rate of the flow, determined by a sediment discharge equation. (For want of an appropriate sediment discharge equation for shallow overland flow, options are available for using two streamflow equations, the Engelund - Hansen total load and Yalin bed load equations, to give the capacity transport rate as a function of the flow conditions.)

Equation 16.17 is applied without consideration of particle size distribution and is solved by an explicit finite difference scheme.

16.6.4. Channel Sediment Transport

For channel flow, sediment transport is simulated as total load, using the one-dimensional mass conservation equation:

$$\frac{\partial(Ac)}{\partial t} + (1-\lambda)W\frac{\partial z}{\partial t} + \frac{\partial(AcV_s)}{\partial x} = q_s \qquad (16.18)$$

where A = flow cross-sectional area; c = volumetric sediment concentration; λ = bed porosity; W = active bed width, over which sediment transport occurs; z = channel bed elevation; and V_s = longitudinal sediment velocity. The term q_s represents the input of sediment from bed erosion and overland flow supply.

Bed erosion is determined as a function of flow shear stress. Noncohesive sediment discharge is determined as a function of the amount of sediment available and the capacity transport rate of the flow (calculated using the Ackers-White or Engelund-Hansen equations). Calculations are carried out for each size fraction in turn and allowance for sediment size distribution enables armour layer development in coarse sediments to be simulated at the bed surface. Cohesive sediment (silt and clay particles) is assumed to move as suspended load and is routed as a single size fraction with the longitudinal sediment velocity approximated by the flow velocity. The flow is assumed to have an unlimited transport capacity for this material. For the larger, noncohesive sediment sizes, moving as a combination of suspended and bed loads, the following expression for sediment velocity is used (Phillips and Sutherland, 1985):

$$V_s = 8.5\, V_* \left[1 - \frac{V_{*c}}{V_*} \right]^{1/2} \qquad (16.19)$$

where V_* = bed shear velocity; and V_{*c} = critical bed shear velocity obtained using the Shields criterion.

Physical dispersion of the sediment is not modelled as it is not considered to be significant for catchment scale modelling. Equation 16.18 is solved by an explicit finite difference scheme.

16.7. MODEL PARAMETER AND DATA PROVISION

16.7.1. Data Needs

SHE simulations require large amounts of time series data and the evaluation of a wide range of parameters, as shown in Table 16.3. In summary the data comprise:

(1) Time series of meteorological inputs (precipitation and data for determining evapotranspiration) needed to drive the simulation;

(2) Time series of output variables (e.g. runoff, sediment yield and well level records) for calibrating and validating the model;

(3) Parameters used to fit the model to a particular river basin (soil, vegetation, topographic and sediment characteristics).

All data and parameters should ideally be spatially distributed.

It is rare for the complete dataset to be available from direct measurements, even for research catchments, and assembling the dataset can be a major task. A range of data collection agencies may need to be visited: for example, in an application of SHE to catchments in India, some 15 agencies representing meteorological, irrigation, agricultural, geological and other professional fields in government, universities and research institutions were approached (Refsgaard et al., 1992). It is also usually necessary to transform the data from the form in which they are provided (on disk, on charts, as maps, in reports) to a computer compatible form in the desired format of the SHE files.

TABLE 16.3.

SHE/SHESED data needs

INPUT DATA AND MODEL PARAMETERS FOR EACH SUBCOMPONENT	
Frame Parameters	Ground surface elevation Impermeable bed elevation Distribution codes for precipitation and meteorological source stations Distribution codes for soil and vegetation types
Water Flow Subcomponents	
Input data	Meteorological and precipitation data
Initial conditions	Phreatic surface levels Overland and channel flow depths Snowpack thickness and temperature
Boundary conditions	Surface flows or water levels at boundaries Man-controlled channel flow diversions and discharges Groundwater flows or potentials at boundaries Groundwater pumping and recharge data
Interception parameters	Drainage parameters Canopy storage capacity (time varying) Ground cover indices (time varying)
Evapotranspiration parameters	Canopy resistance Aerodynamic resistance Ground cover indices (time varying) Ratio between actual and potential evapotranspiration as a function of soil moisture tension Root distribution with depth
Overland and channel flow parameters	Strickler resistance coefficients for overland and channel flows Weir discharge coefficients Channel geometry
Unsaturated zone parameters	Soil moisture tension/content relationship Unsaturated hydraulic conductivity as a function of moisture content
Saturated zone parameters	Porosities or specific yields Saturated hydraulic conductivities
Snowmelt parameters	Degree-day factor Snow zero plane displacement Snow roughness height Snow density

TABLE 16.3. Continued.

TABLE 16.3. Continued.

Sediment Transport Subcomponents	
Input data	Rainfall intensity Overland and channel flow hydraulic data Canopy drainage
Initial conditions	Thickness of loose soil and channel bed material Sediment concentrations in surface flows
Boundary conditions	Sediment concentrations in surface flows at boundaries
Soil erosion parameters	Soil particle size distribution Soil erodibility coefficients Ground/canopy cover indices (time varying) Raindrop size distribution Proportion of drainage falling as leaf drip Leaf drip size Leaf drip fall distance Soil porosity Other parameters supplied by flow component
Channel sediment parameters	Bed material particle size distributions Sediment porosities Other parameters supplied by flow component

16.7.2. Time-Varying Input and Validation Data

The time-varying input data are the most important to obtain for a simulation. They may consist of records of past conditions (the usual case) or be generated for a specified climate scenario (for example for studying the impact of climate change, Bathurst et al., in press). In either case the data must be reliable. Sensitivity analyses carried out as part of SHE applications in India (Refsgaard et al., 1992) and Idaho, USA, (Bathurst and Cooley, in press) suggest that the effect of uncertainty or error in the meteorological data generally outweighs the effect of uncertainty in the parameter values. Thus, if there are significant uncertainties in the precipitation data, then calibrating the model by adjusting the model parameter values is likely to prove a meaningless exercise. The model may be fitted to the calibration period but it is unlikely to represent other periods as accurately.

As the dynamic processes which determine basin response take place over short periods (hours to minutes), the meteorological input data should likewise be supplied at short time intervals, typically one hour or even a few minutes when simulating soil erosion. Wherever possible, they should therefore be derived from continuous recorders or on the basis of frequent sampling. All too often, though, rainfall is found to be measured on a daily basis while autographic gauges are poorly distributed spatially. Daily rainfall data can be disaggregated to finer time resolutions using autographic data from neighbouring areas but this represents an approximation and the simulation results must be treated accordingly (e.g. Bathurst et al., in press).

Potential evapotranspiration is required at a lesser accuracy over short time periods, although its overall contribution to the water balance at, typically, monthly scales should be accurately represented. For application of the Penman-Monteith equation (Eq. 16.2), it is often necessary to derive net radiation from solar radiation measurements and assumptions concerning back radiation.

Calibration and validation data should also be available either on a continuous basis or at intervals sufficient to allow the main variations through time to be monitored. Bulk sediment yields for events can be used but provide a less thorough basis for testing model capability than time-varying sediment discharge. Of particular importance for a distributed model, calibration and validation attain greater worth if data are available not only for the catchment outlet but at points within the catchment. Such data might include snowpack depths, well levels and sediment concentrations along the channel network. Validations based on channel discharge at internal gauging stations are reported by Bathurst (1986) and Bathurst et al. (in press).

16.7.3. Model Parameter Evaluation

The model parameters have physical meaning and can therefore be measured in the field or laboratory. Initial evaluation, especially for those parameters to which the simulation is less sensitive, can be accomplished using information already in the literature. Parallel modelling and field measurement can then be carried out, with the model being used to indicate those parameters and areas of the catchment where greatest accuracy or density of measurement is needed. The fieldwork is thus directed most efficiently at improving the model. This approach was used in the application of the SHE to the Kolar basin in India. An initial calibration was achieved using information from reports and measurements in neighbouring basins. Subsequently a three-week programme of field measurements was carried out to improve the parameter evaluation, concentrating on the soil parameters (Jain et al., 1992).

In theory evaluation of the model parameters could involve an enormous number of measurements. However, in most applications of the SHE to date, the simulated response has been found to vary primarily with just four or five parameters. Measurements can therefore be largely restricted to these parameters. Which parameters these are depends on the type of application and the relative importance of the different catchment processes, although typically the soil hydraulic parameters play an important role in most simulations. The other parameters have a secondary, although not necessarily negligible, impact on the simulation results and may be evaluated from information in the literature.

The soil properties required by the SHE are the saturated hydraulic conductivity, the unsaturated hydraulic conductivity function and the functional relationship between the soil tension and soil water content. Research into means of obtaining these data for SHE applications indicates that the Guelph permeameter provides the most satisfactory basis for direct field measurement. However, the most economical and simple methods, albeit involving greater uncertainty, are predictive models which make use of easily obtained and routinely available soil physical properties such as particle size distribution (Ragab and Cooper, 1993a, b).

16.7.4. Parameter Scale Effects

Within each model grid square, each physical characteristic is represented by one parameter value. As long as the grid square is small compared with the distances over which there is significant spatial variability in catchment properties and hydrological response, this does not compromise the model's ability to represent local variations in response. However, as grid scales increase, the local spatial variability in properties and response becomes subgrid. There are then difficulties in applying the equations of small scale physics (such as the Richards equation, Eq. 16.9) which make up the SHE and evaluating their parameters, at the grid scale (e.g. Beven, 1989). In particular, the field measurements which form the basis of parameter evaluation are most easily carried out at the point or plot scale, which may not be representative of large grid scales. The solution has been to use "effective" parameter values, which represent the subgrid spatial variability, to give a grid scale response. However, this is a pragmatic approach and it is recognized that the concept may not allow an accurate reproduction of the observed response in all circumstances (as shown for example by Binley et al., 1989). There has also not yet been enough research to establish the relationship between the effective parameter value (possibly determined through calibration) and field measurements. Nevertheless, circumstances have been identified in which parameters calibrated at the scale of an experimental plot (area $< 200\ m^2$) can be applied successfully in simulations involving large grid squares. For example, in an application in the Cobres basin, Portugal, parameters evaluated at the 167-m^2 scale of a plot were found to be applicable in a 32-ha subcatchment modelled with a 50-m grid and the 701-km^2 basin modelled with a 2-km grid (Bathurst et al., in press). Simulation of the Kolar basin in India with a 2-km grid was similarly based on parameter values reasonably close to measured values (Jain et al., 1992). However, such close correspondence between effective parameter values and field measurements cannot be assumed in catchments with large topographic variations.

There is always a degree of uncertainty inherent in representing a property with a single parameter value. This is especially so for soil properties since, even in the same soil, there can be significant spatial variability in that property. Consequently, the parameters should be defined in terms of bounds on the physically realistic range of values which they can assume. The range should be kept as narrow as possible and this can be achieved more satisfactorily by field measurement than by literature review. The associated uncertainty in parameter evaluation forms an important aspect of the SHE validation methodology described later.

16.7.5. Data Handling

Considerable manipulation of data on a spatially distributed basis is called for in:

(1) setting up the catchment model (topography, soils, vegetation and sediments);

(2) providing input data (meteorological variables); and

(3) interpreting simulation results (e.g. distributions of soil moisture conditions and sediment concentrations).

Unlike simpler models, which present a single output variable, SHE/SHESED presents multiple outputs covering the entire spectrum of hydrological and sediment variables on a spatially and temporally distributed basis. These may include time-varying records of phreatic surface level, snowpack depth, overland flow depth, soil erosion or any other variable at any grid square or channel link within the catchment. Alternatively, synoptic views of the spatial distribution of any variable across the catchment at any time can be produced (Fig. 16.3). The sheer volume of information which can be incorporated and presented means that manual handling is inefficient and inappropriate and there is instead a heavy dependency on geographical information systems and computer graphics. Thus, for example, within the NERC/ESRC Land Use Programme (NELUP), funded by the UK Natural Environment Research Council (NERC) and Economic and Social Research Council (ESRC) at the University of Newcastle upon Tyne, algorithms have been developed for the automatic generation of catchment characteristics and SHE parameters from routine data (Lunn et al., 1992). The data sources may include, for example, UK Meteorological Office records, Soil Survey databases, digital elevation models and remotely sensed data. Considerable effort has also been expended on the graphical display of time series and synoptic distributions of output data.

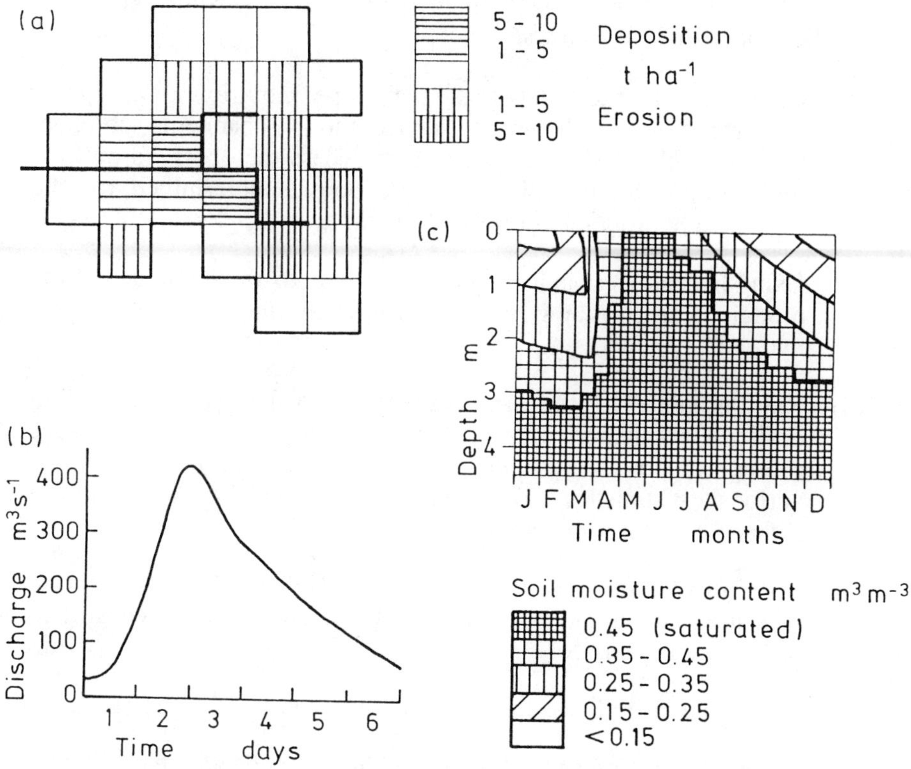

Fig. 16.3. Examples of SHE/SHESED data display options: (a) spatial variation in net soil erosion and deposition for a storm event (channels shown by thick lines); (b) channel flow hydrograph; (c) time varying soil moisture profile at a grid square.

16.8. VALIDATION

Traditional split-sample testing (e.g. Klemeš, 1986) provides a guide to the ability of a model to represent observed catchment behaviour but is incomplete in terms of testing the predictive capability of the model. As the SHE is intended to be used in a predictive mode, for example for predicting the impacts of future land use changes, a more convincing test is required. The WRSRU has therefore developed a "blind" validation methodology, in which the modeller is not allowed sight of the measured output data until the simulations are finished (Ewen and Parkin, in press). This is considered to represent more realistically the conditions in which predictive applications for future states of a basin are likely to be carried out, for which there will be no time series records and an incomplete availability of catchment property data. The procedure is as follows.

(1) Choose a set of test statistics, such as peak outflow discharge or percentage of time the ground adjacent to the channel is saturated; decide on pass/fail criteria, e.g. for success the observed outflow hydrograph should fall within the bounds determined for the simulated hydrograph for 90% of the time.

(2) Using hydrological judgement, define bounds on the values of the model parameters, e.g. maximum and minimum values for soil hydraulic conductivity.

(3) Run the model for the various parameter bound values (including desired combinations of values for different parameters) to produce a range of possible output responses; extract the bounds for the various test statistics. (Up to this point the modeller is not allowed sight of the observed output.)

(4) With all runs completed, compare the predicted bounds with the measured output and assign pass/fail to each statistic, according to the relevant criteria.

(5) Assess the overall pass/fail outcome for the test.

(6) Assess the performance of the model and the modeller, e.g. whether the prediction bounds have been determined with a range narrow enough for meaningful conclusions to be derived from the test.

A first test of the procedure has been carried out, with a team of seven taking part. It was found that very careful judgement is required at steps 1-3, based on past experience and understanding of catchment response mechanisms. In particular, at step 3 an iterative procedure is required in which the results of each run are carefully reviewed for physical realism by the full team, adjustments are made to the parameter bounds and the model is run again. The results of the test are considered to be very positive and are to be published in due course.

16.9. RECENT SHE/SHESED APPLICATIONS

At the time of writing (mid 1993), a number of SHE/SHESED applications had either recently been completed or were nearing completion. These are, or will be, described in detail in other publications. However, the main points from some of them are presented briefly here in order to give a flavour of the way in which the SHE/SHESED combination can be used and the types of application for which it is relevant.

16.9.1. Application in Madhya Pradesh, India

This application, reported in Refsgaard et al. (1992) , Jain et al. (1992) and Lohani et al. (1993), concerned the transfer of the SHE to the National Institute of Hydrology, Roorkee, in Uttar Pradesh, India. Funded by the Commission of the European Communities as a technology transfer project and carried out by the three international SHE partners under the leadership of the Danish Hydraulic Institute, the transfer involved training the Indian scientists and carrying out test applications to six focus basins (800-5000 km^2 in area) in the Narmada basin, Madhya Pradesh. Some particular conclusions from the project are as follows.

(1) The SHE is able to provide a physically reasonable representation of the hydrological processes responsible for basin behaviour in a monsoon environment.

(2) On-site field measurements form an important supplement to data available from existing sources (such as reports from local research institutes), especially for evaluation of the soil hydraulic properties and for an improved assessment of the overall hydrological regime. For a feasibility study the necessary measurements can be made in a few weeks.

(3) The quantity and quality of rainfall data are of the utmost importance for accurate simulation of basin response. Uncertainty in the rainfall input is likely to be the chief constraint on simulation accuracy. It is, therefore, important to assemble and evaluate all rainfall records, and even initiate new measurements, at a very early stage of a project.

The project also provided valuable operational experience in staff training.

16.9.2. Application in Idaho, USA

Snowmelt and soil erosion simulations were carried out at the Reynolds Creek rangeland research basin in Idaho in collaboration with the basin's scientific operators, the Northwest Watershed Research Center of the United States Department of Agriculture's Agricultural Research Service. The snowmelt simulations, carried out at scales of 40 ha, 54 km^2 and 234 km^2 are partly reported in Bathurst and Cooley (in press). The results show the SHE to be capable of successfully simulating hydrograph responses caused by rain falling on snow and by snowmelt generated by heat inputs, including the variation in such responses with elevation and catchment zone. (A simulated variation in the spatial distribution of snowpack thickness through time is illustrated in Fig. 16.4.) They also show how, in the absence of

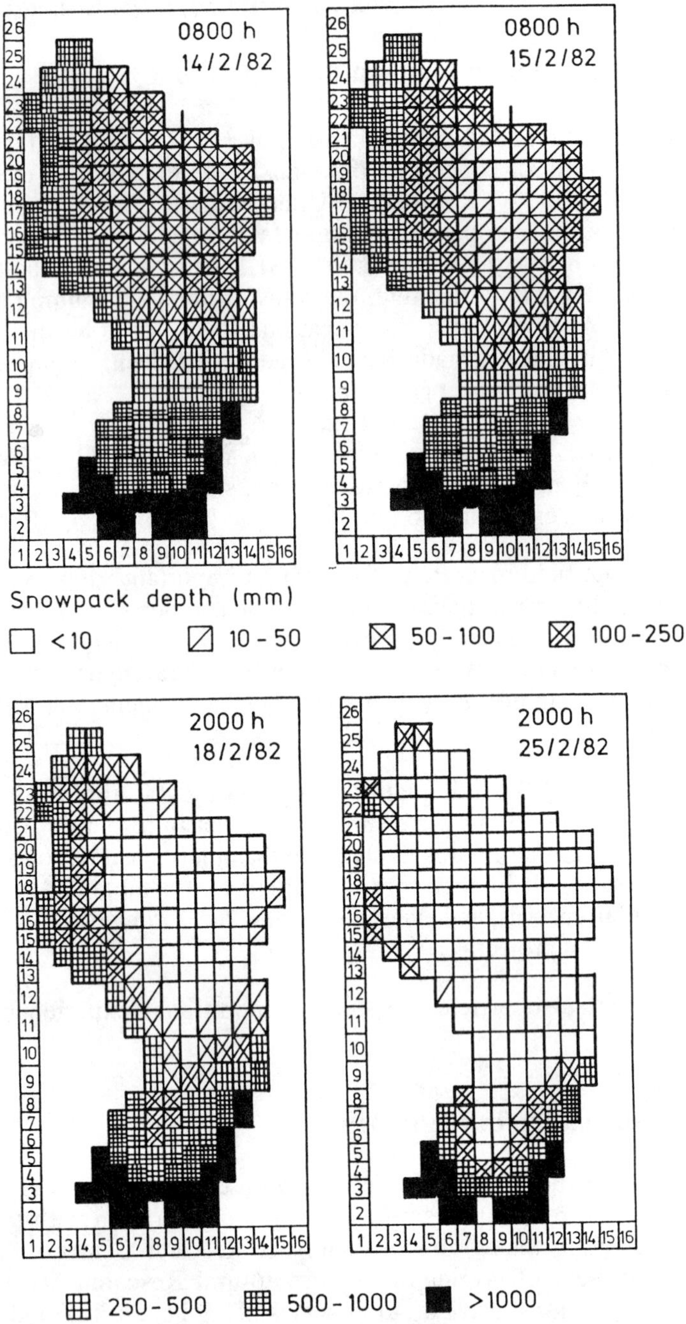

Fig. 16.4. Simulated variation in the spatial distribution of snowpack thickness at the Reynolds Creek catchment, Idaho, for the period 14-25 February 1982. Grid spacings are 1 km and 1.5 km. The channel network and catchment boundary are indicated by heavy lines. Results from J.C. Bathurst (University of Newcastle upon Tyne, UK) and K.R. Cooley (United States Department of Agriculture, Northwest Watershed Research Center, Idaho, USA).

previous field studies of basin response, the SHE can be used to explore different hypotheses of basin behaviour and to identify the one which produces output closest to that observed. The soil erosion simulations are reported in Wicks et al. (1988, 1992). Carried out largely at the scale of rainfall-simulator plots and for a range of land treatments, they indicate a good sensitivity of the soil erodibility coefficients to the major classes of treatment (e.g. Fig. 16.5). Plot calibration of the coefficients in the field is shown to be more reliable than laboratory calibration with disturbed soil samples. Also, sensitivity of the sediment yield simulations to the input water flow data indicates the need for accurate calibration of the water flow component.

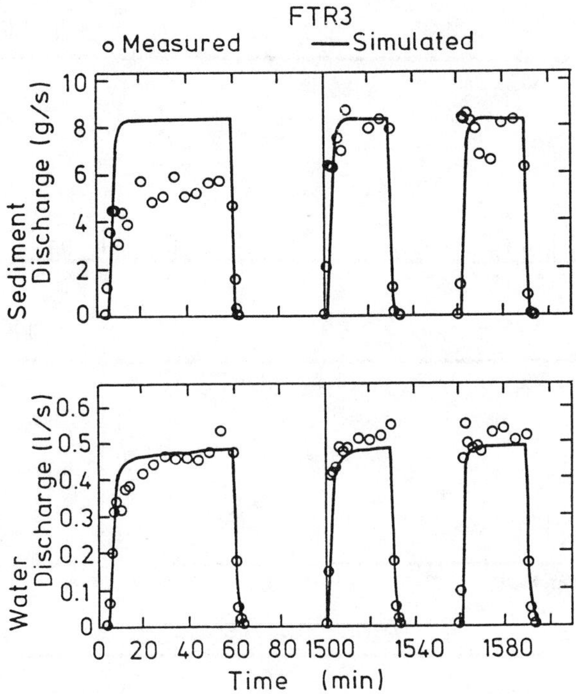

Fig. 16.5. Comparison of simulated water and sediment discharges with measured responses for a 32.54-m² tilled plot subjected to three bursts of artificial rainfall over a 26½-hour period at the Reynolds Creek catchment, Idaho, USA. Reproduced from Wicks et al. (1992) by permission of the American Society of Civil Engineers.

16.9.3. Application in Mediterranean Europe

This application, reported in Bathurst et al. (in press), was part of a 17-organization study of desertification in Mediterranean Europe known as

MEDALUS (Mediterranean Desertification and Land Use) and funded by the Commission of the European Communities. Through collaborations with the New University of Lisbon and the University of Murcia, simulations of water flow, soil erosion and sediment yield were carried out at the plot, subcatchment and full catchment (up to 700 km^2) scales at focus basins in Portugal and Spain, for present day conditions and for specified scenarios for future climate and land use. The principal findings were as follows.

(1) The SHE is able to provide a physically reasonable representation of basins in the Mediterranean area under present and future environmental conditions (e.g. Fig. 16.6).

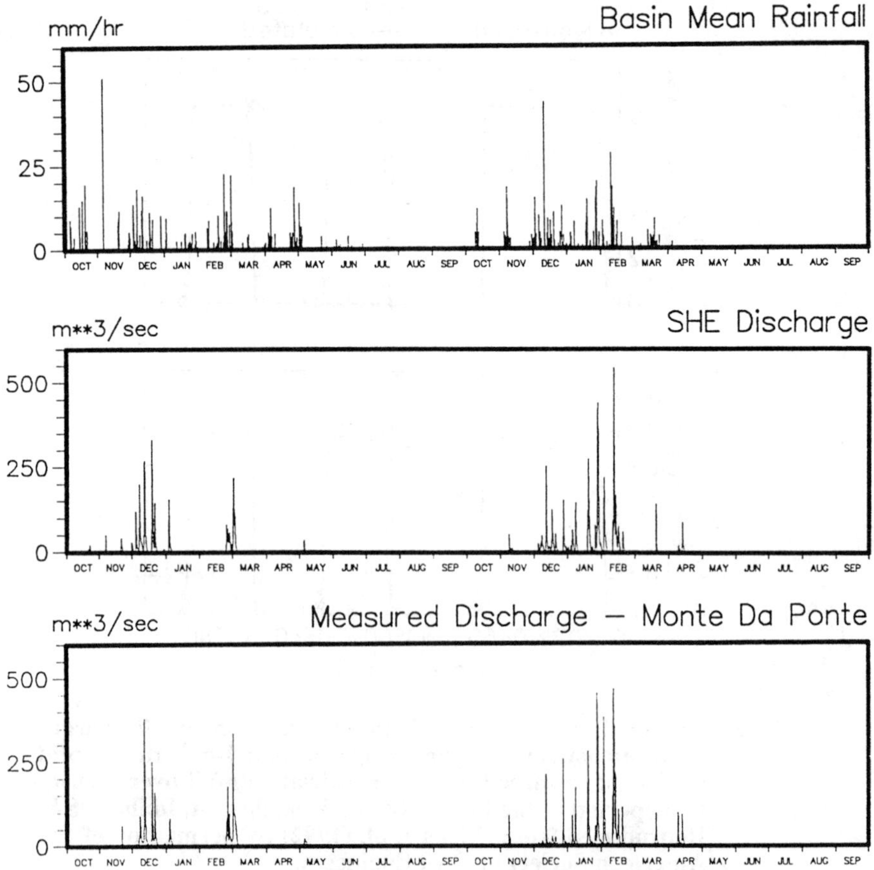

Fig. 16.6. Time series of mean basin rainfall (area weighted) and measured and simulated (SHE) discharge at the outlet to the Cobres catchment (701 km^2), Portugal, for the period October 1977 - September 1979. Reproduced from Bathurst et al. (in press) by permission of John Wiley and Sons Ltd.

(2) The simulations at different scales showed that, for certain conditions, plot-calibrated model parameters can be applied unchanged at model grid scales of up to 2 km, which is of considerable practical importance for river basin modelling.

(3) A powerful method for generating future climate scenarios by combining General Circulation Model predictions with present-day rainfall statistics via a stochastic rainfall model was developed and applied for a future climate based on doubled levels of atmospheric carbon dioxide. The simulation results indicated a reduction in runoff and sediment yield relative to current conditions but the reductions are within the current inter-annual variability.

16.9.4. Application to Radionuclide Migration Modelling

A major upgrading of the SHE/SHESED combination and the addition of a contaminant migration component has been carried out by a team in the WRSRU to create SHETRAN, a general water flow, sediment transport and contaminant migration modelling system. This development has taken place as one project within a large research programme, funded by UK Nirex Ltd and concerned with establishing a safety case for deep underground disposal of low and intermediate level radioactive waste in the UK. The basis of the work is discussed in more detail in Ewen (1990) and Purnama and Bathurst (1991a,b). SHETRAN, although designed with a general capability for modelling contaminant migration, has the specific objective within the Nirex programme of providing a basis for simulating the dispersal of radionuclides in the surface and near-surface regions of the biosphere following their release from a deep radioactive waste repository and upward migration through the geosphere. Results from SHETRAN simulations will contribute to the identification of major migration pathways and processes and the calculation of characteristic migration rates. This information will then support the development and application by other groups of simpler biosphere modelling technologies which can be used for the simulations of thousands to hundreds of thousands of years which will form the core of the safety case assessment. SHETRAN itself is currently limited to simulation periods of a few years to decades. Conditions must be simulated for the possible warmer and colder climates which may evolve at the repository site over the next million years. The SHETRAN flow component has therefore been validated using present day basins from various climatic regions as analogues for future conditions: this work has provided the first test of the blind validation technique referred to earlier.

16.9.5. Application in a Decision Support System

The NERC/ESRC Land Use Programme (NELUP) (Section 16.7.5) is being carried out by the Centre for Land Use and Water Resources Research at the University of Newcastle upon Tyne with the aim of developing a decision support system for planning sustainable patterns of land use, with particular relevance to the UK (O'Callaghan, 1992). An important role of such a system will be the evaluation of the trade-off between environmental quality and economic return referred to in the introduction. Development of the system involves the integration of hydrological, water quality, ecological and economic models interfaced through an information system: the SHE provides the basis for the hydrological and water quality models. Within the project the SHE is being modified for use at large basin scales and is being applied to the 2000-km^2 Tyne catchment in north-east England. Some preliminary results for a subcatchment of the Tyne are described in Dunn et al. (1992) and Lunn et al. (1992) and a more detailed report on the full Tyne applications is to be published as part of a NELUP Handbook. An important modification is SHE-SHELL, an executive code for minimizing the computational cost of performing hydrological simulations of large river basins. For basins of the scale of the Tyne, modelling is carried out using a set of smaller models, each associated with an individual subcatchment of the basin. The subcatchments are coupled at their boundaries by surface and subsurface transfers of water and contaminants. SHE-SHELL determines the subcatchments and manages the sequencing and linking of the subcatchment simulations. It also manages the large data sets and carries out the automatic generation of SHE data from routine data sets. Preliminary results for the Tyne show a good ability to simulate measured time series of river discharge and nitrate concentration.

16.10. FUTURE DEVELOPMENTS

A wider discussion of the possible future development of the SHE is provided in Bathurst and O'Connell (1992). However, three likely developments are the addition of new process components, continued integration within decision support systems and, in the longer term, an expert system for model calibration.

The introduction of a contaminant migration modelling capability has taken place with the development of SHETRAN. Additional process components are needed to represent erosion and sediment yield caused by shallow landslides and gullying, and to represent operation of irrigation systems. Upgrading of existing components to include secondary hydrological processes, for example soil cracking and crusting, may also be required.

Model calibration, at the current level of experience with physically based distributed modelling systems, involves a considerable degree of trial and error. The need to think in three-dimensional terms and to account for the combined effects of different soil, vegetation, topographic and other parameters presents obvious difficulties in deciding how to adjust parameter values to produce good agreement between observed and simulated variations, both at the catchment outlet and internally. Additionally, it is important to ensure that the calibrated model is a physically realistic representation of known basin conditions. However, in the longer term, as experience is accumulated, a more structured approach to integrating these requirements will become possible. Codified into expert systems, it will combine user experience, available data and information on basin response. The result will allow a more efficient approach to calibration and will reduce the length of training required by users of the SHE.

As the multi-faceted nature of environmental problems is recognized, the need is becoming apparent for larger modelling systems which can integrate the relevant and interlinked contributions to such problems. The SHE already provides the basis for water quality, sediment transport and contaminant migration modelling. Also, as noted above, it is being integrated with ecological and economic models to form a decision support system for planning sustainable patterns of land use at the river basin scale, and increasingly sophisticated developments of this nature are likely to continue. More distantly it could be integrated with estuary and coastal models to examine the impact of contaminants and sediments from the land on marine ecosystems and coastline evolution.

16.11. PRACTICAL CONCLUSIONS

From the experience gained in applications of the SHE, generally by the three international SHE partners and more specifically by the WRSRU, a number of practical conclusions can be drawn.

(1) The SHE has been shown to be at home in a wide range of environments. It has been applied in Europe, the USA, India, Thailand and New Zealand, at spatial scales ranging from a rainfall simulator plot (30 m^2) to a catchment of 5000 km^2 and for simulation periods of days to a decade or more. Hydrological responses (triggered by precipitation and snowmelt events) and soil erosion and sediment yield responses have been simulated. Overall the system holds considerable advantages over simpler models for the study of catchment response mechanisms and for applications concerned with the impacts of future land use and climate changes.

(2) Uncertainty in simulation output is reduced as the overall quality of the input data is increased. Precipitation is probably the single item of data to which the results are most sensitive and should be provided with the greatest temporal and spatial resolution possible. The soil parameters are generally the physical property description

to which the results are most sensitive and should be based on direct measurements. Time series data for internal variables are required if a model is to be validated for its capability to represent spatial variability. In any application it is helpful to understand the response mechanisms of the basin being modelled: a relatively short (weeks to months) campaign of field measurements can provide much of the necessary information as well as the basic data from which to evaluate the model parameters. Further research is required on scale dependency in parameter evaluation and the extent to which parameters based on small scale measurements can be used at large grid scales.

(3) Hydrological expertise and an understanding of basin response mechanisms are extremely important in data preparation, model calibration and validation and interpretation of simulation results. The physical basis of the model parameters, the multi-faceted nature of the model process description and the general complexity of SHE applications (involving more than simple rainfall-runoff modelling) mean that a carefully considered approach to model use is required. To a greater extent than is necessary for simpler models, hydrological judgement is required to explain the effects of changing model parameters and to decide the extent to which a calibrated model represents a basin and its response. In effect a more intelligent dialogue is required between modeller and model over the basin behaviour and hydrological reality. This is reflected in the relatively long time (several months) required to train users in the complexities of the SHE and its application. An important task for the future will be to improve the accessibility of the SHE to potential users through the implementation of expert systems, user-friendly graphics and other support.

ACKNOWLEDGEMENTS

The original development of SHESED was carried out by J M Wicks as a PhD project under the supervision of J C Bathurst and P E O'Connell. Subsequent SHE-based development and applications at the Water Resource Systems Research Unit, University of Newcastle upon Tyne, have been carried out by an increasing team of research staff under the leadership of P E O'Connell (the Unit Director). This research forms a basis for some of the discussion in this chapter and its contribution is acknowledged by appropriate publication references in the text. The work of the international SHE team is also gratefully acknowledged. Financial support for SHE-based development and applications at Newcastle has been provided by the Commission of the European Communities, UK Nirex Ltd., the UK Natural Environment Research Council, the UK Economic and Social Research Council and the North Atlantic Treaty Organization.

REFERENCES

Abbott, M.B., Bathurst, J.C., Cunge, J.A., O'Connell, P.E. and Rasmussen, J. 1986a. 'An introduction to the European Hydrological System - Système Hydrologique Européen, "SHE", 1: History and philosophy of a physically-based, distributed modelling system.' J. Hydrol., 87, 45-59.

Abbott, M.B., Bathurst, J.C., Cunge, J.A., O'Connell, P.E. and Rasmussen, J. 1986b. 'An introduction to the European Hydrological System - Système Hydrologique Européen, "SHE", 2: Structure of a physically-based, distributed modelling system.' J. Hydrol., 87, 61-77.

Bathurst, J.C. 1986. 'Physically-based distributed modelling of an upland catchment using the Système Hydrologique Européen.' J. Hydrol., 87, 79-102.

Bathurst, J.C. and Cooley, K.R. In press. 'Use of the SHE hydrological modelling system to investigate basin response to snowmelt at Reynolds Creek, Idaho.' J. Hydrol.

Bathurst, J.C. and O'Connell, P.E. 1992. 'Future of distributed modelling: the Système Hydrologique Européen'. Hydrological Processes, 6, 265-277.

Bathurst, J.C., Kilsby, C. and White, S. In press. 'Modelling the impacts of climate and land use change on basin hydrology and soil erosion in Mediterranean Europe.' In Mediterranean Desertification and Land Use, J. Thornes and J. Brandt (eds), John Wiley & Sons, Chichester, UK.

Beven, K. 1989. 'Changing ideas in hydrology - the case of physically-based models.' J. Hydrol., 105, 157-172.

Binley, A., Beven, K. and Elgy, J. 1989. 'A physically based model of heterogeneous hillslopes 2. Effective hydraulic conductivities.' Wat. Resour. Res., 25, 1227-1233.

Dunn, S., Savage, D. and Mackay, R. 1992. 'Hydrological simulation of the Rede catchment using the Système Hydrologique Européen (SHE).' In Land Use Change: the Causes and Consequences, M.C. Whitby (ed.), Institute of Terrestrial Ecology Symposium No. 27, Her Majesty's Stationery Office, London, pp. 137-146.

Ewen, J. 1990. 'Basis for the subsurface contaminant migration components of the catchment flow, sediment transport, and contaminant migration modelling system SHETRAN-UK.' UK Nirex Ltd Rep. NSS/R229, Harwell, UK.

Ewen, J. and Parkin, G. In press. 'Validation of catchment models for predicting land-use and climate change impacts.' J. Hydrol.

Jain, S.K., Storm, B., Bathurst, J.C., Refsgaard, J.C. and Singh, R.D. 1992. 'Application of the SHE to catchments in India Part 2. Field experiments and simulation studies with the SHE on the Kolar subcatchment of the Narmada River.' J. Hydrol., 140, 25-47.

Klemeš, V. 1986. 'Operational testing of hydrological simulation models.' Hydrol. Sci. J., 31, 13-24.

Lohani, V.K., Refsgaard, J.C., Clausen, T., Erlich, M. and Storm, B. 1993. 'Application of SHE for Irrigation-Command-Area Studies in India.' Proc. Am. Soc. Civ. Engrs., J. Irrig. Drain. Engrg., 119, 34-49.

Lunn, R., Younger, P.L. and Mackay, R. 1992. 'Development of a methodology for hydrological simulation at the basin scale, using the Système Hydrologique Européen (SHE).' In Land Use Change: the Causes and Consequences, M.C. Whitby (ed.), Institute of Terrestrial Ecology Symposium No. 27, Her Majesty's Stationery Office, London, pp. 147-158.

O'Callaghan, J.R. 1992. 'Decision-making in land use.' In Land Use Change: the Causes and Consequences, M.C. Whitby (ed.), Institute of Terrestrial Ecology Symposium No. 27, Her Majesty's Stationery Office, London, pp. 79-87.

Phillips, B.C. and Sutherland, A.J. 1985. 'Numerical modelling of spatial and temporal lag effects in bed load sediment transport.' Proc. 21st Congress Intl. Ass. Hydraul. Res., Melbourne, pp. 571-576.

Purnama, A. and Bathurst, J.C. 1991a. 'Basis for the ground surface and stream channel radionuclide transport components of the catchment modelling system SHETRAN-UK including a review of sorption.' Draft UK Nirex Ltd Rep. NSS/R232, Harwell, UK, 136 pp.

Purnama, A. and Bathurst, J.C. 1991b. 'A review of three features of sediment transport in stream channels: dynamics of cohesive sediment, infiltration of transported sediment into the bed, and stream bank erosion.' Draft UK Nirex Ltd Rep. NSS/R233, Harwell, UK, 124 pp.

Ragab, R. and Cooper, J.D. 1993a. 'Variability of unsaturated zone water transport parameters: implications for hydrological modelling. 1. In situ measurements.' J. Hydrol., 148, 109-131.

Ragab, R. and Cooper, J.D. 1993b. 'Variability of unsaturated zone water transport parameters: implications for hydrological modelling. 2. Predicted vs. in situ measurements and evaluation of methods.' J. Hydrol., 148, 133-147.

Refsgaard, J.C., Seth, S.M., Bathurst, J.C., Erlich, M., Storm, B., Jørgensen, G.H. and Chandra, S. 1992. 'Application of the SHE to catchments in India Part 1. General results.' J. Hydrol., 140, 1-23.

Wicks, J.M. 1988. 'Physically-based mathematical modelling of catchment sediment yield.' PhD thesis, University of Newcastle upon Tyne, Newcastle upon Tyne, UK.

Wicks, J.M. and Bathurst, J.C. In press. 'SHESED: A physically-based, distributed erosion and sediment yield component for the SHE hydrological modelling system.' J. Hydrol.

Wicks, J.M., Bathurst, J.C., Johnson, C.W. and Ward, T.J. 1988. 'Application of two physically-based sediment yield models at plot and field scales.' In Sediment Budgets, Intl. Ass. Hydrol. Sci. Publ. No. 174, Inst. Hydrology, Wallingford, UK, pp. 583-591.

Wicks, J.M., Bathurst, J.C. and Johnson, C.W. 1992. 'Calibrating SHE soil-erosion model for different land covers.' Proc. Am. Soc. Civ. Engrs., J. Irrig. Drain. Engrg., 118, 708-723.

Chapter 17

THE INSTITUTE OF HYDROLOGY DISTRIBUTED MODEL

A. Calver and W. L. Wood

17.1. INTRODUCTION

This chapter describes a physically-based rainfall-runoff modelling system developed in the last decade at the U.K. Institute of Hydrology and generally known as the I.H. Distributed Model. The term 'distributed' in the title refers to the inclusion of spatial dimensions in modelling catchment hydrology.

The model uses established flow equations for surface and subsurface flow and solves these equations by standard numerical methods. Subsurface flow equations used in saturated groundwater modelling practice are extended upward in their zone of application into the less well-tested unsaturated zone. An important aspect of the development of the modelling system is the linkage of the flow equations in a hydrologically appropriate manner and their integration with physical models of interception and evapotranspiration. The aim is, therefore, a physical description of the major processes of the hydrological cycle. As such, this type of modelling is necessarily complex and more appropriate in some circumstances than others. The IHDM has, like any large modelling system, been through various stages of development and continues to do so and to exist in a number of forms for particular purposes.

The chapter is organised on the following lines. The theory and methodology of the surface and subsurface flow equations, their linkages and solution are described, and the practicalities of data and computing requirements outlined. Experience of use of the IHDM in different contexts is discussed, together with guidelines for the efficient use of the model.

17.2. METHODOLOGY

17.2.1. General

Surface water flow in a catchment is described by a kinematic wave formulation solved in the downslope direction by a finite difference method.

This formulation is used for both channel flow and for any overland flow that is generated on the land surface. Subsurface water movement is modelled as Darcian porous medium flow in a formulation which combines saturated and variably unsaturated conditions and is solved in the two-dimensional vertical plane by a finite element method. The third spatial dimension is approximated by a variable width to such planes. Numerical approximation methods are used because of the complexity of the circumstances modelled. Overland flow can be of saturation-excess and/or infiltration-excess type. Channel flow is derived from upstream channel contributions, from channel precipitation, from any overland flow from the ground surface and from lateral throughflow from adjacent saturated soil and/or aquifer material.

The catchment as a whole is modelled by division into channel and hillslope components which are individually modelled and whose flows are appropriately accumulated: Figure 17.1 illustrates the principle. The model produces estimates of flow, water contents and pressure potentials at such locations in the catchment as the user requires. Output is tailored for a particular application: it commonly takes the form of plots of discharge against time and of pressure potential in vertical cross-sections of the

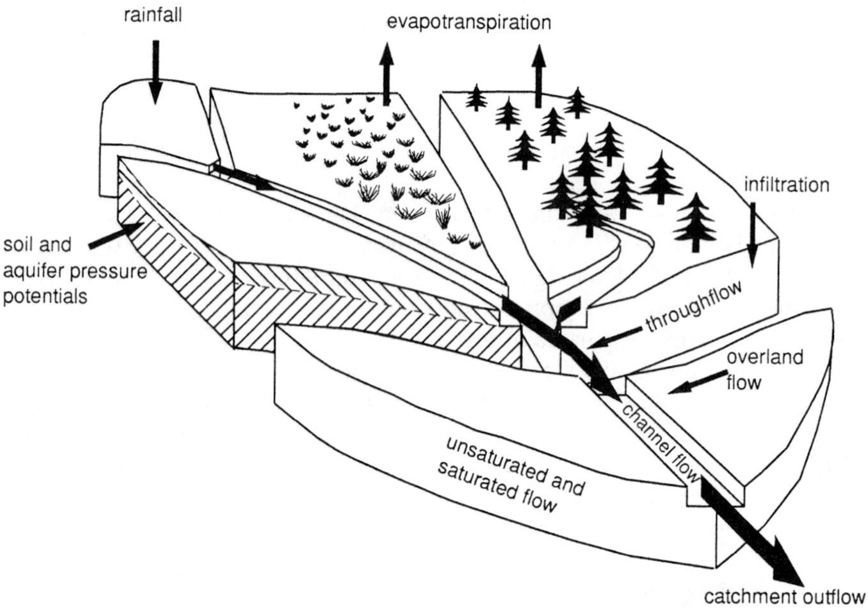

Fig. 17.1. Schematic structure of hillslope and channel components of the IHDM.

catchment. A general description of the methodology is given in an I.H. Report (Beven, Calver and Morris, 1987). Consideration of aspects of water quality is currently being combined with the flow formulation: attention is focussed on nitrogen transport and transformation (Cooper, Ragab, Lewis, Dixon and Whitehead, 1991; Cooper, Ragab, Lewis and Whitehead, 1992).

Catchment components can, of course, be used individually or combined with other formulations for greater or lesser detail as appropriate. Though fundamentally a deterministic approach, the model can be set in a probabilistic framework if required.

17.2.2. Surface Flow

Surface flow, whether sheet flow or channel flow, is represented by the kinematic wave equation

$$\frac{\partial q}{\partial t} + c\frac{\partial q}{\partial X} = cr \tag{17.1}$$

where q is the discharge per unit width, t is time, X is the distance down the slope and r is the effective precipitation (plus any seepage from and minus any infiltration into the ground). The value of c, the wave speed, is given by

$$c = \frac{dq}{dA} \tag{17.2}$$

where A is the cross-sectional area of the flow. Derivations of kinematic wave formulae from the Saint-Venant equations for the dynamics and continuity of unsteady flow (Barré de Saint-Venant, 1848, 1871) are given, for example, by Overton and Meadows (1976) and Wood (1993). The conditions for which the kinematic wave Eq. (17.1) can be used instead of the full St Venant equations have been investigated by Woolhiser and Liggett (1967), Morris and Woolhiser (1980) and Daluz Vieira (1983). Their conclusions depend on the kinematic wave parameter

$$k = \frac{sLg}{HF_L^2} = \frac{sLg}{V_L^2} \tag{17.3}$$

where $\boldsymbol{F}_L = V_L / \sqrt{(gH_L)}$ is the Froude number at $x=L$ where the depth is H_L, the velocity is V_L and g is the acceleration due to gravity. For natural slopes the values of k are expected to be larger than 50 and the kinematic wave approximation can be used. On smooth urban slopes where k is smaller the value of $\boldsymbol{F}_L$ must also be taken into account.

The kinematic wave approximation implicitly takes

$$s = s_f \tag{17.4}$$

where $s = \sin\theta$ is the slope of the ground and s_f is the friction slope (French, 1987). The general formula for the friction slope is given by

$$s_f = Cu^n / H^m \tag{17.5}$$

where u is the average velocity and H is the depth. The parameter C is independent of u and H and the parameters m and n depend on the friction law being used. For example, with the Chézy formula m=1 and n=2 and with the Manning formula m=4/3 and n=2. Substituting into Eq. (17.5) with $q = uHb = uA$, where b is width and A is the cross-sectional area, and rearranging gives a formula for the discharge q assumed to be a single-valued function of A as

$$q = \left(\frac{s}{C}\right)^{1/n} b^{-m/n} A^{1+m/n} \tag{17.6}$$

i.e.

$$q = \alpha A^{\beta} \tag{17.7}$$

where β=5/3 for Manning or β=3/2 for Chézy. Thus

$$\frac{\partial q}{\partial t} = \frac{dq}{dA}\frac{\partial A}{\partial t} = \frac{dq}{dA} b \frac{\partial H}{\partial t} \tag{17.8}$$

In the IHDM the quantities are expressed as per unit width and b=1.

In order to solve the hyperbolic Eq. (17.1) it is required to know the initial value of q at t=0 and a boundary condition. In the IHDM the initial condition is usually taken as q=0 and the boundary condition as $\partial q/\partial x = 0$ at the divideward limit of flow. Equation (17.1) is solved in the IHDM by a version of the finite difference method known as the box scheme using the grid points at the four corners of a box: $[j\Delta X, n\Delta t]$, $[(j+1)\Delta X, n\Delta t]$, $[(j+1)\Delta X, (n+1)\Delta t]$ and $[j\Delta X, (n+1)\Delta t]$. Here ΔX and Δt are the space and time steps and i,j are integers. The box scheme is also known as Preissmann's method (Preissmann, 1961). The dependence of c on q is dealt with by taking

$$\frac{1}{\Delta t}\left(q_{j+1}^{n+1} - q_{j+1}^{n}\right) + \frac{(1-w)}{\Delta X} c^n \left(q_{j+1}^{n} - q_j^n\right) + \frac{w}{\Delta X} c^{n+1}\left(q_{j+1}^{n+1} - q_j^{n+1}\right) = r\left(wc^{n+1} + (1-w)c^n\right) \tag{17.9}$$

where q_j^n is the approximate value of $q(x,t)$ at $(j\Delta X, n\Delta t)$, w is a parameter, $0 \le w \le 1$, and c^n is the kinematic wave speed taken as the mean of the values at the existing and the previous node calculated from the formula

$$q = fs^{1/2} A^{(2+m)/2}, \quad i.e. \quad A = \left(\frac{q}{f\sqrt{s}} \right)^{2/(2+m)} \tag{17.10}$$

f being a roughness parameter (compare Eq. (17.6) with n=2, b=1). This gives

$$c = \frac{dq}{dA} = 1 / \frac{dA}{dq} = (f\sqrt{s})^{2/(2+m)} \left(\frac{2+m}{2} \right) q^{m/(2+m)} \tag{17.11}$$

Equation (17.9) is most accurate when w=0.5 and is locally unconditionally stable for $w \ge 0.5$. Equation (17.9) is also nonlinear because of the dependence of c on q and is solved with iteration until the relative change in discharge is less than some specified tolerance. If this is not reached in a small number of iterations it is possible to reduce the mesh size and/or the time step.

17.2.3. Subsurface Flow

The basic equation of subsurface flow used in the IHDM is

$$S_o \frac{\partial \phi}{\partial t} = \nabla.(K \nabla \phi) \tag{17.12}$$

where ϕ is the piezometric potential or head of water, K is the hydraulic conductivity tensor of the porous medium and S_o is the specific storativity. Bear (1979), for example, gives details of the derivation of this equation which combines Darcy's Law (Darcy, 1856) and a continuity equation and is known as Richards' equation (Richards, 1931). Richards' equation in the IHDM combines flow in the saturated and unsaturated regions up to the surface of the ground. Taking the coordinates as x,y horizontally and z vertically upwards, this general three-dimensional equation can be written in the form

$$C(\psi) \frac{\partial \phi}{\partial t} = \frac{\partial}{\partial x}\left[K_x(\psi) \frac{\partial \phi}{\partial x} \right] + \frac{\partial}{\partial y}\left[K_y(\psi) \frac{\partial \phi}{\partial y} \right] + \frac{\partial}{\partial z}\left[K_z(\psi) \frac{\partial \phi}{\partial z} \right] \tag{17.13}$$

where $C(\psi) = d\theta/d\psi$ is the specific water storativity or the specific water capacity of the saturated/unsaturated medium, θ is the proportional soil moisture content by volume and $\psi = \phi - z$ is the (moisture or) pressure potential. The relationships of θ as a proportion of the saturated value θ_s, and the hydraulic conductivity components as proportions of the saturated value K_s, to the value of ψ vary with the type of soil. In order to solve the parabolic equation (Eq. 17.13) it is necessary to know the initial value of ϕ at time t=0 and also the known head on part of the boundary (S_1) and known flow on the rest of the boundary (S_2).

The IHDM uses two-dimensional vertical sections of the catchment hillslopes with a variable slope width, B (Fig. 17.2). The subsurface flow is thus represented by the equation for $\phi(x,z,t)$:

$$BC(\psi)\frac{\partial \phi}{\partial t} = \frac{\partial}{\partial x}\left[BK_x(\psi)\frac{\partial \phi}{\partial x}\right] + \frac{\partial}{\partial z}\left[BK_z(\psi)\frac{\partial \phi}{\partial z}\right] \qquad (17.14)$$

assuming no variation of material parameters in the y-direction.

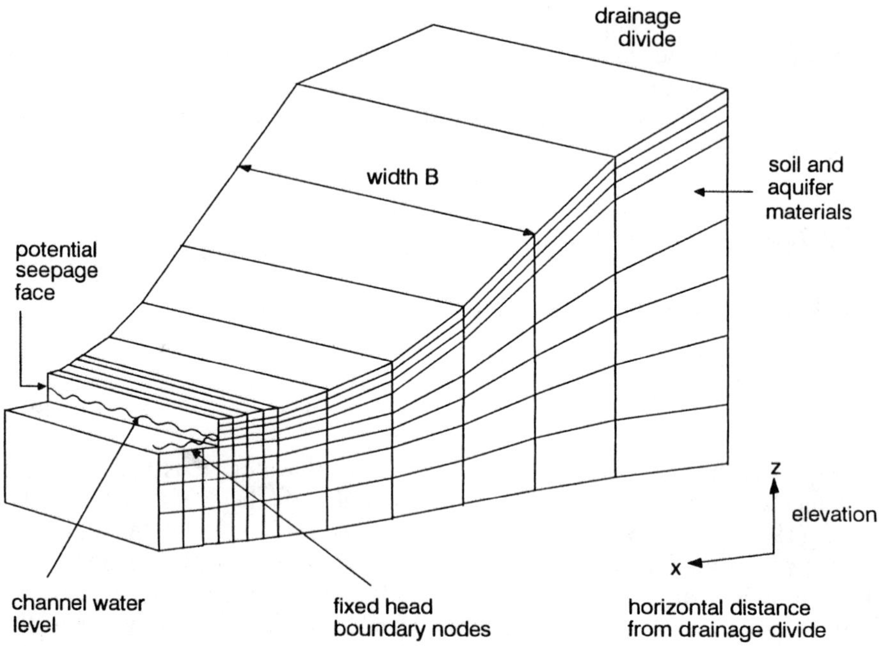

Fig. 17.2. An example of finite element discretisation of an IHDM hillslope.

The boundary conditions for the IHDM hillslope model are most commonly taken as follows, but may be varied in particular applications. There is zero flow at the drainage divide at the top of the slope; the lower boundary of the region commonly also has zero flow across it, or leakage can be included as a known-flow boundary condition. At the surface of the soil the result of effective precipitation is considered; if the soil surface is unsaturated the effective precipitation enters the soil at a flow rate dependent on local conditions; if the soil surface is saturated the effective precipitation is added to the kinematic wave equation (Eq. 17.8) (infiltration-excess overland flow) and the surface boundary condition becomes known-head with the potential at atmospheric i.e $\phi=z$: the small depth of overland flow is neglected. There may be a flow of water out from the saturated surface to join the overland flow (saturation-excess overland flow). Root water uptake from a defined rooting zone can occur depending on local pressure potentials and the evapotranspiration model option used. A known head is used adjacent to the channel at the foot of the slope, depending on 'average' channel water level conditions. This can be replaced by a time-dependent head if necessary. Above this level a potential seepage face occurs where saturated conditions abut this face. Where the soil is unsaturated this is treated as a zero-flow boundary in the solution of the equations. On the active seepage face the soil is saturated hence the pressure is atmospheric and $\phi=z$, and the flow is outwards. The seepage face may be in discrete sections depending on the ground conditions.

The IHDM uses the finite element method for the spatial discretization in the subsurface flow Eq. (17.14) and finite differences for the time-stepping. The way the approximation is constructed in the finite element method means that the numerical solution is readily available anywhere within the region. The IHDM uses piecewise bilinear approximation on quadrilaterals; for convenience these usually have nodes in vertical columns (Fig. 17.2).

The principal ingredients of the finite element program are the formation of element matrices, the assembly into the global matrices, the inclusion of the boundary conditions and the solution of the resultant system of algebraic equations. The bilinear quadrilateral is an example of an isoparametric element where the same basis functions are used for transforming the sides of the element as for approximate solution of the differential equation. The right hand side of Fig. 17.3 shows a bilinear square element with local coordinates referred to axes 0ξ, 0η with origin at the centre of the square whose vertices are (1,1), (-1,1), (-1, -1) and (1, -1), numbered 1,2,3,4 as shown. The bilinear basis functions are

$$N_1 = \frac{1}{4}(1+\xi)\,(1+\eta) \quad \text{etc.} \tag{17.15}$$

With the same basis functions used for transforming the sides of the quadrilateral element:

$$x=\sum_1^4 x_j N_j(\xi,\eta)\,, \quad y=\sum_1^4 y_j N_j(\xi,\eta) \tag{17.16}$$

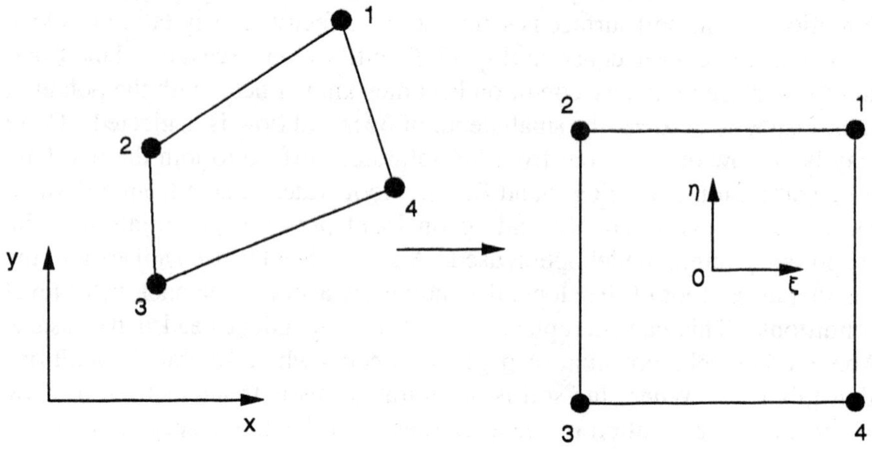

Fig. 17.3. Isoparametric transformation from a general quadrilateral to a square element.

the quadrilateral is transformed into the square as shown in Fig. 17.3.

The element stiffness matrix requires entries of the form

$$\iint_{el}\left(\frac{\partial N_i}{\partial x}\frac{\partial N_j}{\partial x}+\frac{\partial N_i}{\partial y}\frac{\partial N_j}{\partial y}\right)dx\,dy \tag{17.17}$$

where x,y are the global coordinates and el stands for the element. It is necessary to transform the integrand to ξ, η coordinates so as to use the standard square element basis functions and numerical integration formulae. The isoparametric equations (17.16) give x,y in terms of ξ, η but in general they cannot be inverted to give ξ, η in terms of x,y. Hence the chain rule has to be used:

$$\frac{\partial N_i}{\partial \xi}=\frac{\partial N_i}{\partial x}\frac{\partial x}{\partial \xi}+\frac{\partial N_i}{\partial y}\frac{\partial y}{\partial \xi} \tag{17.18}$$

$$\frac{\partial N_i}{\partial \eta}=\frac{\partial N_i}{\partial x}\frac{\partial x}{\partial \eta}+\frac{\partial N_i}{\partial y}\frac{\partial y}{\partial \eta} \tag{17.19}$$

i.e.

$$\begin{bmatrix}\dfrac{\partial N_i}{\partial \xi}\\ \dfrac{\partial N_i}{\partial \eta}\end{bmatrix}=\boldsymbol{J}\begin{bmatrix}\dfrac{\partial N_i}{\partial x}\\ \dfrac{\partial N_i}{\partial y}\end{bmatrix} \tag{17.20}$$

where $\boldsymbol{J}$ is the Jacobian matrix of the transformation given by

$$\boldsymbol{J} = \begin{bmatrix} \dfrac{\partial x}{\partial \xi} & \dfrac{\partial y}{\partial \xi} \\ \dfrac{\partial x}{\partial \eta} & \dfrac{\partial y}{\partial \eta} \end{bmatrix} \tag{17.21}$$

and which can be obtained from equations (17.16). The derivatives required for the integrand in Eq. (17.17) can be obtained from

$$\begin{bmatrix} \dfrac{\partial N_i}{\partial x} \\ \dfrac{\partial N_i}{\partial y} \end{bmatrix} = \boldsymbol{J}^{-1} \begin{bmatrix} \dfrac{\partial N_i}{\partial \xi} \\ \dfrac{\partial N_i}{\partial \eta} \end{bmatrix} = \frac{1}{|\boldsymbol{J}|} \begin{bmatrix} \dfrac{\partial y}{\partial \eta} & -\dfrac{\partial y}{\partial \xi} \\ -\dfrac{\partial x}{\partial \eta} & \dfrac{\partial x}{\partial \eta} \end{bmatrix} \begin{bmatrix} \dfrac{\partial N_i}{\partial \xi} \\ \dfrac{\partial N_i}{\partial \eta} \end{bmatrix} \tag{17.22}$$

where $|\boldsymbol{J}|$ is the determinant of the Jacobian matrix.

Thus

$$\frac{\partial N_i}{\partial x} = \frac{1}{|\boldsymbol{J}|} \left[\frac{\partial N_i}{\partial \xi} \frac{\partial y}{\partial \eta} - \frac{\partial N_i}{\partial \eta} \frac{\partial y}{\partial \xi} \right] \tag{17.23}$$

and

$$\frac{\partial N_i}{\partial y} = \frac{1}{|\boldsymbol{J}|} \left[- \frac{\partial N_i}{\partial \xi} \frac{\partial x}{\partial \eta} + \frac{\partial N_i}{\partial \eta} \frac{\partial x}{\partial \xi} \right] \tag{17.24}$$

where all the right hand side terms in Eqs. (17.23) and (17.24) are obtainable from the isoparametric transformation. It is not necessary to go into details here to see that the result is that

$$\iint_{el} \nabla N_i . \nabla N_j \; dxdy = \iint_{el} \frac{1}{|\boldsymbol{J}|} (\text{polynomial in } \xi, \eta) \; d\xi \, d\eta \tag{17.25}$$

i.e. in general the integrand here is a rational function because $|\boldsymbol{J}|$ is itself an algebraic function of ξ, η. The Gauss numerical integration rules used in the IHDM (Conte and de Boor, 1980) are exact for polynomials but also quite good for rational functions away from singularities, i.e. it is important not to have $|\boldsymbol{J}|$ near zero. $|\boldsymbol{J}|$ is a constant for a parallelogram so that if the elements do not deviate too far from parallelograms the four point Gauss rule should be safe with bilinear quadrilaterals. This samples the integrand at the points $\xi = \pm 1/\sqrt{3}$, $\eta = \pm 1/\sqrt{3}$. The IHDM has provision for the use of the 3×3 and the 4×4 Gauss integration rules also.

The elements and the element nodes are numbered and the element matrix contributions are assembled into the global matrices. The result for the time-dependent problem Eq. (17.14) is

$$\boldsymbol{M}\frac{d\boldsymbol{\phi}^h}{dt}+\boldsymbol{K}\boldsymbol{\phi}^h=\boldsymbol{F} \tag{17.26}$$

where $\boldsymbol{M},\boldsymbol{K}$ are the mass and stiffness matrices respectively and the vector $\boldsymbol{F}$ includes the known terms. The finite element Eqs. (17.26) are the result of multiplying Eq. (17.14) by a test function w taken equal to the $N_i(x,y)$ in turn, integrating by parts and substituting the known flow boundary conditions (see, for example, Wood, 1993, for details). But the known head boundary conditions on S_1 still have to be included. In practice a finite element program assembles the mass and stiffness matrices first ignoring these known head boundary conditions. The values $\partial\phi/\partial n$ on S_1 are not known and the equations of the system, given by taking $w=N_i(x,y)$ where i is the number of a node on S_1, are copied into a reserve to be used later. In the IHDM these equations in the global system are then overwritten so that the diagonal term on the left hand side is multiplied by a large number E and the right hand side is replaced by E times the correct value. The number of equations to be solved is then equal to the total number of nodes. When all the nodal values are available the unknowns in the equations for nodes i on S_1 can be filled in to give the values of

$$\int_{S_1} N_i(x,y)\left[K_x\frac{\partial\phi}{\partial x}n_x+K_z\frac{\partial\phi}{\partial z}n_z\right]dS \tag{17.27}$$

where (n_x, n_z) is the unit outward normal. This gives the flow across the boundary in the neighbourhood of the known head nodes (Lynch, 1984). The overall mass balance figure in the IHDM cannot therefore be used as an independent measure of model performance.

In the solution of Richards' equation allowance has to be made for the fact that $C(\psi)$, K_x, K_z are functions of the unknown piezometric head ø and these variables are included in the integrands of the mass and stiffness matrices. The result is a system of nonlinear equations

$$\boldsymbol{M}(\boldsymbol{\phi})\frac{d\boldsymbol{\phi}}{dt}+\boldsymbol{K}(\boldsymbol{\phi})\,\boldsymbol{\phi}=\boldsymbol{F} \tag{17.28}$$

This system is solved with a backward difference scheme and with iteration to cope with the nonlinearity:

$$\boldsymbol{M}(\overline{\boldsymbol{\phi}})\frac{(\boldsymbol{\phi}_{n+1}-\boldsymbol{\phi}_n)}{\Delta t}+\boldsymbol{K}(\overline{\boldsymbol{\phi}})\boldsymbol{\phi}_{n+1}=\boldsymbol{F} \tag{17.29}$$

where $\boldsymbol{\phi}_n$ is the vector of approximate nodal values at time $n\Delta t$ and $\overline{\phi}$ is a value averaged over the time interval. The backward difference method is unconditionally stable and gives a smooth result.

Equation (17.29) is written in the form

$$\boldsymbol{A}\boldsymbol{\phi}_{n+1} = \boldsymbol{H} \tag{17.30}$$

for each iteration and the system is solved with Gauss elimination adapted for a banded symmetric matrix $\boldsymbol{A}$ as in Zienkiewicz (1971). The iteration is then accelerated by an adaptation of the Cooley algorithm (Cooley, 1983). The equations are solved for the change $\Delta\phi^k_{n+1}$ in the value of the potential from one iteration to the next, where

$$\Delta\,\phi^k_{n+1} = \phi^k_{n+1} - \phi^{k-1}_{n+1} \tag{17.31}$$

The Cooley algorithm assumes that these changes will oscillate in sign from one iteration to the next so that the convergence is accelerated by damping this oscillation. This is equivalent to under-relaxation. If Δ_k is the maximum change in nodal potentials at the kth iteration, the method uses

(a)

$$s = 1 \tag{17.32}$$

for $k=0$

$$s = \Delta_{k+1} / (\omega_k \Delta_k) \tag{17.33}$$

for $k>0$, where ω_k is the relaxation parameter from the previous iteration

(b)

$$\hat{\omega} = (3 + s) / (3 + |s|) \tag{17.34}$$

when $s>-1$, otherwise

$$\hat{\omega} = 0.5 / |s| \tag{17.35}$$

(c) if $\hat{\omega}|\Delta_{k+1}|$ is greater than e_{max}, the specified maximum change in the potential to be allowed in an iteration, then

$$\omega_{k+1} = e_{\max} / |\Delta_{k+1}| \tag{17.36}$$

otherwise

$$\omega_{k+1} = \hat{\omega} \tag{17.37}$$

The iteration stops when $|\Delta_{k+1}|$ is less than the prescribed tolerance. This adaptation of the Cooley algorithm works very efficiently with the IHDM.

More iterations are needed at the start of the run and when rainfall starts but for most time steps only a few iterations are necessary. With an example involving 25 hours without rain, 5 hours with steady rainfall of 4 mm hr^{-1} and another 25 hours without rain, and a tolerance of 0.001m of water, an average of between 2 and 3 iterations per 0.5 hr time step is required and an average of 1 to 2 iterations for a time step of 0.1 hr. The total number of iterations differs little between different grids.

17.3. PRACTICALITIES: DATA AND COMPUTING REQUIREMENTS; CALIBRATION

From the description of the methodology it will be apparent that the model has, theoretically, a large data requirement. In practice it is possible, and indeed necessary, to use the model with considerably less information than that ideally needed.

The data are of three main types, namely a geometric description of the catchment, rainfall data and material hydraulic properties. The cost of data acquisition is increasingly to be considered in distributed model application.

In the U.K., catchment geometry is readily available from Ordnance Survey mapping and, increasingly, from digital terrain systems: in the U.S. topographic maps are available from the U.S. Geological Survey and digital elevation models have been produced by the Defense Mapping agency and (to date a partial cover) by the U.S.G.S. The topographic data have to be divided, with reference also to subsurface conditions, into the catchment components as described above and to be further discretised to produce spatial grids for the flow solution methodologies. A degree of subjectivity inevitably exists in a discretisation algorithm. The discretisation is particularly important (see below) because of its effect on the accuracy of solution achieved.

Rainfall input can be in the form of observed or generated time series or of design storm profiles. The conversion of gross rainfall to net ground surface water input, taking account of interception and evapotranspiration processes, can be made by any suitable method. A preprogram to the IHDM is available (Watts, 1988; Lewis, 1992) which models these processes in a physical way. The methodology is common to a number of other modelling systems: a Rutter interception model (Rutter, Kershaw, Robins and Morton, 1971) and a Penman-Monteith evapotranspiration model (see, for example, Shuttleworth, 1992) are used with a choice of levels of approximation to the soil moisture restraint on root water uptake. Snowmelt routines are also available (Morris and Harding, 1991). The U.K. Meteorological Office produces meteorological data, including rainfall estimates calibrated with respect to radar information, and evapotranspiration estimates (Thompson, Barrie and Ayles, 1981). U.S. climate data are available from the National Climatic Data Center and from private concerns; the weather radar system service commissioned by the National Weather Service is scheduled to expand coverage to most of the U.S. Quantitative hydrological parameters for vegetation are probably still best sought from the research literature; more general divisions of land use classes are available from a number of sources (see, for example, Anderson, Hardy, Roach and Witmer, 1976; Wilson and Henderson-Sellers, 1985; Wyatt, Greatorex Davies, Bunce, Fuller and Hill, 1993); these are, however, more immediately appropriate at larger spatial scales.

Material properties are required for each soil or aquifer material defined in the IHDM. In particular, these are the hydraulic conductivity and water content at saturation and the relationship of these two properties to varying unsaturated pressure potentials for each material: single valued functions are employed, that is, hysteresis is not included. Commonly used sources, since direct data are rarely available, are the semi-empirical functions of Clapp and Hornberger (1978) and van Genuchten (1980). For surface materials, as for channel reaches, roughness coefficients are required. Data are most profitably sought from standard hydraulic texts, research publications, any specific site information and, in the U.K. from databases such as that of the Soil Survey and the HOST (Hydrology of Soil Types) system (Boorman, Hollis and Lilly, 1991). U.S. soil map data are available from the Conservation Service of the Department of Agriculture, and soil water properties, from an Iowa State University database, accompany map data. Remotely sensed data can aid in qualitative landscape division: quantitative information is not readily available, though microwave data (see, for example, Camillo, O'Neill and Gurney, 1986) and its relation to moisture at greater depths (see, for example, Ragab, 1992) may become routinely useful. The British Geological Survey produces extensive hydrogeological mapping, memoirs and borehole records and an aquifer properties database.

Model control information accompanies the physical data to determine the linkage routes through a catchment and to define time steps for calculation. These time steps depend on the expected rate of change of different process variables and should complement the spatial discretisation used.

Early in the development of physically-based models some writers in the field felt that the deterministic nature of the modelling meant that optimisation or calibration would not be required. In practical use, some degree of model calibration is necessary. This arises from inevitable uncertainty over parameter values and from the considerable approximation of the model equations and structure to the real world, despite the nominal physical basis. The case where such a model can be profitably used without optimisation of parameters is when exploring, under the assumptions of the model, ranges of hydrological outcomes from comprehensive ranges of parameter values. For real world applications, the most practical approach is to use such parameter values that are available with reasonable certainty and to calibrate the model for a small number of parameters which have the highest degree of uncertainty, limiting those values to hydrologically acceptable ranges, and focusing attention on those parameters to which model results are most sensitive (see below).

Suggestions have been made for the calibration of models, including the IHDM, using pseudo-likelihood measures (Binley and Beven, 1991; Beven and Binley, 1992; Beven, 1993). Likelihood weights can be successively updated in the light of observations and can provide estimates of the uncertainty of prediction. It is important in such procedures to include, where possible, specific consideration of measurement as well as model errors, and to use a likelihood function which is consistent with subsequent Bayesian updating.

A further problem in model calibration is the question of how much spatial variability to include. The apparent advantage of a discretised catchment in allowing scope for variation of parameters is tempered by how much these distributions can meaningfully be optimised when their precise numerical values are not known. Effects of spatial variability in saturated hydraulic conductivity have been investigated using a three-dimensional version of the IHDM (Binley, Elgy and Beven, 1989) who suggested, under their conditions, more variation in discharge for subsurface (rather than surface) flow dominated responses on a 100 by 150 m slope and, within these subsurface flow cases, less variation for those with low spatial correlation of hydraulic conductivity values.

As it is in practice difficult to acknowledge field variability in detail, the investigation of effective parameters has been sought by a number of authors (see, for example, Freeze, 1975). For the three-dimensional IHDM mentioned above, under similar conditions, Binley, Beven and Elgy (1989) found effective hydraulic conductivities for their subsurface flow dominated cases (though not for the surface dominated cases) but that the values could not be determined explicitly.

In practice model calibration is usually undertaken on a manual basis, relying on hydrological experience in interpreting outcomes and adjusting parameter values accordingly. Automatic parameter optimisation has been undertaken for the IHDM but is not routinely used because of the long run times involved. Manual calibration in particular runs the risk of operator-dependence of the parameter set and its non-uniqueness.

Numerical flow models are comparatively heavy in computing time, even without the lengthier calibration procedures. This is especially so in the case of unsaturated zone modelling, where hydraulic conductivity and water content vary with time and space and require many interpolations on the defining functions.

Advances in computing technology have meant increasingly shorter run times. This should not however be used as a substitute for elegance of modelling solution, whether in the handling of a model like the IHDM or, indeed, in the choice of its use in a particular problem.

In the late nineteen-eighties the use of a Cray (X-MP) was much more practical for the IHDM than, for example, an IBM (4381) mainframe and 386 personal computers. Now, approaching the mid-nineties, workstations are also practical as well as supercomputers. The model offers considerable opportunites for parallel processing. 486 personal computers can be suitable in circumstances where little calibration is required.

17.4. EXPERIENCE

This section deals with the use of the IHDM in a number of contexts: these are grouped into, first, field applications, second methodological investigations, and third general hydrological use unrelated to a particular site.

17.4.1. Field Applications

Early testing of the IHDM was undertaken on Institute of Hydrology instrumented catchments (Roberts, 1989). These sites do not provide all the parameter values ideally required, but streamflow and automatic weather station records were available and were accompanied by experience of field processes. Calibration was undertaken against streamflow in the first instance because of its relative accuracy of measurement and its integration of catchment response.

Tests on small (1-5 km^2) humid temperate upland catchments in central Wales explored the parameters to which the IHDM is most sensitive under those physical circumstances (Calver, 1988). These were found to be hydraulic conductivities, surface roughnesses if there is overland flow and, to a lesser extent, porosity. The initial state of the hillslope water content is also particularly important. Hydraulic conductivity is perhaps the single most important parameter. Its effect on subsurface flow is self-evident: it also serves in the IHDM as an important control on infiltration, which is fully integrated with subsurface conditions and varies over space and time. Conductivity values therefore influence greatly the division between overland flow and throughflow.

The central Wales work also took material parameters optimised for one catchment and transposed them to a nearby physiographically similar catchment, using the topography and rainfall of the latter, with good predictive performance. This type of exercise does not of course remove the danger of non-uniqueness of parameter sets, though it offers a practical degree of confidence in them. The Welsh experimental catchments were also used as a basis for testing optimised parameter values over a range of magnitudes of event, again with good performance: the largest runoff event was estimated to be of 50-100 year return period.

These applications suggested the need to model the fast response of ditched forested catchments (with ditches of some 0.5 m depth and 10 m spacing) as model 'overland flow' rather than as separate entities. This point will be taken up below after further field modelling experience has been described.

Model prediction, in greater detail than solely stream discharge, was compared with field observations of local saturation levels on an experimental hillslope plot in Luxembourg administered by the Vakgroep Fysische Geografie en Bodemkunde, University of Amsterdam: results are published in Calver and Cammeraat (1993). For this site, laboratory and field determinations of material parameters were available and, within these measured ranges, acceptable model flow results were achieved. Seasonal changes in soil structure were acccounted for by changes in hydraulic conductivity values over time. Comparisons of the local within-soil saturated-unsaturated boundary position were made: some were good, some less so. The chief reason for the imprecision of details is likely to be the approximations of a Darcian model to a soil with macropore development.

Unpublished experience on field use of the IHDM suggests a number of points that are dealt with in the remainder of the section. The increasing availablity of measured unsaturated pressure potentials is likely to provide further detailed testing.

Laboratory and field determinations of material parameters may or may not coincide with workable model parameters. Discrepancies can arise because of the scale and conditions of measurement in relation to those of the modelling. This was not so in the Luxembourg example described above. Conductivities from the Soil Survey of England and Wales (determined by the single auger hole method) can, however, be low compared with model 'effective' values. A measurement involving some sort of averaging is often a good indicator of practical model values at larger scales as, for example, with the use of test pumping for aquifer material properties.

The IHDM potentially offers a useful presentation of pore water pressures for the consideration of slope stability. The flexible finite element geometry and the combined consideration of saturated and unsaturated zones are particularly helpful in this context. Unsaturated zone negative pressure potentials offer increased stability. It is unusual to account for this at the design stage but it can help to explain the maintenance of some natural slopes. Figure 17.4 shows an early example of hillslope pressure potential modelling on a southern Italian scarp subject to landsliding believed to be induced, at least in part, by heavy rainfall events. At present the IHDM is used in this context with fixed geometry and the pressure potential fields carried across into standard stability analyses.

An important general point which has emerged in the modelling, and has been noted in a client report (Calver, 1993), is that it appears that flashy catchment responses are not easily modelled by Darcian flow alone. This is not a feature of the IHDM in particular, rather of the basic Darcian formulation, which is incorporated in many models. For flashy hydrographs it appears necessary to invoke an additional source of near-peak flow contribution, whether from macropores, natural soil pipes, artificial field

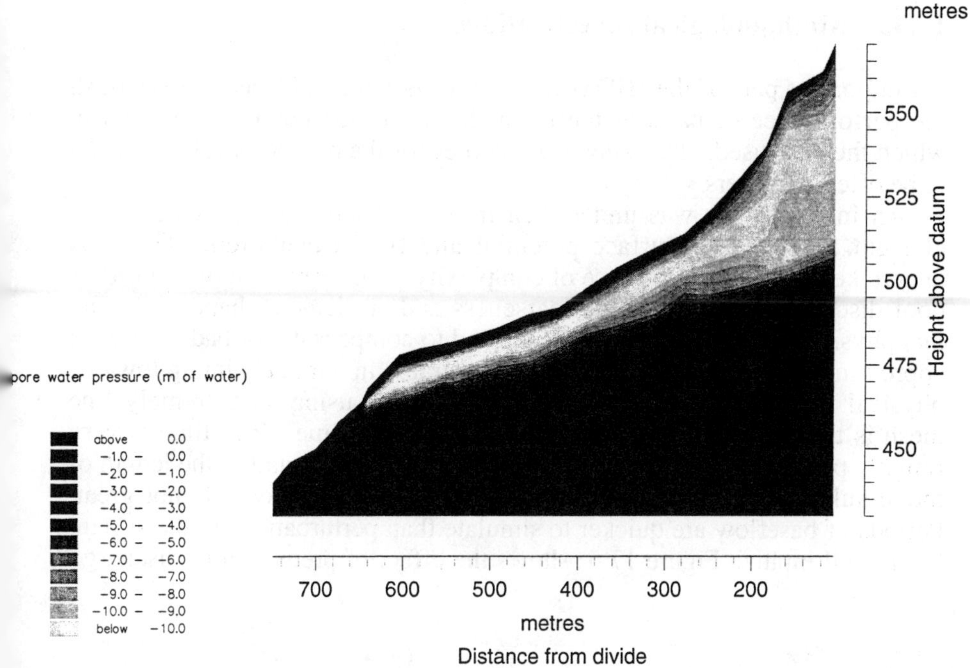

Fig. 17.4. Pressure potential distribution in a vertical section of hillslope, Grassano, southern Italy. Rainfall has entered upper soil horizons of a relatively dry unsaturated zone above the regional saturated zone: the geology is sandstones and interbedded deposits above clays.

drainage or overland flow. Overland flow is a rare occurrence in most areas of IHDM testing work (though is present at the Luxembourg site). Ways of handling and modifying the IHDM, other than using a completely dual porosity formulation, have been suggested (Calver, 1993) to cope in practical ways with flashy response areas which do not show overland flow in the field. In brief, these use spatial or temporal zones of high conductivities or specifically defined non-Darcian rapid water transfer routes.

In field applications the IHDM is generally used most straightforwardly at smaller scales, such as the Welsh and Luxembourg work above and, for example, in the Climex experiment on climatic change on controlled sites in southern Norway (Jenkins, Wright, Berendse, van Breemen, Brussaard, Schulze and Woodward, 1992). At larger scales care is needed with discretised solution models so that the physical basis does not become meaningless. Routines for handling the IHDM at larger scales include classification of catchment components, aggregation/disaggregation procedures and the setting of the model in a probabilistic context. The aim of such methods is to use the physical basis as or where appropriate in an efficient manner.

17.4.2. Methodological Investigations

An important part of the IHDM research programme has been to establish the performance of parts of the methodology in the complex contexts in which they are used. This allows guidelines for the proper use of the model to be offered to users.

An investigation was undertaken into the discretisation of the finite element mesh for subsurface potential and flow calculation. This was undertaken empirically because of complexity of the error determination. A poor discretisation in terms of coarseness and/or element shape can mean that physical parameter values are adjusted to compensate for bad numerical approximation, which is particularly undesirable in a model aiming towards physical reality. The reason for not necessarily using an extremely fine mesh is the accompanying increase in computing time. Run time is very roughly proportional to the number of calculation nodes and to the length of the simulation period: the setting-up overhead of the model is not great. Periods of baseflow are quicker to simulate than perturbations to the system by rainfall inputs. Figure 17.5 shows the effect of prediction of discharge

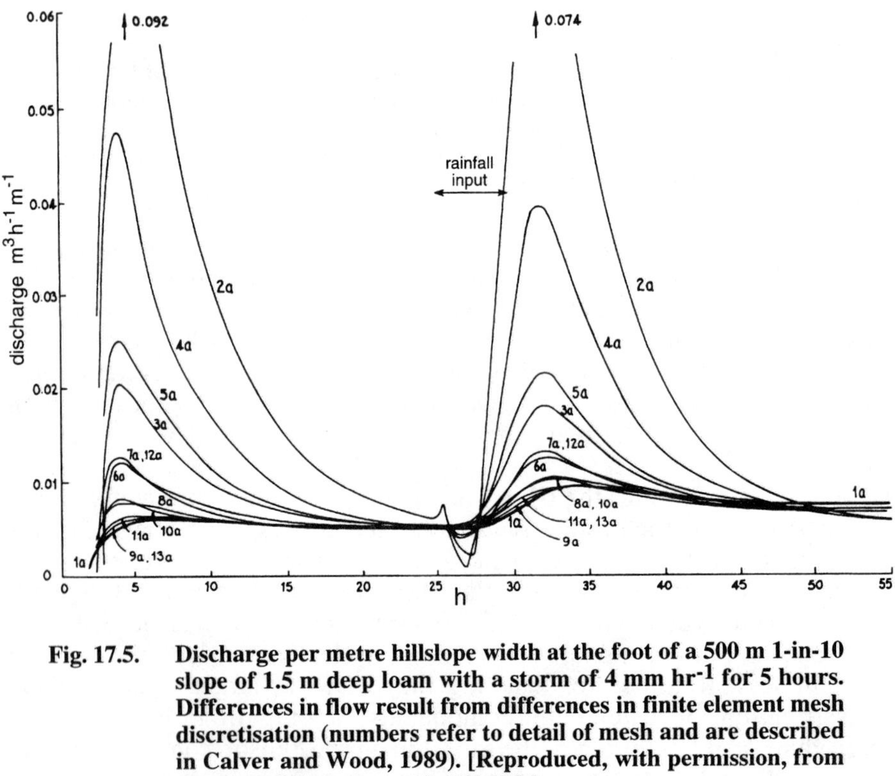

Fig. 17.5. Discharge per metre hillslope width at the foot of a 500 m 1-in-10 slope of 1.5 m deep loam with a storm of 4 mm hr^{-1} for 5 hours. Differences in flow result from differences in finite element mesh discretisation (numbers refer to detail of mesh and are described in Calver and Wood, 1989). [Reproduced, with permission, from Journal of Hydrology, 10, 165-179.]

for cases differing only in the discretisation of the finite element mesh. The long thin shape of many hillslope modelling domains should not be echoed in high aspect ratios in finite element shape. Details of efficient grid densities and aspect ratios are given in Calver and Wood (1989): these vary slightly for pressure potential and for discharge prediction. Figure 17.6 shows data which were the basis of determining grid effciency for discharge prediction.

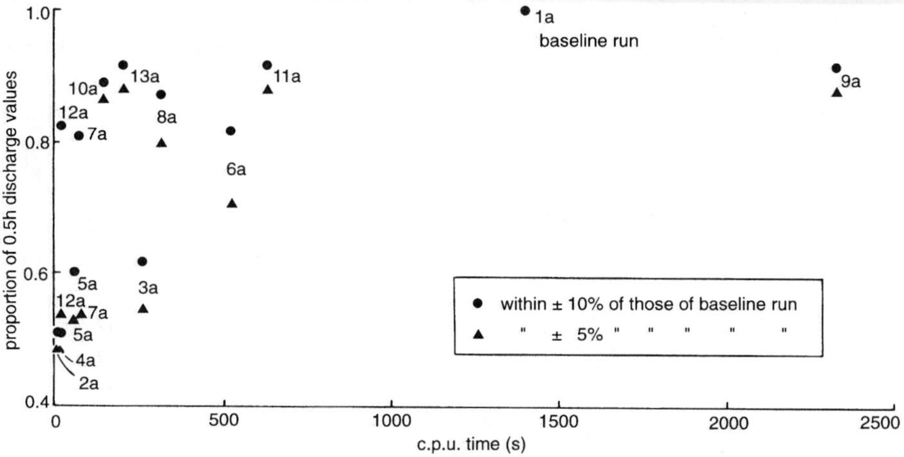

Fig. 17.6. Relationship of IBM mainframe processing time to accuracy of discharge predictions. The 'baseline' run has a 1:1 finite element mesh; the other numbers refer to a variety of discretisations, details of which can be found in Calver and Wood, 1989. [Reproduced, with permission, from Journal of Hydrology, 110, 165-179.]

The question of whether the finite element mass matrix should be lumped (diagonal) or distributed was investigated (Wood and Calver, 1990) because of conflicting advice in the hydrological literature. There was found to be little difference in the rate of convergence of the iteration and the run times between the two methods. Differences in predicted hydraulic potential are very small under slowly varying conditions but approached observation error under rapidly varying conditions. The significant difference is in the discharge from the slope which may be considerably in excess of field measurement error. This suggests the use of a distributed mass matrix, particularly if one is using a good rather than very good discretisation.

An aspect of the uncertainty of model results, namely that resulting from uncertainty of parameter values, was investigated (Binley, Beven, Calver and Watts, 1991) using Monte Carlo and Rosenblueth methods and confining attention to hydraulic conductivity, porosity, surface roughness and (spatially constant) initial pressure potential. The general point emerging was that a considerable band of uncertainty surrounds a model outcome, such that one should be wary of offering a single result without a statement or estimate of this potential source of error. Certainly, too, caution is required in claiming to predict unambiguously an output change resulting from a minor change in physical circumstances in a catchment. Detailed distributions of outcomes are somewhat time-consuming to produce, but it is good practice to offer at least best and worst case outcomes, although these are not necessarily always straightforward to define in advance in a multi-parameter model.

It has been mentioned above that model output is sensitive to the initial water content of the soil and aquifer materials. This effect can last well beyond a single event (see, for example, Fig. 17.7). This type of data is

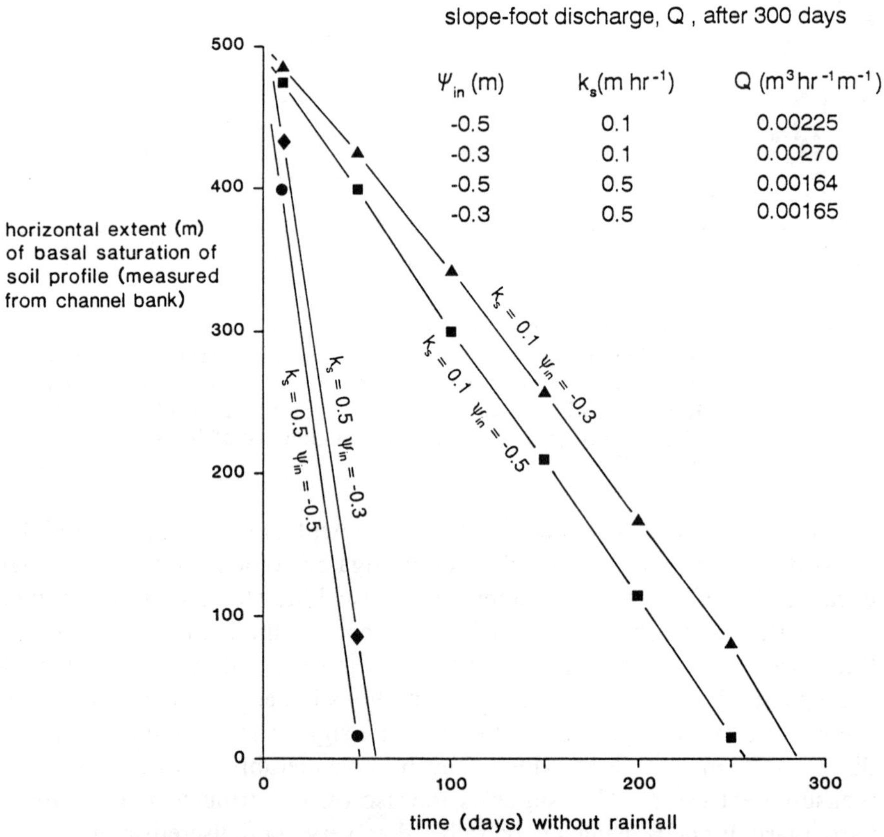

Fig. 17.7. Extent of saturation at base of 1.5 m soil on 500 m 1-in-10 slope over time. The k_s value is the saturated hydraulic conductivity in m hr^{-1}; ψ_{in} denotes the initial (spatially-constant) pressure potential in m water. [Reproduced, with permission, from Journal of Hydrology, 130, 379-397.]

particularly unlikely to be available from measurement, and conventional finite element starting situations are frequently not hydrologically appropriate. Wood and Calver (1992) investigated the detail of mathematical and hydrological conditions at the beginning of model runs. A good mesh is necessary to avoid transient oscillation in solution and, with such a mesh, only a small run-in of a few time steps is numerically necessary. Hydrologically, initial effects are longer lasting and so must be designed, in the absence of observations, to produce the correct background condition of steady or varying baseflow. Figure 17.8 illustrates the effect of different initial conditions and length of model run-in before a storm event is introduced on the magnitude of peak flow predicted. The paper offers guidance on appropriate type of initial condition fields which can be approximated from given stream discharges or be the subject of simple optimisation.

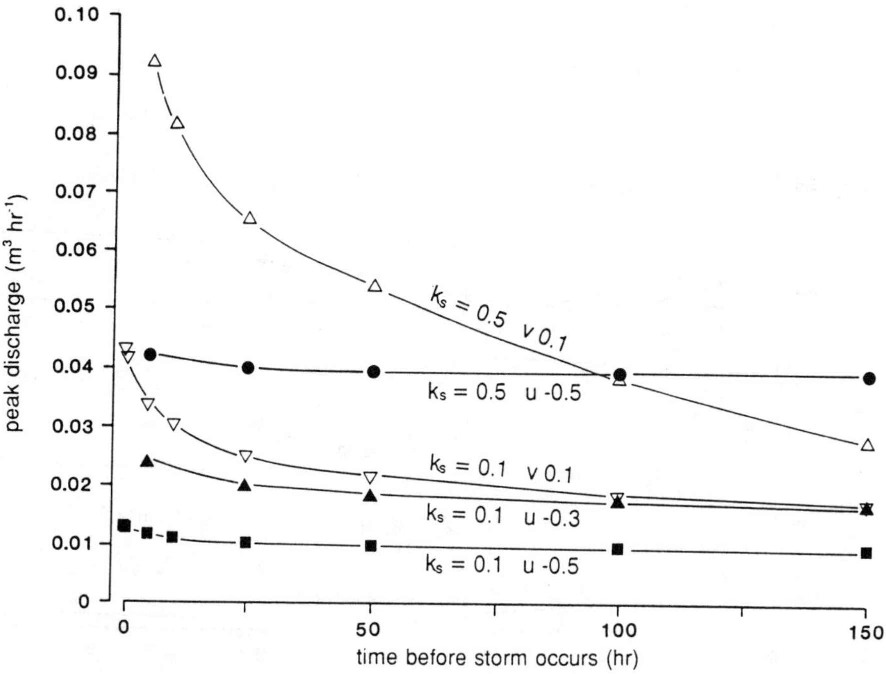

Fig. 17.8. Peak unit-width slope-foot discharge in relation to the time at which storm rainfall is introduced into a simulation under different initial conditions. The k_s value is the saturated hydraulic conductivity in m hr^{-1}: u indicates an initially constant pressure potential at the value given in m of water, and v indicates initial pressure potential at 0.1 of elevation above stream level datum. [Reproduced, with permission, from Journal of Hydrology, 130, 379-397.]

17.4.3. General Hydrological Applications

This section deals with hydrological applications of the IHDM which are general in nature and not related to specific sites. These use the IHDM in predictive mode, exploring outcomes under systematic ranges of input data, as a means of investigation to complement observation and theory, taking advantage of the physical basis of the model.

The IHDM was used to investigate the degree to which the spatial discretisation of rainfall input affected modelled streamflow results (Watts and Calver, 1991). For throughflow conditions on a conceptual 100 km^2 catchment under frontal precipitation, an efficient resolution of rainfall data is around 2.5 km along the storm path: Fig. 17.9 shows the effect of discretisation in relation to predicted peak flow and time to peak. This underlines the potential benefit of radar-derived precipitation fields in

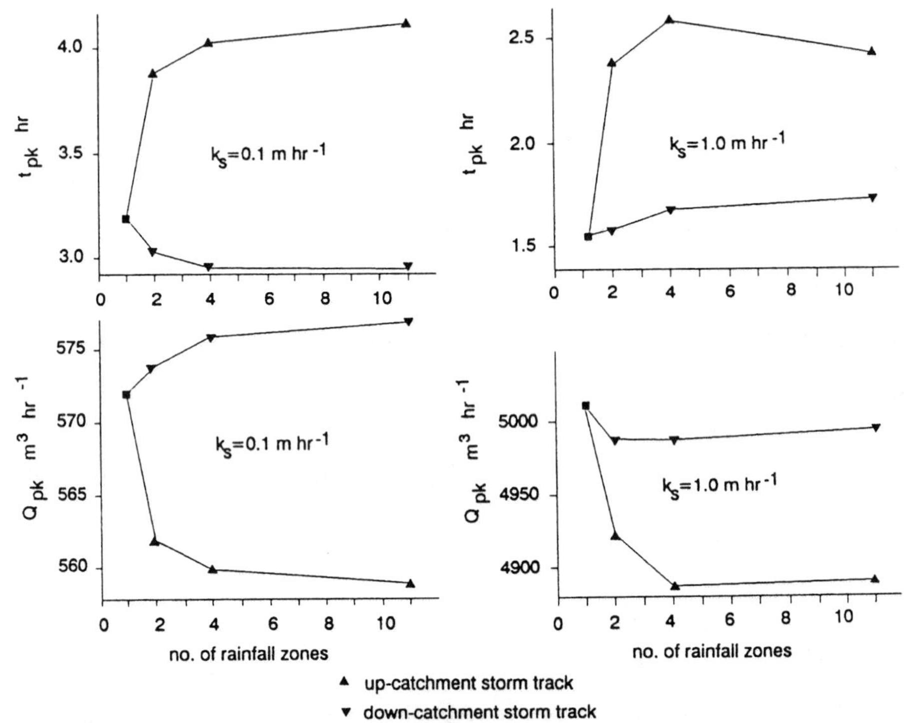

Fig. 17.9. **Effects of number of rainfall zones on peak discharge and time to peak. The k_s value is the saturated hydraulic conductivity; 'zone' indicates division per 10 km distance. [Reproduced, with permission, from Nordic Hydrology, 22, 1-14.]**

detailed distributed modelling. In lumping rainfall, error is most likely to be introduced for lower storm velocities, and the direction of this error can be specified (Fig. 17.10). Peak flow differences between down-catchment and up-catchment storm directions reach a maximum at a storm velocity similar to an averaged peak channel velocity. Such results were qualitatively similar to those reported in the literature for overland flow runoff but the differences were smaller in the case of rainfall modified by a period of throughflow.

Details of pore water velocities and the tracking of water 'particles' through variably unsaturated slope material were investigated under conceptual Darcian conditions by Calver and Binning (1990). This is a particularly computing-heavy application. Figure 17.11 shows examples of

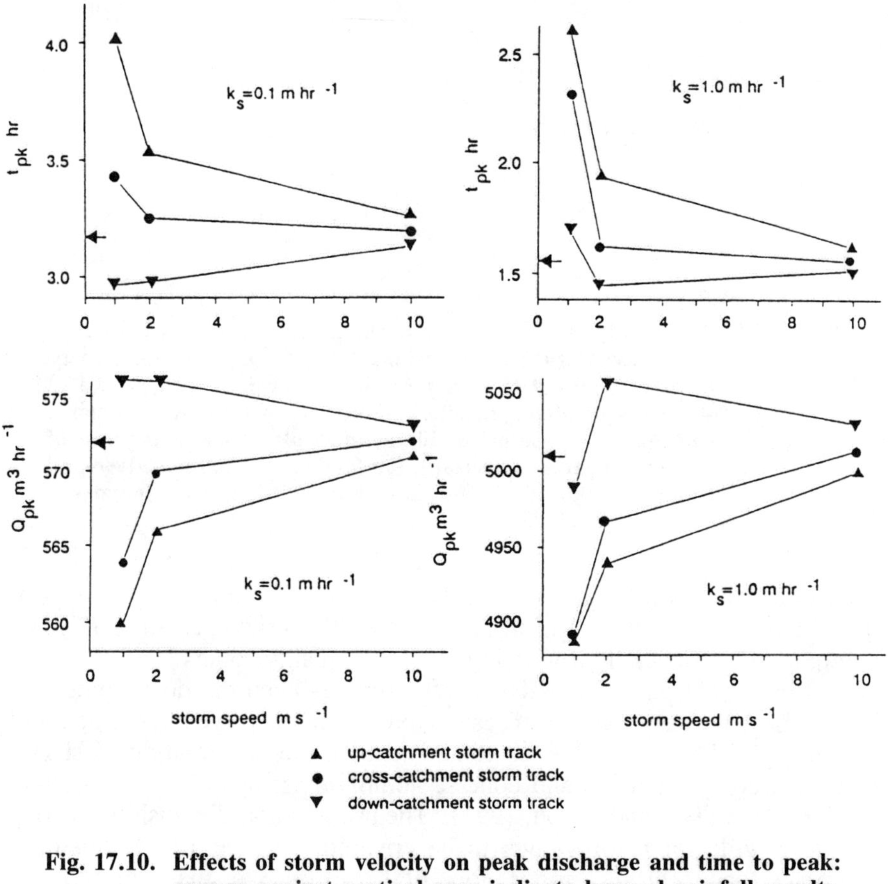

Fig. 17.10. Effects of storm velocity on peak discharge and time to peak: arrows against vertical axes indicate lumped rainfall results. [Reproduced, with permission, from Nordic Hydrology, 22, 1-14.]

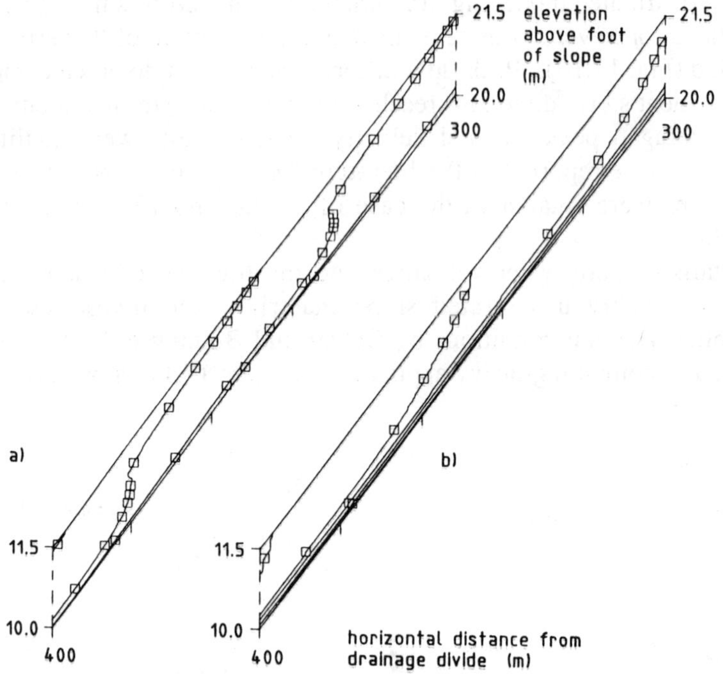

Fig. 17.11. Tracks of water particle movement. (a) Saturated hydraulic conductivity of 0.1 m hr^{-1} with a 1.0 x 10^{-4} m hr^{-1} conductivity layer at 0.50 - 0.75 m depth; single storm. (b) Sequence of storms; saturated hydraulic conductivity of 1.0 m hr^{-1}. Squares show particle positions at 30-day intervals (absent along base of slope where tracks overlap); for further details see Calver and Binning, 1990. [Reproduced, with permission, from Journal of Hydrology, 121, 335-344.]

such tracking, and Fig. 17.12 of hillslope residence times. It is perhaps important here to stress the need to assess the relevance of a Darcian approximation especially if working in a water quality context.

An important part of the IHDM research has been the development of mechanisms for easy transfer of results to those not involved in the detail of modelling. Methods have been established for the presentation of IHDM results in comprehensive and concise nomographs for hillslopes or small catchments (Calver and Wood, 1991). The use of a non-dimensional setting allows a wide range of values to be covered. Figure 17.13 illustrates examples for a 500 m, 1-in-10 slope of 1.5 m of loam material and Fig. 17.14 for one of the Welsh catchments referred to above. These nomographs can be used in calibration or predictive mode. If numerical values of physical parameters are known, discharge can be estimated for a particular

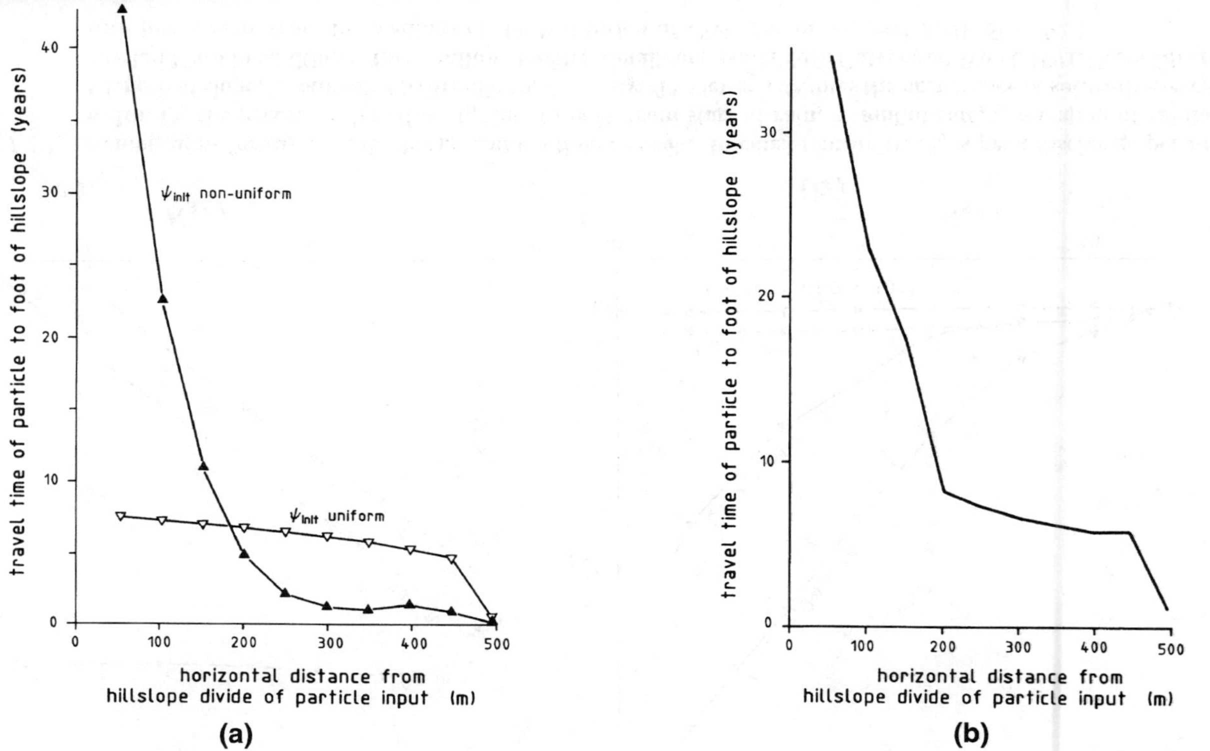

Fig.17.12. **(a) Hillslope travel times for water particles introduced along 500 m hillslope surface under uniform and non-uniform initial pressure potential conditions; saturated hydraulic conductivity 0.1 m hr^{-1}; single rainstorm input. (b) Water particle travel times under sequence of storms; saturated hydraulic conductivity 1.0 m hr^{-1}. For further details see Calver and Binning, 1990. [Reproduced, with permission, from Journal of Hydrology, 121, 335-344.]**

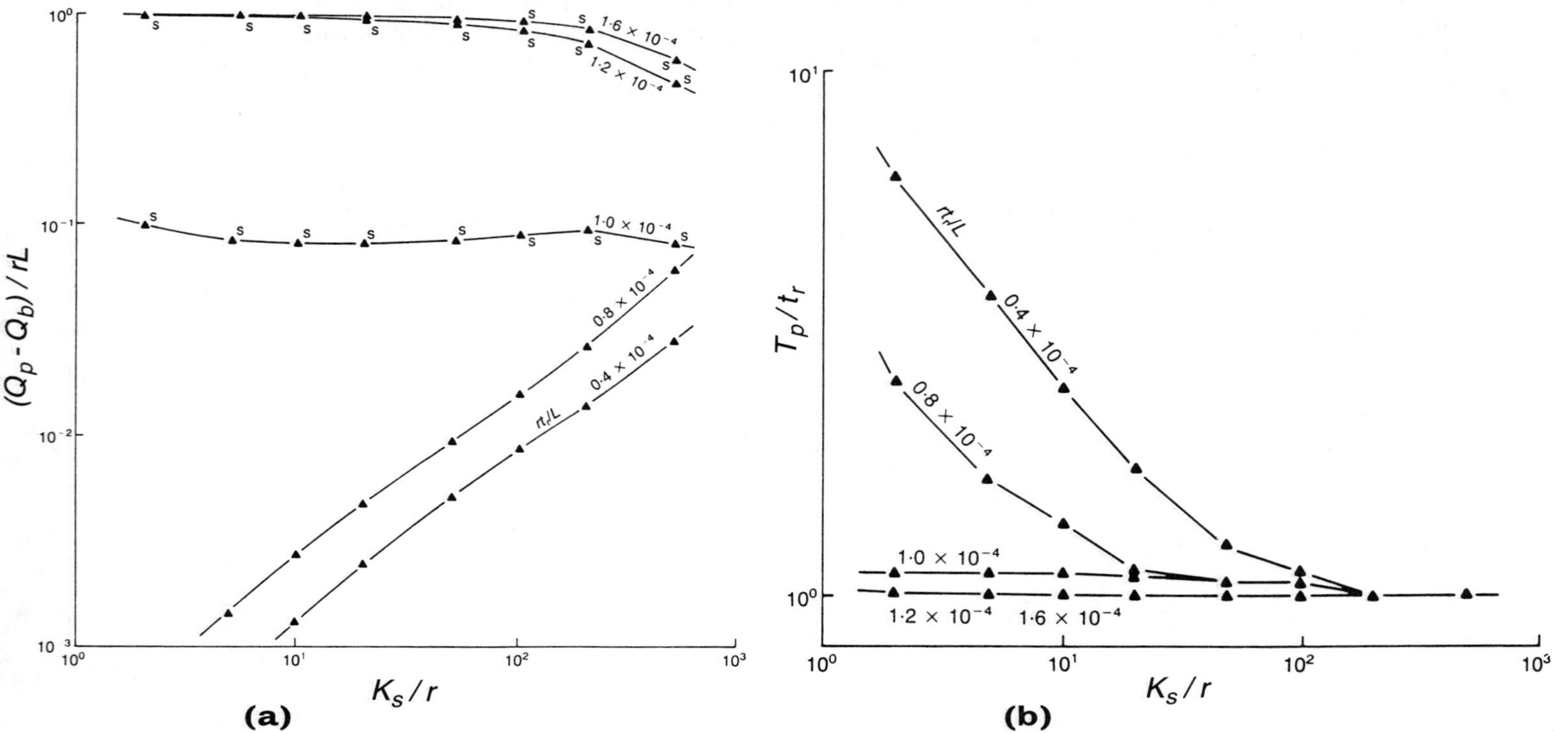

Fig.17.13. Nomographs for (a) peak discharge and (b) time to peak. In consistent units, Q_p is peak discharge per unit width, Q_b the preceding baseflow, t_p time to peak from start of rain, r rainfall rate, t_r duration of rainfall, L length of slope, k_s saturated hydraulic conductivity. The letter s denotes the occurrence of saturation-excess overland flow in addition to throughflow. Further details are available in Calver and Wood, 1991. [Reproduced, with permission, from Proceedings of the Institution of Civil Engineers, part 2, 91, 593-602.]

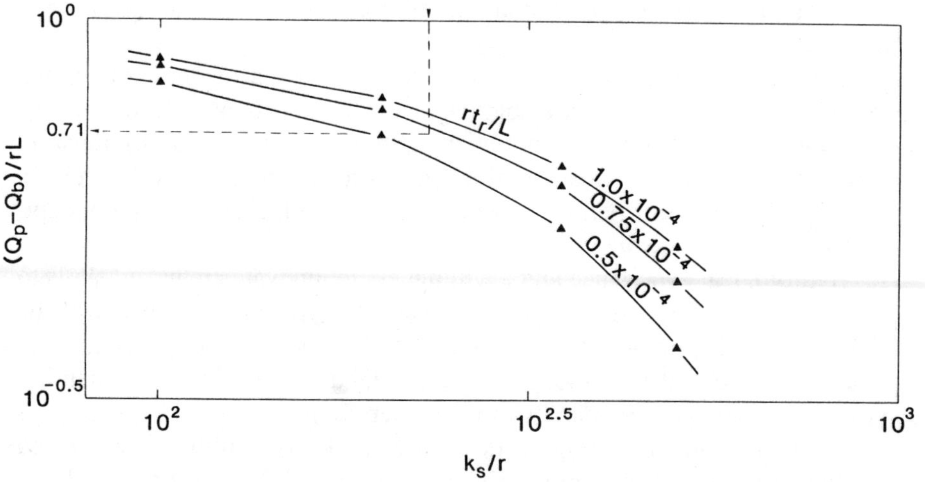

Fig. 17.14. Nomographs for peak discharge for the Tanllwyth catchment. Variables as in Fig. 17.13. Interpolation for event of 12 February 1976 (r=4.35 mm hr^{-1}, t_r=12 hr, L=800 m, Q_b=0.05 m^{3}sec^{-1}, k_s=1.0 m hr^{-1}, width, w=2100 m : gives peak catchment discharge 1.5 m^{3}sec^{-1}; observed 1.7 m^{3}sec^{-1}) .

event of interest or for a design storm: Fig. 17.14 includes an example. If, however, discharge is known but there is uncertainty over a parameter value, this can be estimated from the graph. Convergence on a particular value by interpolation from a number of events increases confidence in the estimation.

17.5. CONCLUDING REMARKS

The IHDM is among the group of models which link established flow equations to simulate catchment hydrological behaviour. Its use of numerical solution methods means its practical evolution has been associated with developments in computing as well as in hydrological research. A realistic idea is available today of what can and what cannot be achieved by such types of model.

The solution methods are reasonably well explored and guidelines have been established for removing as much error as possible from this source in the complex context in which they are used in the model. It remains the case that the equations are not exact for real world, but in many cases are perhaps the best which are standardly available. The biggest question over the generality of such models may be the use of Darcian flow particularly in the unsaturated zone: approximations to meet such problems have been offered.

The establishment of parameter values is an area of concern for the model user. Values should not disguise other problems; 'averaged' or 'effective' values may or may not be workable; there are ways of indicating complexities of real processes albeit in an approximate way by certain usages of parameter values; even where straightforward in concept, parameter values may be unknown.

Hydrological experience and a willingness to innovate are useful skills to bring to developing and applying models of the type and complexity of the IHDM. The degree of spatial variability to include is not only a matter of field knowledge, difficult though that may be, but a question of acceptability of further degrees of complexity of parameter optimisation.

In much field modelling work such a model is unduly complex and inefficient to use. It is of most relevance to field applications in which the *generation* of runoff within the catchment is important because of its position or water quality or because of scope for its practical modification.

Model results should be accompanied where possible by an error estimate, or by consideration of best and worst cases, to convey to the user of results the nature of their reliability and sensitivity.

With time and the wider use of distributed models, a working practical parameter value database can be established in relation to physiographic conditions: this can help future users by cutting down optimisation times through the establishment of good initial parameter estimates.

For those willing to embark on more complex modelling, such as dual porosity formulations in the unsaturated zone, models like the IHDM can offer indications of the circumstances and scales where this may be worthwhile.

In exploratory use, IHDM-type physically-based models offer tools to accompany other sources of information. This is a major role in research hydrology. This is particularly so if they lead, together with field experience and theoretical development, to the establishment of efficient partially-distributed models which become important in the practical sphere for those not wishing to rely on lumped or black-box models. Distributed modelling can be an information route to process approximation in hydrological modelling for contexts in which this is a desirable modelling strategy.

ACKNOWLEDGEMENTS

Financial support for the IHDM research project has been provided by the Flood Defence Division of the U.K. Ministry of Agriculture, Fisheries and Food and the U.K. Natural Environment Research Council. Many I.H. personnel, past and present, have contributed to the model development and testing, particularly, in the formative stages, Keith Beven and Liz Morris. The authors are grateful for information from and discussions with colleagues, particularly David Maidment of the Department of Civil Engineering, University of Texas at Austin, and Frank M. Law and David Jones of the U.K. Institute of Hydrology. Some of the diagrams in this chapter are reproduced, with permission, from the quoted journals.

REFERENCES

Anderson, J.R., Hardy, E.E., Roach, J.T. and Witmer, R.E. 1976. A land use and land cover classification system for use with remote sensor data. U.S. Geological Survey Professional Paper 964.

Barré de Saint-Venant, A.J.C. 1848. Études théoriques et pratiques sur le mouvement des eaux courantes. Paris.

Barré de Saint-Venant, A.J.C. 1871. Théorie du mouvement non permanent des eaux, avec application aux crues des rivières et à l'introduction des marées dans leur lits. Comptes rendus des séances de l'Académie des Sciences, 73, 147-154 and 237-240.

Bear, J. 1979. Hydraulics of Groundwater, McGraw-Hill.

Beven, K. 1993. Prophecy, reality and uncertainty in distributed hydrological modelling. Advances in Water Resources, 16, 41-51.

Beven, K. J. and Binley, A. M. 1992. The future of distributed models: model calibration and predictive uncertainty. Hydrological Processes, 6, 279-298.

Beven, K., Calver, A. and Morris, E. 1987. The Institute of Hydrology Distributed Model. U.K. Institute of Hydrology Report No. 98.

Binley, A. M. and Beven, K. J. 1991. Physically-based modelling of catchment hydrology: a likelihood approach to reducing predictive uncertainty. In: Computer modelling in the environmental sciences, ed. D. G. Farmer and M. J. Rycroft, Clarendon Press, Oxford.

Binley, A.M., Beven, K.J., Calver, A. and Watts, L.G. 1991. Changing responses in hydrology: assessing the uncertainty in physically-based model predictions. Water Resources Research, 27, 1253-1261.

Binley, A., Beven, K. and Elgy, J. 1989. A physically based model of heterogeneous hillslopes 2. Effective hydraulic conductivities. Water Resources Research, 25, 1227-1233.

Binley, A., Elgy, J. and Beven, K. 1989. A physically based model of heterogeneous hillslopes. 1. Runoff production. Water Resources Research, 25, 1219-1226.

Boorman, D.B., Hollis, J. and Lilly, A. 1991. The production of the Hydrology of Soil Types (HOST) data set. British Hydrological Society, Proceedings of the 3rd National Hydrology Symposium, Southampton, 6.7-6.13.

Calver, A. 1993. Darcy and flood generation. Report to UK. Ministry of Agriculture, Fisheries and Food.

Calver, A. 1988. Calibration, sensitivity and validation of a physically-based rainfall-runoff model. Journal of Hydrology, 103, 103-115.

Calver, A. and Cammeraat, L.H. 1993. Testing a physically-based runoff model against field observations on a Luxembourg hillslope. Catena, 20, 273-288.

Calver, A. and Wood, W.L. 1991. Dimensionless hillslope hydrology. Proceedings of the Institution of Civil Engineers, part 2, 91, 593-602.

Calver, A. and Binning, P. 1990. On hillslope water flow paths and travel times. Journal of Hydrology, 121, 335-344.

Calver, A. and Wood, W.L. 1989. On the discretisation and cost-effectiveness of a finite element solution for hillslope subsurface flow. Journal of Hydrology, 110, 165-179.

Camillo, P.J., O'Neil, P.E. and Gurney, R.J. 1986. Estimating soil hydraulic parameters using passive microwave data. International Electrical and Electronics Engineers Transactions on Geoscience and Remote Sensing, GE-24, 930-936.

Clapp, R.B. and Hornberger, G.M. 1978. Empirical equations for some soil hydraulic properties. Water Resources Research, 14, 601-604.

Conte, S.D. and de Boor, C. 1980. Elementary Numerical Analysis, McGraw-Hill.

Cooley, R.L. 1983. Some new procedures for numerical solution of variably saturated flow problems. Water Resources Research 19, 1271-1285.

Cooper, D.M., Ragab, R., Lewis, D.R., Dixon, A.J. and Whitehead, P.G. 1991. Modelling nitrate leachate to surface waters. Report to U.K. Ministry of Agriculture, Fisheries and Food.

Cooper, D.M., Ragab, R., Lewis, D.R. and Whitehead, P.G. 1992. Modelling nitrate leaching to surface waters. Report to U.K. Ministry of Agriculture, Fisheries and Food.

Daluz Vieira, J.H. 1983. Conditions governing the use of approximations for the St Venant equations for shallow surface water flow. Journal of Hydrology, 60, 43-58.

Darcy, H. 1856. Les fontaines publiques de la ville de Dijon. Dalmont. Paris.

Freeze, R.A. 1975. A stochastic-conceptual analysis of one-dimensional groundwater flow in nonuniform homogeneous media. Water Resources Research, 11, 725-741.

French, R.H. 1987. Open-Channel Hydraulics, McGraw-Hill.

van Genuchten, M. Th. 1980. A closed form equation for predicting the hydraulic conductivity of unsaturated soils. Soil Science Society of American Journal, 44, 892-898.

Jenkins, A., Wright, R.F., Berendse, F., van Breemen, N., Brussaard, L., Schulze, E.D. and Woodward, F.I. 1992. The 'Climex' project - climate change experiment. In: Experimental manipulations of biota and biogeochemical cycling in ecosystems, edited by L. Rasmussen, T. Brydges and P. Mathy, Ecosystems Research Report 4, Commission of the European Communities, 71-77.

Lencastre, A. 1987. Handbook of Hydraulic Engineering, English Language Edition, Ellis Harwood.

Lewis, D.R. 1992. The IHDM preprogram: a spatially distributed effective rainfall model. Report to U.K. Ministry of Agriculture, Fisheries and Food.

Lynch, D.R. 1984. Mass conservation in finite element groundwater models. Advances in Water Resources, 7, 67-75.

Morris, E. M. and Harding, R.J. 1991. Parameterisation of turbulent transfers between glaciers and the atmosphere. Proceedings Symposium Glaciers-Ocean-Atmosphere Integration, Leningrad, International Association Scientific Hydrology Publication 208, 543-548.

Morris, E.M. and Woolhiser, D.A. 1980. Unsteady one-dimensional flow over a plane: partial equilibrium and recession hydrographs. Water Resources Research, 16, 355-360.

Overton, D. E. and Meadows, M. E. 1976. Stormwater modelling. Academic Press, New York.

Preissmann, A. 1961. Propagation des intumescences dans les canaux et rivières. First Congress of the French Association for Computation, Grenoble.

Ragab, R. 1992. Assessment of the relationship between remotely sensed topsoil moisture content and profile moisture content. In: Soil moisture modelling, Proceedings of National Hydrology Research Centre Workshop, 141-153.

Richards, L.A. 1931. Capillary conduction of liquids through porous mediums. Physics, 1, 318-333.

Roberts, A. 1989. The catchment research data base at the Institute of Hydrology. U.K. Institute of Hydrology Report No. 106.

Rutter, A.J., Kershaw, K.A., Robins, P.C. and Morton, A.J. 1971. A predictive model of rainfall interception in forests, I: Derivation of the model from observations in a plantation of Corsican pine. Agricultural Meteorology, 9, 367-384.

Shuttleworth, W.J. 1992. Evaporation. In: Handbook of Hydrology, ed. D.R. Maidment, McGraw-Hill, New York.

Thompson, N., Barrie, I.A. and Ayles, M. 1981. The Meterological Office rainfall and evaporation calculation system: MORECS. Hydrological Memorandum 45.

Watts, L.G. 1988. The Preprograms to the Institute of Hydrology Distributed Model. U.K. Institute of Hydrology Report No. 103.

Watts, L.G. and Calver, A. 1991. Effects of spatially-distributed rainfall on runoff using a physically-based model. Nordic Hydrology, 22, 1-14.

Wilson, M. F. and Henderson-Sellers, A. 1985. A global archive of land cover and soils data for use in general circulation climate models. Journal of Climatology, 5, 119-143.

Wood, W.L. 1993. Introduction to Numerical Methods for Water Resources. Oxford University Press.

Wood, W.L. and Calver, A. 1990. Lumped versus distributed mass matrices in the finite element solution of subsurface flow. Water Resources Research, 26, 819-825.

Wood, W.L. and Calver, A. 1992. Initial conditions for hillslope hydrology modelling. Journal of Hydrology, 130, 379-397.

Woolhiser, D.A. and Liggett, J.A. 1967. Unsteady one-dimensional flow over a plane - the rising hydrograph. Water Resources Research, 3, 753-771.

Wyatt, B.K., Greatorex Davies, N., Bunce, R. G. H., Fuller, R. M. and Hill, M.O. 1993. Comparisons of land cover definitions. Institute of Terrestrial Ecology Report to U.K. Department of the Environment.

Zienkiewicz, O.C. 1971. The Finite Element Method in Engineering Science, McGraw-Hill, London.

Chapter 18

TOPMODEL

K. Beven, R. Lamb, P. Quinn, R. Romanowicz, and J. Freer

18.1. MODELLING GOALS.

TOPMODEL is not a hydrological modelling package. It is rather a set of conceptual tools that can be used to reproduce the hydrological behaviour of catchments in a distributed or semi-distributed way, in particular the dynamics of surface or subsurface contributing areas. It is based on some simple approximate hydrological theory but recognises that, because of the lack of measurements of internal state variables and catchment characteristics, the representation of the internal hydrological responses of the catchment must necessarily be functional while introducing the minimal number of parameters to be calibrated (see Kirkby, 1975; Beven, 1989a). The modelling concepts have always been kept simple enough that the model structures used can be modified to bring the predictions closer to the modeller's perceptions of the behaviour of a particular catchment. The distributed nature of the predictions can be a great aid in this respect.

Thus, TOPMODEL may be seen as a product of two objectives. One is the development of a pragmatic and practical forecasting and continuous simulation model. The other is the development of a theoretical framework within which perceived hydrological processes, issues of scale and realism and model procedures may be researched. Parameters are intended to be **physically interpretable** and their number is kept to a minimum to ensure that their values do not become merely the statistical artefacts of a calibration exercise (although, as will be seen, this problem can never be entirely avoided). The model, in practice, represents an attempt to combine the computational and parametric efficiency of a lumped approach with the link to physical theory and possibilities for more rigorous evaluation offered by a distributed model.

18.2. A 'PERCEPTUAL' MODEL OF RUNOFF PRODUCTION

The modelling process involves a sequence of stages of simplification. Beven (1991) has suggested that the hydrologist might have a much more complex qualitative model of catchment hydrology than can be either written down in mathematical form or programmed for a computer. Each hydrologist may have an individual 'perceptual model' of this type and Beven (1994, see also 1989b, 1991) has outlined one such complex model and its representation in distribution function form. The salient features of this model are briefly described here since they form a basis for the TOPMODEL concepts.

Hydrological processes within a catchment are dynamically and heterogeneously distributed. The interaction of vegetation and rainstorm dynamics may lead to very non-uniform inputs to the upper boundary of the soil. Soil and bedrock heterogeneity may further complicate flow paths, causing phenomena such as linked saturated 'pockets', macropore and pipe flow and flow 'fingering'. At the lower boundary of the soil, irregularities in the bedrock interface may affect the pattern of subsurface discharges to streams.

The general tendency of water to flow downhill is, however, amenable to macroscale conceptualisation. For the case of saturated zone flow, for soils that are shallow relative to the hillslope scale, this gravitational control should result in a water table that is nearly parallel to the surface topography over much of the hillslope length, at least when conditions are wet enough for there to be lateral downslope saturated flow. Water may then be expected to flow down from steep to shallow slopes, and into areas of slope convergence.

It is clear, however, that surface and subsurface hillslope flow processes are very complex. Overland flow, generated by either the infiltration-excess or saturation-excess mechanisms, may take place through a vegetated mat, through organic litter, or in ephemeral rills and channels. Flow may also re-infiltrate locally as 'run-on'. Saturation-excess flow, which may have a contribution from 'return flow', may be caused either by the water table rising to the surface or by local filling of near surface soil layers, leading to a form of perched water table. The area where the water table reaches the surface may vary over time, causing the source area for saturation-excess overland flow to be dynamic. Similarly, the area contributing to stormflow via subsurface pathways may vary in time (Cappus, 1960; Dunne and Black, 1970; Hewlett 1974). Unsaturated zone flow may be highly irregular with a complex geometry of preferential and other pathways. It has to be impossible to incorporate all the perceived heterogeneity of unsaturated flows in a set of relatively simple, macroscale concepts. Hydraulic soil properties are also frequently highly variable in the vertical, often with higher conductivities close to the surface.

18.3. A 'CONCEPTUAL' MODEL OF RUNOFF PRODUCTION - TOPMODEL

The complexity of such a complex perceptual model mitigates against the formal conceptualisation of hydrological processes into the functional mathematical structures in a model such as TOPMODEL or any other hydrological model. Every model is an attempt to capture the essence of this complex nature in a manageable way, but it is important to recognise that this conceptualisation also involves a considerable degree of simplification.

In making such a conceptualisation TOPMODEL is premised upon two basic assumptions:

A1 that the dynamics of the saturated zone can be approximated by successive steady state representations;

A2 that the hydraulic gradient of the saturated zone can be approximated by the local surface topographic slope, tanß.

These assumptions lead to simple relationships between catchment storage (or storage deficit) and local levels of the water table (or storage deficit due to drainage) in which the main factor is the topographic index (a/tanß) first proposed by Kirkby (1975) and developed as a complete hydrological model by Beven and Kirkby (1976, 1979). The Kirkby index represents the propensity of any point in the catchment to develop saturated conditions. High values will be caused by high values due to either long slopes or upslope contour convergence, and low slope angles. The assumptions are similar to as those used in the development of the 'wetness' index developed independently by O'Loughlin (1981, 1986). TOPMODEL in its original form, however, takes advantage of the mathematical simplifications allowed by a third assumption:

A3 that the distribution of downslope transmissivity with depth is an exponential function of storage deficit or depth to the water table:

$$T = T_0 e^{-S/m}$$

where T_0 is the lateral transmissivity when the soil is just saturated (m^2/h), S is local storage deficit (m) and m is a model parameter (m). Beven (1984) has provided data to suggest that an exponential decline in (vertical) soil conductivity with depth may be adequate to describe the vertical changes in the hydraulic properties of a wide range of soils and has shown (Beven, 1986a) that the equivalent exponential transmissivity function can be derived under the (possibly dubious) assumption of isotropy.

In terms of water table depth this can be written as

$$T = T_0 e^{-fz}$$

where z is local water table depth (m) and f is a scaling parameter (m^{-1}). The parameters f and m are approximately related by $f = \Delta\theta_1/m$ where $\Delta\theta_1$ is an effective water content change per unit depth in the unsaturated zone due to rapid gravity drainage (down to 'field capacity').

A physical interpretation of the decay parameter m is that it controls the effective depth of the catchment soil profile. This it does interactively with the parameter T_0, which defines the transmissivity of the profile when saturated to the surface. A larger value of m effectively increases the active depth of the soil profile. A small value, especially if coupled with a relatively high T_0, generates a shallow effective soil, but with a pronounced transmissivity decay. This combination tends to produce a well-defined and relatively shallow recession curve response in the model hydrograph. These principles are shown in Fig. 18.1.

Under the assumption A2 of an effective water table gradient and saturated flow parallel to the local surface slope tanß, then at any point i on a hillslope, the downslope saturated subsurface flow rate q_i per unit contour length (m^2/h) may be described by the equation :

$$q_i = T_0 \tan\beta \, e^{-fz_i} \tag{18.1}$$

where T_0 and tanß are local values at point i.

Then under assumption A1 that, at any time step, quasi-steady-state flow exists throughout the soil, then assuming (A4) a spatially homogeneous recharge rate r (m/h) entering the water table, the subsurface downslope flow per unit contour length q_i may also be given by :

$$q_i = ra \tag{18.2}$$

where a is the area of the hillslope per unit contour length (m^2) that drains through point i.

By combining Eqs. (18.1) and (18.2) it is possible to derive a formula for any point relating local water table depth to the topographic index ln(a/tanß) at that point, the parameter f, the local saturated transmissivity and the effective recharge rate, r :

$$z_i = -\frac{1}{f} \ln \frac{ra}{T_0 \tan\beta} \tag{18.3}$$

An expression for the catchment lumped, or mean, water table depth ($\overline{z}$) may be obtained by integrating Eq. (18.3) over the entire area of the catchment (A) that contributes to the water table. In what follows we will

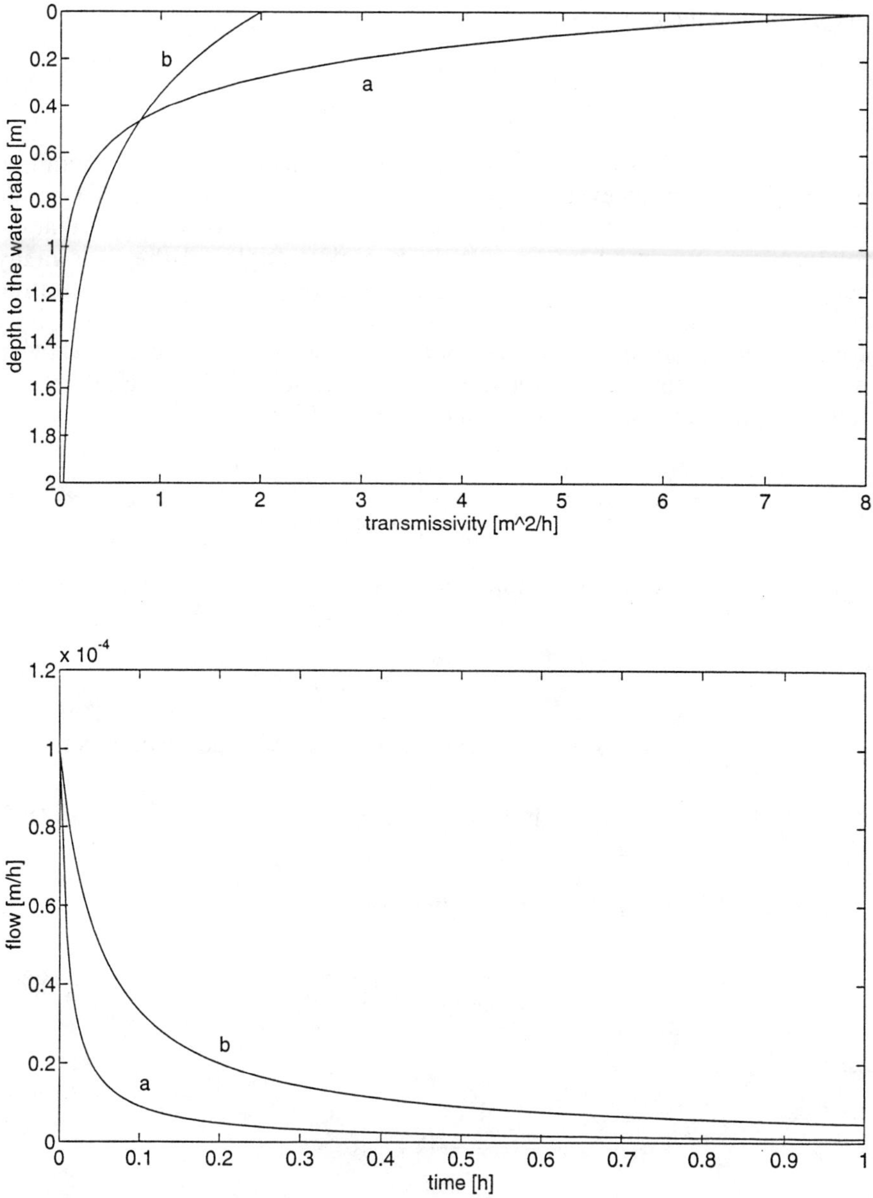

Fig. 18.1. **Exponential relationships between transmissivity and depth to the water table and the related recession curves. Curve a: $T_0 = 8\ m^2/h$, m = 0.02 m; curve b: $T_0 = 2\ m^2/h$, m = 0.05 m.**

express this areal averaging in terms of a summation over all points (or pixels) within the catchment :

$$\bar{z} = \frac{1}{A}\sum_i -\frac{1}{f}\ln\frac{ra}{T_0\tan\beta} \tag{18.4}$$

In spatially integrating the whole catchment, it is also implicitly required that Eq. (18.4) holds even at such locations where water is ponded on the surface ($z_i < 0$). Beven (1991) justifies this assumption on the basis that the relationship expressed by Eq. (18.1) is exponential and that, for many catchments, surface flow is likely to be relatively slow due to vegetation cover. By using Eq. (18.3) in Eq. (18.4), if it is assumed that r is spatially constant, ln r may be eliminated and a relationship found between mean water table depth, local water table depth, the topographic variables and saturated transmissivity. This has the form :

$$\bar{z} = z_i - \frac{1}{f}\left[\gamma - \ln\frac{a}{T_0\tan\beta}\right] \tag{18.5}$$

where $\ln(a/T_0\tan\beta)$ is the soil-topographic index of Beven (1986a), and

$$\gamma = \frac{1}{A}\sum_i \ln\frac{a}{T_0\tan\beta} \tag{18.6}$$

A separate areal average value of transmissivity may be defined, thus :

$$\ln T_e = \frac{1}{A}\sum_i \ln T_0 \tag{18.7}$$

Equation (18.5) may now be rearranged to give :

$$f(\bar{z} - z_i) = \left[\ln\frac{a}{\tan\beta} - \lambda\right] - \left[\ln T_0 - \ln T_e\right] \tag{18.8a}$$

where :

$$\lambda = \frac{1}{A}\sum_i \ln\frac{a}{\tan\beta} \tag{18.9}$$

is a topographic constant for the catchment.

Equation (18.8), may also be written in terms of storage deficit (see for example Beven, 1986a,b) as :

$$\frac{(\bar{S} - S_i)}{m} = \left[\ln\frac{a}{\tan\beta} - \lambda\right] - \left[\ln T_0 - \ln T_e\right] \tag{18.8b}$$

These Eqs. (18.8) express the deviation between the catchment average water table depth (or deficit) and the local water table depth (or deficit) at any point in terms of the deviation of the local topographic index from its areal mean, and the deviation of the logarithm of local transmissivity from its areal integral value. The relationship are scaled by the parameters f or m. Given a value of $\bar{z}$ or $\bar{S}$, Eqs. (18.8) may be used to predict the pattern of water table depths over a catchment, based on knowledge of the spatial distribution of the soil-topographic index a/T_0tanß or, if it is assumed that the soil saturated transmissivity is spatially constant, the topographic index a/tanß. Equation (18.8) implies that every point having the same soil/topographic index value a/T_0tanß behaves functionally in an identical manner. The a/T_0tanß variable is therefore an index of **hydrological similarity** .

The spatial distribution of a/tanß (see Fig. 18.2) may be derived from analysis of a Digital Terrain Model (DTM) of the catchment (Quinn et. al. 1991), although early applications of TOPMODEL had to rely upon manually interpreted maps (Beven and Kirkby, 1979) or analytical hillslope functions (Beven and Wood, 1983).

To calculate the surface (or subsurface) contributing area, the catchment topographic index is expressed in distribution function form (Fig. 18.2b). Discretisation of the a/T_0tanß distribution function brings computational advantages. Given that all points having the same value of a/T_0tanß are assumed to behave in a hydrologically similar fashion, then the computation required to generate a spatially-distributed local water table pattern reduces to one calculation for each a/T_0tanß class; calculations are not required for each individual location in space (Fig. 18.3). This approach should be computationally more efficient than a solution scheme that must make calculations at each of a large number of spatial grid nodes, a potentially significant advantage when iterative calibration and parameter sensitivity procedures are carried out.

Of particular interest is the case in which Eq. (18.8) predicts that the local water table is above the surface ($z_i < 0$), or above a specified near-saturated capillary fringe ($z_i - \psi_o$). Such areas are where saturated overland flow are predicted to occur, and their spatial distributions constitute the variable saturated source areas which generate the modelled surface runoff response. In the procedural model, any rainfall falling upon the saturated source area is taken to be runoff, along with rainfall in excess of that required to top fill areas where local water table depth z_i is small. Equation (18.8) may also be used to predict the pattern of subsurface stormflow contributing areas, or flow through different soil horizons (Robson et al., 1992) if these can be defined by some threshold value of water table depth.

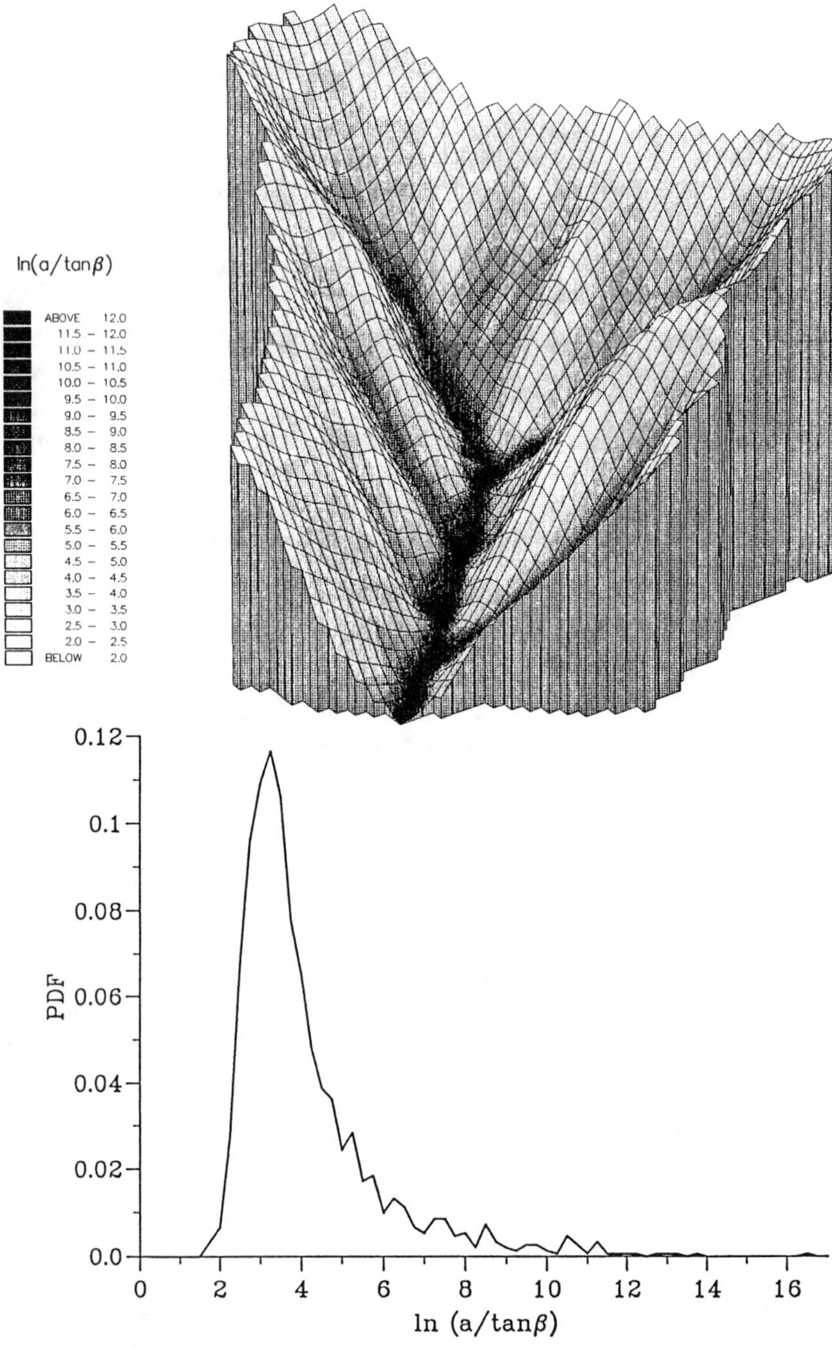

Fig. 18.2. **A. Distribution function of ln (a/tanß) index for the Maimai catchment, New Zealand. B. Pattern of ln (a/tanß) within the topography of the Maimai catchment.**

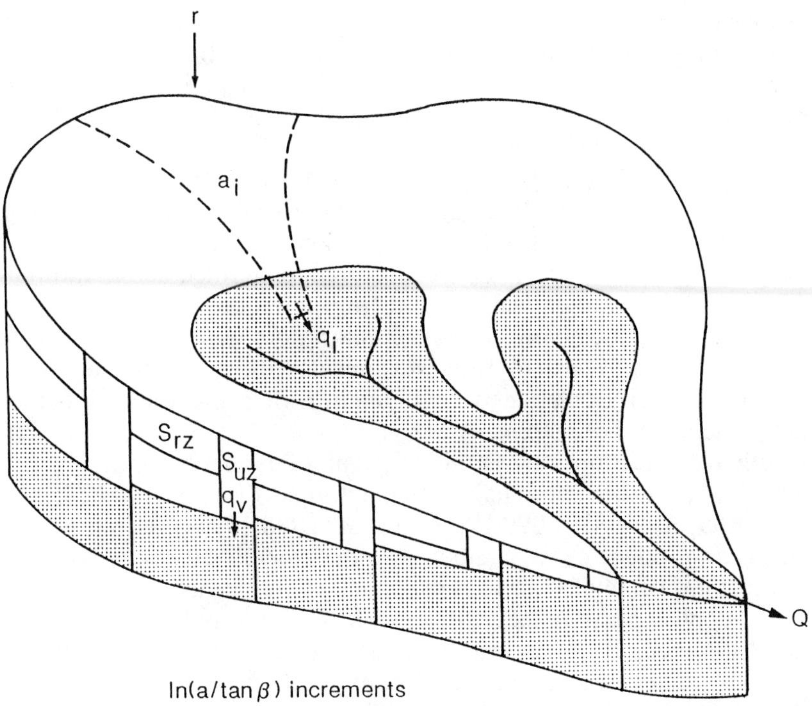

Fig. 18.3. Schematic representation of storage elements within a discrete ln(a/tanß) increment representation of a catchment area, showing the root zone store store S_{rz}, vertical drainage store S_{uz}, and recharge to the saturated zone, q_v, for one increment, and area, a_i, draining through a particlar point i. The shaded area represents the area of surface saturation corresponding, in this case, to values of ln(a/tanß) falling in the highest class.

18.3.1. The Organisation of Stores in TOPMODEL

There are various ways of organising the stores embodied in TOPMODEL. Essentially a simple series of stores is envisaged, routing water from the surface to the saturated zone. It is notable that, in any chain of stores, the form of the aggregate output will be most strongly controlled by the action of the least dynamic store (Kirkby 1975, Beven and Kirkby 1979). This applies to combinations of non-linear stores, such as those perceived to exist in catchment hydrology. Thus, it is important to accurately represent the non-linearity of the most slowly-responding store, whilst more dynamic stores may be approximated by simpler, linear representations. Kirkby (1975) presented a schematic outline of the different response times characteristic of various catchment processes from which it is apparent that the saturated zone is typically the slowest to respond. TOPMODEL therefore adopts a non-linear saturated zone store, but generally adopts simpler, linear representations of other, more dynamic stores. Storage, internal fluxes and inputs/outputs are generally expressed in terms of metres equivalent of water (per unit time).

Beven and Kirkby (1979) applied TOPMODEL to the Crimple Beck catchment in the north of England using three stores :- an interception or depression store, an infiltration store and a saturated zone store. The interception and depression store was required to be filled before infiltration could occur. Evaporation was allowed from this store at full potential rate until the store became empty. Thc infiltration store was formulated to allow for an Hortonian overland flow response, with a catchment average infiltration capacity calculated as a function of a maximum infiltration capacity and upper zone storage.

In other applications of TOPMODEL, more explicitly physically-based approaches to infiltration have been adopted (Beven 1984, 1986a,b; Sivapalan et. al. 1987, 1990; Wood et. al. 1988,1990). However, these approaches introduce extra parameters to the model which may be difficult to calibrate, especially for the case of spatially heterogeneous soils.

In a second application of TOPMODEL to the Crimple Beck site, Beven et. al. (1984) found that the earlier model structure, routing all infiltrated water immediately to the saturated zone, generated an overestimate of discharge which they attributed to an underestimation of evapotranspirative losses. It was considered that this problem reflected the model's inability to delay realistically the vertical unsaturated flow of water between the infiltration store and the water table. A notional 'field capacity' was therefore introduced to the model. This comprised a value of infiltration storage which had to be exceeded before flow to the saturated zone was permitted. The water thus held from entering the saturated zone remained available for evapotranspiration, reducing the model overestimation of storm flows following dry spells.

A revised formulation of the TOPMODEL stores has been presented by Quinn (1991), Beven (1991) and Quinn and Beven (1993) and is shown in Fig. 18.4 for the case of saturation expressed in terms of depth to the water table. There will be one such collection of stores for each $a/T_0\tan\beta$ class (Fig. 18.3), so that a proper account is taken of the predicted variation in depth to the water table in routing water through the unsaturated zone. Here, the 'root zone' storage has the same function as the interception and evaporation stores of earlier TOPMODEL versions, although it may be necessary to add an additional interception and surface storage, particularly for forest canopies. The notion of a zone of non-active soil water reflects the "field capacity" concept. Vertical flux from the root zone to the water table may occur in the 'gravity drainage' zone whenever field capacity is satisfied. The very simplest conversion between a storage deficit due to drainage and water table depth assumes that rapid gravity drainage affects only the largest pores, down to some 'field capacity', and that the difference

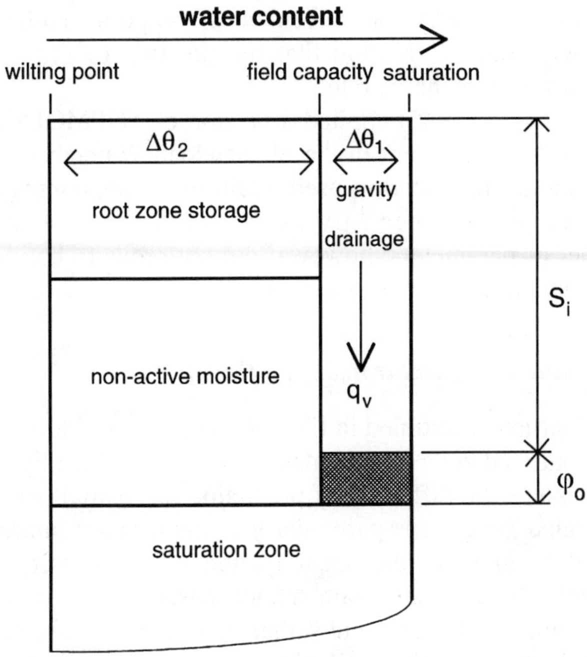

Fig. 18.4. A simple version of the vertical storage elements in TOPMODEL. S_i is the local gravity drainage storage deficit, q_v is local recharge to the saturated zone, and $_o$ is the depth of the "capillary fringe".

in storage between saturation and field capacity does not change with depth. This results in the simple linear scaling

$$s_i = (\theta_s - \theta_{fc})(z_i - \psi_o) = \Delta\theta_1(z_i - \psi_o) \qquad (18.10)$$

where θ_s is moisture content at saturation, θ_{fc} is moisture content at "field caoacity", $\Delta\theta_1$ is an effective drained porosity and ψ_o is an effective depth of capillary fringe assumed to be at saturation. Note that if saturation is expressed only as a storage deficit, it is not necessary to calibrate the parameters $\Delta\theta_1$ and $\psi_{o,}$ but any comparisons with observed water table levels must then be done separately.

A more complex form of conversion from storage deficit to water table depth has been used by Sivapalan et al. (1987, 1990; Wood et al., 1990) in which it is assumed that some soil moisture characteristics are known and that initially a complete gravity drainage holds above the water table. This has been used to define initial conditions prior to single storm simulations but introduces additional parameters associated with the soil moisture characteristic functions (the Brooks-Corey relationships in their case with up to 3 parameters) and, while a convenient assumption mathematically, is inconsistent with the assumption that the position of the water table is controlled by a steady recharge rate.

Hornberger et. al. (1985) applied a version of TOPMODEL with more stores to a forested catchment in the Shenandoah Watershed Acidification Study. This model formulation used additional components to allow for modification of rainfall chemistry by vegetation and forest litter. The greater number of parameters, however, did not prove to be identifiable on the basis of simulating water fluxes alone.

Moisture accounting : unsaturated zone fluxes

The basic soil structure outlined in Fig. 18.4 may be used to accommodate a variety of unsaturated zone process descriptions as defined by the modeller. Beven (1991) notes the difficulty of predicting the spatial pattern of fluxes in the unsaturated zone. It is particularly difficult to account explicitly for the effects of local soil heterogeneity and macroporosity (Beven and Germann, 1992). No adequate mathematical description of unsaturated flow in structured soils with parameters that can be identified at a practical prediction scale is currently available (see discussions in Beven, 1987a, 1989a,b, 1993) and if parameter values are to be determined by calibration then only a minimal parameterisation is allowable.

A minimal parameterisation must account for the changes in unsaturated zone fluxes with local unsaturated zone storage and depth to the water table (or storage deficit), allowing for the fact that any functional relationships may be highly nonlinear. Two formulations that have been adopted in past TOPMODEL applications have assumed that the unsaturated zone flows are essentially vertical and have been expressed in terms of drainage flux from the unsaturated zone. These approaches have used either a simple storage deficit-dependent time delay or a conductivity-based flow equation. Both calculate a vertical drainge flux q_v for each topographic class.

Expressed in terms of storage deficit, Beven and Wood (1983) suggested that a suitable functional form for the vertical flux q_v at any point i is :

$$q_v = \frac{S_{uz}}{S_i t_d} \tag{18.11}$$

where S_{uz} is storage in the unsaturated (gravity drainage) zone, S_i is the local saturated zone deficit due to gravity drainage, and dependent on the depth of the local water table. Parameter t_d is a time constant. Equation (18.11) is the equation of a linear store but with a time constant $S_i\, t_d$ that increases with increasing depth to the water table.

A second form was suggested by Beven (1986a,b) on the basis of the Darcian flux at the base of the unsaturated zone which, for an exponential conductivity function, can be expressed as :

$$q_v = \alpha\, K_0\, e^{-f z_i} \tag{18.12}$$

where α the effective vertical hydraulic gradient is a parameter, K_o the saturated conductivity at the surface and z_i the local water table depth. If the value of α is set to unity, thus assuming that the vertical flux is equal to the saturated hydraulic conductivity just at the water table, it is eliminated as a parameter.

Accounting for evapotranspiration with a minimal number of parameters poses a problem of similar complexity to that of unsaturated zone drainage. TOPMODEL follows generally-adopted practice in calculating actual evapotranspiration (E_a) as a function of potential evaporation (E_p) and root zone moisture storage for cases where E_a cannot be specified directly. In the TOPMODEL description of Beven (1991), evaporation is allowed at the full potential rate for water draining freely in the unsaturated zone and for predicted areas of surface saturation. When the gravity drainage zone is exhausted, evapotranspiration may continue to deplete the root zone store at the rate E_a, given by :

$$E_a = E_p\left(1 - \frac{S_{rz}}{S_{r\max}}\right) \tag{18.13}$$

where the variables S_{rz} and S_{rmax} are, respectively, root zone storage deficit and maximum allowable storage deficit. If some effective root zone depth z_{rz} can be asumed S_{rmax} can be calculated from $z_{rz}\left(\theta_{fc} - \theta_{wp}\right) = z_{rz}\left(\Delta\theta_2\right)$ where 6_{wp} is moisture content at wilting point. For calibration it is only necessary to specify a value for the single parameter S_{rmax}, but the other forms may help in defining this value.

The flux of water entering the water table locally at any time is q_v . This drainage is also a component of the overall recharge of the lumped saturated zone. To account for the catchment average water balance, all the local recharges must be summed. If Q_v is the total recharge to the water table in any time step, then

$$Q_v = \sum_{i=1}^{n} q_{v,i} A_i \tag{18.14}$$

where A_i' is the fractional area associated with topographic index class i as a fraction of total catchment area.

Moisture accounting : saturated zone fluxes

Output from the saturated zone is given by the baseflow term, Q_b. This may be calculated in a distributed sense by the summation of sub-surface flows along each of m stream channel reaches of length l. Recalling Eq. (18.1), we may write :

$$Q_b = \sum_{j=1}^{m} l_j (T_0 \tan \beta) e^{-f z_j} \qquad (18.15)$$

Substituting for z_j using Eq. (18.5) and rearranging, it can be shown that :

$$Q_b = \sum_{j=1}^{m} l_j a_j e^{-\gamma} e^{-f z}$$

Since a_j represents contributing area per unit contour length, then :

$$\sum_{j=1}^{m} l_j a_j = A$$

Therefore :

$$Q_b = A e^{-\gamma} e^{-f z} \qquad [18.16)$$

where A is the total catchment area (m^2). It is therefore possible to calculate baseflow in terms of the average catchment water table ($\bar{z}$):

$$Q_b = Q_0 e^{-f z} \qquad [18.17)$$

where :

$$Q_0 = A e^{-\gamma} \qquad [18.18)$$

is the discharge when $\bar{z}$ or $\bar{S}$ equals zero. This is the same form as that originally assumed by Beven and Kirkby (1979) if the linear relationship between storage deficit and depth to water table of Eq. (18.10) holds. Solution of Eq. (18.17) for a pure recession in which inputs are assumed to be zero shows that discharge has an inverse or first order hyperbolic relationship to time as:

$$\frac{1}{Q_b} = \frac{1}{Q_o} + \frac{t}{m} \qquad [18.19)$$

Thus, if Eq. (18.17) is an appropriate relationship to represent the subsurface drainage of a given catchment a plot of $1/Q_b$ against time should plot as a straight line (eg. Fig. 18.5) with slope 1/m. Thus, given at least some recession curves, not greatly influenced by evapotranspiration or snowmelt processes, it should be possible to set the value of m which will then need minimal calibration.

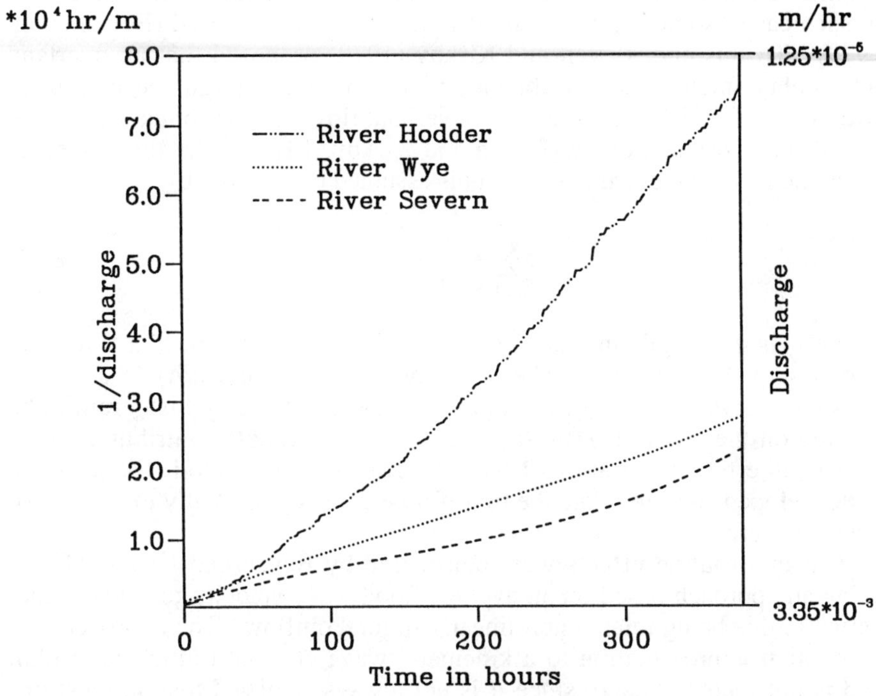

Fig. 18.5. Master recession curves for three different UK catchments plotted as 1/Q versus time. A straight line plot indicates correspondence with Eq. (18.17) which may be an adequate representation for these cases.

The catchment average storage deficit before each time step is updated by subtracting the unsaturated zone recharge and adding the baseflow calculated for the previous time step, thus :

$$\bar{z}_t = \bar{z}_{t-1} + \Delta\theta_1\left[Q_{b_{t-1}} - Q_{v_{t-1}}\right] \qquad [18.20)$$

A similar calculation can be made for a catchment average storage deficit formulation.

Equation (18.17) can be used to initialise the saturated zone of the model at the start of a run. If an initial discharge is known and assumed to be only the result of drainage from the saturated zone, Eq. (18.17) can be inverted to give a value for $\bar{z}_{t=0}$.

18.3.2. Channel Routing and Sub-catchment Structure

For many catchments, especially large ones, it may be inappropriate to assume that all runoff reaches the catchment outlet within a single time step. In such cases, some routing of model output may be required (Beven et. al. 1979). To this end, Beven and Kirkby (1979) proposed that an overland flow delay function and a channel routing function might be employed within the TOPMODEL structure. Overland flow may be routed by the use of a distance-related delay (Beven and Kirkby, 1979). The time taken to reach the basin outlet from any point is assumed to be given by :

$$\sum_{i=1}^{N} \frac{x_i}{\upsilon \tan \beta_i} \qquad [18.21)$$

where x_i is the length and $\tan\beta_i$ the slope of the i th segment of a flow path comprising N segments. The velocity parameter υ (m/h) is assumed constant. Given , Eq. (18.21)) allows a unique time delay histogram to be derived on the basis of basin topography for any runoff contributing area. This is, in effect, a variation of the time-area routing method (Clark 1945), but developed so as to relate the runoff time delay dynamically to the size of the source area.

Channel routing effects were considered by Beven and Kirkby (1979) using an approach based on an average flood wave velocity for the channel network, this being related non-linearly to total outflow. The approach was an explicit approximation to a kinematic wave channel routing algorithm and is not recommended, since it is not always stable. Most applications have been based on a simple constant wave speed routing algorithm, equivalent to the channel width function based algorithms used by Surkan (1969), Kirkby (1976) and Beven (1979, Beven and Wood, 1993), which has the advantage that it introduces only a single wave speed parameter.

18.3.3. Deriving the Topographic Index

An analysis of catchment topography is required in order to derive the a/tanß distribution function (Fig. 18.2). In order to obtain discrete values of a/tanß some sampling of topography is implied. Early development of TOPMODEL relied upon the manual analysis, based on map information, of local slope angles, upslope contributing areas and cumulative areas. Beven and Kirkby (1979) outlined a computerised technique used to derive the topographic distribution function (and the overland flow delay histogram) based on the division of the catchment into sub-basin units. Each unit was then discretised into small 'local' slope elements on the basis of dominant flow paths (inferred from lines of greatest slope). Calculation of a/tanß was carried out for the downslope edge of each element. Although an approximation, this method was felt to be justified by its relative efficiency and because field observations of flow paths could be used in defining the slope elements to be analysed. In particular, the effects of field drains and roads in controlling effective upslope contributing areas could be taken into account. Such human modifications of the natural hydrological flow pathways are not normally included in Digital Terrain Maps (DTM) of catchment areas.

However, given a DTM, more sophisticated computerised methods are now available. Quinn (1991) developed a group of Digital Terrain Analysis (DTA) programs, based on raster elevation data, with the aim of investigating their utility in deriving the topographic information required by TOPMODEL. Application of these techniques to catchment modelling studies has been described by Quinn et. al. (1991) and Quinn and Beven (1993). There, the functional nature of the a/tanß index was discussed in terms of the quality of the representation of hydrologically significant topographic features by DTA methods. There are subjective choices to be made in the DTA. Quinn et al. (1991) describe an multiple flow direction algorithm that is based on the distribution of area to all downslope grid elements; Wolock (1993), on the other hand, uses an algorithm based on a single flow direction with the greatest slope. Other techniques have also been used; most raster DTMs have been derived from contour data digitised from existing maps and Moore et al. (1986) have used DTA techniques based on flowpaths derived directly from contour data to calculate the a/tanß index, a computerised version of the original manual (and exceedingly tedious) Beven and Kirkby (1979) procedure.

Although these techniques offer great time savings in the application of TOPMODEL, they also raise important theoretical questions, of which two are perhaps most immediate. Firstly, how should 'upslope contributing area' be calculated from topographic data and, secondly, what scale of resolution should be adopted for the catchment DTM (see for example the discussion in Quinn et al., 1994)? More generally, it may be noted that DTA methods, in the context of TOPMODEL, may have implications concerning the hydrological processes inferred to be related to topography and on the appropriate effective parameter values. The DTM must have a fine enough resolution to properly reflect the effect of topography on surface and subsurface flow pathways. Coarse resolution DTM data may, for example, fail to represent some convergent slope features. However, too fine a resolution may introduce perturbations to flow directions and slope angles that may not be reflected in the smoother (?) water table surface. The appropriate resolution will depend on the scale of the hillslope features, but 50m or better data is normally suggested. Experience at Lancaster and elsewhere (for example Jordan, 1992; Wendling, 1992; Chairat and Delleur, 1993a) shows that the scale of the DTM used does affect the derived topographic index distribution, in particular inducing a shift in the mean value λ and the calibrated values of parameters in particular applications (Bruneau et al., 1994; Quinn et al., 1994). The location of rivers, and their influence on the DTM cells that contain them, is also problematic. There is a choice of allowing the upslope area to continue to accumulate along the river network, or allowing the river to act as a sink for hillslope flows. The first will tend to overestimate, the second to underestimate, riparian contributing areas if the river width is small compared with the DTM grid scale.

It is worth noting that a parameterisation of the ln(a/tanß) distribution may sometimes be useful. Sivapalan et al. (1990) introduced the use of a gamma distribution in their scaled version of TOPMODEL. Wendling (1992) also used a gamma distribution in a runoff production function for a flood forecasting model and Wolock (1993) gives details of a gamma distribution version for continuous simulation.

18.3.4. The TOPMODEL Concept and Hydrological Similarity

The ln(a/tanß) topographic index or ln(a/T_0tanß) soil-topographic index, given acceptance of the simplifying assumptions of TOPMODEL, are indices of hydrological similarity at a point within the catchment. The response of any individual catchment, as predicted by TOPMODEL, then depends upon the similarity in the distribution of the indices and in the input sequences to which the catchment is subjected including both the time and space variability of rainfall rates and evapotranspiration losses. At small scales, say individual hillslopes, the **pattern** of variability may also be important, but at larger scales it may be possible to have a sufficient sample of the possible variability of topography, soil and vegetation within the catchment to allow representation of that variability in terms of simple distribution functions. The scale at which this simplification may be valid, which they termed the Representative Elementary Area (REA), has been investigated in hypothetical simulations by Wood et al. (1988, 1990). They used realistic topographic data together with stochastic spatial models of soil and rainfall to evaluate the variability of hydrological storm runoff responses predicted at different spatial scales. Both infiltration excess and saturation excess runoff production mechanisms were included (Fig. 18.6).

Their results suggested a minimal REA for the conditions simulated of the order of 1 km^2. It is important to note that they did not suggest that this is the scale at which a **lumped** parameter description could be used; it may still be necessary to represent the heterogeneity of runoff production but at scales greater than the REA this can be done by means of a distribution function or functions representing that variability. Sivapalan et al. (1987, 1990) have taken this concept further in producing a non-dimensional version of TOPMODEL in which different catchments can be compared in terms of various scaling coeffcients. Similarity in these scaling coefficients should imply similarity in runoff production.

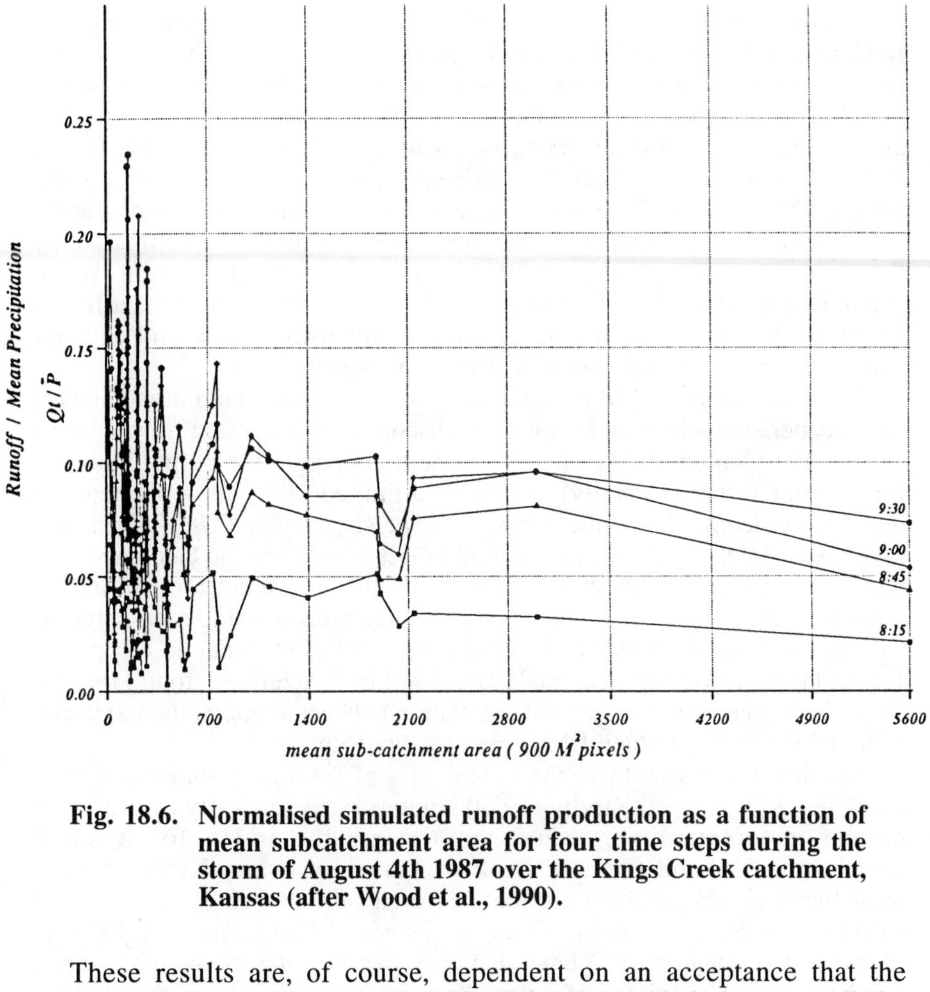

Fig. 18.6. Normalised simulated runoff production as a function of mean subcatchment area for four time steps during the storm of August 4th 1987 over the Kings Creek catchment, Kansas (after Wood et al., 1990).

These results are, of course, dependent on an acceptance that the assumptions that underlie the TOPMODEL topographic index are realistic. These assumptions are, primarily, that the upslope contributing area extends to the hillslope divide and that the water table is parallel to the soil surface. Only then is the index easy to calculate from digital terrain analysis. It has already been shown above how an effective soil transmissivity can be used to modify the index but this then requires additional information that rarely, if ever, is available at the catchment scale. In addition, both assumptions can be relaxed (see below) but also at the cost of requiring additional information.

There have been some tests of the hydrological similarity implied by the index. Troch et al. (1993a) have compared the TOPMODEL conceptualisation of Sivapalan et al. (1987) with field data and the output of a fully dynamic three-dimensional Richards equation based finite element model (although with quite different assumptions about soil conductivities). They showed that the quasi-steady state TOPMODEL representation was a reasonable approximation for the well measurements. In comparisons with field data, Moore et al. (1988) found a good correlation with the topographic index and surface soil moisture measurements in a small fallow catchment, but Ladson and Moore (1992) found very poor correlations with gravimetric and remotely sensed (PBMR) surface soil moistures on the FIFE Konza Prairie site in Kansas, a natural prairie site with a dense grass cover at the surface. We have similarly found little correlation between the topographic index and both gravimetric and surface soil moistures estimated by Synthetic Aperture Radar (SAR) for the agricultural Slapton Wood catchment in Devon, UK. Surface soil moisture is, of course, affected strongly by evapotranspiration as well as downslope flow and a better correlation might be expected with water table levels.

Burt and Butcher (1986) demonstrated that in some humid catchments with deeper subsurface and sometimes disconnected saturated flow systems such as the Slapton Wood catchment, the topographic index (not allowing for a variable transmissivity) may have a limited utility as a predictor of water table depth. To confirm this, we have shown that TOPMODEL can give good simulations of Slapton Wood discharges once the catchment has wetted up in winter, but that it is very difficult to predict the wetting up sequence. Jones (1986) also discusses the limitations of using a static index, such as the $\ln(a/\tan\beta)$ index, for predicting catchment dynamic responses. Jordan (1992), working in a small catchment in Switzerland, found that for some dates there was a very good correlation between topographic index and water table depth, on other dates a poor relationship.

Another test variable is the extent of soil saturation under different conditions. Beven and Kirkby (1979) demonstrated similarity in pattern of measured saturated areas and the topographic index for a small subcatchment of the Crimple Beck in Yorkshire, while Kirkby (1978) presented a similar analysis for the measured saturated areas of the WC-4 catchment at Sleepers River, Vermont (Dunne, 1978). Merot (1988) and Merot and Bruneau (1993) have demonstrated that the topographic index can be a useful predictor of valley bottom wetland areas in the Bocage landscapes of Brittany in France, while Moore et al. (1988) have shown how it can predict zones of potential erosion.

18.3.5. Model Calibration

The parameter values introduced in the description of the TOPMODEL concepts have been given names based on physical descriptions of the flow processes. The descriptions used, however, are simplified physical descriptions, deliberately so since one response to the lack of detailed physical descriptions at the hillslope scale is to minimise the number of parameter values to be calibrated. A number of studies have suggested that there is only enough information in a set of rainfall-runoff observations to calibrate 4 or 5 parameters (Kirkby, 1975; Hornberger et al., 1985). In essence, TOPMODEL provides a nonlinear functional model for fast and slow catchment responses and basic flow routing using a small number of parameters.

It has been argued in recent years that the realism even of the most complex 'physics-based' hydrological models may be illusory (Beven 1987a,1989, Grayson et al., 1992). One implication of this argument is that the physical interpretation of calibrated parameters may be difficult, since calibration of parameter values can compensate for model structural errors and there may be many different parameter sets that can give acceptable simulations and be physically plausible. One consequence of these arguments is that the idea of a set of "optimum" parameter values should be rejected in favour of a concept of equifinality of parameter sets (and possibly also model structures, see Beven, 1993). This concept allows that there may be multiple optima in the parameter response, interactions between parameters, errors in the observed data and, indeed, different "optimal" sets if different calibration periods are used.

In fact, certain TOPMODEL parameters may, in principle, be calibrated on the basis of field measurements and some applications have been made using only measured and estimated values (Beven and Kirkby, 1979; Beven et al., 1983). Each formulation of TOPMODEL may present an individual parameter set to be calibrated; however there are invariably three or four critical parameters that most directly control model response. These are the saturated zone parameter f (or m in the original storage deficit formulation), the saturated transmissivity values T_0, and the root zone parameter S_{rmax}, and in larger catchments a channel routing velocity, v.

It has already been noted that the saturated transmissivity decay parameter, m, may be derived from an analysis of catchment recession curves. Since this is one of the most important model parameters it reinforces the idea that to simulate hydrological responses at the catchment scale, the most useful measurements will be made at the same scale especially in ungauged catchments. The most useful measurement will be stream discharge followed by a good estimate of the rainfall input, especially if it varies spatially.

Other parameters that can (in principle) be estimated in the field include soil transmissivity profiles and root zone available water capacity. For early applications of TOPMODEL, Beven and Kirkby (1979) described a sprinkling infiltrometer method for calibrating the interception, infiltration store and overland flow velocity parameters. These techniques were used by Beven et al. (1984) with reasonable success in applications to three U.K. catchments. These measurements are, however, essentially "point" measurement techniques, whereas the model requires either effective values at the model grid scale, or ideally distributions that reflect the spatial heterogeneity in the catchment. The difficulties of sampling, and the lack of a theory of effective parameter values, mean that such heterogeneity (as well as the simplifications and errors inherent in the model structure) will inevitably introduce uncertainty into the model predictions.

Recent work at Lancaster has focused on investigating model performance in terms of equifinality of parameter sets and uncertainty, rather than optimisation of fit (Binley et al. 1991; Binley and Beven 1991; Beven and Binley 1992; Beven, 1993). To this end, a methodology has been developed (Beven and Binley 1992) whereby the calibration of a hydrological model generates a set of uncertainty bounds defining the range of expected model responses. This procedure has been termed Generalised Likelihood Uncertainty Estimation (GLUE). Many model simulations are performed, each associated with a randomly chosen set of parameter values. The uncertainty methodology accords each parameter **set** a likelihood weight, defined by some subjectively chosen 'likelihood function' or performance index. This likelihood weight may be defined in terms of specific assumptions about the nature of the model residuals (see for example Sorooshian et al., 1983) but could equally use possibilistic measures derived from Fuzzy Set theory (see for example van Straten and Keesman, 1990). Non-acceptable simulations are given a likelihood weight of zero. The cumulative likelihood of all retained simulations can be scaled to sum to unity.

At any time step, it is then possible to derive a probability density function for the whole range of model simulations weighted by their likelihoods. From this function, uncertainty bounds may be drawn (typically the fifth and ninety-fifth percentiles) defining the likely range of model predictions. No 'optimum' simulation is recognised; however, as more calibration data is considered, or as more combinations of parameters are tested, simulation likelihoods may be updated using a Bayesian approach, thus modifying the model's uncertainty bounds. If these bounds become narrower, then model calibration may be deemed to have improved. The case where the predictive uncertainty bounds do not envelope an observed response suggests a fundamental failure of the model (or an inappropriate likelihood function !). If the likelihood distribution changes so as to widen the uncertainty bounds then a reduction in the predictive power of the model is indicated.

An important feature of the uncertainty estimation procedure is that it allows simulation likelihoods to be defined according to different types of calibration data. For instance, borehole levels may be combined with observed discharge in calculating the power of a model to predict both water table depths and streamflow (Binley and Beven, 1991). This should give an improved indication of the realism of a model/parameter set combination. Field experience may be accommodated within this framework. For a catchment where overland flow was thought not to occur, all model simulations exhibiting overland flow could, for example, be given a likelihood of zero. It will be realised that traditional optimisation corresponds here to one set of parameters having a likelihood of one, with all others given zero. One of the consequences of estimating uncertainties associated with predictions is that it raises the question of what type of data would be most valuable in constraining that uncertainty (see Beven and Binley, 1992), especially with respect to the distributed predictions of the model. At present, the use of transmissivity measurements, water table measurements and saturated contributing area measurements are all being explored within this framework.

The GLUE methodology is computationally demanding. Thousands of model realisations may be required in order to explore fully the parameter space defined for a given catchment, each realisation being retained in storage for likelihood updating. Although this may be a significant impediment when complex physics-based distributed models are to be calibrated, the simple structure of TOPMODEL renders it particularly suitable for calibration using such uncertainty estimation techniques. The technique is also well suited to implementation on parallel computers. Some example calculations using TOPMODEL for the Maimai catchment in New Zealand are shown in Fig. 18.7 (see also Beven, 1993).

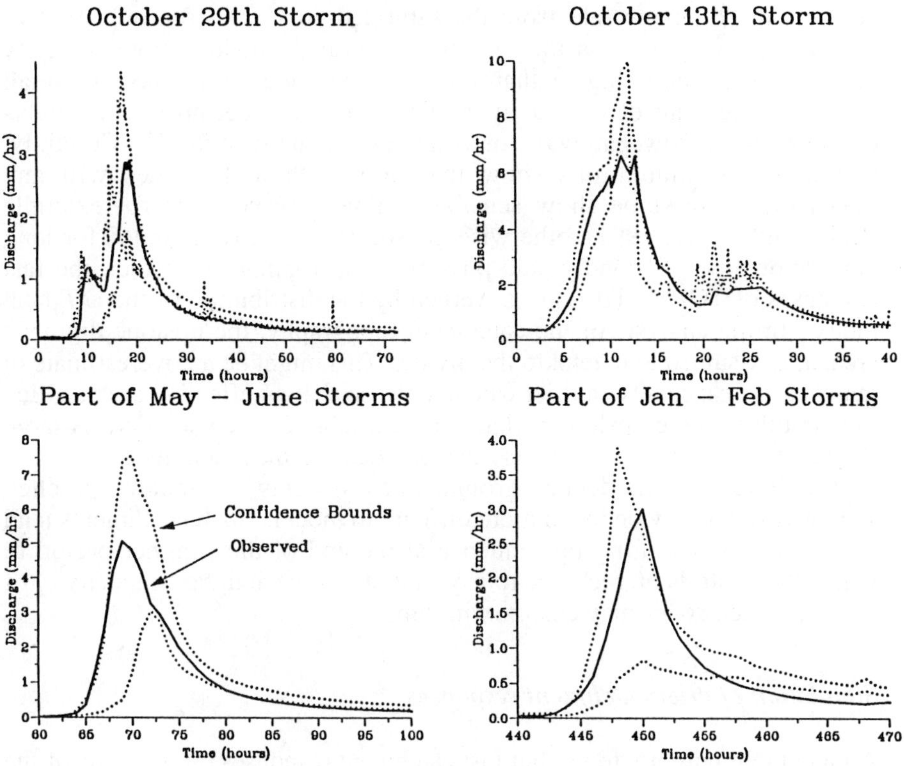

Fig. 18.7. Observed discharges (solid line) and 90% prediction confidence limits (dotted lines) for different storm periods for the Maimai catchment, New Zealand.

18.3.6. Example Applications and Results

Simulation of Humid catchment responses

TOPMODEL was originally developed to simulate small upland catchments in the U.K. (Beven and Kirkby, 1979; Beven et al., 1984). These studies showed that it was possible to get reasonable results with a minimum of calibration of parameter values. Since then there have been applications to a number of other catchments in humid temperate regimes in the eastern USA (Beven and Wood, 1983; Hornberger et al., 1985), New Zealand (Beven, 1993), Scotland (Robson et al., 1993). In all of these cases it has been found that, after calibration of the parameters, TOPMODEL provides good simulations of stream discharges, and broadly believable simulations of variable contributing areas (see Fig. 18.8 and Quinn and Beven, 1993).

Experience with model calibration, however, has shown that the fitted parameter values may be difficult to interpret physically. In particular, calibration of the T_0 transmissivity parameter (to which the simulations tend not to be very sensitive) often yields very high values. This parameter controls the drainage rate from the saturated zone. There could be two reasons for this. One is that effective lateral downslope transmissivity values may be much higher than might be expected on the basis of small scale measurements of vertical hydraulic conductivity because of the effects of preferential flow pathways or zones of fractured regolith. Certainly, Darcian flow simulations show that only with high values will any reasonably long slope show sensible drainage curves (see for example Binley and Beven, 1991), otherwise the soil stays nearly saturated for long periods of time. Secondly, and possibly exacerbating this effect, the fast responses of TOPMODEL are governed by the distribution of the $a/T_0\tan\beta$ index. In the analysis of the catchment topography, the upslope drainage area, a, is assumed to extend to the divide. This might be an overestimate in many cases, especially in drier catchments, so that an effective a value, due to vairability in the upslope recharge rate, might be much smaller. A large T_0 might then compensate for any overestimates in the $a/\tan\beta$ index.

Catchments with deeper groundwater systems, or locally perched saturated zones may be much more difficult to model. Such catchments tend to go through a wetting up sequence at the end of the summer period in which the controls on recharge to any saturated zone and the connectivity of local saturated zones may change with time.

Simulation of drier catchment responses

A model that purports to predict fast catchment responses on the basis of the dynamics of saturated contributed areas may not seem to be a likely contender to simulate the responses of catchments that are often dry, such as in mediterranean or savannah climates. However, Durand et al. (1992) have

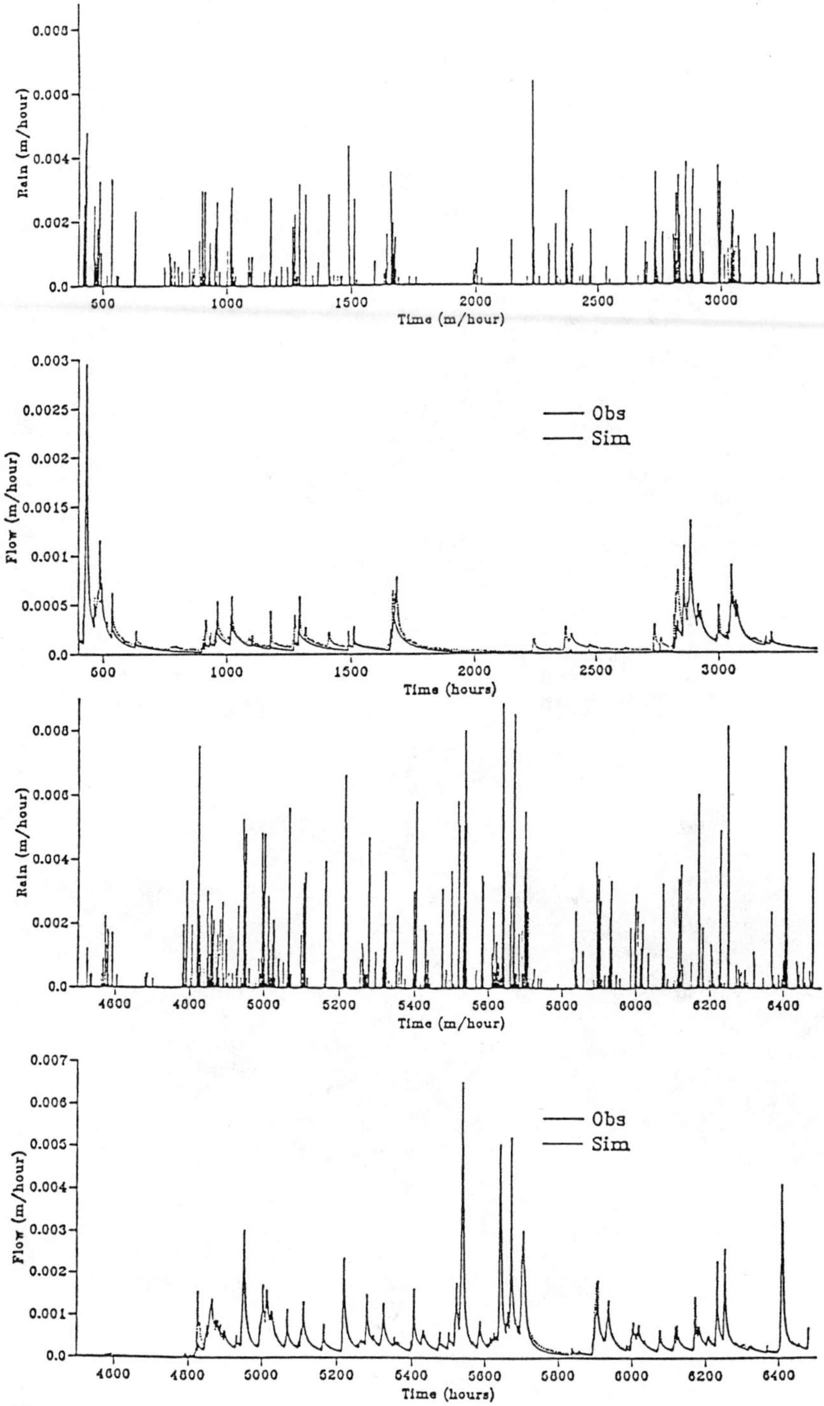

Fig. 18.8A. Rainfalls, observed discharges (solid line) and predicted discharges (dotted line) for a split-record validation simulation for the River Severn catchment at Plynlimon, Wales (after Quinn and Beven, 1993)

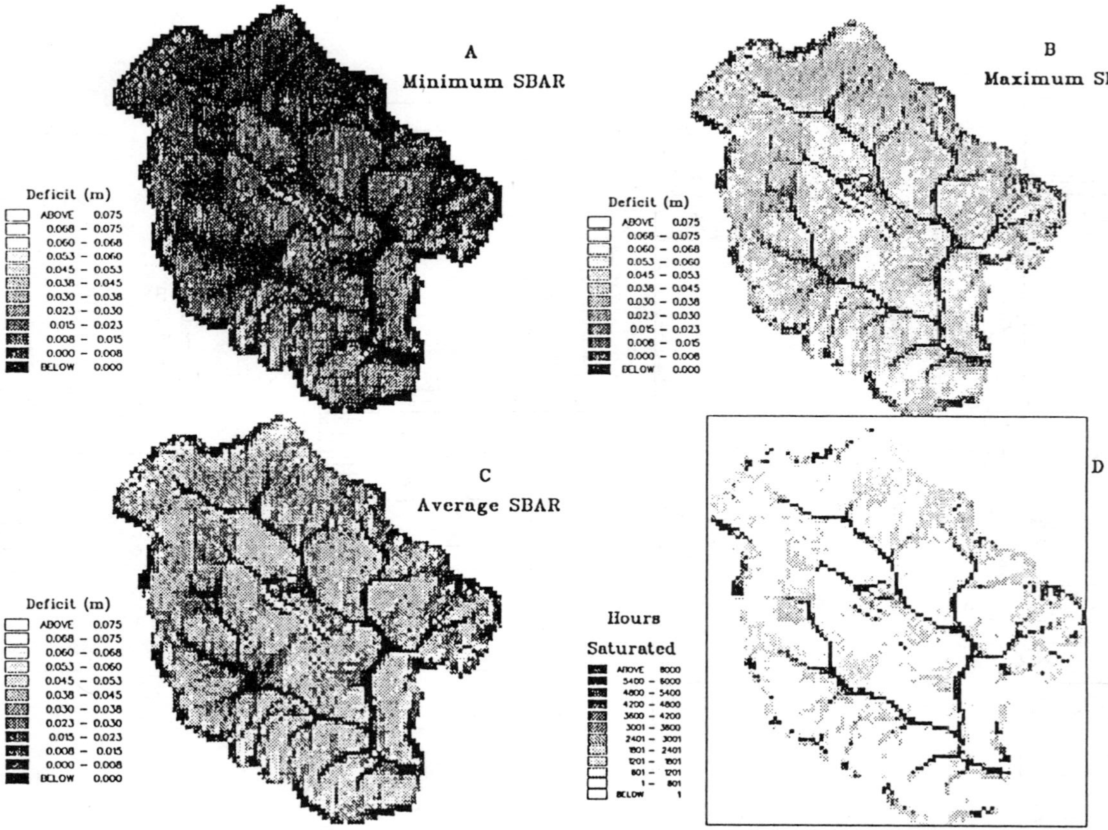

Fig.18.8B. Predicted patterns of storage deficits and saturation for TOPMODEL simulation of the River Wye catchment at Plynlimon, Wales. (A) Moisture deficit map with largest saturated area, 1985. (B) Moisture deficit map with smallest saturated area, 1985. (C) Moisture deficit map with average saturation, 1985. (D) Cumulative frequency of saturation for individual 50 m pixels during nine months in 1985 (after Quinn and Beven, 1993).

shown that TOPMODEL can successfully simulate discharges in such catchments at Mont-Lozère in the Cevennes, southern France. Sempere-Torres (1990) and Wendling (1992) have also used a TOPMODEL based runoff production function to simulate the response of the Gardon D'Anduze and Réal Collobrier catchments in southern France respectively. They show that the runoff production function can be successfully used for flood forecasting purposes after calibration to a small number of storms and proved to be more robust in validation than other functions studied.

However, experience in modelling the Booro-Borotou catchment in the Côte d'Ivoire (Quinn, 1991), and catchments in the Prades mountains of Cataluña, Spain, suggests that TOPMODEL will only provide satisfactory simulations once the catchment has wetted up. In many low precipitation catchments, of course, the soil may never reach a "wetted" state, and the response may be controlled by the connectivity of any saturated downslope flows. Such catchments also tend to receive precipitation in short, high intensity storms. Such rainfalls may lead, at least locally, to the production of infiltration excess overland flow which is not usually included in TOPMODEL (but see Beven, 1986a,b). The underlying assumptions of the TOPMODEL concepts must always be borne in mind relative to the pertinent perceptual model for a particular catchment.

Flood frequency predictions

Beven (1986a,b) linked a version of TOPMODEL to a random rainstorm and interstorm period model to make flood frequency predictions for a number of climates. The simulations made use of soil-topographic index distributions based on hypothetical distributions of hydraulic conductivity, which were also used with the variation of the Green-Ampt infiltration model of Beven (1984) to make predictions of distributed infiltration excess runoff. The model allowed runoff production to be analysed in terms of volumes of infiltration excess, saturation excess and subsurface stormflow, together with the frequency distributions of infiltration excess and saturation excess contributing areas. These were, however, hypothetical simulations.

Beven (1987b) applied a similar model to the Wye catchment at Plynlimon in mid-Wales with storm based simulations and a random initial condition model based on field observations. The model was able to reproduce the observed peak over threshold frequency characteristics of the catchment for the 14 year record available after calibration of a single parameter. Uncertainty in the frequency predictions was evaluated using multiple simulations of 14 year periods (Fig. 18.9).

Sivapalan et al. (1990) produced a scaled flood frequency model based on the TOPMODEL concepts and showed that catchment runoff production could be compared on the basis of eight similarity variables. Their flood frequency curves were derived from storm by storm simulations and showed a transition between saturation excess overland flow dominated flood peaks

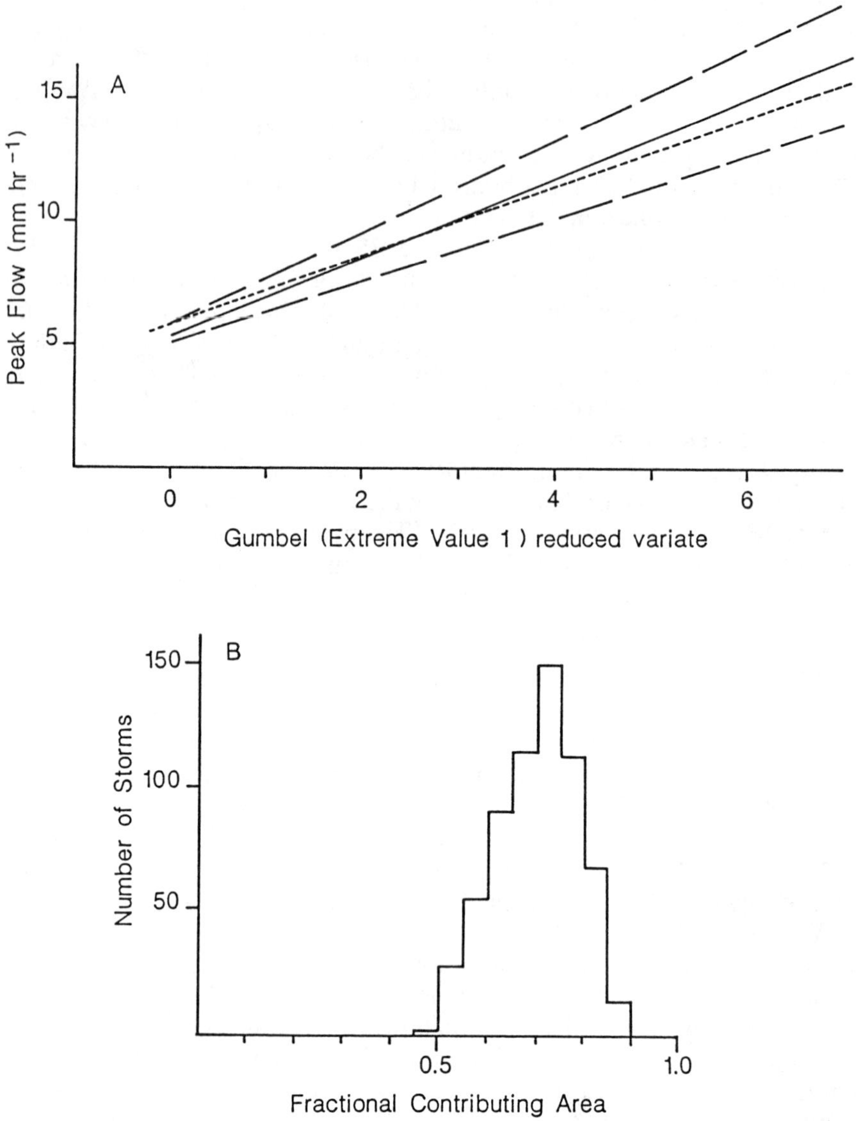

Fig. 18.9. **A. TOPMODEL simulated mean flood frequency curve (± 1 std. dev. derived from 10 realisations each of 14 years of data) for the River Wye, Plynlimon hourly mean discharge data compared with observed peaks over threshold curve for the 14 years 1969-1982.**

B. Simulated maximum contributing area frequency histograms for the peaks over threshold storms of the 10 realisations used in A. (after Beven, 1987).

TOPMODEL in a GIS framework

As a distributed model that can make use of data on topographic, soil and vegetation information TOPMODEL is well suited to implementation within a Geographical Information Systems (GIS) framework. Although many GIS systems have not had the facilities to couple interactive modelling and visualisation of distributed simulation results, a number of such TOPMODEL implementations have been now been made. At Lancaster, we have been involved in linking TOPMODEL to the Water Information System (WIS) developed by the Institute of Hydrology, Wallingford and ICL Ltd. (Romanowicz et al., 1993a,b). WIS can directly access the 50 m topographic database for the UK, together with rainfall and discharge data for gauging sites. The TOPMODEL implementation can, for a selected point on the river network, determine the appropriate topographic index distribution, carry out parameter calibration and sensitivity analysis including the GLUE uncertainty estimation simulations on an external parallel processor, and display the resulting distributed predictions (for example Fig. 18.10).

Other TOPMODEL implementations within a GIS framework have been using the SPANS Modelling Language SML (Stuart and Stocks, 1993); using GRASS routines (Chairat and Delleur, 1993b); with the PVwave modelling and visualisation system (Clapp et al, 1992); in the RHESSys system of Band et al. (1993) which also includes the FOREST-BGC ecological modelling components of Running and Coughlan (1988); the TAPES-G system of Moore et al. (1993) and the Modular Hydrological Modeling System (MHMS) system of Leavesley et al. (1992).

TOPMODEL and Land Surface-Atmosphere interactions

One limitation of the current generation of atmospheric general circulation models (GCM) is that they take litttle account of the variability of surface characteristics and hydrology in their calculation of land surface to atmosphere fluxes. TOPMODEL provides one very simple way of incorporating such heterogeneity into soil-vegetation-atmospheric models in a way in which the surface availability of water for evapotranspiration can be allowed to vary dynamically. This possibility has been explored in a number of papers, notably those of Famiglietti and Wood (1991), Famiglietti et al. (1992) and Quinn et al. (1994). These papers all show that the hydrological variability simulated by TOPMODEL may have an important effect on predicted evapotranspiration fluxes under certain circumstances. However, it must be remembered that the assumptions of TOPMODEL will not be met everywhere within a GCM grid square. A distribution function approach might still be appropriate (see for example Dolman, 1992), but it may be necessary to consider a wider index of hydrological similarity.

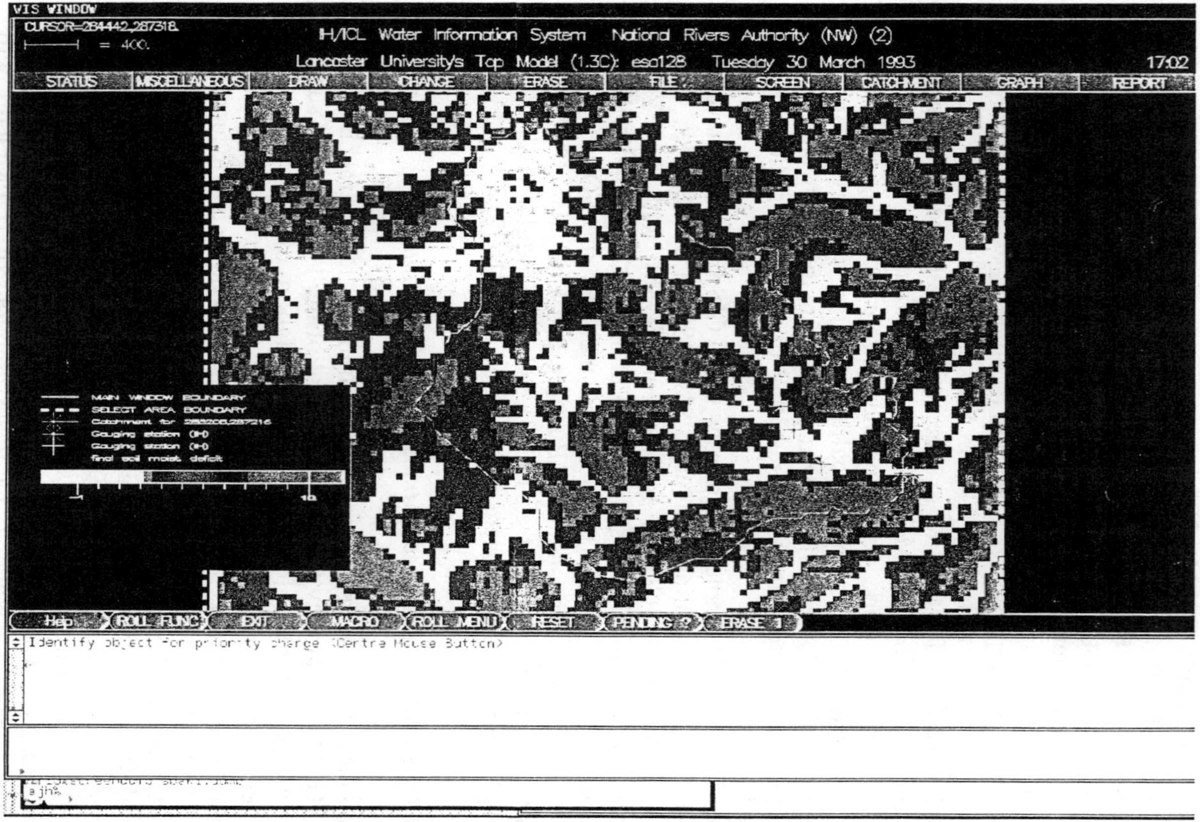

Fig.18.10. **The implementation of TOPMODEL within the Institute of Hydrology's Water Information System (WIS). The pattern shown is soil moisture deficit predictions for the River Severn at Plynlimon, Wales.**

One study has looked at the interaction of longer time climate change and hydrology using TOPMODEL (Wolock and Hornberger, 1991) and has done so in an interesting way. The model used was that calibrated for the White Oak Run catchment in Virginia by Hornberger et al. (1985) but the model was run with stochastic input sequences using a storm generation model similar to that of Beven (1986a,b). Sixty year simulations were made under different climatic scenarios, but unlike many other studies of climate change effects, the use of multiple realisations allowed the calculation of uncertainties in the hydrological effects on peak flows and basin yield for each scenario. The results demonstrate the difficulty in unambiguously detecting the effects of climate change.

TOPMODEL and geochemical predictions

One attraction of TOPMODEL is the possibility of making predictions of the split between surface and subsurface runoff production and the way in which this varies in different parts of a catchment. In fact, by assuming some knowledge about the depths and chemical characteristics of different soil horizons, it is possible to use the predictions of depth to the water table in different locations at different times to examine the mix of soil waters entering the stream channel. This possibility has been explored by Robson et al. (1992) in an application of TOPMODEL to the Hafren catchment, Plynlimon, mid-Wales. They showed that the model results compared well with a two soil component mixing interpretation of chemical signals in the stream, provided that the flow generated on the saturated contributing areas was assumed to have a well mixed composition (either a mix of organic soil and deep waters or the chemistry of mean rainfall). The possibilities of inferring chemical behaviour from such distributed predictions still requires more exploration and internal validation.

The fact that there may be some relationship between the topographic characteristics of a catchment and its chemical characteristics, with the inference that this is due to the effects of topography on flow pathways, has also been explored by Wolock et al. (1989,1990). In studies of catchments in the north-eastern United States and Wales they showed that the mean of the ln(a/tanß) distribution is strongly related to catchment acidification. TOPMODEL predictions of areas susceptible to surface saturation have also been used in Wales to decide areas for liming (Waters et al., 1991).

18.3.7. Restrictive Assumptions and Limitations

TOPMODEL may be expected to perform best when tested against catchments where its assumptions are met, in particular those of an exponential saturated zone store, a quasi-parallel water table and a topographic control on water table depth. Such catchments might be expected to have relatively shallow, homogenous soils and are likely to be generally quite wet, exhibiting the variable source area runoff mechanism which TOPMODEL simulates. Clearly, these conditions may not be met everywhere. Similarly, some TOPMODEL assumptions may be violated over certain ranges of behaviour. The response of the Crimple Beck (Beven and Kirkby 1979), for instance, did not conform to the expected response of an exponential store at the extremes of the recession curve. At high flows such deviance may be due to residual runoff, but at low flows, when it is reasonable to assume that only the saturated zone is contributing to streamflow, a conflict between model assumption and reality is suggested. A similar case will be illustrated below. Another instance where model assumptions may be invalidated is for catchments where an infiltration-excess runoff mechanism is thought to be important; this mechanism is either ignored or modelled only crudely in TOPMODEL. The criticisms suggested here do not necessarily invalidate the TOPMODEL approach, but serve to demonstrate that the model has been somewhat restricted by its specificity to circumstances that accord with its assumptions. There may therefore be restrictions on the geographical scope of TOPMODEL applications.

A fundamental assumption of existing TOPMODEL theory is that the saturated zone behaves as an exponential store (Eqs. (18.1) and (18.17)). Catchment recession behaviour does not always conform to this model. Figure 18.11 shows observed recession curve behaviour during low flow for the Ringelbach, a small catchment in the Vosges, France (Ambroise, 1988, 1990) as the inverse of discharge plotted against time. The marked inflection in the curve deviates from the straight line response expected of an exponential store. Such deviation may lead to baseflow errors in the model response. Some work has now been completed on relaxing the exponential assumption within a TOPMODEL framework and will be subject of a future paper.

It has been shown how the topographic index $a/\tan\beta$ is used to control the position of the water table in TOPMODEL. This procedure requires the assumption that the water table surface is everywhere approximately parallel to the ground, a condition which may not always be met, especially for deep soils. Quinn et. al. (1991) described a method whereby the assumption of a parallel water table may be relaxed. In basic TOPMODEL theory, the depth of the water table is defined with respect to the soil surface. Quinn et. al. (1991) introduced a 'reference level', based on a characteristic water table shape which may then deviate from being parallel to the soil surface. This reference level was used in order to calculate an 'effective' $a/\tan\beta$ distribution function for use in TOPMODEL. Calculated local water table depths were then adjusted according to the local reference level, requiring unsaturated zone calculations for each $a/\tan\beta$ class to be now distributed over a range of water table depth values. Although often likely to be based

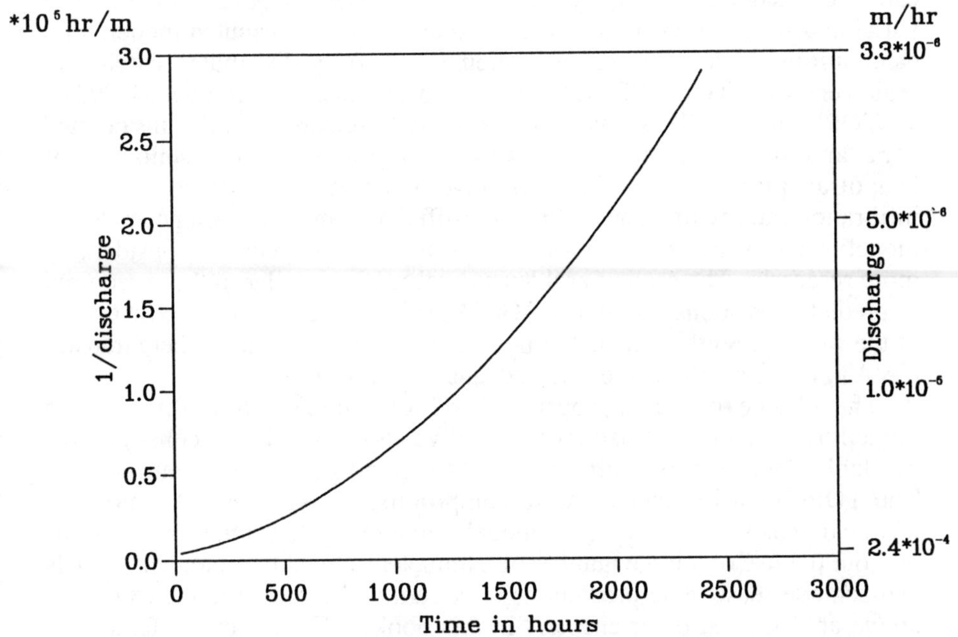

Fig. 18.11. Recession curve plot for the Ringelbach catchment, Vosges, showing the deviation from the behaviour of Eq. (18.17) (after Ambroise, 1988). Compare with Fig. 18.5.

largely on judgement, the specification of a reference level offers a means of applying TOPMODEL more realistically to catchments with a complex water table. There is no mechanism for predicting perching of water tables, however, due to a subsurface "infiltration excess" process. Incorporation of a layered structure is clearly not difficult within the distribution function framework of TOPMODEL but since the development of perched water tables will tend to be local and difficult to observe, such a process will inevitably be hard to quantify.

A further assumption, currently implicit in the use of the topographic index, is that the upslope contributing area a should be constant for any point. This may be a valid assumption in a moist climatic region, but for more arid catchments, there may be considerable evapotranspirative losses from distal areas of the catchment. These may be sufficient to prevent some parts of the catchment from contributing any water at all to the water table, especially during dry periods. The 'effective contributing area' of the catchment may thus be variable. There is evidence of discontinuities in hillslope water tables even in humid catchments (eg Burt and Butcher, 1986) and recent work with the wetting up sequences in Mediterranean catchments suggests that the linking together of local areas of saturation on the hillslopes may be very important in controlling the catchment scale response. TOPMODEL presently assumes that the effective area of the catchment is defined solely by the watershed. One way of allowing for such variable upslope contributing areas is to use a fully distributed model, with calculations made for every grid element. A fully distributed catchment scale version of TOPMODEL has been implemented by Famiglietti (1992).

TOPMODEL still uses very simplified representations of the unsaturated zone and it is difficult to see how to improve the representation of evapotranspiration, the effects of macroporosity and heterogeneity of infiltration rates without adding an unjustifiable number of parameters to the model (see Beven, 1989a, 1994). This is one area that requires considerably more research. The scaling problems involved are similar to those required in predicting responses at the sub-GCM grid scale, and it may be that some of the work currently in hand to improve soil-vegetation-atmosphere transfer (SVAT) models will lead to some advances in this respect.

The choice of an appropriate level of simplification for a given modelling problem is a matter of subjective choice within the constraints of available data, understanding of processes, and parameter identifiability. The TOPMODEL concepts are a compromise, making use of simplified physical reasoning to create a model capable of distributed predictions without the use of a large number of parameters. Most hydrological models can achieve reasonable predictions of stream discharges (see the examples above and in most other chapters in this book). The important feature of TOPMODEL is that the distributed nature of the predictions, the maps that are produced, allow a greater level of scrutiny of the results in relation to perceptions and measurements of the internal responses of the catchment under study. Such predictions may not be correct in spatial detail but major or unexpected differences in the nature of the spatial **pattern** of the response should stand out. This will often be the case; the TOPMODEL concepts will not be appropriate in many catchments, especially those with deeper groundwater systems. However, the simplicity of the model is such that it may be possible to reason as to how to modify the model structure to more closely reproduce the patterns of behaviour, such as in the introduction of the reference level concept by Quinn et al. (1991).

The reference level concept retains the analysis of subsurface drainage as a set of steady state water table configurations but allows for the fact that the local hydraulic gradient may be quite different from the surface slope angle. The index $a/\tan\beta$ is then calculated with a and $\tan\beta$ values derived from some assumed reference water table, for example that which might be calculated under Dupuit-Forchheimer assumptions (see Troch et al. 1993b, but note that they do not allow for the variability of transmissivity with depth or for slope convergence and divergence!). This idea has one further implication for the operation of the model, there may now be a variety of depths to the water table, with different unsaturated zone routing characteristics for each class of $a/\tan\beta$ values. This remains, however, quite simple to implement.

There is also current work at Lancaster aimed at generalising the model to a wider class of recession curve functions and reproduce the changes in effective upslope contributing area associated with the wetting up of a dry catchment. It is hoped that this will allow the application of TOPMODEL to a wider geographical range than is afforded by its original set of assumptions, while retaining the two major advantages of TOPMODEL: parametric parsimony and the ability to map the predictions back into space so that they can be compared with that complex perceptual model of the field hydrologist.

ACKNOWLEDGEMENTS

All work with the TOPMODEL concepts owes a continuing debt to Mike Kirkby who originated the idea of the topographic index. We are also indebted to many colleagues and collaborators over the years, those who have suggested ideas, provided data sets or implemented code, notably Dick Iredale, Nick Schoffield, Andy Tagg, George Hornberger, David Sappington, Roger Clapp, Eric Wood, Murugesu Sivapalan, Dominic Thongs, Jay Famigilietti, David Wolock, Charles Obled, Daniel Sempere-Torres, Jacques Wendling, Pierre Chevallier, Bruno Ambroise, Jean-Louis Perrin, Larry Band, Pep Piñol, Jeff McDonnell, Ross Woods, Philippe Merot, Patricia Bruneau, Sarka Blazkova, Tim Musgrove, Lotta Andersson, Karin Sundblad, Kevin Bishop, Alice Robson, Roger Moore and Jean-Pierre Jordan. The original development of TOPMODEL was started in 1974 at the University of Leeds with funding from the UK NERC. Recent work on the concepts at Lancaster University has been funded by the UK NERC, the NERC TIGER program and the European Commission ENCORE and DM2E projects.

REFERENCES

Ambroise, B, 1988, Interactions eaux souterraines - eaux de surface dans le bassin du Ringelbach à Soultzeren (Hautes Vosges, France): rôle hydrologique des surfaces saturées, in P Dahlblom and G Lindh, Proc. Int. Symp. on Interaction between Groundwater and Surface Water, Ystad, Sweden, 231-238.

Ambroise, B, 1990, Méthodes d'étude de la variabilité du cycle hydrique dans le petit bassin du Ringelbach, in H Lang and A Musy (eds.), Hydrology in Mountain Regions. i. Hydrological Measurements - the Water Cycle. IASH Pubn. No. 193, 327-334.

Band, L E, Patterson, P, Nemani, R, and Running, S W. 1993, Forest ecosystem processes at the watershed scale: incorporating hillslope hydrology, Agric. Forest Meteor., **63**, 93-126

Beven, K J. 1979 On the Generalized Kinematic Routing Method, Water Resour. Res. **15**; 1238-1242

Beven, K J. 1984, Infiltration into a class of vertically non-uniform soils, Hydrol. Sci. J., **29**, 425-434.

Beven, K J. 1986a Hillslope Runoff Processes and Flood Frequency Characteristics. In A. D. Abrahams (ed.) Hillslope Processes, 187-202, Allen and Unwin, Boston.

Beven, K J. 1986b, Runoff production and flood frequency in catchments of order n: an alternative approach, in V K Gupta, I Rodriguez-Iturbe and E F Wood (Eds.), Scale Problems in Hydrology, Reidel, Dordrecht, 107-131.

Beven, K J. 1987a, Towards a New Paradigm in Hydrology, Water for the Future: Hydrology in Perspective, IASH Publication No. **164**; 393-403.

Beven, K J. 1987b, Towards the use of catchment geomorphology in flood frequency predictions, Earth Surf. Process. Landf., **12**, 69-82.

Beven, K J. 1989a Changing Ideas in Hydrology: The Case of Physically-Based Models, J. Hydrol., **105**; 157-172.

Beven, K J. 1989b, Interflow, in H J Morel-Seytoux (Ed.) Unsaturated flow in hydrologic modelling, Proc. NATO ARW, Arles, Reidel, Dordrecht, 191-219

Beven, K J. 1991 Spatially Distributed Modelling: Conceptual Approach to Runoff Prediction. In D.S. Bowles and P.E. O'Connell (eds) Recent Advances in the Modelling of Hydrologic Systems, 373-387, Kluwer Academic.

Beven, K J. 1993 Prophesy, reality and uncertainty in distributed hydrological modelling, Advances in Water Resources, **16**, 41-51.

Beven, K. J. 1994, Process, heterogeneity and scale in modelling soil moisture fluxes, in S. Sorooshian and V K Gupta (Eds.) Global Environmental Change and Land Surface Process in Hydrology: The trials and Tribulations of Modelling and Measuring. Proc. NATO ARW, Tucson, Arizona, to be published by Springer-Verlag.

Beven, K J and Kirkby, M J. 1976, Towards a simple physically-based variable contributing area model of catchment hydrology, Working Paper 154, School of Geography, University of Leeds.

Beven, K J and Kirkby, M J. 1979 A physically based variable contributing area model of basin hydrology Hydrol. Sci. Bull., **24**(1),43-69.

Beven, K, Gilman, K, and Newson, M. 1979 Flow and Flow Routing in Upland Channel Networks, Bull. Hydrol. Sci. **24**, 303-325.

Beven, K J and Germann, P. 1982 Macropores and Water Flow in Soils, Water Resour. Res. **18**(5),1311-1325.

Beven, K J and Wood, E F. 1983 Catchment Geomorphology and the Dynamics of Runoff Contributing Areas, J.Hydrol. **65**, 139-158.

Beven, K J, Kirkby, M J, Schoffield, N, and Tagg, A. 1984 Testing a Physically-based Flood Forecasting Model (TOPMODEL) for Three UK Catchments, J. Hydrol. **69**; 119-143

Beven, K J, Wood, E F, and Sivapalan, M., 1988, On hydrological heterogeneity: catchment morphology and catchment response, J. Hydrology, **100**, 353-375.

Beven, K and Binley, A M. 1992 The Future of Distributed Models: Model Calibration and Uncertainty Prediction, Hydrological Processes, **6,** 279-298.

Beven, K J and Wood, E F, 1993, Flow routing and the hydrological response of chennel networks, in K J Beven and M J Kirkby (eds.), Channel Network Hydrology, Wiley, Chichester, Ch. 5.

Binley, A M and Beven, K J. 1991 Physically-based modelling of catchment hydrology: a likelihood approach to reducing predictive uncertainty. In D. G. Farmer and M .J. Rycroft (eds.) Computer Modelling in the Environmental Sciences, 75-88, The Institute of Mathematics and its Applications Conference Series,Clarendon Press.

Binley, A M, Beven, K J, Calver, A, and Watts, L G. 1991 Changing Responses in Hydrology: Assessing the Uncertainty in Physically Based Model Predictions, Water Resour. Res., **27**(6); 1253-1261.

Bruneau, P, Gascuel-Odoux, C, Robin, P, Merot, Ph, and Beven, K J, 1994, The sensitivity to space and time resolution of a hydrological model using digital elevation data, Hydrological Processes, in press.

Burt, T P and Butcher, D P, 1986 Development of topographic indices for use in semidistributed hillslope runoff models, in D Baltenau and O Slaymaker (Eds.), Geomorphology and Land Management, Zeits. Geomorph. Suppl. Band **58**, 1-19

Cappus, P, 1960, Bassin versant expérimental d'Alrance: études des lois de l'écoulement. Application au calcul et à la prévision des débits. La Houille Blanche, **A**, 493-520.

Chairat, S and Delleur, J W, 1993a, Effects of the topographic index distribution on the predicted runoff using GRASS, Proc. Amer. Water Resour. Assoc. Symp. on Geographic Information Systems in Water Resources, Mobile, Al. March, 1993

Chairat, S and Delleur, J W, 1993b, Integrating a physically based hydrological model with GRASS, in K Kovar and H P Nachtnebel (Eds). HydroGIS 93: Application of Geographical Information Systems in Hydrology and Water Resources, IAHS Pubn. No. **211**, 143-150

Clapp, R B, Timmins, S P and Huston, M A. 1992 Visualising the surface hydrodynamics of a forested watershed, in Proc. IX Internat. Conf. on Computational Methods in Water Resources, Denver, CO, June 1992.

Clark, C O. 1945 Storage and Unit Hydrograph, Trans. Am. Soc. Civ. Engrs., **110**;1416-1446

Dolman, A J, 1992, A note on areally averaged evaporation and the value of effective surface conductance, J. Hydrology, 138, 583-589.

Dunne, T. 1978 Field studies of hillslope flow processes. In M. J. Kirkby (Ed.), Hillslope Hydrology, 227-293, Wiley.

Dunne, T and Black, R D, 1970, An experimental investigation of runoff production in permeable soils, Water Resour. Res. **6**, 478-490.

Durand, P, Robson, A, and Neal, C. 1992 Modelling the hydrology of submediterranean montane catchments (Mont Lozère, France), using TOPMODEL: initial results, J. Hydrol., **139**, 1-14.

Famiglietti, J S. 1992 Aggregation and scaling of spatially-variable hydrological processes: local, catchment-scale and macroscale models of water and energy balance. Unpublished PhD thesis. Princeton University, Princeton, NJ.

Famiglietti, J S and Wood, E F. 1991 Evapotranspiration and runoff from large land areas: land surface hydrology for atmospheric general circulation models. Surv. Geophys., **12**, 179-204.

Famiglietti, J S, Wood, E F, Sivapalan, M and Thongs, D J. 1992 A catchment scale water balance model for FIFE, J. Geophys. Res., **97**(D17), 18997-19007.

Grayson, R B, Moore, I D and McMahon, TA, 1992, Physically-based hydrologic modelling. 2. Is the concept realistic?, Water Resour. Res., 28,2659.

Hewlett, J D. 1974 Comments on Letters Relating to 'Role of Subsurface Flow in Generating Surface Runoff. 2. Upstream Source Areas' by R. Allen Freeze, Water Resour. Res. **10**; 605-607.

Hornberger, G M, Beven, K J, Cosby, B J and Sappington, D E. 1985 Shenandoah Watershed Study: Calibration of a Topography-Based, Variable Contributing Area Hydrological Model to a Small Forested Catchment, Water Resour. Res. **21**; 1841-1850.

Huff, D D, O'Neill, R V, Emanuel, W R, Elwood, J W and Newbold, J D. 1982 Flow Variability and Hillslope Hydrology, Earth Surf. Process Landf., **7**; 91-94.

Jones, A J J, 1986, Some limitations of the a/s index for predicting basin wide patterns of soil water drainage, Z. Geomorph., NF, Suppl. Band **60**, 7-20.

Jordan, J-P, 1992, Identification et modélisation des processus de génération des crues: application au bassin versant de la Haute-Mentue. Thèse de Doctorat XXX. Département de Génie Rural, Ecole polytechnique fédérale de Lausanne, Switzerland.

Kirkby, M J. 1975 Hydrograph Modelling Strategies. In R. Peel, M. Chisholm and P. Haggett (Eds) Process in Physical and Human Geography. Heinemann, 69-90

Kirkby, M J, 1976, Tests of the random network model and its application to basin hydrology, Earth Surface Processes, 1, 197-212.

Kirkby, M J, 1978, Implications for Sediment Transport, in M J Kirkby (ed.) Hillslope Hydrology, Wiley, Chichester, Ch. 9

Ladson, A R and Moore, I D, 1993, Soil water prediction by microwave remote sensing and topographic attributes, J. Hydrology, 144

Leavesley, G H, Restrepo, P, Stannard, L G and Dixon, M. 1992 The Modular Hydrologic Modeling System - MHMS, in Managing Water Resources during Global Change, American Water Resources Association, 263-264

Merot, Ph. 1988, Les zones de sources à surface variable et la question de leur localisation, Hydrol. Continent., **3**, 105-115.

Merot, Ph and Bruneau, P. 1993 Sensitivity of bocage landscapes to surface run-off: application of the Kirkby index, Hydrol. Process., **7**, 167-176.

Moore, I D, Mackay, S M, Wallbrink, P J, Burch, G J and O'Loughlin, E M. 1986 Hydrologic characteristics and modelling of a small forested catchment in southeastern New South Wales, J. Hydrol., **83**, 307-335.

Moore, I D, Burch, G J and Mackenzie, D H. 1988 Topographic effects on the distribution of surface soil water and the location of ephemeral gullies, Trans. Amer. Sco. Agric. Engrs., **31**, 1098-1107.

Moore, I D, Gallant, J C, Guerra, L and Kalma, J D. 1993 Modelling the spatial variability of hydrological processes using Geographical Information Systems, in K Kovar and H P Nachtnebel (Eds). HydroGIS 93: Application of Geographical Information Systems in Hydrology and Water Resources, IAHS Pubn. No. **211**, 161-169.

Obled, Ch, Wendling, J and Beven, K J, 1994, The sensitivity of hydrological models to spatial rainfall patterns: an evaluation using observed data, J. Hydrology, in press.

O'Loughlin, E M. 1981 Saturation regions in cachments and their relation to soil and topographic properties, J. Hydrol., **53**, 229-246.

O'Loughlin, E M. 1986 Prediction of Surface Saturation Zones in Natural Catchments by Topographic Analysis, Water Resour. Res., **22**; 794-804.

Quinn, P F. 1991 The Role of Digital Terrain Analysis in Hydrological Modelling, unpubl. Ph.D., Lancaster University, U.K.

Quinn, P F, Beven, K, Chevallier, P and Planchon, O. 1991 The Prediction of Hillslope Flow Paths for Distributed Hydrological Modelling Using Digital Terrain Models, Hydrological Processes, **5**;59-79.

Quinn, P F and K J Beven, 1993, Spatial and temporal predictions of soil moisture dynamics, runoff, variable source areas and evapotranspiration for Plynlimon, mid- Wales, Hydrol. Process., **7**, 425-448.

Quinn, P F, Beven, K J and Culf, A. 1994, The introduction of macroscale hydrological complexity into land surface-atmosphere transfer function models and the effect on planetary boundary layer development, J. Hydrology, in press

Quinn, P F, Beven, K J and Lamb, R, 1994, The ln(a/tanß) index: how to calculate it and how to use it in the TOPMODEL framework. submitted to Hydrol. Process.

Robson, A, Beven, K J and Neal, C. 1992 Towards identifying sources of subsurface flow: a comparison of components identified by a physically based runoff model and those determined by chemical mixing techniques, Hydro. Process., **6**, 199-214.

Robson, A J, Whitehead, P G and Johnson, R C. 1993. An application of a physically based semi-distributed model to the Balquhidder catchments, J. Hydrol., **145**, 357-370.

Romanowicz, R, Beven, K J and Moore, R V. 1993a, Topmodel as an application module within WIS, in K Kovar and H P Nachtnebel (Eds.) HydroGIS 93: Application of Geographical Information Systems in Hydrology and Water Resources, IAHS Pubn. No. **211**,

Romanowicz, R, Beven, K J and Moore, R V. 1993b, GIS and distributed hydrological models, in P M Mather (Ed.), Geographical Information Handling - Research and Applications, Wiley, Chichester, 197-205.

Running, S W and Coughlan, J C, 1988, A general model of forest ecosystem proceses for regional applications. 1. Hydrologic balance, canopy gas exchange and primary production processes, Ecological Modelling, **42,** 125-154.

Sempere-Torres, D, 1990, Calcul de la lame ruisselée dans la modélisation pluidébit: limitations des approches globales et introduction simplifiée de la topographie et de la variabilité spatiale des pluies. Thèse de Doctorat., Institut de Méchanique de Grenoble, France.

Sivapalan, M, Beven, K J and Wood, E F. 1987 On Hydrological Similarity: 2 A Scaled Model of Storm Runoff Production, Water Resour. Res. **23**(12); 2266-2278.

Sivapalan, M, Wood, E F and Beven, K J. 1990 On Hydrological Similarity: 3. A dimensionless flood frequency distribution, Water Resour. Res., **26**, 43-58.

Sooroshian, S, Gupta, V K and Fulton, J L, 1983, Evaluation of maximum likelihood parameter estimation techniques for conceptual rainfall-runoff models: influence of calibration data variability and length on model credibility, Water Resour. Res., 19, 251-259.

Stuart, N and Stocks, C. 1993, Hydrological modelling within GIS: an integrated approach, in HydroGIS 93: Application of Geographic Information Systems in Hydrology and Water Resources, IAHS Publ. no. **211**, 319-329.

Surkan, A J, 1969, Synthetic hydrographs: effects of network geometry, Water Resour. Res., 5, 112-128.

Troch, P A, Mancini, M, Paniconi, C and Wood, E F, 1993a, Evaluation of a distributed catchment scale water balance model, Water Resour. Res., 29, 1805-1817.

Troch, P A, De Troch, F P and Brutsaert, W, 1993b, Effective water table depth to describe initial conditions prior to storm rainfall in humid regions, Water Resour. Res., 29, 427-434.

van Straten, G and Keesman, K, 1990, Uncertainty propogation and speculation in projective forecasts of environmental change: a lake eutrophication example, J. Forecasting, 10, 163-190.

Waters, D, Jenkins, A, Staples, T and Donald, A P. 1991 The importance of hydrological source areas in terrestrial liming, J. Instn. Water Engrs. Manag., **5**, 336-341.

Wendling, J, 1992, Modélisation pluie-débit: comparaison d'approches conceptuelles/physico-déterministes, globales/semi-distribuées. Essai de prise en compte de la variabilité spatiale des pluies. (Application au bassin versant du Réal Collobrier). Thèse de Doctorat, Institut national polytechnique de Grenoble, France.

Wolock, D M, 1993, Simulating the variable-source-area concept of streamflow generation with the watershed model TOPMODEL, U.S. Geological Survey, Water-Resources Investigations Report 93-4124, Lawrence, Kansas, 33 pp.

Wolock, D M, Hornberger, G M, Beven, K J and Campbell, W G. 1989 The relationship of catchment topography and soil hydraulic characteristics to lake alkalinity in the Northeastern United States, Water Resour. Res., **25**, 829-837.

Wolock, D M, Hornberger, G M and T Musgrove, T. 1990, Topographic controls on episodic streamwater acidification in Wales, J. Hydrol., **115**, 243-259.

Wolock, D M and Hornberger, G M. 1991 Hydrological effects of changes in levels of atmospheric carbon dioxide, J. Forecasting, **10**, 105-116.

Wood, E.F., Sivapalan, M., Beven, K J, and Band, L. 1988 Effects of Spatial Variability and Scale with implications to Hydrological Modelling, J. Hydrol. **102**, 29-47.

Wood, E F, Sivapalan, M and Beven, K J. 1990 Similarity and scale in catchment storm response, Rev. Geophys. **28**, 1-18.

Chapter 19

DISTRIBUTED PARAMETER HYDROLOGIC MODELLING USING VECTOR ELEVATION DATA: THALES AND TAPES-C

R. B. Grayson, G. Blöschl, and I. D. Moore

This chapter presents THALES, a distributed parameter rainfall-runoff model based on the terrain analysis software TAPES-C. THALES is designed as an investigative tool for testing hypotheses, analyzing field data and exploring scale issues. TAPES-C is a set of computer programs that partition a catchment into streamtubes. The main advantage of this approach lies in dynamic modelling as it allows water movement in complex terrain to be represented by a series of one-dimensional equations. This enables much simpler and faster solutions than two-dimensional representations.

THALES can simulate runoff generation from infiltration excess and saturation excess mechanisms. Infiltration is modelled either by the Green and Ampt, the Smith and Parlange or by Philip's equation. Overland flow is based on the kinematic assumption for either sheet, rill, or channel flow and runon infiltration can be handled. In the unsaturated zone, water movement is assumed to occur vertically only and to drain into a saturated layer. Lateral water movement in the saturated layer (subsurface stormflow) is approximated by the kinematic assumption. Exfiltration of subsurface flow is possible.

The model is evaluated for a $2m^2$ laboratory microcatchment, based on outflow hydrographs and saturation patterns. Subsequently, the model is used to analyze hypothetical scenarios in a $17km^2$ catchment. These include moving storms, different assumptions on initial conditions, and investigations into the existence of a Representative Elementary Area.

19.1. INTRODUCTION

Many hydrologic and water quality models crudely represent the three-dimensional nature of natural landscapes and therefore crudely represent spatially distributed hydrologic processes. As transport modelling becomes increasingly important in hydrologic and environmental assessment, this becomes a limiting factor in the predictive power of these models. The effect of topographic convergence and divergence on flow characteristics in natural landscapes has a major impact on hydrologic and hydraulic variables. Without the accurate prediction of terrain characteristics, accurate prediction of spatial hydrological response is not possible.

Digital elevation models (DEMs) provide the basic data for characterizing the topographic attributes of landscapes. Vector DEMs are based on the most common form of elevation data storage, the topographic map. Contour lines are represented in the vector DEM by a series of x, y coordinates (cartesian coordinate system in the horizontal plane) for each isoline of elevation (Moore et al., 1988). Topographic maps are prepared directly from aerial photographs or field surveys so the information has undergone the minimum of manipulation, therefore, minimizing errors.

Vector-DEMs are capable of providing a model structure based on the physics of water movement on the land surface (Moore and Grayson, 1989; Moore et al., 1991). The first vector or contour-based method of partitioning catchments was proposed by Onstad and Brakensiek (1968) who termed it a "stream-path" or "stream-tube" analogy. Speight (1974, 1980) subsequently used the concept to classify landforms, especially specific catchment areas (upslope contributing area per unit contour width) and specific dispersal areas (downslope dispersal area per unit contour width). Once the partitioning of a catchment is complete using this "stream-tube" technique, the data storage requirements are no more onerous than for any other form of DEM because only the element attributes must be stored.

The major disadvantages of such methods are (i) that digitizing is relatively time consuming and the data storage requirements are much greater than for grid-based methods or Triangulated Irregular Networks (e.g. Palacios and Cuevas, 1989); (ii) the information density of vector-DEMs is greater in areas with steep slopes than in flatter areas, where the topography is less well defined; (iii) the use of vector data in the partitioning of terrain into model elements is not straight forward and (iv) the "stream-tube" defined by surface topography may not be appropriate for the representation of subsurface flow, particularly in flat areas.

Despite these problems, there are a number of applications where the vector-based element networks are particularly useful. The computation of static terrain indicies is simple, as is the case with grid data, however the main advantage is in dynamic modelling. With this form of partitioning a catchment, one-dimensional flow can be assumed within each element, allowing water movement in a complex three-dimensional catchment to be represented by a series of coupled one-dimensional equations. The equations can be solved using a one-dimensional finite-difference or finite-element scheme. This is much simpler and faster than most other forms of topographic representation which generally require two-dimensional forms of flow equations to adequately represent topographic effects.

In this chapter, we briefly discuss some general considerations related to model development and model use, then describe the terrain analysis programs which form the basis for the dynamic hydrologic model THALES. The final section provides an illustration of some different applications of THALES ranging from modelling on a small laboratory catchment to investigations into scale effects and the Representative Elementary Area concept.

19.2. MODELLING CONSIDERATIONS

Despite the obvious advantages of distributed parameter hydrologic models for process analysis, many authors have argued that these models must be used with a great deal of caution, if at all for the purpose of prediction (James and Burges, 1982; Klemes, 1982; Beven, 1983; Anderson and Rogers, 1987; Beven, 1989). One of the problems is related to the differences in scale between that at which the algorithms were derived and that at which the models are applied. For example, saturated and unsaturated matrix flow theory was essentially developed from 1 dm^3 laboratory samples, while typical model elements are several orders of magnitude larger. Authors such as Philip (1972, 1975), Klemes (1983), Hillel (1986), Anderson and Rogers (1987), Beven (1989), and Grayson et al. (1992a, b) have indicated that the algorithms used to represent small scale hydrologic process may not be valid at larger scales and when they are, their parameterisation is uncertain and some algorithms are poor representations of reality. For example subsurface flow through macropores is known to occur but is extremely difficult to characterise.

In addition to these problems, the distributed nature of the models complicates proper testing and validation because the detail of information provided by the model is often much greater than that measured in the catchment (James and Burges, 1982; Dunne, 1982; Beven, 1983; Klemes, 1983; Burges, 1986; Anderson and Rogers, 1987). Also a large number of parameter sets can result in similar model output (such as outflow hydrograph) Grayson et al. (1992a, b) Philip (1975) i.e. the model parameters are not unique (Blöschl et al., 1993a).

These problems point to the importance of evaluating **distributed** model behaviour rather than an integrated value such as runoff, when assessing the performance of distributed parameter models (e.g., Blöschl et al., 1991; Moore and Grayson, 1991). Ideally, model development would be in conjunction with, and as an integral part of, carefully planned and executed field studies established for the purpose of advancing our knowledge of the natural system and our capacity to model that system. The measurements should include spatial patterns of hydrologic quantities. Generally this is not the case so model testing and validation is frequently incomplete.

These points illustrate some fundamental constraints on the degree to which models are truly "physically-based" and indicates that in many applications, the models are likely to be used as complex conceptual formulations, i.e. the algorithms become functional relationships (Bergström, 1991). Nevertheless, models like THALES are very useful for certain applications. They are an indispensable tool in the development and testing of theories and hypotheses, the analysis of data and in assisting to determine what data should be collected. However, they are not a substitution for lack of data or lack of understanding of the natural system. They are not an end in themselves, nor should they be static "*In science, and particularly in modelling, no 'last word' (even if it is, for a fleeting moment, the 'latest word') should ever be allowed to become the 'final word'.*" (Hillel, 1986).

It is with this philosophy that THALES was developed and the examples given in the final section of this chapter reflect the type of investigation for which we believe models like THALES are appropriate.

19.3. TERRAIN ANALYSIS PROGRAMS

In using Onstad and Brakensiek's (1968) "stream-tube" concept, contour lines are assumed to be equipotential lines, and pairs of orthogonals to the equipotential lines (streamlines) form the "stream-tubes" (Fig. 19.1). Cayley (1859) called these streamlines "slope lines (*lignes de la plus grande pente*)" and "orthogonal trajectories", and described the general relationship between contour lines and slope lines at summits, immits (depressions), ridges and saddle-points in the landscape. Adjacent contour lines and streamlines define irregularly shaped elements (Figs. 19.1 to 19.3). Surface runoff enters an element orthogonal to the upslope contour line and exits orthogonal to the downslope contour line, with the adjacent streamlines being no-flow boundaries. Flow from one element can then be successively routed to downslope elements formed by the same stream-tubes.

TAPES-C, Topographic Analysis Programs for the Environmental Sciences - Contour, is a set of computer programs that automatically partition a catchment into interconnected, irregularly shaped elements using this "stream-tube" approach. The method is a development of the contour-based terrain analysis methods described by Moore et al. (1988) and O'Loughlin (1986).

Four programs are used in this analaysis:

1) PREPROC

2) PLOTCON

3) EDITCON

4) TAPES-C

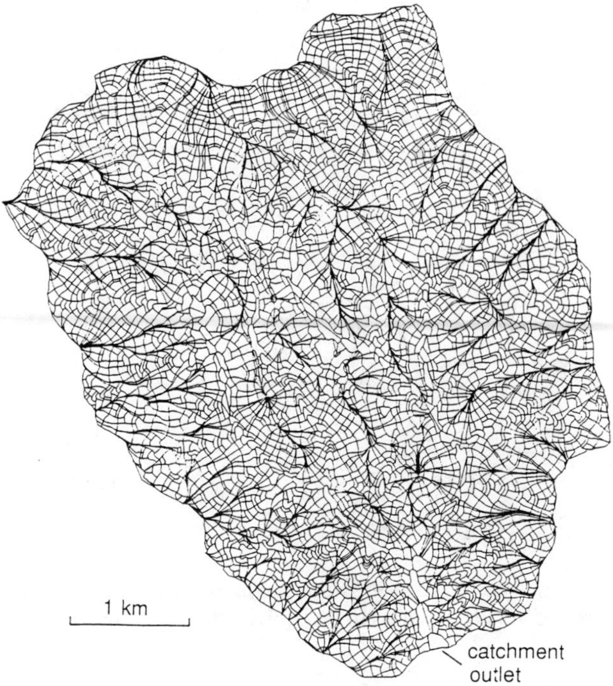

Fig. 19.1. Element network for the Coweeta topography, N.C. The elements are defined by adjacent flowpaths and contours.

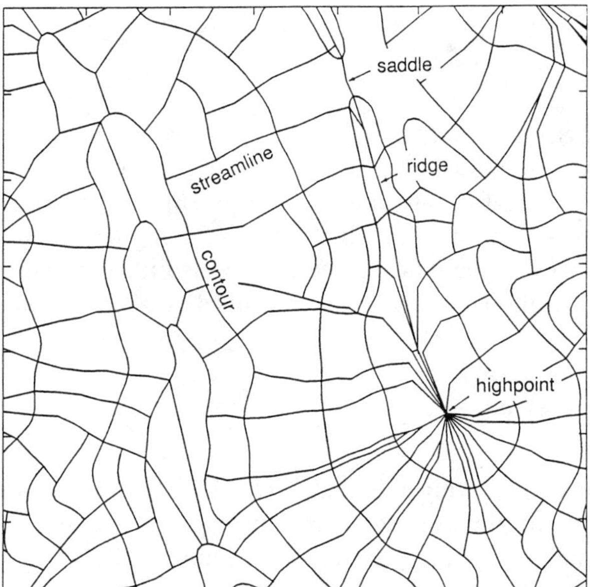

Fig. 19.2. Detail of the element network of Fig. 19.1. The area shown is 700m by 700m.

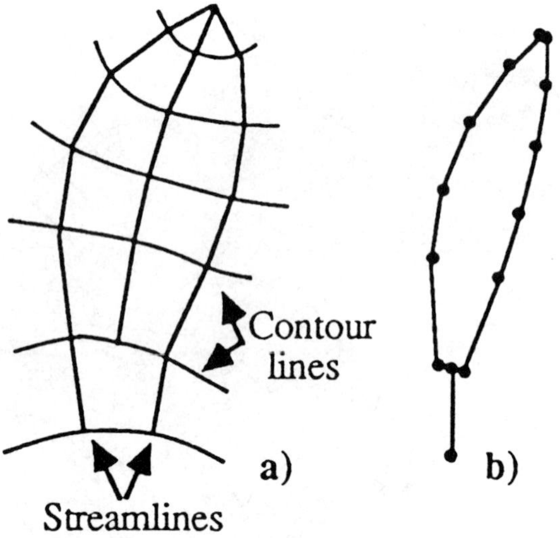

Fig. 19.3. **(a) A hypothetical hillslope segment subdivided into elements using TAPES-C and (b) its equivalent one-dimensional network.**

PREPROC

Preproc includes a basic digitizing program and involves smoothing and resampling of the contour vertices (Moore et al., 1988; Moore and Grayson, 1991).

PLOTCON

Plotcon processes contour segments (such as those produced by ARC-INFO or ANUDEM (Hutchinson, 1992) and includes the following steps: (a) Joining contour lines to continuous strings; (b) clipping contour lines outside a specified catchment area; (c) flipping (=reversing the orientation) of the contour lines so that the downslope direction is on a consistent side of the contour, and (d) smoothing of the contour lines by a moving average filter if required.

EDITCON

Editcon allows for editing the vector DEM. This includes moving and removing contour lines as well as inputing and editing saddlepoints, highpoints, and ridgepoints. *EDITCON* features fully graphic-interactive program control which greatly enhances the speed of data preparation as compared to more traditional methods.

TAPES-C

Tapes-C subdivides a catchment into elements using the "stream-tube" approach and calculates a variety of topographic attributes for each element. It uses Moore's et al. (1988) approximate method of estimating flow trajectories or 'streamlines'. The streamlines between contour lines are approximated by short straight-line segments, which simplifies the computer program and minimizes the computational time required to subdivide a catchment. Streamlines are computed using two criteria: (1) minimum distance between adjacent contour lines (O'Loughlin, 1986); and (2) orthogonals to the downslope contour line. The first criterion is used in ridge areas and the second criterion is used in valley areas. Application of the two criteria is determined by the curvature of the contour lines (i.e., plan curvature in radians per metre). The contour curvature is computed by numerical differentiation of nodes (of x,y-coordinates) defining the contour line.

TAPES-C performs the partitioning of the catchment beginning at the contour line of lowest elevation and ending at the highest contour line, successively determining the elements for each adjacent pair of contour lines. The application of TAPES-C to the Coweeta catchment in North Carolina (Swank and Crossley, 1988) is illustrated in Fig. 19.1.

The contour interval in TAPES-C can be varied and generally a denser sampling of contour lines is used in regions of flat topography. Current topographic maps generally do not define certain types of topographic singularities such as pass or saddle points (Warntz, 1975). Furthermore, because of the large contour interval, typically 20 ft for the 1:24,000 scale United States Geological Survey map series and 10 m or 20 m for the 1:25,000 scale Australian map series, ridges and interfluve areas, particularly at or near the catchment boundary, are often poorly defined. In the regions where such topographic features appear, streamlines are difficult to estimate with the limited information appearing on the topographic maps. TAPES-C estimates the streamline in such cases by projecting an orthogonal from the downslope contour line upslope until it intersects the catchment boundary. The downslope contour line, the two adjacent streamlines and the catchment boundary then define an element.

One of the limitations with contour based catchment partitioning is that the elements tend to be larger (and so information least dense) in valleys and the reverse is true on ridges. Therefore, TAPES-C subdivides elements in the channel areas until the upslope arclength of each element is below a certain treshold. Also, a postprocessing step was introduced which combines upslope/downslope neighbouring elements if the combined flow lengths (flow distances) do not exceed a given value. This splitting and joining produces much more uniform element sizes (Fig. 19.2).

For each element, TAPES-C computes the following topographic attributes: (1) element area; (2) total upslope contributing area; (3) connectivity of upslope and downslope elements; (4) x,y,z-coordinates of the element centroid; (5) x,y,z-coordinates of the midpoint on the downslope contour bounding the element; (6) the average slope of the element orthogonal to the contour; (7) the widths of the element on the upslope and downslope contour lines bounding the element; (8) the flow distance across the element; and (9) aspect or azimuth of the element. The computation of these attributes has been previously decribed by Moore et al. (1988, 1991)

Stream channel networks are defined by TAPES-C using a method based on the upslope contributing area criteria proposed by Band (1986). A stream channel passes through an element if the total upslope contributing area of the element is greater than a critical value. The stream channel is assumed to intersect the upslope and downslope contour lines bounding an element at the points of maximum curvature of the contour lines.

19.4. THALES

THALES is a dynamic hydrological model based on the element network created by TAPES-C. A major consideration in the development of THALES was the ability to assess the model simulations both quantitatively against measured data and qualitatively against field observation in order to ensure that the model is working well (or poorly) for the right reasons. THALES was developed to explore general issues relating to model use and to identify, through example, some barriers to the improvement in modelling of small catchment hydrology with distributed-parameter, process-based models.

The philosophy behind development of the model was to minimise the number of parameters while maintaining the basic components of a "physically-based" model. The rationale was that any additional algorithms simply provided "more knobs to twiddle" and would offer no additional information about the true catchment hydrologic response. Our intention was not to develop yet another model but rather to make use of the structure that TAPES-C provides to investigate some issues related to the modelling of hydrologic response, and particularly those related to topography. By keeping the model as simple as possible for a model of this genre, the "*subtle interdependence between the 'sensitive' and 'insensitive' parameters*"

(Hornberger et al., 1985) is minimised and the effect of the underlying assumptions is easier to identify.

19.5. MODEL DESCRIPTION

Separate surface and subsurface models based on the TAPES-C analysis were presented in Moore and Grayson (1991). THALES is a combination of those two models and is named after Thales of Miletos, a Greek philosopher who recognised the influence of topography on water movement. THALES can simulate kinematic overland flow from saturation excess and infiltration excess mechanisms, as well as subsurface stormflow. A similar development to THALES, also based on the original work of Moore et al. (1988) and O'Loughlin (1986), is known as TOPOG (Vertessy et al.; 1993).

THALES is a relatively simple model. The elements and attributes calculated in TAPES-C define the area within which model parameters are assumed to be constant and provide the element network for the routing of water. The model is based on the assumption that infiltrated water flows downslope in a saturated layer overlying an impermeable base and is modelled as kinematic subsurface flow. If the subsurface flow rate exceeds the capacity of the soil profile to transmit the water, surface saturation occurs and rain falling on the saturated areas becomes direct runoff (saturation excess flow). Runoff can also be generated by infiltration excess overland flow. The infiltrated water is assumed to flow vertically through an unsaturated zone to become part of the saturated layer from where it can flow downslope (Fig. 19.4).

The infiltration rate and volume as well as the rainfall excess are computed for each element using either a relationship based on an equivalence between the Green and Ampt and the Horton infiltration capacity equations (Morel-Seytoux, 1988a, b) or the relationship of Smith and Parlange (1978). It is a simple matter to include other infiltration algorithms such as those of Philip (1957) or the "3-parameter" model by Parlange et al. (1982). Variable rainfall intensity infiltration is simulated, but the rainfall intensity in any time increment Δt is assumed constant.

The rainfall intensity and infiltration behaviour is assumed to be uniform on each element, but varying between elements. Based on an equivalence between the Green and Ampt and the Horton infiltration capacity equations (Morel-Seytoux, 1988a, b), the time from the onset of rainfall to surface ponding, t_p, is given by:

$$t_p = \frac{\kappa(\theta_s - \theta_i) H_c}{i(i - \kappa)} \tag{19.1}$$

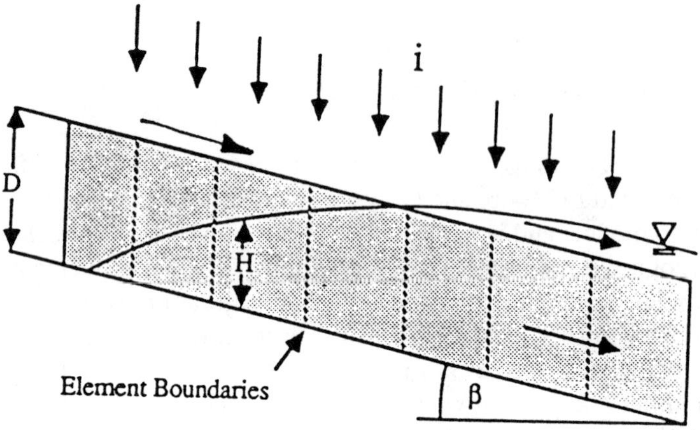

Fig. 19.4. Schematic representation of the flow system modelled by THALES.

where κ is the resaturated hydraulic conductivity, θ_s is the field saturated water content, θ_i is the inital soil water content, H_c is the effective capillary drive, and i is the rainfall intensity (Mein and Larson, 1973; Morel-Seytoux, 1988a). After surface ponding occurs, i.e., for $t > t_p$, the infiltration capacity is calculated by

$$f = \kappa + \frac{1}{k\Delta t}\left(i_p - \kappa\right)\left(e^{k\Delta t} - 1\right)e^{-k\left(t - t_p\right)} \tag{19.2}$$

where f is the average infiltration rate between times t-Δt and t, i_p is the rainfall intensity that induced ponding, and k is a constant. Equation (19.2) is Horton's equation, which Morel-Seytoux (1988b) demonstrated is equivalent to the Green and Ampt equation (Green and Ampt, 1911) if the parameter k is defined in terms of the Green and Ampt parameters κ, θ_s, θ_i, and H_c by

$$k = 8\kappa / \left\{\left(\theta_s - \theta_i\right) H_c\right\} \tag{19.3}$$

In practice, k is generally a fitted parameter. Equation (19.2) is used in the model because it gives an explicit solution for the average infiltration rate during a time interval, whereas the Green and Ampt equation must be solved implicitly. Equations (19.1) and (19.2) assume that the depth of ponding on the soil surface is negligible. Schmid (1989) showed that this assumption produces a maximum error of 11% (but generally less than 5%) in the computed rainfall excess rates and that "the usual neglect of water depth by overland flow modellers appears justified." This infiltration algorithm is computationally efficient and has five parameters κ, θ_s, θ_r, H_c, and μ, and one initial condition θ_i.

The Smith and Parlange (1978) model is applied in a similar way and can be expressed in the form:

$$f = K_s \frac{\exp\left(\frac{F}{B}\right)}{\exp\left(\frac{F}{B}\right) - 1} \tag{19.4}$$

where f is the infiltration rate, K_s is the saturated hydraulic conductivity, F is the volume of water infiltrated and $B = H_c (\theta_s - \theta_i)$ where H_c is the capillary drive, θ_s is the saturated water content and θ_i is the initial water content.

Application of this equation requires an implicit solution if the time interval used in the model is large. For short time intervals (10 to 30 s) a rapid algorithm for the calculation of infiltration rates is provided by an explicit solution based on the infiltrated volume at the previous time step. During periods of no rain, the soil profile drains and for the next rainfall period, a new value of θ_i is used to determine the value of B in Eq. (19.4).

Using either infilration algorithm, during drainage, the soil water profile is assumed to be rectangular and the hydraulic conductivity is related to soil water content by the Brooks and Corey (1964) relationship that has the saturated hydraulic conductivity (K_s), the saturated soil water content (θ_s), and the residual soil water content (θ_r) as parameters.

$$K(\theta) = K_s \left(\frac{\theta - \theta_r}{\theta_s - \theta_r} \right)^{2b+3} \tag{19.5}$$

The water that enters the soil profile is added to the unsaturated zone which then discharges into the underlying saturated zone; i.e. only vertical flow is modelled in the unsaturated zone. The rate at which water enters the saturated zone is given by the hydraulic conductivity computed from the Brooks and Corey (1964) relationship, i.e., it is assumed that the hydraulic gradient is unity. The soil water content is determined from a volume balance of the unsaturated zone. The total unsaturated soil volume for an element changes over time as the saturated depth varies and a water balance is used to redistribute the unsaturated soil water content as a result of a change in total unsaturated soil volume. Lateral flow enters and exits the element via the saturated zone. If the saturated zone reaches the surface, saturated source area runoff occurs and exfiltration of subsurface flow is possible.

Surface runoff may continue after rainfall ceases and this surface runoff can infiltrate in a downslope element if the soil profile of the downslope element is not saturated, i.e. where Hortonian overland flow is occurring. This runoff-runon problem is likely to be important where broad sheet flow occurs, but relatively unimportant when the flow concentrates in defined rills, channels and streams.

19.5.1. Surface Runoff

Water is routed through the channel and rill system and overland flow elements by solging the unsteady continuity equation and the kinematic approximation of the momentum equation which can be expressed, respectively, as:

$$\frac{\delta A}{\delta t} + \frac{\delta Q(A)}{\delta s} = q(t) \tag{19.6a}$$

and

$$Q(A) = \alpha R^{m1} A^{m2} \tag{19.6b}$$

where A is the flow cross-sectional area, Q(A) is the discharge, q(t) is the lateral inflow (or outflow) rate per unit length, s is the distance in the down-slope direction, R is the hydraulic radius, and α, m1 and m2 are constants that depend on the form of the resistance equation used and are functions of the geometry, slope and roughness of the overland flow element, rill or channel, and the fluid viscosity and the type of flow (laminar or turbulent). The value of these constants varies in each flow element or segment. THALES assumes that the flow is fully turbulent and uses Manning's equation for the resistance equation. It is a simple matter to modify the program to incorporate either the Chezy equation (turbulent flow) or the Darcy-Weisbach equation, which can be written to model laminar, transitional or fully turbulent flow (Moore and Foster, 1990).

Equation (19.6) is formulated as a four-point implicit numerical scheme that is solved using the Newton-Raphson and Regula-falsa methods (Moore and Foster, 1990). The Regula-falsa method is only used when the Newton-Raphson method fails to converge. It will always find a solution providing the initial conditions are correctly chosen, but the rate of convergence is slow. The four-point implicit scheme can be expressed as:

$$\phi\frac{\Delta t}{\Delta s}Q_{s+\Delta s,t+\Delta t}+\frac{1}{2}A_{s+\Delta s,t+\Delta t}=\frac{1}{2}\left[A_{s,t}+A_{s+\Delta s,t}-A_{s,t+\Delta t}\right]$$
$$+(1-\phi)\frac{\Delta t}{\Delta s}\left[Q_{s,t}-Q_{s+\Delta s,t}\right]+\phi\frac{\Delta t}{\Delta s}Q_{s,t+\Delta t}+\Delta t\, q(t) \tag{19.7}$$

where ϕ is a weighting coefficient with values ranging from 0.5 to 1.0, but more typically from 0.7 to 0.8 for numerically stable solutions.

Overland Flow

For shallow sheet flow, R = h = A/w, where h is the depth of flow and w is the width of the element on the downstream face, so that the coefficients and exponents of (6b) can be written to give Manning's equation as:

$$\alpha=\frac{\tan^{1/2}\beta}{nw^{2/3}},\ m1=0,\ \ and\ \ m2=5/3 \tag{19.8}$$

where β is the slope angle of the element. The lateral inflow per unit width, q, is:

$$q=(r_e-f)\frac{A_e}{\Delta x_e} \tag{19.9}$$

where r_e - f is the rainfall excess, A_e is the plan area of the element and Δx_e is the flow distance across the element. For infiltration, f is positive and for exfiltration it is negative.

Channel Flow

Channels are assumed to be trapezoidal in shape with the hydraulic radius, R, given by:

$$R=\frac{A}{w_c+2\left(1=z^2\right)\left\{-\frac{w_c}{2z}+\left[\left(\frac{w_c}{2z}\right)^2+\frac{4A}{z}\right]\right\}} \tag{19.10}$$

where w_c and z are the bottom width and sideslope ratio of the channel, respectively. The coefficients and exponents in (6b) are $\alpha=\tan^{1/2}\beta/n$, m1 =

2/3 and m2 = 1. Alternatively, if the shape and characteristics of the channels are unknown or highly variable, the hydraulic radius can be approximated by:

$$R = \xi A^{1/2} \tag{19.11}$$

where ζ is a constant that depends on the shape of the channel. For the most hydraulically efficient cross-sections, ζ is 0.354 for for rectangular channels, 0.352 for trapezoidal channels and 0.364 for parabolic channels (Moore and Burch, 1986; Moore and Foster, 1990). In such cases, $\alpha = \zeta^{2/3} \tan^{1/2} \beta/n$, m1 = 0 and m2 = 4/3. The lateral inflow to the channel, $q = Q_o/\Delta x_e$, where Q_o is the overland flow discharge for the element computed from (6) to (7). Infiltration or exfiltration from or to the channel is not specifically modelled, but is included in the overland flow calculations.

Rill Flow

Rill flow is modelled in the same manner as channel flow, with the hydraulic radius given by (11) and ζ equal to 0.354, 0.352 or 0.364 for rectangular, trapezoidal and parabolic cross-sections, respectively. Rill flow only occurs if the overland flow discharge per unit width exceeds a critical value, i.e. if $Q_o/w_o \geq q_{cr}$. For rill flow, the rill density must be specified,and for agricultural land typically varies from about 1.0 to 2.0 rills per metre (Moore and Foster, 1990).

19.5.2. Subsurface Flow

Lateral, saturated, subsurface flow is approximated by the kinematic assumption (Beven, 1982; Sloan and Moore, 1984). Equation (19.6a) remains the continuity equation, but the momentum equation becomes:

$$Q(A) = K_{sL} \, A Sin\beta \tag{19.12}$$

where K_{sL} is the saturated hydraulic conductivity in the lateral flow direction, A is the cross sectional area of flow (element width X depth of saturated soil) and β is the slope of the impermeable layer. Lateral flow in the unsaturated zone is not represented.

High hydraulic conductivities are indicated by the small response times of many catchments to subsurface flow, particularly forested catchments (e.g. Hewlett and Hibbert, 1967). This may be due to water flowing through interconnected macropores created by roots, other organic matter and cracks between aggregates and rocks rather than through the soil matrix (e.g., Mosley, 1979; Sloan and Moore, 1984). Hence, the hydraulic conductivity used to charaterize the subsurface flow may be different from that measured from small soil cores. For field applications of the model K_{sL} should be considered as an "effective" hydraulic conductivity of the soil profile in the lateral direction rather than the measured value of hydraulic conductivity for the soil matrix.

19.5.3. Using THALES

The input parameters required by the model are water content at field capacity, drainable porosity, effective hydraulic conductivity of the soil profile and the coefficient and exponent of the flow area-discharge relationship (given, for example, by Manning's equation or the Darcy-Weisbach equation) as well as the infiltration parameters. The element structure of THALES, derived from TAPES-C, allows each element to have different infiltration, surface and subsurface flow parameters although this is usually not practical because of the large data requirements. In most applications, the parameters are measured for each soil type or region of different surface condition and it is assumed that these parameters do not vary within each region or soil type. A map of the regions of different parameters can then be numerically overlayed on the element map of the catchment. This approach is based on the assumption that the variation in parameter values within a region is less than that between regions and that the within-element characteristics are uniform. Neither of these assumptions is necessarily true (e.g., Beven, 1989; Sharma et al., 1980; Loague and Gander, 1990).

The solution scheme begins at the uppermost element progressing along a contour and then down to the next element on the lower contour. This continues until the last element is reached after which the calculations for the next time step begin at the uppermost element. For each element the streamlines define no-flow boundaries, the upper contour is an inflow boundary while the lower outflow boundary is the computational "node". The simulation calculates the surface and subsurface flow, infiltration rate, exfiltration rate and rainfall excess for every element at every time step. Any of these data can be written to output files for later analysis. The elements where the conditions for kinematic overland flow are violated, based on the kinematic flow number (Woolhiser and Liggett, 1967) are identified and checks are made of local (element) and global (catchment) water balances. The convergence criteria for the flow routing algorithms as well as the time step, start and finish times of the simulation, are chosen by the user.

THALES includes the option of applying a non-uniform distribution of antecedent soil water conditions. The distribution scales the initial water content of one element to the whole catchment using either a static or quasi dynamic wetness index. The static index is that of Beven and Kirkby (1979):

$$w_i = In(A / b \tan\beta) \tag{19.13}$$

where A/b is the specific catchment area (the upslope contributing area per unit width) and β is the surface slope angle . This index has been developed for humid regions where the soil is relatively wet throughout the whole year.

The quasi dynamic index is given by Barling et al. (1994) and is more realistic to use on catchments where drying of the surface soils occurs. This is because in such catchments, the parameter A becomes a function of time since the last rainfall.

THALES should be thought of as a modelling framework or "library of concepts". It is written in modular form and the algorithms presented here can easily be replaced by others if required. For example, where simulation over long periods is required, interception and evapotranspiration are important and these have been incorporated into research versions of THALES for particular applications. Algorithms for Philip's (1957) infiltration equation as used by Wood et al (1988) were added for another application. Similarly, modifications have been made to incorporate a model that enables the simulation of moving storm cells. Some applications of THALES follow.

19.6. APPLICATIONS

19.6.1. Model Evaluation Based on Saturation Patterns

THALES was evaluated on a $2m^2$ sandbed microcatchment (Grayson, 1990; Moore and Grayson, 1991). The sandbed surface represented topographic convergence and divergence and produced saturated subsurface flow and variable source area runoff. A sculptured impermeable surface was overlain by a layer of wet fine sand of uniform thickness (65±6mm). The saturated hydraulic conductivity of the repacked sandbed was 1530±230mm/h. Rainfall was applied to the microcatchment by an oscillating nozzle rainfall simulator at a rate of 55±5mm/h in two bursts. Details on the instrumental setup and model parameters are given in Grayson (1990) and Moore and Grayson (1991).

Figure 19.5 shows simulated and observed runoff at the outlet of the sandbed. The model performs well on the rising limbs but not as well on the recession limbs. As pointed out earlier, distributed hydrologic models should also be evaluated in terms of their distributed behaviour. Therefore, the

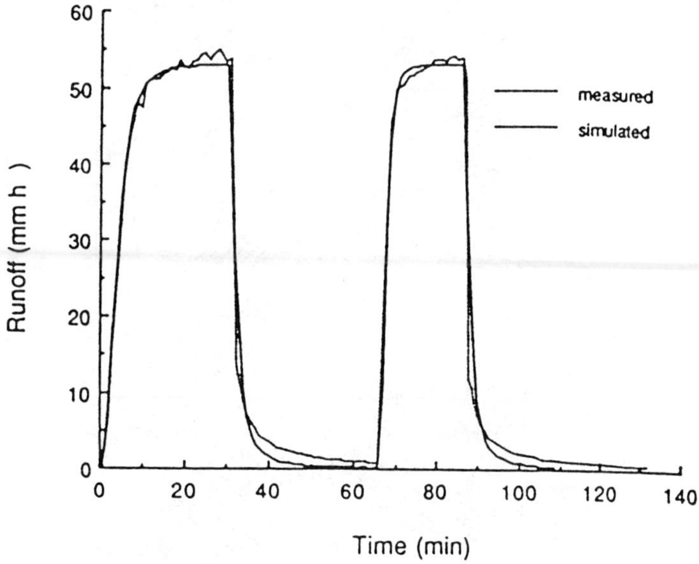

Fig. 19.5. Predicted and observed runoff hydrographs from the sandbed catchments.

predicted areal extent of the zones of surface saturation (=saturation patterns) was compared with the observations.

The observed saturation patterns were derived from stereo photographs in which they showed up as dark areas. Figure 19.6 presents simulated and observed saturation patterns at three different times during the first rainfall period. The general trend of expanding saturation areas with time is modelled well. However, the overall size of the saturated zones is slightly underestimated and the simulated patterns tend to be slimmer (less well rounded) than the observed patterns. There are two reasons for these differences: (1) the effect of the capillary fringe was ignored in this relatively simple model. (2) More importantly, subsurface water ridging occured in the valleys creating large hydraulic gradients away from the valley. This tends to disperse water away from the draining lines, increasing the size of the saturation zones, as observed in the prototype. However, the model assumes that the land surface gradient represents the subsurface hydraulic gradient and so can not mimic the broadening of the saturated zone caused by the ridging effect.

The high saturated hydraulic conductivities in the prototype create a worst case scenario for the occurence of this phenomenon. In most natural landscapes, hydraulic conductivities and slopes are smaller and flow path lengths are longer, thus damping out this effect. Figures 19.5 and 19.6

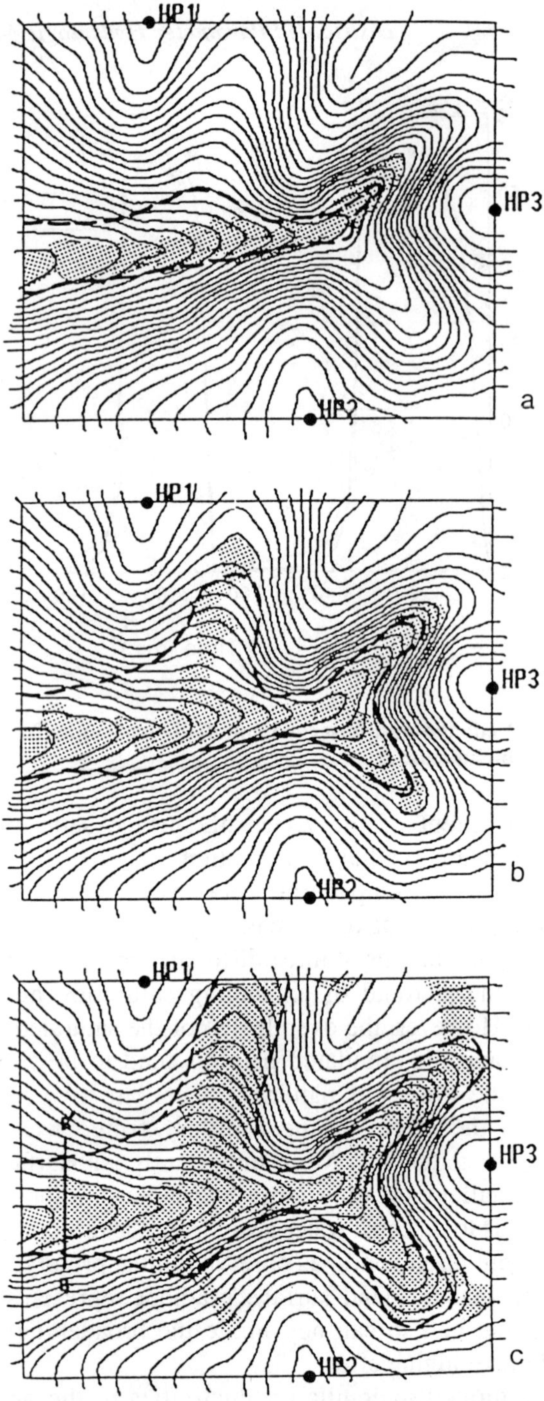

Fig. 19.6. **Predicted (stippled area) and observed (bounded by dashed line) surface saturated zones at three different times during a rainfall event: (a) 2.5 min., (b) 4.25 min., and (c) 23.5 min. Solid lines refer to the contours (contour interval of 10 mm) and HP indicates the highpoints.**

clearly demonstrate that accurate runoff simulations at the catchment outlet do not necessarily imply accurate simulations of distributed catchment response.

19.6.2. Moving Storms

The following simulations were performed using the topography of the $17km^2$ Coweeta catchment (Fig. 1) and hypothetic scenarios for model parameters and rainfall. Specifically, for the moving storm scenario, runoff generation by the infiltration excess mechanism only was assumed using Philip's (1957) equation with a saturated hydraulic conductivity of 5mm/h and a sorptivity of 14 $mmh^{-0.5}$. Channels were defined by a threshold area of 10ha and Mannings n in the channels was set to 0.1. Mannings n on the hillslopes was either n=0.1 or n=0.4. Two storms were moved over the catchment, the first moving towards the catchment outlet (downstream) and the second upstream. The storms were moved at 5km/h, had a Eulerian duration of 24min and an intensity of 45mm/h which corresponds to a storm length of 2km and a total precipitation of 18mm at any point.

Figure 19.7a shows the simulated catchment outflow for the two storms using Mannings n = 0.4 on the hillslopes. As would be expected, the rising limb for the downstream moving storm is steeper and the peak flow is higher (Yen and Chow, 1969). However, the difference in peak flows is only 20%. Figure 19.7b shows a similar scenario, but with smoother surface conditions on the hillslopes (n=0.1). The catchment response is 'flashier' than for the rough case in Fig. 19.7a. Also, the difference in flow peaks between the upstream and the downstream moving storms is much larger (about 50%). The different results for the rough and the smooth scenarios are easily interpreted when examining the simulated velocity distribution within the catchment.

For the smooth cases (Fig. 19.7b) overland flow velocities are similar to those in the channels. These are close to the storm velocity thereby producing 'resonance' and hence the large differences in peaks. Increasing the roughnesses on the hillslopes (Fig. 19.7a) results in much lower velocities on the hillslopes as compared to the channels. Therefore, the velocity distribution is bimodal which dampens the 'resonance' significantly. We believe the scenario of Fig. 19.7a is the more realistic one and illustrates the general tendency of catchments to 'integrate' the internal fluctuations into 'well behaved' hydrographs. In smaller catchments (or on hillslopes) with smaller variations in velocities the effect of the integration will be less pronounced resulting in larger differences in runoff for different storm velocities and/or directions.

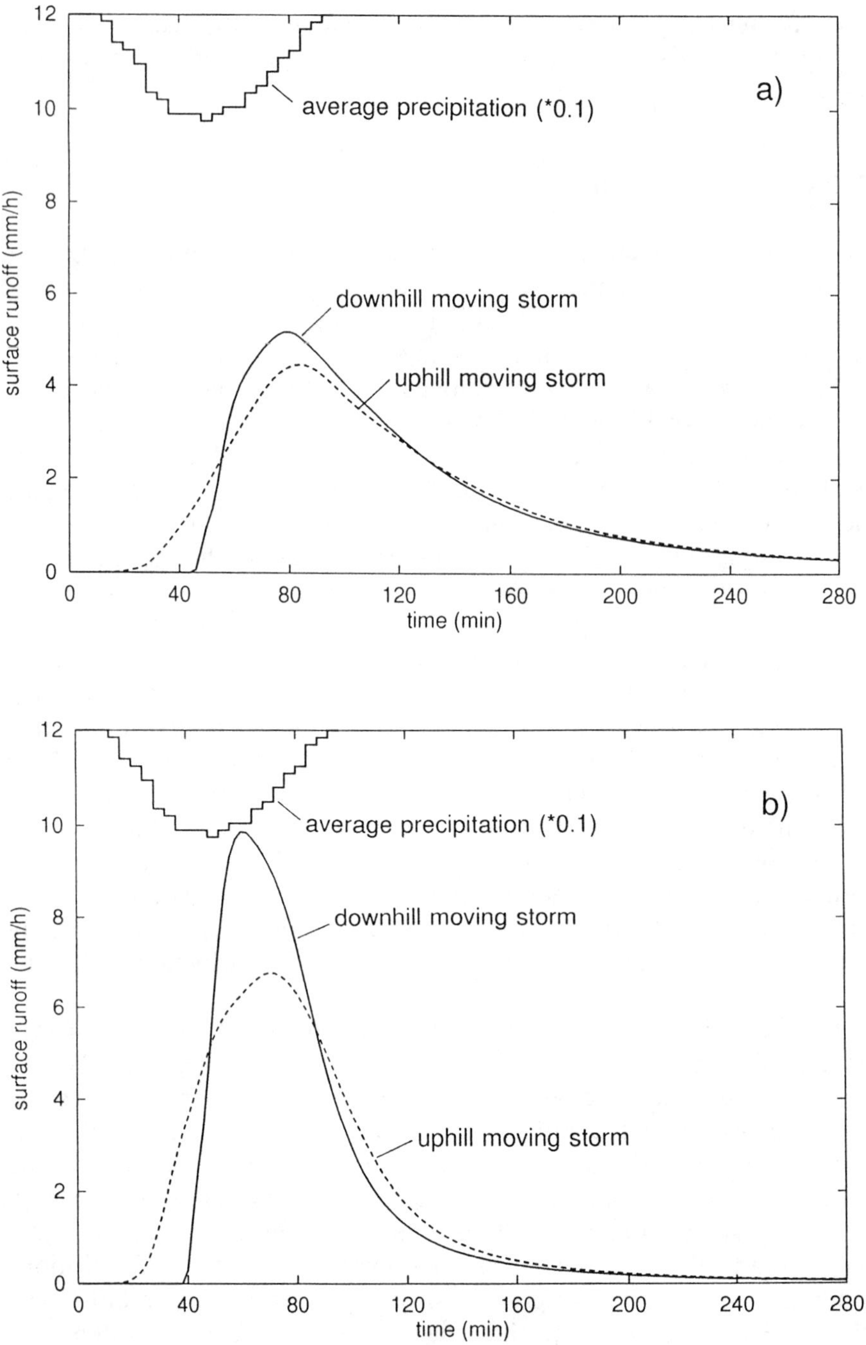

Fig. 19.7. Runoff response for the Coweeta scenario for upstream and down stream moving storms. Mannings n=0.1 in the channels. Mannings n on the hillslopes is (a) n=0.4 and (b) n=0.1.

19.6.3. Effect of Random and 'Organized' Initial Conditions

This scenario, again, is based on the Coweeta topography (Fig. 19.1) and both saturation excess and infiltration excess mechanisms are assumed to be operative. Infiltration parameters and overland flow parameters are those used for Fig. 19.7a. Two different initial conditions are examined. The first case (Fig. 19.8a assumes scaling of the initial saturation deficit based on the static topographic wetness index (Beven and Kirkby, 1979) according to:

$$S_i = \overline{S} + m * \left(\overline{w}_i - w_i\right) \tag{19.14}$$

where S_i is the initial saturation deficit in an element, w_i is the wetness index for the element (Eq. 19.13) and $\overline{w}$ is the wetness index over the catchment.

The mean saturation deficit $\overline{S}$ was set to 15mm and the parameter m was set to 30mm. This is termed the 'organized' case (Blöschl et al., 1993b; Blöschl and Sivapalan, 1994). In the second case (Fig. 19.8b, the initial saturation deficit is assumed to be a stationary random field with the same probability density function and covariance function as the 'organized' case. So, the two cases do not differ in terms of their statistical properties but they differ in terms of their spatial pattern.

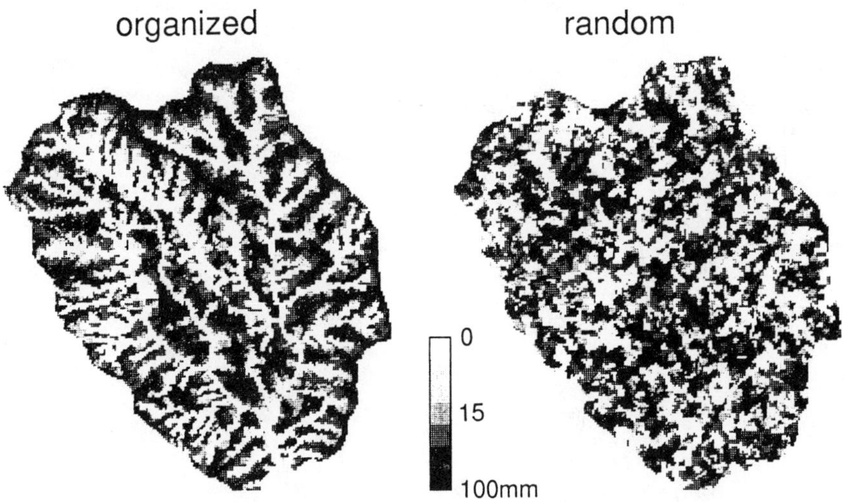

Fig. 19.8. 'Organized' and random initial saturation deficit for the simulations in Fig. 19.9.

Figure 19.9a shows simulated hydrographs at the catchment outlet for a 1hr (spatially uniform) rainblock of 5mm/h intensity. This rainfall intensity involves saturation excess infiltration only (hydraulic conductivity = 5mm/h). The 'organized' and the random case (Fig. 19.9a yield the same

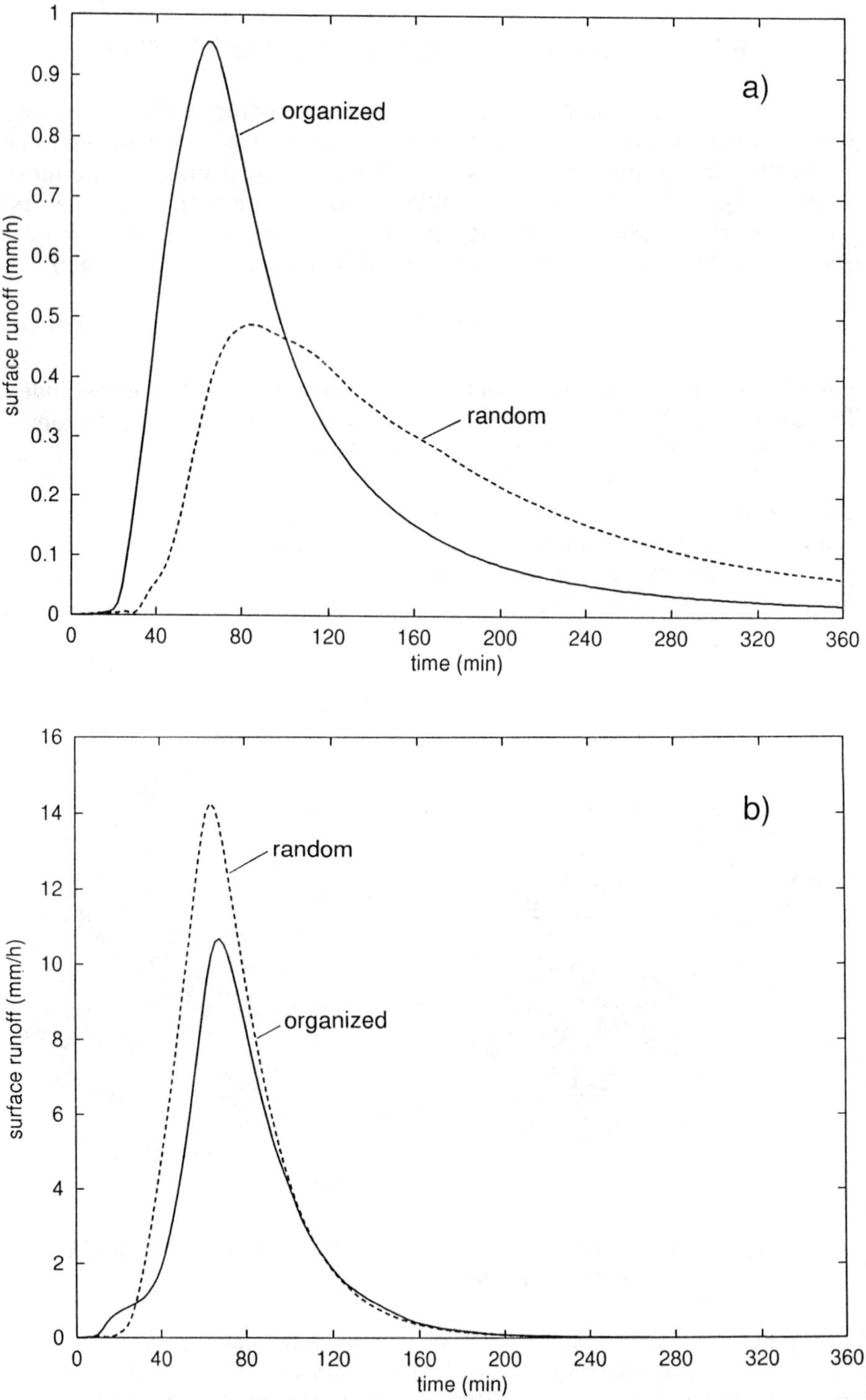

Fig. 19.9. Runoff response for the Coweeta scenario for 'organized' and random initial saturation deficit as in Fig. 19.8. Rainfall is assumed to be a 1hr block with (a) 5mm and (b) 30mm intensity.

runoff volume since the probability density functions of initial saturation deficit are identical. The runoff peaks for the 'organized' case, however, are substantially larger. This is because, in the 'organized' case, the saturated area at the beginning of the storm is 'hydraulically connected' to the outlet while in the random case it is not and runon can reinfiltrate.

Figure 19.9b shows a similar scenario but for a rainfall intensity of 30mm/h. Here, both infiltration excess and saturation excess mechanisms are operative. Figure 19.9b shows a reversed effect to that of Fig. 19.9a with higher peaks for the random case. This is because the hollows (=channel areas) are saturated very quickly thus producing similar conditions in the channel areas for most of the storm. However, the organized case tends to have larger saturation deficits on the hillslopes which translates into smaller peaks at the catchment outlet.

19.6.4. Scale Effects and the Representative Elementary Area

This scenario examines the existence of a Representative Elementary Area (REA). The REA (Wood et al., 1988) promises a spatial scale over which the process representations can remain simple and at which distributed catchment behaviour can be represented without the apparently undefinable complexity of local heterogeneity. This concept is based on the assumption that, at a certain scale (the REA), small scale processes may be parameterized as subgrid variability and large scale processes may be represented explicitly as element-to-element variations. Provided an REA exists for a particular catchment, the same parameterisation of subgrid processes can be used for a range of grid (or element) sizes.

The present simulations involve the topography of the Coweeta catchment (following Wood et al., 1988). Several assumptions on infiltration are examined: - impervious surface conditions, - spatially variable conductivity and topographically defined initial saturation deficit, - spatially invariant conductivity and topographically defined initial saturation deficit, - spatially invariant conductivity and random initial saturation deficit. Rainfall is assumed to consist of a 1hr rainblock and is spatially variable (a random field) with 2500m correlation length and an average intensity of 5mm/h. Details are given in Blöschl et al.(1994).

Runoff peaks were plotted versus upslope contributing area for a set of nested catchments. In these plots, the REA should appear as the scale with a minimum in variability of peak runoff as a function of catchment size (Hubbert, 1956, Blöschl et al, 1994).

Figure 19.10 shows results for one realisation of precipitation and different assumptions on infiltration. There is some variability below catchment areas of 0.5km*0.5km, substantial variability above 2km*2km but little variability in between. Different assumptions about infiltration do not affect the large scale variation but they significantly affect the small scale variability. Clearly, the small scale variability is related to the effect

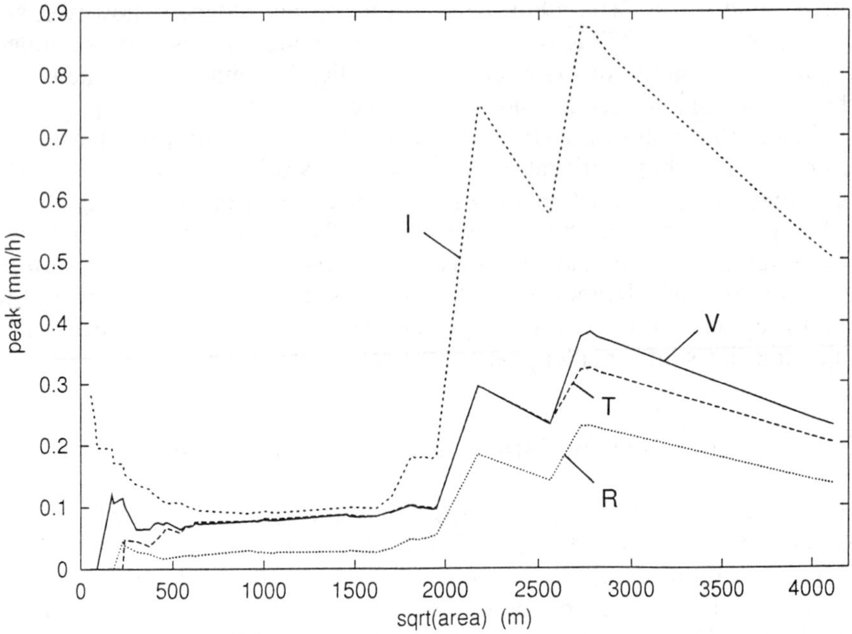

Fig. 19.10. Peak flow vs. subcatchment area for a set of nested subcatchments of the Coweeta topography. Spatially variable precipitation (correlation length of 2500m). Different assumptions about infiltration: I impervious surface conditions, V spatially variable conductivity and topographically defined initial saturation deficit, T spatially invariant conductivity and topographically defined initial saturation deficit, R spatially invariant conductivity and random initial saturation deficit.

of variable soil parameters and topography while the large scale variability is due to precipitation patterns. A minimum in variance in between indicates that an REA does exist for the scenario examined here. In general, the REA will be a function of catchment specific conditions. Nevertheless, in the context of distributed parameter rainfall-runoff modelling the definition of an REA for a particular catchment may be useful when a separation of scales exists such as when large scale frontal precipitation is dominant.

ACKNOWLEDGEMENTS

The authors wish to thank the Fonds zur Förderung der wissenschaftlichen Forschung, Vienna, project no. J0699-PHY for financial support. Günter Blöschl is presently on leave at the Centre for Resource and Environmental Studies, The Australian National University, Canberra, Australia.

REFERENCES

Anderson, M.G. and C.C.M. Rogers, Catchment scale distributed hydrological models - A discussion of research directions, *Progress in Phys. Geogr. 11*, 28-51, 1987.

Band, L.E., Topographic partition of watersheds with digital elevation models, *Water Resour. Res.*, 22, 15-24, 1986.

Barling, R.D., I.D. Moore and R.B. Grayson, A quasi-dynamic wetness index for characterising the spatial distribution of zones of surface saturtion ans soil water content. *Water Resour. Res.,* In press.

Bergström, S., Principles and confidence in hydrological modelling.Nordic Hydrol. 22, 123-136, 1991.

Beven, K. and M.J. Kirkby, A physically based, variable contributing area model of basin hydrology, *Hydrol. Sci. Bull. 24*(1), 43-69, 1979.

Beven, K., A. Calver and E.M. Morris, The Institute of Hydrology Distributed Model, *Inst. Hydrol. Rep. 98*, 1987.

Beven, K., Changing ideas in hydrology - The case of physically-based models, *J. Hydrol. 105*, 157-172, 1989.

Beven, K., Kinematic subsurface stormflow, *Water Resour. Res.*, 17, 1419-1424, 1981.

Beven, K., On subsurface stormflow: Predictions with simple kinematic theory for saturated and unsaturated flows, *Water Resour. Res.*, 18, 1627-1633, 1982b.

Beven, K., Surface water hydrology - Runoff generation and basin structure, *Rev. of Geophys. and Space Phys. 21*(3), 721-729, 1983.

Beven, K.J., and M.J. Kirkby, A physically-based variable contributing area model of basin hydrology, *Hydrol. Sci. Bull.*, 24, 43-69, 1979.

Blöschl, G. and Sivapalan, M. (1994) Scale issues in hydrological modelling- A review. Hydrol. Processes. In press.

Blöschl, G., Grayson, R. B. and Sivapalan, M. (1994) REA. Hydrol. Processes. In press.

Blöschl, G., Grayson, R. B., Sivapalan, M., and Moore, I.D. (1993b) Organisation and randomness in catchments and the verification of distributed hydrologic models. AGU Fall Meeting.

Blöschl, G., Gutknecht, D. and Kirnbauer, R. (1993a) On the evaluation of distributed hydrologic models. In: R. Rosso (ed.) Advances in distributed hydrology. Proceedings of a workshop held in Bergamo, Italy, June 25-26, 1992. Water Resources Publications.

Blöschl, G., Kirnbauer, R. and Gutknecht, D., Distributed snowmelt simulations in an alpine catchment. 1. Model evaluation on the basis of snow cover patterns. *Wat. Resour. Res.* 27(12), 1991, pp. 3171-3179.

Brooks, R.H. and A.T. Corey, Hydraulic properties of porous media, *Hydrol. Pap. no. 3*, Colorado State Univ., Fort Collins, Colorado, 27p, 1964.

Burges, S.J., Trends and directions in hydrology, *Water Resour. Res. 22*(9), 1S-5S, 1986.

Cayley, A., On contour and slope lines, *London, Edinbourgh and Dublin Philosophical Magazine and Journal of Science*, 4th. Series, 18, 264-268, 1859.

Dunne, T., Models of runoff processes and their significance, In: *Scientific Basis of Water Resource Management*, Nat. Academy Press, 17-30, 1982.

Grayson, R. B., I. D. Moore and T. A. McMahon, 1992a.Physically based hydrologic modelling: I. A terrain based model for investigative purposes. Water Res. Res. 26 (10): 2639-2658.

Grayson, R. B., I. D. Moore and T. A. McMahon, 1992b.Physically based hydrologic modelling: II. Is the concept realistic ?. Water Res. Res. 26 (10): 2659-2666.

Grayson, R.B., Terrain Based Hydrologic Modelling for Erosion Studies, Ph.D. thesis, Dept. Civil and Agric. Engr., The University of Melbourne, Parkville, Victoria, Australia, 1990.

Green, N.H., and C.A. Ampt, Flow of air and water through soils, *J. Agric. Sci.*, 4, 1-24, 1911.

Hewlett, J.D., and A.R. Hibbert, Factors affecting the response of small watersheds to precipitation in humid areas, in *Forest Hydrology*, edited by W.E. Sopper and H.W. Lull, pp. 275-290, Pergamon, Oxford, 1967.

Hillel, D, Modelling in soil physics: A critical review. In: *Future Developments in Soil Science Research*, A collection of Soil. Sci. Soc. Am. Golden Anniversary contribtutions presented at Annual Meeting, New Orleans: 35-42, 1986.

Hornberger, G.M., K.J. Beven, B.J. Cosby and D.E. Sappington, Shenandoah watershed study: calibration of a topography-based, variable contributing area hydrological model to a small forested catchment, *Water Resour. Res. 21*(12), 1841-1850, 1985.

Hubbert, M.K. 1956. 'Darcy's law and the field equations of the flow of underground fluids', Trans. Amer. Inst. Min. Met. Eng., 207, 222-239.

Hutchinson, M.F. and T.I. Dowling, A continental hydrologic al assessment of a new grid-based digital elevation model of Australia, Hydrological Processes, 6(3), 1992.

James, L.D. and S.J. Burges, Selection, calibration and testing of hydrologic models, In: C.T. Haan, H.P. Johnson and D.L. Brakenseik (Editors), Hydrological Modeling of small watersheds. *Am. Soc. Agric. Eng. Monograph 5*, 435-472, 1982.

Klemes, V., Conceptualisation and scale in hydrology. *J. Hydrol.* 65, 1983, pp. 1-23.

Klemes, V., Empirical and causal models in hydrology. In: *Scientific Basis of Water Resource Management*, Nat. Academy Press, 1982, pp. 95-104.

Loague, K. and G.A. Gander, R-5 revisited, 1, Spatial variability on a small rangeland catchment, *Water Resour. Res. 26*(5): 957-971, 1990.

Mein, R.G., and C.L. Larson, Modeling infiltration during a steady rain, *Water Resour. Res.*, 9, 384-394, 1973.

Moore, I.D. and G.J. Burch, Sediment transport capacity of sheet and rill flow: Application of unit stream power theory, *Water Resour. Res. 22(8)*, 1350-1360, 1986.

Moore, I.D. and G.R. Foster, Hydraulics and overland flow, In: *Process studies in hillslope hydrology*, M.G. Anderson and T.P. Burt (eds), John Wiley, Chichester, England, 215-254, 1990.

Moore, I.D. and Grayson R.B., Terrain based prediction of runoff with vector elevation data, *Water Resour. Res.*27(6), 1991, pp. 1177-1191.

Moore, I.D. and Mein, R.G. Application of three daily runoff models to four Australian catchments. Monash University Civil Engineering Research Report No. 3/76.

Moore, I.D., and R.B. Grayson, Hydrologic and digital terrain modelling using vector elevation data, *EOS, Trans. Am. Geophys. Union*, 70, 1091, 1989.

Moore, I.D., E.M. O'Loughlin and G.J. Burch, A contour-based topographic model for hydrological and ecological applications, *Earth Surf. Processes and Landf. 13*, 305-320, 1988b.

Moore, I.D., R.B. Grayson, and A.R. Ladson, Digital terrain modelling: a review of hydrological, geomorphological and biological applications, *Hydrological Processes*, 1991 (in press).

Morel-Seytoux, H.J., Equivalence between infiltration parameters in Horton and Morel-Seytoux formulae, *Proc.8th Annual A.G.U. Front Range Branch Hydrol. Days*, April 1988, Fort Collins, 248-259, 1988a.

Morel-Seytoux, H.J., Recipe for simple but physically based modeling of the infiltration and local runoff processes, *Proc.8th Annual A.G.U. Front Range Branch Hydrol. Days,* April 1988, Fort Collins, 226-247, 1988b.

Mosley, M.P., Streamflow generation in a forested watershed, New Zealand, *Water Resour. Res.*, 15, 795-806, 1979.

O'Loughlin, E.M., Prediction of surface saturation zones in natural catchments by topographic analysis, *Water Resour. Res.*, 22, 229-246, 1986.

Onstad, C.A. and D.L. Brakensiek, Watershed simulation by stream path analogy, *Water Resour. Res. 4*(5), 965-971, 1968.

Palacios, O. and Cuevas, B. 1989. 'Transformation of TIN data into a kinematic cascade', *Trans. Am. Geophys. Union*, **70**, 1091.

Parlange, J.-Y., Lisle, I., Braddock, R. D. and Smith, R. E. (1982) The three-parameter infiltration equation. Soil. Sci 133, 337-341.

Philip, J. R. (1957) The theory of infiltration: 1. The infiltration equation and its solution. Soil Sci. 83, 345-357.

Philip, J.R. 1957. 'The theory of infiltration, 1-7', Soil Sci., 83, 84, and 85. Pichler, H. ????. 'Dynamik der Atmosphare', BI-Taschenbuch ???.

Philip, J.R., Future problems in soil water research. *J. Soil Sci.* 116, 1972, pp. 328-335.

Philip, J.R., Some remarks on science and catchment prediction, In: T.G. Chapman and F.X. Dunin (Editors), *Prediction in Catchment Hydrology: a National Symposium on Hydrology*, 25-27 November 1975, Canberra, Austrailian Academy of Science: 23-30, 1975.

Schmid, B.H., On overland flow modelling: Can rainfall excess be treated as independent of flow depth? *J. Hydrol.*, 107, 1-8, 1989.

Sharma, M.L., G.A. Gander, and C.G. Hunt, Spatial variability of infiltration in a watershed, *J. Hydrol.*, 45, 101-122, 1980.

Sloan, P.G., and I.D. Moore, Modelling subsurface stormflow on steeply sloping forested watersheds, *Water Resour. Res.*, 20, 1815-1822, 1984.

Smith, R.E. and J.Y. Parlange, A parameter-efficient hydrologic infiltration model, *Water Resour. Res. 14*(3), 533-538, 1978.

Speight, J.G., A parametric approach to landform regions, *Special Publication, Institute of British Geographers*, 7, 213-230, 1974.

Speight, J.G., The role of topography in controlling through flow generation: A discussion, *Earth Surface Processes and Landforms*, 5, 187-191, 1980.

Swank, W. T. and Crossley, D. A. (ed) (1988) Forest hydrology and ecology at Coweeta. Springer. New York. 512 pp.

Takasao, T., M. Shiiba, and H. Kitamura, A distributed runoff model and lumping scale of a river basin, *Proc. 31st. Annual Conf. Jap. Soc. Civil Engrs. 2-86*, pp. 163 (in Japanese), 1976.

Vertessy, R.A., T.J. Hatton, P.J. O'Shaughnessy andd M.D.A. Jayasuriya, 1993. Predicting water yield from a Mountain Ash forest using a terrain analysis based catchment model. *J of Hydrol.,* 150, 665-700.

Warntz, W., Stream ordering and contour mapping, *J. Hydrol.*, 25, 209-227, 1975.

Wood, E.F., Sivapalan, M., Beven, K., and Band, L. 1988. 'Effects of spatial variability and scale with implications to hydrologic modeling', J. Hydrol., 102, 29-47.

Woolhiser, D.A. and J.A. Liggett, Unsteady one-dimensional flow over a plane: The rising hydrograph, *Water Resour. Res. 3*(3), 753-771, 1967.

Yen, B. C. and Chow, V. T (1969) A laboratory study of surface runoff dueto moving rainstorms. WRR 5(5) 989-1006.

Chapter 20

KINEROS - A KINEMATIC RUNOFF AND EROSION MODEL

R.E. Smith, D.C. Goodrich, D.A. Woolhiser, and C.L. Unkrich

20.1. INTRODUCTION

KINEROS, an acronym for KINematic runoff and EROSion model, has evolved over a number of years primarily as a research tool. However some consulting firms have been attracted by several of its unique features and KINEROS has been used as an engineering tool in the U.S. and abroad.

The logical structure on which KINEROS is based, routing surface runoff over cascades of overland flow planes contributing lateral inflow to channels, was developed in the late 1960's (Woolhiser, et al. 1970). An interactive infiltration component was added by Rovey (1974) and the program, known then as KINGEN, was documented and released (Rovey, et al. 1977). The model was tested with data from the CSU Outdoor Experimental Rainfall-Runoff Facility (Dickinson et al. 1967), agricultural plots and small watersheds, and with urban runoff data. After the release of KINGEN, several modifications were made, including the addition of an erosion and sediment transport component, modification of the infiltration component, addition of a pond element, and modification of the input to allow spatial variability of rainfall. These modifications were incorporated into a model called KINEROS and were documented and released (Woolhiser, et al. 1990a). Since then, KINEROS was modified to account for small scale spatial variation of saturated conductivity. This version of KINEROS was subjected to rigorous testing on split sample data from semiarid watersheds ranging from 1.5 to 630 ha. (Goodrich 1991, Woolhiser et al. 1990b, Michaud, 1992). The results were very encouraging for basins up to about 3 sq. mi. (760 ha.) and demonstrated the necessity of accounting for small scale spatial variability of saturated conductivity and rainfall. These studies also emphasized the importance of carefully checking the experimental rainfall-runoff data to prevent misleading conclusions regarding model applicability.

The following discussion provides an overview of the theory and structure of KINEROS and an analysis of model results. First, the basic theory and assumptions of the processes are presented. Next, the results of testing of KINEROS are presented. Finally, a description of improvements underway will be presented.

20.2. KINEROS BASIC STRUCTURE

20.2.1. Representation of Catchment Geometry

The KINEROS model user must transform the catchment of interest into an equivalent network composed of runoff surfaces or "planes", intercepting channels (or conduits in the case of urban storm drains), and ponds or detention storages. Each of these is oriented such that 1-dimensional flow can be assumed, and each such unit is termed an *element* of the overall network. Figure 20.1 illustrates schematically and conceptually the major

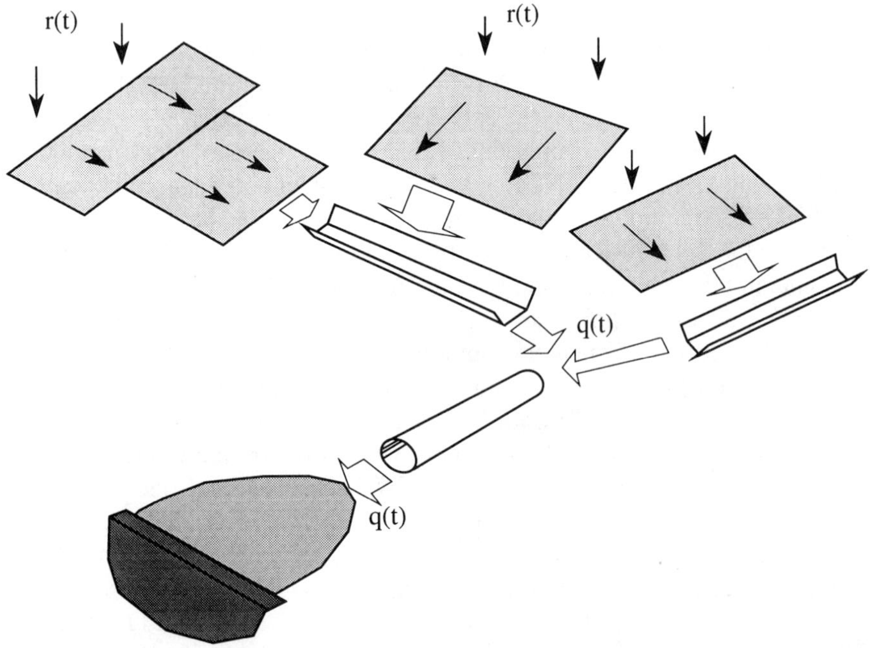

Fig. 20.1. Schematic example of various hydrologic elements used in KINEROS, and some of the ways that they may be interconnected to represent urban or rural watersheds.

types of elements and their connections. The User Manual (Woolhiser et al, 1990a) gives examples to aid in this geometric simplification. Runoff surfaces may be composed of a cascade of rectangular surfaces, which allows the simulation of converging flow areas, or areas of nonuniform slope, hydraulic resistance, or soils. Hortonian runoff is then simulated for the network of elements, culminating in the production of a simulated hydrograph at the outlet. The processes simulated, and the equations involved, will be described in general order of occurrence in the runoff - erosion generation process.

20.2.2. Rainfall Excess Calculation

20.2.2.1. Distributed Rainfall Data. Rainfall information for as many as 20 gage sites within or adjacent to the basin may be provided for use in the model. KINEROS requires information in the form of accumulated depth / time pairs, and converts this data into rainfall rate pulses. The raingages are identified by number, and input data must indicate which gage is associated with each catchment element. A gage weighting factor may be used to adjust the data from a gage for the associated element.

20.2.2.2. Interception. A total depth of interception may be specified for each runoff element, based on the vegetation or other surface condition. This amount is taken from the earliest rainfall pulses until the potential interception depth is filled. The modified rainfall pulse data then becomes input to the soil surface.

20.2.2.3. Infiltration during rainfall. As will be discussed later, each flow element is divided along its flow path into computational finite increments separated by nodes. The rainfall excess rate at a computational node for each surface element is found by the difference between the rainfall rate and the rate at which the soil can absorb or infiltrate rainfall. Soil water infiltration varies considerably in time depending on both soil hydraulic parameters and other time-varying soil conditions. The infiltration model used in KINEROS (Smith and Parlange, 1978) is based on an approximate solution of the basic equation of unsaturated flow. The model requires three parameters to describe the infiltration behavior of the soil during a rainfall event. $f_c = f(I, \theta_i)$

Infiltration capacity, f_c, is the maximum rate at which water can enter the soil surface. It is always determined by soil properties and water content of the soil. In KINEROS, it is described as a function of two variables: $f_c = f(I, \theta_i)$, where I is the cumulative amount of infiltrated water, starting at the beginning of the storm, when initial water in the soil profile, θ_i, is assumed uniform. f_c is also a function of three primary soil parameters, G, S_i, and K_s, which will be discussed below. Unless rainfall rate exceeds the infiltration capacity, the infiltration rate at any time during a rainstorm is limited by the rainfall rate r,

$$f = Min[f_c, r] \tag{20.1}$$

and this definition of f leads to the definition of I;

$$I = \int_o^t f dt \quad . \tag{20.2}$$

I is the independent variable for the infiltration capacity, f_c (Smith and Parlange, 1978),

$$f_c = K_s \frac{\exp(I/B)}{\exp(I/B)-1} \tag{20.3}$$

in which K_S is the surface soil effective saturated hydraulic conductivity, cm/min and B is a term combining a capillary drive parameter and the soil saturation deficit,

$$B = (\theta_s - \theta_i)G \quad . \tag{20.4}$$

These equations describe an implicit relation, but solution is not difficult, as discussed below. Initial water content θ can vary from 0 (in theory) to a maximum value, θ_S, slightly less than the soil porosity. Soils exhibit a rapidly decreasing relative hydraulic conductivity, $k_r = K/K_S$, as soil water tension is increased (or as soil capillary head, ψ, decreases). The effective capillary drive, G, reflects the soil's capillary properties. G is defined by the integral of the unsaturated relative hydraulic conductivity function for the soil.

$$G = \int_{-\infty}^{0} k_r(\psi) d\psi \quad . \tag{20.5}$$

Estimated values for the parameters G and K_S for a variety of soil textural classes have been determined by statistical study of measured data (Rawls et al. 1982) and are furnished in the KINEROS User Manual. It should be noted that the parameters have substantial variability within textural classes, so that values obtained from the tables should be considered as "starting values" only.

The parameter θ_S is usually slightly below the soil porosity ϕ and is not difficult to determine since it has a small range. The saturation deficit (θ_S -

θ_i) is actually one parameter, because θ_s is fixed from storm to storm (except for cultivated soils). For ease of estimation, the KINEROS input parameter for soil water is a scaled water content, $S = \theta/\phi$, (ϕ is the soil porosity) which varies from 0 to 1. Thus initital soil conditions are represented by the variable S_i (= θ_i/ϕ).. In KINEROS, S_{max} is the scaled equivalent of θ_s. S_i cannot be as great as S_{max}, and is usually limited at the lower end not by 0 but by a residual saturation S_r. S_r conceptually represents that water which cannot be removed from the soil by capillary forces, but is often merely a fitting parameter for soil data. The parameter S_r is used in soil water redistribution during breaks in a rainfall sequence. Thus, there are two parameters, K_s, and G to characterize the soil, and the variable S_i to characterize the initial condition. Although in practice, the parameters K_s and G are often treated as constants, they are affected by tillage operations and biological activities between rainfall events.

While Eq. (20.3) is not explicitly a function of time, it can be transformed into a time dependent function. The advantage of I as an independent variable is that Eq. (20.3) describes in one expression both the occurrence of ponding and the decay of f_c after ponding (Smith, 1982).

Figure 20.2 shows the relation f(I) represented by Eq. (20.3). Actual infiltration is the shaded part. As this figure shows, I is accumulated as $\sum rdt$ until the limit described in Eq. (20.3) is reached, after which I is accumulated as described by Eqs. (20.2) and (20.3).

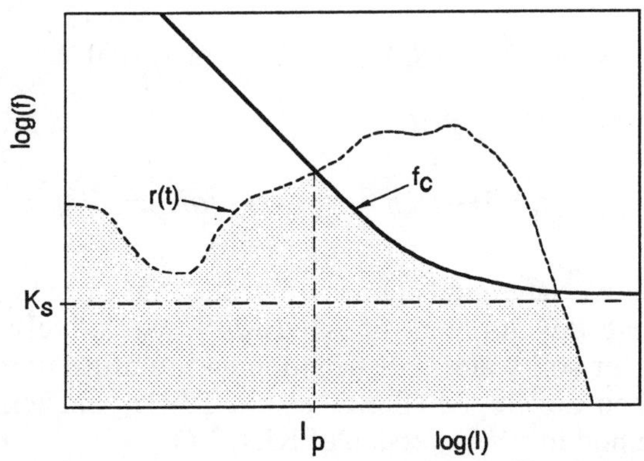

Fig. 20.2. Schematic of the "universal" relation of infiltration rate to infiltrated volume, and to infiltration capacity, f_c.

20.2.2.4. Infiltration during recession. Infiltration calculations should continue after rainfall slows or ceases as long as there is water on the soil surface. This implies a further interaction between the infiltration model and the surface flow model. During recession, however, it is not assumed that the entire surface is covered with water. KINEROS assumes a relatively simple relation between mean surface water depth, h, and the proportion of the surface covered, p_{rc}, at any location along the flow direction, such that $p_{rc} = h/h_c$ [for $h < h_c$]. h_c is accepted as a recession parameter for each surface element.

20.2.2.5. Recovery of infiltration capacity during an interruption. The infiltration theory summarized above may be used for any value of $r > K_s$. When r falls below K_s, but is greater than 0, the surface soil water content will approach a value which is dependent on r, and this asymptotic value $S_*(r)$ can be estimated by a function

$$S_*(r) = S_r + (S_{max} - S_r)\left[\frac{r}{K_s}\right]^p \qquad (r<K_s) \tag{20.6}$$

In theory, the parameter p varies with soil type, but is taken as constant in KINEROS for simplicity. KINEROS assumes that during a break in rainfall S_* is approached asymptotically, and that the value of S_* cannot be less than a user supplied estimate of "field capacity". Estimates are supplied in the User Manual. The asymptote is approached more rapidly for soils with large value of K_s, and more slowly for cases where I is large and when there is water on the surface. Mathematically, these concepts are described as

$$S_i^j = S_i^{j-1} + (S_* - S_i^{j-1})[1 - \exp(-\eta \Delta t)] \tag{20.7}$$

in which

$$\eta = [1 - h/h_c]\exp(-2I)\sqrt{aK_s} \tag{20.8}$$

and S_i^j is the relative saturation at node i for time step j. It should be pointed out that since the development of KINEROS, a relatively efficient and accurate theory for soil water redistribution and reinfiltration during storm intervals has been developed (Smith *et al.*, 1993), and will replace this approximate method in a new version of KINEROS.

20.2.2.6. Numerical Method of Solution. During a computational time interval Δt(i), the appropriate value of *f* is its interval mean value

$$f = [I(t+\Delta t) - I(t)]/\Delta t \quad . \tag{20.9}$$

Prior to ponding, solution is direct, since f = r. After ponding, the value of f may be found by either numerical integration, or series approximation for the exponential function in Eq. (20.3), which leads to a time explicit function. For the period after ponding, KINEROS uses numerical integration, solving the following equation in small subdivisions δt until t+Δt is reached:

$$\Delta t = \int_{I(t)}^{I(t+\Delta t)} \frac{dI}{f_c(I)} \tag{20.10}$$

20.2.3. Hortonian Overland Flow

20.2.3.1. Basic Equations. After sufficient rainfall excess water appears on a surface to overcome surface tension and fill small surface depressions, overland flow begins. Viewed at a microscale, overland flow is an extremely complicated, multidimensional, and heterogenous process. At a larger scale, however, it can be usefully viewed as a one-dimensional flow process in which flux is a power function of the unit storage (storage per unit area, or mean depth of flow);

$$Q = \alpha h^m \tag{20.11}$$

where Q is the discharge per unit width, h is the unit storage of water (or mean depth for smooth surfaces), and α and m are parameters related to slope, surface roughness, and flow condition (laminar or turbulent flow). Equation (20.11) is used in conjunction with the equation of continuity,

$$\frac{\partial h}{\partial t} + \frac{\partial Q}{\partial x} = q(x,t) \tag{20.12}$$

where t is time, x is the distance coordinate, and q(x,t) is the rate of local input, or lateral inflow. For overland flow, Eq. (20.11) may be substituted into Eq. (20.12) to obtain

$$\frac{\partial h}{\partial t} + \alpha m h^{m-1} \frac{\partial h}{\partial x} = q(x,t) \tag{20.13}$$

A definition sketch of one-dimensional surface flow in response to a distributed input is shown in Fig. 20.3. It must be emphasized that KINEROS does not assume overland flow elements to be flat planes, but only assumes that Eq. (20.11) describes the normal flow process. This is all that is required by the kinematic assumption. High slopes or supercritical flow are not required.

The kinematic wave equations are simplifications of the de Saint Venant equations, and do not preserve all of the properties of the more complex equations, such as backwater and diffusive wave attenuation. Attenuation does occur in kinematic routing from shocks or from spatially variable

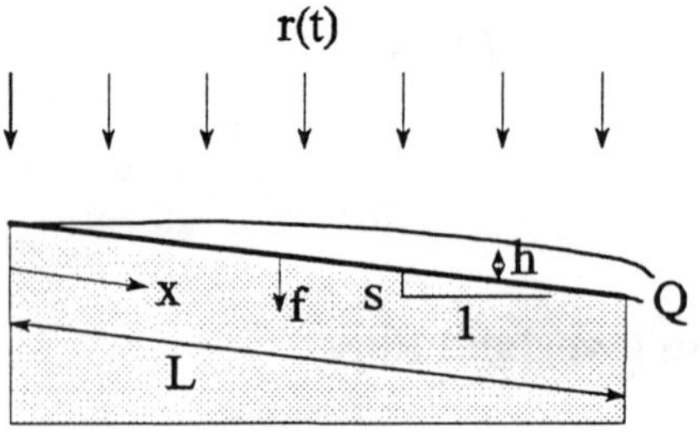

Fig. 20.3. Definition sketch of overland flow on an infiltrating surface.

infiltration. The kinematic routing method is an excellent approximation for most overland flow conditions (Woolhiser and Liggett 1967, Morris and Woolhiser 1980).

20.2.3.2. Boundary Conditions. The depth or unit storage at the upstream boundary must be specified to solve Eq. (20.13). If the upstream boundary is a flow divide, the boundary condition is

$$h(0,t) = 0 \tag{20.14}$$

If another surface is contributing flow at the upper boundary, the boundary condition is

$$h(0,t) = \left[\frac{\alpha_u h_u (L,t)^{m_u} W_u}{\alpha W} \right]^{1/m} \tag{20.15}$$

This merely states an equivalence of discharge, in terms of the values of h and the appropriate parameters at the bottom (length L) of the upstream element (subscript u) and at the upstream end of the element in question

20.2.3.3. Numerical Solution. KINEROS solves the kinematic wave equations using a four-point implicit method. Notation for this method is shown in Fig. 20.4, and the finite difference equation for Eqs. (20.11) and (20.12) is

$$\begin{aligned} &h_{j+1}^{i+1} - h_{j+1}^{i} + h_j^{i+1} - h_j^{i} + \\ &+ \frac{2\Delta t}{\Delta x} \left\{ \theta_w \left[\alpha_{j+1}^{i+1} \left(h^m\right)_{j+1}^{i+1} - \alpha_j^{i+1} \left(h^m\right)_j^{i+1} \right] + \left(1-\theta_w\right) \left[\alpha_{j+1}^{i} \left(h^m\right)_{j+1}^{i} - \alpha_j^{i} \left(h^m\right)_j^{i} \right] \right\} \\ &- \Delta t \left(\bar{q}_{j+1} + \bar{q}_j \right) = 0 \end{aligned} \tag{20.16}$$

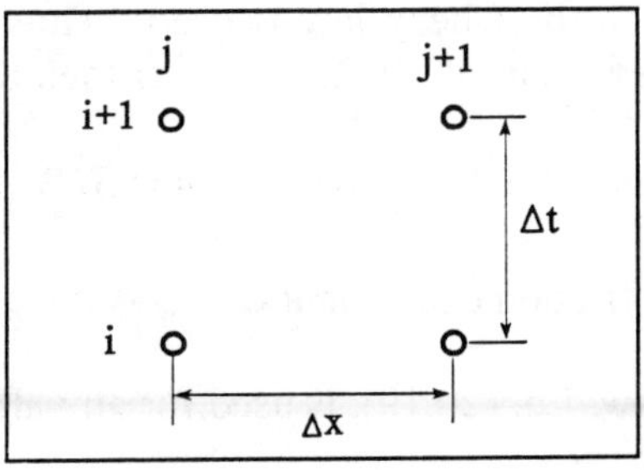

Fig. 20.4. Notation for the space and time dimensions of the finite difference grid.

where θ_w is a weighting parameter (usually 0.6 to 0.8) for the x derivatives at the advanced time step. A solution is obtained by Newton's method (sometimes referred to as the *Newton-Raphson* technique). While the solution is unconditionally stable in a linear sense, the accuracy is highly dependent on the size of Δx and Δt values used. The differencc scheme is nominally of first order accuracy.

20.2.3.4. Roughness Relationships. Four options for α and m in Eq. (20.11) are provided in KINEROS:

1. The Manning hydraulic resistance law is used. In this option

$$\alpha = 1.49\frac{S^{1/2}}{n}, \quad \text{and} \quad m = 5/3 \tag{20.17}$$

where S is the slope, n is a Manning's roughness coefficient for overland flow, and English units are used. A table of values of n for use on surfaces are suggested in the User Manual.

2. A laminar law is used until the Reynolds number (R_e=Q/ν) exceeds a critical value (R_{cr}), and then Manning's law is used. Thus if R_e is smaller than R_{cr},

$$\alpha = \frac{8gS}{k\nu}, \quad \text{and} \quad m = 3 \tag{20.17}$$

where ν is the kinematic viscosity, g is the acceleration of gravity, and k is a dimensionless parameter that has a theoretical value of 24 for a hydraulically smooth surface and may be on the order of 14,000 for turf (Morgali 1970). If R_e is greater than R_{cr}, Eq. (20.17) is used.

3. A laminar law is used until the Reynolds number exceeds a critical value and then the Chezy law is used. Thus if R_e is smaller than R_{cr}, Eq. (20.18) is used. If R_e is greater than R_{cr},

$$\alpha = CS^{1/2}, \qquad \text{and} \qquad m = 3/2 \tag{20.19}$$

where C is the Chezy friction coefficient.

4. The Chezy law, Eq. (20.19), is used for all values of R_e.

Overland flow response characteristics are controlled by the slope, slope length, and hydraulic resistance parameters as well as the rainfall intensity and infiltration characteristics. For example, if the Manning resistance law is used, the time to equilibrium (t_e)[seconds] of a surface of length (L) and slope (S) with a constant rate of lateral inflow (q)[L/T], is

$$t_e = \left[\frac{nL}{1.49S^{1/2}q^{2/3}}\right]^{3/5} \tag{20.20}$$

where n is the Manning resistance coefficient. Therefore, to maintain the time response characteristics of the catchment we must retain the slope, slope length, and hydraulic resistance in our simplified version of a complex catchment. The length, width, and slope of the overland flow elements can be determined from topographic maps. The hydraulic resistance parameter is more difficult to determine. First, a decision must be made regarding the appropriate flow law. It is generally agreed that overland flow on a surface begins as laminar flow and eventually becomes turbulent at large Reynolds numbers. Values of the critical Reynolds number range from 100 to 1,000 (Chow 1959, Yu and McNown 1964, Morgali 1970). However, raindrops impacting on the water film generate local areas of turbulent flow and have an effect similar to increasing the viscosity of the fluid. For example, the relationship between the DarcyWeisbach friction factor (f_D) and Reynolds number for laminar flow over a hydraulically smooth surface is

$$f_D = \frac{24}{R_e} \quad . \tag{20.21}$$

Several laboratory experiments (cf. Glass and Smerdon 1967, Li 1972) have shown that with raindrop impact this relationship becomes

$$f_D = \frac{k}{R_e}; \qquad k > 24 \tag{20.22}$$
$$k = k_o + \chi(r)$$

where k_o is the appropriate value for a given surface geometry without raindrop impact and $\chi(r)$ is some function of the rainfall rate. Although the effects of raindrop impact can be significant on hydraulically smooth surfaces such as laboratory catchments, they are not important for hydraulically rough surfaces usually found in the field. The laminar flow model in KINEROS does not include a relationship between k and rainfall rate. Ranges of k_o for various surfaces are presented in the User Manual (Woolhiser et al. 1990).

Although the range of k_o for each surface is large, the time of equilibrium is proportional to $k_o^{1/3}$, so a 10 percent error in k_o will cause a smaller error in t_e. It should be noted that the relationship $R_e = Q/\nu = uh/\nu$ is used for all surfaces in KINEROS although, for example, the local depth (h) may not be the appropriate characteristic length to use in flow through dense sod. In KINEROS the critical Reynolds number is calculated from the appropriate turbulent Manning's n or Chezy C specified in the input data. Ranges of Manning's n and Chezy C obtained from experiments reported in the literature are presented in the User Manual (Woolhiser et al. 1990a).

20.2.4. Channel Routing

20.2.4.1. Basic Equations. Unsteady, free surface flow in channels is also represented by the kinematic approximation to the equations of unsteady, gradually varied flow. As indicated in Fig. 20.1, channel segments may receive uniformly distributed but time-varying lateral inflow from overland flow elements on either or both sides of the channel, from one or two channels at the upstream boundary, or from an area at the upstream boundary. The dimensions of overland flow units are chosen to completely cover the watershed, so rainfall on the channel is not considered directly. The continuity equation for a channel with lateral inflow is

$$\frac{\partial A}{\partial t} + \frac{\partial Q}{\partial x} = q_c(x,t) \tag{20.23}$$

where A is the cross-sectional area, Q is the channel discharge, and $q_c(x,t)$ is the net lateral inflow per unit length of channel. Under the kinematic assumption, Q can be expressed as a unique function of A and Eq. (20.23) can be rewritten as

$$\frac{\partial A}{\partial t} + \frac{\partial Q}{\partial A}\frac{\partial A}{\partial x} = q_c(x,t) \tag{20.24}$$

The kinematic assumption is embodied in the relationship between channel discharge and cross-sectional area such that

$$Q = \alpha R^{m-1} A \qquad (20.25)$$

where R is the hydraulic radius. If the Chezy relationship is used, $\alpha = CS^{1/2}$ and m = 3/2. If the Manning equation is used, $\alpha = 1.49S^{1/2}/n$ and m = 5/3. Channel cross sections may be approximated as trapezoidal or circular, as shown in Fig. 20.5.

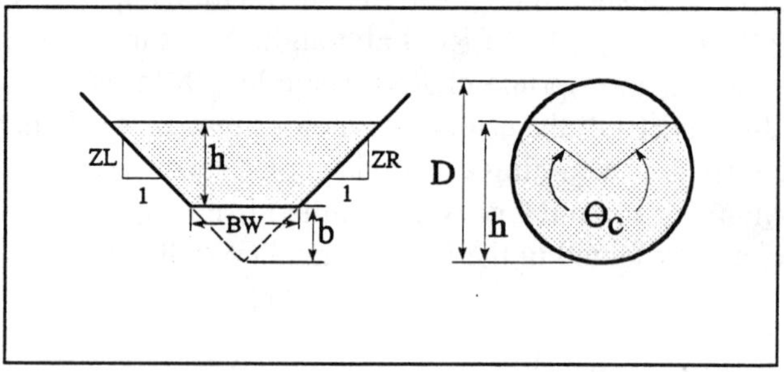

Fig. 20.5. Definition diagrams for a trapezoidal channel (left), and a channel conduit (right).

20.2.4.2. Circular Conduits. In an urban environment, circular conduits must be used to represent storm sewers. To apply the kinematic model, there must be no backwater, and the conduit is assumed to maintain free surface flow conditions at all times--there can be no pressurization. The continuity equation for a circular conduit is the same as Eq. (20.24) with $q_c = 0$. That is, there is no lateral inflow. The upper boundary condition is a specified discharge as a function of time. The most general discharge relationship and the one often used for flow in pipes is the Darcy-Weisbach formula,

$$S_f = \frac{f_D}{4R} \frac{u^2}{2g} \qquad (20.26)$$

where S_f is the friction slope, f_D is the Darcy-Weisbach friction factor, and u is the velocity (Q/A). Under the kinematic assumption, the conduit slope S may be substituted for S_f in Eq. (20.26), so that

$$u = 2\sqrt{\frac{2g}{f_D} RS} \ . \qquad (20.27)$$

Discharge is computed by using Eq. (20.27), and a modification of Eq. (20.25),

$$Q = \frac{\alpha A^m}{p^{m-1}} \tag{20.28}$$

where p is the wetted perimeter, α is $[8gS/f_D]^{1/2}$, and m = 3/2. A schematic drawing of a partially full circular section is shown in Fig. 20.5. Geometric relationships for partially full conduits are further discussed in the documentation [Woolhiser et al. (1990a)].

20.2.4.3. Numerical Method for Channels. The kinematic equations for channels are solved by a four point implicit technique similar to that for overland flow surfaces, except that A is used instead of h, and the geometric changes with depth must be considered:

$$\begin{aligned} &A_{j+1}^{i+1} - A_{j+1}^{i} + A_{j}^{i+1} - A_{j}^{i} + \\ &+\frac{2\Delta t}{\Delta x}\left\{\theta_w\left[\frac{dQ^{i+1}}{dA}\left(A_{j+1}^{i+1} - A_{j}^{i+1}\right)\right] + \left(1-\theta_w\right)\left[\frac{dQ^{i}}{dA}\left(A_{j+1}^{i} - A_{j}^{i}\right)\right]\right\} \\ &\qquad -0.5\Delta t\left(q_{cj+1}^{i+1} + q_{cj}^{i+1} + q_{cj+1}^{i} + q_{cj}^{i}\right) = 0 \end{aligned} \tag{20.29}$$

where θ_W is a weighting factor for the space derivative. The terms dQ/dA are dependent only on channel geometry. Newton's iterative technique is used to solve for the unknown area A_{j+1}^{i+1}. The finite difference equation for circular conduits is the same as Eq. (20.29) except that the lateral inflow terms (q_c) are zero.

The appropriate value of Manning's n or Chezy C to use for channels in KINEROS depends on (1) the channel material (that is, grassed waterway, gravel bedded stream, concrete-lined channel), (2) the degree to which the channel conforms to the idealized trapezoidal cross section, and (3) how straight the channel reach is. Because of these factors, choice of the appropriate parameters is highly subjective, except for artificial channels. Barnes (1967) estimated Manning's n for several streams and presented pictures of the stream reaches. These photographs are very useful in obtaining estimates for natural channels. Values of Manning's n or Chezy C for artificial channels can be obtained from several sources, including Chow (1959).

In arid and semiarid regions, infiltration into channel alluvium may significantly affect runoff volumes and peak discharge. If the channel infiltration option is selected, Eq. (20.3) is used to calculate accumulated infiltration at each computational node, beginning either when lateral inflow begins or when an advancing front has reached that computational node.

Because the trapezoidal channel simplification introduces significant error in the area of channel covered by water at low flow rates (Unkrich and Osborn 1987), an empirical expression is used to estimate an "effective wetted perimeter." The equation used in KINEROS is

$$p_e = \min\left[\frac{h}{0.15\sqrt{BW}}, 1.0\right] p \tag{20.30}$$

where p_e is the effective wetted perimeter for infiltration, h is the depth, BW is the bottom width, and p is the channel wetted perimeter at depth h. This equation states that p_e is smaller than p until a threshold depth is reached, and at depths greater than the threshold depth, p_e and p are identical. The channel loss rate is obtained by multiplying the infiltration rate by the effective wetted perimeter. Further experience may suggest changes in the form and parameters of Eq. (20.30).

20.2.5. Detention Structure Routing

In addition to surface and channel elements, a watershed may contain detention storage elements, which receive inflow from one or two channels and produce outflow from an uncontrolled outlet structure. This element can represent a pond, or a flume or other flow measuring structure with backwater storage. KINEROS accomodates up to three such elements. Rain falling on the pond and infiltration from the pond are not considered. As long as outflow is solely a function of water depth, the dynamics of the storage are well described by the mass balance and outflow equations,

$$\frac{dV}{dt} = q_t - q_o \tag{20.31}$$

and

$$q_o = C_1(h_r - h_z)^{c_2} \tag{20.32}$$

in which $V = V(h_r)$ is storage volume $[L^3]$,

h_r = water surface elevation [L],

q_I = inf low rate $[L^3/T]$,

q_O = outflow rate $[L^3/T]$,

h_z = outflow weir or flume zero flow elevation [L], and

C_1 and c_2 = weir or orifice coefficients.

Water surface elevation (h_r) is measured from some datum below the lowest pond elevation, and $V(h_r)$ is determined from the description of the storage geometry. Equation (20.31) is written in finite difference form over a time interval (Δt) and the stage at time $t + \Delta t$ is determined by the bisection method. Equation (20.32) is general and can represent most control structures such as weirs, overfalls, or orifices. For weirs, C_1 must include the overfall length.

For purposes of water routing, the reservoir geometry may be described by a simple relationship between V, surface area, and h_r. Details for specification of detention pond geometry are given in the User Manual (Woolhiser et al. 1990a).

20.2.6. Surface Erosion and Sediment Transport

As an optional feature, KINEROS can simulate the movement of eroded soil along with the movement of surface water. KINEROS accounts separately for erosion caused by raindrop energy and erosion caused by flowing water, and continues the simulation through channel and pond elements. One necessary limitation is that, because a single mean particle size is used to characterize the eroded material, the effective soil particle size needs to be similar for all the eroding elements.

20.2.6.1. Upland Erosion. The general equation used in KINEROS to describe the sediment dynamics at any point along a surface flow path is a mass balance equation similar to that for kinematic water flow (Bennett 1974):

$$\frac{\partial(AC_s)}{\partial t} + \frac{\partial(QC_s)}{\partial x} - e(x,t) = q_c(x,t) \tag{20.33}$$

in which C_s = sediment concentration $[L^3/L^3]$,

Q = water discharge rate $[L^3/T]$

A = cross sectional area of flow $[L^2]$,

e = rate of erosion of the soil bed $[L^2/T]$, and

q_s = rate of lateral sediment inflow for channels $[L^3/T/L]$.

For upland surfaces, e is assumed to be composed of two major components--production of eroded soil by splash of rainfall on bare soil, and hydraulic erosion (or deposition) due to the interplay between the shearing force of water on the loose soil bed and the tendency of soil particles to settle under the force of gravity. Thus e may be positive (increasing concentration in the water) or negative (deposition). Net erosion is a sum of splash erosion rate as e_s and hydraulic erosion rate as e_h,

$$e = e_s + e_h \tag{20.34}$$

20.2.6.2. Splash Erosion. Based on limited experimental evidence, the splash erosion rate can be approximated as a function of the square of the rainfall rate (Meyer and Wischmeier 1969). This relationship can lead to physically unrealistic concentrations at the upstream boundary, so in KINEROS the splash erosion is somewhat modified as follows:

$$\begin{aligned} e_s &= c_f k(h) r q \quad ; \quad q > 0 \\ &= 0 \qquad\qquad\; ; \quad q < 0 \end{aligned} \tag{20.35}$$

in which c_f is a constant related to soil and surface properties, and k(h) is a reduction factor representing the reduction in splash erosion caused by increasing depth of water. Since q approaches r for high rainfall rates, this relation approximates the r^2 relation found by experiment except at the inception of runoff. k(h) is 1.0 prior to runoff and its minimum is 0 for very deep flow. The function k(h) is given by the empirical expression

$$k(h) = \exp(-c_h h) \tag{20.36}$$

The parameter c_h represents the damping effectiveness of surface water, and does not vary widely. Both c_f and k(h) are always positive, so e_s is always positive when there is rainfall and a positive rainfall excess (q).

20.2.6.3. Hydraulic Erosion. The hydraulic erosion rate e_h represents the rate of exchange of sediment between the flowing water and the soil over which it flows, and may be either positive or negative. KINEROS assumes that for any given surface water flow condition (velocity, depth, slope, etc.), there is an equilibrium concentration of sediment that can be carried if that flow continues steadily. Hydraulic erosion rate (e_h) is estimated as being linearly dependent on the difference between the equilibrium concentration and the current sediment concentration. In other words, hydraulic erosion/deposition is modeled as a kinetic transfer process;

$$e_h = c_g (C_{mx} - C_s) A \tag{20.37}$$

in which C_{mx} is the concentration at equilibrium transport capacity, $C_s = C_s(x,t)$ is the current local sediment concentration, and c_g is a transfer rate coefficient $[T^{-1}]$. Clearly, the transport capacity is important in determining hydraulic erosion, as is the selection of transfer rate coefficient. Conceptually, when deposition is occurring, c_g can be shown to be equal to the particle settling velocity, v_s, divided by the hydraulic depth, h. For erosion conditions on cohesive soils, the value of c_g must be reduced, and v_s/h is used as an upper limit for c_g.

20.2.6.4. Transport Capacity. Many transport capacity relations have been proposed in the literature, but most have been developed and tested for relatively deep, mildly sloping flow conditions, such as streams and flumes. In the absence of a clearly superior relation for shallow surface runoff, KINEROS provides the user a choice between several transport relationships. Julien and Simons (1985) showed that each could be represented by a generalized relation of the type

$$q_m = C_{mx}Q = \omega S^{\beta} Q^{\gamma} r^{\delta} \left[1 - \frac{\tau_c}{\tau_o} \right]^{\varepsilon} \quad ; \; \tau_o \geq \tau_c \tag{20.38}$$

in which q_m = transport capacity $[L^2T^{-1}]$,

τ_o = bed shear stress [L /T],

τ_c = critical shear stress [L /T], and

ω = a coefficient.

Exponents ß, γ, and ε have values of either 0 or between 1 and 2, and ß varies from 0 to -2.24. Although not every transport relation selectable in KINEROS has non-zero values for all the exponents, ß, γ, and ε, all relations use local hydraulic conditions such as slope, velocity, or depth of flow. Some, in addition, use sediment specific gravity and the mean particle size of the soil rather than requiring selection of an empirical coefficient. The equations for each transport relationship are discussed in detail in the User Manual (Woolhiser et al. 1990a).

Particle settling velocity is calculated from particle size and density, assuming the particles have drag characteristics and terminal fall velocities similar to those of spheres (Fair and Geyer 1954). This relation is

$$v_s^2 = \frac{4}{3} \frac{g(S_s - 1)d}{C_D} \tag{20.39}$$

in which g = gravitational acceleration $[LT^{-2}]$,

S_s = particle specific gravity,

C_D = drag coefficient, and

d = particle diameter [L].

The drag coefficient is a function of particle Reynolds number,

$$C_D = \frac{24}{R_n} + \frac{3}{\sqrt{R_n}} + 0.34 \tag{20.40}$$

in which R_n is the particle Reynolds number, defined as

$$R_n = \frac{v_s d}{\nu} \tag{20.41}$$

where ν is kinematic viscosity of water [L^2/T]. Settling velocity of a particle is found by solving Eqs. (20.39, 20.40, and 20.41) for v_s.

20.2.6.5. Numerical Method for Sediment Transport. Equations (20.33 - 20.37) are solved numerically at each time step used by the surface water flow equations. A four-point finite-difference scheme is used; however, iteration is not required since given current and immediate past values for A and Q and previous values for C_s, the finite difference form of this equation is explicit:

$$C_{s\,j+1}^{\,i+1} = f\left(C_{s\,j}^{\,i},\ C_{s\,j+1}^{\,i},\ and\ C_{s\,j}^{\,i+1}\right)$$

The value of C_{mx} is found from current hydraulic conditions.

20.2.6.6. Initial Conditions for Erosion. When runoff commences during a period when rainfall is creating splash erosion, the initial condition on the vector C_s should not be taken as zero. The initial sediment concentration at ponding ($C_s(t=t_p)$), can be found by simplifying Eq. (20.33) for conditions at that time. Variation with respect to x vanishes, and hydraulic erosion is zero. Then,

$$\frac{\partial(AC_s)}{\partial t} = e(x,t) = c_f rq - C_s v_s \tag{20.42}$$

where k(h) is assumed to be 1.0 since depth is zero. Since A is zero at time of ponding, and dA/dt is the rainfall excess rate (q), expanding the left-hand side of Eq. (20.42) results in

$$C_s(t = t_p) = \frac{c_f rq}{q + v_s} \tag{20.43}$$

The sediment concentration at the upper boundary of a single overland flow element, $C_s(0,t)$, is given by an expression identical to Eq. (20.43), and a similar expression is used at the upper boundary of a channel.

20.2.7. Channel Erosion and Sediment Transport

The general approach to sediment transport simulation for channels is nearly the same as that for upland areas. The major difference in the equations is that splash erosion (e_s) is neglected in channel flow, and the term q_s becomes important in representing lateral inflows. Equations (20.33) and (20.37) are equally applicable to either channel or distributed surface flow, but the choice of transport capacity relation may be different for the two flow conditions. For upland areas, q_s will be zero, whereas for channels it will be the important addition that comes with lateral inflow from surface elements. The close similarity of the treatment of the two types of elements allows the program to use the same algorithms for both types of elements.

The erosion computational scheme for any element uses the same time and space steps employed by the numerical solution of the surface water flow equations. In that context, Eqs. (20.33) and (20.37) are solved for $C_s(x,t)$, starting at the first node below the upstream boundary, and from the upstream conditions for channel elements. If there is no inflow at the upper end of the channel, the transport capacity at the upper node is zero and any lateral input of sediment will be subject there to deposition. The upper boundary condition is then

$$C_s(0,t) = \frac{q_s}{q_c + v_s BW} \tag{20.44}$$

where BW is the channel bottom width. A(x,t) and Q(x,t) are assumed known from the surface water solution.

20.2.8. Sediment Routing Through Detention Structures

For shallow rapid flow where erosion is generally more important than deposition, the use of a mean particle diameter can usually be justified for simulation purposes. When reservoirs are important elements in the catchment, however, KINEROS asks the user to specify a particle size distribution, because deposition is the only sedimentation process, and settling velocity is very sensitive to particle size. The distribution is characterized by a mean and a standard deviation and is assumed pseudonormally distributed into a number of particle size classes specified by the user.

The approach for pond sedimentation in KINEROS is similar to that for tank sedimentation, and particle fall velocities and flow-through velocities are used to find the trajectories that intersect the reservoir bottom. Particle fall velocities are calculated for each particle size class using Eqs. (20.39 and 20.41). Particles are assumed distributed uniformly through the reservoir depth in the first section at the inlet end, and the relative fall versus lateral velocities from that point forward determine the proportion of each particle size class that deposits between successive cross sections. At successive sections, each particle size class will have a conceptual depth from the surface to the top of the area still containing sediment of that size.

Suspended and slowly falling particles are subject to molecular diffusion and dispersion, which can sometimes modify the time distribution of outflow concentrations. KINEROS simulates this modification using an effective dispersion coefficient, which is supplied by the user. Typical values suggested are from 10^{-5} to 10^{-4} ft^2/s. In many cases, dispersion is unimportant and can be neglected, such as for very short flow-through situations or for large particles. If a pond is located within a watershed, the mean particle diameter of the outflow will be smaller than that of the inflow. KINEROS cannot account for this phenomenon.

Since the time necessary for a reservoir outflow to approach zero may be significantly longer than that for a channel input hydrograph to return to zero, the user may need to specify a much longer simulation time if ponds are included in the catchment geometry.

20.3. EVALUATION OF KINEROS

20.3.1. Testing of KINEROS on the Colorado State University Outdoor Rainfall-Runoff Experimental Facility

The kinematic overland flow routing component of KINEROS and its precursors has been thoroughly tested using data from the Colorado State University Outdoor Rainfall-Runoff Experimental Facility (ORREF). This unique facility allows relatively precise measurement of rainfall and runoff at a scale comparable to small catchments, and has been described by Dickenson et al. (1967) and Woolhiser et al. (1971). The 0.2 ha (one-half acre) watershed consists of two parallelogram elements 35 m long contributing lateral inflow to a triangular channel 21.3 m long. A conical segment contributes runoff at the head of the channel. The overland flow elements have 5% slopes and the channel slope is 3%. Rainfall rates could be set at 12.5, 25, 50, and 100 mm/hr. The surface was covered with impervious butyl rubber and gravel was placed on the surface in various densities and spatial patterns to increase flow resistance.

Singh (1974) analyzed data from 210 experimental runs with 50 different configurations on the conical section and found that the kinematic wave formulation provided a good description of surface runoff from the facility. Singh (1974) used a numerical solution of the equations describing converging overland flow. Kibler and Woolhiser(1970) have demonstrated that the response of a converging section can be well approximated by the response of a cascade of rectangular surfaces as used in KINEROS.

Additional tests of the kinematic model using data from the ORREF were performed by Lane et al. (1975), Rovey, et al. (1977), and Wu et al. (1978), who also demonstrated that the flow measuring structure caused a lag in the measured rising hydrographs and that when this lag was accounted for, better agreement was achieved between measured and computed hydrographs. It should be pointed out that the researchers cited above used a second-order Lax-Wendroff numerical scheme which is more accurate than the current scheme in KINEROS. However if the guidelines in the KINEROS manual for choice of dx and dt increments are followed, there will be little difference in the numerical results.

20.3.2. Comparison to Other Models and Application on Small Watersheds

Zevenbergen and Peterson (1988) evaluated a variety of storm-event hydrologic and erosion models including a pre-release version of KINEROS (Woolhiser et al., 1990a). Also included in their evaluation were the following models: Areal Nonpoint Source Watershed Environmental Response Simulation (ANSWERS); Watershed and Sediment Runoff Simulation Model for Multiple Watersheds (MULTSED); Precipitation Runoff Modeling System (PRSM); and Hydrology and Sedimentology Watershed Model II (SEDIMOT, SEDCAD version). PRSM was not fully evaluated as it was primarily designed for continuous simulation.

A primary goal of the study was to judge the relative ability these models to predict sediment yield. There were three evaluation categories: "1) degree of physical process simulation; 2) how well the model results compare with observations at an instrumented watershed, and; 3) the sensitivity of model results to variations in input parameters." User manuals of the models were used to assess the type and number of physical processes treated in each model as well as watershed representation capability. They judged ANSWERS and KINEROS as allowing the greatest flexibility in watershed segmentation and MULTSED and KINEROS as simulating the greatest number of channel processes.

Each of the models was tested with runoff and sediment yield observations from nine storm events from a 1.9 ha (4.7 acre) experimental watershed in northwest Mississippi. The events ranged from small to quite large (ten year return period for two- and four-hour durations). Zevenbergen and Peterson (1988) graphically depicted runoff simulation ability of MULTSED, KINEROS and ANSWERS for one event and the agreement of the KINEROS simulation was judged to be "excellent," and slightly better than the simulations of the other two models. To evaluate sediment yield predictability, they calibrated detachment and transport coefficients on the second largest event and computed the coefficient of determination (CD) on the remaining eight events. Using the tractive force transport relation,

KINEROS had the highest CD (0.96) for sediment (CD = 1.0 represents perfect agreement). MULTSED also had a CD = 0.96 using the Meyer-Peter, Muller/Einstein relation. [A discrepancy pointed out by Zevenbergen and Peterson (1988) between ANSWERS and KINEROS using the Yalin transport relationship was due to a coding error in KINEROS which was corrected before the 1990 release.] The general conclusion of this study stated "that all the models except SEDIMOT are quite complementary with respect to hydrology and similar results can be achieved from ANSWERS, KINEROS or MULTSED models for sediment yield."

The KINEROS model has been applied to several semiarid watersheds covering a range of basin scales within the USDA-ARS Walnut Gulch Experimental Watershed operated by the Southwest Watershed Research Center in Tucson, AZ. A location map of the watershed and the specific subwatersheds discussed in this and the following section is presented in Fig. 20.6. Typical Walnut Gulch watershed characteristics include desert brush vegetation, sandy loam soils with significant rock content, and high amounts of soil surface erosion pavement. The subwatersheds are subject to high intensity precipitation from air-mass thunderstorms of limited spatial extent (Renard, 1970).

In February of 1981 the LH-6 (0.36 ha) watershed (see Fig. 20.6) was treated with the herbicide tebuthiuron to kill the woody plants, and was seeded with a mixture of grasses and forbes in June 1984. The adjacent LH-2 (1.4 ha) watershed was not treated with herbicide. KINEROS was used to model post-treatment events from LH-6 using parameters derived from pretreatment conditions (1973-1980) to assess whether the model would be able to detect changes in the runoff regime due to vegetation changes (see Woolhiser et al., 1990b for more detail). LH-2 was used for paired watershed comparisons between treated and untreated conditions.

Split sample calibration and verification of KINEROS using pretreatment runoff data was conducted to see if sufficient confidence could be gained in the model before it was used to attempt detection of vegetation change. Model performance for both calibration and verification was evaluated by inspection of individual hydrographs, examination of scatter plots of measured versus simulated runoff volumes and peak rates, and the coefficient of efficiency, E (Nash and Sutcliffe, 1970). $E = 1$ if the model predicts observed runoff with perfection. If $E < 0$, the model's predictive power is worse than simply using the average of observed values.

For the LH-6 watershed, 1:480 scale base maps indicated a representation with 23 overland flow elements and 7 channel elements (Fig. 20.6). The degree of topographic discretization is a central question in the application of a distributed model such as KINEROS. In many applications the channel elements used are those observed on the best maps available. However, stream network depiction is highly dependent on map scale and

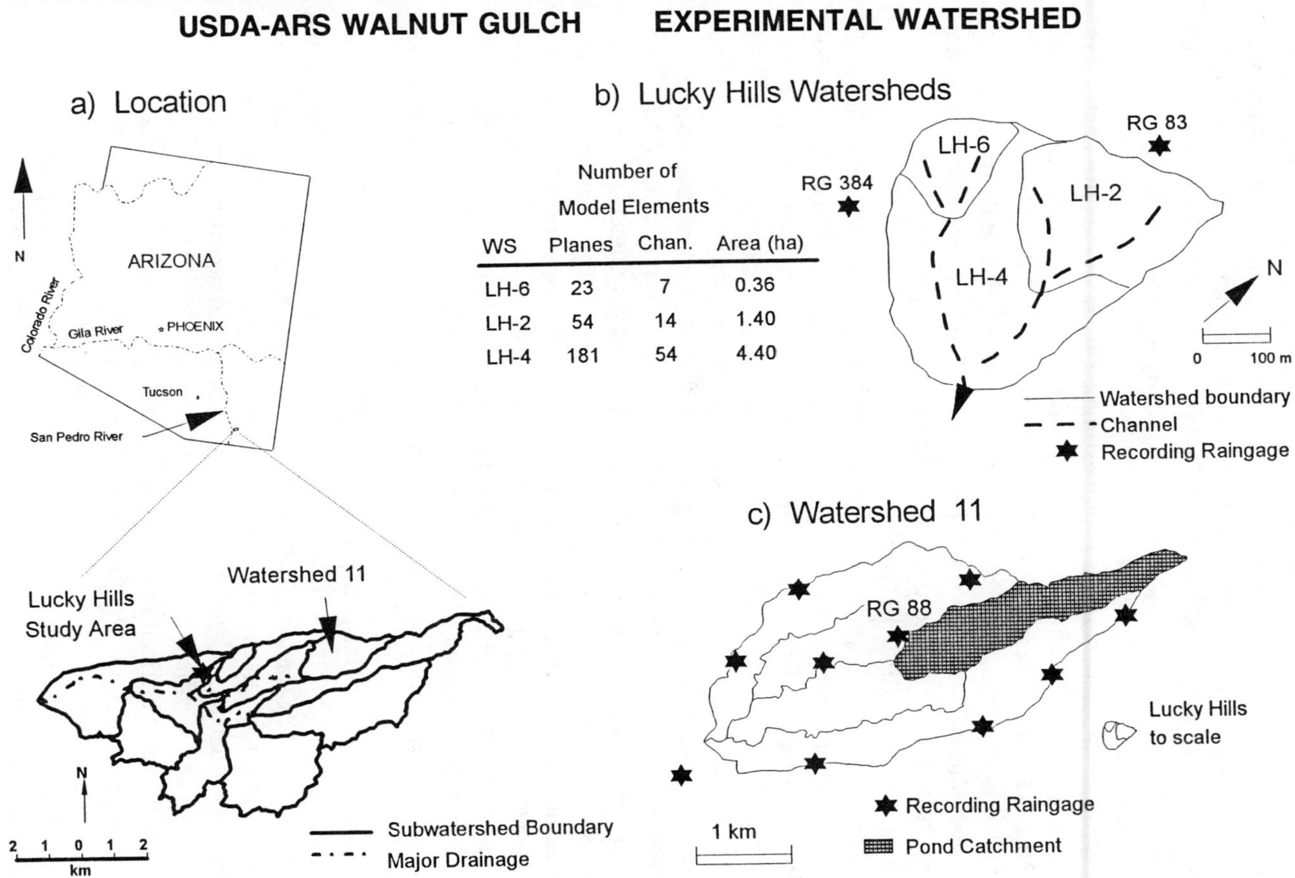

WS	Number of Model Elements: Planes	Chan.	Area (ha)
LH-6	23	7	0.36
LH-2	54	14	1.40
LH-4	181	54	4.40

Fig. 20.6. a) Study area location map; b) Lucky Hills Watersheds; and c) Watershed 11.

cartographic judgement. This problem was addressed in Goodrich et al. (1988) and Goodrich (1991). It was concluded that an average channel support area of 15% (the average area upstream of channel initiation) was adequate to preserve model simulation of surface runoff for watersheds from 1.4 to 631 ha. Along with this it should be emphasized that high levels of discretization are not justified without having spatially distributed rainfall data of a commensurate detail.

Infiltration parameters (K_S) were estimated from soil texture analysis of 17 soil samples. Manning's roughness values (n) were estimated from the KINEROS User's Manual table 5 (Woolhiser et al., 1990a). Input rainfall data were obtained from raingage 83 for both LH-6 and LH-2 for the pretreatment split sample test. Pre-storm initial soil moisture conditions were calculated with the daily water balance component of the chemical transport model CREAMS (USDA, 1980).

For each of the watersheds, 10 pretreatment events were selected covering a large range of event sizes, initial conditions, and complexity of rainfall patterns. To decrease the optimization parameter space, it was assumed that the relative local values of K_s and n estimated from field measurements were fixed, and parameter multipliers were optimized. The Nash-Sutcliffe coefficient of efficiency (E), discussed above was used as an objective function for event runoff volume and peak runoff rate. When final multipliers were applied to initial field estimated parameters, the resulting values were checked to insure that they were physically realistic and were not acting as mere fitting parameters. These calibrated parameter sets were then used to model runoff response for an independent verification event set of 30 pretreatment events. Note that by using nested watersheds LH-6 and LH-2 within watershed LH-4, internal model verification was possible and some degree of interior, or distributed, model confidence could be established.

Using the optimum parameters, the KINEROS model performed very well as judged by the efficiency statistic (E) for both calibration and verification event sets on LH-6 and LH-2 (see columns labelled 1 in Table 20.1). Extrapolation capability is demonstrated by the simulation of runoff from events outside the calibration range of observed runoff volumes (0.8 to 14.5 mm) to a verification range of 0.3 to 47.5 mm. Typical scatter plots for calibration and verification events sets of simulated runoff volumes versus observed data are illustrated in Fig. 20.7 for pretreatment LH-6 events.

The favorable pretreatment modeling results encouraged use of the model for possible detection of hydrologic changes due to post-treatment vegetation changes. The ensuing analysis, described in more detail by Woolhiser et al. (1990b) demonstrated that due to model bias, predicting a

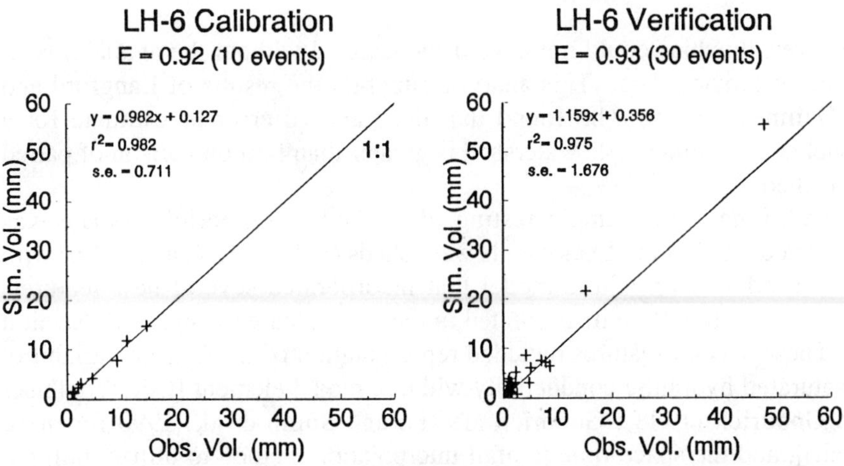

Fig. 20.7. Scatter plots for KINEROS calibration and verification for runoff volume on LH-6 for pretreatment conditions (1973-1980).

TABLE 20.1.

KINEROS calibration and verification coefficient of efficiency (E) for runoff volume (mm) and peak discharge (mm/hr) using optimum multipliers for the pretreatment period (1973-1980).

Basin		Calibration Efficiency				Verification Efficiency			
	Area	Volume		Peak		Volume		Peak	
	ha	1	2	1	2	1	2	1	2
LH-6	0.36	.92	.98	.93	.95	.93	.98	.77	.79
LH-2	1.40	.92	.97	.95	.97	.84	.93	.79	.93
LH-4	4.40	–	.96	–	.88	–	.99	--	.96

1 - one raingage. no heterogeneity

2 - two raingages, rainfall interpolation, small scale K_S heterogeneity

change in hydrologic regime can be misleading if model parameters were estimated from a small subset of pretreatment runoff data, and if the changes in runoff volumes are small. When the model bias for the treated watershed was accounted for by comparing results to the adjacent untreated LH-2 watershed, an apparent, but small, decease in runoff volumes was indicated due to enhanced infiltration on the treated and seeded LH-6 watershed.

However, this bias could not have been detected without the availability of paired watershed data. This analysis supports the results of Langford and McGuinness (1976) who found that the standard error of estimate for a model applied to a single watershed is greater than for comparison of paired watersheds.

Additional split sample testing, described in Goodrich (1991), was conducted on all three Lucky Hills watersheds (LH-2, LH-4, and LH-6) with a modified version of the model that incorporated several new features, some of which will be incorporated in the next release version (see section 4). These special features included representation of small scale variability of saturated hydraulic conductivity within a model element [(see Woolhiser and Goodrich (1988); Goodrich (1991); and Smith et al., 1990 for more detail)], and the space-time rainfall interpolation method to utilize multiple raingage information (described in section 20.4, below). As in the analysis above the calibration event set was selected to cover a range of runoff volume and intitial mositure conditions. The verification set contained events outside the calibration range to test model authenticity. For the Lucky Hills examples, rainfall data were obtained from raingage 83 and 384 which are separated by roughly 300m. An additional parameter, the coefficient of variation of saturated hydraulic conductivity, $C_v(K_s)$, was used during this optimization.

The results are summarized in the column labelled "2" in Table 20.1. Observed best and worst calibration simulation hydrographs for LH-4 are illustrated in Fig. 20.8. Timing errors of approximately eight to ten minutes or less are irrelevant in the Lucky Hills data, owing to normal clock errors. The better simulations occur for the largest events, and the worst occur for the smallest events for all watersheds. For the very small events, uncertainty of measurement error can become a large percentage of the observed runoff. The nonlinearities in the infiltration process will aslo tend to dominate the rainfall-runoff transformation. In all cases, the rainfall interpolation and K_s variability algorithms improved model results. Further analysis demonstrated that the majority of the improvement as measured by E came from using the additional raingage. Sensitivity to changes in C_v were particularly acute for small to medium events where the rainfall rates interact with spatially variable infiltration capacities. However, these events do not influence the E statistic to a large extent. For large events the variation in K_s (and therefore infiltration capacities) is largely overwhelmed by high rainfall intensities. Internal model consistency was demonstrated by successful application of the parameters from LH-4 to watersheds LH-2 and LH6, with E values from 0.86 to 0.97.

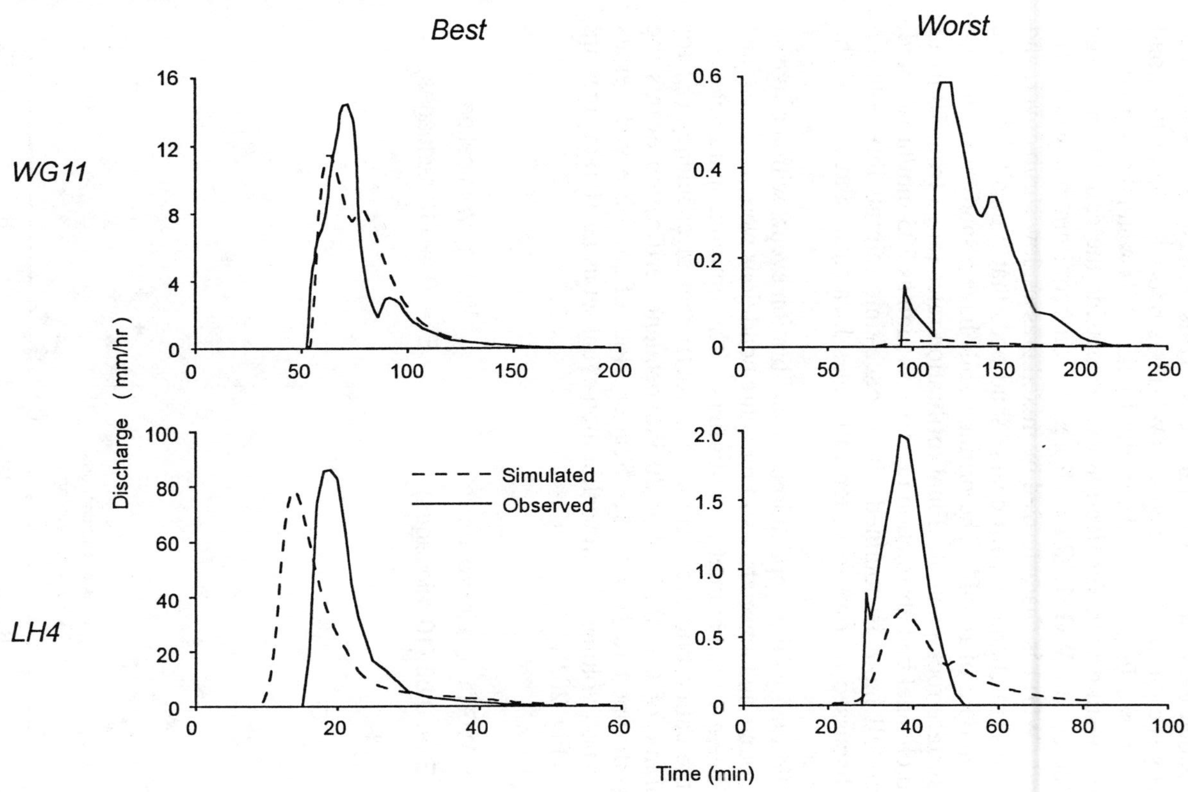

Fig. 20.8. Best and Worst sample hydrographs from calibration of the LH4 and WG11 watershed simulations.

20.3.3. Application to Large Semiarid Watersheds

Application of KINEROS to a larger subwatershed within Walnut Gulch (Watershed 11) and the entire basin has also been undertaken. Goodrich (1991) extended the analysis from the Lucky Hills watersheds discussed in the previous section to Watershed 11 (WG11 is 630 ha without the pond subcatchment, see Fig. 20.6). As in the Lucky Hills watersheds, parallel split sample testing with verification events outside the calibration range was conducted on WG11. Rainfall resolution requirements were also investigated. WG11 is characterized by mixed grasses, low shrub cover, and ephemeral stream channels with coarse sand beds that typically abstract a significant amount of runoff. The primary soils have a sandy loam texture and high surface rock content. Final infiltration rates range from 10 to 13 mm/hr and channel bed abstraction rates are as high as 225 mm/hr (Renard, 1970). Rainfall input is obtained via a space-time rainfall interpolation scheme (discussed below) from ten raingages located in and around the subwatershed.

As in the Lucky Hills application, ten calibration events with a range of rainfall amounts and antecedent soil moisture conditions were used to adjust the same basin-wide parameter multipliers. The optimum parameter set modeled the calibration event set with a coefficient of efficiency (E) for runoff volume of E = 0.86. For 20 independent verification events the efficiency dropped to E = 0.49. Scatter plots of the observed versus simulated runoff volumes for the calibration event set using 10 raingages are presented in Fig. 20.9.

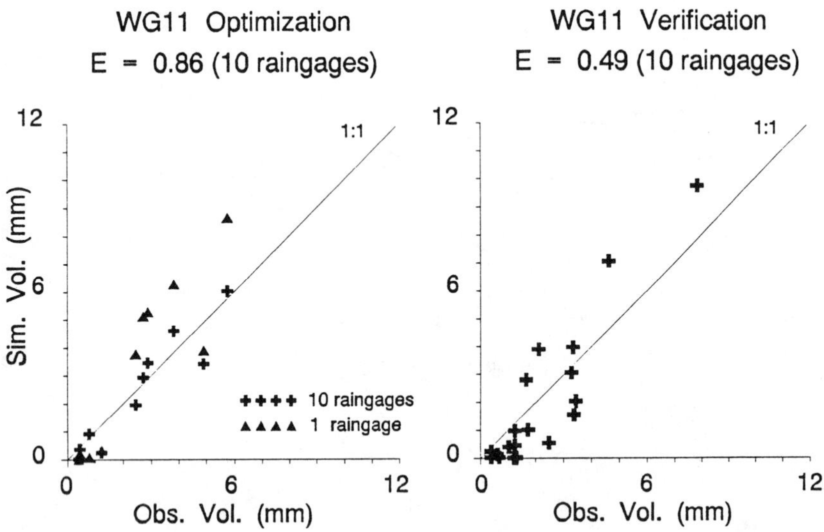

Fig. 20.9. Walnut Gulch Watershed 11 calibration and verification for runoff volume (10 raingages) and calibration set with 1 raingage.

Observed and best and worst simulation hydrographs (based on runoff volume) for the calibration event set are illustrated in Fig. 20.8. As in Lucky Hills the worst simulations occur for the smallest events.

Possible explanations for the model's poorer performance with the verification set include differing channel losses and antecedent soil conditions, and large random rainfall variability that may not have been captured in the small subset of ten calibration events in contrast to Lucky Hills examples. In addition, the same degree of field sampling to estimate initial parameters was not possible in the much larger WG11 watershed. However, the results cannot be dismissed outright as being unacceptable and are good when compared to other modeling efforts with Walnut Gulch data (Hughes and Beater, 1989).

To assess the impacts of spatial rainfall representation, the same events were simulated using data from a single raingage (RG 88) located near the center of WG11 (see Fig. 20.6). Use of a single raingage implies a spatially uniform rainfall representation. In Fig. 20.9, the modeled runoff volumes for the single raingage are also plotted. The figure illustrates that for 6 of 10 of the calibration events, and for all of the medium to large events, the uniform representation overestimated runoff volume (from 14 to 93%). In addition, the coefficient of efficiency for runoff volume dropped to 0.07. Because peak runoff rate is highly correlated with runoff volume in the semi-arid environment of Walnut Gulch (Goodrich, 1991), similar impacts on peak runoff rate occurred when the single raingage was used instead of the ten-gage network. This suggests that at the scale of WG11, the uniform rainfall representation results in poor estimations. Based on records from the dense raingage network on Walnut Gulch, runoff producing rainfall usually is restricted to only portions of the subwatersheds, and is not evenly distributed. This justifies the need for distributed rainfall representation within the model.

Michaud and Sorooshian (1992) applied KINEROS to the entire 15,000 ha Walnut Gulch Watershed. The primary objective of their study was to evaluate the accuracy of KINEROS simulations from a flash flood forecasting perspective under data constraints that are typical of watersheds with "ALERT" (Automated Local Evaluation in Real Time) flood warning systems. This limits the rain data to 1 gage per 2000 ha. KINEROS simulations were also compared to two SCS curve number models, lumped and distributed (McCuen, 1982), for "ungaged" and calibrated cases.

To match typical ALERT raingage density, only 8 recording raingages were used. Six calibration and 24 verification events were selected from the largest storms on record. However, 8 of the 24 verification events were from conventional streamflow measurements before a special supercritical flume was installed and the data from these events are of questionable quality. Instead of the rainfall interpolation scheme described below, they used kriging methods for interpolation. Small scale variability of saturated hydraulic conductivity (K_s) within model elements was also employed.

In general, they found that KINEROS was "reasonably accurate in simulating time to peak, but provides poor simulations of peak flow and runoff volume." This conclusion was supported by the E statistics for 24 verification events presented in Table 20.2 (from table 4-3 of Michaud and Sorooshian, 1992). The spatially-lumped SCS model could not accurately simulate flood peak and time to peak. For the calibration case, the distributed SCS model with channel losses was approximately as accurate as KINEROS. For the ungaged, or uncalibrated case, KINEROS was more accurate than the distributed SCS model. The major reasons cited for poor KINEROS runoff simulations were poor estimation of rainfall and prediction of infiltration. Roughly 50% of the simulation error was attributed to the coarse rainfall sampling.

TABLE 20.2.

KINEROS Nash-Sutcliffe efficiency for 24 verification events for the entire Walnut Gulch Watershed (from table 4-3 of Michaud and Sorooshian, 1992).

Case	Peak Rate	Runoff Volume	Time to Peak
Ungaged	-0.15	-0.52	0.91
Calibrated	-0.68	-0.39	0.78

20.3.4. Discussion of Applications

For the applications examined, KINEROS does a relatively good job simulating runoff and sediment yield at watershed scales of up to approximately 1000 ha (Zevenbergen and Peterson, 1988; Woolhiser et al., 1990b; and Goodrich, 1991). Distributed (interior) model accuracy was demonstrated within small (5 ha, 0.05 km^2) nested subwatersheds. Confidence in model accuracy at this scale highlighted an additional beneficial model use as a tool for data checking. Extremely poor simulation results prompted closer examination of the original rainfall and runoff gage charts. In numerous cases, errors in gage instrumentation or errors in data processing, such as improper time corrections were detected. For watersheds larger then 1000 ha, application of a detailed, process based model, such as KINEROS, may be difficult to justify in the absence of distributed rainfall data, given comparable results from a simpler model which does not entail the costs associated with detailed basin characterization required for KINEROS model inputs (Michaud and Sorooshian, 1992). But this assessment implies the existence of good data for proper calibration, and comparison of calibrated models.

20.4. IMPROVEMENTS UNDERWAY

A comprehensive revision of the FORTRAN program has recently been completed in anticipation of a second release (tentatively called KINEROS2) in 1994. The new code has a more consolidated, modular structure and includes the following additional capabilities:

(1) A composite "urban" element has been developed to mitigate the complexity of modeling an urban environment;

(2) The open channel algorithm has been extended to allow a compound cross section with an overbank section where hydraulic and infiltration parameters can differ from those in the main section;

(3) The new reservoir routing routine accounts for seepage through the wetted area, rainfall on the reservoir itself, and initial storage;

(4) The program will now synthesize rainfall input for individual elements using data from a network of raingages;

(5) The new infiltration model uses a new soil-physics based method for treating redistribution of water during any interruption in the storm, and will treat a soil profile comprised of two soil layers;

(6) A base flow can be specified for each open channel element;

(7) Time varying inflow from sources outside the model, such as measured flows or output from other models, can be injected into the modeled system at any point;

(8) Optional graphical output at the end of simulation showing the rainfall pattern and hydrograph simulated on any element;

(9) Soil and sediment are characterized by a distribution of up to 5 particle size classes rather than a single median particle size.

The composite urban element consists of four components: an open curbside gutter, an impervious area connected directly to the gutter, an impervious area contributing to a pervious area, and the pervious area which empties into the gutter. An interpretation of the element components as they might relate to a residential lot is shown in Fig. 20.10. The fraction of total area assigned to each component defines the element along with the hydraulic and infiltration parameters.

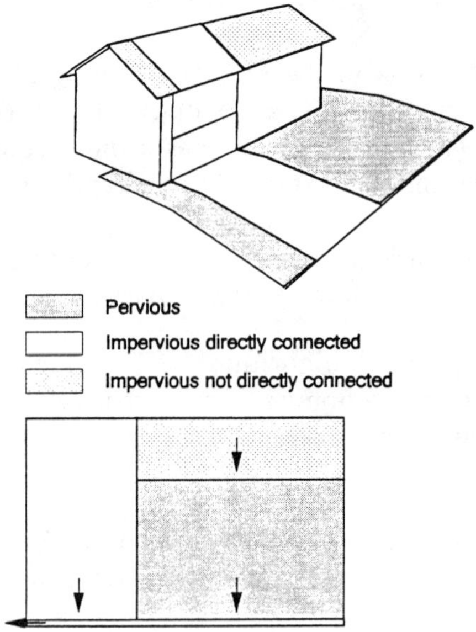

Fig. 20.10. Interpretation of an urban element in the new KINEROS2 model.

The compound channel algorithm is based on two independent kinematic equations - one for the main channel and one for the overbank section - that are written in terms of the same datum for flow depth. In writing the separate equations, it is explicitly assumed that no energy transfer occurs between the two sections, and upon adding the two equations the common datum implicitly requires the water surface elevation to be equal in both sections (Fig. 20.11). Each section has its own set of parameters describing

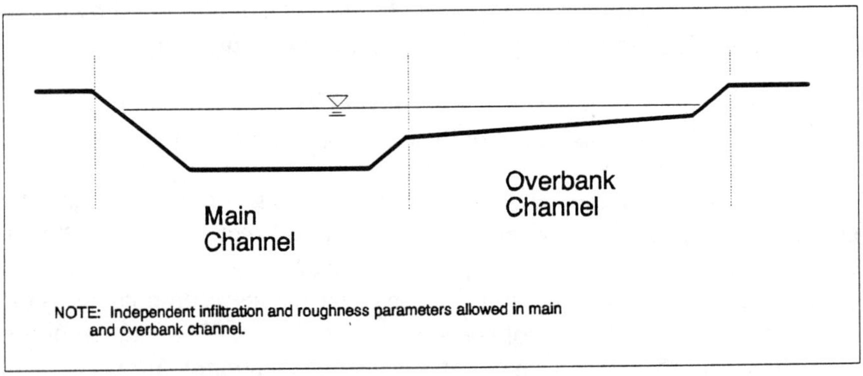

Fig. 20.11. Compound channel cross-section diagram in KINEROS2.

the hydraulic roughness, bed slope, and infiltration characteristics. A compound channel element can be linked with other compound channels or with simple trapezoidal channel elements.

Spatial relationships between raingages and model elements are expressed by rules based on proximity. The position of an element relative to the raingage network is described by a representative point such as the centroid of its area. If this point lies within the triangle formed by the three nearest raingages (in the xy plane), it is projected up to the plane passing through the intensity (z) at each raingage to obtain an interpolated intensity (Fig. 20.12a). If the representative point is not within this triangular region but lies within an area bounded by two parallel lines which pass through the

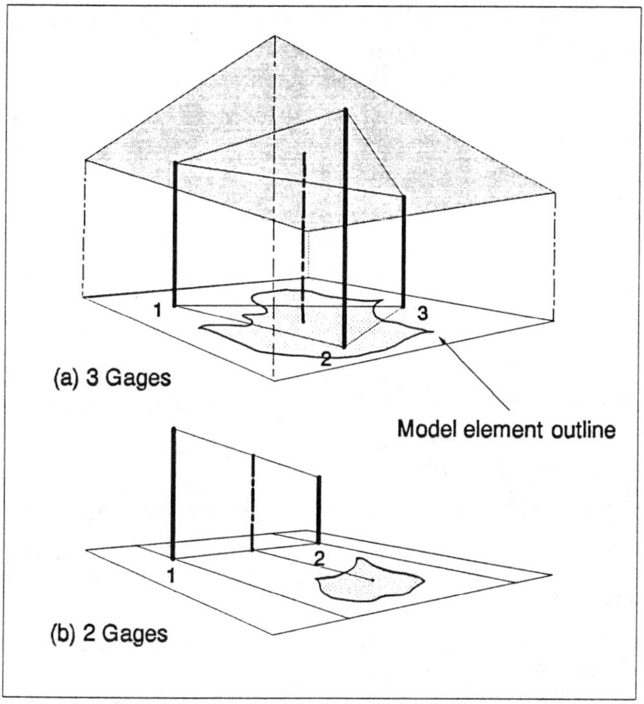

Fig. 20.12. Interpolation of rainfall intensity to element centroid.

two nearest raingages and are perpendicular to a line segment connecting the two raingages, the point is projected to the line segment and the intensity is linearly interpolated according to the relative position of the projected point on the line (Fig. 20.12b). If neither of these two conditions are met breakpoint data from the nearest raingage are used exclusively.

The breakpoint times from the two or three contributing raingage data sets are combined to define the time sequence for computing these intensities. The piece-wise planar (3 gages) or linear (2 gages) approximation of the rainfall intensity field is recomputed each time any raingage intensity changes. The intensity interpolated at the runoff model element centroid is applied uniformly over the entire element. Therefore, the efficacy of the scheme becomes questionable if the model elements are large enough to encompass more than three raingages, because the resulting rainfall input will reflect data from at most three raingages and will neglect rainfall input from the other gages.

The new soil water profile simulation method allows input of a complex pattern of intensity, with multiple storms or significant rainfall hiatus. This method was developed by extending the infiltration model to include very high initial water contents, tracking secondary wetting fronts, and formulating an approximate simulation of redistribution based on soil hydraulic characteristics (Smith, et al, 1993). The two-layer soil model allows study of such natural phenomena as porous channel alluvium overlying a subsoil, and surface layers or crusted soils. Treatment of several sediment size classes allows realistic simulation of a watershed composed of eroding surface soils feeding alluvial channels comprised of different particle sizes, and allows a comparison of particle size distributions at the source and at the watershed outlet.

REFERENCES

Barnes, H.H., Jr. 1967. Roughness characteristics of natural channels. U.S. Geological Survey Water Supply Paper 1849, 213 pp.

Bennett, J.P. 1974. Concepts of mathematical modeling of sediment yield. *Water Resources Research* **10**(3):485-492.

Chow, V.T. 1959. *Open Channel Hydraulics*. 680 pp. McGraw-Hill, New York.

Dickinson, W.T., M.E. Holland, and G.L. Smith. 1967. An Experimental Rainfall-Runoff Facility. Hydrology Paper No. 25, Colorado State University, Fort Collins, CO, 81 pp.

Fair, G.M., and J.C. Geyer. 1954. *Water Supply and Wastewater Disposal.* 973 pp. John Wiley and Sons, New York.

Glass, L.J., and E.T. Smerdon. 1967. Effect of rainfall on the velocity profile in shallow-channel flow. *Transactions of the American Society of Agricultural Engineers* **10**(3):330-332, 336.

Goodrich, D. C., 1991. Basin scale and runoff model complexity. Univ. of Arizona, Dept. of Hydrology and Water Resources Tech. Rep. No. HWR 91-010, 361 p.

Goodrich, D.C., D.A. Woolhiser, and S. Soorooshian. 1988. Model complexity required to maintain hydrologic response. Proceedings, 1988 National Conf. of the Hydraulics Division, American Society of Civil Engineers, Colorado Springs, CO, Aug 8-12, pp.431-436.

Hughes, D. A., and Beater, A. B. (1989). The applicability of two single event models to catchments with different physical characteristics. *Hydrologic Sciences Journal*, **34**(1/2):63-78.

Julien, P.Y., and D.B. Simons. 1985. Sediment transport capacity of overland flow. *Transactions of the American Society of Agricultural Engineers* **28**(3):755-761.

Kibler, D.F., and D.A. Woolhiser, 1970. The kinematic cascade as a hydrologic model. Hydrology Paper No. 39. Colorado State University, Fort Collins, CO. 27 pp.

Lane, L.J., D.A. Woolhiser, and V. Yevjevich. 1975. Influence of simplifications in watershed geometry in simulation of surface runoff. Hydrology paper No. 81, Colorado State University, Fort Collins, CO. 50 pp.

Langford, K. J., and J. L. McGuinness, 1976. A comparison of modeling and statistical evaluation of hydrologic change. *Water Resources Research*, **12**(6):1322-1323.

Li, Ruh-Ming. 1972. Sheet flow under simulated rainfall. M.S. thesis, 111 pp., Colorado State University, Fort Collins.

McCuen, R. H., 1982. A Guide to Hydrologic Methods Using SCS Methods, Prentice Hall, Englewood Cliffs, N.J., 145 p.

Meyer, L.D., and W.H. Wischmeier. 1969. Mathematical simulation of the process of soil erosion by water. *Transactions of the American Society of Agricultural Engineers* **12**(6):754762.

Michaud, J.D. 1992. Distributed rainfall-runoff modeling of thunderstorm generated floods: A case study in a mid-sized semiarid watershed in Arizona. PhD. Dissertation, Univ. of Arizona, Tucson.

Michaud, J. D., and S. Sorooshian, 1992. Rainfall-runoff modeling of flash floods in semi-arid watersheds. Univ. of Arizona, Dept. of Hydrology and Water Resources Tech. Rep. No. HWR 92-030, 319 p.

Morgali, J.R. 1970. Laminar and turbulent overland flow hydrographs. *Journal of the Hydraulics Division*, American Society of Civil Engineers **91**(HY2):441-460.

Morris, E.M., and D.A. Woolhiser. 1980. Unsteady one-dimensional flow over a plane: Partial equilibrium and recession hydrographs. *Water Resources Research* **16**(2):355-360.

Nash, J. E. and J. V. Sutcliffe, 1970. River flow forecasting through conceptual models, I. A discussion of principles. *Journal of Hydrology*. **10**:282-290.

Rawls, W.J., D.L. Brakensiek, and K.E. Saxton. 1982. Estimation of soil water properties. *Transactions of the American Society of Agricultural Engineers* **25**(5):1316-1320, 1328.

Renard, K. G. 1970. The hydrology of semiarid rangeland watersheds. USDA-ARS pub. 41162.

Rovey, E.W. 1974. A kinematic model of upland watersheds. M.S. Thesis, Colorado State University, Fort Collins, 119 pp.

Rovey, E.W., D.A. Woolhiser, and R.E. Smith. 1977. A distributed kinematic model of upland watersheds. Hydrology Paper No. 93, Colorado State University, 52 pp.

Singh, V.P. 1974. A nonlinear kinematic wave model of surface runoff. PhD. Dissertation, Colorado State University, Fort Collins, CO. 282 pp.

Smith, R.E. 1982. Rational models of infiltration hydrodynamics. *In Modeling Components of Hydrologic Cycle,* Water Resources Publications, Littleton, CO, pp. 107-126.

Smith, R.E. 1983. Flux infiltration theory for use in watershed hydrology. *In Proceedings of the National Conference on Advances in Infiltration*, pp. 313-323, American Society of Agricultural Engineers, St. Joseph MI.

Smith, R.E., C. Corradini, and F. Melone. 1993. Modeling infiltration for multistorm runoff events. *Water Resources Research* **29**(1):133-144.

Smith, R.E., and J.-Y. Parlange. 1978. A parameter-efficient hydrologic infiltration model. *Water Resources Research* **14**(3):533-538.

Smith, R. E., D. C. Goodrich, and D. A. Woolhiser, 1990. Areal effective infiltration dynamics for runoff on small catchments. Trans. 14th Inter. Congress of Soil Sci., Volume I: Commission I, p. 22-27, August, Kyoto, Japan.

Unkrich, C.L., and H.B. Osborn. 1987. Apparent abstraction rates in ephemeral stream channels. *Hydrology and Water Resources in Arizona and the Southwest*, Offices of Arid Land Studies, University of Arizona, Tucson, **17**:34-41.

USDA. 1980. CREAMS: A Field-scale model for chemicals, runoff, and erosion from agricultural management systems. Knise, W.G., ed. U.S. Department of Agriculture Conservation Research Report No. 26, 640 pp.

Woolhiser, D. A. and D. C. Goodrich. 1988. Effect of storm rainfall intensity patterns on surface runoff. *Journal of Hydrology*, (102):335-354.

Woolhiser, D.A., C.L. Hanson, and A.R. Kuhlman. 1970. Overland flow on rangeland watersheds. *Journal of Hydrology (New Zealand)* **9**(2):336-356.

Woolhiser, D.A., M.E. Holland, G.L. Smith, and R.E. Smith. 1971. Experimental investigation of converging overland flow. *Trans. ASAE* **14**(4):684-687.

Woolhiser, D.A., R.E. Smith, and D.C. Goodrich. 1990a. KINEROS, A Kinematic Runoff and Erosion Model: Documentation and User Manual. U.S. Dept. of Agriculture, Agricultural Research Service, ARS-77, 130 pp.

Woolhiser, D. A., D. C. Goodrich, W. E. Emmerich, and T. O. Keefer, 1990b. Hydrologic effects of brush to grass conversion. Proc. ASCE Nat. Conf. on Irrig. and Drain., Durango, CO, p. 293-302, July 11-15.

Woolhiser, D.A., and J.A. Liggett. 1967. Unsteady, one-dimensional flow over a plane-the rising hydrograph. *Water Resources Research* **3**(3):753-771.

Wu, Y-H, V. Yevjevich, and D.A. Woolhiser. 1978. Effects of surface roughness and its spatial distribution on runoff hydrographs. Hydrology Paper No. 96, Colorado State University, Fort Collins, CO. 47 pp.

Yu, Y.S., and J.S. McNown. 1964. Runoff from impervious surfaces. *Journal of Hydraulic Research* **2**(1):3-24.

Zevenbergen, L. W., and M. R. Peterson, 1988. Evaluation and testing of storm-event hydrologic models. Proc. ASCE Nat. Conf. on Hydraulic Engr., Colo. Springs, CO, Aug. 6-12., p. 467472.

Chapter 21

MIKE 11 - A GENERALIZED RIVER MODELLING PACKAGE

K. Havnø, M. N. Madsen, and J. Dørge

21.1. INTRODUCTION

The importance of accurate modelling of water-environmental systems is emphasized by the need for more and more refined utilization under normal and extreme conditions. Problems such as drought, river flooding, water pollution, inefficient irrigation, soil erosion and sedimentation, deforestation, poor river basin management, etc., have become major regional and international concerns.

The increasing demand for water resources and environmental protection requires improved techniques which can simulate in an integrated way the overall and local behaviour of the complex systems. Computer based mathematical models can be used to assess the impact of alternative utilization and to optimize, globally and locally, the utilization while minimizing the negative effects.

The assessment of hydroecological and economic impacts of changes in river management strategies builds on a basic understanding of the hydrological, hydrodynamic, morphological and biological/chemical processes. This requires the development of integrated modelling systems or generalized river modelling packages reflecting the highest technical/scientific state of the art within each field.

21.1.1. Introduction to MIKE 11

MIKE 11 is a comprehensive, one-dimensional modelling system for the simulation of flows, sediment transport and water quality in estuaries, rivers, irrigation systems and other water bodies. It is a 4th generation modelling package designed for micro-computers with DOS or UNIX operating systems and provides the user with an efficient interactive menu and graphical support system with logical and systemetic layouts and sequencing of the menus. The package was introduced on the market in 1989; after 4 years the number of installations world-wide exceeds 300. Figure 21.1. shows the MIKE 11 development history.

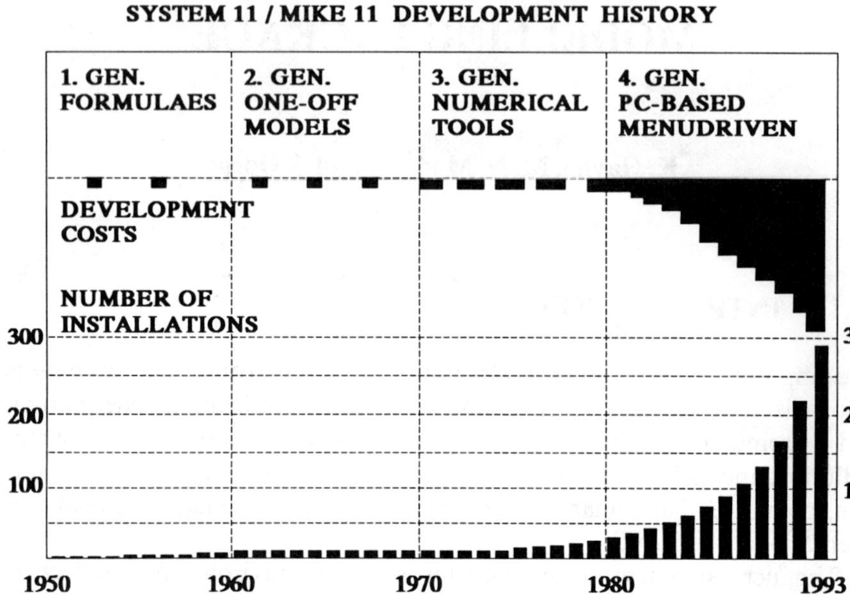

Fig. 21.1. MIKE 11 Development History.

A characteristic feature of the modelling system is its integrated modular structure which is presented in the following sections. The integrated structure allows computational modules to be added if and when required, without complicating model computations when only basic modules are required.

21.1.2. Integrated Modelling Structure

MIKE 11 has been designed to have an integrated modular structure with basic computational modules for hydrology, hydrodynamics, advection-dispersion, water quality and cohesive and non-cohesive sediment transport. Add-on computational modules are available for most of these basic modules. The modular structure of MIKE 11 is illustrated in Fig. 21.2. As shown the model system is equipped with two database systems (from which relevant data are automatically extracted during model simulations) and with a graphical and digital pre- and post processor for presenting results and

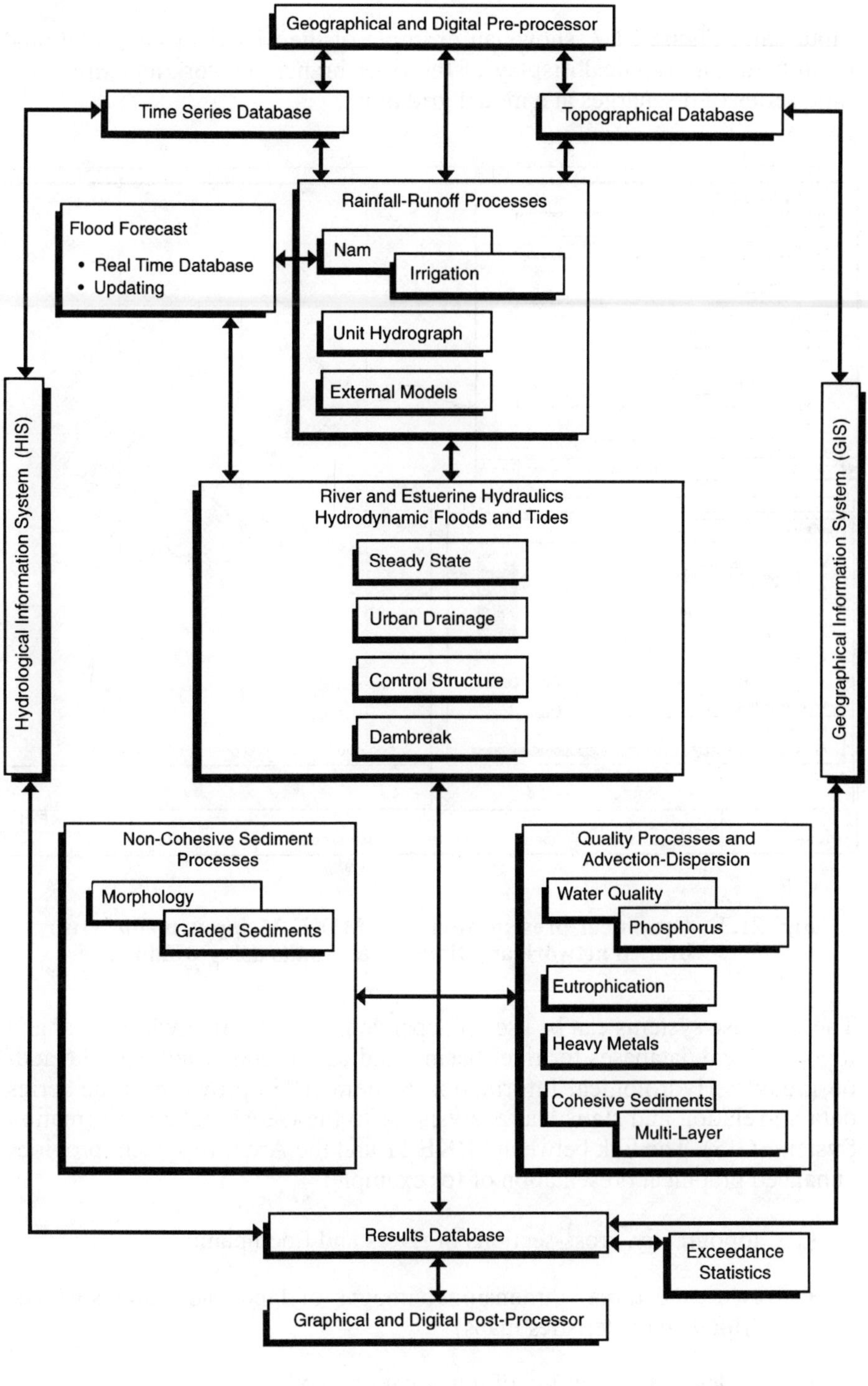

Fig. 21.2. Modular structure of MIKE 11.

input data. Figure 21.3. shows an example of the MIKE 11 computational main menu, a graphical display of the river branch network and simulated time series of discharges at a model grid point.

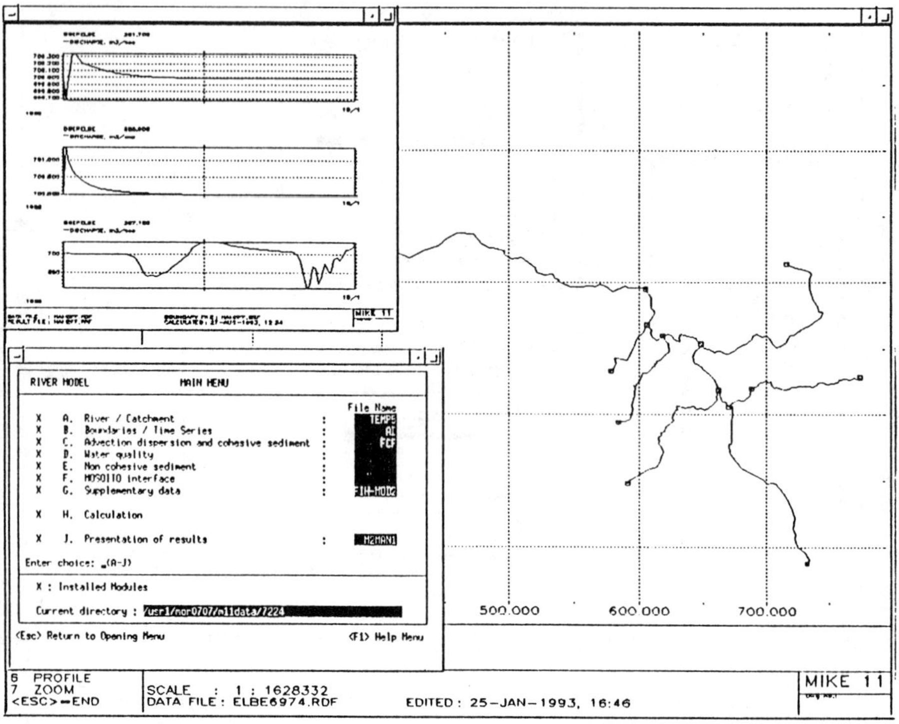

Fig. 21.3. Graphical presentation of a MIKE 11 input menu, river branch network and time series results using windows.

The database systems can be used independently as general hydrological and topographical databases for river basins, and can be connected with the add-on module, Hydrological Information System (HIS) providing time series data correlation and statistical analysis, or to the Geographical Information System (GIS). The link between MIKE 11 and the Arc/Info system provides enhanced graphical presentation of for example:

- topography, cross-sections of rivers and flood plains
- two- and three- dimensional images of land and water surfaces (flood characteristics maps)
- tables and contouring of inundatation depths.

the extent of simulated flood inundation areas.

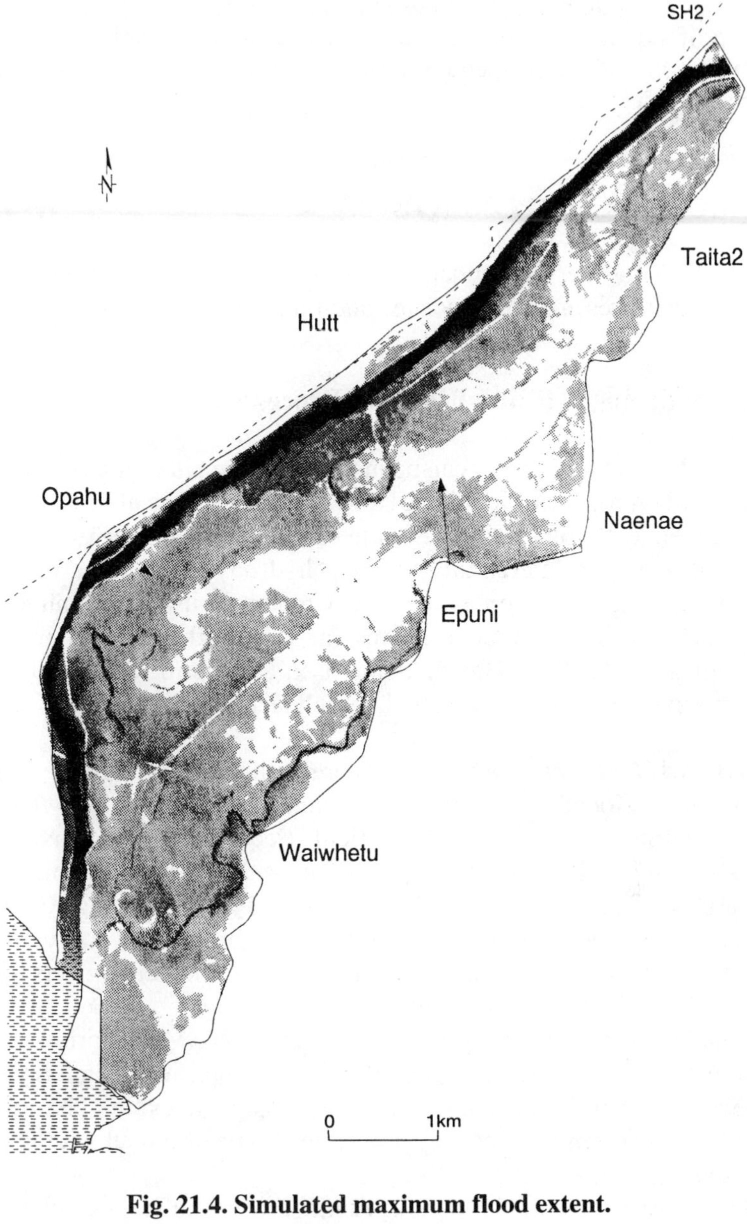

Fig. 21.4. Simulated maximum flood extent.

In the following sections the theoretical background for the basic modules for hydrology, hydrodynamics, sediment transport and standard water quality are presented, together with the add-on modules or more specialized modules associated with each field of the basic modules. Owing to the comprehensive number of modules and hence technical fields only a limited number of equations will be introduced. The interested reader is thus advised to consult the respective references for a detailed discussion on a particular subject.

21.2. MODEL COMPONENTS

In the following the theoretical background for the different basic model components is presented. Associated add-on modules are discussed briefly.

21.2.1. Hydrology (Rainfall-Runoff Processes)

The rainfall-runoff processes can be modelled using either the NAM module or the Unit Hydrograph Module (UHM). Both hydrological modules can be used independently of MIKE 11, or the simulated catchment runoff may be utilized directly as lateral inflows in a hydrodynamic river or channel network simulation. Whereas the NAM model is used to simulate rural catchment runoff based on a representation of the land phase of the hydrological cycle, the UHM model is used to describe the runoff from a single storm event using the unit hydrograph technique.

21.2.1.1. ***NAM Model.*** NAM is an abbreviation of the Danish "Nedbør-Afstrømnings-Model", meaning precipitation-runoff-model. This model has been developed by the Hydrological Section of the Institute of Hydrodynamics and Hydraulic Engineering at the Technical University of Denmark, see Ref. (1). The model is widely used in a large range of climatic regions and has further demonstrated its applicabilty in several workshops organized by WMO for the intercomparison of hydrological models for real-time simulations, see Refs. (2) and (3).

A mathematical hydrological model such as NAM comprises a set of linked mathematical statements describing, in a simplified quantitative form, the behavior of the land phase of the hydrological cycle. The model is defined as a deterministic, conceptual, lumped type of model with moderate input data requirements.

The model simulates the rainfall-runoff process in rural catchments. It operates by continuously accounting the moisture content in four different and mutually interrelated storages; snow storage, surface storage, lower-zone storage and groundwater storage (see Fig. 21.5), representing physical elements of the catchment.

The model area can be divided into a number of subcatchments. As each subcatchment is treated as one unit the parameters and variables are representative of average values for the entire subcatchment.

The input data to the model are: precipitation, potential evapotranspiration and temperature (for the snow routine). The primary results are mean daily values of streamflow and information about other elements of the land phase of the hydrological cycle such as the temporal variation of the soil moisture content and the groundwater recharge.

If the runoff and groundwater recharge of any of the catchments are affected by irrigation schemes, separate calculations can be performed on each irrigation area defined within a catchment. A modified description of the infiltration is used and the time series of irrigation water is specified as additional water supply to the surface storage.

The effect of irrigation on the following can be described:

- overall catchment water balance
- local infiltration and groundwater recharge in irrigated areas
- the distribution of the catchment runoff among overland flow, interflow and baseflow.

21.2.1.2. ***Unit Hydrograph module (UHM).*** This module is used to simulate the runoff from single storm events using unit hydrograph techniques, see Ref (19). The model executes optional excess rainfall calculations dependent on the infiltration loss:

- a fixed initial loss followed by a constant loss rate
- a loss rate proportional to the rainfall intensity (i.e. the 'rational method')
- a loss rate determined by the SCS (US Soil Conservation Service) curve number method

Excess rainfall is routed to the river using unit hydrograph or time area methods including the SCS dimensionless and SCS triangular synthetic unit hydrographs (see Figs. 21.6.), or user defined hydrographs.

The input data required include catchment description data and time series of rainfall.

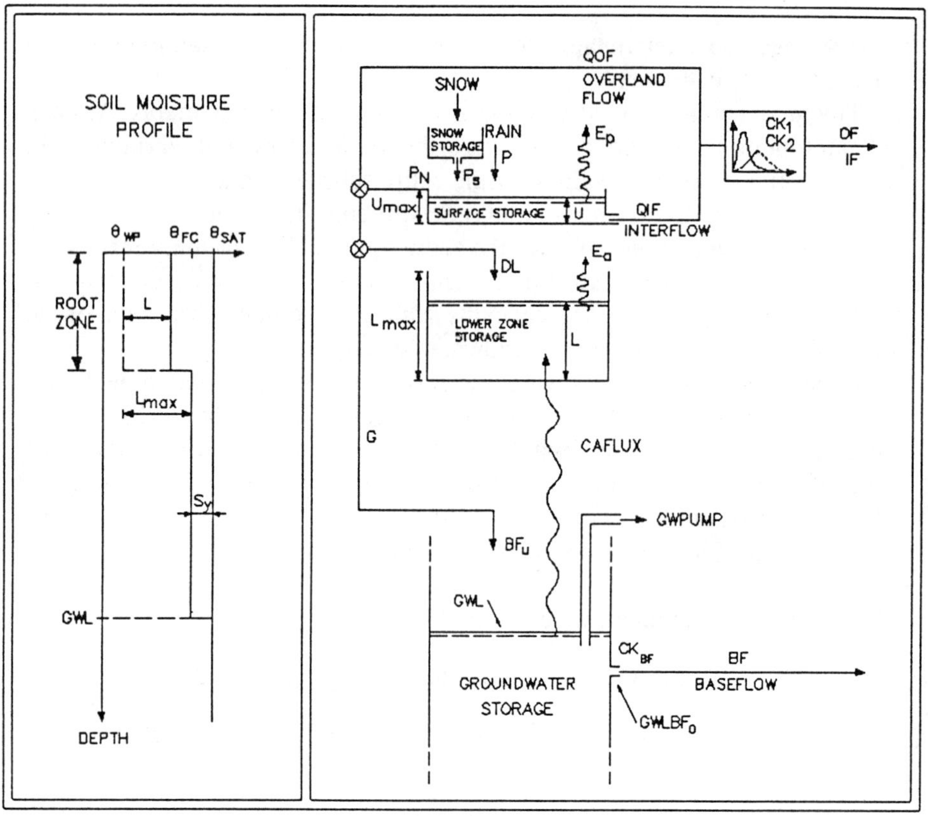

Fig. 21.5. Structure of the NAM Model.

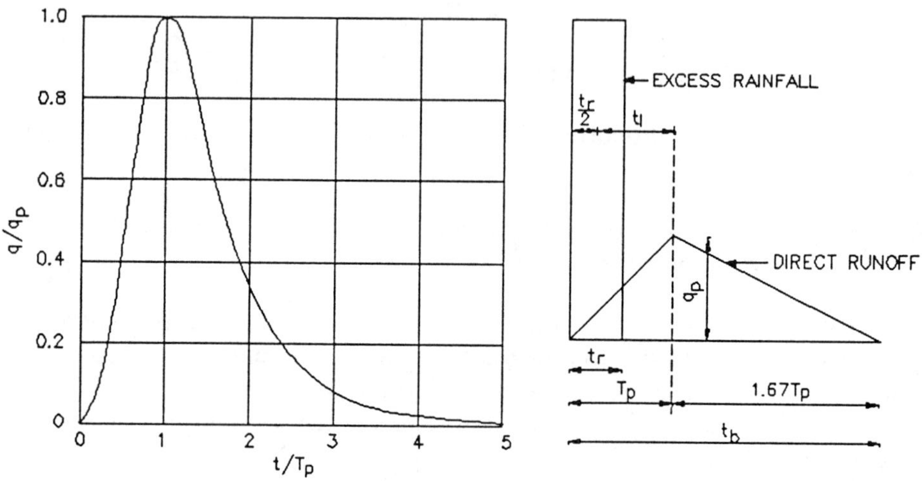

Fig. 21.6. SCS dimensionless and triangular synthetic hydrographs.

21.2.2. Hydrodynamic Modelling

The hydrodynamic computational module forms the basis for the advection-dispersion (AD), water quality (WQ), and non-cohesive sediment transport (NST) modules. Typical applications are:

- drainage
- low flow analyses
- irrigation
- tidal analyses
- flood control and alleviation:
  - fluvial flooding
  - tidally influenced flooding
- impact assessment of proposed river works:
  - flood embankments
  - flood plain modification (eg construction)
  - channel straightening
  - channel diversion
  - dredging
  - upstream ponding / hydraulic structures

Based on topographical- and time series data (model boundary conditions), as stored in the cross-sectional and time series databases respectively, the model computes water levels, discharges and flow velocities at all internal model grid points.

21.2.2.1. ***Basic equations and solution technique.*** The hydrodynamic model is based on the one-dimensional Saint-Venant equations. The two equations representing conservation of mass and momentum are respectively:

$$\frac{\partial Q}{\partial x} + \frac{\partial A}{\partial t} = q \qquad (21.1)$$

$$\frac{\partial Q}{\partial t}+\frac{\partial\left(\alpha\frac{Q^2}{A}\right)}{\partial x}+gA\frac{\partial h}{\partial x}+\frac{gQ|Q|}{C^2AR}=0 \qquad (21.2)$$

where

Q = discharge, m^3s^{-1}

A = flow area, m^2

q = lateral inflow, m^2s^{-1}

h = stage above datum, m

C = Chezy resistance coefficient, $m^{1/2}s^{-1}$

R = hydraulic or resistance radius, m

α = momentum distribution coefficient

Wind friction on the free surface is taken into account by adding a wind friction term to the right hand side of the momentum equation above.

The computational grid comprises alternating Q (discharge) and h (water level) points as illustrated in Fig. 21.7. It is automatically generated with Q-points placed midway between neighbouring h-points and at structures. h-points are placed at locations where cross-sectional data are available.

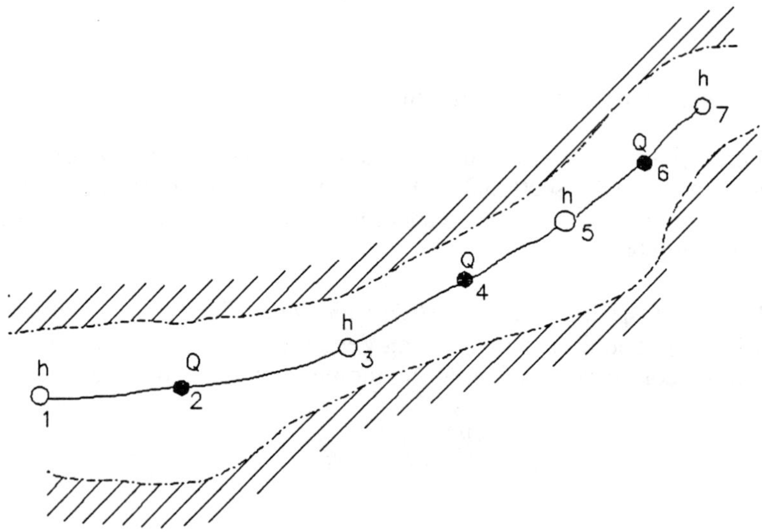

Fig. 21.7. Computational grid.

The differential equations are solved in a 6-point implicit finite difference scheme with alternating Q and H points, known as the Abbott-Ionescu scheme, Ref (4), see Fig. 21.8. The momentum equation is conveniently centered at Q-points whereas the continuity equation is centered at h-points.

The generalized matrix solution procedure applies to simply as well as multiply connected regions (looped and branched networks).

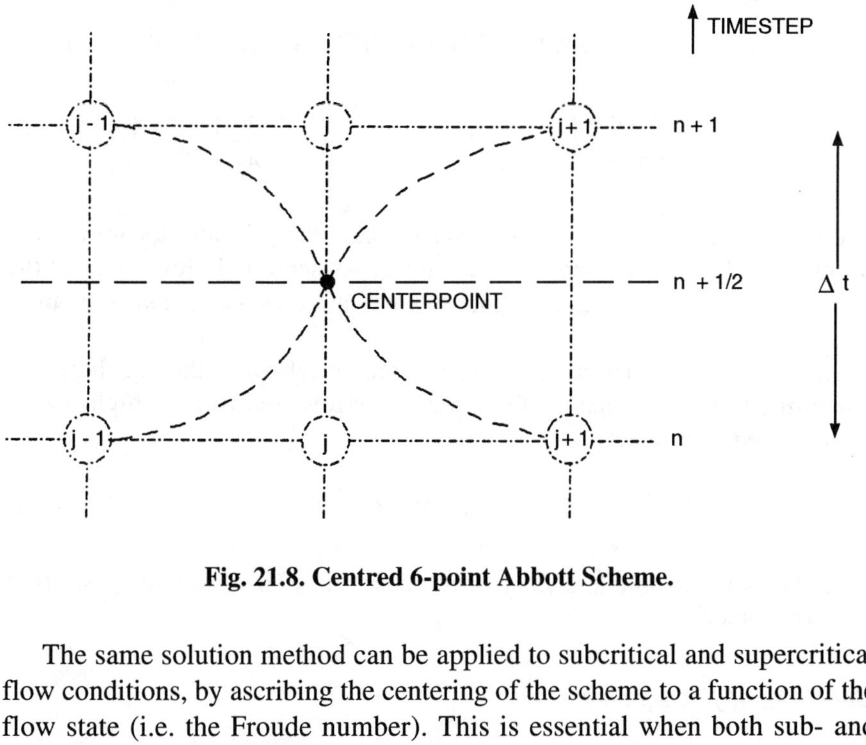

Fig. 21.8. Centred 6-point Abbott Scheme.

The same solution method can be applied to subcritical and supercritical flow conditions, by ascribing the centering of the scheme to a function of the flow state (i.e. the Froude number). This is essential when both sub- and supercritical flow situations have to be described within the same model simulation, as is frequently the case for rivers and flood plains.

For the subcritical flow descriptions, the scheme is used in its fully-time-centered form i.e. at time level $(n+{}^1/2)$ and at the middle gridpoint in space. During transition to supercritical flow the centering is gradually moved upstream and forward in time, along with suppression of the convective momentum term. In the case of supercritical flow the general equations can also be replaced by a kinematic wave formulation allowing for larger timesteps in the computations.

In order to incorporate the effects of local energy losses associated with flows through narrow cross sections or flow at hydraulic structures such as weirs, movable gates, culverts, etc. it is necsessary to substitute the momentum equation with an energy equation. In the case of drowned flow, the discharge can be determined via an energy head loss description:

$$\left(h + \frac{Q^2}{2gA^2}\right)_{j+1} - \left(h + \frac{Q^2}{2gA^2}\right)_{j-1} = \Delta H \tag{21.3}$$

The description generally used for flow through a structure is given by:

$$\Delta H = INFLOW\ HEADLOSS + OUTFLOW\ HEADLOSS \tag{21.4}$$

$$= \zeta_1 \frac{Q_s^2}{2gA_s^2}\left(1 - \frac{A_s}{A_1}\right) + \zeta_2 \frac{Q_s^2}{2gA_s^2}\left(1 - \frac{A_s}{A_2}\right)^2$$

where ΔH is the energy head loss over the structure, ζ_1 and ζ_2 are in- and outlet head loss coefficients, A_s is the cross-sectional flow area at the structure and A_1 and A_2 are up-and downstream cross-sectional flow areas respectively.

In the case of critical flow (e.g. free overflow) the discharge is determined directly using the free overflow relation, which for a broadcrested weir reads:

$$Q_c = \alpha_c 1.705 b H_s^{3/2} \tag{21.5}$$

where H_s is the available energy head above the weir crest and a_c is a free overflow factor.

Modelling of flood-plains

In practice, the simulation of flow over wide flood plains is of a two-dimensional nature. Several descriptions can be formulated with MIKE 11 to simulate areas which become inundated during flooding. One method is to establish a quasi two-dimensional description by dividing the inundated area into a number of cells, or separate flood plains. As shown in Fig. 21.9., this approach corresponds to a looped network of one-dimensional branches.

The division of flood plains may be based on natural features, such as elevated roads, embankments, dykes, etc. The flow between two cells or to the main river may be specified using a broad crested weir, culvert formulation, or river channel description.

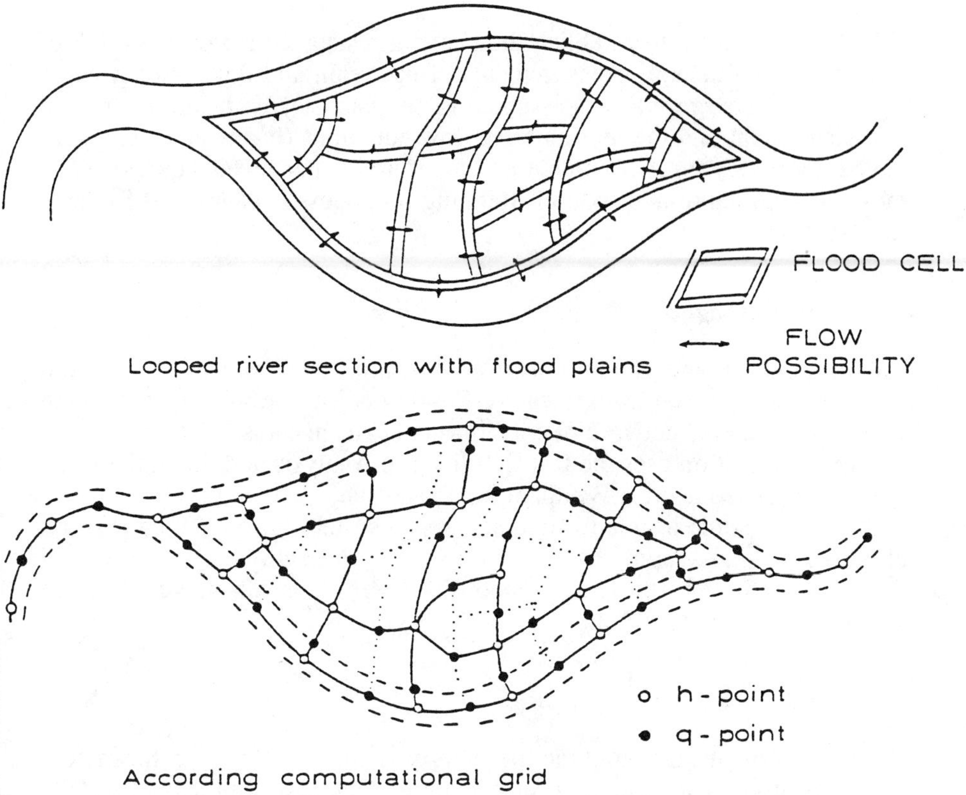

Fig. 21.9. Quasi two-dimensional simulation of flood plains.

21.2.2. Add-on Modules

- *Quasi-Steady State*

This description can be used for the simulation of long-term phenomena e.g. morphological changes or effects of pollutants on water quality.
A nested iteration procedure is used to solve the quasi-steady version of the Saint-Venant equations. An overall iteration is used to obtain the correct discharge distribution in the system (if there are bifurcations) and a backwater iteration calculates the water levels in each branch. The quasi-steady state module is thus able to simulate flow in looped and branched systems and includes a rapid convergence/iteration procedure.

As the calculations are based on steady flow conditions there is no upper stability limit to the computational time step. Thus long-term simulations take only a fraction of the computational time required by the fully dynamic solution.

- *Automatic calibration of bed resistancce*

This facility is used to establish the channel resistance for each stretch of river with simultaneous measurements of upstream and downstream water levels and discharge. Flow is assumed to be quasi-steady during the period of measurement and the non-uniform flow equations (backwater curve) are solved iteratively until computed and measured water levels coincide. The variation with depth and time of Manning roughness is calculated for each cross section.

- *Urban Drainage*

The urban drainage module is used for modelling storm runoff and progressive inundation from river overflows or coastal storm surges draining through a combined network of culverts and open channels.

The computation uses the fast Cell Integral Analysis and the initial flood surge is simulated using a Whitham front analysis.

The UD module is able to simulate looped systems of combined culverts and open channels, rapid changes in cross-section, shallow street flows and control structures. Both dynamic and diffusive waves can be simulated in addition to supercritical flow.

- *Control Structures*

With the control structure facility it is possible to simulate the actual operation of control gates (e.g. in irrigation channels, shiplocks or outlets from reservoirs) as complex rules. It is possible to simulate either underflow (e.g. rectangular sluice gates with a vertical sliding or rotating section) or overflow (movable weirs) structures. Critical flow characteristics are defined automatically.

The gate operation includes features such as:

- gate operation as a function of time or hydraulic variables (water levels and discharges or differences in water levels and discharges) anywhere in the river branch system

- different gate operations may be applied for positive and negative flow direction respectively

- an override option is available for emergency situations, e.g. the gates should always be closed/open if an extreme event occurs.

- *Dambreak*

This module is used to simulate breach development in a dam structure due to overtopping or pipe failure.

Discharge over the dam structure is treated as for a broadcrested weir except that the size and shape of the structure changes with time as the breach develops. Flow over the crest and through the breach are calculated separately (see Fig. 21.10.) using the same equations as for the broadcrested weir.

The breach is initially trapezoidal. Its geometric development is either a known function of time or of erosion based on the sediment transport capacity of the breach flow.
The erosion based failure may be due to overtopping (as above) or to pipe failure in which flow occurs through a pipe in the dam wall. The pipe enlarges due to soil erosion allowing the flow to increase. The dam eventually collapses into the void created by the piping failure after which the failure is simulated as an overtopping process.
The growth of the breach can be limited by specifying a Limiting Cross-Section eg a surface of hard rock or the natural valley bottom.

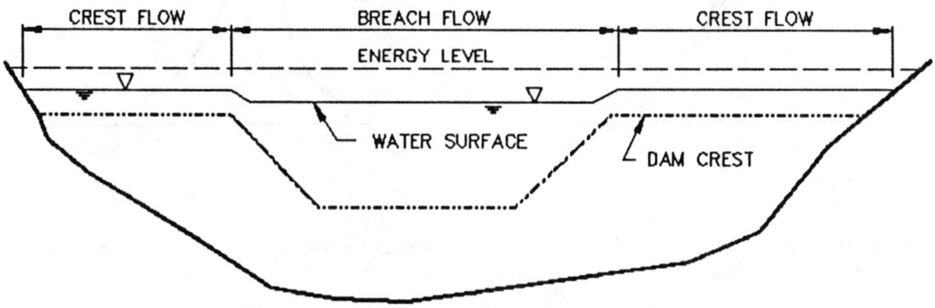

Fig. 21.10. Combined Flow over Breach and Crest

The flow is calculated once the breach geometry has been determined. Free overflow and drowned flow can be simulated including combinations of these eg flow at the breach may be drowned whilst that at the crest may be free.

21.2.3. Flood Forecasting

This module is used for real-time flood or storm surge forecasting. It provides a reliable modelling system for the translation of real-time rainfall, water level, discharge and reservoir operation data into forecasts of river flows and water levels. It can be used with manual input of data or with a real-time data link to a telemetry system in an automatic data collection and processing system.

The main components of the module are:

- a real-time data management system

- NAM rainfall-runoff module
- the hydrodynamic module
- an automatic updating procedure

The updating procedure utilizes measured and/or calculated discharge or water levels to minimize the difference between observed and simulated flow/water levels at the time of forecast. It distinguishes between amplitude and phase errors which are then taken into account in the forecast. The updating can be carried out on the basis of both water levels and discharges. Figure 21.11. shows the principles in the updating routine.

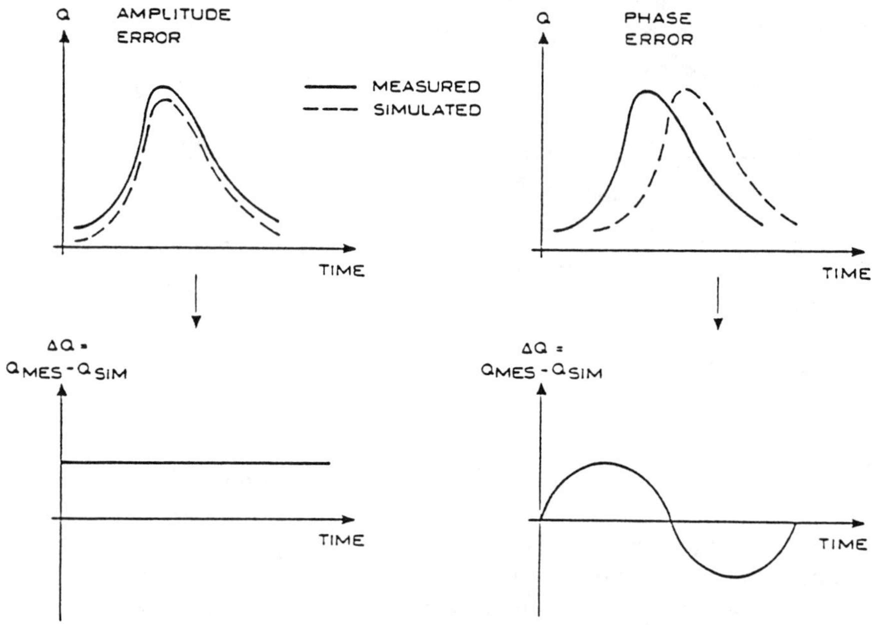

Fig. 21.11. Principles in Flood Forecasting updating routine

21.2.4. Non-Cohesive Sediment Transport

The description of non-cohesive sediment transport can be based on four different models for the calculation of sediment transport rate and alluvial roughness. These are the Engelund-Hansen model and the Ackers-White model for the calculation of the total load, and the Engelund-Fredsøe and van Rijn models for the calculation of bed load and suspended load separately.

The non-cohesive sediment transport can be computed in one of two modes; explicit and morphological.

In the explicit mode, output is required from the hydrodynamic module (HD) in terms of discharge, water levels, (cross-sectional area and hydraulic radius) both in time and space. There is no feedback from the sediment transport calculations to the HD module. Results are in the form of volume transport rates and accumulated volumes of deposition or erosion. The explicit mode is useful where significant morphological changes are unlikely to occur.

In the morphological mode, sediment transport is calculated in parrallel with the HD module. Communication between the sediment transport calculations and the HD module is achieved through solution of the sediment continuity equation and through updating the bed resistance and the following sediment transport calculation. Results are in the form of sediment transport rates, bed level changes, resistance number and dune dimensions (depending on the transport relationship adopted).

Engelund - Hansen Model

The Engelund - Hansen sediment transport formula, Ref. (8), gives the total bed material transport. Although derived for a dune covered bed, the equation was found to be applicable to the upper regime (plane bed and anti-dunes) as well.

Ackers & White Model

Ackers and White (1973), Ref. (5), presented a semi-empirical sediment transport model. The model is partly based on dimensional analysis but physical arguments have been used to derive the form of the function that was tested.

When determining the model parameters, Ackers & White distinguished among three cases, viz. fine grains, coarse grains and a transitional grain size.

Engelund & Fredsøe Model

The sediment transport model presented by Engelund & Fredsøe (1976), Ref. (6), gives a more detailed description of sediment transport and its relation to the flow resistance. The total sediment transport is split into bed load and suspended load and the sediment transport is calculated from the skin friction , i.e. the shear stress which is acting on the surface of the dunes. It is often necessary to distinguish between suspended load and bed load, and to be able to estimate the variation of the sediment transport along the dunes if the river model is to be used to calculate the dimensions of the dunes. Separate computation of suspended load and bed load is also advantageous when calibrating models against measured suspended sediment transport rates.

Dune dimensions

The equilibrium dune hight and dune length can be formulated in terms of the suspended (Φ_s) and bed load (Φ_b) transport (Fredsoe, 1977 and 1982, Refs. (9) and (10)), see Fig. 21.12.

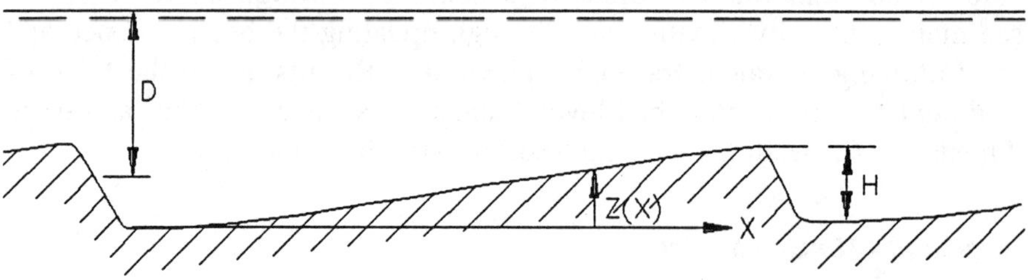

Fig. 21.12. Dune dimensions, definition sketch.

The general equation describing the equilibrium dune height thus reads:

$$\frac{H/D}{1-1/2H/D} = \left. \frac{\Phi_b}{2\theta \dfrac{\partial(\Phi_b + \Phi_s)}{\partial\theta}} \right|_{top} \qquad (21.6)$$

where top refers to the condition at the top of the dune and θ is a dimensionless bed shear stress.

Flow resistance

If the dimensions (height and length) of the dunes are known, the hydraulic resistance (bed shear stress) can be calculated from the water depth and the specific discharge. Furthermore, the portion of the dimensionless bed shear stress, θ', which is acting as skin friction on the gently curved upstream side of the dunes can be determined.

The total bed shear stress, τ, is split up into the skin friction, τ', and the portion which is caused by the form drag on the dunes τ'':

$$\tau = \tau' + \tau'' \qquad (21.7)$$

τ'' is mainly caused by the expansion loss behind each dune and can therefore be described as Carnot losses (Engelund & Hansen (1967)), Ref (8).

Van Rijn Model

In the van Rijn transport models (van Rijn, 1984a,b , Refs. (11) and (12)) the sediment load is divided between bed load and suspended load according to the relative magnitudes of the bed shear velocity, and the particle fall velocity. When the bed shear velocity exceeds the fall velocity then sediment is transported as both suspended and bed load. Bed load is considered to be transported by rolling and saltation and the rate is described as a function of saltation height. The suspended load is determined from the depth-integration of the product of the local concentration and flow velocity. The reference concentration is determined from the bed load transport.

- ***Graded non-cohesive sediments, Add-on module***

This module is used to simulate sediment transport in highly non-uniform sediments or where bed shear stresses are close to the threshold of movement. Fraction sediment transport rates and the variation in time and space of particle size distribution are calculated.

The module can be used with all four sediment transport models discussed above. Bed material comprises an active layer overlying a passive layer (see Fig. 21.13). The module is developed to incorporate a multi-layer bed.

Bed load transport takes place in the active layer. In the calculation of the fraction sediment transport the criterion for the threshold of movement is modified to allow for the non-uniformity of the material.

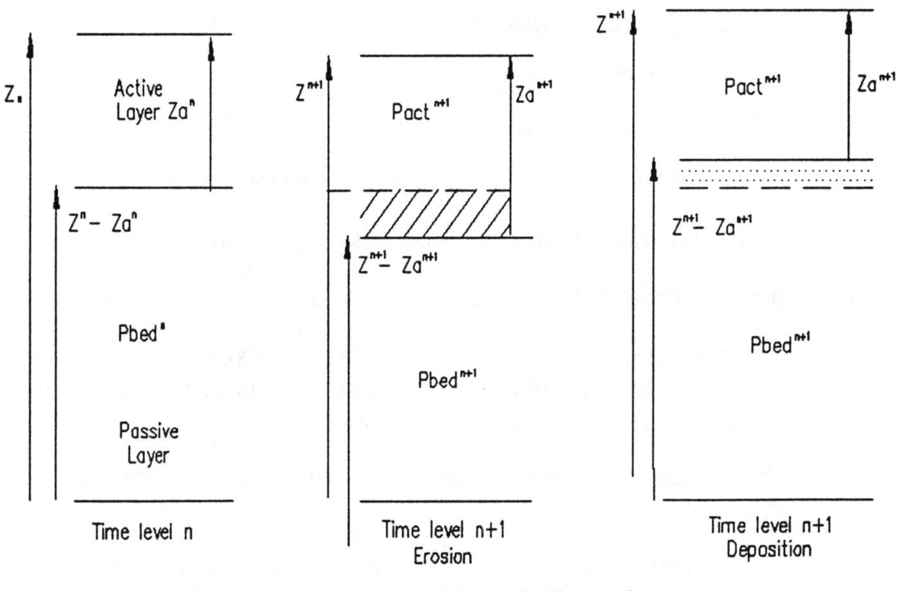

Fig. 21.13. Active and Passive Sediment layers.

21.2.5. Advection-dispersion and Cohesive Sediment Transport

The pure advection-dispersion model, i.e without inclusion of biological and chemical processes, is used to simulate the advection and dispersion of conservative materials or materials with a simple first order decay only. Typical applications include studies of saline intrusion or the behaviour of cohesive sediment. The latter requires the inclusion of erosion and deposition of fine sediments in the advection-dispersion equations.

21.2.5.1. ***Advection-dispersion.*** The one-dimensional (vertically and laterally integrated) equation for the conservation of mass of a substance in solution, ie the one-dimensional advection-dispersion equation reads:

$$\frac{\partial AC}{\partial t} + \frac{\partial QC}{\partial x} - \frac{\partial}{\partial x}\left(AD\frac{\partial C}{\partial x}\right) = -AKC + C_2 \bullet q \qquad (21.8)$$

C = concentration (arbitrary unit)

D = dispersion coefficient (m^2/s)

A = cross-sectional area (m^2)

K = linear decay coefficient (s^{-1})

C_2 = source/sink concentration

q = lateral inflow

x = space coordinate (m)

t = time coordinate (t)

The equation reflects two transport mechanisms:

(i) advective (or convective) transport with the mean flow.

(ii) dispersive transport due to concentration gradients.

The main assumptions underlying the advection-dispersion equation are:

(i) the considered substance is completely mixed over the cross-sections, implying that a source/sink term is considered to mix instantaneously over the cross-section;

(ii) the substance is conservative or subject to a first order reaction (linear decay);

(iii) Fick's diffusion law applies, ie the dispersive transport is proportional to the concentration gradient.

The advection-dispersion equation is solved numerically using an implicit finite difference scheme with negligible numerical dispersion. A correction term has been introduced in order to reduce the third order truncation error which makes it possible to simulate dispersion/convection of concentration profiles with very steep fronts, see Ref (18).

21.2.5.2. *Cohesive sediment transport.* The cohesive sediment transport is based on the advection dispersion (AD) equations by including additional source/sink terms for the description of erosion and deposition respectively. The erosion rate depends on the local hydraulic conditions whereas the deposition rate depends on the concentration of the suspended sediment and on the hydraulic conditions.

The rate of deposition can be expressed by:

$$S = -\frac{w \cdot C}{h*} \bullet \left(\frac{\tau}{\tau_{cd}} - 1 \right), \quad \tau \leq \tau_{cd} \tag{21.9}$$

where

C the concentration of the suspended sediment

w the mean settling velocity of suspended particles

h* the average depth through which the particles settle

τ_{cd} the critical shear stress for deposition

τ the bed shear stress

The resistance against erosion of cohesive sediments is determined by the submerged weight of the individual particles and by the interparticle electro-chemical bonds which must be overcome by the shear forces before erosion starts.

The rate of erosion has been described by the expression (Ariathurai et al. 1977), Ref (14),

$$S = -\frac{M_*}{h} \bullet \left(\frac{\tau}{\tau_{ce}} - 1 \right), \quad \tau \geq \tau_{ce} \tag{21.10}$$

where

τ_{ce} = critical shear stress for erosion

M_* = erodibility of the bed

Applying the cohesive sediment transport model with the erosion/ deposition relations as specified above may be sufficient when evaluating potential erosion deposition patterns in the river system.

A more detailed description of cohesive sediment transport is also available utilizing the general one-dimensional approach of flows and transports with a superimposed two-dimensional description of the erosion / deposition and a full three-dimensional description of the river bed:

- ***Multi-layer bed***

The distribution of sediment at the river bed is often highly non-uniform. The river bed may consist of consolidated mud at the side slopes while there is soft fluid mud in the central deeper parts of river cross-sections. Also vertically, there will be variation in the composition of the mud.

A multi-layer bed model comprising three layers is implemented:

- Weak fluid mud (top layer)
- Fluid mud
- Underconsolidated bed

Each of these layers is associated with different time scales. In tidally influenced reaches of a river, weak fluid mud is the sediment which is depositred around flood slack tide and is taken into suspension once more as the ebb flow starts to pick up.

Each layer is characterized by having its own sediment properties such as: critical shear stress for erosion, erosion function, density and transition rate.

The layered structure is illustrated in Fig. 21.14. below.

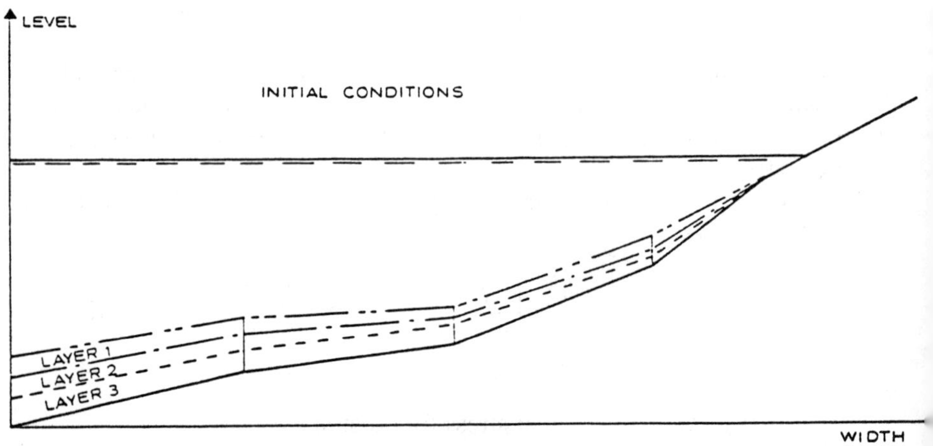

Fig. 21.14. Multi layer structure of river bed in advanced cohesive sediment transport model.

Transfer of sediment from one layer to the next is modelled through first order interaction processes between the layers. Sliding of sediment down the sloping sides of a cross-section is also taken into account.

The river bed will erode when the bed shear exceeds the critical shear stress for erosion of the layer which is exposed to the flow. Two types of erosion are implememted :

Instantaneous Erosion: In this case all the sediment available in a layer will re-intrain instantaneously when the shear stress exceeds the threshold value. This type of erosion is associated with the instability of the sediment deposit, and normally applies to the weak top layer.

Gradual Erosion: When the applied bed shear stress exceeds the critical shear stress for erosion the sediment is assumed to erode gradually as a non-linear function of the excess shear stress.

Deposition is likewise modelled two-dimensionally.

The model distinguishes between two settling regimes: One regime where the fall velocity will increase with the concentration (see van Rijn, 1989, Ref. (15)), and one regime where the hindered settling effect is taken into account.

21.2.6. Water Quality

Water quality models have been used in environmental impact assessment studies since the early seventies. The water quality model in MIKE 11 consists of several modules describing different aspects of water quality in areas influenced by human activities, e.g. pollution with organic matter and subsequent oxygen depletion, pollution with heavy metal or problems with unnaturally high production of algae (eutrophication).

The water quality models are based on the one-dimensional equation of conservation of mass of a dissolved or suspended material, i.e. the advection-dispersion equation, as presented in section 21.2.5., combined with a system of coupled differential equations describing the physical, chemical and biological interactions in the modelled water systems.

21.2.6.1. ***BOD/DO and bacterial fate model.*** The BOD/DO model deals with the basic aspects of river water quality in areas influenced by human activities, e.g. oxygen depletion and ammonia levels as a result of organic matter loadings. Good oxygen conditions are essential for the diversity of the animal community, and ammonia, in its non-ionised form, is highly toxic to fish and fish larvae.

The state variables in the BOD/DO model are dissolved oxygen (DO), water temperature, organic matter (expressed as dissolved, suspended and deposited BOD_5), ammonia/ammonium (NH4-N) and nitrate (NO3-N).

The processes affecting the oxygen levels are degradations of organic matter, respiration, nitrification (oxygen consuming), photosynthesis (oxygen production) and exchange of oxygen with the atmosphere as illustrated in Fig. 21.15.

The BOD/DO model is a relatively complex model with a network of processes among the 7 components (see Fig 21.15.). The mathematical formulations can be found in Refs. (16) and (17).

The water temperature strongly affects the biological processes in the water column. The temperature is computed as the result of the difference between the solar energy input from sunrise to sunset (heating), and the radiation from the water surface (cooling). The diurnal and seasonal variation of the water temperature are calculated from the latitude and other relevant parameters.

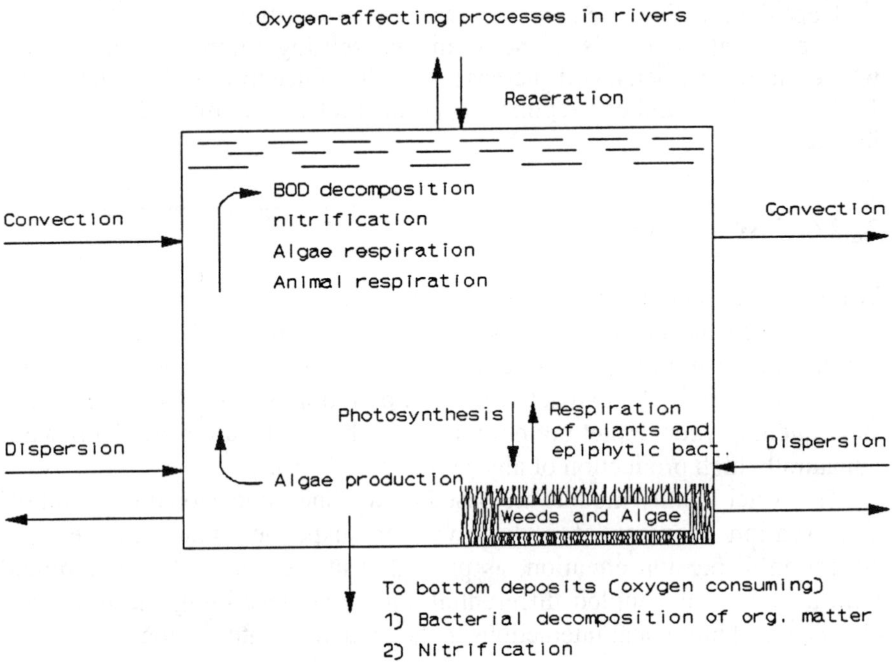

Fig. 21.15. Processes affecting oxygen levels.

The model can be applied at six different levels. The relevant level is determined from a consideration of the study requirements and data availablity. The six model levels describe the following relationships:

- BOD-DO

- BOD-DO including exchange with organic matter from the river bed
- BOD-DO with nitrification
- BOD-DO including exchange with bed, nitrification and denitrification
- BOD-DO including immediate and delayed oxygen demand and exchange with the river bed
- BOD-DO including all the above processes

For the most complex levels, three types of BOD are considered (dissolved, suspended and deposited) as well as oxygen, ammonia, nitrate and temperature.

The decay of bacteria (faecal and total coliform bacteria) are included in the standard water quality model. Eschericia coli (E. coli) is one of the dominant species in faeces from human and warm blooded animals. The organism itself is normally considered non-pathogenic, but is very often used as an indicator of faecal pollution and hence a potential of real pathogenic organisms (e.g. other bacteria and viruses).

Enteric bacteria die-off can be modelled very well by a first order decay reaction. The die-off constant or decay rate is highly variable due to interaction by environmental factors on bacterial die-off. The main factors, accounted in the model are light, temperature and salinity variations.

21.2.6.2. ***Phosphorus model.*** Phosphorus components are modelled using a special version of the BOD/DO model, where BOD/DO relations together with nitrogen and phosphorus dynamics are calculated. The processes included are described below and illustrated in Fig. 21.16.

The processes influencing the concentration of dissolved phosphorus (DP) are: release of inorganic phosphorus, adsorption of dissolved phosphorus, phosphorus uptake by vegetation and degradation of BOD. The processes influencing the concentration of particulate inorganic phosphorus (PP) are its deposition and resuspension, and the release and adsorption of dissolved phosphorus. Particulate organic phosphorus is not described explicitly as a state variable but is related to the concentration of organic material measured as BOD which is itself expressed as a function of the BOD decay rate and the suspension and deposition of sediment with attached BOD.

The total phosphorus concentration is given as the sum of the dissolved inorganic and organic particulate phosphorus.

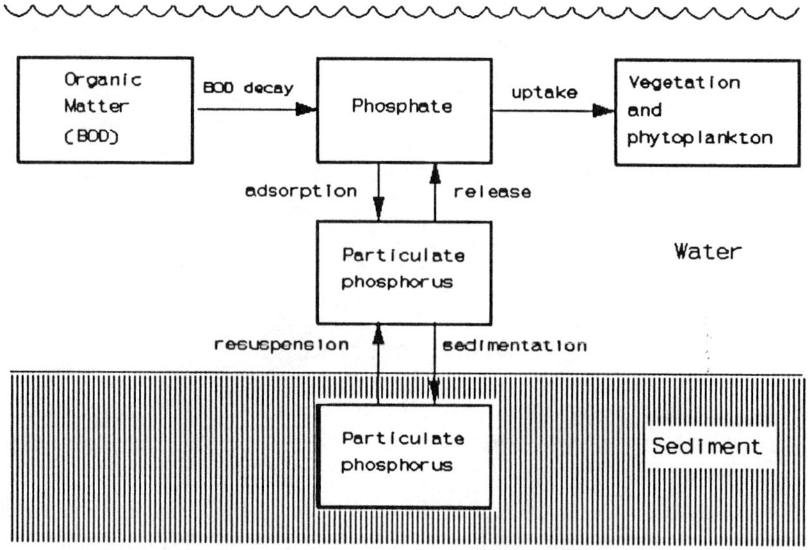

Fig. 21.16. Phosphorus processes.

21.2.6.3. ***Heavy metal model.*** This model is used in investigations of heavy metal pollution (industrial or urban) and dispersion of heavy metal due to its release from polluted sediments. The interaction between metals and particulates, both in the water phase and at the bed, are simulated using a kinetic description of the adsorption/desorption processes.

Heavy metals exist either dissolved in the water or attached to suspended matter in the aquatic environment. The extent to which heavy metal is bound to suspended matter differs from metal to metal and depends also on the nature of the particulates.

Adsorption of heavy metals on to suspended matter leads to their removal from the water column due to the deposition of the suspended material, but transfers the problem to the bed sediments (see Fig. 21.17.).

Differences between the concentration of heavy metal in the porewater of the sediment and in the water column result in a diffusive transport of heavy metal from the polluted sediments to the cleaner water. Another transport mechanism is resuspension of the sediments.

The HM module comprises 6 state variables:

- dissolved metal in the water
- adsorbed metal in the water
- suspended sediment
- dissolved metal in the porewater of the bed

- metal adsorbed on sediment in the bed
- sediment layer thickness

21.2.6.4. ***Eutrophication model.*** This model is used to investigate pollution sources (e.g. domestic and industrial sewage, agricultural run-off) with particular emphasis on the process of eutrophication (nutrient enrichment)

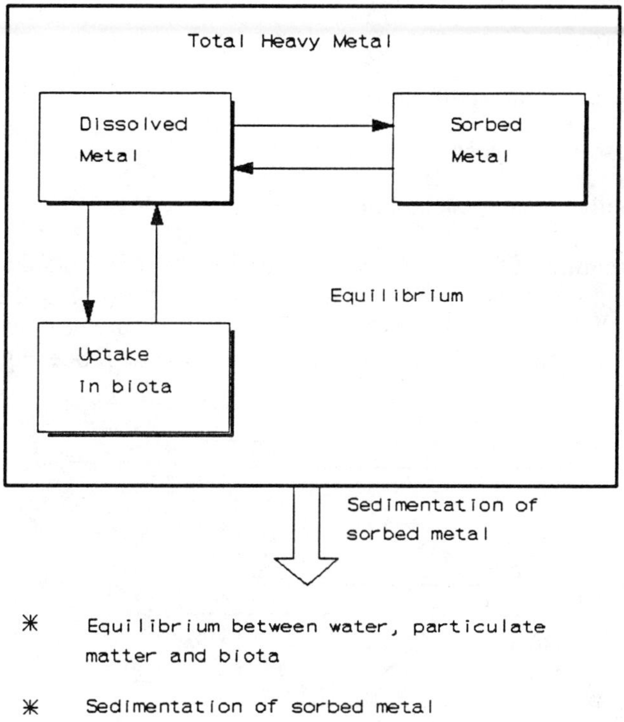

Fig. 21.17. Modelled mechanisms.

and its effects. The model describes nutrient cycling, growth of phytoplankton and zooplankton (including nutrient dynamics) and elements of the oxygen balance.

The module comprises 12 state variables describing biotic and abiotic components:

- phytoplankton carbon
- phytoplankton nitrogen
- phytoplankton phosphorus

- chlorophyll-a
- zooplankton
- detritus (dead organic) carbon
- detritus nitrogen
- detritus phosphorus
- inorganic nitrogen
- inorganic phosphorus
- dissolved oxygen
- benthic vegetation carbon

The dynamics of the primary production by phytoplankton and the grazing on phytoplankton by zooplankton (the first and second steps of the basic food chain in the pelagic part of the ecosystem) are described. Primary production by benthic vegetation can also be modelled. See Fig. 21.18, for illustration.

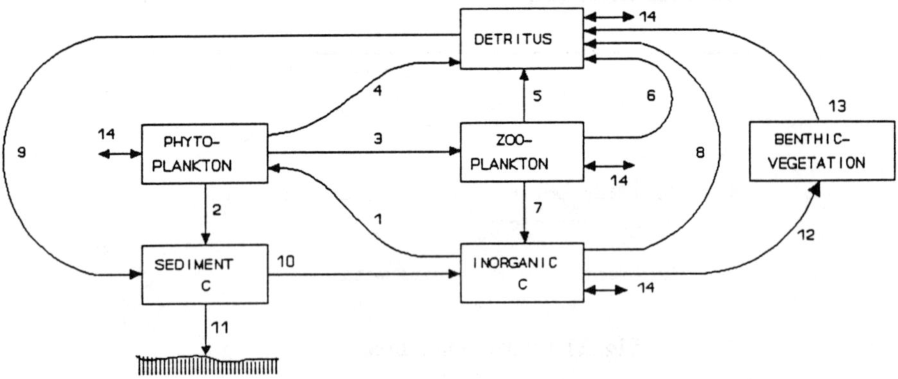

1. production, phytoplankton
2. sedimentation, phytoplankton
3. grazing
4. extinction, phytoplankton
5. excretion, zooplankton
6. extinction, zooplankton
7. respiration, zooplankton
8. mineralisation of suspended detritus
9. sedimentation of detritus
10. mineralisation of detritus
11. accumulation in sediment
12. production, benthic vegetation
13. extinction, benthic vegetation
14. exchange with surrounding waters

Fig. 21.18. General diagram of the eutrophication model.

The production of both the pelagic and the benthic system are calculated as a consequence of BOD loadings, available nutrients and factors such as light intensity at the surface, transparency of the water column and water temperature. The eutrophication model is a complex model with many coupled relations. In the computation of the pelagic system the following processes are included:

- the growth of phytoplankton
- grazing on phytoplankton by zooplankton
- the transformation of phytoplankton and grazers to dead organic material (detritus)
- settling of phytoplankton and detritus
- release of nutrients from organic matter degradation in water and sediment
- the corresponding oxygen demand

The eutrophication model consists of 12 coupled first order differential equations describing the above processes. The transport process for the pelagic components and the biological processes are coupled and solved simultaneously.

CASE STUDIES

21.3.1. The River System of Bangladesh

21.3.1.1. Introduction

The river system of Bangladesh is complex and subjected to flooding with a wide variety of causes and patterns. The application of a single general purpose model requires that the unifying features of the causes of flooding are identified and accounted. The causes may be categorised as follows:

- rainfall over the land area of Bangladesh
- large volumes of water in rivers crossing the international boundary
- cyclonic surges coupled with high (spring) tides in the Bay of Bengal

Flows crossing the international boundary may be further categorised:

- a flood of long duration, e.g. several months, on the major rivers, rising slowly and falling slowly
- a flood of short duration, e.g. several hours, on smaller rivers, rising steeply and falling steeply.

The MIKE 11 model employed is based on the fundamental principals of hydrology and hydraulics, and is able to represent the wide range of phenomena that gives rise to flooding in Bangladesh. The models now available and undergoing further development at the Surface Water Modelling Centre in Dhaka represent the application of state-of-the-art technology to the world's most complex river system. The following provides an overview of the application.

21.3.1.2. Application to the river system

Model Scheme Plan

Bangladesh has innumerable rivers and khals and it is impractical to include them all explicitly in the model. While it would be possible to include all but the smallest of khals in one single model (i.e. there is no practical software or hardware limitation) the sheer problem of managing the huge quantity of data input to and output from such a model would be an impossible task.

A structured approach is therefore adopted. This entails one General Model covering the entire country which includes the main rivers explicitly, while smaller channels are only included implicitly, ie lumped with the larger channels. The scheme plan of the General Model is shown in Fig. 21.19.

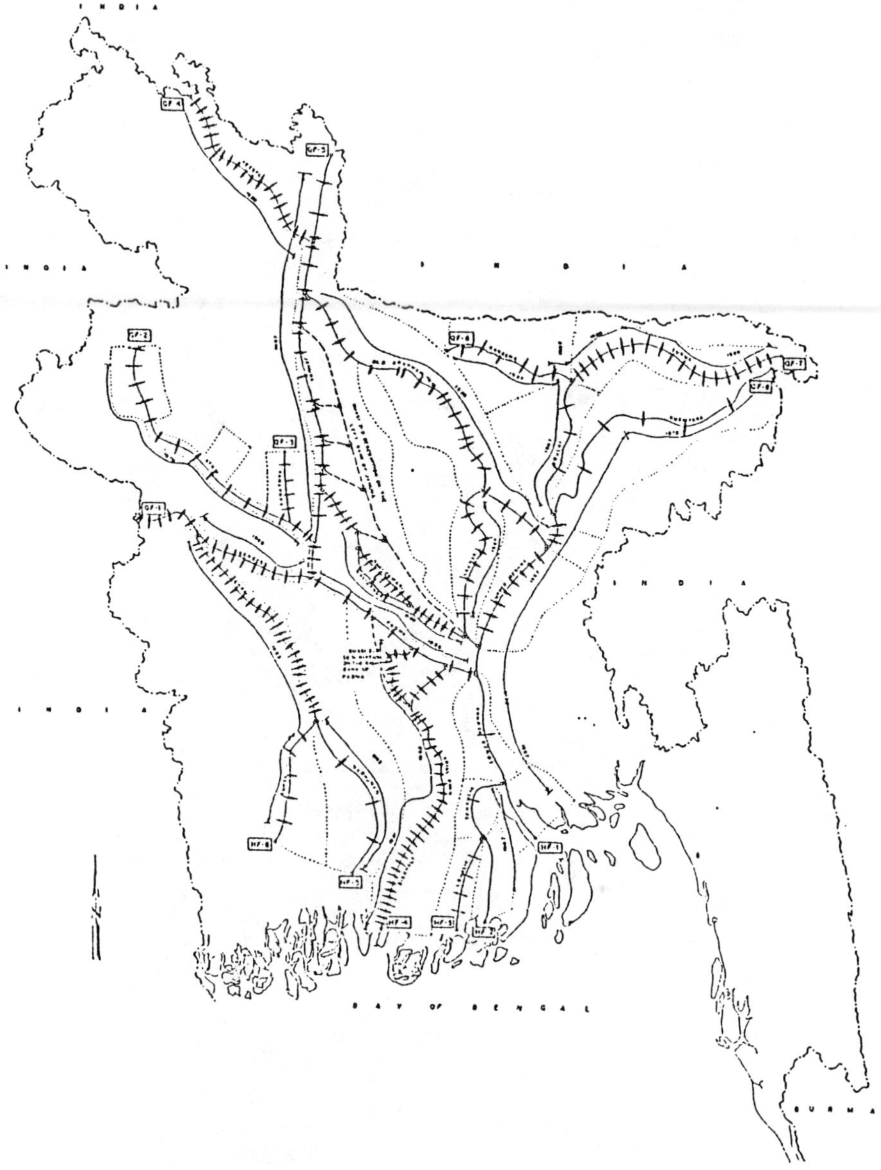

Fig. 21.19. Scheme plan of general model.

The next level in the structured approach is Regional Models. These show a finer level of detail, i.e. include smaller channels explicitly (e.g. Fig. 21.20). As the Regional Models cover a smaller area, their volume of data input and output is of the same order of magnitude as the General Model.

The General and Regional Models have an appropriate level of detail for the planning of regional water management studies and feasibility studies of

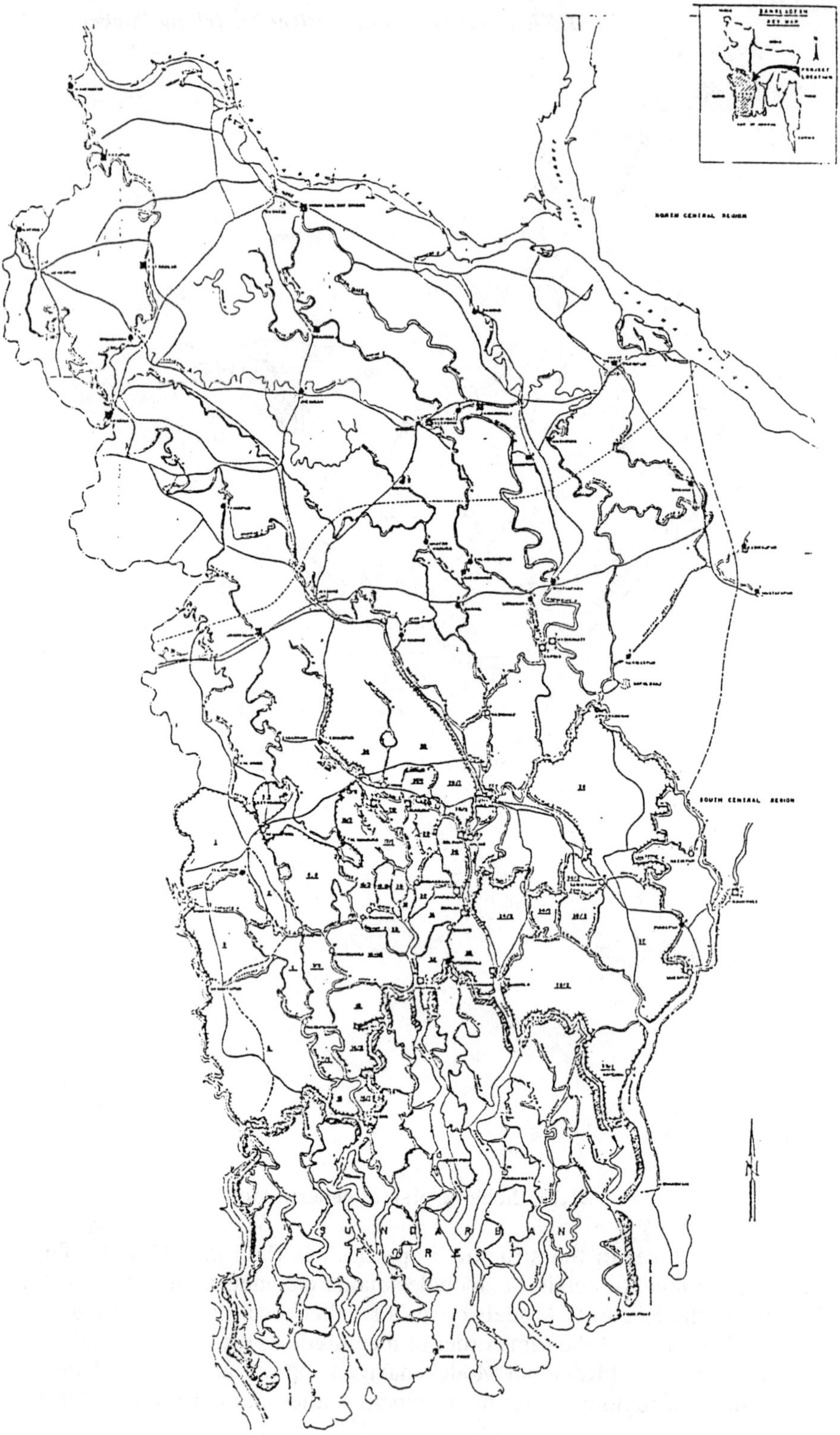

Fig. 21.20. Scheme plan of Southwest region model.

flood control and drainage projects. For detailed design, it may be necessary to develop more detailed models, adding a third level to our structured approach.

Description of River System

The topography is represented by cross sections along the river branches and stage:area relationships for flood plains. Generally, the computation based on average velocity in a cross section works perfectly well for defined river channels.

Where there is significant flow across a flood plain, i.e. not parallel to the flow in the river, a more complex model schematic has to be introduced. By constructing a network of one dimensional flow branches, a two dimensional flow pattern can be simulated. The one/quasi two dimensional modelling approach is appropriate, even to river systems as complex as in Bangladesh. In addition to avoiding unnecessary computational complexities and computer power, a single model represents both the essentially one dimensional dry season flow, and the two dimensional nature of flood plain flow.

The model requires boundary conditions, describing the flow at the boundaries or limits of the model. The General Model requires rainfall from some 120 rainfall stations distributed throughout the country, discharge time series on 8 major rivers crossing the international border and water level time series at 6 stations at or near the Bay of Bengal. In general, daily values are required for these time series. In certain cases, a higher frequency of reading is required, e.g. hourly for tidal water levels and 3 hourly for rainfall stations in areas of intense rainfall.

Model Calibration

Before the model output may be used for particular studies, it is necessary to adjust certain empirically determined parameters, chiefly the hydraulic resistance of the river channels, such that for selected events, the model output matches observed discharges and water levels at a number of key locations distributed throughout the model. This process of calibration may be separated into 4 steps.

Step 1 An initial run is made with assumed global values of the hydraulic resistance; major discrepancies in the model output when compared with the observations are examined. This indicates errors in the model scheme plan, e.g. is the interconnectivity of the river channels correct?, are the flood plains attached to the rivers in the right way?

Step 2 The distribution of the flow among the branches of the river network is scrutinised. By adjusting the relative hydraulic resistance of the individual river branches, it should be possible to obtain an accurate distribution of the flow in time and among the various branches of the river system. Figure 21.21 shows a well calibrated tidal discharge station.

Step 3 The third step is to adjust the hydraulic resistance within each river branch such that the modelled and observed water levels show a good match. The hydraulic resistance can be varied both longitudinally along the branch and vertically within each river cross section.

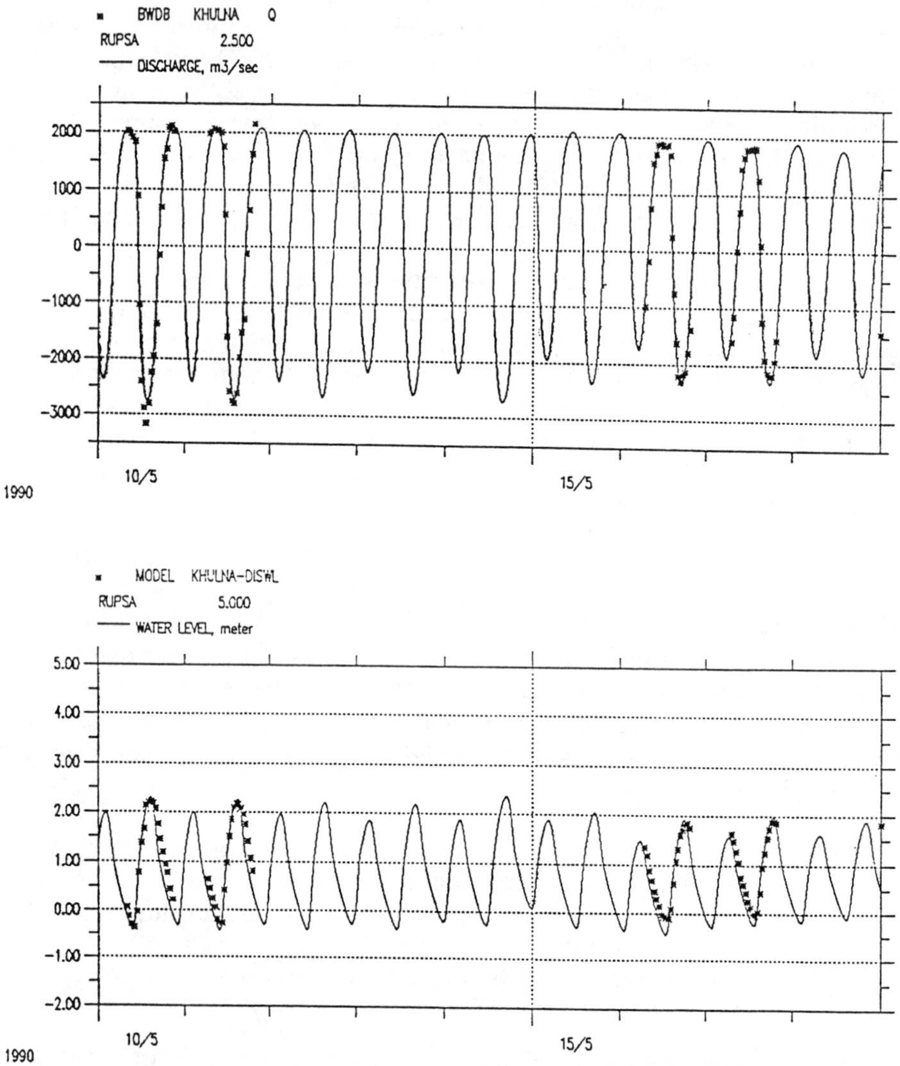

Fig. 21.21. Calibration of tidal discharge station eg Khulna.

Step 4 Having calibrated the model for one event or a series of events, the model should be verified against one or more historic events, preferable exhibiting different patterns of flooding

While the resistance and other hydraulic parameters have an empirical basis, extensive documented measurements have defined the range within which these parameters may be expected to vary for certain conditions. The modeller has to be fully aware of the 'normal range' of values that will be expected.

21.3.1.3. *Model Applications*

Once the model has been satisfactorily calibrated and verified, it may then be used with confidence to simulate events for which no or limited comparison data are available. In the process of calibration, the modeller will have gained a great deal of knowledge about the behaviour of the river system and the database may also have been considerably improved through suspect data either having been checked or discarded.

Analysis of Events

The model has been set up to represent the existing conditions which prevail in the river system, ie with existing structures, embankments, etc. The results from the model simulation of historic events can be studied to reveal more information about the pattern of flooding that can be gained from the observations, i.e. discharges, velocities and water levels along reaches of the river system for which no measurements are available. Figure 21.22 shows the discharge over the left bank flood plain of the Jamuna from 1986 to 1989. This is impossible to measure in practice.

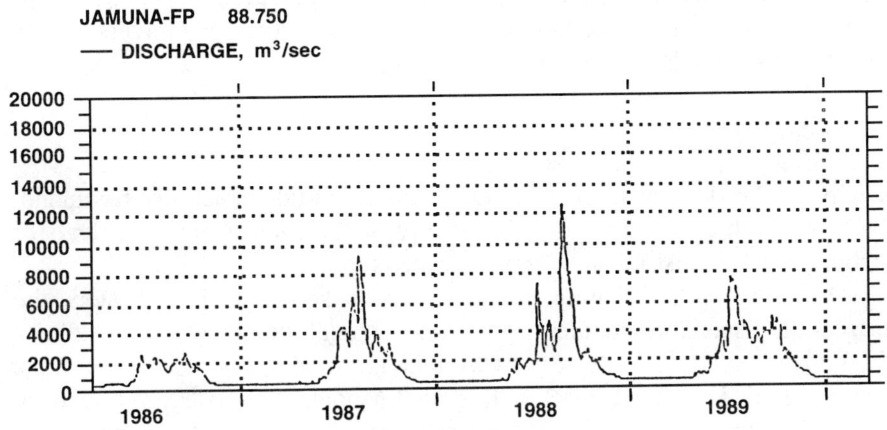

Fig. 21.22. Discharge over Jamuna left bank flood plain, from 1986 to 1989.

The model may be run for a long period, e.g. 25 years, and the local design conditions derived for particular locations through an analysis of the model output. A greater understanding of the existing pattern of flooding can be achieved which may guide the design engineer in selecting design options to alleviate the flooding, which can be tested on the model.

Introducing a Proposed Scheme

The model may be altered to represent a possible future condition of the river system. For instance, channels can be closed off, embanked, dredged to any specified cross section and new channels can be cut. Structures, e.g. barrages, regulators, weirs, culverts, pumps, can be added to the model. Existing embankments and structures can also be removed to represent some past condition of the river system. Analysing the results of the altered model can give the design engineer a comprehensive and accurate picture of the effect of a proposed deign scheme.

Synthesis of Events

Hydrologic events may also be synthesised to represent for example the flood with a specified return period. Owing to the complexity of the river system in Bangladesh, the return period of a historic flood exhibits a high degree of spatial variability over the country, and it is generally more satisfactory to select particular historic event(s) as representative of the design conditions.

Possible changes to the environment outside the model area may also be input to the model to assess their effect on the modelled area, for example a rise in mean sea level (see Fig. 21.23), deforestation of an upstream catchment leading to increased sediment in the river or conversely the construction of a dam lowering the sediment load.

Salinity

Owing to the flat terrain of Bangladesh, the flood tide penetrates far inland, bringing with it saline water. The quality of the river water is critical for irrigation and industrial and domestic water supply.

Once the hydrodynamic model has been set up and calibrated, the advection/dispersion module may be run to model the movement of salinity in the river system. As for the hydrodynamic model, boundary conditions are required. For the General or Regional Models, the upstream boundary would be zero concentration at the fresh water inflow, and the downstream boundary the time varying concentration where the rivers discharge into the Bay of Bengal.

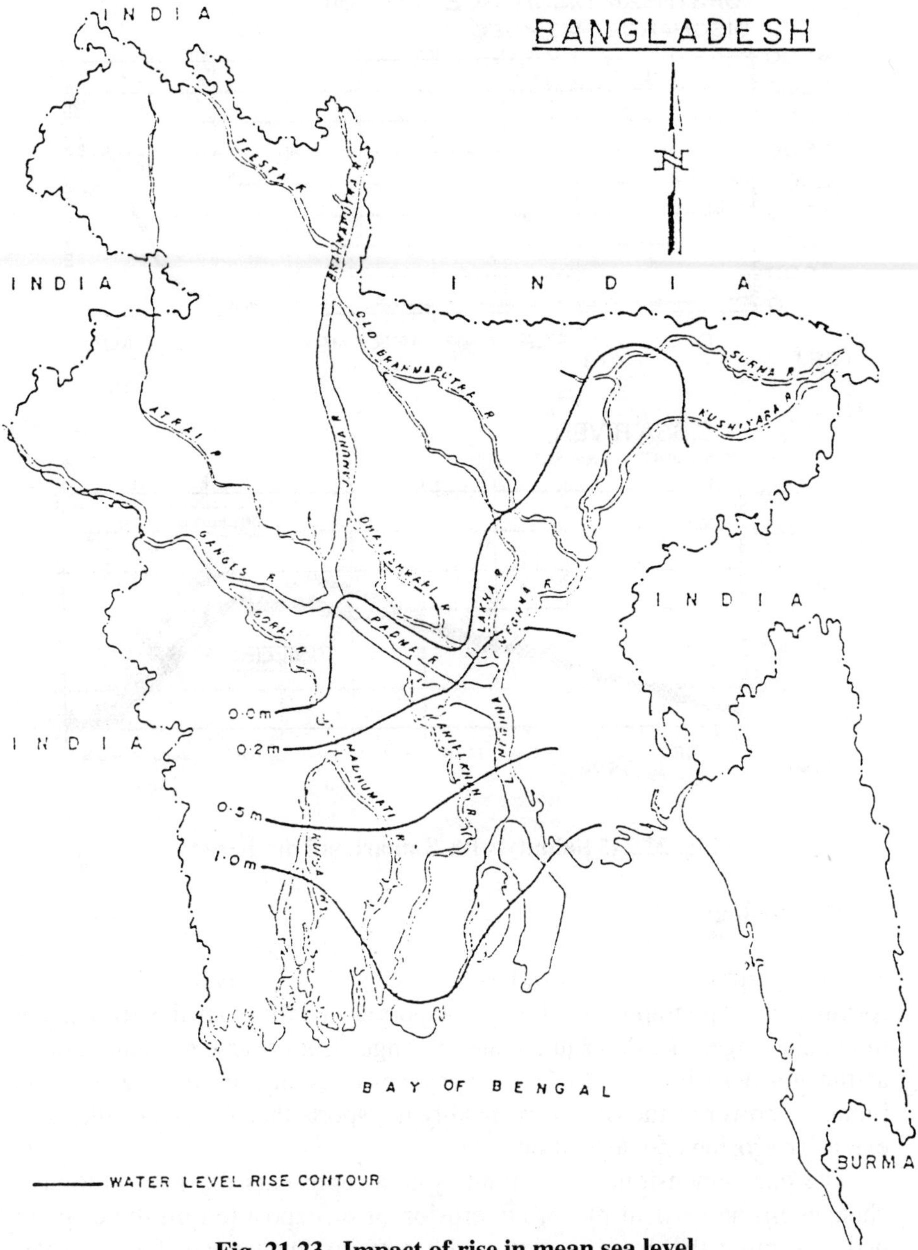

Fig. 21.23. Impact of rise in mean sea level.

Sample results from the model runs show (Fig. 21.24) the tidal variation and the annual variation of salinity. Note that the influence of the upstream discharge on salinity is greater than the influence of the tidal variation.

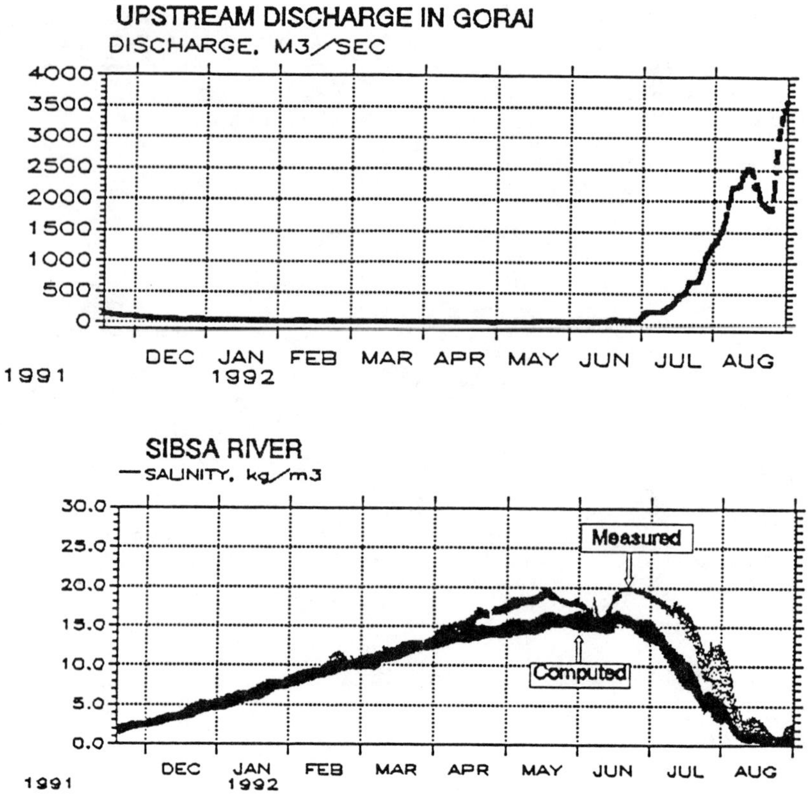

Fig. 21.24. Salinity with Kamarkhali discharge.

Morphology

The topography of Bangladesh has been shaped by its rivers. The material forming the topography has been transported mainly from the Himalayan mountain range. The deposited material ranges from coarse sand and gravel at the northern borders to fine clayey particles in the south west. The immense power of the river flow readily transports the deposits resulting in extensive erosion and deposition.

The one dimensional modelling system represents the morphological changes in the vertical plane, i.e. erosion of or deposition on the channel bed, assuming the cross section width remains unchanged. Over a river reach, patterns of erosion and deposition frequently compensate. Nonetheless, results from the morphological model have to be interpreted in the context of information gained from other sources, e.g. sequential satellite images, and not least practical experience.

Separation of bed and suspended load is essential for calibration against measurements of suspended sediment transport. Bed load cannot practically be measured on large rivers. Sample results of sediment transport rates in the Jamuna and Ganges over several years are shown in Fig. 21.25. The total transported load in the rivers in 1988 was approximately 125 Mm^3 and 80 Mm^3 respectively.

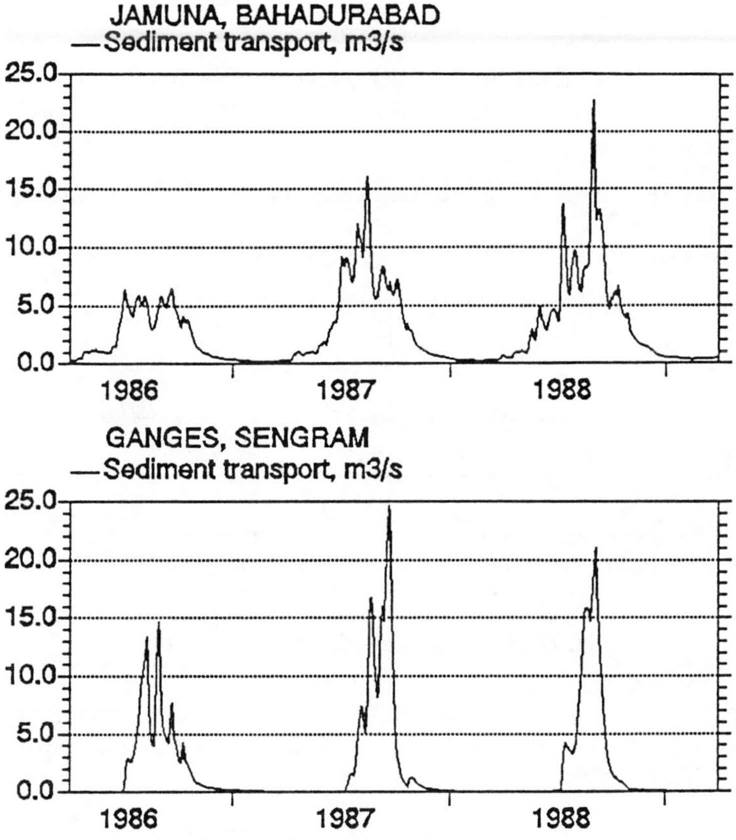

Fig. 21.25. Sediment transport rates Jamuna/Ganges.

Cohesive Transport

The transport of cohesive sediments is described by the advection/dispersion process discussed earlier. Erosion and deposition are influenced by the interparticle electro-chemical bonds which must be overcome by the shear force on the bed before erosion starts.

Deposition occurs when the flow velocity is sufficiently low to allow the particles to settle. The rate of deposition is a function of the hydraulic characteristics as well as the concentration of the suspended sediment. If the deposited particles remain undisturbed for a period of time, the bonding forces increase and consolidation takes place, making erosion more difficult.

A sample plot showing accumulated deposition in a river branch in the southwest of Bangladesh is shown in Fig. 21.26. The plot illustrates the relative magnitude of tidal and upstream discharge transport; i.e. the accumulation rate is larest during the wet season from June to October.

Bed Level Updating

The morphological model updates the bed level changes due to erosion or deposition using the continuity of sediment. The increase/decrease in the bed

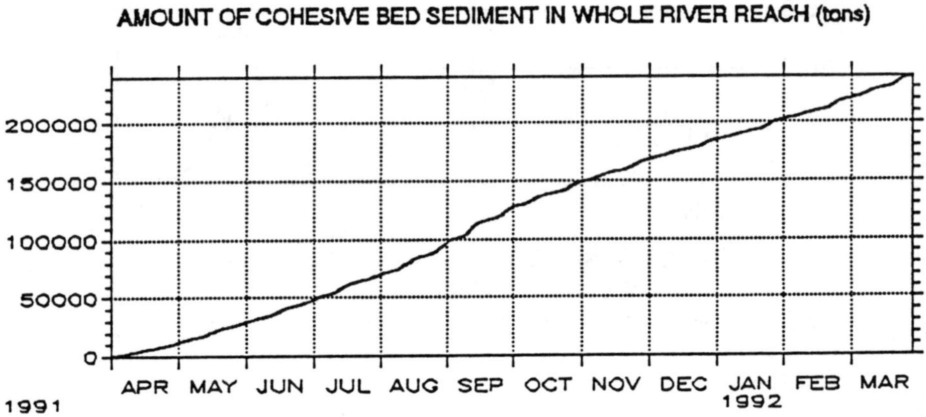

Fig. 21.26. Cohesive sediment deposition.

level of the Jamuna over a period of 25 years under present conditions, and under an embanking and bridge scenario, are shown in Fig. 21.27. The seasonal change in bed level is also recorded.

21.3.2. Cohesive Sediment and Water Quality, River USK, Wales

The River Usk rises in the mountains of south Wales and flows in an incised, alluvial valley to the Severn Estuary at Newport. The tidal influence from the Severn Estuary penetrates approximately 25 km upstream beyond Newbridge-on-Usk, see Fig. 21.28. The mean spring tidal range at the docks in the Severn Estuary is 12 m and the mean neap range is about 5 m.

A prominent feature of the estuary is its high suspended load during spring tides and its much lower suspended load on neap tides. During spring tide the turbid water is advected up and down the estuary. During neap tides, much of the mud settles to the bed to form pools of liquid mud. During the neap to spring tidal cycle these pools of liqued mud are then re-eroded.

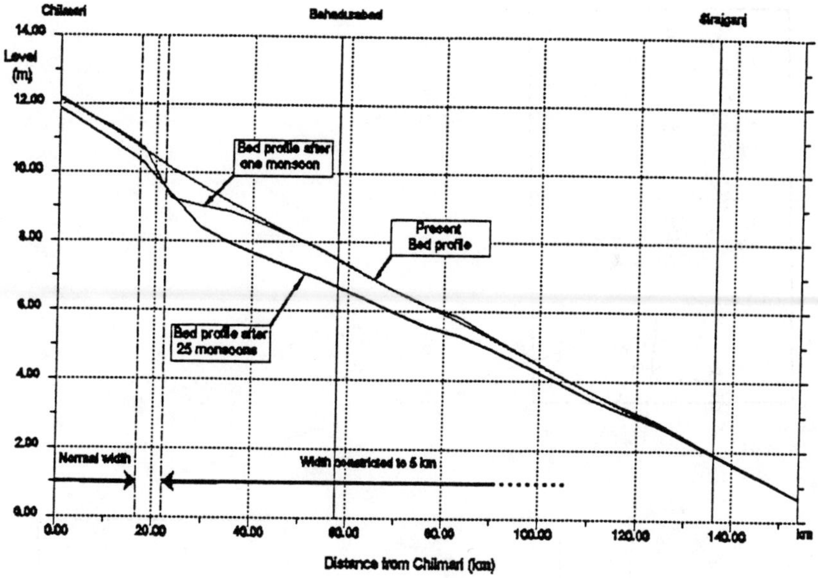

Developed bed Profile on Severely Constricted Jamuna River

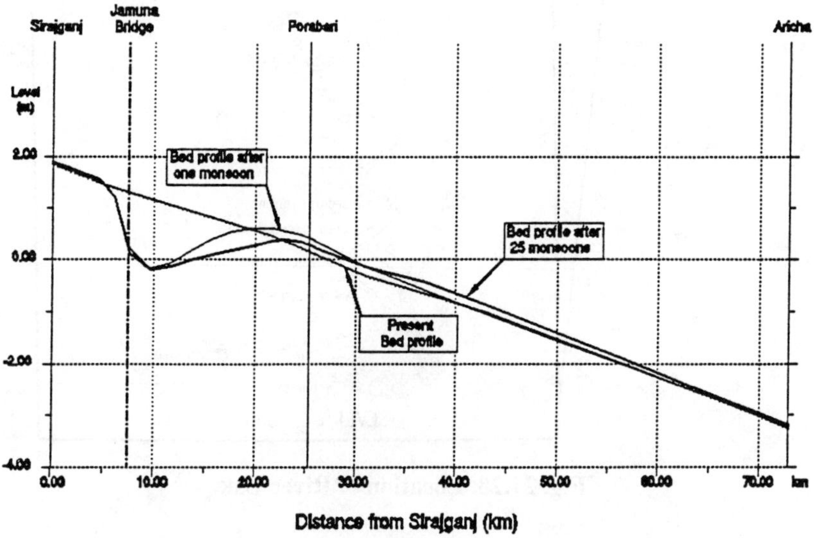

Developed bed Profile after Jamuna Bridge

Fig. 21.27. **Bed-level changes width constriction / present and JMP present.**

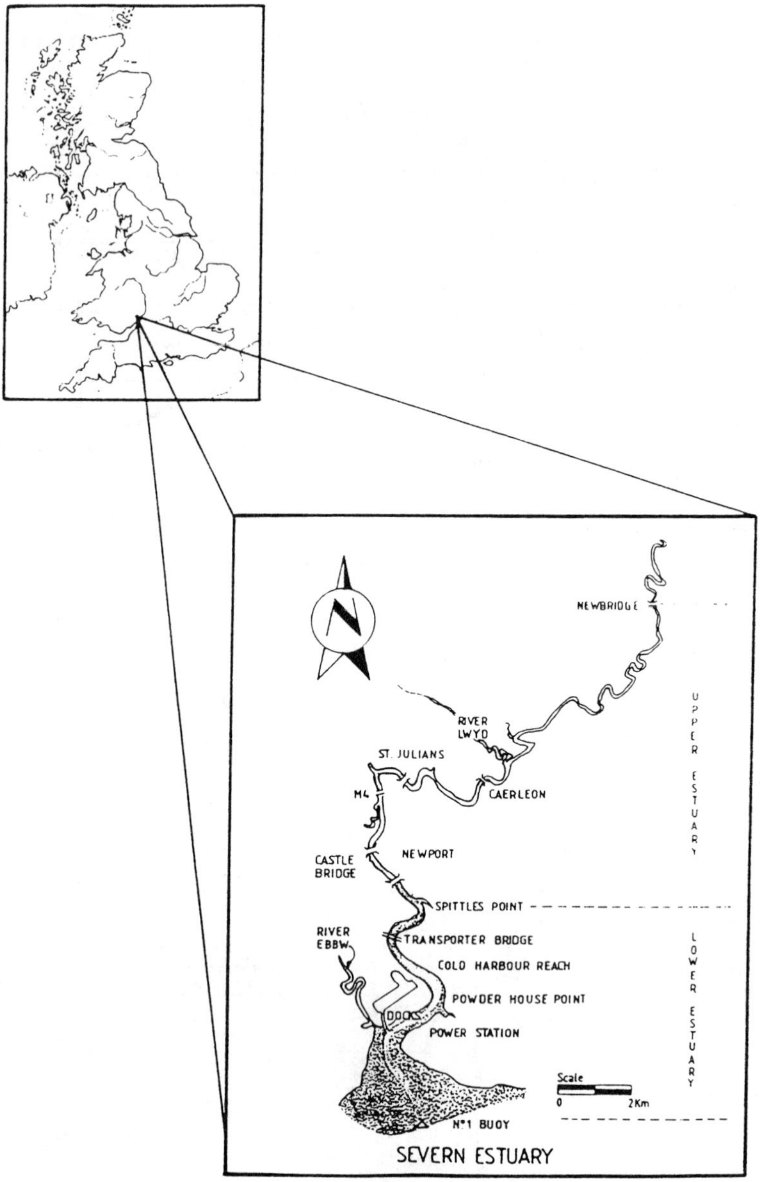

Fig. 21.28. Location of River Usk.

The large tidal ranges in the mouth of River Usk at Newport, Wales, combined with extensive discharges of crude sewage into the river, cause such problems as:

- risk of flooding in the city of Newport

- deteriorated water quality in the river
- aesthetic nuissance due to malodorous sediment at the river banks during low tides.

In order to solve these problems it was proposed to build a barrage across the river in Newport. This would create a constant water level upstream of the barrage and thus exclude some of the above mentioned problems. Additionally, it was proposed to relocate some of the sewage outfalls from the upstream part of the river to the downstream part - otherwise major eutrophication problems in the impounded lake behind the barrage might occur. This relocation would, of course, contribute to an increased load of pollutants to the downstream parts of the river.

The proposed construction of the barrage thus formed the background of this case study:it was necessary to demonstrate that water quality standards set by the national water authorities were met also after the construction. As fisheries interests in the upper parts of the river are of major importance, allowance had to be made for the passage of migratory fish. Thus future levels of dissolved oxygen and ammonia up-and downstream of the barrage had to to be within specified ranges.

The concentration of dissolved oxygen in the river is highly dependent on the oxygenation processes related to the sediment and fluid mud in suspension, which was verified from measurements. The simulation of the present and the future variation in the level of DO require a detailed description of the cyclic behaviour of the sediment and fluid mud as described above.

The MIKE 11 standard water quality model combined with the advance cohesive sediment transport module was selected by the water authorities as a suitable tool to predict future water quality in the downstream reach of the river. Hence both models were set up and calibrated based on the pre-barrage situation and were subsequently used to predict future post-barrage conditions.

Model calibration, Cohesive sediment

Two fluid mud surveys on two consecutive days showed dramatic changes in the amount of fluid mud. The first day (9/11 1989) a total of approximately 20,000 tons of fluid mud was observed in the river. This amount of fluid mud should be compared to a total transport in and out of the river during spring tide conditions of approximately 60,000 tons. The next day (10/10 1989) nearly all the fluid mud had disappeared, only about 3,000 tons of mud were left. The tidal variation and the river flow measured beyond the tidal limit for this period are shown in Figs. 21.29 and 21.30. respectively.

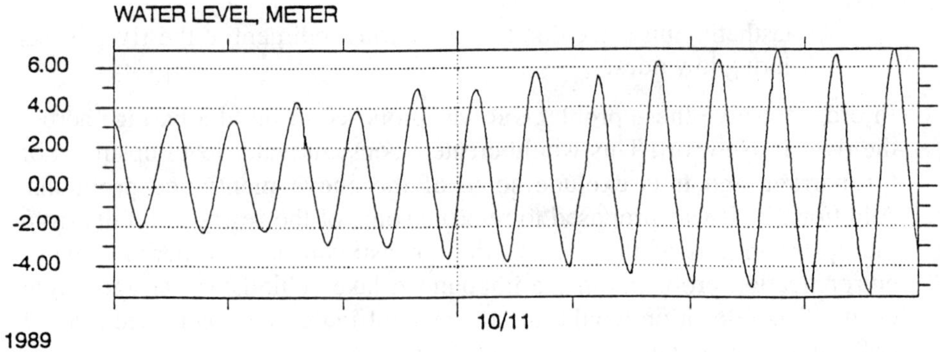

Fig. 21.29. Tidal variation. Neap to Spring.

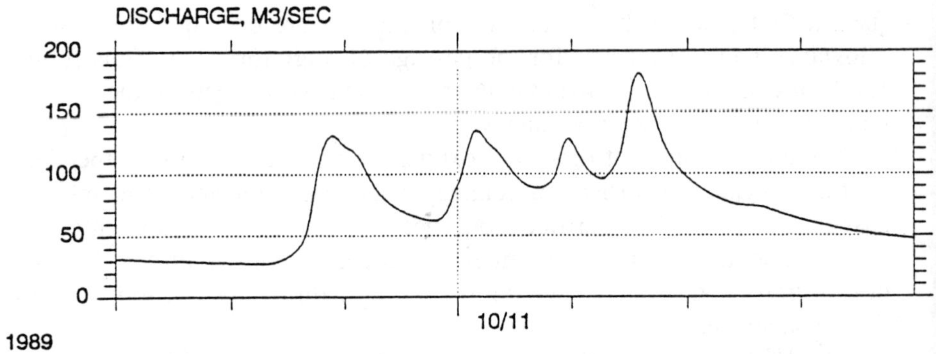

Fig. 21.30. River discharge.

The figures indicate that the removal of the fluid mud coincided with a peak in the river flow.

For calibration of the model the density limits for the three layers were assessed from field data (nuclear transmittance density gauge). The rate at which sediment was transfered from one layer to another was determined from laboratory consolidation experiments and the critical shear stress determined via flume experiments. The sediment fall velocity and the erosion function (erosion coefficient and exponent) were determined via calibration of the model on the above mentioned data and on observed vertical suspension profiles.

The simulated development of weak fluid mud and fluid mud is shown in Fig. 21.31. The frequent peaks in this figure represent the deposition and resuspension of weak fluid mud at each flood slack tide. The simulation has been carried out with an erosion exponent of 3. In Fig. 21.32, a sample

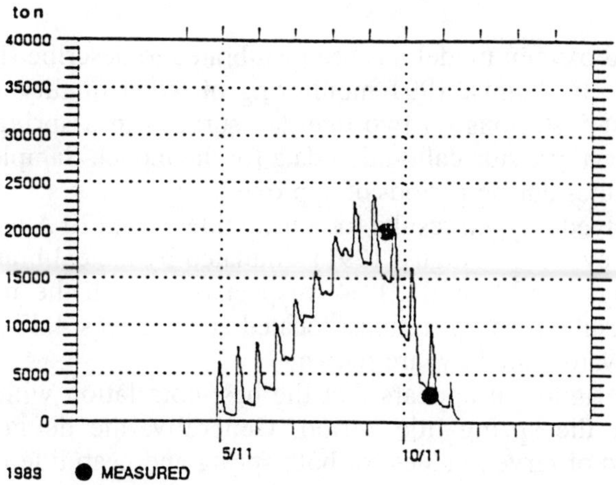

Fig. 21.31. Simulations and observed fluid mud.

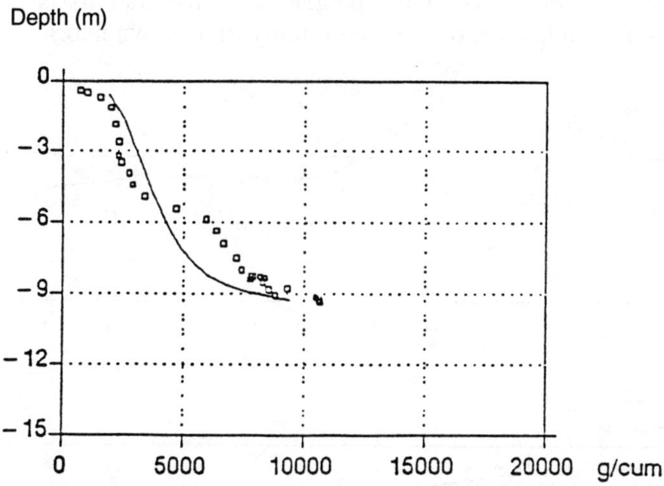

Fig. 21.32. Simulated and observed vertical distribution of suspended solids.

output of the simulated vertical distribution of suspended solid concentration is shown.

The simulated erosion rates and simulated concentrations of suspended sediment produced by the sediment-transport model were used as input to the water quality model.

Model calibration, Water quality

The dissolved oxygen model has been calibrated to describe field data from August 1991. In August 1991 monitoring of water quality variables was carried out at 5 stations on two one-day surveys in a spring-neap-spring cycle in order to provide calibration data for the model. Samples were taken every 30 minutes during periods of approximately 12 hours.

The dissolved oxygen results for a neap tide event (20 August 1991) and a spring tide event (27 August 1991) simulated by the calibrated model are shown in Figs. 21.33. and 21.34. respectively, with the measured data indicated. Results for two stations, located at chainages 4.27 and 14.57 km from the seaward boundary, are shown.

From the figures it appears that the best correlation with field data is achieved for the spring tide period. Generally, the deviation between simulated and observed values for both spring and neap tide is in the range 0-1.5 g/m^3 (corresponding to approx. 0-20%), and the maximum deviation is approx. 3 g/m^3. The underestimation of dissolved oxygen at chainage 14.57 km in the neap simulation (Fig. 21.33b) may be caused by an inaccuracy in the applied fresh water boundary conditions, and has nothing to do with the suspended sediment - dissolved oxygen relationship. This is revealed by inspection of simulated and observed salinity (not shown here).

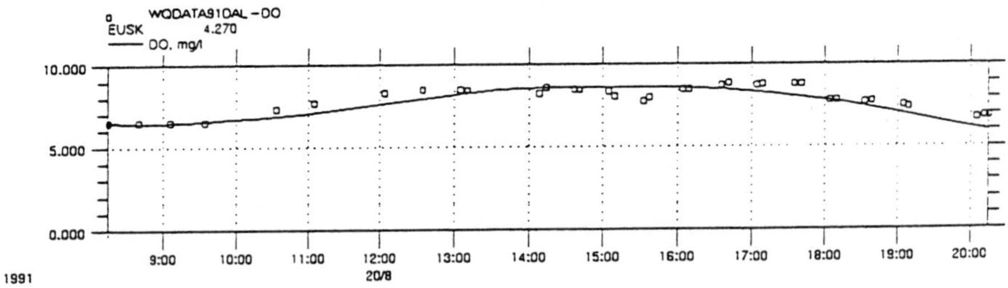

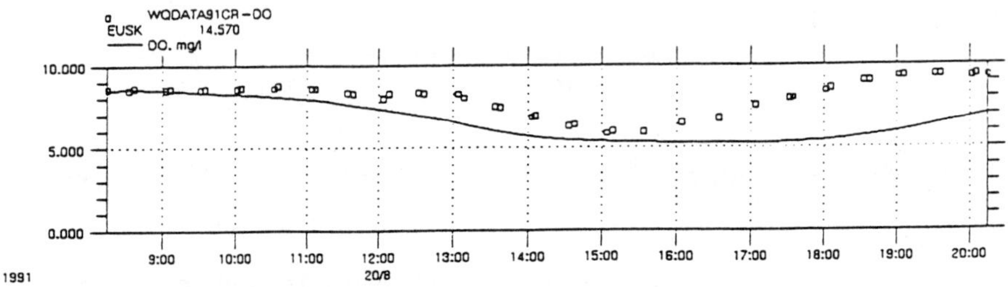

Fig. 21.33. Calibration of water quality model. Simulated and observed dissolved oxygen. Neap tide, August 1991.

a: Chainage 4.27 km from seaward boundary.
b: Chainage 14.57 km from seaward boundary.

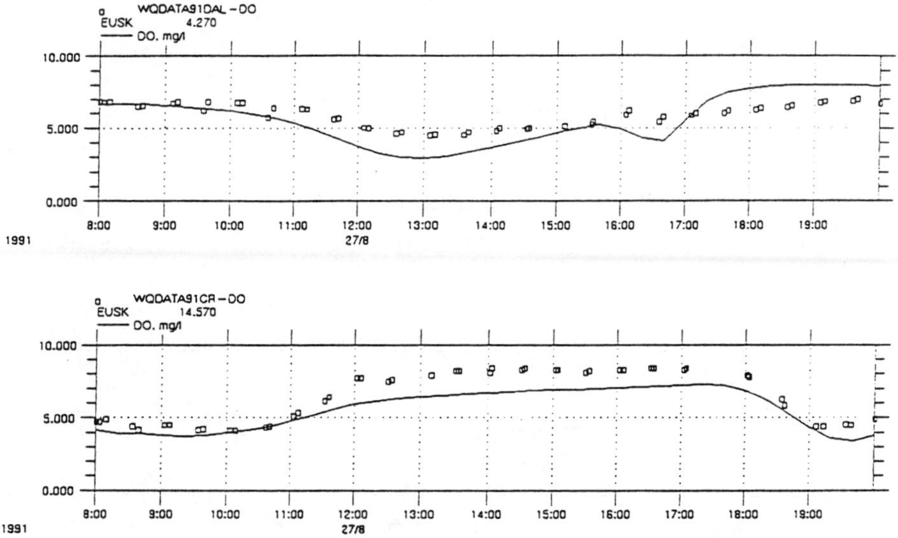

Fig. 21.34. Calibration of water quality model. Simulated and observed dissolved oxygen. Spring tide, August 1991.

a: Chainage 4.27 km from seaward boundary.
b: Chainage 14.57 km from seaward boundary.

Predicted Future River Water Quality Conditions

After construction of the barrage the tidal river will extend only up to approximately 6 km from the mouth. In order to do a worst-case analysis of the future water quality conditions, a so-called 'critical period' was defined. This included, e.g., low fluvial flow, high flushing rates from the impoundment behind the barrage, low oxygen content in the flushing water and high temperatures. Results of dissolved oxygen concentrations from such a scenario are shown in Fig. 21.35. The simulated minimum oxygen concentrations (and maximum ammonia concentrations, not shown here) revealed that the water quality standards set by the authorities were not met by this scenario. Therefore, the procedure for flushing had to be modified. Periodical flushing is necessary in order to flush out sediment accumulated at the downstream side of the barrage. The timing, duration and intensity of flushing are subject to changes and by subsequent modifications it was demonstrated that a flushing procedure that allowed for compliance with the water quality standards was feasible.

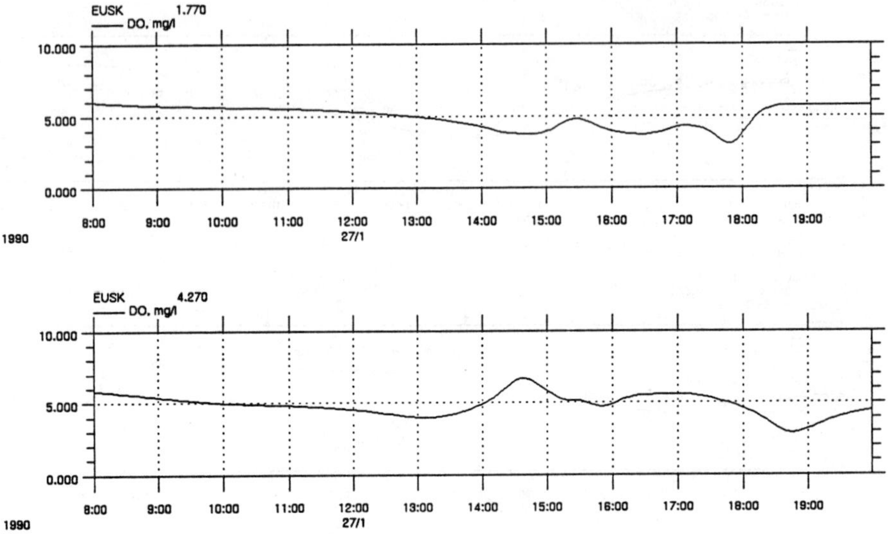

Fig. 21.35. Worst case simulation of future dissolved oxygen conditions. Flushing from impoundment applied.

a: Chainage 1.770 km from seaward boundary
b: Chainage 4.270 km from seaward boundary.

REFERENCES

(1). Nielsen, S.A. and Hansen, E. (1973): Numerical simulation of rainfall runoff process on a daily basis. Nordic Hydrology, 4, pp. 171-190.

(2). WMO, 1986: Intercomparison of Models of Snowmelt Runoff. Operational Hydrology Report No. 23.

(3). WMO, 1988: Real-Time Intercomparison of Hydrological Models, Report of the Workshop held in Vancover, Canada, from 30 July to 8 August 1987. Technical Reports to the Commision for Hydrology -No. 23.

(4). Abbott, M.B., 1979: Computational Hydraulics. Elements of the Theory of Free Surface Flows. Pitman.

(5). Ackers, P. and White, W.R. (1973). "Sediment transport: new approach and analysis", Proc. ASCE, JHD, 99, HY11, pp. 2041-2060.

(6). Engelund, F. and Fredsøe, J., 1976. "A sediment transport model for straight alluvial channels". Nordic Hydrology, Vol. 7, No. 5.

(7). Engelund, F. and Fredsøe, J., 1982. "Hydraulic theory of alluvial rivers", Advances in Hydro Science, Vol. 13, pp. 187-215.

(8). Engelund, F. and Hansen, E., 1967. "A monograph on sediment transport in alluvial streams", Teknisk Forlag, Copenhagen.

(9). Fredsøe, J., 1979. "Dimensions of stationary dunes. Part 1. Low sediment transport rate". Prog. Rep. 49, Institute of Hydrodyn. & Hydr. Engrg., Technical University of Denmark.

(10). Fredsøe, J., 1982. "Shape and Dimensions of stationary dunes in rivers". ASCE, Vol. 108, No. Hy 8.

(11). Rijn, L.C. van, 1984a, "Part I: Bed load transport", J. Hyd. Eng., 110, 10 Oct.

(12). Rijn, L.C. van, 1984b, "Part II: Suspended load transport", J. Hyd. Eng, 110, 11 Nov.

(13). Rouse, H., 1937. "Modern conceptions of the mechanics of fluid turbulence". Trans. ASCE, Vol. 102, pp. 463-543.

(14). Ariathurai, R., MacArthur, R.C. & Krone, R.B. (19719. Mathematical Model of Estuarial Sediment Transport. Department of Civil Engineering, University of California at Davis.

(15). Van Rijn, L.C., "Handbook on Sediment Transport by Current and waves", Delft Hydraulics, report H461, June 1989.

(16). S.E. Jørgensen, M.J.Gromiec, "Mathematical Submodels in Water Quality Systems", Developments in Environmental Modelling,14, Chapter 12, page 217- 240, 1989.

(17). A.M. Hansen, Hanne.K. Bach "Modelling River Water Quality and Impact from Sewers and Storm Sewer overflows", Presented at the Symposium on Water Quality Modelling, Harrogate, 13, Nov. 1991.

(18). K.W. Olesen, K. Havnø, "A Water Quality Modelling Package for Forth-Generation Modelling", Proc. IAMR Congress, Ottowa, Canada, 1989.

(19). V.T. Chow, David R. Maidment, Larry W. Mays. Applied Hydrology. McGraw-Hill Book Company, 1988.

Chapter 22

EPA STORM WATER MANAGEMENT MODEL -- SWMM

W. C. Huber

The U.S. Environmental Protection Agency's Storm Water Management Model has been widely applied in the U.S., Canada and elsewhere for analysis of complex hydrologic, hydraulic and water quality problems associated with urban drainage. This chapter gives a brief description of the program and provides examples of SWMM applications.

22.1. SWMM DESCRIPTION

The U.S. Environmental Protection Agency (EPA) Storm Water Management Model, SWMM, is a comprehensive water quantity and quality simulation model developed primarily for urban areas. The model was developed in 1969-1971 by three groups: Metcalf and Eddy, Inc., the University of Florida, and Water Resources Engineers, Inc. (now Camp, Dresser and McKee, Inc.). The original documentation (Metcalf and Eddy et al., 1971) has been superseded by the current Version 4 documentation (Huber and Dickinson, 1988; Roesner et al., 1988) although some portions of the original reports are useful for understanding program algorithms. The original model was developed primarily to address pollution problems due to combined sewer overflows (CSOs), but the model has been widely used for analysis of hydrologic and hydraulic problems of both combined and separate sewer systems as well as for urban non-point pollution problems. The model has the capability to be used for purposes that include routine drainage design, screening and planning for future drainage issues, water quality problems, complex hydraulic and flow routing issues, and many other engineering tasks. Single event and continuous simulation can be performed for almost all components of the rainfall, runoff and quality cycles for a watershed, including watersheds with a non-urban composition. A somewhat dated bibliography lists many applications of the model through the year 1985 (Huber et al., 1985). An update is provided by James (1993b). The program is written in Fortran and is microcomputer based, although the source code is available for compilation on other machines.

The model is structured in the form of several "blocks" or modules, as shown in Fig. 22.1. The principal computational blocks include the Runoff Block for generation of runoff and quality constituents from rainfall (plus simple routing), the Transport Block for kinematic wave routing and for additional dry weather flow and quality routing, the Storage/Treatment Block for reservoir routing and simulation of treatment and storage quality processes, and the Extended Transport or Extran Block for hydraulic routing of flow (no quality routing) using the complete Saint Venant equations. The dynamic flow routing of the Extran Block permits hydraulic simulation of dynamic backwater conditions, looped drainage networks, surcharging, pressure flow etc. The flow routing characteristics of the various blocks are listed in Table 22.1. Additional blocks shown in Fig. 22.1 perform services such as rainfall pre-processing, statistical analysis, line printer graphing, etc. Further detail on these blocks is provided below.

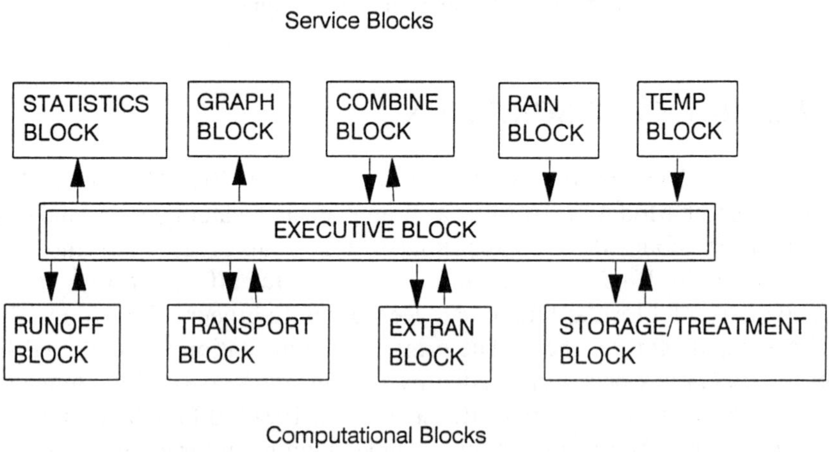

Fig. 22.1. Blocks or modules of the SWMM program (Huber and Dickinson, 1988). Arrows indicate transfer of data by off-line interface files. The Executive Block is only used to call up the other subroutines.

As described in the user's manuals (Huber and Dickinson, 1988; Roesner et al., 1988), SWMM simulates real storm events based on rainfall, other meteorological inputs (e.g., evaporation, temperature), and catchment, transport, storage, and treatment characterization to predict outcomes in the form of quantity (hydrographs and runoff volumes) and quality

TABLE 22.1.

Flow Routing Characteristics of Runoff, Transport and Extran Blocks.

	Runoff Block	Transport Block	Extran Block
1. Flow routing method	Nonlinear reservoir, cascade of conduits	Kinematic wave, cascade of conduits	Complete equations, interactive conduit network
2. Relative computational expense for identical network schematizations	Low	Moderate	High
3. Attenuation of hydrograph peaks	Yes	Yes	Yes
4. Time displacement of hydrograph peaks	Weak	Yes	Yes
5. In-conduit storage	Yes	Yes	Yes
6. Backwater and downstream control effects	No	No[a]	Yes
7. Flow reversal	No	No	Yes
8. Surcharge	Weak	Weak	Yes
9. Pressure flow	No	No	Yes
10. Branching tree network	Yes	Yes	Yes
11. Network with looped connections	No	No	Yes
12. Number of pre-programmed conduit shapes	5	16	8
13. Alternative hydraulic elements (e.g., Pumps, Weirs, Orifices)	WO	PWO	PWO
14. Dry weather flow and infiltration generation (base flow)	Some	Yes	Yes
15. Pollutograph routing	Yes	Yes	No
16. Solids scour/deposition	No	Yes	No
17. Input of hydrographs/pollutographs from external source	No	Yes	Yes

[a]Backwater may be simulated as a horizontal water surface behind a storage element.

(pollutographs and pollutant load) values. Input sources are found in the Runoff Block, which generates surface and subsurface runoff based on precipitation hyetographs, antecedent conditions, land characterization, and topography. The Runoff, Transport, and Extran Blocks route flows and pollutants through the sewer or drainage system to make up the central core of the program (Extran does not perform pollutant routing). The Storage/Treatment Block forms the correctional devices portion of the program that characterizes the effects of control devices upon flow and quality.

SWMM does not include a block for analysis of receiving water quality. Other models available from the U.S. Environmental Protection Agency (EPA) Center for Exposure Assessment Modeling (CEAM) may be interfaced with SWMM and other surface runoff models and used for this purpose (Ambrose and Barnwell, 1989). A direct interface to the CEAM WASP4 model is included in the Transport Block of SWMM.

In addition to the computational blocks mentioned above, SWMM uses six service blocks. The Executive Block assigns logical unit numbers to off-line files and determines the block or sequence of blocks to be executed. Line printer plots of hydrographs, pollutographs and other time series output may be accomplished through the Graph Block. (The power of current PCs for graphical output is woefully under-utilized by Version 4 of SWMM; however, pre- and post-processors available from private vendors for the model partially remedy this situation as does the opportunity to utilize spreadsheets and other microcomputer graphing software.) The Combine Block allows for the manipulation of interface files so that results may be aggregated for input into subsequent blocks. The Rain Block processes U.S. National Weather Service (NWS) and Canadian Atmospheric Environment Service (AES) long-term precipitation series (typically hourly or 15 minute data) for input into the Runoff Block for continuous simulation. Similarly to the Rain Block, the Temp Block processes NWS long term temperature data for input into the Runoff Block for snowmelt calculations. (Long term evaporation and wind speed data may also be entered in the Temp Block.) Finally, the Statistics Block has the capability to process the time step output from a continuous simulation by separating output into discrete storm events, ranking the events according to almost any desired criterion (e.g., peak flow rate, runoff volume, pollutant load, event mean concentration, etc.), assigning empirical frequencies and return periods to runoff and pollutant parameters, tabulating and graphing the results, and calculating statistical moments of the untransformed and log transformed data. The Statistics Block may also be used for analysis of the precipitation time series; the Rain Block also includes a statistical analysis option based on the EPA SYNOP (Synoptic Precipitation) program (EPA, 1976) that performs a similar function.

Blocks are run in a modular fashion, and only the blocks of interest need be utilized. For example, runoff hydrographs generated by an external procedure may be routed using the Extran or Transport Blocks; there is no requirement that the Runoff Block must be used. In addition, output from one block may be used as input to a subsequent run of the same block, providing a means for disaggregation of large systems.

SWMM offers the choice of metric or U.S. customary units. Default values for catchment parameters may be specified by the user; the model does not include such parameters internally, but the model documentation attempts to provide guidance for all required input parameters. While SWMM has tremendous computational power, it is not a "user friendly" model in the conventional sense for a microcomputer. The basic program available from the EPA CEAM is run in a batch mode, and the user must provide an input data file prepared using an ASCII editor. Output is in tabular form only. Line printer graphs of hydrographs and pollutographs can be output, or the tabular listings can be exported into other software (e.g., spreadsheets) for access to conventional PC (personal computer) graphics. Fortunately, private vendor efforts at pre- and post-processors for SWMM alleviate this lack of user friendliness to a large extent. Linkage of SWMM to geographical information systems (GIS) and computer-aided design (CAD) systems has been demonstrated (Huber et al., 1991; Curtis and Huber, 1993).

SWMM is updated every 1-2 years and made available by the U.S. EPA at the following address:

Model Distribution Coordinator
Center for Exposure Assessment Modeling (CEAM)
U.S. Environmental Protection Agency
960 College Station Road
Athens, Georgia 30605-2720 USA

Phone: (706) 546-3549
Bulletin board: (706) 546-3402

Internet address for anonymous ftp: ftp.epa.gov

SWMM and other CEAM programs can be down-loaded from the CEAM computer bulletin board system or anonymous ftp (file transfer protocol) site on the Internet, and may also be obtained on disks. Users must supply the disks, but there is no other charge for processing orders. However, depending upon the software, there may be a nominal charge (typically less than US$150) for documentation. (Some programs include documentation in the form of WordPerfect files on disks, but not SWMM.) The user receives an executable program, the Fortran source code, examples, and limited documentation on the distribution disks. The program is carefully checked by the CEAM in order to eliminate any errors that can be detected by different compilers. A Windows version of SWMM is available from another EPA office (Wu, 1993). SWMM and many EPA and other federal government environmentally-related models are also available on an optical disk or CD-ROM (Walentowicz, 1993).

22.2. COMPUTATIONAL METHODS

Runoff is generated from rainfall in the Runoff Block by a nonlinear reservoir technique (Fig. 22.2) that couples the spatially lumped continuity equation with Manning's equation for overland flow (Huber and Dickinson, 1988; Bedient and Huber, 1992). A conceptual subcatchment width (width of overland flow) is included in Manning's equation to provide a volumetric flow rate. An estimate for this width is the subcatchment area divided by the overland flow path length. Infiltration losses are computed using either the Green-Ampt or integrated Horton equation (Bedient and Huber, 1992). Hence, required input data for each subcatchment include area,

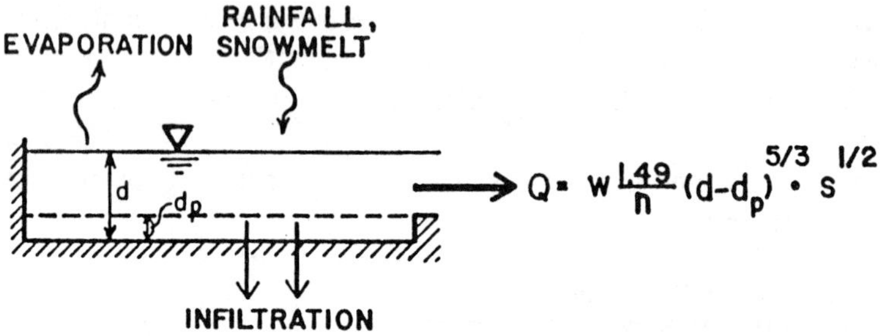

Fig. 22.2. Nonlinear reservoir conceptualization of Runoff Block subcatchment (Huber and Dickinson, 1988). The "1.49" in Manning's equation would be just "1.0" for metric applications. The continuity and Manning's equation are used to solve for the two unknowns, depth and flow.

imperviousness, slope, Manning roughness, a conceptual subcatchment width, depression storage, and infiltration parameters. Water that infiltrates into the soil can be additionally routed (as a user option) through subsurface unsaturated and saturated zone groundwater storages (Fig. 22.3). Infiltration into the soil is cut off if the water table rises to the surface. Snowmelt is optionally simulated using a combination degree-day and energy balance method based on U.S. National Weather Service procedures (Anderson, 1973). Different degree-day coefficients can be used on different urban surfaces (pervious and impervious), and these coefficients can vary seasonally. Urban snow removal and the linkage of quality buildup to the presence of snow may also be simulated. Infiltration into Runoff Block

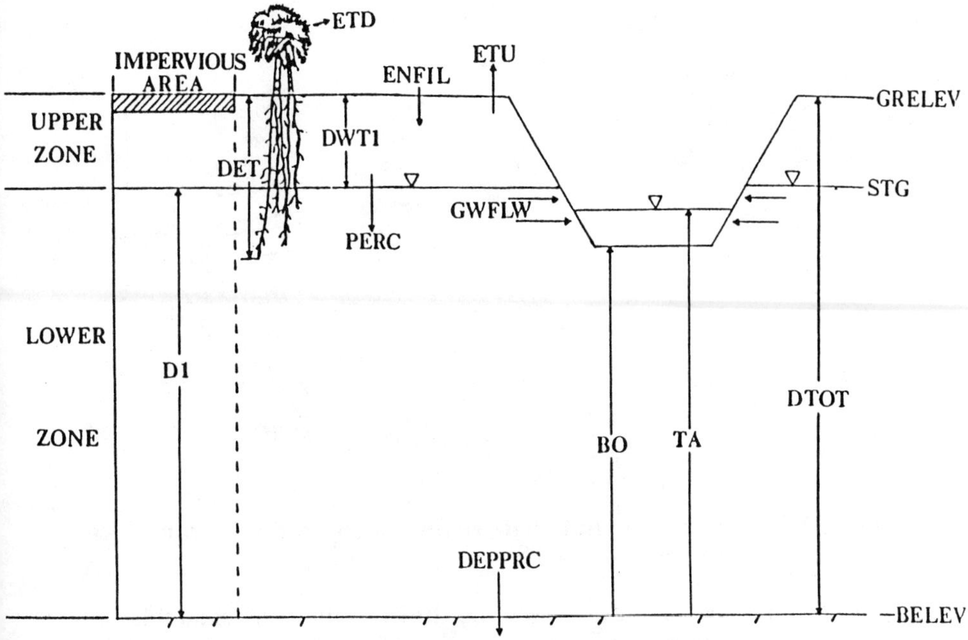

Fig. 22.3. Subsurface flow routing in the Runoff Block (Huber and Dickinson, 1988). The upper zone is unstaurated; the lower zone is saturated.

channel/pipes can be simulated using triangular unit hydrographs that are driven by rainfall excess over the sewered area of each subcatchment. Infiltration may also be entered into a Transport Block sewer network.

The Runoff and Rain Blocks accept an arbitrary input hyetograph on either a storm-event or long-term basis (for continuous simulation). Both historic (measured) and synthetic design storms may be used as model input.

SWMM flow routing options are described in Table 22.1. Depending on the block or method used, input data include conduit or channel dimensions, slope (or invert and ground surface elevations), Manning roughness, and characteristics of hydraulic structures such as weirs, orifices and pumps. The Storage/Treatment and Transport Blocks also provide simple storage-indication (Puls method) reservoir routing. The Transport Block also includes options for temporally and spatially variable input of infiltration and dry-weather flow into the sewer network, as well as a hydraulic design routine that will re-size circular and rectangular pipes to pass maximum flows.

The most sophisticated and easily the most complex and difficult to use flow routing option is provided by the Extran Block (Roesner et al., 1988). The drainage system is conceptualized as a series of links and nodes (Fig. 22.4). The nodes (junctions) have characteristics of storage, as defined by

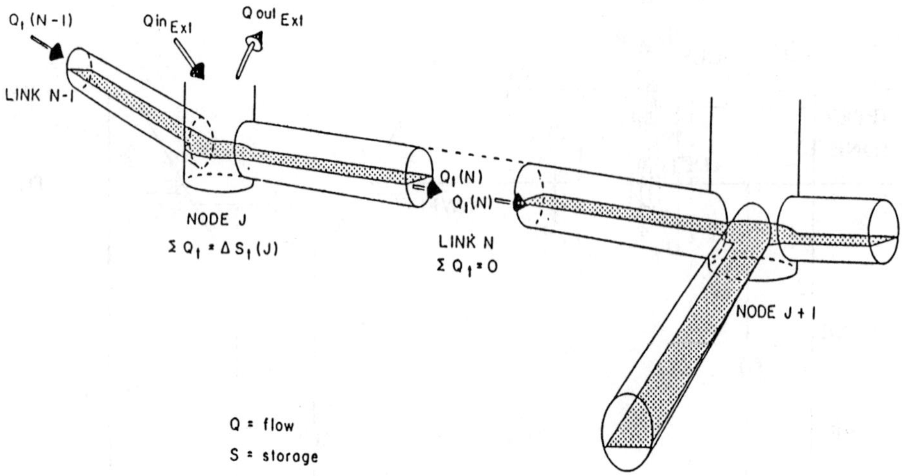

Fig. 22.4. Conceptual link-node configuration in the Extran Block (Roesner et al., 1988).

adjacent pipes or channels, and the links (pipes or channels) have characteristics of transport. That is, the continuity equation is applied at nodes, and the momentum equation is applied along links. Multiple conduits (or open channels) can join at nodes (Fig. 22.5). Thus, the

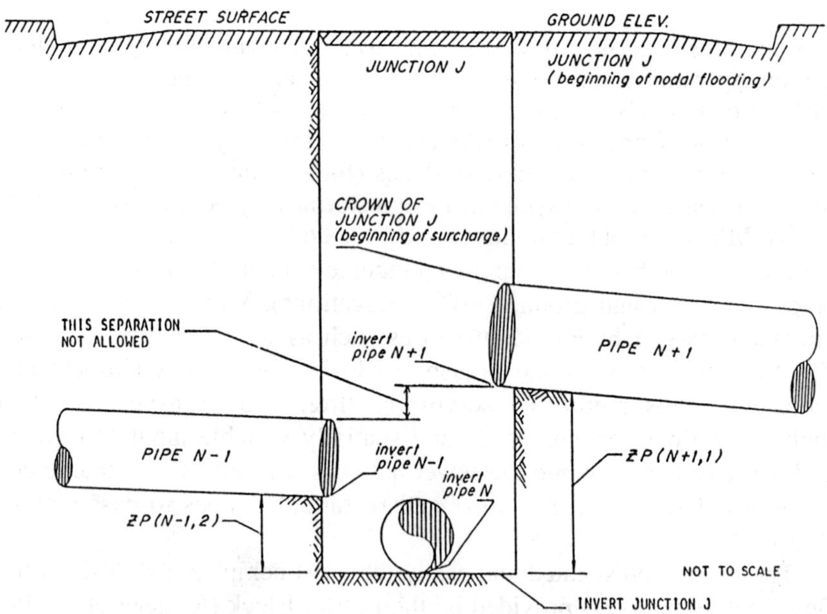

Fig. 22.5. Junction (node) definitions in the Extran Block (Roesner et al., 1988).

continuity equation is solved at nodes simultaneously with the one-dimensional momentum equation along the links (Roesner et al., 1988). The explicit solution technique often requires a short (< 10 sec) time step to maintain the Courant stability criterion, although "equivalent pipes" may be used to alleviate this requirement. User-defined channel cross sections can also be input (Fig. 22.6) so that Extran can be used as a dynamic backwater

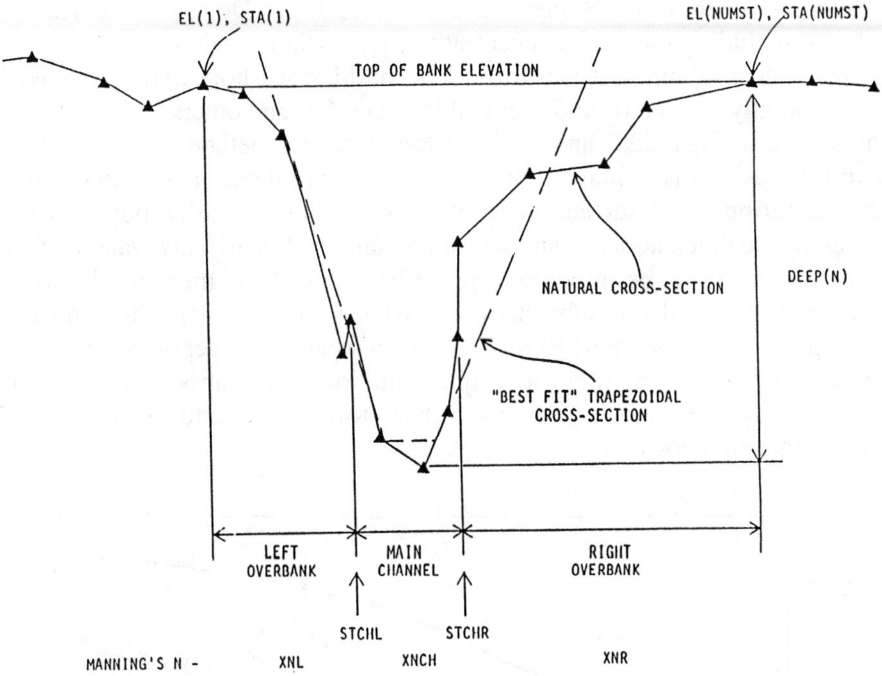

Fig. 22.6. User-defined channel cross section in the Extran Block (Roesner et al., 1988).

model for natural stream channels. For convenience, the required cross-sectional data are in the Hydrologic Engineering Center (HEC) HEC2 model format, and HEC2 cross-sectional data can be inserted directly into Extran (and Transport) Block simulations. Surcharge or pressure-flow conditions in Extran are simulated by assuming that the summation of inflows equals the summation of outflows at surcharged nodes in the absence of significant storage at a node. (Ponds and detention storage may also be simulated.) An iterative procedure is then used to balance the flows within the surcharged portion of the sewer network. (A Priessman slot method for surcharge conditions is available as an option.) Several boundary conditions options are available, including free outfall, constant head, and tidally and other variable heads, with and without flap gates (one-way "valves"). The use of Extran requires hydraulic expertise on the part of the engineer and patience to learn its peculiarities and requirements. However, it has proven to be an extremely powerful tool in the hands of capable users for analysis of complex hydraulic problems in urban areas.

Water quality processes can be simulated by SWMM by a variety of options (Huber, 1985; Huber and Dickinson, 1988; Donigian and Huber, 1991), including constant concentration, regression relationships (load vs. flow), and buildup and washoff. The buildup-washoff option is based conceptually on data from Sartor and Boyd (1972) and others (Huber, 1985) as shown in Figs. 22.7 and 22.8. Parameters and methods can vary with land use. Other quality processes include those associated with precipitation, land-surface erosion, catchbasins (or gully pots), street cleaning, sedimentation, scour, deposition, and treatment. Any water quality constituent may be modeled, provided that it is measurable as a concentration and not subject to excessive scour and deposition during transport. (The Transport Block includes a weak scour-deposition routine, based on Shield's criterion for incipient motion.) Output is in the form of pollutographs (concentration vs. time), total loads and seasonal and statistical summaries.

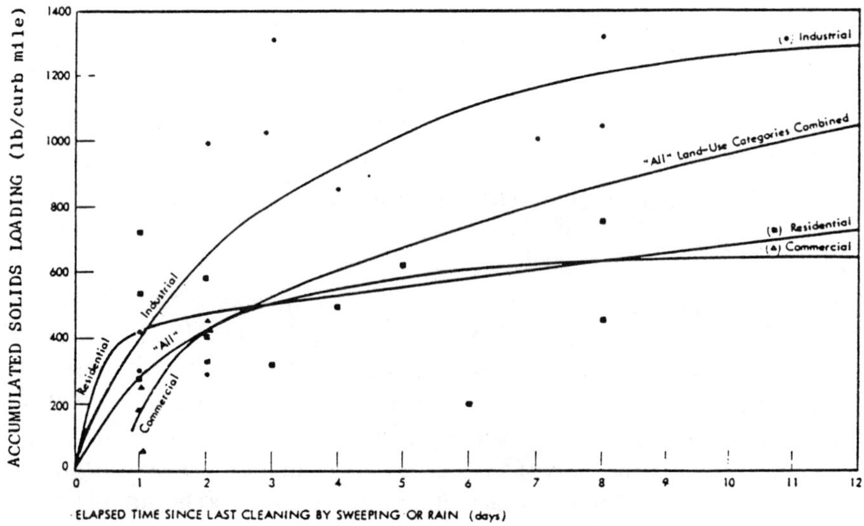

Fig. 22.7. Buildup of street surface solids reported by Sartor and Boyd (1972).

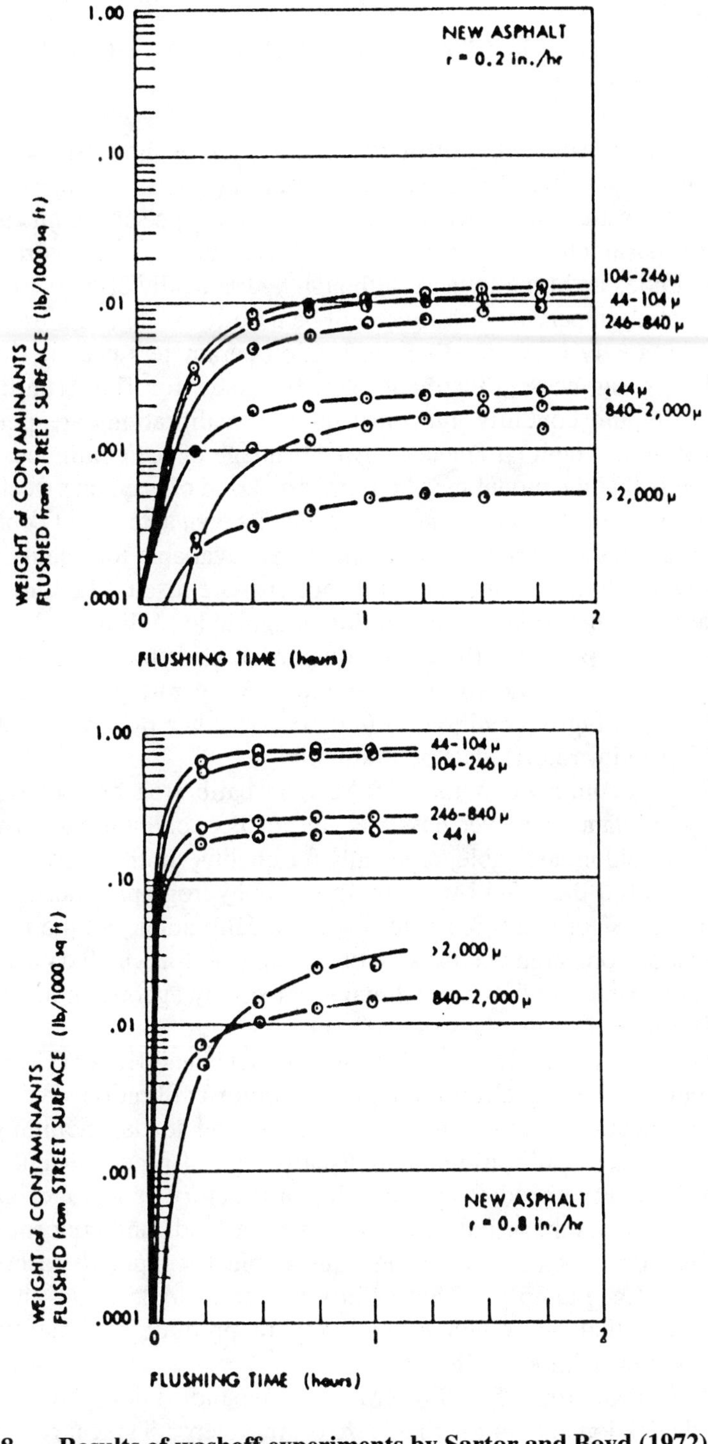

Fig. 22.8. **Results of washoff experiments by Sartor and Boyd (1972). a. Runoff rate = 0.2 in/hr (5 mm/hr). b. Runoff rate = 0/8 in/hr (20 mm/hr).**

22.3. SENSITIVITY, ACCURACY, AND COMPUTATION TIME

Although somewhat difficult to generalize, Runoff Block hydrographs are most sensitive to area and imperviousness. Because these two parameters are determined a priori, in principle, the subcatchment width -- a shape parameter -- is often used for calibration of hydrograph peak magnitudes and time to peak, although hydraulically effective (i.e., directly connected) imperviousness may still be used for volume calibrations. If the area is mostly pervious, then infiltration parameters become important as well. Roughness adjustments can bc made for flow routing through channels and conduits, but most quantity calibrations are performed on runoff hydrograph parameters. An advantage of the continuous simulation option is that the model may be run for a period of, say, one week before the storm of interest to let the model generate its own antecedent conditions.

As discussed earlier, several options are available for quality generation, but when buildup and washoff options are used, the most sensitive parameters are related to buildup magnitude, followed by a washoff exponent. Because there are numerous possible sources of quality parameters in the model, it is important to separate processes as much as possible during the calibration (e.g., dry-weather flow vs. catchbasins vs. buildup and washoff), data permitting.

It is recommended that SWMM be calibrated and verified against measured data whenever possible. In practice, calibration and verification data are seldom available, especially for quality. In their absence and given good rainfall data, SWMM often predicts hydrograph peaks and volumes within ±25% of measured values without calibration, but the model should at least be compared against a nearby catchment for which data are available for corroboration of its predictions -- not simply against another model that suffers from the same lack of calibration.

As described by Huber (1985) and Donigian and Huber (1991), all urban nonpoint source quality simulations require calibration data in order to predict accurate values of concentrations and loads, including SWMM. Without local calibration data (usually end-of-pipe concentrations and flows), relative predictions may be satisfactory (e.g., one management scheme vs. another), but absolute values of loads and concentrations will almost certainly be in error and unsuitable for input to receiving water models, for example. This is in contrast to hydrologic and hydraulic modeling that are much more likely to produce sufficiently accurate hydrographs without calibration.

Execution time for SWMM runs depends upon the detail of the schematization and number and size of time steps. Typical runs with 10-20 subcatchments, 10-20 pipes and a few hundred time steps require 1-5 minutes on a 486/33-type microcomputer. Continuous simulation with hourly precipitation input requires on the order of a 5-10 seconds per year per subcatchment. Complex Extran runs with 200 pipes and thousands of time steps can take several hours of microcomputer time.

22.4. COMPARABLE MODELS

Many models are available for analysis of urban runoff hydrology, although a smaller number also perform quality calculations (Donigian and Huber, 1991; ASCE and WEF, 1993). Among these are the non-proprietary EPA HSPF (Hydrologic Simulation Program Fortran), U.S. Geological Survey DR3M-Qual (Distributed Rainfall Runoff Routing Model - Quality), and Corps of Engineers, HEC STORM (Storage, Treatment, Overflow and Runoff Model) models. In addition, two proprietary European models, the Danish MOUSE and UK Wallingford models are widely used outside the U.S. Examples of useful third-party software for a SWMM graphical user interface (GUI) include XP-SWMM (XP Software, Tampa, FL) and MTVE (10 Brooks Software, Ann Arbor, MI).

22.5. EXAMPLES OF SWMM USAGE

SWMM is just one of several modeling options, but it has been used in hundreds of studies in the U.S., Canada, and elsewhere, most often for analysis of complex drainage systems in highly urbanized areas. Unfortunately, documentation of most of these case studies is in the form of consultants' reports and is difficult to obtain.

Several examples of SWMM applications may be found in readily available journals such as the American Society of Civil Engineers Journal of Hydraulic Engineering, Journal of Environmental Engineering, and Journal of Water Resources Planning and Management. These papers are not listed here in the interest of space. However, a very useful forum for papers on SWMM applications as well as papers on many other models and topics related to urban runoff consists of the several proceedings of the EPA Storm and Water Quality Modeler Users Group Meetings. These proceedings and sources are listed in Table 22.2. Proceedings dated after 1981 will contain references to SWMM and Extran versions similar to the current version, although even the earliest proceedings contain useful examples. James (1993b) provides a very useful index to paper titles from these proceedings.

TABLE 22.2.

Sources of EPA Storm and Water Quality Modeler User's Group Proceedings

Date	Meeting Location	Report No.	NTIS No.	Source[a]
May 1978	Ottawa, ON	EPA-600/9-78-019	PB-285993/2BE	NTIS
Nov. 1978	Annapolis, MD	EPA-600/9-79-003	PB-290742/6BE	NTIS
May 1979	Montreal, PQ	EPA-600/9-79-026	PB80-105663	NTIS
Jan. 1980	Gainesville, FL	EPA-600/9-80-017	PB80-177876	NTIS
Jun. 1980	Toronto, ON	EPA-600/9-80-064	PB81-173858	NTIS
Jan. 1981	Austin, TX	CHI-R83		1
Sep. 1981	Niagara Falls, ON	CHI-R81		1
Mar. 1982	Washington, DC	EPA-600/9-82-015	PB83-145540	NTIS
Oct. 1982	Ottawa, ON			2
Jan. 1983	Gainesville, FL	EPA-600/9-83-015	PB84-118454	NTIS
Sep. 1983	Montreal, PQ	GREMU 83/02		3
Apr. 1984	Detroit, MI	EPA-600/9-85-003	PB85-168003	NTIS
Sep. 1984	Burlington, ON	CHI-R128		1
Jan. 1985	Gainesville, FL	EPA/600/9-85/016	PB85-228302	NTIS
Dec. 1985	Toronto, ON	CHI-R149		1
Mar. 1986	Orlando, FL	EPA/600/9-86/023	PB87-117438	NTIS
Sep. 1986	Toronto, ON			2
Mar. 1987	Denver, CO	EPA/600/9-87/016	PB88-125430	CERI
Oct. 1987	Victoria, BC			4
Oct. 1988	Denver, CO	EPA/600/9-89/001	PB89-195002	NTIS
Apr. 1990	Eatontown, NJ	Not available		
Feb. 1992	Toronto, ON	James (1993a)		5
Feb. 1993	Toronto, ON	James (1994)		5
Nov. 1993	Reno, NV	EPA (in preparation)		NTIS
Feb. 1994	Toronto, ON	In preparation		
Jun. 1994	Denver, CO	EPA (In preparation)		NTIS

[a]Sources:
Note, prices will vary. Contact source.

NTIS:
National Technical Information Service
5285 Port Royal Road
Springfield, Virginia 22161
(703) 487-4650

Table 22.2. Continued.

CERI: (free while supplies last, then from NTIS)
Center for Environmental Research Information
Environmental Protection Agency
Cincinnati, Ohio 45268
(513) 569-7562

1. Computational Hydraulics Inc.
   36 Stuart Street
   Guelph, Ontario N1E 4S5
   (519) 767-0197
   (613) 564-3911

2. IMPSWM
   University of Ottawa
   Department of Civil Engineering
   Ottawa, Ontario K1N 6N5

3. Dr. Patrick Beron
   Groupe de Rescherches su
   L'eau en Mileu Urbain
   Department de Genie Civil
   Ecole Polytechnique de Montreal
   Case Postale 6097, Succursale 'A'
   Montreal, Quebec H3C 3A7
   (514) 340-4802

4. Charles Howard and Assoc., Ltd.
   852 Fort Street
   2nd Floor, West Wing
   Victoria, British Columbia
   V8W 1H8
   (604) 385-0206

5. Lewis Publishers
   CRC Press, Inc.
   2000 Corporate Blvd., NW
   Boca Raton, Florida 33431
   (800) 272-7737

The following discussion will provide a few brief examples to illustrate typical SWMM applications and model input/output. Most applications for cities are much more complex.

22.6. RUNOFF BLOCK QUANTITY EXAMPLE

The U.S. Geological Survey (USGS) monitored the Kings Creek apartment complex in Miami, Florida for flow and quality in 1977-78 (Hardee et al., 1979) and documented hydrologic properties of the catchment for modeling purposes (Miller, 1979). The 5.9 ha (14.7 ac) catchment has an overall imperviousness of 71% and a directly connected imperviousness of 44%. The catchment can be subdivided into 13 subcatchments as shown in Fig. 22.9; the drainage system of concrete pipes is shown in Fig.22.10. A SWMM input file is shown in Table 22.3, and output for four storm events is compared to measured hydrographs in Fig. 22.11.

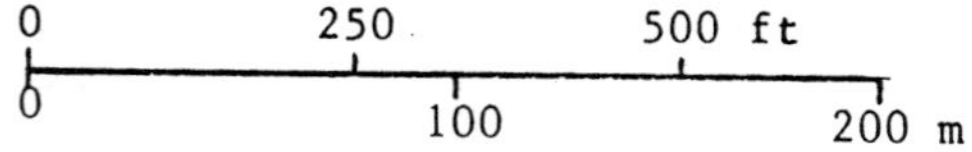

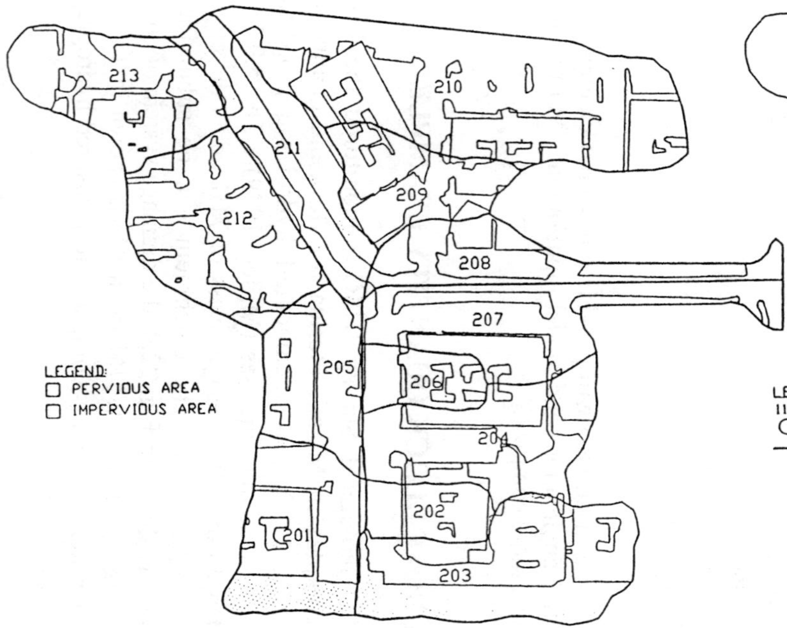

Fig. 22.9. Kings Creek Apartment catchment subcatchments and impreviousness (Huber et al., 1991).

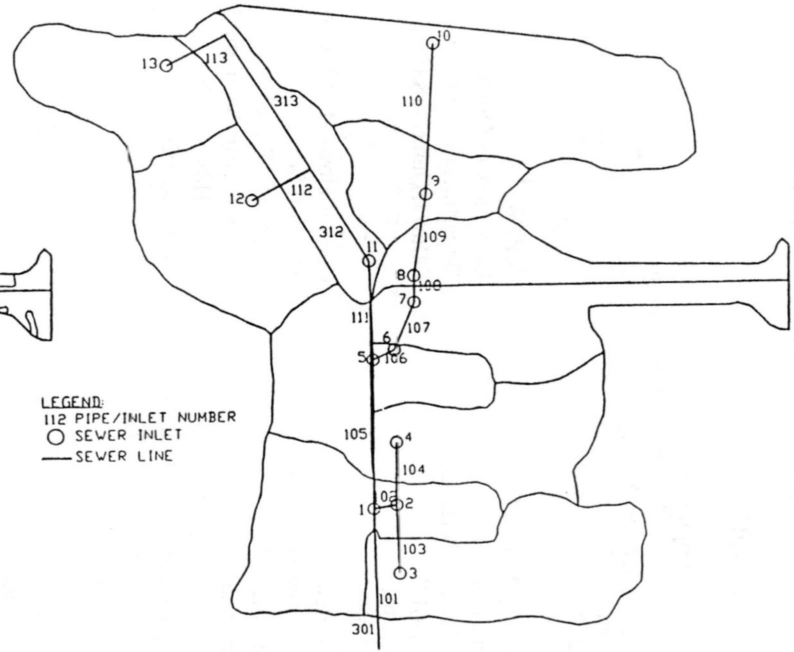

Fig. 22.10. Kings Creek Apartment catchment pipe system (Huber et al., 1991).

TABLE 22.3.

Input Data for Multiple-Family Catchment Runoff Block Example. Note: Rows beginning with asterisks (*) are comment lines.

```
SW 1 0 3
MM 6 11 12 13 14 15 16
$RUNOFF
A1 'MIAMI, KINGS CREEK APARTMENTS, 5/11/77 FIRST STORM'
A1 'HIGH-DENSITY RESIDENTIAL BASIN'
* METRIC ISNOW NRGAG INFILM KWALTY IVAP NHR NMN NDAY MONTH IYRSTR
B1   0      0     1      0       0     0   4   40  11    5      77
B2  1 1 0
*    WET  WETDRY  DRY  LUNIT  LONG
B3  120     600   600    2     1.5
D1  0
*  KTYPE KINC KPRINT KTHIS KTIME KPREP NHISTO THISTO TZRAIN
E1   0    14     1     0     0     1     14     10      280
* RAINFALL 10-MIN DEPTH INCREMENTS (INCHES)
E3 .6662 .4762 .0169 .0091 .0108 .003 0 0 0 0 .0123 .0076 .0062 .0091
* NAMEG NGTO NPG GWIDTH GLEN SLOPE LSLOPE RSLOPE ROUGHN DFULL GDEPTH
G1   113 313   2    1.75  110 .00682    0      0    .022    0     0
G1   313 312   2    2.5   205 .00237    0      0    .022    0     0
G1   112 312   2    1.75  110 .00078    0      0    .022    0     0
G1   312 111   2    2.5   175 .00237    0      0    .022    0     0
G1   111 105   2    2.5   165 .001      0      0    .022    0     0
G1   105 101   2    3     240 .00163    0      0    .013    0     0
G1   101 301   2    4     220 .00505    0      0    .013    0     0
G1   110 109   2    2     210 .00148    0      0    .022    0     0
G1   109 108   2    2.25  165 .00382    0      0    .022    0     0
G1   108 107   2    2.5    40 .02375    0      0    .022    0     0
G1   107 106   2    3      90 .00722    0      0    .022    0     0
G1   106 105   2    3      40 .029      0      0    .022    0     0
G1   104 102   2    1.5   105 .00038    0      0    .022    0     0
G1   103 102   2    1.5   105 .00076    0      0    .022    0     0
G1   102 101   2    1.5    32 .02969    0      0    .022    0     0
* JK NAMEW NGTO WIDTH WAREA IMP SLOPE IMPN PERN ISTOR PSTOR FO  FC  K
H1 1 213   113  1440  1.079 41.1 .04  .01  .2   .05   .15   1 .5 .00115
H1 1 212   112  2360  1.452 59.5 .04  .01  .2   .05   .15   1 .5 .00115
H1 1 211   111   760  1.049 35.7 .04  .01  .2   .05   .15   1 .5 .00115
H1 1 210   110  2520  2.798 49.3 .04  .01  .2   .05   .15   1 .5 .00115
H1 1 209   109  2040   .761 31.7 .04  .01  .2   .05   .15   1 .5 .00115
H1 1 208   108   640  1.006 51.0 .04  .01  .2   .05   .15   1 .5 .00115
H1 1 207   107   720  1.204 48.6 .04  .01  .2   .05   .15   1 .5 .00115
H1 1 206   106   800   .395 31.9 .04  .01  .2   .05   .15   1 .5 .00115
H1 1 205   105  1960   .842 47.1 .04  .01  .2   .05   .15   1 .5 .00115
H1 1 204   104  1240  1.236 37.1 .04  .01  .2   .05   .15   1 .5 .00115
H1 1 203   103  1360  1.412 40.2 .04  .01  .2   .05   .15   1 .5 .00115
H1 1 202   102   800   .352 31.0 .04  .01  .2   .05   .15   1 .5 .00115
H1 1 201   101  1560  1.157 35.9 .04  .01  .2   .05   .15   1 .5 .00115
*  NPRNT  INTERV
M1   1       1
*  NDET  STARTP  STOPPR
M2  1      0       0
*  IPRNT
M3  301
$ENDPROGRAM
```

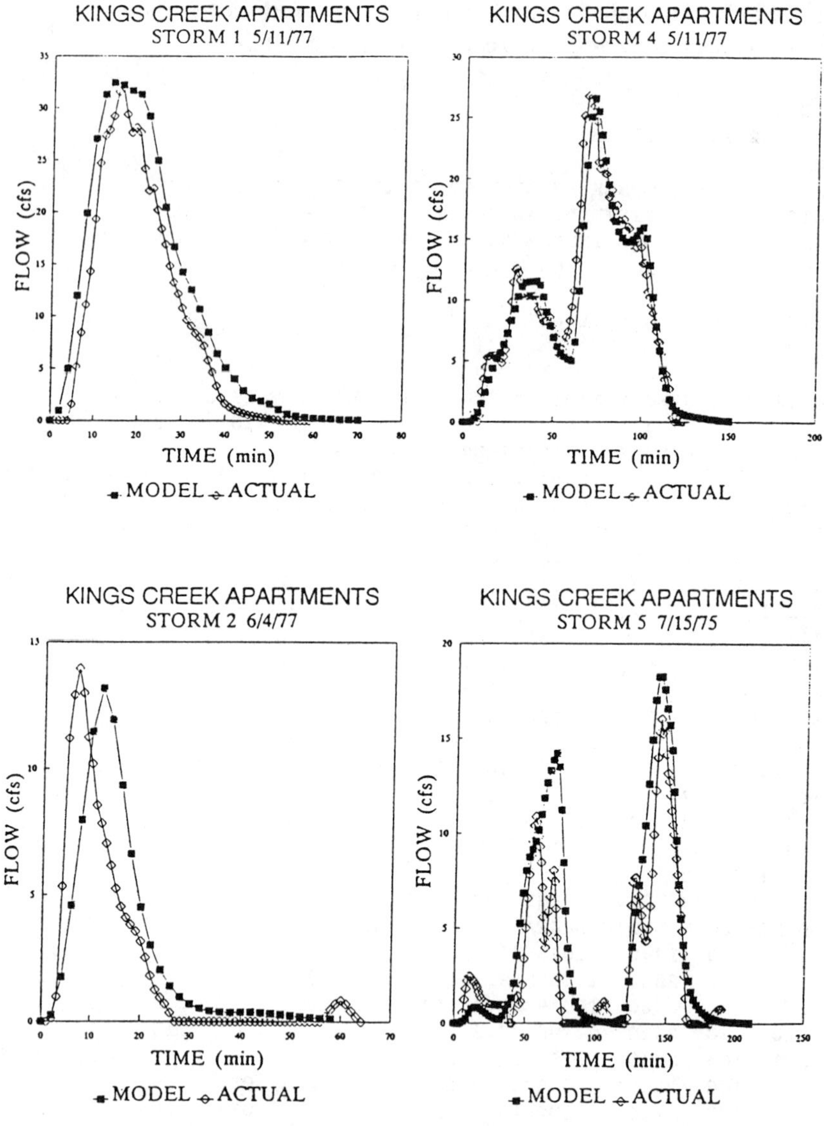

Fig. 22.11. Comparison of measured and predicted hydrographs for four storms.

Note that the input file is "free format" and may be annotated by inserting lines with asterisks in the first column. Five storms were calibrated as a group, and another five were used to verify the calibration for peak flows and for runoff volume. The comparisons shown in Fig. 22.11 are good and not significantly different from an uncalibrated run. The most sensitive input parameters are subcatchment area, imperviousness and width.

22.7. RUNOFF BLOCK QUALITY EXAMPLE

Quality data for the Kings Creek apartment complex are not as plentiful as the flow data but nonetheless may be used for model testing. Example SWMM output is compared for flow and five water quality parameters in Fig. 22.12 using a four-subcatchment and four-pipe catchment schematization. SWMM input data are shown in Table 22.4.

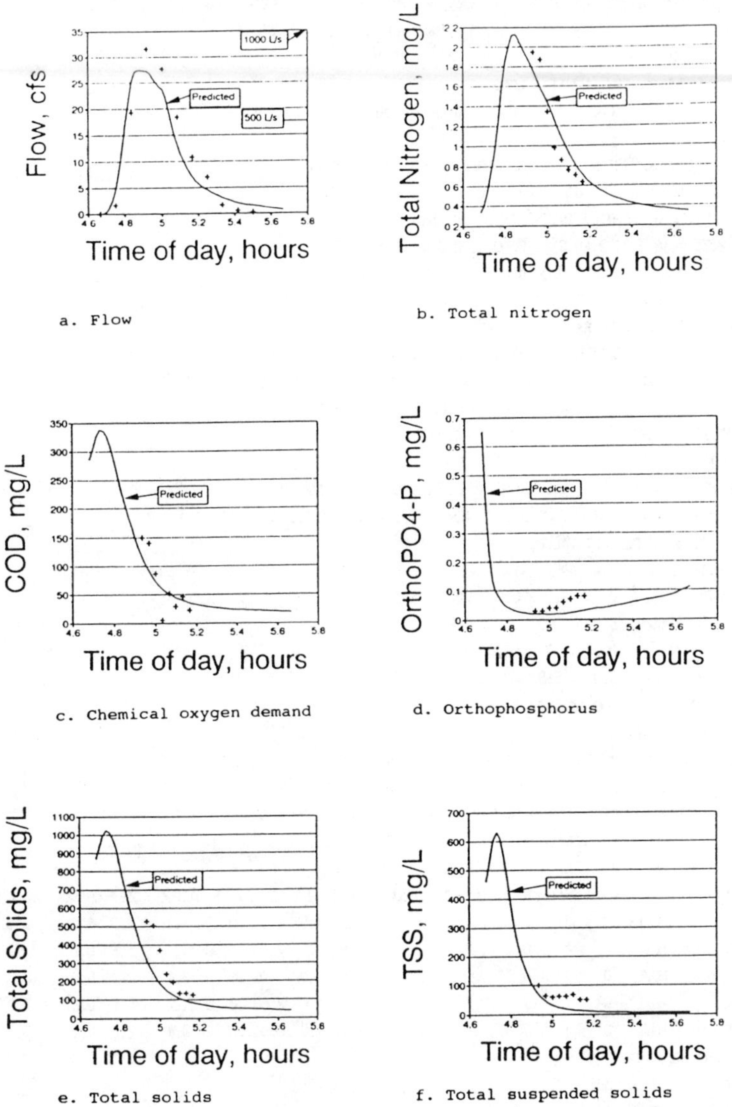

Fig. 22.12. Comparison of predicted and measured flow and quality for storm of May 11, 1977 on Kings Creek Apartment catchment.

TABLE 22.4.

Input Data for Runoff Block Quality Example.

```
SW  1  0  9
MM  4   10 11 12 13
$ANUM        Use alphanumeric labels (in quotes '')
$RUNOFF
A1 'KINGS CREEK APARTMENTS - Runoff Block Example # 44R'
A2 'RUNOFF AND QUALITY MODELING OF USGS STORM # 3, May 5, 1977'
*  METRIC ISNOW NRGAG INFILM KWALTY IVAP NHR NMN NDAY MONTH IYRSTR
B1   0       0   1     1       1      1  04   40  11   5      77
B2   0       0   0
*  WET    WET/DRY  DRY  LUNIT   LONG
B3  60.    60.0   86400.  1     60.0   minutes = simulation length
D1  0
*   KTYPE KINC KPRINT KTHIS KTIME KPREP NHISTO THISTO TZRAIN
E1   0     10    0      0     0     0      24     5.0    280
* RAINFALL INPUT, 5-MIN. INTENSITIES, IN./HR
E2     2.76 5.16 3.36 2.40 0.12 0.00 0.12 0.00 0.00 0.12
E2     0.00 0.00 0.00 0.00 0.00 0.00 0.00 0.00 0.00 0.00
E2     0.00 0.00 0.00 0.00
*  MONTHLY EVAPORATION RATES, IN./DAY
F1   0.09 0.11 0.14 0.17 0.18 0.17 0.17 0.16 0.14 0.13 0.11 0.09
*    NAMEG    NGTO   NPG  GWIDTH GLEN   SLOPE  GS1 GS2  N    DFULL GDEPTH
G1  'P001'  'Outlet' 2    4.     220.    0.0050 0.  0.  0.013  0.     0.
G1  'P005'  'P001'   2    3.     240.    0.0021 0.  0.  0.013  0.     0.
G1  'P009'  'P005'   2    2.25   505.    0.0040 0.  0.  0.022  0.     0.
G1  'P011'  'P005'   2    2.5    655.    0.0020 0.  0.  0.022  0.     0.
* GREEN-AMPT INFIL. PARAMS. USED (SUCTION, HYDR. COND., INIT. MOIS. DEFIC.)
* '-2' LINE SETS DEFAULT VALUES
* JK NAMEW  NGTO  WIDTH  AREA  %IMP  SLP  IMPN PERVN IDS PDS SUCT HYDC IMD
H1 0  '-2'  ' '    0     0.0   0.0  .027 .015 .410 .060 .20 5.0 .30  .06
H1 1 'CA-1' 'P001'  600  4.999  39.0  0.0       /
H1 1 'CA-7' 'P005'  200  1.599  48.6  0.0       /
H1 1 'CA-8' 'P009'  500  4.565  46.7  0.0       /
H1 1 'CA-11' 'P011' 600  3.580  47.0  0.0       /
*  WATER QUALITY DATA
*  NQS JLAND IROS IROSAD DRYDRY CBVOL DRYBSN RAINIT REFFDD KLNBGN KLNEND
J1  5    1    0     0     7.00  0.0   1.00   0.00    0.00    0      0
* LNAME     METHOD JACGUT  DDLIM   DDPOW  DDFACT  CLFREQ AVSWP DSLCL
J2  ' '       -2     0   6000000.   1.0    0.0     0.0    0.0      0
J2 'APARTMTS' 0      1     0         0     0.7      0      0       0
*     PNAME    PUNIT  NDIM KALC KWASH KACGUT LINKUP QFACT1  2    3    4  5 WASHPO RCOEF CBFACT CONCRN REFF
J3 ' COD '    'MG/L'  0    1    0       1     0     1.E05 1.0  2.5  .0 .0  1.1    5.0  0.0      5.   .35
J3 'TOT.SOL'  'MG/L'  0    1    0       1     0     1.E05 1.0  7.00 .0 .0  1.1    5.5  0.0      5.   .40
J3 'TOT.NIT'  'MG/L'  0    1    0       1     0     1.E05 1.0   .05 .0 .0  2.1     .7  0.0      0.3  .45
J3 ' TSS '    'MG/L'  0    1    0       1     0     1.E05 1.0  2.00 .0 .0  1.2   14.0  0.0      5.0  .40
J3 'O-PO4-P'  'MG/L'  0    1    0       1     0     1.E02 1.0 .0006 .0 .0  0.1    2.0  0.0     .002  .25
*     NAMEW  KL   BA    GQ
L1  'CA-1'    1 11.0  36.0  /
L1  'CA-7'    1 11.0  36.0  /
L1  'CA-8'    1 11.0  36.0  /
L1  'CA-11'   1 11.0  36.0  /
M1  1   1
M2  1   0   0
M3 'Outlet'
$ENDPROGRAM
```

The flow prediction is not calibrated, but the quality parameters represent from six to eleven trial calibrations of buildup- washoff parameters, with varying degrees of success. Comparisons of detailed pollutographs (concentration vs. time) are seldom necessary in application, because only the total storm event load is usually of interest. Loads are generally easier to calibrate than detailed pollutographs -- thus the effort expended for Fig. 22.12 may be misleading. Note that concentration measurements are available only for a 16-min period during the middle of the storm and make it difficult to determine the nature of the pollutograph near the beginning.

22.8. EXTRAN EXAMPLE

The Extran Block is commonly applied for flow routing when the drainage system is characterized by backwater, a looped network, surcharging, or other complexities. It can be used for simple drainage networks as well if the user is willing to expend the extra effort, although Runoff or Transport Block routing usually is adequate for uncomplicated hydraulic situations.

An example of Extran Routing is shown in Fig. 22.14 for the network of Fig. 22.13 (Personal communication; J.A. Aldrich; Camp, Dresser and McKee; 1985); input data are shown in Table 22.5.

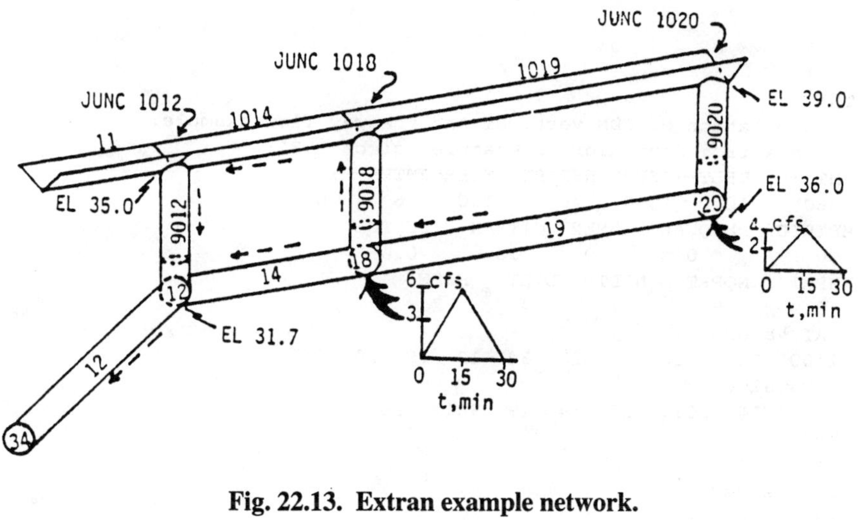

Fig. 22.13. Extran example network.

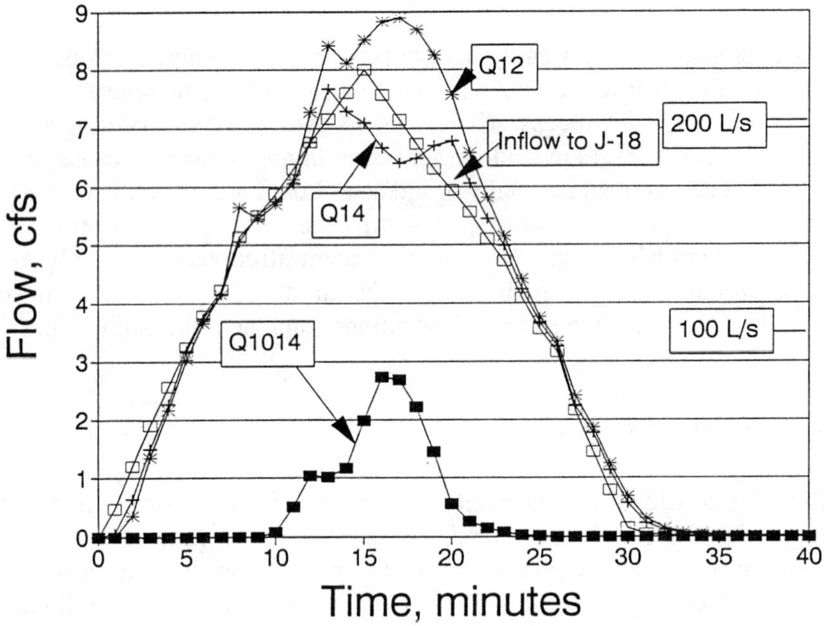

Fig. 22.14. Predicted hydrographs for Extran example.

TABLE 22.5.

Input Data for Extran Example.

```
SW    1      0       0
MM    1      1
$EXTRAN
A1  ' Simulation of CDM vertical and roadway flow example.
A1  ' Data from John Aldrich example, circa 1985.'
*   NTCYC  DELT  TZERO NSTART INTER JNTER REDO
B1  360    10.0   0.0    6      100    6     0
*  METRIC   NEQUAL   AMEN    ITMAX   SURTOL
B2   0         0        0.     30      0.05
*  NHPRT   NQPRT   NPLT   LPLT   NJSW
B3   8        9       1      4      2
*  JPRT values
B4  1100  1012  1018  1020  34  12  18  20
*  CPRT values
B5  11  1014  1019  12  14  19  9012  9018  9020
*  JPLT
B6  18
*  KPLT values
B7  1014  14  12  19
```

TABLE 22.5. Continued.

TABLE 22.5. Continued.

```
* NCOND  NJ1  NJ2  QO NKLASS AFULL DEEP WIDE LEN  ZP1 ZP2 ROUGH STHETA SPHI
C1 1014 1012 1018  0.   6     0    0.5  0.  75.  0.  0.  0.020  60.   40.
C1 1019 1018 1020  0.   6     0    0.5  0. 200.  0.  0.  0.020  30.   30.
C1   11 1100 1012  0.   6     0    2.0  2. 100.  0.  0.  0.020   4.    4.
C1 9012 1012   12  0.   1     0    1.   0. 100.  0.  0.  0.004   0.    0.
C1 9018 1018   18  0.   1     0    1.   0. 100.  0.  0.  0.004   0.    0.
C1 9020 1020   20  0.   1     0    1.   0. 100.  0.  0.  0.004   0.    0.
C1   12   34   12  0.   1     0    1.   0. 100.  0.  0.  0.012   0.    0.
C1   14   12   18  0.   1     0    1.   0.  75.  0.  0.  0.012   0.    0.
C1   19   18   20  0.   1     0    0.67 0. 200.  0.  0.  0.012   0.    0.
*       JUN  GRELEV         Z  QINST  YO
D1     1100   138.       135.    0.   0.
D1     1012   138.       135.    0.   0.
D1     1018   138.       135.2  0.   0.
D1     1020   141.       139.    0.   0.
D1       34   138.       127.2  0.   0.
D1       12   138.       131.5  0.   0.
D1       18   138.       132.2  0.   0.
D1       20   141.       136.   0.   0.
*     JFREE  NBCF
I1     1100   1
I1       34   1
*      NTIDE
J1       1
*      NINC
K1       2
*    JSW1      JSW2
K2    18        20
*    TEO   QCARD(1) QCARD(2)
K3  0.0      0.0      0.0
K3  0.25     6.0      4.0
K3  0.5      0.0      0.0
K3  2.0      0.0      0.0
$ENDPROGRAM
```

The situation is one in which a conduit (no. 14) is unable to accept both the upstream inflow and the local triangular hydrograph inflow at junction 18. Conduits 9012, 9018 and 9020 are in fact manholes that are represented for this simulation by equivalent pipes so that water can flow up (or down), to (and from) the triangular land surface channels. Thus, when conduit 14 cannot accept all the flow, the head builds up at junction 18 until water flows up conduit (manhole) 9018 and into surface channel 1014. Most of the water then flows back down conduit (manhole) 9012 and out conduit 12, while a small amount continues on the surface down conduit 11. While conduit 1014 is serving as a "relief drain," the head increases at junction 12 and briefly reduces the flow in conduit 14. Only a model such as Extran can simulate the surcharging and pressure flow conditions that lead to such a complex flow pattern. However, again it should be noted that there is a considerable "learning curve" before a novice user can readily apply Extran to new studies.

22.9. SUMMARY AND CONCLUSIONS

The U.S. Environmental Protection Agency and several consulting firms and universities have invested large amounts of time and resources into developing the Storm Water Management Model into a useful and widely-accepted package for analysis of urban drainage systems. SWMM is capable of continuous (long-term) and storm event simulation of both quantity and quality processes, and is available from the U.S. EPA at only a nominal cost. However, the program is not user friendly and must be run in a batch mode unless pre- and post- processing software is purchased from private vendors. But the program is in the public domain and may be freely modified and coupled to spreadsheets and other software by the user. SWMM has been used in hundreds of locations in the U.S., Canada, Europe and elsewhere and should continue to prove useful in the future.

22.10. APPENDIX: METRIC CONVERSIONS

Although SWMM is completely metric (as an option), all the examples in this chapter contain U.S. customary units. Conversions are provided below:

1 inch (in) = 2.54 cm
1 foot (ft) = 30.48 cm
1 m = 3.28 ft
1 ha = 2.47 acres (ac)

1 m^3/s = 1000 L/s = 35.3 ft^3/s = 35.3 cfs
1 mile (mi) = 5280 ft = 1609 m = 1.61 km
1 kg = 2.2 pound (lb)

REFERENCES

Ambrose, R.B., Jr. and T.O. Barnwell, Jr., "Environmental Software at the U.S. Environmental Protection Agency's Center for Exposure Assessment Modeling," Environmental Software, Vol. 4, No. 2, pp. 76-93, 1989.

American Society of Civil Engineers and Water Environment Foundation, Design and Construction of Urban Stormwater Management Systems, ASCE Manual of and Reports of Engineering Practice No. 77, ASCE, New York, 1993.

Anderson, E.A., National Weather Service River Forecast System -- Snow Accumulation and Ablation Model, NOAA Tech. Memo NWS HYDRO-17, National Weather Service, National Oceanic and Atmospheric Administration, Silver Spring, MD, 1973.

Bedient, P.B. and W.C. Huber, Hydrology and Floodplain Analysis, Addison-Wesley Publishing Co., Reading, MA, Second Edition, 1992.

Curtis, T.G. and W.C. Huber, "An ARC/INFO Engineering Shell for the Storm Water Management Model (SWMM)," Proc. Runoff Quantity and Quality Modeling Conference, Reno, NV, November 1993, EPA report in press (through NTIS), Environmental Protection Agency, Athens, GA, 1994.

Donigian, A.S., Jr. and W.C. Huber, Modeling of Nonpoint Source Water Quality in Urban and Non-urban Areas, EPA/600/3-91/039, Environmental Protection Agency, Athens, GA, June 1991.

Environmental Protection Agency, Areawide Assessment Procedures Manual, Three volumes, EPA-600/9-76-014, Environmental Protection Agency, Cincinnati, OH, July 1976 et seq.

Hardee, J., Miller, R.A. and H.C. Mattraw, Jr., Stormwater Runoff Data for a Multi-Family Residential Area; Dade County, Florida, USGS Open-File Report 79- 1295, Tallahassee, FL, 1979.

Huber, W.C., "Deterministic Modeling of Urban Runoff Quality," in Urban Runoff Pollution, H.C. Torno, J. Marsalek and M. Desbordes, eds., NATO ASI Series, Series G: Ecological Sciences, Vol. 10, Springer-Verlag, New York, pp. 167-242, August 1985.

Huber, W.C. and R.E. Dickinson, Storm Water Management Model User's Manual, Version 4, EPA/600/3-88/001a (NTIS PB88-236641/AS), Environmental Protection Agency, Athens, GA, 1988.

Huber, W.C., Heaney, J.P. and B.A. Cunningham, Storm Water Management Model (SWMM) Bibliography, EPA/600/3-85/077 (NTIS PB86-136041), Environmental Protection Agency, Athens, GA, 1985.

Huber, W.C., Zollo, A.F., Tarbox, T.W. and J.P. Heaney, Integration of the SWMM Runoff Block with Arc/Info and AutoCAD: A Case Study, Draft Report to the Environmental Protection Agency, Risk Reduction Engineering Laboratory, Contract No. 68-CO-0033, Edison, NJ, 1991.

James, W., ed., New Techniques for Modeling the Management of Stormwater Quality Impacts, Proc. of the Stormwater and Water Quality Management Modeling Conference, Toronto, Ontario, February 1992, Lewis Publishers, Boca Raton, FL, 1993a.

James, W., "Introduction to the SWMM Environment," Chapter 1 in New Techniques for Modeling the Management of Stormwater Quality Impacts, W. James, ed., Proc. of the Stormwater and Water Quality Management Modeling Conference, Toronto, Ontario, February 1992, Lewis Publishers, Boca Raton, FL, pp. 1-28, 1993b.

James, W., ed., Current Practices in Modeling the Management of Stormwater Impacts, Proc. of the Stormwater and Water Quality Management Modeling Conference, Toronto, Ontario, February 1993, Lewis Publishers, Boca Raton, FL, 1994.

Metcalf and Eddy, Inc., University of Florida, and Water Resources Engineers, Inc., Storm Water Management Model, Volume I -- Final Report, EPA Report 11024DOC07/71 (NTIS PB-203289), Environmental Protection Agency, Washington, DC, July 1971.

Miller, R.A., Characteristics of Four Urbanized Basins in South Florida, USGS Open-File Report 79-694, Tallahassee, FL, May 1979.

Roesner, L.A., Aldrich, J.A. and R.E. Dickinson, Storm Water Management Model User's Manual Version 4: Addendum I, EXTRAN, EPA/600/3-88/001b (NTIS PB88- 236658/AS), Environmental Protection Agency, Athens, GA, 1988.

Sartor, J.D. and G.B. Boyd, Water Pollution Aspects of Street Surface Contaminants, EPA-R2-72-081 (NTIS PB-214408), Environmental Protection Agency, Washington, DC, November 1982.

Walentowicz, R., ed., Exposure Models Library and Integrated Model Evaluation System, EPA/600/C-92/002, Revised September 1993, Environmental Protection Agency, Office of Health and Environmental Assessment, Washington, DC, 1993.

Wu, Y., Development of SWMM and SWRRBWQ Windows Interfaces," Proc. Runoff Quantity and Quality Modeling Conference, Reno, NV, November 1993, EPA report in press (through NTIS), Environmental Protection Agency, Athens, GA, 1994.

Chapter 23

MIKE SHE

J. C. Refsgaard and B. Storm

23.1. INTRODUCTION

MIKE SHE is a comprehensive deterministic, distributed and physically-based modelling system for the simulation of all major hydrological processes occurring in the land phase of the hydrological cycle. It simulates water flow, water quality and sediment transport.

MIKE SHE is a further development based on the SHE modelling concept developed by a European consortium of three organizations: the Institute of Hydrology (UK), the French consulting firm SOGREAH and DHI (Abbott et. al. 1986).

MIKE SHE is a fourth-generation, user-friendly modelling package comprising a number of comprehensive pre- and post-processors including digitizing, graphical editing, contouring, grid-averaging and graphical-result presentation with options for display of animations. It has been designed for efficient application on relatively low-cost, high-performance workstations.

MIKE SHE is applicable to a wide range of water resources and environmental problems related to surface water and groundwater systems and the dynamic interaction between these. Typical areas of application are:

- river basin planning
- water supply
- irrigation and drainage
- contamination from waste disposal sites
- impacts of farming practises (including the use of agrochemicals and fertilisers)
- soil and water management
- effects of changes in land use

- effects of changes in climate
- ecological evaluations, including those associated with wetland areas.

MIKE SHE is applicable on spatial scales ranging from a single soil profile (for infiltration studies) to large regions, which mayinclude several river catchments. It has been tested and proved in a large number of research and consultancy projects. The experience record covers a wide range of climatological and hydrological regimes.

23.2. INTEGRATED MODULAR STRUCTURE

The basic MIKE SHE Module is MIKE SHE WM for the description of water movement in the area under study. The WM Module itself has a modular structure with one component dedicated to each hydrological process. Several components include alternative options for describing certain specific processes.

The user can produce his own model configuration adapted specifically to the local hydrological conditions and the purpose of the study.

The following add-on modules are available for water quality, soil erosion and irrigation studies:

- **MIKE SHE AD** - advection and dispersion of solutes
- **MIKE SHE GC** - geochemical processes
- **MIKE SHE CN** - crop growth and nitrogen processes in the root zone
- **MIKE SHE SE** - soil erosion
- **MIKE SHE DP** - dual porosity
- **MIKE SHE IR** - irrigation

In this presentation only the WM module is described in details, while a brief introduction is given to the AD module.

23.3. WATER MOVEMENT MODULE (MIKE SHE WM)

23.3.1. Programme Structure

The overall model structure is illustrated in Fig. 23.1. MIKE SHE WM has been designed with a modular programme structure comprising six process-oriented components, each describing the major physical processes in individual parts of the hydrological cycle and in combination describing the entire hydrological cycle:

- Interception/evapotranspiration (ET)
- Overland- and Channel Flow (OC)
- Unsaturated Zone (UZ)
- Saturated Zone (SZ)
- Snow Melt (SM)
- Exchange between Aquifer and Rivers (EX)

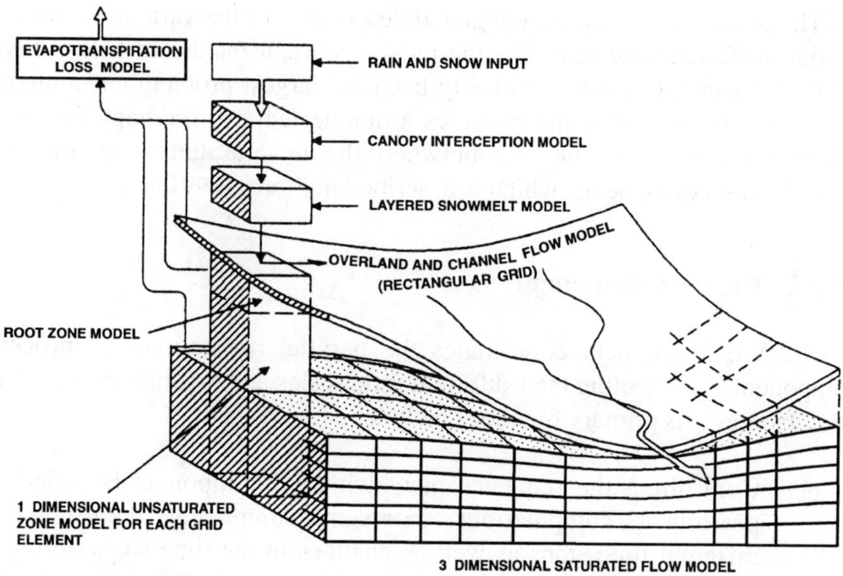

Fig. 23.1. Schematic representation of the components of the MIKE SHE.

The modular form of system structure, or architecture, ensures a great flexibility in the description of the individual physical processes. Data availability or specific hydrological conditions may favour one model description compared to another. By ensuring that the data flow between components are unchanged, alternative methods which are generally accepted in a certain geographical region or a country can be included in the MIKE SHE WM system if required.

The governing partial differential equations for the flow processes are solved numerically by efficient and stable finite difference methods in separate process components. All process descriptions operate at time steps consistent with their own most appropriate temporal scales. Hence the processes may be simulated using different time steps which may be updated during the simulation and coupled with the adjoining processes as and when their time steps coincide. The facility allows for a very efficient operation, making it possible to carry out simulations extending over long periods of time.

Individual components can also be operated separately to describe a single process. This may be relevant in a range of applications, where only rough estimates of data exchange from other parts of the hydrological cycle are required. An example could be a groundwater study where only approximate recharge estimates may be required and a full coupling to the unsaturated zone above the groundwater table is unimportant.

The ability to provide an integrated description of the various processes, despite different time scales, is the most important feature of MIKE SHE WM. This integration has probably been the largest problem encountered during its development and provides a unique feature. Perhaps the most difficult coupling is the one between the unsaturated zone and the groundwater components, which is described in Storm (1991).

23.3.2. Frame Component

A FRAME component coordinates the parallel running of the process components by selecting their different time scales and organizing their data interchanges. Its primary functions include:

(1) Controlling the sequence in which each component is called to perform its computations. For some components there may be different time steps as well as changes in the time steps during a simulation, depending on the rate of hydrological response. If the time steps differ between components, it is necessary to accumulate calculated values over a period before they are transferred to another. In general, the larger time steps are used in the saturated zone component compared to the other components.

(2) Controlling which components are included in the simulation. Dummy versions, which simply provide the necessary boundary variables for the other components may substitute process components if the concerned processes are irrelevant for the application.

(3) Controlling which option of each process description is used if more than one method is available for a component.

(4) Controlling the exchange of data between components. The results provided by one component will in some cases need to be processed into a different form for input to another component.

(5) Controlling the output variables to be stored on disk for postprocessing and the time interval with which each variable is stored.

The data flow between components is illustrated in Fig. 23.2.

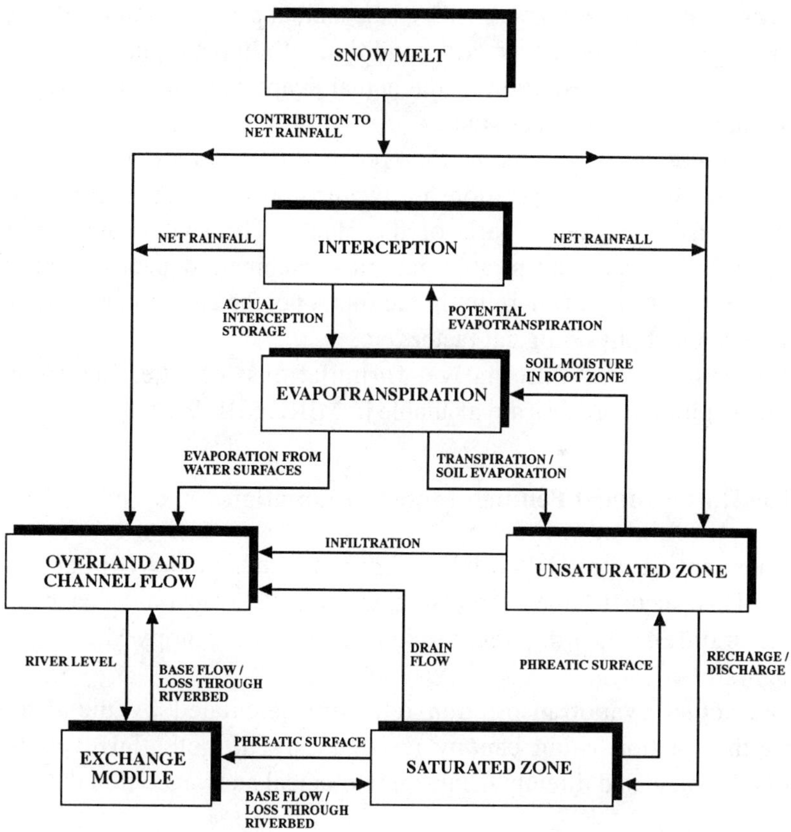

Fig. 23.2. Data flow between the components of the MIKE SHE WM.

23.3.3. Interception and Evapotranspiration Component

During rainfall, part of this will be intercepted by the vegetation and subsequently lost by evaporation.

It is assumed that the interception is either depending on the rainfall rate and/or the interception capacity. The importance of interception depends very much on vegetation type, development stage, density of vegetation and the climatic conditions. Dense forest canopies may account for a considerable interception loss, whereas for shorter and sparser vegetation, such as grass and agricultural crops, the evaporation loss may be much smaller and often insignificant.

Evapotranspiration involves the transfer of large quantities of water. In temperate areas approximately 70% of the annual precipitation is returned to the atmosphere, while under arid conditions it almost equals the rainfall. For this reason the prediction of the actual evapotranspiration plays a key role in many water resources studies.

The evapotranspiration is the total process of evaporation from soil and water surfaces and transpiration, the water uptake by plant roots that is transpired from the leafy parts of the plant. The spatial and temporal variation in the evapotranspiration rate in a catchment depends on multiple factors such as water availability in the rootzone, the aerodynamic transport conditions, plant physiological factors etc.

At present, two alternative formulations of the interception/evapotranspiration process are available in MIKE SHE WM.

(a) The Rutter Model/Penman-Monteith Equation

The interception is modelled by a modified Rutter model (Rutter et. al., 1978). This calculates the evaporation, the actual storage on the canopy, and the net rainfall reaching the ground surface as canopy drainage and throughfall.

The actual evapotranspiration rates are calculated by the Penman-Monteith equation using canopy resistances. The potential evapotranspiration is calculated directly using climatological and vegetation data.

(b) The Kristensen-Jensen Model

The interception storage is calculated based on actual leaf area index and an interception capacity coefficient. The net rainfall is calculated by a simple water balance approach.

The actual evapotranspiration is calculated on the basis of potential rates and the actual soil moisture status in the root zone (Kristensen and Jensen, 1975).

EQUATIONS

(a) The Rutter model/Penman-Monteith equation

Interception

The Rutter model is essentially an accounting procedure for the amount of water stored on the canopy. The rate of change in storage is given by

$$\frac{\partial C}{\partial t} = Q - Ke^{b(C-S)} \tag{23.1}$$

where

$$Q = \begin{cases} P_1\, P_2 \left(P - E_p\, C/S\right) & when\ C \le S \\ P_1\, P_2 \left(P - E_p\right) & when\ C \le S \end{cases} \tag{23.2}$$

with the notation as given at the end of this section.

Evapotranspiration

The Penman-Monteith equation (Monteith, 1965) for predicting the actual evapotranspiration rates is

$$E_a = \frac{\Delta R_n + \dfrac{\rho C_p \delta_e}{r_a}}{\lambda\{\Delta + \gamma\left(1 + r_s / r_a\right)\}} \tag{23.3}$$

The total actual evapotranspiration calculated for each grid square depends on the wetness of the canopy and the degree of ground covered by the canopy:

$$E_t = P_1\, P_2\, E_p\, C/S + E_a\left(1 - C/S\right) P_1\, P_2 + E_s\left(1 - P_1\, P_2\right) \tag{23.4}$$

(b) The Kristensen-Jensen model

Interception

The intercepted water storage capacity is calculated by:

$$S = S_{int} LAI \tag{23.5}$$

The evaporative demand constraint is first applied to the intercepted water and if this is not satisfied the remaining part is applied for transpiration and soil evaporation. If the interception storage capacity is exceeded during rainfall, the surplus of rain will be calculated as throughfall.

Evapotranspiration

The evaporative demand is only met if the soil moisture content in the rootzone is sufficient. For lower moisture contents the actual transpiration is calculated according to a procedure illustrated in Fig. 23.3.

The soil evaporation demand is reduced below the potential value for moisture contents below field capacity, according to the procedure illustrated in Fig. 23.4.

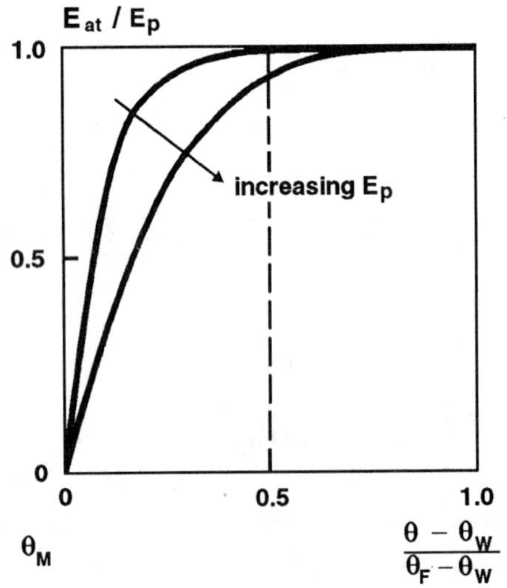

Fig. 23.3. Relationship between actual transpiration and soil moisture content in the Kristensen-Jensen model.

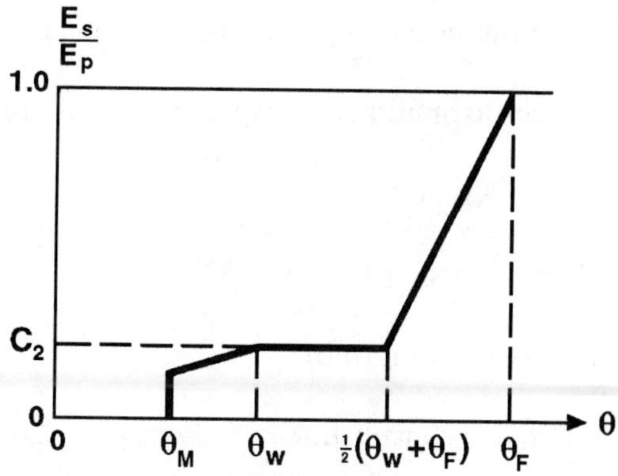

Fig. 23.4. Relationship between actual soil evaporation and the and soil moisture content in the Kristensen-Jensen model.

Symbol List

E_t - total evapotranspiration rate

E_a - actual evapotranspiration

E_{at} - actual transpiration

E_p - potential evapotranspiration

E_s - soil evaporation

C - actual water depth on canopy (mm)

S_{int} - interception storage coefficient (mm)

S - canopy storage capacity (mm)

P - rainfall rate (mm/s)

P_1 - proportion of ground planview hidden by vegetation

P_2 - ratio of total leaf area to area of ground covered by vegetation

K, b - canopy drainage parameters

t - time (s)

R_a - net radiation minus energy flux into the ground (W/m^2)

Δ_e - slope on specific humidity - temperature curve (mb/C°)

ρ - density of air (Kg/m^3)

C_p - specific heat capacity of air (J/kg/C°)

δ_e - vapour pressure deficit (mb)

r_a - aerodynamic resistance (s/m)

r_s - canopy resistance (s/m)

λ - latent heat of vaporization (J/kg)

γ - psychometric constant (mb/C°)

LAI - leaf area index

θ_F - soil moisture content at field capacity

θ_W - soil moisture content at wilting point

θ_M - irreduable soil moisture content

Solution Technique

The Rutter model is solved by analytical integration. All other equations are solved directly.

The transpiration is calculated as sinks at each computational node in the rootzone and the total transpiration is found as a weighted average according to the actual root mass distribution. For the top node, the additional loss due to soil evaporation is also considered.

Examples of Output

The ET component calculates, at each grid square where calculation are performed with the UZ component, time series of e.g. evaporation from interception storage and ponded water, soil evaporation, and transpiration. This is illustrated in Fig. 23.5.

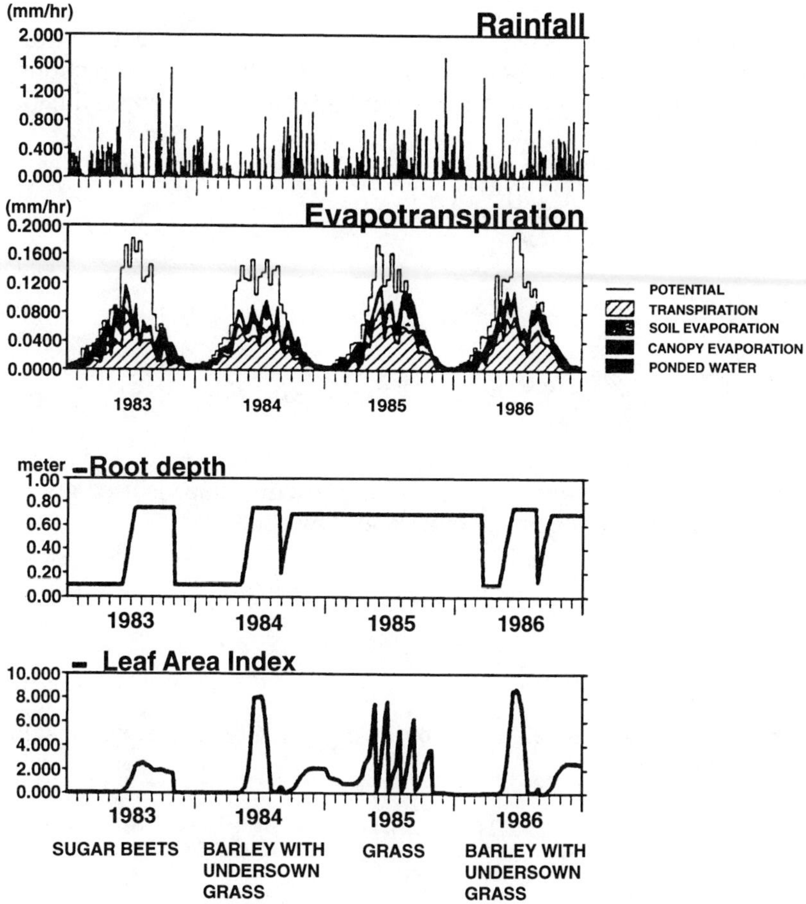

Fig. 23.5. Example of output from ET calculations for a four year period in a single grid square.

23.3.4. Overland - and Channel Flow Component

Water accumulated on the soil surface during heavy rainfall or when the groundwater table rises to the ground surface responds to gravity by flowing down-gradient over the land surface enroute to the stream channel system. From here it discharges through the channels (rivers) to the outlet of the catchment.

During its journey to the streams, the flowing water may diminish because of evaporation or infiltrate in areas with more permeable soils.

The stream channel system is assumed to run along the boundaries of the grid squares.

EQUATIONS

The overland flow process is described by solving the equations of continuity and conservation of momentum in two horizontal directions (the Saint Venant equations). In the latter the diffusive wave approximation is applied.

$$\frac{\partial h}{\partial t}+\frac{\partial (uh)}{\partial x_i}+\frac{\partial (vh)}{\partial x_j}=q$$

$$\frac{\partial h}{\partial x_i}=S_{ox_i}-S_{fx_i} \qquad x_i - direction \qquad (23.6)$$

$$\frac{\partial h}{\partial x_j}=S_{ox_j}-S_{fx_j} \qquad x_j - direction$$

The Strickler/Manning-type law for each friction slope is used giving the following relations between velocities and flow depths.

$$uh=K_{x_i}\; I_{x_i}^{1/2}\; h^{5/3}$$

$$vh=K_{x_j}\; I_{x_j}^{1/2}\; h^{5/3} \qquad (23.7)$$

The channel flow is calculated by an equivalent set of equations, but in one dimension only.

$$\frac{\partial A}{\partial t}+\frac{\partial (Au)}{\partial x}=Q$$

and (23.8)

$$\frac{\partial h}{\partial x}=S_{ox}-S_{fx}$$

Symbol list

$h(x_i, x_j)$	-	local water depth exceeding any detention storage (m)
x_i, x_j	-	space coordinates (m)
t	-	time (s)
$u, v(x_i, x_j, t)$	-	flow velocities (m/s)
$q(x_i, x_j, t)$	-	source/sink per unit horizontal area ($m^3/s/m^2$)

$S_{oi}, S_{oj}\left(x_i, x_j\right)$	-	bed slope in x- and y-directions
$S_{fi}, S_{fj}\left(x_i, x_j\right)$	-	friction slope in x- and y-directions
$K_i, K_j\left(x_i, x_j\right)$	-	Strickler roughness coefficients in x- and y-directions ($m^{1/3}/s$)
$I_i, I_j\left(x_i, x_j\right)$	-	gradients of the water surface levels in x- and y directions
$A(x)$	-	cross-sectional area of the channel (m^2)
$Q(x)$	-	source/sink (lateral in- and outflow from overland flow, drainage flow and exchange with the aquifer) (m/s)

Important Parameters

The Strickler roughness coefficients used both in the overland flow and the river flow descriptions influence the timing and shape of the simulated hydrograph simulations. In particular for the overland flow part these parameters may be subject to calibration.

Solution Technique

Implicit finite difference techniques are used for both the overland flow (Thomas, 1973) and the channel flow (Preissman and Zaoui, 1979). The former uses a modified Gauss-Seidel iterative solution scheme, analogous to that often used for the groundwater flow.

Examples of Output from the OC Component

The OC component calculates the temporal and spatial ponding depths and flows on the ground surface. In the river system the temporal and spatial variations in water levels and discharge are produced. A snapshot from a simulation is shown in Fig. 23.6. The figure illustrates the ponded water on the land surface in a fence diagram of the water level in the channel system.

23.3.5. Unsaturated Zone Component

The unsaturated zone is a crucial part of the hydrological system in a catchment. It plays an important role in many modelling applications, e.g. for recharge estimation, surface-groundwater interaction and agricultural pollution. The unsaturated zone refers here to the mostly-unsaturated soil

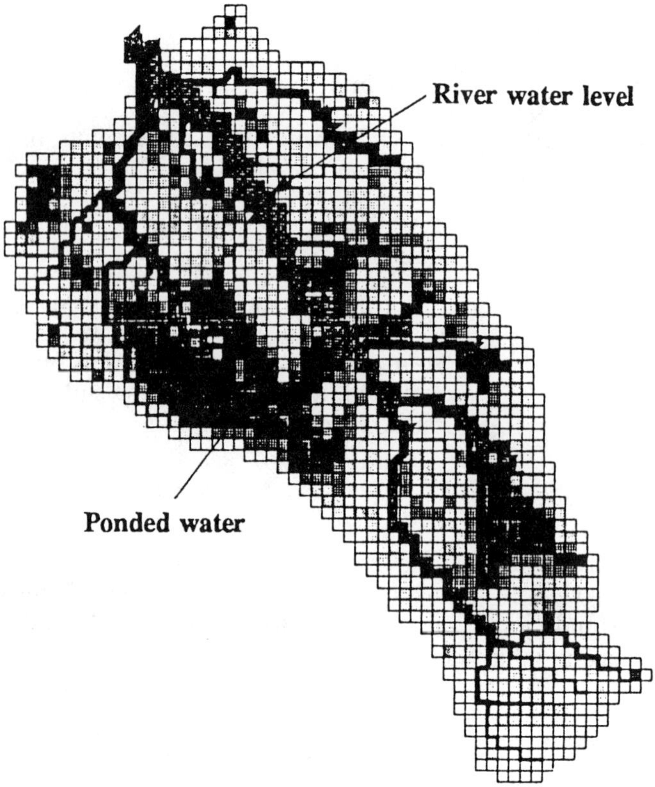

Fig. 23.6. Example of output from the OC calculations. Spatial variation of depths of water levels in river and on surface at a given time.

profile extending from the land surface down to the groundwater table. The profile is usually heterogeneous, consisting of horizons with distinct differences in the physical properties of the soil.

The unsaturated zone is characterized by cyclic fluctuations in the soil moisture as water is removed from the soil profile by evapotranspiration and percolation and replenished by rainfall.

Percolation may introduce a rise in the water table, whereas upward flow from the groundwater due to capillary rise may occur in areas with a high groundwater table and high evaporation demands.

Unsaturated flow can usually be considered vertical since gravity plays a major role during the percolation of water. The unsaturated flow is therefore only represented in MIKE SHE WM by a vertical flow component, which fulfils the requirements of most situations. However, this assumption may limit the validity of the flow description in some special cases, e.g. on very steep hillslopes with contrasting soil properties in the profile.

EQUATIONS

In its most comprehensive mode, MIKE SHE WM solves Richards equation for one-dimensional vertical flow, which includes the effects of:

- gravity
- soil suction
- soil evaporation
- transpiration

in the form:

$$C\frac{\partial\psi}{at}=\frac{\partial}{\partial z}\left(K\frac{\partial\psi}{\partial z}\right)+\frac{\partial K}{\partial z}-S \tag{23.9}$$

The equation is solved over all representative grid squares in the model area.

In many soils pronounced macropore flow is observed. In addition to Richards equation, which only considers the water flow in the micropores (soil matrix), an empirical bypass function is introduced, which calculates a direct instantaneous groundwater recharge as a certain percentage of the net rainfall and the actual soil moisture conditions in the root zone.

Symbol list

$\psi(z,t)$	-	pressure head (m)
t	-	time (s)
z	-	vertical space coordinate (m)
C	-	soil water capacity (m^{-1})
$K(\theta,z)$	-	hydraulic conductivity (m/s)
θ	-	soil moisture content
$S(z,t)$	-	source /sink term (e.g. root extraction) (s^{-1})

Important Parameters

Two non-linear functional relationships are required to solve Richards' equation: the unsaturated hydraulic conductivity function $K(\theta)$ and the soil moisture retention curve, describing the $\theta-\psi$ relation. Both functions should be known for any soil type included in any soil profile within the model area.

In applications made on the catchment scale, these functions are not known in detail. Usually representative values can be obtained for the soil types. The parameters may therefore be subject to calibration.

Solution Technique

Richards' equation is solved numerically by an implicit finite difference technique using the double-sweep algorithm. The variables are defined at every computational node in the vertical.

The time-varying upper boundary at the land surface can shift between a flux-controlled boundary (net rainfall) and a soil-controlled head boundary during ponding. The time-varying lower boundary is usually the groundwater table, specified by a positive value of ψ at the computational node just below the groundwater table. If unsaturated conditions develop in the entire soil profile a zero-flux boundary is used at the impermeable interface until saturated conditions build up from the bottom.

A coupling procedure between the unsaturated and the saturated solutions is included to compute the correct soil moisture and the water table dynamics in the lower part of the soil profile. This overcomes any problems arising from the parallel running of the individual components.

Example of Output

The UZ component computes the temporal variation in soil moisture in all nodes above the groundwater table. The infiltration and recharge rate is also calculated. The temporal variation in soil moisture for two distances to the groundwater table is illustrated in Fig. 23.7.

23.3.6. Groundwater Component

Saturated subsurface flow plays a significant role in the hydrological cycle. During drought periods it provides and sustains streamflow through baseflow, while during storm events it may contribute significantly to the stormflow as well as influence the magnitude of overland flow provided by the rising water table.

Aquifers in catchments can contain large water resources storage which, besides natural changes, may be subject to abstraction for water supply and irrigation. The influence of this human activity may influence the natural recharge and discharge properties and thereby change the flow regime in the catchment.

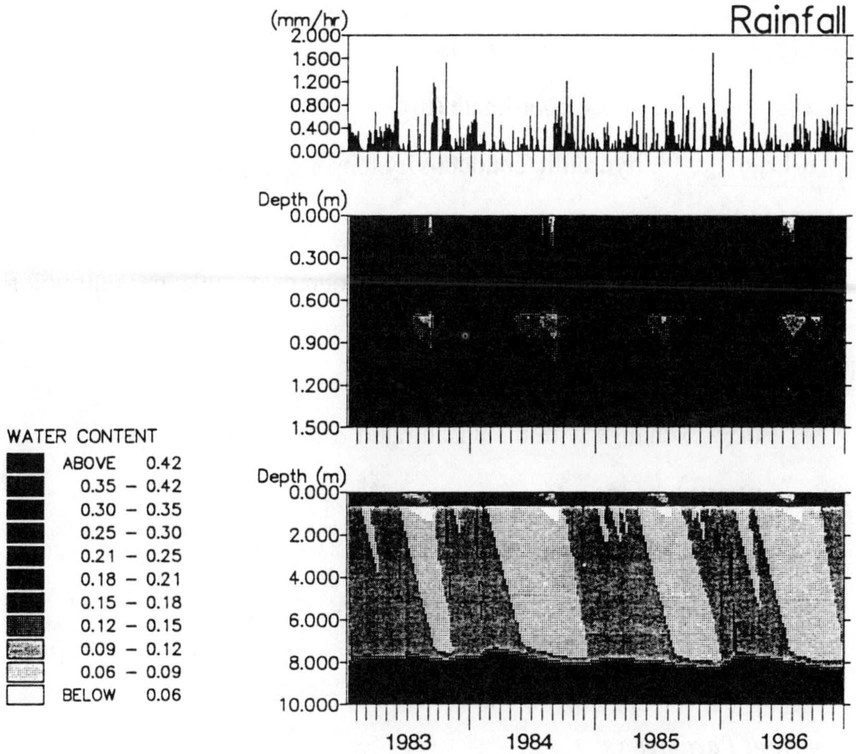

Fig. 23.7. Example of output from the UZ calculations. Temporal variation of soil moisture content at two different UZ-columns.

EQUATIONS

The temporal and spatial variations in the hydraulic head are calculated by solving the partial differential equation of groundwater flow in three dimensions:

$$\frac{\partial}{\partial x_i}\left(K_{ij}\frac{\partial h}{\partial x_j}\right) = S_s\frac{\partial h}{\partial t} + R \quad i,j = 1,2,3 \tag{23.10}$$

The equation is applicable both for a single-layered aquifer as well as a multi-layered aquifer system.

For applications on single aquifer systems, where a two-dimensional flow solution is adequate, the equation may degenerate to two dimensions by integrating over the saturated depth.

Symbol list

$h(x_i)$	-	hydraulic head (m)
x_i	-	space coordinate (m)
$K(x_{ij})$	-	hydraulic conductivity tensor (m/s)
$S_s(x_{ij})$	-	specific storage
$R(x_{ij})$	-	volumetric flow rate via sources or sinks per unit volume ($m^3/s/m^2$)

The source/sink term R includes

- flow exchange with the unsaturated zone
- river exchange rates
- abstraction/injection rates
- evaporation losses
- flow through drain pipes

Important Parameters

A MIKE SHE WM generated model which includes the groundwater component will require information about the hydrogeology in terms of the hydraulic conductivity K (in vertical and horizontal directions), and the storage coefficient for unconfined and confined conditions. Based on a hydrogeological model for the aquifer system, parameter values for each computational node is generated by interpolation with the pre-processor.

Solution Technique

The equation is solved numerically by a finite-difference method using a modified Gauss-Seidel implicit, iterative scheme (Thomas,1973) providing a value for the hydraulic head in time at each computational node. The discretization may either be a rigid network for fully three-dimensional flow or following the geological layering for a quasi-three dimensional flow as illustrated below. In the latter case, each layer (aquifer and aquitard) is represented by one computational node only, and the vertical variations in the groundwater head within each layer is neglected.

Examples of models for a single unconfined aquifer (Karup Å) and a multilayer aquifer system (Langvad Å) respectively, are shown in Fig. 23.8.

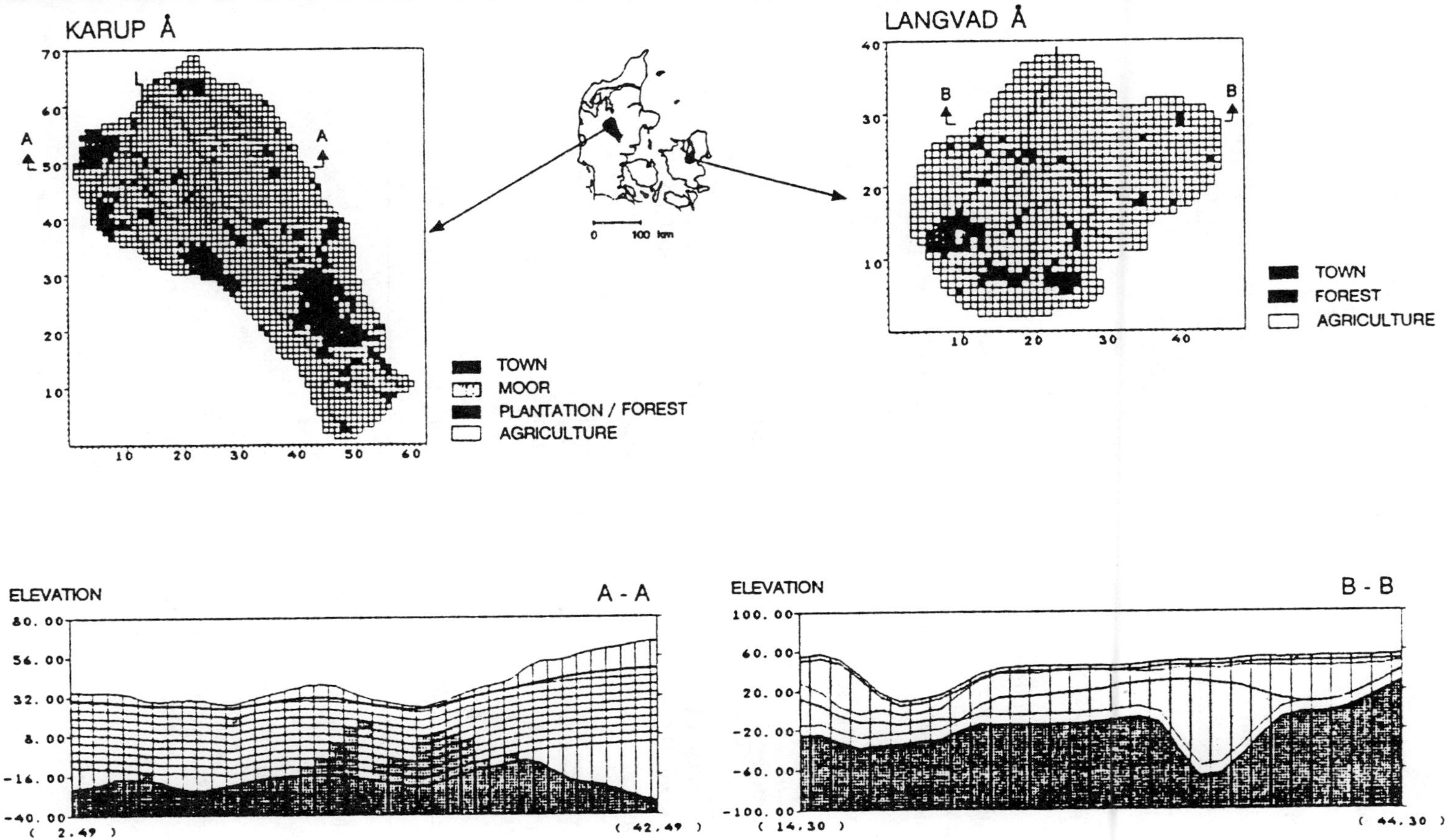

Fig. 23.8. Example of model setup for the SZ component for an unconfined aquifer (a. Karup Å) and multilayer aquifer system (b. Langvad Å).

The main characteristics of the SZ component and the type of regimes for which it is valid for are described in Table 23.1.

TABLE 23.1.

The main characteristics of the SZ component of the MIKE SHE WM

Soil/Rock Characteristics	**Fluid Conditions**
Confined aquifers	Single fluid
Aquitards	
Watertable aquifers	
Multiple aquifer	**Flow Conditions**
Heterogeneous	
Anisotropic	Unsaturated *
Single soil layer	Transient
Layered soils	
	Boundary Conditions
Geometry	*Internal grid squares*:
	Time-varying point source
1-dimensional	Time-varying line source
- horizontal	Time-varying distributed limited
- vertical	area (infiltration to top-layer)
2-dimensional	*At the model boundary*:
- horizontal	Specified time-varying
- vertical	distributed head
	Specified constant flux
3-dimensional	Specified constant head gradient

* includes the unsaturated zone component

Examples of Outputs

The SZ component provides information on the hydraulic heads and flows in time and space. An example of different maps to illustrate simulated hydraulic heads is shown in Fig. 23.9.

Spatial Distribution:

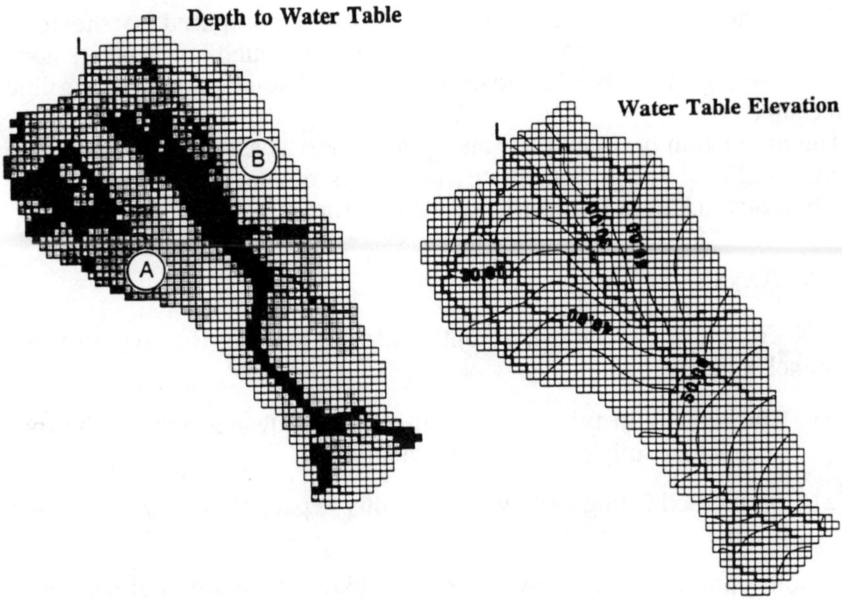

Time series:

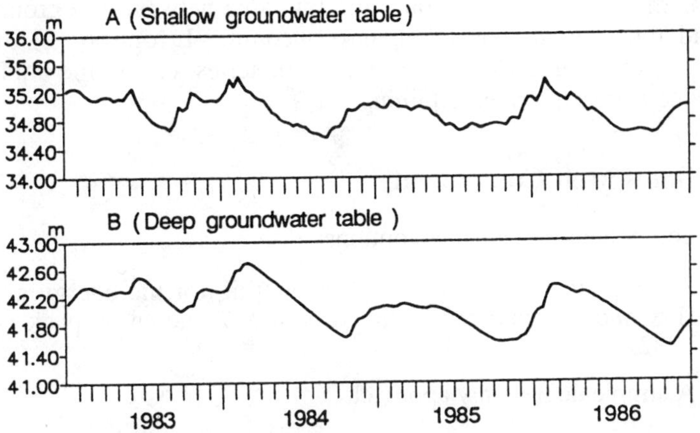

Fig. 23.9. Example of output from the SZ component illustrating both the temporal and spatial variation of ground water tables for an unconfined aquifer.

23.3.7. Aquifer-River Exchange Component

The river system usually influences a large part of the groundwater system at its traverse of the catchment in many directions. The river will control the groundwater head in both the horizontal and vertical directions, defining the recharge and discharge areas.

Since the river surface area is often small compared to the total catchment area, it can in many applications be represented in a separate node system running along the boundaries of the grid squares acting as a line source/sink.

The interaction between river and groundwater is assumed to take place in the middle of the intermediate river links connecting adjacent river computational nodes at the corners of the grid squares.

EQUATIONS

The EX component includes two options for describing the river/groundwater exchange:

(1) The river is in full contact with the groundwater aquifer (but not necessarily fully penetrating the aquifer)

(2) A riverbed lining of low permeability separates the river from the groundwater aquifer.

In both options, Darcy's law is applied taking the additional head loss around the river bottom approximately into account.

In option (1), assuming the degree of river penetration is the limiting factor for the flow exchange, only the low area between the groundwater table and the river bed is taken into account. In option (2) the flow exchange is calculated as two resistances in series, one in the aquifer and one situated across the riverbed lining.

23.3.8. Snow Melt Component

The SM component includes two options:

(1) An energy balance approach accounting for the energy and mass flux and for changes in the structure of the snow pack (Morris, 1982)

(2) A simple degree-day approach

23.3.9. Parameters and Data Requirements

The application of distributed, physically-based models such as MIKE SHE requires the provision of large amounts of parametric and input data. Some of these may be time-dependent.

It is important to notice that MIKE SHE **allows** the user to utilize a large quantity of data, but it does not necessarily restrict the use of MIKE SHE if not all data are available. A model that is set up, or "instantiated", using

MIKE SHE can be simplified according to its users conceptualization of the natural system and the data availability. Although MIKE SHE is an advanced modelling system, it is easy to apply and provides a greater flexibility as compared to many simpler models when describing natural systems.

No matter which mathematical model approach is used, the user should emphasize the reliability of the model output in view of the limiting factors for describing the natural system.

It is important to recognize that MIKE SHE WM requires calibration if it is applied on a different scale from that for which the equations have been developed and the parametric data are representative. In this connection it is important to realize the representativeness of model output compared to measured values.

Table (23.2) describes the parametric and input data for MIKE SHE WM.

TABLE 23.2.

Data and parameter requirements for each grid square or channel link

Frame	
Input data	Horizontal discretization Ground surface elevation Distribution codes for rainfall and meteorological stations
Interception	
Model parameters (Rutter Model) (for each crop type)	Canopy drainage parameters Canopy storage capacity (time varying) Ground cover indices (time varying)
Model parameters (K-J model) (for each crop type)	Leaf area index (time varying) Interception capacity coefficient
Input data	Rainfall rate
Evapotranspiration	
Model parameters (P-M equation)	Canopy resistance Aerodynamic resistance Ground cover indices (time varying) Ratio between actual and potential evapotranspiration as a function of soil moisture tension Root distribution with depth
Model parameters (K-J model) (for each crop type)	Empirical constants describing the ration between actual and potential evapotranspiration as function of soil moisture Leaf area index (time varying) Rooting depth (time varying) Root distribution coefficient
Input data	Meteorological data

Table 23.2 Continued.

TABLE 23.2. Continued.

Overland and channel flow	
Model parameters	Strickler roughness coefficients for overland and river flows
	Detention storage capacity on ground surface
	Coefficients of discharges for weir formulae
Input data	Specified levels and flows at boundaries
	Man-controlled discharges
	Topography of overland flow plane and river cross sections
	Riverbed lining thickness
	Riverbed lining permeability
Unsaturated zone	
Model parameters (for each soil type)	Soil moisture tension/content relationship
	Unsaturated hydraulic conductivity as a function of soil moisture content
	Maximum bypass ratio of net rainfall
Input data	Distribution codes for soil profiles
	Distribution codes for soil types in soil profiles
	Vertical node discretization in UZ
Saturated zone	
Model parameters	Storage coefficients
	Saturated hydraulic conductivities
	Drainage depth
	Time constant for drainage routing
Input data	Specified flows, gradients and heads at boundaries
	Location of abstraction or recharge wells
	Pumping and recharge rates
	Vertical node discretization in SZ
Snow melt	
Model parameters	Degree-day factor
	Snow zero plane displacement
	Snow roughness height
Input data	Meteorological and precipitation data

Data Overlays

A number of input data and parametric data values are required for each grid square in the horizontal grid. This is done as a stack of overlays containing codes describing structural data as well as data types. For each data type (e.g. soil type codes) a number of attributes (soil physical/properties) are attached specifying the actual properties of the data type.

23.4. ADVECTION DISPERSION MODULE (MIKE SHE AD)

MIKE SHE AD simulates the detailed transport and spreading of dissolved conservative solutes from non-point and point sources. It may be coupled with chemical reaction modules to describe also the behaviour of non-conservative solutes.

The module describes the transport and spreading of solutes in the water flowing on the ground surface, in the soil water and in the groundwater by solving the advection- dispersion equation for the respective regimes.

In the following only the equations and the methodology is shown for the groundwater component. The methodology for the other components is very similar. Like for the WM module the largest problem encountered during the development of the AD module has been associated to the couplings between the various components.

EQUATIONS *(shown for groundwater component only)*

The transport of dissolved solutes in groundwater is described by solving the advection-dispersion equation numerically in a fully three-dimensional scheme.

The governing partial differential equation may be written as

$$\frac{\partial c}{\partial t} = -\frac{\partial}{\partial x_i}(cv_i) + \frac{\partial}{\partial x_i}\left(D_{ij}\frac{\partial c}{\partial x_j}\right) + R_c \quad i,j = 1,2,3 \qquad (23.11)$$

Symbol list

$c(x_i)$ - concentration of the solute [mg/l]

R_c - sources or sinks [mg/l/s]

D_{ij} - dispersion coefficients [m^2/s]

v_i - velocity components [m/s]

Important Parameters

The dispersion coefficients are the key parameters. They are determined from a number of dispersivities and the velocity components. The dispersivities have to be provided as input parameters.

Traditionally, the dispersion coefficients are assumed to solely depend on the longitudinal dispersivity, α_L, and the transverse dispersivity, α_T, but this assumption is only valid for isotropic conditions. Instead we suggest that the dispersion coefficients are calculated from four dispersivities so that a more correct description is achieved under anisotropic conditions.

Solution Technique

The equation is solved numerically by an enhancement of the QUICKEST method originally introduced by Leonard (1979) for unsteady flows in one dimension. The scheme has been further developed to two and three dimensions by Justesen et al (1989), Ekebjerg and Justesen (1991) and Vested et al (1992).

The equation is formulated in an explicit scheme using upstream differencing techniques for the advection term and central differencing for the dispersion term. By developing the equations to third order the scheme becomes mass conservative and is computationally very efficient compared to implicit schemes. The numerical method is characterized by generating negligible numerical dispersion.

23.5. PRE- AND POSTPROCESSOR

The user interface of the MIKE SHE module includes powerful pre- and postprocessing facilities with particularly strong GIS capabilities. The software can be applied to the input data and results of all the MIKE SHE modules.

Some examples of the software capabilities are:

- Digitization of contours from maps such as those of the ground surface and geological layers
- Digitization of the river system layout
- Digitization of areally-distributed information, such as land-use, soil types, etc.
- Interpolation routines to provide point values and grid averages
- Graphical editing of 2-D data and river data
- Isoline plots
- Plots of the variations in space of a variable in any layer or along any line through the model
- Plots of time series of any variable

All graphical presentations can be in colour and are produced with a UNIRAS graphics package.

The user can easily design and produce animation of any variable. This provides a unique opportunity to present the dynamic behaviour of the simulated system and adds a new dimension to error handling and interpretation of results.

23.6. APPLICATION EXPERIENCE

The MIKE SHE is today being used operationally by a large number of organisations in different countries ranging from university and research organisations to consulting engineering companies.

The original SHE version has been tested on a number of research catchments and applied to a few other projects, see e.g. Bathurst (1986), Refsgaard et al (1992) and Bathurst (this book) for further details.

MIKE SHE has as an extended version of SHE been applied to a large number of projects during the past few years. A list of applications in which DHI has been directly involved are shown in Tables (23.3) and (23.4) for research and consultancy projects, respectively. These applications illustrate the very wide range of water resources problems for which the MIKE SHE is a suitable tool.

TABLE 23.3.

List of MIKE SHE applications on externally funded research projects

Location	Project Title	Period	Topics
Denmark	Strategic environmental research programme	1993-96	Groundwater pollution
EU	Development of a European Soil Erosion Model	1992-94	Soil Erosion
Sweden	Effects of forestry drainage and clear-cutting on flood conditions	1991-92	Impacts of human activity on floods
EU	Modelling of the nitrogen and pesticide transport and transformation on catchment scale	1991-94	Agricultural pollution
Denmark	Research programme on groundwater pollution from waste disposal site	1988-90	Groundwater pollution
Denmark	Research programme on nitrogen, phosphorous and organic matter	1986-90	Simulation of nitrogen on catchment scale. Coupling with Daisy crop growth and nitrogen model
Denmark	Validation of pesticide models	1994-96	Leaching of pesticides in clayey soils with preferential flow paths

TABLE 23.4.

List of MIKE SHE applications on consultancy projects with DHI involvement

Location	Project Title	Period	Topics
Estonia	Tapa Airbase - Groundwater Model	1993	Groundwater pollution
Slovakia	Danubian Lowland - Groundwater Model	1992 -95	Surface water quality, river and reservoir erosion and sedimentation, groundwater quality and geochemistry (redox), wetland ecology Coupling with ARC/INFO and Informix
Denmark	Four projects on Optimisation of remedial measures for safeguarding groundwater resources from pollution from waste disposal sites	1992 -93	Groundwater pollution
UK	River Management Study, River Avon, Wessex	1992 -93	Effects of groundwater abstraction and augmentation schemes on streamflow
Denmark	Environmental impact assessment of a highway construction	1992 -93	Effects of groundwater drawdown due to tunnel construction
Australia	Irrigation salinity	1991 -93	Process simulation of an irrigation district with focus on flow and salinity transport
Denmark	Identification of new well field for water supply	1991 -92	Groundwater, effects of abstraction on streamflows and wetlands
Hungary	Assessment of pollution hazards in groundwater supplies	1991	Groundwater pollution
Denmark	Water supply planning in Aarhus county	1988 -90	Groundwater resources, effects of abstractions on hydraulic heads and streamflows

In addition to the applications listed in the two tables MIKE SHE is used by other organisations to a large number of projects which are not known to the authors.

The SHE was originally developed with a view to describe the entire landphase of the hydrological cycle in a given catchment with a level of detail sufficiently fine to be able to claim a physically-based concept (Abbott et al. 1986a,b). The equations used in the model are with few exceptions non-empirical and well-known to represent the physical processes at the appropriate scales in the different parts of the hydrological cycle. The parameters in these equations can be obtained from measurements as long as they are compatible with the representative volumes for which the equations are derived.

In most regional catchment studies carried out so far, it has not been possible to represent the spatial variations in catchment characteristic with such a detail that the model could be considered physically-based. In fact, this was realised at an early stage when the applications changed from testing against analytical solutions and small scale research areas to applications on medium sized catchment areas.

However,it is the experience that the spatial resolutions and variations in properties used, provide a very good representation of the conditions in the areas modelled. In practise the spatial variations are derived from maps describing topography, soil and land-use pattern and interpreted geological conditions, combined with information about the general properties of the different map units. The model parameters values are then modified during calibration to match observed conditions at discrete points.

There are a number of fundamental scale problems which needs to be carefully considered in the model applications. This is particular important when describing the interaction between thesurface flow and the subsurface flows. A few areas where scale problems are encountered include:

- The interaction between groundwater and river. Since the flow is based on Darcy's law using the gradient between the river water level and the groundwater heads in the adjacent grid squares, the flow rates and the resultant head changes will depend on the spatial resolution used. This is an important aspect in for example simulating the hydrograph recessions correct.

- In catchments with a dense drainage network it is often not possible to represent entire drainage system (many streams are of ephemeral nature). For such situations subgrid variations in the topography need to be accounted for in order to simulate the hydrograph response in the main streams correctly.

- For modelling of infiltration and vertical unsaturated flow in the soil, the hydraulic parameters used in Richards' equation can be obtained from laboratory measurements on small undisturbed soil samples. However, for grid squares covering large areas (eg. 25 ha) these are seldom representative unless completely homogeneous conditions exist in the horizontal directions. Therefore effective or representative parameters are used, which means that the simulated soil moisture conditions can not be verified directly.

In fact much of the criticism against MIKE SHE arise often from the way the unsaturated flow is simulated and very seldom on how the groundwater conditions are treated.

For most catchment simulations the use of Richards' equation becomes conceptual rather than physically based and simpler approaches could be chosen. Nevertheless, this equation provides a good routing description, and the capability to simulate capillary rise under shallow water table conditions is an important option for studies where eg. wetland areas are included. For situations where Darcy law' does not apply a simple macropore option is included in the solution.

Because representative parameter values are used, the reliability of the results depends very much on the data available for comparison of the simulated spatial and temporal variations with observations. This is well-known from groundwater applications, where the aquifer properties (conductivities or transmissivities) are derived based on calibration against observed head variations in discrete points. For regional catchment studies themodel performance is usually evaluated based on comparisons against river discharges and groundwater heads. Very seldom measured soil moisture data are available, and if they are such comparisons will require that the site specific properties are known.

It is often stated that distributed models requires a large number of data and therefore very time consuming and complicated to set up and calibrate. In fact, a number of short-term screening evaluation projects have been carried out with MIKE SHE for example in connection with studying the contamination risks from waste disposals. In these studies only sparse existing information about the hydrogeological conditions was available. The model was used to obtain an improved knowledge about the possible flow patterns around the waste disposal site based on the existing geological interpretations. These applications could also be used to identify where existing knowledge is lacking and assist in defining appropriate monitoring programme.

Another common argument against distributed models is the risk of over-parameterization. This risk is of course always there. However, the general experience is that if the data are lacking to describe the spatial variations in the catchment it is too time-consuming and not worthwhile to modify a large number of parameter values in order to improve eg. hydrograph predictions. In such cases very few parameters are used in the calibration and the reliability of the results are evaluated with this in mind.

23.7. ONGOING RESEARCH

From the above application records it appears that MIKE SHE already has been used comprehensively both for research studies and for practical routine applications. These applications reflect that for certain types of studies there is no adequate alternative to an integrated, distributed, physically-based modelling approach like the MIKE SHE. Nevertheless, it is realized that MIKE SHE, in its present form, is far from being complete as a tool for advanced hydrological analyses. Many both practical and fundamental problems need to be solved through future research activities. A very significant part of the research carried out in these years in the international scientific community is in fact of direct relevance and most valuable in this context.

At DHI research and developments related to MIKE SHE is carried out in the following fields:

- Improvement of process descriptions. Research work on macropore flow and solute transport is presently undertaken. Other activities such as inclusion of density effects in the groundwater component and description of hysteresis phenomena in the unsaturated zone are planned in the coming years.

- Improvements of numerical efficiency is continuously taking place.

- Interface to geographic information systems (ARC/INFO) is being developed.

- Coupling with DHI's generalized river modelling system (MIKE 11, see Chapter 21 of this book) is going on. A coupled MIKE SHE/MIKE 11 enables description of sediment transport and water quality processes also in the river system as well as description of complex river and canal systems. Typical area of application is irrigation command areas, where networks of both irrigation and drainage canals exist together with a large number of different hydraulic regulating structures.

- Fundamental research on establishment of rigorous methodology on parameterization, calibration and validation is urgently needed. Some first, small steps have been taken as described in Refsgaard et al (1992), Jain et al (1992) and DHI (1993).

- Fundamental research on scale problems related to spatial variability of hydrological parameters is urgently needed. In particular problems related to different scales used for data sampling, process description and model discretization need to be addressed. Although comprehensive international research is carried out in these years no operational results and conclusions are evident.

23.8. CASE STUDY: MODELLING OF NITROGEN TRANSPORT AND TRANSFORMATION ON A CATCHMENT SCALE

23.8.1. Introduction

The present case study is one of the outputs from a comprehensive Danish research and development programme (1986-90), which was carried out with the aim of studying the pollution from nutrients and organic matters in agriculture. The research programme was multidisciplinary and involved a large number of research institutions. It included field investigations, process studies and modelling.

The present case study briefly describes a distributed hydrological modelling of nitrate transport and transformation for the 440 km^2 Karup River catchment. The nitrogen modelling covers the entire land phase of the hydrological cycle - from the source on the soil surface, through the soil zone and the ground water to the streams. The modelling was based on the MIKE SHE CN, which is a combination of the MIKE SHE WM + AD modules and the DAISY model for simulation of the nitrogen dynamics in the root zone.

A more thorough description of the DAISY model is given by Hansen et al (1991), whereas a detailed description of the present case study is presented by Styczen and Storm (1993).

23.8.2. Model setup

The Karup catchment was represented in a three-dimensional network. The discretization is 500 m in the horizontal directions and varies in the vertical from 5 to 40 cm in the unsaturated zone, and 5 m in the permanently saturated zone. Information on soil and vegetation properties were collected, and from the information from a number of wells, a three-dimensional geological map was superimposed on the model grid to provide the hydrogeological parameter values. The topography and the river network have been digitized, and all relevant climatological data collected. The overall land use has been identified. Some of the data and parts of the MIKE SHE model setup are shown in Figs. 23.5 - 23.9.

23.8.3. Results

Discharge and groundwater table hydrographs

The streamflow is simulated for the period 1969 - 1988 at several sites. A comparison with measured discharge at the catchment outlet is shown for four years in Fig. 23.10. In addition the simulated groundwater table is compared with observations in selected wells (Fig. 23.11). The comparison shows that the modelling system simulates the hydrological regime with acceptable accuracy.

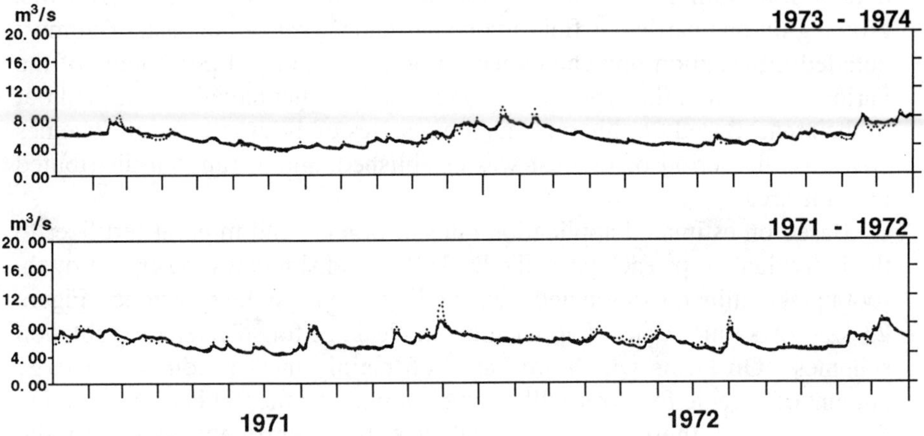

Fig. 23.10. Comparison between simulated and observed river runoff for the period 1971-74.

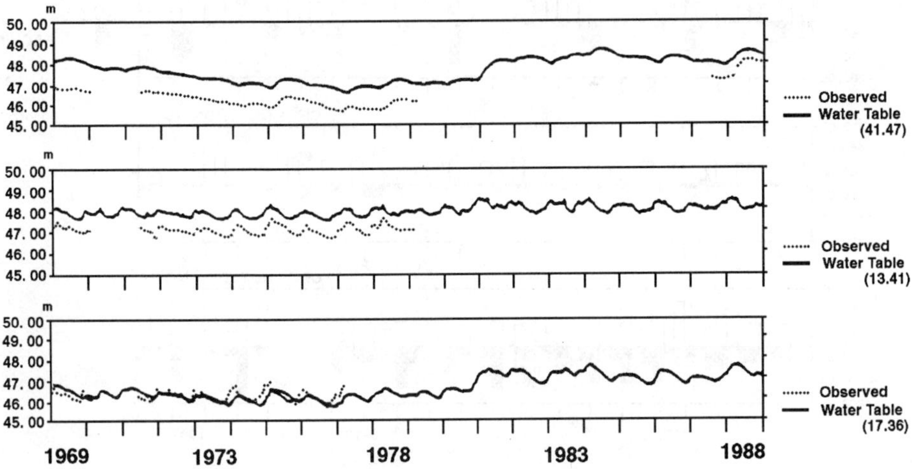

Fig. 23.11. Comparison between simulated and observed groundwater table time series in selected wells.

Leakage from the Root Zone

To simulate the trend in the nitrate concentrations in the ground water and the streams, it is necessary to have information on the history of the fertilizer application in space and time. This information is difficult to obtain in details, for example it is not possible to estimate which type of crop was growing on one particular field in one particular year in the past. The most detailed information one can expect to obtain is an areal percentage of the various crops, and the types of farming practices that have been carried out in the area. Based on this information a series of 14 crop rotation schemes covering the period of interest was established, and at random distributed over the area.

Based on estimated application rates of organic and mineral fertilizer to the individual crops each year, the DAISY model simulates the crop growth, root uptake, mineralization and leakage of nitrate from the root zone. Figure 23.12 shows time series of application and leakage for selected crop rotation schemes. On farms which are based on mainly meat production a large amount of organic fertilizer will often be applied on the fields in the autumn. In this period there is a potential risk for significant losses to the groundwater system.

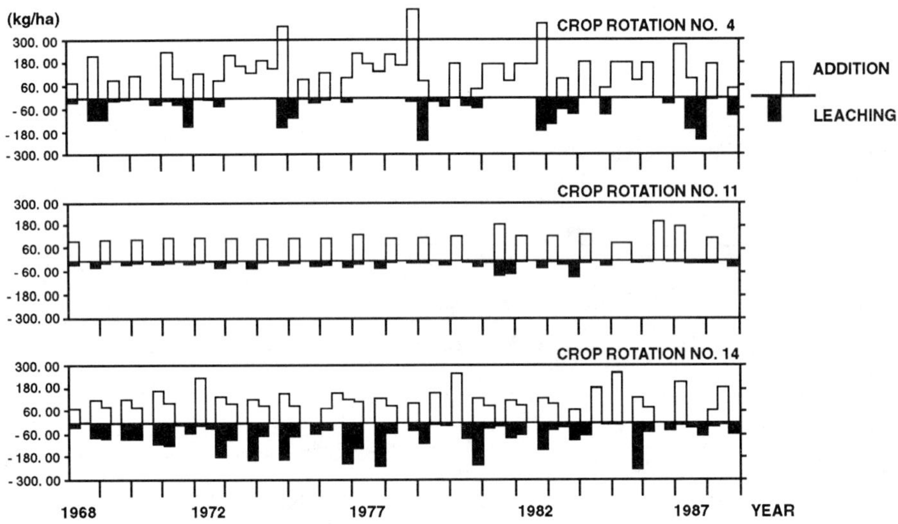

Fig. 23.12. Nitrate leaching ($NO_3^- - N$) from three of the crop rotations calculated by DAISY and summarized over four-months periods. The shown additions of *N* only include mineral fertilizer and the already mineralized part of manure.

Nitrate concentrations in groundwater

While the root zone model simulates one 'soil column' at a time the total model allows studies of the variations in space and time at regional scale. Fig. 23.13 illustrates the variation in simulated NO_3^--concentrations in the upper groundwater layer of the Karup catchment below three selected cropping schemes for two points with different depths of the unsaturated zone. A deep unsaturated zone is seen to dampen the influence of a single year.

Figure 23.14 shows the spatial variation in simulated NO_3^--concentrations in the upper groundwater layer at a specific time. The very large variation of concentration both in space and time is noticed.

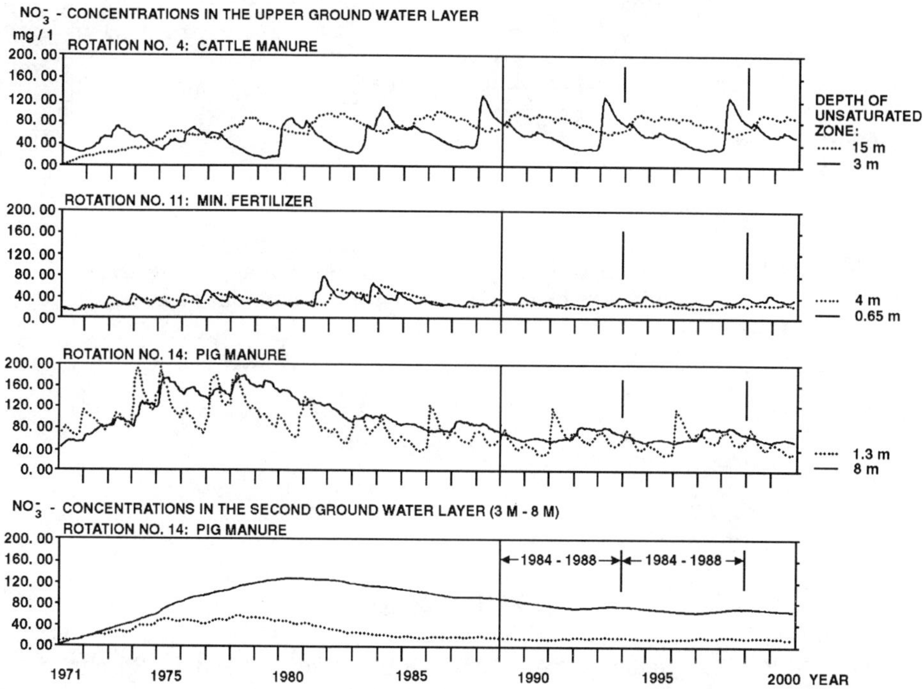

Fig. 23.13. Temporal variation in NO_3^--concentrations in the upper groundwater layer beneath three selected rotation schemes, with two different distances to the groundwater table. The data are extracted from selected grids (not averaged).

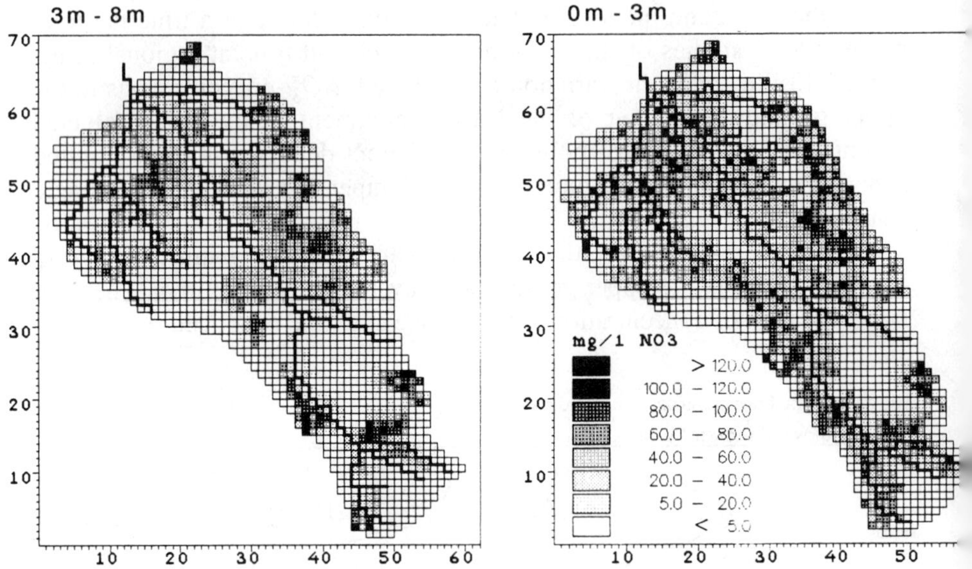

Fig. 23.14. Spatial variation in NO_3^--concentrations in the upper groundwater layer over the entire catchment at a specific time.

REFERENCES

Abbott, M.B., J.C. Bathurst, J.A. Cunge, P.E. O'Connell and J. Rasmussen (1986a): An Introduction to the European Hydrological System - Système Hydrologique Européen "SHE" l: History and philosophy of a physically based distributed modelling system. Journal of Hydrology, 87, 45-59.

Abbott, M.B., J.C. Bathurst, J.A. Cunge, P.E. O'Connell and J. Rasmussen (1986b): An Introduction to the European Hydrological System - Système Hydrologique Européen "SHE" 2: Structure of a physically based distributed modelling system. Journal of Hydrology, 87, 61-77.

Bathurst, J.C. (1986): Physically-based distributed modelling of an upland catchment using the Système Hydrologique Européen. Journal of Hydrology, 87, 79-102.

Carr, R.S., J.F. Punthakey, R. Cooke and B. Storm (1993): Large scale catchment simulation using the MIKE SHE model: 1. Process simulation of an irrigation district. Proceedings of the International conference on Environmental Management, Geo-Water and Engineering aspects. Ed. R.N. Chowalhury and M. Sivakumar, Wollongong, NSW Australia.

Mudgway, L.B. and R.J. Nathan (1993): Process modelling of flow and salt transport between shallow groundwater and surface drainage in the Tragowel Plains. Investigation Branch Report No. 1993/13. Rural Water Cooperation, Victoria, Australia.

DHI (1993): Validation of hydrological models, Phase II. Danish Hydraulic Institute, Hørsholm.

Ekebjærg, L. and P. Justesen (1991): An explicit scheme for advection-diffusion modelling in two dimensions. Computer Methods in Applied Mechanics and Engineering, 88 (3), 287-297.

Hansen, S., H.E. Jensen, N.E. Nielsen and H. Svendsen (1991): Simulation of nitrogen dynamics and biomass production in winter wheat using the Danish simulation model DAISY. Fertilizer Research, 27, 245-259.

Jain, S.K., B. Storm, J.C. Bathurst, J.C. Refsgaard and R.D. Singh (1992): Application of the SHE to catchments in India - Part 2: Field experiments and simulation studies on the Kolar Subcatchment of the Narmada River. Journal of Hydrology, 140, 25-47.

Jensen, K.H. (1983): Simulation of water flow in the unsaturated zone including the root zone. Series paper No. 33. Institute of Hydrodynamics and Hydraulic Engineering. Technical University of Denmark.

Justesen, P., K.W. Olesen and H.J Vested (1989): High accuracy modelling of advection in two and three dimensions. IAHR Congress, Ottawa, Canada.

Kristensen, K.J. and S.E. Jensen (1975): A Model for Estimating Actual Evapotranspiration from Potential Evapotranspiration. Nordic Hydrology, Vol. 6, pp. 70-88.

Leonard, B.P. (1979): Simple high-accuracy resolution program for convective modelling of discontinuities. International Journal for Numerical Methods in Fluids, 8, 1291-1318.

Lohani, V.K., J.C. Refsgaard, T. Clausen, M. Erlich and B. Storm (1993): Application of the SHE for irrigation command area studies in India. Journal of Irrigation and Drainage Engineering, 119 (1), 34-49.

Morris, E.M. (1982): Sensitivity of the European Hydrological System snow models, Proc. Symp. on Hydrological Aspects of Alpine and High-mountain areas. IAHS Publ. 138, 221-231.

Monteith, J.L. (1965): Evaporation and environment, Symp. Soc. Ex Biology. 19, 205-234.

Preismann, A. and Zaoui, J. (1979): Le module "Ecoulement de surface" due Système Hydrologique Européen (SHE). Paper presented at 18th IAHR Congress, Cagliari, Italy.

Refsgaard, J.C., T.H. Christensen and H.C. Ammentorp (1991): A Model for oxygen transport and consumption in the unsaturated zone. Journal of Hydrology, 129, 349-369.

Refsgaard, J.C., S.M. Seth, J.C. Bathurst, M. Erlich, B. Storm, G.H. Jørgensen and S. Chandra (1992): Application of the SHE to catchment in India - Part 1: General results. Journal of Hydrology, 140, 1-23.

Refsgaard, A. and G.H. Jørgensen (1990): Use of three-dimensional modelling in groundwater management and protection. VIII International Conference on Computational Methods in Water Resources, Venice, Italy.

Rutter, A.J., A.J. Morton and P.C. Robins (1975): A predictive model of rainfall interception in forests II. Generalisation of the model and comparison with observations in some coniferous and hardwood stands. J. Applied Ecology, 12, 367-80.

Storm, B., G.H. Jørgensen and M. Styczen (1987): Simulation of water flow and soil erosion processes with a distributed physically-based modelling system. Forest Hydrology and Watershed Management Proceedings of the Vancouver Symposium, August 1987, IAHS-AISH Publ. No. 167, 1987.

Storm, B. (1991): Modelling of saturated flow and the coupling of the surface and subsurface flow. In recent advances in the modelling of hydrologic system. Ed. Bowles D.S. and P.E. O'Connell. Kluwer, The Netherlands.

Storm, B., M. Styczen and T. Clausen (1991): Three-dimensional modelling of nitrate transport in a catchment. Paper presented at the International Conference on Nitrogen, Phosphorus and Organic Matter - May 13-15, 1991, Helsingør, Denmark. National Agency of Environment Protection, Denmark.

Styczen, M. and S.A. Nielsen (1989): A view of soil erosion theory, process, research and model building: Possible interactions and future developments. Quaderni di Scienza del Suolo, Vol. II, Firenze.

Styczen, M. and B. Storm (1993): Modelling of N-movements on catchment scale - a tool for analysis and decision making. 1. Model description & 2. A case study. Accepted for publication in Fertilizer Research, 17 pp.

Thomas, R.G. (1973): Groundwater models. Irrigation and drainage. Spec. Pap. Food Agricultural Organis. No. 21, U.N., Rome.

Vested, H.J, P. Justesen and L. Ekebjærg (1992): Advection-diffusion modelling in three dimensions. Applied Mathematical Modelling.

Chapter 24

SWRRB--A WATERSHED SCALE MODEL FOR SOIL AND WATER RESOURCES MANAGEMENT

J.G. Arnold and J.R. Williams

24.1. INTRODUCTION

A model called SWRRB (Simulator for Water Resources in Rural Basins) was developed for simulating hydrologic and related processes in rural basins (Williams et al., 1985). The objective in model development was to predict the effect of management decisions on water and sediment yields with reasonable accuracy for ungaged rural basins throughout the U.S. To satisfy the objective, the model had to be (a) physically based, using readily available inputs (calibration is not possible on ungaged basins); (b) capable of computing the effects of management changes on outputs; (c) computationally efficient to allow simulation on a variety of management strategies without excessive cost; (d) capable of simulating long periods for use in frequency analysis; and (e) capable of operating on subdivided basins (soils, land use, management, etc. make subdivision necessary).

To create SWRRB, the CREAMS daily rainfall hydrology model was modified for application to large, complex, rural bains. The major changes involved were (a) the model was expanded to allow simultaneous computations on several subbasins to predict the basin water yield; (b) a return flow component was added; (c) a reservoir storage component was added for use in determing the effects of farm ponds and reservoirs on water and sediment yield; (d) a weather simulation model (rainfall, solar radiation, and temperature) was added to provide for longer term simulations and more representative weather inputs, both temporally and spatially; (e) a better method was developed for predicting the peak runoff rate; (f) a crop growth model was added to account for annual variation in growth; (g) a simple flood routing component was added; (h) components were added to simulate sediment movement through ponds, reservoirs, streams, and valleys; and (i) transmission losses were calculated.

Recently, most of the SWRRB model development has been focused on problems involving water quality. Example additions include the GLEAMS (Leonard et al., 1987) pesticide fate component, optional SCS technology for estimating peak runoff rates and newly developed sediment yield equations. These and other less significant developments extend SWRRB's capabilities to deal with a wide variety of watershed management problems. The major processes included in the model are surface runoff, percolation, return flow, evapotranspiration, transmission losses, pond and reservoir storage, sedimentation, pesticide fate, nutrient cycling, and crop growth.

24.2. APPLICATIONS

The SWRRB model is currently being used throughout the world by consulting engineers, government agencies, universities, and chemical companies. Several applications follow to demonstrate the model's potential capabilities.

Hydrology Unit Model of the United States (HUMUS). The SCS is using the EPIC (Erosion-Productivity Impact Calculator), SWRRB, and ROTO models as part of the 1997 Resource Conservation Assessment. The models will be linked to national economic models and used for national planning of water supply and quality on the 18 major river basins in the U.S. This system of models allows water, sediment, and attached pollutants to be tracked from their point of origin to major rivers, reservoirs, and coastal zones. Geographic information systems are utilized to integrate the models with national soils, land use, and digital elevation databases. The GIS automatically extracts model input from the map layers and display model output.

Coastal Pollutant Discharge Inventory. As a part of the National Coastal Pollutant Discharge Inventory, the National Oceanic and Atmospheric Administration (NOAA) is using SWRRB to estimate non-point source loadings from non-urban lands in all coastal counties of the U.S. (Singer et al., 1988). Site-specific data are obtained from the SCS's National Resources Inventory and Soils-5 data bases, NOAA weather stations, U.S. Geological Survey digital land use data tapes, and other local sources. Simulations have been run for cropland, rangeland, and forest land in approximately 770 subwatersheds comprising the Gulf Coast, eastern, and western coastal zones of the U.S. Results are compiled by season and added to a comprehensive data based containing pollutant loadings from all significant discharge sources.

Pesticide Assessment. SWRRB was modified to include simulation of pesticide concentration in the runoff and sediment. The EPA has adopted this version of SWRRB as its pesticide assessment model. They have prepared their own user's manual, "Pesticide Runoff Simulator" (Holst and Kutney, 1987). Many chemical companies and consulting firms are using this version of the model for environmental assessment.

Effect of Urbanization on Reservoir Loadings. White Rock Lake in Dallas was built in 1910. The SCS completed sediment surveys on the lake in 1935, 1956, 1970, 1977, and 1984. Throughout this period, the percentage of urban area has increased from 0 to 77% in 1984. SWRRB was utilized to estimate the effects of urbanization on water and sediment delivery to White Rock Lake (Arnold et al., 1987).

Initially, simulated model results were compared with measured water yields, peak flow rates, and sediment yields. The comparisons showed that the model could do a reasonable job predicting the effect of urbanization on these variables. Also, the effect of urbanization on delivery ratios showed a positive linear correlation.

24.3. HYDROLOGY

The hydrology model is based on the water balance equation

$$SW_t = SW + \sum_{i=1}^{t} (R_i - Q_i - ET_i - P_i - QR_i) \qquad (24.1)$$

where SW is the soil water content minus the 15-bar water content, t is time in days, and R, Q, ET, P, and QR are the daily amounts of precipitation, runoff, evapotranspiration, percolation, and return flow; all units are in mm.

Since the model maintains a continuous water balance, complex basins are subdivided to reflect differences in ET for various crops, soils, etc. Thus, runoff, is predicted separately for each subarea and routed to obtain the total runoff for the basins. This increases accuracy and gives a much better physical description of the water balance.

24.3.1. Surface Runoff

The runoff model simulates surface runoff volumes and peak runoff rates, given daily rainfall amounts. Runoff volume is estimated by using a modification of the Soil Conservation Service (SCS) curve number technique (USDA-SCS, 1972). The technique was selected for use because (a) it is reliable and has been used for many years in the United States; (b) it is computationally efficient; (c) the required inputs are generally available; and (d) it relates runoff to soil type, land use, and management practices. The use of readily available daily rainfall data is a particularly important attribute of the curve number technique because for many locations, rainfall data with time increments of less than 1 day are not available. Also, rainfall data manipulations and runoff computations are more efficient for data taken daily than at shorter intervals.

There are two options for estimating the peak runoff rate -- the modified Rational formula and the SCS TR-55 method (USDA-SCS, 1986). A stochastic element is included in the Rational equation to allow realistic simulation of peak runoff rates, given only daily rainfall and monthly rainfall intensity.

Runoff Volume

Surface runoff is predicted for daily rainfall by using the SCS curve number equation (USDA-SCS, 1972)

$$\begin{aligned} Q &= \frac{(R - 0.2\, s)}{R + 0.8\, s}, && R > 0.2\, s, \\ Q &= 0.0\ , && R \le 0.2\, s) \end{aligned} \tag{24.2}$$

where Q is the daily runoff, R is the daily rainfall, and s is a retention parameter (see "Notations" section). The retention parameter, s, varies (a) among watersheds because soils, land use, management, and slope all vary and (b) with time because of changes in soil water content. The parameter s is related to curve number (CN) by the SCS equation (USDA-SCS, 1972)

$$s = 254 \left(\frac{100}{CN} - 1 \right) \tag{24.3}$$

The constant, 254, in Eq. (24.3) gives s in mm. Thus, R and Q are also expressed in mm. CN_2--the curve number for moisture condition 2, or average curve number--can be obtained easily for any area by using the SCS hydrology handbook (USDA-SCS, 1972). The handbook tables consider soils, land use, and management. Assuming that the handbook CN_2 value is appropriate for a 5% slope, we developed the following equation for adjusting that value for other slopes.

$$CN_{2s} = \frac{1}{3}(CN_3 - CN_2)\ [1 - 2 \exp\ (-13.86\ S)] + CN_2 \tag{24.4}$$

where CN_{2s} is the handbook CN_2 value adjusted for slope, CN_3 is the curve number for moisture condition 3 (wet), and S is the average slope of the watershed. Values of CN_1, the curve number for moisture condition 1 (dry), and CN_3 corresponding to CN_2 are also tabulated in the handbook. For computing purposes, CN_1 and CN_3 were related to CN_2 with the equations

$$CN_1 = CN_2 - \frac{20\,(100 - CN_2)}{100 - CN_2 + \exp[2.533 - 0.0636(100 - CN_2)]} \tag{24.5}$$

$$CN_3 = CN_2 \exp\,[0.00673\quad(100 - CN_2)] \tag{24.6}$$

Fluctuations in soil water content cause the retention parameter to change according to the equation

$$s = s_1\left(1 - \frac{FFC}{FFC + \exp\quad[w_1 - w_2(FFC)]}\right) \tag{24.7}$$

where s_1 is the value of s associated with CN_1, FFC is the fraction of field capacity, and w_1 and w_2 are shape parameters. FCC is computed with the equation

$$FFC = \frac{SW - WP}{FC - WP} \tag{24.8}$$

where SW is the soil water content in the root zone, WP is the wilting point water content (1,500 kPa for many soils), and FC is the field capacity water content (33 kPa for many soils).

Values for w_1 and w_2 are obtained from a simultaneous solution of Eq. (24.7) according to the assumptions that $s=s_2$ when FCC=0.6 and $s=s_3$, when (SW-FC)/(PO-FC)=0.5

$$w_1 = \ln\left(\frac{60.}{1. - s_2 / s_1} - 60.\right) + 60.\ w_2 \tag{24.9}$$

$$w_2 = \frac{\ln\left(\frac{60.}{1. - s_2 / s_1} - 60.\right) - \ln\ \left(POFC\frac{}{1. - s_3 / s_1} - POFC\right)}{POFC - 60.} \tag{24.10}$$

where s_3 is the CN_3 retention parameter and the porosity-field capacity ratio POFC is computed with the equation

$$POFC = 100. + 50.\left(\frac{\sum_{\ell-1}^{M} (PO_\ell - FC_\ell)}{\sum_{\ell-1}^{M} (FC_\ell - WP_\ell)}\right) \tag{24.11}$$

where PO is the porosity of soil layer ℓ. Equations (24.9) and (24.10) assure that CN_1 corresponds with the wilting point and that the curve number cannot exceed 100.

The FFC value obtained in Eq. (24.8) represents soil water uniformly distributed through the top 1.0 m of soil. Runoff estimates can be improved if the depth distribution of soil water is known. For example, water distributed near the soil surface results in more runoff than the same volume of water uniformly distributed throughout the top meter of soil. Also, a soil surface associated with a uniform distribution of soil water results in more runoff than a soil surface that is dry. Since SWRRB estimates water content of each soil layer daily, the depth distribution is available. The effect of depth distribution on runoff is expressed in the depth weighting function

$$FFC^* = \frac{\sum_{\ell=1}^{M} FFC_\ell \frac{Z_\ell - Z_{\ell-1}}{Z_\ell}}{\sum_{\ell=1}^{M} \frac{Z_\ell - Z_{\ell-1}}{Z_\ell}}, \qquad Z_\ell \le 1.0m \tag{24.12}$$

where FFC^* is the depth weighted FFC value for use in Eq. (24.7), Z is the depth in m to the bottom of soil layer ℓ, and M is the number of soil layers. Equation (24.12) performs two functions: (1) it reduces the influence of lower layers because FFC_ℓ is divided by Z_ℓ and (2) it gives proper weight to thick layers relative to thin layers because FFC is multiplied by the layer thickness.

There is also a provision for estimating runoff from frozen soil. If the temperature in the second soil layer is less than 0°C, the retention parameter is reduced by using the equation

$$s_f = 0.1s \tag{24.13}$$

where s_f is the retention parameter for frozen ground. Equation (24.13) increases runoff for frozen soils but allows significant infiltration when soils are dry.

Peak Runoff Rate

SWRRB contains two methods for estimating peak runoff rate--the modified Rational formula and the SCS TR-55 method (USDA-SCS, 1986).

Rational Equation Method. The rational equation can be written in the form

$$q_\rho = (\rho)\,(r)\,(A)/360 \tag{24.14}$$

where q_ρ is the peak runoff rate in $m^3 \bullet s^{-1}$, ρ is a runoff coefficient expressing the watershed infiltration characteristics, r is the rainfall intensity in mm•h^{-1} for the watershed's time of concentration, and A is the drainage area in ha. The runoff coefficient can be calculated for each storm if the amount of rainfall and runoff are known

$$\rho = \frac{Q}{R} \tag{24.15}$$

Since R is input and Q is computed with Eq. (24.1), ρ can be calculated directly. Rainfall intensity can be expressed with the relationship

$$r = \frac{R_{tc}}{t_c} \tag{24.16}$$

where R_{tc} is the amount of rainfall in mm during the watershed's time of concentration, t_c in h. The value of R_{tc} can be estimated by developing a relationship with total R. The Weather Service's TP-40 (Hershfield, 1961) provides accumulated rainfall amounts for various durations and frequencies. Generally, R_{tc} and R_{24} (24-h duration is appropriate for the daily time step model) are proportional for various frequencies. Thus,

$$R_{tc} = \alpha\ R_{24} \tag{24.17}$$

where α is a dimensionless parameter that expresses the proportion of total rainfall that occurs during t_c.

The peak runoff equation is obtained by substituting Eqs. (24.15), (24.16), and (24.17) into Eq. (24.14).

$$q_\rho = \frac{(\alpha)\ (Q)\ (A)}{360\ (t_c)} \tag{24.18}$$

The time of concentration can be estimated by adding the surface and channel flow times

$$t_c = t_{cc} + t_{cs} \tag{24.19}$$

where t_{cc} is the time of concentration for channel flow and t_{cs} is the time of concentration for surface flow in h. The t_{cc} can be computed by using the equation

$$t_{cc} = \frac{L_c}{v_c} \tag{24.20}$$

where L_c is the average channel flow length for the watershed in km and v_c is the average channel velocity in m•s^{-1}. The average channel flow length can be estimated by using the equation

$$L_c = \sqrt{(L)(L_{ca})} \tag{24.21}$$

where L is the channel length from the most distant point to the watershed outlet in km and L_{ca} is the distance from the outlet along the channel to the watershed centroid in km. Average velocity can be estimated by using Manning's equation and assuming a trapezoidal channel with 2:1 side slopes and a 10:1 bottom width/depth ratio. Substitution of these estimated and assumed values gives

$$t_{cc} = \frac{\sqrt{(L)(L_{ca})}(n)^{0.75}}{0.489(q_c)^{0.25}(\sigma)^{0.375}} \tag{24.22}$$

where n is Manning's n, q_c is the average flow rate in $m^3 \bullet s^{-1}$, and s is the average channel slope in $m \bullet m^{-1}$. Assuming that L_{ca}=0.5L and converting units (L from m to km, t_{cc} from s to h, and q_c from $m^3 \bullet s^{-1}$ to $mm \bullet h^{-1}$) gives the equation

$$t_{cc} = \frac{1.75\ (L^*)\ (n)^{0.75}}{(q_c^* \bullet \Sigma A)^{0.25}\ (\sigma)^{0.375}} \tag{24.23}$$

where L^* is the channel length in km and q_c^* is the average flow rate in $mm \bullet h^{-1}$. The average flow rate is obtained from the estimated average flow rate from a unit source in the watershed (1 ha area) and the relationship

$$q_c^* = q_o^*\ A^{-0.5} \tag{24.24}$$

where q_o^* is the average flow rate from a 1 ha area in $mm \bullet h^{-1}$. The unit source flow rate is estimated with the equation

$$q_o^* = \frac{Q}{DUR} \tag{24.25}$$

where DUR, the rainfall duration in h is calculated using the equation

$$DUR = \frac{4.605}{-2\ln(1.-\alpha_{.5})} \tag{24.26}$$

where $\alpha_{.5}$ is computed with Eq. (24.17) using $R_{.5}$ instead of R_{tc}. Equation (24.26) is derived assuming rainfall intensity is exponentially distributed. Details of the procedure for estimating $\alpha_{.5}$ and DUR are given in the water erosion section of this chapter.

Substituting Eq. (24.25) into Eq. (24.23) gives the final equation for t_{cc}

$$t_{cc} = \frac{1.75\ (L^*)\ (n)^{0.75}}{(q_o^*)^{0.25}\ (A)^{0.125}\ (\sigma)^{0.375}} \tag{24.27}$$

A similar approach is used to estimate t_{cs}:

$$t_{cs} = \frac{\lambda}{v_s} \qquad (24.28)$$

where λ is the surface slope length in m and v_s is the surface flow velocity in m•s^{-1}.

Applying Manning's equation to a strip 1 m wide down the slope length, assuming flow is concentrated into a small trapezoidal channel with 1:1 side slopes and 5:1 bottom width-depth ratio gives the velocity equation

$$v_S = \frac{0.8375 \; d^{0.666} \; s^{0.5}}{n} \qquad (24.29)$$

where v_s is the flow velocity in m^3•s^{-1}, d is flow depth in m, s is the land surface slope in m•m^{-1}, and n is Manning's roughness coefficient for the surface.

The average flow depth, d, can be calculated from Manning's equation as a function of flow rate

$$d = \left(\frac{q_o \bullet n}{5.025 \; s^{0.5}} \right)^{0.375} \qquad (24.30)$$

where q_o is the average flow rate in m^3•s^{-1}. Substituting Eqs. (24.29) and (24.30) into Eq. (24.28) gives

$$t_{cs} = \frac{0.0216 \; (\lambda \bullet n)^{0.75}}{(q_o^*)^{0.25} \; (s)^{0.375}} \qquad (24.31)$$

To properly evaluate α, variation in rainfall patterns must be considered. For some short duration storms, most or all the rain occurs during tc causing α to approach its upper limit of 1.0. Other storms of uniform intensity cause α to approach a minimum value. All other patterns cause higher α values than the uniform pattern, because r_{tc} is greater than r_{24} for all patterns except the uniform. By substituting the products of intensity and time into Eq. (24.17), an expression for the minimum value of α, α_{mn}, is obtained

$$\alpha_{mn} = \frac{t_c}{24} \qquad (24.32)$$

Thus, a ranges within the limits

$$t_c/24 \leq \alpha \leq 1.0$$

Although confined between limits, the value of a is assigned with considerable uncertainty when only daily rainfall and simulated runoff amounts are given. Thus, α is generated from a triangular distribution with the base ranging from $t_c/24$ to 1.0.

The peak of the α distribution changes monthly because of seasonal differences in rainfall intensities. The Weather Service (U.S. Department of Commerce, 1979) provides information on monthly maximum rainfall intensities that can be used to estimate the peak α for each month.

TR-55 Method. The SCS TR-55 Method for estimating peak runoff rate has been described in detail (USDA-SCS, 1986). Only the essential equations are presented here. The peak runoff rate is dependent on the rainfall distribution and amount, the runoff curve number, and the time of concentration as expressed in the equation

$$q_p = (q_p^*)\ (R) \tag{24.33}$$

where q_p is the peak rate in mm•h^{-1}, q_p^* is the peak rate per unit of rainfall in mm•h^{-1}•mm^{-1}, and R is the storm rainfall amount in mm. A set of curves are available (USDA-SCS, 1986) for estimating q_p^* given rainfall distribution (SCS Type I, IA, II, or III), the runoff curve number, and the watershed time of concentration. A map of the United States is given (USDA-SCS, 1986) to determine appropriate storm type. The curves relating q_p^*, t_c, and CN have been fitted with a seventh degree polonomial for use in computer solutions.

Time of concentration is estimated with the equation

$$t_c = t_{cc} + t_{cs} + t_{sf} \tag{24.34}$$

where t_c is the watershed time of concentration in h and t_{cc}, t_{cs}, and t_{sf} are the t_c components attributed to channel flow, surface flow, and shallow channel flow. The channel component is computed with the equation

$$t_{cc} = \frac{(L - \lambda - L_{sf})\ n}{3.6\ d^{0.67}\ \sigma^{0.5}} \tag{24.35}$$

where t_{cc} is the channel time of concentration in h, L is the channel length from the most distant point to the watershed outlet in km, λ is the surface slope length in km, L_{sf} is the shallow flow length in km, n is Manning's roughness coefficient, d is the average channel flow depth in m, and σ is the channel slope in m•m^{-1}. The shallow flow component of t_c is estimated with the equation

$$t_{sf} = \frac{L_{sf}}{v_{sf}} \tag{24.36}$$

where v_{sf} is the average shallow flow velocity in km•h^{-1} is estimated with the equation

$$v_{sf} = 17.7\ S^{0.5} \leq 2.19 km \bullet h^{-1} \tag{24.37}$$

where S is surface flow slope in m•m^{-1}. The length of shallow flow, L_{sf} is estimated with the equations

$$L_{sf} = 0.05 \qquad L > 0.1km$$
$$= L - 0.05 \qquad 0.05 < L < 0.1\ km \qquad (24.38)$$
$$= 0.0 \qquad L < 0.05km$$

Note that t_{sf} is constrained to less than 0.1h even with S=0.001. Thus, t_{sf} is generally insignificant. The surface flow component of t_c is estimated with the equation

$$t_{cs} = \frac{0.0913\ (\lambda \bullet n)^{0.8}}{S^{0.4}\ \ R^{0.5}} \qquad (24.39)$$

Percolation

The percolation component of SWRRB uses a storage routing technique combined with a crack-flow model to predict flow through each soil layer. Once water percolated below the rootzone, it is lost from the watershed (becomes groundwater or appears as return flow in downstream basins). The storage routing technique is based on the equation

$$SW_i = SW_{oi} \exp(-\Delta t \ / \ TT_i) \qquad (24.40)$$

where SW_O and SW and the soil water contents at the beginning and end of the day in mm, Δt is the time interval (24 h), and TT is the travel time through layer i in h. Thus, the percolation can be computed by subtracting SW from SW_O.

$$O_i = SW_{Oi}[1. - \exp(-\Delta t \,/\, TT_i)] \qquad (24.41)$$

where O is the percolation rate in mm•d^{-1}.

The travel time, TT_i, is computed for each soil layer with the linear storage equation

$$TT_i = (SW_i - FC_i) \,/\, H_i \qquad (24.42)$$

where H_i is the hydraulic conductivity in mm•h^{-1} and FC is the field capacity minus wilting point water content for layer i in mm. The hydraulic conductivity is varied from the saturated conductivity value at saturation to near zero at field capacity.

$$H_i = SC_i \left(\frac{SW_i}{UL_i} \right)^{\beta_i} \qquad (24.43)$$

where SC_i is the saturated conductivity for layer i in mm•h^{-1} and is a parameter that causes H_i to approach zero as SW_i approaches FC_i.

The equation for estimating β is

$$\beta_i = \frac{-2.655}{\log_{10} (FC_i / UL_i)} \tag{24.44}$$

The constant (-2.655) in Eq. (24.44) was set to assure H_i=0.002SC_i at field capacity.

Flow through a soil layer may be reached by a saturated lower soil layer. If the layer immediately below the layer being considered is saturated, no flow can occur. The effect of lower layer water content is expressed in the equation

$$O_{ci} = O_i \sqrt{1 - \frac{SW_i + 1}{Ul_i + 1}} \tag{24.45}$$

where O_{ci} is the percolation rate for layer i in mm•d^{-1} corrected for layer i+l water content and O_i is the percolation computed with Eq. (24.41).

The crack-flow model allows percolation of infiltrated rainfall even though the soil water content is less than field capacity. When the soil is dry and cracked, infiltrated rainfall can flow through the cracks of a layer without becoming part of the layer's soil water. However, the portion that does become part of a layer's stored water cannot percolate until the storage exceeds field capacity.

Crack-flow is simulated with the relationships similar to those used to estimate percolation above field capacity. The amount of percolate caused by crack flow is estimated with the equation

$$O_i = O_{i-1} \left[\exp(-\Delta t / TT_{ci})\right] \tag{24.46}$$

where O_{i-1} is the flow from the layer above in mm•d^{-1} (R-Q for the top layer) and TT_{ci} is the crack flow travel time in h. Crack flow travel time is estimated with the equation

$$TT_{ci} = I_{i-1} / (\zeta_i) \tag{24.47}$$

where ζ_i is a dimensionless (0-1) soil parameter that expressed degree of cracking and is a function of soil water and clay content.

$$\zeta_i = 0.01(CLA)_i \left(1 - \frac{SW}{SW + \exp(7.0 - 0.11SW)}\right) \tag{24.48}$$

where CLA_i is the clay content of soil layer i (0-1).

Percolation is also affected by soil temperature. If the temperature in a particular layer is O°C or below, no percolation is allowed from that layer.

Water can, however, percolate into the layer as dictated by Eqs. (24.41) and (24.46).

Since the one-day time interval is relatively low for routing flow through soils, it is desirable to divide the water into several parts for routing. This is necessary because flow rates are dependent upon soil water content which is continuously changing. For example, if the soil is extremey wet, Eqs. (24.41), (24.42), and (24.43) might greatly over estimate percolation if ony one routing was performed using the entire amount (SW_i-FC_i). To overcome this problem, SWRRB divides each layer's inflow into 4-mm slugs for routing. Also, by dividing the inflow into 4-mm slugs and routing each slug individually through all layers, the lower layer water content relationship (Eq. (24.43)) is allowed to function. The lower layer water content effect cannot be calculated realistically by routing all of the input through one layer before proceeding to the next layer.

24.3.2. Lateral Subsurface Flow

The kinematic storage model developed by Sloan et al. (1983) uses the mass continuity equation with the entire soil profile as the control volume. The mass continuity equation in finite difference form for the kinematic storage model is

$$\frac{S_2 - S_1}{t_2 - t_1} = iL - \frac{q_{lat1} + q_{lat2}}{2} \tag{24.49}$$

where S is the drainable volume of water stored in the saturated zone m•m^{-1} (water above field capacity), t is time in h, q_{lat} is the lateral flow in m^3•h^{-1}, i is the rate of water input to the saturated zone in m^2•h^{-1}, L is the hillslope length in m, and subscripts 1 and 2 refer to the beginning and end of the time step, respectively. The saturated thickness normal to the hillslope, H_o, is

$$H_o = \frac{2S}{\Theta_d L} \tag{24.50}$$

where Θ_d is the drainable porosity of the soil. The drainable volume of water stored, S, is updated daily in the watershed model. The lateral flow at the hillslope outlet is given by

$$q_{lat} = H_o v w \tag{24.51}$$

where v is the velocity of flow at the outlet and w is the hillslope width. Velocity at the outlet is estimated from

$$v = K_s \sin(\alpha) \tag{24.52}$$

where K_s is the saturated conductivity. Combining Eqs. (24.50) and (24.52) into 51 yields

$$q_{lat} = 0.024 \frac{2\ S\ K_s \sin(\alpha)}{\Theta_d L} \tag{24.53}$$

where q_{lat} is in mm•d^{-1}, S in m•h^{-1}, a in m•m^{-1}, Θ_d in m•m^{-1}, and L in m. If the saturated zone rises above the soil later, water is allowed to flow to the layer above (back to the surface for the upper soil layer). The amount of flow upward is estimated from K_s and the saturated slope length.

$$q_{sat} = 24\ K_s\ L_s / L \tag{24.54}$$

where q_{sat} is the upward flow in mm•d^{-1}, and L_s is the saturated slope length in m.

Most comprehensive hydrologic models divide the soil profile into multiple layers, and allow for percolation from one soil layer to the next and percolation from the bottom soil layer past the soil profile (as rechage to the shallow aquifer). To account for multiple layers, the model is applied to each soil layer independently starting at the upper layer.

24.3.3. Evapotranspiration

The model offers two options for estimating potential evaporation-Hargreaves and Samani (1985), and Priestley-Taylor (1972). The Priestley-Taylor method requires solar radiation and air temperature as input. The Hargreaves method requires air temperature only.

The model computes evaporation from soils and plants separately, as described by Ritchie (1972). Potential soil water evaporation is estimated as a function of potential evaporation and leaf area index (LAI, area of plant leaves relative to the soil surface area). Actual soil water evaporation is estimated by using exponential functions of soil depth and water content. Plant water evaporation is simulated as a linear function of potential evaporation and leaf area index.

Potential Evaporation

The Priestley-Taylor (1972) method provides estimated of potential evaporation based only on temperature and radiation is

$$E_o = 1.28 \left(\frac{h_o}{HV}\right) \left(\frac{\delta}{\delta + y}\right) \tag{24.55}$$

where E_o is the potential evaporation in mm, δ is the slope of the saturation vapor pressure curve in kPa°C^{-1}, γ is a psychrometer constant in kPa°C^{-1}, h_o is the net radiation in MJ•m^{-2}, and HV is the latent heat of vaporization in

$MJ \cdot kg^{-1}$. The latent heat ov vaporization is estimated with the temperature function

$$HV = 2.50 - 0.0022\ T \tag{24.56}$$

where T is the mean daily air temperature in °C. The saturation vapor pressure is also estimated as a function of temperature by using the equation

$$e_a = 0.1 \exp\left(54.88 - 5.03\ \ln\ (T+273) - \frac{6791}{T+273}\right) \tag{24.57}$$

The slope of the saturation vapor pressure curve is estimated with the equation

$$\delta = \left(\frac{e_a}{T+273}\right)\left(\frac{6791}{T+273} - 5.03\right) \tag{24.58}$$

The psychrometer constant is computed with the equation

$$\gamma = 6.6\ X\ 10^{-4}\ PB \tag{24.59}$$

where PB is the barometric pressure (kPa). The barometric pressure is estimated as a function of elevation by using the equation

$$PB = 101 - 0.0115\ ELEV + 5.44\ X\ 10^{-7} ELEV^2 \tag{24.60}$$

where ELEV is the elevation of the site in m.

The maximum possible solar radiation is computed with the equations

$$RAMX = 30\left(1.0 + 0.0335\ \sin\left[\frac{2\pi}{365}\ (i + 88.2)\right]\right)$$

$$XT\ \sin\left(\frac{2\pi}{360} LAT\right)\ \sin(SD) + \cos\left(\frac{2\pi}{360} LAT\right)\ \cos(SD)\ \sin(XT)) \tag{24.61}$$

$$XT = \cos^{-1}\left(-\tan\left(\frac{2\pi}{360} LAT\right)\tan(SD)\right), \quad 0 \le XT \le \pi \tag{24.62}$$

where LAT is the latitude of the site in degrees, SD is the sun's declination angle (radians), and i is the day of the year. The sun's declination angle is calculated with the equation

$$SD_i = 0.4102\ \sin\left[\frac{2p}{365}\ (i - 80.25)\right] \tag{24.63}$$

The net radiation is estimated with the equation

$$h_{oi} = RA_i\,(1 - AB_i) \tag{24.64}$$

where RA is the solar radiation in MJ•m^{-2} and AB is albedo.

The Hargreaves and Samani (1985) method estimates potential evaportranspiration as a function of extraterrestrial radiation and air temperature. Hargreaves' method was modified for use in SWRRB by increasing the temperature difference exponent from 0.5 to 0.6. Also, extraterrestrial radiation is replaced by RAMX (maximum possible solar radiation at the earth's surface) and the coefficient is adjusted from 0.0023 to 0.0032 for proper conversion. The modified equation is

$$E_o = 0.0032\left(\frac{RAMX}{HV}\right)(T + 17.8)\,(T_{mx} - T_{mn})^{0.6} \tag{24.65}$$

where T_{mx} and T_{mn} are the daily maximum and minimum air temperatures in °C.

Both methods estimate albedo by considering the soil, crop, and snow cover. If a snow cover exists with 5 mm or greater water content, the value of albedo is set to 0.6. If the snow cover is less than 5 mm and no crop is growing, the soil albedo is the appropriate value. When crops are growing, albedo is determined by using the equation

$$AB = 0.23\,(1.0 - EA) + (AB_s)\,(EA) \tag{24.66}$$

where 0.23 is the albedo for plants, AB_s is the soil albedo, and EA is a soil cover index. The value of EA ranges from 0 to 1.0 according to the equation

$$EA = \exp(-0.05CV) \tag{24.67}$$

where CV is the sum of the above ground biomass and crop residue in t•ha^{-1}.

Soil and Plant Evaporation

The model computes evaporation from soils and plants separately by an approach similar to that of Ritchie (1972).

$$E_p = \frac{(E_o)(LAI)}{3.0}, \qquad 0 \le LAI \le 3.068 \tag{24.68}$$

$$E_p = E_o, \qquad LAI > 3.0 \tag{24.69}$$

where E_p is the predicted plant water evaporation rate in mm•d^{-1}. If soil water is limited, plant water evaporation will be reduced as described in the plant growth section of this chapter.

Potential soil water evaporation is simulated by considering soil cover according to the following equation

$$E_s = (E_o)(EA) \tag{24.70}$$

where E_s is the potential soil water evaporation rate in mm•d^{-1}. Potential soil water evaporation is reduced during periods of high plant water use with the equation

$$E_s^* = \min\left[ES, \frac{(E_s)(E_o)}{E_s + E_p}\right] \tag{24.71}$$

When E_p is low $E_s^* \rightarrow E_s$ but as E_p approaches E_o, $E_s^* \rightarrow E_s/(1.+EA)$.

Actual soil water evaporation is estimated on the basis of the top 0.2 m of soil and snow cover, if any. If 5 mm or more (water content) of snow is present albedo is set to 0.6 and EA to 0.5 for estimating E_o and snow is evaporated at that rate. When all snow is evaporated, soil water evaporation begins. Such evaporation is governed by soil depth and water content according to the equation

$$EV_Z = E_s^*\left[\frac{Z}{Z + \exp\ (2.374 - 0.00713\ Z)}\right] \tag{24.72}$$

EV is the total soil water evaporation in mm from soil of depth Z in mm. The coefficients of Eq. (24.72) are set to give EV=0.5 E_s^* when Z=10 mm and EV=0.95 E_s^* when Z=100 mm.

Potential soil water evaporation for a layer is estimated by taking the difference between EV's at the layer boundaries

$$SEV_\ell = EV_{Z(\ell)} - EV_{Z(\ell-1)} \tag{24.73}$$

where SEV is the potential soil evaporation for layer ℓ in mm.

The depth distributed estimate of soil water evaporation may be reduced according to the following equation if soil water is limited in a layer

$$SEV_\ell^* = SEV_\ell \exp\left(\frac{2.5(SW_\ell - FC_\ell)}{FC_\ell - WP_\ell}\right), \qquad SW_\ell < FC_\ell \tag{24.74}$$

where SEV_ℓ^* is the adjusted soil water evaporation estimate in mm.

$$SEV_\ell^* = SEV_\ell, \qquad SW_\ell \geq FC_\ell \tag{24.75}$$

The final step in adjusting the evaporation estimate is to assure that the soil water supply is adequate to meet the demand

$$SEV_\ell^* = \min\ (SEV_\ell^*\ , SW_\ell - b_w\ WP_l) \tag{24.76}$$

where b_w may range from 0. to 1.0 in the top 0.5 m of soil and is set to 1.0 below 0.5 m. Thus, SWRRB can be adjusted to allow the top 0.5 m to dry down to any fraction of wilting point.

24.3.4. Snowmelt

If snow is present, it may be melted on days when the second soil layer temperature exceeds 0°C. Snow is melted as a function of the snow pack temperature using the equation

$$SML = T\,(1.52 + 0.54\ SPT) \qquad 0. \le SML \le SNO \tag{24.77}$$

where SML is the snowmelt rate in mm•d^{-1}, SNO is the snow present in mm of water, T is the mean daily air temperature in °C, and SPT is the snow pack temperature in °C. The snow pack temperature is estimated with the equation

$$SPT = \min\ (T_s\ , T(2)) \tag{24.78}$$

where T_s is the temperature at the top of the snow pack and T(2) is the temperature at the center of soil layer 2. The equations for estimating T_s and T(2) are presented in the soil temperature section. Melted snow is treated the same as rainfall for estimating runoff volume and percolation, but rainfall energy is set to 0.0 and peak runoff rate is estimated by assuming uniformly distributed rainfall for a 24-h duration.

24.3.5. Transmission Losses

Many semiarid watersheds have alluvial channels that abstract large quantities of streamflow (Lane, 1982). The abstractions, or transmission losses, reduce runoff volumes as water is lost as the flood wave travels downstream. Chapter 19 of the SCS Hydrology Handbook (USDA, 1983 update) by Leonard J. Lane describes a procedure for estimating transmission losses for ephemeral streams. This procedure is used in SWRRB when channel data (width, depth, Manning's "n") are available. It is based on the method for estimating transmission losses in the absence of observed inflow-outflow data assuming no lateral inflow or out-of-bank flow. The procedure is based on derived regression equations and enables the user to estimate transmission losses for similar channels of arbitrary length and width. Analysis of observed data resulted in regression equations of the form

$$a_r = -\ 0.001831\ \ (CHK)\ \ DUM \tag{24.79}$$

$$k_r = -1.09 \ln(1.0 - 0.2649)\ \ (CHK)\ \ DU\,/\,P \tag{24.80}$$

$$b_r = e^{-k_r} \tag{24.81}$$

where a_r is the unit channel intercept in m^3, CHK is the effective hydraulic conductivity of the channel alluvium in mm/hr (Lane, 1982; USDA, 1983 update), DU is the duration of streamflow in h, k_r is the decay factor in $m{\bullet}km^{-1}$ P_{in} is inflow volume of m^3, and b_r is the unit channel regression slope. The regression parameters are

$$a_x = [a_r(1 - b_r)] \ \ (1.0 - b_r L_c) \tag{24.82}$$

$$b_x = e^{-2.04k_r} \ \ L_c w \tag{24.83}$$

$$P_o = -a_x \ / \ b_x \tag{24.84}$$

where a_x is the regression intercept in $m{\bullet}km^{-1}$, b_x is the regression slope, w is average width of flow in m, L_c is length of reach in km, and P_o is the threshold volume for a unit channel in m^3. The prediction equation for volume after losses, Q_l, is

$$\begin{aligned} Q_\ell &= 0 & P_{in} &< P_o \\ Q_\ell &= -a_x + b_x P_{in} & P_{in} &> P_o \end{aligned} \tag{24.85}$$

Transmission losses are calculated twice for each subbasin. First, losses occuring within the subbasin are estimated. The inflow volume, P is assumed to be equal the surface runoff from the subbasin. The flow duration, is estimated from

$$DU = (Q) \ (A) \ /(1.8 \ q_p) \tag{24.86}$$

where Q is the surface runoff volume in mm, A is the drainage area in ha, and q_p is the peak flow rate in $m^3{\bullet}s^{-1}$. Then losses occurring from the subbasin outlet to the basin outlet are estimated. The prediction equation for peak discharge after losses q_{pl}, is

$$\begin{aligned} q_{pl} &= 0 & Q &= 0 \\ q_{pl} &= 12.1a_x \ / \ DU - (1.0 - b_x)P_{in} + b_x q_p & Q &> 0 \end{aligned} \tag{24.87}$$

Now duration of flow from subbasin to basin outlet is estimated with Eq. (24.86) using Q_l and q_{pl}, and the procedure is repeated to estimate losses from the subbasin outlet to the basin outlet. The volume and peak rate after losses are

$$Q_l = Q - TL \tag{24.88}$$

where TL is the transmission loss in mm.

24.4. WEATHER

The weather variables necessary for driving SWRRB are precipitation, air temperature, and solar radiation. If daily precipitation data are available, they can be input directly to SWRRB. If not, the weather component of the model can simulate daily rainfall and temperature. Solar radiation is always simulated. One set of weather variables may be simulated for the entire basin, or they can be simulated for each subbasin. Descriptions of the models used for simulating precipitation, temperature, and solar radiation follow.

24.4.1. Precipitation

The SWRRB precipitation model developed by Nicks (1974) is a first-order Markov chain model. Thus, input for the model must include monthly probabilities of receiving precipitation. On any given day, the input must include information as to whether the previous day was dry or wet. A random number (0-1) is generated and compared with the appropriate wet-dry probability. If the random number is less than or equal to the wet-dry probability, precipitation occurs on that day. Random numbers greater than the wet-dry probability give no precipitation. Since the wet-dry state of the first day is established, the process can be repeated for the next day and so on throughout the simulation period.

If wet-dry probabilities are not available, the average monthly number of rainy days may be substituted. The probability of a wet day is calculated directly from the number of wet days:

$$PW = NWD / ND \tag{24.89}$$

where PW is the probability of a wet day, NWD is the number of rainy days, and ND is the number of days, in a month. The probability of a wet day after a dry day can be estimated as a fraction of PW.

$$P(W / D) = \beta PW \tag{24.90}$$

where P(W/D) is the probability of a wet day following a dry day and where β is a fraction usually in the range of 0.6 to 0.9. The probability of a wet day following a wet day can be calculated directly by using the equation

$$P(W / W) = 1.0 - \beta + P(W / D) \tag{24.91}$$

where P(W/W) is the probability of a wet day after a wet day. When $\beta \rightarrow 1.0$, wet days do not affect probability of rainfall--P(W/D)=P(W/W)=PW. Conversely, low β values give strong wet day effects--$\beta \rightarrow 0.0$, P(W/D) $\rightarrow 1.0$. Thus, β controls the interval between rainfall events but has no effect on the number of wet days. For many locations, β=0.75 gives satisfactory estimates of P(W/D). Although Eqs. (24.90) and (24.91) may give slightly different probabilities than those estimated from rainfall records, they do guarantee correct simulation of the number of rainfall events.

When precipitation event occurs, the amount is generated from a skewed normal daily precipitation distribution

$$R_i = \left(\frac{\left(\left(SND_i - \frac{SCF_k}{6.0} \right) \left(\frac{SCF_k}{6.0} \right) + 1 \right)^3 - 1.}{SCF_k} \right) RSDV_k + \overline{R}_k \quad (24.92)$$

where R is the amount of rainfall for day i, in mm, SND is the standard normal deviate for day i, SCF is the skew coefficient, RSDV is the standard deviation of daily rainfall in mm, and R is the mean daily rainfall in month k.

If the standard deviation and skew coefficient are not available, the model simulates daily rainfall by using a modified exponential distribution.

$$R_i = \frac{(-\ln\ (\mu))^{\zeta}\, \overline{R_k}}{\int_{0.0}^{1.0} (-\ln\ (\chi))^{\zeta}\, dx} \quad (24.93)$$

where μ is a uniform random number (0.0-1.0) and ζ is a parameter usually in the range of 1.0 to 2.0. The larger the ζ value, the more extreme the rainfall events. A value of 1.3 gives satisfactory results at many locations in the United States. The denominator of Eq. (24.93) assures that the mean long-term simulated rainfall is correct. The modified exponential is usually a satisfactory substitute and requires only the monthly mean daily rainfall as input.

If precipitation is to be simulated for each subbasin, the amount computed using Eqs. (24.92) or (24.93) is assumed to be the mean for all gages for the day. The center of the storm is located in a rectangle with boundaries set at a distance of 100 km from the basins' maximum and minimum x and y rain gage coordinates. Thus, the storm center could be located in the basin or as much as 100 km in any direction from the basin. Each storm center is defined by drawing two random numbers-one for the x scale and one for the y scale. Rainfall at each gage is computed using an area reduction function (Nicks and Igo, 1980).

$$RF = \frac{1 - (AO)\ (DUR)^{-0.148}}{337.5 + 1.09\ AO} \quad (24.94)$$

where RF is a reduction ratio (0-1), DUR is the rainfall duration in h, and AO is area represented by the rain gage in km^2. The circular area, AO, is calculated using the distance between the gage and the storm center to compute the diameter.

$$AO = 0.7854\ (x_c - x_i)^2 + (y_c - y_i)^2 \quad (24.95)$$

where x and y are coordinates of the rain gages in km and subscripts c and i refer to the storm center and the gage for which RF is being computed.

Finally, rainfall is generated for each gage using the equation

$$R_i = \frac{(RF_i)\ (R)\ (N)}{\sum_{i=1}^{N} RF_i} \tag{24.96}$$

where R_i is the daily rainfall for rain gage i, RF_i is the reduction factor, R is the mean daily rainfall, and N is the number of subbasins.

Daily precipitation is partitioned between rainfall and snowfall using a combination of maximum daily air temperature (T_{mx}) and surface layer soil temperature (T(1)). If the average of T_{mx} and T(1) is 0°C or below, the precipitation is snowfall, otherwise, it is rainfall.

24.4.2. Air Temperature and Solar Radiation

The model developed by Richardson (1981) was selected for use in SWRRB because it simulates temperature and radiation, which are mutually correlated with rainfall. The residuals of daily maximum and minimum air temperature and solar radiation are generated from a multivariate normal distribution.

The multivariate generation model used implies that the residuals of maximum temperature, minimum temperature, and solar radiation are normally distributed and that the serial correlation of each variable may be described by a first-order linear autoregressive model. Details of the multivariate generation model were described by Richardson (1981). The dependence structure of daily maximum temperature, minimum temperature, and solar radiation was described by Richardson (1982).

The temperature model requires monthly means of maximum and minimum temperatures and their standard deviations as inputs. If the standard deviations are not available, the long-term observed extreme monthly minimums and maximums may be substituted. The model estimates standard deviation as 0.25 of the difference between the extreme and the mean for each month. For example:

$$SDTMX_k = 0.25\ (TE_{mx,k} - \overline{T}_{mx,k}) \tag{24.97}$$

where SDTMX is the standard deviation of the daily maximum temperature, TE is the extreme daily maximum temperature, and $\overline{T}$ is the average daily maximum temperature for month k.

The solar radiation model uses the extreme approach extensively. Thus, only the monthly means of daily solar radiation are required as inputs. The equation for estimating standard deviation is

$$SDRA_k = 0.25\ (RAMX_k - RA_k) \tag{24.98}$$

where SDRA is the standard deviation of daily solar radiation in MJ•m^{-2}, RAMX is the maximum daily solar radiation at midmonth, and RA is the mean daily solar radiation for month k.

Maximum temperature and solar radiation tend to be lower on rainy days. Thus, it is necessary to adjust the mean maximum temperature and solar radiation downward for simulating rainy day conditions. For T_{mx} this is accomplished by assuming that wet day values are less than dry day values by some fraction of T_{mx}-T_{mn}:

$$TW_{mx,k} = TD_{mx,k} - \Omega_T(T_{mx}, k - T_{mn,k}) \qquad (24.99)$$

where TW is the daily mean maximum temperture for wet days in °C in month k, TD is the daily mean maximum temperature for dry days, Ω_T is a scaling factor ranging from 0.0 to 1.0, T_{mx} is the daily mean maximum temperature, and T_{mn} is the daily mean minimum temperature. Choosing Ω_T=1.0 provides highest deviations on wet days and Ω_T=0.0 ignores the wet day effect. Observed data indicate that Ω_T usually lies between 0.5 and 1.0.

Since Eq. (24.99) gives lower mean maximum temperature values for wet days, a companion equation is necessary to slightly increase mean maximum temperature for dry days. The development is taken directly from the continuity equation

$$(T_{mx,k})(ND_k) = (TW_{mx,k})(NWD_k) + (TD_{mx,k})\ (NDD_k) \qquad (24.100)$$

where ND is the number of days in a month, NWD is the number of wet days, and NDD is the number of dry days. The desired equation is obtained by substituting Eq. (24.99) into Eq. (24.100) and solving for TD

$$TD_{mx,k} = T_{mx,k} + \left(\frac{NWD_k}{ND_k}\right) \Omega_T\ (T_{mx,k} - T_{mn,k}) \qquad (24.101)$$

Use of the continuity equation guarantees that the long-term simulated value for mean maximum temperture agrees with the input value of T_{mx}.

The method of adjusting solar radiation for wet and dry days is similar to that of adjusting maximum temperature. The radiation on wet days is a fraction of the dry day radiation

$$RAW_k = \Omega_R RAD_k \qquad (24.102)$$

where RAW is the daily mean solar radiation on wet days in MJ•m^{-2}, Ω_R is a scaling factor ranging from 0.0 to 1.0, and RAD is the daily mean solar radiation on dry days. An Ω_R value of 0.5 gives satisfactory results for many locations. The dry day equation is developed by replacing temperature with radiation in Eq. (24.101) and substituting Eq. (24.102) for RAW. Then,

$$RAD_k = \frac{(RA_k)\ (ND_k)}{\Omega_R(NWD_k) + NDD_k} \qquad (24.103)$$

where RA is the daily mean solar radiation for month k in MJ•m^{-2}

24.5. SEDIMENT YIELD

Sediment yield is computed for each subbasin with the Modified Universal Soil Loss Equation (MUSLE) (Williams and Berndt, 1977).

$$Y = 11.8\ (V\ q_p)^{0.56}\ (K)\ (C)\ (PE)\ (LS) \tag{24.104}$$

where Y is the sediment yield from the subbasin in t, V is the surface runoff column for the subbasin in m^3, q_p is the peak flow rate for the subbasin in $m^3 \cdot s^{-1}$, K is the soil erodibility factor, C is the crop management factor, PE is the erosion control practice factor, and LS is the slope length and steepness factor.

The LS factor is computed with the equation (Wischmeier and Smith, 1978)

$$LS = (\lambda / 22.1)^{\xi}(65.41\ S^2 + 4.565\ S + .065) \tag{24.105}$$

The exponent ξ varies with slope and is computed in SWRRB with the equation

$$\xi = 0.6\ [1 - \exp(-35.835\ S)] \tag{24.106}$$

The crop management factor, C, is evaluated for all days when runoff occurs using the equation,

$$C = \exp[(-0.2231 - CVM)\ \exp(-0.00115\ \ CV) + CVM] \tag{24.107}$$

where CM is the soil cover (above ground biomass+residue in $kg \cdot ha^{-1}$ and CVM is the minimum value of C. The value of CVM is estimated from the average annual C factor using the equation

$$CVM = 1.463 \ln\ (CVA) + 0.1034 \tag{24.108}$$

The value of CVA for each crop is determined from tables prepared by Wischmeier and Smith (1978). Values of K are contained in the SCS Soils-5 database, and PE factors can be estimated for each subbasin using information contained in Wischmeier and Smith (1978).

24.6. NUTRIENTS

Subbasin nutrient yield and nutrient cycling were taken from the EPIC model (Williams et al., 1984) and modified as necessary for inclusion into the SWRRB model. SWRRB allows for simultaneous computations on each subbasin and routes the water, sediment, and nutrients from the subbasin outlets to the basin outlet.

24.6.1. Nitrogen

Nitrate Loss in Surface Runoff

The amount of NO_3-N in runoff is estimated for each subbasin for considering the top soil layer (10 mm thickness) only. The total amount of water leaving the layer is the sum of runoff, lateral subsurface flow, and percolation.

$$QT = Q + O_1 + QR_1 \tag{24.109}$$

where QT is the total water lost from the first layer in mm, Q is the runoff volume in mm, O_1 is the percolation from the first layer in mm, and QR_1 is the lateral flow from the first layer in mm. The amount of NO_3-N lost with QT is

$$VNO3 = (QT)\ (C_{NO3}) \tag{24.110}$$

where VNO3 is the amount NO_3-N lost from the first layer and C_{NO3} is the concentration of NO_3-N in the first layer. At the end of the day, the amount of NO_3-N left in the layer is

$$WNO3 = WNO3_o - (QT)\ (C_{NO3}) \tag{24.111}$$

where $WNO3_O$ and WNO3 are the weights of NO_3-N contained in the layer at the beginning and ending of the day. The NO_3-N concentration can be estimated by dividing the weight of NO_3-N by the water storage volume:

$$C'_{NO3} = C_{NO3} - C_{NO3}\left(\frac{-QT}{PO_1 - WP_1}\right) \tag{24.112}$$

where C'_{NO3} is the concentration of NO_3-N at the end of the day, PO is the soil porosity, and WP is the wilting point water content for soil layer one in mm. Equation (24.112) is a finite different appoximation for the exponential equation

$$C'_{NO3} = C_{NO3} - \exp\left(\frac{-QT}{PO_1 - WP_1}\right) \tag{24.113}$$

thus, VNO3 can be computed for any QT value by integrating Eq. (24.113).

$$VNO3 = WNO3\left(1 - \exp\left(\frac{-QT}{PO_1 - WP_1}\right)\right) \tag{24.114}$$

The average concentration of QT for the day is

$$C_{NO3} = VNO3\,/\,QT \tag{24.115}$$

Amounts of NO_3-N contained in runoff, lateral flow, and percolation are estimated as the products of the volume of water and the concentration from Eq. (24.115).

Nitrate Leaching

Leaching and lateral subsurface flow in lower layers are treated with the same approach used in the upper layer except surface runoff is not considered.

Organic N Transport by Sediment

A loading function developed by McElroy et al. (1976) are modified by Williams and Hann (1978) for application to individual runoff events is used to estimate organic N loss for each subbasin. The loading function

$$YON = 0.001\ (Y)\ \ (CON)\ \ (ER) \qquad (24.116)$$

where YON is the organic N runoff loss at the subbasin outlet in kg•ha^{-1}, CON is the concentration of organic N in the top soil layer in g•t^{-1}, Y is the sediment yield in t•ha^{-1}, and ER is the enrichment ratio. The value of CON is input to the model and is constant throughout the simulation. The enrichment ratio is the concentration of organic N in the sediment divided by that of the soil. Enrichment ratios are logarithmically related to sediment concentration as described by Menzel (1980). An individual event enrichment sediment concentration relationship was developed for SWRRB considering upper and lower bounds. The upper bound of enrichment ratio is the inverse of the sediment delivery ratio. Exceeding the inverse of the delivery ratio implied that more organic N leaves the watershed than is dislodged from the soil. The delivery ratio is estimated for each runoff event using the equation

$$DR = (q_p\ /\ r_{ep})^{0.56} \qquad (24.117)$$

where DR is the sediment delivery ratio (subbasin sediment yield divided by gross sheet erosion), q_p is the peak runoff rate in mm•h^{-1}, and r_{ep} is the peak rainfall excess rate in mm•h^{-1}. Equation (24.117) is based on sediment yield estimated using MUSLE (Williams, 1975). The rainfall excess rate cannont be evaluated directly because the hydrology model only predicts the total daily runoff volume. An estimate of the rate can be obtained, however, using the equation

$$r_{ep} = r_p - f \qquad (24.118)$$

where r_p is the peak rainfall rate in mm•h^{-1} and f is the average infiltration rate in mm•h^{-1}. The average infiltration rate can be computed from the equation

$$f = R - Q\ /\ DUR \qquad (24.119)$$

where DUR is the rainfall duration in h and R is rainfall in mm. The procedure for estimating the rainfall duration is given in Williams et al. (1984).

$$DUR = 4.605R / r_p \tag{24.120}$$

The lower limit of enrichment ratio is 1.0-sediment particle size distribution is the same as that of the soil. thus, 1≤ER≤1/DR. The logarithmic equation estimating enrichment ratio is

$$ER = x_1 c_a^{x_2} \tag{24.121}$$

where c_a is the sediment concentraion in g•m^{-3} and x_1 and x_2 are parameters set by the upper and lower limits. To approach an enrichment ratio is 1.0, the sediment concentration would be extremely high. Conversely, a very low sediment concentration would cause the enrichment ratio to approach 1/DR. The simultaneous solution of Eq. (24.121) at the boundaries assuming sediment concentrations range from 500 to 250,000 g•m^{-3} gives

$$x_2 = \log(1 \;/\; DR) \;/\; 2.699 \tag{24.122}$$

$$x_1 = 1 /(0.25)^{x_2} \tag{24.123}$$

Crop Uptake

Crop use of N is estimated using a supply and demand approach. The daily (day i) crop N demand can be computed using the equation

$$UND_i = (C_{NB})_i \; B_i - (C_{NB})_{i-1} \; B_{i-1} \tag{24.124}$$

where UND_i is the N demand of the crop in kg•ha^{-1}, C_{NB} is the optimal N concentration of the crop, and B is the accumulated in kg•ha^{-1}. The optimal crop N concentration is computed as a function of growth stage using the equation

$$C_{NB} = 4.0 \;(bn) + 1.54 \;(bn) \;\exp(-bn \;\; B_1) \tag{24.125}$$

where bn is a crop parameter expressing N concentration and B_1 is the fraction of the growing season. The value of B_1 is estimated as a function of heat units

$$B_{1,i} = \sum_{k=1}^{i} \frac{HUI}{PHU} \tag{24.126}$$

where HU is the daily heat units in °C above the crop's base temperature and PHU is the potential heat units to mature the crop in °C.

The crop is allowed to take N from any soil layer that has roots. Uptake starts at the upper layer and proceeds downward until the daily demand is met or until all N has been depleted. If the soil cannont supply the daily N demand for legumes, the deficit is attributed to N fixation.

24.6.2. Phosphorus

Soluble P Loss in Surface Runoff

The SWRRB approach is based on the concept of partitioning pesticides into the solution and sediment phases as described by Leonard and Wauchope (Knisel, 1980). Because P is mostly associated with the sediment phase, the soluble P runoff equation can be expressed in the simple form

$$YSP = 0.01(C_{LPP})\ (Q) / k_d \tag{24.127}$$

where YSP is the soluble P in $kg{\bullet}ha^{-1}$ lost in runoff volume Q in mm, C_{LPP} is the concentration of AP in soil layer P in $g{\bullet}t^{-1}$, and k_d is the P concentration in the sediment divided by that of the water in $m^3{\bullet}t^{-1}$. The value of CLP is input to the model and remains constant. The value of k_d used in SWRRB is 175.

P Transport by Sediment

Sediment transport of P is simulated with a loading function as described in organic N transport. The P loading function is

$$YP = 0.01\ \ (Y)\ \ (C_p)\ \ (ER) \tag{24.128}$$

where YP is the sediment phase of P loss in runoff in $kg{\bullet}ha^{-1}$ and c_p is the concentration of P in the top soil layer in $g{\bullet}t^{-1}$.

24.6.3. Fertilizer Application

The date and rate of N and P application is input to the model. The entire amount of N and P is added to the first soil layer and is available for water and sediment transport, leaching, and crop uptake.

24.7. PESTICIDES

GLEAMS (Leonard et al., 1987) technology for simulating pesticide transport by runoff, percolate, soil evaporation, and sediment was added to SWRRB. Pesticides may be applied at any time and rate to plant foliage or below the soil surface at any depth. When pesticide is applied, there is a loss to the atmosphere. Thus, the amount that reaches the ground or plants is expressed by the equation

$$PAPE = (PAPR)\ \ (PAEF) \tag{24.129}$$

where PAPE is the effective amount of pesticide applied in kg•ha^{-1} PAPR is the actual amount applied in kg•ha^{-1}, and PAEF is an application efficiency factor.

To determine how much pesticide reaches the ground, the amount of ground cover provided by plants is estimated with the equation

$$GC_{\ell} = (1. - ERFC\ \ (1.33LAI_{\ell,m} - 2.))\ /\ 2. \tag{24.130}$$

where GC is the fraction of the ground that is covered by plants, and LAI is the leaf area index on day, ℓ. Therefore, the amount of pesticide that reaches the plants is computed with the equation

$$FP = (GC)\ \ (PAPE) \tag{24.131}$$

where FP is the amount of pesticide that is intercepted by plants. The remaining pesticide falls to the ground and is simply the difference between the effective amount applied and the amount intercepted by plants.

$$GP = PAPE - FP \tag{24.132}$$

where GP is the amount of pesticide that reaches the ground.

Pesticide that remains on the plant foliage can be washed off by rain storms. It is assumed that the fraction of pesticide that is potentially dislodgeable is washed off the plants once a threshold rainfall amount is exceeded. The model uses a threshold value of 2.5 mm and potential washoff fractions for various pesticides have been estimated (Leonard et al., 1987). The appropriate equations for computing washoff and adding to the ground amount are:

$$\begin{aligned} WO &= (WOF)\ \ (FP), & R &\geq 2.5mm \\ WO &= 0., & R &< 2.5mm \end{aligned} \tag{24.133}$$

$$GP = GP + WO \tag{24.134}$$

$$FP = FP - WO \tag{24.135}$$

where WO is the amount of pesticide washed off the plants by a rainstorm of R in mm and WOF is the washoff fraction for the particular pesticide.

Pesticide on the plants and in the soil is lost from the system based on the decay equations

$$GP = (GP_o)\ \exp\ (-0.693\ /\ HLS) \tag{24.136}$$

$$FP = (FP_o) \ \exp\ (-0.693 / HLP) \qquad (24.137)$$

where GP_o and GP are the initial and final amounts of pesticide on the ground, FP_o and FP are the initial and final amounts of pesticide on the plants, HLS is the half life for pesticide in the soil in d, and HLP is the half life of the foliar residue in d. Values of HLP and HLS have been established for various pesticides (Leonard et al., 1987).

Another way that pesticide can be lost from the zone considered in computing runoff (top 10 mm of soil) is through leaching. The GLEAMS leaching component is used here with slight modification. The change in the amount of pesticide contained in the top 10 mm zone is expressed as a function of time, concentration, and amount of infiltration using the equation

$$\frac{dGP}{dt} = (PSTC_w) \ (f) \qquad (24.138)$$

where GP is the amount of pesticide in the top 10 mm zone at time t, $PSTC_w$ is the pesticide concentration in the water in $g \bullet t^{-1}$, and f is the water flow rate through the zone in $mm \bullet h^{-1}$. The total amount of pesticide contained in the top 10 mm zone is the sum of the adsorbed and mobile phases.

$$GP = 0.01 \ (PSTC_w) \ (SW) + 0.1 \ (PSTC_s) \ (BD) \qquad (24.139)$$

where SW is the amount of water stored in the top 10 mm of soil in mm, PSTC is the concentration of adsorbed pesticide in $g \bullet t^{-1}$, and BD is the soil bulk density in $t \bullet m^{-3}$. The ratio of the concentration of pesticide adsorbed to the concentration of pesticide in the water has been estimated for various pesticides (Leonard et al., 1987) and is expressed by the equation

$$K_d = \frac{c_s}{c_w} \qquad (24.140)$$

where K_d is the partitioning constant. Substituting Eq. (24.140) into Eq. (24.139) gives

$$GP = 0.01 \ (PSTC_w) \ (SW) + 0.1 \ (PSTC_w) \ (K_d) \ (BD) \qquad (24.141)$$

Solving Eq. (24.141) for $PSTC_w$ gives

$$PSTC_w = \frac{GP}{0.01 \ \ SW + 0.1 (K_d) \ \ (BD)} \qquad (24.142)$$

Substituting $PSTC_w$ from Eq. (24.142) into Eq. (24.138) yields

$$\frac{dGP}{dt} = \frac{(GP)(f)}{0.01 \ \ SW + 0.1 \ \ (K_d) \ \ (BD)} \qquad (24.143)$$

Rearranging Eq. (24.143) and integrating gives the equation expressing the amount of pesticide as a function of the amount of water flowing through the zone.

$$GP = GP_o \exp\left(\frac{-f}{0.01SW + 0.01(K_d)(BD)}\right) \qquad (24.144)$$

where GP_0 is the initial amount of pesticide in the top 10 mm zone in kg• ha^{-1}, GP is the amount that remains after the amount of flow (*f* in mm) passes through the zone, SW is the initial water storage in mm, k_d is the partitioning coefficient in m^3•t^{-1}, and BD is the soil bulk density in t•m^{-3}. To obtain the amount of pesticide leached by the amount of water *f*, GP is subtracted from GP_0 using the equation

$$PSTL = GP_o\left(1. - \exp\left(\frac{-f}{0.01\ SW + 0.1\ (K_d)\ (BD)}\right)\right) \qquad (24.145)$$

where PSTL is the amount of pesticide leached by *f*.

Pesticide concentration in percolate is computed with the equation

$$PSTC_L = \min\begin{cases} PSTL\,/\,f \\ PSO1 \end{cases} \qquad (24.146)$$

where PSO1 is the pesticide solubility and $PSTC_L$ is the pesticide concentration in the percolate in g•m^{-3}. Finally, PSTL is the product of O and $PSTC_L$.

Pesticide loss in surface runoff is estimated with a modification of Eq. (24.145) that includes an abstraction coefficient

$$PSTQ = GP_o\left(1. - \exp\left(\frac{-(Q)\ (ab)}{0.01\ SW + 0.1\ (ab)\ (K_d)\ (BD)}\right)\right) \qquad (24.147)$$

where PSTQ is the pesticide loss in surface runoff in kg•ha^{-1}, Q is the surface runoff volume in mm, and ab is the abstraction coefficient.

The calculation sequence is: (1) Soil layer 1--Leaching is estimated first by substituting 0 for *f* in Eqs. (24.145) and (24.146). Then PSTL is removed from layer 1 and placed into layer 2. Next runoff and lateral flow losses are estimated simultaneously by substituting Q+QR for Q in Eq. (24.147). The average concentration is obtained by dividing the resulting PSTQ by Q+QR. Individual losses are estimated for runoff (Q•$PSTC_{Q+QR}$) and lateral flow (QR•$PSTC_{Q+QR}$). Finally, these losses are removed from layer 1. (2) Soil layers 2-M--Leaching and lateral flow losses are estimated simultaneously by substituting O_ℓ+QR_ℓ for *f* in Eqs. (24.145) and (24.146). Individual losses are calculated as in (1) (QR_ℓ•$PSTC_L$) and (O_ℓ•$PSTC_L$), subtracted from layer ℓ and $O_\ell \bullet PSTC_L$ is added to layer $\ell+1$. The process is repeated layer by later to the bottom of the soil profile.

The total amount of pesticide lost in the runoff is estimated by adding the soluble fraction computed with Eq. (24.147) to the amount that is adsorbed to the sediment. Pesticide yield from the adsorbed phase is computed with an enrichment ratio approach. The enrichment ratio equation is

$$PSTY = (Y)\ (PSTC_s)\ (ER)\ (0.001) \tag{24.148}$$

where PSTY is the pesticide yield adsorbed to the sediment in kg•ha^{-1}, Y is the sediment yield in t•ha^{-1}, and ER is the enrichment ratio (concentration of pesticide in the sediment divided by the pesticide concentration in the top 10 mm of soil) computed with Eq. (24.121). The pesticide concentration in the soil is calculated by substituting Eq. (24.140) into Eq. (24.142) and solving for $PSTC_s$.

$$PSTC_s = \frac{(K_d)\ (GP)}{0.01\ SW + 0.1\ (K_d)\ (BD)} \tag{24.149}$$

Soil layers with low storage volumes have high leaching potentials not only because percolation is greater, but also because storage volume displacement is greater (higher concentration). Pesticides with low K_d values and high solubility are transported rapidly with water. Conversely, high K_d value pesticides are adsorbed to soil particles and travel largely with sediment.

24.8. SOIL TEMPERATURE

Daily average soil temperature is simulated at the center of each soil layer. The basic soil temperature equation is

$$T(Z,t) = \overline{T} + \frac{AM}{2} \exp(-Z\ /\ DD)\ \cos\left[\frac{2\pi}{365}(t-200) - Z\ /\ DD\right] \tag{24.150}$$

where Z is depth from the soil surface in mm, t is time d, T is the average annual air temperature in °C, AM is the annual amplitude in daily average temperature in °C, and DD is the damping depth for the soil in mm. Equation (24.150) provides estimates of air temperature (Z=0) as well as soil temperature. Since air temperature is provided by the weather componenet of SWRRB, the soil temperature model should be capable of using these air temperatures as drivers. Otherwise, Eq. (24.150) would predict the same temperatures for a given day each year. To allow simulated air temperature to be used as the soil temperature driver, equations were developed to estimate soil surface temperature.

$$TGB_{IDA} = (z_j)\ (T_{mx}) + (1 - z_j)(T_{mn}), \qquad R = 0 \tag{24.151}$$

$$TGB_{IDA} = T_{mn}, \qquad R > 0 \qquad (24.152)$$

where TGB_{IDA} is the bare soil surface temperature in °C, T_{mx} and T_{mn} are the maximum and minimum daily air temperature in °C, z is a weighting factor for estimating soil surface temperature on dry days, and j specifies the month. Obviously, Eq. (24.152) uses the minimum air temperature to estimate surface temperature on rainy days. Higher temperatures are estimated on dry days using Eq. (24.151). The value of z is determined by considering the average T_{mx}, T_{mn}, and number of rainy days in this month.

$$z_j = \frac{\dfrac{\left(T_{mxj} + T_{mnj} / 2\right) - \left(T_{mnj}\right)\left(NRD_j\right)}{NDD_j}}{T_{mxj} - T_{mnj}} - T_{mnj} \qquad (24.153)$$

where ND is the number of days in a month, NRD is the number of rainy days, and NDD is the number of dry days. Equation (24.153) assures that the long-term average soil surface temperature and air temperature are equal.

The soil surface temperature is also affected by residue and snow cover. This effect can be simulated by lagging the predicted base surface temperature using the equation

$$TG_{IDA} = (bcv)(TGB_{IDA-1}) + (1 - bcv)\ (TGB_{IDA}) \qquad (24.154)$$

where bcv is a lagging factor for simulating residue and snow cover effects on surface temperature. The value of bcv is 0 for base soil and approaches 1.0 as cover increased as expressed in the equation

$$bcv = \max\begin{cases} \dfrac{CV}{CV + \exp(7.563 - 1.297\ X\ 10^{-4} CV)} \\ \dfrac{SNO}{SNO + \exp(6.055 - 0.3022\ SNO)} \end{cases} \qquad (24.155)$$

where CV is the sum of above ground biomass and crop residue in kg•ha^{-1} and SNO is the water content of the snow cover in mm.

The soil temperature at any depth is estimated with Eq. (24.150) by substituting TG for T(0,t). TG is a better estimate of the surface temperature then T(0,t) because current weather and cover conditions are considered. At soil surface (Z=0), the proper substitution can be accomplished by adding TG and subtracting T(0,t) from Eq. (24.150). Differences between TG and T(0,t) are damped as Z increases. Thus, the final equation for estimating soil temperature at any depth is

$$T(Z,t) = T + \left[\left(\frac{AM}{2} + TG - T(0,t)\right)\right] \cos\left[\frac{2\pi}{365}\left((t - 200) - \frac{Z}{DD}\right)\right] \exp\left(\frac{-Z}{DD}\right) \quad (24.156)$$

The damping depth (DP) is a function of soil bulk density and water content as expressed in the equations

$$DP = 100 + \left(\frac{2500 \; BD}{BD + 686 \; \exp(-5.63 \; BD)} \right) \tag{24.157}$$

$$S_p = \frac{SW}{(0.356 - 0.144 \; BD) \; ZM} \tag{24.158}$$

$$DD = DP \exp\left(\ln\left(\frac{500}{DP}\right)\left[\frac{(1 - S_p)}{(1 + S_p)}\right]^2 \right) \tag{24.159}$$

where DP is the maximum damping depth for the soil mm, BD is the soil bulk density in t•m^{-3}, ZM is the distance from the bottom of the lowest soil layer to the surface in mm, and S_p is a scaling parameter.

24.9. CROP GROWTH

The crop model in SWRRB is a simplification of the EPIC crop model (Williams et al., 1984). SWRRB uses EPIC concepts of phenological crop development based on daily accumulated heat units, harvest index for partitioning grain yield, Monteith's approach (Monteith, 1977) for potential biomass, and water and temperature stress adjustments. However, the detailed EPIC root growth and nutrient cycling models are not included. A single model is used for simulating all the crops considered SWRRB is capable of simulating crop growth for both annual and perennial plants. Annual crops grow from planting date to harvest date or until the accumulated heat units equal the potential heat units for the crop. Perennial crops maintain their root systems throughout the year, although the plant may become dormant after frost.

Phenological development of the crop is based on daily heat unit accumulation. It is computed using the equation

$$HU_i = (T_{mx,i} + T_{mn,i} \; / \; 2) - T_{bj}, \qquad HU_k \geq 0 \tag{24.160}$$

where HU, T_{mx} and T_{mn} are the values of heat units, maximum temperature, and minimum temperature in °C on day i and T_b is the crop-specific base temperature in °C (no growth occurs at or below T_b) of crop j. A heat unit index (HUI) ranging from 0 at planting to 1 at physiological maturity is computed as follows.

$$HU_i = \frac{\sum_{k=1}^{i} HU_k}{PHU_j} \tag{24.161}$$

where HUI is the heat unit index for day i and PHU is the potential heat units required for maturity of crop j. The value of PHU is calculated by the model from normal planting and harvest dates.

24.9.1. Potential Growth

Interception of solar radiation is estimated with Beer's law equation (Monsi and Saeki, 1953)

$$PAR_i = 0.02092\ (RA_i)[1. - \exp(-0.65\ LAI_i)] \qquad (24.162)$$

where PAR is photosynthetic active radiation in MJ•m^{-2}, RA is solar radiation in Ly, LAI is the leaf area index, and subscript i is the day of the year. Using Monteith's approach (Monteith, 1977), potential increase in biomass for a day can be estimated with the equation

$$DB_{p,i} = (BE_j)\ (PAR_i) \qquad (24.163)$$

where B_p is the daily potential increase in total biomass in kg•ha^{-1} and BE is the crop parameter for converting energy to biomass in kg•m^{-2}•ha^{-1}•MJ.

LAI is simulated as a function of heat units and biomass. LAI is estimated with the equations

$$LAI_i = \frac{(LAI_{mx})\ (B_{AG})}{B_{AG} + \exp(9.5 - .0006\ B_{AG})}, \qquad HIU_i \le DLAI \qquad (24.164)$$

$$LAI_i = (16)(LAI_{mx})(1 - HIU_i)^2, \qquad HIU_i > DLAI \qquad (24.165)$$

where LAI_{mx} is the maximum LAI potential for the crop, B_{AG} is above ground biomass in kg•ha^{-1}, and DLAI is the fraction of the growing season when LAI starts declining (≈.75).

The fraction of total biomass partitioned to the root system normally decreases from 0.3 to 0.5 in the seedling to 0.05 to 0.20 at maturity (Jones, 1985). The model estimates the root fraction to range linearly from 0.4 at emergence to 0.2 at maturity. Thus, the daily root fraction is computed with the equation

$$RWT_i = (0.4 - 0.2\ HUI_i) \qquad (24.166)$$

where RWT is the fraction of total biomass partitioned to the root system on day i. Thus, B_{AG} is calculated from the equation

$$B_{AG} = (1 - RWT_i)\ (B_{TOTi}) \qquad (24.167)$$

where B_{TOT} is total biomass in kg•ha^{-1} on day i.

24.9.2. Crop Yield

The economic yield of most grain, pulse, and tuber crops is a reproductive organ. Crops have a variety of mechanisms which insure that their production is neither too great to be supported by the vegetative components nor too small to insure survival of the species. As a result, harvest index (economic yield/above-ground biomass) is often a relatively stable value across a range of environmental conditions. Crop yield is estimated using the harvest index concept.

$$YLD_j = (HI_j)\ (B_{AG}) \tag{24.168}$$

where YLD is the amount of the crop removed from the field in kg•ha^{-1}, HI is the harvest index at harvest, and B_{AG} is the above-ground biomass in kg•ha^{-1} for crop j. Harvest index increases non-linearly from zero at planting using the equation

$$HIA_i = HIO_j(HUFH_i - HUFH_{i-1}) \tag{24.169}$$

where HLA is the harvest index on day i, HIO is the harvest index under favorable growing conditions, and HUFH is the heat unit factor that affects harvest index on day i and on the previous day i-1.

The harvest index heat unit is computed with the equation

$$HUFH_i = \frac{HUI_i}{HUI_i + \exp(6.50 - 10.0\ \ HUi_i)} \tag{24.170}$$

The constants in Eq. (24.170) are set to allow $HUFH_i$ to increase from 0.1 at HUI_i=0.5 to 0.92 at HUI_i=0.9. This is consistent with economic yield development of grain crops which produce most economic yield in the second half of the growing season.

Most grain crops are particularly sensitive to water stress from shortly before until shortly after anthesis, when major yield components are determined (Doorenbos and Kassam, 1979). Optimum conditions for growth may reduce harvest index slightly if dry matter accumulation is large and economic yield is limited by sink size. The harvest index is affected by water stress using the equation

$$HIA_i = \frac{HIA_i}{1. + (WSYF_j)\ (FHU_i)\ (0.9 - WS_i)} \tag{24.171}$$

where HIA is the adjusted harvest index, WSYF is a crop parameter expressing drought sensitivity, FHU is a function of crop stage, and WS is the water stress factor for day i. Notice that harvest index may increase slightly on days with WS values greater than 0.9. The crop stage factor, FHU, is estimated with the equation

$$FHU_i = \sin\left(\frac{\pi}{2}\left(\frac{HUI_i - 0.3}{0.3}\right)\right), \qquad 0.3 \le HUI_i \le 0.9 \tag{24.172}$$

$$FHU_i = 0., \qquad HUI < 0.3 HUI > 0.9 \tag{24.173}$$

Thus, water stress only affects harvest index between 0.3 and 0.9 of maturity with the greatest effect occurring at 0.6.

24.9.3. Growth Constraints

The potential biomass predicted with Eq. (24.163) is adjusted daily if one of the plant stress factors is less than 1.0 using the equation

$$\Delta B = (\Delta B_p)\ (REG) \tag{24.174}$$

where REG is the one crop growth regulating factor (the minimum stress factor).

Water Stress

The water stress factor is computed by considering supply and demand in the equation

$$WS = \frac{\sum_{i=1}^{N_\ell} U_i}{E_P} \tag{24.175}$$

where WS is the water stress factor (0-1) and U_i is plant water use in layer i in mm. The value of E_p is predicted in the evapotranspiration model.

A model (Williams and Hann, 1978) for simulating plant water uptake is used to estimate the E_p distribution. Root depth is calculated with the equation

$$RD_n = \frac{2.5\sum_{i=1}^{n} E_{pi}}{\sum_{i=1}^{N_g} E_{pi}} \tag{24.176}$$

where RD_n is the fraction of the root zone that contains roots on day n and N_g is the number of days in the growing season.

The potential water use in each soil layer is estimated with the equation

$$u_{pi} = \frac{E_p}{1-\exp(-\Lambda)}\left[-\Lambda\left(\frac{RD_i}{RZ}\right)\right]\sum_{j=1}^{i-1} u_j \qquad (24.177)$$

where u_{pi} is the potential water use rate from layer i in mm•d^{-1}, RZ is the root zone depth in mm, and Λ is the use rate-depth parameter. The details of evaluating Λ are given by Williams and Hann (1978). The value of Λ used in SWRRB (3.065) was determined assuming that about 30% of the total water use comes from the top 10% of the root zone. Equation (24.177) allows roots to compensate for water deficits in certain layers by using more water in layers with adequate supplies. The potential water use must be adjusted for water deficits to obtain the actual use for each layer.

$$u_i = u_{pi}, \qquad SW_i > 0.25\ FC_i \qquad (24.178)$$

$$u_i = u_{pi}\left(\frac{SW_i}{0.25\ FC_i}\right), \qquad SW_i \le 0.25\ FC_i \qquad (24.179)$$

Note that the actual use for the layers above the i'th layer in Eq. (24.177) must be known This presents no problems, however, since u_j=0 when i=1.

Temperature Stress

The temperature stress factor is computed with the equation

$$TS = \exp\delta[(T_o - T)/T]^2 \qquad (24.180)$$

where TS is the temperature stress factor (0-1) is the temperature stress parameter for the crop, δ is the optimal temperature for the crop in °C (25°C for summer and 13°C for winter crops), and T is the daily average air temperature in °C. The stress parameter is evaluated by appropriate substituting and rearranging of Eq. (24.180).

$$\delta = \frac{\ln\ (0.9)}{\left(\frac{T_o - [(T_o + T_b)/2]}{(T_o + T_b)/2}\right)^2} \qquad (24.181)$$

where T_b is the base temperature for the crop in °C (8°C for summer and 0°C for winter crops). Equation (24.181) sets TS=0.9 when the air temperature is half way between T_b and T_o.

24.10. TILLAGE AND RESIDUE

The SWRRB tillage component was designed to partition the above ground biomass at harvest. A portion of the biomass is incorporated into the soil while the remainder is left on the soil surface as residue. Once the residue is incorporated, it has no impact on the model. Also no change is made in bulk density due to tillage.

The residue is set at harvest according to one of the four tillage practices chosen.

Fall plow:

$$RSD = 0.5\ WLV \tag{24.182}$$

Spring plow:

$$RSD = 0.25\ \ WLV \tag{24.183}$$

Conservation tillage:

$$RSD = 0.50\ \ WLV \tag{24.184}$$

Zero tillage:

$$RSD = 0.95\ \ WLV \tag{24.185}$$

where RSD is the residue left after harvest in kg•ha^{-1} and WLV is the above ground biomass minus crop yield at harvest in kg•ha^{-1}.

The residue decays throughout the remainder of the year (harvest to next harvest) as a function of soil water content and soil temperature.

$$DECR = 0.05\ \min(CDG, SUT) \tag{24.186}$$

where DECR is the daily decrease in residue, and

$$RSD = RSD\ (1 - DECR) \tag{24.87}$$

where CDG and SUT are soil water and soil temperature residue decay factors.

$$CDG = \frac{0.9\ ST_2}{ST_2 + \exp(9.93 - 0.312\ ST_2)} \tag{24.188}$$

and

$$SUT = \frac{SW_2}{FC_2} \tag{24.189}$$

where ST_2, SW_2, and FC_2 are the soil temperature, soil water content, and field capacity of the second soil layer.

24.11. IRRIGATION

The SWRRB user has the option to simulate dryland or irrigated agricultural areas. If irrigation is indicated, he must also specify the runoff ratio (volume of water leaving the field/volume applied) and a plant water stress level to start irrigation. The plant water stress factor ranges from 0 to 1.0 (1 means no stress and 0 means no growth). These stress factors are described in the Crop Growth section. When the user-specified stress level is reached, water is applied according to the equation

$$AIR = \frac{FC - SW}{1 - EFI} \tag{24.190}$$

where FC is the root zone field capacity in mm, SW is the root zone water content before irrigation in mm, EFI is the runoff ratio, and AIR is the volume of irrigation water applied in mm.

24.12. FLOOD ROUTING

Since SWRRB is primarily a long-term water and sediment yield simulator, a high degree of accuracy in predicting hydrographs is not as necessary as for other applications like flood control planning and flood forecasting. Also, SWRRB must operate as efficiently as possible to be useful in water resources planning that requires long-term simulations of numerous management strategies. Therefore, flood routing is not performed in SWRRB and daily bains outflow is estimated by summing subbasin outflows. All surface runoff reaches the basin outlets on the day of the event. Subsurface flow is lagged as a function of basin lag time. Accurate prediction of subbasin peak rates is important since they are used in estimating subbasin sediment yields.

24.13. PONDS AND RESERVOIRS

24.13.1. Ponds

This component of SWRRB was designed to account for the effects of farm ponds on water yield. The water balance equation is

$$VM = VM_O + QI - QO - EV - SEP \tag{24.191}$$

where VM_O is the volume of the water stored in all ponds within a subbasin at the beginning of the day, VM is the volume at the end of the day, QI is the inflow during the day, QO is the outflow, EV is the evaporation, SEP is the seepage, and all units are m^3. The inflow, QI is composed of surface runoff from the total pond drainage area and rainfall on the water surface area.

Outflow occurs when the volume exceeds the permanent pool storage capacity and is described with the equation

$$\begin{aligned} QO &= VM - VM_{mx}, \qquad && VM > VM_{mx} \\ QO &= 0, && VM \le VM_{mx} \end{aligned} \tag{24.192}$$

where VM_{mx} is the maximum permanent pool storage of all ponds in the subbasin in m^3.

The evaporation is computed with the equation

$$EV = 10\ (\eta)\ (E_o)\ (SA) \tag{24.193}$$

where η is an evaporation coefficient (0.6) and SA is the surface area of all ponds in the subbasin in ha.

Seepage from the ponds is computed with the equation

$$SEP = 240\ (SC)\ (SA) \tag{24.194}$$

where SC is the saturated conductivity of the pond bed in $mm \cdot h^{-1}$.

Since pond surface area is required for computing evaporation and seepage, a relationship between surface area and volume is necessary. Data from a large number of stock ponds and small reservoirs in Texas and Oklahoma (USDA-SCS, 1957) indicate that surface area can be calculated with the equation

$$SA = \omega_1\ (VM)^{\omega_2} \tag{24.195}$$

where ω_1 is a parameter ($1.3 \cdot 10^{-4}$) and ω_2 is a fairly constant parameter (0.9). The SWRRB model assumes ω_2=0.9 and determines ω_1 for each subbasin using SA_{mx} and VM_{mx}.

24.13.1. Reservoirs

Although this component was mainly designed to simulate flow through small reservoirs like those constructed on SCS PL566 projects, it can also be used on larger reservoirs. The reservoir water balance component is similar to the pond component except it allows flow from principal and emergency spillways. The reservoir outflow function is expressed in the equation

$$\begin{aligned} QOR &= VR - VR_F, \qquad && VR > VR_F \\ QOR &= (RR)(Dt), && VR_S < VR \le VR_F \\ QOR &= 0. && VR < VR_s \end{aligned} \tag{24.196}$$

where QOR is the daily outflow in m^3, VR is the volume of water in the reservoir in m^3, VR_F is the reservoir capacity at the emergency spillway crest in m^3, RR is the principal spillway release rate in $m^3 \cdot s^{-1}$, and VR_S is the reservoir capacity at the principal spillway crest in m^3.

The relationships (Eqs. (24.193), (24.194), and (24.195)) used to estimate evaporation and seepage from ponds are also applicable to reservoirs. However, the method for estimating ω_1 and ω_2 is slightly different. Since the surface areas and volumes for the principal and emergency spillway crest elevations are generally readily available, those values can be used for a simultaneous solution of Eq. (24.195). The resulting equations are

$$\omega_1 = \frac{\log SA_F - \log SA_S}{\log VR_F - \log VR_S} \tag{24.197}$$

$$\omega_2 = SA_F \,/\, VR^{\omega_2} \tag{24.198}$$

where SA is the reservoir surface area and subscripts F and S refer to emergency and principal spillway crest elevations, respectively.

24.14. SEDIMENT ROUTING

24.14.1. Channel Sediment Routing

The sediment routing model consists of two components operating simultaneously (deposition and degradation). Deposition in the stream channel is based on the fall velocity of the sediment particles (Arnold et al., 1990). With a temperature of 22°C and a sediment density of 1.2 $t{\bullet}m^{-3}$, Stokes' Law for fall velocity becomes:

$$V_f = 411\ (d^2) \tag{24.199}$$

where V_f is the fall velocity in $m{\bullet}h^{-1}$ and d is the sediment particle diameter. The depth (y_f) that sediment of particle size d will fall during time, TT, is

$$y_f = (V_f)\ (TT) \tag{24.200}$$

The sediment delivery ratio (DR) through the reach is estimated with the equations:

$$DR = \frac{1 - 0.5\ (y_f)}{d_q} \qquad y_f < d_q \tag{24.201}$$

$$DR = \frac{0.5\ (d_q)}{y_f} \qquad y_f > d_q \tag{24.202}$$

where d_q is the depth of flow.

Finally, deposition is calculated with the equation:

$$DEP = SED_{IN}\ (1 - DR) \tag{24.203}$$

Stream power is used to predict degradation in the routing reaches. Williams (1980) used Bagnold's (1977) definition of stream power to develop a methods for determining degradation in channels. Bagnold defined stream power, SP, with the equation:

$$SP = \gamma\ q\ S_w \tag{24.204}$$

where γ is the density of the water, q is the flow rate, and S_W is the water surface slope. By applying stream power to bed load predictions (Bagnold, 1977) and estimating model parameters (Williams, 1980), the equation for sediment reentrained, DEG_r, is

$$_R = \alpha_{sp}\ \gamma^{1.5}\ (dur)\ (w)\ \left(d_q\ S_W\ V_c\right)^{1.5} \tag{24.205}$$

where α_{sp} is a parameter dependent on maximum stream power for the reach and V_c is the velocity in the channel.

The parameter α_{sp} can be estimated with the equation:

$$\alpha_{sp} = (\gamma_{wq}\ S_c)_{mx}^{-0.5} \tag{24.206}$$

where S_c is the slope of the channel and the subscript mx refers to the maximum flow expected in the reach for extreme events. The value of q is assumed to equal some maximum rainfall intensity (250 mm/hr) and α_{sp} becomes:

$$\alpha_{sp} = \left(69.44\ \gamma\ DA\ S_c\right)^{-0.5} \tag{24.207}$$

where DA is the drainage area into the reach in km^2.

All of the stream power is used for reentrainment of loose and deposited material until all of the material has been removed. When this occurs, degradation of the bed material, DEG_B, begins and is calculated by:

$$DEG_B = K\ C\ DEG_R \tag{24.208}$$

where K and C are MUSLE (Williams and Berndt, 1977) factors for the stream channel. Total degredation, DEG, is the sum of the reentrainment and bed degredation components. This amount is also allowed to be redeposited before reaching the basin outlet.

$$DEG = (DEG_R + DEG_B)\ (1 - DR) \tag{24.209}$$

Finally, the amount of sediment reaching the basin outlet, SED_{out}, is:

$$SED_{out} = SED_{in} - DEP + DEG \tag{24.210}$$

where SED_{in} is the sediment entering the reach.

24.14.2. Reservoir Sediment Routing

The sediment balance equation for reservoirs is:

$$SR_i = SR_{i-1} + SR_{in} - SR_{out} - SR_{DEP} \tag{24.211}$$

where SR_i is the total sediment in the reservoir, SR_{i-1} is the total sediment in the reservoir on the previous day, SR_{in} is the incoming sediment, SR_{out} is the sediment transported in the sediment outflow, and SR_{DEP} is the amount of sediment deposited in the reservoir. Sediment outflow from reservoirs is calculated with the equation

$$SR_{out} = c_o q_o \qquad q_o > 0 \tag{24.212}$$

$$SR_{out} = 0 \qquad q_o = 0 \tag{24.213}$$

where c_o is the outflow sediment concentration. The outflow concentration is a function of the reservoir concentration at the beginning and end of the day

$$c_o = (cs_1 + cs_2) / 2 \tag{24.214}$$

where cs_1 and cs_2 are the reservoir sediment concentrations at the beginning and end of the day, respectively.

The initial reservoir concentration is input to the model. The inflow concentration can be calculated since q_i and SR_{in} are simulated, but the final reservoir concentration is unknown. It can be computed using the continuity equation

$$V_2 \; cs_2 = V_1 \; cs_1 + q_i \; c_i - q_o \; c_o \tag{24.215}$$

where V_1 and V_2 are the storage volumes at the beginning and end of the day, and c_i is the inflow sediment concentration. Substituting Eq. (24.214) into Eq. (24.215) and rearranging yields an expression for the final concentration

$$cs_2 = \frac{V_i \; cs_i + q_i \; c_i - (q \frac{o}{2}) \; cs_1}{V_2 + (\frac{q_o}{2})} \tag{24.216}$$

Between storms the final reservoir concentration decreases to an equilibrium concentration according to the equation

$$c_s = (cs_2 - cs_e)\ \exp(-k_s\ t\ d_{50}) + cs_{eR} \tag{24.217}$$

where cs is the resevoir concentration t days after the value of cs_2 is obtained, k_s is the decay constant, d_{50} is the median particle size of the inflow sediment, and cs_e is the equilibrium sediment concentration (intput to the model). A value of k_s is evaluated by assuming that 99% of the 1 µm particles are settled within 25 days (k_s=0.184).

24.15. NUTRIENT AND PESTICIDE ROUTING

24.15.1. Nitrate Routing

Once NO_3-N enters a stream it is considered a conservative material for the duration of an individual runoff event (Williams, 1980). Thus, NO_3-N routing is simply a matter of adding the yields from all subbasins to determine the basin yield.

24.15.2. Organic N Routing

The loading function approach is also used in routing organic N from the subbasin outlets to the basin outlet.

$$YON_{Bj} = 0.01\ (Y_B)_j\ (CONSB)_j\ (ER_R)_j \tag{24.218}$$

where YON_B is the organic N runoff loss at the basin outlet in kg•ha^{-1}, Y_B is the sediment yield reaching the basin outlet from subbasin j in t•ha^{-1}, CONSB is the concentration of organic N in the sediment reaching the subbasin j outlet in g•t^{-1}, and ER_R is the enrichment ratio for the channel routing from subbasin j to the channel outlet. The estimate of ER_R is calculated by Eqs. (24.121), (24.122), and (24.123) with the delivery ratio for the channel routing calculated from

$$DR = \frac{(Y_{SB})_j}{(Y_B)_j} \tag{24.219}$$

where Y_{SB} is the sediment yield at the subbasin outlet in t•ha^{-1}, and Y_B is that sediment yield from subbasin j after it has been routing to the basin oulet in t•ha^{-1}. Arnold et al. (1991) describe the sediment routing procedure in detail.

As with NO_3-N routing, once soluble P enters a stream it is considered a conservative material and routing is accomplished by adding the yields from all subbasins to determine the basin yield.

24.15.3. P Routing

Again, the loading function approach is used in routing P from the subbasin outlets to the basin outlet.

$$YP_{Bj} = 0.01\ (Y_B)_j\ (C_{PSB})_j\ (ER_R)_j \tag{24.220}$$

where YP_B is the P yield at the basin outlet in kg•ha^{-1}, and C_{PSB} is the P concentration in the sediment reaching the subbasin j outlet in g•t^{-1}.

24.15.4. Pesticide Routing

Soluble pesticide runoff is considered conservative in the stream channels and no decay or volatilization is considered. Adsorbed pesticides are allowed to be deposited if sediment deposition occurs. The delivery ratio/enrichment ratio approach used for nutrient routing is also used for adsorbed pesticides in the stream channels.

24.16. LAKE WATER QUALITY

Few models predict the effect of land management on off-site concerns such as lake water quality often environmental engineers assume a non-point source loading to lake water quality models and cannot directly predict the effects of land management decisions. At the same time, agricultural engineers and scientists model the non-point source loading and do not consider the off-site effects.

Here an integrated model was developed by linking SWRRB with a lake toxic model developed by Chapra (1989). The integrated model operates on a daily time step and is efficient enough to run for multiple years on a microcomputer. The model tracks the fate of pesticides from their initial application on the land to their final fate in the lake. This allows decision makers to directly predict the influence of upland agricultural management decisions on lake water quality.

24.16.1. Toxic Balance

The lake toxic balance is taken from Chapra (1989) and assumes well-mixed conditions. The system is partitioned into a well-mixed surface water layer underlain by a well-mixed sediment layer. The toxic is partitioned into dissolved and particulate in both the surface-water and sediment layers. The major processess simulated by the model are loading, outflow, reactions, volatilization, settling, diffusion, resuspension, and burial. Loading to the lake is estimated by non-point source model components, and toxic outflow is simply the water outflow from the lake water balance multiplied by the concentraion. Volatilization is the exchange of toxic across the air-water interface and acts only one the dissolved fraction while settling acts only on the particulate-associated fraction. Toxic movement due to random water motion of mixing is referred to as diffusion, which is the mechanism by which contaminated sediment pore water can seep back into the lake. Resuspension occurs when flow suspends particles that have been previously deposited. In the burial process, the toxic is covered to a depth where it can no longer react with the lake and is considered lost from the system. The mass balance can be described mathematically as

$$V_1 \frac{\Delta C_1}{\Delta t} =$$

$$Q_1 C_1 - Q_o C_1 - K_i V_i C_i - V_v A F_{dl} C_1 - V_s F_{pl} C_i + v_i A C_2 + V_C A \left(F_{d2} c_2 - F_{d1} c_1 \right) \quad (24.221)$$

$$V_2 \frac{\Delta C_2}{\Delta t} =$$

$$K_2 V_2 C_2 + V_g A F_{dl} C_1 - V_r A C_2 - V A C_2 + V_d A \left(F_{dl} C_1 - F_{d2} C_2 \right) \quad (24.222)$$

where subscripts 1 and 2 indicate the water layer and sediment layer, respectively, V is volume in m^3, C is total (dissolved and particulate) toxic concentration in mg• m^3, Δ is the time step (one day), Q is water volume in m^3 (subscripts 1 and o indicate inflow and outflow, respectively), K is a first order reaction coefficient in day^{-1}. V_V is the volatilization mass transfer coefficient in m•day, A is area m^2, V_S is settling velocity in m•day, V_r is resuspension velocity in m•day, V_d is diffusion mixing velocity in m•day, V_b is burial velocity in m•day, and F_d and F_p are the fraction of total contaminant in dissolved and particulate form, respectively.

Soil-Liquid Partitioning

The total toxic concentration is separated into two components

$$C = C_d + C_p \quad (24.223)$$

where C_d is the dissolved component and C_p is the particulate component. These components represent fraction of the total concentration as in

$$C_d = F_d \ C \tag{24.224}$$

$$C_p = F_p \ C \tag{24.225}$$

The fractions F_d and F_p are a function of the contaminant's partition coefficient and the suspended solids concentration in the lake.

$$F_d = \frac{1}{1 + K_d \ S} \tag{24.226}$$

$$F_p = \frac{K_d \ S}{1 + K_d \ S} \tag{24.227}$$

where K_d is a partition coefficient in $m^3 \bullet g$ and S is the suspended solids concentration in the lake in $g \bullet m^3$. The partition coefficient quantifies the tendency of the toxic to associate the solid matter.

24.16.2. Phosphorus Mass Balance

A simple model for phosphorus mass balance was taken from Thomann and Mueller (1987). The model assumes: 1) completely mixed lake, 2) phosphorus limited, and 3) total phophorus can be used as a measure of trophic status. The first assumption ignores lake stratification and intensification of phytoplankton in the eplimnion. The second assumption is generally valid when non-point sources dominated and the third assumption implies that a relationship exists between total phosphorus and biomass.

Despite the severity of the assumptions, the following model has been shown to produce very useful results (Thomann and Mueller, 1987). The mass balance equation for total phosphorus is

$$V \frac{\Delta p}{\Delta t} = W - K_s \ p \, V - Q_p \tag{24.228}$$

where V is volume of the lake m^3, p is total phosphorus concentration in the lake in $mg \bullet l$, Q is outflow volume in l, K_s is overall loss rate of total phosphorus in day^{-1}, W is total phosphorus inflow in mg, and Δt is the time increment (one day). Vollenwelder (1975) and Chapra and Tarapchak (1976) discuss estimates of K_s and methods of estimating K_s from readily available parameters.

24.16.3. Relationship Between Total Phosphorus and Trophic Status

Empirical relationships from measured data have been postulated by several researchers including Bartsch and Gakstatter (1978), Rast and Lee (1978), Dillon and Rigler (1975), and Smith and Shapiro (1981). The relationships are similar and the Rast and Lee (1978) relationship was selected for use in the model.

$$\log_{10}(chl\ a) - 0.76\ \log_{10}(p) - 0.259 \qquad (24.229)$$

where chl is a chlorophyll a concentration in mg^{-1}.

Another measure of lake trophic status is secchi depth. Since water clarity is easily perceived by the general public, secchi depth is an important measure of water quality. Table 24.1 shows that relationship between secchi depth and public perception of water quality. The relationship between secchi depth and chlorophyll concentration is (Rast and Lee, 1978).

$$\log_{10}(S_d) = 0.473\ \log_{10}(chl\ a) + 0.803 \qquad (24.230)$$

where S_d is the secchi depth in m.

TABLE 24.1

Relationship between secchi depth and public perception of water quality of Annebessacook Lake, Maine (from USEPA, 1980).

Secchi depth (mm)	Public Perception of Water Quality
0 - 0.9	Gross pollution; lake is totally unusable for recreation
1 - 1.9	Algae blooms still evident; quality unacceptable for most uses
2 - 2.9	Some complaints of declining water quality; some impairement of water use
3 - 3.9	Satisfactory quality; no impairment of water use
4 - 4.9	Excellent water quality; a positive factor encouraging lake use
5 +	Exceptional quality for this lake

NOTATIONS

α	=	dimensionless parameter that expresses the proportion of total rainfall that occurs during t_c
α_{sp}	=	parameter dependent on maximum stream power for the reach
ρ	=	runoff coefficient expressing the watershed infiltration characteristics
σ	=	average channel slope ($m \cdot m^{-1}$)
λ	=	surface slope length in (m)
λ	=	surface slope length (km)
Δt	=	time interval (24 h)
ζ_i	=	dimensionless soil parameter (0-1)
θ_d	=	drainable porosity of the soil
δ	=	slope of the saturation vapor pressure curve (kPa°C^{-1})
γ	=	psychrometer constant (kPa°C^{-1})
β	=	fraction usually in the range of 0.6 to 0.9
μ	=	uniform random number (0.0-1.0)
ζ	=	parameter usually in the range of 1.0 to 2.0
Ω_R	=	scaling factor (0.0 to 1.0)
Ω_T	=	scaling factor (0.0 to 1.0)
f	=	water flow rate through the zone ($mm \cdot h^{-1}$)
Λ	=	use rate-depth parameter
δ	=	optimal temperature for the crop (°C)
η	=	evaporation coefficient
ω_1	=	parameter ($1.3X10^{-4}$)
ω_2	=	fairly constant parameter (0.9)
γ	=	density of water
Δ	=	time step (one day)
Δt	=	time increment (one day)
A	=	drainage area (ha)
A	=	area (m^2)
ab	=	abstraction coefficient
AB	=	albedo
AB_s	=	soil albedo

AIR = volume of irrigation water applied (mm)

AM = annual amplitude in daily average temperature (˚C)

AO = area represented by the rain gage (km^2)

a_r = unit channel intercept (m^3)

a_x = regression intercept ($m \bullet km^{-1}$)

B_1 = fraction of the growing season

B_{AG} = above ground biomass ($kg \bullet ha^{-1}$)

bcv = lagging factor for simulating residue and snow cover effects on surface

BD = soil bulk density ($t \bullet m^{-3}$)

BE = crop parameter for converting energy to biomass ($kg \bullet m^{-2} \bullet ha^{-1} \bullet MJ$)

bn = crop parameter expressing N concentration

B_p = daily potential increase in total biomass ($kg \bullet ha^{-1}$)

b_r = unit channel regression slope

B_{TOT} = total biomass on day i ($kg \bullet ha^{-1}$)

b_x = regression slope

C = crop management factor

C = total (dissolved and particulate) toxic concentration ($mg \bullet m^3$)

C = MUSLE factor for stream channel

c_a = sediment concentraion ($g \bullet m^{-3}$)

C_d = dissolved component

CDG = soil water residue decay factors

CHK = effective hydraulic conductivity of the channel alluvium (mm/hr)

chl = chlorophyll *a* concentration (mg^{-1})

c_i = inflow sediment concentration

CLA_i = clay content of soil layer i (0-1)

C_{LPP} = concentration of AP in soil layer P ($g \bullet t^{-1}$)

CM = soil cover (above ground biomass+residue) ($kg \bullet ha^{-1}$)

CN_1 = curve number for moisture condition 1 (dry)

CN_2 = curve number for moisture condition 2 (average)

CN_{2s} = handbook CN_2 value adjusted for slope

CN_3 = curve number for moisture condition 3 (wet)

C_{NB} = optimal N concentration of the crop

C_{NO3}	=	concentration of NO_3-N in the first layer
c_o	=	outflow sediment concentration
CON	=	concentration of organic N in the top soil layer ($g \cdot t^{-1}$)
CONSB	=	concentration of organic N in the sediment reaching the subbasin j outlet
C_p	=	concentration of P in the top soil layer ($g \cdot t^{-1}$)
C_p	=	particulate component
C_{PSB}	=	P concentration in the sediment reaching the subbasin j outlet ($g \cdot t^{-1}$)
cs	=	resevoir concentration t days after the value of cs_2 is obtained
cs_1, cs_2	=	reservoir sediment concentrations at the beginning and end of the day
cs_e	=	equilibrium sediment concentration (input to the model)
CV	=	sum of above ground biomass and crop residue ($kg \cdot ha^{-1}$)
CVM	=	minimum value of C
C'_{NO3}	=	concentration of NO_3-N at the end of the day
d	=	flow depth (m)
d	=	sediment particle diameter
d_{50}	=	median particle size of the inflow sediment
DA	=	drainage area into the reach (km^2)
DD	=	damping depth for the soil (mm)
DECR	=	daily decrease in residue
DEG	=	total degradation
DLAI	=	fraction of the growing season when LAI starts declining
DP	=	maximum damping depth for the soil (mm)
d_q	=	depth of flow
DR	=	sediment delivery ratio (subbasin sediment yield divided by gross sheet erosion)
DU	=	duration of streamflow (h)
DUR	=	rainfall duration (h)
EA	=	soil cover index
EFI	=	runoff ratio
ELEV	=	elevation of the site (m)
E_o	=	potential evaporation (mm)
E_p	=	predicted plant water evaporation rate ($mm \cdot d^{-1}$)
ER	=	enrichment ratio

ER_R	=	enrichment ratio for the channel routing from subbasin j to the channel outlet
E_s	=	potential soil water evaporation rate (mm•d^{-1})
ET	=	evapotranspiration (mm)
EV	=	evaporation (m^3)
FC	=	field capacity water content (mm)
FC_2	=	field capacity of the second soil layer
F_d	=	fraction of total contaminant in dissolved form
FFC	=	fraction of field capacity
FFC^*	=	depth weighted FFC value
FHU	=	function of crop stage
F_p	=	fraction of total contaminant in particulate form
FP	=	final amount of pesticide on the plant
FP_o	=	initial amount of pesticide on the plant
GC	=	fraction of the ground that is covered by plants
GP	=	final amount of pesticide on the ground
GP_o	=	initial amount of pesticide on the ground
H_i	=	hydraulic conductivity (mm•h^{-1})
HI	=	harvest index at harvest
HIA	=	adjusted harvest index
HIO	=	harvest index under favorable growing conditions
HLA	=	harvest index on day i
HLP	=	half life of the foliar residue (d)
HLS	=	half life for pesticide in the soil (d)
h_o	=	net radiation (MJ•m^{-2})
H_o	=	saturated thickness normal to the hillslope
HU	=	heat units (°C)
HUFH	=	heat unit factor that affects harvest index on day i and on the previous day i-1
HUI	=	heat unit index for day i
HV	=	latent heat of vaporization (MJ•kg^{-1})
i	=	rate of water input to the saturated zone (m^2•h^{-1})
i	=	day of the year

K	=	MUSLE factor for stream channel
K	=	soil erodibility factor
K	=	first order reaction coefficient (day^{-1})
k_d	=	P concentration in the sediment divided by that of the water ($m^3 \cdot t^{-1}$)
k_d	=	partitioning coefficient ($m^3 \cdot t^{-1}$)
k_r	=	decay factor ($m \cdot km^{-1}$)
k_s	=	decay constant
K_s	=	saturated conductivity
K_s	=	overall loss rate of total phosphorus (day^{-1})
L	=	channel length from the most distant point to the watershed outlet (km)
L	=	hillslope length (m)
L^*	=	channel length (km)
LAI	=	leaf area index
LAI_{mx}	=	maximum LAI potential for the crop
LAT	=	latitude of the site (degrees)
L_c	=	average channel flow length for the watershed (km)
L_{ca}	=	distance from the outlet along the channel to the watershed centroid (km)
L_s	=	saturated slope length (m)
LS	=	slope length and steepness factor
M	=	number of soil layers
n	=	Manning's roughness coefficient
N	=	number of subbasins
ND	=	number of days in a month
NDD	=	number of dry days
N_g	=	number of days in the growing season
NRD	=	number of rainy days
NWD	=	number of wet days
O	=	percolation rate ($mm \cdot d^{-1}$)
O_1	=	percolation from the first layer (mm)
O_{ci}	=	percolation rate for layer i ($mm \cdot d^{-1}$)
O_{i-1}	=	flow from the layer above ($mm \cdot d^{-1}$)
p	=	total phosphorus concentration in the lake ($mg \cdot l$)

P	=	percolation (mm)
P(W/D)	=	probability of a wet day following a dry day
P(W/W)	=	probability of a wet day after a wet day
PAEF	=	application efficiency factor
PAPE	=	effective amount of pesticide applied ($kg \cdot ha^{-1}$)
PAPR	=	actual amount of pesticide applied ($kg \cdot ha^{-1}$)
PAR	=	photosynthetic active radiation ($MJ \cdot m^{-2}$)
PB	=	barometric pressure (kPa)
PE	=	erosion control practice factor
PHU	=	potential heat units to mature the crop (°C)
P_{in}	=	inflow volume (m^3)
P_o	=	threshold volume for a unit channel (m^3)
PO	=	soil porosity
POFC	=	porosity-field capacity ratio
PSO1	=	pesticide solubility ($g \cdot m^{-3}$)
PSTC	=	concentration of adsorbed pesticide ($g \cdot t^{-1}$)
$PSTC_L$	=	pesticide concentration in the percolate ($g \cdot m^{-3}$)
$PSTC_w$	=	pesticide concentration in the water ($g \cdot t^{-1}$)
PSTL	=	amount of pesticide leached by *f*
PSTQ	=	pesticide loss in surface runoff ($kg \cdot ha^{-1}$)
PSTY	=	pesticide yield adsorbed to the sediment ($kg \cdot ha^{-1}$)
PW	=	probability of a wet day
q	=	flow rate
Q	=	surface runoff volume (mm)
q^*_c	=	average flow rate ($mm \cdot h^{-1}$)
q^*_p	=	peak rate per unit of rainfall ($mm \cdot h^{-1} \cdot mm^{-1}$)
q_ρ	=	peak runoff rate ($m^3 \cdot s^{-1}$)
q_c	=	average flow rate ($m^3 \cdot s^{-1}$)
QI	=	inflow during the day
q_{lat}	=	lateral flow ($m^3 \cdot h^{-1}$)
q_o	=	average flow rate ($m^3 \cdot s^{-1}$)
QO	=	outflow (m^3)

q_o^*	=	average flow rate from a 1 ha area ($mm \cdot h^{-1}$)
QOR	=	daily outflow (m^3)
q_p	=	peak flow rate ($m^3 \cdot s^{-1}$)
QR	=	return flow (mm)
QR_1	=	lateral flow from the first layer (mm)
q_{sat}	=	upward flow ($mm \cdot d^{-1}$)
QT	=	total water lost from the first layer (mm)
r	=	rainfall intensity for the watershed's time of concentration ($mm \cdot h^{-1}$)
R	=	daily rainfall in month k
RA	=	solar radiation ($MJ \cdot m^{-2}$)
RAD	=	daily mean solar radiation on dry days
RAMX	=	maximum daily solar radiation at midmonth
RAW	=	daily mean solar radiation on wet days ($MJ \cdot m^{-2}$)
RD_n	=	fraction of the root zone that contains roots on day n
REG	=	crop growth regulating factor
r_{ep}	=	peak rainfall excess rate ($mm \cdot h^{-1}$)
RF	=	reduction ratio (0-1)
RF_i	=	reduction factor
R_i	=	daily rainfall for rain gage i
r_p	=	peak rainfall rate ($mm \cdot h^{-1}$)
RR	=	principal spillway release rate ($m^3 \cdot s^{-1}$)
RSD	=	residue left after harvest ($kg \cdot ha^{-1}$)
RSDV	=	standard deviation of daily rainfall (mm)
R_{tc}	=	amount of rainfall during the watershed's time of concentration (mm)
RWT	=	fraction of total biomass partitioned to the root system on day i
RZ	=	root zone depth (mm)
s	=	land surface slope ($m \cdot m^{-1}$)
S	=	suspended solids concentration in the lake ($g \cdot m^3$)
S	=	drainable volume of water stored in the saturated zone ($m \cdot m^{-1}$)
S	=	average slope of the watershed
s_1	=	value of s associated with CN_1
s_3	=	CN_3 retention parameter

SA	=	reservoir surface area
S_c	=	slope of the channel
SCF	=	skew coefficient
SC_i	=	saturated conductivity for layer i ($mm{\bullet}h^{-1}$)
S_d	=	secchi depth (m)
SD	=	sun's declination angle (radians)
SDRA	=	standard deviation of daily solar radiation ($MJ{\bullet}m^{-2}$)
SDTMX	=	standard deviation of the daily maximum temperature
SED_{in}	=	sediment entering the reach
SEP	=	seepage (m^3)
SEV	=	potential soil evaporation for layer ℓ (mm)
SEV^*_ℓ	=	adjusted soil water evaporation estimate (mm)
s_f	=	retention parameter for frozen ground
SML	=	snowmelt rate ($mm{\bullet}d^{-1}$)
SND	=	standard normal deviate for day i
SNO	=	water content of snow cover (mm)
S_p	=	scaling parameter
SPT	=	snow pack temperature (°C)
SR_{DEP}	=	amount of sediment deposited in the reservoir
SR_i	=	total sediment in the reservoir
SR_{i-1}	=	total sediment in the reservoir on the previous day
SR_{in}	=	incoming sediment
SR_{out}	=	sediment transported in the sediment outflow
ST_2	=	soil temperature of the second soil layer
SUT	=	soil temperature residue decay factors
SW	=	amount of water stored in the top 10 mm of soil (mm)
S_W	=	water surface slope
SW	=	soil water content in the root zone (mm)
SW_2	=	soil water content of the second soil layer
SW_O	=	soil water contents at the beginning of the day (mm)
t	=	time (d)
T	=	mean daily air temperature (°C)

T(2)	=	temperature at the center of soil layer 2
T_b	=	crop-specific base temperature (°C)
t_c	=	watershed's time of concentration (h)
t_{cc}	=	time of concentration for channel flow (h)
t_{cs}	=	time of concentration for surface flow (h)
TD	=	daily mean maximum temperature for dry days
TE	=	extreme daily maximum temperature
TGB_{IDA}	=	bare soil surface temperature (°C)
TL	=	transmission loss (mm)
T_{mx}, T_{mn}	=	daily maximum and minimum air temperatures (°C)
T_s	=	temperature at the top of the snow pack
t_{sf}	=	t_c components attributed to shallow channel flow
TT	=	travel time (h)
TT_{ci}	=	crack flow travel time (h)
TW	=	daily mean maximum temperture for wet days (°C)
U_i	=	plant water use in layer i (mm)
UND_i	=	N demand of the crop (kg•ha^{-1})
u_{pi}	=	potential water use rate from layer i (mm•d^{-1})
v	=	velocity of flow at the outlet
V	=	volume (m^3)
V_f	=	fall velocity (m•h^{-1})
V_1, V_2	=	storage volumes at the beginning and end of the day
V_b	=	burial velocity (m•day)
v_c	=	average channel velocity (m•s^{-1})
V_c	=	velocity in the channel
V_d	=	diffusion mixing velocity (m•day)
VM	=	volume of the water stored in all ponds within a subbasin at the end of the day
VM_{mx}	=	maximum permanent pool storage of all ponds in the subbasin (m^3)
VM_O	=	volume of the water stored in all ponds within a subbasin at the beginning of the day
VNO3	=	amount NO_3-N lost from the first layer
V_r	=	resuspension velocity (m•day)

VR	=	volume of water in the reservoir (m^3)
VR_F	=	reservoir capacity at the emergency spillway crest (m^3)
VR_S	=	reservoir capacity at the principal spillway crest (m^3)
v_s	=	surface flow velocity ($m^3 \bullet s^{-1}$)
V_S	=	settling velocity (m•day)
v_{sf}	=	average shallow flow velocity ($km \bullet h^{-1}$)
V_V	=	volatilization mass transfer coefficient (m•day)
w	=	average width of flow (m)
W	=	total phosphorus inflow (mg)
w_1, w_2	=	shape parameters
WLV	=	above ground biomass minus crop yield at harvest ($kg \bullet ha^{-1}$)
WNO3	=	weight of NO_3-N contained in the layer at the end of the day
$WNO3_O$	=	weight of NO_3-N contained in the layer at the beginning of the day
WO	=	amount of pesticide washed off the plants by a rainstorm (mm)
WOF	=	washoff fraction for the particular pesticide
WP	=	wilting point water content for soil layer one (mm)
WS	=	water stress factor for day i
WSYF	=	crop parameter expressing drought sensitivity
x, y	=	coordinates of the rain gages (km)
x_1, x_2	=	parameters set by the upper and lower limits
Y	=	sediment yield ($t \bullet ha^{-1}$)
Y_B	=	sediment yield reaching the basin outlet from subbasin j ($t \bullet ha^{-1}$)
YLD	=	amount of the crop removed from the field ($kg \bullet ha^{-1}$)
YON	=	organic N runoff loss at the subbasin outlet ($kg \bullet ha^{-1}$)
YON_B	=	organic N runoff loss at the basin outlet ($kg \bullet ha^{-1}$)
YP	=	sediment phase of P loss in runoff ($kg \bullet ha^{-1}$)
YP_B	=	P yield at the basin outlet ($kg \bullet ha^{-1}$)
Y_{SB}	=	sediment yield at the subbasin outlet ($t \bullet ha^{-1}$)
YSP	=	soluble P lost in runoff ($kg \bullet ha^{-1}$)
z	=	weighting factor for estimating soil surface temperature on dry days
Z	=	depth to the bottom of soil layer ℓ (m)
ZM	=	distance from the bottom of the lowest soil layer to the surface (mm)

REFERENCES

Arnold, J.G. and C.O. Stockle. 1992. Simulation of supplemental irrigation from on-farm ponds. ASCE J. Irrig. Drainage Div. 117(3):408-424.

Arnold, J.G., M.D. Bircket, J. R. Williams, W.R. Smith, and H.N. McGill. 1987. Modeling the effects of urbanization on basin water yield and reservoir sedimentation. Water Resources Bull. 23(6):1101-1107.

Arnold, J.G., J.R. Williams, A.D. Nicks, and N.B. Sammons. 1990. SWRRB: A Basin Scale Simulation Model for Soil and Water Resources Management. Texas A&M Univ. Press, College Station.

Arnold, J.G., J.R. Williams, R.H. Griggs, and N.B. Sammons. 1991. SWRRBWQ - A basin scale model for assessing management impacts on water quality. (Model Documentation / User Manual - Draft)

Bagnold, R.A. 1977. Bedload transport in natural rivers. Water Resources Res. 13(2):303-312.

Bartsch, A.F. and J. H. Gakstatter. 1978. Management decisions for lake systems on a survey of trophic status, limiting nutrients, and nutrient loadings. pp. 371-394 *In* American-Soviet Symp. on Use of Mathematical Models to Optimize Water Quality Management. USEPA A-600/9-78-024.

Chapra, S.C. 1989. Water quality modeling of toxic organics in lakes. CADSWES Working Paper No. 4. Univ. Colorado, Boulder.

Chapra, S.C. and S.J. Tarpchak. 1976. A chlorophyll a model and its relationship to phosphorus loading plots for lakes. Water Resources Res. 12(6):1260-1264.

Dillon, D.J. and F.H. Rigler. 1975. The phosphorus-chlorophyll relationship in lakes. Limnol. Oceanogr. 19(4):767-773.

Doorenbos, J. and A.H. Kassam. 1979. Yield response to water. Irrigation and Drainage Paper 33. Food Agric. Org. United Nations, Rome.

Hargreaves, G.H. and Z.A. Samani. 1985. Reference crop evapotranspiration from temperature. Applied Engr. Agric. 1:96-99.

Hershfield, D.M. 1961. Rainfall frequency atlas of the United States for durations from 30 minutes to 24 hours and return periods from 1 to 100 years. U.S. Dept. Commerce Tech. Paper No. 40.

Holst, R. and L. Kutney. 1978. Pesticide runoff simulator user's manual. Contract No. 68-01-3840.

Knisel, W.G. 1980. CREAMS, A field scale model for chemicals, runoff, and erosion from agricultural management systems. U.S. Dept. Agric. Conserv. Res. Rept. No. 26.

Lane, L.J. 1982. Distributed model for small semi-arid watersheds. ASCE J. Hydr. Div. 108(HY10):1114-1131.

Leonard, R.A., W.G. Knisel, and D.A. Still. 1987. GLEAMS: Groundwater loading effects on agricultural management systems. Trans. ASAE 30(5):1403-1428.

McElroy, A.D., S.Y. Chiu, J.W. Nebgen, and others. 1976. Loading functions for assessment of water pollution from nonpoint sources. Environ. Prot. Tech. Serv., EPA 600/2-76-151.

Menzel, R.G. 1980. Enrichment ratios for water quality modeling. pp. 486-492 In W.G. Knisel, ed., CREAMS, A field scale model for chemicals, runoff, and erosion from agricultural management systems. U.S. Dept. Agric. Conserv. Res. Rept. No. 26.

Monsi, M. and T. Saeki. 1953. Uber den Lictfaktor in den Pflanzengesellschaften und sein Bedeutung fur die Stoffproduktion. Japan J. Bot. 14:22-52.

Monteith, J.L. 1977. Climate and the efficiency of crop production in Britain. Phil. Trans. Res. Soc. London Ser. B. 281:277-329.

Nicks, A.D. 1974. Stochastic generation of the occurrence, pattern, and location of maximum amount of daily rainfall. pp. 154 - 171 *In* Proc. Symp. Statistical Hydrology, Aug. - Sept. 1971, Tucson, AZ. U.S. Dept. Agric., Misc. Publ. No. 1275.

Nicks, A.D. and F.A. Igo. 1980. A depth-area-duration model of storm rainfall in the southern plains. Water Resources Res. 16(5):939-945.

Priestley, C.H.B. and R.J. Taylor. 1972. On the assessment of surface heat flux and evaporation using large-scale parameters. Mon. Weather Rev. 100:81-92.

Rast, W. and G.F. Lee. 1978. Summary analysis of the North American OECD eutrophication project: Nutrient loading-lake response relationships and trophic state idices. USEPA-600/3-78-008.

Richardson, C.W. 1981. Stochastic simulation of daily precipitation, temperature, and solar radiation. Water Resources Res. 17(1):182-19O.

Richardson, C.W. 1982. Dependence structure of daily temperature and solar radiation. Trans. ASAE 25(3):735-739.

Ritchie, J.T. 1972. A model for predicting evaporation from a row crop with incomplete cover. Water Resources Res. 8:1204-1213.

Singer, M.P., F.D. Arnold, R.H. Cole, J.G. Arnold, and J.R. Williams. 1988. Use of SWRRB computer model for the National Pollutant Discharge Inventory. pp. 119-131 *In* W.L. Lyle and T.J. Hoban (eds.) Proc. Symp on Coastal Water Resources.

Sloan, P.G., I.D. Morre, G.B. Coltharp, and J.D. Eigel. 1983. Modeling surface and subsurface stormflow on steeply-sloping forested watersheds. Water Resources Inst. Report 142. Univ. Kentucky, Lexington.

Smith, V.H. and J. Shapiro. 1981. Chlorophyll - phosphorus relations in indivdual lakes, their importance to lake restoration studies. Environ. Sci. Tech. 15:444-451.

Thormann, R.V. and J.A. Mueller. 1987. Principles of Surface Water Quality Modeling and Control. Harper & Row, Inc., New York.

U.S. Department of Agriculture, Soil Conservation Service. 1957. Farm Pond Survey--Texas.

U.S. Department of Agriculture, Soil Conservation Service. 1972. National Engineering Handbook, Hydrology Section 4, Chapters 4-1O.

U.S. Department of Agriculture, Soil Conservation Service. 1983. National Engineering Handbook, Hydrology Section 4, Chapters 19.

U.S. Department of Agriculture, Soil Conservation Service. 1986. Urban hydrology for small watersheds. Tech. Release 55.

U.S. Department of Commerce. 1979. Maximum short duration rainfall. National Summary, Climatic Data.

Vollenwelder, R.A. 1975. Input-output models with special reference to the phosphorus loading concept in limnology. Schweiz. Z. Hydrol. 37:53-83.

Williams, J.R. 1975. Sediment yield prediction with universal equation using runoff energy factor. U.S. Dept. Agric., Agric. Res. Serv., ARS-S-40.

Williams, J.R. 1980. SPNM, a model for predicting sediment, phosphorus, and nitrogen from agricultural basins. Water Resources Bull. 16(5):843-848.

Williams, J.R. and H.D. Berndt. 1977. Sediment yield prediction based on watershed hydrology. Trans. ASAE 20(6):1100-1104.

Williams, J.R. and R.W. Hann. 1978. Optimal operation of large agricultural watersheds with water quality constraints. Texas Water Resources Institute, Texas A&M Univ., Tech. Rept. No. 96.

Williams, J.R., C.A. Jones, and P.T. Dyke. 1984. A modeling approach to determining the relationship between erosion and soil productivity. Trans. ASAE 27:129-144.

Williams, J.R., A.D. Nicks, and J.G. Arnold. 1985. Simulator for water resources in rural basins. ASCE J. Hydr. Engr. 111(6):970-986.

Wischmeier, W.H. and D.D. Smith. 1978. Predicting rainfall erosion losses, a guide to conservation planning. U.S. Dept. Agric., Agric. Handbook No. 537.

Chapter 25

THE EPIC MODEL

J.R. Williams

25.1. INTRODUCTION

The Erosion-Productivity Impact Calculator (EPIC) (Williams et al., 1984) model was developed to assess the effect of soil erosion on soil productivity. It was used for that purpose as part of the 1985 RCA (1977 Soil and Water Resources Conservation Act) analysis. Since the RCA application, the model has been expanded and refined to allow simulation of many processes important in agricultural management (Sharpley and Williams, 1990).

EPIC is a continuous simulation model that can be used to determine the effect of management strategies on agricultural production and soil and water resources. The drainage area considered by EPIC is generally a field-sized area, up to 100 ha (weather, soils, and management systems are assumed to be homogeneous). The major components in EPIC are weather simulation, hydrology, erosion-sedimentation, nutrient cycling, pesticide fate, plant growth, soil temperature, tillage, economics, and plant environment control.

Recently, most of the EPIC model development has been focused on problems involving water quality and global climate/CO_2 change. Example additions include the GLEAMS (Leonard et al., 1987) pesticide fate component, nitrification and volatilization submodels, a new more physically based wind erosion component, optional SCS technology for estimating peak runoff rates, newly developed sediment yield equations, and mechanisms for simulating CO_2 effects on crop growth and water use. These and other less significant developments extend EPIC's capabilities to deal with a wide variety of agricultural management problems.

Example applications include: (1) 1985 RCA analysis; (2) 1988 drought assessment; (3) soil loss tolerance tool; (4) Australian sugarcane model (AUSCANE); (5) pine tree growth simulator; (6) global climate change analysis (effect of CO_2, temperature, and precipitation change on runoff and crop yield); (7) farm level planning; (8) five-nation EEC assessment of environmental/agricultural policy alternatives; (9) Argentine assessment of erosion/productivity; (10) USDA-Water Quality Demonstration Project Evaluation; (11) N leaching index national analysis.

The purpose here is to fully describe the new more comprehensive EPIC model and to present example applications. Descriptions of the EPIC components and the mathematic relationships used to simulate the processes involved follow.

25.2. HYDROLOGY

25.2.1. Surface Runoff

The runoff model simulates surface runoff volumes and peak runoff rates, given daily rainfall amounts. Runoff volume is estimated by using a modification of the Soil Conservation Service (SCS) curve number technique (U.S. Department of Agriculture, Soil Conservation Service 1972). The technique was selected for use because (a) it is reliable and has been used for many years in the United States; (b) it is computationally efficient; (c) the required inputs are generally available; and (d) it relates runoff to soil type, land use, and management practices. The use of readily available daily rainfall data is a particularly important attribute of the curve number technique because for many locations, rainfall data with time increments of less than 1 day are not available. Also, rainfall data manipulations and runoff computations are more efficient for data taken daily than at shorter intervals.

There are two options for estimating the peak runoff rate -- the modified Rational formula and the SCS TR-55 method (USDA-SCS, 1986). A stochastic element is included in the Rational equation to allow realistic simulation of peak runoff rates, given only daily rainfall and monthly rainfall intensity.

Runoff Volume

Surface runoff is predicted for daily rainfall by using the SCS curve number equation (U.S. Department of Agriculture, Soil Conservation Service 1972)

$$\begin{aligned} Q &= \frac{(R-0.2s)^2}{R+0.8s}, & R &> 0.2s \\ Q &= 0.0, & R &\le 0.2s \end{aligned} \tag{25.1}$$

where Q is the daily runoff, R is the daily rainfall, and s is a retention parameter (see "Notations" section). The retention parameter, s, varies (a) among watersheds because soils, land use, management, and slope all vary and (b) with time because of changes in soil water content. The parameter s is related to curve number (CN) by the SCS equation (U.S. Department of Agriculture, Soil Conservation Service 1972)

$$s = 254\left(\frac{100}{CN} - 1\right) \tag{25.2}$$

The constant, 254, in Eq. (25.2) gives s in mm. Thus, R and Q are also expressed in mm. CN_2--the curve number for moisture condition 2, or average curve number--can be obtained easily for any area by using the SCS hydrology handbook (U.S. Department of Agriculture, Soil Conservation

Service 1972). The handbook tables consider soils, land use, and management. Assuming that the handbook CN_2 value is appropriate for a 5% slope, we developed the following equation for adjusting that value for other slopes.

$$CN_{2s} = \frac{1}{3}(CN_3 - CN_2)[1 - 2\exp(-13.86S)] + CN_2 \tag{25.3}$$

where CN_{2s} is the handbook CN_2 value adjusted for slope, CN_3 is the curve number for moisture condition 3 (wet), and S is the average slope of the watershed. Values of CN_1, the curve number for moisture condition 1 (dry), and CN_3 corresponding to CN_2 are also tabulated in the handbook. For computing purposes, CN_1 and CN_3 were related to CN_2 with the equations

$$CN_1 = CN_2 - \frac{20(100 - CN_2)}{100 - CN_2 + \exp[2.533 - 0.0636(100 - CN_2)]} \tag{25.4}$$

$$CN_3 = CN_2 \exp[0.00673(100 - CN_2)] \tag{25.5}$$

Fluctuations in soil water content cause the retention parameter to change according to the equation

$$s = s_1\left(1 - \frac{FFC}{FFC + \exp[w_1 - w_2(FFC)]}\right) \tag{25.6}$$

where s_1 is the value of s associated with CN_1, FFC is the fraction of field capacity, and w_1 and w_2 are shape parameters. FCC is computed with the equation

$$FFC = \frac{SW - WP}{FC - WP} \tag{25.7}$$

where SW is the soil water content in the root zone, WP is the wilting point water content (1,500 kPa for many soils) and FC is the field capacity water content (33 kPa for many soils).

Values for w_1 and w_2 are obtained from a simultaneous solution of Eq. (25.6) according to the assumptions that $s=s_2$ when FCC=0.6 and $s=s_3$, when (SW - FC)/(PO - FC) = 0.5

$$w_1 = \ln\left(\frac{60.}{1.-s_2/s_1} - 60.\right) + 60.w_2 \tag{25.8}$$

$$POFC = 100. + 50. \left(\frac{\sum_{\ell-1}^{M} (PO_\ell - FC_\ell)}{\sum_{\ell-1}^{M} (FC_\ell - WP_\ell)} \right) \tag{25.9}$$

where s_3 is the CN_3 retention parameter and the porosity-field capacity ratio POFC is computed with the equation

$$w_2 = \frac{\ln\left(\frac{60.}{(1. - s_2/s_1)} - 60. \right) - \ln\left(\frac{POFC}{1. - s_2/s_1} - POFC \right)}{POFC - 60.} \tag{25.10}$$

where PO is the porosity of soil layer ℓ.

Equations (25.8) and (25.9) assure that CN_1 corresponds with the wilting point and that the curve number cannot exceed 100.

The FFC value obtained in Eq. (25.7) represents soil water uniformly distributed through the top 1.0 m of soil. Runoff estimates can be improved if the depth distribution of soil water is known. For example, water distributed near the soil surface results in more runoff than the same volume of water uniformly distributed throughout the top meter of soil. Also, a soil surface associated with a uniform distribution of soil water results in more runoff than a soil surface that is dry. Since EPIC estimates water content of each soil layer daily, the depth distribution is available. The effect of depth distribution on runoff is expressed in the depth weighting function

$$FFC^* = \frac{\sum_{\ell-1}^{M} FFC_\ell \frac{(Z_\ell - Z_{\ell-1})}{Z_\ell}}{\sum_{\ell=1}^{M} \frac{Z_\ell - Z_{\ell-1}}{Z_\ell}}, \quad Z_\ell \leq 1.0\text{m} \tag{25.11}$$

where FFC* is the depth weighted FFC value for use in Eq. (25.6), Z is the depth in m to the bottom of soil layer ℓ, and M is the number of soil layers. Equation (25.11) performs two functions: (1) it reduces the influence of lower layers because FFC_ℓ is divided by Z_ℓ and (2) it gives proper weight to thick layers relative to thin layers because FFC is multiplied by the layer thickness.

There is also a provision for estimating runoff from frozen soil. If the temperature in the second soil layer is less than 0°C, the retention parameter is reduced by using the equation

$$s_f = 0.1s \tag{25.12}$$

where s_f is the retention parameter for frozen ground. Equation (25.12) increases runoff for frozen soils but allows significant infiltration when soils are dry.

The final step in estimating the runoff volume is an attempt to account for uncertainty. The retention parameter or curve number estimate is based on land use, management, hydrologic soil group, land slope, and soil water content and distribution and is adjusted for frozen soil. However, many complex natural processes and artificial diversions that affect runoff are not accounted for in the model. Thus, the final curve number estimate is generated from a triangular distribution to account for this uncertain variation. The mean of the triangle is the best estimate of curve number based on using Eqs. (25.2), (25.3), (25.6), (25.7), (25.11), and (25.12). The extremes are ±5 curve numbers from the mean. The generated curve number is substituted into Eq. (25.2) to estimate runoff with Eq. (25.1).

Peak Runoff Rate

EPIC contains two methods for estimating peak runoff rate--the modified Rational formula and the SCS TR-55 method (USDA-SCS, 1986).

Rational Equation Method. The rational equation can be written in the form

$$q_{\rho} = (\rho)(r)(A)/360 \qquad (25.13)$$

where q_ρ is the peak runoff rate in $m^3 \cdot s^{-1}$, $_\rho$ is a runoff coefficient expressing the watershed infiltration characteristics, r is the rainfall intensity in $mm \cdot h^{-1}$ for the watershed's time of concentration, and A is the drainage area in ha. The runoff coefficient can be calculated for each storm if the amount of rainfall and runoff are known

$$\rho = \frac{Q}{R} \qquad (25.14)$$

Since R is input and Q is computed with Eq. (25.1), ρ can be calculated directly. Rainfall intensity can be expressed with the relationship

$$r = \frac{R_{tc}}{t_c} \qquad (25.15)$$

where R_{tc} is the amount of rainfall in mm during the watershed's time of concentration, t_c in h. The value of R_{tc} can be estimated by developing a relationship with total R. The Weather Service's TP-40 (Hershfield 1961) provides accumulated rainfall amounts for various durations and frequencies. Generally, R_{tc} and R_{24} (24-h duration is appropriate for the daily time step model) are proportional for various frequencies. Thus,

$$R_{tc} = \alpha R_{24} \qquad (25.16)$$

where α is a dimensionless parameter that expresses the proportion of total rainfall that occurs during t_c.

The peak runoff equation is obtained by substituting Eqs. (25.14), (25.15), and (25.16) into Eq. (25.13).

$$q_p = \frac{(\alpha)(Q)(A)}{360(t_c)} \tag{25.17}$$

The time of concentration can be estimated by adding the surface and channel flow times

$$t_c = t_{cc} + t_{cs} \tag{25.18}$$

where t_{cc} is the time of concentration for channel flow and t_{cs} is the time of concentration for surface flow in h. The t_{cc} can be computed by using the equation

$$t_{cc} = \frac{L_c}{v_c} \tag{25.19}$$

where L_c is the average channel flow length for the watershed in km and v_c is the average channel velocity in $m \cdot s^{-1}$. The average channel flow length can be estimated by using the equation

$$L_c = \sqrt{(L)(L_{ca})} \tag{25.20}$$

where L is the channel length from the most distant point to the watershed outlet in km and L_{ca} is the distance from the outlet along the channel to the watershed centroid in km. Average velocity can be estimated by using Manning's equation and assuming a trapezoidal channel with 2:1 side slopes and a 10:1 bottom width/depth ratio. Substitution of these estimated and assumed values gives

$$t_{cc} = \frac{\sqrt{(L)(L_{ca})}(n)^{0.75}}{\left(0.489(q_c)^{0.25}(\sigma)^{0.375}\right)} \tag{25.21}$$

where n is Manning's n, q_c is the average flow rate in $m^3 \cdot s^{-1}$, and σ is the average channel slope in $m \cdot m^{-1}$. Assuming that L_{ca}=0.5L and converting units (L from m to Km, t_{cc} from s to h, and q_c from $m^3 \cdot s^{-1}$ to $mm \cdot h^{-1}$) gives the equation

$$t_{cc} = \frac{1.75(L^*)(n)^{0.75}}{(q_c^* \bullet A)^{0.25}(\sigma)^{0.375}} \tag{25.22}$$

where L^* is the channel length in Km and q^*_c is the average flow rate in $mm \cdot h^{-1}$. The average flow rate is obtained from the estimated average flow rate from a unit source in the watershed (1 ha area) and the relationship

$$q_c^* = q_o^* A^{-0.5} \tag{25.23}$$

where q_o^* is the average flow rate from a 1 ha area in mm·h^{-1}. The unit source flow rate is estimated with the equation

$$q_o^* = \frac{Q}{DUR} \tag{25.24}$$

where DUR, the rainfall duration in h is calculated using the equation

$$DUR = \frac{4.605}{-2\ln(1.-\alpha_{.5})} \tag{25.25}$$

where $\alpha_{.5}$ is computed with Eq. (25.16) using $R_{.5}$ instead of R_{tc}. Equation (25.25) is derived assuming rainfall intensity is exponentially distributed. Details of the procedure for estimating $\alpha_{.5}$ and DUR are given in the water erosion section of this chapter.

Substituting Eq. (25.24) into Eq. (25.22) gives the final equation for t_{cc}

$$t_{cc} = \frac{1.75(L^*)(n)^{0.75}}{(q_o^*)^{0.25}(A)^{0.125}(\sigma)^{0.375}} \tag{25.26}$$

A similar approach is used to estimate t_{cs}:

$$t_{cs} = \frac{\lambda}{v_s} \tag{25.27}$$

where λ is the surface slope length in m and v_s is the surface flow velocity in m·s^{-1}.

Applying Manning's equation to a strip 1 m wide down the slope length, assuming flow is concentrated into a small trapezoidal channel with 1:1 side slopes and 5:1 bottom width-depth ratio gives the velocity equation

$$v_S = \frac{0.8375 d^{2/3} s^{0.5}}{n} \tag{25.28}$$

where v_s is the flow velocity in m^3·s^{-1}, d is flow depth in m, s is the land surface slope in m·m^{-1}, and n is Manning's roughness coefficient for the surface.

The average flow depth, d, can be calculated from Manning's equation as a function of flow rate

$$d = \left(\frac{q_o \bullet n}{5.025 s^{0.5}}\right)^{0.375} \tag{25.29}$$

where q_o is the average flow rate in m^3·s^{-1}. Substituting Eqs. (25.28) and (25.29) into Eq. (25.27) gives

$$t_{cs} = \frac{0.0216(\lambda \bullet n)^{0.75}}{(q_o^*)^{0.25}(s)^{0.375}} \tag{25.30}$$

To properly evaluate α, variation in rainfall patterns must be considered. For some short duration storms, most or all the rain occurs during tc causing α to approach its upper limit of 1.0. Other storms of uniform intensity cause α to approach a minimum value. All other patterns cause higher α values than the uniform pattern, because r_{tc} is greater than r_{24} for all patterns except the uniform. By substituting the products of intensity and time into Eq. (25.16), an expression for the minimum value of α, α_{mn}, is obtained

$$\alpha_{mn} = \frac{t_c}{24} \tag{25.31}$$

Thus, α ranges within the limits

$$\frac{t_c}{24} \leq \alpha \leq 1.0$$

Although confined between limits, the value of α is assigned with considerable uncertainty when only daily rainfall and simulated runoff amounts are given. Thus, α is generated from a triangular distribution with the base ranging from $t_c/24$ to 1.0.

The peak of the α distribution changes monthly because of seasonal differences in rainfall intensities. The Weather Service (U.S. Department of Commerce, 1979) provides information on monthly maximum rainfall intensities that can be used to estimate the peak α for each month.

TR-55 Method. The SCS TR-55 Method for estimating peak runoff rate has been described in detail (USDA-SCS, 1986). Only the essential equations are presented here. The peak runoff rate is dependent on the rainfall distribution and amount, the runoff curve number, and the time of concentration as expressed in the equation

$$q_p = (q_p^*)(R) \tag{25.32}$$

where q_p is the peak rate in $mm{\cdot}h^{-1}$, q^*_p is the peak rate per unit of rainfall in $mm{\cdot}h^{-1}{\cdot}mm^{-1}$, and R is the storm rainfall amount in mm. A set of curves are available (USDA-SCS, 1986) for estimating q^*_p given rainfall distribution (SCS Type I, IA, II, or III), the runoff curve number, and the watershed time of concentration. A map of the U.S. is given (USDA-SCS, 1986) to determine appropriate storm type. The curves relating q^*_p, t_c, and CN have been fitted with a seventh degree polonomial for use in computer solutions.

Time of concentration is estimated with the equation

$$t_c = t_{cc} + t_{cs} + t_{sf} \tag{25.33}$$

where t_c is the watershed time of concentration in h and t_{cc}, t_{cs}, and t_{sf} are the t_c components attributed to channel flow, surface flow, and shallow channel flow. The channel component is computed with the equation

$$t_{cc} = \frac{(L - \lambda - L_{sf})n}{3.6 d^{0.67} \sigma^{0.5}} 34 \tag{25.34}$$

where t_{cc} is the channel time of concentration in h, L is the channel length from the most distant point to the watershed outlet in Km, λ is the surface slope length in Km, L_{sf} is the shallow flow length in Km, n is Manning's roughness coefficient, d is the average channel flow depth in m, and σ is the channel slope in m·m^{-1}. The shallow flow component of t_c is estimated with the equation

$$t_{sf} = \frac{L_{sf}}{v_{sf}} \tag{25.35}$$

where v_{sf} the average shallow flow velocity in Km·h^{-1} is estimated with the equation

$$v_{sf} = 17.7S^{0.5} \le 2.19 Km \bullet h^{-1} \tag{25.36}$$

where S is surface flow slope in m·m^{-1}. The length of shallow flow, L_{sf} is estimated with the equations

$$\begin{aligned} L_{sf} &= 0.05 && L > 0.1\ Km \\ &= L - 0.05 && 0.05 < L < 0.1\ Km \\ &= 0.0 && L < 0.05\ Km \end{aligned} \tag{25.37}$$

Note that t_{sf} is constrained to less than 0.1h even with S = 0.001. Thus, t_{sf} is generally insignificant. The surface flow component of t_c is estimated with the equation

$$t_{cs} = \frac{0.0913(\lambda \bullet n)^{0.8}}{S^{0.4} R^{0.5}} \tag{25.38}$$

25.2.2. Percolation

The EPIC percolation component uses a storage routing technique to simulate flow through soil layers. Flow from a soil layer occurs when soil water content exceeds field capacity. Water drains from the layer until the storage returns to field capacity. The reduction in soil water is simulated with the routing equation

$$SW_\ell = (SW_{o\ell} - FC_\ell)\exp(-\Delta t/TT_\ell) + FC_\ell \tag{25.39}$$

where SW and SW_0 are the soil water contents at the end and the start of time interval Δt (24 h) and TT is travel time through layer ℓ in h.

Thus, daily percolation can be computed by taking the difference between SW and SW_0.

$$0_\ell = (SW_{o\ell} - FC_\ell)[1.0 - \exp(-\Delta t/TT_\ell)] \tag{25.40}$$

where O is the percolation rate for layer ℓ in mm·d^{-1}.

Travel time through a layer is computed with the linear storage equation

$$TT_\ell = \frac{PO_\ell - FC_\ell}{SC_\ell} \tag{25.41}$$

where PO is the porosity in mm, FC is field capacity in mm, and SC is saturated conductivity in mm·h^{-1}.

The routing process is applied from the soil surface layer by layer through the deepest layer. Since the saturated conductivity of some layers may be much lower than that of others, the routing scheme can lead to an impossible situation (porosity of low saturated conductivity layers may be exceeded). For this reason, a back pass is executed from the bottom layer to the surface. If a layer's porosity is exceeded, the excess water is transferred to the layer above. This process continues through the top layer.

There is also a provision for upward movement when a lower layer exceeds field capacity. Movement from a lower layer to an adjoining upper layer is regulated by soil water to field capacity ratios using the equation

$$UF_\ell = \left(SW_\ell - FC_\ell\right)\left(1. - \exp\left(0.05\left(\frac{SW_{\ell-1}}{FC_{\ell-1}} - \frac{SW_\ell}{FC_\ell}\right)\right)\right) \tag{25.42}$$

Saturated conductivity may be input or estimated for each soil layer by using the equation

$$SC_\ell = \frac{12.7\ (100 - CLA_\ell)(SS_\ell)}{100 - CLA_\ell + \left[11.45 - 0.097\ (100 - CLA_\ell)\right]} \tag{25.43}$$

where CLA is the percentage of clay in soil layer ℓ and SS is the soil strength factor (described in the Growth Constraints section of this chapter).

Percolation is also affected by freezing temperature. Water can flow into a frozen layer but is not allowed to percolate from the layer.

25.2.3. Lateral Subsurface Flow

Lateral subsurface flow is calculated simultaneously with percolation. The lateral flow function (similar to Eq. (25.40)) is expressed in the equation

$$QH_\ell = (SW_{o\ell} - FC_\ell)[1.0 - \exp(-24./TT_{H\ell})] \tag{25.44}$$

where QH is the lateral flow rate for soil layer ℓ in mm·d^{-1} and $TT_{H\ell}$ is the lateral flow travel time in d.

The lateral flow travel time is estimated for each soil layer by using the equation

$$TT_{H\ell} = (TT_\ell)/(S) \tag{25.45}$$

where S is the land surface slope in m·m^{-1}.

Equations (25.40) and (25.44) must be solved simultaneously to avoid one process dominating the other, simply because the solution occurs first. Thus, an equation for the sum of percolation and lateral flow is written as

$$0_{\ell} + QH_{\ell} = (SW_{o\ell} - FC_{\ell})(1.0 - \exp\left(\frac{-24.}{TT_{\ell}}\right)\exp\left(\frac{-24.}{TT_{H\ell}}\right) \tag{25.46}$$

Taking the ratio of QH/0 and substituting the resulting QH into Eq. (25.46) leads to the equation

$$0 + 0\left(\frac{1.0 - \exp\left(\frac{-24.}{TT_{H\ell}}\right)}{1.0 - \exp\left(\frac{-24.}{TT_{\ell}}\right)}\right) = (SW_{0\ell} - FC_{\ell})\left(1.0 - \exp\left(\frac{-24.}{TT_{\ell}}\right)\exp\left(\frac{-24.}{TT_{H\ell}}\right)\right) \tag{25.47}$$

Solving for 0 gives the final percolation equation

$$0 = \frac{\left(SW_{0\ell} - FC_{\ell}\right)\left(1. - \exp\left(\frac{-24.}{TT_{\ell}}\right)\exp\left(\frac{-24.}{TT_{H\ell}}\right)\right)}{1. + \left(\frac{1. - \exp\left(\frac{-24.}{TT_{H\ell}}\right)}{1. - \exp\left(\frac{-24.}{TT_{\ell}}\right)}\right)} \tag{25.48}$$

The calculated 0 value is substituted into Eq. (25.46) to obtain the final estimate of QH.

Equations (25.46) and (25.48) are used to partition flow between vertical and lateral components. The lateral flow component is estimated by considering subsurface flow from a hillside. This is generally adequate for small areas composed of a hillside landscape. However, for larger more complex areas, there is need to lag lateral flow to simulate return flow. EPIC uses an additional routing equation to accomplish the proper lag.

$$QR = \left(1. - \exp\left(\frac{-1.}{TT_R}\right)\right)\sum_{\ell=1}^{M} QH_{\ell} \tag{25.49}$$

where QR is the return flow volume in mm and TT_R is the return flow travel time in d (Input to EPIC).

25.2.4. Evapotranspiration

The model offers four options for estimating potential evaporation--Hargreaves and Samani (1985), Penman (1948), Priestley-Taylor (Priestley-

Taylor 1972), and Penman-Monteith (Monteith, 1965). The Penman and Penman-Monteith methods require solar radiation, air temperature, wind speed, and relative humidity as input. If wind speed, relative humidity, and solar radiation data are not available, the Hargreaves or Priestley-Taylor methods provide options that give realistic results in most cases.

The model computes evaporation from soils and plants separately, as described by Ritchie (1972). Potential soil water evaporation is estimated as a function of potential evaporation and leaf area index (LAI, area of plant leaves relative to the soil surface area). Actual soil water evaporation is estimated by using exponential functions of soil depth and water content. Plant water evaporation is simulated as a linear function of potential evaporation and leaf area index.

Potential Evaporation

The Penman (1948) option for estimating potential evaporation is based on the equation

$$E_0 = (\frac{\delta}{\delta+\gamma})(\frac{h_0 - G}{HV}) + (\frac{\gamma}{\delta+\gamma}) f(V)(e_a - e_d) \tag{25.50}$$

where E_0 is the potential evaporation in mm, δ is the slope of the saturation vapor pressure curve in kPa°C^{-1}, γ is a psychrometer constant in kPa°C^{-1}, h_0 is the net radiation in MJ·m^{-2}, G is the soil heat flux in MJ·m^{-2}, HV is the latent heat of vaporization in MJ·kg^{-1}, f(V) is a wind speed function in mm·d^{-1}·kPa^{-1}, e_a is the saturation vapor pressure at mean air temperature in kPa, and e_d is the vapor pressure at mean air temperature in kPa. The latent heat of vaporization is estimated with the temperature function

$$HV = 2.50 - 0.0022T \tag{25.51}$$

where T is the mean daily air temperature in °C. The saturation vapor pressure is also estimated as a function of temperature by using the equation

$$e_a = 0.1\exp\left(54.88 - 5.03\ \ln(T+273) - \frac{6791}{T+273}\right) \tag{25.52}$$

The vapor pressure is simulated as a function of the saturation value and the relative humidity:

$$e_d = (e_a)(RH) \tag{25.53}$$

where RH is the relative humidity expressed as a fraction. The slope of the saturation vapor pressure curve is estimated with the equation

$$\delta = \left(\frac{e_a}{T+273}\right)\left(\frac{6791}{T+273} - 5.03\right) \tag{25.54}$$

The psychrometer constant is computed with the equation

$$\gamma = 6.6X10^{-4}\, PB \tag{25.55}$$

where PB is the barometric pressure (kPa). The barometric pressure is estimated as a function of elevation by using the equation

$$PB = 101 - 0.0115ELEV + 5.44X10^{-7}\, ELEV^2 \tag{25.56}$$

where ELEV is the elevation of the site in m. The soil heat flux is estimated by using air temperature on the day of interest plus 3 days prior.

$$G = 0.12\left(T_i - \left(\frac{T_{i-1} + T_{i-2} + T_{i-3}}{3}\right)\right) \tag{25.57}$$

where T is the mean daily air temperature on day i in °C. Since G is usually relatively small, it is assumed zero in EPIC. Solar radiation is adjusted to obtain net radiation by using the equation

$$h_{oi} = RA_i(1.0 - AB_i) - RAB_i\left(\frac{0.9RA_i}{(RAMX_i)} + 0.1\right) \tag{25.58}$$

where RA is the solar radiation in $MJ \cdot m^{-2}$, AB is albedo, RAB is the net outgoing long wave radiation in $MJ \cdot m^{-2}$ for clear days, and RAMX is the maximum solar radiation possible in $MJ \cdot m^{-2}$ for the location on day i. The RAB value is estimated with the equation

$$RAB_i = 4.9X10^{-9}(0.34 - 0.14\sqrt{e_d})(T_i + 273)^4 \tag{25.59}$$

The maximum possible solar radiation is computed with the equations

$$RAMX = 30\left(1.0 + 0.0335\sin\left[\frac{2\pi}{365}(i + 88.2)\right]\right. \tag{25.60}$$

$$\ldots.\left(XT\sin\left(\frac{2\pi}{360}LAT\right)\sin(SD) + \cos\left(\frac{2\pi}{360}LAT\right)\cos(SD)\sin(XT)\right)\Bigg)$$

$$XT = \cos^{-1}\left(-\tan\left(\frac{2\pi}{360}LAT\right)\tan(SD)\right), \qquad 0 \le XT \le \pi \tag{25.61}$$

where LAT is the latitude of the site in degrees, SD is the sun's declination angle (radians), and i is the day of the year. The sun's declination angle is calculated with the equation

$$SD_i = 0.4102\sin\left(\frac{2\pi}{365}(i - 80.25)\right) \tag{25.62}$$

Finally, the wind function of the Penman equation is approximated with the relationship

$$f(V) = 2.7 + 1.63V \tag{25.63}$$

where V is the mean daily wind speed at a 10-m height in m·s^{-1}.

The Penman-Montieth method (Monteith, 1965) was recently added to EPIC to provide a means for estimating the effects of CO_2 changes (Stockle et al., 1992). The Penman-Montieth equation is expressed as

$$E_o = \frac{\delta(h_o - G) + 86.7AD(e_a - e_d)/AR}{(HV)(\delta + \gamma)} \tag{25.64}$$

and

$$E_p = \frac{\delta(h_o - G) + 86.7AD(e_a - e_d)/AR}{HV(\delta + \gamma(1. + CR/AR))} \tag{25.65}$$

where AD is the air density in g·m^{-3}, AR is the aerodynamic resistance for heat and vapor transfer in s·m^{-1}, and CR is the canopy resistance for vapor transfer in s·m^{-1}. Air density is estimated with the equation

$$AD = 0.01276PB/(1 + 0.0367T) \tag{25.66}$$

The aerodynamic resistance is computed with the equation

$$AR = \frac{6.25\left(\ln\left(\frac{10. - ZD}{ZO}\right)\right)^2}{V} \tag{25.67}$$

where ZD is the displacement height of the crop in m, ZO is the surface roughness parameter in m, and V is the daily mean wind speed in m·s^{-1}. The surface roughness is estimated with the equation

$$ZO = 0.131CHT^{0.997} \tag{25.68}$$

and the crop displacement height is estimated with the equation

$$ZD = 0.702CHT^{0.979} \tag{25.69}$$

where CHT is crop height in m. When no crop is growing the aerodynamic resistance is estimated with the equation

$$AR = \frac{350.}{V} \tag{25.70}$$

The canopy resistance is computed with the equation

$$CR = \frac{p_1}{(LAI)(g_o^*)(1.4 - 0.00121CO_2)} \tag{25.71}$$

where p_l is a parameter ranging from 1.0 to 2.0, LAI is the leaf-area-index of the crop, g_o^* is the leaf conductance in $m \cdot s^{-1}$, and CO_2 is the carbon dioxide level in the atmosphere in ppm. Leaf conductance is estimated from the crop input rate adjusted for vapor pressure deficit (VPD).

$$g_o^* = (g_o)(FV) \tag{25.72}$$

where g_o the crops leaf resistance when the VPD is less than the crops threshold VPD and FV is the VPD correction factor.

$$FV = 1. - b_v(VPD - VPD_t) \geq 0.1 \tag{25.73}$$

where b_v is a crop coefficient and VPD_t is threshold VPD for the crop.

The Priestly-Taylor (Priestly and Taylor 1972) method provides estimates of potential evaporation without wind and relative humidity inputs. The simplified equation based only on temperature and radiation is

$$E_o = 1.28\left(\frac{h_o}{HV}\right)\left(\frac{\delta}{\delta + \gamma}\right) \tag{25.74}$$

The net radiation is estimated with the equation

$$h_{oi} = RA_i(1 - AB_i) \tag{25.75}$$

instead of Eq. (25.58) which is used in the Penman method.

The Hargreaves method (Hargreaves and Samani, 1985) estimates potential evaportranspiration as a function of extraterrestrial radiation and air temperature. Hargreaves' method was modified for use in EPIC by increasing the temperature difference exponent from 0.5 to 0.6. Also, extraterrestrial radiation is replaced by RAMX (maximum possible solar radiation at the earth's surface) and the coefficient is adjusted from 0.0023 to 0.0032 for proper conversion. The modified equation is

$$E_o = 0.0032\left(\frac{RAMX}{HV}\right)(T + 17.8)(T_{mx} - T_{mn})^{0.6} \tag{25.76}$$

where T_{mx} and T_{mn} are the daily maximum and minimum air temperatures in °C.

All four methods estimate albedo by considering the soil, crop, and snow cover. If a snow cover exists with 5 mm or greater water content, the value of albedo is set to 0.6. If the snow cover is less than 5 mm and no crop is growing, the soil albedo is the appropriate value. When crops are growing, albedo is determined by using the equation

$$AB = 0.23(1.0 - EA) + (AB_s)(EA) \tag{25.77}$$

where 0.23 is the albedo for plants, AB_s is the soil albedo, and EA is a soil cover index. The value of EA ranges from 0 to 1.0 according to the equation

$$EA = \exp(-0.05CV) \tag{25.78}$$

where CV is the sum of the above ground biomass and crop residue in $t \cdot ha^{-1}$.

Soil and Plant Evaporation

The model computes evaporation from soils and plants separately by an approach similar to that of Ritchie (1972). For all methods except Penman-Monteith, potential plant water evaporation is computed with the equations

$$E_p = \frac{(E_o)(LAI)}{3.0}, \qquad 0 \leq LAI \leq 3.0 \tag{25.79}$$

$$E_p = E_o, \qquad LAI > 3.0 \tag{25.80}$$

where E_p is the predicted plant water evaporation rate in mm·d^{-1}. If soil water is limited, plant water evaporation will be reduced as described in the plant growth section of this chapter.

Potential soil water evaporation is simulated by considering soil cover according to the following equation

$$E_s = (E_o)(EA) \tag{25.81}$$

where E_s is the potential soil water evaporation rate in mm·d^{-1}. Potential soil water evaporation is reduced during periods of high plant water use with the equation

$$E_s^* = \min[ES, \frac{(E_s)(E_o)}{E_s + E_p}] \tag{25.82}$$

When E_p is low $E^*_s \rightarrow E_s$ but as E_p approaches E_o, $E^*_s \rightarrow E_s/(1.+EA)$.

Actual soil water evaporation is estimated on the basis of the top 0.2 m of soil and snow cover, if any. If 5 mm or more (water content) of snow is present albedo is set to 0.6 and EA to 0.5 for estimating E_o and snow is evaporated at that rate. When all snow is evaporated, soil water evaporation begins. Such evaporation is governed by soil depth and water content according to the equation

$$EV_Z = E_s^* \left[\frac{Z}{Z + \exp\,(2.374 - 0.00713\,Z)} \right] \tag{25.83}$$

EV is the total soil water evaporation in mm from soil of depth Z in mm. The coefficients of Eq. (25.83) are set to give EV=0.5 E^*_s when Z=10 mm and EV=0.95 E^*_s when Z=100 mm.

Potential soil water evaporation for a layer is estimated by taking the difference between EV's at the layer boundaries

$$SEV_\ell = EV_{Z(\ell)} - EV_{Z(\ell-1)} \tag{25.84}$$

where SEV is the potential soil evaporation for layer ℓ in mm.

The depth distributed estimate of soil water evaporation may be reduced according to the following equation if soil water is limited in a layer

$$SEV_\ell^* = SEV_\ell \exp\left(\frac{2.5(SW_\ell - FC_\ell)}{FC_\ell - WP_\ell}\right), \qquad SW_\ell < FC_\ell \qquad (25.85)$$

where SEV_ℓ^* is the adjusted soil water evaporation estimate in mm.

$$SEV_\ell^* = SEV_\ell, \qquad SW_\ell \geq FC_\ell \qquad (25.86)$$

The final step in adjusting the evaporation estimate is to assure that the soil water supply is adequate to meet the demand

$$SEV_\ell^* = \min\,(SEV_\ell^*, SW_\ell - b_w WP_\ell) \qquad (25.87)$$

where b_w may range from 0. to 1.0 in the top 0.5 m of soil and is set to 1.0 below 0.5 m. Thus, EPIC can be adjusted to allow the top 0.5 m to dry down to any fraction of wilting point.

25.2.5. Snowmelt

If snow is present, it may be melted on days when the second soil layer temperature exceeds 0°C. Snow is melted as a function of the snow pack temperature using the equation

$$\begin{gathered} SML = T(1.52 + 0.54SPT) \\ 0. \leq SML \leq SNO \end{gathered} \qquad (25.88)$$

where SML is the snowmelt rate in $mm \cdot d^{-1}$, SNO is the snow present in mm of water, T is the mean daily air temperature in °C, and SPT is the snow pack temperature in °C. The snow pack temperature is estimated with the equation

$$SPT = \min\left(T_S, T(2)\right) \qquad (25.89)$$

where T_s is the temperature at the top of the snow pack and T(2) is the temperature at the center of soil layer 2. The equations for estimating T_s and T(2) are presented in the soil temperature section. Melted snow is treated the same as rainfall for estimating runoff volume and percolation, but rainfall energy is set to 0.0 and peak runoff rate is estimated by assuming uniformly distributed rainfall for a 24-h duration.

25.2.6. Water Table Dynamics

The water table height is simulated without direct linkage to other soil water processes in the root zone to allow for offsite water effects. The model drives the water table up and down between input values of maximum and minimum depths from the surface. The driving mechanism is a function of rainfall, surface runoff, and potential evaporation, as given in the equation

$$WTBL_i = WTBL_{i-1} - W1(WTBL_{i-1} - WTL) \tag{25.90}$$

where WTBL is the depth in m from the surface to the water table on day i, W1 is the driving function, and WTL is the appropriate limit. The driving equations are

$$W1 = \min(0.1, |W2|) \tag{25.91}$$

$$W2 = \frac{RFS - QS - EOS}{EOS} \tag{25.92}$$

where RFS, QS, and EOS are the sums of rainfall, runoff, and potential evaporation for 30 days before day i and W2 is a scaling factor. Equation (25.92) causes the water table to rise faster than it falls because the denominator is larger during recession.

The maximum water table depth, WTMX, is substituted into Eq. (25.90) for WTL when the water table is falling. Conversely, WTL is set to the minimum water table depth, WTMN, on the rising side.

$$WTL = WTMX, \quad W2 \le 0.0 \tag{25.93}$$

$$WTL = WTMN, \quad W2 > 0.0 \tag{25.94}$$

Obviously, Eq. (25.90) gives highest rising rates when W2 is large and when WTBL≈WTMN the rate of rise approaches zero. The reverse is true on the falling side.

25.3. WEATHER

The weather variables necessary for driving the EPIC model are precipitation, air temperature, and solar radiation. If the Penman methods are used to estimate potential evaporation, wind speed and relative humidity are also required. Of course, wind speed is also needed when wind-induced erosion is simulated. If daily precipitation, air temperature, and solar radiation data are available, they can be input directly into EPIC. Rainfall and temperature data are available for many areas of the United States, but solar radiation, relative humidity, and wind data are scarce. Even rainfall and temperature data are generally not adequate for the long-term EPIC simulation (100 years+). Thus, EPIC provides options for simulating various combinations of the five weather variables. Descriptions of the models used for simulating precipitation, temperature, radiation, relative humidty, and wind follow.

25.3.1. Precipitation

The EPIC precipitation model developed by Nicks (1974) is a first-order Markov chain model. Thus, input for the model must include monthly probabilities of receiving precipitation. On any given day, the input must include information as to whether the previous day was dry or wet. A random number (0-1) is generated and compared with the appropriate wet-dry probability. If the random number is less than or equal to the wet-dry probability, precipitation occurs on that day. Random numbers greater than the wet-dry probability give no precipitation. Since the wet-dry state of the first day is established, the process can be repeated for the next day and so on throughout the simulation period.

If wet-dry probabilities are not available, the average monthly number of rainy days may be substituted. The probability of a wet day is calculated directly from the number of wet days:

$$PW = NWD \,/\, ND \tag{25.95}$$

where PW is the probability of a wet day, NWD is the number of rainy days, and ND is the number of days, in a month. The probability of a wet day after a dry day can be estimated as a fraction of PW.

$$P(W/D) = \beta PW \tag{25.96}$$

where P(W/D) is the probability of a wet day following a dry day and where β is a fraction usually in the range of 0.6 to 0.9. The probability of a wet day following a wet day can be calculated directly by using the equation

$$P(W/W) = 1.0 - \beta + P(W/D) \tag{25.97}$$

where P(W/W) is the probability of a wet day after a wet day. When β→1.0, wet days do not affect probability of rainfall--P(W/D)=P(W/W)=PW. Conversely, low β values give strong wet day effects--β→0.0, P(W/D)→1.0. Thus, β controls the interval between rainfall events but has no effect on the number of wet days. For many locations, β=0.75 gives satisfactory estimates of P(W/D). Although Eqs. (25.96) and (25.97) may give slightly different probabilities than those estimated from rainfall records, they do guarantee correct simulation of the number of rainfall events.

When precipitation event occurs, the amount is generated from a skewed normal daily precipitation distribution

$$R_i = \left(\frac{\left(\left(SND_i - \dfrac{SCF_k}{6.0} \right)\left(\dfrac{SCF_k}{6.0} \right) + 1 \right)^3 - 1.}{SCF_k} \right) RSDV_k + \bar{R}_k \tag{25.98}$$

where R is the amount of rainfall for day i, in mm, SND is the standard normal deviate for day i, SCF is the skew coefficient, RSDV is the standard deviation of daily rainfall in mm, and R is the mean daily rainfall in month k.

If the standard deviation and skew coefficient are not available, the model simulates daily rainfall by using a modified exponential distribution.

$$R_i = \frac{(-1n(\mu))^{\zeta} \ \overline{R_k}}{\int_{0.0}^{1.0} (-1n\ (\chi))^{\zeta} dx} \qquad (25.99)$$

where μ is a uniform random number (0.0-1.0) and ζ is a parameter usually in the range of 1.0 to 2.0. The larger the ζ value, the more extreme the rainfall events. A value of 1.3 gives satisfactory results at many locations in the United States. The denominator of Eq. (25.99) assures that the mean long-term simulated rainfall is correct. The modified exponential is usually a satisfactory substitute and requires only the monthly mean daily rainfall as input.

Daily precipitation is partitioned between rainfall and snowfall using a combination of maximum daily air temperature (T_{mx}) and surface layer soil temperature (T (1)). If the average of T_{mx} and T (1) is 0°C or below, the precipitation is snowfall, otherwise, it is rainfall.

25.3.2. Air Temperature and Solar Radiation

The model developed by Richardson (1981) was selected for use in EPIC because it simulates temperature and radiation, which are mutually correlated with rainfall. The residuals of daily maximum and minimum air temperature and solar radiation are generated from a multivariate normal distribution.

The multivariate generation model used implies that the residuals of maximum temperature, minimum temperature, and solar radiation are normally distributed and that the serial correlation of each variable may be described by a first-order linerar autoregressive model. Details of the multivariate generation model were described by Richardson (1981). The dependence structure of daily maximum temperature, minimum temperature, and solar radiation was described by Richardson (1982).

The temperature model requires monthly means of maximum and minimum temperatures and their standard deviations as inputs. If the standard deviations are not available, the long-term observed extreme monthly minimums and maximums may be substituted. The model estimates standard deviation as 0.25 of the difference between the extreme and the mean for each month. For example:

$$SDTMX_k = 0.25(TE_{mx,k} - \overline{T}_{mx,k}) \qquad (25.100)$$

where SDTMX is the standard deviation of the daily maximum temperature, TE is the extreme daily maximum temperature, and T is the average daily maximum temperature for month k.

The solar radiation model uses the extreme approach extensively. Thus, only the monthly means of daily solar radiation are required as inputs. The equation for estimating standard deviation is

$$SDRA_k = 0.25(RAMX_k - RA_k) \tag{25.101}$$

where SDRA is the standard deviation of daily solar radiation in $MJ \cdot m^{-2}$, RAMX is the maximum daily solar radiation at midmonth, and RA is the mean daily solar radiation for month k.

Maximum temperature and solar radiation tend to be lowr on rainy days. Thus, it is necessary to adjust the mean maximum temperature and solar radiation downward for simulating rainy day conditions. For T_{mx} this is accomplished by assuming that wet day values are less than dry day values by some fraction of T_{mx} - T_{mn}:

$$TW_{mx,k} = TD_{mx,k} - \Omega_T(T_{mx,k} - T_{mn,k}) \tag{25.102}$$

where TW is the daily mean maximum temperture for wet days in °C in month k, TD is the daily mean maximum temperature for dry days, Ω_T is a scaling factor ranging from 0.0 to 1.0, T_{mx} is the daily mean maximum temperature, and T_{mn} is the daily mean minimum temperature. Choosing Ω_T=1.0 provides highest deviations on wet days and Ω_T=0.0 ignores the wet day effect. Observed data indicate that Ω_T usually lies between 0.5 and 1.0.

Since Eq. (25.102) gives lower mean maximum temperature values for wet days, a companion equation is necessary to slightly increase mean maximum temperature for dry days. The development is taken directly from the continuity equation

$$(T_{mx,k})(ND_k) = (TW_{mx,k})(NWD_k) + (TD_{mx,k})(NDD_k) \tag{25.103}$$

where ND is the number of days in a month, NWD is the number of wet days, and NDD is the number of dry days. The desired equation is obtained by substituting Eq. (25.102) into Eq. (25.103) and solving for TD

$$TD_{mx,k} = T_{mx,k} + \left(\frac{NWD_k}{ND_k}\right)\Omega_T(T_{mx,k} - T_{mn,k}) \tag{25.104}$$

Use of the continuity equation guarantees that the long-term simulated value for mean maximum temperature agrees with the input value of T_{mx}.

The method of adjusting solar radiation for wet and dry days is similar to that of adjusting maximum temperature. The radiation on wet days is a fraction of the dry day radiation

$$RAW_k = \Omega_R RAD_k \tag{25.105}$$

where RAW is the daily mean solar radiation on wet days in $MJ \cdot m^{-2}$, Ω_R is a scaling factor ranging from 0.0 to 1.0, and RAD is the daily mean solar radiation on dry days. An Ω_R value of 0.5 gives satisfactory results for many locations. The dry day equation is developed by replacing temperature with radiation in Eq. (25.103) and substituting Eq. (25.105) for RAW. Then,

$$RAD_k = \frac{(RA_k)(ND_k)}{\Omega_R(NWD_k) + NDD_k} \quad (25.106)$$

where RA is the daily mean solar radiation for month k in $MJ \cdot m^{-2}$.

25.3.3. Wind

The original EPIC wind model (Richardson and Wright, 1984) simulated mean daily wind speed and daily direction. The new EPIC wind erosion model, WECS, (Wind Erosion Continuous Simulation) requires wind speed distribution within the day and the dominant direction. Daily wind speed distribution is simulated using the modified exponential equation

$$v_j = a_1 V_i (-\ln(f_j))^{a_2} \quad (25.107)$$

where v_j is the wind speed at f fraction of the day, V_i is the mean wind speed for day i, and a_1 and a_2 are parameters. Equation (25.107) is an expression of wind speed probabilities rather than wind speed as a function of time during the day. The daily time distribution of wind speed is not required because all other EPIC components operate on a daily time step. Experimental work with data from Bushland, TX and Temple, TX indicates that $0.1 < a_2 \quad 0.6$. Values of a_2 are generated daily from a triangular distribution with base ranging from 0.1 to 0.6 and peak at 0.35. The value of a_1 can be closely appointed with the equation

$$a_1 = 1.5567(a_2)^{0.1508} \exp(-0.4336 a_2) \quad (25.108)$$

Equation (25.108) assures that

$$V_i = \int_0^{1.0} v df \quad (25.109)$$

Values of v_j are simulated with f starting with the threshold windspeed for erosion. The threshold f value (f_o) is determined using the wind erosion Eq. (25.141) and Eq. (25.107) in the form

$$f_o = \exp\left[\left(\frac{v_{*\tau}}{0.0408 a_1 V_i}\right)^{1/a_2}\right] \quad (25.110)$$

where $v_{*\tau}$ is the threshold friction velocity in $m \cdot s^{-1}$. Larger f values produce non-erosive wind speeds. Windspeeds greater than $v_{*\tau}$ are produced as f is reduced.

The mean daily wind speed is simulated using the modified exponential equation

$$V_i = b_{1,k} \quad V_k[-\ln(RN)]^{b_{2,k}} \tag{25.111}$$

where V_k is the mean wind speed for month k, RN is a random number, b_2 is a parameter for month k, and b_1 is calculated directly using an approach similar to that described in Eq. (25.108). Experimental work with data from Bushland, TX and Temple, TX indicates that $b_2 \approx 0.3$.

Wind direction expressed as radians from north in a clockwise direction is generated from an empirical distribution specific for each location. The empirical distribution is simply the cumulative probability distribution of wind direction. The "Climatic Atlas of the United States" (U.S. Department of Commerce, 1968) gives monthly percentages of wind from each of 16 directions. Thus, to estimate wind direction for any day, the model draws a uniformly distributed random number and locates its position on the appropriate monthly cumulative probability distribution.

25.3.4. Relative Humidity

The relative humidity model simulates daily average relative humidity from the monthly average by using a triangular distribution. As with temperature and radiation, the mean daily relative humidity is adjusted to account for wet- and dry-day effects. The assumed relation between relative humidity on wet and dry days is

$$RHW_k = RHD_k + \Omega_H(1.0 - RHD_k) \tag{25.112}$$

where RHW is the daily mean relative humidity on wet days for month k, RHD is the daily mean relative humidity on dry days, and Ω_H is a scaling factor ranging from 0.0 to 1.0. An Ω_H value of 0.9 seems appropriate for many locations. Using the continuity equation as described in the temperature and radiation sections produces the equation

$$RHD_k = \frac{RH_k - \Omega_H\left(\frac{NWD}{ND}\right)}{1.0 - \Omega_H\left(\frac{NWD}{ND}\right)} \tag{25.113}$$

where RH is the long-term average relative humidity for month k.

The appropriate value (RHW or RHD) is used as the peak of a triangular distribution to generate daily relative humidity. The upper limit of the triangular distribution is set with the equation

$$RHU_i = RHP_i + (1.0 - RHP_i)\exp(RHP_i - 1.0) \tag{25.114}$$

where RHU is the largest relative humidity value that can be generated on day i and RHP is the peak of the triangular distribution (RHW or RHD). The lower limit is set with the equation

$$RHL_i = RHP_i[1.0 - \exp(-RHP_i)] \qquad (25.115)$$

where RHL is the lowest relative humidity value that can be generated on day i.

To assure that the simulated long-term value for mean relative humidity agrees with input RH, the generated value is adjusted by using the equation

$$RHG_i^* = RHG_i \left(\frac{RHP_i}{RH_i}\right) \qquad (25.116)$$

where RHG* is the generated relative humidity on day i adjusted to the mean of the triangle, RHG is the relative humidity generated from the triangle, and RH is the mean of the triangle.

If relative humidity data is not available, dew point temperature may be substituted and EPIC estimates the monthly mean relative humidity using Eqs. (25.52) and (25.53). If dew point temperature is not available EPIC estimates monthly mean relative humidity using the equation

$$RH_K = 0.9 - \frac{0.8\Delta T_K}{\Delta T_K + \exp(5.12 - 0.127\Delta T_K)} \qquad (25.117)$$

where $\Delta T_k = T_{mx} - T_{mn}$ for month k.

25.4. EROSION

25.4.1. Water

Rainfall/Runoff

The EPIC component for water-induced erosion simulates erosion caused by rainfall and runoff and by irrigation (sprinkler and furrow). To simulate rainfall/runoff erosion, EPIC contains six equations--the USLE (Wischmeier and Smith, 1978), the Onstad-Foster modification of the USLE (Onstad and Foster, 1975), the MUSLE (Williams, 1975), two recently developed variations of MUSLE, and a MUSLE structure that accepts input coefficients. Only one of the equations (user specified) interacts with other EPIC components. The six equations are identical except for their energy components. The USLE depends strictly upon rainfall as an indicator of erosive energy. The MUSLE and its variations use only runoff variables to simulate erosion and sediment yield. Runoff variables increased the prediction accuracy, eliminated the need for a delivery ratio (used in the USLE to estimate sediment yield), and enables the equation to give single storm estimates of sediment yields. The USLE gives only annual estimates. The Onstad-Foster equation contains a combination of the USLE and MUSLE energy factors.

Thus, the water erosion model uses an equation of the form

$$Y = \chi(K)(CE)(PE)(LS)(ROKF)$$

$$\chi = EI \qquad \textit{for USLE} \qquad (25.118)$$

$$\chi = 0.646EI + 0.45(Q \cdot q_p)^{0.33} \qquad \textit{for Onstad-Foster}$$

$$\chi = 1.586(Q \cdot q_p)^{0.56} A^{0.12} \qquad \textit{for MUSLE}$$

$$\chi = 2.5(Q \cdot q_p)^{0.5} \qquad \textit{for MUST}$$

$$\chi = 0.79(Q \cdot q_p)^{0.65} A^{0.009} \qquad \textit{for MUSS} \qquad (25.119)$$

$$\chi = by_1 Q^{by_2} q_p^{by_3} A^{by_4} \qquad \textit{for MUSI}$$

where Y is the sediment yield in $t \cdot ha^{-1}$, k is the soil erodibility factor, CE is the crop management factor, PE is the erosion control practice factor, LS is the slope length and steepness factor, ROKF is the coarse fragment factor, Q is the runoff volume in mm, q_p is the peak runoff rate in $mm \cdot h^{-1}$, A is the watershed area in ha, MUST is a new equation theoretically developed from sediment concentration bases, MUSS is a new equation developed by fitting small watershed data (no channel erosion), and MUSI allows user input of four coefficients ($b_{y1}....b_{y4}$). The PE value is determined initially by considering the conservation practices to be applied. The value of LS is calculated with the equation (Wischmeier and Smith, 1978)

$$LS = \left(\frac{\lambda}{22.1}\right)^{\xi} (65.41S^2 + 4.56S + .065) \qquad (25.120)$$

where S is the land surface slope in $m \cdot m^{-1}$, λ is the slope length in m, and ξ is a parameter dependent upon slope. The value of ξ varies with slope and is estimated with the equation

$$\xi = 0.3S/[S + \exp(-1.47 - 61.09S)] + 0.2 \qquad (25.121)$$

The crop management factor is evaluated for all days when runoff occurs by using the equation

$$CE = \exp[(\ln 0.8 - \ln CE_{mn,j})\exp(-1.15CV) + \ln CE_{mn,j}] \qquad (25.122)$$

where $CE_{mn,j}$ is the minimum value of the crop management factor for crop j and CV is the soil cover (above ground biomass plus residue) in $t \cdot ha^{-1}$.

The soil erodibility factor, K, is evaluated for the top soil layer at the start of each year of simulation with the equation

$$K = \left(0.2 + 0.3\exp[-0.0256SAN(1 - SIL/100)]\right)\left(\frac{SIL}{CLA + SIL}\right)^{0.3} \quad (25.123)$$

$$\ldots\ldots \left(1.0 - \frac{0.25\,C}{C + \exp(3.72 - 2.95\,C)}\right)\left(1.0 - \frac{0.7\,SN1}{SN1 + \exp(-5.51 + 22.9\,SN1)}\right)$$

where SAN, SIL, and CLA, and C are the sand, silt, clay, and organic carbon contents of the soil in % and SN1=1-SAN/100. Equation (25.123) allows K to vary from about 0.1 to 0.5. The first term gives low K values for soils with high coarse-sand contents and high values for soils with little sand. The fine sand content is estimated as the product of sand and silt divided by 100. The expression for coarse sand in the first term is simply the difference between sand and the estimated fine sand. The second term reduces K for soils that have high clay to silt ratios. The third term reduces K for soils with high organic carbon contents. The fourth term reduces K further for soils with extremely high sand contents (SAN>70%).

The runoff model supplies estimates of Q and q_p. To estimate the daily rainfall energy in the absence of time-distributed rainfall, it is assumed that the rainfall rate is exponentially distributed:

$$r_t = r_p \exp(-t/\kappa) \quad (25.124)$$

where r is the rainfall rate at time t in mm◊h^{-1}, r_p is the peak rainfall rate in mm◊h^{-1}, and k is the decay constant in h. Equation (25.124) contains no assumption about the sequence of rainfall rates (time distribution). The USLE energy equation in metric units is

$$RE = \Delta R\left(12.1 + 8.9\log\frac{\Delta R}{\Delta t}\right) \quad (25.125)$$

where RE is the rainfall energy for water erosion equations and ΔR is a rainfall amount in mm during a time interval Δt in h. The energy equation can be expressed analytically as

$$RE = 12.1\int_0^\infty r\,dt + 8.9\int_0^\infty r\log r\,dt \quad (25.126)$$

Substituting Eq. (25.124) into Eq. (25.126) and integrating gives the equation for estimating daily rainfall energy:

$$RE = R\left[12.1 + 8.9(\log r_p - 0.434)\right] \quad (25.127)$$

where R is the daily rainfall amount in mm. The rainfall energy factor, EI, is obtained by multiplying Eq. (25.127) by the maximum 0.5-h rainfall itensity ($r_{.5}$) and converting to the proper units:

$$EI = R\left[12.1 + 8.9(\log r_p - 0.434)\right](r_{.5})/1000 \quad (25.128)$$

To compute values for r_p, Eq. (25.124) is integrated to give

$$R = (r_p)(\kappa) \tag{25.129}$$

and

$$R_t = R[1 - \exp(-t/\kappa)] \tag{25.130}$$

The value of $R_{.5}$ can be estimated by using $\alpha_{.5}$, as mentioned in the Hydrology section of this chapter:

$$R_{.5} = \alpha_{.5} R \tag{25.131}$$

To determine the value of r_p, Eqs. (25.131) and (25.129) are substituted into Eq. (25.130) to give

$$r_p = -2R\ln(1 - \alpha_{.5}) \tag{25.132}$$

Since rainfall rates vary seasonally, $\alpha_{.5}$ is evaluated for each month by using Weather Service information (U.S. Department of Commerce, 1979). The frequency with which the maximum 0.5-h rainfall amount occurs is estimated by using the Hazen plotting position equation (Hazen, 1930).

$$F = \frac{1}{2\tau} \tag{25.133}$$

where F is the frequency with which the largest of a total of τ events occurs. The total number of events for each month is the product of the number of years of record and the average number of rainfall events for the month. To estimate the mean value of $\alpha_{.5}$, it is necessary to estimate the mean value of $R_{.5}$. The value of $R_{.5}$ can be computed easily if the maximum 0.5-h rainfall amounts are assumed to be exponentially distributed. From the exponential distribution, the expression for the mean 0.5-h rainfall amount is

$$R_{.5,k} = \frac{R_{.5F,k}}{-\ln F_k} \tag{25.134}$$

where $R_{.5,k}$ is the mean maximum 0.5-h rainfall amount, $R_{.5F,k}$ is the maximum 0.5-h rainfall amount for frequency F, and subscript k refers to the month. The mean $\alpha_{.5}$ is computed with the equation

$$\alpha_{.5,k} = \frac{R_{.5,k}}{R_k} \tag{25.135}$$

where R is the mean amount of rainfall for each event (average monthly rainfall/average number of days of rainfall) and subscript k refers to the month. Daily values of $\alpha_{.5}$ are generated from a triangular distribution. The lower limit determined by a uniform rainfall rates gives $\alpha_{.5}$ equal to 0.5/24 or 0.0208. The upper limit of $\alpha_{.5}$ is set by considering a large rainfall event. In a large event, it is highly unlikely that all the rainfall occurs in 0.5 h (α=1). The upper limit of $\alpha_{.5}$ can be estimated by substituting a high value for r_p (250 mm·h^{-1} is generally near the upper limit of rainfall intensity) into Eq. (25.130)

$$\alpha_{.5u} = 1 - \exp(-125/R) \qquad (25.136)$$

where $\alpha_{.5u}$ is the upper limit of $\alpha_{.5}$. The peak of the $\alpha_{.5}$ triangular distribution is $\alpha_{.5,K}$ from Eq. (25.135).

The coarse fragment factor is estimated with the equation (Simanton et al. 1984)

$$ROKF = \exp(-.03 ROK) \qquad (25.137)$$

where ROK is the percent of coarse fragments in the surface soil layer.

Irrigation

Erosion caused by applying irrigation water in furrows is estimated with MUST

$$Y = 2.5(Q \cdot q_p)^{.5}(K)(CE)(PE)(LS) \qquad (25.138)$$

where CE, the crop management factor, has a constant value of 0.5. The volume of runoff is estimated as the product of the irrigation volume applied and the irrigation runoff ratio.

The peak runoff rate is estimated for each furrow by using Manning's equation and assuming that the flow depth if 0.75 of the ridge height and that the furrow is triangular. If irrigation water is applied to land without furrows, the peak runoff rate is estimated by dividing the runoff volume by the duration (24 h).

25.4.2. Wind

The original EPIC wind erosion model (WEQ) required daily mean wind speed as a driving variable. The new EPIC wind erosion model WECS (Wind Erosion Continuous Simulation) requires the daily distribution of wind speed to take advantage of the more mechanistic erosion equation. The new approach estimates potential wind erosion for a smooth bare soil by integrating the erosion equation through a day using the wind speed distribution. Potential erosion is adjusted using four factors based on soil properties, surface roughness, cover, and distance across the field in the wind direction.

Basic Equation

The basic WECS wind erosion equation is

$$YW = (FI)(FR)(FV)(FD)\int_{0}^{DW} YWR\,dt \tag{25.139}$$

where YW is the wind erosion in kg·m^{-1}, FI is the soil erodibility factor, FR is the surface roughness factor, FV is the vegetative cover factor, FD is the mean unsheltered travel distance of wind across a field, DW is the duration of wind greater than threshold velocity in s, and YWR is the wind erosion rate in kg·m^{-1}·s^{-1} at time t.

The wind erosion rate is calculated with the equation of Skidmore (1986).

$$YWR = C\left(\frac{\rho_a}{g}\right)\left[v_{*o}^2 - v_{*\tau}^2 - 0.5\left(\frac{SW}{WP}\right)^2\right]^{3/2} \tag{25.140}$$

where C is a parameter ≈2.5, ρ_a is the air density in kg·m^{-3}, g is the acceleration of gravity in m·s^{-2}, v_{*o} is the friction velocity in m·s^{-1}, and $v_{*\tau}$ is the threshold friction velocity in m·s^{-1}. SW and WP are the actual and 1500 kPa water contents of the top soil layer (10 mm thick), respectively. The soil water term of Eq. (25.140) was developed by Chepil (1956) and Skidmore (1986). Substituting acceleration of gravity (9.8 m·s^{-2}) and assuming air density is 1 kg·m^{-3} gives

$$YWR = 0.255\left[v_{*o}^2 - v_{*\tau}^2 - 0.5(SW/WP)^2\right]^{3/2} \tag{25.141}$$

The friction velocity is estimated with the equation

$$v_{*o} = \frac{(K)(v)}{\ln\left(\frac{Z}{Z_o}\right)} \tag{25.142}$$

where K is Von Karman's constant ≈0.4, v is the wind speed in m·s^{-1} at height Z (10 m for WECS input), and Z_o is the aerodynamic roughness (0.00055 m used in WECS). Substituting the WECS values into Eq. (25.142) gives

$$v_{*o} = 0.0408\ v \tag{25.143}$$

The threshold friction velocity is estimated with the equation

$$v_{*\tau} = 0.1\sqrt{g\left(\frac{\Delta\rho}{\rho_a}\right)D} \tag{25.144}$$

where $\Delta\rho$ is the mineral soil density (2650 kg·m^{-3}) and D is the soil particle size diameter in m. Substituting constants into Eq. (25.144) and expressing D in μm gives

$$v_{*_\tau} = 0.0161\ \sqrt{D} \qquad (25.145)$$

Soil Erodibility Factor

WECS uses the soil erodibility concept of WEQ expressed in dimensionless form with the equation

$$FI = I/695. \qquad (25.146)$$

where I is the soil erodibility factor of the Woodruff and Siddoway (1965) model in t·ha^{-1} and FI is the dimensionless soil erodibility factor of the new model.

Roughness Factor

The surface roughness factor (FR) is based upon the shelter angle concept developed by Potter et al. (1990). This roughness index calculates the erodible fraction of the soil surface by estimating the portion susceptible to abrasion by saltating particles. The shelter angle index incorporates both roughness due to random cloddiness and oriented roughness (ridges) due to tillage operations. The effect of oriented roughness varies as a function of wind direction, which is selected each day so that the statistical distribution of wind direction approaches that of the simulation site. FR is estimated with the equation (Potter and Zobeck, 1990)

$$FR = 1 - \exp\left[-(wn_1/RFB)^{RFC}\right] \qquad (25.147)$$

where wn_1 is the descent angle of saltating sand grains (about 15° from horizontal). A 15° impact angle has also been shown to cause maximum aggregate abrasion (Hagen et al., 1988). The coefficient RFC is calculated with the equation

$$RFC = 0.77\left(1.002^{RHTT}\right) \qquad (25.148)$$

where RHTT is the ridge height in mm. The coefficient RFB is estimated with the equations

$$RFB = RRF + RIF \qquad (25.149)$$

$$RRF = 11.9\left[1 - \exp(-(RRUF/9.8)^{1.3})\right] \qquad (25.150)$$

$$RIF = \left|\left[\sin(wn_2)\right]\ [1.27(RHTT^{0.52})\right]\right| \qquad (25.151)$$

where RRF is the clod roughness factor, RRUF is the random roughness in mm, RIF is the ridge roughness factor, and wn_2 is the angle of the wind relative to ridges. Both RRUF and RHTT are altered by wind and water erosion and tillage.

Vegetative Cover Factor

The vegetative cover factor is based on the approach used in the original EPIC model. A vegetative cover equivalent factor is simulated daily as a function of standing live biomass, standing dead residue, and flat crop residue.

$$VE = 0.253(\omega_1 SB + \omega_2 SR + \omega_3 FR)^{1.36} \tag{25.152}$$

where VE is the vegetative cover equivalent factor, SB is the standing biomass in $t \cdot ha^{-1}$, SR is the standing crop residue in $t \cdot ha^{-1}$, FR is the flat residue in $t \cdot ha^{-1}$, and ω_1, ω_2, and ω_3 are crop specific coefficients. The vegetative equivalent is converted to a vegetative factor for use in the new model.

$$FV = \frac{VE}{VE + \exp(0.48 - 1.32VE)} \tag{25.153}$$

Equation (25.153) assures that $0.0 \le FV \le 1.0$.

Unsheltered Distance Factor

Field length along the prevailing wind direction is calculated as in the original model (Cole et al., 1982) by considering the field dimensions and orientation and the wind direction.

$$WL = \frac{(FL)(FW)}{FL\left[\cos\left(\frac{\pi}{2} + \theta - \phi\right)\right] + FW\left[\sin\left(\frac{\pi}{2} + \theta - \phi\right)\right]} \tag{25.154}$$

where WL is the unsheltered field length along the prevailing wind direction in m, FL is the field length in m, FW is the field width in m, θ is the wind direction clockwise from north in radians, and ϕ is the clockwise angle between field length and north in radians. The new model distance factor (FD) is calculated as described by Stout (1990) using the equation

$$FD = 1.0 - \exp[-(WL/wn_3)] \tag{25.155}$$

where wn_3 is a parameter determined experimentally to lie in the range $50. < wn_3 < 90$. A value of 70. is used in EPIC. The integration of Eqs. (25.141) and (25.107) is accomplished numerically using variable Δf steps.

Large steps of Δf=0.1 are taken initially in simulating low wind speeds. The step size is reduced as f is reduced to gain better definition of the wind speed distribution at high speeds. At the end of each integration step, Δf is cut in half if

$$\left(\frac{v_j + v_{j-1}}{V_i}\right)(f_{j-1} - f_j) > 0.1 \qquad (25.156)$$

25.5. NUTRIENTS

25.5.1. Nitrogen

Nitrate Losses--Leaching, Surface Runoff, Lateral Subsurface Flow

The amount of NO_3-N lost when water flows through a layer is estimated by considering the change in concentration. Thus, the equation

$$VNO3 = (QT)(c_{NO3}) \qquad (25.157)$$

where VNO3 is the amount of NO_3-N lost from a soil layer and c_{N03} is the average concentration of NO_3-N in the layer during the percolation of volume QT through the layer. At the end of the day, the amount of NO_3-N left in the layer is

$$WN03 = WN03_0 - (QT)(c_{N03}) \qquad (25.158)$$

where $WNO3_0$ and WNO3 are the weights of NO_3-N contained in the layer at the beginning and ending of the day. The No_3-N concentration can be calculated by dividing the weight of NO_3-N by the water storage volume:

$$c'_{NO3} = c_{NO3} - c_{NO3}\left(\frac{QT}{(bl)(PO)}\right) \qquad (25.159)$$

where c'_{NO3} is the concentration of NO_3-N at the end of a day, PO is soil porosity, and bl is a fraction of the storage PO occupied by percolating water. Equation (25.159) is a finite difference approximation for the exponential equation

$$c'_{NO3} = c_{NO3} \exp\left(\frac{-QT}{bl \cdot PO}\right) \qquad (25.160)$$

Thus, VNO3 can be computed for any volume, QT, by integrating Eq. (25.160)

$$VNO3 = WNO3\left(1. - \exp\left(\frac{-QT}{(bl)(PO)}\right)\right) \qquad (25.161)$$

The average concentration during the percolation of QT (1 day since EPIC is a daily time step model) is

$$c_{NO3} = \frac{VNO3}{QT} \qquad (25.162)$$

Amounts of NO_3-N contained in runoff, lateral flow, and percolation are estimated as products of the volume of water and the concentration from Eq. (25.162). The calculation sequence is: (1) Soil layer 1--Leaching is estimated first by substituting 0 for QT in Eq. (25.161) and solving for VNO3. Then VNO3 is removed from $WNO3_1$ and placed into $WNO3_2$. Next, runoff and lateral flow losses are estimated simultaneously by substituting Q+QR, for QT in Eq. (25.161) and the resulting VNO3 into Eq. (25.162). Individual losses are estimated for runoff ($Q \cdot c_{NO3}$) and lateral flow ($QR_1 \cdot c_{NO3}$). Finally, these losses are removed from $WNO3_1$. (2) Soil layers 2-M--Leaching and lateral flow losses are estimated simulataneously by substituting $QR_\ell + O_\ell$ for QT in Eq. (25.161) and the resulting VNO3 into Eq. (25.162). Individual losses are calculated as in (1) ($QR_\ell \cdot c_{NO3}$) and ($O_\ell \cdot c_{NO3}$), subtracted from $WNO3_\ell$ and added to $WNO3_{\ell+1}$. The process is repeated layer by layer to the bottom of the soil profile.

NO_3-N Transport by Soil Water Evaporation

When water is evaporated from the soil, NO_3-N is moved upward into the top soil layer by mass flow. The equation for estimating this NO_3-N transport is

$$ENO3 = \sum_{\ell=2}^{M} \text{SEV}_\ell^* (c_{NO3})_\ell \qquad (25.163)$$

where ENO3 is the amount of NO_3-N in $kg \cdot ha^{-1}$ moved from lower layers to the top layer by soil water evaporation E_s in mm, subscript ℓ refers to soil layers, and M is the number of layers contributing to soil water evaporation (maximum depth is 0.2m).

Organic N Transport by Sediment

A loading function developed by McElroy et al. (1976) and modified by Williams and Hann (1978) for application to individual runoff events is used to estimate organic N loss. The loading function is

$$YON = 0.001(Y)(c_{ON})(ER) \qquad (25.164)$$

where YON is the organic N runoff loss in $kg \cdot ha^{-1}$, Y is the sediment yield in $t \cdot ha^{-1}$, c_{ON} is the concentration of organic N in the top soil layer in $g \cdot t^{-1}$, and ER is the enrichment ratio. The enrichment ratio is the concentration of organic N in the sediment divided by that in the soil. Enrichment ratios are logarithmically related to sediment concentration as described by Menzel (1980). An individual event enrichment-sediment concentration relationship was developed for EPIC considering upper and lower bounds. The upper bound of enrichment ratio is the inverse of the sediment delivery ratio. Exceeding the inverse of the delivery ratio implies that more organic N leaves the watershed than is dislodged from the soil. The delivery ratio is estimated for each runoff event by using the equation

$$DR = \left(\frac{q_p}{r_{ep}}\right)^{0.5} \tag{25.165}$$

where DR is the sediment delivery ratio (sediment yield divided by gross sheet erosion), q_p is the peak runoff rate in mm·h^{-1}, and r_{ep} is the peak rainfall excess rate in mm·h^{-1}. Equation (25.165) is based on sediment yield estimated by using MUST. The rainfall excess rate cannot be evaluated directly because the hydrology model predicts only the total daily runoff volume. An estimate of the rate can be obtained, however, using the equation

$$r_{ep} = r_p\left(\frac{Q}{R}\right)^{0.1} \tag{25.166}$$

where r_p is the peak rainfall rate in mm·h^{-1} and the runoff-rainfall ratio Q/R is used to account for infiltration. The Q/R is raised to the 0.1 power to account for intensity effects on infiltration -- Q/R is greater than average near r_p.

The lower limit of enrichment ratio is 1.0--sediment particle size distribution is the same as that of the soil. Thus, 1≤ER≤1/DR. The logarithmic equation for estimating enrichment ratio is

$$ER = x_1 c_s^{x_2} \tag{25.167}$$

where c_s is the sediment concentration in g·m^{-3} and x_1 and x_2 are parameters set by the upper and lower limits. For the enrichment ratio to appraoch 1.0, the sediment concentration must be extremely high. Conversely, for the enrichment ratio to approach 1/DR, the sediment concentration must be low. The simultaneous solution of Eq. (25.167) at the boundaries assuming that sediment concentrations range from 500 to 250,000 g·m^{-3} gives

$$x_2 = -\log\left(\frac{1}{DR}\right)/2.699 \tag{25.168}$$

$$x_1 = \frac{1.}{(0.1)^{x_2}} \tag{25.169}$$

Denitrification

As one of the microbial processes, denitrification is a function of temperature and water content. The equation used to estimate the denitrification rate is

$$DN_\ell = WNO3_\ell(1 - \exp[-1.4(TF_{N\ell})(C_\ell)]), SWF \geq 0.95 \qquad (25.170)$$

$$DN = 0., \qquad SWF < 0.95$$

DN is the denitrification rate in layer ℓ in kg·ha^{-1}·d^{-1}, TF_n is the nutrient cycling temperature factor, C is the organic carbon content in %, and SWF is the soil water factor. The temperature factor is expressed by the equation

$$TF_{N\ell} = \max\begin{cases} 0.1 \\ \dfrac{T_\ell}{T_\ell + \exp(9.93 - 0.312T_\ell)} \end{cases} \qquad (25.171)$$

where T is soil temperature in °C and subscript ℓ refers to the layers. The soil water factor considers total soil water in the equation

$$SWF_\ell = \frac{SW_\ell}{FC_\ell} \qquad (25.172)$$

where SW is the soil water content in layer ℓ and FC is the field capacity in mm.

Mineralization

The N mineralization model is a modification of the PAPRAN mineralization model (Seligman and van Keulen, 1981). The model considers two sources of mineralization: fresh organic N pool, associated with crop residue and microbial biomass, and the stable organic N pool, associated with the soil humus. Mineralization from the fresh organic N pool is estimated with the equation

$$RMN_\ell = (DCR_\ell)(FON_\ell) \qquad (25.173)$$

where RMN is the N mineralization rate in kg·ha^{-1}·d^{-1} for fresh organic N in layer ℓ, DCR is the decay rate constant for the fresh organic N, and FON is the amount of fresh organic N present in kg·ha^{-1}. The decay rate constant is a function of C:N ratio, C:P ratio, composition of crop residue, temperature, and soil water:

$$DCR_\ell = 0.05 \quad (CNP_\ell) \sqrt{\left(\frac{SW_\ell}{FC_\ell}\right) \cdot TF_{N\ell}} \qquad (25.174)$$

where CNP is a C:N and C:P ratio factor and FC is the soil water content in mm at field capacity. The value of CNP is calculated with the equation

$$CNP_\ell = \min\begin{cases}\exp\,[-0.693\,(CNR - 25)/25] \\ \exp\,[-0.693\,(CPR_\ell - 200)/200] \\ 1.0\end{cases} \tag{25.175}$$

where CNR is the C:N ratio and CPR is the C:P ratio in layer ℓ. The C:N and C:P ratios of crop residue are computed for each soil layer with the equations

$$CNR_\ell = \frac{0.58FR_\ell}{FON_\ell + WNO3_\ell} \tag{25.176}$$

$$CPR_\ell = \frac{0.58FR_\ell}{FOP_\ell + AP_\ell} \tag{25.177}$$

where FON is the amount of fresh organic N in kg·ha^{-1} , FOP is the amount of fresh organic P in kg·ha^{-1}, and AP is the amount of labile P in kg·ha^{-1} for layer ℓ.

Organic N associated with humus is divided into two pools--active and stable--by using the equation

$$ON_{a\ell} = (RTN_\ell)(ON_\ell) \tag{25.178}$$

where ON_a is the active or readily mineralizable pool in kg·ha^{-1}, RTN is the active pool fraction, ON is the total organic N in kg·ha^{-1}, and the subscript ℓ is the soil layer number. The active pool fraction in the plow layer depends on the number of years the soil has been cultivated and is estimated with the equation

$$RTN_\ell = 0.4\exp(-0.0277YC) + 0.1 \tag{25.179}$$

where YC is period of cultivation before the simulation starts in yr. The concepts expressed in Eq. (25.179) are based on work of Hobbs and Thompson (1971). Below the plow layer the active pool fraction is set at 40% of the plow layer value, based on work of Cassman and Munns (1980). Organic N flux between the active and stable pools is goverened by the equilibrium equation

$$RON_\ell = BKN\left(ON_{a\ell}\left(\frac{1}{RTN_\ell}\right) - ON_{s\ell}\right) \tag{25.180}$$

where RON is the flow rate in kg·ha^{-1}·d^{-1} between the active and stable organic N pools, BKN is the rate constant (≈1 X 10^{-5}·d^{-1}), ON_s is the stable organic N pool, and subscript ℓ is the soil layer number. The daily flow of humus related organic N (RON) is added to the stable pool and subtracted from the active pool.

Only the active pool of organic N is subjected to mineralization. The humus mineralization equation is

$$HMN_\ell = (CMN)(ON_{a\ell})(SWF_\ell \cdot TF_{N\ell})^{0.5}(BD_\ell)^2/(BDP_\ell)^2 \quad (25.181)$$

where HMN is the mineralization rate in kg·ha^{-1}·d^{-1} for the active organic N pool in layer ℓ, CMN is the humus rate constant ($\approx$0.0003 d^{-1}), BD is the settled bulk density of the soil in t·m^{-3}, and BDP is the current bulk density as affected by tillage in t·m^{-3}. To maintain the N balance at the end of the day, the humus mineralization is subtracted from the active organic N pool; the residue mineralization is subtracted from the FON pool; 20% of RMN is added to the active ON pool; and 80% of RMN is added to WNO3 pool.

Immobilization

The daily amount of immobilization is computed by subtracting the amount of N contained in the crop residue from the amount assimilated by the microorganisms:

$$WIM_\ell = (DCR_\ell)(FR_\ell)(0.016 - c_{NFR}) \quad (25.182)$$

where WIM is the N immobilization rate in layer ℓ in kg·ha^{-1}·d^{-1}; 0.016 is the result of assuming that C=0.4 FR, that C:N of the microbial biomass and their labile products = 10, and that 0.4 of C in the residue is assimilated; and c_{NFR} is the N concentration in the crop residue in g·g^{-1}. Immobilization may be limited by N or P availability. If the amount of N available is less than the amount of immobilization predicted from Eq. (25.182), the decay rate constant is adjusted with the relationship

$$DCR'_\ell = \frac{0.95WNO3_\ell}{FR_\ell(0.016 - c_{NFR})} \quad (25.183)$$

where DCR′ allows 95% use of the available NO_3-N in soil layer ℓ. A similar adjustment is made if P is limiting. The crop residue is reduced by using the equation

$$FR_\ell = FR_{o\ell} - (DCR'_\ell)(FR_{o\ell}) \quad (25.184)$$

where FR_o and FR are the amounts of residue in soil layer ℓ at the start and end of a day in kg·ha^{-1}. Finally, the immobilized N is added to the FON pool and subtracted from the WNO3 pool.

Nitrification

Nitrification, the conversion of ammonia N to nitrate N is estimated using a combination of the methods of Reddy et al. (1979) and Godwin et al. (1984).

The approach is based on the first-order kinetic rate equation of Reddy et al. (1979)

$$RNV_i = WNH3_\ell\left(1.-\exp\left(-AKN_\ell - AKV_\ell\right)\right) \qquad (25.185)$$

where RNV is the combined nitrification and volatilization kg·ha^{-1}·d^{-1}, WNH3 is the weight of NH_3 in kg·ha^{-1}, AKN is the nitrification regulator, and AKV is the volatilization regulator for soil layer ℓ. The nitrification regulator is a function of temperature, soil water content, and soil pH as expressed in the equation

$$AKN_\ell = (TF)_\ell (SWF)_\ell (PHF)_\ell \qquad (25.186)$$

where temperature factor, TF, is estimated with the equation

$$TF_\ell = 0.41(T_\ell - 5.)/10., \qquad T > 5. \qquad (25.187)$$

where T_ℓ is the temperature of soil layer ℓ in °C. The soil water factor, SWF, is computed with the equations

$$SWF_\ell = \frac{SW_\ell - WP_\ell}{SW25_\ell - WP_\ell}, \qquad SW_\ell < SW25_\ell \qquad (25.188)$$

$$SWF_\ell = 1., \qquad SW25_\ell < SW_\ell < FC_\ell \qquad (25.189)$$

$$SWF_\ell = 1. - \frac{SW_\ell - FC_\ell}{PO_\ell - FC_\ell}, \qquad SW_\ell > FC_\ell \qquad SWF_\ell \geq 0. \qquad (25.190)$$

where SW is the soil water content, WP is the wilting point soil water content, FC is the field capacity soil water content, and SW25 is the water content at WP+0.25(FC-WP) all in mm for soil layer ℓ. The soil pH factor PHF is computed with the equations

$$PHF_\ell = 0.307PH_\ell - 1.269, \qquad PH_\ell < 7. \qquad (25.191)$$

$$PHF_\ell = 1., \qquad 7. < PH_\ell < 7.4 \qquad (25.192)$$

$$PHF_\ell = 5.367 - 0.599PH_\ell, \qquad PH_\ell > 7.4 \qquad (25.193)$$

where PH is the soil pH value for soil layer ℓ.

Volatilization

Volatilization, the loss of ammonia to the atmosphere, is estimated simultaneously with nitrification. Volatilization of surface-applied ammonia is estimated as a function of temperature and wind speed using the equation

$$AKV_1 = (TF)_1 (WNF) \qquad (25.194)$$

where AKV is the volatilization regulator from Eq. (25.185) and WNF is the wind speed factor for surface application (soil layer ℓ).

$$WNF = 0.335 + 0.16\ln(V_i) \tag{25.195}$$

where V is the mean wind speed in m·s^{-1} for day i. Depth of ammonia within the soil, cation exchange capacity of the soil, and soil temperature are used in estimating below surface volatilization.

$$AKV_\ell = (TF)_\ell (CECF)_\ell (DPF)_\ell, \qquad \ell > 1 \tag{25.196}$$

where the cation exchange capacity factor, CECF, is estimated with the equation

$$\begin{aligned} CECF_\ell &= 1. - 0.038\ CEC_\ell \\ CECF_\ell &\geq 0. \end{aligned} \tag{25.197}$$

where CEC is the cation exchange capacity for soil layer ℓ. The depth factor, DPF, is computed with the equation

$$DPF_\ell = 1. - \frac{ZZ}{ZZ + \exp(4.706 - 0.0305ZZ)} \tag{25.198}$$

where ZZ is the depth to the middle of soil layer ℓ in mm.

$$ZZ = 1000.\left(\frac{Z_\ell + Z_{\ell-1}}{2}\right) \tag{25.199}$$

where Z is the depth to the bottom of layer ℓ in mm. The parameters in Eq. (25.198) are set to give DPF=0.95 at ZZ=5 mm and DPF=0.05 at ZZ=100 mm.

To partition nitrification and volatilization rates appropriately, Eq. (25.195) is solved for each process separately.

$$RNIT = 1. - \exp(-AKN) \tag{25.200}$$

$$RVOL = 1. - \exp(-AKV) \tag{25.201}$$

where RNIT and RVOL are the separate estimates of nitrification and volatilization. However, to satisfy Eq. (25.185), Eqs. (25.200) and (25.201) must be adjusted. Thus, the final equations

$$RVOL^* = \frac{(RNV)(RVOL)}{RVOL + RNIT} \tag{25.202}$$

and

$$RNIT^* = RNV - RVOL^* \tag{25.203}$$

Rainfall

To estimate the N contribution from rainfall, EPIC uses an average rainfall N concentration for a location for all storms. The amount of N in rainfall is estimated as the product of rainfall amount and concentration.

25.5.2. Phosphorus

Soluble P Loss in Surface Runoff

The EPIC approach is based on the concept of partitioning pesticides into the solution and sediment phases as described by Leonard and Wauchope (Knisel, 1980). Because P is mostly associated with the sediment phase, the soluble P runoff equation can be expressed in the simple form

$$YSP = 0.01(c_{LP\ell})(Q)/k_d \tag{25.204}$$

where YSP is the soluble P in kg·ha^{-1} lost in runoff volume Q in mm, $c_{LP\ell}$ is the concentration of AP in soil layer ℓ in g·t, and k_d is the P concentration of the sediment divided by that of the water in m^3·t^{-1}. The value of k_d used in EPIC is 175.

P Transport by Sediment

Sediment transport of P is simulated with a loading function as described in organic N transport. The P loading function is

$$YP = 0.001(Y)(c_p)(ER) \tag{25.205}$$

where YP is the sediment phase P lost in runoff in kg·ha^{-1} and c_p is the concentration of P in the top soil layer in g·t^{-1}.

Mineralization

The P mineralization model developed by Jones et al. (1984) is similar in structure to the N mineralization model. Mineralization from the fresh organic P pool is estimated for each soil layer with the equation

$$RMP_\ell = (DCR_\ell)(FOP_\ell) \tag{25.206}$$

where RMP is the mineralization rate of fresh organic P in layer ℓ in kg·ha^{-1}·d^{-1} and FOP is the fresh organic P in crop residue in kg·ha^{-1}. Mineralization of organic P associated with humus is estimated for each soil layer by using the equation

$$HMP_\ell = \frac{(CMN)(ON_{a\ell})(OP_\ell)(SWF_\ell \cdot TF_{N\ell})^{0.5}(BD_\ell)^2}{(ON_\ell)(BDP_\ell)^2} \tag{25.207}$$

where HMP is the humus P mineralization rate in kg·ha^{-1}·d^{-1} and OP is the organic P content of soil layer ℓ in kg·ha^{-1}. The ratio of ON_a to ON is used in Eq. (25.207) to calculate the active portion of the OP pool. This eliminates the need for maintaining two OP pools corresponding to the ON_a and ON_s pools.

To maintain the P balance at the end of a day, humus mineralization is subtracted from the organic P pool; residue mineralization is subtracted from the FOP pool; 20% of RMP is added to the OP pool; and 80% of RMP is added to labile P pool.

Immobilization

The P immobilization model, also developed by Jones et al. (1984), is similar in structure to the N immobilization model. The daily amount of immobilization is computed by subtracting the amount of P contained in the crop residue from the amount assimilated by the microorganisms:

$$WIP_\ell = (DCR'_\ell)(FR_\ell)(0.16LF_{I\ell} - c_{PFR}) \tag{25.208}$$

where WIP is the P immobilization rate in layer ℓ kg·ha^{-1}·d^{-1}; 0.16 is the result of assuming that C=0.4 FR and that 0.4 of the C in FR is assimilated by soil microorganisms; LF_I is the labile P immobilization factor allowing the P:C ratio of soil microorganisms to range from 0.01 to 0.02 as a function of labile P concentration, and c_{PFR} is the P concentration in the crop residue. The labile P immobilization factor is computed with the equations

$$LF_{I\ell} = 0.01 + 0.001c_{LP\ell}, \qquad c_{LP\ell} \le 10 \tag{25.209}$$

$$LF_{I\ell} = 0.02, \qquad c_{LP\ell} > 10 \tag{25.210}$$

The immobilized P is added to the FOP pool and subtracted from the labile P pool.

Mineral P Cycling

The mineral P model was developed by Jones et al. (1984). Mineral P is transferred among three pools: labile, active mineral, and stable mineral. Fertilizer P is labile (available for plant use) at application but may be quickly transferred to the active mineral pool. Flow between the labile and active mineral pools is governed by the equilibrium equation

$$MPR_\ell = AP_\ell - MP_{a\ell}\left(\frac{PSP_\ell}{1 - PSP_\ell}\right) \tag{25.211}$$

where MPR is the mineral P flow rate for layer ℓ in kg·ha^{-1}·d^{-1}, MP_a is the amount in the active mineral P pool in kg·ha^{-1}, and PSP is the P sorption coefficient defined as the fraction of fertilizer P remaining in the labile pool after the initial rapid phase of P sorption is complete. The daily amount of P computed with Eq. (25.211) flows to the active mineral P pool and is, therefore, added to that pool and subtracted from the labile pool. Obviously, the flow reverses when labile P is less than $MP_{a\ell}\ PSP_\ell / (1-PSP_\ell)$. Since reverse flow is much slower, Eq. (25.211) is multiplied by 0.1 when the resulting MPR_ℓ is negative. The P sorption coefficient is a function of chemical and physical soil properties as described by the following equations (Jones et al., 1984).

In calcareous soils

$$PSP_\ell = 0.58 - 0.0061 CAC_\ell \tag{25.212}$$

In noncalcareous, slightly weathered soils

$$PSP_\ell = 0.02 + 0.0104 AP_\ell \tag{25.213}$$

In noncalcareous, moderately weathered soils

$$PSP_\ell = 0.0054 BSA_\ell + 0.116 PH_\ell - 0.73 \tag{25.214}$$

In noncalcareous, highly weathered soils

$$PSP_\ell = 0.46 - 0.0916 \ln CLA_\ell \tag{25.215}$$

where PSP is the P sorption coefficient for soil layer ℓ, CAC is the $CaCO_3$ concentration in $g{\cdot}t^{-1}$, and BSA is the base saturation by the ammonium acetate (NH_4OA_c) method in %. PSP is constrained within the limits of $0.05 \leq PSP \leq 0.75$. At equilibrium the stable P pool is assumed to be four times as large as the active mineral P pool. Flow between the P pools is governed by the equation

$$ASPR_\ell = \omega_\ell (4 MP_{a\ell} - MP_{s\ell}) \tag{25.216}$$

where ASPR is the flow rate between the active and stable mineral P pools in $kg{\cdot}ha^{-1}{\cdot}d^{-1}$ for soil layer ℓ, ω is the flow coefficient in d^{-1}, and MP_s is the amount of stable mineral P in $kg{\cdot}ha^{-1}$. The daily amount of P computed with Eq. (25.215) flows into the stable pool and is subtracted from the active pool. Obviously, the flow reverses when $MP_{s\ell} > 4\,MP_{a\ell}$ Since reverse flow is much slower, Eq. (25.216) is multiplied by 0.1 when the resulting ASPR is negative. The flow coefficient, ω, is a function of PSP as expressed by the equations (Jones et al., 1984)

$$\omega_\ell = \exp(-1.77 PSP_\ell - 7.05) \tag{25.217}$$

for noncalcareous soils, and

$$\omega_\ell = 0.0076 \tag{25.218}$$

for calcareous soils.

25.6. PESTICIDE FATE

GLEAMS (Leonard et al., 1987) technology for simulating pesticide transport by runoff, percolate, soil evaporation, and sediment was added to EPIC. Pesticides may be applied at any time and rate to plant foliage or below the soil surface at any depth. When pesticide is applied, there is a loss to the atmosphere. Thus, the amount that reaches the ground or plants is expressed by the equation

$$PAPE=(PAPR)(PAEF) \tag{25.219}$$

where PAPE is the effective amount of pesticide applied in kg·ha^{-1} PAPR is the actual amount applied in kg·ha^{-1}, and PAEF is an application efficiency factor.

To determine how much pesticide reaches the ground, the amount of ground cover provided by plants is estimated with the equation

$$GC_{\ell}=(1.-ERFC(1.33LAI_{\ell,m}-2.))/2. \tag{25.220}$$

where GC is the fraction of the ground that is covered by plants, and LAI is the leaf area index on day, l. Therefore, the amount of pesticide that reaches the plants is computed with the equation

$$FP=(GC)(PAPE) \tag{25.221}$$

where FP is the amount of pesticide that is intercepted by plants. The remaining pesticide falls to the ground and is simply the difference between the effective amount applied and the amount intercepted by plants.

$$GP=PAPE-FP \tag{25.222}$$

where GP is the amount of pesticide that reaches the ground.

Pesticide that remains on the plant foliage can be washed off by rain storms. It is assumed that the fraction of pesticide that is potentially dislodgeable is washed off the plants once a threshold rainfall amount is exceeded. The model uses a threshold value of 2.5 mm and potential washoff fractions for various pesticides have been estimated (Leonard et al., 1987). The appropriate equations for computing washoff and adding to the ground amount are:

$$\begin{aligned} WO &= (WOF)\ (FP), && R \geq 2.5mm \\ WO &= 0., && R < 2.5mm \end{aligned} \qquad 25.223)$$

$$GP=GP+WO \tag{25.224}$$

$$FP=FP-WO \tag{25.225}$$

where WO is the amount of pesticide washed off the plants by a rainstorm of R in mm and WOF is the washoff fraction for the particular pesticide.

Pesticide on the plants and in the soil is lost from the system based on the decay equations

$$GP = (GP_o)\exp(-0.693/HLS) \tag{25.226}$$

$$FP = (FP_o)\exp(-0.693/HLP) \tag{25.227}$$

where GP_o and GP are the initial and final amounts of pesticide on the ground, FP_o and FP are the initial and final amounts of pesticide on the plants, HLS is the half life for pesticide in the soil in d, and HLP is the half life of the foliar residue in d. Values of HLP and HLS have been established for various pesticides (Leonard et al., 1987).

Another way that pesticide can be lost from the zone considered in computing runoff (top 10 mm of soil) is through leaching. The GLEAMS leaching component is used here with slight modification. The change in the amount of pesticide contained in the top 10 mm zone is expressed as a function of time, concentration, and amount of infiltration using the equation

$$\frac{dGP}{dt} = (PSTC_w)(f) \tag{25.228}$$

where GP is the amount of pesticide in the top 10 mm zone at time t, $PSTC_w$ is the pesticide concentration in the water in $g \cdot t^{-1}$, and f is the water flow rate through the zone in $mm \cdot h^{-1}$. The total amount of pesticide contained in the top 10 mm zone is the sum of the adsorbed and mobile phases.

$$GP = 0.01(PSTC_w)(SW) + 0.1(PSTC_s)(BD) \tag{25.229}$$

where SW is the amount of water stored in the top 10 mm of soil in mm, PSTC is the concentration of adsorbed pesticide in $g \cdot t^{-1}$, and BD is the soil bulk density in $t \cdot m^{-3}$. The ratio of the concentration of pesticide adsorbed to the concentration of pesticide in the water has been estimated for various pesticides (Leonard et al., 1987) and is expressed by the equation

$$K_d = \frac{c_s}{c_w} \tag{25.230}$$

where K_d is the partitioning constant. Substituting Eq. (25.230) into Eq. (25.229) gives

$$GP = 0.01(PSTC_w)(SW) + 0.1(PSTC_w)(K_d)(BD) \tag{25.231}$$

Solving Eq. (25.231) for $PSTC_w$ gives

$$PSTC_w = \frac{GP}{0.01SW + 0.1(K_d)(BD)} \tag{25.232}$$

Substituting $PSTC_w$ from Eq. (25.232) into Eq. (25.228) yields

$$\frac{dGP}{dt} = \frac{(GP)(f)}{0.01SW + 0.1(K_d)(BD)} \tag{25.233}$$

Rearranging Eq. (25.233) and integrating gives the equation expressing the amount of pesticide as a function of the amount of water flowing through the zone.

$$GP = GP_o \exp\left(\frac{-f}{0.01SW + 0.01(K_d)(BD)}\right) \tag{25.234}$$

where GP_o is the initial amount of pesticide in the top 10 mm zone in kg·ha^{-1}, GP is the amount that remains after the amount of flow (f in mm) passes through the zone, SW is the initial water storage in mm, k_d is the partitioning coefficient in m^3·t^{-1}, and BD is the soil bulk density in t·m^{-3}. To obtain the amount of pesticide leached by the amount of water f, GP is subtracted from GP_o using the equation

$$PSTL = GP_o\left(1. - \exp\left(\frac{-f}{0.01SW + 0.1(K_d)(BD)}\right)\right) \tag{25.235}$$

where PSTL is the amount of pesticide leached by f.

Pesticide concentration in percolate is computed with the equation

$$PSTC_L = \min\begin{cases} PSTL/f \\ PSO1 \end{cases} \tag{25.236}$$

where PSO1 is the pesticide solubility and $PSTC_L$ is the pesticide concentration in the percolate in g·m^{-3}. Finally, PSTL is the product of O and $PSTC_L$.

Pesticide loss in surface runoff is estimated with a modification of Eq. (25.235) that includes an abstraction coefficient

$$PSTQ = GP_o\left(1. - \exp\left(\frac{-(Q)(ab)}{0.01SW + 0.1(ab)(K_d)(BD)}\right)\right) \tag{25.237}$$

where PSTQ is the pesticide loss in surface runoff in kg·ha^{-1}, Q is the surface runoff volume in mm, and ab is the abstraction coefficient.

The calculation sequence is: (1) Soil layer 1--Leaching is estimated first by substituting 0 for f in Eqs. (25.235) and (25.236). Then PSTL is removed from layer 1 and placed into layer 2. Next runoff and lateral flow losses are estimated simultaneously by substituting Q+QR for Q in Eq. (25.237). The average concentration is obtained by dividing the resulting PSTQ by Q+QR. Individual losses are estimated for runoff ($Q \cdot PSTC_{Q+QR}$) and lateral flow ($QR \cdot PSTC_{Q+QR}$). Finally, these losses are removed from layer 1. (2) Soil layers 2-M--Leaching and lateral flow losses are estimated simultaneously by substituting $O_\ell + QR_\ell$ for f in Eqs. (25.235) and (25.236). Individual losses are calculated as in (1) ($QR_\ell \cdot PSTC_L$) and ($O_\ell \cdot PSTC_L$), subtracted from layer ℓ and $O_\ell \cdot PSTC_L$ is added to layer ℓ+1. The process is repeated layer by later to the bottom of the soil profile.

The total amount of pesticide lost in the runoff is estimated by adding the soluble fraction computed with Eq. (25.237) to the amount that is adsorbed to the sediment. Pesticide yield from the adsorbed phase is computed with an enrichment ratio approach. The enrichment ratio equation is

$$PSTY = (Y)(PSTC_s)(ER)(0.001) \tag{25.238}$$

where PSTY is the pesticide yield adsorbed to the sediment in $kg \cdot ha^{-1}$, Y is the sediment yield in $t \cdot ha^{-1}$, and ER is the enrichment ratio (concentration of pesticide in the sediment divided by the pesticide concentration in the top 10 mm of soil) computed with Eq. (25.167). The pesticide concentration in the soil is calculated by substituting Eq. (25.230) into Eq. (25.232) and solving for $PSTC_S$.

$$PSTC_s = \frac{(K_d)(GP)}{0.01SW + 0.1(K_d)(BD)} \tag{25.239}$$

Soil layers with low storage volumes have high leaching potentials not only because percolation is greater, but also because storage volume displacement is greater (higher concentration). Pesticides with low K_d values and high solubility are transported rapidly with water. Conversely, high K_d value pesticides are adsorbed to soil particles and travel largely with sediment.

25.7. SOIL TEMPERATURE

Daily average soil temperature at the center of each soil layer is simulated for use in nutrient cycling and hydrology. The basic soil temperature equation is

$$T_{\ell,i} = LAG(T_{\ell,i-1}) + (1.0 - LAG)\ (FZ_\ell(\bar{T} + TG_i) + TG_i) \tag{25.240}$$

where T is the soil temperature at the center of layer ℓ on day i in °C, LAG is a coefficient ranging from 0.0 to 1.0 that allows proper weighting of yesterday's temperature, $\bar{T}$ is the long-term average annual air temperature at the site, TG is the soil surface temperature, and FZ is a depth factor. Thus, given yesterday's temperature, Eq. (25.240) estimates today's temperature as a function of soil surface temperature, depth, and a lag coefficient. It is assumed that the temperature remains almost constant at some depth called damping depth and is approximately $\bar{T}$. The depth weighting factor governs temperature changes between the soil surface and the damping depth according to the equation

$$FZ_\ell = \frac{ZD}{ZD + \exp(-0.867 - 2.08ZD)} \tag{25.241}$$

where

$$ZD = \frac{Z_\ell + Z_{\ell-1}}{2.0DD} \tag{25.242}$$

where Z is soil depth from the surface in m and DD is the damping depth in m. Obviously, Eqs. (25.241) and (25.242) make near surface temperatures a strong function of TG. As depth increases, $\overline{\text{T}}$ has more influence until finally at the damping depth, the temperature is within 5% of $\overline{\text{T}}$.

The damping depth is a function of soil bulk density and water content as expressed in the equation

$$DP = 1.0 + \frac{2.5BD}{BD + \exp(6.53 - 5.63BD)} \tag{25.243}$$

$$\S = \frac{0.001(SW)}{(0.356 - 0.144BD)Z_M} \tag{25.244}$$

$$DD = DP \exp(\ln \left(\frac{0.5}{DP}\right)\left(\frac{1-\S}{1+\S}\right)^2) \tag{25.245}$$

where DP is the maximum damping depth for soil in m, BD is the soil bulk density in $t{\cdot}m^{-3}$, Z_M is the distance from the bottom of the lowest soil layer to the surface in m, and § is a scaling parameter.

To complete the solution of Eq. (25.240), the soil surface temperature must be estimated. The first step is to estimate the bare soil surface temperature (TGB). Of course, TGB is usually closely related to the air temperature. Other important factors that also influence TGB are precipitation, solar radiation, and previous soil temperature. When precipitation occurs, the soil surface temperature usually decreases. Thus, the appropriate air temperature for estimating TGB is near the daily minimum.

$$TGBW_i = T_{mn,i} + \Omega_s(T_{mx,i} - T_{mn,i}) \tag{25.246}$$

where TGBW is the bare soil surface temperature on wet day, i, and Ω_s is a scaling factor to adjust for wet days. The value of Ω_s ranges from 0.0 to 1.0, but more realistic results can be obtained using $\Omega_s \approx 0.1$. The companion equation for dry days is derived from the continuity equation (Eq. (25.103).

$$TGBD_i = \frac{T_i - PW_k(T_{mn,i} + \Omega_s(T_{mx,i} - T_{mn,i}))}{1. - PW_k} \tag{25.247}$$

where TGBD is the bare soil surface temperature on dry day, i, and PW is the probability of a wet day in month k.

The soil surface temperature is also affected by solar radiation. Therefore, a net radiation term is added to Eqs. (25.246) and (25.247).

$$TGBW_i = T_{mn,i} + \Omega_s(T_{mx,i} - T_{mn,i}) + RAH_i \tag{25.248}$$

$$TGBD_i = \frac{T_i - PW_k(T_{mn,i} + \Omega_s(T_{mx,i} - T_{mn,i}))}{1. - PW_k} + RAH_i \qquad (25.249)$$

where the radiation term RAH is estimated with the equation

$$RAH_i = 2.\left(\left(\frac{RA_i(1. - AB_i)(EA_i)}{15.}\right)^2 - 1.\right) \qquad (25.250)$$

where solar radiation, RA; albedo, AB; and the cover factor, EA are discussed and evaluated in the Evapotranspiration section. Equation (25.250) increases soil surface temperatures when radiation is high but also makes adjustments for albedo and cover. Low radiation can also cause up to a two °C drop in surface temperature.

If the soil surface is not bare, the surface temperature can be affected considerably by the amount of cover (crop residue or snow). This effect can be simulated by combining the estimated bare surface temperature for the day with the previous day's temperature in the second soil layer (the top 10 mm layer is considered too thin for this purpose).

$$TG_i = (bcv)(T_{2,i}) + (1. - bcv)(TGB^*{}_i) \qquad (25.251)$$

where TG is the final estimate of soil surface temperature in °C and bcv is a lagging factor for simulating residue and snow cover effects on surface temperature. The value of bcv is 0 for bare soil and approaches 1.0 as cover increases, as expressed in the equation

$$bcv = \max\begin{cases} \dfrac{CV}{CV + \exp\ (5.34 - 2.40CV)} \\ \dfrac{SNO}{SNO + \exp\ (2.303 - 0.2197SNO)} \end{cases} \qquad (25.252)$$

where CV is the sum of above ground biomass and crop residue in $t \cdot ha^{-1}$ and SNO is the water content of the snow cover in mm.

25.8. CROP GROWTH MODEL

A single model is used in EPIC for simulating all the crops considered (corn, grain sorghum, wheat, barley, oats, sunflower, soybean, alfalfa, cotton, peanuts, potatoes, durham wheat, winter peas, faba beans, rapeseed, sugarcane, sorghum hay, range grass, rice, casava, lentils, and pine trees). Of course, each crop has unique values for the model parameters. EPIC is capable of simulating growth for both annual and perennial crops. Annual crops grow from planting date to harvest date or until the accumulated heat units equal the potential heat units for the crop. Perennial crops maintain their root systems throughout the year, although they may become dormant after frost. They start growing when the average daily air temperature exceeds their base temperature.

Phenological development of the crop is based on daily heat unit accumulation. It is computed by using the equation

$$HU_k = \left(\frac{T_{mx,k} + T_{mn,k}}{2}\right) - T_{b,j}, \qquad HU_k \geq 0 \tag{25.253}$$

where HU, T_{mx}, and T_{mn} are the values of heat units, maximum temperature, and minimum temperature in °C on day k, and T_b is the crop-specific base temperature in °C (no growth occurs at or below T_b) of crop j. A heat unit index (HUI) ranging from 0 at planting to 1 at physiological maturity is computed as follows:

$$HUI_i = \frac{\sum_{k=1}^{i} HU_k}{PHU_j} \tag{25.254}$$

where HUI is the heat unit index for day i and PHU is the potential heat units required for the maturation of crop j. The value of PHU may be inputted or calculated by the model from normal planting and harvest dates. Date of harvest, leaf area growth and senescence, optimum plant nutrient concentrations, and partition of dry matter among roots, shoots, and economic yield are affected by HUI.

25.8.1. Potential Growth

Interception of solar radiation is estimated with a Beer's law equation (Monsi and Saeki, 1953)

$$PAR_i = 0.5(RA)_i[1. - \exp(-0.65 LAI_i)] \tag{25.255}$$

where PAR is intercepted photosynthetic active radiation in $MJ \cdot m^{-2}$, RA is solar radiation in $MJ \cdot m^{-2}$, LAI is the leaf area index, and subscript i is the day of the year. The constant 0.5 is used to convert solar radiation to photosynthetically active radiation (Monteith, 1973). Experimental studies indicate that the extinction coefficient varies with foliage characteristics, sun angle, row spacing, row direction, and latitude (Thornley, 1976). The value used in EPIC (0.65) is representative of crops with narrow row spacings (Uchijima et al., 1968). A somewhat smaller value (0.4-0.6) might be appropriate for tropical areas in which average sun angle is higher and for wide row spacings (Begg et al., 1964; Bonhomme et al., 1982; Muchow et al., 1982). Using Monteith's approach (Monteith, 1977), potential increase in biomass for a day can be estimated with the equation

$$\Delta B_{p,i} = 0.001(BE)_j(PAR)_i \quad (25.256)$$

where ΔB_p is the daily potential increase in biomass in t·ha^{-1}, BE is the crop parameter for converting energy to biomass in kg·ha^{-1}·MJ^{-1}·m^2.

Biomass energy conversion is affected by vapor pressure deficit (VPD) and by atmospheric CO_2 level. The biomass conversion factor BE is adjusted using the equations of Stockle et al. (1992)

$$BE^* = \frac{100.CO_2}{CO_2 + \exp(bc_1 - bc_2(CO_2))} \quad (25.257)$$

where CO_2 is the atmospheric CO_2 level in ppm and bc_1 and bc_2 are crop parmeters. The VPD correction is accomplished in the equation

$$BE' = BE^* - bc_3(VPD - 1.) \qquad VPD > 0.5 \quad (25.258)$$

where VPD is the vapor pressure deficit in kPa and bc_3 is a crop parameter. The value of VPD is e_a (saturation vapor pressure at mean air temperature) minus e_d (vapor pressure at mean air temperature). Finally BE' is substituted into Eq. (25.256) to obtain the corrected biomass estimate.

In most crops, leaf area index (LAI) is initially zero or very small. It increases expoentially during early vegetative growth, when the rates of leaf primordia development, leaf tip appearance, and blade expansion are linear functions of heat unit accumulation (Tollenaar et al., 1979; Watts, 1972). In vegetative crops such as sugarcane and some forages, LAI reaches a plateau, at which time the rates of senescence and growth of leaf area are approximately equal. In many crops, LAI decreases after reaching a maximum and approaches zero at physiological maturity. In addition, leaf expansion, final LAI, and leaf duration are reduced by stresses (Acevedo et al., 1971; Eik and Hanway, 1965).

LAI is simulated as a function of heat units, crop stress, and crop development stages. From emergence to the start of leaf decline, LAI is estimated with the equations

$$LAI_i = LAI_{i-1} + \Delta LAI \quad (25.259)$$

$$\Delta LAI = (\Delta HUF)(LAI_{mx})(1.- \exp[(5.0(LAI_{i-1} - LAI_{mx})]) \sqrt{REG_i} \quad (25.260)$$

where LAI is the leaf area index, HUF is the heat unit factor, and REG is the value of the minimum crop stress factor discussed in more detail below. Subscript mx is the maximum value possible for the crop and Δ is the daily change. The exponential function of Eq. (25.260) prevents LAI from exceeding LAI_{mx} when HUF is adjusted for vernalization of certain crops. The heat unit factor is computed using the equation

$$HUF_i = \frac{HUI_i}{HUI_i + \exp[ah_{j,1} - (ah_{j,2})(HUI_i)]} \quad (25.261)$$

where $ah_{j,1}$ and $ah_{j,2}$ are parameters of crop j, and HUI is the heat unit index.

From the start of leaf decline to the end of the growing season, LAI is estimated with the equation

$$LAI_i = LAI_0 \left(\frac{1 - HUI_i}{1 - HUI_0} \right)^{ad_j} \tag{25.262}$$

where ad is a parameter that governs LAI decline rate for crop j and subscript o is the day of the year when LAI starts declining.

Crop height is estimated with the relationship

$$CHT_i = HMX_j \sqrt{HUF_i} \tag{25.263}$$

where CHT is the crop height in m and HMX is the maximum height for crop j.

The fraction of total biomass partitioned to the root system normally decreases from 0.3 to 0.5 in the seedling to 0.05 to 0.20 at maturity (Jones, 1985). The model simulates this partitioning by decreasing the fraction linearly from emergence to maturity. Thus, the potential daily change in root weight is computed with the equation

$$\Delta RWT_i = \Delta B_{P,i}(ar_1 - ar_2 HUI_i) \tag{25.264}$$

where ΔRWT is the change in root weight in $t \cdot ha^{-1}$ and ar_1 and ar_2 are crop parameters with typical values of ≈0.4 and 0.2. The potential change in root weight through the root zone is simulated as a function of plant water use in each layer of soil with the equation

$$\Delta RW_{i,\ell} = (\Delta RWT_i) \left(\frac{u_{i,\ell}}{\sum_{\ell=1}^{M} u_{i,\ell}} \right) \tag{25.265}$$

where RW is the root weight in soil layer ℓ in $t \cdot ha^{-1}$, M is the total number of soil layers, and u is the daily water use rate in layer ℓ in $mm \cdot d^{-1}$.

Rooting depth normally increases rapidly from the seeding depth to a crop-specific maximum. In many crops, the maximum is usually attained well before physiological maturity (Borg and Grimes, 1986). Rooting depth is simulated as a function of heat units and potential root zone depth:

$$\Delta RD_i = 2.5(RDMX_j)(\Delta HUF_i), \quad RD_i \le RZ \tag{25.266}$$

where RD is the root depth in m, RDMX is the maximum root depth in m for crop j in ideal soil, and RZ is the soil profile depth in m.

The economic yield of most grain, pulse, and tuber crops is a reproductive organ. Crops have a variety of mechanisms which ensure that their production is neither too great to be supported by the vegetative components nor too small to ensure survival of the species. As a result, harvest index (economic yeild/above-ground biomass) is often a relatively stable value across a range of environmental conditions. In EPIC, crop yield is estimated by using the harvest index concept:

$$YLD_j = (HI_j)(B_{AG})_j \tag{25.267}$$

where YLD is the amount of the crop removed from the field in t·ha^{-1}, HI is the harvest index, and B_{AG} is the above-ground biomass in t·ha^{-1} for crop j. For non-stressed conditions harvest index increases nonlinearly from 0 at planting to HI at maturity according to the equation

$$HIA_i = HI_j(\sum_{k=1}^{i} \Delta HUFH_k) \tag{25.268}$$

where HIA is the harvest index on day i and HUFH is the heat unit factor that affects harvest index.

The harvest index heat unit factor is computed with the equation

$$HUFH_i = \frac{HUI_i}{HUI_i + \exp(11.1 - 0.1 HUI_i)} \tag{25.269}$$

The constants in Eq. (25.269) are set to allow $HUFH_i$ to increase from 0.1 at HUI_i=0.5 to 0.95 at HUI_i=0.95. This is consistent with the economic yield development of grain crops, which produce the greatest economic yield in the second half of the growing season.

25.8.2. Water Use

The potential water use, E_p, is estimated as described in the evapotranspiration section. The potential water use from the soil surface to any root depth is estimated with the function

$$U_{Pi} = \frac{E_{Pi}}{1. - \exp(-\Lambda)}\left(1. - \exp\left[-\Lambda\left(\frac{Z}{RZ}\right)\right]\right) \tag{25.270}$$

where U_p is the total water use in rate in mm·d^{-1} to depth Z in m on day i, RZ is the root zone depth in m, and Λ is a water use distribution parameter. The amount used in a particular layer can be calculated by taking the difference between U_{Pi} values at the layer boundaries:

$$u_{P\ell} = \frac{E_{Pi}}{1.-\exp(-\Lambda)}\left(\left(1.-\exp\left[-\Lambda\left(\frac{Z_\ell}{RZ}\right)\right]\right) - \left(1.-\exp\left[-\Lambda\left(\frac{Z_{\ell-1}}{RZ}\right)\right]\right)\right) \tag{25.271}$$

where $u_{p\ell}$ is the potential water use rate for layer ℓ in mm·d^{-1}. Equation (25.271) applies to a soil that provides poor conditions for root development when Λ is set to a high value like 10. The high Λ value gives high water use near the surface and very low use in the lower half of the root zone. Since there is no provision for water deficiency compensation in any layer, considerable water stress may be incorrectly indicated if Eq. (25.271) is used. To overcome this problem, Eq. (25.271) was modified to allow plants to compensate for water deficiency in a layer by using more water from other layers. Total compensation can be accomplished by taking the difference between U_{Pi} at the bottom of a layer and the sum of water use above a layer:

$$u_{P\ell} = \frac{E_{Pi}}{1. - \exp(-\Lambda)}\left(1. - \exp\left[-\Lambda\left(\frac{Z_\ell}{RZ}\right)\right]\right) - \sum_{k=1}^{\ell-1} u_k \qquad (25.272)$$

where u_k is the actual water use rate in mm·d^{-1} for all layers above layer ℓ. Thus, any deficit can be overcome if a layer that is encountered has adequate water storage. Neither Eq. (25.271) (no compensation) nor Eq. (25.272) (total compensation) is satisfactory to simulate a wide range of soil conditions. A combination of the two equations, however, provides a very general water use function:

$$u_{P\ell} = \frac{E_{Pi}}{1.- \exp(-\Lambda)}\left(1.-\exp\left[-\Lambda\left(\frac{Z_\ell}{RZ}\right)\right]-(1.-UC)\left(1.-\exp\left[-\Lambda\left(\frac{Z_{\ell-1}}{RZ}\right)\right]\right)\right)-UC\sum_{k=1}^{\ell-1} u_k \qquad (25.273)$$

where UC varies over a range (0.-1.) and is the water deficit compensation factor. In soils with a good rooting environment, UC=1. gives total compensation. The other extreme, poor conditions, allow no compensation (UC=0). The procedure for estimating UC is described in the Growth Constraints section of this chapter.

The potential water use in each layer calculated with Eq. (25.273) is reduced when the soil water storage is less than 25% of plant-available soil water (Jones and Kiniry, 1986) by using the equation

$$u_\ell = u_{P\ell} \exp\left(5.\left(\frac{4.(SW_{\ell i} - WP_\ell)}{(FC_\ell - WP_\ell)} - 1.\right)\right), \; SW_\ell < \frac{FC_\ell - WP_\ell}{4} + WP_\ell \qquad (25.274)$$

$$u_\ell = u_{P\ell}, SW_\ell \geq \frac{FC_\ell - WP_\ell}{4.} + WP_\ell \qquad (25.275)$$

where SW is the soil water content in layer ℓ on day i in mm and FC and WP are the soil water contents at field capacity and wilting point for layer ℓ.

25.8.3. Nutrient Uptake

Nitrogen

Supply and Demand. Crop use of N is estimated by using a supply and demand approach. The daily crop N demand is the difference between the crop N content and the ideal N content for that day. The demand is estimated with the equation

$$UND_i = (c_{NB})_i (B)_i - \sum_{k=1}^{i-1} UN_k \tag{25.276}$$

where UND is the N demand rate of the crop in kg·ha^{-1}·d^{-1}, c_{NB} is the optimal N concentration of the crop in kg·t^{-1}, B is the accumulated biomass in t·ha^{-1} for day i, and UN is the actual N uptake rate in kg·ha^{-1}·d^{-1}. The optimal crop N concentration declines with increasing growth stage (Jones, 1983a) and is computed as a function of growth stage by using the equation

$$c_{NBi} = bn_1 + bn_2 \exp(-bn_3 HUI_i) \tag{25.277}$$

where bn_1, bn_2, and bn_3 are crop parameters expressing N concentration and HUI (heat unit index) is the fraction of the growing season.

Soil supply of N is assumed to be limited by mass flow of NO_3-N to the roots

$$UN_{\ell,i} = u_{\ell,i} \left(\frac{WNO3_\ell}{SW_\ell} \right)_i \tag{25.278}$$

where UN is the rate of N supplied by the soil in kg·ha^{-1}·d^{-1}, WNO3 is the amount of NO_3-N in kg·ha^{-1}, SW is the soil water content in mm, u is water use rate in mm·d^{-1}, and subscript ℓ refers to the soil layers. The total mass flow supply is estimated by summing the layer supplies:

$$UNS_i = \sum_{\ell=1}^{M} UN_{\ell,i} \tag{25.279}$$

where UNS is the N supply rate from soil to plants in kg·ha^{-1}·d^{-1}. Since mass flow uptake can produce questionable results when N concentrations are extremely high or low, UN values obtained from Eq. (25.278) are adjusted:

$$UNa_{\ell,i} = UN_{\ell,i} \left(\frac{UND_i}{UNS_i} \right), \qquad UNa_{\ell,i} \le WNO3_{\ell,i} \tag{25.280}$$

Equation (25.280) assures that actual N uptake cannot exceed the plant demand when mass flow estimates are too large. It also provides for increased N supply when mass flow estimates are too low despite the availability of NO_3.

Fixation. Daily N fixation is estimated as a fraction of daily plant N uptake for legumes:

$$WFX_i = FXR_i \cdot UN_i, \qquad WFX \leq 6.0 \tag{25.281}$$

where WFX is the amount of N fixation in $kg \cdot ha^{-1}$ and FXR is the fraction of uptake for day i. The fraction, FXR, is estimated as a function of soil NO_3 and water content and plant growth stage.

$$FXR = \min(1.0, FXW, FXN) \cdot FXG \tag{25.282}$$

where FXG is the soil growth stage factor, FXW is the soil water content factor, and FXN is the soil NO_3 content factor. The growth stage factor inhibits N fixation in young plants prior to development of functional nodules and in old plants with senescent nodules (Patterson and LaRue, 1983).

$$FXG_i = 0.0, \qquad HUI_i \leq 0.15, \quad HUI \geq 0.75 \tag{25.283}$$

$$FXG_i = 6.67 HUI_i - 1.0, \qquad 0.15 < HUI_i \leq 0.3 \tag{25.284}$$

$$FXG_i = 1.0, \qquad 0.3 < HUI_i \leq 0.55 \tag{25.285}$$

$$FXG_i = 3.75 - 5.0 HUI_i, \qquad 0.55 < HUI_i \leq 0.75 \tag{25.286}$$

where HUI is the heat unit index for day i. The soil water content factor reduces N fixation when the water content at the top 0.3 m is less than 85% of field capacity (Albrecht et al., 1984; Bouniols et al., 1991) using the equation

$$FXW_i = \frac{SW3_i - WP3}{0.85(FC3 - WP3)}, \qquad SW3 < 0.85(FC3 - WP3) + WP3 \tag{25.287}$$

where SW3, WP3, and FC3 are the water contents in the top 0.3 m of soil on day i, at wilting point, and at field capacity.

The amount of NO_3 in the root zone can affect N fixation (Harper, 1976; Bouniols et al., 1985) determines the soil NO_3 factor, FXN.

$$FXN = 0., \qquad WNO3 > 300. kg \cdot ha^{-1} \cdot m^{-1} \tag{25.288}$$

$$FXN = 1.5 - 0.005(\frac{WNO3}{RD}), \quad 100. < WNO3 \leq 300. \tag{25.289}$$

$$FXN = 1.0, \quad WNO3 \le 100. kg \cdot ha^{-1} \cdot m^{-1} \tag{25.290}$$

where WNO3 is the weight of NO_3-N in the root zone in kg·ha^{-1}·m^{-1} and RD is the root depth in m. This approach reduces N fixation when the NO_3-N content of the root zone is greater than 100 kg·ha^{-1}·m^{-1} and prohibits N fixation at N contents greater than 300 kg·ha^{-1}·m^{-1}.

Phosphorus

Crop use of P is estimated with the supply and demand approach described in the N model. The daily demand is computed with Eq. (25.276) written in the form

$$UPD_i = (c_{PB})_i (B)_i - \sum_{k=1}^{i-1} UP_k \tag{25.291}$$

where UPD is the P demand for the plant in kg·ha^{-1}, UP is the actual P uptake in kg·ha^{-1}, and c_{PB} is the optimal P concentration for the plant. As in the case of N, the optimal plant P concentration is computed with Eq. (25.277) written in the form

$$c_{PBi} = bp_1 + bp_2 \exp(-bp_3 HUI_i) \tag{25.292}$$

where bp_1, bp_2, and bp_3 are parameters calculated from crop-specific optimum P concentration at the seedling stage, halfway through the season and at maturity. Soil supply of P is estimated using the equation

$$UPS_i = 1.5 UPD_i \sum_{\ell=1}^{M} (LF_u)_\ell \left(\frac{RW_\ell}{RWT_i} \right) \tag{25.293}$$

where UPS is the amount of P supplied by the soil in kg·ha^{-1}, LF_u is the labile P factor for uptake, RW is the root weight in layer ℓ in kg·ha^{-1}, and RWT is the total root weight on day i in kg·ha^{-1}. The constant 1.5 allows 2/3 of the roots to meet the P demand of the plant if labile P is not limiting This approach is consistent with studies suggesting that roots of P-deficient plants (or plants whose root systems have been pruned) can absorb P faster than the roots of normal plants (Andrews and Newman, 1970; DeJager, 1979; Jungk and Barber, 1974).

The labile P factor for uptake ranges from 0.1 to 1.0 according to the equation

$$LF_{u\ell} = 0.1 + \frac{0.9 c_{LP\ell}}{c_{LP\ell} + \exp(8.01 - 0.360 C_{LP\ell})} \tag{25.294}$$

where c_{LP} is the labile P concentration in soil layer ℓ in g·t^{-1}. Equation (25.294) allows optimum uptake rates when c_{LP} is above 20 g·t^{-1}. This is consistent with critical labile P concentrations for a range of crops and soils (Sharpley et al., 1990). Sharpley et al. (1984, 1985) described methods of estimating c_{LP} from soil test P and other soil characteristics.

25.8.4. Growth Constraints

Potential crop growth and yield are usually not achieved because of constaints imposed by the plant environment. The model estimates stresses caused by water, nutrients, temperature, aeration, and radiation. These stresses range from 0.0 to 1.0 and affect plants in several ways. In EPIC, the stresses are considered in estimating constraints on biomass accumulation, root growth, and yield. The biomass constraint is the minimum of the water, nutrient, temperature, and aeration stresses. The root growth constraint is the minimum of soil strength, temperature, and aluminum toxicity. Though topsoil aluminum toxicity can have a direct effect on shoot growth, EPIC simulates only its indirect effects through its inhibition of root growth and water use. A description of the stress factors involved in determining each constraint follows.

Biomass

The potential biomass predicted with Eq. (25.256) is adjusted daily if any of the five plant stress factors is less than 1.0 using the equation

$$\Delta B = (\Delta B_p)(REG) \tag{25.295}$$

where REG is the crop growth regulating factor (the minimum stress factor).

Water Stress. The water stress factor is computed by considering supply and demand in the equation

$$WS_i = \frac{\sum_{\ell=1}^{M} u_{i,\ell}}{E_{Pi}} \tag{25.296}$$

where WS is the water stress factor, u is the water use in layer ℓ, and E_P is the potential plant water use on day i. This is consistent with the concept that drought stress biomass production in proportion to transpiration reduction (Hanks, 1983).

Temperature Stress. The plant temperature stress is estimated with the equation

$$TS_i = \text{sine}\left(\frac{\pi}{2}\left(\frac{TG_i - T_{bj}}{T_{oj} - T_{bj}}\right)\right), \quad 0 \le TS_i \le 1 \tag{25.297}$$

where TS is the plant temperature stress factor, TG is the average daily soil surface temperature in °C, T_b is the base temperature for crop j, and T_o is the optimal temperature for crop j. Equation (25.297) produces symmetrical plant growth stress about the optimal temperature and it is driven by average daily soil surface temperature. This approach allows growth of small plants to respond realistically to low soil surface temperature found in temperate regions in the spring. The presence of soil residues can retard simulated soil warming and reduce crop growth. As the crop canopy develops, it shades the soil surface, and simulated average soil surface temperature approaches average air temperature.

In addition to the temperature stress imposed by Eq. (25.297) plant growth does not occur on days when the average daily temperature exceeds 1.5 (T_{oj}).

Nutrient Stress. The N and P stress factors are based on the ratio of simulated plant N and P contents to the optimal values. The stress factors vary non-linearly from 1.0 at optimal N and P contents to 0. when N or P is half the optimal level (Jones, 1983a). In the case of N, the scaling equation is

$$SN_{S,i} = 200\left(\frac{\sum_{k=1}^{i} UN_k}{(c_{NB})_i (B)_i} - 0.5\right) \tag{25.298}$$

where SN_S is a scaling factor for the N stress factor, c_{NB} is the optimal N concentration of the crop on day i, B is the accumulated biomass in $kg \cdot ha^{-1}$, and UN is the crop N uptake on day k in $kg \cdot ha^{-1}$. The N stress factor is computed with the equation

$$SN_i = \frac{SN_{S,i}}{SN_{S,i} + \exp(3.52 - 0.026 SN_{S,i})} \tag{25.299}$$

where SN is the N stress factor for day i. The P stress factor, SP, is computed with Eqs. (25.298) and (25.299) written in P terms.

Aeration Stress. When soil water content approaches saturation, plants may suffer from aeration stress. The water content of the top 1 m of soil is considered in estimating the degree of stress:

$$SAT = 100\left(\frac{SW1}{PO1} - CAF_j\right)/(1. - CAF_j) \tag{25.300}$$

$$AS_i = 1. - \frac{SAT}{SAT + \exp(5.10 - .082 SAT)}, \quad SAT > 0.0 \tag{25.301}$$

where SAT is the saturation factor, SW1 is the water content of the top 1 m of soil in mm, PO1 is the porosity of the top 1 m of soil in mm, CAF is the critical aeration factor for crop j (≈0.85 for many crops), and AS is the aeration stress factor. Finally, the value of REG is determined as the lowest of the stress factors, WS, TS, SN, SP, and AS.

Root Growth

As described in Eq. (25.265), root growth is proportional to water use. Water use from a soil layer is estimated as a function of soil depth, water content, and a compensation factor according to Eqs. (25.273) and (25.274). Soil strength, temperature, and aluminum toxicity stress factors are calculated from soil properties. The lowest of these three stress factors constrains root growth by governing the water use compensation factor.

Cold soil temperatures may limit root growth, especially when subsoil layers warm slowly in the spring (Taylor, 1983). The temperature stress for each soil layer is computed by substituting soil temperature at the center of the layer for soil surface temperature in Eq. (25.297). Numerous studies have shown that root growth is affected by soil strength. Three important strength determinants are bulk density, texture, and water content (Eavis, 1972; Monteith and Bonath, 1965; Taylor et al., 1966). All three variables are considered in estimating the EPIC soil strength stress factor by using the following equations:

$$SS_{\ell} = \frac{BD_{\ell}}{BD_{\ell} + \exp[bt_1 + bt_2 + (BD_{\ell})]} \qquad (25.302)$$

where SS is the soil strength factor in layer ℓ, BD is the soil bulk density in $t \cdot m^{-3}$ adjusted for water content, and bt_1 and bt_2 are parameters dependent upon soil texture. The values of bt_1 and bt_2 are obtained from a simultaneous solution of Eq. (25.302) by substituting boundary conditions for stress. The lower boundary, where essentially no stress occurs, is given by the equation (Jones, 1983b)

$$BDL = 1.15 + 0.00445 SAN \qquad (25.303)$$

where BDL is the bulk density near the lower boundary (SS=1.) for a particular percentage of sand, SAN. The upper boundary is given by the equation (Jones, 1983b)

$$BDU = 1.5 + 0.005 SAN \qquad (25.304)$$

where BDU is the bulk density near the upper boundary (SS≈0.2) for a particular percentage of sand, SAN. The equations for estimating bt_1 and bt_2 are

$$bt_2 = \frac{\ln(0.0112 BDL) - \ln(8.BDU)}{BDL - BDU} \qquad (25.305)$$

$$bt_1 = \ln(0.0112 BDL) - (bt_2)(BDL) \qquad (25.306)$$

Equations (25.305) and (25.306) assure that Eq. (25.302) gives SS values of 1.0 and 0.2 for BD=BDL and BD=BDU.

The water-content-adjusted bulk density is estimated with Grossman's equation (Grossman et al., 1985)

$$BD_{\ell,i} = BD3 + (BDD - BD3)\left(\frac{FC_\ell - SW_{\ell,i}}{FC_\ell - WP_\ell\left(4.083 - 3.33BDD^{1/3}\right)}\right) \quad (25.307)$$

where BD is the water-content-adjusted bulk density of day i, BD3 is the bulk density of 33 kPa water content, BDD is the bulk density of the oven dry soil, FC is the field capacity, WP is the wilting point, and SW is the soil water content for layer ℓ on day i.

Aluminum (Al) toxicity can limit root growth is some acid soil layers, and Al saturation is a widely used index of its effects (Abruna et al., 1982; Brenes and Pearson, 1973; Pavan et al., 1982). Because crops and cultivars differ in sensitivity to Al toxicity (Foy et al., 1974; Mugwira et al., 1980), EPIC expresses Al toxicity as a function of this sensitivity. The Al toxicity stress factor associated with root growth is estimated with the equations

$$ATS_\ell = \frac{100 - ALS_\ell}{100 - ALO_j}, \qquad ALS_\ell > ALO_j \quad (25.308)$$

$$ATS_\ell = 1.0, \qquad ALS_\ell \le ALO_j \quad (25.309)$$

where ATS is the Al toxicity stress factor (0-1) for soil layer ℓ, ALS is the Al saturation in %, and ALO is the maximum ALS value crop j can tolerate without stress in %. Crop specific values of ALO are determined from the equation

$$ALO_j = 10 + 20(ALT_j - 1) \quad (25.310)$$

where ALT_j is the Al tolerance index number for crop j. Values of ALT range from 1 to 5 (1 is sensitive; 5 is tolerant) for various crops. Finally, the root growth constraint, RGF, is the lowest of the stress factors SS, ATS, and TS.

Water Use. Plant water is governed by the root growth stress factor and the water deficit compensation factor of Eq. (25.273). Recall that the water deficit compensation factor, UC, allows total compensation if the value is 1.0 and no compensation at 0.0. The value of UC for any layer is estimated as the product of the root growth stress factors for the layer and all layers above:

$$UC_\ell = \prod_{k=1}^{\ell} RGF_k \quad (25.311)$$

Thus, a low RGF_k greatly reduces water compensation for layer k and all layers below k.

The final estimates of water use for each layer are obtained by multiplying the u_l values in Eqs. (25.274) and (25.275) by RGF:

$$u_{\ell}^{*} = (u_{\ell})(RGF)_{\ell} \tag{25.312}$$

25.8.5. Crop Yield

Crop yield may be reduced through water-stress-induced reductions in the harvest index. Most grain crops are particularly sensitive to water stress from shortly before until shortly after anthesis, when major yield components are determined (Doorenbos and Kassam, 1979). Optimum conditions for growth may reduce harvest index slightly if dry matter accumulation is large and economic yield is limited by sink size. The harvest index is affected by water stress according to the equation

$$HIA^{*} = (HIA - HIA_{o})\frac{WUR}{WUR + \exp(6.13 - 0.0883WUR)} + HIA_{o} \tag{25.313}$$

where HIA^* is the final estimate of harvest index applied to above-ground biomass to estimate yield, HIA is the simulated potential harvest index, HIA_o is the minimum harvest index for the crop, and WUR is the simulated water use ratio. The value of HIA is simulated daily using the equation

$$HIA_{i} = HI_{j}\left(\frac{100HUI_{i}}{100HUI_{i} + \exp(11.1 - 10.HUI_{i})}\right) \tag{25.314}$$

The exponential parameters are set to give HIA=0.1 when HUI=0.5 and HIA=0.95 when HUI=0.95. Thus, if the growing season is shortened by frost or for other reasons, the potential harvest index HI is not attained. The water use ratio WUR is estimated at harvest using the equation

$$WUR = 100\left(\frac{\sum_{i=1}^{K} U_{i}}{\sum_{i=1}^{K} E_{pi}}\right) \tag{25.315}$$

where U and E_p are the actual and potential plant water use rates in $mm \cdot d^{-1}$ for day i and K is the number of days in the growing season. The exponential parameters in Eq. (25.313) are set to give $0.05(HIA-HIA_o)$ when WUR=0.10 and $0.90(HIA-HIA_o)$ when WUR=0.50.

Thus, if the actual to potential water use ratio is greater than 0.5 there is little reduction to harvest index.

25.8.6. Winter Dormancy

The day length growth constraint is used to simulate a winter dormant period for fall planted crops. This constraint is only imposed for areas that have a growing season of fewer than 12 months. A 12-month growing season is defined in the model as having no month with mean minimum temperature of lower than 5°C. If there is a dormant winter period, it is defined as the time when day length is within 1 h of the location's minimum day length.

If a crop becomes dormant in winter, the heat unit summation is set to zero. This provides for rapid new growth when temperatures increase in the spring. During the dormant period, the plants are not allowed to grow. The standing live biomass is actually reduced during this period because of frost and short day length. The day length reduction factor is estimated with the equation

$$FHR_i = 0.35\left(1.0 - \frac{HRLT_i}{HRLT_{mn} + 1}\right) \quad (25.316)$$

where FHR is the day length reduction factor, $HRLT_i$ is the day length on day i in h, and $HRLT_{mn}$ is the minimum day length for the location. Day length is a function of the time of year and latitude as expressed in the equation

$$HRLT_i = 7.64 \cos^{-1}\left(\frac{-\sin\left(\frac{2\pi}{360} LAT\right)\sin(SD)_i - 0.044}{\cos\left(\frac{2\pi}{360} LAT\right)\cos(SD)_i}\right) \quad (25.317)$$

where LAT is the latitude of the watershed in degrees and SD, the sun's declination angle, is defined by the equation

$$SD_i = 0.4102 \sin\left(\frac{2\pi}{365}(i - 80.25)\right) \quad (25.318)$$

The frost reduction factor is estimated with the equation

$$FRST_i = \frac{-T_{mn,i}}{-T_{mn,i} - \exp(af_{j,i} + af_{j,2} \cdot T_{mn,i})}, T_{mn,j} < -1.°C \quad (25.319)$$

where $FRST_i$ is the frost damage factor, T_{mn} is the minimum temperature on day i in °C, and $af_{j,1}$ and $af_{j,2}$ are parameters expressing the crop's frost sensitivity. The reduction in standing live biomass is estimated with the equation

$$\Delta B_{AG,i} = 0.5 \cdot B_{AG,i}\,(1.0 - HUI_i) \cdot \max(FHR_i, FRST_i) \quad (25.320)$$

where ΔB_{AG} is the reduction in above ground biomass in $t \cdot ha^{-1}$ on day i, HUI is the heat unit index, and B_{AG} is the above ground biomass in $t \cdot ha^{-1}$ on day i.

Note that frost damage is greater when plants are small (HUI≈0) and approaches 0 as the plants near maturity.

There is also a provision for frost kill of annual plants. Kill occurs if

$$FRST_i(1.-SNOF) \geq 0.9 \tag{25.321}$$

where SNOF is a snow cover factor calculated with the equation

$$SNOF = \frac{SNO}{SNO + \exp(2.303 - 0.220SNO)} \tag{25.322}$$

where SNO is the water content of snow cover in mm.

25.9. TILLAGE

The EPIC tillage component was designed to mix nutrients and crop residues within the plow depth, simulate the change in bulk density, and convert standing residue to flat residue. Other functions of the tillage component include simulating ridge height and surface roughness.

Each tillage operation is assigned a mixing efficiency (0-1). The tillage mixing equation is

$$X_\ell = (1 - EF)X_{o\ell} + \left(\frac{Z_\ell - Z_{\ell-1}}{PD}\right) EF \sum_{k=1}^{M} X_{ok} \tag{25.323}$$

where X is the amount of the material in layer ℓ after mixing in kg·ha^{-1}, EF is the mixing efficiency of the tillage operation (0-1), X_o is the amount of the material before mixing kg·ha^{-1}, and M is the number of soil layers in the plow depth, PD, in m.

The change in bulk density in the plow layer is simulated for each tillage operation by using the equation

$$BDP_\ell = BDP_{o\ell} - (BDP_{o\ell} - \frac{2}{3} BD_{o\ell})(EF) \tag{25.324}$$

where BDP is the bulk density after tillage, BDP_o is the bulk density in soil layer ℓ before tillage, and BD_o is the bulk density of the soil when it has completely settled after tillage. Between tillage operations, the soil settles with each rainfall event according to the equations

$$SZ_\ell = \frac{0_{\ell-1}}{Z_\ell^{0.6}} \left(1. + \frac{2.SAN_\ell}{SAN_\ell + \exp(8.597 - 0.075SAN_\ell)} \right) \tag{25.325}$$

$$BDP_{\ell,i} = BDP_{\ell,i-1} + (BD_\ell - BDP_{\ell,i-1}) \left(\frac{SZ_\ell}{SZ_\ell + \exp(3.92 - 0.0226SZ_\ell)} \right) \tag{25.326}$$

where SZ_ℓ is a scaling factor for soil layer ℓ, $O_{\ell-1}$ is the percolation rate into the layer in mm·d^{-1} (R-Q for the top layer) and SAN is the percentage of sand in the layer. The exponential parameters of Eq. (25.326) are set to give $0.10(BD_\ell - BDP_{\ell i-1})$ when $SZ_\ell=5$ and $0.95(BD_\ell - BDP_{\ell i-1})$ when $SZ_\ell=100$. Thus, near the surface soils with little sand will almost completely settle with 100 mm percolation. Equations (25.325) and (25.326) cause fast settling when rainfall is large and soils are sandy and have been tilled recently. Also, settling is much faster near the surface (this allows simulation of long-term deep chiseling effects). Of course, settling is relatively slow for soils low in sand content, especially in low rainfall areas.

Another important function of the tillage model, converting standing residue to flat residue, is accomplished with the equation

$$SR=(SR_o)\exp[-56.9(PD)(EF)] \qquad (25.327)$$

where SR_0 and SR are the standing residue weights before and after tillage in t·ha^{-1} and PD is the plow depth in m.

Other functions of the tillage component include simulating ridge height and interval and surface roughness. These variables are specified for each tillage implement. However, the ridge interval and height are computed after each tillage operation to reflect the combined effects of the current and previous operations. The ridge height is estimated by using the equations

$$HR=HT_k+(HT_{k-1}-HT_k)\exp(-PD_k/PD_{k-1}), HT_k<HT_{k-1} \qquad (25.328)$$

$$HR=HT_k, \qquad HT_k \geq HT_{k-1} \qquad (25.329)$$

where HR is the ridge height after the tillage operation k in m, HT is the input ridge height for the tillage operation in m, and k refers to the sequence of operations. After each tillage operation, the ridge interval is set to the input ridge interval of the operation with the greater HT.

The user specifies the date and depth for each tillage operation. The tillage operation is carried out on the specified date if the soil is dry enough. If not, the operation occurs on the next suitable day.

It is also possible to schedule operations by fraction of heat unit accumulations. The heat unit schedule may be user input or automatically developed by EPIC. Various combinations of scheduling (by date, heat units input, or automatic heat units) are also permitted.

The harvest index and harvest efficiency provide adequate flexibility to accommodate almost any harvest strategy. The harvest index (HI) is input for each crop and adjusted during each year of simulation as described in the Crop Yield section. Normally, the adjusted HI dictates the fraction of the above ground biomass removed from the crop. Thus, for a grain crop like corn, about 40-50% is removed. However, if corn is cut for silage, the input HI would be about 0.95. An option to override HI allows single crops to be harvested in two different ways. For example, oats could be harvested for grain by using the model adjusted value of HI≈0.4 and then the straw could be baled by using the appropriate override value (0.5-0.95). The harvest efficiency (HE) indicates what portion of the harvested material actually leaves the field. For most operations, HE may range between 0.7 to 0.95. However, it can be set as low as 0.0 to simulate the plowing under of cover crops.

25.10. PLANT ENVIRONMENT CONTROL

The plant environment control component provides mechanisms for applying irrigation water, fertilizer, lime, and pesticide or for simulating grazing or drainage systems.

25.10.1. Drainage

Drainage via underground drainage systems is treated as a modification of the natural lateral subsurface flow of the area. Drainage is simulated by indicating which soil layer contains the drainage system and the time required for the drainage system to reduce plant stress. The drainage time in d replaces the travel time in Eq. (25.44) for the layer containing the system.

25.10.2. Irrigation

The EPIC user has the option to simulate dryland or irrigated agricultural areas. Sprinkler or furrow irrigation may be simulated and the applications may be scheduled by the user or automatically. As implied, the user scheduled option allows application dates and rates to be inputted. With the automatic option, the model decides when and how much water to apply.

Required inputs for the automatic version include a signal to trigger applications (the three trigger choices include: plant water stress level (0-1), plow layer soil water tension in kPa, or root zone soil water deficit in mm); the maximum volume applied to each crop in mm; the runoff fraction; minimum and maximum single application volumes in mm; and the minimum time interval between applications in d.

Two modes of application, rigid and flexible, are available.

Rigid mode:

(1) User schedule--The exact input volumes are applied on specified dates.

(2) Automatic option--Maximum single application volumes are applied when triggered.

Flexible mode:

(1) User schedule--The application volume is the minimum of the specified volume, the maximum single application volume, and the volume required to fill the root zone to field capacity.

(2) Automatic option--The application volume is the minimum of the maximum single application volume and the volume required to fill the root zone to field capacity.

Also, irrigation does not occur when the application volume derived from the appropriate mode and option (except for rigid, user-scheduled) is less than the input minimum single application volume.

The application mode (rigid or flexible) is fixed for the entire crop rotation. However, the trigger value and criterion (plant water stress level, soil water tension, or root zone water deficit) and the runoff fraction may be changed at any time during the operation schedule. Also, a combination of user and automatic scheduling is permitted.

25.10.3. Fertilization

Fertilizer application is similar to irrigation--scheduling may be input or automatic and rigid and flexible modes are available. Required inputs for the automatic version include a trigger (plant N stress level (0-1)); maximum annual N applied to a crop in $kg \cdot ha^{-1}$; and minimum time between applications in d. Automatic fertilizing at planting is also optional.

Rigid mode:

(1) User schedule--The exact input rates of N and P are applied at specified depths on scheduled dates.

(2) Automatic option--A fraction of the annual maximum N rate for the crop is applied when triggered. The application fraction and the maximum rate are inputs. Also P is applied at a rate necessary to bring the plow layer (0.2 m) P concentration to a level specified at the start of a simulation. All automatic applications are placed in the second soil layer.

Flexible mode:

(1) User schedule--The model samples N and P concentration in the root zone and compares with user preset N and P concentrations. Applications occur on schedule at specified depths and rates if the root zone N and P concentrations do not exceed user standards; otherwise, applications are delayed until N and P concentrations are depleted below user standards.

(2) Automatic option--The N application rate is the difference between the preset rate (application fraction times the maximum annual rate) and the root zone N content. The P application strategy is the same as in the rigid mode.

Other features and limitations include only mineral N (in NO_3 form) and P may be applied automatically. Organic N and P and ammonia are applied by user scheduling. The maximum annual N application for a crop can be changed at planting. A combination of user and automatic scheduling is permitted. Automatic applications occur only when N is the active crop growth constraint even though the trigger value is reached. Thus, the annual N and P application rates vary according to the crop's needs, the soil's ability to supply those needs, and the magnitude of the N stress relative to water and temperature stresses.

25.10.4. Liming

EPIC simulates the use of lime to neutralize toxic levels of Al and/or to raise soil pH to near-optimum levels. Different algorithms are used to estimate lime requirements of "highly weathered" soils (Oxisols, Ultisols, Quartzipsamments, Ultic subgroups of Alfisols, and Dystric suborders of Inceptisols) (Sharpley et al., 1985) and other soils. The highly weathered soils have large amounts of variable-charge clays. Moderate amounts of lime are required to increase their pH to about 5.5 and convert extractable Al to more inactive forms. However, the pH of these soils is highly buffered above pH 5.5, and very large amounts of lime are required to raise the pH to near 7.0. As a result, soils with variable charge clays are usually limed only to reduce Al saturation to acceptable levels.

The Al saturation of each soil layer is estimated with the equations (Jones, 1984)

$$ALS_\ell = 154.2 - 1.017 BSA_\ell - 3.173 C_\ell - 14.23 PH_\ell, \quad PH_\ell \leq 5.6 \tag{25.330}$$

$$ALS_\ell = 0., \qquad PH_\ell > 5.6 \tag{25.331}$$

where ALS is the Al saturation of soil layer ℓ in % calculated as KCl-extractable Al divided by effective cation exchange capacity (ECEC), BSA is the base saturation calculated from cation exchange capacity (CEC) determined by the NH_4OAc (pH=7.0) method in %, C is the organic carbon content in %, and PH is the soil pH. For highly weathered soils, the lime required to neutralize toxic Al in the plow layer is estimated with the equation

$$RLA = 0.1(ALS)(ECEC)(PD)(BD) \tag{25.332}$$

where RLA is the lime required to neutralize Al in $t \cdot ha^{-1}$, ECEC is the effective cation exchange capacity in $cmol(p+) \cdot kg^{-1}$, BD is the soil bulk density in $t \cdot m^{-3}$, and PD is the plow depth in m.

ECEC is calculated as SMB/ALS (Soil Survey Staff, 1982), where SMB in $cmol \cdot kg^{-1}$ is the sum of the bases extracted by NH_4OAc (pH=7.0). The constant 0.1 converts cmol(p+) kg^{-1} extractable aluminum to equivalent $CaCO_3$ in $t \cdot ha^{-1}$, assuming 2 cmol(p+) $CaCO_3$ are required to completely neutralize 1 cmol(p+) extractable Al (Kamprath, 1970). At the end of each year, enough lime is applied to meet the lime requirement (RLA) if RLA≥1 t·ha^{-1}. If RLA<1 t·ha^{-1} no lime is applied. When lime is applied, the plow layer PH is raised to 5.4 and ALS is reduced to 0.

For EPIC, soil acidification and decreasing base saturation are caused by addition of fertilizer N and symbiotic N fixation by legumes. All fertilizer N is assumed to derive from anhydrous amonia, urea, amonium nitrate, or mixtures of these with equivalent acidifying effects. The $CaCO_3$ equivalent of fertilizer or fixed N is assumed to be 1.8 kg $CaCO_3$·kg^{-1}N (Pesek et al., 1971). This is within the range of variation reported by Pierre et al. (1971) for fertilized corn and by Nyatsanga and Pierre (1973) and Jarvis and Robson (1983) for legumes.

At the end of each year of simulation, the plow layer PH is reduced to reflect the change in base saturation caused by N fertilizer and N fixation. The change in base saturation is computed with the equation

$$\Delta BSA = \frac{0.036(FN + WFX)}{(PD)(BD)(CEC)} \qquad (25.333)$$

where FN is the amount of N fertilizer added during the year in kg·ha^{-1} and WFX is the amount of N fixation by legumes in kg·ha^{-1}. The PH value is reduced by using the equation

$$PH = PH_0 - 0.05\ \Delta BSA \qquad (25.334)$$

where the constant 0.05 approximates the slope of the relationship between PH and ΔBSA for several soils when the values of BSA are between 60 and 90 (Peech, 1965).

For other soils, the lime requirement is the amount of lime needed to raise soil pH to 6.5 according to the equation

$$RLA = 0.01(PD)(BD)(CEC)(\Delta BSA) \qquad (25.335)$$

where ΔBSA is the change in base saturation needed to raise soil pH to 6.5. The constant 0.05 converts ΔBSA in % to equivalent $CaCO_3$ in t·ha^{-1}, assuming that applied $CaCO_3$ reacts with equivalent unsaturated CEC. The ΔBSA is estimated with the relation

$$\Delta BSA = \min\begin{cases}(6.5 - PH)/0.023 \\ 90 - BSA\end{cases} \qquad (25.336)$$

For soils that are not highly weathered, lime application is simulated if at the end of the year, RLA>2.0 t·ha^{-1}. When lime is applied, PH is changed to 6.5, base saturation is increased by ΔBSA, and ALS is set to 0.

25.10.5. Pests

The three pests considered by EPIC are insects, weeds, and plant diseases. The effects of all three pests are expressed in the EPIC pest factor. Crop yields are estimated at harvest as the product of simulated yield and pest factor. The pest factor ranges from 0.0 to 1.0--1.0 means no pest damage and 0.0 means total crop destruction by pests. The pest factor is simulated daily as a function of temperature, moisture, and ground cover.

$$PSTI_K = \sum_{i=1}^{K} T_{mn,i}(1.+RFS_i - RFS_T), \qquad (25.337)$$

$$RFS_i > RFS_T$$

$$CV_i > CV_T$$

$$T_{mn,i} > 0.0$$

where PSTI is the accumulated pest index for day k, T_{mn} is the minimum temperature for day i in °C, RFS_i is the accumulated rainfall for 30 days preceeding day i in mm, RFS_T is the threshold 30-day rainfall amount in mm, CV is the ground cover (live biomass and crop residue) on day i in $t{\cdot}ha^{-1}$, and CV_T is the threshold cover value in $t{\cdot}ha^{-1}$. When TMN is less than 0.0, the pest index is reduced using the equation

$$PSTI_K = PSTI_{K-1} + T_{mn} \qquad (25.338)$$

Thus, the pest index grows rapidly during warm moist periods with adequate ground cover and is reduced by cold temperatures. This general pest index is an attempt to account for major differences in pest problems related to climate variability.

When pesticides are applied, the pest index is reduced using the equation

$$PSTI_K = PSTI_{K-1} - 1000.(PSTE) \qquad (25.339)$$

where PSTE is the pesticide kill fraction ranging from near 0.0 to near 1.0. Thus, if the kill fraction approaches 1.0, the pest index is reduced nearly 1000 units.

At harvest, the pest factor is computed from the pest index using the equation

$$PSTF = 1.0 - (1.0 - PSTM)\left(\frac{PSTI^*}{PSTI^* + exp(2.70 - 0.499\, PSTI^*)}\right) \qquad (25.340)$$

$$PSTI^* = \frac{PSTI}{k}$$

where PSTF is the pest factor used to adjust crop yield, PSTM is the minimum pest factor value for a crop, and k is time since last harvest in d.

25.10.6. Furrow Diking

Furrow diking is the practice of building small temporary dikes across furrows to conserve water for crop production. Since they reduce runoff, they may also aid in erosion control. The EPIC furrow diking model allows construction of dikes for any ridge spacing and at any interval down the furrows. Dikes may be constructed or destroyed mechanically on any day of the year. If estimated runoff for a particular event exceeds the dike storage volume, overtopping occurs and all of the estimated runoff is lost. If not, all of the rainfall infiltrates and is available for plant use. When runoff destroys the dikes, the model rebuilds them automatically. Rainstorms that do not overtop the dikes cause settling and, thus, reduce storage volume. Settling is esimated with the equation

$$H = H_o \exp(-0.1Y) \tag{25.341}$$

where H_0 is the dike height in m before settling, H is the dike height after settling, and Y is the USLE estimate of soil loss in $t \cdot ha^{-1}$. Ridge height is also reduced with the settling function contained in Eq. (25.341). The dikes are rebuilt automatically when $H/H_0 < 0.7$.

The dike storage volume is estimated by assuming that the furrow and the dike are triangular and that the dike side slopes are 2:1. Given the dike and ridge heights, the dike interval, and the slope down the furrow, the volume can be calculated directly. There are two possible dike configurations that require slightly different solutions. Normally, the dike interval is relatively short (1.0-3.0 m) and the slope along the furrow is relatively flat (<1.0%). When the dike is full, water extends from the top of the downslope dike up the furrow to a point above the toe of the upslope dike. The volume is calculated by using cross-sectional areas at the toes of the two dikes. This approach computes the volume in three parts (between the top and the toe of the downslope dike; between the toes of the two dikes; and between the toe and the waterline on the upslope dike). Beginning at the centerline of the downslope dike, the volume equations are

$$DV_I = \frac{1}{2}(H)(D_2)(W_2) \tag{25.342}$$

$$DV_{II} = \frac{1}{4}(DI - 4H)[(D_2)(W_2) + (D_3)(W_3)] \tag{25.343}$$

$$DV_{III} = \frac{1}{4}(XD - DI + 2H)(D_3)(W_3) \tag{25.344}$$

where DV is the dike volume between cross sections in m^3, H is the dike height in m, D is the water depth in m, W is the water surface width in m, DI is the dike interval in m, XD is the distance from the center of the downslope dike to the waterline on the upslope dike in m, and subscripts 2 and 3 refer to cross sections 2 and 3. Cross section 2 is a the toe of the downslope dike and cross section 3 is at the toe of the upslope dike. Water depth is calculated with the equations

$$D_2 = H - 2(S)(H) \tag{25.345}$$

$$D_3 = H - S(DI - 2H) \tag{25.346}$$

where S is the slope in m·m^{-1} along the furrow. Water surface width is a function of depth and ridge spacing, RS in m.

$$W = RS\left(\frac{D}{H}\right) \tag{25.347}$$

The distance XD is computed with the equations

$$XD = DI - 2(H - DZ) \tag{25.348}$$

$$DZ = H - (S)(XD) \tag{25.349}$$

where DZ is the water line elevation on the upslope dike. The constant 2 in Eq. (25.348) comes from the assumed 2:1 dike sideslopes. Simultaneous solution of Eqs. 25.348 and 25.349 yields

$$XD = \frac{DI}{1+2S} \tag{25.350}$$

Substituting D, W, and XD into Eqs. 25.342, 25.343, and 25.344 and summing gives

$$DV = \frac{1}{4}\left(\frac{RS}{H}\right)\left(H^2(1-2S)^2(DI-2H) + [H - S(DI-2H)]^2\left(\frac{DI}{1+2S} - 2H\right)\right) \tag{25.351}$$

Equation (25.351) is divided by the total surface area of a furrow dike to convert volume from m^3 to mm.

$$DV = \frac{250}{(DI)(H)}\left(H^2(1-2S)^2(DI-2H) + [H - S(DI-2H)]^2\left(\frac{DI}{1+2S} - 2H\right)\right) \tag{25.352}$$

In the simpler and more unusual dike configuration, the upslope waterline does not extend to the toe of the upslope dike. Only one cross section is involved and the volume is computed into two parts. Equation (25.342) is used to calculate the most downslope volume, and the upslope volume is calculated with the equation

$$DV_2 = \frac{1}{4}(D_2)(W_2)(H)\left(\frac{1}{S} - 2\right) \tag{25.353}$$

Adding Eqs. 25.342 and 25.353, substituting D and W, and converting from m^3 to mm gives

$$DV = \frac{250H^2(1-2S)^2}{(S)(DI)} \tag{25.354}$$

Thus, the average dike volume of a field is estimated with Eq. (25.352) or (25.354) as dictated by slope and dike height and interval. However, no field is exactly uniform in slope; dike and ridge heights vary, and furrow and dike side slopes may not be triangular. Therefore, the model provides a user-controlled dike efficiency factor to allow for varying conditions across a field. The dike efficiency factor also provides for conservative or optimistic dike system design.

25.10.7. Grazing

Livestock grazing is simulated as a daily harvest operation. Users specify daily grazing rate in $kg \cdot ha^{-1}$, minimum grazing height in mm, harvest efficiency, and date grazing begins and ends. Harvest efficiency is used to estimate the fraction of grazed plant material used by animals--not returned as manure, etc. Any number of grazing periods may occur during a year and the grazing schedule may vary from year to year within a rotation. Grazing ceases when forage height is reduced to the user-specified cutoff value and resumes automatically when new growth exeeds the cutoff height if the grazing period has not expired.

25.11. ECONOMICS

The economic component of EPIC is more accurately represented as a crop budget and accounting subsystem. The algorighms keep track of the costs of producing and marketing the crops. Costs (and income) are divided into two groups: those costs which do not vary with yield and those that do. These groups will be addressed in turn. All cost registers are cleared at harvest. All operations after harvest are charged to the next crop in the cropping sequence.

Tillage and (preharvest) machine operation costs are assumed to be independent of yield. These operation costs must be calculated outside of EPIC and are inputted as one variable into the tillage file. This cost cell contains all costs associated with the single operation or activity (e.g., a chiseling activity includes fuel, labor, depreciation, repair, interest, etc., for both the tractor and the chisel). A budget generator program like the Micro Budget Management System (MBMS) (McGrann et al., 1986) is convenient for making these calculations. This is an updated interaction program developed from the Enterprise Budget Generator (Kletke, 1979). The

MBMS is more compatible with EPIC in that it has output capabilities to itemize cost by machine operation. This information (when converted to metric units) can be inputted directly into the equipment file in EPIC. Farm overhead, land rent, and other fixed costs can be charged to the crop by first creating null operations in the equipment file with machine number and cost information only and then triggering the cost in EPIC with a null activity. Government payments can be credited by using negative cost entries in the same way.

Costs which are yield and management dependent are entered into EPIC in two regions of the input data. Seed costs, seeding rates, and crop prices are entered in the crop parameter file for each crop code. Seed costs are calculated as the produce of seeding rate and cost per kilogram. Amendment costs are calculated similarly. The amendments include elemental N and P, irrigation water, and lime. Total cost per hectare is based on the sum of costs for machinery operations, seed, and amendments. Market value per hectare is based on the product of crop yield and net crop price. Net crop price is the market price minus the harvest, hauling, and other processing costs which are yield dependent. The net price must be determined outside EPIC.

When valid cost figures are entered into these EPIC input cells, the model will return annual cost and returns by crop. EPIC budget information is valuable not only for profit analyses but also risk analyses, since the annual distributions of profits and costs can be captured. Risk analyses capability greatly enhances the analytical value of EPIC for economic studies.

The greatest value of EPIC to economic analyses is not its internal economic accounting, but the stream of physical outputs on daily, monthly, annual, or multi-year periods that can be input into economic models, budget generators, and risk analysis systems. EPIC estimates crop yields, movement of nutrients and pesticides, and water and sediment yields. Changes in inputs necessary to respond to changes in management, soil quantity and quality, climate (i.e. global warming), droughts, etc., are also estimated. These outputs become inputs into economic and natural resource models facilitating comprehensive analyses of alternative policies and programs.

NOTATIONS

α	=	ratio of the maximum rainfall amount during a period equal the watershed time of concentration to the total rainfall for the storm
α_{mn}	=	minimum value of α
$\alpha_{.5}$	=	ratio of the maximum rainfall amount during 0.5 h to the total rainfall for the storm
$\alpha_{.5p}$	=	$\alpha_{.5}$ value at the peak of gamma distribution
$\alpha_{.5u}$	=	upper limit of $\alpha_{.5}$
β	=	probability of a wet day following a dry day divided by probability of a wet day
κ	=	decay constant in exponential rainfall rate distribution (h)
Λ	=	plant-water-use-rate/soil-depth parameter
σ	=	average channel slope ($m \cdot m^{-1}$)
ρ	=	runoff coefficient expressing the watershed infiltration characteristics
λ	=	land surface slope length (m)
δ	=	slope of saturated vapor pressure curve (kPa/°C)
γ	=	psychrometer constant (kPa/°C)
μ	=	uniform random number (0.0-1.0)
ζ	=	power parameter in modified exponential rainfall distribution
Ω_H	=	parameter relating mean relative humidity for wet and dry days
Ω_R	=	ratio of mean solar radition on wet days to mean solar radiation on dry days
Ω_s	=	scaling factor to adjust soil surface temperature on wet days
Ω_T	=	parameter relating mean maximum air temperature for wet and dry days
η	=	gamma distribution shape parameter for estimating wind velocity
ξ	=	slope length power parameter (dependent upon slope) in the Universal Soil Loss Equation (USLE)
ν	=	gamma distribution shape parameter for estimating $\alpha_{.5}$
θ	=	wind direction clockwise from north (radians)
ϕ	=	clockwise angle between field length and north (radians)
ψ_1, ψ_2	=	wind erosion parameters related to vegetative cover factor
ω	=	flow coefficient governing mineral P flow between active and stable pools (d^{-1})

§	=	scaling parameter for soil temperature damping depth
χ	=	dimensionless variable (0-1) expressing frequency with which a specified wind velocity occurs ($m \cdot s^{-1}$)
af_1, af_2	=	crop parameters for frost sensitivity
ah_1, ah_2	=	crop parameters that determine the shape of the leaf - area - index - development curve
ad	=	crop parameter that governs leaf-area-index-decline rate
A	=	drainage area (ha)
AB	=	albedo
AB_s	=	soil albedo
AD	=	air density ($g \cdot m^{-3}$)
AE	=	average annual wind energy ($kWh \cdot m^{-2}$)
AIR	=	volume of irrigation water applied (mm)
AL0	=	maximum ALS value a crop can tolerate without stress (%)
ALS	=	aluminum saturation of the soil (%)
ALT	=	aluminum tolerance index number (1-5)
AKN	=	nitrification regulator
AKV	=	volatilization regulator
AP	=	labile P in soil ($kg \cdot ha^{-1}$)
AR	=	aerodynamic resistance for heat and vapor transfer ($s \cdot m^{-1}$)
AS	=	aeration root growth stress factor (0-1)
ASPR	=	flow rate of P between active and stable mineral P pools ($kg \cdot ha^{-1} \cdot d^{-1}$)
ATS	=	root growth stress factor due to aluminum toxicity (O- 1)
b	=	TP-40 rainfall distribution parameter
bcv	=	lagging factor for simulating residue and snow cover effects on soil surface temperature (0-1)
bn_1, bn_2, bn_3	=	crop parameters for plant N concentration equation
bp_1, bp_2, bp_3	=	crop parameters for plant P concentration equation
bt_1, bt_2	=	soil strength parameters related to soil texture and bulk density
B	=	accumulated plant biomass (above ground and roots) ($t \cdot ha^{-1}$)
B_{AG}	=	above ground biomass of growing crop ($t \cdot ha^{-1}$)
B_P	=	potential crop biomass ($t \cdot ha^{-1}$)
BD	=	soil bulk density ($t \cdot m^{-3}$)
BD_o	=	bulk density of soil completely settled after tillage ($t \cdot m^{-3}$)

BDD	=	bulk density of oven dry soil ($t \cdot m^{-3}$)
BD3	=	bulk density at 33 kPa water content ($t \cdot m^{-3}$)
BDL	=	bulk density near lower boundary of stress (causes no root growth stress) ($t \cdot m^{-3}$)
BDP	=	current bulk density as affected by tillage ($t \cdot m^{-3}$)
BDP_o	=	bulk density in soil layer before tillage ($t \cdot m^{-3}$)
BDU	=	bulk density near upper stress boundary (root growth stress≈0.2 ($t \cdot m^{-3}$)
BE	=	crop parameter--converts energy to biomass ($kg \cdot MJ^{-1}$)
BKN	=	rate constant for N flow between pools of active and stable organic N (d^{-1})
BSA	=	base saturation as determined by ammonium acetate (NH_4OAc) method (Z)
C_{LP}	=	labile P concentration in the soil ($g \cdot t^{-1}$)
c_{NB}	=	optimal N concentration for a plant ($kg \cdot t^{-1}$)
c_{NFR}	=	N concentration in crop residue ($g \cdot g^{-1}$)
c_{N03}	=	$N0_3$ concentration in water ($g \cdot m^{-3}$)
c_{ON}	=	concentration of organic N in the soil ($g \cdot t^{-1}$)
c_p	=	P concentration in the soil ($g \cdot t^{-1}$)
c_{PB}	=	optimal P concentration for a plant ($g \cdot t^{-1}$)
c_{PFR}	=	P concentration in crop residue ($kg \cdot t^{-1}$)
c_s	=	sediment concentration in runoff ($g \cdot m^{-3}$)
C	=	organic carbon content of soil (%)
CAC	=	$CaCO_3$ concentration in soil ($g \cdot t^{-1}$)
CAF	=	fraction of porosity containing water when aeration starts limiting root growth
CE	=	USLE crop management factor
CE_{mn}	=	minimum value of crop management factor
CEC	=	cation exchange capacity as determined by the NH_4OA_c method ($cmol \cdot kg^{-1}$)
CHT	=	crop height (m)
CLA	=	clay content of the soil (%)
CNN	=	rate constant for N mineralization from humus (d^{-1})
CN	=	Soil Conservation Service (SCS) runoff curve number
CN_1	=	SCS runoff curve number for moisture condition 1 (dry)
CN_2	=	SCS runoff curve number for moisture condition 2 (average)

CN_3 = SCS runoff curve number for moisture condition 3 (wet)

CNP = C:N and C:P rate parameter for crop residue decay

CNR = C:N ratio of crop residue

CPR = C:P ratio of crop residue

CR = canopy resistance for vapor transfer ($s \cdot m^{-1}$)

CV = soil cover (above ground biomass plus crop residue) ($t \cdot ha^{-1}$)

D = water depth in furrow dike system (m)

DCR = decay rate constant for fresh organic material (crop residue) (d^{-1})

DD = soil temperature damping depth (m)

DE = daily wind energy ($kWh \cdot m^{-2}$)

DI = furrow dike interval (m)

DN = denitrification rate ($kg \cdot ha^{-1} \cdot d^{-1}$)

DP = maximum damping depth for soil temperature (m)

DR = sediment delivery ratio (sediment yield divided by gross erosion)

DUR = rainfall duration (h)

DV = furrow dike volume (m^3)

DW = duration of wind greater than threshold velosity (s)

DZ = waterline elevation on upslope furrow dike (m)

e_a = saturation vapor pressure at mean air temperature (kPa)

e_d = vapor pressure at mean air temperature (kPa)

E = evaporation (mm)

E_o = potential evaporation (mm)

E_P = potential evaporation rate of plant water ($mm \cdot d^{-1}$)

E_s = potential evaporation rate of soil water ($mm \cdot d^{-1}$)

EA = soil cover index (0-1)

ECEC = effective cation exchange capacity ($cmol \cdot kg^{-1}$)

EF = mixing efficiency of tillage operation (0-1)

EI = USLE rainfall energy factor

EIR = runoff ratio (volume of irrigation runoff/volume of water applied)

ELEV = elevation of site (m)

EN03 = amount of NO_3-N moved from lower layers to the top layer by soil evaporation ($kg \cdot ha^{-1}$)

EOS = 30-day sum of potential evaporation (mm)

ER = enrichment ratio (concentration of organic N or total P in the sediment/soil concentrations)

EV	=	daily total evaporation from the soil (mm)
f	=	average infiltration rate during a storm ($mm \cdot h^{-1}$)
F	=	frequency with which the largest of a total of τ events occurs
FC	=	field capacity (33 kPa for many soils) water content (mm)
FD	=	mean unsheltered travel distance of wind across a field
FFC	=	ratio of soil water content above wilting point to difference between field capacity and wilting point
FFC*	=	depth weighted FFC value
FHU	=	crop stage factor governing water stress on harvest index
FHR	=	winter dormancy day length factor for reducing standing live biomass (0-1)
FI	=	soil erodibility factor
FL	=	field length (m)
FN	=	amount of N fertilizer added ($kg \cdot ha^{-1}$)
FON	=	fresh organic N in crop residue ($kg \cdot ha^{-1}$)
FOP	=	fresh organic P in crop residue ($kg \cdot ha^{-1}$)
FP	=	amount of pesticide intercepted by plants
FR	=	flat crop residue ($t \cdot ha^{-1}$)
FRST	=	factor for estimating frost damage to crop (0-1)
FV	=	vegetative cover factor
FW	=	field width (m)
FXG	=	crop growth-stage factor regulating N fixation
FXN	=	soil NO_3 content factor governing N fixation
FXR	=	fraction of N uptake provided by N fixation
FXW	=	soil water content factor governing N fixation
FZ	=	depth factor in soil temperature equation
g_1, g_2, g_3	=	crop coefficients relating grouing biomass and residue to small grain equivalent
G	=	soil heat flux ($MJ \cdot m^{-2}$)
GC	=	fraction of ground covered by plants
GP	=	amount of pesticide reaching the ground
h_o	=	net radiation ($MJ \cdot m^{-3}$)
H	=	furrow dike height (m)
HI	=	potential harvest index--ratio of crop yield to above ground biomass
HIA	=	actual harvest index (harvest index adjusted for water stress)

HLP	=	half life of foliar residue (d)
HMN	=	mineralization rate for active organic N pool ($kg \cdot ha^{-1} \cdot d^{-1}$)
HMP	=	humus P mineralization rate ($kg \cdot ha^{-1} \cdot d^{-1}$)
HMX	=	maximum crop height (m)
HR	=	ridge height (m)
HRLT	=	day length (h)
$HRLT_{mn}$	=	minimum day length (h)
HT	=	input ridge height for tillage operation (m)
HU	=	daily heat units--average daily temperature minus base temperature of crop (°C)
HUF	=	heat unit factor for driving leaf-area-index development (0-1)
HUFH	=	heat unit factor for estimating harvest index development (0-1)
HUI	=	heat unit index--ratio of accumulated to potential heat units (0-1)
HV	=	latent heat of vaporization ($MJ \cdot kg^{-1}$)
I	=	wind erosion soil erodibility index ($t \cdot ha^{-1}$)
IR	=	ridge interval (m)
k_d	=	P concentration in sediment divided by that of water ($m^3 \cdot t^{-1}$)
K	=	USLE soil erodibility factor
KR	=	ridge roughness for wind erosion (mm)
L	=	channel length from most distant point to watershed outlet (km)
L_c	=	average channel flow length for watershed (km)
L_{ca}	=	distance along channel to watershed centroid (km)
LAG	=	lag parameter for estimating soil temperature (0-1)
LAI	=	leaf area index--area of plant leaves relative to the soil surface
LAT	=	latitude of watershed (degrees)
LF_I	=	labile P immobilization factor
LF_u	=	labile P factor for crop uptake (0-1)
LS	=	USLE slope length and steepness factor
MP_a	=	amount in active mineral P pool ($kg \cdot ha^{-1}$)
MP_s	=	amount in stable mineral P pool ($kg \cdot ha^{-1}$)
MPR	=	flow rate from labile to active mineral P pool ($kg \cdot ha^{-1} \cdot d^{-1}$)
n	=	Manning's roughness coefficient for water flow
ND	=	number of days in a month
NDD	=	number of dry days in a month

NWD	=	number of wet days in a month
O	=	percolation rate (mm$\cdot$d^{-1})
ON	=	organic N content of the soil (kg$\cdot$ha^{-1})
ON_a	=	active organic N pool (kg$\cdot$ha^{-1})
ON_s	=	stable organic N pool (kg$\cdot$ha^{-1})
OP	=	organic P content of the soil (kg$\cdot$ha^{-1})
PAEF	=	application efficiency factor
PAPE	=	effective amount of pesticide applied (kg$\cdot$ha^{-1})
PAPR	=	actual amount of pesticide applied (kg$\cdot$ha^{-1})
PAR	=	photosynthetic active radiation (MJ$\cdot$m^{-2})
PB	=	barometric pressure (kPa)
PD	=	plow depth (m)
PE	=	factor for uater erosion control practice
PH	=	soil pH
PHU	=	potential heat units for crop maturity (°C)
PO	=	soil porosity (mm)
PO1	=	porosity of top 1 m of soil (mm)
PSO1	=	pesticide solubility (g$\cdot$m^{-3})
PSP	=	P sorption coefficient
PSTC	=	pesticide concentration (g$\cdot$t^{-1})
PSTE		pesticide kill fraction (0.0-1.0)
PSTF	=	pest factor used to adjust crop yield
PSTI	=	accumulated pest index
PSTL	=	amount of pesticide leached by *f*
PSTM	=	minimum pest factor value
PSTQ	=	pesticide loss in surface runoff (kg$\cdot$ha^{-1})
PSTY	=	pesticide yield adsorbed to sediment (kg$\cdot$ha^{-1})
PW	=	probability of wet day
P(W/D)	=	probability of a wet day follouing a dry day
P(W/W)	=	probability of a wet day following a wet day
q_c	=	average flow rate in channel (used to estimate watershed time of concentration) (m$^3\cdot$s^{-1})
q_p	=	peak runoff rate (m$^3\cdot$s^{-1})
q^*_p	=	peak runoff rate (mm$\cdot$h$^{-1}\cdot$mm^{-1})

q_s	=	average surface flow rate ($m^3 \cdot s^{-1}$)
Q	=	runoff volume (mm)
Q*	=	runoff volume (m^3)
QR	=	lateral flow rate ($mm \cdot d^{-1}$)
QS	=	30-day sum of runoff (mm)
QT	=	total water loss from a soil layer (used to transport NO_3) (mm)
r	=	rainfall intensity ($mm \cdot h^{-1}$)
r_{ep}	=	peak rainfall excess rate ($mm \cdot h^{-1}$)
r_p	=	peak rainfall rate ($mm \cdot h^{-1}$)
R	=	daily rainfall (mm)
$\overline{R}$	=	mean daily rainfall (mm)
R_t	=	rainfall amount during time t (mm)
$\overline{R}_{.5}$	=	mean maximum 0.5-h rainfall amount (mm)
$R_{.5F}$	=	maximum 0.5-h rainfall amount for frequency F (mm)
RA	=	solar radiation ($MJ \cdot m^{-2}$)
$\overline{RA}$	=	mean daily solar radiation ($MJ \cdot m^2$)
RAB	=	net outgoing long wave radiation for clear days ($MJ \cdot m^{-2}$)
RAD	=	daily mean solar radiation on dry days ($MJ \cdot m^{-2}$)
RAMX	=	maximum solar radiation possible at a site on a given day ($MJ \cdot m^{-2}$)
RAW	=	daily mean solar radiation on wet days ($MJ \cdot m^{-2}$)
RC	=	residue composition factor
RD	=	root depth (m)
RDMX	=	maximum root depth (m)
RE	=	rainfall energy for water erosion equations
REG	=	crop growth constraint (lowest value from among the stress factors estimated for water, nutrients, aeration, and temperature) (0-1)
RFS	=	30-day sum of rainfall (mm)
RGF	=	root growth constraint (lowest value from among the stress factors estimated for soil strength, temperature, and aluminum toxicity) (O-1)
RH	=	relative humidity
$\overline{RH}$	=	mean of the triangular relative humidity distribution
RHD	=	daily mean relative humidity on dry days
RHG	=	relative humidity generated from triangular distribution
RHG*	=	generated relative humidity adjusted to the mean of the triangular distribution

RHL	=	lowest relative humidity value that can be generated
RHP	=	peak of the triangular relative humidity distribution (RHW or RHD)
RHU	=	largest relative humidity value that can be generated
RHW	=	daily mean relative humidity on wet days
RIF	=	ridge roughness factor
RLA	=	lime required to neutralize aluminum in the plow layer ($t \cdot ha^{-1}$)
RMN	=	mineralization rate for fresh organic N ($kg \cdot ha^{-1} \cdot d^{-1}$)
RMP	=	mineralization rate for fresh organic P ($kg \cdot ha^{-1} \cdot d^{-1}$)
RNV	=	combined nitrification and volatilization ($kg \cdot ha^{-1} \cdot d^{-1}$)
ROK	=	percent of coarse fragments in the surface soil layer
ROKF	=	coarse fragment factor
RON	=	flow rate between active and stable organic N pools ($kg \cdot ha^{-1} \cdot d^{-1}$)
RRF	=	clod roughness factor
RRUF	=	random roughness (mm)
RS	=	ridge spacing (m)
RSDV	=	standard deviation on daily rainfall (mm)
RTN	=	ratio of N in active pool to total organic N
RW	=	root weight in a soil layer ($t \cdot ha^{-1}$)
RWT	=	total root weight ($t \cdot ha^{-1}$)
RZ	=	soil profile depth (m)
s	=	SCS runoff curve number retention parameter (m)
s_f	=	SCS runoff curve number retention parameter for frozen soil (mm)
S	=	land surface slope ($m \cdot m^{-1}$)
SAN	=	sand content of a soil layer (%)
SAT	=	saturation index for top 1 m of soil
SC	=	saturated conductivity of a soil layer--that is, rate of water drainage through a saturated layer ($mm \cdot h^{-1}$)
SCF	=	skew coefficient for skewed normal rainfall distribution
SD	=	sun's declination angle (radians)
SDRA	=	standard deviation of daily solar radiation ($MJ \cdot m^{-2}$)
SDTMX	=	standard deviation of daily maximum temperature (°C)
SDV	=	standard deviation of daily wind velocity ($m \cdot s^{-1}$)
SEV	=	potential evaporation from a soil layer (mm)
SEV*	=	soil-water-adjusted evaporation from a soil layer (mm)

SIL = silt content of soil (%)

SMB = sum of bases ($cmol \cdot kg^{-1}$)

SML = snowmelt rate ($mm \cdot d^{-1}$)

SN = N stress factor for crop growth (0-1)

SN_S = N stress scaling factor

SND = standard normal deviate

SNO = water content of snow (mm)

SPT = snow pack temperature (°C)

SR = standing residue from previous crop ($t \cdot ha^{-1}$)

SS = root growth stress factor due to soil strength (0-1)

SW = soil water content (mm)

SWF = soil water factor, associated uith water content of soil pores, for a soil layer

t = time (h)

t_c = watershed's time of concentration (h)

t_{cc} = time of concentration for channel flow (h)

t_{cs} = time of concentration for surface flow (h)

T = temperature (°C)

$\overline{T}$ = long-term average annual air temperature (°C)

T_b = base temperature for a crop (plants start growing) (°C)

T_{mn} = daily minimum air temperature (°C)

T_{mx} = daily maximum air temperature (°C)

T_o = optimal temperature for a crop (°C)

T_s = temperature at top of snow pack (°C)

TD = daily mean maximum temperature for dry days (°C)

TE = long-term extreme maximum and minimum monthly temperatures

TF_N = temperature factor for nutrient cycling

TG = soil surface temperature (°C)

TGB* = final estimate of bare soil surface temperature (°C)

TGBD = bare soil surface temperature on dry days (°C)

TGBW = bare soil surface temperature on wet days (°C)

TH = temperature stress factor affecting harvest index (0-1)

TS = temperature stress factor for crop growth (0-1)

TT = travel time through a soil layer (h)

TT_R	=	lateral flow travel time (d)
TW	=	daily mean maximum temperature for wet days (°C)
u	=	plant water use rate in a soil layer ($mm \cdot d^{-1}$)
UP	=	potential plant water use rate for the entire root zone ($mm \cdot d^{-1}$)
UC	=	plant-use/soil-water-deficit compensation factor (0-1)
UN	=	actual N uptake rate from a soil layer by a plant ($kg \cdot ha^{-1} \cdot d^{-1}$)
UND	=	plant N demand rate ($kg \cdot ha^{-1} \cdot d^{-1}$)
UNS	=	N supply rate from soil to plants ($kg \cdot ha^{-1} \cdot d^{-1}$)
U_p	=	actual P uptake rate from a soil layer by a plant ($kg \cdot ha^{-1} \cdot d^{-1}$)
UPD	=	plant P demand rate ($kg \cdot ha^{-1} \cdot d^{-1}$)
UPS	=	P supply rate from a soil layer to plants ($kg \cdot ha^{-1} \cdot d^{-1}$)
v_c	=	average channel flow velocity ($m \cdot s^{-1}$)
v_j	=	wind speed at *f* fraction of day
v_s	=	surface flow velocity ($m \cdot s^{-1}$)
V	=	mean daily wind speed ($m \cdot s^{-1}$)
$\overline{V}$	=	annual average wind speed ($m \cdot s^{-1}$)
V_L	=	lower limit of erosive wind speed ($m \cdot s^{-1}$)
V_p	=	wind speed at the peak frequency ($m \cdot s^{-1}$)
V_u	=	upper limit of wind speed ($m \cdot s^{-1}$)
VE	=	quantity of vegetative cover expressed as small grain equivalent ($kg \cdot ha^{-1}$)
VN03	=	amount of flow related NO_3-N loss from a soil layer ($kg \cdot ha^{-1}$)
w_1, w_2	=	shape parameters for soil water-runoff curve number retention parameter relationship
W	=	water surface width in furrow dike system (m)
WC	=	climatic factor for wind-induced erosion
WE	=	soil loss from wind-induced erosion ($t \cdot ha^{-1}$)
WF	=	field length factor for wind-induced erosion
WFX	=	nitrogen fixation by legumes ($kg \cdot ha^{-1}$)
WIM	=	N immobilization rate ($kg \cdot ha^{-1} \cdot d^{-1}$)
WIP	=	P immobilization rate ($kg \cdot ha^{-1} \cdot d^{-1}$)
WK	=	soil ridge roughness factor
WL	=	field length along the prevailing wind direction (m)
WNF	=	wind speed factor for surface application
WNO3	=	weight of NO_3-N in a soil layer ($kg \cdot ha^{-1}$)

WO	=	amount of pesticide washed off plants in rainfall (mm)
WP	=	soil water content of layer at wilting point (mm)
WS	=	water stress factor for crop growth (0-1)
WSYF	=	crop parameter expressing drought sensitivity
WTBL	=	depth from surface to water table (m)
WTL	=	limits of water table--WTMN or WTMX (m)
WTMN	=	minimum water table depth from the surface (m)
WTMX	=	maximum water table depth from the surface (m)
x_1, x_2	=	nutrient enrichment ratio parameters
X	=	material mixed by tillage (nutrients, crop residue, soil properties) ($kg \cdot ha^{-1}$)
XD	=	distance from center of downslope furrow dike to water line on upslope dike (m)
Y	=	sediment yield from water-induced erosion ($t \cdot ha^{-1}$)
YC	=	period of cultivation before simulation starts (yrs)
YLD	=	crop yield ($t \cdot ha^{-1}$)
YON	=	organic N runoff loss ($kg \cdot ha^{-1}$)
YP	=	sediment phase P lost in runoff ($kg \cdot ha^{-1}$)
YWR	=	wind erosion rate at ($kg \cdot m^{-1} \cdot s^{-1}$)
YSP	=	soluble P lost in runoff volume ($kg \cdot ha^{-1}$)
YW	=	wind erosion ($kg \cdot m^{-1}$)
Z	=	soil depth from the surface (m)
ZD	=	displacement height of crop (m)
ZO	=	surface roughness parameter (m)

REFERENCES

Abruna, F., J. Rodriquez, and S. Silva. 1982. Crop response to soil acidity factors in Ultisols and Oxisols in Puerto Rico. VI. Grain sorghum. J. Agric. Univ. P.R. 61:28-38.

Acevedo, E., T.C. Hsiao, and D.W. Henderson. 1971. Immediate and subsequent growth responses of maize leaves to changes in water status. Plant Physiol. 48:631-636.

Albrecht, S.L., S.M. Bennett, and K.J. Boote. 1984. Relationship of nitrogenase activity to plant water stress in field-grown soybeans. Field Crops Res. 8:61-71.

Andrews, R.E. and E.I. Newman. 1970. Root density and competition for nutrients. Oecol. Plant. 5:319-334.

Begg, J.E., J.F. Bierhuizen, E.R. Lemon, D.K. Misra, R.O. Slatyer, and W.R. Stern. 1964. Diurnal energy and water exchanges in bulrush millet in an area of high solar radiation. Agric. Meteorol. 1:294-312.

Bonhomme, R., F. Ruget, M. Derieux, and P. Vincourt. 1982. Relations entre production de matiere seche aerienne et energie interceptee chez differents genotypes de mais. C.R. Adad. Sci. Paris 294:393-398.

Borg, H. and D.V. Grimes. 1986. Depth development of roots with time: An empirical description. Trans. ASAE 29:194-197.

Bouinols, A., M. Cabelguenne, C.A. Jones, A. Chalamet, J.L. Charpenteau, and J.R. Marty. 1991. Simulation of soybean nitrogen nutrition for a silty clay soil in southern France. Field Crops Res. 26:19-34.

Brenes, E. and R.W. Pearson. 1973. Root responses of three Gramineae species to soil acidity in an Oxisol and an Ultisol. Soil Sci. 116:295-302.

Cassman, K.G. and D.N. Yunns. 1980. Nitrogen mineralization as affected by soil moisture, temperature, and depth. Soil Sci. Soc. Am. J. 44:1233-1237.

Chepil, W.S. 1956. Influence of moisture on erodibility of soil by wind. Soil Sci. Soc. Am. Proc. 20:288-292.

Cole, G.W., L. Lyles, and L.J. Hagen. 1982. A simulation model of daily wind erosion soil loss. 1982 ASAE Winter Meeting, Paper No. 82-2575.

DeJager, A. 1979. Localized stimulation of root growth and phosphate uptake in *Zea mays* L. resulting from restricted phosphate supply. pp. 391-403 *In* J.L. Harley and R. S. Russell (eds.) The Soil-Root Interface. Academic Press, New York.

Doorenbos, J. and A.H. Kassam. 1979. Yield response to water. Irrigation and Drainage Paper 33. Food Agric. Org. United Nations, Rome.

Eavis, B.W. 1972. Soil physical conditions affecting seedling root growth. I. Mechanical impedance, aeration, and moisture availability as influenced by bulk density and moisture levels in a sandy loam soil. Plant Soil 36:613-622.

Eik, K. and J.J. Hanway. 1965. Some factors affecting development and longevity of leaves of corn. Agron. J. 57:7-12.

Foy, C.D., H.N. Lafever, J.W. Scuartz, and A.L. Fleming. 1974. Aluminum tolerance of wheat cultivars related to region of origin. Agron. J. 66:751-758.

Godwin, D.C., C.A. Jones, J.T. Ritchie, P.L.G. Vlek, and L.G. Youngdahl. 1984. The water and nitrogen components of the CERES models. pp. 95-100 *In* Proc. Intl. Symp. on Minimum Data Sets for Agrotechnology Transfer, March 1983, Patancheru, India. Intl. Crops Research Instittue for the Semi-Arid Tropics.

Grossman, R.B., V.D. Nettleton, and B.R. Brasher. 1985. Application of pedology to plant response prediction for tropical vertisols. *In* Proc. Fifth Intl. Soil Classification Workshop, Sudan, November 1982.

Hagen, L.J., E. L. Skidmore, and J.B. Layton. 1988. Wind erosion abrasion: Effects of aggregate moisture. Trans. ASAE 31:725-728.

Hanks, R.J. 1983. Yield and uater-use relationships: An overview. pp. 393-411 *In* H.M. Taylor, W.R. Jordan, and T.R. Sinclair, eds., Limitations to Efficient Water Use in Crop Production. Am. Soc. Agron., Crop Sci. Soc. Am., Soil Sci. Soc. Am., Madison, WI.

Hargreaves, G.H. and Z.A. Samani. 1985. Reference crop evapotranspiration from temperature. Applied Engr. Agric. 1:96-99.

Harper, J.E. 1976. Contribution of dinitrogen and soil or fertilizer nitrogen to soybean production. pp. 101-107 *In* L.D. Hill (ed.) Proc. World Soybean Research Conf.

Hazen, Allen. 1930. Flood Flows, A Study of Frequencies and Magnitudes. John Wiley & Sons, Inc., New York.

Hershfield, D.M. 1961. Rainfall frequency atlas of the United States for durations from 30 minutes to 24 hours and return periods from 1 to 100 years. U.S. Dept. Commerce Tech. Paper No. 40.

Hobbs, J.A. and C.A. Thompson. 1971. Effect of cultivation on the nitrogen and organic carbon content of a Kansas argiustoll [Chernozem]. Agron. J. 63:66-68.

Jarvis, S.C. and A.D. Robson. 1983. The effect of nitrogen nutrition of plants on the development of acidity in western Australian soils. I. Effects with subterranean clover grown under leaching conditions. Aust. J. Agric. Res. 34:341-353.

Jones, C.A. 1983a. A survey of the variability in tissue nitrogen and phosphorus concentrations in maize and grain sorghum. Field Crops Res. 6:133-147.

Jones, C.A. 1983b. Effect of soil texture on critical bulk densities for root growth. Soil Sci. Soc. Am. J. 47:1208-1211.

Jones, C.A. 1984. Estimation of percent aluminum saturation from soil chemical data. Commun. Soil Sci. Plant Anal. 15:327-335.

Jones, C.A. 1985. C-4 Grasses and Cereals. John Wiley & Sons, Inc., New York.

Jones, C.A. and J.R. Kiniry, eds. 1986. CERES-Maize. Texas A&M Univ. Press, College Station.

Jones, C.A., C.V. Cole, A.N. Sharpley, and J.R. Williams. 1984. A simplified soil and plant phosphorus model. I. Documentation. Soil Sci. Soc. Am. J. 48(4):800-805.

Jungk, A. and S.A. Barber. 1974. Phosphate uptake rate of corn roots as related to the proportion of the roots exposed to phosphate. Agron. J. 66:554-557.

Kamprath, E.J. 1970. Exchangeable aluminum as a criterion for liming leached mineral soils. Soil Sci. Soc. Am. Proc. 34:252-254.

Kletke, D.D. 1979. Operation of the enterprise budget generator. Oklahoma State Univ., Agric. Exp. Sta. Res. Rept. P-790.

Knisel, W.G. 1980. CREAMS, A field scale model for chemicals, runoff, and erosion from agricultural management systems. U.S. Dept. Agric. Conserv. Res. Rept. No. 26.

Leonard, R.A., W.G. Knisel, and D.A. Still. 1987. GLEAMS: Groundwater loading effects on agricultural management systems. Trans. ASAE 30(5):1403-1428.

McElroy, A.D., S.Y. Chiu, J.W. Nebgen, and others. 1976. Loading functions for assessment of water pollution from nonpoint sources. Environ. Prot. Tech. Serv., EPA 600/2-76-151.

McGrann, J.M., K.D. Olson, T.A. Powell, and T.R. Nelson. 1986. Microcomputer budget management system user manual. Dept. Agric. Econ., Texas A&M Univ., College Station.

Menzel, R.G. 1980. Enrichment ratios for water quality modeling. pp. 486-492 In W.G. Knisel, ed., CREAMS, A field scale model for chemicals, runoff, and erosion from agricultural management systems. U.S. Dept. Agric. Conserv. Res. Rept. No. 26.

Monsi, M. and T. Saeki. 1953. Uber den Lictfaktor in den Pflanzengesellschaften und sein Bedeutung fur die Stoffproduktion. Japan J. Bot. 14:22-52.

Monteith, J.L. 1965. Evaporation and environment. Symp. Soc. Exp. Biol. 19:205-234.

Monteith, J.L. 1973. Principles of Environmental Physics. Edward Arnold, London.

Monteith, J.L. 1977. Climate and the efficiency of crop production in Britain. Phil. Trans. Res. Soc. London Ser. B. 281:277-329.

Monteith, N.H. and C.L. Banath. 1965. The effect of soil strength on sugarcange growth. Trop. Agric. 42:293-296.

Muchow, R.C., D.B. Coates, G.L. Wilson, and M.A. Foale. 1982. Growth and productivity of irrigated *Sorghum bicolor* (L. Moench) in Northern Australia. I. Plant density and arrangement effects on light interception and distribution, and grain yield, in the hybrid Texas 610SR in low and medium latitudes. Aust. J. Agric. Res. 33:773-784.

Mugwira, L.M., S.J. Patel, and A.L. Fleming. 1980. Aluminum effects on growth and Al, Ca, Mg, K, and P levels in triticale, wheat, and rye. Plant Soil 57:467-470.

Nicks, A.D. 1974. Stochastic generation of the occurrence, pattern, and location of maximum amount of daily rainfall. pp. 154-171 *In* Proc. Symp. Statistical Hydrology, Aug.-Sept. 1971, Tucson, AZ. U.S. Dept. Agric., Misc. Publ. No. 1275.

Nyatsanga, T. and W.H. Pierre. 1973. Effect of nitrogen fixation by legumes on soil acidity. Agron. J. 65:936-940.

Onstad, C.A. and G.R. Foster. 1975. Erosion modeling on a watershed. Trans. ASAE 18(2):288-292.

Patterson, T.G. and T.A. Larue. 1983. Nitrogen fixation (C_2H_2) by soybeans: Cultivars and seasonal effects and comparison of estimates. Cro. Sci. 23:488-492.

Pavan, M.A., F.T. Bingham, and P.F. Pratt. 1982. Toxicity of aluminum to coffee in Ultisols and Oxisols amended with $CaC0_3$, $MgC0_3$, and $CaS0_4/2H_20$. Soil Sci. Soc. Am. J. 46:1201-1207.

Peech, M. 1965. Lime requirement. Agronomy 9:927-932.

Penman, H.L. 1948. Natural evaporation from open, bare soil and grass. Proc. Soc. London Ser. A193:120-145.

Pesek, J., G. Stanford, and N.L. Case. 1971. Nitrogen production and use. pp. 217-269 *In* R.A. Olson, ed., Fertilizer Technology and Use. Soil Sci. Soc. Am., Madison, WI.

Pierre, W.H., J.R. Webb, and W.D. Shrader. 1971. Quantitative effects of nitrogen fertilizer on the development and downward movement of soil aciditiy in relation to level of fertilization and crop removal in a continuous corn cropping system. Agron. J. 63:291-297.

Potter, K.N. and T.M. Zobeck. 1990. Estimation of soil microrelief. Trans. ASAE 33(1):156-161.

Potter, K.N., T.M. Zobeck, and L.J. Hagen. 1990. A microrelief index to estimate soil erodibility by wind. Trans. ASAE 33(1):151-155.

Priestley, C.H.B. and R.J. Taylor. 1972. On the assessment of surface heat flux and evaporation using large-scale parameters. Mon. Weather Rev. 100:81-92.

Reddy, K.R., R. Khaleel, M.R. Overcash, and P.W. Westerman. 1979. A nonpoint source model for land areas receiving animal wastes: II. Ammonia volatilization. Trans. ASAE 22(6):1398-1404.

Richardson, C.W. 1981. Stochastic simulation of daily precipitation, temperature, and solar radiation. Water Resources Res. 17(1):182-19O.

Richardson, C.W. 1982. Dependence structure of daily temperature and solar radiation. Trans. ASAE 25(3):735-739.

Richardson, C.W. and D.A. Wright. 1984. WGEN: A model for generating daily veather variables. U.S. Dept. Agric., Agric. Res. Ser., ARS-8.

Ritchie, J.T. 1972. A model for predicting evaporation from a row crop with incomplete cover. Water Resources Res. 8:1204-1213.

Seligman, N.G. and H. van Keulen. 1981. PAPRAN: A simulation model of annual pasture production limited by rainfall and nitrogen. pp. 192-221 *In* M.J. Frissel and J.A. van Veen, eds., Simulation of Nitrogen Behaviour of Soil-Plant Systems, Proc. Workshop, Wageningen, Jan.-Feb. 1980.

Sharpley, A.N., C.A. Jones, and J.R. Williams. 1990. The nutrient component of EPIC. Chapter 7, pp. 152-166 In A.N. Sharpley and J.R. Williams (eds.) EPIC--Erosion/Productivity Impact Calculator: 1. Model Documentation. U.S. Dept. Agric. Tech. Bull. No. 1768.

Sharpley, A.N., C.A. Jones, C. Gray, and C.V. Cole. 1984. A simplified soil and plant phosphorus model: II. Prediction of labile, organic, and sorbed phosphorus. Soil Sci. Soc. Am. J. 48:800-805.

Sharpley, A.N., C.A. Jones, C. Gray, and others. 1985. A detailed phosphorus characterization of seventy-eight soils. U.S. Dept. Agric., Agric. Res. Ser., ARS-31.

Sharpley, A.N. and J.R. Williams, eds. 1990. EPIC--Erosion/Productivity Impact Calculator: 1. Model Documentation. U.S. Dept. Agric. Tech. Bull. No. 1768.

Simanton, J.R., E. Rawitz, and E.D. Shirley. 1984. Effects of rock fragments on erosion of semiarid rangeland soils. Chapter 7, pp. 65-72 *In* Erosion and Productivity of Soils Containing Rock Fragments. Soil Sci. Soc. Am., Madison, WI.

Skidmore, E.L. 1986. Wind-erosion climatic erosivity. Climate Change 9:195-208.

Soil Survey Staff. 1982. Procedures for collecting soil samples and methods of analysis for soil survey. U.S. Dept. Agric., Soil Conserv. Ser., Soil Survey Investigations Rept. No. 1 (revised August 1982). Washington, DC.

Stockle, C.O., J.R. Williams, N.J. Rosenberg, and C.A. Jones. 1992. A method for estimating the direct and climatic effects of rising atmospheric carbon dioxide on growth and yield of crops: Part I--Modification of the EPIC model for climate change analysis. Agric. Systems 38:225-238.

Stout, J.E. 1990. Wind erosion with a simple field. Trans. ASAE 33:1597-1600.

Taylor, H.M. 1983. A program to increase plant available water through rooting modification. pp. 463-472 *In* Root Ecology and Its Practical Application. Intl. Symp., Gumpenstein, September 1982. Budndesanstalt fur alpenlandische Landwirtschalf, A-8952 Irding.

Taylor, H.M., G.M. Robertson, and J.J. Parker, Jr. 1966. Soil strength--root penetration relations for medium to coarse-textured soil materials. Soil Sci. 102:18-22.

Tollenaar, M., T.B. Daynard, and R.B. Hunter. 1979. Effect of temperature on rate of leaf appearance and flowering date of maize. Crop Sci. 19:363-366.

Thornley, J.H.M. 1976. Mathematical Models in Plant Physiology. Academic Press, New York.

Uchijima, Z., T. Udagawa, T. Horie, and K. Kobayashi. 1968. The penetration of direct solar radiation into corn canopy and the intensity of direct radiation on the foliage surface. J. Agron. Meteorol. Tokyo 3:141-151.

U.S. Department of Agriculture, Soil Conservation Service. 1972. National Engineering Handbook, Hydrology Section 4, Chapters 4-1O.

U.S. Department of Agriculture, Soil Conservation Service. 1986. Urban hydrology for small watersheds. Tech. Release 55.

U.S. Department of Commerce. 1968. Climatic Atlas of the United States. Environmental Science Services Administration, Environmental Data Service.

U.S. Department of Commerce. 1979. Maximum short duration rainfall. National Summary, Climatic Data.

Watts, W.R. 1972. Leaf extension in *Zea mays*. II. Leaf extension in response to independent variation of the temperature of the apical meristem, of the air around the leaves, and of the rootzone. J. Exp. Bot. 23:713-721.

Williams, J.R. 1975. Sediment yield prediction with universal equation using runoff energy factor. U.S. Dept. Agric., Agric. Res. Serv., ARS-S-40.

Williams, J.R. and R.W. Hann. 1978. Optimal operation of large agricultural watersheds with water quality constraints. Texas Water Resources Institute, Texas A&M Univ., Tech. Rept. No. 96.

Williams, J.R., C.A. Jones, and P.T. Dyke. 1984a. The EPIC model and its application. pp. 111-121 *In* Proc. ICRISAT-IBSNAT-SYSS Symp. on Minimum Data Sets for Agrotechnology Transfer, March 1983, Hyderabad, India.

Williams, J.R., C.A. Jones, and P.T. Dyke. 1984b. A modeling approach to determining the relationship between erosion and soil productivity. Trans. ASAE 27:129-144.

Wischmeier, W.H. and D.D. Smith. 1978. Predicting rainfall erosion losses, a guide to conservation planning. U.S. Dept. Agric., Agric. Handbook No. 537.

Woodruff, N.P. and F.H. Siddoway. 1965. A wind erosion equation. Soil Sci. Soc. Am. Proc. 29(5):602-608.

Chapter 26

AGNPS: AN AGRICULTURAL NONPOINT SOURCE MODEL

R.A. Young, C.A. Onstad, and D.D. Bosch

Nutrients and sediment in runoff, seepage, or percolation resulting from land management activities have a major impact on water quality. Sources of such pollution include soil erosion and sedimentation on rural and urban land and eroding streambanks, nutrients and organic materials from livestock wastes and agricultural land, and stormwater from urban areas.

Managing nonpoint sources of pollution, in addition to being politically, economically, and socially difficult, is technically complex. Pollutant sources often are located over a large geographic area and are not readily identifiable. By locating and targeting specific areas within a watershed that have high potential for soil and nutrient losses, available federal and state cost-sharing funds can be used more efficiently to alleviate potential pollution problems and protect water quality.

The AGNPS (Agricultural NonPoint Source) model was developed by the Agricultural Research Service (ARS) in cooperation with the Minnesota Pollution Control Agency (MPCA) and the Soil Conservation Service (SCS). The model was developed to analyze and provide estimates of runoff water quality from agricultural watersheds ranging in size from a few acres to upwards of 50,000 acres. It is relatively simple to use and runs on an IBM-compatible personal computer. Estimates can be used to compare the performance of selected watersheds experiencing similar runoff events. After a watershed has been identified as having a potential pollution problem, remedial measures can be recommended on the basis of an assessment of the effects of applying alternative management practices. This is accomplished by varying input data consistent with alternative management practices being investigated and analyzing the resulting watershed responses.

26.1. MODEL STRUCTURE

AGNPS is a distributed parameter, event-based model. It simulates the behavior of runoff, sediment, and nutrient transport from watersheds that have agriculture as their prime use. The nutrients considered include nitrogen (N) and phosphorus (P), both of which are essential plant nutrients and major contributors to surface water pollution. It also considers chemical oxygen demand (COD). COD is a measure of the oxygen required to oxidize organic and oxidizable inorganic compounds in water. As such, it is used as an indicator of degree of pollution. Basic model components include hydrology, erosion, and sediment and chemical transport. In addition, the model considers point sources of sediment from gullies, streambanks, construction sites, etc., and inputs of water, sediment, nutrients, and COD from animal feedlots, springs, and other point sources. Water impoundments, such as tile-outlet terraces, also are considered as depositional areas of sediment and sediment-associated nutrients.

The model operates on a cell basis. Cells are uniformly square areas subdividing the watershed, allowing analyses at any point within the watershed. Potential pollutants are routed through cells from the watershed divide to the outlet in a stepwise manner so that flow at any point between cells may be examined. All watershed characteristics and inputs are expressed at the cell level. Figure 26.1 shows a typical small watershed of 1040 acres containing 26 cells, each of which is 40 acres in size. Arrows in each cell depict the major drainage pathways.

For watersheds exceeding 2,000 acres, cell sizes of 40 acres are generally recommended. Smaller cell sizes are recommended for smaller watersheds. Accuracy of results can be increased by reducing the cell size, but this increases the time and labor required to run the model. Conversely, enlarging the cell size reduces time and labor, but the savings must be balanced against the loss of accuracy resulting from treating larger areas as homogeneous units.

26.1.1. Hydrology

Runoff

Runoff volume and peak flow rate are calculated in this part of the model. Runoff volume estimates are based on the SCS curve number method (USDA, 1972). The basic equation is as follows:

$$Q = \frac{(P - 0.2S)^2}{P + 0.8S} \tag{26.1}$$

where Q is the runoff volume, P is the rainfall, and S is a retention parameter, all expressed in uniform dimensions of length. Except where

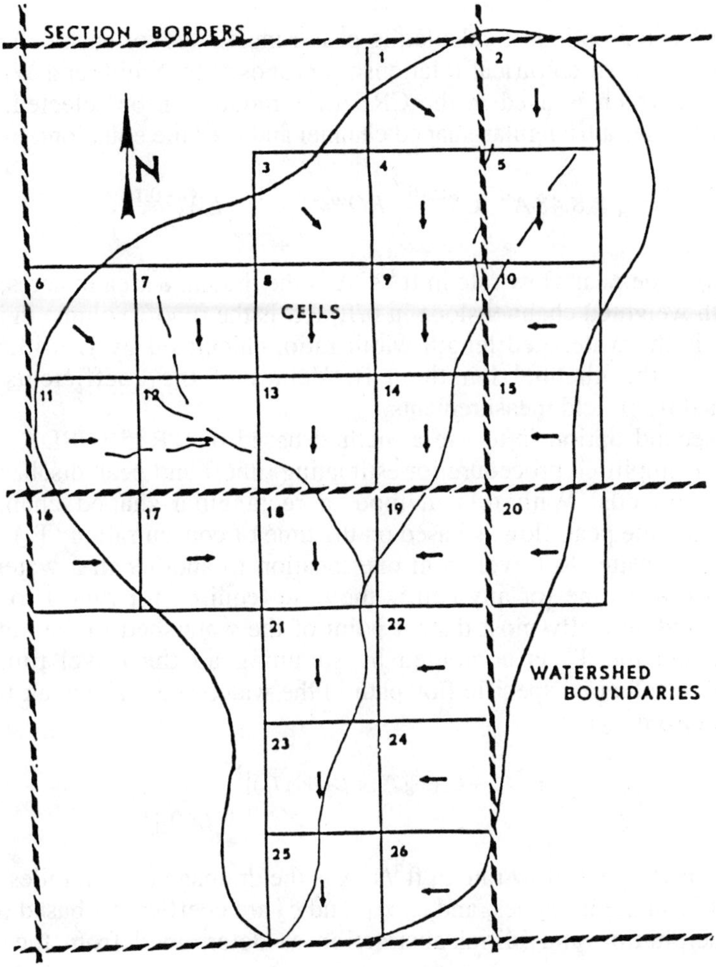

Fig. 26.1. Sample watershed showing subdivision into cells.

otherwise indicated, equations are generalized relationships valid in any system of units. The retention parameter is defined in terms of a curve number (CN), as follows:

$$S = \frac{1000}{CN} - 10 \tag{26.2}$$

The curve number depends upon land use, soil type, and hydrologic soil condition. This method was chosen because of its simplicity and widespread use among the principal user agencies for which the model was developed.

Peak Runoff Rate

The user has the option of calculating peak runoff rates for each cell in two different ways. An empirical relationship proposed by Smith and Williams (1980) and which is used in the CREAMS model can be selected. This method assumes a triangular-shaped channel and uses the equation:

$$Q_p = 8.48\,A^{0.7}\,CS^{0.16}\left(RO^{0.82A^{0.017}}\right)LW^{-0.19} \tag{26.3}$$

where Q_p is the peak flow rate in ft^3/s, A is the drainage area in acres, CS is the length weighted channel slope in ft/ft, RO is the runoff volume in inches and LW is the watershed length-width ratio, calculated by $L^2/(43{,}560A)$ where L is the channel length in ft. Values of the coefficients were determined from field measurements.

The second option is to use a method based on TR55 (USDA, 1986) which is a simplified procedure for estimating runoff and peak discharges in small watersheds. With this method, a rectangular-shaped channel is assumed and the peak flow is based on the time of concentration (T_c), or the time it takes water to travel from one location to another in a watershed. The total travel time for any cell is the time required for runoff to travel from the hydraulically most distant point of the watershed to the outlet of that cell and the T_c is computed by summing all the travel times for consecutive cells in a specific flowpath in the watershed. The peak flow is calculated from T_c by:

$$Q_p = 10^{\log\left[C_o + C_1(\log T_c) + C_2(\log T_c)\right]^2}\left[\frac{A}{640}\right]Q \tag{26.4}$$

where Q_p is the peak flow rate in ft^3/s, A is the drainage area in $miles^2$, Q is the runoff volume in inches, and c_0, c_1, and c_2 are coefficients based on 24-hour precipitation and initial abstraction as determined from the curve number.

With either method of calculating peak flow, the user has the option of entering known channel characteristics, i.e. channel length, width, and depth, or using hydrogeomorphic relationships to determine channel geometry. The hydraulic geometry predicted by geomorphic calculations is a data-smoothing approach that predicts the downstream trend of increasing channel widths, depths, and lengths within well-defined geomorphic regions. This allows the user to estimate the channel dimensions as a function of total drainage area into the cell.

Hydrograph Development

The peak flow rates calculated from either method are used to generate a triangular hydrograph for each cell. The shape of the hydrograph can be user defined. The hydrograph is partitioned into uniform increments with at least three increments on the rising limb. Flow rates are calculated for each increment using the average time to peak, the flow duration, and the increment duration.

26.1.2. Erosion and Sediment Transport

Erosion

A modified form of the universal soil loss equation (USLE) (Wischmeier and Smith, 1978) is used to estimate upland erosion for single storms as follows:

$$SL = (EI)\, KLSCP[SSF] \tag{26.5}$$

where SL is the soil loss, EI is the product of the storm total kinetic energy and maximum 30-minute intensity, K is the soil erodibility factor, LS is the topographic factor, C is the cover and management factor, P is the supporting practice factor, and SSF is a factor to adjust for slope shape within the cell. Erosion equation factors are calculated using procedures found in **Agricultural Handbook 537** (Wischmeier and Smith, 1978). Soil loss is calculated for each cell of the watershed. Eroded soil and sediment yield are subdivided into five particle size classes--clay, silt, small aggregates, large aggregates, and sand.

Sediment Transport

After runoff and upland erosion are calculated for each cell, detached sediment is routed from cell to cell through the watershed to the outlet. The procedure used involves sediment transport and depositional relations described by Foster et al. (1981) and Lane (1982). The basic routing equation is derived from the steady-state continuity equation as follows:

$$Q_s(x) = Q_s(o) + Q_{sl}(x / L_r) - {}_o\int^{x} D(x)\, w dx \tag{26.6}$$

where $Q_s(x)$ is the sediment discharge at the downstream end of the channel reach, $Q_s(o)$ is the sediment discharge into the upstream end of the channel reach, Q_{sl} is the lateral sediment inflow rate, x is the reach length, w is the channel width, and D(x) is the deposition rate. Deposition rate is estimated as follows:

$$D(x) = [V_{ss} / q(x)][q_s(x) - g_s'(x)] \tag{26.7}$$

where V_{ss} is the particle fall velocity, q(x) is the discharge per unit width, $q_s(x)$ is the sediment load per unit width, and $g'_s(x)$ is the effective transport capacity per unit width. Effective transport capacity is computed using a modification of the Bagnold stream power equation (Bagnold, 1966), as follows:

$$g'_s(x) = \eta g_s = \eta k \tau v^2 / V_{ss} \tag{26.8}$$

where g'_s is the transport capacity, η is an effective transport factor, k is the transport capacity factor, τ is the shear stress, and v is the average channel flow velocity determined by Manning's equation. Values for the effective transport capacity are described by Young et al. (1986). Sediment load for each of the five particle size classes leaving a cell is calculated as follows:

$$Q_s(x) = \left[\frac{2q(x)}{2q(x) + \Delta x V_{ss}}\right]\left[Q_s(o) + Q_{sl}\frac{x}{L} - \frac{w\Delta x}{2}\left[\frac{V_{ss}}{q(o)}\left[q_s(o) - g'_s(o)\right] - \frac{V_{ss}}{q(x)}g'_s(x)\right]\right] \tag{26.9}$$

where symbols are as defined above. Equation (26.9) is the basic routing equation that drives the sediment transport model.

The sediment transport calculations allow for deposition and/or scouring of all particle sizes during channel flow based on transport capacity and sediment availability. However, the user has the option of not allowing any scouring, as in the case of a non-erodible channel bed, or allowing only specific particle sizes to scour, as might be the case in a relatively stable channel bed.

26.1.3. Chemical Transport

Nutrients

The chemical transport calculations estimate transport of N, P, and COD throughout the watershed. Relationships used to calculate chemical transport have been adapted from CREAMS (Frere et al., 1980) and a feedlot evaluation model (Young et al., 1982), with modifications for the effects of variations in soil texture. Chemical transport calculations are divided into soluble and sediment adsorbed phases. Nutrient yield in the sediment
adsorbed phase is calculated using total sediment yield from a cell, as follows:

$$Nut_{sed} = \left(Nut_f\right) Q_s(x) E_r \tag{26.10}$$

where Nut_{sed} is N or P transported by sediment, Nut_f is N or P content in the field soil, and E_r is the sediment enrichment ratio, calculated as follows:

$$E_r = 7.4 Q_s(x)^{-0.2} T_f \tag{26.11}$$

where $Q_s(X)$ is sediment yield and T_f is an adjustment to correct sediment adsorbed nutrient enrichment ratios for sand and clay soils (Young et al., 1987).

Soluble nutrient estimates consider the effects of nutrient levels in rainfall, fertilization, and leaching. Soluble nutrients contained in runoff are estimated as follows:

$$Nut_{sol} = C_{nut}\ Nut_{ext} Q \qquad (26.12)$$

where Nut_{sol} is the concentration of soluble N or P in the runoff, C_{nut} is the mean concentration of soluble N and P at the soil surface during runoff, Nut_{ext} is an extraction coefficient of N and P for movement into runoff, and Q is the total runoff.

Chemical Oxygen Demand

COD in the model is assumed soluble. Estimates of COD in runoff are based on calculated runoff volumes and average concentration of COD in that volume. Background concentrations of COD available in the literature are used as a basis for predicting COD concentrations from each cell.

All soluble nutrients and COD are allowed to decay with time once they enter a channel according to either an exponential decay function incorporated within the model or to a user input rate of reduction.

26.1.4. Point Source Inputs

The model treats nutrient and COD contributions from animal feedlots as point sources and routes them with contributions from diffuse sources. Chemical contributions from feedlots are estimated using the feedlot pollution model developed by Young et al. (1982) as a subroutine. The feedlot model estimates nutrient concentrations and mass at both the feedlot edge and at a receiving body of water.

Other point-source inputs of water and nutrients, such as springs, wastewater treatment plant discharges, etc. are accounted for by inputting incoming flow rates and concentrations of N, P, and COD to the cells in which they occur.

Erosion from streambanks, streambed, gullies, and other sediment sources are accounted for using estimated values as point sources. Sediment from these sources is added to upland sediment and considered in the transport phase of the model. Nutrients from these sources is added to other nutrient inputs to a cell and routed accordingly.

26.1.5. Impoundments

Sediment and runoff routing through impoundments is accomplished using relationships described by Laflen et al. (1978). These relations were developed for impoundment terrace systems having pipe outlets. The fraction of each particle class passing through the impoundment is a function of surface area, depth, diameter of the pipe outlet, and infiltration rate. In terms of water quality, impoundments reduce peak discharges, sediment yield, and yield of sediment-attached chemicals.

26.2. MODEL INPUT

Table 26.1 lists the inputs for AGNPS. Parameter values may be obtained from published data, available farmstead and watershed records, or on-site

TABLE 26.1.

Input information required to run AGNPS.

Hydrologic
- Amount of precipitation
- Rainfall energy-intensity value (EI) or storm duration and type
- Runoff curve number

Soils
- Soil texture
- Soil erodibility

Upland drainage
- Slope length
- Slope steepness
- Slope aspect
- Slope shape

Channel drainage
- Occurrence of defined channel in cell
- Channel slope gradient
- Channel side slope or channel dimensions
- Manning's roughness coefficient

Agricultural management
- Fertilizer application level
- Fertilizer incorporation level
- Cropping management factor (C)
- Conservation practice factor (P)
- Surface condition constant

Other information
- Estimated erosion from other sources (i.e. gullies, streambank, etc.)
- Terrace impoundments - number, size, and size of outlet pipe
- Feedlots - number, size, and type of animal
- Other point sources (i.e., springs, treatment plant discharge, etc.)

inspection. The first items, under hydrologic input, include information that applies to the watershed as a whole, including the type of rainfall, the N contribution from rainfall, and the method to be used for the hydrology calculations. The remaining items have to be determined for each cell. The model's user manual (Young et al., 1987) contains tables listing standard variables for many of the required parameters. Default values for some parameters are also incorporated into the model. These default values will be used for a cell whenever no other value is input by the user.

26.3. MODEL OUTPUT

Various output options are available with the model. Preliminary output given for all watersheds being analyzed includes watershed area and cell size, storm precipitation and erosivity (EI), estimates of runoff volume and peak flow rate at the watershed outlet, and area-weighted erosion, both upland and channel. Also given are estimates of sediment delivery ratio, sediment enrichment ratio, mean sediment concentration, and total sediment yield for each of five sediment particle size classes. Also available is a nutrient analysis, which includes N, P, and COD mass per unit area for both soluble and sediment adsorbed nutrients, and N, P, and COD concentrations in the runoff.

A more detailed output can be obtained for any or all cells if desired, and includes runoff volume, peak flow, upland erosion, channel erosion, sediment yield by particle size, deposition, soluble N, P, and COD (both concentration and mass load), and sediment attached N and P (mass load). Cell selection for output depends on the problem. For example, if a stream reach needs to be protected for fish habitat, those cells having outlets to the stream may be the only ones that need to be examined. If an impoundment at the watershed outlet needs protection, then only the output at the outlet of the watershed may need to be examined. If a potential problem is shown to exist, upland cells may be selected and examined to identify and prioritize problem areas where management alternatives can be implemented in order to correct the problem.

If the chosen cell size is too large to adequately reflect conditions in a certain area of the watershed or, if upon examination of the output, a cell is considered to have a problem which requires more detailed examination, a cell division capability allows for the division of any cell from the original grid into fourths. This division process can be repeated three times, resulting in cells 1/64th the size of the original cell. This process enables a more detailed analysis of critical areas within a watershed.

Table 26.2 lists a summary of output parameters that can be obtained for each cell or all cells if desired. The output can be displayed in either tabular

TABLE 26.2.

Output at the watershed outlet or for any cell

Hydrology output	
	Runoff volume (inches)
	Peak runoff rate (cubic feet/second)
	Fraction of runoff generated within the cell
Sediment output	
	Sediment yield (tons)
	Sediment concentration (ppm)
	Sediment particle size distribution
	Upland erosion (tons/acre)
	Amount of deposition (%)
	Sediment generated within the cell (tons)
	Enrichment ratios by particle size
	Delivery ratios by particle size
Chemical output	
	Nitrogen
	Sediment associated mass (pounds/acre)
	Concentration of soluble material (ppm)
	Mass of soluble material (pounds/acre)
	Phosphorus
	Sediment associated mass (pounds/acre)
	Concentration of soluble material (ppm)
	Mass of soluble material (pounds/acre)
	Chemical Oxygen Demand
	Concentration (ppm)
	Mass (pounds/acre)

or graphics format. The graphics output will outline the watershed and display up to 13 different input and/or output variables simultaneously. This enables the user to quickly assess environmental conditions within the watershed and to locate critical areas needing further examination. An example of a graphics output is shown in Fig. 26.2.

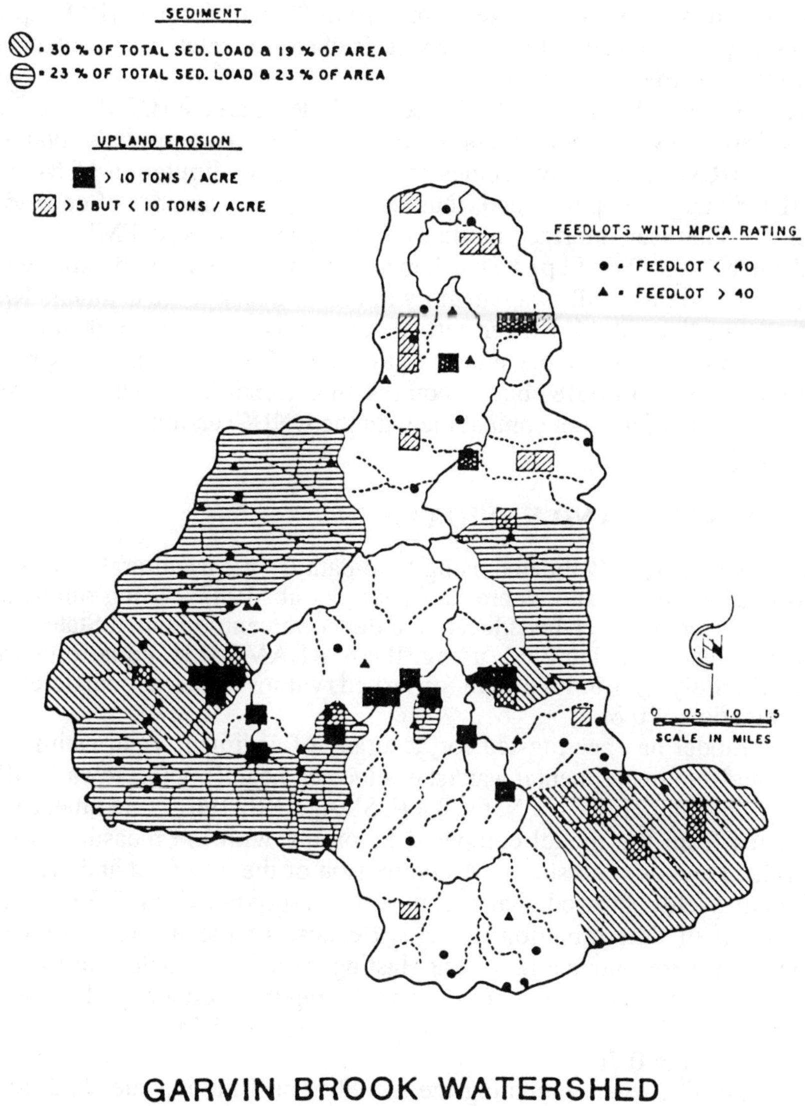

Fig. 26.2. Estimated sediment discharge from Garvin Brook watershed, Winona County, Minnesota.

26.4. SYSTEM REQUIREMENTS

AGNPS is written in C and turbo-pascal and is designed to run on any IBM compatible personal computer with MS-DOS version 3.0 or later. It requires 2 Mb available of either extended or expanded memory, a hard disk with 3 Mb or more of free space, an 80286 or later processor, and a graphics adapter and monitor (CGA minimum). An 8087 math co-processor is highly recommended for speed of execution. AGNPS will support IBM, Epson, and Okidata dot matrix printers, Hewlett Packard LaserJet laser printers, and Hewlett Packard compatible plotters.

AGNPS uses both the DOS Protected Mode Interface (DPMI) and Phar Lap's 286 DOS Extender in order to use all of a computer's available memory (RAM). This overcomes the normal space limitation of 640K of RAM for many computers and, thus, allows a larger number of cells to be entered. The user interface modules of the program use DPMI while the model itself uses Phar Lap[1]. Phar Lap also allows for virtual memory which means the program will swap memory to the hard disk if the available RAM memory is exhausted. However, virtual memory is much slower than RAM.

A UNIX version of the model is available for use with a geographic information system (GIS) but it requires an external input data file. Also, the graphics output is not compatible with the UNIX version.

26.5. TESTING AND VALIDATION

The model has been validated using field data from agricultural watersheds in Minnesota, Iowa, and Nebraska (Young et al., 1987). Using runoff data from 20 different watersheds located in the north central United States, peak runoff rates as calculated using the CREAMS [3] equation, were approximately 1.6% less than observed values with a coefficient of determination of 0.81.

The model has been tested for accuracy of sediment yield estimations with data from experimental watersheds located near Treynor, Iowa (USDA, ARS, 1970) and Hastings, Nebraska (USDA, ARS, 1967). Sediment yield estimates from the model compared favorably with the measured values from the Treynor watersheds. A comparison of the observed and predicted sediment yields showed that the model overpredicted by 2.3%, with a coefficient of determination of 0.95. Because of the limited number of values that were obtainable for the Hastings watershed, determining if the model was consistently under or over predicting the sediment yields was not possible. Available data for this watershed resulted in an coefficient of determination of 0.76.

Lee (1987), when validating the model in an Illinois watershed, found that the simulated and observed data for runoff volume and sediment yield were well represented when compared with observed data. Koelliker and Humbert (1989) found close agreement between observed and estimated water and sediment yields from 5 watersheds in northeast Kansas.

While the chemical components of AGNPS had undergone some basic testing on land slopes during the development of the CREAMS model, generally there are few good chemical data from watersheds. Data from measureable rainfall runoff events in seven Minnesota watersheds over a 3

---

[1] Mention of a trade name, proprietary product, or specific equipment does not constitute a guarantee or warranty by the U.S. Department of Agriculture and does not imply its approval to the exclusion of other products that may be suitable.

year period were used to test the chemical components of the model. Unfortunately, during this test period only relatively small (1-year, 24-hour) runoff events were measured. Due to the extremely low sediment concentrations from such small events, the data were insufficient to adequately test either the sediment yield estimates or the sediment particle-size relationships of AGNPS. However, a comparison of measured versus estimated N and P concentrations from 20 different sampling points indicated that AGNPS generally provided realistic estimations of nutrient concentrations in runoff water. Comparisons of total nitrogen (TN) and total phosphorus (TP) values yielded coefficients of determination of 0.63 and 0.78, respectively. The model tends to overpredict nutrient concentrations, but estimated values are generally within the range of acceptability and errors in estimation would be conservative. The lack of data points from larger rainfall events tends to lend an element of uncertainty to the relationships shown. Further testing of all components of the model is being done as additional data becomes available.

26.6. EXAMPLE

The AGNPS model has been used in many states and several countries to prioritize watersheds for severity of water quality problems, to pinpoint critical areas within a watershed contributing to pollution, and to evaluate the effects of applying alternative management practices. An example of using AGNPS for assessing problems within a watershed is the Garvin Brook Watershed, Winona County, Minnesota. The Garvin Brook watershed is a 31,640-acre watershed in southeastern Minnesota that drains into the Mississippi River near the city of Winona. It is an area of hilly, rolling topography in which approximately one-third of the land is forested and two-thirds is agricultural. The principal drainage way in the watershed is a designated trout stream which is thought to be degrading due to runoff from agricultural areas. The purpose in using AGNPS in this case was to determine the sediment and nutrient loads being delivered to the trout stream. The model was used to evaluate the watershed response to a 25-year frequency, 24-hour duration storm. Output was screened to locate feedlots having a high potential pollutant rating and individual cells having excessively high upland erosion rates. Then, by examining the output of every cell bordering the trout stream, total loads of sediment and nutrients to the stream were estimated. Subwatersheds were determined for those cells bordering the stream that contributed excessively high quantities of either sediment or nutrients to the waterway.

Figure 26.2 shows that over 30% of the total sediment load entering the stream probably originated in three subwatersheds comprising about 19% of the total area. Another 23% of the sediment load to the stream came from subwatersheds making up about 23% of the total area. While contributions of TN and TP were apparently more uniformly distributed over the watershed, those areas producing the majority of these nutrients corresponded closely with those areas having the highest contributions of sediment, with 23% of the total area producing 28% of the TN and TP entering the stream.

Output from the AGNPS model was used to pinpoint "critical" areas within the Garvin Brook watershed where control measures could be used to best advantage in reducing the input of potential pollutants to the trout stream. For example, two of the high sediment yielding subwatersheds had cells with very high erosion rates located directly above their outlets (Fig. 26.2). Yields from these subwatersheds might be sufficiently reduced by controlling upland erosion rates within these few cells. On the other hand, the third subwatershed contributing high sediment loads to the stream contained no specifically high sediment yielding cells. Instead, general land use and topographic factors for the whole area contributed to the high sediment yields. In this case, alternative management practices to reduce sediment yield over the entire area would have to be employed. This output information was used to encourage local farmers to sign up for cost sharing to implement appropriate control measures.

A second example showing a different type of application is the Salmonson Creek Catchment in Big Stone County, Minnesota. Big Stone Lake is 26 miles long, covering 12,360 acres astride the South Dakota-Minnesota border. The lake is a potentially significant recreational resource for the area. However, in recent years, problems of poor water quality, excessive algal blooms and aquatic vegetation, and increased sedimentation have led to reduced recreational use and declining land values around the lake. The lake has a drainage area of approximately 730,000 acres consisting primarily of rolling, highly erodible farmland with a wide variety of soil types. While municipal sewage systems and lakefront septic systems contribute significantly to the hyper-eutrophic conditions in the lake, most of the problems lie with drainage and runoff from the primarily agricultural areas in the watershed. A cooperative effort between several federal, state, and local agencies and groups was begun to try and substantially improve the lake water quality and reduce its hyper-eutrophic condition (German and Dingels, 1984). The nonpoint source portion of this effort has concentrated on selecting certain targeted subwatersheds for initial analysis and treatment. One of these, the 15,640-acre Salmonson Creek subwatershed (Dingels, 1986), is shown in Fig. 26.3.

The AGNPS model was used to identify critical areas within this subwatershed where excessive upland erosion and runoff have resulted in high contributions of sediment and nutrients at the outlet where it entered Big Stone Lake. Once these critical areas were identified, appropriate best management practices (BMP's) were recommended to reduce erosion/pollution problems.

In order to estimate the potential benefits of implementing the BMP's, inputs to the model were altered to reflect the application of the

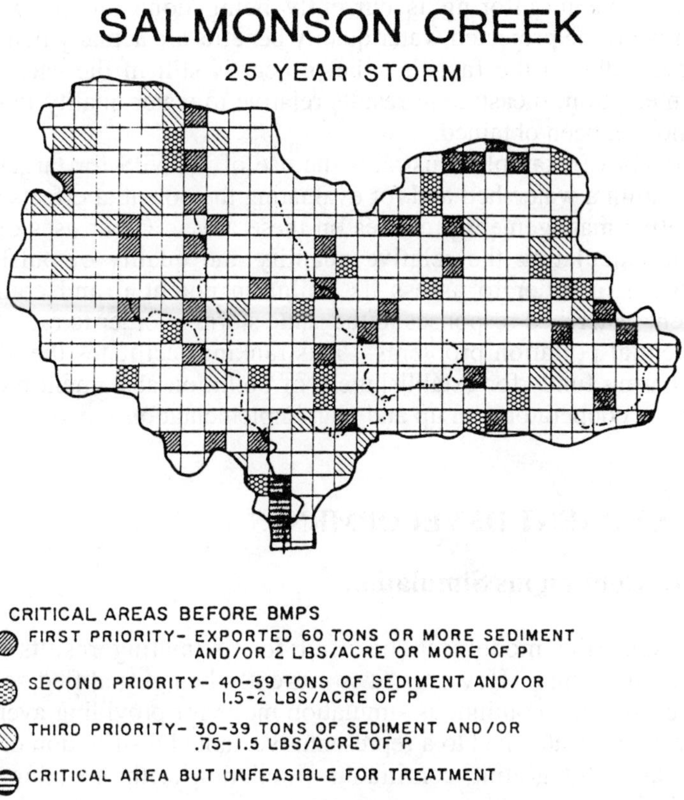

Fig. 26.3. Salmonson Creek Watershed, Big Stone County, MN.

recommended practices. The model was then rerun and the new output used to estimate the projected differences in sediment and nutrient yields at the watershed outlet. Table 26.3 shows the predicted sediment and nutrient yields from the Salmonson Creek subwatershed before and after the BMP's were implemented.

TABLE 26.3.

Estimated sediment and nutrient yields from a 25-yr, 24-hr storm before and after best management practices on the Salmonson Creek catchment.

	Before	After	% Reduction
Sediment (tons)	4460	2647	41
Total N (lb/a)	2.7	1.9	30
Total P (lb/a)	0.9	0.6	33

Extensive monitoring is currently being done on the watershed to determine if the projected water quality benefits are actually being achieved. However, due to the fact that the project is still in the earlier stages of implementation, measurable results relative to water quality improvements have not yet been obtained.

The above example illustrates the use of AGNPS for targeting critical areas within a watershed and for evaluating the potential effects of applying alternative management practices on these areas. Other uses of the model include calculating the relative quantity and quality of outflow from a watershed in order to assess its pollution potential and comparison of different watershed responses to the same storm in order to rank the severity of potential pollution problems. This ranking facilitates the allocation of cost-sharing funds for establishment of pollution abatement measures and can aid in cooperator sign up and plan implementation.

26.7. CURRENT DEVELOPMENTS

26.7.1. Continuous Simulation

At present, the model can be used for estimating results of specific hydrologic events. However, it is currently being modified so that it can also be used in a continuous simulation mode for providing average annual response of a watershed to a representative annual distribution of rainstorms for an area. Integrating continuous simulation to the existing distributed spatial abilities enables an even more representative description of the stochastic physical processes controlling the movement of sediment and agricultural chemicals from the field into the surrounding environment.

While the basic components of the model will remain the same, the continuous simulation mode involves several modifications to the model as it now exists. These include incorporation of a weather generator (Nicks and Lane, 1989) to generate a representative sequence of storms for the area of interest and the annualization of most of the localized time dependent factors in the model representing specific vegetative and soil conditions. Input parameters will be initialized for a certain point in time. The weather generator generates daily weather for the desired period of time and the model, operating on a daily time step, calculates daily changes in soil, vegetative, and soil moisture conditions and keeps a running balance of water, sediment and nutrient movement in each cell. Also, upland erosion is estimated using techniques adapted from the revised universal soil loss equation (RUSLE) (USDA, In press). Output results can be viewed for any single cell within the watershed, a sub-basin within the watershed, or for the entire watershed in either tabular or graphical mapped format. Runoff events taking place within the simulation can be viewed individually, as a monthly or yearly accumulation of storm results, or as an average annual value derived from a multiple-year scenario.

Because the continuous simulation model requires calculations to be made over an entire year, a set of winter routines has been incorporated into AGNPS to deal with snowmelt runoff and runoff from frozen soil conditions. The snowmelt-frozen soil procedure is composed of three separate components which interact with each other on a daily basis. These components deal with soil frost, snowmelt, and snowdrifting. The frost component estimates the extent of frost development and thawing over the winter period as well as changes in soil water content and infiltration capacity. The snowmelt component estimates the amount of snowmelt occurring and how much snowmelt water is available for runoff in the spring. The snowdrift component estimates the depth, density, and distribution of snow cover over the watershed. Prediction of the effects of soil frost and snowmelt on runoff and soil erosion at any point within a watershed is achieved by allowing the three components of the snowmelt-frozen soil process to interact on a daily basis.

26.7.2. Urban Component

AGNPS, in its current form, does not address nutrient and sediment loads from urban areas. It is common for a predominantly agricultural watershed draining into a lake to have low density urban fringe areas which can contribute significant pollutant loads to the lake. It is important to integrate an urban component into the model. This component has been developed for use in the continuous simulation version of AGNPS. The procedure is based on concepts of the USEPA SWMM model (Huber and Dickinson, 1988), the US Corp of Engineers' STORM model (U.S. Corps of Engineers, 1977), and a rainfall accumulation-washoff model. Appropriate techniques for estimating heavy metal loading are also included.

26.7.3. Water Body Component

In many instances, watersheds will contain lakes, reservoirs, and/or wetlands, as well as streams or rivers. A procedure has been developed to link lake processes with watershed processes and to simulate lake and watershed responses to land management and weather conditions. The procedure links runoff, sediment, and chemical input variables from a watershed to a lake model, FARMPOND (Gallegos et al., 1985). This one-dimensional lake model simulates temperature stratification, mixing by wind, sedimentation, inflow density currents, and algal growth. Unsteady advection-diffusion equations characterize the dynamics of suspended sediment, nitrogen and phosphorus, and chlorophyll. The same random generation of weather conditions that drives AGNPS on a daily basis are used to drive the water body model. Resulting impacts of alternative management plans are simulated by changing agricultural practices and land use, thereby modifying inflow characteristics to lakes.

26.7.3. Pesticide Component

A menu-driven interactive pesticide submodel has been developed which allows the evaluation of different management practices on pesticide runoff and leaching. Pesticides are divided into 6 categories: herbicides, insecticides, fungicides, nematicides, growth regulators, and desiccants/defoliants. Pesticides are selected from a database containing both common and trade names. Default inputs are suggested depending upon the pesticide selected and the method of application. Inputs include the time of pesticide application, application rate, application efficiency, percent ground cover, soil and foliar pesticide residues, soil and foliar pesticide half-life, incorporation depth and efficiency, water solubility, and organic carbon sorption coefficient (K_{oc}). An enrichment ratio approach is used to estimate pesticide yield from the adsorbed phase. The submodel outputs the mass (lb/acre) of pesticide lost in the runoff and sediment yield. In addition to surface runoff, the amount of pesticide leached from the surface is estimated. Outputs are both tabular and graphical, indicating areas of pesticide losses within the watershed.

26.7.4. Economic Component

An economic component is being developed which will allow the application of cost/benefit ratios to AGNPS output. This process will help to provide reasonable estimates of the typical benefits that could be achieved by applying a selected management plan within a rural or urban watershed in terms of typical costs associated with the repair and/or maintenance of existing man-made capital improvements.

26.8. SUMMARY

The AGNPS model is a distributed parameter, event-based model developed to analyze and provide estimates of runoff water quality from primarily agricultural watersheds. The model is relatively easy to use and is flexible in its capability to address a wide range of problematic conditions within watersheds. The outputs help describe:

1) The extent of NPS problems--how much sediment and nutrients are being generated and transported throughout a watershed.

2) The source of such problems--"hot spot" cells and/or drainages delivering substantial levels of NPS pollutants to waterways.

3) The potential environmental impacts of alternative land uses, conservation treatments, and management scenarios involving crops, tillages, and various management practices and control measures.

These outputs allow the user to evaluate the impact of applying alternative management practices at any point within the catchment. Such a procedure enables the user to rate upland catchment areas in priority order to remedy a downstream problem.

REFERENCES

Bagnold, R. A. 1966. An approach to the sediment transport problem from general physics. Prof. Paper 422-J. U.S. Geological Survey, Reston, VA.

Dingels, S. P. 1987. Big Stone Lake Watershed Project, Minnesota, South Dakota. In: Cooperating for Clean Water. Estimating water quality improvements from management of agricultural nonpoint source pollution. Briefing paper No. 2. Nat. Assoc. of State Depts. of Agriculture Res. Foundation. p. 7.

Foster, G. R., L. J. Lane, J. D. Nowlin, J. M. Laflen, and R. A. Young. 1981. Estimating erosion and sediment yield on field-sized areas. Trans. ASAE 24(50):1253-1262.

Frere, M. H., J. D. Ross, and L. J. Lane. 1980. The nutrient submodel. In: CREAMS, A field scale model for chemicals, runoff, and erosion from agricultural management systems. USDA Cons. Res. Report 26(1):65-85.

Gallegos, C. L., F. R. Schiebe, and H. G. Stefan. 1985. Modeling water movement and thermal structure in agricultural impoundments. In: USDA-ARS-SCS, Natural Resources Modeling Symposium, Pingree Park, CO. Oct. 16-21, 1983. Rep. ARS-30. USDA-ARS, Fort Collins, CO. pp. 371-378.

German, D. and S. Dingels. 1984. Big Stone Lake Restoration Project. U.S. EPA 315 Clean Lakes Program. Unpublished Report.

Huber, W. C. and R. E. Dickinson. 1988. Storm Water Management Model Version 4, User's Manual. EPA/600/3-88/001a (NTIS PB88-236641/AS). Environmental Protection Agency. Athens, GA. 569 pp.

Knisel, W. G., [ed]. 1980. CREAMS: A field-scale model for chemicals, runoff, and erosion from agricultural management systems. Conservation Research Report No. 26 USDA-SEA, Washington, D.C. 643 pp.

Laflen, J. M., H. P. Johnson, and R. O. Hartwig. Erosion modeling on impoundment terraces. Trans. ASAE 21(6):1131-1135.

Lane, L. J. 1982. Development of a procedure to estimate runoff and sediment transport in ephemeral streams. In: Recent developments in the explanation and prediction of erosion and sediment yield. Proceedings of the Exeter Symposium. IAHS Publ. No. 137. p. 275-282.

Lee, M. T. 1987. Verification and applications of a nonpoint source pollution model. In: Proceedings of the National Engineering Hydrology Symposium. ASCE, New York, NY.

Nicks, A. D. and L. J. Lane. 1989. Weather generator. In: USDA-Water Erosion Prediction Project: Hillslope Profile Model Documentation. NSERL Rpt. N. 2. USDA-ARS, National Soil Erosion Research Laboratory. W. Lafayette, IN.

Smith, R. E. and J. R. Williams. 1980. Simulation of surface water hydrology. In: CREAMS, A field scale model for chemicals, runoff, and erosion from agricultural management systems. USDA Conserv. Res. Report 26(1):15.

U.S. Corps of Engineers. 1977. Storage, treatment, overflow, runoff model, STORM, user's manual, generalized computer program. 723-S8-L7520. Hydrologic Engineering Center. Davis, CA.

Central Great Plains Experimental Watershed. A summary report of 30 years of hydrologic research. USDA Hydrograph Laboratory through contract with Univ. of Nebraska Statistical Laboratory, Hastings, Nebraska. Undated Mimeo Report (probably late 1967).

USDA, 1967, 1970. Annual Research Report, North Central Watershcd Research Center, Columbia, MO.

USDA, 1972. Hydrology. Sec. 4, Chap. 10, p. 10.5-10.6. Soil Conservation Service National Engineering Handb. Washington, D.C.

USDA, 1986. Urban Hydrology for small watersheds. Soil Conservation Service-Engineering Div. tr-55.

Williams, J. R., A. D. Nicks, and J. G. Arnold. 1985. SWRRB, a simulator for water resources in rural basins. J. Hydraulic Eng. 111(6):970-986.

Williams, J. R., C. A. Jones, and P. T. Dyke, 1984. A modelling approach to determining the relationship between erosion and soil productivity. Trans. ASAE 27(1):129-144.

Wischmeier, W. H. and D. D. Smith. 1978. Predicting rainfall erosion losses. USDA Handb. 537. Washington, D. C. 58 pp.

Young, R. A., C. A. Onstad, D. D. Bosch, and W. P. Anderson. 1987. AGNPS: An agricultural nonpoint source pollution model: A watershed analysis tool. USDA Conservation Research Report 35. 77 pp.

Young, R. A., M. A. Otterby, and A. Roos. 1982. A technique for evaluating feedlot pollution potential. J. Soil Water Conserv. 37:21-23.

Young, R. A. , M. A. Otterby, and A. Roos. 1982. An evaluation system to rate feedlot pollution potential. ARM-NC-17. April.

Young, R. A., C. A. Onstad, D. D. Bosch, and W. P. Anderson. 1989. AGNPS: A non-point source pollution model for evaluating agricultural watersheds. J. Soil and Water Conserv. 44(2):164-172.

Young, R. A., C. A. Onstad, and D. D. Bosch. 1986. Sediment transport capacity in rills and small channels. In: Proc., Fourth Federal Interagency Sediment Conf., Subcomm. on Sedimentation of the Interagency Advisory Comm. on Water Data, Vol. 2. Washington, D.C. pp. 6-25 to 6-33.

Chapter 27

SPUR-91: SIMULATION OF PRODUCTION AND UTILIZATION OF RANGELANDS

D.H. Carlson, T. L. Thurow, J.R. Wight

27.1. DESIGN INTENT AND EVOLUTION OF SPUR-91

The ability of a computer model to simulate the complexity of ecosystem function and predict ecosystem response to various management practices has great potential as a research tool and as a decision aid to resource managers. In 1981 the Agricultural Research Service of the U.S. Department of Agriculture initiated a project to create a model capable of simulating the complex climate, hydrology, plant, and animal interactions on rangeland ecosystems. Specific objectives and uses for the model include:

1) Predict changes in water quality and quantity resulting from management practices;

2) Simulate rates and amounts of runoff and sediment occurring in stream channels;

3) Predict by species or species-groups the daily biomass and nitrogen content of live and dead rangeland vegetation;

4) Simulate the spatial heterogeneity of rangeland communities;

5) Predict the response of vegetation to herbivory, nitrogen availability, and water availability;

6) Simulate herbage removal by cattle and wildlife from rangelands;

7) Provide sufficient detail to allow simulation of sub-divided watersheds;

8) Provide useful information for environmental impact analyses.

This complex modeling task was approached by assigning development of individual submodels to various specialty research groups within the USDA-ARS organization. Existing technologies were incorporated into the model, and when necessary, were modified for rangeland application. Submodels were then integrated to form a unified ecosystem simulation model.

The product of this six year effort was the Simulation of Production and Utilization of Rangelands model (SPUR) (Wight and Skiles, 1987). When the SPUR model was released, its status regarding predictive capability and user documentation was not adequate for wide-spread practical use in resource management applications (Carlson and Thurow, 1992). Limited validation of submodel output had been performed (Wight and Skiles, 1987) but validation of the entire model was not conducted using independent long-term data sets with concurrent hydrology, plant and animal data. Thus, the effectiveness of the hydrology-plant-animal interface had not been assessed prior to the model's release. The many critical decisions (e.g., time steps, prioritization of process order, accounting procedures, importance of scale) that must be made by model developers when integrating various submodels into a complete working model requires that after the submodels are linked together, refinement and validation must focus on performance of the complete model.

Problems encountered by SPUR users and new information in the form of extensive data sets that would enable comprehensive validation of the model prompted Texas A&M University in cooperation with the USDA-Soil Conservation Service to upgrade the model. Validation of the field-scale version of the original SPUR model began with a 4-year data set comprised of simultaneous measurements of hydrology, vegetation, livestock, and meteorological parameters (Carlson et al., 1990; Heitschmidt and Dowhower, 1991) collected at the Texas Agricultural Experiment Station near Throckmorton on the Rolling Plains rangelands of north Texas. Consistent deviations between predicted and observed values for this validation indicated serious weaknesses within the model. For example, the shrub component at the site was lost from the system after a six-year simulation due to competition from annual forbs and shortgrasses. To determine the source of numerous unrealistic output anomalies, a sensitivity analysis was conducted. Generally, the sensitivity analysis revealed that hydrology variables had little influence over plant outputs even though they represented processes of key importance within the system. MacNeil et al. (1987) concluded, in a separate sensitivity analysis, that the plant and hydrology components "seemed too independent." It became apparent that submodel integration was a major limitation of the original model.

Initial validation results were analyzed to identify specific problem areas within SPUR. Submodel integration problems within the existing model structure were corrected using an iterative evaluation/modification process to ensure that changes provided a logical, consistent improvement in correspondence between observed and predicted data. The refined model was then tested using long-term simultaneous watershed/plant/livestock data sets from three ecologically diverse locations in Texas. Based on the results of each validation, additional shortcomings of the model were highlighted and corrected. The result of these refinement efforts was SPUR-91 (Carlson and Thurow, 1992). A fourth data set, collected near the site of data used in the first validation, was then conducted. This validation, along with retests of the other data sets, confirmed that the SPUR-91 model was operating well across a diverse range of soil, climate, and vegetation which characterize these sites. A final sensitivity analysis was conducted on SPUR-91 to complete the evaluation process. The performance results of these tests are discussed later in this chapter.

A "final" model version is not likely to be reached as long as added insight into biophysical processes improve. An iterative process of design, implementation, and evaluation is the key to development of models which have the capability of accurately predicting the effects of natural resource management and solving problems in a consistent manner (Carlson et al., 1993).

27.2. STRUCTURE

SPUR-91 is a physically-based model designed to provide biophysical simulation capability for rangeland ecosystems. Some empirical functions were included in the model where physical processes were not well understood or would require inclusion of many detailed field measurements. The complexity of biological, ecological and physical processes which comprise natural systems makes modeling a potentially valuable tool for anticipating responses to management actions. One way to simplify complex models is to partition design into interactive modules. In that way a scientific advance allowing better predictive capability for one module will improve the interactive product of the combined modules. This approach to simulation of complex ecosystem function and management response was used in development of the SPUR-91 model (Wight and Skiles, 1987; Carlson et al., 1993).

The SPUR-91 model is composed of five basic components:

1) Climate,

2) Hydrology,

3) Plant,

4) Animal,

5) Economics.

The interlinkage of the components is illustrated in Fig. 27.1. Two versions of SPUR-91 address different levels of landscape resolution. The field-scale version was designed to simulate plant and animal interactions at a pasture

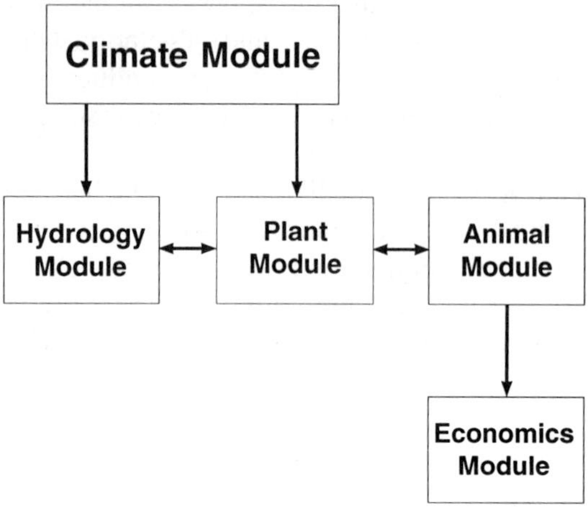

Fig. 27.1. Interlinkage of components within SPUR-91.

or field level. The basin-scale version was designed to simulate small watershed processes (Table 27.1). A generalized flowchart of the main program for the field- and basin-scale versions of SPUR-91 is presented in Fig. 27.2 and Fig. 27.3, respectively (Carlson and Thurow, 1992).

The UTIL system, a Universal Text Integration Language (USDA-ARS, 1991) is used to ease data entry. UTIL is a data file editor that has been developed to help users of large computer models and other data intensive programs. A description of each variable or field and its range limits is provided. UTIL was designed to be used with many different computer models. This versatility saves users the burden of learning a different editor for each model. UTIL is currently used in several other prominent computer models such as WXGEN, SWRRB, and EPIC (USDA-ARS, 1991). File format is consistent from one run to the next. UTIL file formats have been developed for input files required by the climate generator for SPUR-91 and for those required by both the field-scale and basin-scale versions of the model. A simplified user-guide for UTIL specific to its SPUR-91 application is presented in the user guide for SPUR-91 (Carlson and Thurow, 1992).

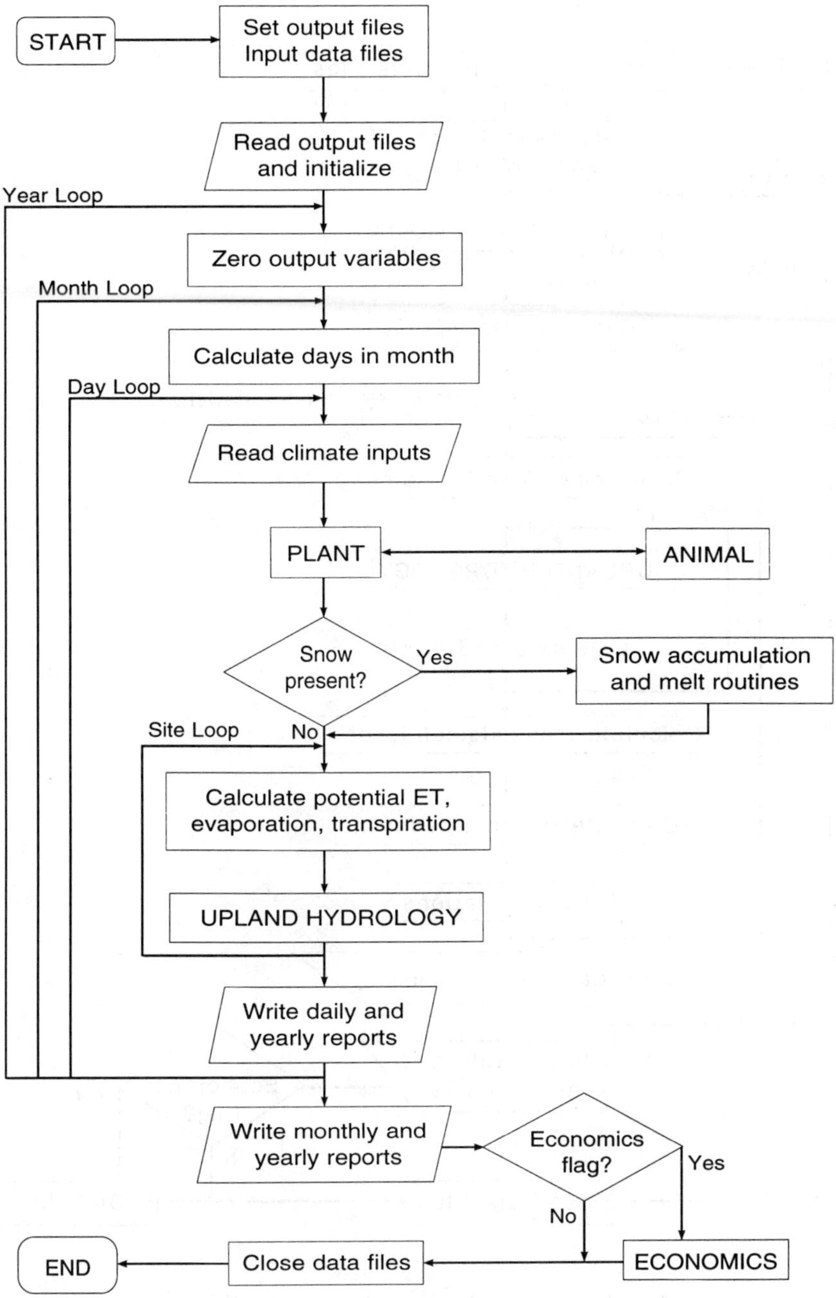

Fig. 27.2. Flowchart for the field-scale version of SPUR-91.

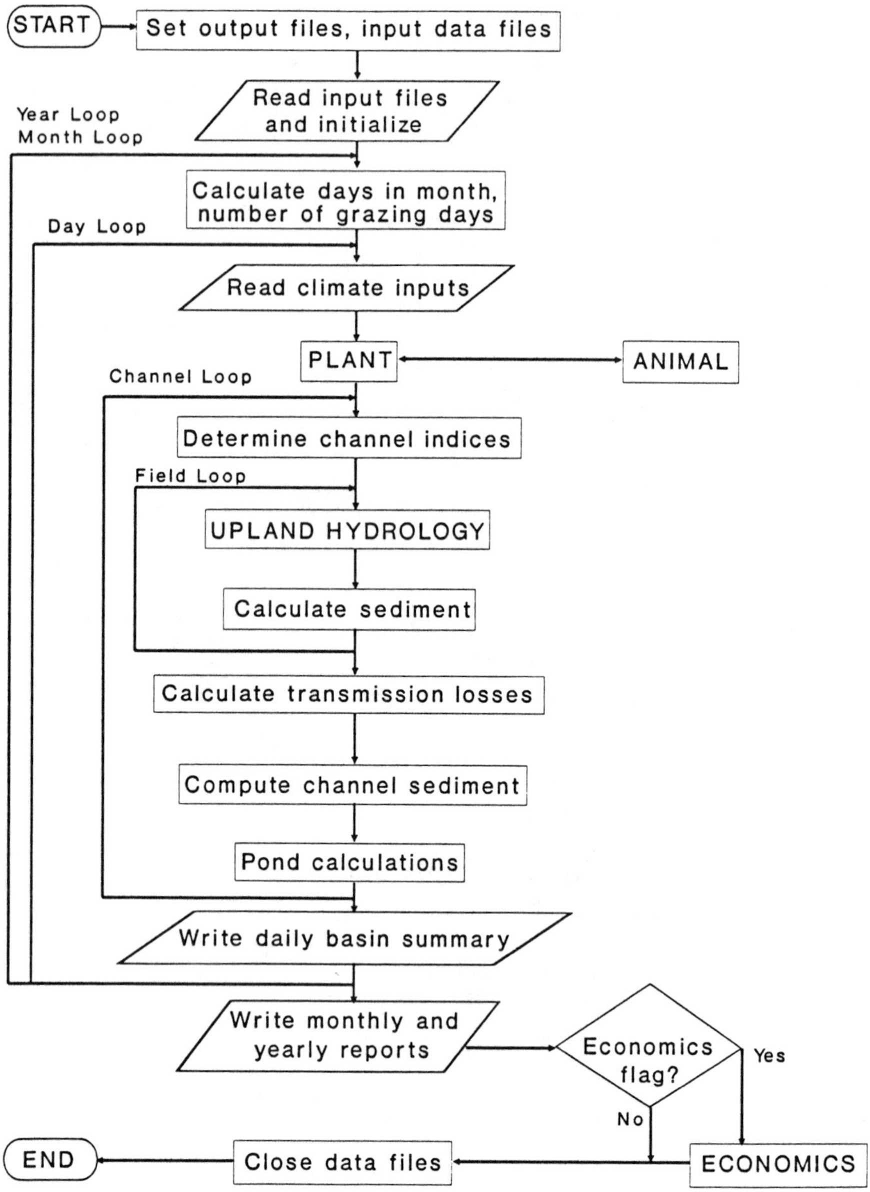

Fig. 27.3. Flowchart for the basin-scale version of SPUR-91.

TABLE 27.1.

Characteristics, scope, and input requirements of the two versions of SPUR-91.

FIELD-SCALE VERSION			BASIN-SCALE VERSION	
		Emphasis		
Plant-animal processes and interactions.			Hydrology of small watersheds.	
		Intended Use		
Simulate effects of field management.			Simulate effects of land-scape management.	
		Scope		
1 field			27 fields	
9 sites			0 sites	
8 soil layers per site			8 soil layers per field	
7 plant species			7 plant species	
1 steer variety			1 steer variety	
10 wildlife species			10 wildlife species	
			9 channels and ponds	
		Inputs Required[1]		
Minimum	**Maximum**		**Minimum**	**Maximum**
48	491	< HYDROLOGY FILE >	81	2154
52	556	< PLANT FILE >	54	1132
24	436	< ANIMAL FILE >	24	562
8	8	< ECONOMICS FILE >	8	8

1 Number of inputs required for each submodel depends on how many sites, plant species, fields, etc. that the user chooses to simulate.

27.3. SUBMODEL DISCUSSION

The emphasis of SPUR-91 is on predicting water fate and plant production, therefore the emphasis of this chapter will be on a technical discussion of these two submodels. A brief overview of the other model components is also provided. The animal and economic components are fairly simplistic and will require substantial improvements to be responsive to rangeland management conditions (Carlson and Thurow, 1992).

27.4. CLIMATE SUBMODEL

The climate component operates independently of the rest of the model and provides daily values for precipitation, maximum and minimum air temperatures, solar radiation, and wind run. These inputs drive a range of model processes including water yield, erosion, plant growth, and species composition. Climate data can be entered from weather records or can be generated stochastically. The climate generator model (CLIMGN-91) is separate from SPUR-91 and is a modification of the WGEN model (Richardson and Wright, 1984). The climate generator is designed to develop a realistic climate pattern for a site, and can be used to fill in data gaps or create an entire data set of the required climatic variables. CLIMGN-91 provides three options:

1) All variables generated,

2) Daily precipitation data provided by the user which is used by the model to generate the other four variables,

3) Generation of wind run only (user provides the other climatic variables).

The model is designed to preserve the dependence in time, the internal correlation, and the seasonal characteristics which exist in actual weather data for the location. The precipitation-generation component of CLIMGN-91 is a Markov chain-gamma model. The probability of rain on a given day is conditioned on the wet or dry status of the previous day. Precipitation and wind run are generated independently of the other variables. Maximum and minimum temperatures and solar radiation are generated depending on whether the day is wet or dry. Procedures have been developed which provide the user with the option of altering generated precipitation or temperature variables by inputting actual mean monthly values. The basic structure of the model is uncomplicated and is structured to enable general application (Richardson et al., 1987).

27.5. ANIMAL SUBMODEL

The animal component simulates herbage removal by grazing animals. Herbivore categories are limited to designation as either livestock or other herbivores (wildlife). Range sites are used to characterize production potential of different portions of the landscape. The range sites (up to nine for the field-scale version, up to 27 for the basin-scale version) and plant species (seven different species possible) may be assigned selection values that will influence the degree of herbivory. Also, within the vegetation groups, the forage is subdivided as live or dead to further influence herbivore selection. The relative abundance of each vegetation group has no effect on diet composition. Daily consumption by wildlife species precedes consumption by livestock. Wildlife are only a sink for the forage they harvest. No other plant-animal interactions (e.g., trampling and nutrient recycling) are incorporated in the wildlife portion of the animal component. In contrast, livestock are not only a sink for the forage they harvest, but they also influence vegetation through trampling effects and return of nitrogen and organic matter to the system via deposition of feces and urine. Livestock growth is simulated on a steer-equivalent basis. The basic structure of this growth submodel is adapted from the Texas A&M University Beef Simulation Model (Sanders, 1977; Notter, 1977; Sanders and Cartwright, 1979). The animal component has a narrow scope and only simulates steer-equivalent growth under moderate grazing intensity.

27.6. ECONOMICS SUBMODEL

The economic component is a simple application of cost-benefit analysis. The only benefits included in SPUR-91 are attributable to weight gain of steers calculated in the animal component section. Gross returns are calculated as the weight gain which is used to determine the final sale weight. The sale weight is multiplied by a sale price to determine the value of the livestock produced. The costs included in the analysis are the cost of the stockers, the fee and non-fee costs of using the pastures and other costs (e.g., interest) that must be incurred to obtain the benefits. Net benefits at a point in time are determined by subtracting costs from gross benefits. The economic component has a narrow scope and does not factor in the many multi-product benefits and costs of rangeland management. Benefits and costs other than animal production, such as value of water yield or cost of poor water quality (siltation or extra downstream purification costs) are not considered (Carlson and Thurow, 1992).

27.7. HYDROLOGY SUBMODEL

27.7.1 Upland Hydrology

Daily Water Balance

The upland hydrology submodel is based primarily on the SWRRB model (Williams et al., 1985) with modifications made for rangeland application.

The submodel maintains a daily water balance (all units are in inches) using the equation:

$$SW = SW_0 + P - Q - ET - PL - QR$$

where:

SW	=	Current soil water content,
SW_0	=	Initial (current day - 1) soil water content,
P	=	Precipitation,
Q	=	Surface runoff,
ET	=	Evapotranspiration,
PL	=	Percolation loss below the root zone (deep drainage) to groundwater storage,
QR	=	Return flow (basin-scale version only).

The soil profile of each site of the field is divided into layers (maximum of eight) according to differences in soil physical properties and vegetation rooting patterns. Because range vegetation can typically operate at soil water tensions below -1.5 MPa, Campbell's equation (Campbell, 1974) is used to compute a -5.0 MPa water content, which is defined as the wilting point for vegetation within the SPUR-91 model. The total storage, field capacity and initial water storage are all expressed, within SPUR-91, in terms of plant available water (i.e. total soil water content minus soil water content at -5.0 MPa) (Carlson and Thurow, 1992).

The percolation routine is similar to that used in CREAMS (Knisel, 1980) and SWRRB. Percolation can occur in two ways: storage routing and crack flow. Storage routing to the next layer occurs only when the soil water content of the layer exceeds field capacity. The travel time through each soil layer is based on the hydraulic conductivity of the soil layer. Restricted flow (due to impermeable caliche layers, etc.) can be simulated by setting the hydraulic conductivity to zero. The crack flow model permits percolation when soils are not saturated. Crack flow occurs only on days when water enters a layer, and the magnitude of flow is greatest when the layer below is dry. Percolation of water below the root zone is defined as deep drainage. For the basin-scale version, a portion of deep drainage is used to estimate subsurface flow from the basin interior, based on the time required to reach the basin outlet.

SPUR-91 computes soil evaporation and transpiration separately, and they are dependent on potential evapotranspiration. Potential evapotranspiration (PET) is calculated using Ritchie's (1972) formula:

$$PET = (0.0504 * H_0 * B) / (PC + B)$$

where H_o is the net solar radiation (langleys), B is the slope of the saturation vapor pressure curve at mean air temperature, and PC is the psychometric constant. Net solar radiation is calculated from daily solar radiation and the albedo. For soil evaporation, the potential evaporation (ESO) is the minimum of either:

$$ESO = PET * e^{-0.4\,LAI}$$

or

$$ESO = PET * GR$$

where LAI is the leaf area index, and GR is the mulch factor. The actual soil evaporation is calculated using Ritchie's technique (Ritchie, 1972). Potential transpiration (EPO) is calculated as: PET - ES (actual soil evaporation) if the LAI is greater than three. This rarely occurs on rangelands. If LAI is less than three, then EPO = (PET * LAI)/3. The sum of EPO and ES is not allowed to exceed PET; in instances where this would occur the ES is reduced accordingly. If the soil water content in a layer is less than 0.35 * FC, then plant transpiration is reduced from EPO by (EPO * SW)/(0.35 * FC), where SW is the current soil water content and FC is field capacity. Transpiration is distributed down through the soil layers where roots occur using an exponential function to describe a generalized root distribution. If one layer's transpirational demand cannot be met because the soil water content is too low, then that demand is added to the demand of the next soil layer (Renard et al., 1987; Carlson and Thurow, 1992).

The SPUR-91 model uses the curve number technique (USDA-SCS, 1972) to determine daily runoff. Soil water content is updated daily and a daily curve number is computed based on this soil water storage (Williams and LaSeur, 1976), rather than using the three curve number (dry (I), normal (II), wet (III)) approach. Daily runoff is then calculated based on daily rainfall and the storage retention parameter calculated from soil water content. The runoff calculations have greater dependence on upper soil layers.

Sediment Yield

The SPUR-91 model computes peak flow of surface runoff in order to calculate upland sediment yield. The peak flow rate is calculated from the equation:

$$Q_p = (C5 * Q)/D$$

where:

Q_p = Peak flow rate (in/h)

Q = Daily runoff volume (in)

D = Duration of runoff (h)

C5 = Watershed parameter representing hydrograph shape.

Runoff duration is calculated from the equation:

$$D = C1 * A^{C2}$$

where:

A = Watershed area (acres)

C1 = Watershed input parameter representing slope and size relationships

C2 = Watershed input parameter representing climatic characteristics

Watershed parameters are derived from a study by Murphey et al. (1977). Sediment yield is then calculated using the modified MUSLE (Williams and Berndt, 1977). The MUSLE includes factors which address soil erodibility, vegetation cover, management practices, erosion control practices, and the length and steepness of slope. These parameters do not affect any other component or process in the model.

Pond Water Balance and Sedimentation

The pond water balance equation used in SPUR-91 is based on SWRRB. Inflow into the pond consists of runoff from the areas draining into the pond plus precipitation received on the pond surface. Water loss from the pond can occur in three ways. Outflow is assumed to occur through a spillway when the maximum pond storage is exceeded. Evaporation from the pond surface is dependent on the surface area and an evaporation coefficient. Seepage from the pond bottom is dependent only on the conductivity of the pond bed material (assumed homogenous) (USDA-SCS, 1957; Hanson et al., 1975). The final pond volume at the end of the day is calculated as inflow less outflow, evaporation and seepage. A pond can only be located at the end of a channel section. Sediment contained in the inflow is assumed to be retained within the pond. Outflow, therefore, is free of sediment (Renard et al., 1987).

Validation Results

Renard et al. (1987) evaluated the upland hydrology submodel of SPUR using data sets from the Lucky Hills Watershed in southeast Arizona.

Evapotranspiration, an important parameter determining antecedent moisture, closely followed patterns of precipitation for most years. The correlation between predicted and actual annual runoff for the upland area was strong (r=0.94), but the correlation between predicted and observed runoff for the entire watershed was slightly weaker (r=0.81). Over the 17-year period, simulated cumulative runoff over-predicted actual runoff by 35% on the upland area and by 30% over the entire watershed. The curve number used to estimate runoff is **very** sensitive. Wilcox et al. (1989) validated the original SPUR submodel on sagebrush rangelands and obtained fairly weak correlations between short-term predicted and observed runoff. The magnitude of total runoff, however, was accurately predicted to be small. Results for the SPUR-91 validation conducted at the Throckmorton site in Texas showed that:

1) There was good correspondence between actual and simulated weekly soil water content on vegetated sites over four years (r=0.90),

2) Runoff was poorly simulated on a monthly basis (r=0.37 to 0.58) but the proportion of runoff in the 4-year water budget was accurately predicted (within 7%),

3) Monthly evapotranspiration was the major outflow of water and was simulated well by SPUR-91 (r=0.62 to 0.82),

4) Estimates of the proportion of deep drainage in the 4-year water budget were very good (within 4%).

More details of the validation are presented in the "Plant-Hydrology Interface" section of this chapter.

27.7.2. Basin Hydrology

The basin hydrology submodel combines streamflow routing procedures and sediment transport equations to predict rates and amounts of runoff and sediment from small basins. The streamflow routing component uses runoff and sediment from adjacent upland and lateral areas, along with features of the channel, as inputs to estimate transmission losses and predict runoff volume, peak rate, and flow duration for each channel section. The sediment transport component computes hydraulic parameters such as velocity and flow depth and uses these, along with channel features and the runoff hydrograph, to predict sediment yield. Channel transport algorithms were developed under the assumption of ephemeral streamflow, which occurs as a result of rainfall events. The hydrograph model was not intended for application in perennial streams or where runoff is dominated by snow melt. In addition, channels are assumed to be alluvial, composed of non-cohesive materials (Lane, 1987).

Streamflow

The submodel first calculates amount of runoff and sediment which is produced on adjacent upland or lateral areas for each channel section (maximum of nine sections allowed). Runoff volume is calculated using the procedures described in the "Upland Hydrology" section. A double triangle hydrograph is used to model the runoff event (Ardis, 1973). Peak discharge is assumed to be a function of the runoff volume and time characteristics of the hydrograph (Lane, 1987). Peak discharge (Q) is then obtained from the equation:

$$Q = (C5 * V)/D$$

where V is runoff volume, D is flow duration, and C5 is a coefficient representing hydrograph shape. Mean flow duration (D) is then calculated from the equation:

$$D = C1 * A^{C2}$$

where A is the drainage area, C1 is a coefficient representing the drainage slope and size characteristics, and C2 is a parameter representing climatic characteristics (Murphey et al., 1977). Transmission loss for a channel section is calculated on the basis of a regression equation established by Lane et al. (1971). Outflow from a channel reach only occurs when inflow surpasses a threshold volume. Inflow volumes less than this threshold are lost via infiltration into channel alluvium. A differential equation approximating the rate of change in runoff volume with distance (Lane, 1983) is then used to develop the basic transmission equation:

$$V(x,w) = a(x,w) + b(x,w) * V_{up} + F(x,w) * (V_{lat} / x)$$

where V(x,w) is the outflow volume, V_{up} is the upstream inflow volume, V_{lat} is the lateral inflow volume, a(x,w) is the regression intercept, b(x,w) is the regression slope, F(x,w) is the channel section parameter, x is the length of the channel section, and w is the average width of flow. Based on data from 14 channel sections in Arizona, Texas, Kansas, and Nebraska, equations to determine the transmission loss parameters were developed (Lane, 1987). They are dependent on the unit channel intercept, effective hydraulic conductivity, mean duration of flow, a decay factor, and mean volume of flow. Mean volume of flow (V) is determined from the equation:

$$V = C3 * A^{C4}$$

where A is the drainage area and C3 and C4 are hydrograph parameters (Murphey et al., 1977). Runoff volume following transmission losses is used along with mean flow duration to determine the peak discharge at the end of the channel section. The model is set up to allow a channel system whereby downstream channel sections can receive upstream inflow as well as inflow from lateral contributing areas (Lane, 1987).

Sediment Yield

The Manning equation is the basis for the hydraulic calculations of the sediment transport component, assuming normal flow occurs in channels with a rectangular cross section. Bed, bank and grain resistance to flow are estimated to determine hydraulic radius. Piecewise normal approximation (Lane, 1982a) to the double triangle hydrograph is used (over nine intervals) to estimate the hydraulic variables for sediment transport calculations. The sediment transport rate is assumed to be equal to the sediment transport capacity. Particles larger than 0.062 mm are transported as bed load while smaller particles are transported as suspended load. Bed load calculations are a modification of the DuBoys-Straub formula (Graf, 1971), dependent on effective shear stress, sediment transport coefficient, proportion and diameter of particles in a given size class, and critical shear stress for particles in a given size class, summed over all size classes. The suspended load model is based on the concept of stream power (Bagnold, 1966). This equation utilizes average velocity, proportion of particles smaller than 0.062 mm, effective shear stress, and a coefficient which incorporates efficiency parameters and the settling velocity of suspended particles to determine suspended load. Sediment yield is then determined from the piecewise normal approximating hydrograph and the particle size distribution of bed sediments.

Validation Results

Water and sediment routing submodels were validated for watersheds in Nebraska and Arizona. The model produced reasonable estimates of mean runoff volume and peak rates for 222 runoff events from eight watersheds (Lane, 1982b). Lane (1987) observed a good fit ($r^2 = 0.97$) for predicted vs. sediment discharge (calibrated simulation) but a poorer performance ($r^2 = 0.78$) for predicted vs. sediment yield (uncalibrated simulation).

Snow Hydrology

The snow hydrology submodel is based on the HYDRO-17 model used by the National Weather Service (Anderson, 1973) and adapted for the 24-hour time step used in SPUR-91. Because air temperature is a readily available climatic parameter, the model is structured as a set of empirical equations relating air temperature with fluxes in the energy balance which represent the physical processes affecting snow accumulation and snow melt (Cooley et al., 1987). The submodel keeps a daily accounting of the snowpack present on the site. Snow accumulation is dependent on air temperature and the temperature which the user sets to differentiate rain from snow. Air temperature is used as an index of the energy exchange across the snow-air interface. Snow melt is divided into two categories (rain-on-snow and non-rain periods) because of the seasonal variation in the melt rates for these two melt processes, as well as differences in the energy transfer relations (Cooley et al., 1987).

Rain-on-snow melt occurs at the snow surface. The energy balance of the snowpack is calculated as the sum of net radiation transfer, advected heat transfer, latent heat transfer, sensible heat transfer, and heat transfer across the soil-snow interface (Anderson, 1976). Snowmelt is then determined from an equation relating air temperature, water equivalent of precipitation, atmospheric pressure, saturation vapor pressure, and a wind function. For rain-on-snow, calculations assume that conditions exist whereby the wet-bulb and air temperatures are equal.

Because of the wide variety of meteorological conditions which can exist during dry conditions, non-rain snowmelt is not calculated using energy balance equations. Instead, non-rain snowmelt is determined from an empirical relationship with air temperature. A sinusoidal relationship between the melt factor and season was developed to account for seasonal variation of data from the contiguous United States (Cooley et al., 1987). Groundmelt is calculated as a small but continuous process at the bottom of the snowpack.

Snow hydrology submodel evaluations by Cooley et al. (1987) demonstrated a strong correlation between simulated and observed snow water equivalent (amount of water in the snowpack in mm) for the first year (1979, $r=0.998$) with the calibrated model. The snow water equivalent was underpredicted, however, by 15 to 40% three out of four test years (using the model calibrated for 1979) due to premature snowmelt (30 to 90 days). This problem was attributed to the use of temperature as an index of the complex physical processes controlling snow accumulation and melt. More accurate simulation of snow accumulation and melt would require inclusion of many variables which are not widely available (Cooley et al., 1987).

27.8. NECESSITY OF HYDROLOGY-PLANT INTERLINKAGE FOR RANGELAND SIMULATION

Hydrologic condition of a site is determined by a complex interaction of vegetation, soils and climatic variables (Ponce and Meiman, 1983). The great diversity of these factors on rangelands complicates modeling efforts.

To model runoff, biomass production, plant succession, or any complex process on rangelands, it is important to clearly understand the mechanisms of how the water, soil, and plant components interact. The type of vegetation present on rangelands strongly influences a host of parameters which have great temporal and spatial impacts on the hydrology of the site. For short-term simulation, the seasonal dynamics of plant growth and cover have important influences on site hydrology, especially where a small proportion of annual storms have a major impact on soil water content and runoff volume. Over the long-term, hydrologic condition is often tied to plant succession patterns and various land management practices. Therefore, a sophisticated plant component which can simulate these changes must be incorporated into models designed to simulate rangeland hydrology.

27.8.1. Vegetation Effects on Infiltration

The rate at which water can enter the soil is dependent on a variety of vegetation parameters. The type of vegetation affects the amount and structure of associated cover. The amount of cover is usually greatest on sites dominated by trees and shrubs. For tree interspaces and grasslands, cover is greater on sites dominated by bunchgrasses, followed by shortgrasses (Blackburn, 1975; Thurow et al., 1986). Infiltration rates on many rangelands are strongly related to this cover relationship (Blackburn, 1975; Wood and Blackburn, 1981a; Knight et al., 1984; Thurow et al., 1986; Thurow et al., 1988). Different rangeland vegetation types are also associated with different amounts of litter. Bunchgrasses and shrubs generally produce more litter than do shortgrasses and forbs. Litter helps to increase soil organic matter content and create a more consistent temperature and moisture microenvironment for microorganism activity which aids soil structure (Thurow, 1991). Quantity of living plant material and associated litter is also strongly correlated with infiltration. Dee et al. (1966) found that infiltration rates in rangeland plant communities, on the same soil series, were directly correlated with amount of standing vegetation, accumulated litter, and state of succession of the community.

Infiltration rates also vary seasonally because of the variation in growth dynamics of different plant species (Thurow et al., 1988; Blackburn et al., 1992). On rangelands in Texas, Thurow et al. (1988) found little seasonal fluctuation in infiltration rates on bunchgrass sites. However, where shortgrasses and forbs dominated, seasonal infiltration patterns varied. Cover on sites dominated by annual species is generally highly variable over time because amount of cover rapidly increases during warm moist periods that favor growth but decrease during dormant seasons. Therefore the soil is well covered during portions of the year and quite bare during other times. This is in contrast to cover dynamics of perennial shrubs and grasses were there is less fluctuation in the standing cover throughout the year (Thurow, 1991).

27.8.2. Vegetation Effects on Erosion

Runoff varies with the type and amount of vegetation. Sediment production is closely related to amount of runoff, and is therefore significantly correlated with vegetation parameters. The amount of interrill erosion varies depending on plant growth form. Interrill erosion is generally greater from tree interspace sites than from under trees and shrubs (Blackburn, 1975; Knight et al., 1984; Thurow et al., 1986). Sediment loss is generally greater from shortgrass than bunchgrass- or shrub-dominated rangeland (Blackburn, 1975; Wood and Blackburn, 1981b; Blackburn et al., 1982; Knight et al., 1984; McCalla et al., 1984; Warren et al., 1986; Pluhar et al., 1987; Thurow et al., 1988;), even when cover is equal. The bunchgrass growth form and accumulated litter at the base provide an effective obstruction to overland flow. By slowing and diverting runoff, organic cover reduces the kinetic energy and reduces transport capacity (Thurow, 1991).

27.8.3. Vegetation Effects on Evapotranspiration

Evapotranspiration is usually the largest component of the water budget for rangelands, and therefore reliable estimates of evapotranspiration are required to accurately model rangeland hydrology. Unfortunately, reliable estimates are hard to obtain, especially in arid and semi-arid regions (Lemeur and Zhang, 1990). The magnitude of evapotranspiration is dependent in part on vegetation characteristics. Plants differ in their ability to withdraw soil water. When sufficient soil water is available, plants tend to transpire at a rate dominated by the energy available at the transpiring surface. But when the supply becomes limited, as is the usual case on rangelands, plants begin to regulate transpirational flux through stomata via physiological and morphological mechanisms (Beneke, 1976). These mechanisms vary depending on plant species. For example, there is a consistent difference in the water use efficiencies between C_3 and C_4 grasses (different photosynthetic pathways) due to inherent physiological, chemical and morphological characteristics (Hsaio and Acevedo, 1974; Christie, 1978; Holmstead, 1989; Hicks et al., 1990). Some species have the ability to withdraw soil water even at very negative soil water potentials due to physiological adaptations.

Differences in rooting depth between plant species is also an important determinant of the extent of evapotranspirational loss. Sites dominated by trees and shrubs (as opposed to sites dominated by herbaceous species) generally have greater evapotranspiration because their more extensive root systems provide access to a larger volume of soil and therefore soil water.

Differences in the timing of soil water depletion is an additional factor explaining lower evapotranspiration from herbaceous vs. tree- or shrub-dominated sites. Herbaceous species can be subject to stress-induced dormancy due to drought or temperature extremes, thus periodically reducing amount of live leaf (transpiring) cover. Conversion of vegetation types on rangelands can reduce evapotranspirational loss where deep-rooted species are replaced by shallow-rooted species (Johnston, 1970; Davis and Pase, 1977; Hibbert, 1983). While the amount of live leaf tissue is related to amount of evapotranspiration, consumptive water use by different plant communities is not necessarily proportional to the stand density or standing crop (Branson et al., 1976; Sosebee, 1976).

27.8.4. Herbivore Effects on Hydrology

Livestock and wildlife, which are an inherent part of rangeland communities, have a significant indirect impact on the runoff and erosion of a site. Herbivore impacts are an important component in rangeland hydrology models because herbivores affect the hydrology of a site by their consumption of cover and transpiring leaf tissue and by trampling of soil and standing vegetation. Plant species composition on a site is a major determinant of hydrologic condition. Because livestock are selective eaters, they remove the most desirable plants first. These desirable species are often bunchgrasses which provide the best vegetative cover. Therefore, less desirable species, which may be woody vegetation, shortgrasses or annuals, are likely to increase under heavy grazing (Archer and Smeins, 1991; Briske, 1991). In this way, long-term grazing can shift the vegetation on a site. The degree to which grazing changes species composition and reduces standing vegetation and litter is dependent on grazing frequency and intensity (Wood and Blackburn, 1981; Thurow et al., 1988a; Thurow, 1991). Heavy year-long grazing usually reduces hydrologic condition of grasslands because the species composition shifts from bunchgrass-dominated to shortgrass-dominated communities and because total cover and litter is reduced. Rotational short-duration grazing systems reduce this selective grazing through brief, high intensity use. High intensity grazing, however, removes great amounts of vegetation for short periods during the season. This can also have profound effects on hydrologic response depending on the timing of precipitation and grazing. Moderate rotational grazing fosters favorable species composition and hydrologic response.

27.9. PLANT SUBMODEL

The plant component of SPUR-91 simulates the flows of carbon and nitrogen through the soil-plant-animal interface. There are seven carbon and eight nitrogen state variables between which the flows take place. The initial values for state variables are set by the user depending on site characteristics and available field data for total (carbon) phytomass. The nitrogen state variables are then set within SPUR-91 dependent on their carbon phytomass counterparts. The gains to and losses from each of these state variables are listed in Tables 27.2 and 27.3.

TABLE 27.2.

Gains to and losses from the seven carbon state variables of the plant submodel which can occur.

VARIABLE	GAINS	LOSSES
PHYTM1	Photosynthesis Root to shoot transfer Germinating shoots	Trampled green shoots Green shoot death Shoot to seed transfer Shoot to root transfer Respiration Maximum leaf area shoot death
PHYTM2	Germinating roots Shoot to root transfer	Root respiration Root mortality Root to shoot transfer
PHYTM3	Shoot to seed transfer	Germinating phytomass Seed mortality
PHYTM4	Green shoot death Maximum leaf area shoot death	Dead to litter transfer
DROOTS	Total dead roots Root mortality	Dead root to organic matter transfer
ALIT	Total litter Trampled green shoots Dead to litter transfer Seed mortality	Litter to organic matter transfer
AORG	Dead root to organic matter transfer Litter to organic matter transfer	Loss to atmosphere

TABLE 27.3.

Gains to and losses from the eight nitrogen state variables of the plant submodel which can occur.

VARIABLE	GAINS	LOSSES
PNC1	Root to shoot transfer	Transfer due to trampling
	Germinating shoots	Nitrogen in dying shoots
		Propagule nitrogen
PNC2	Nitrogen in germinating roots	Nitrogen lost to dead roots
		Root to shoot transfer
PNC3	Propagule nitrogen	Nitrogen in germinating shoots
		Nitrogen in dead propagules
PNC4	Nitrogen in dying shoots	Dead shoot to litter transfer
DROOTN	Total nitrogen in dead roots	Total organic nitrogen in dead roots
	Nitrogen lost to dead roots	
ALITN	Total nitrogen in litter	Total organic nitrogen in litter
	Dead to litter transfer	
	Transfer from trampling	
	Nitrogen in dead propagules	
AORGN	Litter to organic matter transfer	Mineralized nitrogen
	Dead root to organic matter transfer	
SNIO	Fixation	Denitrification
	Mineralized nitrogen	Immobilized nitrogen in litter
		Immobilized nitrogen in dead roots
		Nitrogen uptake by plants

Abiotic variables which influence plant processes include a and soil temperatures, soil water potential, adjusted solar radiation, accumulated wind run, and soil bulk density. Process rates in the model are calculated by reducing maximum rates according to non-dimensional scalar multipliers ranging from 0 to 1. The four basic response curves used by the model are:

1) ATANF, which is a sigmoidal curve used primarily for soil water potential effects,

2) BELL, which is a bell curve used for temperature responses,

3) HYP, which is a rectangular hyperbola function representing the effect of nitrogen on photosynthesis,

4) THRESH, which is a threshold-type curve primarily used for the effect of soil water on photosynthesis (Fig. 27.4).

The model is unique in that it has the capability of simulating competition between different species or functional groups (maximum of seven) for limiting resources. To do this, the user selects appropriate values for a variety of coefficients which adjust the curves to match a particular species' typical response. In this way, different species have variable response to controlling variables. The plant submodel simulates the following basic processes:

1) Carbon assimilation,

2) Translocation of assimilates,

3) Respiration,

4) Germination and seed production,

5) Mortality and senescence,

6) Nitrogen cycling and nitrogen flows,

7) Decomposition,

8) Plant-animal interactions (Carlson and Thurow, 1992).

27.9.1. Carbon Assimilation

In order to determine daily photosynthesis, daily patterns of net carbon dioxide assimilation rates are generated and then integrated using Simpson's Rule (Detling et al., 1979). Photoperiod is first determined (Parton and Logan, 1981), and then the time of sunrise is derived. Eight integration

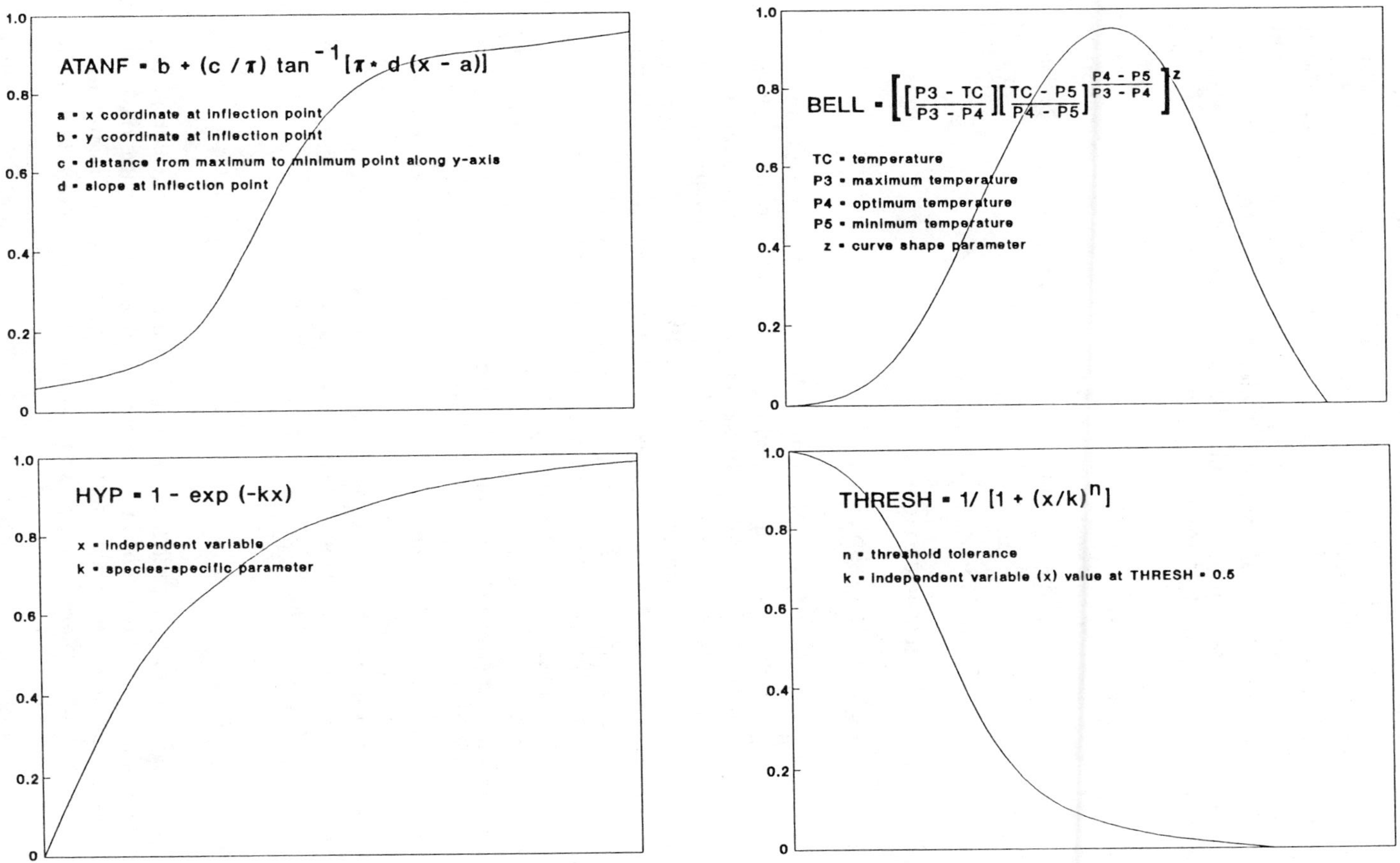

Fig. 27.4. Controlling function options of the plant growth submodel.

intervals are used to determine net photosynthetic rate. Total daily radiation is used to calculate daily maximum radiation and then maximum photosynthetic rate. The maximum net photosynthetic rate is reduced according to temperature, soil water content, nitrogen and leaf age to obtain the expected net photosynthetic rate. The daily assimilation rate is then dependent on the current leaf area of the species. The submodel provides the option of reducing late-season photosynthesis according to leaf age (calendar day) (Hanson et al., 1987).

27.9.2. Translocation

There are three basic translocation processes simulated by SPUR-91. These include:

1) Translocation of biomass from roots to shoots,

2) Transfer from shoots to roots,

3) Translocation from shoots to propagules.

Translocation from roots to shoots is primarily involved in growth initiation (perennials) in temperature-controlled climates. This process is controlled by both temperature and surface soil water content. These two factors not only act as on/off switches, but also moderate the rate of translocation. Root-to-shoot translocation occurs in spring during growth initiation or at times when some process (e.g., removal of forage by herbivores) has removed a critical amount of shoot biomass, thus simulating how plants utilize energy stored in roots to continue growth when aboveground biomass is not sufficient to generate the required energy.

Translocation from shoots to roots occurs as a shunt to prevent shoot biomass from exceeding the amount which can be supported by the root biomass, according to the root-to-shoot ratio specified by the user. After senescence begins, an additional proportion of shoot biomass is translocated to roots to simulate storage processes in temperature-controlled climates. Translocation from shoots to propagules is set to a specific daily rate, and begins at the outset of flowering and seed production, which is controlled by calendar day (Hanson et al., 1987; Carlson and Thurow, 1992).

27.9.3. Respiration

Both root and shoot respiration are calculated within the model. For shoots, only the night-time maintenance respiration is calculated because day-time respiration is assumed to be accounted for in calculations of net assimilation. A value representing daily expected proportion of recent photosynthate to be respired is initially chosen by the user, and is then automatically adjusted during the simulation run based on total leaf area and night-time hours calculated by the photoperiod function. Since respiration is based on the daily photosynthesis, it is affected by the same climatic factors which affect photosynthesis. Root respiration is set to a maximum daily rate and reduced according to soil temperature (Hanson et al., 1987; Carlson and Thurow, 1992).

27.9.4. Seed Production and Germination

SPUR-91 contains a state variable to deal with a seed bank component of plant establishment. Germination from this seed bank occurs when specific conditions of temperature and surface soil water potential are met for a given species (set by user). A maximum daily germination rate is set and reduced according to temperature effects. When germination occurs, biomass of seeds is partitioned to roots and shoots depending on the root: shoot ratio set by the user. A specified portion of photosynthate can be transferred to the seed bank after flowering. The seed bank is subject to losses at a daily rate specified by the user. There is a complex interaction of factors controlling germination and survival. Therefore, the annual flushes that sometimes occur on rangelands generally cannot be simulated by SPUR-91, but a general simulation of average performance is obtained (Carlson and Thurow, 1992).

27.9.5. Mortality and Senescence

Shoots, roots and propagules are all subject to mortality losses. The timing of senescence processes in the model are controlled primarily by dates (julian days) provided by the user. This reliance on dates poses a problem, especially in areas of more moderate temperature regimes or where the potential exists for multiple growing seasons. In the first case, narrow ranges of temperature increase problems with simulation because most on/off controls are no longer very meaningful. In the second case, there is much confusion about when senescence actually begins and ends. The model could be improved if senescence processes were controlled biotically and abiotically (Carlson and Thurow, 1992).

Shoots are subject to a variety of losses. Shoots are subject to water-stress death at any time. A maximum daily rate is set and then reduced according to the current-day soil water status in those layers where a given species' roots are located. Also, if the minimum temperature established for a species' survival is reached, then 10% of shoots for that day are removed from live aboveground to standing dead compartments. Once senescence begins, an additional proportion of shoots die at a set daily rate. This rate is increased after senescence ends. Whenever shoots exceed the maximum leaf area, these additional shoots are also removed from the live aboveground compartment (two check-points within the model). Trampling of green shoots by livestock can also lead to shoot death. Root mortality is set to a maximum daily rate which is reduced dependent on both temperature and soil water content level. Propagules die at a daily set rate which is independent of temperature and soil water content (Hanson et al., 1987; Carlson and Thurow, 1992).

27.9.6. Nitrogen Processes and Nitrogen Flows

Nitrogen compartmentation and flows in SPUR-91 parallel those which occur for total biomass (carbon model). The daily flow rate between nitrogen compartments is pre-set in the source code, and is based on biomass (carbon) transfers, as follows:

$$PNC1 = PHYTM1 * 0.03$$

$$PNC2 = PHYTM2 * 0.01$$

$$PNC3 = PHYTM3 * 0.04$$

$$PNC3 = PHYTM4 * 0.05$$

where PHYTM1 to PHYTM4 are live standing phytomass, live roots, propagules, and standing dead phytomass compartments, respectively. PNC1 to PNC4 are the nitrogen counterparts of the respective PHYTM1 to PHYTM4 carbon compartments.

Nitrogen uptake occurs only if the plant is photosynthesizing, indicated by presence of transpiration. The expected uptake rate is based on the root biomass, a maximum rate set by the user, the size of the available soil inorganic nitrogen pool, and the nitrogen-use efficiency. This expected maximum uptake rate is then reduced by the effects of soil temperature and soil water content. Actual uptake is then distributed between species depending on uptake demand vs. available soil nitrogen. If the nitrogen content of roots exceeds 1%, then uptake ceases. The nitrogen taken up by the plant then "flows" between live tissue compartments, dependent on phytomass totals for each compartment. A nitrogen root: shoot balance is also maintained, and during senescence processes nitrogen is conserved. Similarly, nitrogen is removed from dead roots and litter depending on the nitrogen content of the respective donor pools, and then passed to organic matter. Immobilization of nitrogen from decomposition of litter and dead roots occurs in order to maintain the proper carbon: nitrogen ratios. Other below ground processes simulated by SPUR-91 include:

1) Nitrogen fixation, defined as 1% of precipitation (does not account for symbiotic fixation),

2) Denitrification, which is controlled by the soil nitrogen content, maximum rate, and soil water potential,

3) Mineralization, which is 4% of soil organic matter undergoing decomposition (Hanson et al., 1987; Carlson and Thurow, 1992).

The nitrogen flow and decomposition processes are based on very limited data; therefore the cycling details within the ecosystem are not well understood. More attention needs to be devoted to this area of research.

27.9.7. Decomposition

Decomposition processes begin in SPUR-91 when standing dead phytomass is knocked down and transferred to the litter compartment. Standing dead can be knocked down by wind, precipitation, or can be trampled by livestock. Litter and dead roots decay at a used-designated rate to form soil organic matter. All species lose their identity once they enter the litter or dead root compartments. Both soil temperature and soil water potential affect decomposition rates. As with most plant processes, a maximum rate is set by the user and then reduced on a daily basis. Decomposition will not occur unless some soil inorganic nitrogen is available. Only 40% of litter and dead root biomass subject to decomposition is transferred to organic matter. The rest is lost to the atmosphere as carbon dioxide (Hanson et al., 1987; Carlson and Thurow, 1992). The nitrogen cycling aspects of decomposition were outlined above.

27.9.8. Plant-Animal Interactions

Livestock grazing on rangelands impacts plant species by removing plant tissue and also through trampling of live and dead phytomass. The model attempts to simulate both of these effects. In SPUR-91, removal of plant tissue is dependent on amount of phytomass available, animal demand, and animal preferences. The animal component allows the user to input a preference vector which describes how grazing pressure is distributed between the available forage plants and maturity classes. The fact that some species (e.g., bunchgrasses) are more susceptible to livestock damage than other species (e.g., shortgrasses) because of morphological characteristics (Briske 1991), is simulated in SPUR-91. Several parameters establish:

1) Proportion of phytomass susceptible to trampling (differentiates those species which are subject to trampling (herbaceous) from those which are not (trees and shrubs),

2) Susceptibility of standing dead to trampling,

3) Susceptibility of green shoots to trampling (separates those species which are susceptible to actual damage (shoot death) caused by trampling from those plant species which are structurally resistant to damage from trampling) and is dependent on stocking rate.

Once livestock consume vegetation, organic matter and nitrogen can be cycled back to the system via animal feces and urine. Wildlife are only a sink for tissue removal and so do not contribute to nutrient cycling. The fraction of time spent grazing each site is computed after the diet selection process is complete. Indigestible carbon and nitrogen are excreted and distributed over the field in proportion to the time spent grazing each site (MacNeil et al., 1987).

27.9.9. Validation Results of the Plant Submodel

Hanson et al. (1987) state that, for a two-species system, SPUR produced results typical of shortgrass prairie phytomass dynamics. An independent validation by Stout et al. (1990) found that the SPUR-simulated plant biomass did not agree with observed plant biomass based on the results of regression analysis. However, simulated values were within the range of observed biomass values for most sampling dates. Results of the SPUR-91 validation conducted at the Throckmorton site in Texas showed that:

1) There was a good agreement between observed and predicted total herbaceous standing crop (r=0.72),

2) Time series simulation of individual perennial grass species was fair to excellent (r=0.39 to 0.68),

3) Time series simulation of the large erratic variations typical of seasonal and annual production of annual grasses and forbs was poor,

4) A determinant-type growth curve could be generated to approximate long-term response of woody species (e.g., mesquite).

More details of the validation are presented in the "Plant-Hydrology Interface" of this chapter.

27.10. PLANT-HYDROLOGY INTERFACE

27.10.1. Specific Linkages in SPUR-91

The original SPUR developers expressed concerns about poor intercommunication between the plant and hydrology submodels (MacNeil et al., 1987). Because rangeland hydrology and plant processes are closely intertwined, this was a major limitation of SPUR. Due to several key modifications (Carlson and Thurow, 1992), the linkage between SPUR-91 hydrology and plant submodels has been enhanced. These linkages are designed to mirror the key interrelationships between the two components which exist in natural rangeland ecosystems.

Vegetation characteristics have a great effect on evapotranspiration. The major linkage between plant and hydrology components naturally involves evapotranspiration and its effect on soil water content via the daily water balance. These interconnections are detailed in the equations presented in the "Upland Hydrology" section of this chapter. Potential evaporation is dependent on potential evapotranspiration and on either 1) the live leaf area of the vegetation cover, or 2) a mulching factor which represents relative litter cover on the site, whichever provides the lowest estimate of potential evaporation. To calculate actual soil evaporation, Ritchie's technique (Ritchie, 1972) was slightly modified so that the soil evaporation parameter could be increased during periods of no or low vegetative cover. The required daily cover data (combined leaf area of both live and dead standing vegetation) is provided by the plant submodel. Ritchie's evapotranspiration model tends to underpredict evaporation losses from bare soil conditions because wind speed and vapor pressure deficit were not included in the equations to obtain potential evaporation. The modification, therefore, helps to correct this underestimation under low cover conditions. The depth to which soil evaporation occurs in SPUR-91 is also dependent on the daily vegetation cover as provided by the plant submodel. Depth of evaporation ranges from 76 mm under extensive cover conditions down to 157 mm under bare soil conditions.

Potential transpiration calculations are dependent on live leaf area. If the live leaf area index (LAI) is less than three (which is common on rangelands), then the potential transpiration is dependent on the potential evapotranspiration and one-third of LAI. If LAI is greater than three, then potential transpiration is the difference between potential evapotranspiration and actual soil evaporation. In this way, transpirational demand is based on total live leaf area and potential transpiration. Depth to which transpiration occurs is dependent on vegetation rooting depth. This depth is closely tied to type of vegetation present on the site. Transpirational demand is distributed through the soil layers based on an exponential function which describes typical rooting patterns. The daily loss of soil water to evapotranspiration is monitored by the hydrology component's daily water balance calculations.

The daily soil water content (converted to soil water potential) which is maintained by the hydrology submodel is a major controlling variable for many plant processes. A species-specific plant parameter designates how many soil layers a particular plant species has access to. A weighted mean soil water potential for each species, based on individual rooting depth and parametric root distribution, is then derived from the soil water content information provided by the hydrology submodel. This mean soil water potential is used to determine the effect of soil water content on various plant growth processes. By determining the effect of soil water content on photosynthesis, for example, competition between different plant species for soil water can be simulated. Species response is therefore controlled by soil water contents only where their roots actually occur within the soil profile.

Therefore, shallow-rooted species will be affected differently than will deep-rooted species. Some processes, such as seed germination and translocation from roots to shoots, are controlled by the soil water potential in the surface soil only (upper 15 cm). Individual species responses are further affected by soil water potential by assigning values to other plant parameters (e.g., soil water potential occurring at 1/2 maximum photosynthesis, drought tolerance coefficient, water potential when translocation begins) which are used in response curve equations. These differential responses are the basis for plant composition changes and succession due to grazing and other disturbances.

The curve number technique for estimating runoff (and hence the amount of water entering the soil) is a weak link in the hydrology-plant interface. The curve number technique has poor runoff prediction capability on rangelands, due in part to the poor curve number data base applicable to rangeland soils (Wood and Blackburn, 1984). This method does not take into account storm intensity or duration for a particular rainfall event. This limitation introduces problems when using the method for discontinuous storms or for storms of duration longer than one day (Rallison, 1980). Storm intensity is an indication of raindrop kinetic energy and is thus an important determinant of runoff and erosion. The curve number technique only incorporates the effects of "average" plant composition and cover. Consequently, infiltration is unresponsive to **current** plant cover characteristics, which act to dissipate raindrop energy. While the curve number is a limitation of the model in simulating runoff on a short-term basis, it gives reasonable estimates of runoff over the long-term. The need for accurate simulation of **each** runoff event is important, however, because a poor estimation of daily runoff affects the amount and timing of water available for evapotranspiration and deep percolation. In turn, this affects soil water available for plant growth which determines production and succession (Carlson and Thurow, 1992).

27.10.2. Sensitivity Analysis

Ongoing evaluation and refinement of simulation models is a necessary activity which enables modeling tools to improve by evolving to respond to user experience as well as new information. A critical part of this evaluation process is a detailed sensitivity analysis which provides valuable information on how the magnitude of output response is associated with variation of single input parameters and can provide some superficial indication as to how the parameters interrelate (Carlson et al., 1993). A sensitivity analysis was conducted for SPUR-91 by increasing and decreasing one input parameter at a time by 20% from its base value and comparing the 10-year mean of each output variable with a base-run 10-year mean. A value of 20% was selected as the reasonable range of variance for selection of input parameter values by model users. For some input parameters, a range different than 20% was chosen to better represent the expected or likely range of values. The annual output variables selected for comparison were runoff, evapotranspiration, deep drainage, plant available water, and sediment production. In addition, the following plant component state variables, averaged over 24 dates per year, were selected for comparison: standing live phytomass, live root phytomass, standing dead phytomass, dead roots, litter, organic matter, and soil inorganic nitrogen. This was done for three cover types (bare ground, herbaceous vegetation only, and herbaceous vegetation with a mesquite overstory) for which long-term data was available.

The curve number was the parameter which most influenced the various hydrology output variables. The phytomass-to-leaf area conversion factor, theoretical maximum net photosynthetic rate, maximum temperature for positive plant activity, species-specific rooting depth, julian day that senescence begins, and additional shoot death after senescence were among the ten most important parameters influencing hydrology output variables. The resultant change from the base value, however, was generally less than 20%. There were only a few hydrology parameters that significantly influenced plant output variables (soil evaporation parameter, curve number, and -0.3 and -15 bar volumetric water content of the soil layer). These parameters were sensitive (5 to 50% change in base value) and therefore require accurate parameterization for optimum model performance.

By comparing the results of the sensitivity analyses for different plant cover types, the user can begin to understand how parameter interactions affect model results. For example, the curve number was a very sensitive parameter in all three cover types as it affected hydrology output variables. However, the curve number was more sensitive on the herbaceous covered sites than on the herbaceous+mesquite covered sites in terms of effects on plant component output variables. Clearly, other variables describing plant growth processes were interacting with variables affected directly by the curve number (hydrology variables) to give this variation in sensitivity. Detailed results of the sensitivity analyses for the three different vegetation cover types are presented in Carlson and Thurow (1992).

27.10.3. Performance Results

SPUR-91 was tested on four different sites in Texas (Fig. 27.5) which had quite different soil, climate, and vegetation communities. For all simulations, precipitation data from the site or obtained from the nearest weather station (within 40 km of study site) were used in the climate data input file. All other climatic variables (maximum and minimum temperature, solar radiation, and wind run) were generated using guidelines from the SPUR-91 workbook and user guide (Carlson and Thurow, 1992). Initial parameterization of hydrology, plant, and animal input files was based on all available field data, from information provided by prior validations, and

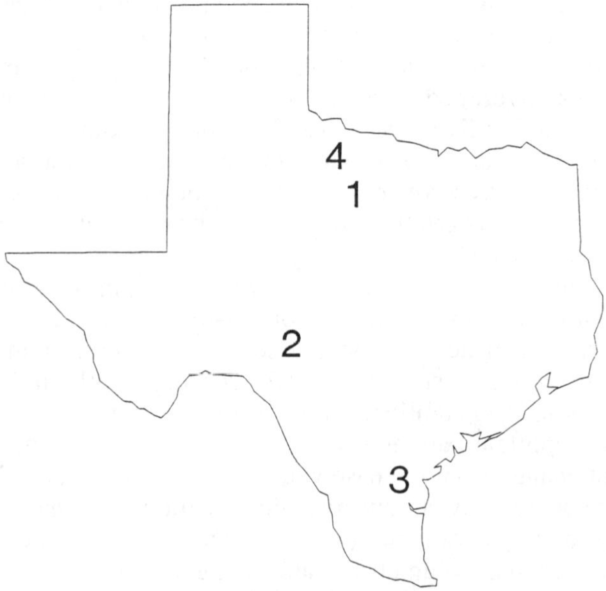

Fig. 27.5. Texas A&M University experiment station sites from which long-term water/plant/animal simultaneously-measured data sets were procured to validate SPUR-91.
1= Throckmorton in the Rolling Plains, 2= Sonora on the Edwards Plateau, 3= Alice in the South Texas Plains, and 4=Vernon in the Rolling Plains.

from information and insight provided by the SPUR-91 workbook and user guide. When soil data were not provided, information was obtained for the specific soil series from the Soil Conservation Service soil guides available for most counties in Texas. SPUR-91 was calibrated for each site using the first year of data. Remaining years of each data set were then used to test the model.

Throckmorton Site

The Throckmorton site had the most extensive data set and was used for the primary validation of SPUR-91. This site was located on the eastern edge of the Rolling Plains resource region approximately 22 km north of Throckmorton, TX (33°20'N, 99°14'W). Climate was semi-arid continental with average annual precipitation (646 mm) distributed with peaks in May and September. Soils were silty clay loams (slopes less than 3%) that were deep, well-drained, and slowly permeable. Dominant midgrasses on the site were Texas wintergrass (*Stipa leuchotricha* Trin. and Rupr.) and sideoats grama (*Bouteloua curtipendula* (Michx.) Torr.). Perennial shortgrasses included buffalograss (*Buchloe dactyloides* (Nutt.) Engelm.) and curly mesquite (*Hilaria belangeri* (Steud.) Nash). A virgin stand of honey mesquite (*Prosopis glandulosa* Torr. *var. glandulosa*) was the dominant woody species, providing about 30% cover. Three different cover treatments (mesquite with an herbaceous understory (H+M), herbaceous vegetation only (H), and bare ground (BG)) were evaluated from 1986 to 1989 using large (25m^2) non-weighable lysimeters. A variety of soil, vegetation, climatic and water balance parameters were monitored throughout the study. The model was parameterized using information provided by this data set.

Hydrology

Observed 4-year mean soil water content was predicted within 1%, 8%, and 6% on the H+M (574 mm), H (638 mm), and BG (758 mm) cover types, respectively. The temporal fluctuation of weekly soil water contents generated by SPUR-91 were weakly correlated with observed soil water content on the BG treatment (r=0.23) but strongly correlated on the H (r=.91) and H+M (r=.90) cover types (Fig. 27.6). The deviation between simulated and observed soil water content which occurred at the end of 1987 on the H lysimeter was due primarily to inaccuracies in the prediction of soil water content of lower layers. Initialized rooting depth calibrated for 1986 apparently was not deep enough to accommodate deeper soil water extraction by herbaceous species which began at the start of a dry period (fall of 1987). SPUR-91 does not simulate rooting depth changes over time. This may cause inaccuracies when simulating periods of drought or heavy stocking situations which lead to severe defoliation (with subsequent changes in rooting depth). The 10-year simulations of weekly soil water content for all cover types demonstrate the robustness of the soil water routing component (Fig. 27.6).

SPUR-91 did not provide output which accurately reflected observed monthly runoff for any cover type (r=0.37 to 0.58). The model underpredicted the observed 4-year total runoff on the BG site by 30% and on the H+M site by 8%, but overpredicted 4-year total runoff on the H treatment by a factor of two (Table 27.4). The model was calibrated using data for 1986, and SPUR-91 closely predicted runoff from 1986 to 1988 (within 1%). Because 1989 had an unusually large amount of runoff, the curve number technique was not adequate. SPUR-91 accurately predicted the proportion of runoff in the total water budget for all cover types (Table 27.4).

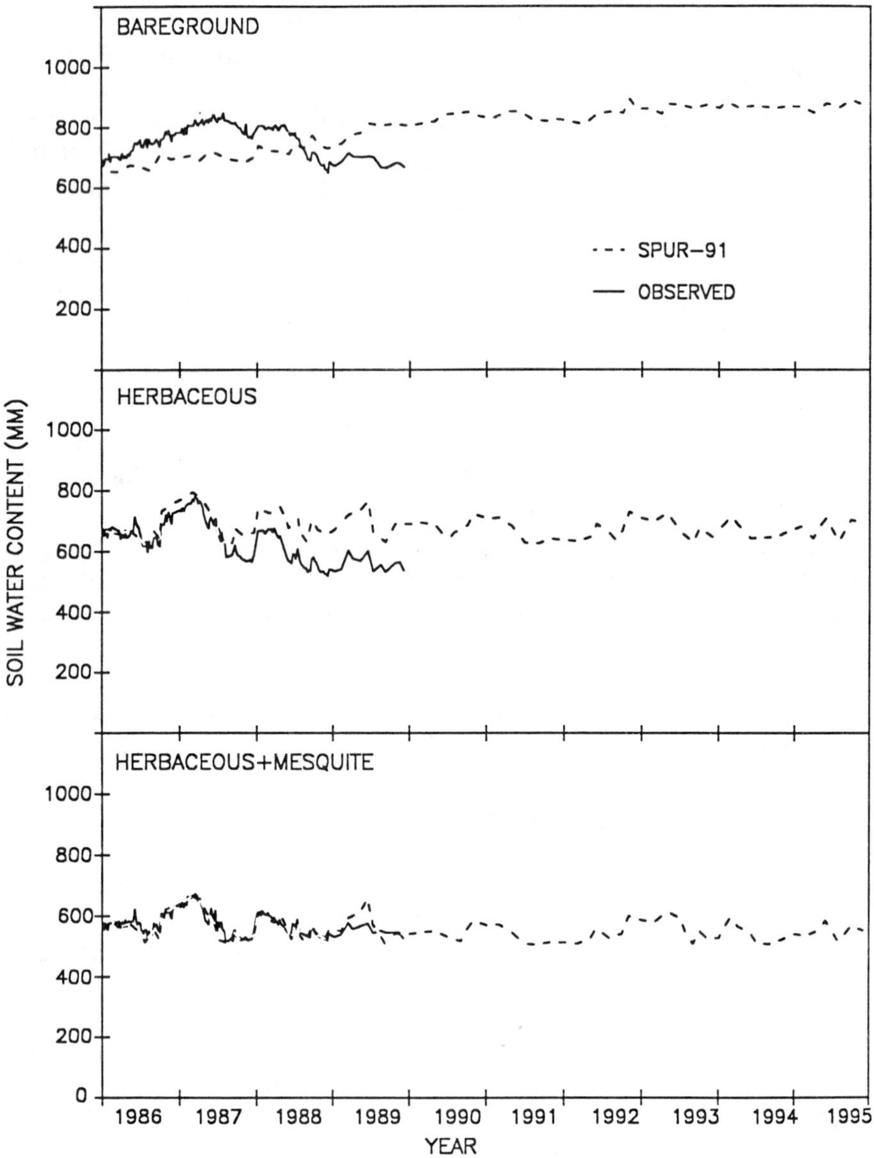

Fig. 27.6. Simulated and observed weekly water content (mm) of the soil profile in bare ground (BG), herbaceous (H), and herbaceous plus mesquite (H+M) sites from 1986-1989 and predicted weekly water content for 1990-1995 at Throckmorton, TX.

TABLE 27.4.

Observed and predicted 4-year water balance in mm (and as a percent of precipitation received) for bare ground, herbaceous, and herbaceous+mesquite cover types for three replicates of each treatment at the Throckmorton site.

	OBSERVED	SPUR-91
BARE GROUND		
Runoff	665 (24.8)	467 (17.6)
Evapotranspiration	2057 (76.7)	2066 (77.7)
Deep Percolation	36 (1.3)	112 (4.5)
HERBACEOUS		
Runoff	44 (1.6)	101 (3.8)
Evapotranspiration	2710 (101.1)	2539 (95.5)
Deep Percolation	27 (1.0)	21 (0.8)
HERBACEOUS+MESQUITE		
Runoff	182 (6.8)	168 (6.3)
Evapotranspiration	2550 (95.1)	2544 (95.7)
	17 (0.6)	0 (0.0)

The 4-year total evapotranspiration was closely predicted by SPUR-91 on the BG (within 1%), H (6%), and H+M (1%) sites (Table 27.4). There were no consistent seasonal patterns of under- or over-prediction of monthly evapotranspiration. The correspondence of temporal fluctuation between observed and SPUR-91-predicted evapotranspiration was $r=0.62$ for BG and $r=0.82$ for H and H+M sites (Fig. 27.7). SPUR-91 accurately predicted that ET accounted for over 95% of total precipitation leaving the vegetated sites.

Less than 5% of precipitation received on this site was lost to deep drainage for any cover type. SPUR-91 adequately predicted the extent of deep drainage on both vegetated and bare soil sites (Table 27.4).

Vegetation

The correlation of temporal fluctuation between observed and SPUR-91-predicted total live standing crop ($r=0.72$) and total standing dead ($r=0.74$) were close, as were the live standing crop simulations for the dominant individual species and species groups (Fig. 27.8). A determinant-type growth pattern of mesquite could be generated only by setting specific parameters outside the suggested ranges. These perturbations were made because the model does not include shrub/tree growth response algorithms

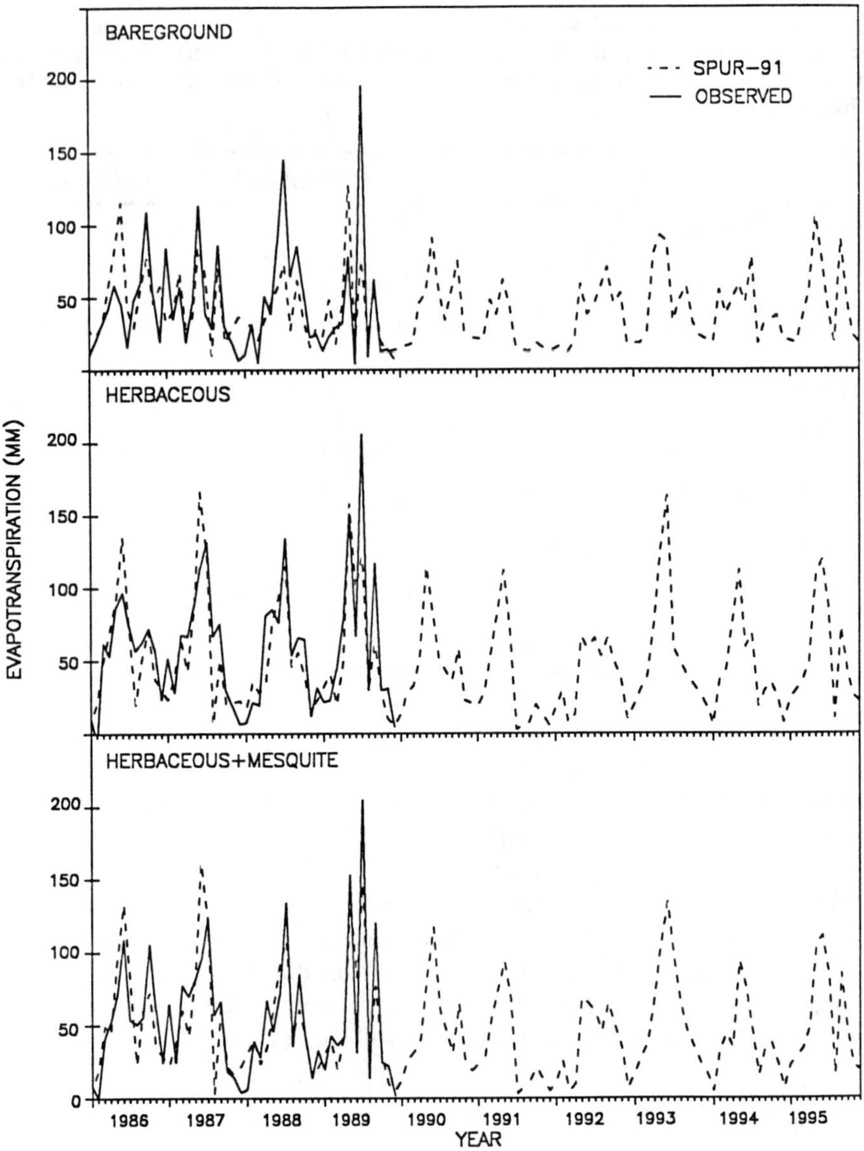

Fig. 27.7. Simulated and observed monthly evapotranspiration (mm) from bare ground (BG), herbaceous (H), and herbaceous plus mesquite (H+M) sites from 1986-1989 and predicted monthly evapotranspiration from 1990-1995 at Throckmorton, TX.

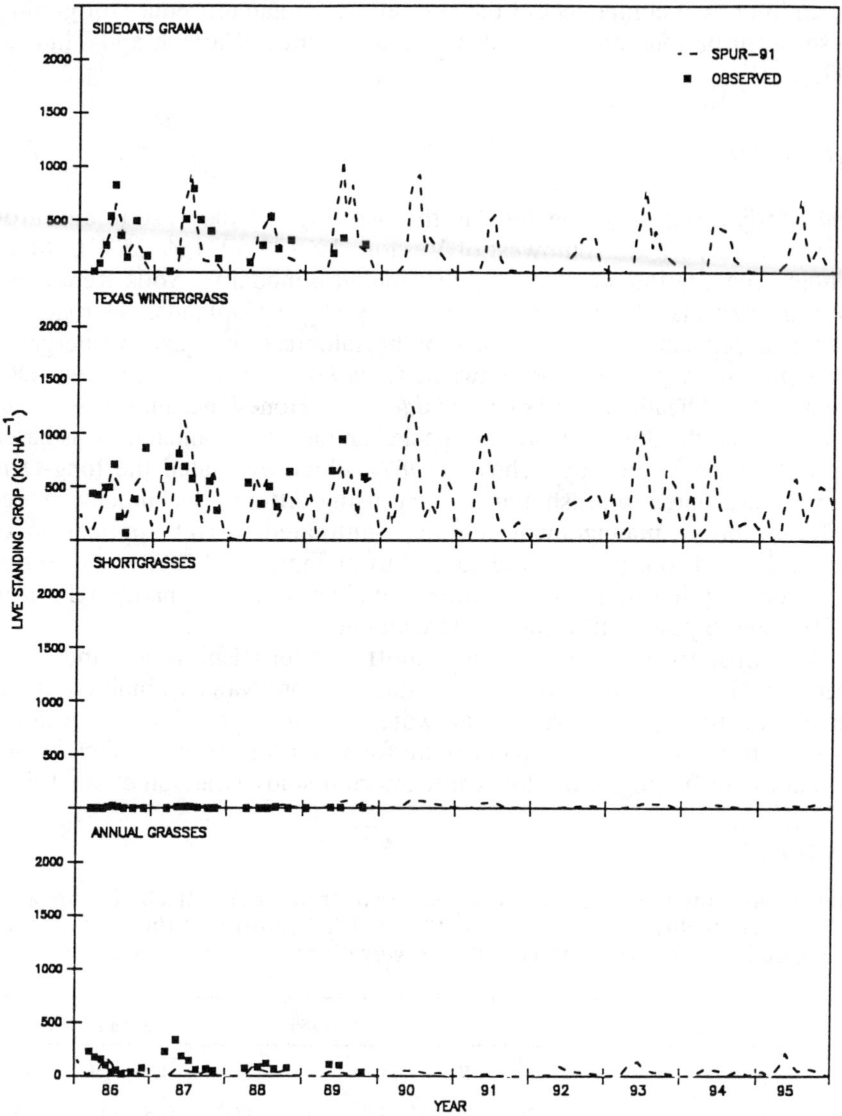

Fig. 27.8. Simulated and observed monthly live standing vegetation (kg ha^{-1}) for dominant plant species and species groups from 1986-1989 and predicted monthly live standing vegetation from 1990-1995 at Throckmorton, TX.

pertaining to the woody component or the proper timing of the effect of environmental factors on growth processes. For cases where trees or shrubs are an important component of the system, a stop-gap procedure for guiding these perturbations are outlined in the user guide (Carlson and Thurow, 1992).

Vernon Site

The study area was located in the Rolling Plains resource region approximately 37 km southwest of Vernon, TX (33°50'N,99°25'W). Mean annual precipitation was 652 mm distributed bimodally. Soils were deep, well-drained clay loams located on gently sloping uplands. Herbaceous perennial vegetation was dominated by buffalograss and Texas wintergrass. Predominant forbs were broomweed (*Xanthocephalum texacanum* (DC) Shinners) and heath aster (*Aster ericoides* L.). Honey mesquite was multi-stemmed and the dominant woody species on the site. Validation was based on a study conducted by Schultz (1982) which evaluated the long-term changes attributed to brush management using different techniques. Three different brush management regimes (untreated, aerial application of herbicide, and root-plowed and seeded to kleingrass (*Panicum coloratum* L.)) were applied in 1974. Treatment evaluations were made in 1979 to 1980, 5 and 6 years following brush treatment.

The proportion of simulated runoff, evapotranspiration, and deep drainage (Table 27.5) were very similar to those values obtained at the Throckmorton site where actual water balance data were available. Evapotranspiration was the major route for water loss from all sites, which concurs with findings from the Throckmorton study (Carlson et al., 1990).

TABLE 27.5.

Simulated water balance components for untreated (U), herbicide-sprayed (S), and root-plowed (R) sites in 1979 and 1980 (mm) and the 2-year mean water budget (percent of precipitation received) at the Vernon site.

	1979			1980			2-year Mean		
	U	S	R	U	S	R	U	S	R
Runoff	17	26	7	85	97	69	9.5	11.5	7.1
Evapotranspiration	677	667	699	373	354	366	97.8	95.4	99.2
Deep percolation	0	0	0	0	0	0	0	0	0

Evapotranspiration was 2% to 5% less on the sprayed site as compared to the untreated site. This substantiates Schultz' conclusion that a long-term increase in production was not realized on the sprayed sites due to the limited production potential of grasses present on the site following brush treatment. No deep drainage was predicted beyond 150 cm.

The model results accurately reflected the observed species composition on all treatments. Both actual and simulated data showed a greater abundance of cool-season grasses and forbs on the sprayed as compared to the untreated site. The general growth response of mesquite generated by SPUR-91 in the untreated site seemed realistic and conformed to the measured mesquite cover (36%). Mesquite regrowth was a relatively small proportion of the system 6 years after spraying; measured cover was about 3%, which corresponded to the SPUR-91 estimate on the sprayed site. No mesquite was present on the root-plowed sites. Schultz' conclusions were supported by the SPUR-91 validation. Whether honey mesquite control can enhance long-term production of the herbaceous component is an important question which many managers will need to address. SPUR-91 appears to have the potential to aid in this decision process.

La Copita Site

The site was located in the South Texas Plains resource region, about 24 km southwest of Alice, TX (27°40'N,98°12'W). Mean annual precipitation averaged 716 mm with peaks in May and September. The soils were fine sandy loams located on slopes less than 3%. The area has been described as a landscape of discrete shrub clusters in a grassland matrix (Scifres and Koerth, 1987). Shrub clusters were dominated by honey mesquite. Subdominant shrubs included limepricklyash (*Zanthoxylum fagara* (L.) Sarg.), lotebush (*Ziziphus obtusifolia* (T. and G.) Gray) and Texas colubrina (*Colubrina texensis* (T. and G.) Gray). Herbaceous species consisted of a mix of warm-season mid- and shortgrasses and cool season grasses and forbs. Major warm-season grasses were red lovegrass (*Eragrostis secundiflora* Presl.), red grama (*Bouteloua trifida* Thurb.), and threeawns (*Aristida* spp.). Validation was based on hydrology data obtained from a lysimeter study by Weltz (1987) and vegetation data was obtained from a study by Scanlon (1988) which evaluated plant growth dynamics on herbaceous interspace and shrub cluster sites.

For all treatments (bare soil, herbaceous interspace, and shrub cluster), SPUR-91 consistently underestimated soil water content, although the seasonal fluctuations were very closely simulated. This discrepancy points out the necessity of providing actual field data for soil input parameters. Volumetric soil water content parameters are sensitive and therefore even reasonable estimates made using guidelines presented in the workbook and user guide (Carlson and Thurow, 1992) may not be accurate enough. SPUR-91 does predict that the soil water content on bare ground is significantly greater than that on vegetated sites, and the soil water content of interspaces is greater than that of shrub clusters. SPUR-91 predicted less runoff for bare soil sites and greater runoff for vegetated sites compared to actual data (Table 27.6). Weltz (1987) stated that more than half of total runoff came from two large storms. Therefore, the inability of the curve number technique to simulate single storm events can cause simulation problems with both hydrology and plant growth processes. Observed evapotranspiration for both interspace and shrub cluster sites was estimated within 3%, and within 15% for bare soil sites (Table 27.6). No deep drainage was predicted from the vegetated sites. Species composition was more accurately predicted on the interspace than on shrub clusters. The conclusions drawn by Weltz (1987) and Scanlon (1988) concerning the hydrological differences of interspace vs. shrub clusters were all supported by the SPUR-91 validation.

TABLE 27.6.

Two-year observed and simulated water budget in mm (and as a percent of precipitation received) on bare soil, interspace, and shrub cluster lysimeters at La Copita, Texas.

	OBSERVED		SPUR-91	
BARE SOIL				
Runoff	115	(9.6)	65	(4.9)
Evapotranspiration	850	(72.5)	1167	(87.9)
Deep Percolation	78	(6.5)	145	(10.9)
INTERSPACE				
Runoff	31	(2.6)	116	(8.5)
Evapotranspiration	1130	(94.4)	1288	(94.6)
Deep Percolation	22	(1.8)	0	(0.0)
SHRUB CLUSTER				
Runoff	22	(1.8)	109	(8.0)
Evapotranspiration	1211	(101.2)	1337	(98.2)
Deep Percolation	0	(0.6)	0	(0.0)

Sonora Site

The study site was located on the Edwards Plateau resource region approximately 57 km south of Sonora (31° N, 100° W). Annual median precipitation was 438 mm. Soils were shallow silty clays overlaying fractured limestone. The oak-grassland community was dominated by live oak (*Quercus virginiana* Mill.) clusters which occupied about 27% of the pastures. Predominant midgrasses present on the site were sideoats grama, Texas wintergrass, and threeawns. Curly mesquite and hairy grama

(*Bouteloua hirsuta* Lag.) were the dominant shortgrasses. Validation was based on a study by Thurow (1985) which examined vegetative and hydrologic responses to livestock grazing. Three grazing treatments were examined from 1978 to 1982: heavy continuous, moderate continuous, and livestock excluded (1980 to 1982 only). Within each grazing treatment, three vegetation cover types were examined: oak mottes, bunchgrass, and shortgrass sites.

Measured runoff from heavy continuous (HC) and moderate continuous (MC) pastures was 10% and 3% of precipitation received, respectively. The simulated total runoff and sediment for both the MC and HC pastures accurately predicted the proportion of runoff in the water balance of the sites (Table 27.7). Runoff was consistently greater on shortgrass sites as compared to midgrass sites, with the least runoff occurring from oak mottes.

TABLE 27.7.

Simulated five-year (1978-1982) fate of water expressed in mm (and as a percent of precipitation received) on midgrass, shortgrass, and oak motte sites and averaged over all sites (percent of precipitation received) for three different grazing regimes at Sonora, TX.

	MIDGRASS		SHORTGRASS		OAK MOTTES		ALL SITES
MODERATE CONTINUOUS							
Runoff	251	(9.8)	387	(15.2)	135	(5.3)	10.9
Deep Percolation	86	(3.4)	14	(0.5)	0	(0.0)	1.3
Evapotranspiration	2306	(90.3)	2275	(89.1)	2565	(100.5)	92.9
Sediment Production (tons/acre)							0.19
HEAVY CONTINUOUS							
Runoff	300	(11.8)	81	(16.3)	158	(6.2)	12.2
Deep Percolation	213	(8.3)	81	(3.2)	0	(0.0)	3.9
Evapotranspiration	2011	(78.8)	2024	(79.0)	2535	(99.3)	84.5
Sediment Production (tons/acre)							0.38
LIVESTOCK EXCLOSURE[1]							
Runoff	153	(9.8)	240	(15.3)	82	(5.2)	10.8
Deep Percolation	5	(0.3)	0	(0.0)	0	(0.0)	0.1
Evapotranspiration	1520	(97.1)	1413	(90.2)	1591	(101.6)	94.9

1 Three year (1980-1982) total.

This agrees with the actual data (Table 27.7). Thurow noted that the temporal irregularity of storms strongly influenced the growth of warm-season grasses, thereby making infiltration rate and water storage availability very important. Since SPUR-91 cannot accurately predict infiltration for individual storms, this can reduce accuracy of plant growth. On the MC pastures, observed 5-year mean total live standing crop was estimated within 34% on midgrass sites and within 1% on shortgrass sites. The observed 5-year mean total live standing crop was underestimated by 83% on midgrass sites and by 62% on shortgrass sites in the HC pastures, although production on both sites was very low for most years. Midgrasses were lost after the third simulation year in HC pastures, as occurred in the actual data set. SPUR-91 demonstrated the capability of effectively modeling the results of midgrasses being more susceptible to grazing pressure. Growth response curves of live oak were realistic for all livestock treatments. Thurow concluded that livestock management systems which decrease soil stability by removing cover and increasing soil surface disturbance, such as heavy continuous grazing, degrade hydrologic condition. This conclusion is supported by the SPUR-91 validation.

Performance Implications

The results of the multiple validations of SPUR-91 in Texas demonstrate two points:

1) Extensive validation can highlight model strengths and point out weaknesses within the model which need to be addressed,

2) SPUR-91 appears to have the potential for aiding in the assessment of various management practices to accomplish specific objectives, such as vegetation manipulation for increased herbaceous production and off-site water yield.

The model could prove to be a valuable tool for managers in making supply forecasts, economic analyses, etc. Currently, however, the information provided by SPUR-91 simulations, at least for certain applications, is more reliably used to predict general trends of management responses rather than absolute values.

SUMMARY

SPUR-91 is a comprehensive rangeland ecosystem model composed of five basic submodels (climate, hydrology, plant, animal and economics). The climate component operates independently of the rest of the model and provides daily inputs for precipitation, maximum and minimum air temperature, solar radiation, and wind run. The hydrology component maintains a daily water balance and simulates snow accumulation and melt. This information is used in conjunction with watershed characteristics to determine streamflow and sediment transport. The plant submodel simulates carbon and nitrogen flows through the ecosystem and simulates competition between plant species for limited resources. The animal submodel simulates herbage removal by livestock and wildlife and determines a productivity index of livestock growth. The economic component uses cost-benefit analysis to determine the value of livestock produced.

Hydrologic condition of a site is determined by a complex interaction of vegetation, soils, and climatic variables. The great diversity of these factors on rangelands complicates modeling efforts. The type of vegetation present on rangelands strongly influences a host of parameters which have great temporal and spatial impacts on the hydrology of the site. Inclusion of a detailed plant submodel is therefore essential for more accurate simulation of hydrologic processes on rangelands.

REFERENCES

Anderson, E.A. 1973. National Weather Service river forecast system-- snow accumulation and ablation model. United States Department of Commerce, National Oceanic and Atmospheric Administration, Technical Memorandum NWS Hydro-17.

Anderson, E.A. 1976. A pint energy and mass balance model for snow cover. United States Department of Commerce, National Oceanic and Atmospheric Administration, Technical Memorandum, National Weather Service HYDRO-17.

Archer, S. and F.E. Smeins. 1991. Ecosystem-level processes. pp. 109-140, IN: R.K. Heitschmidt and J.W. Stuth (eds.). Grazing management: an ecological perspective. Timber Press, Portland, OR.

Ardis, C.V. 1973. A double triangular model for generating storm hydrographs. First World Congress on Water Resources Proceedings 4:350-360.

Bagnold, R.A. 1966. An approach to the sediment transport problem from general physics. U.S. Geological Survey Professional Paper 422-I.

Beneke, P. 1976. Soil water relations and water exchange of forest ecosystems. pp. 101-131, IN: O.L. Lange, L. Kappen, and E.D. Schulze (eds.). Water and plant life. Ecological Studies 19, Springer-Verlag, New York, NY.

Blackburn, W.H. 1975. Factors influencing infiltration and sediment production of semiarid rangelands in Nevada. Water Resources Research 11:929-937.

Blackburn, W.H., R.W. Knight, and M.K. Wood. 1982. Impact of grazing on watershed: a state of knowledge. Texas Agricultural Experiment Station MP-1496.

Blackburn, W.H., F.B. Pierson, C.L. Hanson, T.L. Thurow, and A.L. Hanson. 1992. The spatial and temporal influence of vegetation on surface soil factors in semiarid rangelands. Transactions of ASAE 35:479-486.

Branson, F.A., R.F. Miller, and I.S. McQueen. 1976. Moisture relationships in twelve northern desert shrub communities near Grand Junction, Colorado. Ecology 57:176-197.

Briske, D.D. 1991. Developmental morphology and physiology of grasses. pp. 85-108, IN: R.K. Heitschmidt and J.W. Stuth (eds.). Grazing management: an ecological perspective. Timber Press, Portland, OR.

Campbell, G.S. 1974. A simple method for determining unsaturated conductivity from moisture retention data. Soil Science 117:311-314.

Carlson, D.H., and T.L. Thurow. 1992. SPUR-91: Workbook and user guide. Texas Agricultural Experiment Station MP-1743.

Carlson, D.H., T.L. Thurow, and C.A. Jones. 1993. Biophysical simulation models as a foundation for decision support systems. pp. 37-67 IN: J.W. Stuth and B.G. Lyons (eds.). Decision support systems for the management of grazing lands: emerging issues. Man and the Biosphere Series Vol. 11, Parthenon Publishing Group, London.

Carlson, D.H., T.L. Thurow, R.W. Knight, and R.K. Heitschmidt. 1990. Effect of honey mesquite on the water balance of Texas Rolling Plains rangeland. J. Range Manage. 43:491-496.

Christie, E.K. 1978. Ecosystem processes in semiarid grasslands. I. Primary production and water use of two communities possessing different photosynthetic pathways. Australian Journal of Agricultural Research 29:773-787.

Davis, E.A., and C.P. Pase. 1977. Root systems of shrub live oak: implications for water yield in Arizona chaparral. Journal of Soil and Water Conservation 32:174-180.

Dee, R.F., T.W. Box, and E. Roberston, Jr. 1966. Influence of grass vegetation on water intake of Pullman silty clay loam. Journal of Range Management 19:77-79.

Cooley, K.R., E.P. Springer, and A.L. Huber. 1987. Hydrology component: snowmelt. pp. 31-40, IN: J.R. Wight and J.W. Skiles (eds.). SPUR -- Simulation of production and utilization of rangelands : Documentation and user guide. United States Department of Agriculture, Agricultural Research Service, ARS-63.

Detling, J.K., W.J. Parton. and H.W. Hunt. 1978. An empirical model for estimating CO_2 exchange of Bouteloua gracilis biomass (HBK) Lag. in shortgrass prairie. Oecologia 33: 137-147.

Detling, J.K., W.J. Parton, and H.W. Hunt. 1979. A simulation model of Bouteloua gracilis biomass dynamics on the North American shortgrass prairie. Oecologia 38: 157-191.

Graf, W.H. 1971. Hydraulics of sediment transport. McGraw-Hill Book Co., New York, NY.

Hanson, C.L., E.L. Neff, and D.A. Woolhiser. 1975. Hydrologic aspects of water harvesting in the northern Great Plains. Proceedings of Water Harvesting Symposium, United States Department of Agriculture, Agricultural Research Service, ARS-W-22.

Hanson, J.D., J.W. Skiles, and W.J. Parton. 1987. Plant component. pp. 57-74, IN: J.R. Wight and J.W. Skiles (eds.). SPUR -- Simulation of production and utilization of rangelands : Documentation and user guide. United States Department of Agriculture, Agricultural Research Service, ARS-63.

Heitschmidt, R.K., and S.L. Dowhower. 1991. Herbage response following control of honey mesquite within single tree lysimeters. Journal of Range Management 44:144-149.

Hibbert, A.R. 1983. Water yield improvement potential by vegetation management on western rangelands. Water Resources Bulletin 19:375-381.

Hicks, R.A., D.D. Briske, C.A. Call, and R.J. Ansley. 1990. Co-existence of a perennial C_3 bunchgrass in a C_4 dominated grassland: an evaluation of gas exchange characteristics. Photosynthetica 24:63-74.

Holmstead, G.L. 1989. Water-use and growth of three C_4 bunchgrasses: evaluation under field and controlled environment conditions. MS Thesis, Texas A&M University, College Station, TX.

Hsiao, T.C., and E. Acevedo. 1974. Plant responses to water deficits, water use efficiency, and drought resistance. Agricultural Meteorology 14:59-84.

Innis, G.S. (ed.). 1978. Grassland Simulation Model. Springer-Verlag, New York, NY.

Johnston, R.S. 1970. Evapotranspiration from bare, herbaceous, and aspen plots: a check on a former study. Water Resources Research 6:324-327.

Knight, R.W., W.H. Blackburn, and L.B. Merrill. 1984. Characteristics of oak mottes, Edwards Plateau, Texas. Journal of Range Management 37:534-537.

Knisel, W.G. (ed). 1980. CREAMS: A field-scale model for chemicals, runoff, and erosion from agricultural management systems. United States Department of Agriculture, Conservation Research Report No. 26.

LaMeur, R. and L. Zhang. 1990. Evaluation of three evapotranspiration models in terms of their applicability for an arid region. Journal of Hydrology 114:395-411.

Lane, L.J. 1982a. Development of a procedure to estimate runoff and sediment transport in ephemeral streams. pp. 275-282, IN: Proceedings of the Symposium of the International Association of Hydrological Sciences, Exeter, England, July, 1982. IAHS Pub. No. 137.

Lane, L.J. 1982b. A distributed model for small semiarid watersheds. Proceedings, American Society of Civil Engineers, Journal of Hydraulics Division 108 (HY10):1111-1131.

Lane, L.J. 1983. Transmission losses. pp. 191-194, IN: United States Department of Agriculture, Soil Conservation Service, National engineering handbook, Hydrology Section 4, Chapter 9.

Lane, L.J. 1987. Hydrology component: water routing and sedimentation. pp. 41-56, IN: J.R. Wight and J.W. Skiles (eds.). SPUR -- Simulation of production and utilization of rangelands : Documentation and user guide. United States Department of Agriculture, Agricultural Research Service, ARS-63.

Lane, L.J., M.H. Diskin, and K.G. Renard. 1971. Input-output relationships for an ephemeral stream channel system. Journal of Hydrology 13:22-40.

MacNeil, M.D., T.G. Jenkins, R.W. Rice, and L.J. Koong. 1987. Animal component. pp. 75-84, IN: J.R. Wight and J.W. Skiles (eds.). SPUR -- Simulation of production and utilization of rangelands : Documentation and user guide. United States Department of Agriculture, Agricultural Research Service, ARS-63.

MacNeil, M.D., J.W. Skiles, and J.D. Hanson. 1987. Sensitivity analysis. pp. 89-98, IN: J.R. Wight and J.W. Skiles (eds.). SPUR -- Simulation of production and utilization of rangelands : Documentation and user guide. United States Department of Agriculture, Agricultural Research Service, ARS-63.

McCalla, G.R. II, W.H. Blackburn, and L.B. Merrill. 1984. Effects of livestock grazing on sediment production, Edwards Plateau of Texas. Journal of Range Management 37:291-294.

Murphey, J.B., D.E. Wallace, and L.J. Lane. 1977. Geomorphic parameters predict hydrograph characteristics in the Southwest. Water Resources Bulletin 13:25-38.

Notter, D.R. 1977. Simulated efficiency of beef production for a cow-calf feedlot management system. PhD. dissertation, University of Nebraska, Lincoln, NB.

Parton, W.J., and J.A. Logan. 1981. A model for diurnal variation in soil and air temperature. Agricultural Meteorology 23:205-216.

Parton, W.J., J.S. Singh, and D.C. Coleman. 1978. A model of production and turnover of roots in shortgrass prairie. Journal Applied Ecology 47: 515-542.

Pluhar, J.J., R.W. Knight, and R.K. Heitschmidt. 1987. Infiltration rates and sediment production as influenced by grazing systems in the Texas Rolling Plains. Journal of Range Management 40:240-243.

Ponce, S.L., and J.R. Meiman. 1983. Water yield augmentation through forest and range management-issues for the future. Water Resour. Bull. 19:415-419.

Rallison, R.E. 1980. Origin and evolution of the SCS runoff equation. Proceedings of the Watershed Management Symposium, American Society of Civil Engineers, Boise, ID.

Renard, K.E., E.D. Shirley, J.R. Williams, and A.D. Nicks. 1987. Hydrology component: upland phases. pp. 17-30, IN: J.R. Wight and J.W. Skiles (eds.). SPUR -- Simulation of production and utilization of rangelands : Documentation and user guide. United States Department of Agriculture, Agricultural Research Service, ARS-63.

Richardson, C.W., C.L. Hanson, and A.L. Huber. 1987. Climate generator. pp. 3-16, IN: J.R. Wight and J.W. Skiles (eds.). SPUR -- Simulation of production and utilization of rangelands : Documentation and user guide. United States Department of Agriculture, Agricultural Research Service, ARS-63.

Richardson, C.W., and D.A. Wright. 1984. WGEN: A model for generating daily weather variables. United States Department of Agriculture, Agricultural Research Service, ARS-8.

Ritchie, J.T. 1972. A model for predicting evaporation from a row crop with incomplete cover. Water Resources Research 8:1204-1213.

Sanders, J.O. 1977. Application of a beef cattle production model to the evaluation of genetic selection criteria. PhD. dissertation, Texas A&M University, College Station, TX.

Sanders, J.O., and T.C. Cartwright. 1979. A general cattle production systems model. I. Description of the model. Agricultural Systems 4: 217-227.

Scanlon, J.C. 1988. Spatial and temporal vegetation patterns in a subtropical *Prosopis* savanna woodland, Texas. PhD. Dissertation, Texas A&M University, College Station, TX.

Schultz, R.D. 1982. Net primary production and seasonal herbaceous dynamics following brush management. PhD. Dissertation, Texas A&M University, College Station, TX.

Scifres, C.J., and B.H. Koerth. 1987. Climate, soils, and vegetation of the La Copita Research Area. Texas Agricultural Experiment Station Publication MP-1626.

Sosebee, R.E. 1976. Hydrology: the state of the science evapotranspiration. pp. 95-104, IN: H.F. Heady, D.H. Falkenburg, and J.P. Riley. Watershed Management on range and forest lands. Utah State University, Logan, UT.

Stout, W.L., L.C. Vona, J.W. Skiles, J.A. Shaffer, G.A. Jung, and R.L. Reid. 1990. Evaluating SPUR model for predicting animal gains and biomass on eastern hill land pastures. Agricultural Systems 34:169-178.

Thurow, T.L. 1985. Hydrologic interrelationships with vegetation and soil as affected by selected grazing systems and climate on the Edwards Plateau. PhD. Dissertation, Texas A&M University, College Station, TX

Thurow, T.L. 1991. Hydrology and erosion. pp. 141-160, IN: R.K. Heitschmidt and J.W. Stuth (eds.). Grazing management: an ecological perspective, Timber Press Inc., Portland, OR.

Thurow, T.L., W.H. Blackburn, and C.A. Taylor Jr. 1986. Hydrologic characteristics of vegetation types as affected by livestock grazing systems, Edwards Plateau, Texas. J. Range Manage. 39:505-509.

Thurow, T.L., W.H. Blackburn, and C.A. Taylor Jr. 1988. Infiltration and interrill erosion responses to selected livestock grazing strategies, Edwards Plateau, Texas. J. Range Manage. 41: 296-302.

United States Department of Agriculture, Agricultural Research Service (USDA-ARS). 1991. Installing and running the UTIL program on an IMB-PC or compatible. Feb. 1 version, Temple, TX.

United States Department of Agriculture, Soil Conservation Service (USDA-SCS). 1957. Farm pond survey-Texas.

United States Department of Agriculture, Soil Conservation Service (USDA-SCS). 1972. National engineering handbook, Hydrology Section 4, Chapters 4-10.

Warren, S.D., W.H. Blackburn, and C.A. Taylor, Jr. 1986. Effects of season and stage of rotation cycle on hydrologic condition of rangeland under intensive rotation grazing. Journal of Range Management 39:486-490.

Weltz, M.A. 1987. Observed and estimated (ERHYM-II model) water budgets for south Texas rangelands. Ph.D. Dissertation, Texas A&M University, College Station, TX.

Wight, J.R. and J.W. Skiles (eds.). 1987. SPUR -- Simulation of production and utilization of rangelands : Documentation and user guide. United States Department of Agriculture, Agricultural Research Service, ARS-63.

Wilcox, B.P., C.L. Hanson, J.R. Wight, and W.H. Blackburn. 1989. Sagebrush rangeland hydrology and evaluation of the SPUR hydrology model. Water Resources Bulletin 25:1-14.

Williams, J.R., and H.D. Berndt. 1977. Sediment yield prediction based on watershed hydrology. American Society of Agricultural Engineers Transactions 20:1100-1104.

Williams, J.R., and W.V. La Seur. 1976. Water yield model using SCS curve numbers. Journal of the Hydraulics Division, American Society of Civil Engineers, 102(HY9):1241-1253.

Williams, J.R., A.D. Nicks, and J.G. Arnold. 1985. Simulator for water resources in rural basins. Journal Hydraulic Engineering, American Society of Civil Engineers 111: 970-986.

Wood, M.K., and W.H. Blackburn. 1981a. Grazing systems: their influence on infiltration rates in the Rolling Plains of Texas. Journal of Range Management 34:331-335.

Wood, M.K., and W.H. Blackburn. 1981b. Sediment production as influenced by livestock grazing on the Texas Rolling Plains. Journal of Range Management 34:228-231.

Wood, M.K., and W.H. Blackburn. 1984. An evaluation of the hydrologic soil groups as used in the SCS runoff method on rangelands. Water Resources Bulletin 20:379-389.

Chapter 28

HYDROLOGY COMPONENTS OF CREAMS AND GLEAMS MODELS

W. G. Knisel and J. R. Williams

The hydrology components of the CREAMS and GLEAMS water quality loading models are described. A daily rainfall option using a modification of the USDA-SCS curve number method is the preferred runoff-estimating procedure of CREAMS, but the modified Green-Ampt infiltration option is also described. A revised soil layering procedure to better represent the soil profile was incorporated in the daily rainfall procedure of CREAMS that is described as the principal component of GLEAMS.

28.1. INTRODUCTION

Methods of estimating direct runoff or total streamflow were relatively simple rainfall-runoff relationships prior to the development of high-speed digital computers. Very few models generated hydrographs beyond the simple triangular shape with linear rising and falling sides. Only limited laborious and time-consuming applications were made that involved hydrograph routing procedures. In the early 1950's, the USDA, Soil Conservation Service (1972)[1] developed a simple procedure for estimating direct runoff from storm rainfall, generally applied for design-type 24-hour rainfall. The procedure was a simple graphical method of estimating runoff volume rather than a storm hydrograph. The method related runoff to rainfall as a function of a curve number which represented combined soil-cover complex and 5-day antecedent rainfall. Soils were grouped by potential hydrologic response, i. e. runoff potential. Cover included type of crop (row crop, small grain, close-seeded crops, meadow, and etc.), tillage

---

[1] Earlier versions of the hydrology handbook were published in 1960 and 1964. The 1972 version is readily available and was repeatedly referenced in literature of the late 1970's and 1980's. The latest available version is dated 1988.

(straight row, contour, residue management, and etc.), and hydrologic condition or receptiveness of rainfall for infiltration. The procedure became known as the "SCS curve number method".

Development of digital mainframe computers in the 1950's and 1960's, and associated development of the FORTRAN computer language enabled engineers to formulate processes into programs for iterative calculations. This made it feasible for calculations on a continuous basis to estimate runoff from all storms rather than just design storms. Williams and LaSeur (1976) used the "SCS curve number method" to develop a computerized water yield model for mixed land-use agricultural watersheds. Their modification of the SCS (USDA-SCS, 1972) procedure included a daily soil moisture accounting procedure using a climatic index based upon lake evaporation. The modified procedure explained 95% of the variation in monthly runoff from a mixed land-use watershed in Texas compared with 65% of the variation explained by the original method.

Williams and Hann (1978) further modified the water yield model to develop a watershed water-quality loading model. Their revisions in the hydrology component consisted of replacing the climatic index with a procedure for estimating daily consumptive water use for individual crops. They retained the "SCS curve number method" as the basis for runoff computations. Further, a routine was incorporated for routing infiltrated water into, within, and through the root zone in the water accounting method.

Williams (1982) extended the modified SCS curve number runoff method into a problem-oriented computer language when he developed the SPNM model--Sediment, Phosphorus, and Nitrogen Movement. SPNM hydrology was essentially like that of Williams and Hann (1978).

In 1978, the USDA, Agricultural Research Service (then Science and Education Administration) initiated an effort to develop a mathematical model for assessing nonpoint source pollution from field-size areas. That effort resulted in the CREAMS model (Knisel, 1980)--Chemicals, Runoff, and Erosion from Agricultural Management Systems. CREAMS provided model users with two options for simulating hydrologic response from field-size areas. Hydrology option one of CREAMS was the modified SCS curve number method (Williams and Nicks, 1982) using daily rainfall as input. Option two was the Green and Ampt (1911) infiltration equation modified by Smith and Parlange (1978). The infiltration option required breakpoint or hourly rainfall as input.

Since development of CREAMS, other models have been developed that resulted in a continuum of models that have the daily rainfall (modified SCS curve number) method in common. They include SWRRB (Williams et al., 1985; Arnold et al., 1990), GLEAMS (Leonard et al., 1987; Nutter et al., 1993; Knisel et al., 1993), EPIC (Williams and Renard, 1985; Sharpley and

Williams, 1990), PRZM (Carsel et al., 1985), and SWRRB-WQ (Arnold and Williams, 1989).

All of the above named models included components other than hydrology, i. e. erosion, plant nutrients, and pesticides. They were developed to assess water quality loadings and impact of erosion on long-term sustained productivity. This chapter describes the hydrologic component of two of those models: CREAMS and GLEAMS.

28.2. CREAMS

CREAMS was developed to assess edge-of-field nonpoint source pollutant loadings for alternate management systems. Since crops may change within a rotation cycle, fertilization and pest control practices vary from year to year. Time of fertilizer and pesticide application relative to runoff-producing rainfall each year may result in highly variable runoff losses. Therefore, CREAMS was designed to make long-term (up to 20 years) daily simulations to assess the storm-by-storm differences from year-to-year. Models are available to simulate runoff and sediment yield from single design-type storms, for example, ANSWERS (Beasley et al., 1977). They are entirely satisfactory for the purpose they were developed. CREAMS, too, can be applied for a single event, but that was not the purpose for which it was developed.

Several simplifying assumptions were made to facilitate model development. The field unit was considered to consist of a natural drainage that has:

1) a relatively homogenous soil,

2) a single crop at any one time (not to be confused with crop rotation),

3) a single management practice over the entire area, and

4) uniform rainfall over the entire area. Homogeneity of soil and rainfall is the most strenuous, but the assumption is not restrictive for simulation purposes.

Perhaps a more critical factor for model application is the availability of data. Time-intensity rainfall data are not readily available except at intensively monitored research sites. At the time of the CREAMS model development, mainframe computers were in wide-spread use with data input by punch cards or magnetic tape, either via remote terminals or on site at computer centers. Thus, data requirements and their associated input were significant considerations in model development.

FORTRAN computer language was widely used in scientific applications. Due to the urgency of the modelling project, existing models and components had to be used where possible. As a result, most of the available components were programmed in FORTRAN IV, and this is the basic language used in the CREAMS model. By the time of publication, the source code was compiled with a FORTRAN-77 compiler.

The time step in the hydrology component is one day between storms. The daily time increment is used in the SCS curve number procedure which utilizes daily rainfall as input. The infiltration option uses hourly or irregular time increments as small as one minute during rainfall events with the same daily time increment between storm events.

Precipitation, snowmelt, infiltration and redistribution of soil water, percolation, evaporation, transpiration, and surface runoff were considered to be the important processes to be formulated. These processes are described below in sufficient detail to provide insight to the model and its operations.

28.2.1. Runoff Processes

Daily Rainfall Option

Hydrologists in the USDA-Soil Conservation Service developed a procedure in the mid-1950's to estimate runoff from storm rainfall as a function of soil, land use and treatment, and antecedent conditions (Mockus, 1949). The method later became known as the SCS curve number procedure (USDA-SCS, 1972). The curve number represents the soil-cover complex and 5-day antecedent rainfall. The runoff calculation is expressed as

$$Q = \frac{(P - 0.2S)^2}{(P + 0.8S)} \tag{28.1}$$

where Q is daily runoff, cm, P is precipitation, cm, and S is available soil water storage or retention, cm, on the day that rainfall occurs. The SCS antecedent rainfall index was referenced to dry (condition I), normal (condition II), or wet (condition III). Williams and LaSeur (1976) modified the SCS procedure for storm-by-storm computations by replacing the antecedent rainfall condition with a soil water accounting procedure.

The retention parameter, S, in Eq. (28.1) is related to the soil water content as

$$S = S_{mx} \frac{(UL - SM)}{(UL)} \tag{28.2}$$

where SM is the soil moisture storage in the root zone on the day of rainfall, cm, and UL is the upper limit of plant available and drainable soil water storage in the root zone defined as

$$UL = (\phi - BR15)\, RD \tag{28.3}$$

where ø is the soil porosity in the root zone, BR15 is the volumetric soil water content at 1500 kPa, cm/cm, and RD is the effective rooting depth, cm. SM in Eq. (28.2) is

$$SM = (\theta - BR15)\, RD \tag{28.4}$$

where - is the volumetric soil water content, cm/cm.

The maximum value of storage, S_{mx} in Eq. (28.2), is estimated for antecedent moisture condition I (USDA-SCS, 1972) as

$$S_{mx} = 2.54\left(\frac{100}{CN_1} - 1\right) \tag{28.5}$$

where CN_I is curve number for moisture condition I calculated from the normally-cited CN_{II} with the polynomial

$$CN_1 = 1.348(CN_{II}) - 0.01379(CN_{II})^2 + 0.0001177(CN_{II})^3 - 16.91 \tag{28.6}$$

The curve number for moisture condition II, CN_{II} can be obtained for any hydrologic soil, and land use and treatment (USDA-SCS, 1972).

Equation (28.1) should give a good estimate of runoff if the storage, or retention parameter, S, is well estimated. If soil water is uniformly distributed in the root zone, Eq. (28.2) should give a good estimate of S. If water content is greater near the soil surface, S in Eq. (28.2) would be under-estimated, and likewise runoff calculation by Eq. (28.1), would be low. If water content is greater near the bottom of the root zone, for example a high water-table soil, then runoff would be S and Q would be over-estimated. A depth-weighted retention parameter is calculated in CREAMS as

$$S = S_{mx}\left[1.0 - \sum_{i=1}^{7}(WF_i)\left(\frac{SM_i}{UL_i}\right)\right] \tag{28.7}$$

where WF_i is the weighting factor for the i-th computational layer defined by

$$WF_i = 1.016\left[\exp\left(-4.16\frac{D_{i-1}}{RD}\right) - \exp\left(-4.16\frac{D_i}{RD}\right)\right] \tag{28.8}$$

where D is the depth from the soil surface to the bottom of the computational soil layer, cm. Equation (28.8) assures that the sum of the weighting factors sum to 1.0.

As indicated by Eq. (28.7), there are seven computational soil layers in the CREAMS daily rainfall option (Smith and Williams, 1980). The layer thicknesses are as follows: layer 1 = 1/36 RD, layer 2 = 5/36 RD, and layers 3-7 = 1/6 RD.

Runoff hydrographs are not generated in the CREAMS model, but an instantaneous peak rate of runoff, q_p, cm/hr, for the field area is estimated by

$$q_p = 160(A)^{0.3}(SL)^{0.159}\left(\frac{Q}{2.54}\right)^{0.836(A)^{0.0166}}(LWR)^{-0.187} \tag{28.9}$$

where A is drainage area, ha, SL is the average field slope, m/m, and LWR is the length width ratio. The average field slope is the maximum difference in elevation from the field outlet, EL_o, to the highest point on the edge of the field, EL_h, divided by the length of the flow path, FPL, m, between the two points

$$SL = \frac{EL_h - EL_o}{FPL} \tag{28.10}$$

The length width ratio is estimated as

$$LWR = \frac{(FPL)^2}{1000\,A} \tag{28.11}$$

The variable exponents of A and Q in Eq. (28.9) prevent unreasonably high estimates of peak runoff rates for small drainage areas, i. e. less than 1 ha. Model applications have been made for rainfall simulator plots without obvious questionable peak rates.

Infiltration Option

The Green and Ampt infiltration equation (1911) modified by Smith and Parlange (1978) is included as an option in CREAMS hydrology (Smith and Williams, 1980). If hourly or breakpoint rainfall data are available, this option provides the user the capability to use the specific storm characteristics. The infiltration procedure partitions rainfall amount into infiltration and runoff for each time interval during the storm. Runoff for each time interval is summed and routed to the field outlet to estimate storm runoff volume, Q, and peak rate of runoff, q_p. The time interval can range from 1 minute to several hours, for high intensity and low intensity storms, respectively. This time interval should not be confused with the daily time step between storm events. Time-intensity data can accommodate multiple storms in one 24-hr period, and the total summed for the day and maximum peak rate of runoff provided.

Infiltration during a rainfall event consists of two phases: infiltration, F_p, to the time of ponding, t_p, when rainfall rate, r_p, exceeds infiltration rate, f_t; and infiltration, F', that occurs at a rate declining in time as root zone soil water content increases. At the beginning of rainfall, the soil has some initial degree of saturation, ST_i, defined as

$$ST_i = \frac{\theta_i}{\phi} \tag{28.12}$$

where θ is the volumetric water content, and the subscript i denotes initial condition. The variables in Eq. (28.12) represent the surface infiltration-control depth, D_s, usually in the range of 2-5 cm.

In the early stage of rainfall, the near-surface saturation increases from ST_i to a maximum theoretical value $ST_o = 1$ (100% saturation) if rainfall lasts long enough and rainfall volume is sufficient. For $ST \geq ST_o$, the soil controls surface water flux, and the time when this begins is called time of ponding, t_p. The infiltration to time of ponding, F_p, is approximately equal to the initial abstraction of rainfall, I_a, in the SCS curve number procedure (USDA-SCS, 1972), and is given in Eq. (28.1) as 0.2S. I_a is a function of soil water storage, S, and F_p is a function of rainfall rate.

The Green and Ampt (1911) infiltration relation assumes piston flow, and is given by

$$K_s t = F - \phi H_c (ST_o - ST_i)\, 1n\left[1 + \frac{F}{\phi H_c (ST_o - S_i}\right] \tag{28.13}$$

where K_s is effective saturated conductivity, cm/hr, t is time from start of ponding, hr, H_c is effective capillary tension, cm, and F is cumulative depth of infiltration, cm. Cumulative infiltration, F, is given by

$$F = L\phi(ST_o - ST_i) \tag{28.14}$$

where L is the depth of wetting. The total head across the wetting front is $-(H_c+L)$, and by Darcy's law, infiltration rate, f, cm/hr, is

$$f = K_s\left(\frac{H_c + L}{L}\right) \tag{28.15}$$

Solving Eq. (28.14) for depth of wetting, L, and substituting into Eq. (28.15) gives

$$f = K_s\left[\frac{H_c\phi(ST_o - ST_i) + F}{F}\right] \tag{28.16}$$

and solving for F gives

$$F = \frac{H_c\phi(ST_o - ST_i)K_s}{r_f - K_s} \tag{28.17}$$

The ponding depth, or cumulative infiltration at the time of ponding, F_p, is

$$F_p = \frac{H_c\phi(ST_o - ST_i)K_s}{r_p - K_s} \tag{28.18}$$

where r_p is the rainfall rate at the time of ponding.

Since infiltration rate, f, represents the change in the change in infiltration depth, F, in time t, then f = dF/dt, and with substitution into Eq. (28.16) and integration for a finite difference gives

$$K_s(t+\Delta t) = (F+\Delta F) - H_c\phi(ST_o - ST_i)\ln\left[1+\frac{F+\Delta F}{H_c\phi(ST_o - ST_i)}\right] \tag{28.19}$$

Subtracting Eq. (28.18) for Eq. (28.19) and rearranging gives

$$K_s t = \Delta F - H_c\phi(ST_o - ST_i)\ln\left[1+\frac{\Delta F}{H_c\phi(ST_o - ST_i)+F}\right] \tag{28.20}$$

Using the first term of the series approximation for the natural logarithm, Eq. (28.20) can be solved for ΔF as

$$\Delta F = \left[2K_s\Delta t\left[H_c\phi(ST_o - ST_i)+F\right]+\left[F-\left(\frac{K_s}{2}\right)\Delta t\right]^2\right]^{0.5} - \left[F-\left(\frac{K_s}{2}\right)\Delta t\right] \tag{28.21}$$

The mean infiltration rate for the i-th time interval is

$$f_i = \frac{\Delta F_i}{\Delta t_i} \tag{28.22}$$

Runoff during the i-th interval is

$$q_i = r_i - f_i \tag{28.23}$$

where r_i is rainfall rate for the i-th interval, cm/hr, for $t > t_p$. The cumulative infiltration is updated for each time interval as

$$F_{i+1} = F_i + \Delta F_i \tag{28.24}$$

If $r_i\ \Delta t_i < \Delta t_i\ \Delta F_i$,

$$F_{i+1} = F_i + r_i \Delta t_i \tag{28.25}$$

Total storm runoff, Q, cm, is the sum for n-intervals is

$$Q = \sum_{i=1}^{n} q_i \Delta t_i \tag{28.26}$$

Smith and Williams (1980) suggested that if breakpoint rainfall is not available, the more readily available hourly rainfall data can be used in hydrology option 2. Since rainfall intensity may vary significantly in an hour, r_p in Eq. (28.18) is estimated as 133% of the hourly rate. However, for fields the size normally used in CREAMS model applications, the time of concentration is much less than 1 hr, and the use of hourly data is grossly inadequate for an infiltration-based model.

Time lapse greater than 3 hr without rainfall signals a separate storm event for the infiltration option even though it occurs on the same day. The lapse allows some infiltration recovery, but the soil remains wet and is estimated in the model as

$$(ST_o - ST_i) = 0.9 - \left(\frac{t_e}{180}\right) 0.05 \tag{28.27}$$

where t_e is the time elapsed without rainfall, min. Runoff is summed for each separate storm during the day to give the total daily runoff.

Infiltration simulated from breakpoint rainfall produces a histogram of excess rainfall, r_e, for each breakpoint interval, Δ. However, the excess rainfall rate for each interval cannot be summed directly to calculate peak rate of storm runoff, and each histogram element must be routed to the field outlet. If the excess rainfall rate above the infiltration rate is uniform and lasts long enough, for example a rainfall simulator, equilibrium can occur, and the runoff rate would be equal to the rainfall excess rate. This rarely, if ever, occurs with natural rainfall even for field-size areas. Shallow water flow hydraulics allows calculation by simulating movement of characteristic "waves" over the flow plane from the upstream point to the outlet. Beginning at the origin, i. e. distance x = 0 and time t = 0, an upstream characteristic can be described for equilibrium conditions. On one side of the characteristic, the flow is uniform unsteady (rising), and on the other side the flow is nonuniform steady. Along any characteristic the depth of flow, h, can be estimated for any breakpoint interval, i, with its respective rainfall excess from the relation

$$h_i = h_{i-1} + (r_e)_i \, \Delta t_i \tag{28.28}$$

The velocity (celerity) of the characteristic, change in distance with respect to time (dx/dt), is

$$v = m\alpha h^{m-1} \tag{28.29}$$

where m is the uniform flow exponent (1.5 for the Chezy roughness law), and a is the uniform flow coefficient equal to the Chezy roughness times the square root of the slope ($CS^{0.5}$). Combining Eqs. (28.28) and (28.29) and integrating for the characteristic beginning at time t_{i-1} gives the distance as

$$x_i = \frac{\alpha}{(r_e)_i}\left[h_{i-1} + (r_e)_i \Delta t\right]^m + x_{i-1} \tag{28.30}$$

Eq. (28.30) gives the position of the "wave" from the upstream edge at any time, and can be applied successively for all increments of time to obtain the distance the peak moves. The peak runoff rate can be estimated from the greatest depth h reached at x = L (L being the total length of flow plane) for the fastest characteristic since, at all points, for kinematic flow

$$q = \alpha h^m \tag{28.31}$$

To estimate the peak outflow from a complex rainfall pattern, the characteristic flow path is chosen along which the largest rates of rainfall excess occur, and thus the largest depth at the watershed outlet. This characteristic would usually include the time interval for the largest rainfall excess, plus the intervals of the largest excess before and after, necessary for the characteristic to travel the total distance, L. Thus, the peak rate can be estimated by taking the depth h (Eq. (28.28)) for the peak rainfall excess interval, and adding adjacent intervals on each side of the peak interval. This gives the fastest characteristic for which x, Eq. (28.30), equals or exceeds L.

Complex slope shapes of natural landscapes can be represented by hydraulically equivalent uniform slopes for solving kinematic surface water flow equations. If the slope length is broken into N segments of different slope and roughness, the equivalent single plane values can be determined. For each segment, j, for j = 1,N, there is a flow coefficient, a_j, such that

$$\alpha_j = C_j \sqrt{S_j} \tag{28.32}$$

The Chezy roughness coefficient, C in Eq. (28.32), is the same as 1.49/n where n is the Manning roughness coefficient. It is necessary to estimate a composite flow coefficient, a_j, for the set of N planes which can be done by taking the average or composite slope, S_c, and solving Eq. (28.32) with a composite C_c.

Detention storage, D, on the composite plane of length L is

$$D = \int_0^L h\,dx \tag{28.33}$$

Equilibrium discharge for any plane of length x and rainfall excess r_e, is

$$r_e x = \alpha h(x)^m \tag{28.34}$$

from which the equilibrium depth h is

$$h = \left(\frac{r_e x}{\alpha}\right)^{1/m} \tag{28.35}$$

Then detention storage on the equivalent plane is the sum of storage on the N planes

$$D = \sum_{j=1}^{N} D_j \tag{28.36}$$

Combining Eqs. (28.33), (28.35), and (28.36) gives

$$\int_0^L \left(\frac{r_e x}{\alpha_c}\right)^{1/m} dx = \int_0^{x_1} \left(\frac{r_e x}{\alpha_1}\right)^{1/m} dx + \int_{x_1}^{x_2} \left(\frac{r_e x}{\alpha_2}\right)^{1/m} dx + \ldots + \int_{x_{N-1}}^{x_N} \left(\frac{r_e x}{\alpha_N}\right)^{1/m} dx \tag{28.37}$$

Integrating Eq. (28.37) results in

$$\frac{1}{b}\left(\frac{r_e}{\alpha_c}\right)^{1/m} L^b = \left[\frac{1}{b}\left(\frac{r_e}{\alpha_1}\right)^{1/m} x_1{}^b + \left(\frac{r_e}{\alpha_2}\right)^{1/m} (x_2{}^b - x_1{}^b) + \ldots + \left(\frac{r_e}{\alpha_N}\right)^{1/m} (L^b - x_{N-1}{}^b)\right] \tag{28.38}$$

where b = (m+1)/m. Dividing both sides of Eq. (28.38) by $(1/b)(r_e)^{1/m}(L)^b$ gives

$$\left(\frac{1}{\alpha_c}\right)^{1/m} = \sum_{j=1}^{N} \left(\frac{1}{\alpha_j}\right)^{1/m} \left[\frac{(x_j{}^b - x_{j-1}{}^b)}{L^b}\right] \tag{28.39}$$

and rearranging gives

$$\alpha_c = \left[\frac{L^b}{\sum_{j=1}^{N} \left[\frac{\left(x_j{}^b - x_{j-1}{}^b\right)}{\alpha_j{}^{1/m}}\right]}\right]^m \tag{28.40}$$

where $x_N = L$.

28.2.2. Percolation

The daily rainfall option of CREAMS hydrology uses seven computational soil layers to represent the root zone as specified in the previous section. The infiltration option uses only two computational layers: a surface infiltration control layer, D_s, and the root zone, D_p. These two different representations of the root zone result in different routines for percolation of infiltration in excess of field capacity.

Daily Rainfall Option

The daily rainfall option of CREAMS uses a storage-routing technique to simulate movement of excess infiltration through the root zone. In this option, infiltration volume, F, is estimated as the difference between daily precipitation, P, and daily runoff, Q, as

$$F = P - Q \tag{28.41}$$

The soil water routing equation for outflow, 0, from a layer is

$$O = \sigma\left(F + \frac{ST}{\Delta t}\right) \quad for \; \left(F + \frac{ST}{\Delta t}\right) > FC \tag{28.42}$$

where FC is field capacity, approximated by water retention at 10 kPa, or 33 kPa depending upon soil texture, s is the storage coefficient, and Δt is the routing interval (1 day). If infiltration plus storage does not exceed field capacity, percolation is not predicted. The storage coefficient is a function of the travel time through the soil layer expressed as

$$\sigma = \frac{2\,\Delta t}{2tt + \Delta t} \tag{28.43}$$

where tt is the travel time through the storage layer. Travel time is estimated by

$$tt = \frac{SM - FC}{r_c} \tag{28.44}$$

where SM is soil moisture storage, cm, and r_c is saturated conductivity of the soil profile, cm/hr. CREAMS uses a single value of r_c, the most restrictive value for the profile.

Infiltration Option

The surface infiltration control layer is important in its initial conditions for partitioning rainfall into runoff and infiltration. Also, the upper soil zone,

D_s, is subject to soil evaporation from the evapotranspiration model (to be presented below). The lower soil zone, D_p, is subject to root extraction of water, i. e. plant transpiration, during the growing season.

Water moves from the upper soil zone to the root zone as a function of the positive difference in saturation between the two zones as

$$O_s = C_s(S_s)^3 (S_s - S_p) \phi D_s \quad for \quad (S_s > S_p) \tag{28.45}$$

where O_s is the outflow from the surface layer, cm, C_s is a storage coefficient, normally 0.1, S_s is the volumetric saturation in the surface zone, cm/cm, S_p is the volumetric saturation in the root zone, cm/cm, ø is effective porosity (total porosity minus immobile water content, i. e. water content at 1500 kPa tension), cm^3/cm^3, and D_s is depth of surface zone, generally 2 to 5 cm.

Percolation from the root zone is predicted to occur when S_p exceeds field capacity, and is estimated as the daily excess of S_s over field capacity. Water uptake by crop roots occurs from both surface and root zones in proportion to the relative root depth which varies with leaf area index up to the maximum depth, RD. Thus, if depth of rooting is twice the surface infiltration control depth, evapotranspiration is taken equally from the surface layer, D_s, and the root zone, D_p, if sufficient water is available in both layers.

28.2.3. Soil Water Accounting

Water entering the soil from precipitation, snowmelt, or irrigation, as infiltration F, becomes soil water storage, SM, evapotranspiration, ET, or percolation, O, below the root zone. Then soil moisture, SM, on any day, i, can be expressed as

$$SM_i = SM_{i-1} + F_i + M_i + IR_i - ET_i - O_i \tag{28.46}$$

where M is snowmelt, and IR is irrigation.

28.2.4. Snowmelt

Snow accumulation in the CREAMS model occurs when mean daily air temperature is below freezing on a day when precipitation occurs. The water equivalent depth of snowfall, as reported by the National Weather Service (NWS), is shown in the precipitation file. The method of accumulation and melt allows no sublimation.

When mean daily air temperature is above freezing, daily snowmelt, M_i, cm, is estimated by the relation (Stewart et al., 1975)

$$M_i = 0.457 \;\; TO_i \tag{28.47}$$

where T0 is the temperature above 0°C.

The snowmelt model is very simple, but it allows the user to input only mean monthly temperature and mean monthly solar radiation. Kothandaraman and Evans (1972) showed that a Fourier series could be fitted to mean monthly temperature and radiation data, and the computed coefficients used to interpolate daily values. The fitting technique does not allow a mid-winter thaw that normally occurs in most cold climatic regions. Also, snowmelt generally occurs at sufficiently low daily amounts that little or no runoff occurs even from significant snowpack.

Knisel et al. (1985) modified CREAMS hydrology to give an approximate procedure for simulating frozen soil and to reduce available storage depending upon soil type. Infiltration characteristics of a frozen soil change with soil type and water content at the time of freezing. For example, a clay soil with a water content near field capacity when it freezes will have little remaining pore space for water movement. Conversely, expansion on freezing of water in a sandy soil near field capacity may result in increased infiltration rate. This effect is as difficult to estimate as the water retention characteristics of soils when only the texture is known. In order to partially account for soil differences, Knisel et al. (1985) used the input curve number to adjust the available storage for frozen conditions. If curve number is equal to or less than 80, a value of 95 is used during frozen soil, and for curve numbers greater than 80, a value of 98 is used. The new curve number is used only during periods of frozen soil, and a new maximum available storage is calculated (S_{mx} in Eq. (28.5)). When frozen soil ends in the spring, the original S_{max} is returned.

Frozen soil is assumed to occur if mean monthly temperature for any month is below freezing. Frost remains in the soil for some lag time after mean daily temperatures rise above freezing in the spring. The duration is based upon the December solar radiation in CREAMS since that generally is the lowest monthly value in the northern hemisphere. The number of days with frozen soil in the spring, beyond the first day with mean daily temperature above freezing, is estimated as

$$DFS = 35.4 - 0.154\ R_D \tag{28.48}$$

where R_D is the December radiation, gm-cal/cm^2/day. The prolonged frozen soil using DFS results in reduced infiltration from snowmelt for an extended period in the spring.

28.2.5. Evapotranspiration

Potential Evaporation

The soil water accounting procedure given by Eq. (28.46) includes estimation of evapotranspiration. Soil evaporation and plant transpiration are calculated separately in CREAMS since transpiration is required to simulate plant uptake of nutrients. The model calculates growth of plant leaf area and growth of roots during the growing season to estimate root extraction of water. The method of Ritchie (1972) is used to calculate evaporation and transpiration for agricultural crops. The method estimates potential evaporation by the Priestly-Taylor (1972) method as

$$E_o = \frac{1.28 \Delta H_o}{\Delta + \gamma} \tag{28.49}$$

where Δ is the slope of the saturation vapor pressure curve at the mean air temperature, H_o is net solar radiation, gm-cal/cm²/day, and γ is the psychrometric constant (0.68). Δ is calculated as

$$\Delta = \frac{5304}{T_k^2} e^{(21.255 - 5304/T_k)} \tag{28.50}$$

where T_k is mean daily air temperature, °K. Net solar radiation, H_o, in Eq. (28.49) is

$$H_0 = \frac{(1 - \beta) R}{58.3} \tag{28.51}$$

where ß is albedo (used as a constant 0.23 in CREAMS).

The constant 1.28 in Eq. (28.49) has been deliberated by several people. Stemmler (1982) found that a value of 0.86 gave a better estimate of potential evaporation in western Europe.

Soil Evaporation

Potential soil evaporation, E_{so}, cm, is estimated from E_o as

$$E_{so} = E_o e^{(-0.4 LAI)} \tag{28.52}$$

where LAI is the leaf area index, m^2/m^2, of the crop (Ritchie, 1972). LAI is defined as the sum of leaf area per unit area of soil.

Actual soil evaporation is computed in CREAMS in two stages. In the first stage, soil evaporation is limited only by the evaporative energy at the soil surface, and it is equal to the potential evaporation for unlimited water availability. When the accumulated soil evaporation exceeds the stage one upper limit, the stage two evaporative process begins. The stage one upper limit, U, mm, is estimated as

$$U = 9(\alpha_s - 3)^{0.42} \tag{28.53}$$

where a_s is a soil evaporation parameter dependent upon soil water transmission characteristics, ranging from 3.3 to 5.5 mm/d$^{0.5}$. Ritchie (1972) suggests a value of 4.5 loamy soils, 3.5 for clays, and 3.3 for sands.

The stage two daily soil evaporation, mm, is estimated as

$$E_s = \alpha_s\left[t^{0.5} - (t-1)^{0.5}\right] \tag{28.54}$$

where t is the time in days since stage two evaporation began.

Plant Evaporation

Plant evaporation, or transpiration, is estimated by Ritchie's method (1972). When soil water is not limited, plant evaporation is

$$E_p = \frac{E_o(LAI)}{3} \quad for \quad 0 \le LAI \le 3 \tag{28.55}$$

and

$$E_p = E_o - E_s \qquad for \quad LAI > 3 \tag{28.56}$$

When soil water is limited, plant evaporation is reduced by the availability of water expressed as

$$E_{pl} = \frac{E_p(ST)}{0.25PAW} \qquad for \quad ST \le 0.25\ PAW \tag{28.57}$$

where PAW is the plant available water, cm, equal to the field capacity minus wilting point of the soil.

Evapotranspiration is the sum of soil and plant evaporation

$$ET = E_s + E_p \qquad for \quad ST > 0.25\ PAW \tag{28.58}$$

$$ET = E_s + E_{pl} \qquad for \quad ST \le 0.25\ PAW$$

Limited soil water availability results in reduced plant evaporation and also results in reduced plant growth. Plant growth is represented in CREAMS as "idealized" leaf area index values throughout the growing season. Crop growth traces the idealized conditions when soil water is not limiting. However, with limited soil water, actual LAI deviates from the idealized data, representing moisture stress in the plants. If rainfall or irrigation raises the soil water content at some later date, the actual LAI does not return to the idealized curve, but resumes the idealized growth rate at the reduced level. This is characteristic of domestic crops, but is not entirely applicable for native or climax vegetation.

Soil Water Depletion by ET

Daily evapotranspiration calculated by Eq. (28.58) must be distributed properly through the computational soil layers. A root growth component is used for this purpose, and is expressed as

$$FRD_n = \frac{\sum_{i=1}^{n} LAI_i}{\sum_{i=1}^{N} LAI_i} \tag{28.59}$$

where FRD_n is the fraction of the effective rooting depth RD that contains crop roots on day n, and N is the number of days in the growing season. Then the active root depth on day n is

$$RDN = FRD_n\ (RD) \tag{28.60}$$

RDN is adjusted to the bottom of the computational soil layer within which RDN occurs, i. e. roots are assumed to fully penetrate a soil layer. For example, if R = 90 cm, the bottom depths of each of the seven layers are: 2.5, 15, 30, 45, 60, 75, and 90 cm. If RDN from Eq. (28.60) is 40 cm, then RDN is adjusted to 45 cm.

The water use rate, u, cm, is a function of the active root depth expressed as

$$u = u_o\ e^{(-3.065\, RDN)} \tag{28.61}$$

where u_o is the water use rate at the surface. Total water use within any depth can be estimated by integrating Eq. (28.60) to obtain

$$ET = \frac{u_o}{3.065}\left[1.0 - e^{(-3.065\, RDN)}\right] \tag{28.62}$$

The water use rate at the surface, u_o, is calculated each day from Eq. (28.62) to distribute ET by computational soil layer as

$$wu_i = \frac{u_o}{3.065}\left[e^{(-3.065\, RDN_{i-1})} - e^{(-3.065\, RDN_i)}\right] \tag{28.63}$$

where wu_i is water uptake (ET), cm, in soil layer i, and RDN_{i-1} and RDN_i are depths in cm to the top and bottom, respectively, of soil layer i. Eq. (28.63) is solved for all layers to a depth of RDN.

Irrigation

Specific irrigation dates and amounts can be entered into the CREAMS precipitation file, in which case irrigation will be considered exactly like rainfall. This is also how irrigation is considered when comparing model simulation with observed data for validation purposes.

An irrigation option in CREAMS allows the model to irrigate automatically, based upon specified threshold and upper limit of soil water content. The soil water content in the root zone is compared each day with a specified fraction of plant available water, for example the soil water content for stage two water use (0.25 PAW in Eq. (28.57)). When this threshold value is reached, the model takes the difference between the threshold and the specified upper limit, for examplc 1.0 PAW, to determine the irrigation depth to be "applied". The procedure allows water deficit management (WDM) by specifying an upper limit less than field capacity, i. e. 0.90 PAW, a commonly recommended practice in humid regions. Some unfilled storage is left in case of thunderstorm occurrence. WDM is not a common practice in dryer regions or where water availability is limited.

The automatic irrigation option assumes that the rate is equal to or less than the infiltration rate, and runoff is not allowed. In addition to the threshold and upper limits, the model user must specify by code the beginning and ending dates for considering automatic irrigation. For example, irrigation of most crops is discontinued 30 to 60 days prior to harvest.

28.2.6. Sensitive Model Parameters

There are 30 input variables in the CREAMS model hydrology component including codes, considering the 12 values of mean monthly temperature as one variable, for example. In the daily rainfall option, the following parameters are sensitive in the water balance computations: RD--effective root depth; CN2--curve number for moisture condition II; POROS--soil porosity; UL(1) - UL(7)--upper limit of storage in computational layers 1-7; FUL--fraction of upper limit of storage filled at field capacity. The upper limit of storage for each computational layer is estimated as

$$\begin{aligned} UL_1 &= \left(POROS_1 - BR15_1\right)0.028\,RD \\ UL_2 &= \left(POROS_2 - BR15_2\right)0.139\,RD \\ UL_i &= \left(POROS_i - BR15_i\right)0.167\,RD \qquad for\quad i=3,7 \end{aligned} \tag{28.64}$$

where UL is the upper limit of storage (drainable and plant available water), cm, POROS is porosity, cm^3/cm^3, BR15 is volumetric water content at wilting point (1500 kPa), cm/cm, and RD is the effective rooting depth, cm.

Fraction of upper limit filled at field capacity, FUL, is defined as

$$FUL = \frac{FC - BR15}{POROS - BR15} \tag{28.65}$$

where FC is volumetric water content at field capacity (10 or 33 kPa, depending upon soil texture).

Water balance calculations using the infiltration option of CREAMS are sensitive to the same parameters except CN2. GA, the effective capillary tension (H_c in Eq. (28.13)), used in estimating the soil infiltration characteristics, is moderately sensitive (Lane and Ferreira, 1980). In the infiltration option, the surface infiltration control layer (DS) and root depth (DP) are only slightly sensitive compared with RD for the daily rainfall option.

Peak rate of runoff is sensitive to daily rainfall option parameters: DACRE--drainage area; CHS--hydraulic slope (SL in Eq. (28.10)); and WLW--length width ratio (LWR in Eq. (28.11)). Sensitive parameters in the infiltration option for peak runoff rate include: RMN--Manning's "n" (resistance to flow) in the overland flow plane (a_c in Eq. (28.40)); SLOPE--effective hydrologic slope; and XLP--effective hydrologic slope length.

28.2.7. Model Validation

The CREAMS model hydrology component was validated with observed runoff data from several locations representing a wide range of climatic and soil regions (Smith and Williams, 1980). Comparisons of monthly and annual runoff volumes were made for each location. Very limited data are available for validation of the percolation component. Lysimeter data from Coshocton, Ohio, and interceptor drain tile effluent from an instrumented watershed at Tifton, Georgia were compared with monthly and annual simulated percolation volumes from CREAMS (Smith and Williams, 1980).

Knisel et al. (1985) used observed runoff volumes from research watersheds in Iowa, Michigan, Montana, Ohio, and South Dakota to validate the frozen soil modification of CREAMS. Runoff volumes during the winter and spring months of snowmelt were compared with model simulations.

McKenna and Nutter (1982) used runoff data from wooded watersheds in North Carolina, Ohio, and Oklahoma to validate the CREAMS hydrology component for application on forested areas. Two major modifications were needed to adequately represent forests: rainfall interception by forest canopy and subsequent evaporation, and potential and actual evaporation. Evaporative flux under a mature forest is less than that for agricultural crops due to the lack of wind movement. The "as published" model version was found to be satisfactory for pine seed orchards with a grass understory. The grass cover dominated soil evaporation and plant transpiration, and rainfall intercepted by the tree canopy was not a significant component.

Other CREAMS model validations have been performed by numerous model users worldwide, but most studies have been part of more comprehensive projects related to nonpoint pollution by sediment and chemicals.

28.2.8. Model Applications

There have been many applications of the CREAMS model to compare management alternatives for nonpoint source pollution assessment. It is impossible to know or to recount all of the applications. Those mentioned here are cited merely to indicate some of the varied applications. In most applications, emphasis was not on hydrology alone.

Two publications provide a compilation of several model applications. Svetlosanov and Knisel (1982) describe several applications in various European countries. An example case study in the U.S. was included also. A symposium was held in 1989 (Beasley et al., 1991) that brought together several papers describing applications of CREAMS and GLEAMS.

Nicks et al. (1984) used the CREAMS model to assess alternative management practices to reduce sediment-associated pesticide transport into a lake. Representative topographic regimes were selected in the critical area of the Louisiana water quality project, and management alternatives were assessed with CREAMS.

Hakonson et al. (1982) used CREAMS to evaluate landfill cover characteristics on low level radioactive waste disposal sites. The utility of CREAMS was demonstrated, and Lane (1984) developed a user guide to aide model users in similar applications.

Crowder and Young (1985) made many CREAMS model runs to assess the costs of achieving water quality improvements using BMP's in a Rural Clean Water Project in Pennsylvania. Most of the BMP's were associated with degrees of tillage. Young et al. (1985) used CREAMS to analyze nutrient losses from heavy fertilization with dairy manure in the highly intense dairying area of Lancaster County, Pennsylvania. The applications included nitrate leaching with percolate below the root zone as well as surface runoff and sediment-associated losses of nitrogen and phosphorus.

Dumper et al. (1986) used the evapotranspiration component of CREAMS hydrology to assess the potential survival of windbreaks in the midwest. Farmstead protection by windbreaks is dependent upon survival of a mixture of trees and/or tall grasses in early stages of development. Seasonal precipitation and soil water availability were the principal concerns.

Gburek et al. (1991) demonstrated the utility of CREAMS percolation simulation is estimating recharge to groundwater. Adjustment of soil porosity and water retention characteristics for high rock content was necessary to attain the estimated annual groundwater recharge volumes.

Simulation of the hydrologic regime in the high water table conditions of peninsular Florida resulted in modification and the development of CREAMS-WT (Heatwole et al., 1987). When the fluctuating water table rises to near the ground surface during the rainy season, runoff volumes may be high due to relatively high soil water content.

Leonard et al. (1989) compared hydrologic response of Coastal Plain soils to verify groupings based upon soil taxonomy. Soil characteristics from published pedon data provided the input for the comparative simulation results.

Thomas et al. (1991) used CREAMS model simulations to evaluate soil protection by an alternate crop in the Coastal Plain. Winter rape was found to give adequate protection from soil erosion in the winter while providing a cash oilseed crop.

Even though a frozen soil representation was implemented in CREAMS (Knisel et al., 1987), Kallio et al. (1991) found that modifications of the hydrology component to adequately represent snowmelt conditions in Finland. The temperature regime from fitting mean monthly temperature was found inadequate for estimating precipitation state (rainfall or snow) during the winter season. These changes were implemented along with a modified snowmelt model to improve estimates of daily snowmelt and associated runoff.

These are only a few of the many known CREAMS model applications, both domestic and foreign. Model shortcomings have been identified in a number of instances, and these have been considered in the development of the GLEAMS model. Discussion of these improvements will be given in the following discussion of the GLEAMS model.

28.3. GLEAMS

In the mid-1980's, several pesticides had been found in aquifer systems in Georgia, Florida, New York, Wisconsin, and Nebraska (Cohen, 1986). Laboratory instruments and methodology had developed to the extent that very low concentrations of pesticides could be detected, and there was growing concern that models were needed to assess pesticide fate. The CREAMS model was primarily a surface response, edge-of-field loading model with a percolation component in hydrology to consider the total water balance in the root zone. Major modifications of CREAMS were needed to consider the vertical flux of pesticides.

Modifications of CREAMS began in 1984, and the first results were reported in 1985 (Leonard et al., 1985). The revisions were quite extensive in the hydrology component, and the pesticide component was completely modified. The revised model was named GLEAMS, and acronym for Groundwater Loading Effects of Agricultural Management Systems (Leonzard et al., 1987).

It was anticipated that the GLEAMS model would be linked with an unsaturated (vadose) zone model and possibly a saturated zone (aquifer) model. Application of the infiltration option in CREAMS with breakpoint rainfall was quite time consuming, and it was thought that a daily time step was more feasible for GLEAMS and its linkages. Therefore, the infiltration option was deleted from the hydrology component when initial work began on GLEAMS.

Major modifications in the hydrology component of GLEAMS included:

1) change of computational soil layering with input of water retention and transmission characteristics by soil horizon;
2) capability of changing root depth from crop-to-crop within a rotation;
3) include an option to input mean daily temperature;
4) change in irrigation routine to consider
   a) runoff from excess irrigation,
   b) pre-plant irrigation, and
   c) model-applied irrigation based upon soil water content in the active root zone;
5) idealized leaf area data for 78 crops are included in data statements within the model;
6) include an option for simulating hardwood and pine forests; and
7) added the Penman-Monteith method as an option for estimating potential evapotranspiration.

Each of these changes are discussed below with association to the CREAMS hydrology option.

28.3.1. Computational Soil Layers

In CREAMS, computational soil layers were represented as fixed fractions of the total effective rooting depth. This oftentimes resulted in a computational layer including characteristics of two diverse characteristics. Computational layers in GLEAMS as delineated totally within a soil genetic horizon. Also, since most root activity is in the plow layer, and organic matter (organic carbon) is the adsorbing medium for pesticides, computational layers in the plow horizon are thinner than those in the lower soil horizons. To be consistent with the chemistry components of GLEAMS, the surface layer is fixed at 1 cm thickness regardless of total effective root depth.

GLEAMS model users can specify the number of genetic soil horizons for which characteristics are to be entered with a maximum of five. This is generally more than adequate, but extreme situations may require combining of similar horizons.

The surface 1 cm thickness is subtracted from the top horizon thickness, and the remainder is divided into layers of equal thickness not to exceed 10 cm. For example, if the plow layer is 25 cm thick, layer 1 thickness (1 cm) is subtracted to leave 24 cm. Three layers of 8 cm thickness are needed to satisfy the criteria of not exceeding 10 cm per layer.

The thicknesses of computational layers in the remaining horizons are a maximum of 15 cm, except that a maximum of 12 computational layers is allowed. The same criterion is used as above, that is, layers of equal thickness within a horizon. In the lower horizon(s), if the total number of layers exceeds 12, the layers in the lower horizon are allowed to exceed 15 with all layers within a horizon being of equal thickness. Soil layering is performed in the model with the user only specifying the number of genetic horizons and the depth from the soil surface to the bottom of each horizon.

Soil characteristics are user supplied for each horizon. These include: porosity (POR); volumetric water content at field capacity (FC) and wilting point (WP) represented by 33 kPa and 1500 kPa, respectively; and saturated conductivity (SATK). These properties are needed for water balance computations. Other soil properties not needed in the hydrology component but are needed in the erosion and chemistry components are also user supplied as input in hydrology. They include organic matter content (OM), clay content (CLAY), and silt content (SILT).

Characteristics of each computational soil layer within a horizon are assigned the same values within the model. This is entirely appropriate since the horizon characterization is assumed to be uniform.

Saturated conductivity is user supplied for each soil horizon. In addition, the saturated conductivity is also specified for the horizon immediately below the root depth. This horizon oftentimes has characteristics that delineate the effective root zone, for example, a clay pan, a high-organic spodic layer, or a clay layer. Such restricting layers generally have reduced water transmission characteristics that result in seasonal perched water tables within the effective rooting depth. Some clay-pan and plow-pan soils have very shallow root zones, 20-30 cm, that result in high runoff production. Management practices that remove the pan conditions are effective in reducing runoff and increase evapotranspiration. Therefore, it is important to include these soil characteristics in model representation.

28.3.2. Change of Root Depth within a Crop Rotation

Root growth of an annual crop is represented in CREAMS and GLEAMS as a function of the ratio of daily accumulated leaf area index to the total cumulative leaf area for the growing season. This results in root growth to the maximum effective root depth at the end of the growing season. Two additional conditions are provided in GLEAMS.

Perennial crops such as most grasses, alfalfa, clover, fruit trees, and forests maintain their total root depth from year-to-year and do not re-grow roots from the soil surface each year. Admittedly, those crops develop a new mass of feeder roots each year, but they are generated throughout the total root zone. In GLEAMS, perennial crops are designated, and effective root depth is maintained throughout the simulation period and not re-grown each year.

Perennial crops grown in cold climates, such as meadow grasses in rotation, pasture grasses, and alfalfa, have non-transpiring leaf area at the first killing frost in the fall and resume growth again in the early spring. The leaf area, or canopy, is effective as an evaporation suppressant, but does not transpire. Hardwood forests that drop their leaves in the fall and grow new leaves in the spring have some equivalent stem and limb basal area effective as evaporation suppressant but do not transpire. These growth characteristics are represented in GLEAMS by specifying the "end growth" and "beginning growth" dates. The leaf area values on those dates are used to estimate the equivalent evaporation control without resulting in transpiration.

Some annual crops grown in rotation have significantly different root depths, for example, most vegetable crops grown in rotation with field crops. Also, small grain crops grown as a winter cover crop or green manure crop plowed under before maturity do not root to their maximum depth. Silage crops such as corn or small grain that are chopped well before maturity have different root growth than their "full season" counterparts. Both of these conditions can be represented in GLEAMS.

Idealized leaf area values by fraction of the growing season are included as data statements by crop in the model. Model users can input idealized (non-stressed) leaf area data for crops other than the 78 included in the model data statements. The leaf area data are given for the entire growth period from planting to harvest. Planting and harvest dates are specified, and the model translates the fraction of growing season to Julian date between the planting and harvest dates. If a silage crop, hay crop, or a winter cover crop are harvested or plowed down before maturity, a "truncation" date is specified to terminate growth. In the case of silage or a winter cover crop, root growth does not progress to the total effective rooting depth since the cumulative LAI is not achieved at truncation. The water accounting procedure is the same, but soil water uptake by the crop is restricted to the depth of actual root growth simulated by the model.

In the case of shallow-rooting vegetable crops grown in rotation with field crops, for example radishes, celery, etc., a "current crop rooting depth" (CCRD) is specified that is different from the maximum effective rooting depth (RD). In those seasons of vegetable crop production, root growth is simulated only to the CCRD, and water uptake is not simulated below that depth. The water accounting procedure is the same as that for a truncated crop except the root depth at maturity is specified as CCRD.

The root growth simulation, truncation, and shallow-rooted crop representations are more sensitive in chemistry simulation results than in the hydrology component. They do have an effect in water uptake by crops, and as such have some effect on the water balance computations.

28.3.3. Climatic Data Input

Climatic data required for the CREAMS hydrology component, in addition to daily or breakpoint rainfall, included mean monthly temperature and radiation data. Monthly values are fitted with a Fourier series (Kothandaraman and Evans, 1972) to interpolate daily values needed in the model. Mean daily temperatures are used to determine precipitation state. That is, precipitation that occurs when daily temperature is below freezing is considered to be snow, and is accumulated as snow pack to be melted when temperatures rise above freezing. Fitted temperature data are smoothed, and short periods above freezing during the winter cannot be distinguished. Even in cold regions, prolonged mid-winter warm periods occur and the entire snowpack may melt.

Kallio et al. (1991) found that it was necessary to input mean daily temperature to adequately represent the precipitation state and enable better estimates of snowmelt. These affect runoff and percolation occurrence during the winter and early spring months. Temperature is not sensitive in simulation of water balance components during the growing season. An option was included in GLEAMS for user input of mean daily temperature data. Knisel et al. (1991) showed that use of mean daily temperature did not improve the simulation of runoff and percolation volumes in a warm climatic region.

Most chemical transformations in the soil are temperature dependent, that is rate constants are functions of temperature. Since the principle objective of GLEAMS is to simulate effects of management practices on water, sediment, and chemical losses for field size areas, the soil temperature algorithm in the EPIC model (Sharpley and Williams, 1990) was incorporated into the plant nutrient component of GLEAMS. Daily values of maximum and minimum temperature are needed to simulate soil temperature. Therefore, mean monthly maximum and minimum temperature data are required as input in the hydrology component of GLEAMS. The method of Kothandaraman and Evans (1972) is used to interpolate daily max and min temperature.

GLEAMS model users in some irrigated regions report that potential evapotranspiration estimated by the Priestly-Taylor method (Priestly and Taylor, 1972) is lower than observed lysimeter data. Jensen et al. (1990) state that the Priestly-Taylor method assumes high soil moisture conditions, and it should not be used in irrigated areas. However, if irrigation is practiced, then soil water content is maintained well above the non-irrigated condition.

Jensen et al. (1990) compared evapotranspiration estimates by 15 methods with observed lysimeter measurements at 11 locations, ranging from humid to arid, around the world. The methods included the data-intensive Penman method (Penman, 1948) and numerous simplifications with less intense data requirements.

A modification by Monteith (1965) was shown to give extremely good agreement with measured monthly data for all locations examined (Jensen et al., 1990). The method, referred to as Penman-Monteith, requires wind speed and vapor pressure deficit. Jensen et al. (1990) gave four alternative methods of estimating vapor pressure deficit of the Penman method (Penman, 1948). One alternative uses maximum and minimum daily relative humidity, and this gave very acceptable estimates of monthly evapotranspiration.

Climatic data input for GLEAMS include:

1) mean monthly maximum temperature;

2) mean monthly minimum temperature; and

3) mean monthly solar radiation.

Optional data include:

1) mean daily temperature;

2) mean monthly wind movement; and

3) mean monthly dew-point temperature.

The five mean monthly climatic variables (Richardson and Nicks, 1990) have been included in a data base for over 1,000 locations in the United States (Knisel et al., 1993). The data are included as a supplemental data base with import into the hydrology parameter editor as an option to the model user.

28.3.4. Penman-Monteith Option for Potential Evapotranspiration

Jensen et al. (1990) gave excellent treatment of the concepts and methods for estimating evapotranspiration and irrigation requirements. The Priestly-Taylor method (1972) was recognized as an effective method for estimating potential evapotranspiration in humid areas, and also under a nearly complete canopy in an arid or semiarid environment. However, with little or no crop canopy, vapor diffusion from crop leaves and bare soil require more intense treatment for adequate estimation. Thus, methods incorporating wind movement and vapor pressure deficit are generally more effective estimators of potential ET for these conditions.

Methods of calculating ET for the short term to determine timing and amount of irrigation admittedly require hourly measurements of estimators (Jensen et al., 1990). This is not feasible for extended periods of simulation such as that desired for GLEAMS applications. Simplifications and averaging are required for computational feasibility. Since methods cannot be changed periodically during a long-term simulation, the most appropriate method must be selected by the model user to give the best representation for the duration of the simulation.

The Penman-Monteith method (Monteith, 1965) of estimating ET was shown to accurately represent ET measured in lysimeters at 11 locations around the world (Jensen et al., 1990). Therefore, the Penman-Monteith method is included in GLEAMS as an alternative option to the Priestly-Taylor method.

Potential ET is described by Jensen et al. (1990) as the evapotranspiration for a "well watered reference crop such as grass or alfalfa" and is designated by E_{tr}. Actual ET is estimated by a crop coefficient which is a ratio of the reference crop ET. In GLEAMS, the method developed by Ritchie (1972) for an incomplete cover is used to estimate the actual ET for a specific crop.

The Penman-Monteith (1965) expression for daily ET, mm/day, is

$$\lambda E_{tr} = \frac{\Delta}{\Delta + \gamma^*}\left(R_n - G_h\right) + \frac{\gamma}{\Delta + \gamma^*} K_1 \frac{0.622\,\lambda\rho}{P}\frac{1}{r_a}\left(e_z{}^o - e_z\right) \qquad (28.66)$$

where Δ is the slope of the saturation vapor pressure-temperature curve, kPa/°C, γ is the psychrometric constant, kPa/°C, γ^* is the psychrometric constant modified by the ratio of canopy resistance to atmospheric resistance, kPa/°C, R_n is net radiation, MJ/m^2/day, G is the heat flux density to the ground, MJ/m^2/day, λ is latent heat of vaporization, MJ/kg, ρ is density of air, kg/m^3, P is atmospheric pressure, kPa, r_a is diffusion resistance of the air layer (aerodynamic resistance), sec/m, $e_z{}^o$ is saturation vapor pressure of air at height z, kPa, e_z is saturation vapor pressure of air, kPa, and K_1 is a dimension coefficient to assure correct units. Since E_{tr} is defined as potential ET (Jensen et al., 1990), it is equivalent to E_o defined by Eq. (28.49) for the Priestley-Taylor method (Priestley and Taylor, 1972).

The slope of the saturation vapor pressure-temperature curve, Δ in Eq. (28.66) is defined as

$$\Delta = 0.20\left(0.00738\,T_c + 0.8072\right)^7 - 0.000116 \qquad (28.67)$$

where T_c is the mean daily air temperature in °C. The psychrometric constant, γ in Eq. (28.66) is defined as

$$\gamma = \frac{c_p\,P}{0.622\,\lambda} = \frac{1.013\,P}{0.622\,\lambda} = \frac{1.629\,P}{\lambda} \qquad (28.68)$$

where c_p is the specific heat of moist air at constant pressure and is equal to 1.013 kJ °C/kg. P in Eqs. (28.66) and (28.68) is

$$P = P_o\left(\frac{T_o - \alpha\,ELEV}{T_o}\right)^{g/\alpha R} = 101.3\left(\frac{2.88 - 0.0065\,ELEV}{288}\right)^{5.257} \qquad (28.69)$$

where P_o is atmospheric pressure at mean sea level, T_o is absolute temperature of a standard atmosphere, a adiabatic lapse rate, and ELEV is the mean sea level elevation, m, of the location of interest. The latent heat of vaporization, λ, in Eq. (28.68) is estimated from the mean daily temperature as

$$\lambda = 2.501 - 0.002361\, T_c \tag{28.70}$$

The density of air, ρ, in Eq. (28.66) is a function of elevation (Jensen et al., 1990), expressed as

$$\rho = 1.23 - 0.112\ ELEV\ x\ 10^{-3} \tag{28.71}$$

The procedure for estimating net solar radiation from total incoming radiation given in Eq. (28.51) for the Priestly-Taylor method is not satisfactory for R_n in Eq. (28.66). Jensen et al. (1990) cited the works of several researchers for simplified methods of estimating R_n. It is not the purpose of this paper to review the literature, individual citations will not be given. It will suffice to say that Jensen et al. (1990) made an exhaustive survey of all the methods and literature, and provided accurate representation of the different authors.

Net solar radiation, R_n, MJ/m^2, is estimated from incoming solar radiation, RAD, MJ/m^2, as

$$R_n = ((1.0 - ALB)\ R_s - R_b \tag{28.72}$$

where ALB is albedo (used as a constant in GLEAMS, ALB = 0.23), and R_b is the net outgoing long-wave radiation, MJ/m^2. R_b can be estimated as

$$R_b = \left[0.9\,\frac{R_s}{R_{so}} - 0.1\right] R_{bo} \tag{28.73}$$

where R_{so} is radiation, MJ/m^2, that is expected on a day without clouds, and R_{bo} is the net outgoing long-wave radiation, MJ/m^2, on a clear day. R_{so} in Eq. (28.73) can be estimated as a function of latitude and elevation as

$$R_{so} = Ap + BP\cos\left[\left(2\pi\,\frac{d}{365}\right) - CP\right] \tag{28.74}$$

where AP, BP, and CP are coefficients, and d is the Julian day of the year. Coefficient AP is estimated as

$$AP = 31.54 - 0.73\ LAT + 0.00078\ ELEV \tag{28.75}$$

and

$$z_{ov} = 0.1\, z_{om} \tag{28.89}$$

Since both terms appear in the denominator of the logarithmic functions of Eq. (28.85), the Penman-Monteith method of estimating ET was not conceptualized to simulate potential soil evaporation alone. Sharpley and Williams (1990) reported a slight adjustment by always adding 0.01 cm to h_c, resulting in a negligible value preventing division by zero.

Wind movement in Eq. (28.85) is needed for a height 2 m above the crop height. Thus, during the growing season, daily wind movement is needed for varying heights, and can be estimated from the relation

$$U_z = U_{rh} \left(\frac{z}{rh} \right)^{0.2} \tag{28.90}$$

where z is the desired height, cm (h_c + 200), and rh is the reference height, cm (height of the anemometer cup).

Jensen et al. (1990) gave five methods for estimating vapor pressure deficit used in Eq. (28.66). In the absence of hourly data, the method that appeared to give the best results entails averaging vapor pressure at the daily maximum and minimum temperatures and subtracting the vapor pressure at the dew point temperature as

$$\left(e_z^{\,o} - e_z \right) = \frac{e^o (T_x) + e^o (T_n)}{2} - e^o (T_d) \tag{28.91}$$

where T_x, T_n, and T_d are daily maximum, minimum, and dew point temperatures, °C, respectively. Vapor pressure at a specific Centigrade temperature, T_c, is estimated as

$$e^o (T_c) = \exp \left[\frac{16.78\, T_c - 116.9}{T_c + 237.3} \right] \tag{28.92}$$

The Penman-Monteith method of estimating potential ET is recommended for semi-arid and arid areas if wind movement and dew-point temperature data are available, or can be used from the tabulated locations in GLEAMS. The Priestly-Taylor method should be used for forested areas of semi-arid and arid regions because of the relatively low evaporative flux under forest canopy as will be explained below.

28.3.5. Irrigation

The "automatic" irrigation option in CREAMS and early versions of GLEAMS assumed that irrigation water was applied at a rate equal to or less than the minimum infiltration rate of the soil. Any irrigation water in excess of soil water storage was assumed to percolate through the root zone. Also, irrigation demand was calculated as the difference between threshold level and upper level for the entire root zone irrespective of the depth to which crop roots had grown. These assumptions were changed in the present version of GLEAMS

When automatic irrigation is specified in the present version, soil water content is summed in the active root depth and compared with the threshold level. Then the amount of irrigation is calculated as the difference between the threshold level and the user-specified upper limit for that active root depth. This procedure may result in very small amounts of irrigation required to supply the deficit when root depth is shallow. It is assumed that a minimum of 0.63 cm of water can be applied in a single irrigation.

Irrigation water applied on the soil surface infiltrates just as rainfall, and the surface layer is filled to field capacity first and calculations proceed with depth. For those soils with low saturated conductivity, runoff may occur from irrigation. In the present version of GLEAMS, the runoff procedure is invoked with irrigation to determine if and how much runoff occurs. This is a more realistic concept, especially if the upper limit of irrigation is at or above field capacity.

In semi-arid and arid climatic regions, salt accumulations may occur within the root zone. It is a common practice in those regions to apply irrigation water in excess of the field capacity for the maximum root zone to leach salts out of the root zone before a crop is planted, i. e. a pre-plant irrigation. The pre-plant irrigation is applied at least one month before planting in order for the seed zone to warm adequately for seed germination without rooting. This is especially true where irrigation water is obtained from surface (lake) water supplies.

A pre-plant irrigation is user-supplied in GLEAMS, both the amount and the date of irrigation. As in the case of automatic model-applied irrigation, the pre-plant irrigation is subjected to the potential runoff routine of the model.

28.3.6. Forest Option

Potential Evaporation.

CREAMS and the early versions of GLEAMS hydrology were developed for agricultural crops. However, forests represent a significant land use that has potential impact on water quality, both beneficial and harmful impact. An option was added to the GLEAMS model to simulate the hydrologic response of forests and forest management practices (Nutter et al., 1993). The algorithms are based primarily on the work of McKenna and Nutter (1983) to evaluate the hydrology component of CREAMS.

Potential soil evaporation is determined from potential evaporation for agricultural crops as a function of leaf area (Eq. (28.52)), exp(-0.4 LAI), but McKenna and Nutter (1983) found a better relation for forests to be

$$E_{so} = E_o \, e^{(-0.6\,LAI)} \tag{28.93}$$

The reduced soil evaporation results from a reduced evaporative flux due to lower energy input under the forest canopy with a dense understory. Likewise, if potential soil evaporation is reduced, then potential plant evaporation (transpiration) should be increased. Agricultural crops are assumed to transpire at potential when LAI is greater than 3 m^2/m^2. McKenna and Nutter (1983) found another modification is needed for forested areas, and potential transpiration is given by

$$E_p = E_o\left[1 - e^{(-0.6LAI)}\right] \tag{28.94}$$

Eqs. (28.93) and (28.94) provide consistence between the components of potential evapotranspiration.

In the two-stage soil evaporation process for agricultural crops, the upper limit of stage one uses a soil evaporation parameter dependent upon soil texture as given by Eq. (28.53). Ritchie (1972) suggests values of the soil evaporation parameter ranging from 3.3 to 4.5 $mm/d^{0.5}$. Forest hydrologists have shown that the mor humus on forest floors reduce soil evaporation, and McKenna and Nutter (1983) suggested a value of $a_s = 3.0$, which results in an upper limit of stage-one evaporation U = 0. This means that for all practical purposes, stage-one evaporation does not exist, and only stage-two evaporation occurs in a forest environment.

Crop canopy and surface residue for agricultural crops provide a range of evaporation suppressant from 0.5 to 1.0 (1.0 being no suppressant). McKenna and Nutter (1983) recommended a relatively low value of 0.25 for forest applications. This represents an effective control by the heavy forest litter to break the capillary rise of water from the mineral soil.

As discussed in the next section, trunks and limbs of hardwood trees have a capacity to intercept rainfall during the dormant season. McKenna and Nutter (1983) represented the trunks and limbs by LAI in dormant-season (Nutter et al., 1993). This is not a transpiring leaf area, and is so distinguished by a date of beginning growth in the spring (BEGGRO) and a date of ending growth in the fall (ENDGRO) for the perennial trees. From ENDGRO to BEGGRO, the LAI serves other purposes but transpiration is not simulated during that time. This concept is used for perennial agricultural crops such as alfalfa, grasses, and sugarcane, also. Also like perennial agricultural crops, root growth for trees does not begin at zero each year, but remains constant the effective root depth throughout model simulation.

Canopy Interception

Considerable amounts of rainfall, particularly at low intensities, can be intercepted by the forest canopy. Even trunks and limbs of deciduous trees may intercept significant amounts of low-intensity winter-type rainfall after leaf fall. The counterpart of rainfall interception is throughfall, that portion of rainfall reaching the ground to become potential runoff and/or infiltration (Nutter et al., 1993).

The type of leaf is significant in rainfall interception. McKenna and Nutter (1983) designated three groups for throughfall determinations as follows: group 1--pine and mixed hardwood-pine; group 2--short-leafed conifers (spruce/fir/hemlock/cedar); and group 3--mixed hardwoods. The relations were developed from published data (Nutter et al., 1993), and incorporated into GLEAMS. Throughfall, TF, cm, is given as

$$\begin{aligned} &Group\,1: && TF = (0.35\,P) - 0.015 \\ &Group\,2: && TF = (0.31\,P) - 0.020 \\ &Group\,3: && TF = (0.37\,P) - 0.015 && \textit{for growing season} \\ & && TF = (0.38\,P) - 0.008 && \textit{for dormant season} \end{aligned} \tag{28.95}$$

here P is precipitation in cm. Although small amounts of precipitation can be totally intercepted, some interception will become throughfall, depending upon LAI, and add to soil water. An adjustment is made for interception as follows:

$$\begin{aligned} &I = (P - TF)\,r && \textit{for } TF > 0 \\ &I = P\,r && \textit{for } TF \le 0 \end{aligned} \tag{28.96}$$

where I is interception, cm, and r is a ratio defined as

$$\begin{aligned} &r = LAI/3 && \textit{for } LAI \le 3 \\ &r = 1.0 && \textit{for } LAI > 3 \end{aligned} \tag{28.97}$$

McKenna and Nutter (1983) obtained good comparisons of interception loss from Eqs. (28.95-28.97) compared with published values for a southern hardwood forest.

Intercepted rainfall on the forest canopy is available for evaporation. Since it is a "free water surface", the evaporative flux acts on the canopy storage first. There is some potential evaporative demand, and the interception storage is consumed by that demand before soil evaporation is calculated in GLEAMS. After the canopy storage is depleted in the model, soil evaporation is considered in a normal manner.

Frozen Soil

The frozen soil component in CREAMS and GLEAMS (Knisel et al., 1985) assumes that saturated conductivity is markedly reduced in mineral soils. However, published works reviewed by Nutter et al. (1993) indicated that litter on the forest floor prevents formation of solid or "concrete" frost that restricts conductivity. Also, insulation effects of the forest litter impedes fast freezing, and thus delays or prevents solid frost formation, and only granular-type frost that occurs may increase pore size and resultant conductivity. Therefore, frozen soil is not considered in the forest option of GLEAMS.

28.3.7. Sensitive Model Parameters

Sensitive parameters in the GLEAMS model include those in CREAMS plus others. Effective root depth (RD) and current crop root depth (CCRD) are sensitive for runoff and percolation calculation. Soil porosity (POR), field capacity (FC), and SCS curve number (CN2) are sensitive, also. If GLEAMS over-estimates runoff and under-estimates percolation, FC can be decreased and the result will be less runoff and more percolation.

Effective saturated conductivity of the horizon immediately below the root zone, RC, and saturated conductivity by soil horizon, SATK, are sensitive only for high water table conditions. RC and SATK are not used to estimate percolation volume. However, when values are very low, infiltrated water in excess of field capacity may not move freely through the soil resulting in higher ET and runoff and lower percolation. RC and SATK are not sensitive for well drained soils in hydrologic groups A and B.

Mean daily temperature may be sensitive in cold regions that receive large amounts of snow resulting in large volumes of snowmelt in the spring when the soil may be frozen. Mean daily temperature input into GLEAMS is not sensitive in warm regions, but it is significant to determine state of precipitation (rain or snow) in cold regions.

The most sensitive parameters in water balance calculations are: RD, POR, FC, and CN2. Less sensitive parameters include: CCRD, RC, SATK, and mean daily temperature in cold regions.

28.3.8. Model Validation

Little additional validation has been done for the GLEAMS model hydrology component beyond that done for CREAMS. Some validations have been made by graduate students at institutions where some data may have been available from selected studies.

Knisel et al. (1991) validated the water balance components with data from a small field size area in the Georgia Coastal Plain. Reyes et al. (1992) modified the GLEAMS hydrology component to allow interaction between a shallow water table and soil water in a field study in Louisiana. The modifications were used to formulate the GLEAMS-WT model which was validated with 7 years of field data.

LIST OF SYMBOLS

A	Drainage area, ha
AP	Coefficient
BP	Coefficient
BR15	Volumetric soil water content at wilting point (1500 kPa), cm/cm
C	Chezy roughness coefficient
C_s	Infiltration storage coefficient
CCRD	Current crop root depth, cm
CLAY	Clay content of the soil, percent
CN_I, CN_{II}	SCS curve number for antecedent moisture condition I, and II, respectively
CP	Coefficient, Julian date of the longest day of the year
D	Depth of soil from the surface to the bottom of a soil layer, cm
D_s	Depth of the soil surface infiltration control layer, cm
D_p	Depth of the root zone below surface control layer, cm
DFS	Number of days in the spring with frozen soil, days
E_o	Potential evaporation, cm
E_p	Plant evaporation, cm
E_{pl}	Plant evaporation for limited soil water, cm
E_s	Soil evaporation, cm
E_{so}	Potential soil evaporation, cm
E_{tr}	Evapotranspiration for reference crop, mm/day
EL	Elevation, m

ELEV Elevation above mean sea level, m

ET Evapotranspiration, cm

F Infiltration depth, cm

FC Soil water content at field capacity (33 kPa), cm

F_p Infiltration depth to the time of ponding, cm

FPL Length of flow path, m

FRD Fraction of effective root depth

FUL Fraction of upper limit of storage filled at field capacity

G_h Heat flux density to the ground, $MJ/m^2/day$

H_c Effective capillary tension, cm

H_o Net solar radiation, gm-cal/cm^2/day

I Rainfall interception, cm

IR Irrigation depth, cm

K_s Effective saturated conductivity, cm/hr

K_1 Dimension coefficient

L Depth of wetting, cm

LAI Leaf area index, m^2/m^2

LAT Latitude, degrees

LWR Field length-to-width ratio

M Snowmelt, cm

N Number of days in the growing season, days

O Outflow (percolation), cm

O_s Outflow from soil surface infiltration control layer, cm

OM Organic matter content of the soil, percent

P	Precipitation, cm
P	Atmospheric pressure, kPa
P_o	Atmospheric pressure at mean sea level
PAW	Plant available water, cm
POR	Soil porosity, cm^3/cm^3
POROS	Soil porosity, cm^3/cm^3
Q	Runoff volume, cm
R	Total solar radiation, gm-cal/cm^2/day
R_b	Net out-going long-wave radiation, MJ/m^2/day
R_{bo}	Out-going (back) radiation on cloudless day, MJ/^{2}
R_n	Net radiation, MJ/m^2/day
R_s	Observed daily radiation, MJ/m^2
R_D	December mean daily solar radiation, gm-cal/cm^2/day
RD	Effective root depth, cm
RDN	Active root depth on day n, cm
S	Available soil water storage (retention), cm
S_{mx}	Maximum available soil water storage, cm
S_p	Volumetric saturation in the root zone, cm/cm
S_s	Volumetric saturation in the surface infiltration control layer, cm/cm
SATK	Saturated hydraulic conductivity, cm/hr
SILT	Silt content of the soil, percent
SL	Average hydraulic slope of the field, m/m
SM	Soil moisture storage, cm

ST	Volumetric saturation, cm/cm
T_c	Temperature, °C
T_d	Dew point temperature, °C
T_k	Temperature, °K
T_n	Minimum temperature, °C
T_o	Absolute temperature, °K
T_x	Maximum temperature, °C
TF	Throughfall of rainfall, cm
TO	Temperature above 0°C
U	Upper limit of stage-one soil evaporation, mm
U_{rh}	Wind speed at reference height, m/sec
U_z	Wind speed at height z, m/sec
UL	Upper limit of plant-available and drainable soil water storage, cm
WF	Soil water weighting factor
WP	Volumetric soil water content at wilting point, cm/cm
Z_{om}	Momentum roughness length, cm
Z_{ov}	Roughness length for heat and water vapor, cm
Z_p	Height of psychrometer and thermometer, cm
Z_w	Height of anemometer, cm
c_p	Specific heat of moist air at constant pressure, kJ °C/kg
d	Crop height parameter, cm
e	Exponent, 2.713
e_z	Saturation vapor pressure of air, kPa

e_z^o	Saturation vapor pressure of air at height z, kPa
exp	Exponent, 2.713
d	Calendar day
f	Infiltration rate, cm/hr
h	Depth of flow (infiltration option), cm
h	Subscript denoting highest point in the field
h_c	Crop height, cm
i	Soil layer index; subscript denoting initial condition
j	Subscript denoting segment number
ln	Natural logarithm, base e
m	Exponent for uniform flow (1.5 for Chezy law)
n	Day index number
o	Subscript denoting field outlet
q_i	Runoff volume during the i-th interval, cm
q_p	Peak rate of runoff, cm/hr
r	Ratio of LAI for adjusting interception
r_a	Diffusion resistance of the air layer, sec/m
r_c	Saturated conductivity, cm/hr
r_{cc}	Crop canopy resistance, sec/m
r_i	Rainfall rate for the i-th interval, cm/hr
r_p	Rainfall rate at the time of ponding, cm/hr
rh	Reference height, cm
t	Time, (infiltration) hr, (time step) day
t_e	Elapsed time, min

tt	Travel time through a soil computational layer, hr
u	Water use rate, cm
u_o	Water use rate at the soil surface
v	Velocity (celerity) of the characteristic, m/sec
wu	Water uptake, cm
x	Distance, m
z	Height above ground, m
α	Uniform flow coefficient
α	Adiabatic lapse rate
α_c	Resistance to flow in overland flow plane
α_s	Soil evaporation constant, $mm/d^{0.5}$
β	Albedo
γ	Psychrometric constant, kPa/°C
γ^*	Psychrometric constant modified by the ratio of canopy resistance to atmospheric resistance, kPa/°C
Δ	Slope of saturation vapor pressure curve at the mean air temperature
Δ	Increment
ε	Emmittance
θ	Volumetric soil water content, cm/cm
λ	Latent heat of vaporization, MJ/kg
ρ	Density of air, kg/cm^3
ϕ	Soil porosity, cm^3/cm^3
σ	Soil water storage coefficient
σ'	Constant, $MJ/m^2/day/°K^4$

REFERENCES

Arnold, J. G., J. R. Williams, A. D. Nicks, and N. B. Sammons. 1990. SWRRB: A basin scale simulation model for soil and water resources management. Texas A & M University Press, College Station. 241 pp.

Beasley, D. B., E. J. Monke, and L. F. Huggins. 1977. 'ANSWERS: A model for watershed planning'. Purdue Agricultural Experiment Station Journal Paper No. 7038. 34 pp.

Beasley, D. B., W. G. Knisel, and A. P. Rice (Eds.) 1991. Proceedings of the CREAMS- GLEAMS Symposium. Publication No. 4, Agricultural Engineering Department, University of Georgia, Athens. 220 pp.

Cohen, S. Z., C. Eiden, and M. N. Lorber. 1986. 'Monitoring ground water for pesticides in the U.S.A'. Proceedings of the American Chemical Society Symposium, Washington, D. C. Symposium preprint. 15 pp.

Crowder, B. M., and C. E. Young. 1985. 'Modeling agricultural non-point source pollution for economic evaluation of the Conestoga headwaters RCWP project'. U.S. Department of Agriculture, Economic Research Service, Natural Resources Economics Division, Washington, D.C. ERS Staff Report No. AGES850614. 70 pp.

Dumper, T. A., W. G. Knisel, D. L. Hintz, and D. Plachy. 1986. 'Simulating the evapotranspiration of windbreak species with the CREAMS model'. (Extended Abstract.) International Symposium on Windbreak Technology, Lincoln, Nebraska, June 23-27.

Gburek, W. J., R. R. Schnabel, and S. T. Potter. 1991. 'CREAMS as a driver for an aquifer model'. In: D. B. Beasley, W. G. Knisel, and A. P. Rice (Eds.) Proceedings, CREAMS-GLEAMS Symposium. Sept. 27-29, 1989, Athens, Georgia, Agricultural Engineering Department, University of Georgia-Coastal Plain Experiment Station, Publication No. 4, pp. 78-96.

Green, W. A., and G. A. Ampt. 1911. 'Studies on soil physics, I. The flow of air and water thru soils'. Journal of Agricultural Science, 4:1-24.

Hakonson, T. E., L. J. Lane, J. G. Steger, and G. L. DePoorter. 1982. 'Some interactive factors affecting trench cover integrity on low-level waste sites'. Symposium on Site Characterization and Monitoring, Arlington, Virginia, June 15-17, 1982. Nuclear Regulatory Commission Report NUREG/CP-0028 CONF-820674, Vol. 2.

Heatwole, C. D., K. L. Campbell, and A. B. Bottcher. 1987. 'Modified CREAMS hydrology model for Coastal Plain Flatwoods'. Transactions of the American Society of Agricultural Engineers, 30(4):1014-1022.

Jensen, M. E., R. D. Burman, and R. G. Allen (Eds.). 1990. Evapotranspiration and Irrigation Water Requirements. American Society of Civil Engineers, Manuals and Reports on Engineering Practice, No. 70. 332 pp.

Kallio, K., S. Rekolainen, M. Posch, and E. Turtola. 1991. 'Testing and modifying the CREAMS model for Finnish Conditions'. In: D. B. Beasley, W. G. Knisel, and A. P. Rice (Eds.) Proceedings, CREAMS/GLEAMS Symposium. Sept. 27-29, 1989, Athens, Georgia, Agricultural Engineering Department, University of Georgia-Coastal Plain Experiment Station, Publication No. 4, pp. 179-191.

Knisel, W. G. (Ed.) 1980. CREAMS: A Field-Scale Model for Chemicals, Runoff, and Erosion from Agricultural Management Systems. U. S. Department of Agriculture, Science and Education Administration, Conservation Research Report No. 26. 643 pp.

Knisel, W. G., R. A. Leonard, F. M. Davis, and A. D. Nicks. 1993. 'GLEAMS Version 2.10. Part III. User manual'. (Accepted for publication in USDA-ARS Conservation Research Report series.) 205 pp.

Knisel, W. G., R. A. Leonard, F. M. Davis, and J. M. Sheridan. 1991. 'Water balance components in the Georgia Coastal Plain: A GLEAMS model validation and simulation'. Journal of Soil and Water Conservation, 46(6):450-456.

Knisel, W. G., D. C. Moffitt, and T. A. Dumper. 1985. 'Representing seasonally frozen soil with the CREAMS model'. Transactions of the American Society of Agricultural Engineers, 28(5):1487-1493.

Kothandaraman, V., and R. L. Evans. 1972. 'Use of air-water relationships for predicting water temperature'. Illinois State Water Survey, Urbana. 14 pp.

Lane, L. J. 1984. 'Surface water management: A user's guide to calculate a water balance using the CREAMS model'. Los Alamos National Laboratory, Los Alamos, New Mexico. LA-10177-M Manual. 49 pp.

Lane, L. J., and V. A. Ferreira. 1980. 'Sensitivity analysis'. In: W. G. Knisel (Ed.)

CREAMS: A Field-Scale Model for Chemicals, Runoff, and Erosion from Agricultural Management Systems. U.S. Department of Agriculture, Science and Education Administration, Conservation Research Report No. 26, Vol. I, Chap. 6. pp. 113-158.

Leonard, R. A., W. G. Knisel, and D. A. Still. 1987. 'GLEAMS: Groundwater Loading Effects of Agricultural Management Systems'. Transactions of the American Society of Agricultural Engineers 30(5):1403-1418.

Leonard, R. A., H. F. Perkins, and W. G. Knisel. 1989. 'Relating agrichemical runoff and leaching to soil taxonomy: A GLEAMS model analysis'. Proceedings, Georgia Water Re-sources Conference, Athens, Georgia, May 15-16. pp. 158-160.

McKenna, R., and W. L. Nutter. 1983. 'Some modifications to CREAMS for forested applications'. Unpublished final report of a research project on "Runoff Simulation Models for Water Quantity and Quality from Forested Areas", School of Forest Resources, University of Georgia, Athens. 46 pp.

Mockus, V. 1949. 'Estimation of total and peak rates of surface runoff for individual storms'. Interim Survey Report, Grand Neosho River Watershed, USDA, Soil Conservation Service, Lincoln, Nebraska.

Monteith, J. L. 1965. 'Evaporation and the environment'. The State and Movement of Water in Living Organisms, XIXth Symposium. Society for Experimental Biology, Swansea, Cam-bridge University Press. pp. 205-234.

Nicks, A. D., J. B. Burt, W. G. Knisel, and G. A. Gander. 1984. 'Application of a field-scale water quality model for evaluation of a watershed land treatment project'. American Society of Agricultural Engineers, St. Joseph, Michigan, ASAE Paper No. 84-2633. 33 pp.

Nutter, W. L., F. M. Davis, R. A. Leonard, and W. G. Knisel. 1993. 'Simulating forest hydrologic response with GLEAMS'. Submitted for publication in Transactions of the American Society of Agricultural Engineers.

Penman, H. L. 1948. 'Natural evaporation from open water, bare soil and grass'. Proceedings of the Royal Society of London. A193:120-146.

Priestley, C. H. B., and R. J. Taylor. 1972. 'On the assessment of surface heat flux and evaporation using large-scale parameters'. Monthly Weather Review 100:81-92.

Reyes, M. R., R. L. Bengtson, J. L. Fouss, and J. S. Rogers. 1992. 'The GLEAMS hydrology submodel modified for shallow water table conditions'. American Society of Agricultural Engineers, Winter Meeting, Nashville, Tennessee, Dec. 15-18, 1992. Paper No. 92-2564. 20 pp.

Richardson, C. W., and A. D. Nicks. 1990. 'Weather generator description'. In: A. N. Sharpley and J. R. Williams (Eds.) EPIC--Erosion/Productivity Impact Calculator: 1. Model Documentation. U. S. Department of Agriculture, Technical Bulletin, No. 1768. pp. 93-104.

Ritchie, J. T. 1972. 'A model for predicting evaporation from a row crop with incomplete cover'. Water Resources Research 8(5):1204-1213.

Sharpley, A. N., and J. R. Williams (Eds.) 1990. EPIC--Erosion/Productivity Impact Calculator: 1. Model documentation. U. S. Department of Agriculture, Technical Bulletin, No. 1768. 235 pp.

Smith, R. E., and J. Y. Parlange. 1978. 'A parameter-efficient hydrologic infiltration model'. Water Resources Research 14(3):533-538.

Smith, R. E., and J. R. Williams. 1980. 'Simulation of the surface water hydrology'. In: W. G. Knisel (Ed.), CREAMS: A Field-Scale Model for Chemicals, Runoff, and Erosion from Agricultural Management Systems. U. S. Department of Agriculture, Science and Education Administration, Conservation Research Report No. 26. pp. 13-35.

Stemmler, S. 1982. 'Environmental effects of nitrogen fertilization exemplified by groundwater pollution as simulated by CREAMS'. In: V. Svetlosanov and W. G. Knisel (Eds.) European and U. S. Case Studies in Application of the CREAMS Model. International Institute for Applied Systems Analysis, Laxenburg, Austria. CP-82-S11. pp. 49-62.

Stewart, B. A., D. A. Woolhiser, W. H. Wischmeier, J. H. Caro, and M. H. Frere. 1975. 'Control of water pollution from cropland. Vol. I. - A manual for guideline development'. U.S. Department of Agriculture, Agricultural Research Service, Headquarters, Report No. ARS-H-5-1. 111 pp.

Svetlosanov, V., and W. G. Knisel (Eds.) 1982. European and U. S. Case Studies in Application of the CREAMS Model. International Institute for Applied Systems Analysis, Laxenburg, Austria. CP-82-S11. 144 pp.

Thomas, D. L. , M. C. Smith, M. A. Breve, and R. A. Leonard. 1991. 'Alternate production systems'. In: D. B. Beasley, W. G. Knisel, and A. P. Rice (Eds.) Proceedings, CREAMS-GLEAMS Symposium. Sept. 27-29, 1989, Athens, Georgia, Agricultural Engineering Department, University of Georgia-Coastal Plain Experiment Station, Publication No. 4, pp. 149-160.

USDA, Soil Conservation Service. 1972. National Engineering Handbook, Hydrology, Section 4. U. S. Government Printing Office, Washington, D.C. 548 pp.

Williams, J. R. 1979. 'SPNM--A problem-oriented computer language for Sediment, Phosphorus, and Nitrogen Movement'. Mathematics and Computers in Simulation XXIV, North Holland Publishing Company.

Williams, J. R., and R. W. Hann. 1978. 'Optimal operation of large agricultural watersheds with water quality constraints'. Technical Report No. 96, Texas Water Resources Institute, Texas A & M University, College Station, Texas. 152 pp.

Williams, J. R., and W. V. LaSeur. 1976. 'Water yield model using SCS curve numbers'. American Society of Civil Engineers, Journal of Hydraulics Division 102(HY9):1241-1253.

Williams, J. R., and A. D. Nicks. 1982. 'CREAMS hydrology model--Option 1'. In: V. P. Singh (Ed.) Applied Modeling Catchment Hydrology. Proceedings, International Symposium Rainfall-Runoff Modeling. Mississippi State University, Mississippi State. pp. 69-86.

Williams, J. R., and K. G. Renard. 1985. 'Assessments of soil erosion and crop productivity with process models (EPIC)'. In: R. F. Follett and B. A. Stewart (Eds.) Soil Erosion and Crop Productivity. American Society of Agronomy, Madison, Wisconsin. pp. 68-103.

Williams, J. R., A. D. Nicks, and J. G. Arnold. 1985. 'Simulator for water resources in rural basins'. American Society of Civil Engineers, Journal of Hydraulic Engineering 111(6):970-986.

Young, C. E., B. M. Crowder, J. S. Shortle, and J. R. Alwang. 1985. 'Nutrient management on dairy farms in southeastern Pennsylvania. Journal of the Soil and Water Conservation 40(5):443-445.

SUBJECT INDEX